HUMAN BIOLOGY

THIRD EDITION

BOOKS IN THE WADSWORTH BIOLOGY SERIES

Biology: Concepts and Applications, Third, Starr

Biology: The Unity and Diversity of Life, Eighth, Starr and Taggart

Human Biology, Third, Starr/McMillan

Laboratory Manual for Biology, Perry/Morton

Plant Biology, Rost et al.

Introduction to Biotechnology, Barnum

General Ecology, Krohne

Introduction to Microbiology, Ingraham/Ingraham

Living in the Environment, Eleventh, Miller

Environmental Science, Seventh, Miller

Sustaining the Earth, Fourth, Miller

Environment: Problems and Solutions, Miller

Introduction to Cell and Molecular Biology, Wolfe

Molecular and Cellular Biology, Wolfe

Cell Ultrastructure, Wolfe

Marine Life and the Sea, Milne

Essentials of Oceanography, Garrison

Oceanography: An Invitation to Marine Science, Third, Garrison

Oceanography: An Introduction, Fifth, Ingmanson/Wallace

Plant Physiology, Fourth, Salisbury/Ross

Plant Physiology Laboratory Manual, Ross

Plants: An Evolutionary Survey, Second, Scagel et al.

Psychobiology: The Neuron and Behavior, Hoyenga/Hoyenga

Sex, Evolution, and Behavior, Second, Daly/Wilson

Dimensions of Cancer, Kupchella

Evolution: Process and Product, Third, Dodson/Dodson

Genetics: The Continuity of Life, Fairbanks

HUMAN BIOLOGY

THIRD EDITION

CECIE STARR

Belmont, California

BEVERLY McMILLAN

Gloucester, Virginia

Brooks/Cole • Wadsworth

I(T)P® An International Thomson Publishing Company

Pacific Grove, CA • Albany, NY • Belmont, CA • Boston • Cincinnati • Detroit • Johannesburg
London • Madrid • Melbourne • Mexico City • New York • Scottsdale, AZ
Singapore • Tokyo • Toronto

Biology Publisher: Jack C. Carey
Project Development Editor: Kristin Milotich
Editorial Assistant: Susan Lussier
Marketing Manager: Tami Cueny
Project Editor: John Walker
Print Buyer: Karen Hunt
Permissions Editor: Susan Walters
Production: Lachina Publishing Services, Inc.
Interior Designer: Gary Head
Copy Editor: Nancy Tenney
Artists: Artemis (Betsy Palay), Lewis Calver, Raychel
  Ciemma, Robert Demarest, Hans & Cassady, Inc. (Hans
Neuhart), Darwen Hennings, Vally Hennings, Carlyn
Iverson, John W. Karapelou, Leonard Morgan, Palay/
Beaubois, Precision Graphics (Jim Gallagher, Jan Troutt),
Nadine Sokol, Kevin Somerville, Lloyd Townsend
Cover Designer: Cuttris & Hambleton
Cover Image: DNA torus-complex © Mark Barounos;
  Seven-week-old human embryo © CNRI; Fertilized
  human egg © Myriam Wharman; DNA strand
  © Dan McCoy
Compositor: Lachina Publishing Services, Inc.
Color Separator: H&S Graphics
Printer: R.R. Donnelley, Willard

Printed in the United States of America
1 2 3 4 5 6 7 8 9 10

For more information, contact Brooks/Cole Publishing Company, 511 Forest Lodge Road, Pacific Grove, California 93950,
or electronically at http://www.brookscole.com

International Thomson Publishing Europe
Berkshire House
168-173 High Holborn
London, WC1V 7AA, United Kingdom

Nelson ITP, Australia
102 Dodds Street
South Melbourne, Victoria 3205 Australia

Nelson Canada
1120 Birchmount Road
Scarborough, Ontario, Canada M1K 5G4

International Thomson Publishing Southern Africa
Building 18, Constantia Square
138 Sixteenth Road, P.O. Box 2459
Halfway House, 1685 South Africa

International Thomson Editores
Seneca, 53
Colonia Polanco, 11560 México D.F. México

International Thomson Publishing Asia
60 Albert Street
#15-01 Albert Complex
Singapore 189969

International Thomson Publishing Japan
Hirakawa-cho Kyowa Building, 3F
2-2-1 Hirakawa-cho, Chiyoda-ku, Tokyo 102 Japan

Library of Congress Cataloging-in-Publication Data
Starr, Cecie.
  Human biology / Cecie Starr, Beverly McMillan.—3rd ed.
    p. cm.
  Includes bibliographical references and index.
  ISBN 0-534-55089-4
  1. Human biology. I. McMillan, Beverly. II. Title
QP34.5.S727 1998
612.8—dc21   98-43320

CONTENTS IN BRIEF

I FOUNDATIONS

Introduction   Concepts in Human Biology 2

1   The Chemistry of Life 14

2   Cells 38

II BODY SYSTEMS AND FUNCTIONS

3   Tissues, Organ Systems, and Homeostasis 66

4   Musculoskeletal System: Support and Movement 86

5   Digestion and Nutrition 110

6   Blood and Circulation 134

7   Immunity 162

8   The Respiratory System 184

9   The Internal Environment 200

10   Nervous System 218

11   Sensory Reception 246

12   Integration and Control: Endocrine System 266

13   Reproductive Systems 286

14   Development 304

15   Sexually Transmitted Diseases 330

III PRINCIPLES OF INHERITANCE

16   Cell Reproduction 350

17   Observable Patterns of Inheritance 372

18   Chromosome Variations and Medical Genetics 386

19   DNA Structure and Function 404

20   Cancer: A Case Study of Genes and Disease 422

21   Recombinant DNA and Genetic Engineering 438

IV EVOLUTION AND ECOLOGY

22   Principles of Evolution 456

23   Human Evolution 474

24   Ecosystems 486

25   Impacts of the Human Population 508

DETAILED CONTENTS

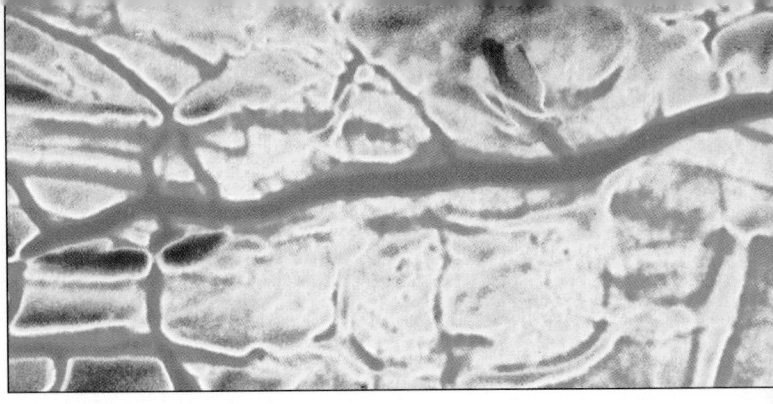

I FOUNDATIONS

INTRODUCTION
CONCEPTS IN HUMAN BIOLOGY

HUMAN BIOLOGY REVISITED 2

Key Concepts 3

I.1 **Cells: The Basic Units of Life** 4
From Non-Life to Life 4
Metabolism 4
Homeostasis 5

I.2 **More Life Characteristics: Reproduction and Inheritance** 6
What Do We Mean by Inheritance? 6
Adaptive Traits 6

I.3 **Organization and Interdependency in the Living World** 7
Organisms Are Interdependent 7

I.4 **Scientific Methods** 8
What Do We Mean by "Theory"? 8
The Limits of Science 9

I.5 **Case Study: Using a Scientific Method in Cancer Research** 10
Moving from Hypothesis to Prediction 10
Testing a Prediction by Experimentation 10

I.6 CHOICES: BIOLOGY AND SOCIETY: HOW DO YOU DEFINE DEATH? 11

Summary 12

1 THE CHEMISTRY OF LIFE

CHEMISTRY IN AND AROUND YOU 14

Key Concepts 15

1.1 **How Matter Is Organized** 16
The Structure of Atoms 16
Isotopes—Varying Atoms of an Element 16
Combinations of Atoms 17

1.2 SCIENCE COMES TO LIFE: USING RADIOISOTOPES TO TRACK CHEMICALS AND SAVE LIVES 18

1.3 **What Is a Chemical Bond?** 19

1.4 **Important Bonds in Biological Molecules** 20
Ionic Bonds: Electrons Gained or Lost 20
Covalent Bonds: Atoms Share Electrons 20
Hydrogen Bonding 21
Free Radicals: Stealing Electrons 21

1.5 **Water, Dissolved Ions, and the World of Cells** 22
Properties of Water That Support Life 22
H_2O Interactions: Acids, Bases, and Salts 22
The pH Scale 23
Buffers and the pH of Body Fluids 23

1.6 **Organic Compounds: Built on Carbon Atoms** 24
Carbon Bonding: The Key to Versatility 24
Hydrocarbons and Functional Groups 24
How Cells Use Organic Compounds—Five Classes of Reactions 25
Condensation and Hydrolysis Reactions 25
The Molecules of Life 25

1.7 **Carbohydrates** 26
Monosaccharides: A Single Sugar Unit 26
Oligosaccharides: Short Chains of Sugar Units 26
Polysaccharides: Many Sugar Units 27

1.8 **Lipids** 28
Fatty Acids 28
Neutral Fats (Triglycerides) 28
Phospholipids 29
Waxes 29
Sterols and Substances Derived from Them 29

1.9 **Amino Acids and Proteins** 30
Primary Structure of Proteins 31
Three-Dimensional Structure of Proteins 31

1.10 **Some Examples of Final Protein Structure** 32
Hemoglobin and Other Complex Proteins 32
Denaturation and Protein Structure 33

1.11 **Nucleotides and Nucleic Acids** 34
Nucleotides with Roles in Metabolism 34
Nucleic Acids—DNA and RNA 34

1.12 FOCUS ON OUR ENVIRONMENT: MAKING NONPOLLUTING FUEL 35

Summary 36

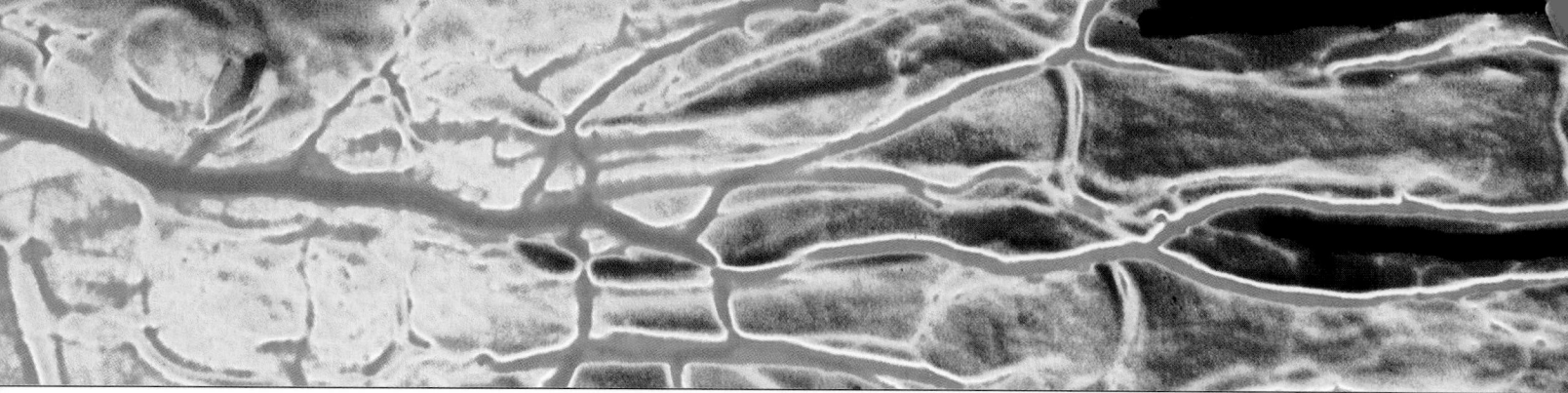

2   CELLS

CELLS FILLED WITH JUICES *38*

Key Concepts *39*

2.1   Basic Cell Structure and Function *40*
The Size and Shape of Cells *40*
Cell Membranes *40*

2.2   How Water Crosses Cell Membranes *42*
Concentration Gradients and Diffusion *42*
Osmosis *43*

2.3   How Dissolved Substances Cross Cell Membranes *44*

2.4   FOCUS ON OUR ENVIRONMENT: WATER POLLUTION AND DISEASE *45*

2.5   Cells and Their Organelles *46*

2.6   SCIENCE COMES TO LIFE: MICROSCOPES: GATEWAYS TO THE CELL *48*

2.7   The Nucleus *49*
The Nucleolus *49*
The Nuclear Envelope *49*
Chromosomes *49*

2.8   The Cytomembrane System *50*
Endoplasmic Reticulum *50*
Golgi Bodies *50*
A Variety of Vesicles *50*

2.9   Mitochondria *52*

2.10   The Cytoskeleton *53*

2.11   Cell Metabolism *54*
Enzymes *54*
Cofactors: Enzyme "Helpers" *55*
How Are Enzymes Controlled? *55*

2.12   Capturing Energy in ATP: The First Two Stages in Human Cells *56*
Overview: How Human Cells Make ATP *56*
The Breakdown of Glucose Begins *56*
The Krebs Cycle and Preparatory Steps *57*

2.13   The Final Stage of Aerobic Respiration *58*
Net ATP Yield of Aerobic Respiration *58*

ATP from Anaerobic Pathways *58*
The ADP/ATP Cycle *59*

2.14   Alternative Energy Sources *60*
Energy from Fats *60*
Energy from Proteins *60*

Summary *62*

II   BODY SYSTEMS AND FUNCTIONS

3   TISSUES, ORGAN SYSTEMS, AND HOMEOSTASIS

THE BODY IN BALANCE *66*

Key Concepts *67*

3.1   Epithelial Tissue *68*
Glands: Derived from Epithelium *68*
Membranes *69*

3.2   SCIENCE COMES TO LIFE: SPARE PARTS FOR THE BODY *70*

3.3   Cell-to-Cell Contacts *72*

3.4   Overview of Connective Tissues *73*
General Characteristics of Connective Tissue *73*
Connective Tissue Proper *73*

3.5   Cartilage and Bone *74*

3.6   Blood and Adipose Tissue *75*

3.7   Muscle Tissue *76*

3.8   Nerve Tissue *77*
Neuron Structure *77*
Neuron Functions *77*

3.9   Organ Systems *78*

3.10   The Skin: Case Study of an Organ System *80*
Skin Functions *80*
Skin Structure *80*

3.11   FOCUS ON OUR ENVIRONMENT: SUN, SKIN, AND THE OZONE LAYER *81*

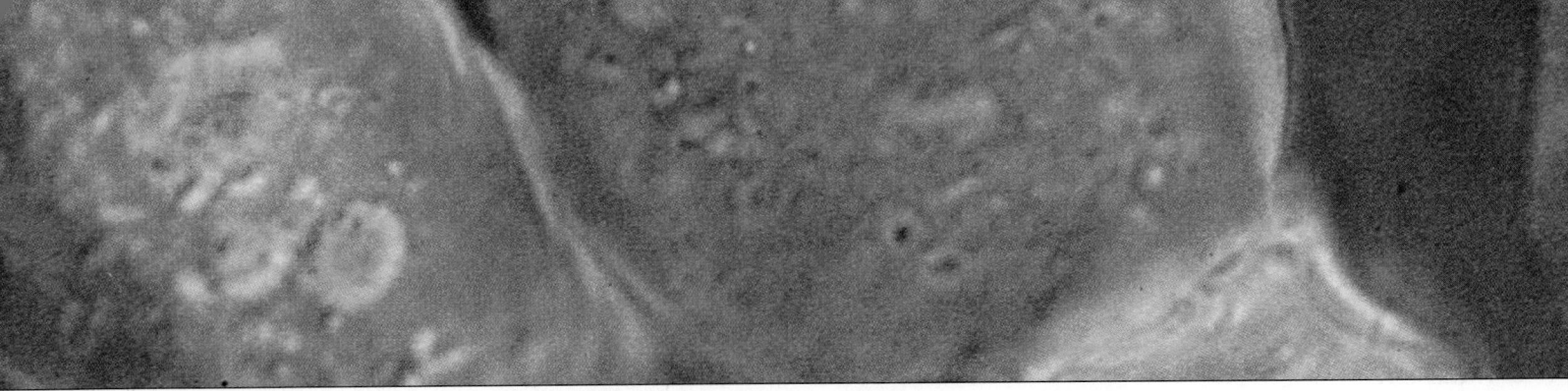

3.12 **Homeostasis and Systems Control** *82*

The Internal Environment *82*

Mechanisms of Homeostasis *82*

Summary *84*

---

4 **MUSCULOSKELETAL SYSTEM: SUPPORT AND MOVEMENT**

ROW, ROW, ROW YOUR BOAT *86*

Key Concepts *87*

4.1 **Bone** *88*

Basic Bone Functions and Structure *88*

How Do Bones Develop? *89*

4.2 **How the Skeleton Grows and Is Maintained** *90*

How Bone Remodeling Works *90*

Bone Remodeling over Time *90*

The Skeleton: A Preview *90*

4.3 **The Axial Skeleton** *92*

The Skull *92*

Facial Bones *92*

Vertebral Column: The Backbone *93*

The Ribs and Sternum *93*

4.4 **The Appendicular Skeleton** *94*

The Pectoral Girdle and Upper Limbs *94*

The Pelvic Girdle and Lower Limbs *95*

4.5 **Joints** *96*

4.6 SCIENCE COMES TO LIFE: THE VULNERABLE KNEE *97*

4.7 **The Musculoskeletal System** *98*

How Skeletal Muscles and Bones Interact *98*

4.8 **Muscle Structure and Function** *100*

How Skeletal Muscle Is Organized *100*

Sliding-Filament Model of Contraction *100*

4.9 **Controlling the Contraction of a Skeletal Muscle** *102*

Neuromuscular Junctions *102*

ATP Formation and Levels of Exercise *103*

4.10 **Properties of Whole Muscles** *104*

Muscle Tension *104*

How Motor Units Function *104*

"Fast" and "Slow" Muscle *105*

4.11 FOCUS ON YOUR HEALTH: MUSCLE MATTERS *106*

Summary *108*

---

5 **DIGESTION AND NUTRITION**

TALES OF A TUBE *110*

Key Concepts *111*

5.1 **Overview of the Digestive System** *112*

Structure of the Digestive Tube *113*

Motility *113*

5.2 **Chewing and Swallowing** *114*

Mouth, Teeth, and Salivary Glands *114*

Swallowing *115*

5.3 **Digestion in the Stomach and Small Intestine** *116*

The Stomach *116*

The Small Intestine *117*

5.4 **Absorption in the Small Intestine** *118*

Mechanisms of Absorption *118*

5.5 **Liver Functions in Digestion** *120*

5.6 **The Large Intestine** *121*

5.7 **Digestion Controls and Nutrient Turnover** *122*

Controls over Digestion *122*

Nutrient Turnover and Organic Metabolism *122*

Disorders That Disrupt Absorption *122*

5.8 **Nutrition** *124*

A Food Pyramid *124*

Carbohydrates *124*

Lipids *125*

Proteins *125*

5.9 **Vitamins and Minerals** *126*

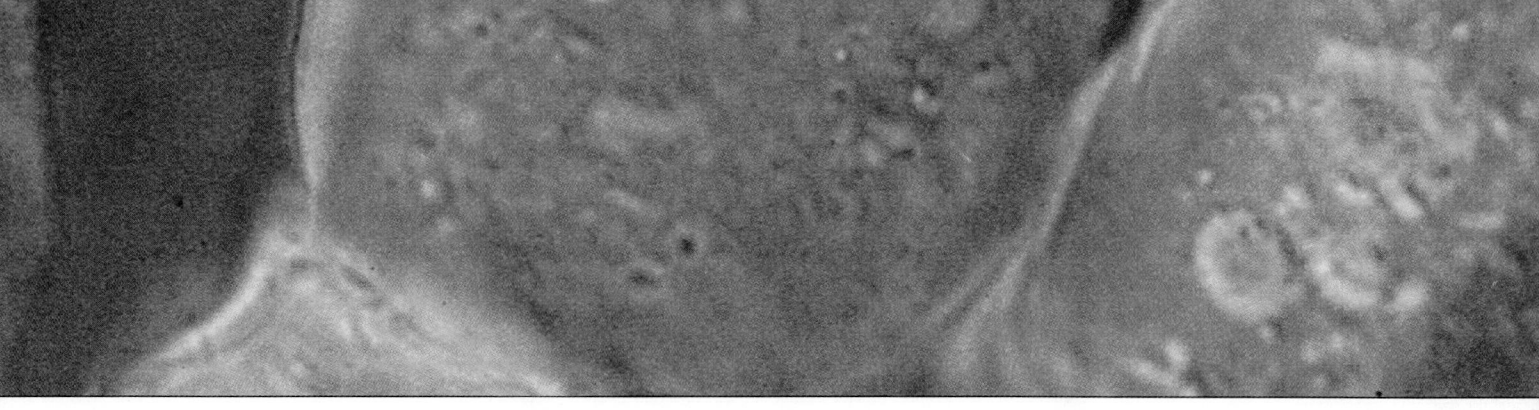

5.10 **Food Energy and Body Weight** *128*
Assessing Body Weight *128*
My Genes Made Me Do It *129*
Dieting and Exercise *129*

5.11 CHOICES: BIOLOGY AND SOCIETY: MALNUTRITION AND UNDERNUTRITION *130*

Summary *132*

---

6 **BLOOD AND CIRCULATION**

HEARTWORKS *134*

Key Concepts *135*

6.1 **Blood** *136*
Plasma *136*
Red and White Blood Cells *136*
Platelets *137*

6.2 **Oxygen Transport in Blood** *138*
Hemoglobin, The Oxygen Carrier *138*
Factors That Affect Oxygen Binding *138*

6.3 **Life Cycle of Red Blood Cells** *139*

6.4 **Blood Typing** *140*
ABO Blood Typing *140*
Rh Blood Typing *140*
Other Applications of Blood Typing *141*

6.5 **Overview of the Cardiovascular System** *142*
Cardiovascular System Components *142*
Links with the Lymphatic System *142*

6.6 **The Heart: A Durable Pump** *144*

6.7 SCIENCE COMES TO LIFE: CAN WE MAKE ARTIFICIAL BLOOD? *145*

6.8 **Blood Circulation** *146*
The Two Circuits *146*
Heartbeat: The Cardiac Cycle *147*

6.9 **How the Heart Contracts** *148*
The Cardiac Conduction System *148*
Neural Controls over Heart Rate *148*

6.10 **Blood Pressure and Velocity in the Cardiovascular System** *149*
Blood Pressure *149*
Blood Velocity *149*

6.11 **How Vessel Structure Affects Blood Pressure** *150*
Resistance at Arterioles *150*
Capillaries *150*
Venules and Veins *150*
How Vessels Help Control Blood Pressure *151*

6.12 **Exchanges of Fluid and Solutes at Capillaries** *152*

6.13 **Hemostasis and Blood Clotting** *153*
Overview of Hemostasis *153*
Clotting Mechanisms *153*

6.14 **Cardiovascular Disorders** *154*
Cholesterol and Cardiovascular Disease *154*
Other "Healthy Heart" Issues *154*

6.15 FOCUS ON YOUR HEALTH: ATHEROSCLEROSIS, HYPERTENSION, AND ARRHYTHMIAS *156*

6.16 **The Lymphatic System** *158*
The Lymph Vascular System *159*
Lymphoid Organs and Tissues *159*

Summary *160*

---

7 **IMMUNITY**

RUSSIAN ROULETTE, IMMUNOLOGICAL STYLE *162*

Key Concepts *163*

7.1 **Three Lines of Defense** *164*
Surface Barriers to Invasion *164*
Nonspecific and Specific Responses *164*

7.2 **Complement Proteins** *165*

7.3 **Inflammation** *166*
The Roles of Phagocytes and Their Kin *166*
The Inflammatory Response *166*

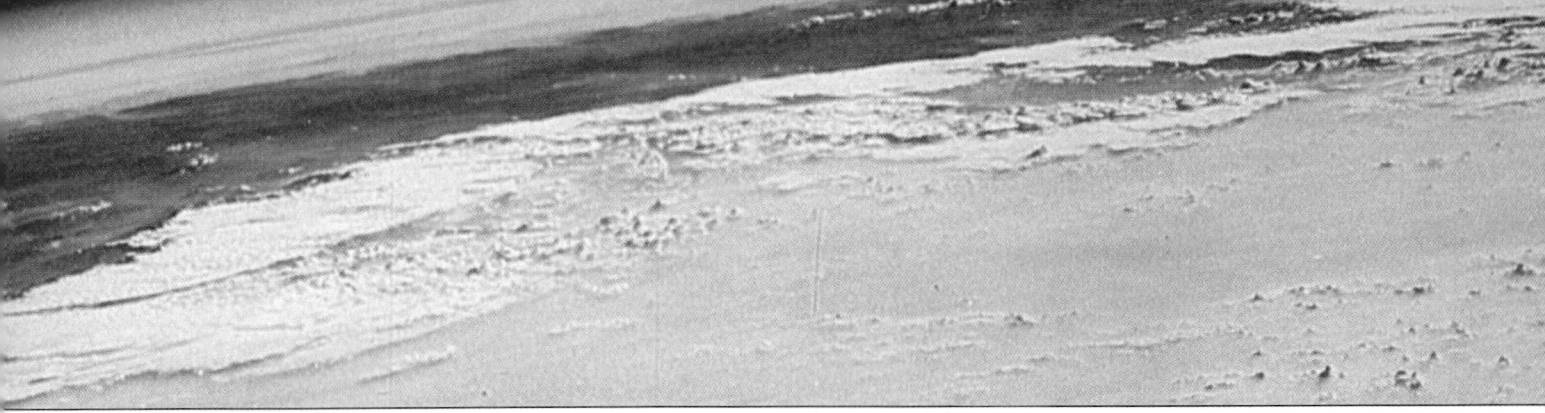

7.4 The Immune System *168*
Defining Features *168*
Antigen-Presenting Cells—The Triggers for Immune Responses *168*
Key Players in Immune Responses *168*
Control of Immune Responses *169*

7.5 Lymphocyte Battlefields *170*

7.6 Cell-Mediated Responses *170*
T Cell Formation and Activation *170*
Functions of Effector T Cells *170*
Natural Killer Cells *171*

7.7 Antibody-Mediated Responses *172*
B Cells and the Targets of Antibodies *172*
The Immunoglobulins *173*

7.8 FOCUS ON YOUR HEALTH: IMMUNE RESPONSES AND ORGAN TRANSPLANTATION *173*

7.9 Immune Specificity and Memory *174*
Formation of Antigen-Specific Receptors *174*
Immunological Memory *175*

7.10 Immunization and Other Practical Applications of Immunology *176*
Immunization *176*
Monoclonal Antibodies *176*
Cytokines *177*

7.11 Abnormal or Deficient Immune Responses *178*
Allergies *178*
Autoimmune Disorders *178*
Deficient Immune Responses *179*

7.12 Immunity Defied: Infectious Disease *180*
Modes of Transmission *180*
Patterns of Occurrence *180*
Virulence *180*
What Antibiotics Can and Cannot Do *180*
Antibiotic Resistance *181*
Case Study: The Resurgence of Tuberculosis *181*

Summary *182*

8   THE RESPIRATORY SYSTEM
CONQUERING CHOMOLUNGMA *184*
Key Concepts *185*

8.1 Overview of the Respiratory System *186*
Airways to the Lungs *186*
Sites of Gas Exchange in the Lungs *187*

8.2 Factors That Influence Gas Exchange *188*
Pressure Gradients and Transport Pigments *188*
Gas Exchange in Unusual Environments *188*

8.3 FOCUS ON OUR ENVIRONMENT: BREATHING AS A HEALTH HAZARD *189*

8.4 Breathing—Cyclic Reversals in Air Pressure Gradients *190*
The Respiratory Cycle *190*
Lung Volumes *191*
Breathing and Sound Production *191*

8.5 Gas Exchange and Transport *192*
Gas Exchange in Alveoli *192*
Gas Transport between Blood and Metabolically Active Tissues *192*

8.6 Controls over Gas Exchange *194*
Matching Air Flow with Blood Flow *194*
Local Chemical Controls *195*
Disrupted Controls over Gas Exchange *195*

8.7 FOCUS ON YOUR HEALTH: TOBACCO AND OTHER THREATS TO RESPIRATORY HEALTH *196*

Summary *198*

9   THE INTERNAL ENVIRONMENT
SURVIVAL AT STOVEPIPE WELLS *200*
Key Concepts *201*

9.1 The Challenge: Shifts in Extracellular Fluid *202*
Water Gains and Losses *202*
Solute Gains and Losses *202*

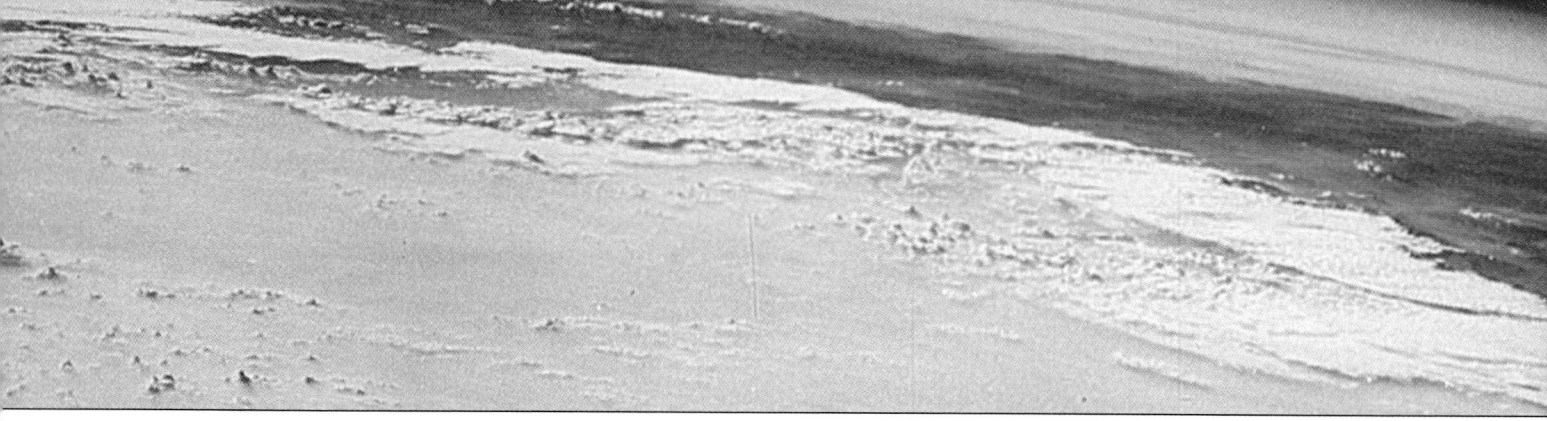

9.2   FOCUS ON OUR ENVIRONMENT: IS YOUR DRINKING WATER POLLUTED? *203*

9.3   **The Urinary System** *204*

9.4   **The Formation of Urine** *206*
Filtration, Reabsorption, and Secretion *206*
Factors That Influence Blood Filtration *206*

9.5   FOCUS ON YOUR HEALTH: AVOIDING TROUBLE IN THE URINARY TRACT *207*

9.6   **Reabsorption of Water and Sodium** *208*
Role of the Proximal Tubule *208*
Urine Concentration and Dilution *209*

9.7   **Hormonal Adjustments of Reabsorption** *210*
How ADH Influences Water Reabsorption *210*
How Aldosterone Influences Sodium Reabsorption *210*
Salt–Water Balance and Thirst *211*

9.8   **Acid–Base Balance** *212*

9.9   SCIENCE COMES TO LIFE: KIDNEY DISORDERS *213*

9.10   **Maintaining the Body's Core Temperature** *214*
Responses to Cold Stress *214*
Responses to Heat Stress *214*

Summary *216*

---

10   **NERVOUS SYSTEM**

WHY CRACK THE SYSTEM? *218*

Key Concepts *219*

10.1   **Neurons—The Communication Specialists** *220*
Functional Zones of a Neuron *220*
A Neuron at Rest, Then Moved to Action *220*
Restoring and Maintaining Readiness *221*

10.2   **Action Potentials** *222*
An All-or-Nothing Spike *222*
Propagation of Action Potentials *223*

10.3   **High-Speed Signals Along Sheathed Axons** *224*

10.4   **Chemical Synapses** *226*
A Smorgasbord of Signals *226*
Synaptic Integration *227*
Removing Neurotransmitter from the Synaptic Cleft *227*

10.5   **The Nervous System: An Overview** *228*
General Paths of Information Flow *229*

10.6   **The Peripheral Nervous System** *230*
Somatic and Autonomic Subdivisions *230*
Sympathetic and Parasympathetic Nerves *230*

10.7   **The Central Nervous System** *232*
The Spinal Cord *232*
The Brain *232*

10.8   **Other Aspects of CNS Structure** *234*
The Cerebral Hemispheres *234*
Brain Cavities and Canals *234*

10.9   FOCUS ON OUR ENVIRONMENT: AN ENVIRONMENTAL ASSAULT ON THE NERVOUS SYSTEM *235*

10.10   **A Closer Look at the Cerebral Cortex** *236*
States of Consciousness *236*
Connections with the Limbic System *237*

10.11   SCIENCE COMES TO LIFE: SPERRY'S SPLIT-BRAIN EXPERIMENTS *238*

10.12   **Memory** *239*

10.13   **How Psychoactive Drugs Affect the Central Nervous System** *240*
Drug Effects *240*
Drug Action and Interactions *241*

10.14   **Commonly Abused Drugs** *242*
Stimulants *242*
Drugs That Reduce Brain Activity *242*
Analgesics *243*
Psychedelics and Hallucinogens *243*
Deliriants *243*

Summary *244*

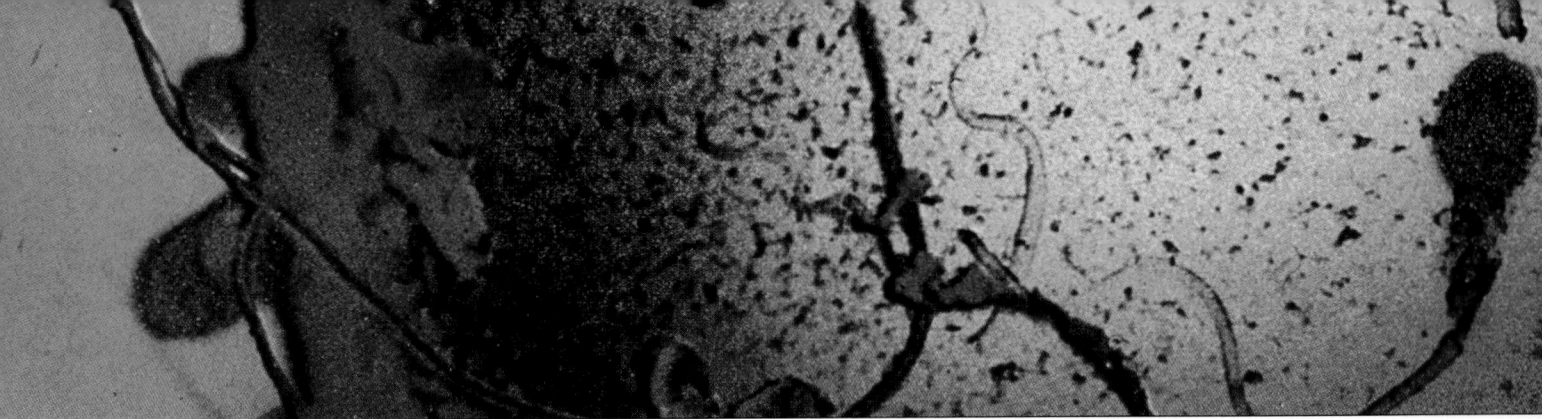

11	**SENSORY RECEPTION**	

NOSING AROUND *246*

Key Concepts *247*

11.1   **Sensory Pathways and Receptors** *248*

Types of Sensory Receptors *248*

Sensory Pathways *248*

11.2   **Somatic Sensations** *250*

Receptors Near the Body Surface *250*

Muscle Sense *250*

A Closer Look at the Sensation of Pain *250*

11.3   **Taste and Smell** *252*

Gustation: The Sense of Taste *252*

Olfaction: The Sense of Smell *252*

11.4   SCIENCE COMES TO LIFE: A TASTE OF SOME SENSORY SCIENCE *253*

11.5   **Hearing** *254*

The Ear *254*

The "How" of Hearing *254*

11.6   **Balance** *256*

11.7   FOCUS ON OUR ENVIRONMENT: NOISE POLLUTION: AN ATTACK ON THE EARS *257*

11.8   **Vision: An Overview** *258*

Eye Structure *258*

Focusing Mechanisms *259*

11.9   FOCUS ON YOUR HEALTH: DISORDERS OF THE EYE *260*

11.10   **From Neuron Signaling to Visual Perception** *262*

Organization of the Retina *262*

Neuronal Responses to Light *263*

Summary *264*

12	**INTEGRATION AND CONTROL: ENDOCRINE SYSTEM**	

RHYTHMS AND BLUES *266*

Key Concepts *267*

12.1   **The Endocrine System** *268*

Hormones and Other Signaling Molecules *268*

Discovery of Hormones and Their Sources *268*

12.2   **Signaling Mechanisms** *270*

Steroid Hormones *271*

Amines and Peptide Hormones *271*

12.3   **The Hypothalamus and Pituitary Gland** *272*

Posterior Lobe Secretions *272*

Anterior Lobe Secretions *272*

12.4   **Examples of Abnormal Pituitary Output** *274*

12.5   **Sources and Effects of Other Hormones** *275*

12.6   **Pancreatic Islets** *276*

Secretions from Pancreatic Islets *276*

Diabetes—Faulty Regulation of Blood Glucose *276*

12.7   **The Adrenal Glands** *278*

Hormones of the Adrenal Cortex *278*

Adrenal Medulla Hormones *279*

12.8   FOCUS ON OUR ENVIRONMENT: ARE "ENDOCRINE DISRUPTERS" HARMING HUMAN HEALTH? *279*

12.9   **Hormones from the Thyroid, Parathyroid, and Some Other Sources** *280*

The Thyroid and Parathyroids—and the Importance of Feedback Control *280*

Thymus Hormones *281*

The Pineal Gland: Melatonin *281*

Hormones from the Heart, GI Tract, and Elsewhere *281*

12.10   **Some Final Examples of Integration and Control** *282*

Local Signaling Molecules *282*

Prostaglandins *282*

Growth Factors *282*

Pheromones Revisited *282*

12.11   SCIENCE COMES TO LIFE: COMMUNICATION MOLECULES PROVIDE HOPE FOR HEALING *283*

Summary *284*

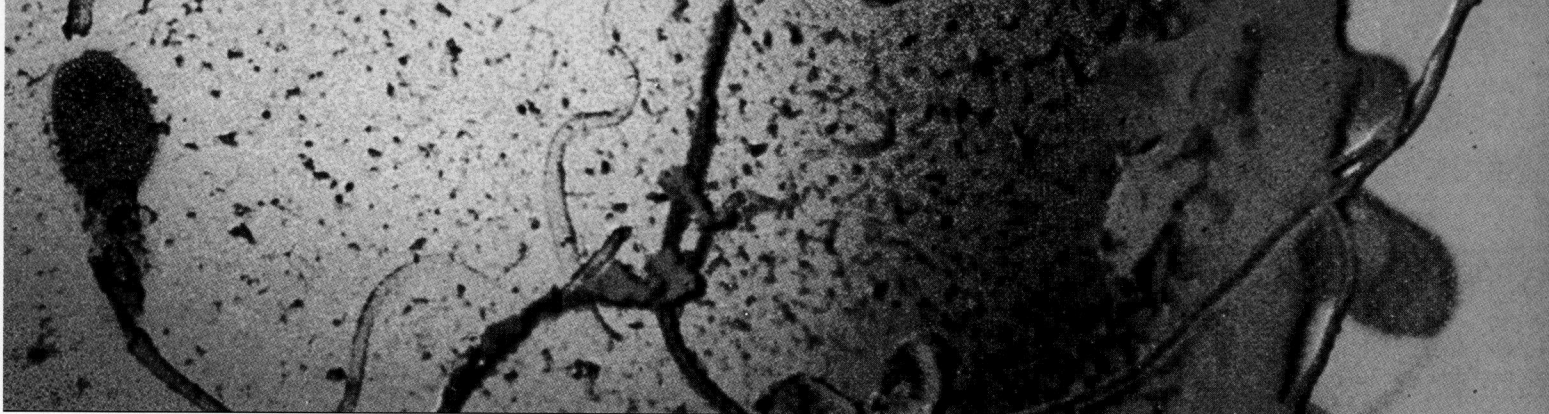

13 REPRODUCTIVE SYSTEMS

GIRLS AND BOYS 286

Key Concepts 287

13.1 The Male Reproductive System 288
Where Sperm Form 288
Where Semen Forms 288

13.2 Male Reproductive Function 290
Sperm Formation 290
Hormonal Controls 290

13.3 The Female Reproductive System 292
The Reproductive Organs 292
Overview of the Menstrual Cycle 293

13.4 Female Reproductive Function 294
Cyclic Changes in the Ovary 294
Cyclic Changes in the Uterus 295

13.5 Visual Summary of the Menstrual Cycle 296

13.6 Sexual Intercourse and Fertilization 297
Sexual Intercourse 297
Fertilization: A Preview 297

13.7 Control of Fertility 298
Birth Control Options 298
Future Options for Fertility Control 299

13.8 Coping with Infertility 300
In Vitro Fertilization 300
Intrafallopian Transfers 300
Artificial Insemination 300

13.9 CHOICES: BIOLOGY AND SOCIETY: DILEMMAS OF FERTILITY CONTROL 301

Summary 302

14 DEVELOPMENT

HOW YOUR JOURNEY BEGAN 304

Key Concepts 305

14.1 Stages of Development 306
Gastrulation: Primary Tissues Form 306
Organogenesis, Growth, and Tissue Specialization 306

14.2 The Beginnings of You—Early Events in Development 308
Fertilization and Cleavage 308
Implantation 308

14.3 Extraembryonic Membranes 310
Yolk Sac, Amnion, Allantois, and Chorion 310
The Importance of the Placenta 310

14.4 The Making of an Early Embryo—A Closer Look 312
Gastrulation Revisited 312
Morphogenesis Revisited 313
"Marching Orders" for Cells 313

14.5 Emergence of Distinctly Human Features 314
Gonad Development 314
Miscarriage 314

14.6 Fetal Development 316
From the Seventh Month to Birth 316
Fetal Circulation 316

14.7 From Birth Onward 318
Birth 318
Three Stages of Labor 318
Nourishing the Newborn 319

14.8 CHOICES: BIOLOGY AND SOCIETY: SAVING BABIES AND MAKING CLONES 319

14.9 How the Mother's Lifestyle Affects Early Development 320
Maternal Nutrition 320
Risk of Infections 320
Prescription Drugs, Illegal Drugs, and Alcohol 321
Effects of Cigarette Smoke 321

14.10 SCIENCE COMES TO LIFE: PRENATAL DIAGNOSIS: DETECTING BIRTH DEFECTS 322

14.11 Summary of Developmental Stages 323
Birth to Adulthood 323
Adult Life 323

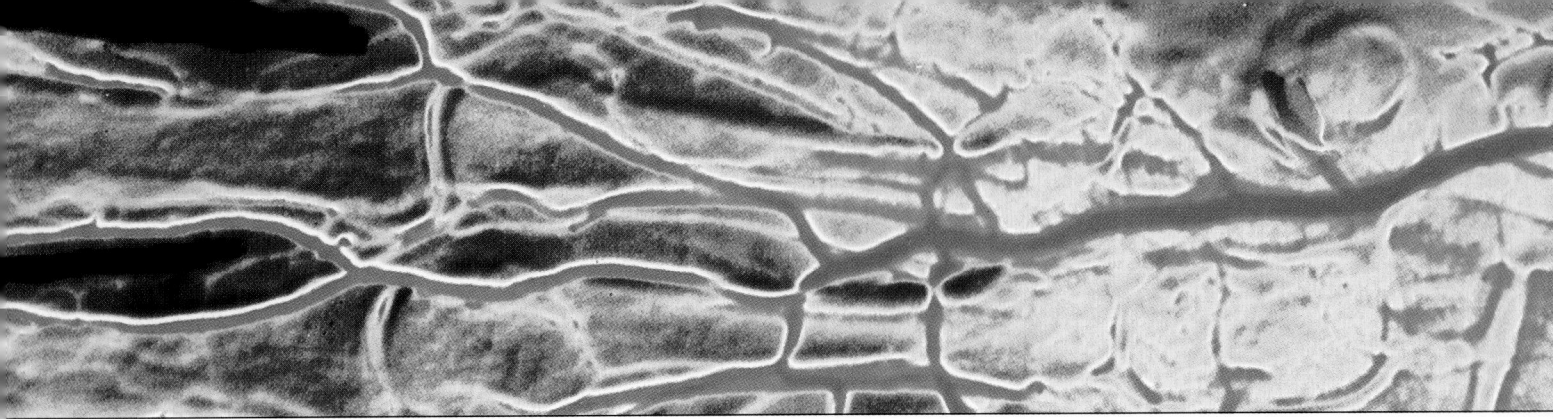

14.12 The Body As It Ages *324*

What Causes Aging? *324*

14.13 Aging of Skin, Muscle, the Skeleton, and Internal Transport Systems *325*

Skin, Muscles, and the Skeleton *325*

Aging in the Cardiovascular and Respiratory Systems *325*

14.14 Age-Related Changes in Some Other Body Systems *326*

The Nervous System and Senses *326*

Reproductive Systems and Sexuality *327*

Immunity, Nutrition, and the Urinary System *327*

Summary *328*

---

15 SEXUALLY TRANSMITTED DISEASES

SAFER SEX *330*

Key Concepts *331*

15.1 Viruses and Bacteria: The Unseen Multitudes *332*

15.2 How Viruses and Bacteria Cause Disease *334*

Several Routes to Disease *334*

Another Kind of Infectious Particle *334*

15.3 HIV and AIDS: A Global Perspective *335*

AIDS—An International Epidemic *335*

15.4 A Closer Look at HIV *336*

HIV Structure and Replication *336*

How HIV Is Transmitted *336*

15.5 HIV Infection *338*

The Initial Attack by HIV *338*

From HIV Infection to AIDS *338*

Possibilities for Treatment *339*

Vaccines: Problems and Prospects *339*

AIDS Prevention *339*

15.6 A Trio of Common STDs *340*

Gonorrhea *340*

Syphilis *340*

Chlamydial Infections *341*

15.7 Other Prevalent STDs *342*

Pelvic Inflammatory Disease *342*

Nongonoccocal Urethritis (NGU) *342*

Genital Herpes *342*

Human Papillomavirus *343*

Type B Viral Hepatitis *343*

Chancroid *343*

15.8 STDs Caused by Parasites or Fungi *344*

Pubic Lice and Scabies *344*

Vaginitis *344*

15.9 FOCUS ON YOUR HEALTH: PROTECTING YOURSELF— AND OTHERS—FROM STDs *345*

Summary *346*

III PRINCIPLES OF INHERITANCE

---

16 CELL REPRODUCTION

TRILLIONS FROM ONE *350*

Key Concepts *351*

16.1 Dividing Cells: The Bridge Between Generations *352*

Overview of Division Mechanisms *352*

Some Key Points about Chromosomes *352*

Mitosis and the Chromosome Number *352*

16.2 Mitosis and the Cell Cycle *354*

The Wonder of Interphase *354*

Summary of the Cell Cycle Phases *355*

16.3 SCIENCE COMES TO LIFE: HENRIETTA'S IMMORTAL CELLS *355*

16.4 A Visual Tour of the Stages of Mitosis *356*

16.5 A Closer Look at Mitosis *358*

Prophase: Mitosis Begins *358*

Transition to Metaphase *358*

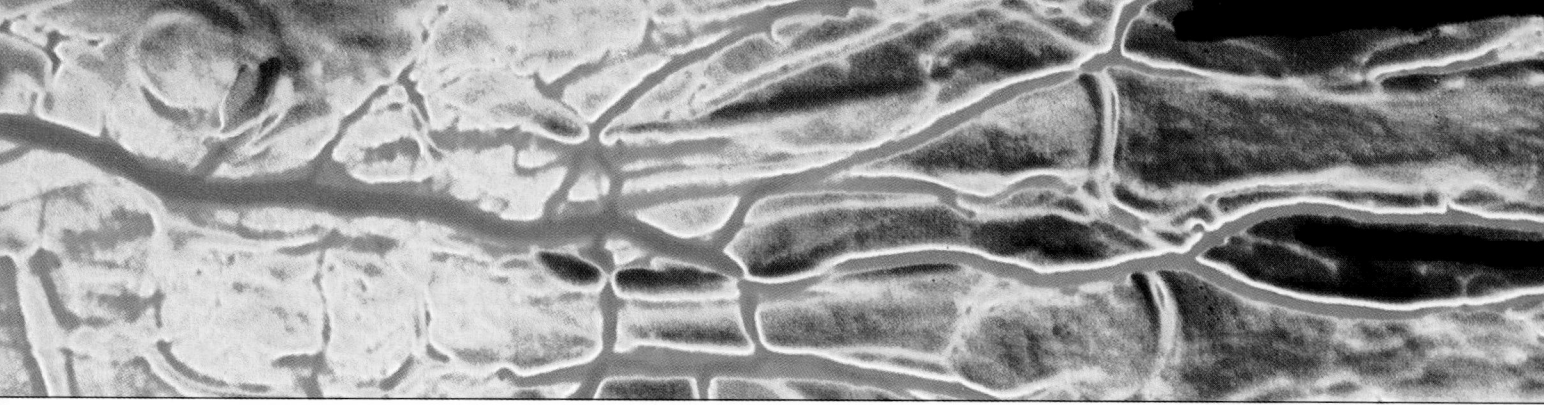

Anaphase *358*

Telophase *359*

16.6 **Division of the Cytoplasm** *360*

16.7 **Overview of Meiosis** *361*

Making Haploid Gametes *361*

Two Divisions, Not One *361*

16.8 **A Visual Tour of the Stages of Meiosis** *362*

16.9 **Key Events during Meiosis I** *364*

Metaphase I Alignments *365*

Separating the Sister Chromatids *365*

16.10 **Meiosis and the Life Cycle** *366*

16.11 FOCUS ON OUR ENVIRONMENT: IONIZING RADIATION: INVISIBLE THREAT TO CELLS *367*

16.12 **Meiosis and Mitosis Compared** *368*

Summary *370*

---

17 **OBSERVABLE PATTERNS OF INHERITANCE**

A SMORGASBORD OF EARS AND OTHER TRAITS *372*

Key Concepts *373*

17.1 **Patterns of Inheritance** *374*

The Origins of Genetics *374*

Some Terms Used in Genetics *374*

17.2 **Mendel's Theory of Segregation** *375*

17.3 **Doing Genetic Crosses and Figuring Probabilities** *376*

17.4 **The Testcross: A Tool for Discovering Genotypes** *378*

17.5 **Independent Assortment** *378*

17.6 **A Closer Look at Independent Assortment** *380*

17.7 **Variations on Mendel's Themes** *381*

Dominance Relations *381*

Multiple Effects of Single Genes *381*

17.8 **Less Predictable Variations in Traits** *382*

Penetrance and Expressivity *382*

Polygenic Traits *383*

Do Genes "Program" Behavior? *383*

17.9 SCIENCE COMES TO LIFE: MOM AND POP GENES: GENOMIC IMPRINTING *383*

Summary *384*

---

18 **CHROMOSOME VARIATIONS AND MEDICAL GENETICS**

AT THE MERCY OF OUR GENES *386*

Key Concepts *387*

18.1 **The Chromosomal Basis of Inheritance** *388*

Genes and Their Locations on Chromosomes *388*

Autosomes and Sex Chromosomes *388*

Gene Mutations on Chromosomes *388*

18.2 **Sex Determination and X Inactivation** *390*

18.3 **Linkage Groups and Crossing Over** *391*

18.4 **Human Genetic Analysis** *392*

Pedigrees: Genetic Connections *392*

Genetic Counseling *393*

18.5 CHOICES: BIOLOGY AND SOCIETY: PROMISE— AND PROBLEMS—OF GENETIC SCREENING *393*

18.6 **Patterns of Autosomal Inheritance** *394*

Autosomal Recessive Inheritance *394*

Autosomal Dominant Inheritance *394*

18.7 **Patterns of X-Linked Inheritance** *396*

X-Linked Recessive Inheritance *396*

X-Linked Inheritance of Dominant Mutant Alleles *397*

Another Rare X-Linked Abnormality *397*

A Few Qualifications *397*

18.8 **Sex-Influenced Inheritance** *398*

18.9 **Changes in Chromosome Structure** *398*

Deletion *398*

Duplications *399*

Inversions and Translocations *399*

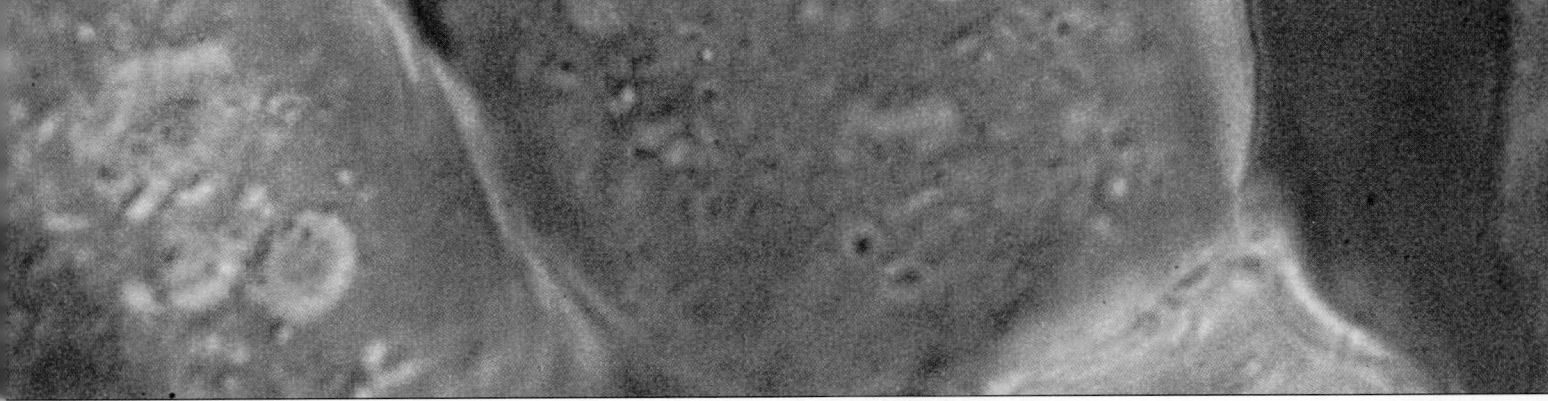

18.10 **Changes in Chromosome Number** *400*
Categories and Mechanisms of Change *400*
Changes in the Number of Autosomes *400*
Changes in the Number of Sex Chromosomes *401*

Summary *402*

---

19 **DNA STRUCTURE AND FUNCTION**

CARDBOARD ATOMS AND BENT-WIRE BONDS *404*

Key Concepts *405*

19.1 **DNA Structure and Replication** *406*
DNA Replication *406*
DNA Repair *407*

19.2 SCIENCE COMES TO LIFE: DNA DETECTIVES: DISCOVERING THE CONNECTION BETWEEN GENES AND PROTEINS *408*

19.3 **Organization of DNA in Chromosomes** *409*

19.4 **DNA into RNA—Protein Synthesis Begins** *410*
A Brief Overview of Protein Synthesis *410*
Transcription *410*

19.5 **From mRNA to Proteins** *412*
The Genetic Code *412*
Roles of tRNA and rRNA *412*

19.6 **Translation** *414*

19.7 **How Mutation Affects Protein Synthesis** *416*

19.8 **Regulating Gene Action** *417*
Regulatory Proteins *417*
Hormones as Regulatory Agents *417*

19.9 FOCUS ON YOUR HEALTH: A TALE OF NUCLEOTIDE REPEATS *417*

Summary *418*

---

20 **CANCER: A CASE STUDY OF GENES AND DISEASE**

THE BODY BETRAYED *422*

Key Concepts *423*

20.1 **Cancer Defined** *424*
Characteristics of Tumors *424*
Characteristics of Cancer Cells *424*

20.2 **The Genetic Triggers for Cancer** *426*
Oncogene Activation *426*
Other Routes to Carcinogenesis *426*

20.3 FOCUS ON OUR ENVIRONMENT: CANCER AND AGRICULTURAL CHEMICALS *428*

20.4 **Diagnosing Cancer** *429*

20.5 **Cancer Treatments and Prevention** *430*

20.6 **Some Major Types of Cancer** *431*

20.7 **Cancers of the Breast and Reproductive System** *432*
Breast Cancer *432*
Cancers of the Reproductive System *433*

20.8 **A Survey of Other Common Cancers** *434*
Oral and Lung Cancers *434*
Cancers of the Digestive System and Related Organs *434*
Urinary System Cancers *434*
Cancers of the Blood and Lymphatic System *434*
Skin Cancer *435*

Summary *436*

---

21 **RECOMBINANT DNA AND GENETIC ENGINEERING**

INGENIOUS GENES *438*

Key Concepts *439*

21.1 **Recombination in Nature—and in the Laboratory** *440*
Plasmids, Restriction Enzymes, and the New Technology *440*
Producing Restriction Fragments *441*

21.2 **Working with DNA Fragments** *442*
Amplification Procedures *442*
Sorting Out DNA Fragments *442*

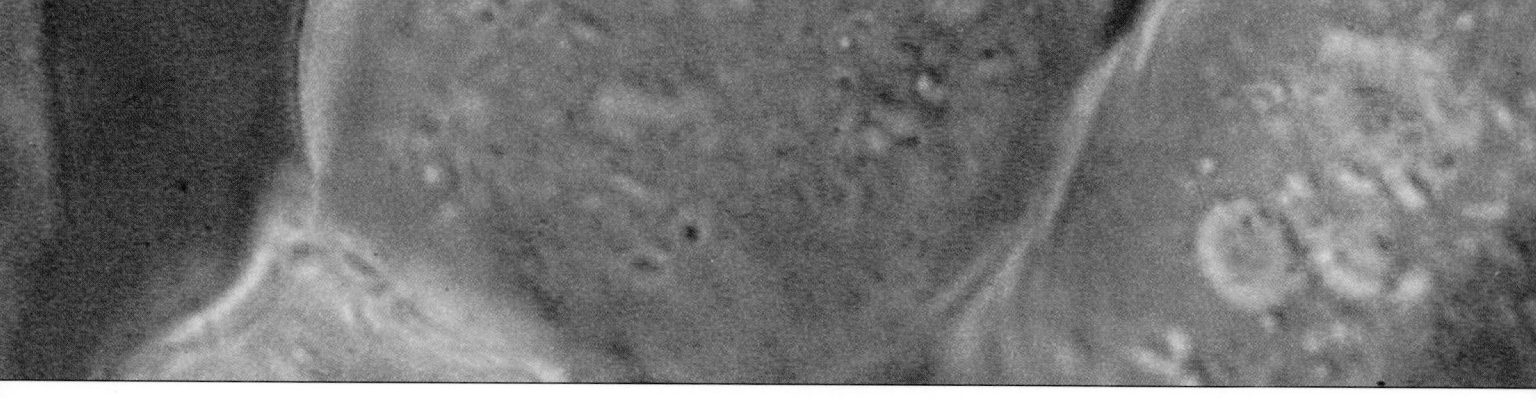

21.3 SCIENCE COMES TO LIFE: DNA FINGERPRINTS *444*

21.4 Modified Host Cells *445*
Blue-White Screening *445*
Use of cDNA *445*

21.5 Bacteria, Plants, and the New Technology *446*
Genetically Engineered Bacteria *446*
Genetic Modification of Plants *446*

21.6 FOCUS ON OUR ENVIRONMENT: BACTERIA THAT CLEAN UP POLLUTANTS *447*

21.7 Genetic Engineering of Animals *448*
Super Mice and Biotech Barnyards *448*
Applying the New Technology to Humans *448*
Gene Therapy *449*
Altering Gene Expression *449*

21.8 Methods and Prospects for Gene Therapy *450*
Strategies for Transferring Genes *450*
Examples of Gene Therapy in Action *451*
Gene Therapy and Your Future *451*

21.9 CHOICES: BIOLOGY AND SOCIETY: ISSUES FOR A BIOTECHNOLOGICAL SOCIETY *452*

Summary *453*

IV EVOLUTION AND ECOLOGY

22 PRINCIPLES OF EVOLUTION
FLOODS AND FOSSILS *456*
Key Concepts *457*

22.1 A Little Evolutionary History *458*

22.2 Some Basic Principles of Evolution *459*
Variation in Populations *459*
Where Does Variation Come From? *459*

22.3 Processes of Microevolution *460*
Mutation: Sole Source of New Alleles *460*
Natural Selection *460*

Genetic Drift and Gene Flow *460*
Reproductive Isolation and Speciation *461*
Rates of Evolutionary Change *461*

22.4 An Introduction to Macroevolution *462*
Biogeography *462*
Fossils: Evidence of Ancient Life *462*
Interpreting the Fossil Record *463*

22.5 Comparisons of Body Form *464*
Comparative Embryology *464*
Vestigial Structures *464*
Homologous and Analogous Structures *465*

22.6 Comparing Biochemistry *466*

22.7 Evolutionary Trees and Their Branchings *466*
Extinctions and Adaptive Radiations *467*
Organizing the Evidence: Classification *467*

22.8 FOCUS ON OUR ENVIRONMENT: GENES OF ENDANGERED SPECIES *468*

22.9 Evolution and Earth History *468*
Origin of the Earth *468*
Conditions on the Early Earth *469*

22.10 The Origin of Life *470*
Synthesis of Biological Molecules *470*
The First Metabolic Pathways *470*
Origin of Self-Replicating Systems *471*
Enter the First Cells *471*

Summary *472*

23 HUMAN EVOLUTION
THE CAVE AT LASCAUX AND THE HANDS OF GARGAS *474*
Key Concepts *475*

23.1 Human Evolution in Perspective *476*
The Mammalian Heritage *476*
Primates *476*

23.2 From Primate to Human: Key Evolutionary Trends *478*
Upright Walking *478*

Precision Grip and Power Grip *478*

Enhanced Daytime Vision *478*

Changes in Dentition *478*

Changes in the Brain and Behavior *479*

Primate Origins *479*

23.3   The First Hominids *480*

23.4   Emergence of Early Humans *482*

Defining "Human" *482*

Early *Homo* and the First Stone Tools *482*

From *Homo erectus* to *H. sapiens* *483*

23.5   SCIENCE COMES TO LIFE: OUT OF AFRICA—
ONCE, TWICE, OR . . . *484*

Summary *485*

---

24   ECOSYSTEMS

CRÊPES FOR BREAKFAST, PANCAKE ICE FOR
DESSERT *486*

Key Concepts *487*

24.1   Introduction to Principles of Ecology *488*

24.2   Ecosystem Organization *490*

Feeding Relationships in Ecosystems *490*

Food Webs *490*

24.3   Energy Flow through Ecosystems *492*

Primary Productivity *492*

Major Pathways of Energy Flow *492*

Ecological Pyramids *492*

**24.4   Case Study: Energy Flow at Silver Springs,
Florida** *494*

24.5   Biogeochemical Cycles—An Overview *495*

24.6   The Hydrologic Cycle *496*

24.7   The Carbon Cycle *498*

24.8   SCIENCE COMES TO LIFE: FROM GREENHOUSE
GASES TO A WARMER PLANET *500*

24.9   The Nitrogen Cycle *502*

Cycling Processes *502*

Nitrogen Scarcity *502*

Human Impacts on the Nitrogen Cycle *502*

24.10   The Phosphorus Cycle *504*

24.11   FOCUS ON OUR ENVIRONMENT: TRANSFER OF
HARMFUL COMPOUNDS THROUGH ECOSYSTEMS *505*

Summary *506*

**25   IMPACTS OF THE HUMAN
POPULATION**

TALES OF NIGHTMARE NUMBERS *508*

Key Concepts *509*

25.1   Introduction to Population Dynamics *510*

Characteristics of Human Populations *510*

Population Size and Patterns of Growth *510*

25.2   A Closer Look at Growth Patterns *512*

Biotic Potential *512*

Limiting Factors and Carrying Capacity *512*

Checks on Population Growth *512*

Life History Patterns *513*

25.3   Human Population Growth *514*

25.4   Air Pollution *516*

Smog *516*

Acid Deposition *516*

Damage to the Ozone Layer *517*

25.5   A Global Water Crisis *518*

Impact of Large-Scale Irrigation *518*

Problems with Water Quality *518*

**25.6   Coping with Solid Wastes and Problems
of Land Use** *519*

Solid Wastes *519*

Conversion of Marginal Lands for
Agriculture *519*

25.7   Destruction of Forests *520*

25.8   Concerns about Energy Use *522*

Fossil Fuels *522*

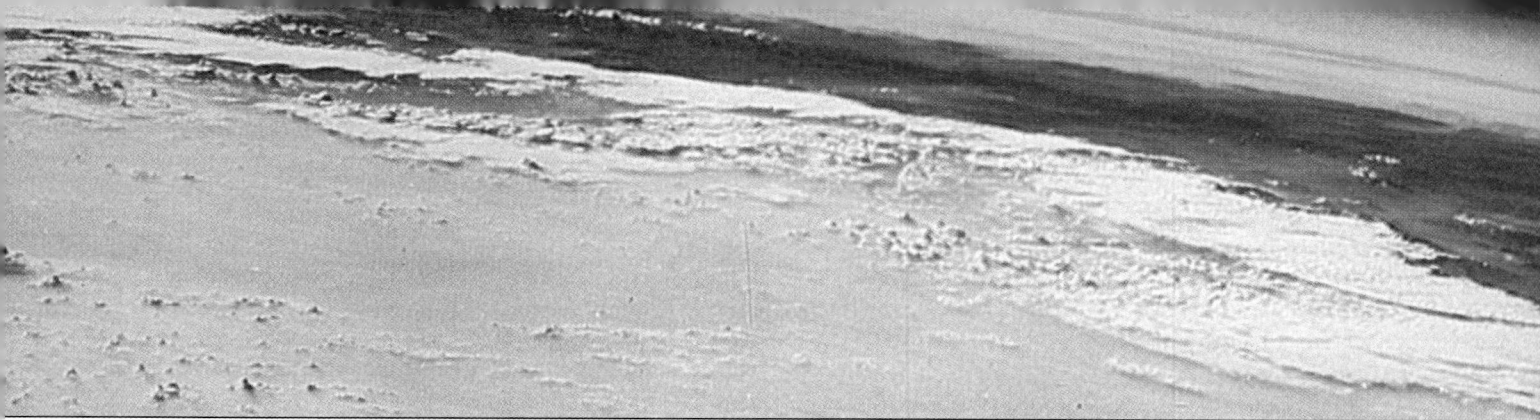

Nuclear Energy *522*

Alternative Energy Sources *523*

25.9 SCIENCE COMES TO LIFE: A PLANETARY EMERGENCY: LOSS OF BIODIVERSITY *524*

25.10 The Idea of Sustainable Living *525*

Summary *526*

APPENDIX I   PERIODIC TABLE OF THE ELEMENTS

APPENDIX II   UNITS OF MEASURE

APPENDIX III   ANSWERS TO GENETICS PROBLEMS

APPENDIX IV   ANSWERS TO SELF-QUIZZES

APPENDIX V   KEY TO CRITICAL THINKING QUESTIONS

GLOSSARY OF BIOLOGICAL TERMS

CREDITS AND ACKNOWLEDGMENTS

INDEX

APPLICATIONS INDEX

PREFACE

As human beings collectively, and as individual people, we are crossing the threshold of a new millennium, a rite of passage that invites reflection on where human biology might be heading. About 500 years ago, early naturalists began to systematically record observations about the living world, although even before them some thinkers had begun to formulate ideas about the operation of the human body and its component parts. About 150 years ago, the monk Gregor Mendel blazed an intellectual trail that helped lead to our modern knowledge of the genetic underpinnings of life, and naturalists such as Charles Darwin began building a framework for understanding the grand scheme of life on earth. These individuals were generalists, seeking answers to broad, sweeping questions about the living world.

About 50 years ago, biologists caught their first glimpse of life's workings—and its fundamental unity—at the level of the DNA molecule. Since then, the study of human biology, like biology generally, has grown to encompass dozens of specialized fields, each focused on one narrow aspect of life and yielding volumes of information about it. For some the focus is how each person develops from a fertilized egg, for others how and why the human heart, brain, or immune system—or some part of it—functions as it does, for still others why we fall prey to cancers. And these are only a tiny sample of the specialized fields of inquiry that contribute to our understanding of the human body and how it functions.

Can an introductory textbook keep up with the rapid splintering of biological inquiry? Given the sometimes daily or weekly announcements of advances in medicine, genetics, research on cancer and nutrition, and other topics, doing so often is nigh unto impossible despite our best efforts. Given this state of affairs, several years ago James Bonner, a teacher and researcher at the California Institute of Technology, wisely predicted that authors and instructors in introductory biology courses would become the new generalists, the ones who give each generation of students broad perspective on what we know about life—and the human body—and what we have yet to learn.

And we must do this, for the biological perspective on human life remains one of education's most valuable gifts. With it, students can cut their own intellectual paths through the thickets of medical, environmental, and social issues that confront us as our world becomes an ever-more complex place in which to live.

CONCERNING THE THIRD EDITION

Like previous editions, this book starts with an overview of the basic concepts and scientific methods, followed by chapters providing an overview of underlying principles of biochemistry and cell structure and function. Next come core chapters on body systems and functions, followed by a unit on the principles of inheritance. A final unit introduces and discusses principles of evolution, including the fascinating, still-unfolding story of the emergence of modern *Homo sapiens*, and principles of ecology that are essential background for an informed citizenry.

As before, we identify and highlight the key concepts, current understandings, and research trends for the major fields of inquiry. We also explain the structure and functioning of tissues, organs, and organ systems—the anatomy and physiology of the human body—in enough detail that students can build a working vocabulary about life's parts and processes. Moreover, the theme that structure follows function is reinforced, chapter after chapter. Working with a cadre of dedicated users and reviewers, we refined and updated throughout, including the review material at the end of each chapter and the book's glossary.

CONCEPT SPREADS

In the introductory chapter, an overview of the living world's level of organization kicks off a story that continues through the rest of the book. Telling such a big, complex story might be daunting unless you remind yourself of the question "How do you eat an elephant?" and its answer, "One bite at a time." We who have told the story again and again know how the parts fit together, but many students need help to keep the story line in focus within and between chapters. And they need to chew on concepts one at a time.

In every chapter we focus each concept on its own two-page spread. That is, we organize the descriptions, art, and supporting evidence for it on two facing pages, at most. Think of this as a concept spread, as in Figure A. Each starts with a numbered tab and ends with boldfaced statements to summarize the key points. Students can use these cues as reminders to digest one topic before starting on another. Well-crafted transitions between spreads help students focus on where topics fit in the larger story and gently discourage memorization for its own sake. The clear demarcation also gives instructors greater flexibility in assigning or skipping topics within a chapter.

By restricting the space available for each concept, we force ourselves to clear away the clutter of superfluous detail. Within each concept spread, we block out headings and subheadings to rank the importance of its various parts. Without a hierarchy, information has all the excitement, flow, and drama of an encyclopedia. Where details are useful as expansions of concepts, we integrate them into suitable illustrations to keep them from disrupting the text flow.

Many students, especially those who are not biology majors, approach biology textbooks with apprehension. If the words don't engage them, they sometimes end up hating both the book *and* the subject. So, our writing aims for a balance between core material and interesting applications. Interrupting the description of muscle function, say, with an anecdote may simply distract the student. Plunking a humorous aside into a chapter on immunity trivializes the story of a body system whose remarkable workings we are only just beginning to comprehend. By contrast, it is appropriate and timely to follow the concept spreads on digestion and absorption with a brief look at energy requirements and strategies for weight loss. It is fine to entertain students *if* doing so reinforces a key concept. Thus, for example, we include the story of how a bulldog named Jimmie helped give an early physiologist key insights into the heart's electrical activity.

BALANCING CONCEPTS WITH APPLICATIONS

Each chapter begins with a lively or sobering application that leads into an adjoining list of key concepts and a brief outline giving the main chapter sections. The list and outline are advance organizers for the chapter as a whole. At strategic points, examples of applications parallel the core material—not so many as to be distracting, but enough to keep minds perking along with the conceptual development. Many brief applications are integrated into the text. Others are in boxed essays, which provide more depth on health, medical, environmental, and social issues for interested students but do not interrupt the text flow. A separate index on the book's last few pages (tinted *green*) lists all applications for quick reference.

FOUNDATIONS FOR CRITICAL THINKING

To help students develop a capacity for critical thinking, we selected certain text discussions and chapter introductions to show students how biologists use critical thinking to solve problems. Among these are the introductions to the chapters on immunity and DNA structure and function, and the discussions of HIV and AIDS in Chapter 15 and of Mendelian genetics in Chapter 17. *Science Comes to Life* essays provide optional, more detailed examples. For example, one of these describes efforts to develop artificial skin, bone, and other tissues (Section 3.2). Another (Section 14.10) helps convey that human biology is not a closed book. Even when research brings an emerging story into sharp focus—in this case, increasingly sophisticated techniques of prenatal diagnosis—it also poses new questions.

Continuing in this edition are *Critical Thinking* questions at the end of chapters. Katherine Denniston of Towson State University developed the new items among these thought-provoking questions. Chapters 17 and 18 also

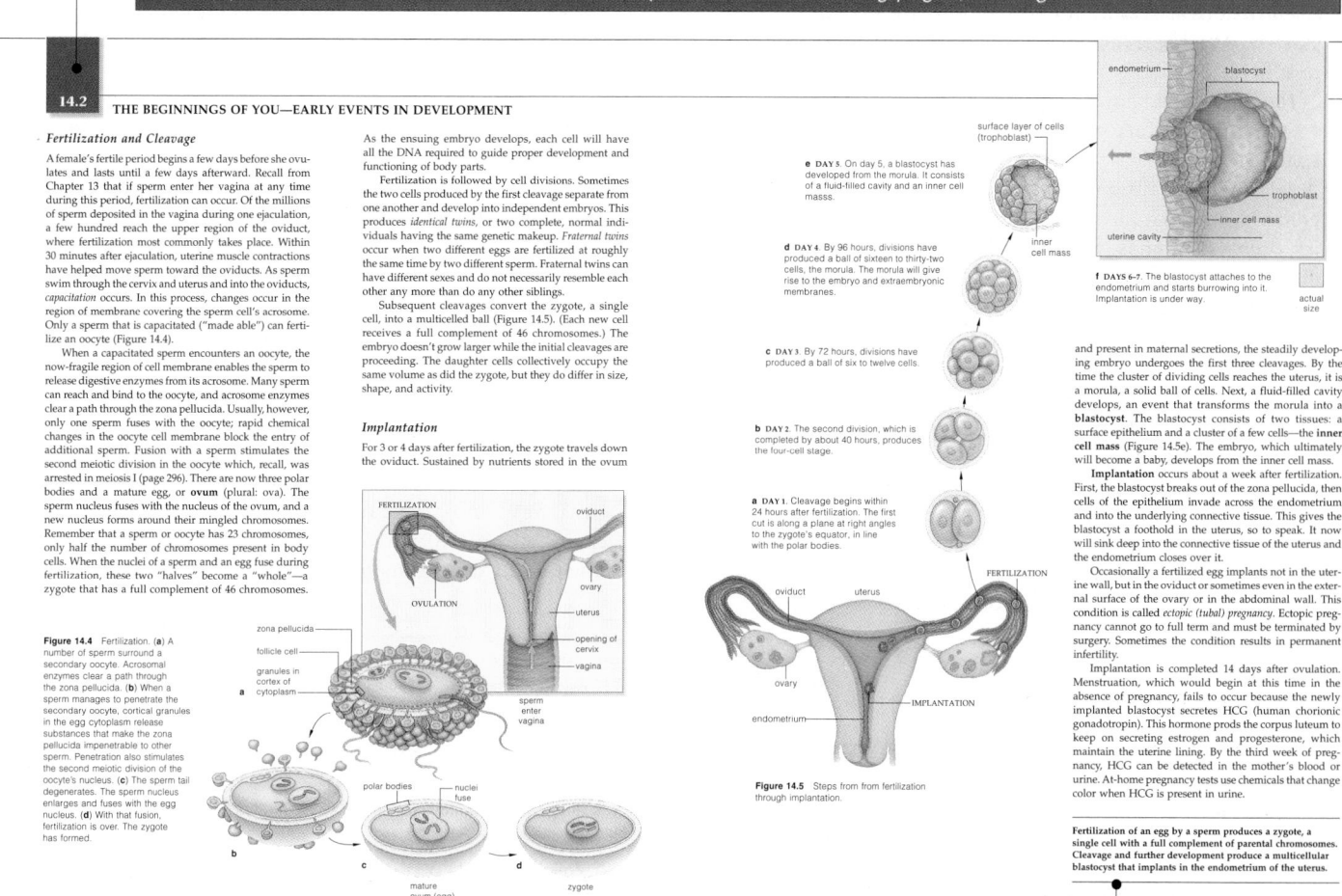

FIGURE A   *A concept spread from this edition.*

include many *Genetics Problems* that help students grasp the principles of inheritance.

VISUAL PREVIEWS OF MAJOR CONCEPTS

While writing the text, we simultaneously develop the illustrations as inseparable parts of the same story. This integrative approach appeals to students who are visual learners. When they can first work their way through a visual overview of some process, then reading through the corresponding text becomes less intimidating. Over the years, students have repeatedly thanked us for our many preview illustrations, which contain step-by-step, written descriptions of biological parts and processes. We break down the information into a series of illustrated steps that are much more inviting than complex, "wordless" diagrams. Figure B is a sample. Notice how simple descriptions, integrated within the art, walk students through the stages by which nutrient molecules are absorbed into the body, one step at a time.

Similarly, we continue to create visual previews for anatomical drawings. These previews integrate structure and function. Students need not jump back and forth from the text, to tables, and to illustrations in order to mentally construct an overview of how an organ system is put together and what its assorted parts do. Even individual descriptions are hierarchically arranged to reflect the organ system's structural and functional organization.

COLOR CODING

In diagrams, we consistently use the same colors for molecules and other structures. Visual consistency makes it easier for students to track complex parts and processes. For instance, amino acids and proteins are always shaded green, carbohydrates pink, lipids yellow and gold, DNA blue and RNA orange. Figure C is the color-coding chart.

New to this edition are icons that invite students to use multimedia. One icon directs them to the Interactive Concepts CD-ROM enclosed with each student copy, another to supplemental material on the Web, and a third to InfoTrac:

CD-ROM ICON: WEB ICON: INFOTRAC ICON:

ZOOM SEQUENCES

Where appropriate, selected illustrations progress from macroscopic to microscopic views of the same subject. For example, Figure D shows an overview illustration of skeletal muscle structure and function that begins with the biceps muscle in a human arm and culminates with the organiza-

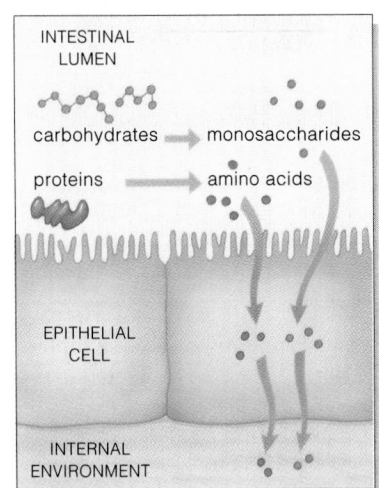

1 Digestion of carbohydrates to monosaccharides, and proteins to amino acids, by pancreatic enzymes.

2 Active transport of monosaccharides and amino acids across plasma membrane of cells; then nutrients move by facilitated diffusion into blood.

1 Emulsification. Wall movements break up fat globules into small droplets. Bile salts keep globules from re-forming. Pancreatic enzymes digest droplets to fatty acids and monoglycerides.

2 Formation of micelles as bile salts combine with digested products and phospholipids. Products readily slip into and out of the micelles.

3 Concentration of monosaccharides and fatty acids in micelles enhances gradients that lead to their diffusion across lipid bilayer of cells' plasma membranes.

4 Reassembly of products into triglycerides inside cells. These join with proteins, then are expelled (by exocytosis) into the internal environment.

Step-by-step art with simple descriptions helps students visualize a process before reading text about it.

FIGURE B *A visual summary from this edition.*

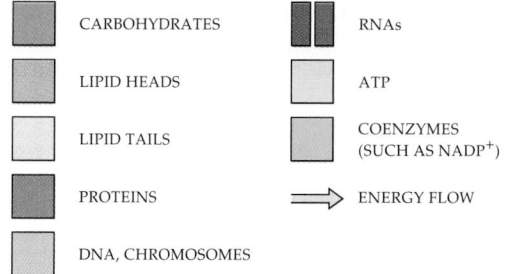

	CARBOHYDRATES		RNAs
	LIPID HEADS		ATP
	LIPID TAILS		COENZYMES (SUCH AS NADP$^+$)
	PROTEINS	⟹	ENERGY FLOW
	DNA, CHROMOSOMES		

FIGURE C *Color-coding chart for the diagrams of biological molecules and cell structures.*

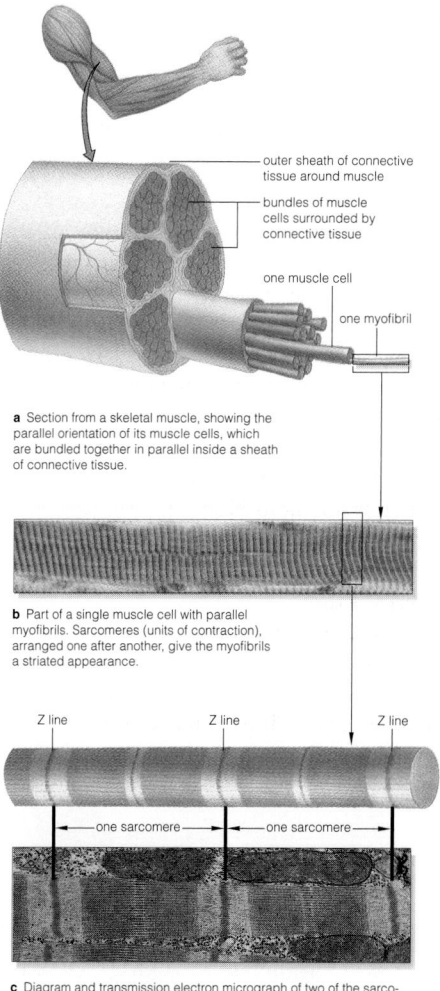

outer sheath of connective tissue around muscle

bundles of muscle cells surrounded by connective tissue

one muscle cell

one myofibril

a Section from a skeletal muscle, showing the parallel orientation of its muscle cells, which are bundled together in parallel inside a sheath of connective tissue.

b Part of a single muscle cell with parallel myofibrils. Sarcomeres (units of contraction), arranged one after another, give the myofibrils a striated appearance.

Z line   Z line   Z line

one sarcomere   one sarcomere

c Diagram and transmission electron micrograph of two of the sarcomeres from a myofibril. This closer view shows how dark "bands," called Z lines, define the two ends of each sarcomere. Mitochondria (the oval-shaped organelles) adjacent to the sarcomeres provide ATP energy for muscle action.

FIGURE D *Example of a zoom sequence.*

tion of a sarcomere, the basic unit of muscle contraction. Another example is the diagram of the organization of DNA in chromosomes in Chapter 19, on DNA structure and function.

END-OF-CHAPTER STUDY AIDS AND APPENDIXES

Figure E shows a sampling of our end-of-chapter study aids, which reinforce the key concepts. Each chapter ends with a summary in list form, review questions, a self-quiz, critical thinking questions, selected key terms, and a list of readings. Italicized page numbers tie the review questions and key terms to relevant text pages.

At the book's end, a glossary includes the boldfaced terms in the text, with pronunciation guides and word origins when such information can make formidable words less so. Our index is detailed enough to help students find a door to the text more quickly.

Students can use the periodic table of the elements in Appendix I for reference purposes. Appendix II includes metric–English conversion charts. Appendix III has detailed answers to genetics problems, and Appendix IV has answers to self-quizzes. Appendix V is a key containing suggestions for students who work the critical thinking questions, and here we have departed from texts that include full answers in the book itself. Instead of being temptingly available at the back of the book, suggested answers to these questions are provided to instructors. The in-text key suggests an approach to thinking through a plausible answer for each question. We hope it will stimulate students to analyze and pull together different facets of what they have learned, and so make their grasp of concepts that much stronger.

SIGNIFICANT CONTENT REVISIONS

Instructors who have been using the second edition of *Human Biology* may wish to evaluate the following overview of significant modifications in content. Overall, each chapter's *conceptual development* is more focused. A major objective for the revision was to make the writing crisp but not too brief, because certain frequently confusing topics become even more so when they are not treated in enough detail. Great thought went into the selective expansions and condensations of every chapter.

Also, we revised all chapters wherever new research findings demanded shifts in presentation. For example, the chapter on immunity (7) incorporates current models of immune responses and the roles of various types of lymphocytes; Chapter 15 presents the current state of affairs with respect to AIDS treatments and our understanding of the biology of nonprogressors. Users of previous editions found the discussions of infectious disease, drug abuse, endocrine disrupters, human genetic engineering, and other timely topics both relevant and useful, and we have updated them to incorporate advances in knowledge. New thinking on diet, obesity, and weight control has been added to Chapter 5.

SUMMARY

Control of Extracellular Fluid

1. Within the body, the cellular environment consists of certain types and amounts of substances dissolved in water. This extracellular fluid fills tissue spaces and (in the form of blood plasma) blood vessels. Its volume and composition are maintained only when the daily intake of water and solutes are in balance. The following processes maintain the balance:

   a. Water is gained by absorption (from the GI tract) and by metabolism. It is lost by urinary excretion, evaporation from the lungs and skin, sweating, and elimination in feces.

   b. Solutes are gained by absorption from the gut, secretion, respiration, and metabolism. They are lost by excretion, respiration, and sweating.

   c. Losses of water and solutes are controlled mainly by adjusting the volume and composition of urine.

2. The human urinary system consists of two kidneys, two ureters, a urinary bladder, and a urethra.

3. Blood is filtered and urine forms in kidney nephrons. Each nephron interacts closely with two sets of blood capillaries (glomerular and peritubular).

   a. A nephron has a cup-shaped beginning (glomerular capsule), then three tubelike regions (proximal tubule, loop of Henle, and distal tubule, which empties into a collecting duct).

   b. The glomerular (Bowman's) capsule surrounds a set of highly permeable capillaries. Together, they are a blood-filtering unit, the glomerulus.

   c. Blood pressure forces water and small solutes out of the capillaries, into the cup. Most of the filtrate is reabsorbed by the tubules and returned to the blood. A portion is excreted as urine.

4. Urine forms in the nephron by three processes:

   a. Filtration of blood at the glomerulus of a nephron, which puts water and small solutes into the nephron.

   b. Reabsorption. Water and solutes to be retained leave the nephron's tubular parts and enter the peritubular capillaries that thread around them. A small volume of water and solutes remains in the nephron.

   c. Secretion. Some ions and a few other substances leave the peritubular capillaries and enter the nephron, for disposal in urine.

5. Reabsorption of many solutes may occur passively, following concentration gradients. In other instances, active transport is required. Sodium reabsorption occurs by active transport. Water reabsorption is always passive, occurring along its osmotic gradient.

6. Urine becomes more or less concentrated by the action of two hormones, ADH and aldosterone. These act on cells of distal tubules and collecting ducts as follows:

   a. ADH is secreted when the body must conserve water; it enhances reabsorption from the distal nephron tubule and collecting ducts. Inhibition of ADH allows more water to be excreted.

   b. Aldosterone conserves sodium by enhancing its reabsorption in the distal tubule. Inhibition of aldosterone allows more sodium to be excreted. Because "water follows salt," aldosterone indirectly influences water reabsorption.

7. Together with the respiratory system and other mechanisms, the kidneys also help maintain the body's overall acid–base balance.

Control of Body Temperature

1. Body temperature is determined by the balance between metabolically produced heat and the heat absorbed from and lost to the environment.

2. Humans are endotherms. Their body temperature is controlled largely by metabolic activity and by precise controls over heat produced and heat lost.

Review Questions

1. Label the regions of the nephron, and identify where filtration, reabsorption, and secretion occur: 206

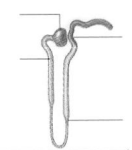

2. How does urine formation help maintain the body's internal environment? 202–203

3. Label the kidney's component parts: 204

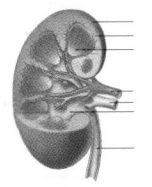

4. Which hormone influences water reabsorption and conservation? Sodium reabsorption and conservation? 210

5. Fatty tissue holds the kidneys in place. Extremely rapid weight loss may cause this tissue to shrink so that the kidneys slip from their normal position. On rare occasions, the slippage may put a kink in one or both ureters and block urine flow. Speculate on what might then happen to the kidneys. 203, 213

Critical Thinking: You Decide (Key in Appendix V)

1. A urinalysis reveals that the patient's urine contains glucose, hemoglobin, and white blood cells (pus). Are any of these substances abnormal in urine? Explain.

2. As a person ages, nephron tubules lose some of their ability to concentrate urine. What is the effect of this change?

3. What homeostatic mechanism is visibly operating in the photograph below? What are its physiological effects?

Self-Quiz (Answers in Appendix IV)

1. The body gains water by _____.
  a. gastrointestinal absorption   c. both a and b
  b. metabolism   d. neither a nor b

2. The body loses water by _____.
  a. evaporation from lungs and skin
  b. elimination in feces
  c. urinary excretion
  d. all of the above

3. Each human kidney contains about _____ nephrons.
  a. 1,000   c. 100,000
  b. 10,000   d. 1,000,000

4. Water and small solutes return to the blood during_____.
  a. filtration   c. secretion
  b. reabsorption   d. both a and b

5. Water and small solutes leave blood during _____.
  a. filtration   c. secretion
  b. reabsorption   d. both a and c

6. A few substances move out of the peritubular capillaries and are moved into the nephron during _____.
  a. filtration   c. secretion
  b. reabsorption   d. both a and c

7. A nephron's reabsorption mechanism depends on _____.
  a. osmosis across the nephron wall
  b. active transport of sodium across the nephron wall
  c. a steep solute concentration gradient starting at the kidney cortex and descending into the medulla
  d. all of the above

8. Water is conserved and urine becomes more concentrated by the action of _____.
  a. ADH   c. aldosterone
  b. renin   d. both a and c

9. Sodium is conserved and urine becomes more concentrated by the action of _____.
  a. ADH   c. aldosterone
  b. renin   d. both a and c

10. Match the following salt-water balance concepts:
  ____ aldosterone   a. blood filter of a nephron
  ____ nephron   b. controls sodium reabsorption
  ____ thirst mechanism   c. occurs at nephron tubules
  ____ reabsorption   d. site of urine formation
  ____ glomerulus   e. controls water gain

Selected Key Terms

ADH 210	peripheral vasodilation 214
aldosterone 210	peritubular capillary 205
core temperature 214	pilomotor response 214
distal tubule 205	proximal tubule 208
endotherm 214	reabsorption 206
extracellular fluid 202	renal corpuscle 205
filtration 206	secretion 206
glomerular (Bowman's) capsule 205	thirst center 211
glomerulus 205	ureter 204
juxtaglomerular apparatus 211	urethra 204
kidney 204	urinary bladder 204
loop of Henle 205	urinary excretion 202
nephron 204	urinary system 202
peripheral vasoconstriction 214	urine 206

Readings

Flieger, Ken. March 1990. "Kidney Disease: When Those Fabulous Filters Are Foiled." *FDA Consumer.* Excellent, nontechnical survey of kidney functions and treatments for kidney disease.

Sherwood, L. 1997. *Human Physiology.* 2d. ed. Belmont, California: Wadsworth.

Smith, H. 1961. *From Fish to Philosopher.* New York: Doubleday. Paperback. This entertaining, classic book traces the evolutionary path that produced the human kidney.

Vander, A., J. Sherman, and D. Luciano. 1994. "The Kidneys and Regulation of Water and Inorganic Ions," in *Human Physiology.* 6th ed. New York: McGraw-Hill, Chapter 15.

For additional readings, go to InfoTrac College Edition, your online library at:
http://www.infotrac-college.com/wadsworth

FIGURE E   *Example of end-of-chapter study aids.*

UNIT I. FOUNDATIONS Our advisors worked closely with us to refine and add clarity to the treatment of several complex topics in this unit. For instance, the sections on acids, bases and buffers, organic compounds, and the four major groups of biological molecules are rewritten, as is the *Science Comes to Life* essay on uses of radioisotopes.

UNIT II. BODY SYSTEMS AND FUNCTIONS The chapters in this core unit provide prime examples of how concept spreads can help move students step by step through individually complex yet interrelated topics. We rewrote throughout to enhance the flow of the biological story of body structures and their functions. Chapter 10, on the nervous system, is now a more integrated presentation and contains a wholly new section on the mechanisms by which memories are created and stored. And we reworked illustrations, such as those for the muscular system (4), the structure of the stomach and small intestine (5), the heart and cardiovascular system (6), and the respiratory cycle (8)— to give just a few examples.

UNIT III. PRINCIPLES OF INHERITANCE Rapid advances in genetics research demanded line-by-line evaluations for adjustments in these chapters. Careful rewriting also made the fascinating although potentially difficult sections— including cell reproduction, protein synthesis, and recombinant DNA technology—even more accessible to the introductory student. In Chapter 16 we devoted more space to explaining mitosis and the cell cycle—always tough concepts for beginning students. In Chapter 17, on chromosome variations and medical genetics, we included more student-friendly illustrations, such as the wonderful photograph of college students that illustrates continuous variation on page 382. We also rewrote the section on X-linked inheritance, adding new examples and improved illustrations. In Chapter 21 on genetic engineering, we added discussions of cloning, micro- and minisatellites, and the now-common method of blue-white screening.

UNIT IV. EVOLUTION AND ECOLOGY   Once again, we worked with leading evolutionists to update the evolution chapters. This included a comprehensive reworking of the chapter on human evolution. The comments of Paul Hertz helped us update Chapter 25, which examines the impacts of human populations on the biosphere.

SUPPLEMENTAL COURSE MATERIAL

MULTIMEDIA SUPPLEMENTS

1. *Interactive Concepts in Biology.* Packaged free with all student copies, this is the first CD-ROM to address the full sweep of biology. The cross-platform CD-ROM covers all concept spreads in the book, and has interactive exercises and quizzes. Because students can learn by doing, it encourages them to manipulate the book's art. Text, graphics, photographs, animations, video, and audio enhance each book chapter.

2. *InfoTrac College Edition.* This online library is available FREE with each copy of *Human Biology.* It gives students access to full articles—not abstracts—from more than 700 scholarly and popular periodicals dating back as far as four years. The 700,000 articles are available through Info-Trac's impressive database that has such periodicals as *Discover, Audubon,* and *Health,* and many pamphlets on health concerns such as AIDS, STDs, and diabetes.

3. *Student Guide to InfoTrac College Edition.* This guide is on the Brooks/Cole Biology Resource Center site on the World Wide Web. It has a critical thinking question and a set of electronic readings for each text chapter.

4. *Biology Resource Center.* All information is arranged by the third edition's chapters. Every month it has new BioUpdates on relevant applications, and hyperlinks and an average of 40 practice quiz questions per chapter. It includes a student feedback site, cool clip art, ideas for teaching on the Web, and a forum where instructors can share ideas on teaching courses. It also includes flashcards for all glossary terms, critical thinking exercises, news groups, a variety of search engines, a BioTutor, and a final Blitz set of practice questions. It also has Internet exercises for each chapter to guide students to doing more than randomly browse sites. A cool event of the quarter will have an ongoing experiment in which students and instructors can participate. The address for the Brooks/Cole Biology Resource Center is:

http://www.brookscole.com/biology

5. *Thomson World Class Course.* This software enables instructors to create their own Web site. Instructors can post course information, office hours, related Internet links, downloaded materials, assignments, and sample tests or quizzes. More information is available at:

http://www.worldclasslearning.com

6. *Internet Activities for General Biology.* This booklet includes online interactive dissections, surveys, genetic crosses, lab experiments, notice postings, and other diverse activities. It has tear-out worksheets that may be handed in for evaluation.

7. *An Introduction to the Internet.* This 100-page booklet helps students learn how to get around the Internet and lists useful biology sites on the Net that correspond to book chapters.

8. *BioLink 3.* With this presentation tool, instructors can easily assemble art and database files with lecture notes to create a fluid lecture that may help stimulate even the least-engaged students. It includes all illustrations in the book, animations and films from the student CD, and art from other Brooks/Cole biology textbooks. Biolink 3 also has a Kudo Browser with an easy drag-and-drop feature that allows file export into such presentation tools as PowerPoint. A file or lecture with BioLink 3 can be posted to the Web, where students can access it.

9. *Overheads.* All the micrographs and diagrams in the book are available as overheads that are reproduced in vivid color with large, bold-lettered labels. All of the diagrams are on CD-ROM, Biolink 3.

10. *Animations and Films from Cycles of Life Telecourse.* Tape One has animations for cell structure and function, and for principles of inheritance and evolution. Tape Two has more on diversity, plant structure and function, animal structure and function, and ecology and behavior.

11. *CNN Videos.* Produced by Turner Learning, nine videos can stimulate and engage students. They cover general biology, anatomy-physiology, and environmental science. New tapes are offered every year for these topics.

12. *Life Science Video Library.* Films for the Humanities and CNN created this library.

13. *Wadsworth Biology Videodisc.* This videodisc has line art with large, boldface labels and often step-by-step or full-motion animation. There are 3,500 still photographs, films, a correlation directory, and bar code guide.

14. *West's Biology Videodisc.* This videodisc provides thousands of additional images.

15. *STELLA II.* This modeling software has 23 simulations to develop critical thinking skills. The textbook's critical thinking questions and 150 additional ones are arranged by chapter in *Critical Thinking Exercises.*

16. *Electronic Study Guide.* This has an average of 40 multiple-choice questions per book chapter that differ from those in the test-item booklet. Students respond to each question, and then an on-screen prompt allows them to review their answers and learn why they are correct or incorrect.

17. *West Nutrition CD.* This interactive learning tool has animations, video, hands-on exercises, and a glossary with pronunciation guides. In-depth sections allow students to learn more about the biochemistry of particular topics.

ADDITIONAL SUPPLEMENTS

Seven respected test writers created the *Test Items* booklet The booklet has more than 3,000 questions in electronic form for IBM and Macintosh in a test-generating data manager. Questions also are available in Microsoft Word.

For each book chapter, an *Instructor's Resource Manual* has chapter objectives, a detailed lecture outline, ideas for lectures, classroom and lab demonstrations, discussion questions, research paper topics, and more. This resource manual is also available electronically on the Biology Resource Center.

An interactive *Study Guide and Workbook* lets students write answers to questions, which are arranged by chapter section with references to specific text pages. For those who wish to modify or select parts, *chapter objectives* are available on the Biology Resource Center.

A 100-page *Answer Booklet* answers the textbook's review questions. (The answers to the self-quizzes and genetics problems are given in appendixes to the book itself.)

Building Your Life Science Vocabulary helps students learn biological terms by explaining root words and their applications.

Study Skills for Science Students. Daniel Chiras's guide explains how to develop good study habits, sharpen memory, prepare for tests, and write term papers.

Strategies for Success: Learning Skills Booklets. Gardner/Jewler's best-selling college success text can be customized and bundled with the text. Topics include managing time, test taking, writing and speaking, and note taking.

Jim Perry and David Morton's *Laboratory Manual* has 20 experiments and exercises, with 300+ full-color labeled photographs and diagrams. Many experiments are divided in parts for individual assignment. Each consists of objectives, discussion (introduction, background, and relevance), list of materials, procedures, pre-lab questions, and post-lab questions. An *Instructor's Manual* for the lab manual lists quantities, procedure for preparing reagents, time requirements for each part of an exercise, hints to make the lab a success, and vendors of materials with item numbers. It has more investigative exercises that can be copied for lab use.

Customized Laboratory Manuals by Phillip Shelp and its accompanying instructor's manual can be tailored for individual courses. The *Photo Atlas for Biology* and *Photo Atlas for Anatomy and Physiology* each have 700 full-color, labeled photographs and micrographs of the cells and organisms that students typically deal with in the lab.

Additional readings supplements are available:
• *Contemporary Readings in Biology* is a collection of articles on applications of interest. • *Current Perspectives in Biology* is another collection of articles • *A Beginner's Guide to Scientific Method* is a supplement for those who wish to treat this topic in detail. • *The Game of Science* gives students a realistic view of what science is and what scientists do. • *Environment: Problems and Solutions* provides a brief 120-page introduction to environmental concerns.

A COMMUNITY EFFORT

Authors can write accurately and often well about their field of interest, but it takes more than this to deal with the full breadth of biology as it relates to the human body. For us, it takes an educational network of more than 2,000 teachers, researchers, and photographers. We list those reviewers whose recent contributions continue to shape our thinking. There simply is no way to describe the thoughtful effort that these individuals and other reviewers before them gave to our books.

G. Tyler Miller graciously contributed three of the *Focus on Our Environment* essays, and he and Steve Wolfe were the guardians of accuracy and teachability in the chapters on chemistry and cell biology.

Gary Head designed the book. Jeff Lachina and his staff at Lachina Publishing Services, Inc., diligently shepherded the book from its manuscript stage through production. Jeff, in particular, spent countless hours ensuring that this project happened when it needed to. We thank him for his warmth and unstinting professionalism. Our Brooks/Cole editor, Kristin Milotich, once again demonstrated that she is an editorial professional with endurance, a big heart, and a sense of humor.

This third edition of *Human Biology* also is much better for the careful copyediting of Nancy Tenney. Lachina Publishing Services organized, analyzed, and finally put together the remarkable illustration program.

We are grateful to all those people whose dedication and competence helped bring this edition to fruition. One of us, Bev McMillan, once again expresses her profound appreciation to biologist *extraordinaire* Jack Musick for his blend of scientific acumen, common sense, and marital support. Finally, immense credit goes to our publisher, Jack Carey. Jack consistently reminded us of our vision of what this new edition could achieve as an educational tool. Jack gives the most, the best, all the time. No authors could ask for more.

REVIEWERS

ALCOCK, JOHN, *Arizona State University*
ALFORD, DONALD K., *Metropolitan State College of Denver*
ANDERSON, D. ANDY, *Utah State University*
ARMSTRONG, PETER, *University of California, Davis*
AULEB, LEIGH, *San Francisco State University*
BAKKEN, AIMÉE H., *University of Washington, Seattle*
BARNUM, SUSAN R., *Miami University*
BEDECARRAX, EDMUND E., *City College of San Francisco*
BENNETT, JACK, *Northern Illinois University*
BOHR, DAVID F., *University of Michigan, Ann Arbor*
BOOTH, CHARLES E., *Eastern Connecticut State University*
BRAMMER, J.D., *North Dakota State University*
BROADWATER, SHARON T., *College of William and Mary*
BROWN, MELVIN K., *Erie Community College–City Campus*
BURKS, DOUGLAS J., *Wilmington College*
CHRISTENSEN, A. KENT, *University of Michigan*
CHRISTENSEN, ANN, *Pima Community College*
CHOW, VICTOR, *City College of San Francisco*
CONNELL, JOE, *Leeward Community College*
COX, GEORGE W., *San Diego State University*
COYNE, JERRY, *University of Chicago*
CROSBY, RICHARD M., *Treasure Valley Community College*
DANKO, LISA, *Mercyhurst College*
DELCOMYN, FRED, *University of Illinois at Urbana-Champaign*
DENNER, MELVIN, *University of Southern Indiana*
DENNISTON, KATHERINE J., *Towson State University*
DLUZEN, DEAN, *Northeastern Ohio University's College of Medicine*
EDLIN, GORDON J., *University of Hawaii*
FAIRBANKS, DANIEL J., *Brigham Young University*
FALK, RICHARD H., *University of California, Davis*
FREY, JOHN E., *Mankato State University*
FROEHLICH, JEFFREY, *University of New Mexico*
FULFORD, DAVID E., *Edinboro University of Pennsylvania*
GENUTH, SAUL M., *Case Western Reserve University*
GIANFERRARI, EDMUND A., *Keene State College of the University of New Hampshire*
GOODMAN, H. MAURICE, *University of Massachusetts Medical School*
GORDON, SHELDON R., *Oakland University*
HAHN, MARTIN, *William Paterson College*
HALL, N. GAIL, *Trinity College*
HASSAN, ASLAM S., *University of Illinois*
HENRY, MICHAEL, *Santa Rosa Junior College*
HERTZ, PAUL E., *Barnard College*
HILLE, MERRILL B., *University of Washington*
HOHAM, RONALD W., *Colgate University*
HOSICK, HOWARD L., *Washington State University*
HUCCABY, PERRY, *Elizabethtown Community College*
HUPP, EUGENE W., *Texas Woman's University*
JOHNS, MITRICK A., *Northern Illinois University*
JOHNSON, LEONARD R., *University of Tennessee College of Medicine*
JOHNSON, VINCENT A., *St. Cloud State University*
JONES, CAROLYN K., *Vincennes University*
KAREIVA, PETER, *University of Washington*
KAYE, GORDON I., *Albany Medical College*
KENYON, DEAN H., *San Francisco State University*
KEYES, JACK, *Linfield College, Portland Campus*
KLEIN, KEITH K., *Mankato State University*
KENNEDY, KENNETH A.R., *Cornell University*
KIMBALL, JOHN W.
KROHNE, DAVID T., *Wabash College*
KRUMHARDT, BARBARA, *Des Moines Area Community College, Urban Campus*
KUPCHELLA, CHARLES E., *Southeast Missouri State University*
KUTCHAI, HOWARD, *University of Virginia*
LAMMERT, JOHN M., *Gustavus Adolphus College*

LAPEN, ROBERT, *Central Washington University*
LASSITER, WILLIAM E., *University of North Carolina–Chapel Hill School of Medicine*
LEVY, MATTHEW N., *Mt. Sinai Hospital*
LITLE, ROBERT C., *Medical College of Georgia*
LUCAS, CRAN, *Louisiana State University–Shreveport*
MANN, ALAN, *University of Pennsylvania*
MANN, NANCY J., *Cuesta College*
MATHIS, JAMES N., *West Georgia College*
MATTHEWS, PATRICIA, *Grand Valley State University*
MAYS, CHARLES, *DePauw University*
McCUE, JOHN F., *St. Cloud State University*
McMAHON, KAREN A., *University of Tulsa*
McNABB, F.M. ANNE, *Virginia Polytechnic Institute and State University*
MILLER, G. TYLER
MITCHELL, JOHN L.A., *Northern Illinois University*
MITCHELL, ROBERT B., *Pennsylvania State University*
MOHRMAN, DAVID E., *University of Minnesota, Duluth*
MOISES, HYLAN, *University of Michigan*
MORK, DAVID, *St. Cloud State University*
MORTON, DAVID, *Frostburg State University*
MOTE, MICHAEL I., *Temple University*
MOWBRAY, ROD, *University of Wisconsin–LaCrosse*
MURPHY, RICHARD A., *University of Virginia Health Sciences Center*
MYKLES, DONALD L., *Colorado State University*
NORRIS, DAVID O., *University of Colorado at Boulder*
PARSON, WILLIAM, *University of Washington*
PETERS, LEWIS, *Northern Michigan University*
ROBERTS, JANE C., *Creighton University*
SCALA, ANDREW M., *Dutchess Community College*
SAPP OLSON, SALLY
SHEPHERD, GORDON M., *Yale University School of Medicine*
SHERMAN, JOHN W., *Erie Community College–North*
SHERWOOD, LAURALEE, *West Virginia University*
SHIPPEE, RICHARD H., *Vincennes University*
SLOBODA, ROGER D., *Dartmouth College*
SMIGEL, BARBARA W., *Community College of Southern Nevada*
SMITH, ROBERT L., *West Virginia University*
SOROCHIN, RON, *State University of New York; College of Technology at Alfred*
STEELE, CRAIG W., *Edinboro University of Pennsylvania*
STEUBING, PATRICIA M., *University of Nevada, Las Vegas*
STEWART, GREGORY J., *West Georgia College*
STONE, ANALEE G., *Tunxis Community College*
SULLIVAN, ROBERT J., *Marist College*
TAUCK, DAVID, *Santa Clara University*
TIZARD, IAN, *Texas A&M University*
TROTTER, WILLIAM, *Des Moines Area Community College Sciences Center*
TUTTLE, JEREMY B., *University of Virginia Health Sciences Center*
VALENTINE, JAMES W., *University of California, Berkeley*
VAN DE GRAAFF, KENT M., *Weber State University*
VAN DYKE, PETE, *Walla Walla Community College*
VARKEY, ALEXANDER, *Liberty University*
WALSH, BRUCE, *University of Arizona*
WARNER, MARGARET R., *Purdue University*
WEISBROT, DAVID R., *William Paterson College*
WEISS, MARK L., *Wayne State University*
WHIPP, BRIAN J., *St. George's Hospital Medical School*
WILLIAMS, ROBERTA B., *University of Nevada, Las Vegas*
WISE, MARY, *Northern Virginia Community College*
WOLFE, STEPHEN L., *University of California, Davis (Emeritus)*
YONENAKA, SHANNA, *San Francisco State University*

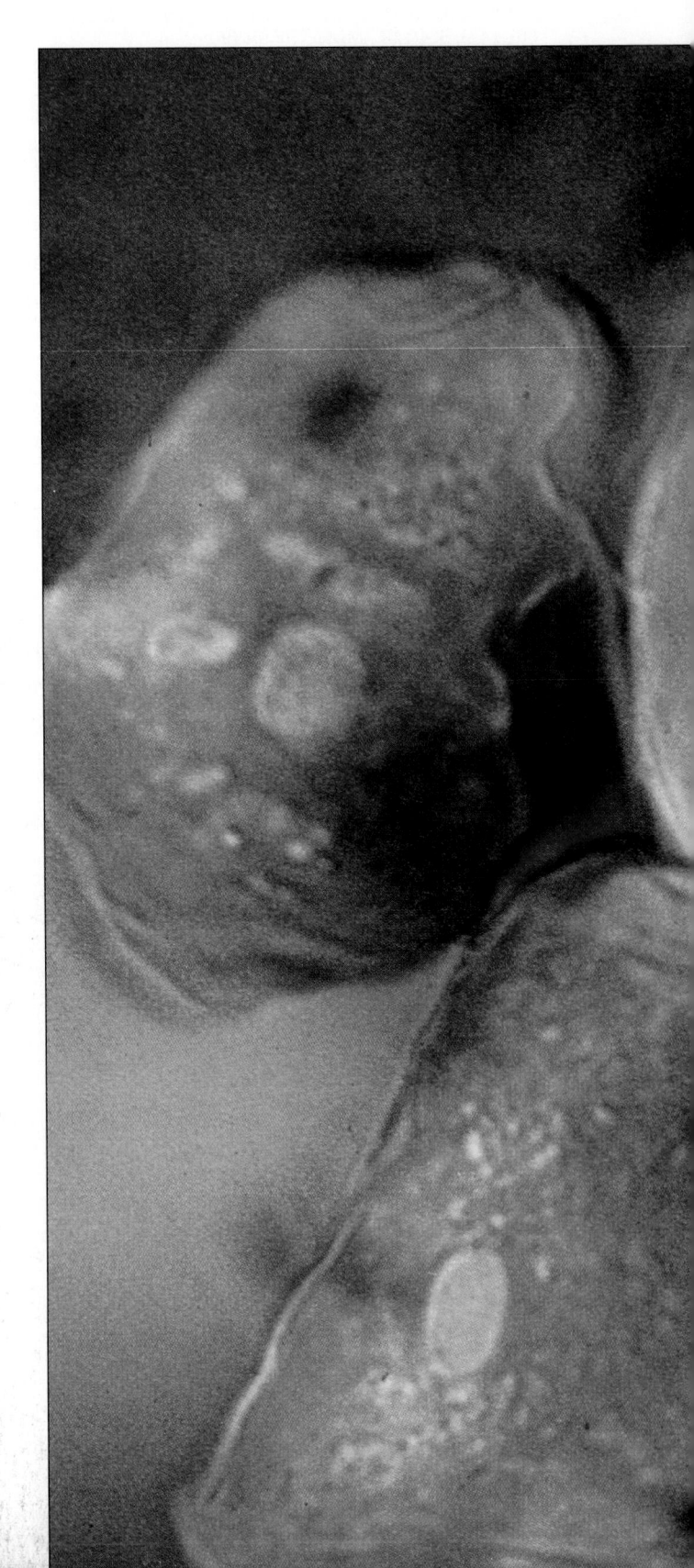

Human cheek cells magnified 200 times. Each cell is a tiny but highly ordered unit, the fundamental unit of life.

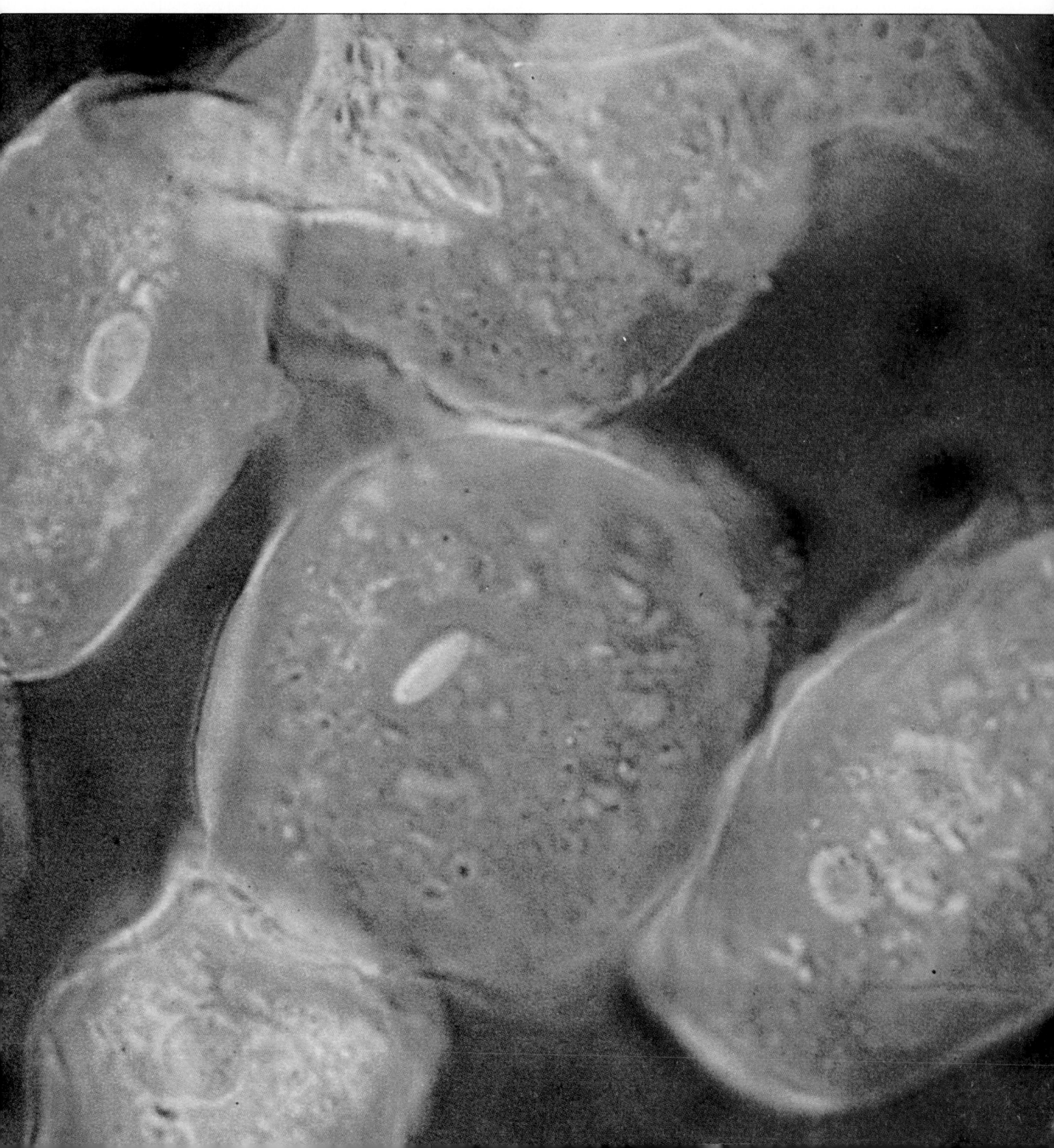

Introduction CONCEPTS IN HUMAN BIOLOGY

Human Biology Revisited

Buried somewhere in your brain are memories of discovering your own hands and feet, your family, the change of seasons, the smell of grass. The memories include early introductions to a great, disorganized parade of insects, flowers, frogs, and furred things—most of them living, some of them dead. There are memories of questions: "What is life?" and "Where do I fit in the world around me?" There are memories of answers, some satisfying, others less so.

By observing, asking questions, and accumulating answers, you have built up a store of knowledge about the world of life (Figure I.1). Experience and education have been refining your questions, and no doubt some answers are difficult to come by. Think of a young man whose brain is functionally dead as a result of a motor-cycle accident. If breathing and other basic functions proceed only as long as he remains hooked up to mechanical support systems, is he "alive"? Think of an embryo, a cluster of cells growing inside a pregnant woman. At what point in its development is it a definably "human" life? If questions like these cross your mind, your thoughts about life obviously run deep.

The point is, this book isn't your introduction to human biology, for you have been studying yourself

and the world around you ever since information began penetrating your brain. This book is simply human biology *revisited* in ways that may help carry your thoughts to more organized levels.

To biologists, the question "What is life?" opens up a story that has been unfolding for several billion years. "Life" is an outcome of ancient events by which nonliving materials became assembled into the first living cells. With the emergence of living cells came biological **evolution**—change in details of the body plan and functions of organisms through successive generations. In the course of evolution, broad groups of life forms, including animals, emerged.

Humans, apes, and some other closely related animal species are primates. As primates we are also mammals, and all mammals, including humans, are *vertebrates*—animals with "backbones." We share our planet with millions of other animal species, as well as with plants, fungi, bacteria, and other organisms. Figure I.2 provides a general picture of the place our human species, *Homo sapiens*, occupies in the living world.

"Life" is also a way of capturing and using energy and materials. "Life" is a way of sensing and respond-ing to specific changes in the environment. "Life" is a capacity to grow, develop, and reproduce. In fact, much of this text explores aspects of anatomy and functioning through which your body maintains the living state. A partial list of those life-support systems includes your brain and an elaborate network of nerve cells, a diges-tive system that can extract nutrients from thousands of foods, respiratory and circulatory systems that keep your body supplied with oxygen, a urinary system, an immune system, a hormone-based communication system, a skeleton and muscles for support and move-ment, and organs for reproduction.

This chapter introduces some basic concepts that provide a foundation for understanding the material to come. Here, too, we introduce several themes that are cornerstones of the study of human biology. One theme is our evolutionary heritage, which includes cultural as well as biological evolution. Another is the state of internal constancy called *homeostasis*. Living cells can function properly only within narrow environmental limits, and each of the life-support systems we will study in succeeding chapters makes an essential contribution to maintaining a constant internal environment. Finally, throughout the text you will also find boxed features that relate basic biological concepts to health topics, environmental concerns, and social issues.

Figure I.1 Think back on all you have known and seen; this is the foundation for your deeper study of human biology.

1. Humans exhibit the basic characteristics of life. They
consist of substances put together according to the same
laws that govern matter and energy. Humans obtain
and use energy and materials from their environment.
They sense and respond to changing conditions in their
environment. And they grow and reproduce, based on
instructions contained in DNA.

2. Life processes depend on a stable internal state called
homeostasis.

3. As with other forms of life, the structures of the
human body, and their functions, have been shaped by
evolution.

4. Biology, like other branches of science, is based on
systematic observations, hypotheses, predictions, and
relentless testing. The external world, not internal
conviction, is the testing ground for scientific theories.

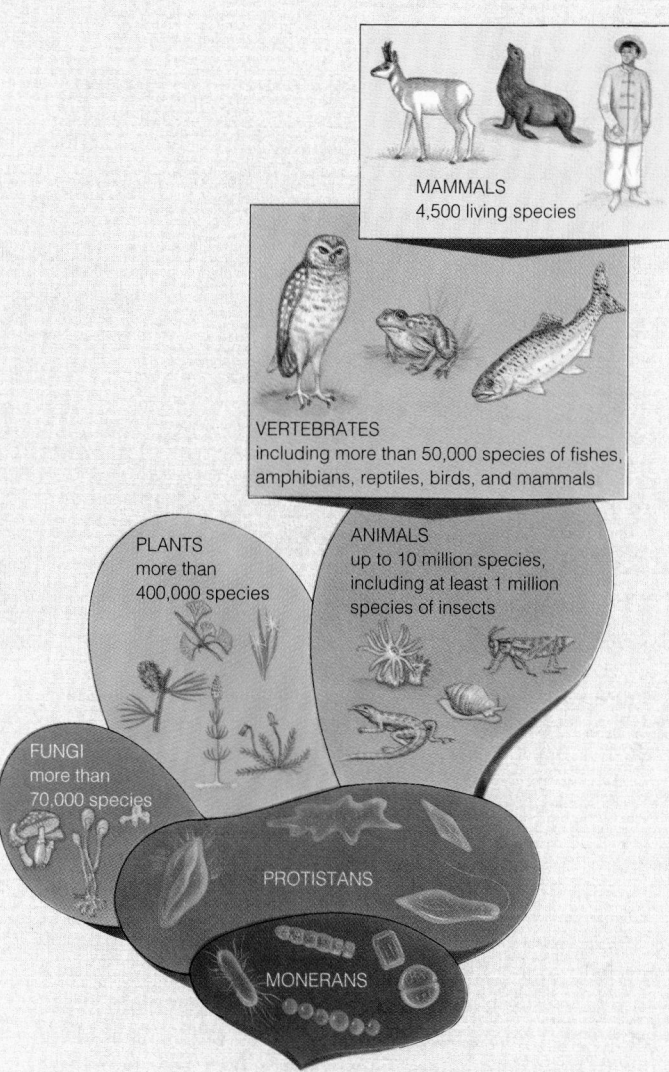

CHAPTER AT A GLANCE

I.1   Cells: The Basic Units of Life

I.2   More Life Characteristics: Reproduction and
      Inheritance

I.3   Organization and Interdependency in the
      Living World

I.4   Scientific Methods

I.5   Case Study: Using a Scientific Method in Cancer
      Research

I.6   *How Do You Define Death?*

MAMMALS
4,500 living species

VERTEBRATES
including more than 50,000 species of fishes,
amphibians, reptiles, birds, and mammals

PLANTS
more than
400,000 species

ANIMALS
up to 10 million species,
including at least 1 million
species of insects

FUNGI
more than
70,000 species

PROTISTANS

MONERANS

Figure I.2   KINGDOMS OF LIFE

Human beings are only one species among millions
inhabiting the Earth. Among the organisms with which
we share our world are at least 1 million species of
insects, more than 400,000 species of plants, and
vast numbers of single-celled organisms, including
amoebas (Protistans) and bacteria (Monerans).

CELLS: THE BASIC UNITS OF LIFE

Picture a climber on a rock face, inching cautiously toward the top (Figure I.3). Without even thinking about it, you know the climber is alive and the rock is not. Yet the climber, the rock, and everything else in the universe are composed of the same fundamental components—atoms and their parts—about which you will read much more in Chapter 1. These components are all held together in predictable patterns, according to basic physical laws. At the heart of those laws is **energy**, which is most simply defined as a capacity to do work. Energy-based interactions hold the components of a rock together—and they hold a human together.

What sets living things apart from the nonliving world? An important part of the answer is a remarkable molecule called **DNA**—short for deoxyribonucleic acid. No chunk of granite or quartz has it. DNA contains the instructions for building and operating each human being from carbon, hydrogen, and a few other kinds of "life-less" substances. A simple analogy is to think of DNA as being like the instructions you follow to turn a disordered heap of ceramic tiles—even just two kinds of tile—into ordered patterns such as these:

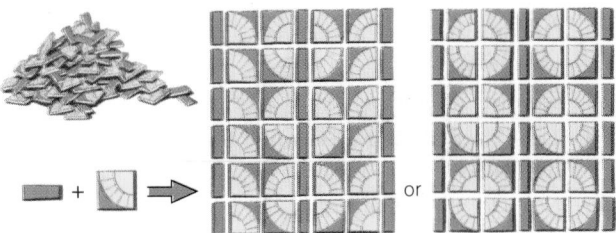

Figure I.3 A climber cautiously chooses her route up a vertical rock face. Although she is alive and the rock is not, both are composed of the same fundamental components.

From Non-Life to Life

Life requires DNA's "directions," certain raw materials, and an input of energy. The **cell** is the smallest unit of matter having the capacity for life. A cell is an organized unit that can survive and reproduce by itself, given DNA instructions and the necessary energy and materials. A nonliving object can never, ever meet these demanding specifications.

Every cell is a tiny wonder, with some fairly amazing attributes that you will only see in living things. One of those attributes is called metabolism, and another is the ability to maintain a state of dynamic equilibrium called *homeostasis*. The rest of this section will introduce you to these very special cell abilities, which receive closer scrutiny in later chapters.

Metabolism

Another basic characteristic of life is **metabolism**—the capacity of a cell to extract energy from its surroundings and to convert and use that energy to maintain itself, grow, and make more cells. Simply put, metabolism means energy transfers. Those energy transfers, which are accomplished by an energy carrier called ATP, power the building of cell parts and specialized cellular "work" such as the contraction of a muscle cell. Energy transfers also make possible other biochemical reactions that are vital to survival. The cell even stores some energy. Like other living things, humans take in energy for metabolism from the external world. This is one reason why a person such as the child in Figure I.4 must consume food. You will learn much more about how the body extracts energy from food in Chapter 5.

Figure I.4 This child is taking in energy from food. Food energy is transferred to ATP, which in turn makes that energy available for metabolic functions required to maintain cells in the living state.

Homeostasis

Cells of your body can sense changes in the environment and make controlled responses. They manage this vital monitoring task with the help of **receptors**, which are molecules and structures that can detect specific information about the environment. When cells receive signals from receptors, they adjust their activities in ways that bring about an appropriate response.

For example, how fast your heart beats is controlled by different nerves that either stimulate a more rapid beat or slow the heart rate down—all depending on some "external" factor, such as whether you are resting or exercising. Similarly, your body can withstand only so much heat or cold, must rid itself of wastes and harmful substances, and requires a regular supply of specific nutrients. Yet temperature shifts, wastes, and harmful substances are unavoidable, and there may be times when you gorge on one type of food or skip a meal entirely.

Suppose you skip breakfast, then lunch, and the level of the sugar glucose in your blood—"blood sugar"—falls. Glucose is the body's main fuel molecule. A hormone then signals cells in your liver to dig into their stores of energy-rich molecules. Those complex molecules (glycogen) are broken down to the sugar glucose, which is released into the bloodstream, and your blood-sugar level returns to normal. This information feeds back to the hormone-secreting cells. Their activity then slows, and your liver stops breaking down glycogen. After you eat, glucose enters your bloodstream, blood sugar rises, and cells of the pancreas increase their secretion of the hormone insulin. Most body cells have receptors for insulin, which prompts them to take up glucose—and your blood-sugar level again returns to normal.

Blood is part of the "internal environment," the fluid environment bathing your cells. Usually, the internal environment of your body is kept fairly constant. When the internal environment is maintained within tolerable ranges, a state called **homeostasis** exists. Homeostasis means "staying the same," and with it healthy cells can survive; without it, they die. And the body systems that maintain homeostasis are built of cells. This fundamental, interdependent relationship can be summarized thus:

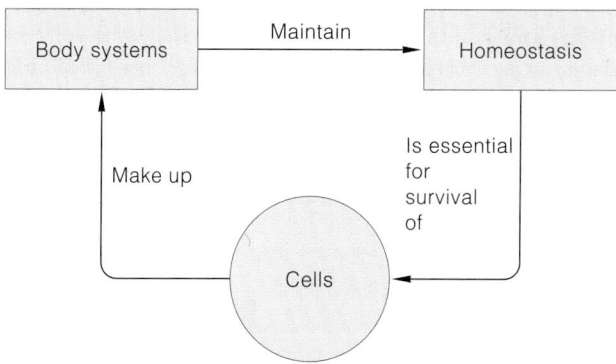

In Chapter 3 we'll take up the topic of homeostasis once more, giving special attention to the two types of feedback mechanisms by which it operates.

The structure and organization of living things begin with instructions contained in DNA molecules. Life emerges at the level of the cell.

Only living things show metabolic activity. Cells take in and use energy to stockpile, tear down, and dispose of materials in ways that promote survival and reproduction.

Cell responses to changes in the environment help maintain the stable internal state called homeostasis.

When homeostasis is maintained, operating conditions inside the cell or the whole body are favorable for survival.

We humans tend to think we enter the world rather abruptly and are destined to leave it the same way. Yet we and all other organisms are part of an immense, on-going journey that began billions of years ago. Think of the first cell produced when a human egg and sperm join. The cell would not even exist if the sperm and egg had not formed earlier according to DNA instructions that were passed down through countless generations. With the time-tested instructions in DNA, a new human body develops in ways that will prepare it, ultimately, for helping to produce individuals of a new generation. With **reproduction**—the production of offspring by parents—the journey of life continues.

What Do We Mean by Inheritance?

Reproduction involves **inheritance**. This word means that parents transmit to their offspring instructions for duplicating their body form and other traits. Molecules of DNA contain the required instructions.

DNA instructions assure that offspring will resemble their parents. The instructions also permit *variations* in the details of traits (Figure I.5). For example, although having five fingers on each hand is a human trait, some people are born with six fingers on each hand. How do such variations come about? In the final analysis, variations in traits arise through **mutations**, which are heritable changes in the structure or number of DNA molecules.

Most mutations are harmful, for the separate bits of information in DNA are part of a coordinated whole. For example, a single mutation in human DNA may lead to hemophilia, a genetic disorder in which blood cannot clot properly after the body is cut or bruised (Chapter 18).

Adaptive Traits

Even though mutations usually cause problems, some mutations may prove harmless, or even beneficial, under prevailing conditions. For example, the disease sickle-cell anemia is caused by a DNA change that results in a defective form of the blood protein hemoglobin. The sickle-cell trait is most prevalent in parts of the world where malaria is common. People who inherit the mutation from both parents suffer the debilitating disease. But, for complex reasons, people who inherit the mutation from only one parent are resistant to malaria. For them, the mutation is adaptive; it increases their chances of survival.

An **adaptive trait** helps an organism survive and reproduce under a given set of environmental conditions. In the long course of human evolution, countless DNA mutations, tested in the environments of our ancestors, have given rise to an elaborate nervous system, efficient mechanisms for taking in and distributing oxygen and food molecules, and other characteristics that enable each of us to live the biologically complex life of a human being. Later in the book we will consider the mechanisms by which evolution occurs.

DNA is the molecule of inheritance. Its instructions for reproducing traits are passed on from parents to offspring.

Mutations introduce variations in heritable traits.

Figure I.5 Reproduction is a life characteristic. Instructions in DNA assure that offspring will resemble their parents—and they also permit variations in the details of traits, as the photograph demonstrates.

ORGANIZATION AND INTERDEPENDENCY IN THE LIVING WORLD

Take a look at Figure I.6, which summarizes the levels of organization in nature. Beyond the levels we have already discussed—nonliving, subcellular components and living cells—Figure I.6 also shows several more inclusive levels of organization, from populations (such as the world's human population) on through whole ecosystems and the biosphere. **Biosphere** refers to all regions of the Earth's waters, crust, and atmosphere in which organisms live. We return to all of these concepts in Chapters 24 and 25, which discuss principles of ecology and the many human impacts on ecosystems.

Organisms Are Interdependent

In general, a flow of energy from the sun maintains the pattern of organization in nature. Plants and some other organisms that capture solar energy are the entry point

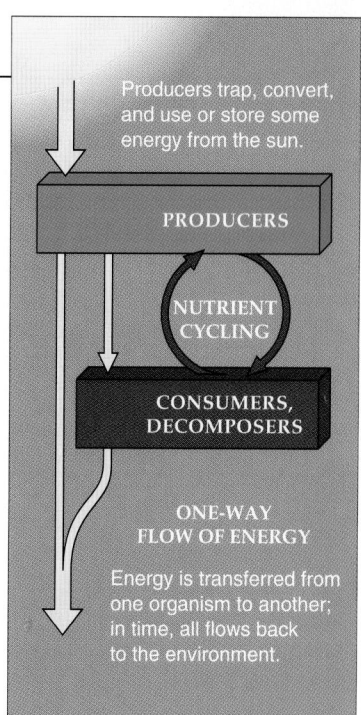

Figure I.7 The flow of energy and the cycling of materials in the biosphere.

for this energy. They are food producers for the living world. Animals, including humans, are consumers; directly or indirectly, they feed on energy stored in plant parts. Thus you tap directly into this stored energy when you eat a banana, and you tap into it indirectly when you eat hamburger made from a steer that fed on grass or grain. Bacteria and fungi are decomposers. When they feed on tissues or the remains of other organisms, they break down biological molecules to simple raw materials, which can be recycled back to producers. In time, all the energy that the producers initially captured from the sun's rays returns to the environment (as heat). For the moment, the simple, powerful message to remember is: Every part of the living world, including each of us, is linked to every other part.

For now, you only need to keep in mind that living things connect with one another by a one-way flow of energy through them and a cycling of materials among them, as Figure I.7 diagrams. The interconnectedness of organisms on our planet affects the structure, size, and composition of the Earth's populations and communities. It affects ecosystems, even the whole biosphere. Once you understand the extent of these interactions and how they are affected by the activities of roughly 6 billion human beings, you will gain insight into the thinning of the ozone layer, global warming, acid rain, and many other modern-day problems.

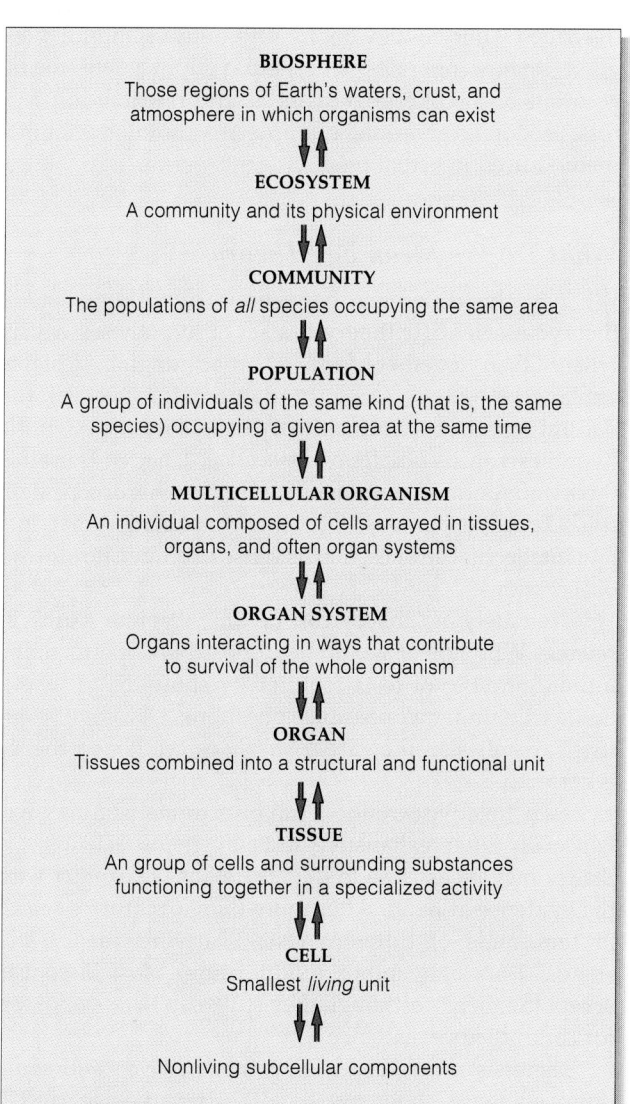

Figure I.6 An overview of levels of organization in nature.

Levels of organization in the living world begin with subatomic particles, atoms, and molecules. Cells, multicellular organisms, and whole ecosystems are part of this continuum of increasing complexity.

All organisms, including humans, are part of webs of organization in nature, in that they depend directly or indirectly on one another for energy and raw materials.

Biology, like science generally, is an ongoing record of discoveries arising from methodical inquiries into the natural world. Human biology focuses, naturally enough, on the workings of the human body and closely related topics.

Our fascination with ourselves is probably as ancient as our beginnings. Great thinkers such as Hippocrates, Leonardo da Vinci, and Charles Darwin pursued their curiosity under the umbrella of "natural history." Modern biologists are just as curious, if not more so, but they now investigate complex topics ranging from the molecular structure of HIV, the virus responsible for AIDS, to the impact on human health of a hole in the stratospheric ozone. In fact, the range of possible specialization within "human biology" is so broad that no single "scientific method" can be used to study all the relevant topics and issues.

Even so, scientists everywhere still have practices in common. *Scientists ask questions, make educated guesses about possible answers, and then devise ways to test their predictions, which will hold true if their guesses are good ones.* The following list describes the steps scientists generally follow when they proceed with an investigation:

1. Ask a question or identify a problem.

2. Develop a **hypothesis**, a testable idea or guess, about what the answer (or solution) might be. This might involve sorting through what has been learned already about related phenomena.

3. With a hypothesis as a guide, make a **prediction**— that is, a statement of what you should be able to observe if you were to go looking for it. This is often called the "if-then" process. (*If* something in cigarette smoke is a cancer-causing agent in human lungs, *then* we should be able to detect a higher rate of lung cancer among smokers than among nonsmokers.)

4. Devise ways to test the accuracy of predictions. You might do this by making observations, developing models, and doing experiments. By definition, an **experiment** is a test in which some phenomenon in the natural world is manipulated in controlled ways to gain insight into its function, structure, operation, or behavior. An essential step in any experiment is establishing a **control group**. Control groups are used to evaluate possible side effects of a test being performed on an experimental group. If an experiment involves laboratory rats, then the control group will be rats; if it involves college students, the controls will be students, and so on. Ideally, members of a control group should be identical to those of an experimental group in every respect— except for the key factor, or **variable**, under study. Both groups also must be large enough so the results won't be due to chance alone. Generally, experiments are

devised to disprove a hypothesis, not to prove it. Why? Because it would be impossible to prove beyond a shadow of a doubt that a hypothesis is correct, for it would take an infinite number of experiments to demonstrate that it holds under all possible conditions.

5. If the test results are not as expected, check for what may have gone wrong. A procedure may have been performed improperly, something may have been overlooked, or the hypothesis may not be a good one.

6. Repeat or devise new tests—the more the better. Hypotheses supported by many different tests are more likely to be correct.

7. Objectively report the test results and conclusions drawn from them.

In broad outline, a scientific approach to studying nature is that simple. Figure I.8 diagrams the steps. You can use this approach to pick your way logically through environmental, medical, and social issues of the sort described later in the book. And understanding how good science operates will help you evaluate media accounts of discoveries, advances, and research in progress. Section I.5 gives one example of a common scientific method used in actual research.

What Do We Mean by "Theory"?

What is the difference between a hypothesis and a theory? A **scientific theory** is a set of hypotheses which, when taken together, form a broad-ranging, testable explanation about some fundamental aspect of the natural world. A scientific theory differs from a scientific hypothesis in its *breadth of application*. Charles Darwin's theory about the evolution of species fits this description: It is a broad, encompassing "Aha!" explanation that, in a few intellectual strokes, makes sense of a huge number of observations.

Ultimately, no theory can be an "absolute truth" in science. Why not? It would be impossible to perform the infinite number of tests required to show that a theory holds true under *all* possible conditions. Objective scientists say only that they are *relatively* certain that a theory is correct.

Such "relative certainty" can be extremely impressive. Especially after exhaustive tests by many scientists, a theory may be as close to the "truth" as we can get with the evidence at hand. After more than a century's worth of thousands of different tests, Darwin's theory still stands, with only minor modification. Most biologists accept the theory, although they still keep their eyes open for contradictory evidence.

Scientists must keep asking themselves, "Will some other evidence show my hypothesis to be incorrect?"

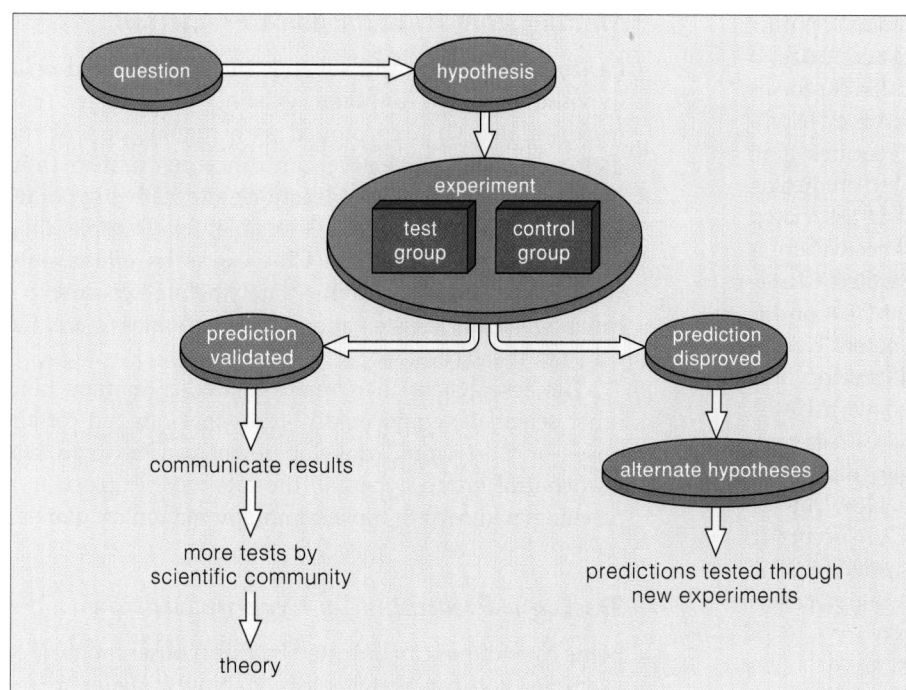

Figure I.8 Example of a scientific method.

They are expected to put aside pride or bias by testing their hypothesis. If an individual scientist doesn't (or won't) do this, *others will*—for science is conducted in a community that is both cooperative and competitive. Ideas are shared and examined with the understanding that it is just as important to expose errors as it is to applaud insights. Scientists can change their minds when presented with new evidence—and this is a strength of science, not a weakness.

The Limits of Science

The call for objective testing strengthens the theories that emerge from scientific studies. Yet it also puts limits on the kinds of studies scientists can carry out. Beyond the realm of scientific analysis, some events remain unexplained. Why do we exist, for what purpose? Why does any one of us have to die at a particular moment and not another? Should one person aid another incurably ill person in the act of suicide? Answers to such questions are subjective. This means they come from within us, as an outcome of all the experiences and mental connections that shape our consciousness. Because people differ so enormously in this regard, subjective answers do not readily lend themselves to scientific analysis. The *Choices* essay on page 11 considers some wrenching personal and societal dilemmas in which scientific information is only one factor in a complex decision-making process.

Outside the scientific arena, subjective answers can have great value. For example, no human society can function without a shared commitment to standards for making judgments, even if the judgments are subjective. Moral, aesthetic, economic, and philosophical standards vary from one society to the next. But all guide their members in deciding what is important and good and what is not. All attempt to give meaning to what we do.

Every so often, scientists stir up controversy when they question or explain a part of our world that was considered beyond natural explanation—that is, belonging to the supernatural. On occasion, a new, natural explanation runs counter to supernatural belief or to some other widely held nonscientific view. This doesn't mean the scientists who raise the questions are any less moral, less law-abiding, less sensitive, or less caring than anyone else. It simply means one more standard guides their work: *The external world, not internal conviction, must be the testing ground for scientific beliefs.*

A scientific approach to studying human biology is based on asking questions, formulating hypotheses, making predictions, devising tests, and then objectively reporting the results.

A scientific theory is a testable explanation about the cause or causes of a broad range of related phenomena. It remains open to tests, revision, and tentative acceptance or rejection.

CASE STUDY: USING A SCIENTIFIC METHOD IN CANCER RESEARCH

To get a feel for how researchers use a common scientific method, put yourself in the shoes of Michael Pariza, a biochemist who has probed the effects of different forms of linoleic acid (LA) on cancer (Figure I.9). LA is a "Jekyll-and-Hyde" substance: The human body requires it in small amounts as a building block for fats, but studies on mice and rats show that in large doses it is associated with the development of certain types of cancer. There is also an alternate form of the compound, called CLA—and here the plot thickens. When you paint CLA on the skin of mice that have been exposed to a potent cancer-causing agent—a **carcinogen**—those mice develop only half as many skin cancers as mice painted with LA. A second experiment shows that force-feeding CLA to mice a few days before administering a carcinogen inhibits the development of cancerous stomach tumors. Now you are getting excited, because you know that CLA occurs in some common human foods, including grilled ground beef and some cheese products. Could CLA inhibit cancer when it is consumed as an additive to food?

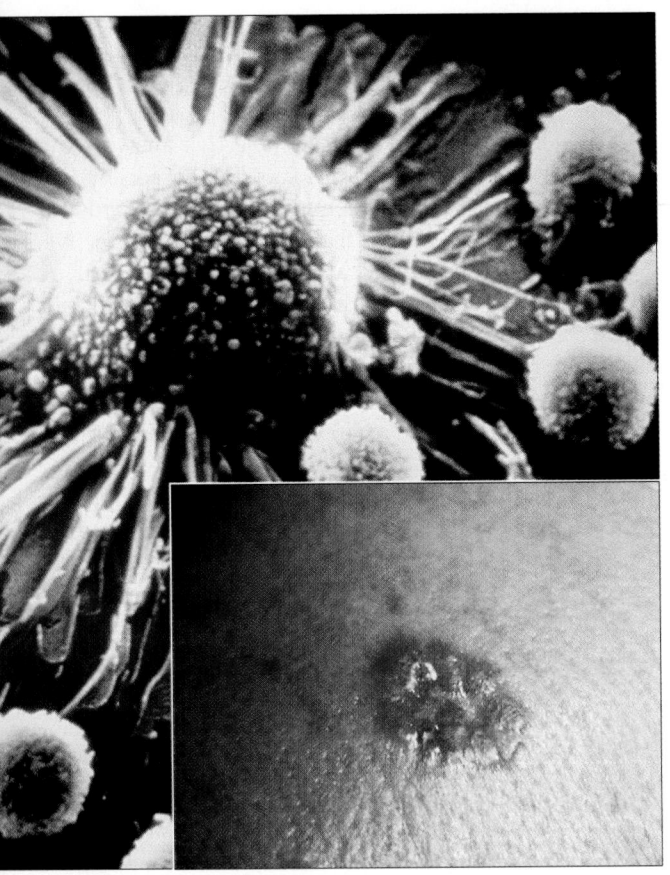

Figure I.9 A skin cancer (basal cell carcinoma) superimposed over a micrograph of a single cancer cell. The smaller cells in the micrograph are weapons of the immune system, which can often—but not always—detect and eliminate cancerous cells.

Moving from Hypothesis to Prediction

Pariza, Clement Ip, and several colleagues agreed on a hypothesis: Based on their existing knowledge, they reasoned that CLA consumed as a regular part of the diet could help prevent certain cancerous tumors from developing. Notice that at least one alternative hypothesis would also have been reasonable; for example, they might have proposed that CLA exerts its effects only when it is administered immediately before exposure to a carcinogen. In science, alternative hypotheses are the rule, not the exception.

The researchers then made a prediction they (and other scientists) could readily test—that rats fed certain doses of CLA would develop fewer cancers than rats in a control group, when all the rats were exposed to a carcinogen known to cause tumors in mammary glands.

Testing a Prediction by Experimentation

Some hypotheses can be tested by direct observation. (For example, you can watch birds visiting a feeder to see whether one type of bird feeds more aggressively.) In this instance, Pariza and his coworkers conducted a series of experiments, using the experimental method described in the previous section. They began with five groups of 30 healthy rats. One group, the control, received a normal diet; the other four groups were fed the same diet, except that each group's food contained a certain amount of CLA, which was added for several weeks before all the animals were exposed to the carcinogen. What was the outcome? The team reported in the prestigious journal *Cancer Research* that, under the conditions of their experiments, regular feeding of CLA in the diet "is effective in cancer prevention."

Pariza, Ip, and the rest of the team have continued to investigate exactly how CLA deters cancer. Based on knowledge they and others have gained, one hypothesis is that CLA may interfere with steps in a pathway that promotes the runaway multiplication of cancer cells—a topic you will read more about in Chapter 20. No one yet knows whether the results of these experiments will apply to humans. If it turns out that they do, a CLA-like substance could become a routine additive to some human foods.

Commonly, scientists develop alternative hypotheses as possible answers to a particular question about the natural world.

Hypotheses often are tested by experimentation. Proper methods, including the uses of controls, are essential to obtaining reliable results.

Choices: Biology and Society

HOW DO YOU DEFINE DEATH?

In an austere hospital room, a young mother and father face the most anguished moment of their lives. Their baby has been born with no brain, except for a small portion of brain stem. For this child, none of the qualities that we associate with human life—the potential to think, feel joy or pain, learn, and speak—will ever be possible. For the moment, however, that bit of brain tissue controls lung and heart functions, so the baby's heart beats sporadically and its lungs occasionally take in air. Within hours or days, even those halting functions will cease.

In the United States, about one baby in a thousand is born with this condition, called *anencephaly*. For parents and physicians alike, the situation is agonizing. During the short period while the heart and lungs minimally function, other organs (such as the liver and kidneys) receive enough oxygen-carrying blood to keep them reasonably healthy. If the child is declared legally dead during that time, those organs can be transplanted and can bring the gift of life to others. If doctors wait until the nubbin of brain stem gives out, potentially transplantable organs will be irreversibly damaged by lack of oxygen and will be useless.

If this were your child, what course would you follow? If the matter went to court, you might not have a choice.

Many states have adopted a strict legal standard for such cases. A person may be declared legally dead only if *all* of the brain, including the brain stem, no longer functions. Some ethicists prefer this approach because it does not put society in the position of determining that some parts of the brain—but not others—make a person truly alive. Other people disagree strongly in cases in which the patient is in a "persistent vegetative state" (has no higher brain function) and there is no hope of recovery. In particular, advocates for a less strict definition of brain death point to the shortage of organs for transplants and the potential for saving other lives. They believe that doctors should be allowed to terminate life support for "hopeless" patients and remove usable tissues and organs.

As individuals, we deal with this kind of issue in many ways. Some people hold religious beliefs requiring medical care as long as the heart can beat (Figure I.10). Others draw up "living wills" designed to convey their wishes that life-support equipment, such as respirators and other high-tech machinery, not be used to prolong life artificially. Many people ignore the issue altogether, hoping that they or their loved ones will never have to grapple with it. Unfortunately life and death are not always so simple.

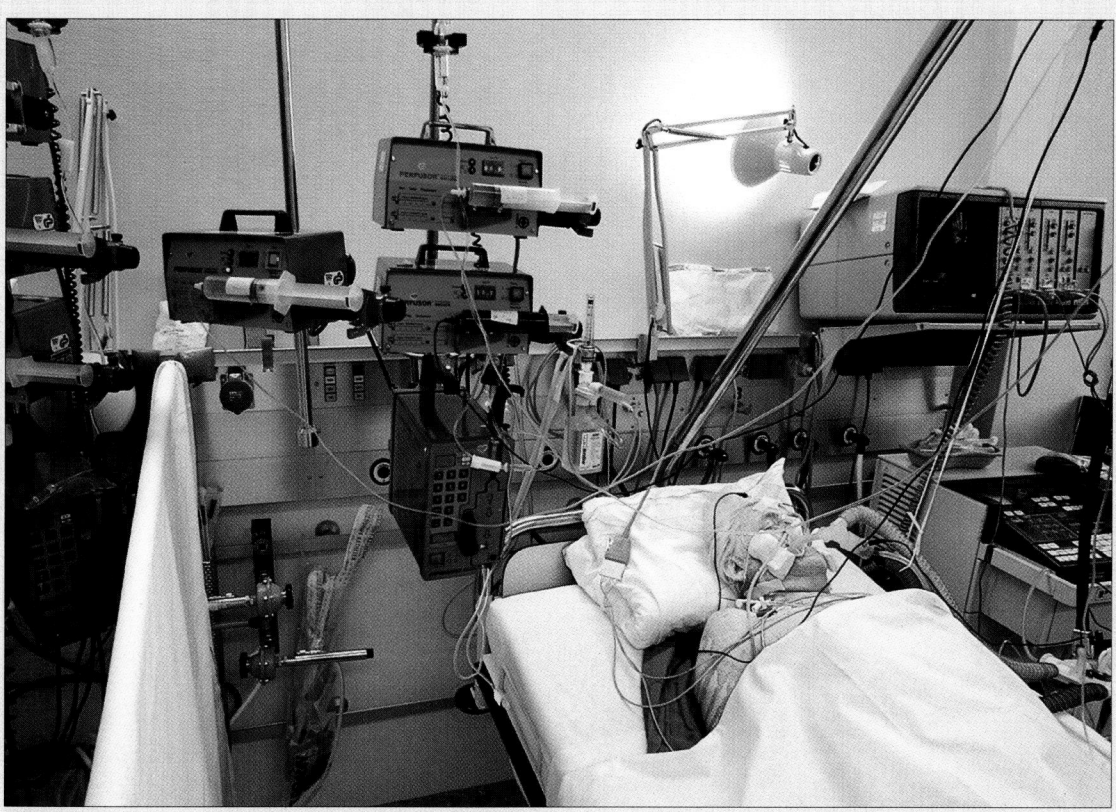

Figure I.10 A critically ill hospital patient. Families of such individuals often find themselves confronting wrenching issues about the type of care their loved one should receive.

SUMMARY

1. As living organisms, humans have the following characteristics:

   a. Their structure, organization, and interactions arise from the basic properties of matter and energy.

   b. Processes of metabolism and homeostasis maintain the living state.

   c. They have the capacity for growth, development, and reproduction, based on instructions contained in their DNA.

2. Diversity in features and characteristics arises through mutation. Mutations introduce changes in the DNA. The changes may lead to variations in the form, functioning, or behavior of individual offspring.

3. Individuals vary in their heritable traits (the traits that parents transmit to offspring). Such variations influence an organism's ability to survive and reproduce. Under prevailing conditions, some varieties of a given trait may be more adaptive than others; such traits will be "selected," whereas others will be eliminated through successive generations. Thus the population changes over time; it evolves. These points are central to the theory of evolution by natural selection.

4. There are many scientific methods, corresponding to many different fields of inquiry. The following key terms are important in all of those fields:

   a. Theory: an explanation of a broad range of related phenomena. An example is the theory of evolution by natural selection.

   b. Hypothesis: a possible explanation of a specific phenomenon; sometimes called an "educated guess."

   c. Prediction: a claim about what an observer can expect to see in nature if a theory or hypothesis is correct.

   d. Test: an effort to gather actual observations that may (or may not) match predicted or expected observations.

   e. Conclusion: a statement about whether a hypothesis (or theory) should be accepted, rejected, or modified, based on tests of the predictions derived from it.

5. A scientific theory is based on systematic observations, hypotheses, predictions, and tests. The external world, not internal conviction, is the testing ground for scientific theories.

6. Science has clear limits, because scientific methods can be applied only to objectively testable hypotheses. Issues and questions having subjective answers based on values are outside the realm of science.

Review Questions

1. For this and subsequent chapters, make a list of the boldface terms that occur in the text. Write a definition next to each, and then check it against the one in the text.

2. Why is it difficult to give a simple definition of life? (For this and subsequent chapters, *italic numbers* following review questions indicate the pages on which the answers may be found.) *2*

3. As a human, you are a living organism. List the characteristics of life that you exhibit. *4*

4. What is energy? Why is the concept of energy relevant to the study of human biology? *4*

5. Define metabolic activity and give an example of a metabolic event. *4*

6. Summarize what is meant by biological organization. *7*

7. Describe the one-way flow of energy and the cycling of materials through the biosphere. *6*

8. Which life characteristic does this diagram depict? *6*

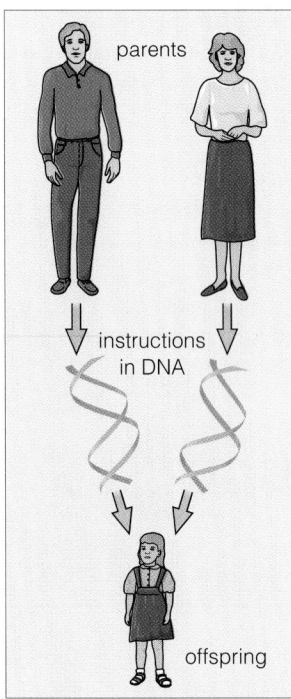

9. List the main elements of the scientific method described in this chapter. *8–9*

10. What is the difference between a hypothesis and a scientific theory? *8–9*

Critical Thinking: You Decide (Key in Appendix V)

1. Court witnesses are asked "to tell the truth, the whole truth, and nothing but the truth." What are some problems inherent in the question? Is there a better alternative?

2. Design a test (or series of tests) to support or refute this hypothesis: A diet high in salt is associated with hypertension (high blood pressure), but hypertension is more common in people with a family history of the condition.

3. In popular magazine articles on diet, exercise, and other health-related topics, the authors often recommend a specific diet or dietary supplement. What kinds of evidence should the articles describe so you can decide whether you can accept their recommendations?

Self-Quiz (Answers in Appendix IV)

1. The complex patterns of structural organization characteristic of life are based on instructions contained in _____.

2. _____ is the ability of cells to extract and transform energy from the environment and use it to maintain themselves, grow, and reproduce.

3. _____ is a state in which the body's internal environment is being maintained within a tolerable range. This state depends on _____, which are cells or structures that detect specific aspects of the environment.

4. Diverse structural, functional, and behavioral traits are considered to be _____ to changing conditions in the environment.

5. The capacity to evolve is based on variations in traits, which originally arise through _____.

6. Each of us has some traits that also were present in our great-great-great-great-grandmothers and -grandfathers. This is an example of _____.
   a. metabolism   c. a control group
   b. homeostasis   d. inheritance

7. A scientific approach to explaining some aspect of the natural world includes all of the following except _____.
   a. a hypothesis   c. faith and public consensus
   b. testing   d. systematic observations

8. A related set of hypotheses that collectively explain some aspect of the natural world is a scientific _____.
   a. prediction   d. authority
   b. test   e. observation
   c. theory

Selected Key Terms

adaptive trait 6	evolution 2	mutation 6
ATP 4	experiment 8	prediction 8
cell 4	homeostasis 5	reproduction 6
control group 8	hypothesis 8	theory 8
DNA 4	inheritance 6	variable 8
energy 4	metabolism 4	

Readings

Alberts, B., and K. Shine. December 4, 1994. "Scientists and the Integrity of Research." *Science*.

Committee on the Conduct of Science. 1989. *On Being a Scientist*. Washington, D.C.: National Academy of Sciences. Paperback.

Larkin, Tim. June 1985. "Evidence vs. Nonsense: A Guide to the Scientific Method." *FDA Consumer*.

Raloff, J. February 15, 1992. "Cancer-Fighting Food Additives." *Science News*.

Rosenthal, Elizabeth. October 1992. "Dead Complicated." *Discover*. Is a person with no functional brain, maintained by life-support machinery, alive or dead? The author, a physician, discusses this and other questions society faces as a result of modern, high-technology medical practices.

 For additional readings, go to InfoTrac College Edition, your online library at:

http://www.infotrac-college.com/wadsworth

1 THE CHEMISTRY OF LIFE

Chemistry In and Around You

Right now you are breathing in oxygen. You would die without it. Two centuries ago, no one had a clue to what oxygen is, where it comes from, and how it helps keep people alive. Then researchers started unlocking the secrets of this chemical substance and others. As their knowledge of chemistry grew, they began to develop such amazing things as fertilizers, nylons, cosmetics, aspirin, antibiotics, and plastic—which, in turn, is used in products ranging from cars and refrigerators to computers, television sets, and jet airplanes.

Today our chemical "magic" brings tremendous benefits *and* some problems. Pesticides are just one example. These powerful chemicals kill unwanted insects and vegetation, and thereby help increase the productivity of farmland—and keep prices down (Figure 1.1). In part because of pesticides and fertilizers, you and other consumers can buy inexpensive, virtually flawless fruits and vegetables. On the other hand, people also can inhale pesticides, ingest trace amounts with food, or absorb them through the skin. In rare cases, a person may be exposed to doses of certain pesticides that are high enough to trigger hives, rashes, asthma, and other moderate allergic reactions.

Today, in the United States and some other places, the application of agricultural chemicals is much more closely regulated than in the past. There is also much closer scrutiny of potentially toxic chemical by-products of industrial processes, some of which have seriously polluted water and soil and, in some instances, have poisoned people exposed to them. Despite the gains, however, government agencies and ordinary citizens in many localities continue to cope with the legacy of our collective use—and, sometimes, abuse—of chemicals.

Everything in and around you is chemistry. Most of the chemical reactions that keep you alive involve four

Figure 1.1a Crop duster applying pesticides, which can dramatically increase the yield of many agricultural crops. About two-thirds of the pesticides used on U.S. croplands are applied this way.

types of biological molecules: proteins, lipids (fats and their relatives), carbohydrates, and nucleic acids (components of DNA, the genetic material, and some other key substances). Those molecules are the foundation for the structure and function of each of your cells. The body uses them as building materials, as enzymes (catalysts) that speed up chemical reactions, and as energy storehouses. Each type of biological molecule is built upon a skeleton of linked carbon atoms.

The basic concepts in this chapter will help you understand processes that go on in your own body, every minute of your life. They provide a foundation of knowledge for understanding mechanisms of homeostasis that will be discussed in subsequent chapters. Beyond this, you owe it to yourself and others to gain understanding of chemical substances. By demystifying chemistry's "magic," you will be better equipped to assess its benefits and risks.

Figure 1.1b Pesticides, fungicides, herbicides, and other chemicals make their way to consumers through a variety of routes, including the supermarket.

KEY CONCEPTS

1. Matter is made up of elements—pure substances that cannot be broken down to simpler substances by ordinary means. Combinations of elements are called compounds.

2. Atoms of each element have subunits, which are arranged in particular ways. These subunits are protons, electrons, and, in nearly all types of atoms, neutrons.

3. Atoms do not have an overall electric charge. However, an atom can acquire a charge if it becomes ionized—that is, if it loses or gains electrons.

4. When the electron structures of two or more atoms join together, this union is a chemical bond. Chemical bonds are the basis of the physical organization and activities of living things, including the human body.

5. There are three types of chemical bonds: ionic bonds, covalent bonds, and hydrogen bonds.

6. Life on Earth began in water and is adapted to water's properties. Two of these properties that are especially important in human biology are water's ability to stabilize temperature and its capacity to mix with some substances and not others.

7. An organic compound has a backbone of one or more carbon atoms to which other atoms are attached. Cells can assemble the organic compounds known as carbohydrates, lipids, proteins, and nucleic acids.

8. Carbohydrates and lipids are the cell's main sources of energy and building blocks. Two nucleic acids, DNA and RNA, are the basis of inheritance and cell reproduction.

9. Many proteins are structural materials; many others are enzymes, which speed up specific metabolic reactions. Still other proteins transport substances, contribute to movements, trigger changes in cell activities, or help the body defend itself against disease.

CHAPTER AT A GLANCE

1.1   How Matter Is Organized
1.2   *Using Radioisotopes to Track Chemicals and Save Lives*
1.3   What Is a Chemical Bond?
1.4   Important Bonds in Biological Molecules
1.5   Water, Dissolved Ions, and the World of Cells
1.6   Organic Compounds: Built on Carbon Atoms
1.7   Carbohydrates
1.8   Lipids
1.9   Amino Acids and Proteins
1.10 Some Examples of Final Protein Structure
1.11 Nucleotides and Nucleic Acids
1.12 *Making Nonpolluting Fuel*

HOW MATTER IS ORGANIZED

All matter is composed of one or more elements. An **element** is a fundamental form of matter that occupies space and has mass. An element cannot be broken apart into something else, at least not by ordinary means. Ninety-two elements occur naturally on Earth; a handful of artificial ones have been created in the laboratory. As is true of all organisms, most of your own body consists of four elements—oxygen, carbon, hydrogen, and nitrogen (Figure 1.2). Your body also contains some calcium, phosphorus, potassium, sulfur, sodium, and chlorine, as well as tiny but important amounts of so-called trace elements. Chemists organize the elements into groups, with the members of each group having similar chemical properties. The periodic table in Appendix II shows these groups.

The Structure of Atoms

The smallest unit that retains the properties of a given element is an **atom**. Each atom is composed of subatomic particles: protons, electrons, and (except for hydrogen)

neutrons (Figure 1.3). **Protons** carry a positive charge (p^+). Together with **neutrons**, which have no charge, protons make up the atom's core region, the atomic nucleus. **Electrons** carry a negative charge (e^-), but they have no mass. They move rapidly around the nucleus, and they occupy most of the atom's volume.

Regardless of the element, atoms have just as many electrons as protons. This means they carry no *net* charge, overall. As you will read in Section 1.3, electrons tend to occur in specific energy levels, or "shells," around the nucleus (Table 1.1).

Each element has a unique *atomic number*, which refers to the number of protons in its atoms. As Table 1.1 indicates, that number is 1 for the hydrogen atom (with one proton) and 6 for the carbon atom (with six protons). Each element also has a *mass number*, which is the number of protons *and* neutrons in the nucleus. A carbon atom with six protons and six neutrons has a mass number of 12. You may have heard the term "atomic weight." It is used in chemistry to refer to the relative masses of atoms, even though, technically speaking, mass is not quite the same thing as weight.

Why be concerned with atomic numbers and mass numbers? This information gives us an idea of whether and how substances will interact—and, as you will discover, *it helps us predict how the substances of life will behave in cells, in the human body, and in the environment.*

Isotopes—Varying Atoms of an Element

All atoms of an element have the same number of protons and electrons, but they may not have the same number of neutrons. If an atom of a given element has more or fewer

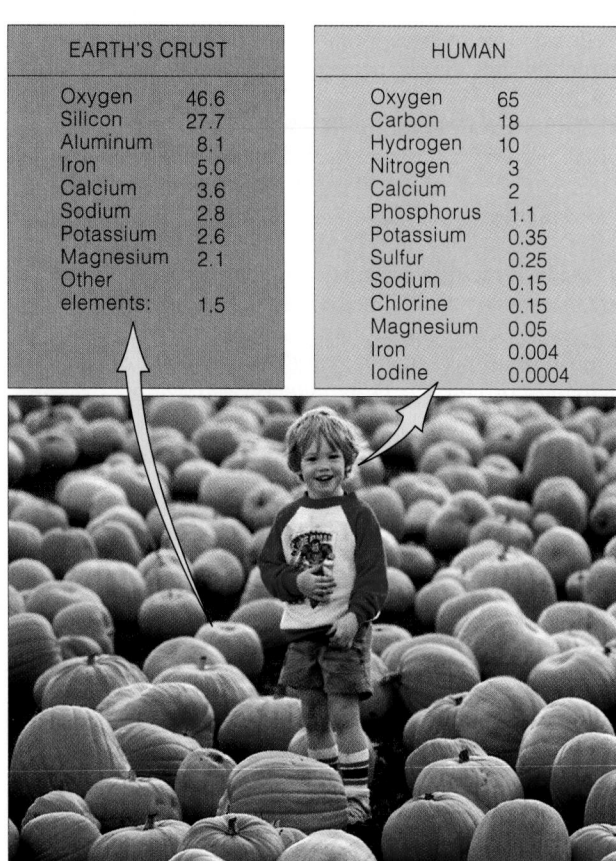

EARTH'S CRUST		HUMAN	
Oxygen	46.6	Oxygen	65
Silicon	27.7	Carbon	18
Aluminum	8.1	Hydrogen	10
Iron	5.0	Nitrogen	3
Calcium	3.6	Calcium	2
Sodium	2.8	Phosphorus	1.1
Potassium	2.6	Potassium	0.35
Magnesium	2.1	Sulfur	0.25
Other		Sodium	0.15
elements:	1.5	Chlorine	0.15
		Magnesium	0.05
		Iron	0.004
		Iodine	0.0004

Figure 1.2 Proportions of different elements in the Earth's crust and in the human body, as percentages of the total weight of each.

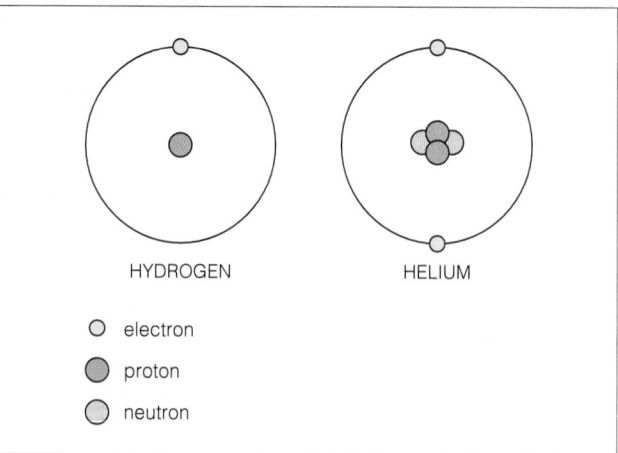

HYDROGEN   HELIUM

○ electron
● proton
○ neutron

Figure 1.3 One model of atomic structure, showing a hydrogen atom and a helium atom. In all types of atoms except hydrogen, the nucleus has at least one neutron as well as protons. Electrons occupy orbitals around the nucleus. At the scale of this drawing, the nucleus actually would be an invisible speck at the atom's center. ✳

Element	Symbol	Atomic Number	Most Common Mass Number	Number of Electrons in Outermost Shell
Hydrogen	H	1	1	1
Carbon	C	6	12	4
Nitrogen	N	7	14	5
Oxygen	O	8	16	6
Sodium	Na	11	23	1
Magnesium	Mg	12	24	2
Phosphorus	P	15	31	5
Sulfur	S	16	32	6
Chlorine	Cl	17	35	7
Potassium	K	19	39	1

neutrons than those having the most common number, it is an **isotope**. As an example, "a carbon atom" might be carbon 12 (six protons, six neutrons) or carbon 14 (six protons, eight neutrons). These are written as ^{12}C and ^{14}C. Radioactive isotopes, or *radioisotopes*, are unstable and tend to lose subatomic particles and energy at a known rate. This process, called **radioactive decay**, transforms an unstable atom into an atom of a different element. Carbon 14, for instance, decays to nitrogen 14. *Science Comes to Life* on the following page describes some uses of radioisotopes in research and medicine.

Combinations of Atoms

When two or more atoms join together, the result is a **molecule**. Many molecules contain atoms of only one element. Molecular nitrogen (N_2), with its two nitrogen atoms, is an example. Figure 1.4 explains how to read the notation used in representing chemical reactions that occur between atoms and molecules.

Many other kinds of molecules are **compounds**, meaning that they consist of two or more elements in proportions that never vary. Water is an example. Every water molecule has one oxygen atom bonded with two hydrogen atoms. Molecules of water in rain clouds or in a lake or in your bathtub *always* have twice as many hydrogen as oxygen atoms.

By contrast, in a **mixture**, two or more elements simply mingle, in proportions that can vary. For example, the sugar sucrose is a compound of carbon, hydrogen, and oxygen. Swirl sucrose molecules and water molecules together, and you get a mixture.

Only living organisms put together and use the "biological molecules" such as carbohydrates, proteins, lipids, and nucleic acids. These and other combinations of atoms come about by way of chemical bonding, the topic you will read about in Section 1.3.

We use symbols for elements when writing *formulas*, which identify the composition of compounds. For example, water has the formula H_2O. Symbols and formulas are used in *chemical equations*, which are representations of reactions among atoms and molecules.

In written chemical reactions, an arrow means "yields." Substances entering a reaction (reactants) are to the left of the arrow. Reaction products are to the right. For example, the reaction between hydrogen and oxygen that yields water is summarized this way:

$$2H_2 \ + \ O_2 \ \longrightarrow \ 2\,H_2O$$

4 hydrogens   2 oxygens   4 hydrogens, 2 oxygens

Note that there are as many atoms of each element to the right of the arrow as there are to the left. Although atoms are combined in different forms, none is consumed or destroyed in the process. The total mass of all products of any chemical reaction equals the total mass of all its reactants. All equations used to represent chemical reactions, including reactions in cells, must be balanced this way.

Figure 1.4   Chemical bookkeeping.

An atom is the smallest unit of matter that retains the properties of a particular element. Each atom consists of subatomic particles: neutrons and positively charged protons, which make up the atomic nucleus, and electrons, which have a negative charge.

Atoms of an element have the same number of electrons and protons, but they can vary in the number of neutrons. Variant forms of atoms of the same element are called isotopes.

In molecules of an element, all of the atoms are of the same kind. In molecules of a compound, atoms of two or more elements are combined (by way of chemical bonding) in unvarying proportions.

USING RADIOISOTOPES TO TRACK CHEMICALS AND SAVE LIVES

To a physician, radioisotopes can be vital tools. Used with care, they can permit the precise diagnosis of disease without subjecting a patient to exploratory surgery. In other circumstances, radioisotopes are part of the actual treatment of certain cancers. For safety's sake, clinicians use only radioisotopes with extremely short half-lives. **Half-life** is the time it takes for half of a quantity of a radioisotope to decay into a different, more stable daughter isotope.

TRACKING TRACERS Various devices can detect radio-isotope emissions. Thus, radioisotopes can be employed in tracers. A *tracer* is a substance with a radioisotope attached to it, rather like a shipping label, that a physician can administer to a patient. The tracking device then follows the tracer's movement through a pathway or pinpoints its destination.

SAVING LIVES In nuclear medicine, radioisotopes are used under carefully controlled conditions to diagnose and treat diseases. Think about your thyroid, the only gland of ours that takes up iodine. After a tiny amount of ^{123}iodine is injected into a patient's bloodstream, the thyroid can be scanned, producing the kinds of images you see in Figure 1.5a. Some cancer treatments rely on the cell-destroying capacities of radioisotopes. In radiation therapy, a cancer that has not spread is bombarded with ^{226}radium or ^{60}cobalt. Patients with irregular heartbeats use pacemakers, which are powered by energy emitted from ^{238}plutonium. (This dangerous radioisotope is sealed in a case to prevent emissions from damaging body tissues.) With the imaging technology called positron-emission tomography (PET), radioisotopes provide information about abnormalities in the metabolic functions of specific tissues. The radio-isotopes are attached to the sugar glucose or to some other biological molecule, then injected into a patient, who is moved into a PET scanner (Figure 1.5b,c). When cells in a target tissue absorb glucose, energy emitted by the radioisotope can be used to produce a vivid image of changes in metabolic activity. PET has been tremendously useful in studying events in the brain (Figure 1.5d).

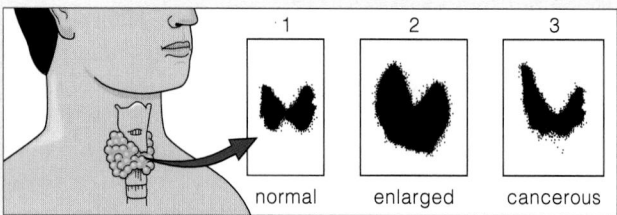

1	2	3
normal	enlarged	cancerous

a

Figure 1.5 (**a**) Scans of the thyroid gland from three patients. (**b**) Patient being moved into a PET scanner. Inside (**c**), a ring of detectors intercepts the radioactive emissions from labeled molecules that were injected into the patient. Computers analyze and color-code the number of emissions from each location in the scanned body region. (**d**) Brain scan of a child who has a neurological disorder. Different colors in a brain scan signify differences in metabolic activity. The right half of this brain shows very little activity. By comparison, cells of the left half absorbed and used the labeled molecule at expected rates.

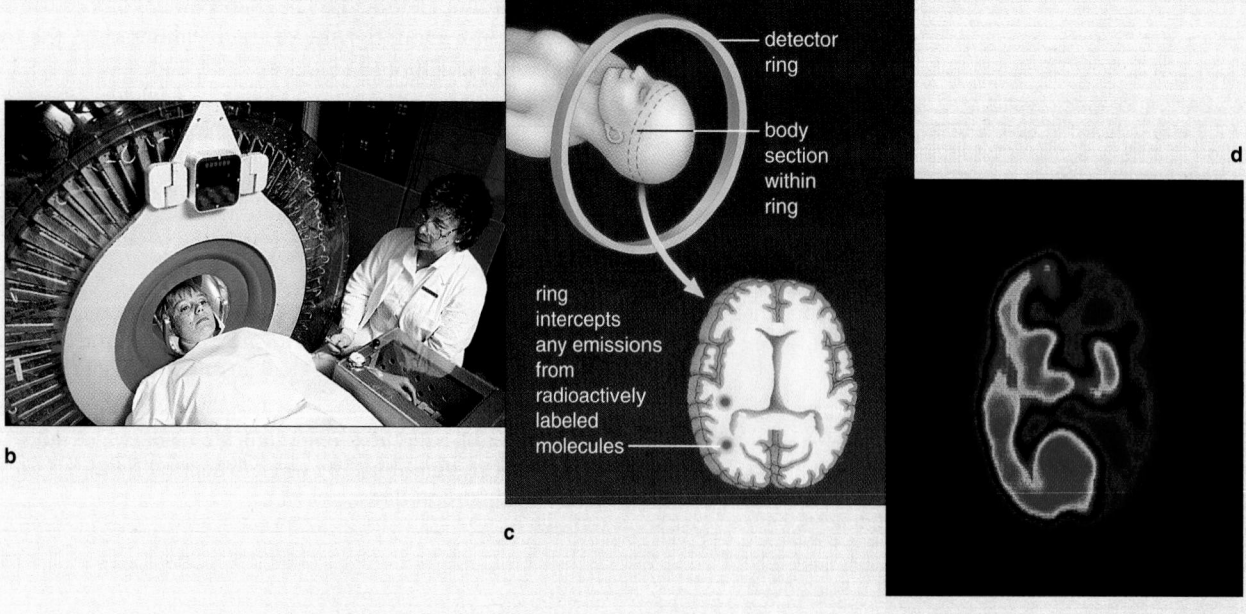

detector ring

body section within ring

d

ring intercepts any emissions from radioactively labeled molecules

b

c

We turn now to **chemical bonds**, which are unions between the electron structures of atoms. The chemical reactions that sustain living cells involve the formation and breaking of chemical bonds. Whether one atom will bond with another atom depends on how many electrons it has and how those electrons are arranged.

Because all of an atom's electrons carry a negative charge, they repel each other. On the other hand, the electrons are *attracted* to the positive charge of the protons. The electrons spend as much time as possible near the protons and far away from each other by moving in **orbitals**. You can think of orbitals as *volumes of space* around the atomic nucleus in which electrons are likely to be at any instant. Just as the number of electrons differs among atoms, so does the number of orbitals. Only one or *two electrons at most* can occupy any orbital.

In hydrogen, the simplest atom, the lone electron occupies a spherical orbital closest to the nucleus, which corresponds to the *lowest available energy level* (Figure 1.6). In all other atoms two electrons fill this orbital.

In larger atoms, the next two electrons occupy a second orbital that surrounds the first. As many as six more electrons may be distributed in three dumbbell-shaped orbitals, and so on. These electrons spend more of their time farther away from the nucleus; we say that they are at *higher energy levels*.

The **shell model** shown in Figure 1.7 is a simple, fairly accurate way to think about how electrons are distributed in atoms. In this model, a series of shells encompasses the orbitals that are available to electrons. The first spherical orbital fits inside the first shell. The next four orbitals fit inside the second shell (at a higher energy level), which encloses the first. More orbitals fit in a third shell, and so on, up to the large, complex atoms of the heavier elements listed in Appendix II.

Having a filled outer shell is the most stable state for atoms. However, many atoms lack enough electrons to

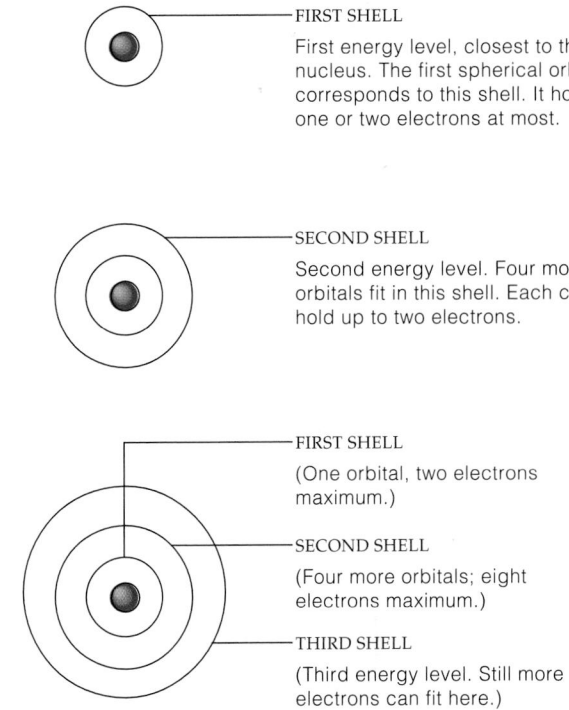

FIRST SHELL

First energy level, closest to the nucleus. The first spherical orbital corresponds to this shell. It holds one or two electrons at most.

SECOND SHELL

Second energy level. Four more orbitals fit in this shell. Each can hold up to two electrons.

FIRST SHELL

(One orbital, two electrons maximum.)

SECOND SHELL

(Four more orbitals; eight electrons maximum.)

THIRD SHELL

(Third energy level. Still more electrons can fit here.)

Figure 1.7 Shell model of electron distribution in atoms. Only three of the many possible energy levels (shells) are shown.

completely fill the outermost shell. For instance, atoms of hydrogen, oxygen, carbon, and nitrogen are this way. *Atoms with an unfilled outer shell tend to form chemical bonds with other atoms in order to fill their outer shell* (Figure 1.7). The single shell of hydrogen and helium atoms is full when it contains two electrons. Other kinds of atoms that have unfilled outer shells follow the *octet rule*—they take part in chemical bonds that result in a filled outer shell containing eight electrons. As you will read in the next section, in different types of chemical bonds electrons are either added to an atom's outer shell, removed from it, or shared between atoms.

Biologically important atoms have unfilled orbitals in their outermost shell. To fill the outermost shell (two or eight electrons), atoms tend to form chemical bonds with other atoms. They do this by gaining, losing, or sharing one or more of their electrons.

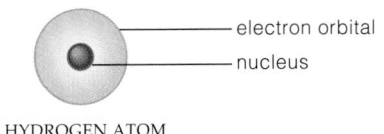

electron orbital

nucleus

HYDROGEN ATOM

Figure 1.6 The electron structure of a hydrogen atom.

Ionic Bonds: Electrons Gained or Lost

Whenever an atom loses or gains one or more electrons, the balance between its protons and electrons shifts and the atom becomes positively or negatively charged. In this state the atom is called an **ion**.

In cells, some neighboring atoms accept or donate electrons among one another. When one atom loses an electron and one gains, *both* become ionized. Depending on cellular conditions, the ions may be separated, or they may remain together through the mutual attraction of their opposite charges. An association of two oppositely charged ions is an **ionic bond**.

Figure 1.8 gives an example of ionic bonding. It shows a portion of a crystal of table salt, or NaCl, in which ions of sodium (Na^+) and chloride (Cl^-) interact. The sodium fluoride in "anti-cavity" toothpastes is another example of a compound held together by ionic bonds.

Covalent Bonds: Atoms Share Electrons

In a **covalent bond**, two atoms *share* electrons. Such bonds form when an atom strongly attracts one or more electrons of a neighboring atom—but not strongly enough to pull them completely from their orbital.

Two interacting hydrogen atoms provide an example of a covalent bond:

MOLECULAR
HYDROGEN
(H_2)

In structural formulas, a single line between two atoms represents a single covalent bond, as in H—H. In a double covalent bond, two atoms share two pairs of electrons. This happens in an O_2 molecule, expressed as O=O. In a triple covalent bond, two atoms share three pairs of electrons, as in N≡N.

Covalent bonds are polar or nonpolar. In a *nonpolar* covalent bond, the participating atoms exert the same pull on electrons and share them equally. The term "nonpolar" refers to the fact that there is no difference in charge at the two ends of the bond. Molecular hydrogen is an example. Its two H atoms, each with one proton, attract the shared electrons equally.

In a *polar* covalent bond, atoms of different elements don't exert the same pull on shared electrons. One exerts a stronger pull, and it ends up with a slight negative charge; the atom is "electronegative." Its effect is balanced by the other atom, which exerts less pull and ends up with a slight positive charge. In other words, in a polar covalent bond the charge is distributed unevenly between the bond's two ends—but taken together the two interacting atoms have no *net* charge.

A water molecule has two polar covalent bonds (Figure 1.9). Its electrons are less attracted to the hydrogen than to the oxygen, which has more protons. The molecule carries no net charge, but its polarity makes it weakly attract neighboring polar molecules and ions.

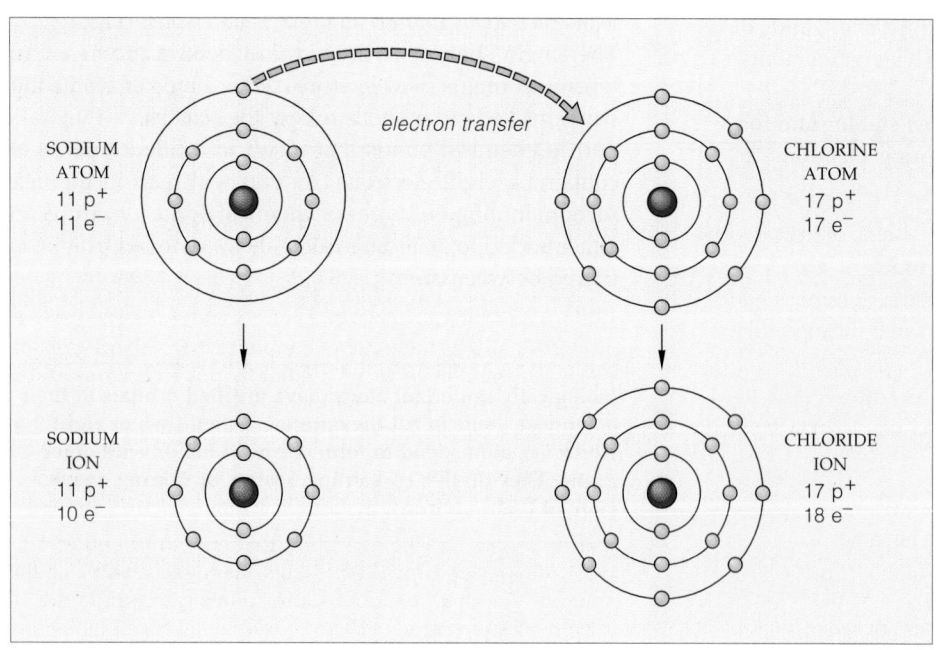

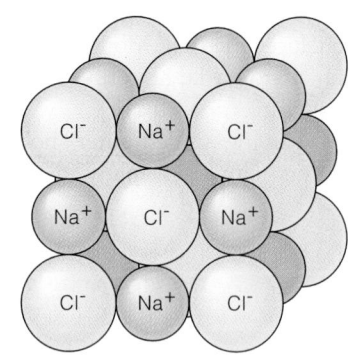

Figure 1.8 Ionic bonding in sodium chloride. (**a**) A sodium atom donates the lone electron in its outermost shell to a chlorine atom, which has an unfilled outer shell. This interaction results in a sodium ion (Na^+) and a chloride ion (Cl^-). (**b**) In each crystal of table salt, or NaCl, many sodium and chloride ions remain together because of the mutual attraction of their opposite charges. Their interaction is an example of ionic bonding.

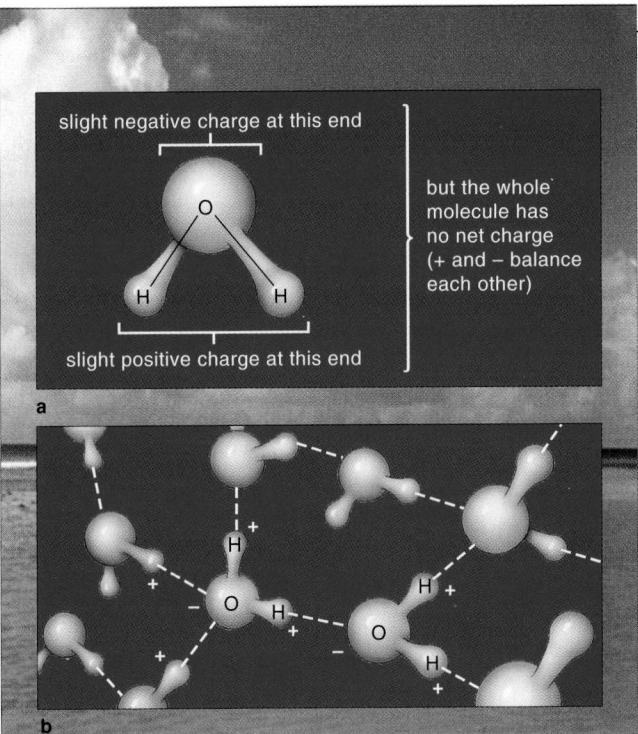

slight negative charge at this end

but the whole molecule has no net charge (+ and − balance each other)

slight positive charge at this end

a

b

Figure 1.9 Water is essential for life. (**a**) Polarity of a water molecule. (**b**) Hydrogen bonds between molecules in liquid water, signified by dashed lines.

Hydrogen Bonding

In a **hydrogen bond**, a small, highly electronegative atom in a molecule interacts weakly with a hydrogen atom that is already part of a polar covalent bond. When participating in such a bond, hydrogen has a slight positive charge, and this attracts other atoms that carry a slight negative charge.

Although hydrogen bonds are a good deal weaker than ionic bonds or covalent bonds, they are common in cells. For example, DNA, the genetic material, consists of two parallel strands of chemical units. Hydrogen bonds link the strands, and collectively they help stabilize DNA molecules. Hydrogen bonds also hold water molecules together (Figure 1.9) and give water many of the life-sustaining properties you will learn about in Section 1.5.

We conclude this discussion of chemical bonding with a look at a type of electron transfer that has become the basis for wholesale lifestyle changes for many people because of the damage such transfers can cause to many of the body's vital biological molecules.

Free Radicals: Stealing Electrons

Doubtless you have read about antioxidants—substances touted for slowing aging, preventing heart disease, and a host of other benefits. An **antioxidant** is a chemical that prevents an atom or molecule from losing one or more electrons to another atom or molecule, an event called *oxidation*. Many normal cell processes require oxidation reactions. However, in certain cases of oxidation, when a transferred electron is "stolen" by a molecule called a free radical, the result can be devastating damage.

A **free radical** is a molecule (such as O_2^-) that includes an oxygen atom that lacks a full complement of electrons in its outer shell. As a result, a free radical is extremely reactive: It will oxidize another, stable molecule, "stealing" an electron to fill the empty slot—and disrupt or destroy that molecule in the process. Potential targets include DNA and molecules that are key components of cell membranes. Free radicals result from normal cell operations. They are also produced by the effects of cigarette smoke, ultraviolet radiation in sunlight, and some other chemical assaults on the body.

Antioxidants combine with free radicals before they do damage. Known antioxidants include vitamins C and E and some carotenoids, which are pigments found in orange and leafy green vegetables, among other foods. Beta carotene is one of these; so are alpha carotene (found in foods such as carrots and pumpkin) and beta cryptoxanthin. Some food and vitamin manufacturers tout the benefits of beta carotene, but studies suggest that in some circumstances other antioxidants may be more effective free-radical scavengers. Melatonin, a hormone produced naturally in the body, also may be an efficient antioxidant, and scientists in many laboratories are exploring its possible effects on normal cell aging.

Researchers also are investigating whether free-radical activity may account for changes associated with lung damage from cigarette smoke and smog, with some degenerative diseases such as rheumatoid arthritis, with the coronary artery disease called atherosclerosis (Chapter 6), and even with damage to sperm cells.

In an ionic bond, two ions of opposite charge attract each other and remain together. Ions form when atoms gain or lose electrons, acquiring a net positive or negative charge.

In a covalent bond, atoms share electrons. If the atoms share electrons equally, the bond is nonpolar. If the sharing is not equal, the bond itself is polar (slightly positive at one end, slightly negative at the other).

In a hydrogen bond, a small, highly electronegative atom or molecule interacts weakly with a neighboring hydrogen atom that is already taking part in a polar covalent bond.

Free radicals are rogue molecules that can oxidize other molecules, stripping electrons from them and disrupting the molecule's normal function. Antioxidants can counter the activity of free radicals.

Properties of Water That Support Life

Life on our planet began in water. By weight, your own body is about two-thirds water, which is essential to maintaining the shape and internal structure of all of your cells. Water also makes up most of the fluid in your blood. Many of the chemical reactions that sustain life require water as a reactant or can occur only after other substances dissolve in water. Water's pivotal role in life processes is linked to properties of the remarkable water molecule.

The polarity of water molecules allows them to form hydrogen bonds with one another and with other polar substances, such as sugars. All polar molecules are attracted to water; they are **hydrophilic** (water-loving). By contrast, water repels nonpolar substances, such as oil. All nonpolar molecules are **hydrophobic** (water-dreading). As you will read in Chapter 2, hydrophobic interactions help organize the rather oily, sheetlike layers of cell membranes.

Water is a *solvent*, in that ions and polar molecules readily dissolve in it. Dissolved substances are called **solutes**. A substance dissolves as clusters of water molecules surround its individual ions or molecules (Figure 1.10). This happens to solutes in cells, between cells, in blood, and in all other body fluids. Most of the chemical reactions in the body occur in water-based solutions.

Hydrogen bonds between water molecules enable water to absorb a great deal of heat energy before it heats up significantly or evaporates. Why is this so? The answer is actually quite simple: In a given volume of water, multiple hydrogen bonds link individual H_2O molecules, and a large amount of heat must be applied to break the bonds. Water's ability to absorb a great deal of heat before becoming hot is the reason it can be used to cool a hot automobile engine. In the same way, water helps stabilize the temperature inside cells, which are mostly water. The chemical reactions in cells constantly produce heat, yet cells must stay relatively cool because cell proteins can function properly only within narrow temperature limits.

When enough heat energy is present, hydrogen bonds between water molecules break apart and do not re-form. Then water evaporates—molecules at the water's surface escape into the air. When many water molecules evaporate, heat energy is lost. This is why sweating helps cool you off on a hot, dry day. Sweat is 99 percent water, and when it evaporates from the more than 2.5 million sweat glands in your skin, heat leaves with it.

H₂O Interactions: Acids, Bases, and Salts

An **acid** is a substance that releases protons—H^+ ions—when it dissolves in water. By contrast, a **base** is a substance that accepts protons. The concentration of H^+ in a solution is measured on the pH scale (Figure 1.11). *Acidic* solutions contain more H^+ than OH^-, and *alkaline* (basic) solutions contain more OH^- than H^+.

When you consume food, stomach cells are stimulated to secrete hydrochloric acid (HCl), which separates into H^+ and Cl^- in water. The H^+ ions make the fluid in your stomach more acidic, and the increased acidity switches on enzymes that can digest (chemically break down) food particles. It also helps kill harmful bacteria. However, eating a meal that contains too much of certain kinds of foods can lead to "acid stomach." Antacids, such as milk of magnesia, are strong bases. When milk of magnesia dissolves, it releases magnesium ions and hydroxyl ions (OH^-). When the OH^- ions combine with excess hydrogen ions in stomach fluid, the reaction raises the pH.

Salts are compounds that release ions *other than* H^+ and OH^- in solutions. Salts and water often form when a strong acid and a strong base interact. Depending on a solution's pH value, salts can form and dissolve easily. Consider how sodium chloride (table salt) forms, then dissolves:

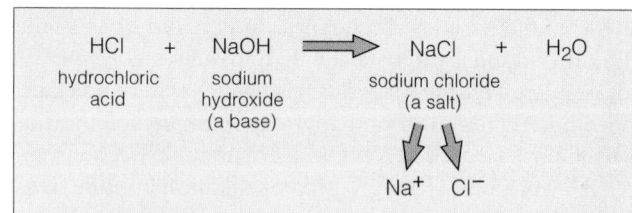

Many other salts also dissolve into ions that have important roles in cells. For instance, sodium ions (Na^+) and potassium ions (K^+) are involved in the transmission of impulses through your nervous system, a topic we will look at closely in Chapter 10. Na^+ and K^+ also are components of *electrolytes*, substances that dissociate into ions in solution and that can conduct an electric current.

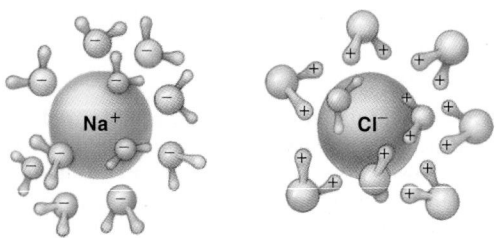

Figure 1.10   Clusters of water molecules around charged ions.

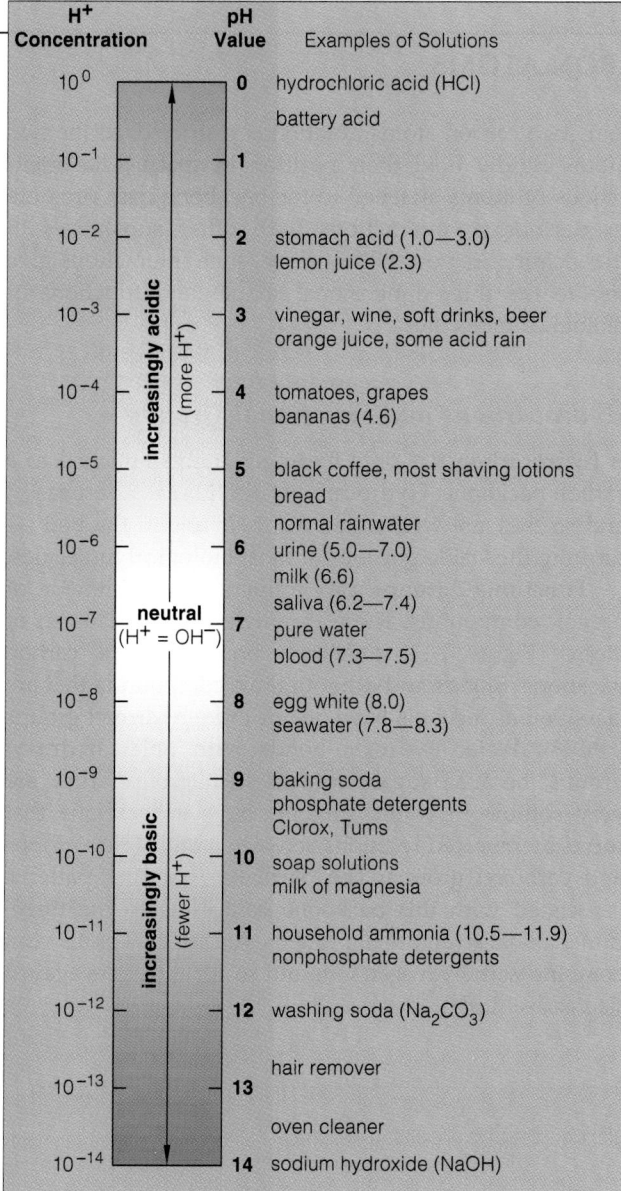

Figure 1.11 The pH scale, in which a liter of fluid is assigned a number according to the number of hydrogen ions in it. The scale ranges from 0 (most acidic) to 14 (most basic).

Figure 1.12 A coal-burning power plant emitting sulfur dioxide.

Figure 1.11 shows that each unit of change on the pH scale corresponds to a tenfold increase or decrease in H^+ concentration. The pH of human blood and most tissue fluids ranges between 7.35 and 7.45. The normal pH inside cells ranges from about 7.0 to 7.2, or from neutral to slightly basic.

Compared to that of cells, the pH of another environment may be much higher or lower. River water ranges between 6.8 and 8.6. Some airborne industrial wastes, such as the sulfur dioxide (SO_2) emissions pictured in Figure 1.12, can react with water vapor and lower its pH so much that the result is acid rain—a serious ecological threat discussed in Chapter 25.

Buffers and the pH of Body Fluids

Chemical reactions in cells are sensitive to even slight shifts in pH. Yet many reactions continually use and produce hydrogen ions. Helping to maintain internal pH within narrow limits are **buffers**—compounds that can combine with hydrogen ions or release them, or both, and so help stabilize pH. Buffers respond to acidic conditions by taking up H^+ and to basic conditions by releasing H^+. Bicarbonate (HCO_3^-) is an important buffer in blood. When the blood becomes too acidic, bicarbonate combines with H^+ to form carbonic acid (H_2CO_3):

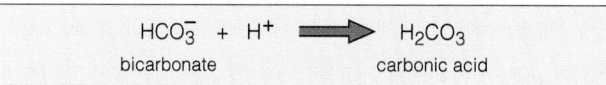

$$HCO_3^- + H^+ \longrightarrow H_2CO_3$$
bicarbonate   carbonic acid

When a person's blood is too basic, carbonic acid can release H^+. This buffering action is crucial. For instance, some lung diseases interfere with the elimination of carbon dioxide, so the blood level of carbonic acid (hence H^+) increases. This abnormal condition, a form of *acidosis*, makes breathing difficult and weakens the body.

Water helps stabilize temperature in the body and dissolves many substances. Ions dissolved in body fluids affect the structure and behavior of biological molecules. The pH of a fluid refers to its concentration of hydrogen ions. Buffers counter shifts in pH by releasing or combining with hydrogen ions.

The pH Scale

Chemists use the **pH scale** to measure the concentration of hydrogen ions in water, blood, and other solutions. (Hydrogen ions are the same thing as unbound, or "free," protons.) Figure 1.11 shows the scale, which ranges from 0 (most acidic) to 14 (most basic). A solution with a pH value of 7 is neutral; it has as many hydrogen ions (H^+) as hydroxide ions (OH^-). An acid solution has more hydrogen ions than hydroxide ions, and a basic solution has proportionately more hydroxide ions. A solution that is neutral becomes more acidic when H^+ ions are added to it and more basic when H^+ ions are removed from it.

ORGANIC COMPOUNDS: BUILT ON CARBON ATOMS

The molecules of life are **organic compounds**, meaning that they consist of atoms of carbon and one or more other elements, held together by covalent bonds. The term comes from a time when chemists thought "organic substances" were obtained only from plants and animals, as opposed to "inorganic" substances obtained from minerals. Although researchers now synthesize organic compounds in laboratories, organic compounds may have been present on the Earth before organisms were.

Carbon Bonding: The Key to Versatility

Living organisms are mostly oxygen, hydrogen, and carbon. Much of the oxygen and hydrogen is in the form of water. Remove the water, and carbon makes up more than half of what's left.

Carbon's importance in life arises from its versatile bonding behavior. A carbon atom can form covalent bonds with as many as four other atoms. Each covalent bond is quite stable because the carbon atoms share pairs of electrons equally. Such bonds link carbon atoms in chain and rings, and these serve as a backbone to which hydrogen, oxygen, and other elements become attached. Although the structural formulas in Figure 1.13 make the molecules they represent look flat, in fact these molecules have three-dimensional shapes, which arise from bonding arrangements in the carbon backbone.

Some carbon atoms can rotate freely around a single covalent bond. However, where a double covalent bond joins two carbon atoms, rotation is restricted, so the two atoms rigidly hold their position in space. The orientations of atoms attached to the backbone may promote or discourage interactions with other substances in the vicinity. As you will soon see, such interactions give rise to the three-dimensional shapes and functions of biological molecules.

Hydrocarbons and Functional Groups

A **hydrocarbon** has only hydrogen atoms attached to a carbon backbone. Hydrocarbons don't break apart easily, and so they are well suited for their role in the body—forming the stable portions of most biological molecules.

Functional groups can influence the behavior of organic compounds. These are single atoms or clusters of atoms (Figure 1.14) covalently bonded to the carbon backbone. Sugars and other organic compounds that are classified as alcohols have one or more hydroxyl groups (—OH). Water hydrogen-bonds with polar hydroxyl groups; because sugars contain such groups, they are very soluble in water. Proteins have a backbone that forms by repeated reactions between many amino groups and carboxyl groups. The particular bonding patterns associated with this backbone contribute to the three-dimensional structure of proteins. Amino groups also can combine with hydrogen ions and so act as buffers against decreases in pH.

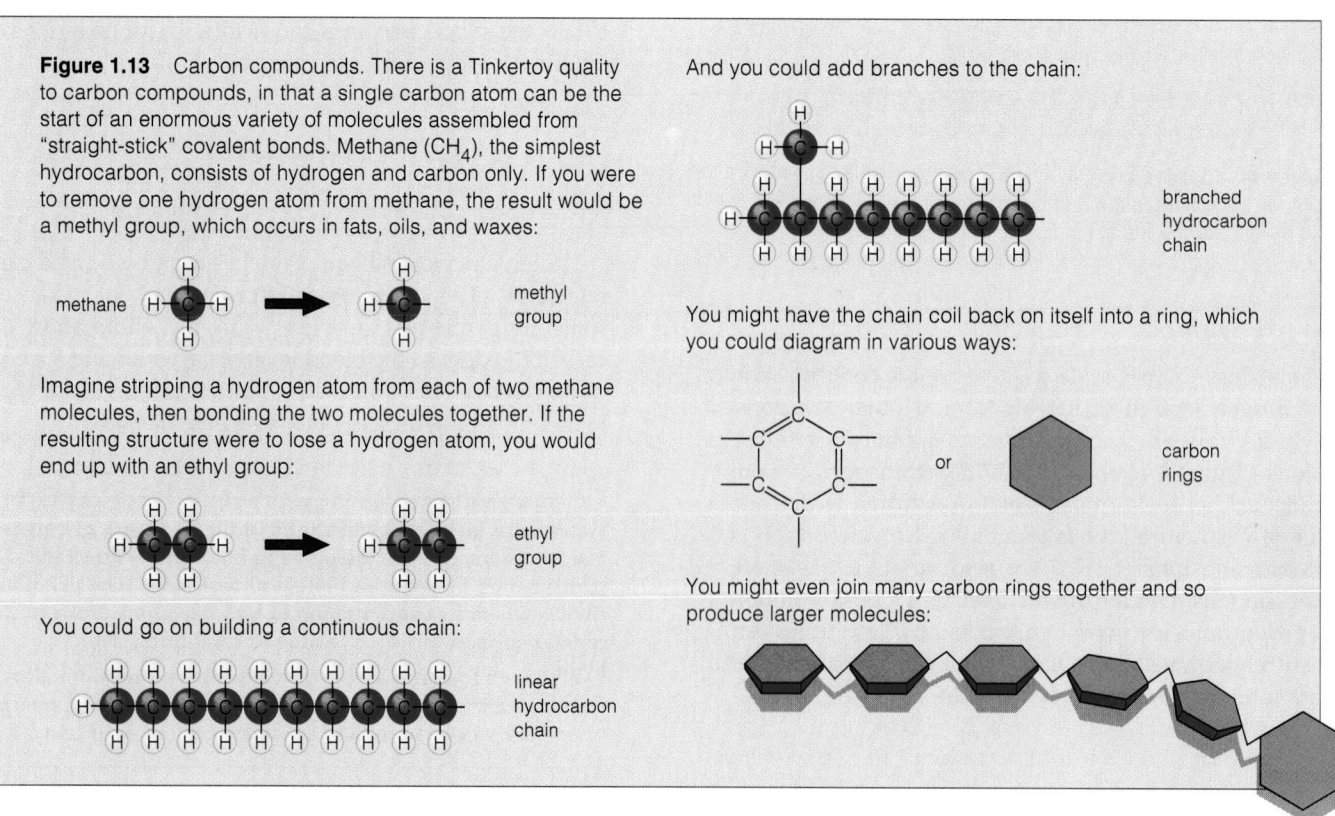

Figure 1.13 Carbon compounds. There is a Tinkertoy quality to carbon compounds, in that a single carbon atom can be the start of an enormous variety of molecules assembled from "straight-stick" covalent bonds. Methane (CH_4), the simplest hydrocarbon, consists of hydrogen and carbon only. If you were to remove one hydrogen atom from methane, the result would be a methyl group, which occurs in fats, oils, and waxes:

methane   methyl group

Imagine stripping a hydrogen atom from each of two methane molecules, then bonding the two molecules together. If the resulting structure were to lose a hydrogen atom, you would end up with an ethyl group:

ethyl group

You could go on building a continuous chain:

linear hydrocarbon chain

And you could add branches to the chain:

branched hydrocarbon chain

You might have the chain coil back on itself into a ring, which you could diagram in various ways:

or   carbon rings

You might even join many carbon rings together and so produce larger molecules:

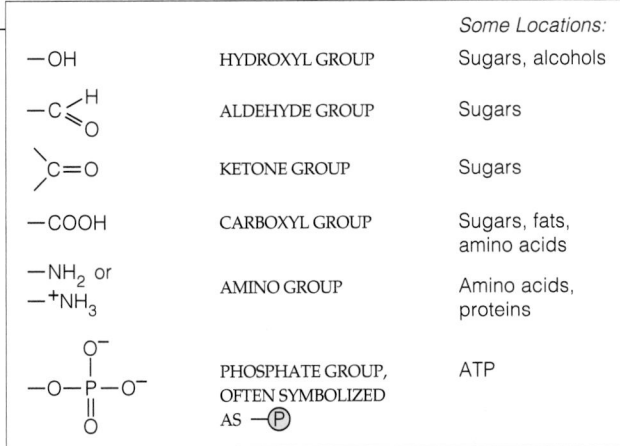

		Some Locations:
—OH	HYDROXYL GROUP	Sugars, alcohols
aldehyde structure	ALDEHYDE GROUP	Sugars
ketone structure	KETONE GROUP	Sugars
—COOH	CARBOXYL GROUP	Sugars, fats, amino acids
—NH₂ or —⁺NH₃	AMINO GROUP	Amino acids, proteins
phosphate structure	PHOSPHATE GROUP, OFTEN SYMBOLIZED AS —Ⓟ	ATP

Figure 1.14 A few examples of functional groups.

How Cells Use Organic Compounds— Five Classes of Reactions

Enzymes are proteins that enable metabolic reactions to proceed rapidly under the chemical and temperature conditions inside cells. You will read much more about them in Chapter 2. For now, it is enough to know that they function in five types of reactions by which most biological molecules are assembled, rearranged, and broken apart:

1. **Functional-group transfer**. One molecule gives up a functional group, which another molecule accepts.

2. **Electron transfer**. One or more electrons stripped from one molecule are donated to another molecule.

3. **Rearrangement**. Internal bonds are rearranged, converting one type of organic molecule into another.

4. **Condensation**. Through covalent bonding, two molecules combine to form a larger molecule.

5. **Cleavage**. A molecule splits into two smaller ones.

Condensation and Hydrolysis Reactions

To get a sense of what goes on, think about just two of these events. In many **condensation** reactions, enzymes remove a hydroxyl group from one molecule and an H atom from another. A covalent bond then forms at the exposed sites (Figure 1.15a). The discarded hydrogen and oxygen atoms may combine to form a water molecule (H_2O). Because this kind of reaction often forms water as a by-product, condensation is sometimes called *dehydration synthesis*. Cells assemble various polymers by repeated condensation reactions. A **polymer** is a large molecule composed of three to millions of subunits, which may or may not be identical. The individual subunits may be called **monomers**.

Hydrolysis, a type of cleavage reaction, is like condensation in reverse. It splits a molecule into two or more

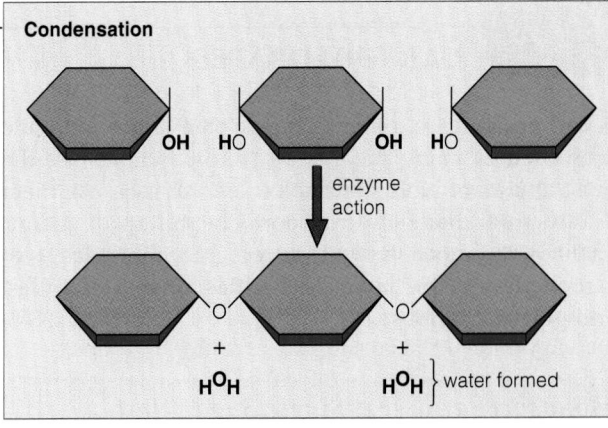

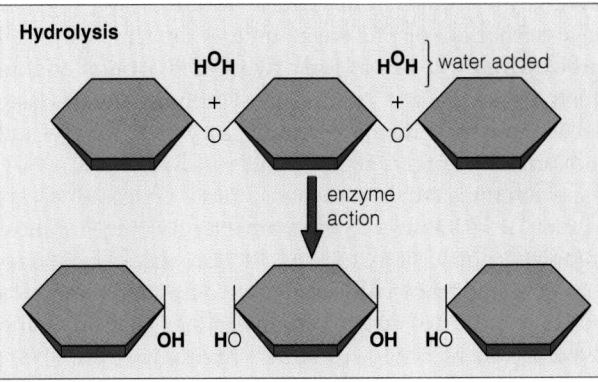

Figure 1.15 (**a**) Condensation. In this generalized example, three molecules covalently bond into a larger molecule and two water molecules are formed. (**b**) Hydrolysis, a water-requiring cleavage reaction. Covalent bonds in a molecule are broken, and H^+ and OH^- derived from water molecules bond to the molecular fragments.

parts. Simultaneously, —H and OH^- derived from water become attached to the exposed sites (Figure 1.15b).

The Molecules of Life

In nature, only living cells can synthesize the organic compounds known as carbohydrates, lipids, proteins, and nucleic acids. Together, *these are the molecules characteristic of life*. Different kinds function as packets of instant energy, energy stores, structural materials, or libraries of hereditary information; or they may play other basic roles.

Some biological molecules, such as simple sugars, fatty acids, amino acids, and nucleotides, are rather small organic compounds. As described next, they can serve as subunits for the synthesis of complex carbohydrates and other large biological molecules.

The diverse three-dimensional shapes and functions of organic compounds begin with flexible and rigid bonding arrangements in their carbon backbones. Functional groups covalently bonded to the backbone add to the diversity.

CARBOHYDRATES

A **carbohydrate** is a simple sugar or a larger molecule composed of many sugar units. Carbohydrates are the most abundant biological molecules. All cells use them as structural materials, transportable packets of energy, and stored forms of energy. We recognize three classes of carbohydrates: the **monosaccharides**, **oligosaccharides**, and **polysaccharides**.

Monosaccharides: A Single Sugar Unit

"Saccharide" comes from a Greek word meaning sugar. A *mono*saccharide, or one sugar unit, is the simplest of all carbohydrates. It has at least two —OH groups and an aldehyde or ketone group. Most simple sugars taste sweet, and they dissolve readily in water. Ribose and deoxyribose (sugar components of RNA and DNA, respectively) have five carbon atoms. Glucose has six (Figure 1.16a). Glucose is the main energy source for most organisms, including humans. It also serves as a precursor (parent molecule) of many compounds and as a building block for larger carbohydrates. Vitamin C and glycerol (an alcohol with three hydroxyl groups) are other examples of compounds that are derived from sugar monomers.

Oligosaccharides: Short Chains of Sugar Units

An *oligo*saccharide is a short chain of two or more covalently bonded sugar units. (*Oligo*- means few.) The simplest kinds, the *di*saccharides, have two sugar units. Lactose, sucrose, and maltose are examples. Lactose (a glucose and a galactose unit) is present in milk. Sucrose, the most plentiful sugar in nature, consists of a glucose and a fructose unit (Figure 1.16c). By crystallizing extracts of sucrose from sugar cane and sugar beets we get table sugar.

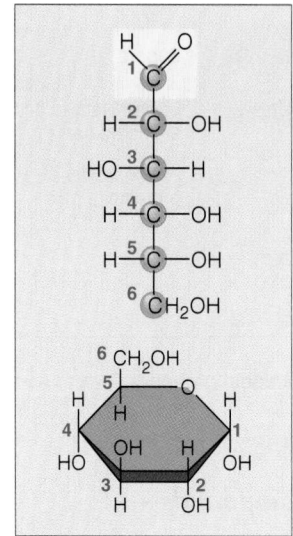
a Glucose

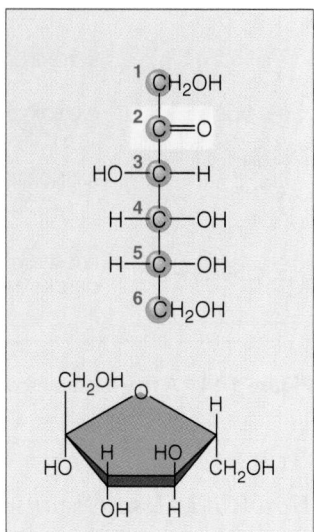
b Fructose

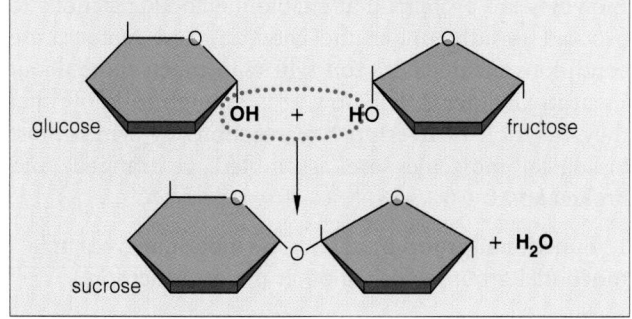
c Formation of sucrose

Figure 1.16 Straight-chain and ring forms of glucose (**a**) and fructose (**b**). For reference purposes, the carbon atoms of sugars are often numbered in sequence, starting at the end of the molecule closest to the aldehyde or ketone group (blocked in *yellow*). (**c**) Condensation of two monosaccharides (glucose and fructose) into a disaccharide (sucrose).

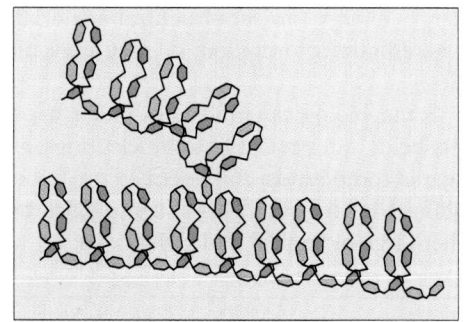

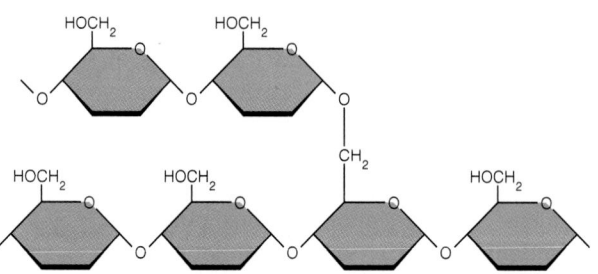

Figure 1.17 Glycogen, a form in which sugars are stored in some animal tissues, including human muscle and liver. A glycogen molecule has a branched structure, as the drawings show.

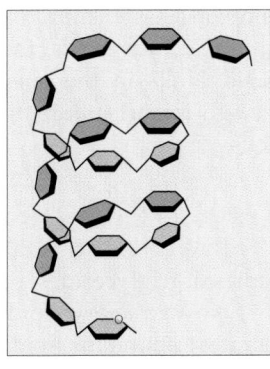
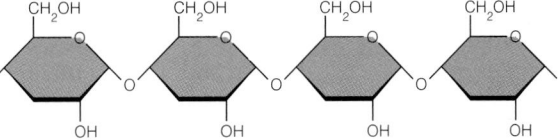

a Starch, a sugar storage form in plants. In amylose, the type of starch shown here, oxygen bridges link glucose subunits.

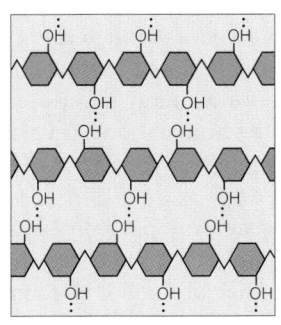
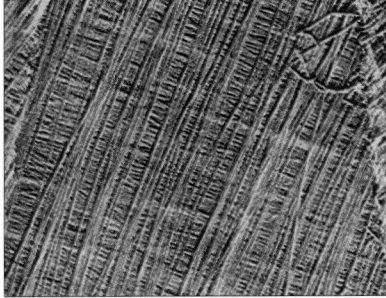

Structural formula for cellulose

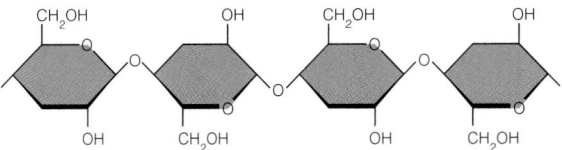

b In cellulose, the fine strands may twist together and then coil to form cellulose threads. The micrograph shows such threads.

Figure 1.18 Starch and cellulose.

Proteins and other large molecules often have oligosaccharides attached to their carbon backbone, as side chains. Some chains have roles in membrane function; others have roles in the body's responses to agents of disease.

Polysaccharides: Many Sugar Units

A *poly*saccharide is a straight or branched chain of many sugar units—often hundreds or thousands of the same or different kinds. Most of the carbohydrates you eat are in the form of polysaccharides. The most common—glycogen, starch, and cellulose—consist only of glucose.

When you eat meat you are consuming protein, and also glycogen. Glycogen is a sugar storage form in animals—notably in muscle tissues and the liver. When blood-sugar levels fall, liver cells break down glycogen and release glucose units to the blood. During exercise, muscle cells tap into their glycogen stores for quick access to energy. Figure 1.17 shows just a few of a typical glycogen molecule's numerous branchings.

Foods such as potatoes, rice, wheat, and corn are all rich in starch (Figure 1.18a), which is a storage form of glucose in plant cells. In starch the glucose subunits are arranged in a linear fashion. Cellulose is a tough, insoluble structural material in plant cell walls (Figure 1.18b). In cellulose, the links between neighboring glucose monomers have a different orientation than they do in glycogen and starch. Also, links form between —OH groups of adjacent cellulose molecules; the result is a fine strand. Humans do not have digestive enzymes that can break down (hydrolyze) the cellulose in vegetables, whole grains, and other plant tissues. We benefit from it, however, as undigested "fiber" that adds bulk and so helps move wastes through the lower digestive tract. You will read more about the health benefits of dietary fiber in Chapter 5.

Carbohydrates are either simple sugars (such as glucose) or molecules composed of many sugar units.

In order of their structural complexity, the three types of carbohydrates are monosaccharides, oligosaccharides, and polysaccharides.

All cells require carbohydrates, which the body uses for structural materials, to store energy, and as transportable packets of energy.

The greasy or oily compounds called **lipids** dissolve readily in one another, but they show very little tendency to dissolve in water. In most organisms, certain lipids are the main reservoirs of stored energy. Others are structural materials in cell components (such as membranes) and products (such as surface coatings). Here we focus on the neutral fats, phospholipids, and waxes, all of which have fatty acid components. We also consider the sterols, each with a backbone of four carbon rings.

Fatty Acids

Hydrocarbons called **fatty acids** have a backbone of up to 36 carbon atoms, a carboxyl (—COOH) group at one end, and hydrogen atoms occupying most or all of the remaining bonding sites. When combined with other molecules, fatty acids typically stretch out, like flexible tails. Tails with one or more double bonds in their backbone are *unsaturated*. Tails having only single bonds are *saturated*. Figure 1.19 shows examples of both types.

When many saturated fatty acid tails occur in a substance, weak attractions make them nestle in parallel, and this gives the substance a rather solid consistency.

The visible fat in a strip of bacon consists of saturated fatty acids. By contrast, in unsaturated fatty acids, the double and triple bonds put rigid kinks in the tails. The packing arrays are less stable and impart fluidity to substances—for example, many oils.

Neutral Fats (Triglycerides)

Butter, lard, and oils are all examples of **triglycerides**, or neutral fats—the most common type of dietary lipid. Triglycerides also are the body's most abundant lipids and its richest source of energy. They have fatty acid tails attached to a glycerol backbone (Figure 1.20). Gram for gram, triglycerides yield more than twice as much energy as carbohydrates. They have many more removable electrons than carbohydrates do—and energy is released when electrons are removed. In the human body, cells of fat-storing (adipose) tissue stockpile triglycerides as fat droplets.

The so-called *trans fatty acids* are only partially saturated ("hydrogenated"). These molecules often are the main ingredient in margarines that are solid at room temperature. There is now a growing body of evidence

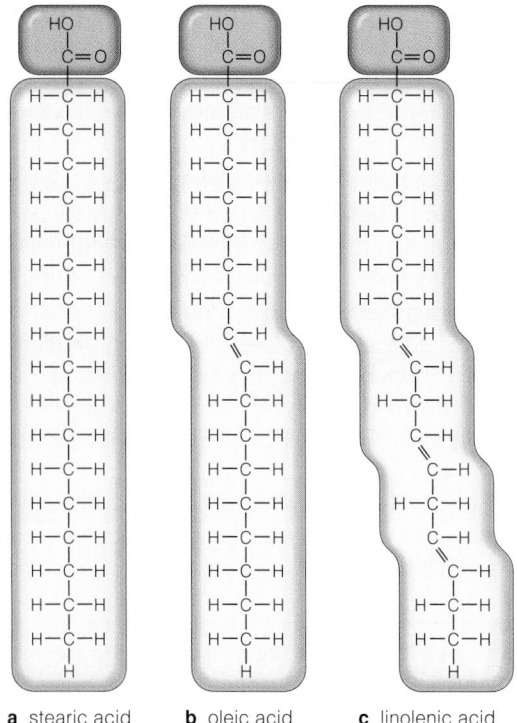

a stearic acid   **b** oleic acid   **c** linolenic acid

Figure 1.19 Structural formulas for three fatty acids. (**a**) Stearic acid's carbon backbone is fully saturated with hydrogens. (**b**) Oleic acid, with its double bond in the carbon backbone, is unsaturated. (**c**) Linolenic acid, with three double bonds, is a "polyunsaturated" fatty acid.

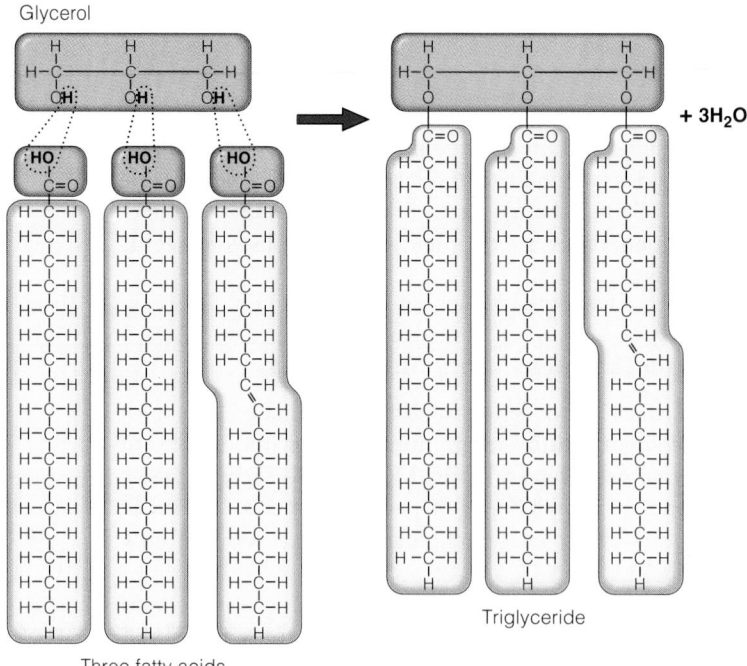

Figure 1.20 Condensation of fatty acids and glycerol into a triglyceride.

that trans fatty acids are linked to biochemical reactions associated with the development of some types of heart disease, a subject we will return to in Chapter 6.

Phospholipids

A **phospholipid** has a glycerol backbone, two fatty acid tails, and a hydrophilic "head" that includes a phosphate group and another polar group (Figure 1.21). Phospholipids are the main components of cell membranes, which have two layers of lipids. Heads of one layer are dissolved in the cell's fluid interior, and heads of the other layer are dissolved in the surroundings. All fatty acid tails, which are hydrophobic, are sandwiched between the two. If you care to look ahead, you can see this arrangement diagrammed in Section 2.1.

Waxes

The lipids called **waxes** have long-chain fatty acids, tightly packed and linked to long-chain alcohols or to carbon rings. Waxes have a firm consistency and repel

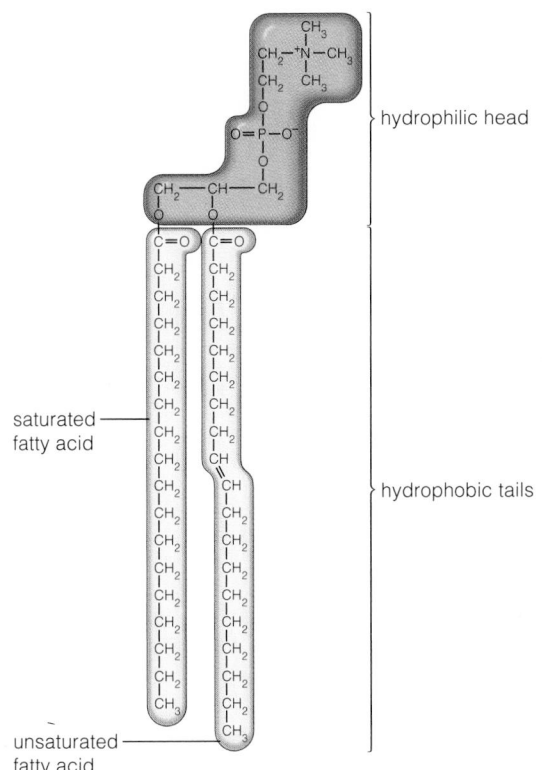

Figure 1.21 Structural formula of a typical phospholipid of animal cell membranes.

water. In humans, glands in the ear canal produce earwax, a yellow-brown substance that protects tissues from drying out. Sometimes the secretions are so prolific that wax obstructs the canal, causing discomfort and impairing a person's ability to hear. Because of the danger of damaging the eardrum, which lies at the inner end of the ear canal, only medical personnel should remove a severe earwax blockage.

Sterols and Substances Derived from Them

Sterols are among the lipids that have no fatty acid tails. Sterols differ in the number, position, and type of their functional groups, but they all have a rigid backbone of four fused-together carbon rings:

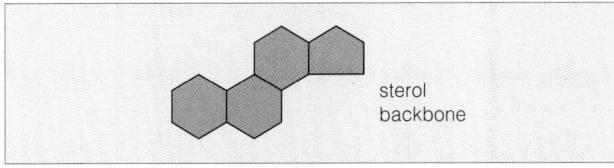

Many people think of the sterol cholesterol as a chemical villain because it is associated with heart and artery disease. As you will read in Chapter 6, serious cardiovascular disorders can arise when there is excess cholesterol in the blood. However, normal amounts of this sterol are not only *not* harmful, they are also crucial to the structure and proper functioning of cells. For instance, cholesterol is a vital component of membranes of every cell in your body. Cells also remodel cholesterol into other important substances. The derivatives include vitamin D (essential for bone and tooth development), bile salts (which assist fat digestion in the small intestine), and steroid hormones such as estrogen and testosterone.

Later chapters will tell more of the fascinating story of how steroid hormones influence reproduction, development, growth, and some other functions—and how artificial steroids used to increase muscle mass can lead to serious health problems.

Lipids are greasy or oily hydrophobic compounds. They include:

- **Neutral fats (triglycerides), major reservoirs of energy**

- **Phospholipids, the main components of cell membranes**

- **Waxes, firm yet pliable components of water-repelling and lubricating substances such as earwax**

- **Sterols (such as cholesterol), components of membranes and precursors of steroid hormones and other vital molecules.**

1.9 AMINO ACIDS AND PROTEINS

Of all the large biological molecules, **proteins** are the most diverse. The proteins called enzymes make metabolic reactions proceed much faster than they otherwise would. Structural proteins are major components of your bones, muscles, hair, and other body elements. Transport proteins help move substances across cell membranes or through body fluids. Regulatory proteins, including the protein hormones, help alter body functions by serving as signals for change in cell activities. Without them, many bodily transitions and functions—waking, sleeping, puberty, and sex, to cite just a few—would be impaired or impossible. Many other proteins function as weapons against harmful bacteria and other invaders. Amazingly, your cells build all these diverse proteins from their pools of only 20 or so kinds of amino acids.

An **amino acid** is a small organic compound. It is built of an amino group, a carboxyl group (an acid), a hydrogen atom, and one or more atoms called its R group. These parts are covalently bonded to the same carbon atom:

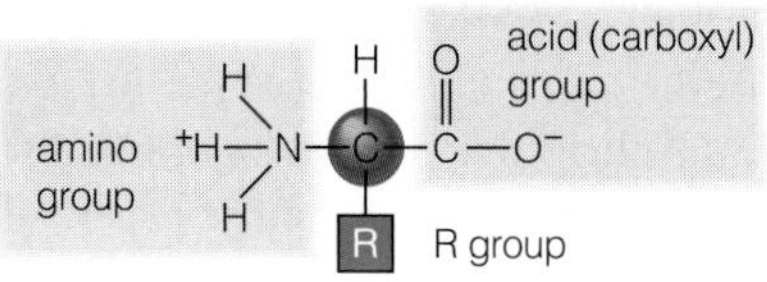
amino group
acid (carboxyl) group
R group

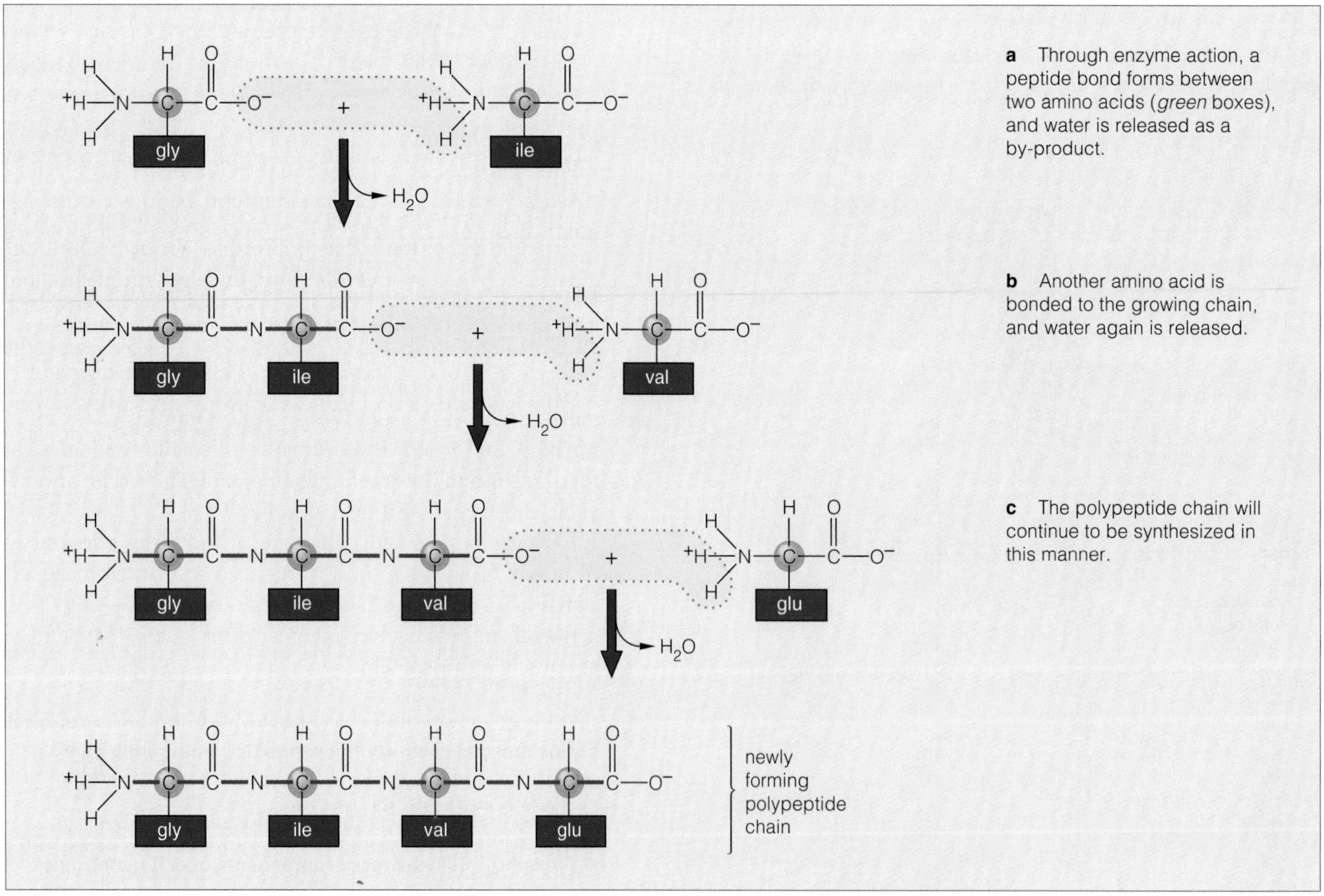
a Through enzyme action, a peptide bond forms between two amino acids (*green* boxes), and water is released as a by-product.
b Another amino acid is bonded to the growing chain, and water again is released.
c The polypeptide chain will continue to be synthesized in this manner.
gly
ile
val
glu
H_2O
newly forming polypeptide chain

Figure 1.22 Peptide bond formation during protein synthesis. This is a condensation reaction.

Primary Structure of Proteins

When cells synthesize proteins, amino acids become linked, one after the other, by *peptide* bonds. As Figure 1.22 shows, this type of covalent bond forms between the amino group of one amino acid and the acid group (carboxyl) of another. Three or more amino acids joined this way form a **polypeptide chain**. The backbone of each chain incorporates nitrogen atoms in this regular pattern: —N—C—C—N—C—C—.

For each particular kind of protein, different amino acid units are selected in a specific order, one at a time, from the 20 kinds available. As you will learn in a later chapter, the order in which amino acids are "chosen" is determined by your genes. Overall, the resulting sequence of linked amino acids is unique for each kind of protein. The sequence represents the protein's *primary* structure. The primary structure of the largest known protein, which is a component of human muscle, is a string of some 27,000 amino acids!

Three-Dimensional Structure of Proteins

Cells make thousands of different proteins. Many are fibrous, with polypeptide chains organized as strands or sheets. Collectively, many such molecules contribute to the shape, internal organization, and movement of your body cells. Many other proteins are globular, with one or more polypeptide chains folded into compact, rounded shapes. Most enzymes are globular proteins.

Each protein's shape and function arise from its primary structure—that is, from information built into its amino acid sequence. That information dictates which parts of a polypeptide chain will coil, bend, or even interact with other chains nearby. And the type of arrangement of atoms in the coiled, stretched-out, or folded regions determines whether a protein will act as, say, an enzyme or a receptor. In such ways, a protein's primary structure dictates both its final structure and its chemical behavior.

Exactly how does primary structure give rise to a protein's shape? *First*, it allows hydrogen bonds to form between different amino acids in a polypeptide chain. *Second*, it puts R groups in positions that allow them to interact. The interactions induce the chain to bend and twist. In many chains, hydrogen bonds form between every third amino acid. The bonding pattern twists the peptide groups into a helical coil like a spiral staircase (Figure 1.23a). In other cases, the hydrogen-bonding pattern holds two or more chains side by side, in a sheet-like array (Figure 1.23b).

Thus proteins also have *secondary* structure, a coiled or extended pattern that is brought about by hydrogen bonds at regular intervals along their polypeptide chains. Most coiled chains undergo further folding. The folding happens when certain R groups interact with other R groups some distance away, with the backbone of the chain, or with cell substances. Folding that arises through interactions among R groups of a polypeptide chain is called the *tertiary* structure of proteins. The following section gives examples of the resulting three-dimensional shapes.

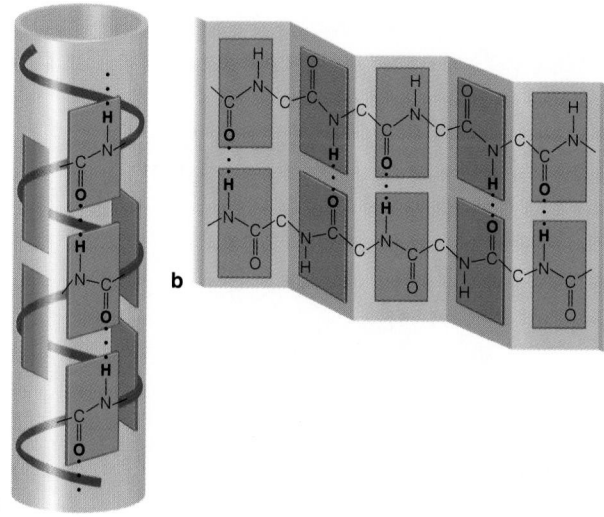

Figure 1.23 Hydrogen bonds (dotted lines) in a polypeptide chain can give rise to a coiled chain (**a**) or a sheetlike array of chains (**b**).

A protein consists of one or more chains of amino acids. During protein synthesis the amino acids are joined together by peptide bonds.

The sequence of amino acids (which kind follows another) is unique for each kind of protein; this sequence is the protein's primary structure. It dictates the protein's unique final structure and its function.

Proteins also have secondary structure, a coiled or extended pattern brought about by hydrogen bonding at intervals along polypeptide chains. Most chains also undergo further folding into a three-dimensional tertiary structure.

SOME EXAMPLES OF FINAL PROTEIN STRUCTURE

Hemoglobin and Other Complex Proteins

As you read this, every single mature red blood cell in your body is transporting about a billion molecules of oxygen, bound to 250 million molecules of **hemoglobin**. Hemoglobin is an example of a protein with *quaternary* structure. In such proteins, numerous hydrogen bonds and some other weak interactions hold two or more polypeptide chains together tightly.

Hemoglobin has four polypeptide chains (Figure 1.24). Each chain is a separate protein, of a type called globin. Hydrogen bonding makes the polypeptide chain of each globin molecule coil and fold into a compact shape. At the chain's center is a ringlike, iron-containing group of molecules called heme. In some other proteins with quaternary structure, disulfide bridges help hold the structure together. Each bridge links the sulfur atoms of two units of a particular amino acid on the chain.

Hemoglobin is a fine example of a globular protein. Keratin, a structural protein of hair (Figure 1.25), is a fibrous one. So is collagen, the most common protein in the human body. Skin, bones, corneas, blood vessels, heart valves—these and other parts of your body depend on the strength inherent in collagen.

One of the rules that applies to biological systems is that there is no separating structure and function. Proteins are a perfect example. To function properly, a given protein must fold properly into its final shape. If some event disrupts or alters the folding, the affected protein molecules can form tangled clumps. Evidence is mounting that misfolded, malfunctioning proteins underlie some symptoms of certain diseases, including Alzheimer's disease (Chapter 10).

In later chapters you will encounter physiological processes in which lipoproteins and glycoproteins are key players. Lipoproteins form when certain proteins that circulate freely in the blood encounter and combine with cholesterol, triglycerides, and phospholipids that were absorbed from the gastrointestinal tract after a

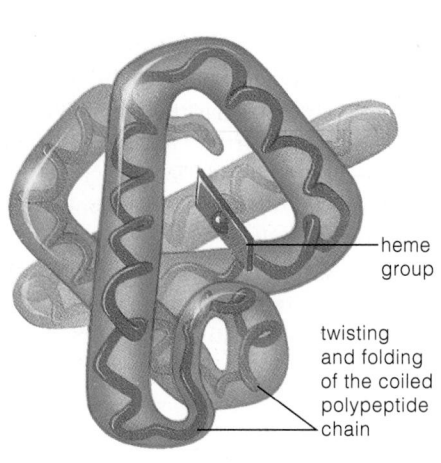

heme group

twisting and folding of the coiled polypeptide chain

a

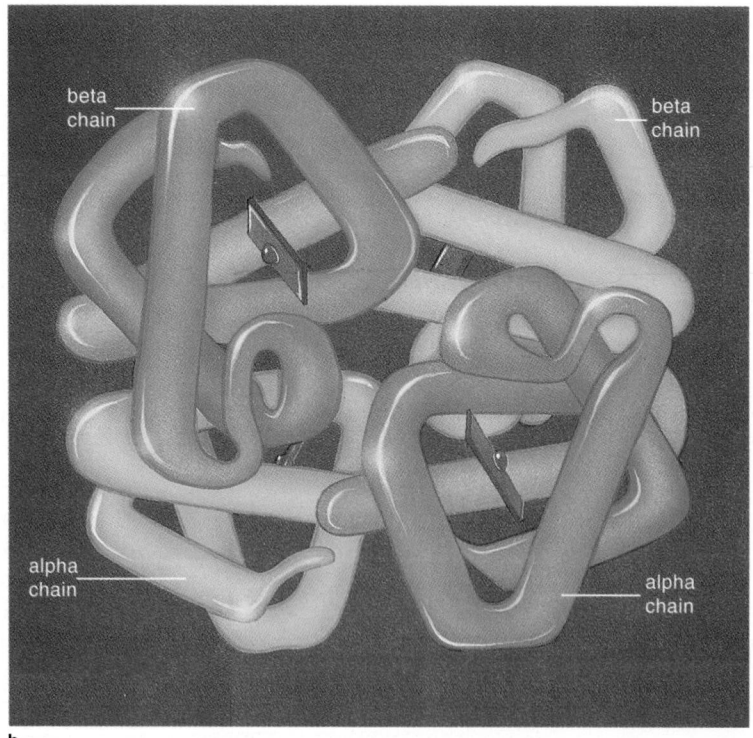

beta chain

beta chain

alpha chain

alpha chain

b

Figure 1.24 (**a**) One of the four polypeptide chains in hemoglobin, an oxygen-transporting protein in blood. The dark green "ribbon" represents the polypeptide chain. Heme, an iron-containing group, binds the oxygen. (**b**) Quaternary structure of human hemoglobin. Many weak bonds hold the four chains tightly together.

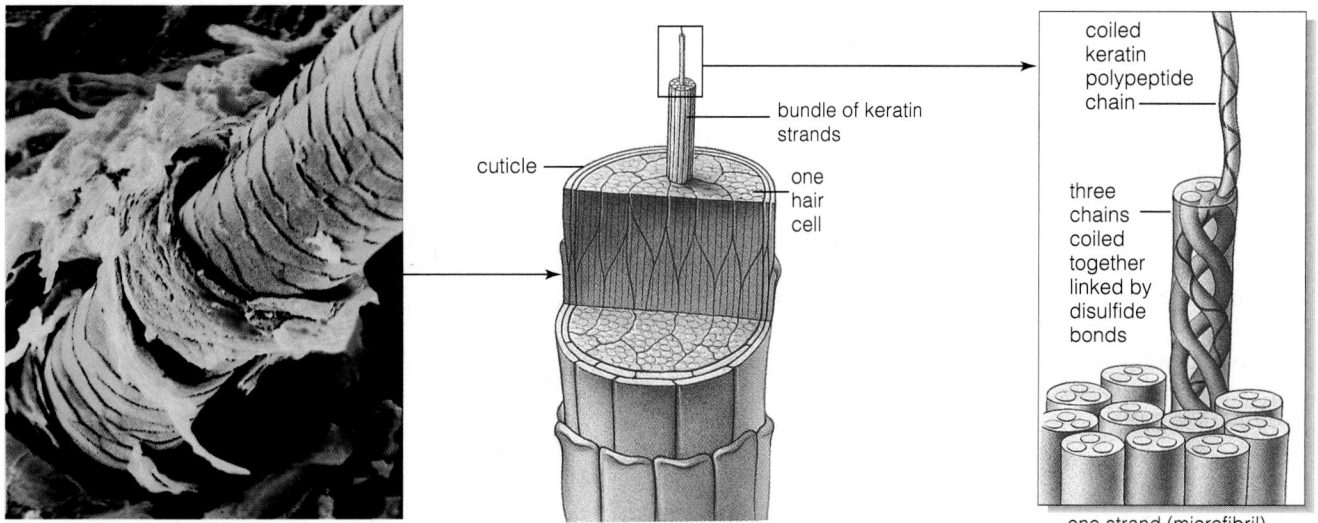

Figure 1.25 Structure of hair. Polypeptide chains of the protein keratin are synthesized inside hair cells, which are derived from cells of the skin. The chains become organized into fine fibers (microfibrils), which in turn become bundled into larger, cablelike fibers (macrofibrils). These practically fill the cells, which eventually die. Dead, flattened cells form a tubelike cuticle around the developing hair shaft.

meal. Not all lipoproteins are the same; they differ from one another in their combinations of protein and lipid components.

Most **glycoproteins** form when new polypeptide chains are being modified into mature proteins. These proteins have oligosaccharides covalently bonded to them. Some of the attached oligosaccharides are linear chains; others are branched. Nearly all of the proteins on the outer surface of animal cells, including those of the human body, are glycoproteins. So are the majority of proteins secreted from body cells (including protein hormones), and most proteins in blood.

Denaturation and Protein Structure

Breaking the weak bonds of a protein or any other large molecule disrupts its three-dimensional shape. We call this event **denaturation**. For example, individually, hydrogen bonds are weak—and they are sensitive to changes in temperature and pH. If the temperature or pH exceeds a given protein's range of tolerance, then its polypeptide chains will unwind or change shape, and the protein will no longer function.

Consider albumin, a protein that is concentrated in the egg white of chicken eggs. When you cook eggs, the heat doesn't disrupt the strong covalent bonds of albumin's primary structure. But it destroys weaker bonds contributing to the three-dimensional shape.

For some proteins, denaturation can be reversed when normal conditions are restored. For instance, you can curl your hair using a curling iron, which temporarily denatures the protein keratin in hair, but within a few hours the curled hairs resume their former shape. With albumin the story is different. There is no way to uncook a cooked egg white.

Many proteins incorporate two or more polypeptide chains, held together by numerous hydrogen bonds. The result is a fourth level of protein structure, called quaternary structure. The blood protein hemoglobin is an example.

Denaturation occurs when some event—such as a change in temperature or pH—breaks the weak bonds of a protein (or other large molecule) and so disrupts its three-dimensional shape. Denaturation sometimes is reversible, but in other cases the structural and functional change in the affected protein is permanent.

NUCLEOTIDES AND NUCLEIC ACIDS

Nucleotides with Roles in Metabolism

The small organic compounds called **nucleotides** have three components: a five-carbon sugar, a phosphate group, and a nitrogen-containing base. The sugar is either ribose or deoxyribose. The base has a single- or a double-carbon ring structure (Figure 1.26).

One nucleotide is central to metabolism. We call it **ATP** (adenosine triphosphate). In a living cell, molecules of ATP couple chemical reactions that *release* energy with other reactions that *require* energy. ATP is so versatile that it can deliver energy from the site of one reaction to virtually any other reaction site in cells. Chapter 2 will discuss in more detail how this crucial molecule functions.

Other nucleotides are subunits of **coenzymes**, which are "go-between" molecules. Some coenzymes accept hydrogen atoms and electrons that are being removed from other molecules and transfer the hydrogens and electrons to other locations where they can be used for other reactions. Examples of coenzymes include certain vitamins and compounds abbreviated NAD$^+$ and FAD.

Still other nucleotides function as chemical messengers within and between cells. One messenger nucleotide (cyclic adenosine monophosphate, or cAMP) is of major importance in processes you will learn about later in the book.

Nucleic Acids—DNA and RNA

In **nucleic acids** four kinds of nucleotides are bonded together in large single- or double-stranded molecules. Each strand's backbone is made up of sugars that are joined to the phosphate groups of adjacent nucleotides by covalent bonds. The nucleotide bases stick out to the side, as you can see in Figure 1.27a. Each kind of nucleic acid has strand regions in which the sequence of particular bases is unique.

Two major nucleic acids are **RNA** (ribonucleic acid) and **DNA** (deoxyribonucleic acid). RNA is usually a single nucleotide strand. As you will see, RNA functions include roles in the processes by which genetic instructions from DNA are used to build the body's proteins. DNA is usually a double-stranded molecule that twists helically, very much like a spiral staircase (Figure 1.27b). Hydrogen bonds hold the two strands together. Genetic instructions for building and operating body structures are encoded in the sequence of bases in DNA.

You will be reading a good deal more about all of the carbon-based molecules introduced in this chapter, including nucleotides and nucleic acids. On the facing page we conclude with *Focus on Our Environment*, an essay that describes ways in which some organic molecules may become valuable sources of nonpolluting fuel in our energy-hungry world.

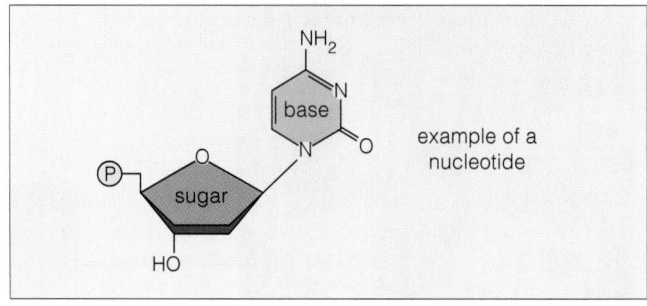

Figure 1.26 Structure of a nucleotide.

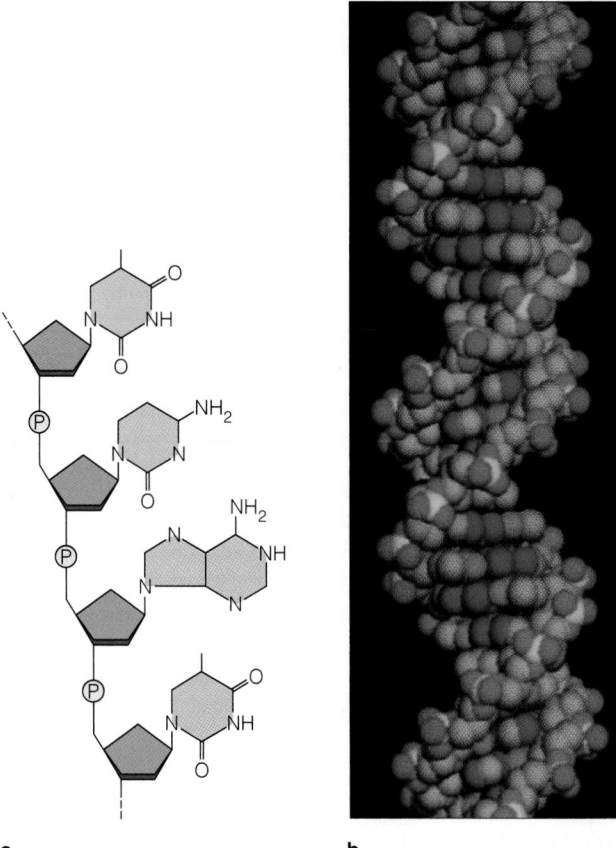

a   b

Figure 1.27 (**a**) Examples of bonds between nucleotides in a nucleic acid molecule. (**b**) Model of DNA. The molecule consists of two strands of nucleotides joined by hydrogen bonds and twisted into a double helix. The nucleotide bases are *blue*.

In nucleic acids, four kinds of nucleotides are bonded together in large single-stranded molecules (RNA) or in double-stranded ones (DNA).

Genetic information is encoded in the particular order of nucleotide bases that follow one another in DNA.

1.12 MAKING NONPOLLUTING FUEL

Hydrogen gas is an environmentally friendly fuel. When it burns, the by-product is pure water. Large-scale use of hydrogen gas for fuel—to operate motor vehicles, for example—would eliminate air and water pollution caused by extracting, transporting, and burning fossil fuels such as oil and coal. Chapter 25 looks in depth at the strong links between fossil-fuel burning and global warming.

Hydrogen gas can be produced by passing an electric current through water, breaking it down into its oxygen and hydrogen components. Once formed, hydrogen gas can be stored in a pressurized tank or in metal hydrides—metal powders that absorb gaseous hydrogen and release it when heated for use as a fuel in a car, furnace, or electricity-producing fuel cell. Unlike gasoline, solid metallic hydrogen compounds will not explode or burn if a vehicle's tank is ruptured in an accident. By weight, hydrogen has about 2.5 times the energy of gasoline, making it an especially attractive fuel for aircraft. However, there's a catch in this scenario. Because present methods of generating hydrogen require electricity or some other form of energy, using hydrogen gas for fuel is too expensive to be cost-effective.

The clear environmental gains associated with using hydrogen as fuel have prompted researchers to look for a thrifty way of making it. One answer may reside with glucose, the simple sugar molecule that contains ample hydrogen. Glucose is one of nature's ultimate renewable resources. For instance, as described in Section 1.7, plants manufacture starch and cellulose, both of which consist of strings of glucose molecules. Plant parts such as tubers (for example, potatoes) are packed with starch. Discarded newsprint and cardboard, wood chips, grass clippings and other garden trimmings—to name just a few items— all are rich sources of cellulose (Figure 1.28).

Proponents say that the key to launching a "hydrogen revolution" is finding a cheap way to liberate the hydrogen in cellulose and starch. As it happens, certain bacteria already do just that. They make enzymes (the catalytic proteins described in Section 1.6) that break down cellulose or starch in steps that liberate hydrogen. When the proper mix of raw materials is present, the hydrogen-liberating reactions take place in a cycle that can be repeated over and over again. Considerable additional research must be done before such a process could be implemented on a commercial scale, but the possibilities are exciting. By one estimate, simply converting the cellulose in waste newspaper could provide enough hydrogen to meet the annual energy needs of a city of one million people.

Other nations are actively exploring the possibilities of various hydrogen-based systems. A test fleet of 200 cars, a bus, and even a large jet airplane have been running on hydrogen fuel for several years in the former Soviet Union. In Japan, one automaker unveiled a prototype car that runs on hydrogen released slowly from metal hydrides heated by the car's radiator coolant. In Germany, researchers are developing small hydrogen-producing systems that could be used by individuals for home heating, cooking, and other power needs. Germany and Saudi Arabia have jointly built a large solar-hydrogen plant, in which sunlight provides the energy to split water into hydrogen and oxygen.

Figure 1.28 Discarded newspapers—one possible source of cellulose, a carbohydrate in plant tissues from which hydrogen may be extracted and used as fuel.

SUMMARY

1. Protons, neutrons, and electrons are the basic constituents of atoms. An element's atoms have the same number of protons and electrons but may vary in the number of neutrons. The variant forms are isotopes. Whether an atom interacts with others depends on the number and arrangement of its electrons.

2. By one model, electrons occupy orbitals (volumes of space) inside a series of shells around the nucleus of an atom. Atoms with one or more unfilled orbitals in their outermost shell tend to participate in chemical bonds.

3. Hydrogen, oxygen, carbon, and nitrogen are the most abundant elements in organisms. They all tend to form bonds with other atoms.

4. Atoms have no net charge, but may gain or lose one or more electrons and become ions, with an overall positive or negative charge.

5. In ionic bonds, positive and negative ions remain together simply by the mutual attraction of their opposite charges. In covalent bonds, atoms share one or more electrons. In hydrogen bonds, a small electronegative atom interacts weakly with a hydrogen atom that is already part of a polar covalent bond.

6. A pH value refers to the concentration of hydrogen ions in some fluid. Acids release hydrogen ions (H^+); bases release hydroxide ions (OH^-) that can combine with H^+. At pH 7, the H^+ and OH^- concentrations in a solution are equal. Buffering mechanisms maintain pH values of blood, tissue fluids, and the fluid inside cells.

7. Water is crucial for the metabolic activity, shape, and internal organization of cells. Because of hydrogen bonds between its polar molecules, water has a range of special properties. Among these are its ability to resist temperature changes and its remarkable capacity to dissolve other polar substances.

8. A carbon atom forms up to four covalent bonds with other atoms. Carbon atoms bonded together in linear or ring structures are the backbone of organic compounds. The chemical and physical properties of many of those compounds depend largely on functional groups (atoms attached to the backbone).

9. Cells assemble, rearrange, and break apart most organic compounds by five kinds of enzyme-mediated reactions: functional-group transfers, electron transfer, internal rearrangements, condensation reactions, and cleavages (including hydrolysis).

10. Cells have pools of dissolved sugars, fatty acids, amino acids, and nucleotides, all of which are small organic compounds that include no more than 20 or so carbon atoms. They are building blocks for the larger biological molecules—the polysaccharides, lipids, proteins, and nucleic acids (Table 1.2).

Review Questions

1. Carbohydrates, lipids, proteins, and nucleic acids are assembled from four main families of small organic molecules. What are these families? 25

2. Which of the following is the carbohydrate, the fatty acid, the amino acid, and the polypeptide? 26, 28, 30
   a. $^+NH_3 \diagdown CHR \diagdown COO^-$
   b. $C_6H_{12}O_6$
   c. $(glycine)_{20}$
   d. $CH_3(CH_2)_{16}COOH$

3. Is the following statement true or false? Most enzymes are proteins, but only some proteins are enzymes. 25

4. Describe the four levels of protein structure. How do a protein's side groups influence its interactions with other substances? What is denaturation? 30–33

5. Distinguish among the following:
   a. monosaccharide, polysaccharide, disaccharide 26–27
   b. peptide bond, polypeptide 30
   c. glycerol, fatty acid 28
   d. nucleotide, nucleic acid 34

Critical Thinking: You Decide (Key in Appendix V)

1. A new high-carbohydrate diet claims to rid the body of sugars. Without sugars, the diet says, your body will burn away fat. Given what you know of biological molecules, how would you critique this claim?

2. Black coffee has a pH of 5, whereas milk of magnesia has a pH of 10. Is coffee twice as acidic as milk of magnesia?

3. Your cotton shirt has stains from whipped cream and strawberry syrup, and your dry cleaner says that two separate cleaning agents will be needed to remove the stains. Explain why two agents are needed and what different chemical characteristic each would have.

4. A store clerk tells you that "natural" vitamin C extracted from rose hips is better for you than synthetic vitamin C. Based on what you know of the structure of organic compounds, does this claim seem credible? Why or why not?

Self-Quiz (Answers in Appendix IV)

1. The backbone of organic compounds forms when _____ atoms are covalently bonded into chains and rings.

2. A carbon atom can form up to _____ bonds with other atoms.
   a. four
   b. six
   c. eight
   d. sixteen

3. All of the following *except* _____ are small organic molecules that serve as the main building blocks or energy sources in cells.
   a. fatty acids
   b. simple sugars
   c. lipids
   d. nucleotides
   e. amino acids

4. Which of the following is *not* a carbohydrate?
   a. glucose molecule
   b. simple sugar
   c. fat
   d. polysaccharide

5. _____, a class of proteins, make metabolic reactions proceed much faster than they would on their own.
   a. DNA molecules
   b. Amino acids
   c. Fatty acids
   d. Enzymes

6. Examples of nucleic acids, the basis of inheritance, are _____.
   a. polysaccharides
   b. DNA and RNA
   c. proteins
   d. simple sugars

7. Which of the following best describes the role of functional groups?
   a. assembling large organic compounds
   b. influencing the behavior of organic compounds

Table 1.2 Summary of the Main Carbon Compounds In Living Things

Category	Main Subcategories	Some Examples and Their Functions	
Carbohydrates *contain an aldehyde or a ketone group and one or more hydroxyl groups*	**Monosaccharides (simple sugars)** **Oligosaccharides**	Glucose Sucrose (a disaccharide)	Structural roles, energy source Form of sugar transported in plants
	Polysaccharides (complex carbohydrates)	Starch Cellulose	Energy storage Structural roles
Lipids *are largely hydrocarbon, generally do not dissolve in water but dissolve in nonpolar solvents*	**Lipids with fatty acids:** *Glycerides:* one, two, or three fatty acid tails attached to glycerol backbone *Phospholipids:* phosphate group, another polar group, and (often) two fatty acids attached to glycerol backbone	Fats (e.g., butter) Oils (e.g., corn oil) Phosphatidylcholine	Energy storage Key component of cell membranes
	Waxes: long-chain fatty acid tails attached to alcohol	Earwax	Helps tissues retain moisture
	Lipids with no fatty acids: *Steroids:* four carbon rings; the number, position, and type of functional groups vary	Cholesterol	Component of animal cell membranes, can be rearranged into other steroids (e.g., vitamin D, sex hormones)
Proteins *are polypeptides (up to several thousand amino acids, covalently linked)*	**Fibrous proteins:** Individual polypeptide chains, often linked into tough, water-insoluble molecules	Keratin Collagen	Structural element of hair, nails Structural element of bones and cartilage
	Globular proteins: One or more polypeptide chains folded and linked into globular shapes; many roles in cell activities	Enzymes Hemoglobin Insulin Antibodies	Increase in rates of reactions Oxygen transport Control of glucose metabolism Tissue defense
Nucleic Acids (and Nucleotides) *are chains of units (or individual units) that each consist of a five-carbon sugar, phosphate, and a nitrogen-containing base*	**Adenosine phosphates** **Nucleotide coenzymes**	ATP NAD^+, $NADP^+$	Energy carrier Transport of protons (H^+) and electrons from one reaction site to another
	Nucleic acids: Chains of thousands to millions of nucleotides	DNA, RNAs	Storage, transmission, translation of genetic information

c. splitting molecules into two or more parts
d. speeding up metabolic reactions

8. In _____ reactions, small molecules become covalently linked, and water can also form.
   a. symbiotic   c. condensation
   b. hydrolysis   d. ionic

9. Match each type of molecule with the correct description.

   _____ chain of amino acids   a. carbohydrate
   _____ energy carrier   b. phospholipid
   _____ glycerol, fatty acids,   c. protein
          phosphate   d. DNA
   _____ chain of nucleotides   e. ATP
   _____ one or more sugar units

10. The shape of large molecules such as proteins is often controlled by what kinds of bonds?
    a. hydrogen   c. covalent   e. single
    b. ionic   d. inert

Readings

Goodsell, D. September–October 1992. "A Look inside the Living Cell." *American Scientist*. Current models of biological molecules.

 For additional readings, go to InfoTrac College Edition, your online library at:

http://www.infotrac-college.com/wadsworth

Selected Key Terms

acid 22
amino acid 30
atom 16
ATP 34
base 22
buffer 23
carbohydrate 26
chemical bonds 19
coenzyme 34
compound 17
condensation 25
covalent bond 20
denaturation 33
DNA 34
electron 16
element 16
enzyme 25
fatty acid 28
functional group 24
glycoprotein 33
hydrocarbon 24
hydrogen bond 21
hydrophilic 22
hydrophobic 22
ion 20
ionic bond 20
isotope 16
lipid 28
molecule 17
monomer 25
monosaccharide 26
neutron 16
nucleic acid 34
nucleotide 34
oligosaccharide 26
organic compound 24
pH scale 23
phospholipid 29
polymer 25
polypeptide chain 31
polysaccharide 26
protein 30
proton 16
RNA 34
salt 22
solute 22
sterol 29
triglyceride 28
wax 29

2 CELLS

Cells Filled with Juices

Around the middle of the 17th century, Robert Hooke, Curator of Instruments for the Royal Society of England, stood at the forefront of a new scientific enterprise—using a magnifying instrument to explore aspects of the living world that are invisible to the unaided eye. When Hooke first peered into a crude microscope to view thinly sliced cork from a tree, he observed tiny compartments (Figure 2.1a). He gave them the Latin name *cellulae*, meaning small rooms—hence the origin of the biological term "cell." In other tissues he discovered cells "fill'd with juices," but could not imagine what they represented.

Given the simplicity of their instruments, it is truly amazing that Hooke and other pioneers in microscopy saw as much as they did. Antony van Leeuwenhoek, a Dutch shopkeeper (Figure 2.1b), had great skill in constructing lenses and possibly the keenest vision. By the late 1600s he was observing wonders such as the "many very small animalcules" in scrapings of tartar from his own teeth. Van Leeuwenhoek's specimens even included sperm cells, which he sketched (Figure 2.1c). By the end of the 19th century, microscopic analysis had advanced to the point where scholars could generally agree upon three generalizations, which together constitute the **cell theory**: First, all organisms, including humans, are composed of one or more cells. Second, the cell is the smallest unit having the properties of life. Third, the continuity of life arises directly from the growth and division of single cells. These insights still hold true.

Today we know that each living cell is a highly organized bit of life in a world that is usually less organized and predictable. In this chapter you will discover that cells generally are built in such a way that they can bring in certain substances, release or keep out

Figure 2.1 Early glimpses into the world of cells. (**a**) Robert Hooke's compound microscope and his drawing of cell walls from cork tissue. (**b**) Antony van Leeuwenhoek, microscope in hand, and (**c**) one of his early sketches of sperm cells.

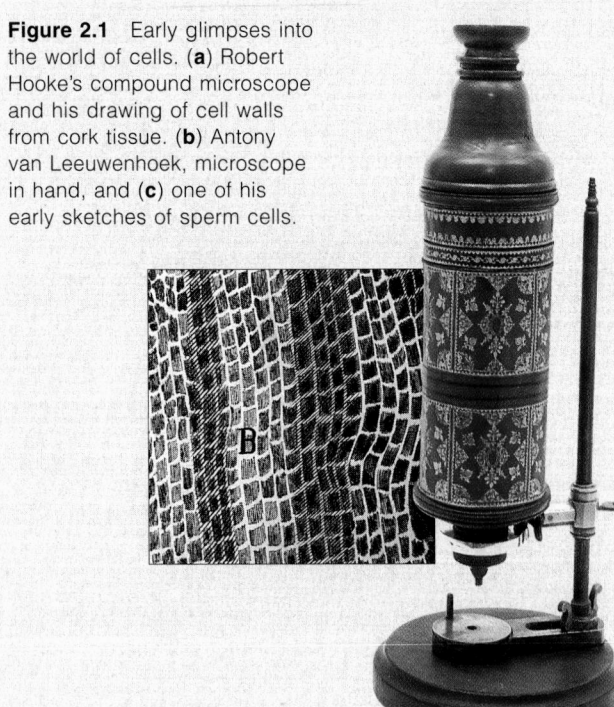

a

b

others, and conduct their internal activities with great precision—all at a breakneck pace. Every cell also is adapted to function under particular environmental conditions, which the body's homeostatic mechanisms must maintain.

The precision and control necessary to keep each cell alive begin at the plasma membrane, a seemingly flimsy surface layer of little more than lipids and proteins. Across this membrane, water molecules and other materials are selectively exchanged between the cell's surroundings and its interior. Further exchanges are made across internal cell membranes, which form compartments called organelles. Organelles perform a great variety of functions that help cells survive and reproduce. The sum total of these activities is called metabolism. We'll also consider in this chapter how cells obtain the energy that keeps the metabolic fires burning.

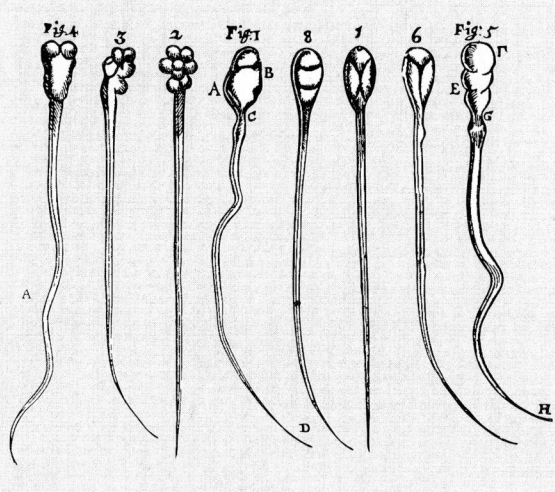

KEY CONCEPTS

1. The cell is the smallest unit having the characteristics of life.

2. Cells have an outer plasma membrane that helps their interior remain distinct from the surroundings. Cells contain cytoplasm, a region that is organized for energy conversions, protein synthesis, movements, and other activities. The cells of humans and other complex (eukaryotic) organisms contain a nucleus in which DNA is located.

3. The nucleus is one of many organelles. Organelles are membrane-bound compartments inside eukaryotic cells. They physically separate different metabolic reactions and allow them to take place in an orderly way.

4. Cell membranes consist mostly of phospholipids and proteins. The phospholipids form two adjacent layers that give the membrane its basic structure and prevent water-soluble substances from freely crossing it. The proteins perform most other membrane functions.

5. Whether a given molecule or ion crosses a membrane depends partly on its concentrations on either side of the membrane. Substances tend to diffuse from regions of higher to lower concentration. Membrane proteins work with or against this tendency.

6. Cells harness and use energy for building, storing, breaking apart, and eliminating substances. These activities are called metabolism.

7. Inside cells, enzymes greatly increase the rate of specific biochemical reactions.

8. ATP transports usable energy in chemical form from one reaction site to another. Most of the ATP used by human cells is generated by aerobic respiration.

CHAPTER AT A GLANCE

2.1   Basic Cell Structure and Function

2.2   How Water Crosses Cell Membranes

2.3   How Dissolved Substances Cross Cell Membranes

2.4   *Water Pollution and Disease*

2.5   Cells and Their Organelles

2.6   *Microscopes: Gateways to the Cell*

2.7   The Nucleus

2.8   The Cytomembrane System

2.9   Mitochondria

2.10 The Cytoskeleton

2.11 Cell Metabolism

2.12 Capturing Energy in ATP: The First Two Stages in Human Cells

2.13 The Final Stage of Aerobic Respiration

2.14 Alternative Energy Sources

BASIC CELL STRUCTURE AND FUNCTION

Cells are amazingly diverse in size, shape, and activities. A bacterium is a complete organism even though it is just a single cell. A cell of your liver is not only more complex structurally than a bacterium, but it is also highly specialized for a specific function within one of the most complex life forms on Earth. Yet all cells share three basic features. All have an outer membrane and a semifluid region of cytoplasm. All cells also have the genetic material DNA concentrated in their interior. Bacteria are *prokaryotic*, a term loosely meaning "before nucleus." Nothing separates their DNA from other internal cell parts. Cells that are more complex than bacteria, including the trillions that make up your body, are *eukaryotic* ("true nucleus"); their DNA is enclosed within a nucleus. Here are the three fundamental structural features of eukaryotic cells and their basic functions:

1. **Plasma membrane**. This thin, outermost membrane maintains the cell as a distinct entity. By doing so, it allows metabolic events to proceed apart from random events in the environment. A plasma membrane does not *isolate* the cell interior. Substances and signals continually move across it, in highly controlled ways.

2. **Nucleus**. This membrane-bound compartment contains the DNA, as well as other molecules that can copy or read its hereditary instructions.

3. **Cytoplasm**. The cytoplasm is everything enclosed by the plasma membrane, *except* for the nucleus. It is a semifluid substance in which particles, filaments, and often compartments are organized.

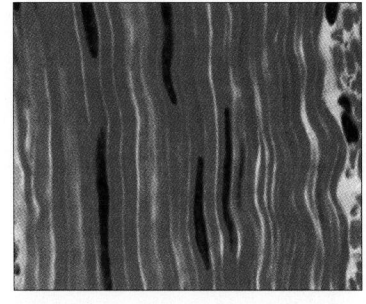
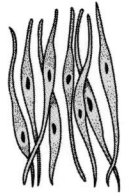

a

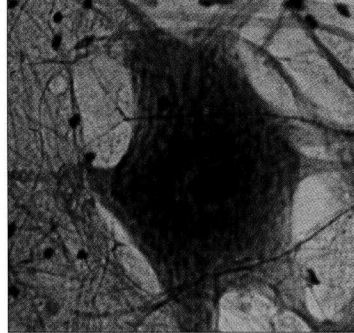
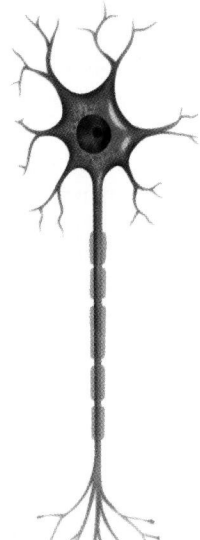

b

Figure 2.2 Two types of cells in the human body. (**a**) Smooth muscle cells, found in the walls of hollow organs such as the stomach, have an elongated oval shape. (**b**) A motor neuron, a type of nerve cell, has threadlike extensions.

The Size and Shape of Cells

A few cells can be seen with the unaided human eye. The list includes the "yolks" of bird eggs and the fish eggs we call caviar. However, most cells can only be observed with a microscope. For instance, each of your red blood cells is about 8 millionths of a meter across. You could fit a string of 2,000 of them across your thumbnail!

If a cell isn't small, it probably is long and thin or has outfoldings and infoldings that increase its surface area relative to its volume. The smaller or narrower or more frilly-surfaced the cell, the more efficiently materials can cross its surface and become distributed throughout the interior. Figure 2.2 shows just two of the many shapes of cells in your own body.

Cell Membranes

Cell membranes consist mostly of phospholipids. A phospholipid, recall, has a hydrophilic (water-loving) head and two fatty acid tails, which are hydrophobic (water-dreading). When phospholipids are immersed

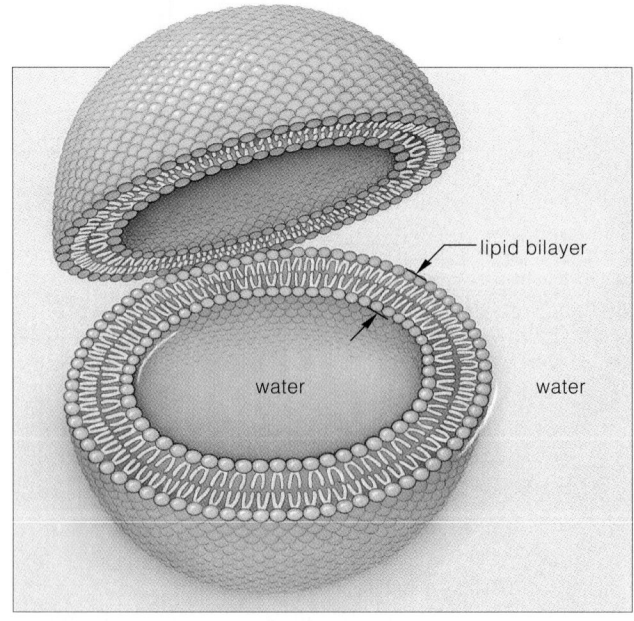

Figure 2.3 Phospholipids arranged in a lipid bilayer.

lipid bilayer

water   water

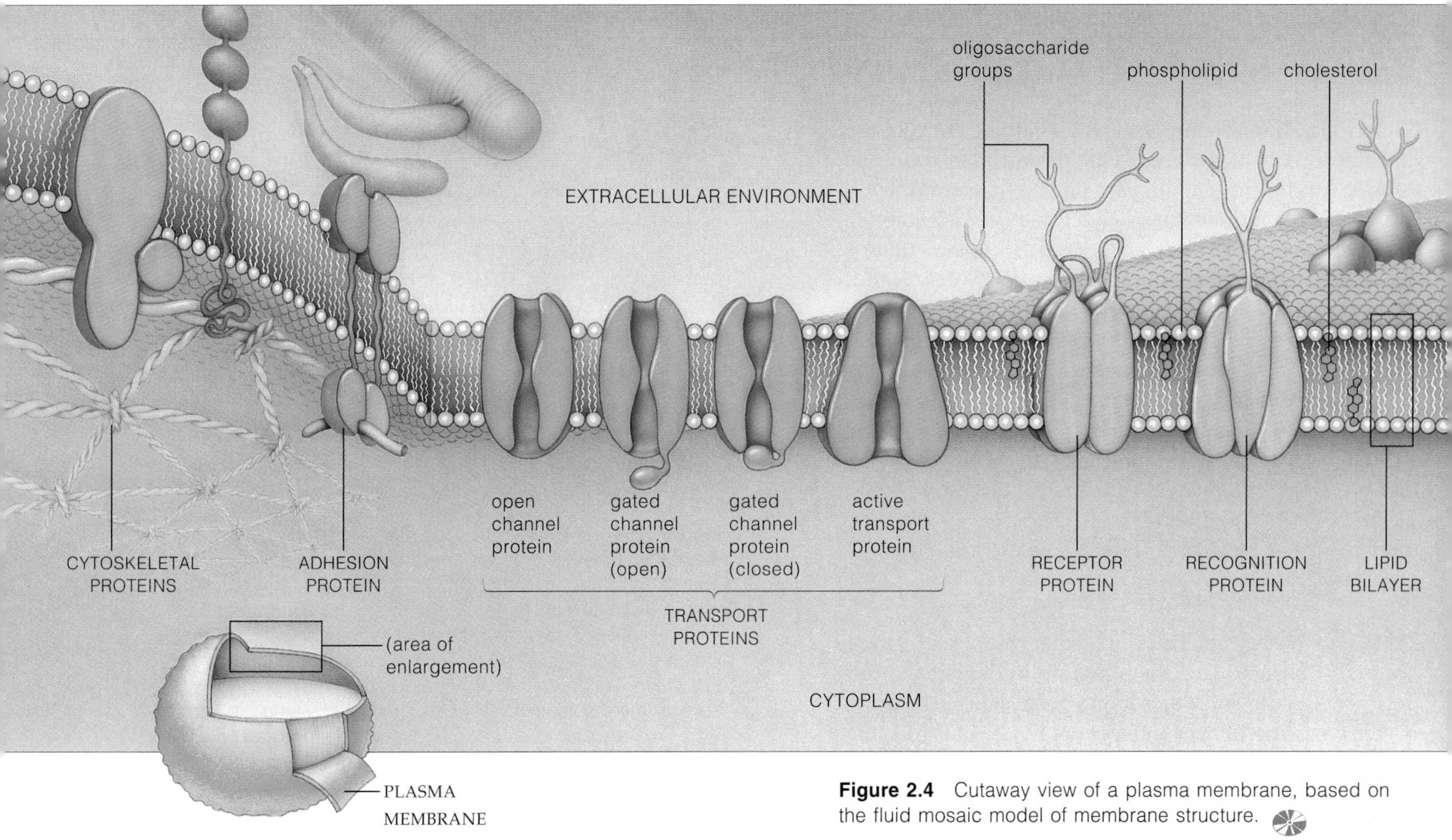

EXTRACELLULAR ENVIRONMENT

oligosaccharide groups   phospholipid   cholesterol

open channel protein | gated channel protein (open) | gated channel protein (closed) | active transport protein

CYTOSKELETAL PROTEINS

ADHESION PROTEIN

TRANSPORT PROTEINS

RECEPTOR PROTEIN

RECOGNITION PROTEIN

LIPID BILAYER

(area of enlargement)

CYTOPLASM

PLASMA MEMBRANE

Figure 2.4 Cutaway view of a plasma membrane, based on the fluid mosaic model of membrane structure.

in water, the molecules cluster together. They may become arranged in two layers, with the fatty acid tails sandwiched between the hydrophilic heads. This arrangement, called a **lipid bilayer**, is the structural basis of cell membranes (Figure 2.3).

Figure 2.4 shows a bit of membrane that corresponds to the **fluid mosaic model**. By this model, cell membranes consist of lipids *and* proteins. The membranes incorporate a variety of phospholipids, glycolipids, and, in the cells of humans (and other animals), cholesterol. Many proteins are embedded in the bilayer or positioned at its outer or inner surface. Different events take place at the two surfaces, which have different numbers and kinds of lipids and proteins, arranged in different ways.

The term "mosaic" refers to the mixed composition of lipids and proteins in the cell membrane. "Fluid" refers to the motions of membrane lipids and their interactions. Phospholipid molecules spin about their long axis, they move sideways, and their tails flex—all of which help keep neighboring molecules from packing into a solid layer. Lipids that have short tails or unsaturated (kinked) tails contribute to the membrane's fluidity.

The proteins associated with the bilayer carry out most membrane functions. Various **transport proteins** allow or encourage water-soluble substances to move through their interior, which opens on both sides of the

bilayer. These proteins bind molecules or ions on one side of the membrane, then release them on the other. **Receptor proteins** bind hormones and other extracellular substances that trigger changes in cell activities. Different cells have different combinations of receptors. Diverse **recognition proteins** at the cell surface are like molecular fingerprints; they identify a cell as being of a specific type. Finally, **adhesion proteins** help particular kinds of cells stick together and thereby remain positioned in their proper tissues.

Human cells are eukaryotic: They have an outermost plasma membrane and an internal region of cytoplasm that contains membrane-bound compartments, including the nucleus.

Each cell membrane has a lipid bilayer. The hydrophobic parts of its lipid molecules are sandwiched between the hydrophilic parts, which are dissolved in the fluid surroundings.

The lipid bilayer gives the membrane its basic structure and serves as a barrier to water-soluble substances.

Proteins associated with the bilayer carry out most membrane functions. Many types transport substances across the bilayer or serve as receptors for extracellular substances. Other types function in cell-to-cell recognition or adhesion.

Picture a cell membrane, with water bathing both sides of its bilayer. Lots of substances are dissolved in the water, but the kinds and amounts are not the same on the two sides. The membrane itself helps establish and maintain those differences, which are essential for cell functioning. How does the cell membrane accomplish this feat? It shows **selective permeability**. *Because of its molecular structure, the membrane allows some substances but not others to cross it in certain ways, at certain times.*

Carbon dioxide, molecular oxygen, and other small, nonpolar solutes can readily cross a cell membrane's lipid bilayer, and so can water molecules:

Although water molecules are polar, they might move through transient gaps that open up in the bilayer, then reclose. Glucose and other large, polar molecules almost never cross the bilayer independently. Neither do ions:

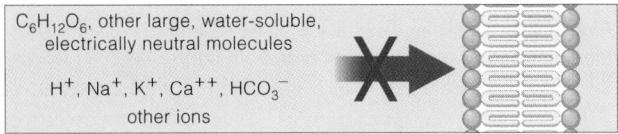

Such water-soluble substances must cross the bilayer through the interior of transport proteins, as you will read in Section 2.3. Why does a solute move one way or another at any given time? The answer starts with concentration gradients.

Concentration Gradients and Diffusion

"Concentration" refers to the number of molecules of a substance in a specified volume of fluid. "Gradient" means that the number in one region is not the same as it is in another. Thus a **concentration gradient** is a difference in the number of molecules or ions of a given substance in two adjoining regions.

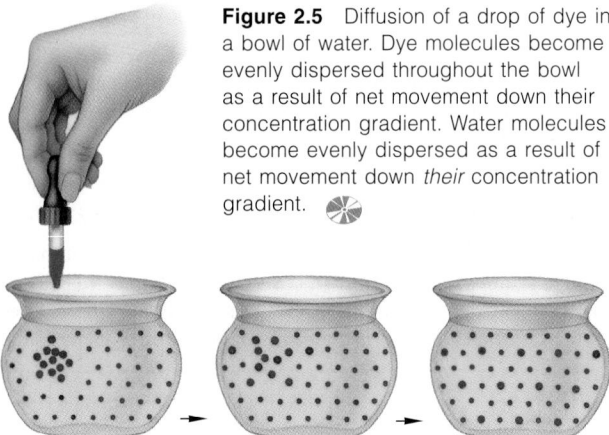

Figure 2.5 Diffusion of a drop of dye in a bowl of water. Dye molecules become evenly dispersed throughout the bowl as a result of net movement down their concentration gradient. Water molecules become evenly dispersed as a result of net movement down *their* concentration gradient.

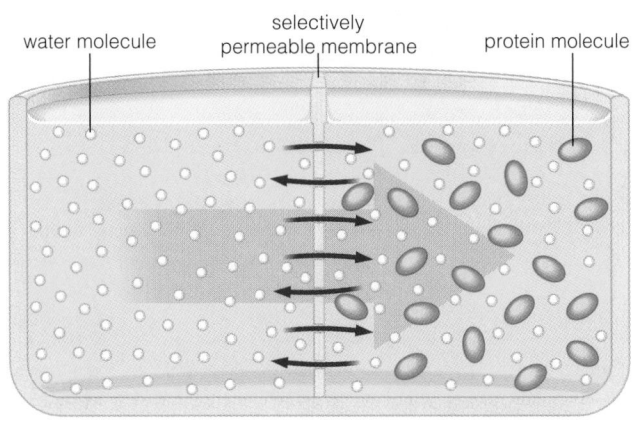

Figure 2.6 Osmosis. In this diagram, a membrane-like barrier divides a container into two compartments. Pure water was poured into the left compartment. An equal amount of a protein-rich solution was poured into the one on the right. Water molecules can move across this membrane. The protein molecules, which are larger, cannot, so they occupy some of the available space in their compartment. Thus, the right compartment contains fewer water molecules than the left. Water molecules move down this concentration gradient. There is a net osmotic movement from left to right (large *blue* arrow).

In the absence of other forces, molecules move down their concentration gradient—that is, *molecules move from a region where they are more concentrated to a region where they are less concentrated.* They do so because they constantly collide with one another at random, millions of times a second. Such collisions send the molecules back and forth, but the *net* movement is away from the place of greater concentration. When the concentration difference is large, molecules move rapidly down the gradient. When the difference is not so great, the net rate of movement is slower. Even when the gradient no longer exists, individual molecules are still in motion; then, however, there is no net movement in any direction.

The net movement of solutes from a region of higher concentration to a region of lower concentration is called **diffusion**. Diffusion is a key factor in the movement of substances across cell membranes and through fluid regions of the cytoplasm.

If a solution contains more than one solute, each solute still diffuses according to its *own* concentration gradient. For example, if you put a drop of dye in one side of a bowl of water, the dye molecules diffuse in one direction: to the region where they are less concentrated. On the other hand, the water molecules move in the opposite direction, to the region where *they* are less concentrated (Figure 2.5). Ultimately both are distributed equally throughout the solution.

The principles of diffusion apply generally to uncharged solutes. When a solute is charged, as ions are, it may diffuse down an **electric gradient**—a difference in charge between two adjoining regions. This movement can occur *against* the solute's concentration gradient.

Many processes, such as the flow of information through your nervous system, depend on the combined influences of electric and concentration gradients to attract specific substances across cell plasma membranes. A **pressure gradient**—a difference in pressure between two adjoining regions—also can influence the direction in which molecules of a solute will diffuse and how rapidly the diffusion will occur.

Osmosis

Because the plasma membrane of a cell is selectively permeable, the concentrations of solutes can increase on one side of the membrane and not the other. For example, the cytoplasm of most of your cells typically contains solutes (such as proteins) that cannot cross the plasma membrane by simple diffusion. Such solutes may become

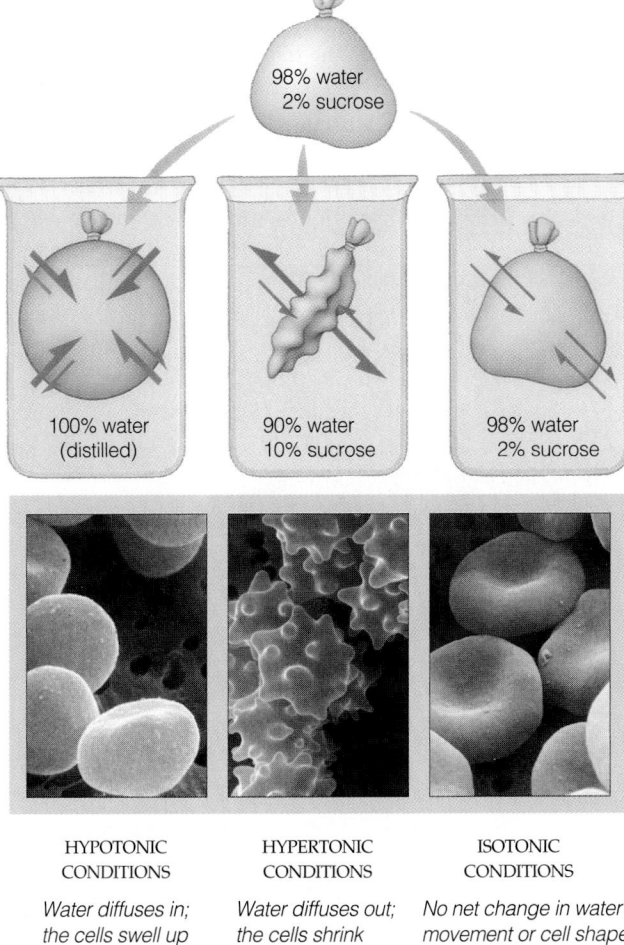

HYPOTONIC CONDITIONS	HYPERTONIC CONDITIONS	ISOTONIC CONDITIONS
Water diffuses in; the cells swell up	*Water diffuses out; the cells shrink*	*No net change in water movement or cell shape*

Figure 2.7 Tonicity and the diffusion of water. In the sketches, membranous bags through which water but not sucrose can move are placed in hypotonic, hypertonic, and isotonic solutions. In each container, arrow width represents the relative amount of water movement. The micrographs correspond to the sketches, which explain what happens to red blood cells that are placed in comparable solutions. Red blood cells have no mechanisms to actively take in or expel water.

quite concentrated within a cell, and the resulting solute concentration gradients affect the diffusion of water across the plasma membrane. The diffusion of water across a selectively permeable membrane in response to solute concentration gradients is called **osmosis** (Figure 2.6).

An important factor called **tonicity** influences osmotic water movements. Tonicity refers to the relative concentrations of the solutes in two fluids, including ones separated by a semipermeable membrane (Figure 2.7). When the two fluids contain equal concentrations of solutes, they are said to be *isotonic* ("iso-" means same) and there is no net movement of water in either direction across the intervening membrane. When the solute concentrations are not equal, one fluid is *hypotonic* (has fewer solutes) and the other is *hypertonic* (has more solutes). Because water moves down its concentration gradient, it always tends to move from a hypotonic solution to a hypertonic one.

Osmotic water movements across a cell's plasma membrane produce *osmotic pressure*. This term refers to the tendency of water to pass from a hypotonic solution to a hypertonic one when a selectively permeable membrane separates the two. The greater the difference in solute concentration between the solutions, the more osmotic pressure builds because of the greater tendency of water to move into the region where solutes are more concentrated. In a few situations osmotic pressure can be counterbalanced by *hydrostatic pressure*. In a cell, this is a mechanical pressure that various factors (such as the plasma membrane's resistance to stretching) exert on the cytoplasm. If the opposing hydrostatic pressure becomes great enough, it can stop or even reverse the osmotic flow of water. Blood is largely water and also exerts hydrostatic pressure in the body. Chapter 6 explores some physiological implications of this fact.

Moment to moment, cell activities and other events alter the factors that influence the solute concentrations of body fluids and water movements between those fluids. If cells did not have mechanisms for adjusting to such differences, they would shrivel or burst, as Figure 2.7 illustrates. In subsequent chapters we will return to some key ways in which osmotic water movements help maintain the body's proper water balance.

Diffusion is the net movement of like molecules or ions from a region of higher concentration to a region of lower concentration.

Osmosis is the diffusion of water across a selectively permeable membrane in response to concentration gradients, fluid pressure, or both.

Small molecules with no net charge can diffuse across the lipid bilayer of cell membranes. Large or charged molecules generally cross membranes through the interior of transport proteins.

HOW DISSOLVED SUBSTANCES CROSS CELL MEMBRANES

Figure 2.8 is an overview of the mechanisms that move solutes across all cell membranes. Combinations of these mechanisms supply cells and organelles with raw materials and rid them of wastes, at controlled rates. The mechanisms also help maintain the volume and pH of a cell or organelle with the ranges that are necessary for optimal functioning.

Again, water and small nonpolar solutes (such as oxygen) diffuse across the lipid bilayer. By contrast, in a mechanism called **passive transport**, a polar substance diffuses down its concentration gradient through the interior of transport proteins spanning the bilayer. This directional movement is also called "facilitated diffusion." Like other passive mechanisms, it does not require an energy boost from ATP, the energy-delivering nucleotide you may recall reading about in Section 1.11.

When a transport protein binds a solute on one side of a membrane, it undergoes a reversible change in shape that shunts the solute through the interior to the other side. Transport proteins are selective about which solutes travel through them; for example, the protein that transports amino acids will not transport glucose.

In **active transport**, a solute also crosses the membrane through the interior of transport proteins, but in this case the net movement is *against* the concentration gradient. Active transport can operate only when transport proteins are activated, as by ATP energy. Active transport can continue until the solute is *more* concentrated on the side of the membrane to which it is being pumped. For example, one active transport system, the calcium pump, helps keep the calcium concentration inside cells at least a thousand times lower than the concentration outside. You will be able to apply this information on active transport in later chapters that describe muscle contraction and some other essential physiological processes.

To move materials in bulk across membranes, cells rely on vesicles that form by way of mechanisms called exocytosis and endocytosis. A vesicle is a small membrane-bound sac that forms

at the plasma membrane *or* that buds from an internal cell membrane. In **exocytosis**, a vesicle in the cytoplasm moves to the plasma membrane and fuses with it. As the vesicle membrane becomes incorporated into the plasma membrane, the vesicle's contents are released to the surroundings. In **endocytosis**, part of the plasma membrane sinks inward and then balloons around particles or liquid droplets. It then seals back on itself, forming a vesicle that transports its contents or stores them within the cytoplasm. When endocytosis brings organic matter into the cell, the process is called *phagocytosis*, which means "cell eating."

Diseases that upset the normal balance of water and certain solutes can be deadly, as this chapter's *Focus on Our Environment* explains.

In passive transport, a solute diffuses through a transport protein; its net movement is down its concentration gradient. In active transport, the net movement of a solute is against its concentration gradient. The transporting protein must be activated, as by ATP energy.

Exocytosis and endocytosis move materials across the membrane in bulk.

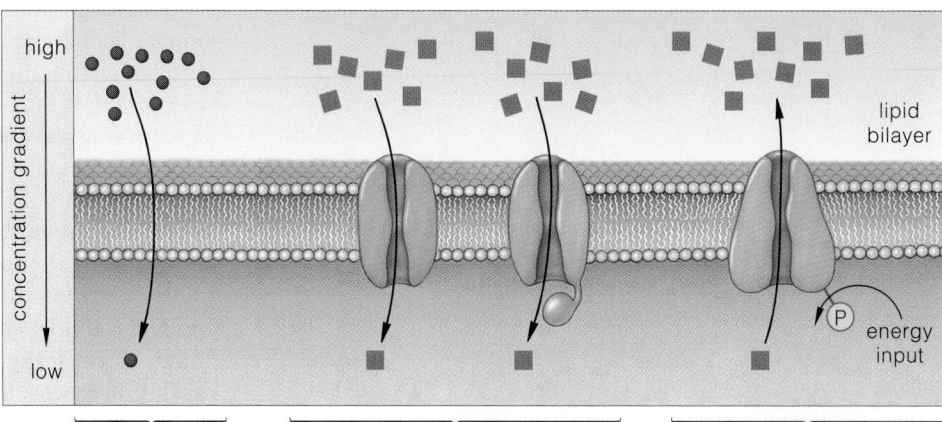

DIFFUSION ACROSS LIPID BILAYER	PASSIVE TRANSPORT	ACTIVE TRANSPORT
Lipid-soluble substances as well as water diffuse across.	*Water-soluble substances, and water, diffuse through interior of transport proteins. No energy boost required. Also called facilitated diffusion.*	*Specific solutes are pumped through interior of transport proteins. Requires energy boost.*

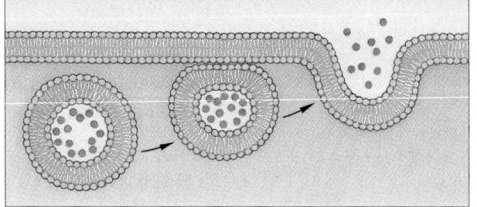

EXOCYTOSIS
Vesicle in cytoplasm moves to plasma membrane, fuses with it; contents released to the outside.

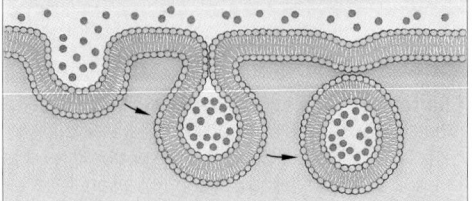

ENDOCYTOSIS
Vesicle forms from a patch of inward-sinking plasma membrane, enters cytoplasm.

Figure 2.8 Overview of the major mechanisms by which dissolved substances cross cell membranes. Exocytosis and endocytosis occur only at the plasma membrane.

2.4 WATER POLLUTION AND DISEASE

For every living cell, survival is a risky business. The risk is easy to see in parts of Asia, Africa, and the Americas, where cholera is a severe health threat to inhabitants—and to international travelers. Caused by the bacterium *Vibrio cholerae*, cholera spreads by way of water or food contaminated by human sewage (Figure 2.9). Every year, cholera kills more people than AIDS does.

The most devastating symptom of cholera is severe diarrhea. Over a period of several days, body tissues rapidly lose water and essential salts. Treatment involves giving fluids that replenish these substances, as well as antibiotics to kill the microbe. Sometimes, however, the body loses so much water so rapidly that blood pressure plunges and the volume of blood in blood vessels falls drastically. The cells that make up tissues become starved for oxygen and nutrients because not enough blood is available to deliver those substances. Other complications can occur as well.

High-quality water is a necessity of life that is easily taken for granted. Increasingly, however, various parts of the world are plagued by periodic drought and pollution of groundwater supplies. For example, in 1993 nearly a quarter of a million people in Milwaukee, Wisconsin, developed diarrhea and other intestinal problems when the municipal water supply became contaminated with agricultural runoff carrying *Cryptosporidium* (a protozoan) from infected dairy cows. Mercury and other industrial pollutants now are threats in fresh waters and the oceans. We return to the interrelationship of human activities and water pollution in Chapter 25.

Figure 2.9 In many places where public sanitation is primitive or lacking altogether, people run the risk of contracting cholera because drinking water and some food sources (especially fish) are contaminated by human sewage that carries the *Vibrio cholerae* bacterium. Cholera can be lethal if it so rapidly depletes the body of water and needed solutes that the internal balance we call homeostasis is fatally disrupted.

CELLS AND THEIR ORGANELLES

In the tiny space of a cell, a vast number of chemical reactions go on simultaneously. Many of the reactions would be incompatible if they occurred within the same compartment of the cell. For example, a molecule of fat can be assembled by some reactions and taken apart by others, but a cell gains nothing if both sets of reactions proceed at the same time on the same fat molecule.

In eukaryotic cells such as those of the human body, organelles are the solution to this dilemma. An **organelle** is an internal, membrane-bound sac or compartment that serves one or more specialized functions. Beyond physically separating incompatible reactions, organelles also permit compatible, interconnected reactions to proceed at different times.

Most types of human cells contain the following kinds of organelles or structures:

Organelle or Structure	Main Function
Nucleus	*Localizing the cell's DNA*
Ribosomes	*Assembling polypeptide chains*
Endoplasmic reticulum	*Routing and modifying newly formed polypeptide chains; also synthesizing lipids*
Golgi body	*Modifying polypeptide chains into mature proteins; sorting and shipping proteins and lipids for secretion or for use inside cell*
Various vesicles	*Transporting or storing substances; digesting substances and structures within the cell; other functions*
Mitochondria	*Producing ATP molecules*
Cytoskeleton	*Imparting shape and internal organization to cell; moving the cell and its internal structures*

Figures 2.10 and 2.11 show where organelles and other structures might be located in a cell in your own body. Keep in mind, though, that there are dramatic differences in the structures and functions of body cells. *Science Comes to Life* (page 48) will give you some insight into the various ways modern researchers can use microscopy to study cells.

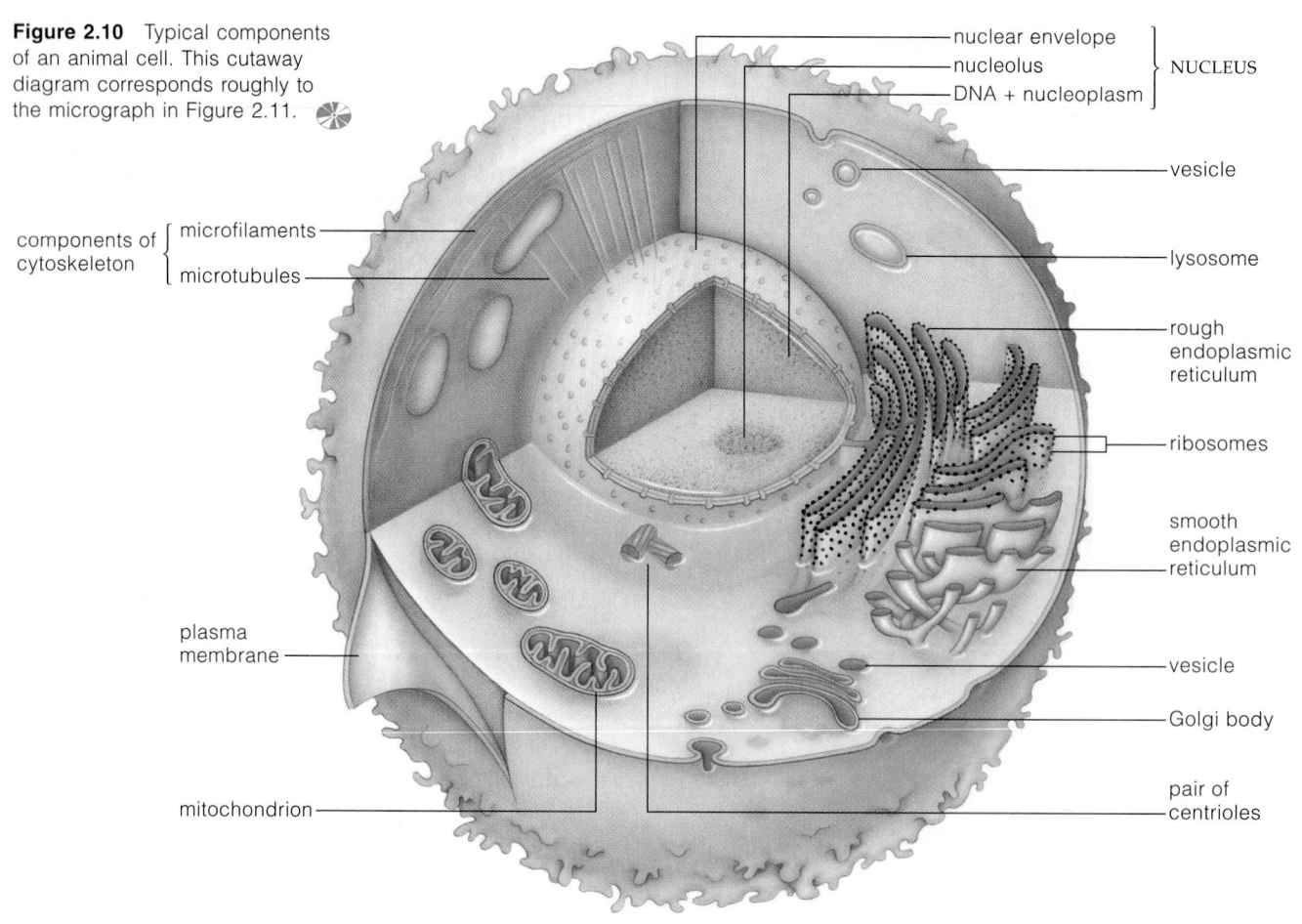

Figure 2.10 Typical components of an animal cell. This cutaway diagram corresponds roughly to the micrograph in Figure 2.11.

nuclear envelope
nucleolus
DNA + nucleoplasm
NUCLEUS

vesicle

lysosome

rough endoplasmic reticulum

ribosomes

smooth endoplasmic reticulum

vesicle

Golgi body

pair of centrioles

components of cytoskeleton — microfilaments — microtubules

plasma membrane

mitochondrion

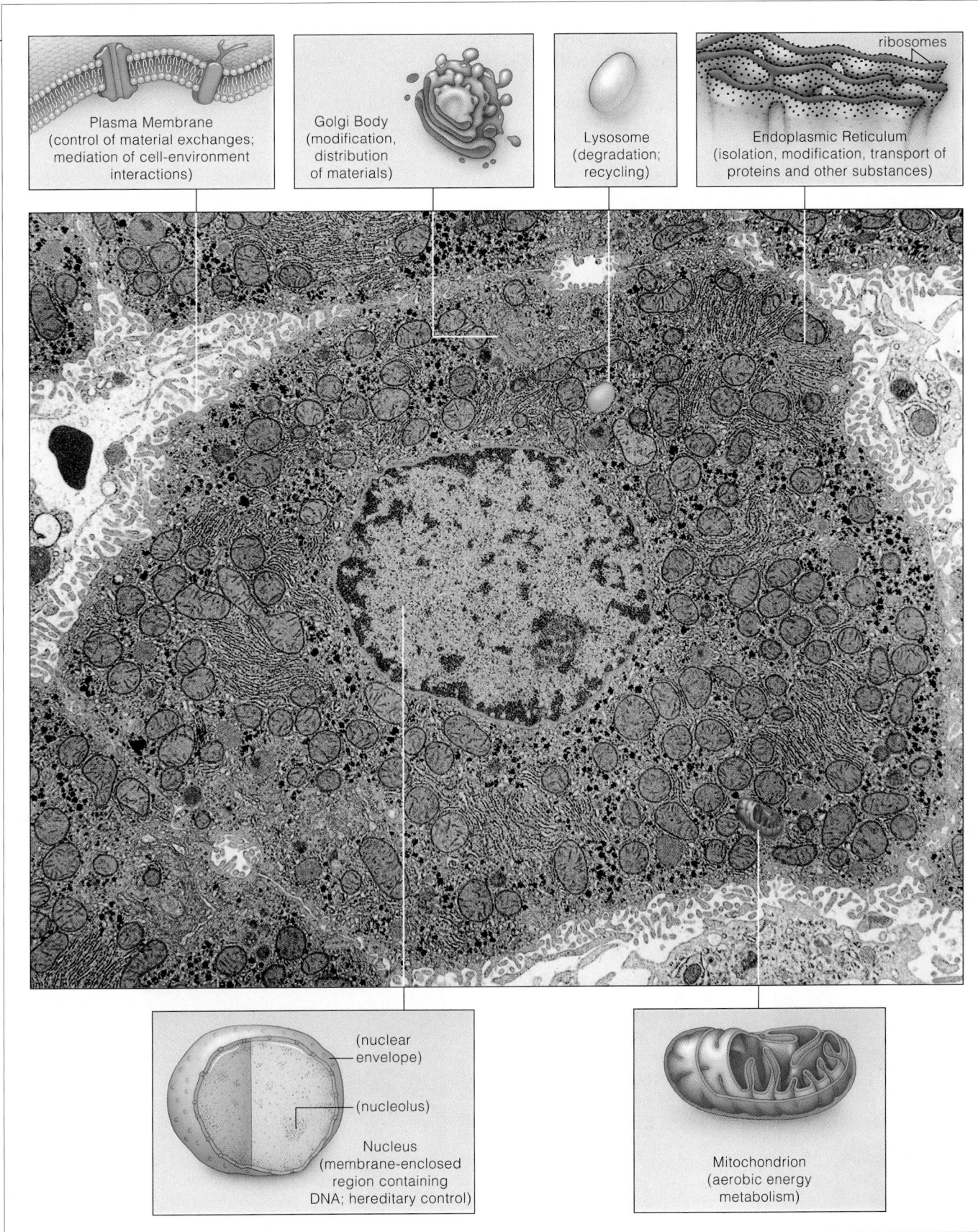

Figure 2.11 Transmission electron micrograph of a liver cell in cross section. Although this is a rat liver cell, it is representative of liver cells in other mammals, including humans.

2.6 MICROSCOPES: GATEWAYS TO THE CELL

Figure 2.12 shows one common type of modern electron microscope. A photograph formed using a microscope is called a *micrograph*. The micrographs in Figure 2.13 compare the sorts of detail different types of microscopes can reveal. The specimen is a red blood cell.

LIGHT MICROSCOPES   Light microscopes work by bending (refracting) light rays. Any rays directly entering a glass lens, except at the very center, are bent. In a *compound light microscope*, two or more glass lenses bend incoming rays to form an enlarged image of a specimen. If you wish to observe a living cell, the cell must be small or thin enough for light to pass through. Also, to be visible, cell parts must differ in color or optical density from their surroundings. Unfortunately, most cell parts are nearly colorless and are uniform in density. For this reason, prior to viewing cells through a light microscope, researchers expose the cells to dyes that react with some cell parts but not others. Although useful, this "staining" can alter cell parts and kill the cell. Dead cells begin to break down at once, so they are sometimes preserved before staining.

No matter how good a glass lens system is, when the diameter of the object being viewed is magnified by 2,000 times or more, cell parts appear larger but are not clearer. The properties of light waves passing through glass lenses limit the resolution of smaller details.

ELECTRON MICROSCOPES   Electrons are particles, but they behave like waves. Electron microscopy is based on accelerating the flow of streams of electrons that have wavelengths about 100,000 times shorter than those of visible light. As a practical matter, this means that electron microscopes can achieve much greater resolution of smaller details than even the best light microscopes can, allowing the viewer to obtain remarkably sharp images of cells or other specimens.

A *transmission electron microscope* uses a magnetic field as the "lens" that bends the electron stream and focuses it into an image, which then is magnified.

With a *scanning electron microscope*, a narrow beam of electrons is directed back and forth across a specimen thinly coated with metal.

The metal responds by emitting some of its own electrons. A detector connected to electronic circuitry transforms the electron energy into an image of the specimen's surface on a television screen. Most of the images have fantastic depth.

Figure 2.13
Comparison of how different types of microscopes reveal cellular details. The cells are human red blood cells.

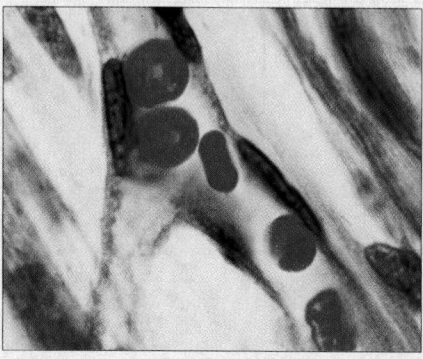

a Red blood cells flowing through a blood capillary as revealed by a light microscope.

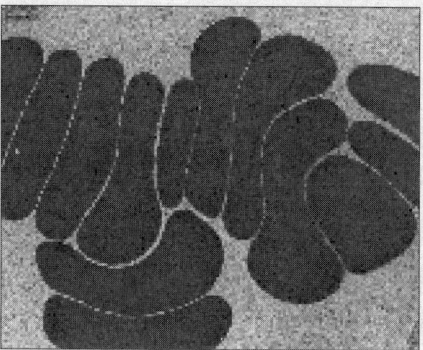

b Transmission electron micrograph showing the inside of mature red blood cells. Such cells lack nuclei and are packed with hemoglobin.

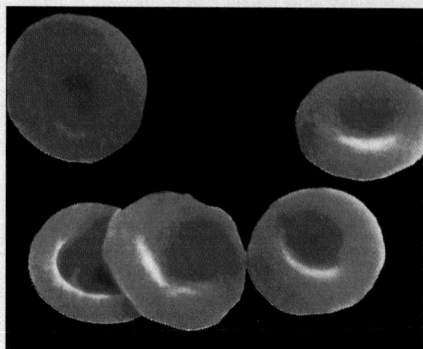

c Scanning electron micrograph (SEM) with color added clearly shows the "concave disk" shape of red blood cells.

Figure 2.12   Microscopy. An electron microscope in a modern laboratory.

THE NUCLEUS

The **nucleus** encloses the DNA in a cell. DNA contains instructions for building all of a cell's proteins and, through those proteins, for determining a cell's structure and function. Figure 2.14 shows the distinctive structure of the nucleus, and Table 2.1 lists its components. The nucleus has two key functions. *First*, it physically separates DNA from the intricate metabolic machinery of the cytoplasm. Each of your body cells contains 46 DNA molecules. Localization of DNA makes it easier to sort out hereditary instructions when it comes time for a cell to divide. The DNA molecules can be sorted into parcels—one parcel for each new cell that forms. *Second*, the membranous boundary of the nucleus helps control the exchange of signals and substances between the nucleus and the cytoplasm.

The Nucleolus

As a cell grows, one or more dense masses appear within the nucleus. Each mass is a **nucleolus** (plural: nucleoli)—a site where the protein and RNA subunits of ribosomes are assembled. Later the subunits are shipped from the nucleus into the cytoplasm. When genetic instructions in the cell order the synthesis of proteins, the ribosome subunits join together and become functional ribosomes.

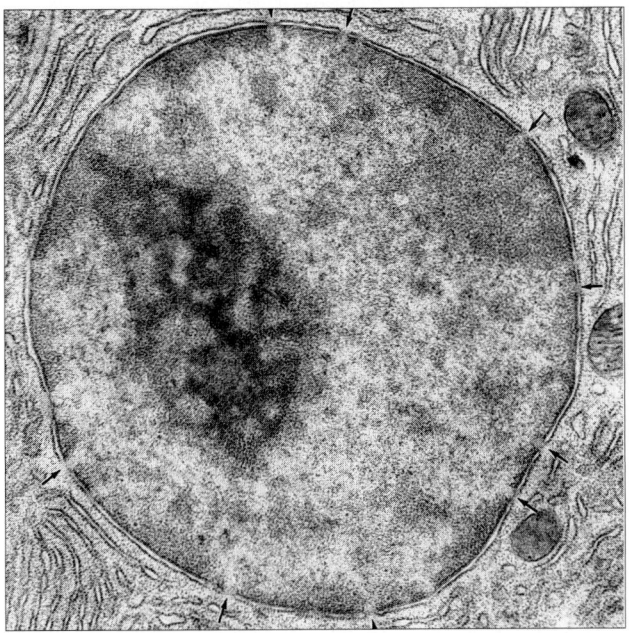

Figure 2.14 Animal cell nucleus, thin section. Arrows point to pores in the nuclear envelope.

Table 2.1	Components of the Nucleus
Nuclear envelope	Double-membraned, pore-riddled boundary between cytoplasm and interior of nucleus
Nucleolus	Dense cluster of the RNA and proteins used to assemble ribosomal subunits
Nucleoplasm	Fluid portion of the nuclear interior
Chromosomes	DNA molecules and numerous proteins attached to them

The Nuclear Envelope

The outermost component of the nucleus, the **nuclear envelope**, has two lipid bilayers, one outside the other. This double-membrane system surrounds the fluid portion of the nucleus. It is continuous with the endoplasmic reticulum, which is a component of the cell's cytomembrane system (Section 2.8). As with all cell membranes, the lipid bilayers of the nuclear membrane act as a barrier to water-soluble substances. Clusters of proteins form pores that span both bilayers. They allow ions and small, water-soluble molecules to move freely across the nuclear envelope, but they control the passage of ribosomal subunits, proteins, and other large molecules.

Chromosomes

Between cell divisions, DNA is threadlike, with many proteins attached to it. At intervals the DNA "thread" loops around the protein molecules, producing a combined structure that resembles beads on a string. Except at extreme magnification, the beaded threads look grainy, as you see in Figure 2.14. Before a cell divides, however, it duplicates its DNA molecules (so that each new cell will get all the required hereditary instructions). Before the DNA molecules are sorted into two sets, they are folded and twisted into condensed structures, proteins and all.

Early microscopists named the threadlike material *chromatin* and the condensed structures *chromosomes*. We now define **chromatin** as a cell's total collection of DNA, together with all of the proteins associated with it. Each **chromosome** is an individual DNA molecule and its associated proteins, whether it is threadlike or condensed. In subsequent chapters, keep in mind that a chromosome doesn't always look the same during the life of a cell.

The nucleus keeps the cell's DNA molecules separated from cytoplasmic machinery, and this serves two functions:

• When a cell divides, the separation makes it easier to parcel out hereditary instructions to new cells.

• Pores across the nuclear envelope help control the passage of larger molecules between the nucleus and cytoplasm.

THE CYTOMEMBRANE SYSTEM

Some polypeptide chains assembled on ribosomes pass through the **cytomembrane system** (Figure 2.15). This series of organelles includes the endoplasmic reticulum, Golgi bodies, and vesicles. In the system, many proteins take on final form and are packaged in vesicles. Also, lipids are assembled and packaged here. Some vesicles deliver their contents to regions where new membranes must be built. Others store proteins or lipids for specific uses. Others move to the plasma membrane and release their contents outside the cell.

Endoplasmic Reticulum

The membranes of the endoplasmic reticulum (ER) begin at the nucleus and curve through the cytoplasm. ER has rough and smooth regions due largely to the presence or absence of ribosomes on the side of the membrane facing the cytoplasm. *Rough ER* is often arranged as stacked, flattened sacs which have many ribosomes attached (Figure 2.16a). Polypeptide chains are assembled on the ribosomes. Newly forming chains enter spaces inside rough ER, where they may acquire carbohydrate chains. Many specialized cells secrete proteins, and rough ER is abundant in them. For example, some ER-rich cells of the pancreas produce and secrete enzymes that enter the small intestine. There, they aid in food digestion.

Smooth ER has no ribosomes (Figure 2.16b). In many cells, smooth ER is the main site of lipid synthesis. In your liver cells, smooth ER inactivates certain drugs and harmful by-products of metabolism. In skeletal muscle cells a type of smooth ER (sarcoplasmic reticulum) stores and releases calcium ions essential in muscle contraction.

Golgi Bodies

In Golgi bodies enzymes put the finishing touches on lipids and proteins, sort them out, and package them in vesicles, which will move them to specific locations.

For example, an enzyme in one Golgi region might alter the carbohydrate chain of a glycoprotein and so flag the protein for delivery to its proper destination.

The flattened membrane sacs of a Golgi body vaguely resemble a stack of pancakes (Figure 2.17). Vesicles form at the final (uppermost) region of a Golgi body when parts of the membrane bulge, then break away. In animal cells, this is the region closest to the plasma membrane.

A Variety of Vesicles

Various types of vesicles move through the cytoplasm or take up positions within it. **Lysosomes**, which are organelles in which

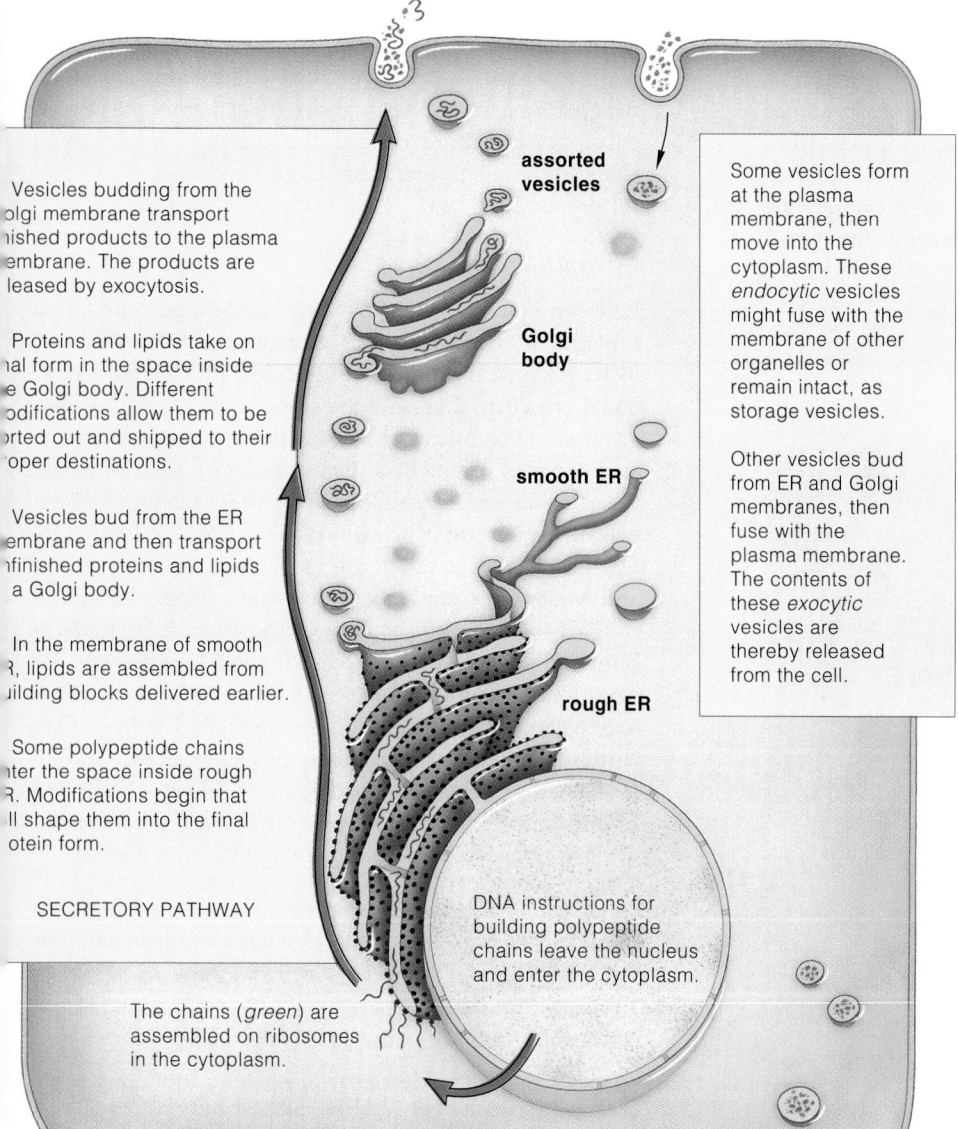

Vesicles budding from the Golgi membrane transport finished products to the plasma membrane. The products are released by exocytosis.

Proteins and lipids take on final form in the space inside the Golgi body. Different modifications allow them to be sorted out and shipped to their proper destinations.

Vesicles bud from the ER membrane and then transport unfinished proteins and lipids to a Golgi body.

In the membrane of smooth ER, lipids are assembled from building blocks delivered earlier.

Some polypeptide chains enter the space inside rough ER. Modifications begin that will shape them into the final protein form.

SECRETORY PATHWAY

The chains (*green*) are assembled on ribosomes in the cytoplasm.

assorted vesicles

Golgi body

smooth ER

rough ER

DNA instructions for building polypeptide chains leave the nucleus and enter the cytoplasm.

Some vesicles form at the plasma membrane, then move into the cytoplasm. These *endocytic* vesicles might fuse with the membrane of other organelles or remain intact, as storage vesicles.

Other vesicles bud from ER and Golgi membranes, then fuse with the plasma membrane. The contents of these *exocytic* vesicles are thereby released from the cell.

Figure 2.15 The cytomembrane system. This system functions in the assembly, modification, packaging, and shipment of proteins and lipids. The *green* arrows highlight a secretory pathway by which certain proteins and lipids are packaged and released from many cells.

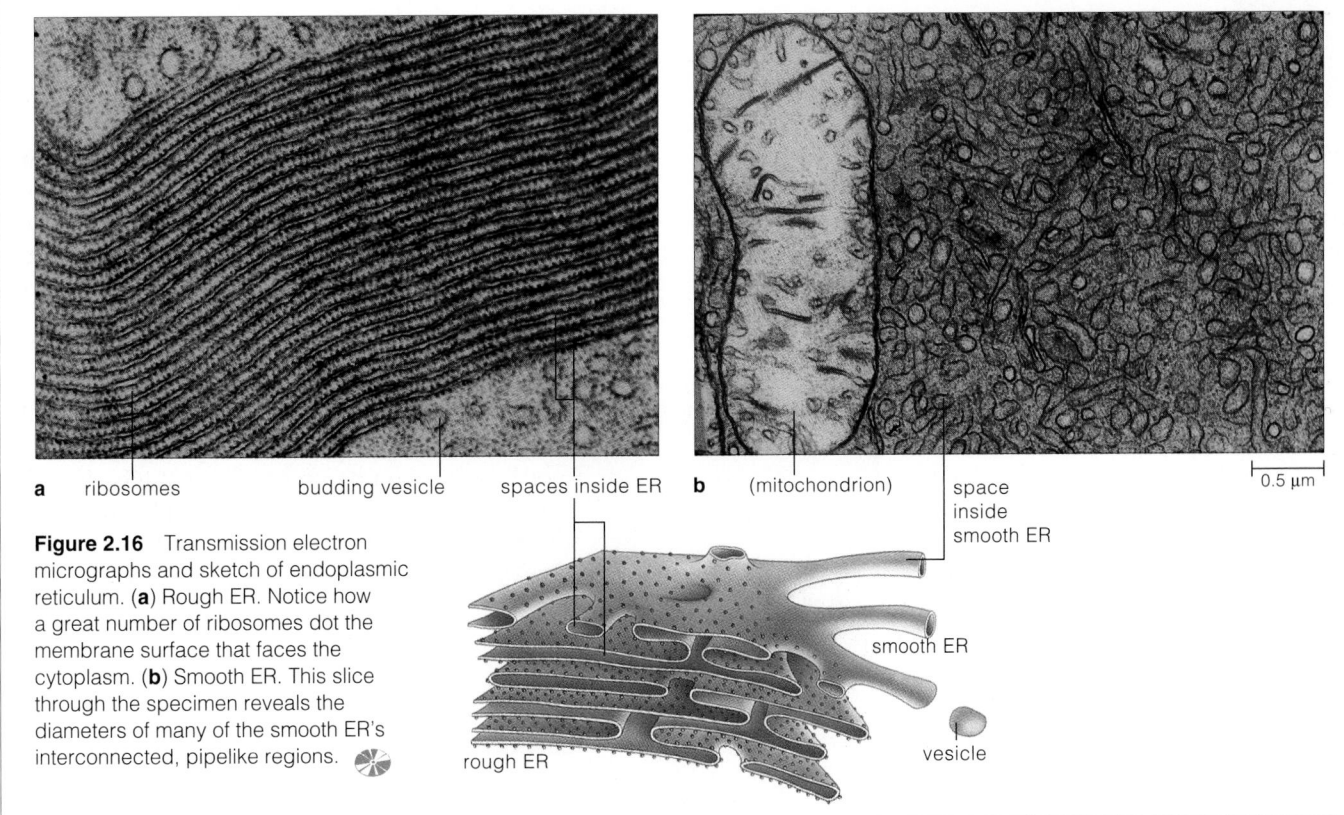

a ribosomes   budding vesicle   spaces inside ER   **b** (mitochondrion)   space inside smooth ER

0.5 µm

smooth ER

vesicle

rough ER

Figure 2.16 Transmission electron micrographs and sketch of endoplasmic reticulum. (**a**) Rough ER. Notice how a great number of ribosomes dot the membrane surface that faces the cytoplasm. (**b**) Smooth ER. This slice through the specimen reveals the diameters of many of the smooth ER's interconnected, pipelike regions.

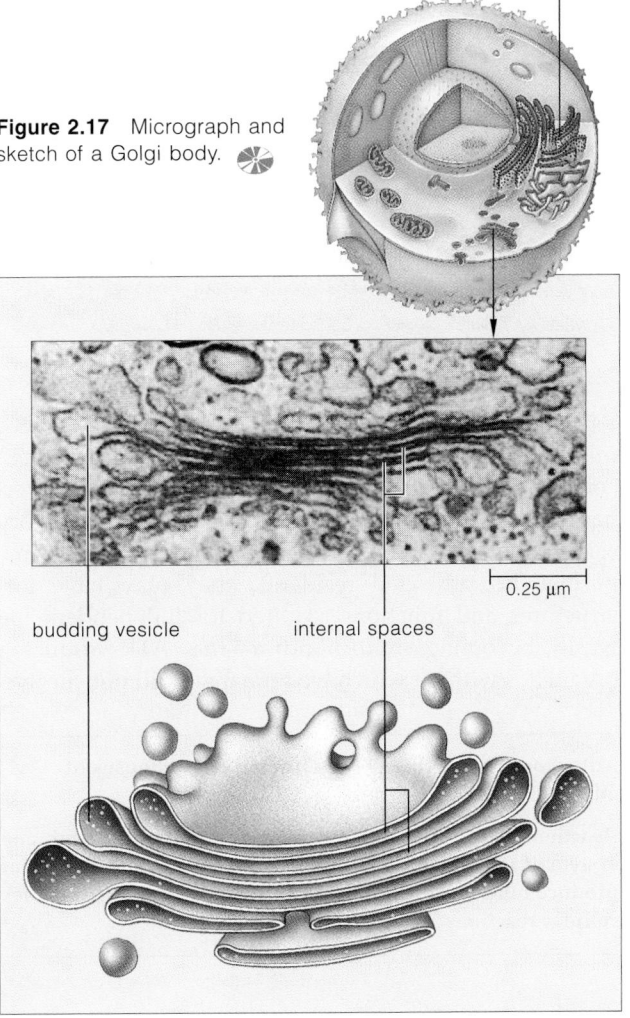

Figure 2.17 Micrograph and sketch of a Golgi body.

0.25 µm

budding vesicle   internal spaces

intracellular digestion takes place, bud from the Golgi membranes. Enzymes they contain can speed the breakdown of proteins, complex sugars, nucleic acids, and some lipids. Often, lysosomes fuse with vesicles that formed at the plasma membrane. They may even help digest some of the cell's own parts. Certain white blood cells of the immune system take in foreign material by phagocytosis and dispose of it in just this manner.

Peroxisomes are sacs of enzymes that break down fatty acids and amino acids. The reactions produce hydrogen peroxide, a potentially harmful substance. But before hydrogen peroxide can injure the cell, another enzyme within peroxisomes converts it to water and oxygen or uses it to break down alcohol. If you drink alcohol, nearly half of it is broken down in peroxisomes of your liver and kidney cells.

In the ER and Golgi bodies of the cytomembrane system, many proteins take on final form and lipids are synthesized.

Lipids, proteins, and other substances become packaged in vesicles destined for export, storage, membrane building, intracellular digestion, and other cell activities.

2.9 MITOCHONDRIA

Energy that ATP molecules carry from one reaction site to another drives nearly all cell activities. In organelles called **mitochondria** (singular: mitochondrion), enzymes break apart glucose and other organic compounds, and energy released in the reactions is used to form *many* ATP molecules. These reactions, which require oxygen, extract far more energy than can be extracted by any other means. When you breathe in, you are taking in oxygen for your mitochondria. Exhale, and you are removing carbon dioxide produced in mitochondria. Some cells have a sprinkling of these organelles; others, such as energy-demanding muscle cells, may contain thousands.

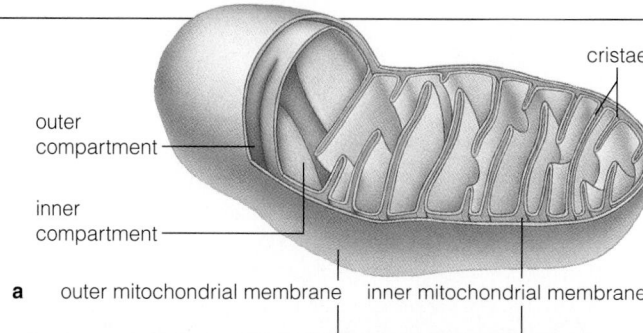

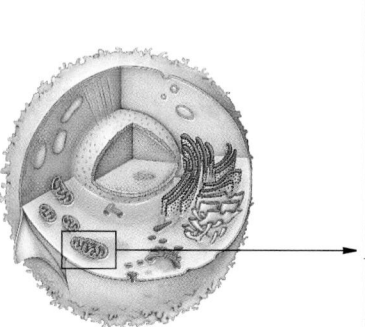

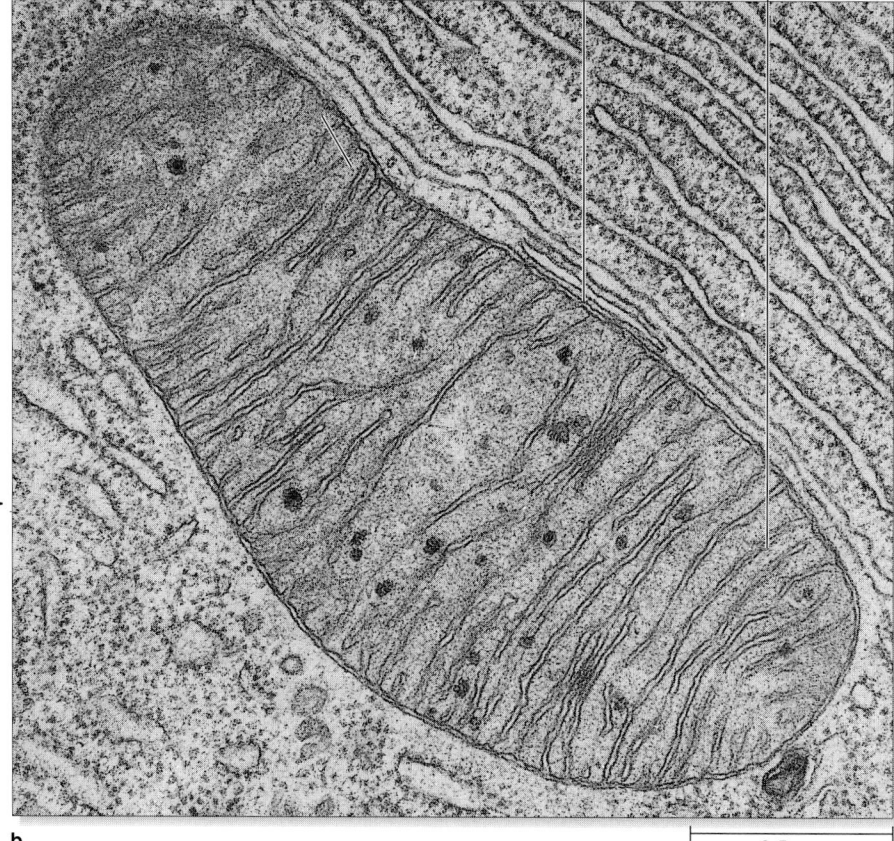

Figure 2.18 (**a**) Sketch and (**b**) transmission electron micrograph of a thin slice through a typical mitochondrion. Reactions inside this organelle produce quantities of ATP, which is the major energy carrier in cells.

0.5 μm

A mitochondrion has two membranes, as you can see in Figure 2.18. Its outer membrane faces the cytoplasm. In most cases, its inner membrane folds back on itself repeatedly, a bit like the folds of an accordion. Each fold is a crista (plural: cristae). Spaces between the cristae form the *matrix*. The double-membrane system creates two compartments, which are used in ATP formation—a subject we will consider in short order.

All eukaryotic cells have at least one mitochondrion. If you look back at Figure 2.11, you will see mitochondria in profusion. The micrograph in the figure shows only one thin slice from one liver cell. Its message is clear: the liver is an exceptionally active, energy-demanding organ!

Mitochondria have intrigued biologists because they resemble bacteria in size and in many details of their biochemistry. They even have their own DNA and some ribosomes, and they divide on their own.

Based on persuasive research, many biologists believe mitochondria evolved from ancient bacteria that were consumed by another ancient cell, yet did not die. Perhaps they were able to reproduce inside the predatory cell and its descendants. If the engulfed bacteria became permanent, protected residents, they may have lost structures and functions required for independent life while becoming mitochondria—the ATP-producing organelles without which we humans could not survive.

Mitochondria are the ATP-producing powerhouses of eukaryotic cells.

Reactions that release energy occur in the internal membrane system of mitochondria. The reactions, which require oxygen, produce much more ATP than can be produced by any other cellular reactions.

THE CYTOSKELETON

A system of fibers, threads, and lattices extends between the nucleus and the plasma membrane. This system, the **cytoskeleton**, gives cells their shape, internal organization, and ability to move. Certain elements of the cytoskeleton reinforce the plasma membrane and hold its proteins in place; others perform the same task for the nuclear envelope. Still other elements serve as supportive scaffolds for specific regions of the cytoplasm.

Microtubules are the largest structural element of the cytoskeleton. Each one is a hollow cylinder made up of parallel rows of subunits of the protein tubulin (Figure 2.19a). Large numbers of microtubules arise from dense areas of material known as **microtubule organizing centers** (**MTOCs**).

Other cytoskeletal elements are delicate, fiberlike **intermediate filaments** and thin, twisted chains of the protein actin called **microfilaments**:

tubulin units

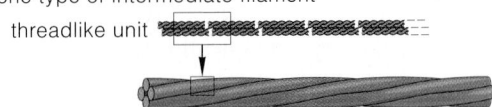

a Part of one microtubule

actin subunit

Part of one type of microfilament

Part of one type of intermediate filament

threadlike unit

Intermediate filaments are the most stable elements of the cytoskeleton. The six known types mechanically strengthen cells or cell parts and help maintain their shape. Microfilaments, the thinnest cytoskeletal elements, take part in movements, especially the kind that take place at the cell surface. They also contribute to the development and maintenance of cell shape.

A **flagellum** (plural: flagella) or **cilium** (plural: cilia) is an example of cytoskeletal organization. Nine pairs of microtubules ring a central pair (Figure 2.19b). A system of spokes and links holds this "9 + 2 array" together. The flagellum or cilium bends when pairs of microtubules in the ring slide over each other. Like whiplike tails, flagella propel human sperm through fluid environments.

Cilia are shorter than flagella, and usually there are more of them per cell. In your respiratory tract, thousands of ciliated cells whisk out mucus laden with dust or other undesirable material.

The microtubules of a flagellum or cilium arise from **centrioles**, which remain at the base of the completed structure, as a **basal body**. A centriole is a type of MTOC, and you will be reading about it and some others later in this book.

The cytoskeleton is the basis of a eukaryotic cell's shape, internal structure, and capacity for movement.

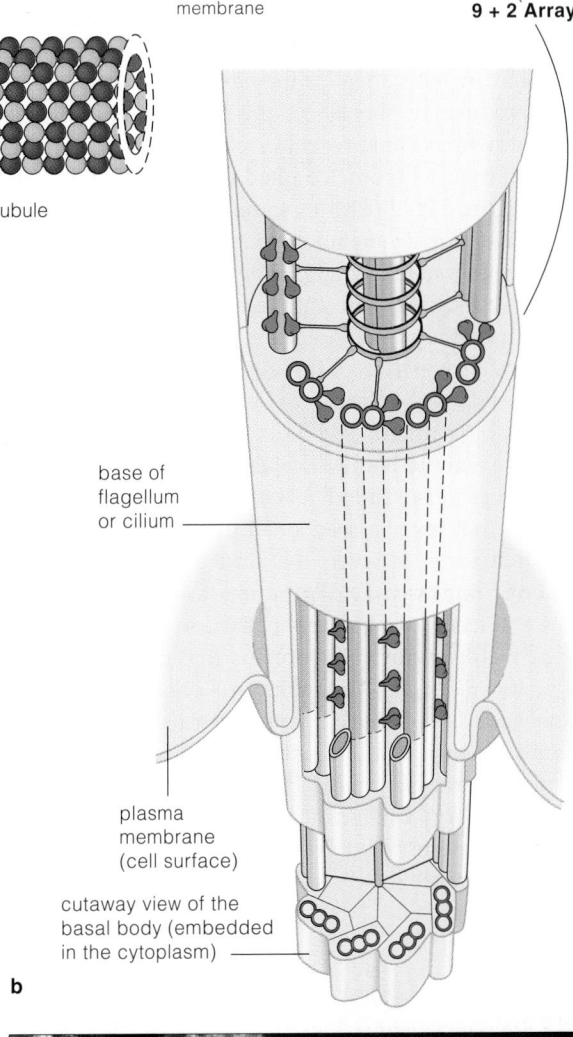

one of nine pairs of microtubules of the outer ring

two central microtubules

central sheath

spokes and links of the connective system

plasma membrane

9 + 2 Array

base of flagellum or cilium

plasma membrane (cell surface)

cutaway view of the basal body (embedded in the cytoplasm)

b

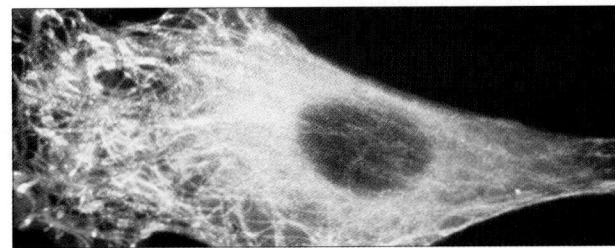

Figure 2.19 Internal organization of flagella and cilia. The micrograph shows the cytoskeleton of an animal cell.

CELL METABOLISM

The 65 trillion living cells in your body have the capacity to acquire energy and use it to build, store, break apart, and eliminate substances in controlled ways. This dynamic activity is called **metabolism**.

Cells normally maintain, increase, and decrease the concentrations of substances by coordinating different metabolic pathways. At any instant, thousands of reactions are transforming thousands of substances in the confines of a cell. Most of the reactions are organized as **metabolic pathways**, in which reactions proceed one after another in orderly steps that enzymes mediate. Often the steps proceed in linear sequence. Often they proceed in a circle, with one of the end products becoming the starting substance for the next cycle. Many times, the end products or substances produced at earlier steps of one pathway become reactants elsewhere, so that two or more pathways become coupled in linear or branching fashion. Figure 2.20 illustrates some of the possibilities.

Overall, the main metabolic pathways are either biosynthetic or degradative. In the *biosynthetic* pathways, small molecules are assembled into complex carbohydrates, proteins, and other large molecules of higher energy content. In the *degradative* pathways, large molecules are broken down to products of lower energy content. Sometimes biosynthetic activity is called **anabolism** and degradative activity is called **catabolism**.

Substrates (also called reactants or precursors) are substances that enter a reaction. **Intermediates** are substances that form between the start and conclusion of a metabolic pathway. **End products** are the substances present at the end of a metabolic reaction or pathway.

Nearly all **enzymes** are proteins that catalyze (speed up) specific reactions. **Cofactors** are organic molecules or metal ions that assist enzymes or transport electrons or atoms. **Energy carriers** activate (donate energy to) substances by transferring functional groups to them. ATP is the main energy carrier in cells. Finally, transport proteins adjust concentration gradients at cell membranes in ways that influence the direction of metabolic reactions.

Quite a few metabolic pathways involve **oxidation-reduction reactions**, or "redox" reactions. Whenever a substance taking part in a biochemical reaction gives up one or more electrons, it is said to be *oxidized*. Conversely, when a substance accepts electrons, it is *reduced*. In general, the phrase "oxidation-reduction reaction" merely means an electron transfer. Shortly, you will see how energy released during electron transfers in mitochondria is captured in the cell's energy currency, ATP.

Enzymes

Enzymes are *catalytic* molecules. They greatly enhance the rate at which specific reactions occur. Most are proteins, although some RNA molecules also show catalytic activity. Enzymes have four features in common. *First*, they do not make anything happen that couldn't happen on its own. But they often make it happen at least a million times faster. *Second*, they are not permanently altered or used up in a reaction; the same enzyme may act over and over again. *Third*, the same enzyme usually works for the forward *and* reverse directions of a reaction. *Fourth*, each type of enzyme is highly selective about its substrates.

In most cases, an enzyme's substrates are specific molecules that it can chemically recognize, bind, and modify in certain ways. For example, thrombin, an enzyme involved in blood clotting, recognizes a side-by-side arrangement of two amino acids, arginine and glycine, and cleaves a peptide bond between them:

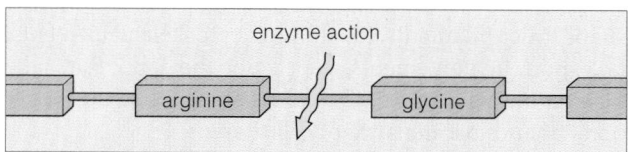

As Figure 2.21 shows, an enzyme has one or more **active sites**. These are crevices in its surface where substrates interact with the enzyme and where a specific biochemical reaction is catalyzed.

Each type of enzyme functions best within a certain temperature range. For example, reaction rates decrease sharply when body temperature becomes too high. The increased heat energy disrupts weak bonds holding an enzyme in its three-dimensional shape. This alters the active site, and substrates cannot bind to it. Even high fevers can alter enzyme structure significantly enough to disrupt metabolism and impair body functions. Normal core body temperature is about 37°C (98.6°F). A person will usually die when it reaches 44°C (112°F).

Enzymes also function best within a certain pH range. In the human body, this range typically is from pH 7.35 to 7.4. Higher or lower pH values generally disrupt enzyme structure and activity.

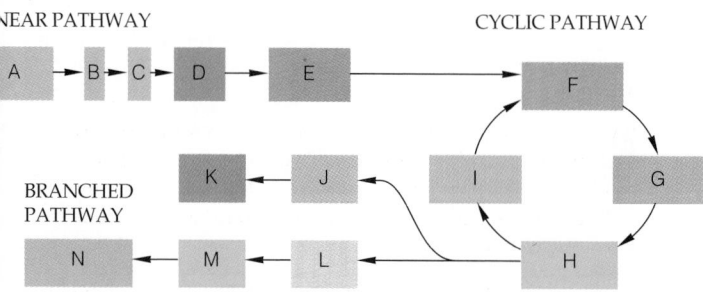

Figure 2.20 Types of reaction sequences of metabolic pathways.

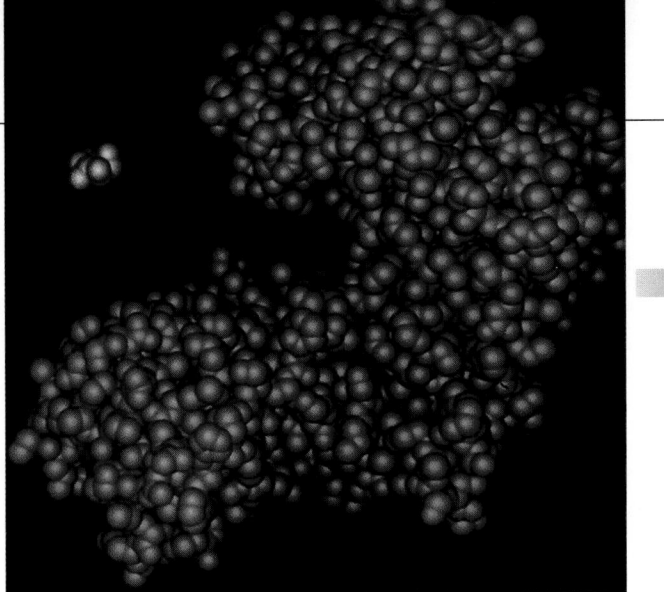

a

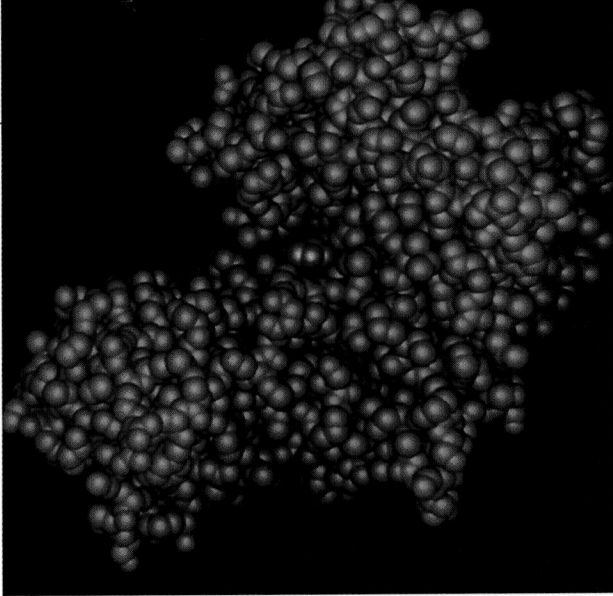

b

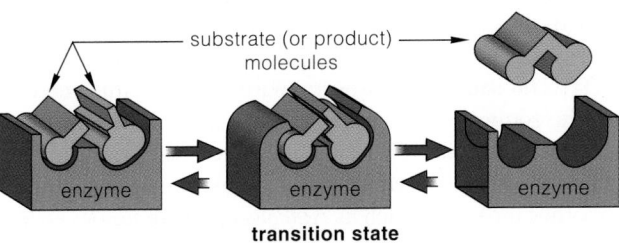

substate (or product) molecules

transition state
(tightest binding but least stable)

c

Figure 2.21 An enzyme at work. (**a**) Model of an enzyme (hexokinase) and its substrate (a glucose molecule, shown here in *red*). The cleft into which the glucose is heading is the enzyme's active site. (**b**) When the substrate makes contact with this active site, the enzyme temporarily changes shape. The upper and lower parts close in around the substrate, making a closer fit. The most precise fit between enzyme and substrate occurs during a transition state that is part of the reaction. (**c**) The enzyme-substrate complex is short-lived, partly because it is usually held together by weak bonds. The enzyme returns to its prebinding shape as a product molecule is released.

Cofactors: Enzyme "Helpers"

During many metabolic reactions, enzymes speed the transfer of electrons, atoms, or functional groups from one substrate to another. "Cofactors" help with the reactions or briefly act as transfer agents.

Cofactors include **coenzymes**. Many of these complex organic molecules are derived from vitamins. Two cofactors, **NAD⁺** (nicotinamide adenine dinucleotide) and **FAD** (flavin adenine dinucleotide), interact with hydrogen atoms liberated from glucose as it is broken down. They transfer electrons from the hydrogens to other reaction sites. The leftover protons—hydrogen ions or H⁺—are attracted to the opposite charge of the electrons and follow along. When loaded with electrons and H⁺, these coenzymes are abbreviated NADH and FADH₂. Some metal ions also serve as cofactors. In mitochondria, for example, ferrous iron (Fe⁺⁺) assists certain membrane proteins (cytochromes) active in ATP production.

How Are Enzymes Controlled?

Cells maintain, increase, and decrease concentrations of substances mostly by controlling enzymes. Some controls govern enzyme synthesis and so reduce or increase the number of enzyme molecules that are operating at any time at a key step in a pathway. Other controls stimulate or inhibit enzymes that are already formed.

In humans, the signaling molecules called hormones have major enzyme-regulating effects. As you will read in Chapter 12, specialized cells release hormones into the bloodstream. Any other cell having receptors for a given hormone will take it up, and the cell's instructions for a specific activity, such as constructing a particular protein, will be altered. This is because the hormone stimulates internal controls that either promote or inhibit the action of specific enzymes. For example, in adults certain regulatory hormones secreted from the pituitary gland (in the brain) trigger periodic increases and decreases in the synthesis of sex hormones. The regulatory hormones do this by triggering an increase or decrease in the amounts of specific enzymes in cells of reproductive organs where a person's sex hormones are synthesized.

Most chemical reactions in cells are organized in the orderly steps of metabolic pathways, which proceed with the help of enzymes.

An enzyme speeds up the rate at which a specific reaction occurs. Enzymes act only on specific substrates, and a given enzyme may catalyze the same reaction repeatedly as long as substrates are available.

Enzymes function best within limited ranges of temperature and pH. Enzyme function also depends on cofactors (coenzymes and metal ions), which help catalyze reactions or transfer electrons, atoms, and functional groups from one substrate to another.

Control mechanisms influence the synthesis of new enzymes and stimulate or inhibit existing enzymes.

CAPTURING ENERGY IN ATP: THE FIRST TWO STAGES IN HUMAN CELLS

Like all other living things, you stay alive by taking in energy. Plants and some other organisms get energy from the sun. Animals, including humans, get energy by eating plants and one another—or food derived from those sources. No matter where it comes from, energy must be put into a form that can drive chemical reactions that sustain life. Energy carried from one reaction site to another by adenosine triphosphate—ATP—serves that function. Recall that an ATP molecule is a nucleotide. It consists of adenine (a nitrogen-containing compound), ribose (a sugar with five carbon atoms), and a string of three phosphate groups. The array of enzymes that take part in thousands of different metabolic reactions catalyze the transfer of ATP's phosphate groups to many kinds of molecules. Such transfers are called **phosphorylations**, and they prime the molecules to enter reactions.

Overview: How Human Cells Make ATP

Cells make ATP by breaking covalent bonds of carbohydrates (especially glucose), lipids, and proteins. During the breakdown reactions, electrons are removed from intermediate compounds—and energy associated with those liberated electrons drives the formation of ATP, as you'll soon read.

Cells of the human body typically form ATP by way of **aerobic respiration**. "Aerobic" means the pathway can't be completed without oxygen. With every breath, you are providing your respiring cells with oxygen. Other energy-releasing pathways are *anaerobic*, for they can be completed without using oxygen. The most common are called **fermentation pathways**. Cells of skeletal muscle and some other human tissues can use a fermentation pathway for a short while when oxygen supplies run low, but they rely mainly on the aerobic pathway.

Both types of energy-releasing pathways start with the same stage of reactions, called **glycolysis**. Enzymes of these reactions split and rearrange each glucose molecule into two molecules of a substance called **pyruvate**. Glycolysis takes place in the cytoplasm, and oxygen has no role in it. In the aerobic pathway, the end of glycolysis marks the beginning of additional reaction stages that take place inside mitochondria—and successful completion of those reactions *does* require oxygen. Breaking the bonds in glucose puts electrons up for grabs, so to speak. In mitochondria, enzymes and coenzymes operate as electron transport systems that have a central role in the formation of ATP. As ATP forms, enzymes catalyze each reaction step. The intermediates produced at a given reaction step serve as substrates for the next enzyme in the pathway. With this overview in mind, let's now get a fuller picture of the biochemical events by which each and every human cell obtains the ATP energy for life.

The Breakdown of Glucose Begins

Glucose, remember, is classified as a simple sugar. Each glucose molecule consists of six carbon, twelve hydrogen, and six oxygen atoms, joined to one another by covalent bonds. In glycolysis, a glucose molecule is broken down to two molecules of pyruvate. The first steps of glycolysis are *energy-requiring*. As you can see in Figure 2.22, the steps proceed only when two ATP molecules each transfer a phosphate group to glucose—and in so doing, donate energy to it. The phosphorylations raise the energy content of glucose to a level high enough so that the *energy-releasing* steps of glycolysis can begin.

First the primed glucose molecule is split into two molecules of **PGAL** (phosphoglyceraldehyde). Each PGAL is converted to an unstable intermediate. This intermediate donates a phosphate group to ADP, forming ATP. The same events occur with the next intermediate in the sequence. Thus, four ATP form by **substrate-level phosphorylation**. This metabolic event is the *direct* transfer of a phosphate group from a substrate of a reaction to another molecule, such as ADP. Remember, however, that two ATP were invested to start the reactions, so the net energy yield at this point is only two ATP.

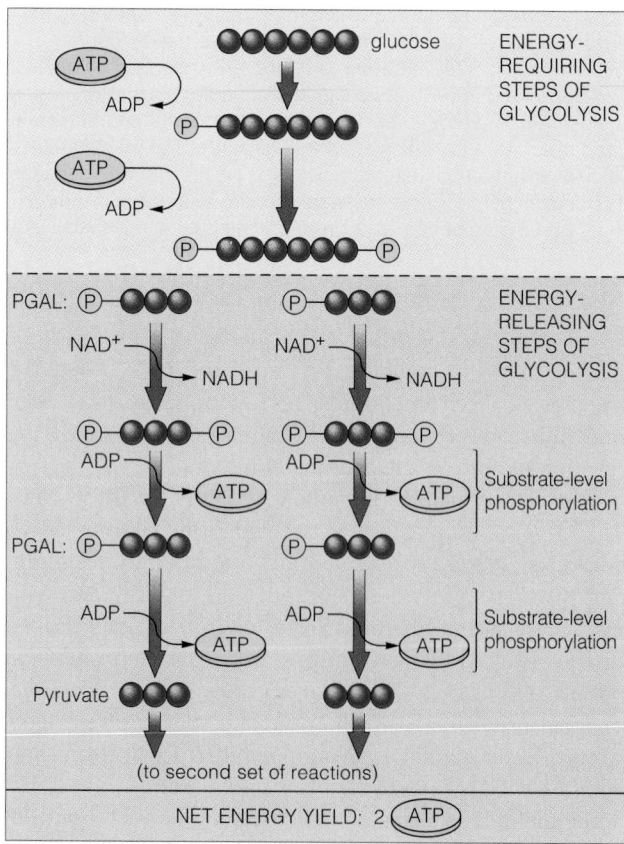

Figure 2.22 Reaction steps of glycolysis. The *red* circles represent the six carbon atoms of the glucose molecule.

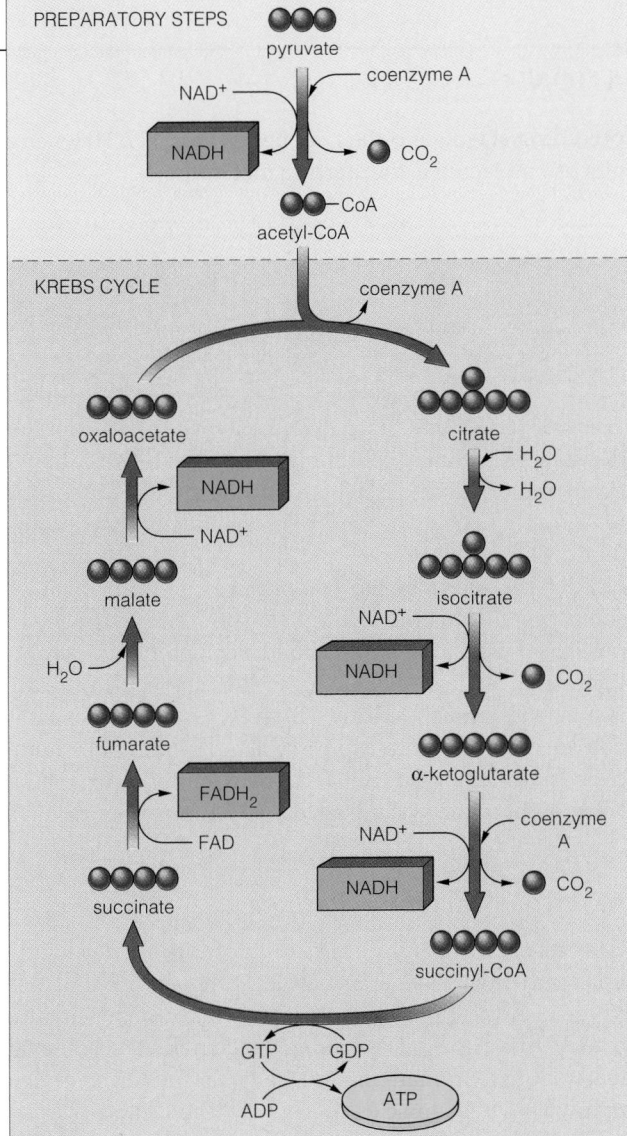

PREPARATORY STEPS

pyruvate

coenzyme A

NAD⁺

NADH

CO_2

CoA

acetyl-CoA

KREBS CYCLE

coenzyme A

oxaloacetate

citrate

H_2O

H_2O

NADH

NAD⁺

malate

isocitrate

NAD⁺

H_2O

NADH

CO_2

fumarate

α-ketoglutarate

FADH₂

NAD⁺

coenzyme A

FAD

NADH

CO_2

succinate

succinyl-CoA

GTP   GDP

ADP   ATP

Figure 2.23 The Krebs cycle and a few reactions that precede it. For each three-carbon pyruvate molecule entering the cycle, three CO_2, one ATP, four NADH, and one FADH₂ molecules form. The steps shown proceed *twice* (remember, the glucose molecule was broken down initially to *two* pyruvate molecules).

Meanwhile, electrons and hydrogen ions released from each PGAL are transferred to the coenzyme NAD⁺. Like all coenzymes, NAD⁺ is reusable. It picks up electrons and hydrogen ions stripped from substrate, becoming NADH. When the coenzyme later gives the electrons and H⁺, it becomes NAD⁺ again.

The Krebs Cycle and Preparatory Steps

After pyruvate forms in the cytoplasm, it may enter a mitochondrion. There, in the mitochondrion's inner compartment (the matrix), the second stage of the aerobic pathway continues. During the second stage, a bit more ATP forms. Carbon and oxygen atoms depart, in the

form of carbon dioxide and water. And coenzymes accept electrons and hydrogen stripped from substrates during the reactions (Figure 2.23).

During a few preparatory steps, an enzyme strips a carbon atom from each pyruvate molecule. A coenzyme picks up a two-carbon fragment (to form acetyl-CoA) and transfers it to oxaloacetate—the entry point of the **Krebs cycle,** a cyclic pathway in which more steps in the breakdown of glucose take place. The cycle was named after Sir Hans Krebs, the biologist who first worked out many of its details.

Notice that six carbon atoms, three in each pyruvate backbone, enter this second stage of reactions. Six also leave (in six carbon dioxide molecules) during the preparatory reactions and the Krebs cycle itself. The CO_2 is transported in blood to the lungs and exhaled.

Second-stage reactions serve three major functions. *First,* NAD⁺ and FAD pick up H⁺ and electrons to become NADH and FADH₂. *Second,* two molecules of ATP are produced, by substrate-level phosphorylations. *Third,* intermediates of the Krebs cycle are rearranged into oxaloacetate.

Cells have only so much oxaloacetate, and it must be regenerated to keep the cyclic reactions going. The two ATP that form do not add much to the small yield from glycolysis. But the reactions load *many* coenzymes with hydrogen and electrons for transport to the sites of the third and final stage of the aerobic pathway:

Glycolysis:	2 NADH
Pyruvate conversion preceding Krebs cycle:	2 NADH
Krebs cycle:	2 FADH₂ + 6 NADH
Coenzymes sent to the third stage:	2 FADH₂ +10 NADH

Cells get some usable energy from glycolysis, a stage of reactions that partially break down glucose or some other carbohydrate to two pyruvate molecules.

Two coenzymes (NADH) and four ATP molecules form at certain steps of glycolysis. After subtracting the two ATP required to start the reactions, the net energy yield of glycolysis is two ATP.

During the second stage of the aerobic pathway, the two pyruvate molecules from glycolysis enter a mitochondrion. Each pyruvate gives up a carbon atom and the remnant enters the Krebs cycle. All of the pyruvate molecule's carbon atoms end up in carbon dioxide.

The preparatory steps and the Krebs cycle itself yield two ATP. Oxaloacetate, the entry point for the cycle, is regenerated. Many coenzymes pick up hydrogen and electrons removed from substrates, for delivery to the final stage of the pathway.

ATP production goes into high gear during the third stage of the aerobic pathway. Electron transport systems and neighboring transport proteins serve as the production machinery. They are embedded in the inner membrane that divides the mitochondrion into two compartments. The actual ATP-making mechanism is called **oxidative phosphorylation**. As it proceeds, electron transport systems interact with electrons and unbound hydrogen (H^+ ions), which are delivered to them by coenzymes that were loaded down during the first two stages of the pathway.

Briefly, electrons are transferred from one molecule of each transport system to the next in line. As certain molecules accept and then donate electrons, they also pick up H^+ ions in the inner compartment, then release them to the outer compartment.

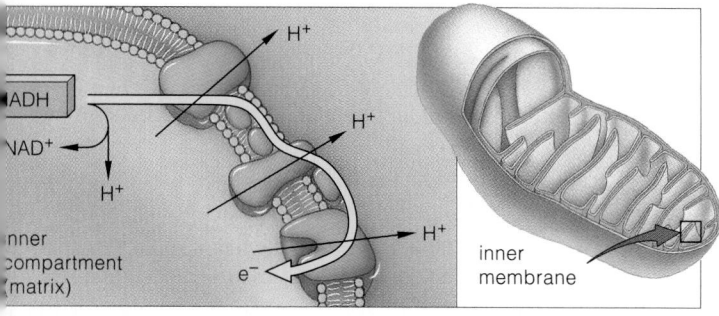

Their action sets up H^+ concentration and electric gradients across the inner mitochondrial membrane. Nearby in the membrane are transport proteins (ATP synthases) that allow the H^+ to follow the gradients back into the inner compartment:

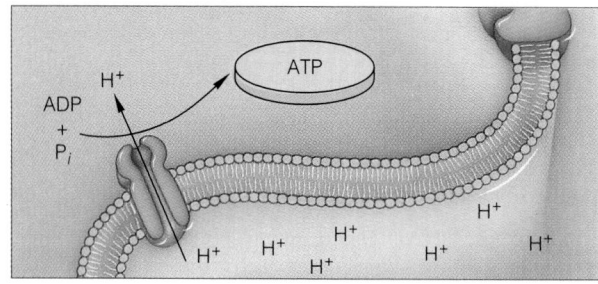

This H+ flow drives the formation of ATP from ADP and unbound phosphate. Free oxygen keeps ATP production going. It withdraws electrons from the end of the transport systems and combines with H^+ to form water. Without oxygen, ATP production stops and cells die.

Net ATP Yield of Aerobic Respiration

In many types of cells, 32 ATP form in the third stage of aerobic respiration. Add these to the net yield from the preceding stages, and the total net yield is 36 ATP when a glucose molecule is the starting material:

$$C_6H_{12}O_6 \; 1 \; 6O_2 \longrightarrow 6CO_2 \; 1 \; 6H_2O$$

glucose   carbon dioxide

TYPICAL NET YIELD = 36 ATP

Think of 36 ATP as a typical yield only. The actual amount depends on cellular conditions, as when cells require a given intermediate elsewhere and so pull it out of the reaction sequence. The yield also depends on how particular cells use the NADH that formed in glycolysis. Any NADH produced in the cytoplasm can't enter a mitochondrion. It can only transfer electrons and hydrogen to proteins in the outer mitochondrial membrane—which then shuttles them to NAD^+ or to FAD molecules already inside the mitochondrion. Both types of coenzymes deliver electrons to transport systems of the inner mitochondrial membrane. But FAD delivers them to a *lower* entry point in the system, so less ATP forms (Figure 2.24).

ATP from Anaerobic Pathways

Cells also can use glucose as a substrate for fermentation pathways. These anaerobic pathways do not use oxygen as the final acceptor of the electrons that ultimately drive the ATP-forming machinery. As you will read shortly, human muscle cells use a fermentation pathway when muscles are highly active and oxygen becomes scarce. Various microorganisms also use such pathways exclusively. Examples include bacteria that live in the animal intestine and those that cause botulism, tetanus, and some other diseases.

Glycolysis is the first stage of fermentation pathways. Here again, enzymes split and rearrange the fragments of a glucose molecule into two pyruvate molecules. Two NADH form, and the net energy yield is two ATP. However, the reactions do not fully break down glucose to carbon dioxide and water; they produce no more ATP beyond the glycolysis yield. The final steps serve to regenerate NAD^+—a coenzyme with a central role in the breakdown reactions.

Lactate fermentation is the pathway that occurs in muscle cells. In this case, the pyruvate formed during glycolysis accepts hydrogen and electrons from NADH. With this transfer, each pyruvate becomes converted to lactate, a three-carbon compound. (Lactate is also called lactic acid.)

Some types of animal cells can switch to lactate fermentation for a quick fix of ATP. As already noted, when a muscle's demands for energy are intense but

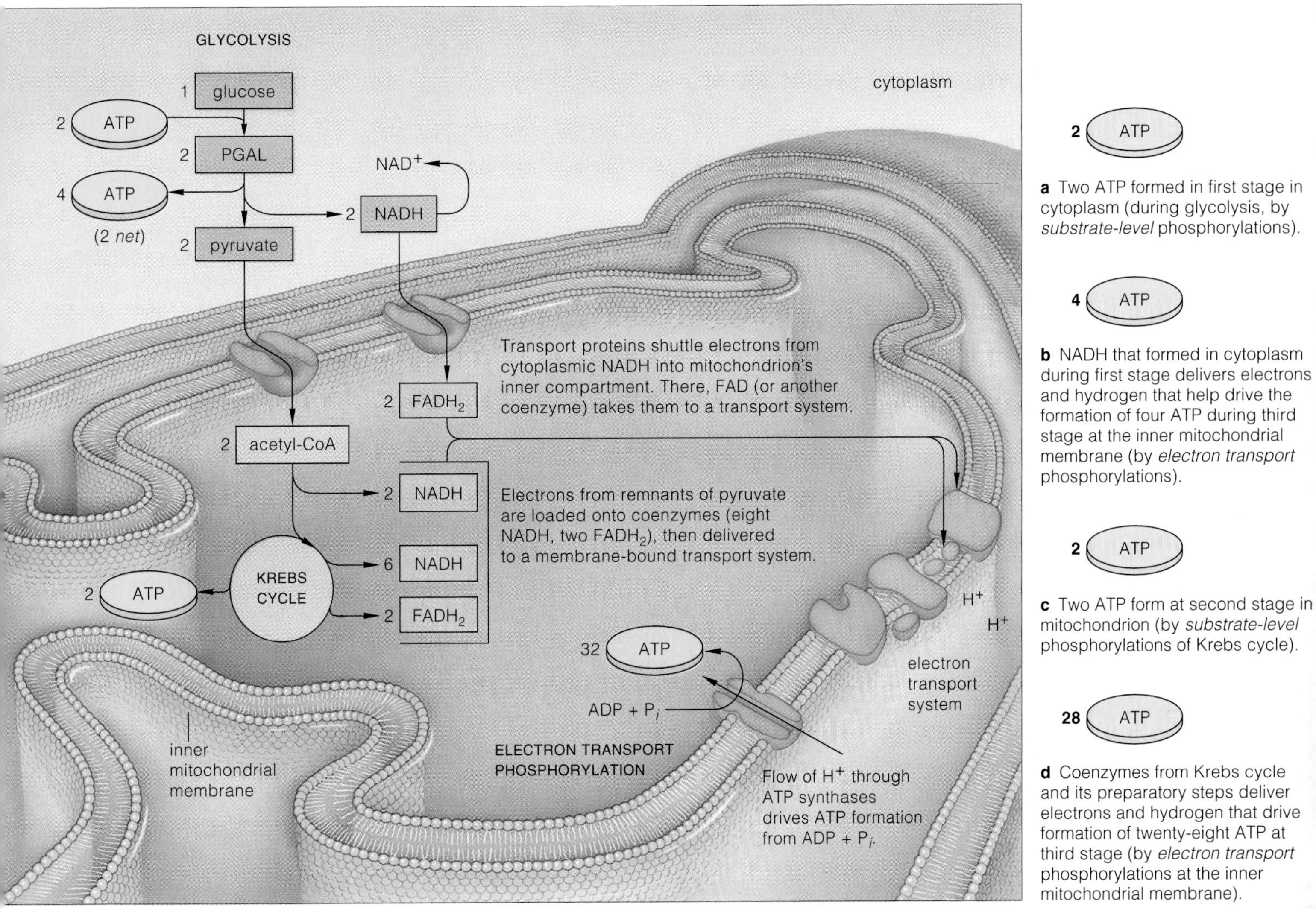

GLYCOLYSIS

1 glucose

2 ATP

2 PGAL

4 ATP

(2 net)

2 pyruvate

NAD+

2 NADH

cytoplasm

a Two ATP formed in first stage in cytoplasm (during glycolysis, by *substrate-level* phosphorylations).

2 FADH₂

Transport proteins shuttle electrons from cytoplasmic NADH into mitochondrion's inner compartment. There, FAD (or another coenzyme) takes them to a transport system.

2 acetyl-CoA

2 NADH

6 NADH

KREBS CYCLE

2 FADH₂

2 ATP

Electrons from remnants of pyruvate are loaded onto coenzymes (eight NADH, two FADH₂), then delivered to a membrane-bound transport system.

32 ATP

ADP + P$_i$

ELECTRON TRANSPORT PHOSPHORYLATION

inner mitochondrial membrane

Flow of H⁺ through ATP synthases drives ATP formation from ADP + P$_i$.

H⁺

H⁺

electron transport system

b NADH that formed in cytoplasm during first stage delivers electrons and hydrogen that help drive the formation of four ATP during third stage at the inner mitochondrial membrane (by *electron transport* phosphorylations).

c Two ATP form at second stage in mitochondrion (by *substrate-level* phosphorylations of Krebs cycle).

d Coenzymes from Krebs cycle and its preparatory steps deliver electrons and hydrogen that drive formation of twenty-eight ATP at third stage (by *electron transport* phosphorylations at the inner mitochondrial membrane).

36 ATP

TYPICAL NET ENERGY YIELD

Figure 2.24 Summary of aerobic respiration. The net yield of this energy-releasing pathway varies according to shifting concentrations of reactants, intermediates, and end products of the reactions.

brief—say, during a short footrace—the muscle cells use this anaerobic pathway. They cannot do so for long, though. When glucose stores are depleted, muscles fatigue and lose their ability to contract.

The ADP/ATP Cycle

ATP activates hundreds of different molecules and drives hundreds of activities, such as synthesizing or degrading organic compounds, actively transporting substances across cell membranes against concentration gradients, and making muscle cells contract. Given its central role in metabolism, it comes as no surprise that cells have a mechanism, called the **ADP/ATP cycle**, for renewing the supply of ATP.

The cycle is simple. In aerobic respiration and many other metabolic pathways, an energy input drives the attachment of unbound phosphate (P$_i$) or a phosphate

group to ADP (adenosine diphosphate) to form an ATP molecule. Later, when the ATP molecule donates a phosphate group somewhere, ATP reverts back to ADP:

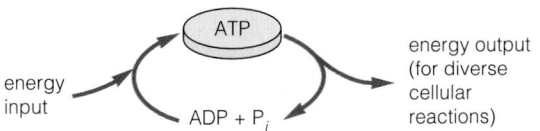

energy input

ATP

ADP + P$_i$

energy output (for diverse cellular reactions)

In the final stage of the aerobic pathway, coenzymes deliver electrons to transport systems at the inner mitochondrial membrane. Electrons and H⁺ move through the systems. Oxygen is the final electron acceptor. ATP forms when H⁺ flows through enzymes that also span the inner membrane.

From start (glycolysis in the cytoplasm) to finish (in the mitochondrion) aerobic respiration commonly has a net yield of 36 ATP for every glucose molecule.

So far, you've looked at what happens after a single glucose molecule enters an energy-releasing pathway. Now you can start thinking about what cells do when they have too many or too few of these molecules.

When you eat, your body absorbs a great deal of glucose and other small organic molecules, which your bloodstream delivers to tissues throughout the body. This is true of mammals generally. Our cells use glucose for ATP production for as long as it is available. Brain cells can use almost nothing else. When a person's food intake exceeds cellular demands for energy, ATP-producing machinery goes into high gear. The ATP concentration may increase to such an extent that it inhibits glycolysis. (This is an example of a negative feedback loop, a major type of mechanism for maintaining homeostasis and one that we will look at closely in Chapter 3.) Then, instead of being broken down to pyruvate, glucose molecules are diverted into a biosynthesis pathway in liver cells and a few other cell types. This pathway yields glycogen, a storage form of glucose in the animal body.

When the level of glucose in your blood decreases slightly, the body breaks into its glycogen stores and breaks down glycogen to glucose-6-phosphate. This compound enters the pathway of glycolysis. Liver cells also convert it back to free glucose and release it. The bloodstream then delivers the glucose to energy-demanding cells of muscles and other organs that have depleted their own glycogen stores.

It's important to note that not all cells store large amounts of glycogen. Liver and muscle cells maintain the largest stores. Even then, glycogen represents only about 1 percent of the total stored energy in an adult. On the average, 78 percent is stored in fats and another 21 percent in proteins.

Table 2.2 gives you a summary of how different types of starting molecules, whether they are complex carbohydrates, fats, or proteins, feed into the aerobic pathway.

The subunits of fats and proteins enter at points beyond glycolysis, as you'll now read.

Energy from Fats

Maybe you avoid butter, ice cream, and other fatty foods, thinking it is better to fill up on carbohydrates and proteins. This is a good idea, as long as you don't stuff yourself with these organic compounds—because your body will convert excess amounts to fats. Remember from Section 1.8 that a fat molecule has a glycerol head and one, two, or three fatty acid tails. Most of the fats stored in the body are triglycerides, which accumulate inside the fat cells in certain tissues (adipose tissues) of the buttocks and other strategic locations beneath the skin.

Between meals or during exercise, the body taps triglycerides as energy alternatives to glucose. At such times, enzymes in fat cells cleave the bonds holding glycerol and fatty acids together. Then the breakdown products enter the bloodstream. When glycerol reaches the liver, enzymes convert it to PGAL—an intermediate of glycolysis. Most cells take up the circulating fatty acids. Enzymes cleave the carbon backbone of fatty acid tails and convert the fragments to acetyl-CoA, which can enter the Krebs cycle (Figure 2.25). Each fatty acid tail has many more carbon-bound hydrogen atoms than glucose, so its breakdown yields much more ATP. When this pathway is operating, fatty acid conversions can supply about half the ATP that your muscle, liver, and kidney cells require.

Energy from Proteins

Eat more proteins than your body requires to grow and sustain itself, and your cells will not store them. Enzymes split the proteins into amino acid units; then they remove the amino group (—NH_3) from each unit. Depending on conditions in the cell, the carbon backbones that remain may be converted to fats or carbohydrates. Or they may enter the Krebs cycle. There hydrogen and electrons from the hydrocarbon backbones are transferred to the cycle's coenzymes. The amino groups undergo conversions that produce urea, a nitrogen-containing waste product that is excreted from the body in urine.

Subunits of complex carbohydrates, fats, and proteins all can serve as energy sources in the human body.

The entry of glucose or another organic compound into an energy-releasing pathway depends on its concentration inside and outside the cell, as well as on the type of cell.

Table 2.2 Summary of Energy Sources in the Human Body

Starting Molecule	Subunit	Entry Point into the Aerobic Pathway
Complex carbohydrate	Simple sugars (e.g., glucose)	Glycolysis
Fat	Fatty acids	Preparatory reactions for Krebs cycle
	Glycerol	Raw material for key intermediate in glycolysis (PGAL)
Protein	Amino acids	Carbon backbones enter Krebs cycle or preparatory reactions

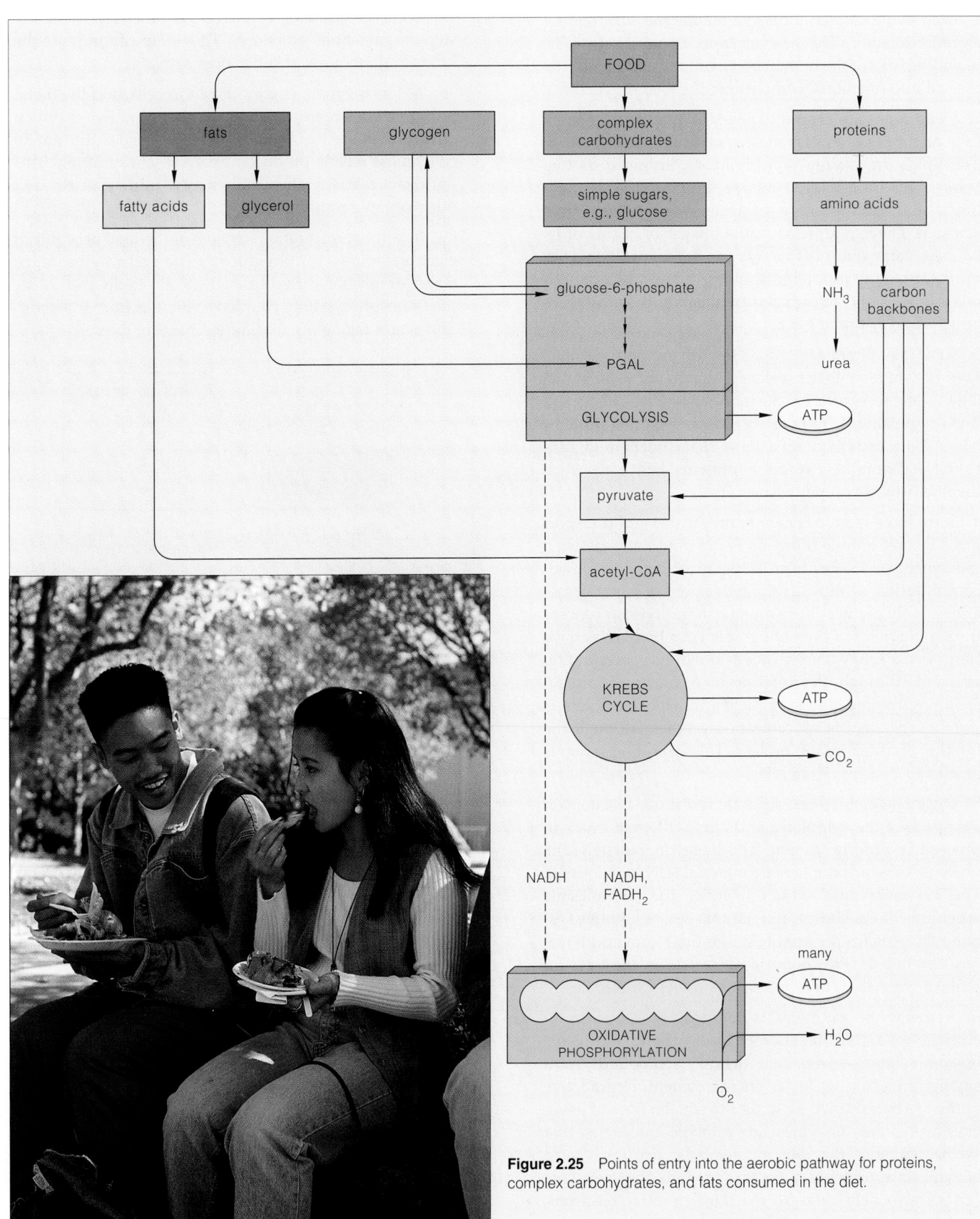

Figure 2.25 Points of entry into the aerobic pathway for proteins, complex carbohydrates, and fats consumed in the diet.

SUMMARY

1. Each living cell has a plasma membrane surrounding an inner region called the cytoplasm. In eukaryotic cells, including cells of the human body, cell membranes divide the cell into functional compartments called organelles.

2. Membranes are crucial to cell structure and function. Cell membranes consist mostly of lipids and proteins. The lipids are mainly phospholipids, arranged as two layers. This lipid bilayer imparts structure to the membrane and bars passage of water-soluble substances across it. Diverse proteins in the bilayer or at one of its surfaces carry out most membrane functions. Properties of membrane phospholipids and proteins make the cell membrane only selectively permeable to substances outside and within the cell.

3. Some membrane proteins allow or promote the passage of water-soluble substances across the membrane. Others are receptors for hormones or other substances that trigger alterations in cell behavior. Still other proteins have short carbohydrate chains that function in cell-to-cell recognition. Adhesion proteins help cells stick together in their proper tissues.

4. Substances cross cell membranes by several transport mechanisms. Simple diffusion is the random, unassisted movement of solutes from a region of higher to lower concentration. Osmosis is the movement of water across a selectively permeable membrane in response to concentration gradients, a pressure gradient, or both. In passive transport, a solute simply moves down its concentration gradient, through the interior of a transport protein spanning the membrane. In active transport, a solute is pumped against its concentration gradient through the interior of a membrane protein. This operation requires an energy boost, as from ATP.

5. Organelle membranes separate metabolic reactions in the space of the cytoplasm and allow different kinds to proceed in orderly fashion. The largest organelle is the nucleus, in which the genetic material (DNA) is located. The cytomembrane system includes the endoplasmic reticulum, Golgi bodies, and various vesicles. In this system new proteins are modified into final form and lipids are assembled. Mitochondria specialize in the oxygen-requiring reactions that produce many ATP molecules.

6. A metabolic pathway is an orderly sequence of enzyme-mediated reactions. Biosynthetic pathways assemble large, energy-rich organic compounds from smaller molecules of lower energy content. Degradative pathways break down molecules to smaller ones of lower energy content. Enzymes are catalytic molecules that greatly enhance the rate of metabolic reactions; most enzymes are proteins. Cofactors, such as the coenzyme NAD^+, also help catalyze reactions or carry electrons,

hydrogen, or functional groups stripped from a substrate to other sites.

7. Controls stimulate or inhibit enzyme activity at key steps in metabolic pathways. They help cooordinate the kinds and amounts of substances available.

8. ATP is the main energy carrier in cells, and the metabolic reactions of nearly all biosynthetic pathways run on energy delivered to them by ATP molecules. In human cells, most ATP is produced by aerobic respiration. This pathway releases chemical energy from glucose and other organic compounds.

9. In aerobic respiration, oxygen is the final acceptor of electrons removed from glucose. The pathway has three stages of reactions: glycolysis, the Krebs cycle, and oxidative phosphorylation coupled with electron transport. The stages take place in different parts of the cell:

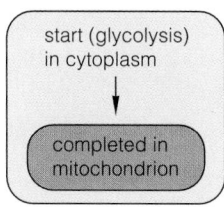

AEROBIC PATHWAY

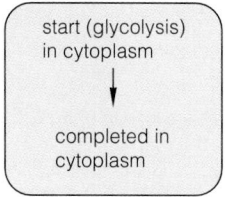

ALL OTHER ENERGY-RELEASING PATHWAYS

The net energy yield of aerobic respiration is commonly 36 ATP molecules.

10. Compared with aerobic respiration, the anaerobic pathway of fermentation has a small net yield (two ATP, from glycolysis) because glucose is not completely degraded. Following the initial reactions (glycolysis), the pathway simply regenerates the NAD^+ required for glycolysis.

Review Questions

1. All cells have a plasma membrane, cytoplasm, DNA, and ribosomes. Eukaryotic cells also have many organelles and a cytoskeleton. Describe the general functions of components. *40, 46*

2. Which organelles are part of the cytomembrane system? Sketch their arrangement in a cell, from the nuclear envelope to the plasma membrane. *50*

3. Distinguish between the following pairs of terms:
   a. diffusion; osmosis *42, 43*
   b. passive transport; active transport *44*
   c. endocytosis; exocytosis *44*

4. What is an enzyme? Describe its role in metabolic reactions. *54*

5. Think about the various energy-releasing pathways. Which reactions occur only in the cytoplasm? Which occur only in a cell's mitochondria? *56–58*

6. State the function of coenzymes in the energy-releasing pathways. *55*

7. Describe the key events of each stage of aerobic respiration:
   a. glycolysis *56*
   b. the Krebs cycle (and preparatory conversions) *57*
   c. oxidative phosphorylation/electron transport *58*

8. Describe the functional zones of a mitochondrion. Explain where the electron transport systems and carrier proteins required for ATP formation are located. *58*

Critical Thinking: You Decide *(Key in Appendix V)*

1. Using the *Science Comes to Life* as a reference, suppose you want to observe the surface of a microscopic section of bone. Would you benefit most from using a compound light microscope, a transmission electron microscope, or a scanning electron microscope?

2. Jogging is considered aerobic exercise because the cardiovascular system (heart and lungs) can adjust its activity to supply the oxygen needs of working cells. In contrast, sprinting the 100-meter dash might be called "anaerobic" exercise and playing golf "nonaerobic" exercise. Explain these last two observations.

Self-Quiz *(Answers in Appendix IV)*

1. The plasma membrane _____.
   a. surrounds an inner region of cytoplasm
   b. separates the nucleus from the cytoplasm
   c. separates the cell interior from the environment
   d. acts as a nucleus in prokaryotic cells
   e. only a and c are correct

2. The _____ is responsible for cell shape, internal structural organization, and cell movement.

3. Cell membranes consist mainly of a _____.
   a. carbohydrate bilayer and proteins
   b. protein bilayer and phospholipids
   c. phospholipid bilayer and proteins
   d. nucleic acid bilayer and proteins

4. Most membrane functions are carried out by _____.
   a. proteins   c. nucleic acids
   b. phospholipids   d. hormones

5. The passive movement of a solute through a membrane protein as it follows its concentration gradient is an example of _____.
   a. osmosis   c. diffusion
   b. active transport d. facilitated diffusion

6. Match each organelle with its correct function.
   ____protein synthesis   a. mitochondrion
   ____movement   b. ribosome
   ____intracellular digestion   c. smooth ER
   ____modification of new proteins d. rough ER
   ____lipid synthesis   e. nucleolus
   ____ATP formation   f. lysosome
   ____ribosome subunit assembly g. flagellum

7. Which of the following statements is *not* true? Metabolic pathways _____.
   a. occur in stepwise series of chemical reactions
   b. are speeded up by enzyme activity
   c. may degrade or assemble molecules
   d. always produce energy (i.e., ATP)

8. Enzymes _____.
   a. enhance reaction rates   c. act on specific substrates
   b. are affected by pH   d. all of the above are correct

9. Match each substance with its correct description.
   ____a coenzyme or metal ion   a. reactant
   ____substance formed at end of b. enzyme
   a metabolic pathway   c. cofactor
   ____mainly ATP   d. energy carrier
   ____substance entering a reaction e. end product
   ____protein that catalyzes a
   reaction

10. Glycolysis starts and ends in the _____.

    a. nucleus   c. plasma membrane
    b. mitochondrion   d. cytoplasm

11. The pathway of aerobic respiration is completed in the ____.

    a. nucleus   c. plasma membrane
    b. mitochondrion   d. cytoplasm

12. Match each type of metabolic reaction with its function:
    ____glycolysis   a. ATP, NADH, and CO_2 form
    ____Krebs cycle   b. glucose to two pyruvate molecules
    ____oxidative   c. H^+ flows through channel
    phosphorylation   proteins, ATP forms

Selected Key Terms

active site *54*
active transport *44*
ADP/ATP cycle *59*
aerobic respiration *56*
centriole *53*
chromosome *49*
cilium *53*
coenzyme *55*
concentration gradient *42*
cytomembrane system *50*
cytoplasm *40*
cytoskeleton *53*
diffusion *42*
endocytosis *44*
endoplasmic reticulum (ER) *50*
exocytosis *44*
FAD *55*
fermentation pathway *56*
flagellum *53*
fluid mosaic model *41*
glycolysis *56*
Golgi body *50*
intermediate filament *53*
Krebs cycle *57*
lipid bilayer *41*

lysosome *50*
metabolic pathway *54*
metabolism *54*
microfilament *53*
microtubule *53*
mitochondrion *52*
NAD^+ *55*
nuclear envelope *49*
nucleolus *49*
nucleus *40*
organelle *46*
osmosis *43*
oxidation-reduction
   reaction *54*
oxidative
   phosphorylation *58*
passive transport *44*
peroxisome *51*
PGAL *56*
plasma membrane *40*
ribosome *46*
substrate *54*
substrate-level
   phosphorylation *56*
vesicle *46*

Readings

Kedersha, N. March 1992. "Psychedelic Cells." Discover. Text plus vivid color photographs of different types of human cells, obtained using the technique of immunofluorescence.

Rothman, J. E., and L. Orci. March 1996. *Scientific American*. "Budding Vesicles in Living Cells." This article explains how researchers on two continents worked together to unveil how vesicles operate as the cell's internal transportation system.

Trefil, James. June 1992. "Seeing Atoms." *Discover*. How researchers use microscopes to study cell organelles, molecules, and even atoms.

 For additional readings, go to InfoTrac College Edition, your online library at:

http://www.infotrac-college.com/wadsworth

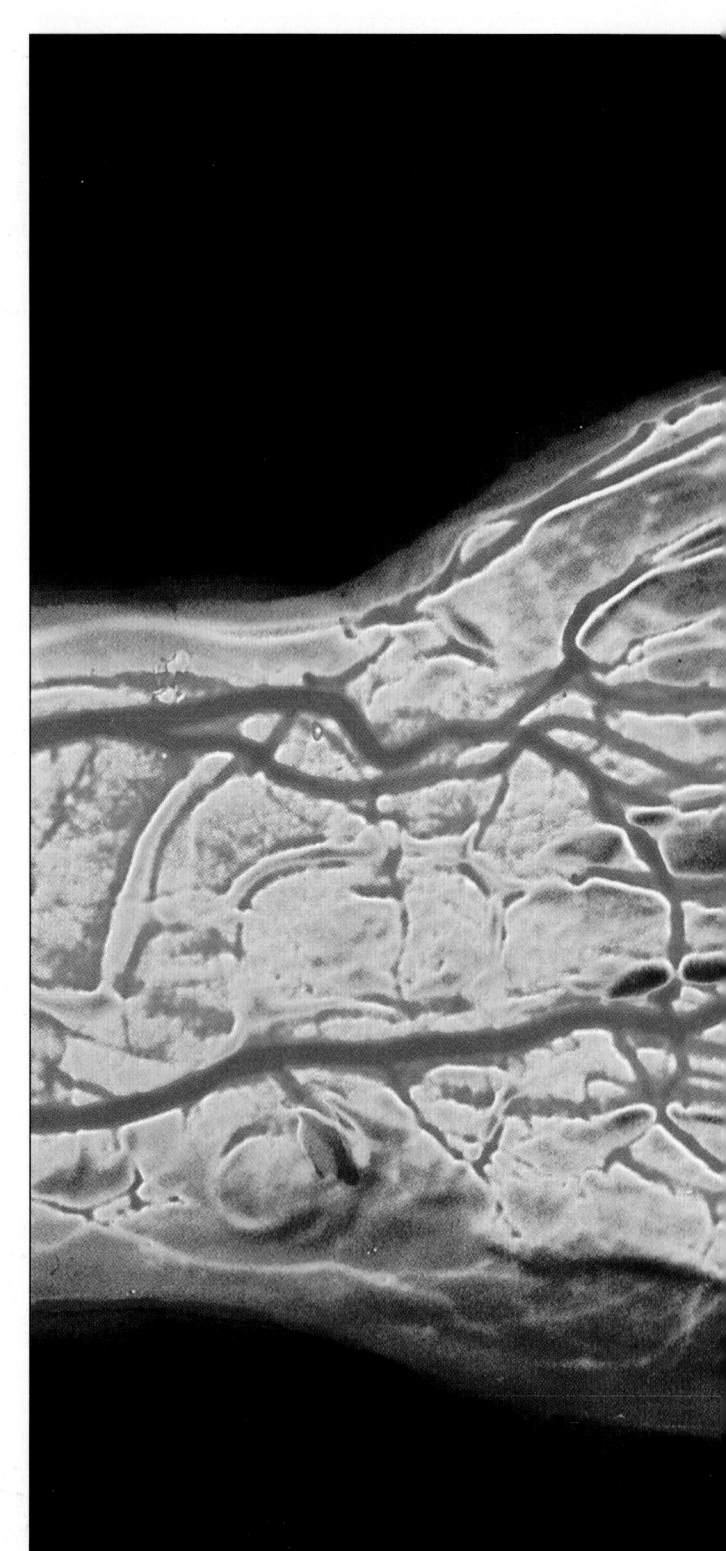

Intricate network of blood vessels in a human hand.

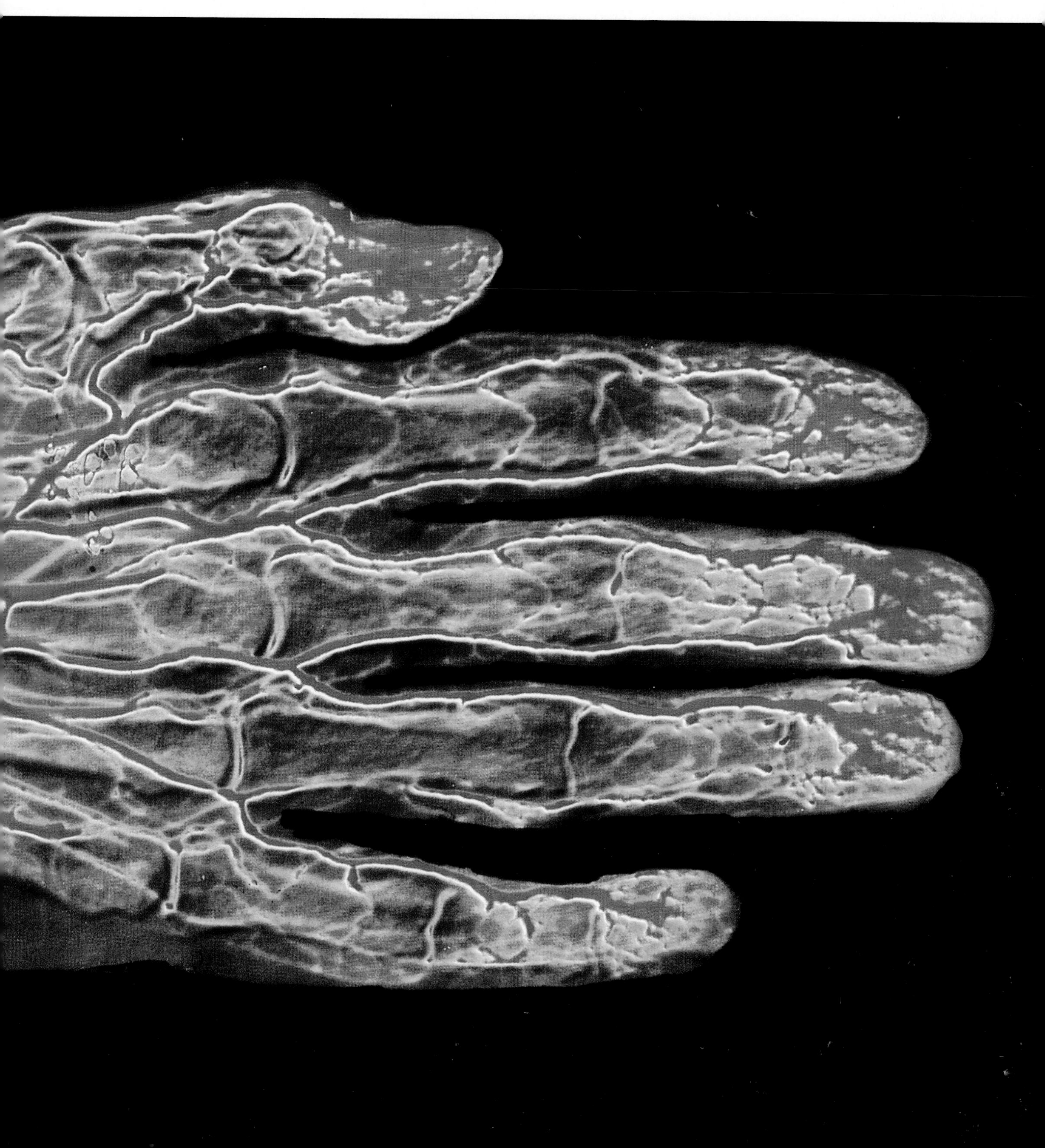

TISSUES, ORGAN SYSTEMS, AND HOMEOSTASIS

The Body in Balance

What does it mean to be physically fit? An exercise physiologist—someone who studies how exercise affects body functions—would loosely define physical fitness as a state of well-being in which the body can meet the demands of daily life, as well as perform adequately in occasional, more challenging activities. For many of us, maintaining a fit body is no easy task—it requires us to juggle a hectic schedule, make changes in a sedentary lifestyle, and resist the temptation to load our plates with foods high in fat or sugar. When we do make a consistent effort to stay in shape, however (Figure 3.1), the result usually is improved functioning of our body's cells, tissues, and organs—including the heart, lungs, and muscles.

Just as "no man is an island," no organ operates in isolation. Each one functions as part of an organ system—a group of interacting organs that perform related functions. These systems are highly specialized.

For instance, your digestive system brings needed nutrients into your body, and your circulatory system—the heart and blood vessels—efficiently transports them to body cells. Together, the circulatory and respiratory systems supply your cells with the oxygen they need for metabolism and ATP production. Those systems also swiftly remove the wastes produced by all that cellular activity.

The coordinated activity of cells, tissues, organs, and organ systems goes on every moment of your life. And as this "business of life" proceeds, it produces changes in the body's internal environment—the bloodstream and fluids bathing body cells. For instance, biochemical events may alter the volume of blood and tissue fluids, add or deplete ions, and so forth. Drastic changes in blood and tissue fluid would kill cells—except that your kidneys and other elements of the urinary system work to keep this disaster from happening.

Governing all this activity is the role of the body's command posts—the nervous system and an endocrine system, which operates by way of hormones. These two control systems work together and mobilize the body as a whole for everything from simple housekeeping, such as basic waste removal, to meeting the dramatically increased physical demands of vigorous exercise.

In this unit we focus squarely on two basic topics—the structure of the human body, or its **anatomy**, and how the body functions—its **physiology**. This chapter surveys the tissues and organ systems that you will be considering in greater depth in Chapters 4–13. It also explores more fully the concept of homeostasis—the all-important stability in the internal environment that is brought about by the coordinated activity of the body's cells, tissues, organs, and organ systems.

Even though humans are extraordinarily complex animals, our body has only four basic types of tissues. These are epithelial, connective, muscle, and nerve tissues. A **tissue** is a group of interacting cells and intercellular substances that each in its own way contributes to the specialized tasks just listed. An **organ**, such as the heart, consists of different tissues organized in specific proportions and patterns. An **organ system** consists of two or more organs that interact physically, chemically, or both in performing a common task, such as blood circulation. Cells, tissues, organs, and organ systems split up the work, so to speak, in ways that contribute to the survival of the body as a whole.

As you work through this chapter, keep in mind that the tissues and other structures you are reading about are the result of a truly amazing developmental journey that begins shortly after the moment a new human being is conceived. Chapter 14 describes the remarkable developmental mechanisms involved. Now, let's turn our attention to some of the features of the four types of specialized tissues present in the adult human body.

Figure 3.1 Exercising to stay fit. As we'll see in this chapter, physical fitness is in part a reflection of the body's ability to maintain homeostasis, keeping internal operating conditions within life-sustaining limits.

KEY CONCEPTS

1. Cells interact to form *tissues*, many of which are combined in *organs*, which are components of *organ systems*.

2. Organs are constructed of four basic classes of tissues: epithelial, connective, muscle, and nerve tissues.

3. Each cell engages in basic metabolic activities that assure its own survival. At the same time, cells, tissues, and organs perform activities that contribute to the survival of the body as a whole.

4. The combined contributions of cells, tissues, organs, and organ systems help maintain a stable internal environment, which is required for our body's survival. This concept helps us understand the functions of any organ system, regardless of its complexity.

CHAPTER AT A GLANCE

3.1   Epithelial Tissue
3.2   *Spare Parts for the Body*
3.3   Cell-to-Cell Contacts
3.4   Overview of Connective Tissues
3.5   Cartilage and Bone
3.6   Blood and Adipose Tissue
3.7   Muscle Tissue
3.8   Nerve Tissue
3.9   Organ Systems
3.10   The Skin: Case Study of an Organ System
3.11   *Sun, Skin, and the Ozone Layer*
3.12   Homeostasis and Systems Control

Skin. The rosy tissue lining your mouth. Each of these is an example of epithelial tissue, or **epithelium** (plural: epithelia). Epithelium always has a free surface, which faces a body cavity or the outside environment. Epithelial cells nestle together, with little material in between them. They are arranged in one or more layers, and they are linked by specialized junctions that provide both structural and functional connections between the cells. Sometimes cells at an epithelium's free surface have slim protrusions of cytoplasm—cilia or microvilli (singular: microvillus). These surface modifications have special roles, as you will see in later chapters. The other surface of an epithelium adheres to a **basement membrane**—a cell-free layer packed with proteins and polysaccharides that lies between the epithelium and the underlying connective tissue (Figure 3.2a).

Basically, there are only two types of epithelia, simple and stratified. *Simple* epithelium lines the body's cavities, ducts, and tubes (Figure 3.2b–d). It is a single layer of cells, which typically function in the diffusion, secretion, absorption, or filtering of substances across the layer. *Stratified* epithelium has two or more cell layers (see the micrograph in Figure 3.2a); protection is its typical function—for example, at the skin surface. *Pseudostratified* epithelium actually is a simple epithelium in disguise. It consists of a single cell layer that looks like a multiple layer in side view because the nuclei of adjacent cells are staggered. Most of the cells bear cilia. Pseudostratified epithelium occurs in the throat, nasal passages, reproductive tract, and other body locations where cilia sweep mucus or some other fluid across the epithelial surface.

The basic types of epithelium are subdivided into several categories, depending on the shape of cells at the free surface. A *squamous epithelium* has flattened cells, a *cuboidal epithelium* has cube-shaped cells, and a *columnar epithelium* has elongated cells. Oxygen and carbon dioxide easily diffuse across the simple squamous epithelium making up the walls of fine blood vessels, as shown in Figure 3.2b. Table 3.1 summarizes the various types of epithelium and their roles. *Science Comes to Life* on page 70 describes experimental efforts to grow epithelium and other tissues outside the body.

Glands: Derived from Epithelium

Glands are secretory cells or multicellular structures that are derived from epithelium and often stay connected to it. Mucus-secreting goblet cells, for instance, are embedded in pseudostratified epithelium that lines the trachea (your windpipe) and other tubes leading to the lungs. Stomach epithelium is densely peppered with glandular cells that secrete mucus and other substances.

Biologists often classify glands according to how their secretions reach the site where they are used. **Exocrine glands** secrete substances onto a free epithelial surface through ducts or tubes. The ducts have only a few branches in *simple* exocrine glands, and many branches in *compound* glands. Mucus, saliva, earwax, oil, milk, and digestive enzymes all are exocrine secretions.

We can also assort exocrine glands into groups according to the makeup of their secretions. The secretions of *apocrine* glands (including some sweat glands and epithelial tissue in milk-secreting mammary glands) include bits of the gland cells that produced them. In *holocrine glands*, such as sebaceous (oil) glands in skin, whole cells full of the material to be secreted are actually shed into the duct, where they burst and release the secretion. Most exocrine glands are *merocrine* glands; their secretions do not include elements of gland cells. Salivary glands and most sweat glands are in this category.

Endocrine glands have no ducts. Their secretions, hormones, pour directly into the extracellular fluid bathing the glands. Typically, the bloodstream picks up hormones and distributes them to target cells elsewhere in the body. Examples of endocrine glands include the thyroid, adrenals, and the pituitary gland. These and other endocrine glands, and the activities of various hormones, are the main topics of Chapter 12.

Table 3.1   Major Types of Epithelium

Type	Shape	Typical Locations
Simple (one layer)	Squamous	Linings of blood vessels, lung alveoli (sites of gas exchange)
	Cuboidal	Glands and their ducts, surface of ovaries, pigmented epithelium of eye
	Columnar	Stomach, intestines, uterus
Stratified (two or more layers)	Squamous	Skin (keratinized), mouth, throat, esophagus, vagina (nonkeratinized)
	Cuboidal	Ducts of sweat glands
	Columnar	Male urethra, ducts of salivary glands
Pseudostratified	Columnar	Throat, nasal passages, sinuses, trachea, male genital ducts

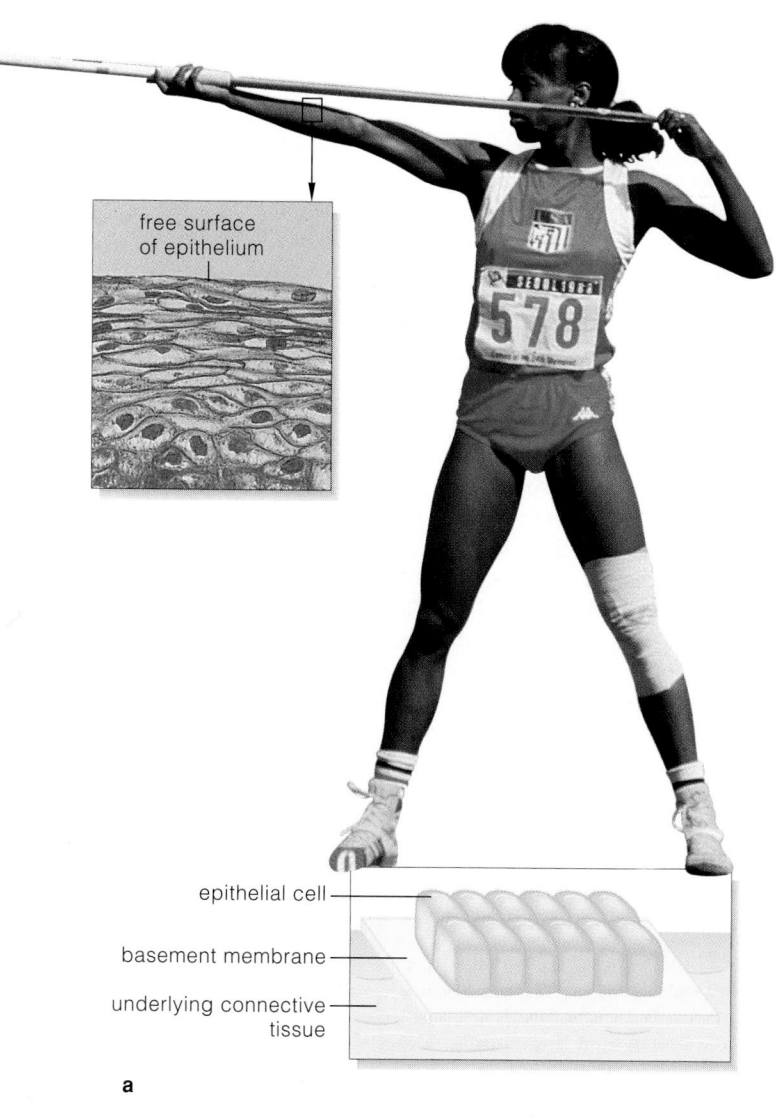

free surface
of epithelium

epithelial cell
basement membrane
underlying connective
tissue

a

Membranes

Various types of thin, sheetlike membranes cover or line organs. They consist of epithelium together with underlying connective tissue. Some, like **mucous membranes** that line the tubes and cavities of the digestive, respiratory, and reproductive systems, contain glands including mucous glands that secrete mucus. In chapters to come, you will read about many examples of how mucous membranes offer protection and secrete or absorb various substances. *Serous membranes*, such as those that line the thoracic cavity and enclose the heart and lungs, do not contain glands. Among other functions, they help hold internal organs in place and provide lubricated smooth surfaces to prevent chafing or abrasion between adjacent organs or between organs and the body wall.

Epithelia are sheetlike tissues with one free surface. Simple epithelium lines body cavities, ducts, and tubes. Stratified epithelium, like that at the skin surface, typically functions in protection.

Glands are secretory cells or multicellular structures derived from epithelium and often remain connected to it.

Figure 3.2 Epithelial tissue. (**a**) All epithelia have a free surface. There is a basement membrane between the opposite surface and underlying connective tissue. The diagram shows simple epithelium, which consists of a single layer of cells. The micrograph shows the upper portion of stratified squamous epithelium, which has more than one layer of cells. The cells become more flattened toward the surface. (**b–d**) Examples of simple epithelium, showing the three basic cell shapes in this type of tissue.

b

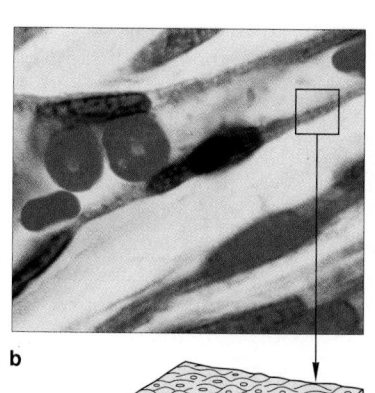

TYPE: Simple squamous

DESCRIPTION: Single layer of flattened cells

COMMON LOCATIONS: Blood vessel walls; air sacs of lungs

FUNCTION: Diffusion

c

TYPE: Simple cuboidal

DESCRIPTION: Single layer of cubelike cells; free surface may have microvilli (absorptive structures)

COMMON LOCATIONS: Salivary glands, sweat glands, and ducts of glands and kidney tubules

FUNCTION: Secretion, absorption

d

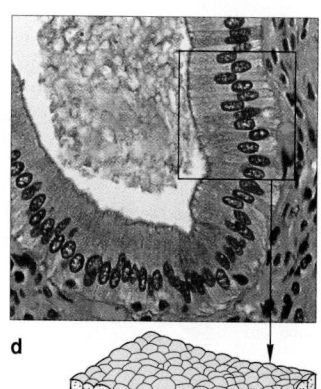
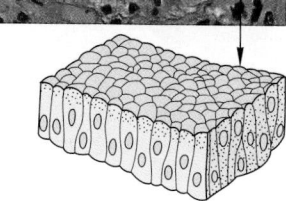

TYPE: Simple columnar

DESCRIPTION: Single layer of tall, slender cells; free surface may have microvilli

COMMON LOCATIONS: Part of lining of gut and respiratory tract

FUNCTION: Secretion, absorption

3.2 SPARE PARTS FOR THE BODY

Need a new swatch of skin or a piece of cartilage? How about a new heart valve grown especially for you? Researchers have spent decades trying to grow tissues—even whole human organs—in the laboratory. Their goal is to help the millions of people who each year lose large amounts of tissue or organs to accidents or disease, and the thousands of children born with certain birth defects. Sooner than you think, "designer" spare parts for the body may be a reality.

GROWING SKIN A new breed of "tissue engineers" have already devised ways to grow whole sheets of new skin, using as their raw material foreskin tissue donated after newborn baby boys are circumcised. The lab-grown tissue does not provoke rejection by the recipient's immune system, and it can be used to replace skin lost due to severe burns, ulcers, or other damage. At one Massachusetts biotechnology firm, a team of scientists are developing a method for growing such designer skin, and they have made major progress. They begin with collagen extracted from cattle. As described in Section 3.4, collagen is one of the body's major structural proteins. After it is purified, this material will provide a structural base for the next step, in which cells from the skin's inner layer (the dermis) are "seeded" on the collagen and allowed to proliferate. After a week or so, cells from skin's upper layer (epidermis) are added. About 14 days later, the mix of skin cells and collagen has become transformed into delicate pink sheets of "custom-made" skin. When layers of the new skin are placed over, say, a leg ulcer of a diabetes patient, blood vessels grow into the new tissue and the wound heals.

SPACE AGE TISSUES A "bioreactor" developed for space shuttle missions (Figure 3.3) can grow heart, liver, and lung cells, among other cell types. All behave much as they would in intact tissues. The device reduces physical shear stresses that can separate cells proliferating in a culture dish. Instead, the cells grow in extremely close proximity to one another—and presumably can more easily exchange intercellular signals that are essential for tissues to perform their normal functions.

Although bioreactor-grown cells are just a beginning—and no one has yet succeeded in growing an entire organ—the future does seem promising. For instance, much work has already been done on molding ultrapure plastic into the correct configuration (using computer-aided design) for a chunk of tissue or a replacement organ. When the design is complete, the plastic framework is seeded with cells that, given the proper chemical cues, can grow into the desired organ. Another scenario is to grow cells of the desired type in a bioreactor, using a dissolvable mesh as the framework. Once the replacement tissue has formed and has been implanted into a patient, the mesh "skeleton" dissolves, in the same way that dissolving stitches do.

ARTIFICIAL BONE Laboratory-synthesized materials called *organoapatites* now are being used to repair bone damaged by disease or accident. Derived from mineral crystals that occur naturally in living bone (and teeth), organoapatites rapidly become integrated into the patient's healthy bone tissue. Blood vessels will even grow into the artificial material, helping to speed up natural processes of bone tissue replacement.

Figure 3.3 A rotating cell culture device developed by NASA researchers creates laboratory conditions that encourage cells to differentiate into specialized cell types.

The Challenge of Replacing Blood

Every three seconds, someone in the United States receives a blood transfusion (Figure 3.4). And the demand for blood for surgery patients and the treatment of severe injuries is growing dramatically. Why? A key reason is that the human population is aging, and older people are more likely to require surgical procedures—such as joint replacement or heart surgery—in which significant blood loss is normal. Blood is transfused in "units," each about one-half liter (roughly a pint). Worldwide, the demand for transfusable blood is exploding, with an annual *increase* of over 7 million units.

Blood donations aren't even close to keeping up with this incessant demand—in the United States or anywhere else. Another problem also lingers. In the 1980s the world came to grips with the horrifying fact that donated blood, and products derived from it, could transmit HIV, the virus responsible for AIDS, as well as some other viral diseases. Although sensitive methods for testing donated blood have since become routine, many people deeply fear receiving transfused blood.

Enter researchers who are racing to develop synthetic blood—substances that can be administered to patients and can perform blood's functions. Since the body has mechanisms for renewing its blood supply (described in Chapter 6), a substitute would need only to fill the gap, so to speak, for a limited time. But the basic functions such a substance would have to carry out are absolutely crucial to survival of the body's cells, tissues, and organs. Most vital is transporting the oxygen that cells need to make ATP during aerobic respiration. This means that, at a minimum, a blood substitute must perform the tasks of hemoglobin— the iron-containing transport protein described in Section 1.10. In addition, it must not be recognizable by the body as "foreign," and so trigger an immune response. Finally, it must be free of dangerous side effects.

EXAMPLES OF "SUBSTITUTE BLOOD" One major type of whole blood substitute is a solution containing large carbohydrate molecules that help maintain the proper water content of the fluid, along with synthetic molecules that contain atoms of carbon and fluorine. This mixture, called a perfluorocarbon, or PFC, can bind roughly two-thirds the amount of oxygen that would be found in normal whole blood. Unfortunately, PFCs can have serious side effects, generally are eliminated from the body quickly, and are difficult to store for long periods. However, they may provide an alternative in cases where an individual's religious beliefs forbid transfusion of real blood.

Another approach is to create a blood substitute using actual human hemoglobin. The hemoglobin molecules are extracted from cow's blood or from donated blood that has outlived its shelf life and is no longer usable. The hemoglobin then is chemically modified in various ways.

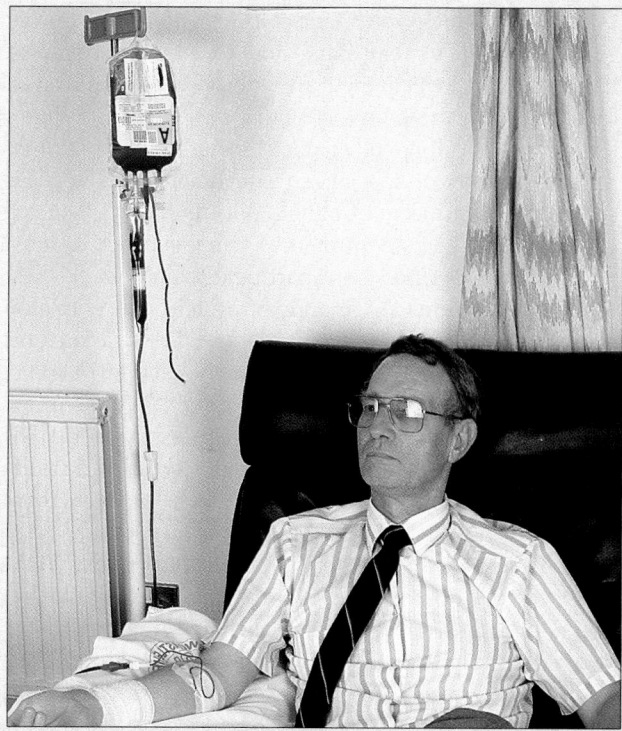

Figure 3.4 Transfusing blood.

For example, hemoglobin molecules extracted from human blood are linked to other substances, mimicking conditions inside a red blood cell that allow hemoglobin to bind oxygen (in the lungs), then release it to respiring cells. This alteration also increases the size of hemoglobin molecules. Hence they remain in the recipient's body longer than they otherwise would, allowing more time for the person's own red blood cells to be regenerated. The hemoglobin molecules also may be reshaped so that they can carry more oxygen than usual. This property has applications in treating some kinds of cancer tumors, which can be rendered more susceptible to radiation by adding oxygen to the cancerous tissue.

Genetic engineers have inserted the gene for human hemoglobin into pigs, which then produce the protein. The goal is to obtain quantities of purified hemoglobin, which can be used to improve the oxygen-carrying capacity of the blood in people suffering from a severe blood loss.

At this writing, efforts to develop blood substitutes all have pitfalls, ranging from high manufacturing costs to short retention in the body to the danger of transmitting disease. For example, products made from cow's blood could carry *bovine spongiform encephalitis*—otherwise known as "mad cow disease." Given their potential for improving health care and saving lives, however, it is only a matter of time before one or more safe, effective blood substitutes become available.

Look at the skin of your hand. Now try to imagine what would happen to that skin—and to all your tissues—if the cells of which they are composed did not stick together. Quite literally, you'd fall apart. Figure 3.5 shows three types of cell-to-cell contacts in epithelium and other tissues. By way of specialized attachment sites at these contacts, cells adhere strongly to one another. Cell-to-cell contacts are especially numerous when substances must not leak from one body compartment to another.

Tight junctions are strands of protein that help stop substances from leaking across a tissue. The strands form gasketlike seals that block the free movement of molecules. In epithelium, for example, tight junctions allow the epithelial cells to control what enters the body. For instance, during food digestion various types of nutrient molecules can diffuse into epithelial cells or enter them selectively by active transport, but tight junctions keep those needed molecules from slipping *between* cells. Tight junctions also prevent the highly acidic gastric fluid in your stomach from leaking out and digesting proteins of your own body instead of those you consume in food.

Adhering junctions cement cells together. One type, called *maculae adherens* junctions or *desmosomes,* are like spot welds at the plasma membranes of two adjacent cells. They are anchored to the cytoskeleton in each cell and help hold cells together in tissues that are subject to stretching, such as epithelium of the skin and stomach. Another type of adhering junction, called *zonula adherens* junctions, form a tight collar around epithelial cells.

Gap junctions help cells communicate by promoting the rapid transfer of ions and small molecules between them. This type of junction contains small, open channels that directly link the cytoplasm of adjacent cells.

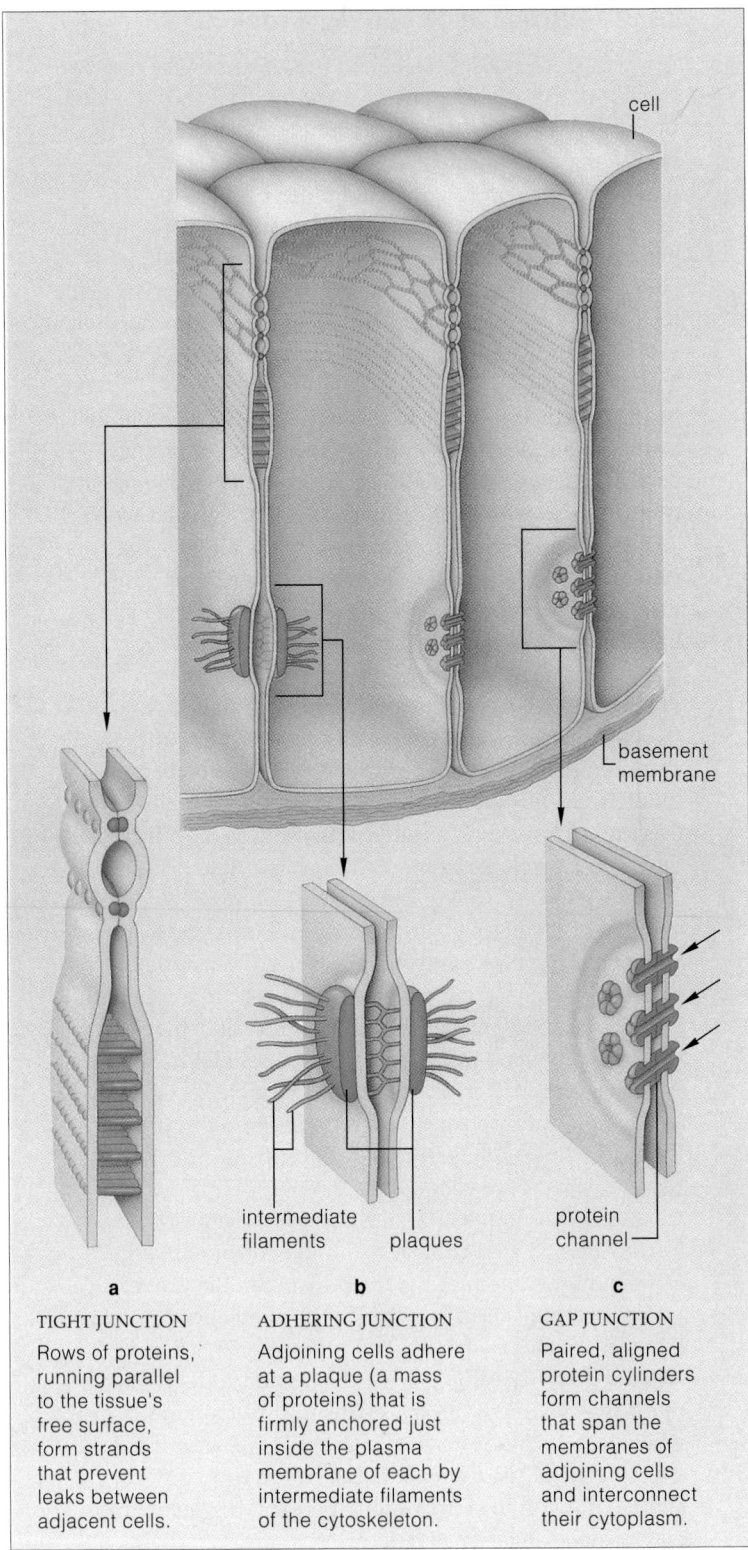

intermediate
filaments   plaques

protein
channel

a

TIGHT JUNCTION

Rows of proteins, running parallel to the tissue's free surface, form strands that prevent leaks between adjacent cells.

b

ADHERING JUNCTION

Adjoining cells adhere at a plaque (a mass of proteins) that is firmly anchored just inside the plasma membrane of each by intermediate filaments of the cytoskeleton.

c

GAP JUNCTION

Paired, aligned protein cylinders form channels that span the membranes of adjoining cells and interconnect their cytoplasm.

Figure 3.5 Examples of cell junctions. (**a**) In some epithelia, protein strands form tight seals that ring each cell and seal it to its neighbors. The seals prevent substances from leaking across the free epithelial surface. The only way they can reach tissues below is to pass *through* the epithelial cells, which have built-in mechanisms that can control their passage. (**b**) Adhering junctions (desmosomes) are like spot welds that "cement" cells of epithelium (and all other tissues) together so that they function as a unit. They are abundant in the skin's surface layer and other tissues subjected to abrasion. (**c**) Gap junctions promote diffusion of ions and small molecules from cell to cell. They are abundant in the heart and other organs in which cell activities must be rapidly coordinated.

OVERVIEW OF CONNECTIVE TISSUES

Of all the tissues in the human body, connective tissues are the most abundant and widely distributed. They also are structurally specialized to serve a variety of functions. As shown in Table 3.2, they include the several types of "connective tissue proper" we consider in this section, as well as specialized connective tissues—cartilage, bone, blood, and adipose tissue—discussed in Sections 3.5 and 3.6.

General Characteristics of Connective Tissue

In **connective tissue**, a relatively small number of cells are dispersed through a "ground substance," rather like bits of fruit in a gelatin salad. In all types except blood, fibroblasts and other kinds of cells secrete collagen fibers and elastic fibers (containing molecules of a protein called elastin), which are structural proteins. Plastic surgeons inject the strong, whitish collagen fibers to "plump up" skin wrinkles, lips, and sunken acne scars. Yellowish elastic fibers are stretchable and help impart the same property to connective tissue. Fibroblasts also secrete the jellylike ground substance, which consists of proteins and polysaccharides. The ground substance and fibers make up an **extracellular matrix**. The fiber-reinforced matrix fills spaces between connective tissue cells and helps the tissue as a whole withstand physical stresses.

Connective Tissue Proper

Tissues of this group have mostly the same components, but in different proportions. **Loose connective tissue** has the most types of cells, and fewer fibers (Figure 3.6a). The fibers, both collagen and elastic, are loosely arranged. Cells in loose connective tissue include fibroblasts and infection-fighting cells (macrophages), as well as other cells that migrate through tissues or take up residence in

Table 3.2 Types of Connective Tissue
Connective tissue proper:
Loose connective tissue
Dense, irregular connective tissue
Dense, regular connective tissue (ligaments, tendons)
Specialized connective tissue:
Cartilage
Bone
Blood
Adipose tissue

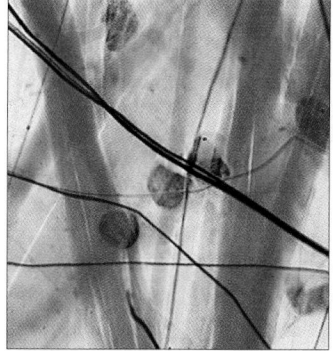

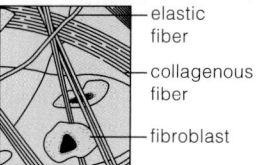

TYPE: Loose connective tissue
COMMON LOCATIONS: Under the skin and most epithelia
FUNCTION: Support, elasticity

a

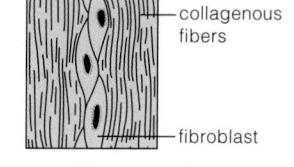

TYPE: Dense, regular
COMMON LOCATIONS: Tendons, skin, kidney capsule
FUNCTION: Support, elasticity

b

Figure 3.6 Examples of loose connective tissue (**a**) and dense connective tissue (**b**).

them. As part of the body's defense arsenal, macrophages can engulf and destroy potential infectious agents. When small wounds allow bacteria to cross the skin or the lining of the intestine, respiratory tract, or urinary tract, macrophages mount an early counterattack. Loose connective tissue often supports epithelia and many organs, and it surrounds blood vessels and nerves.

Dense, irregular connective tissue consists of fibers, mostly collagen-type, and fibroblasts. Its fibers interweave with one another, but not in a regular orientation. This tissue forms protective capsules around organs that are not stretched much. It also is present in the deeper portion of the skin (the dermis).

Dense, regular connective tissue is a "specialized" connective tissue in which the fibers occur in a parallel orientation (Figure 3.6b). The parallel bundles of fibers have rows of fibroblasts between them. In this arrangement, collagen fibers strongly resist being pulled along the axis of their parallel orientation when the tissue is stretched. Examples include collagen-rich tendons, which attach muscle to bone, and elastic ligaments, which attach bone to bone.

Connective tissues bind together, support, strengthen, protect, and insulate other body tissues. Most consist of protein fibers and a variety of cells in a ground substance.

3.5 CARTILAGE AND BONE

Cartilage is a specialized connective tissue that cushions and helps maintain the shape of body parts and serves as a transition tissue in growing bones. Although there are several types of cartilage, all of them have a solid but pliable matrix through which substances diffuse to and from nearby blood vessels. This pliability allows cartilage to resist compression and stay resilient, like a piece of solid rubber. However, the lack of blood vessels in cartilage makes it difficult for this tissue to heal when it is injured. Cells called chondroblasts produce the cartilage matrix. They mature into chondrocytes, which occur in small cavities in the matrix called *lacunae* (Figure 3.7).

The most common type of cartilage in the human body is *hyaline cartilage*, in which the matrix is laced with many small collagen fibers. In freely movable mature bones, hyaline cartilage provides a friction-reducing cover where the bone ends articulate in joints, and it makes up parts of the nose, ribs, and windpipe (trachea). In an embryo, hyaline cartilage is the forerunner of bone in the developing skeleton.

Elastic cartilage occurs in places where a flexible yet rigid structure is required. In addition to collagen fibers, it also has fibers containing the protein elastin. You can bend your outer ear because it is built of elastic cartilage. So is the epiglottis, a flexible flap of tissue that folds down over the opening of the larynx whenever you swallow. As a result, food and liquids enter the tube leading to the stomach, not the passageway leading to the lungs.

Sturdy and resilient *fibrocartilage* is packed with collagen fibers arranged in thick bundles. It can withstand tremendous pressure, and it forms the cartilage "cushions" in joints such as the knee (page 97) and in the disks that separate the vertebrae in the spinal column.

Bone is the weight-bearing tissue of the skeleton. It is the main tissue in bones, organs that support or protect softer body structures. Limb bones interact with attached muscles to bring about movement. Bones also store the mineral calcium, and some produce blood cells.

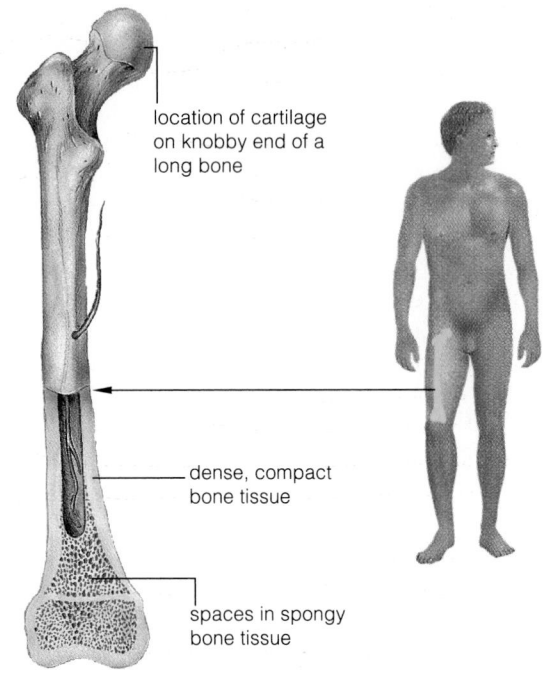

Figure 3.8 Cartilage and bone tissue in a human leg bone. Spongy bone tissue has tiny, needlelike hard parts with spaces between. Compact bone tissue is much more dense.

Unlike other connective tissues, the bone matrix is mineralized and hardened; its collagen fibers and ground substance are loaded with calcium salts. Bone tissue is not completely solid, however; as with cartilage, lacunae occur within the ground substance. And inside lacunae are living bone cells called osteocytes (Figure 4.3). Unlike cartilage, bone also has spaces that contain blood vessels.

Figure 3.8 shows the two basic types of bone tissue. Dense **compact bone** makes up the shafts of long bones and the outer regions of all bones. To the unaided eye, compact bone appears solid. Using a light microscope, you can see that it is laced with tunnel-like channels that contain blood vessels and nerves. By contrast, the *spongy bone* inside the ends of long bones (and in the interior of bones of the skull, pelvis, and breastbone) looks a little like Swiss cheese. Hard, platelike struts separate large, marrow-filled spaces. This structure provides strength without the weight of compact bone. Chapter 4 considers bone structure and functions more fully.

Cartilage and bone are connective tissues. The functions of cartilage include cushioning body parts and helping maintain their shape. Bone is the weight-bearing tissue of the skeleton and functions in movement, calcium storage, and blood cell production.

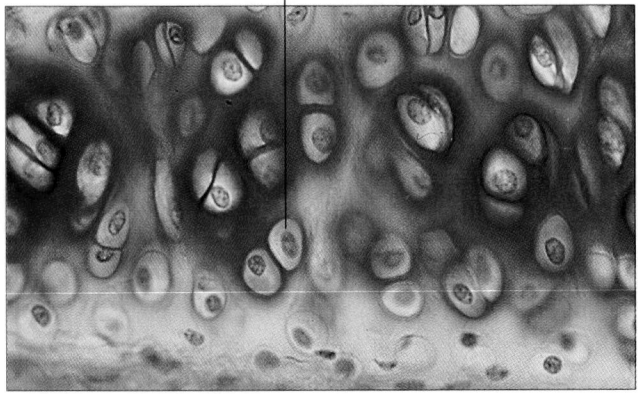

Figure 3.7 Cartilage cells (chondrocytes).

3.6 BLOOD AND ADIPOSE TISSUE

By weight, just under 10 percent of a person's body is the unusual connective tissue called **blood**. Though it might seem strange to include blood in the connective tissue category, many biologists do so. One reason is that blood is derived mainly from connective tissue; the various cells it contains arise from stem cells in bone marrow. Blood also has a fluid portion that contains many dissolved ions and molecules such as the protein fibrin—an essential player in the process of blood clotting (see Section 6.12).

Cells and cell fragments make up the nonfluid portion of blood (Figure 3.9). The most abundant of these are red blood cells. These cells lose their nucleus and organelles as they mature and become loaded with large amounts of the oxygen-binding protein hemoglobin. This structural specialization accords with a key function of red blood cells, which is to transport oxygen throughout the body.

About 5 percent of blood consists of white blood cells, which function in immunity, and platelets, which have a major role in blood clotting. Platelets are fragments that have broken off from large cells (called *megakaryocytes*) in bone marrow. Males and females differ in the relative amounts of different blood components (Figure 3.10). Males typically have a bit more of all types of blood cells; because this includes red blood cells, males generally have more hemoglobin overall than do females.

In addition to oxygen, blood transports nutrients to cells and carries wastes away from them; it also carries hormones and enzymes. Blood is moved through the body by the cardiovascular system—the heart and blood vessels. Chapter 6 returns to the cardiovascular system and the varied services blood performs in the body. Those functions range from helping maintain body temperature to stabilizing the pH in body tissues—all key requirements for homeostasis.

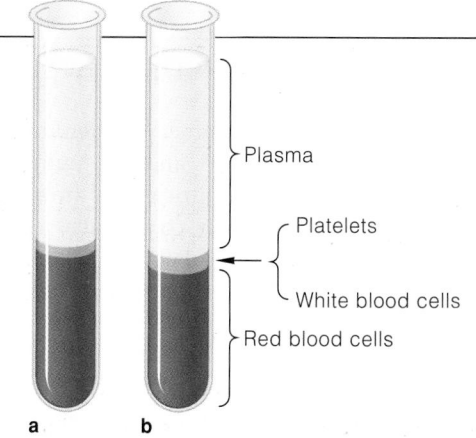

Figure 3.10 Plasma, blood cells, and cell fragments in blood samples from a male (**a**) and a female (**b**). The percentage of cellular components in a given volume of blood is the *hematocrit*. However, health care givers commonly use "hematocrit" to refer to the fraction of a blood sample occupied by red blood cells.

The connective tissue called **adipose tissue** has large, densely clustered cells that are specialized for fat storage (Figure 3.9b). The body can store only limited amounts of carbohydrate and protein. Excess carbohydrates and proteins are converted to fats that are stored in adipose cells. This reserve can fuel the body's basic energy needs during prolonged fasts.

Adipose tissue is laced with blood vessels, which serve as highways for the movement of fats (or their components) to and from individual adipose cells. Adipose tissue usually develops around the kidneys, behind the eyeballs, and under the skin, where it helps insulate the body from heat loss and absorbs shocks. Fat deposits also accumulate in the hips, abdomen, or thighs; the exact location is determined by a person's genetic makeup.

Blood is a transport medium and has key roles in maintaining the constancy of the internal environment.

Adipose tissue, another connective tissue, has densely clustered cells specialized for fat storage.

Figure 3.9 Some components of human blood (**a**) and cells of adipose tissue (**b**).

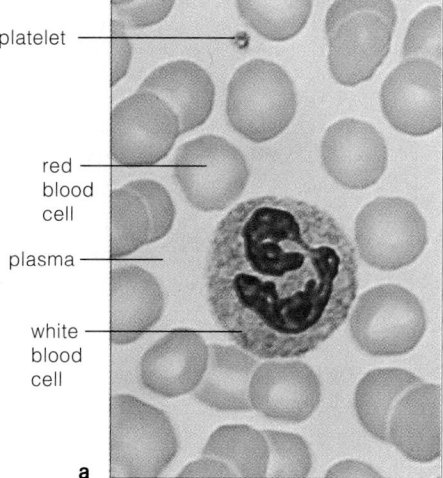

platelet

red blood cell

plasma

white blood cell

a

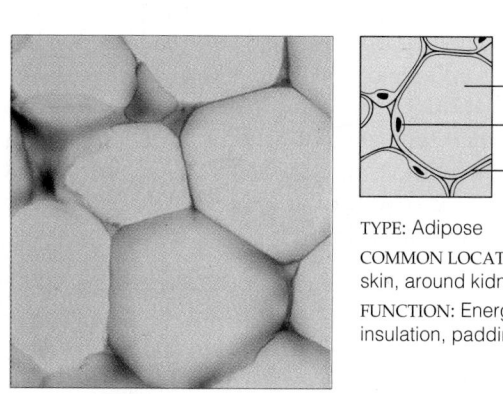

fat droplet

nucleus

plasma membrane

TYPE: Adipose

COMMON LOCATIONS: Under skin, around kidneys, heart

FUNCTION: Energy reserve, insulation, padding

b

MUSCLE TISSUE

In **muscle tissue** cells can contract (shorten) in response to stimulation, then lengthen and so return to their original state. Many long, cylindrical cells are arranged in parallel in these tissues. Their coordinated contraction and relaxation help move the body and its individual parts. There are three types, called skeletal, smooth, and cardiac muscle tissues.

The muscles attached to the bones of your skeleton are composed primarily of **skeletal muscle tissue** (Figure 3.11a). The individual cells of this tissue often are called *muscle fibers*. Each elongated cell has several nuclei, which are located not in the center of the cell but near the outer edges. This multinuclear arrangement comes about when several precursor cells fuse into a single cell, which then matures into a muscle fiber. As the fiber develops, contractile proteins (actin and myosin) accumulate within it. Actin and myosin eventually become organized into a banded pattern that makes skeletal muscle appear *striated* when it is viewed with the light microscope.

Typically, skeletal muscle cells form bundles (called *fascicles*). Then several bundles are enclosed in a tough connective tissue sheath to form "a muscle." Chapter 4 describes how skeletal muscle tissue functions.

The contractile cells of **smooth muscle tissue** are smaller than skeletal muscle fibers, and each has a single nucleus in the center of the cell. There are fewer contractile proteins, and they are organized differently than in skeletal muscle, so smooth muscle does not appear striated. For the same reasons, it also does not contract as rapidly as skeletal muscle, but it can maintain constant tension over longer periods of time. The contractile cells of smooth muscle are tapered at both ends (Figure 3.11b). Cell junctions hold them together and allow coordinated contraction. These cells, too, are enclosed in connective tissue. Smooth muscle tissue occurs in walls of blood vessels, the stomach, the bladder, the uterus, and other internal organs. It is said to be "involuntary" because a person usually cannot directly control its contraction.

Cardiac muscle is the contractile tissue of the heart (Figure 3.11c). It is striated like skeletal muscle, but each cell usually has only a single nucleus. The plasma membranes of adjacent cardiac muscle cells are bound tightly together by adhering junctions at regions called *intercalated disks*. Gap junctions in the disks are passageways for signals that stimulate contraction, allowing individual heart muscle cells to contract as a unit. When one muscle cell receives a signal to contract, its neighbors are also stimulated into contracting.

Muscle tissue, which can contract (shorten) in response to stimulation, helps move the body and specific body parts.

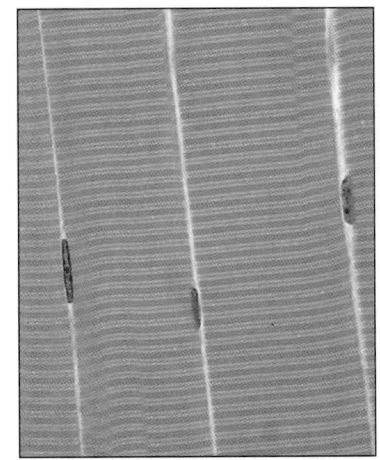
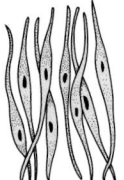

a

TYPE: Skeletal muscle

DESCRIPTION: Long, striated cells with multiple nuclei

COMMON LOCATIONS: In skeletal muscles

FUNCTION: Contraction for voluntary movements

width of one muscle cell

cell nucleus

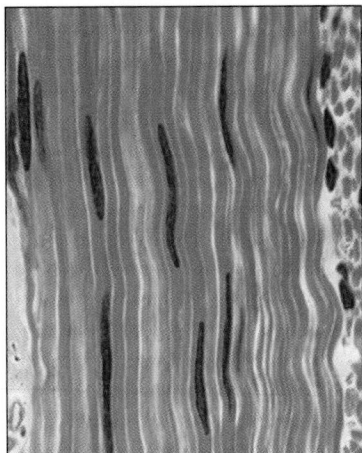

b

(cells, teased apart for clarity here)

TYPE: Smooth muscle

DESCRIPTION: Long, spindle-shapted cells, each with a single nucleus

COMMON LOCATIONS: In hollow organs (e.g., stomach)

FUNCTION: Movement of internal organs

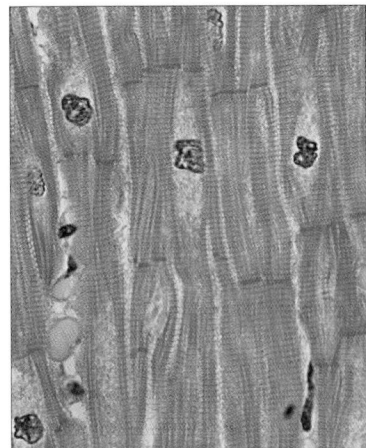

c

junction between adjacent cells

TYPE: Cardiac muscle

DESCRIPTION: Branching striated cells fused at plasma membranes

COMMON LOCATION: Wall of heart

FUNCTION: Pumping of blood in the circulatory system

Figure 3.11 Examples of skeletal, smooth, and cardiac muscle tissues.

NERVE TISSUE

Of all tissues, **nerve tissue** exerts greatest control over the body's responsiveness to changing conditions. In this tissue, cells called *neurons* are organized as communication lines that extend through the body. Tens of thousands of neurons, as well as many accessory cells called *neuroglia* (or simply glia), are centrally located in the brain. Others extend throughout the body.

Neuron Structure

Each neuron consists of a cell body (which contains the nucleus and cytoplasm) and two types of cell processes (Figure 3.12). Branched processes called *dendrites* pick up incoming chemical messages, in the form of molecules of neurotransmitters that are released from other neurons. (As you will read in Chapter 10, neurotransmitters include a range of substances, such as acetylcholine.) Outgoing messages are conducted by an axon. Depending on the type of neuron, its axon may extend outward a few millimeters or more than a meter.

A **nerve** is a cluster of processes from several neurons. Nerves conduct messages from the central nervous system (CNS, consisting of brain and spinal cord) to muscles and glands, and from sensory receptors back to the CNS. Neuroglia and various other types of accessory cells physically support or insulate neurons, help shuttle nutrients to them, and scavenge debris, microorganisms, or other foreign matter.

Neuron Functions

Some types of neurons detect specific changes in both internal and external environmental conditions. For example, as Chapter 11 describes, millions of specialized sensory neurons in epithelium lining the upper reaches of the nose respond to odor molecules. They are key to our senses of smell and taste. Sensory neurons in the retina of the eye are linked with receptors that respond to light, while different types in skin and muscle tissue are linked to receptors that detect pressure, stretching, and other stimuli.

Other neurons receive sensory input, integrate it with additional information, and then trigger responses by influencing the activity of still other neurons. In this way, they coordinate the body's immediate and long-term responses to change. Such cells occur mainly in the brain and spinal cord and make up the central nervous system. Yet other neurons are part of a peripheral nervous system that relays signals to muscles and glands that can carry out responses. The constant flow of information coordinates and regulates the activities of the body's

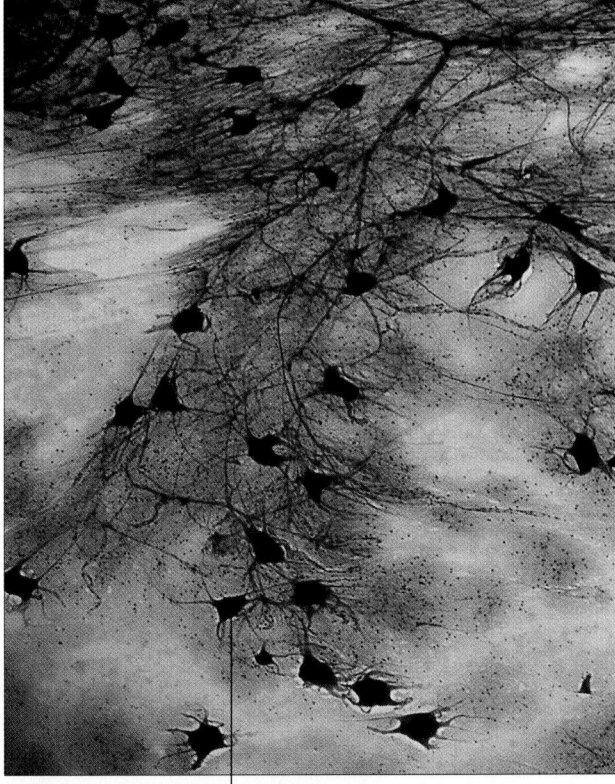

cell body of neuron

Figure 3.12 A sampling of the millions of neurons that form communication lines within and between different regions of the human body. Shown here are motor neurons, which relay signals from the brain or spinal cord to muscles and glands. The large, dark regions are neuron cell bodies. Collectively, the body's neurons sense environmental change, integrate signals about those changes, and initiate appropriate responses.

billions of cells and the tissues and organs they make up. With their essential role in homeostasis, neuron structure and functioning will be key topics in Chapter 10.

As a human embryo develops, epithelial tissues, connective tissues, nerve tissue, and muscle tissue begin to form. The various tissues eventually become organized into organs, and then into organ systems, that are much the same in all vertebrates. These are the levels of structure and function in the human body that we consider next.

Neurons are the basic units of communication in nerve tissue. Different kinds detect specific stimuli, integrate information, and issue or relay commands for response.

ORGAN SYSTEMS

Figure 3.13 provides an overview of the human body's organ systems. Section 3.10 considers the integumentary system—your skin—as an example, and other systems will be the focus in subsequent chapters.

It might seem far-fetched to say that each organ system contributes to the survival of tissues in the body. After all, what could the body's bones and muscles have to do with the life of the cells in tissues? Well, think about it. Interactions between your skeletal and muscular systems allow you to move about—toward sources of nutrients and water, for example. Parts of those systems assist blood circulation, as when contractions of your leg muscles help move blood in veins back to the heart. The bloodstream itself carries nutrients and other substances to cells and transports secreted products and wastes away from them.

Throughout this unit we will be using some standard terms for describing the locations of organs and organ systems in the human body. To get your bearings, take a moment to study Figure 3.14a, which shows the location of some major body cavities in which organs occur. The **cranial cavity** and **spinal cavity** house the central nervous system—your brain and spinal cord. Your heart and lungs reside in the **thoracic cavity**— essentially, inside your chest. The diaphragm muscle separates the thoracic cavity from the **abdominal cavity**, which holds your stomach, liver, most of the intestine, and other organs. Your reproductive organs, bladder, and rectum are located within the **pelvic cavity**. Figure 3.14b defines some other anatomical terms that apply to most animals, including humans.

Figure 3.13 Organ systems of the human body.

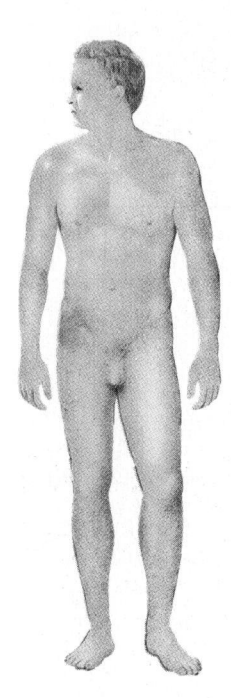
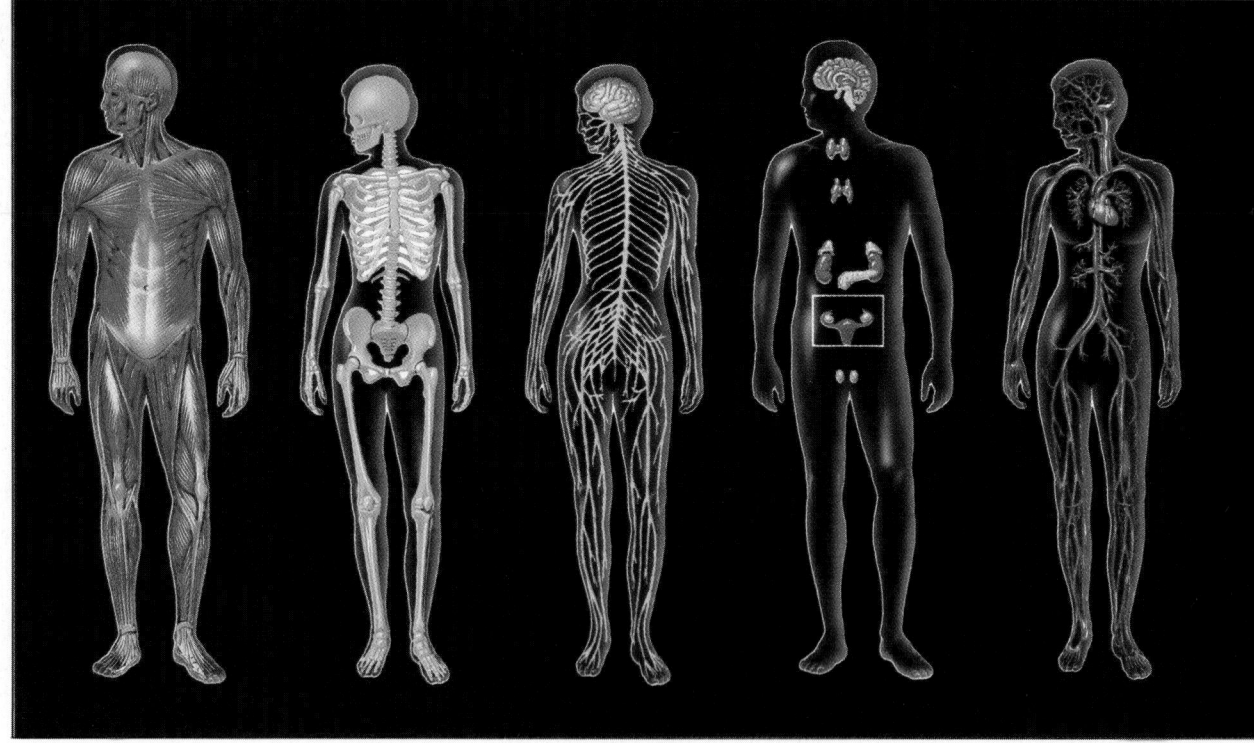

INTEGUMENTARY SYSTEM	MUSCULAR SYSTEM	SKELETAL SYSTEM	NERVOUS SYSTEM	ENDOCRINE SYSTEM	CIRCULATORY SYSTEM
Protection from injury and dehydration; body temperature control; excretion of some wastes; reception of external stimuli; defense against microbes	Movement of internal body parts; movement of whole body; maintenance of posture; heat production	Support, protection of body parts; sites of muscular attachment, blood cell production, and calcium and phosphate storage	Detection of external and internal stimuli; control and coordination of responses to stimuli; integration of activities of all organ systems	Hormonal control of body functioning; works with nervous system in integrative tasks	Rapid internal transport of many materials to and from cells; helps stabilize internal temperature and pH

Figure 3.14 (a) Directional terms and planes of symmetry for the human body. The midsagittal plane divides the body into right and left halves. The transverse plane divides it into superior (top) and inferior (bottom) parts. The frontal plane divides it into anterior (front) and posterior (back) parts. (b) Major cavities in the human body.

SUPERIOR
(of two body parts, the one closer to head)

distal (farthest from trunk or from point of origin of a body part)

frontal plane (*aqua*)

midsagittal plane (*green*)

proximal (closest to trunk or to point of origin of a body part)

ANTERIOR (at or near front of body)

POSTERIOR (at or near back of body)

transverse plane (*yellow*)

INFERIOR
(of two body parts, the one farthest from head)

a

cranial cavity
spinal cavity
thoracic cavity
abdominal cavity
pelvic cavity

b

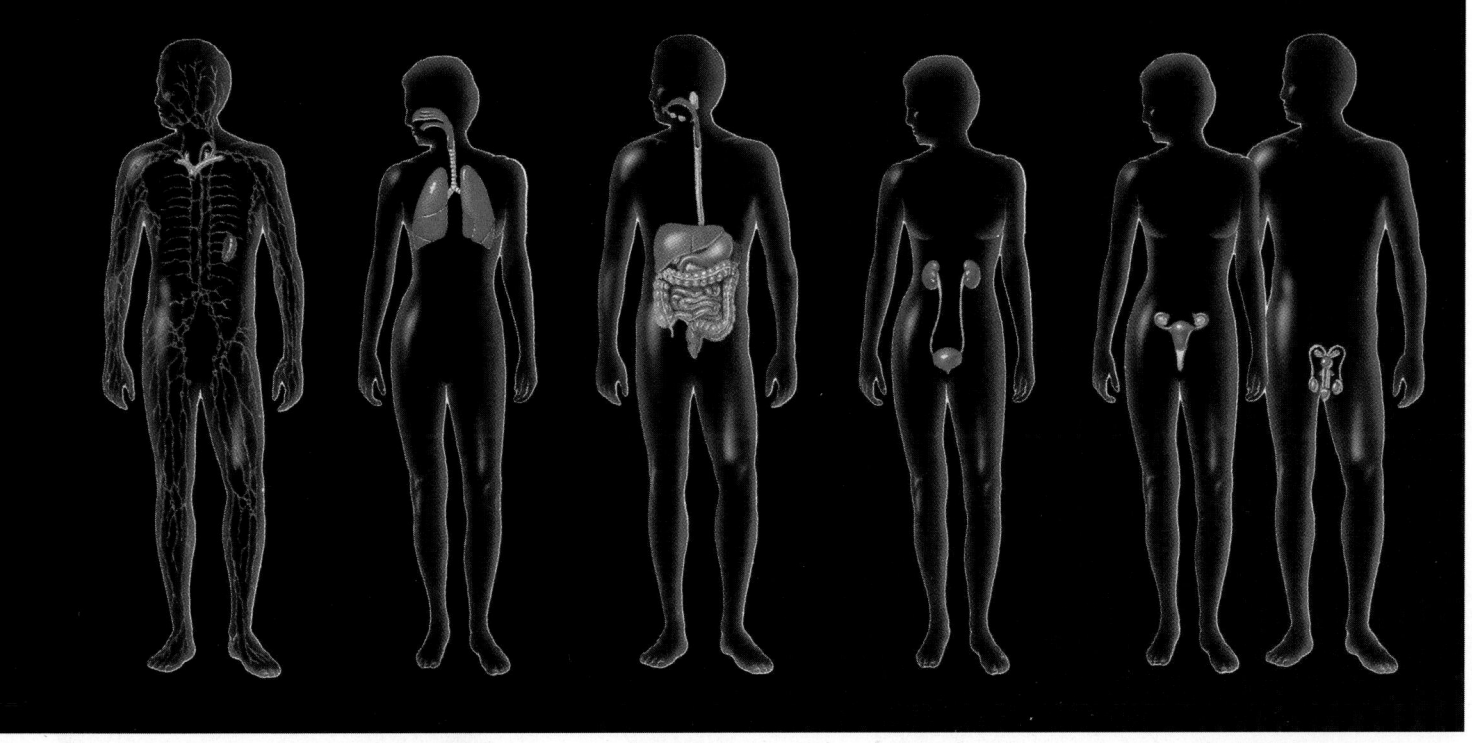

LYMPHATHIC SYSTEM

Return of some tissue fluid to blood; roles in immunity (defense against specific invaders of the body)

RESPIRATORY SYSTEM

Delivery of oxygen to cells; removal of carbon dioxide wastes produced by cells; pH regulation

DIGESTIVE SYSTEM

Ingestion of food, water; preparation of food molecules for absorption; elimination of food residues from the body

URINARY SYSTEM

Maintenance of the volume and composition of extracellular fluid; excretion of blood-borne wastes

REPRODUCTIVE SYSTEM

Male: production and transfer of sperm to the female. Female: production of eggs; provision of a protected, nutritive environment for developing embryo and fetus. Both systems have hormonal influences on other organ systems.

THE SKIN: CASE STUDY OF AN ORGAN SYSTEM

Technically, your body's outer cover—its skin—is an **integument** (after the Latin *integere*, meaning "to cover"). The tough, pliable integument is a barrier against a great variety of environmental insults. It includes the skin and the structures (such as hair and nails) derived from epidermal cells of its outer tissue layers. By weight and size, the integument is the body's largest organ, and it is a good example of an organ with a wide range of functions.

Skin Functions

No garment ever made comes close to skin's qualities. Skin protects the rest of the body from dehydration, abrasion, and bacterial attack. It helps control the body's internal temperature. Its many small blood vessels are a reservoir for blood that can be shunted to metabolically active regions, such as leg muscles during strenuous exercise. Skin produces the vitamin D required for calcium metabolism. And signals from skin's sensory receptors help the brain assess what's going on in the outside world.

Skin Structure

Assuming you are an average-sized adult, your skin weighs about 9 pounds. Stretched out, it would have a surface area of 15 to 20 square feet. Except for places subjected to regular pounding or abrasion (such as the palms of the hands and soles of the feet), human skin is generally not much thicker than a paper towel. As Figure 3.15a shows, skin has two distinct regions—an outermost **epidermis** and an underlying **dermis**. Beneath the dermis is a subcutaneous ("under the skin") layer, the *hypodermis*, a loose connective tissue that anchors the skin and yet allows it some freedom of movement. Fat stored in the hypodermis insulates the body and helps cushion some body parts.

EPIDERMIS Epidermis consists of stratified squamous epithelium. Abundant cell junctions knit the epithelial cells together. The cells arise deep within the epidermis and are pushed toward its free surface as cell divisions produce new cells beneath them.

Most cells of the epidermis are keratinocytes. Each is a tiny factory for manufacturing *keratin*, a tough, water-insoluble protein. Those cells start producing keratin when they are in mid-epidermal regions. By the time they reach the skin's free surface, they are dead and flattened. All that remain are fibers of keratin, packed inside plasma membranes. This is the composition of the outermost layer of skin—the tough, waterproof *stratum corneum* (Figure 3.15b). Millions of the flattened keratin packages wear off daily, but rapid, ongoing cell divisions in the epidermis continually push up replacements. The rapid divisions also contribute to skin's capacity to mend itself quickly after cuts or burns.

In the deepest layer of epidermis, cells called **melanocytes** produce the brownish-black pigment called melanin. This accumulates inside keratin-producing cells, forming a shield against ultraviolet radiation. Skin color variations are due partly to differences in the distribution and metabolic activity of melanocytes. Suntanning increases melanocyte activity (see *Focus on Our Environment*, page 81). In addition, skin color is influenced by hemoglobin and by the yellow-orange pigment carotene in the dermal layer. For example, pale skin does not have much melanin, so the presence of hemoglobin is not masked and the skin looks pinkish.

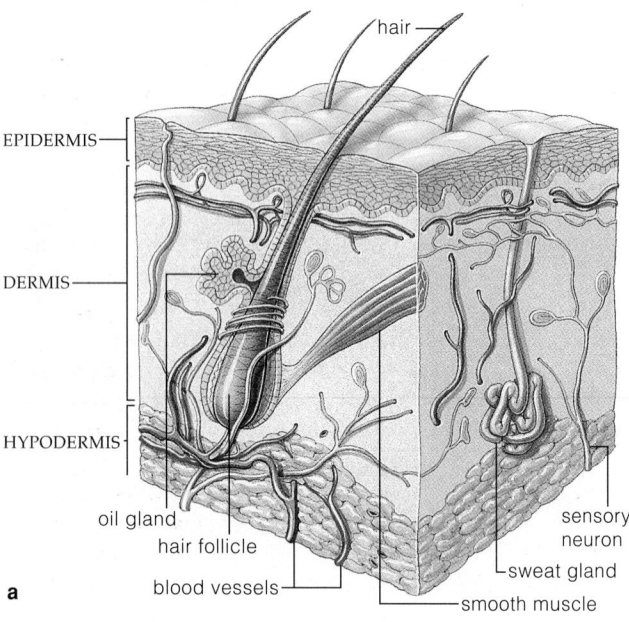

EPIDERMIS

DERMIS

HYPODERMIS

hair

oil gland

hair follicle

blood vessels

sensory neuron

sweat gland

smooth muscle

a

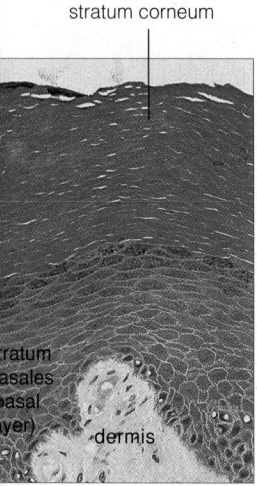

stratum corneum

stratum basales (basal layer)

dermis

Figure 3.15 (**a**) Two-layered structure of human skin. The hypodermis is a subcutaneous layer; it is beneath the skin, not part of it. (**b**) Section through human skin, showing the uppermost layer of epidermis (stratum corneum); deeper epidermal layers, including the stratum basales (basal layer) and the underlying dermis. The dark spots in the basal layer are epidermal cells to which melanocytes have passed pigment. The most common skin cancer, basal cell carcinoma (page 10), arises in this region.

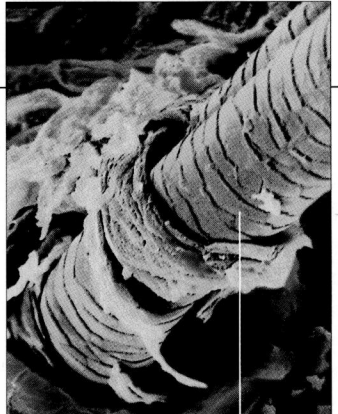

Figure 3.16 Close look at a hair. This scanning electron micrograph shows overlapping cells of the outer layers of a hair shaft, here emerging from the surface of the skin. The average scalp has about 100,000 hairs, but genes, nutrition, and hormones influence hair growth and density.

dead, flattened cells around a developing hair shaft

DERMIS Dense connective tissue makes up most of the dermis, and it fends off damage from everyday stretching and other mechanical insults. There are limits to this protection. For example, the dermis tears when skin over the abdomen is stretched too much during pregnancy, leaving white scars ("stretch marks"). With persistent abrasion, the epidermis separates from the dermis, the gap fills with a watery fluid, and you get a blister.

Blood vessels, lymph vessels, and the receptor endings of sensory nerves thread throughout the dermis. Nutrients from the bloodstream reach epidermal cells by diffusing through the dermal tissue. Sweat glands, oil glands, and the husklike structures called hair follicles reside mostly in the dermis, even though they are derived from epidermal tissue. So are fingernails and toenails.

The fluid secreted from *sweat glands* is 99 percent water; it also contains dissolved salts, traces of ammonia and other metabolic wastes, vitamin C, and other substances. The body has about 2.5 million sweat glands. One type, which abounds in the palms of the hands, soles of the feet, forehead, and armpits, functions mainly in temperature regulation. Another type of sweat gland prevails in skin around the sex organs. Their secretion steps up during stress, pain, and sexual foreplay.

Oil glands (also called *sebaceous glands*) are everywhere except on the palms and soles of the feet. Their oily secretions soften and lubricate the hair and the skin; other secretions kill surface bacteria. In *acne*, oil gland ducts become infected by bacteria, and the glands become inflamed. *Hairs* are mostly keratinized cells. Each has a root embedded in skin and a shaft that projects above the skin surface. As cells divide near the base of the root, older cells are pushed upward, then flatten and die. The outermost layer of the shaft consists of flattened cells that overlap like roof shingles (Figure 3.16).

With its multiple layers of keratinized and melanin-shielded epidermal cells, skin helps the body conserve water, avoid damage from ultraviolet radiation, and resist mechanical stress.

Hairs, oil glands, sweat glands, and other structures associated with skin are derived from epidermal cells.

SUN, SKIN, AND THE OZONE LAYER

The melanin-producing cells of the epidermis are stimulated by exposure to ultraviolet (UV) radiation. With prolonged sun exposure, melanin levels increase and light-skinned people become tanned. Tanning gives some protection against UV radiation, but over the years, tanning causes elastin fibers in the dermis to clump together. The skin loses its resiliency and begins to look leathery.

Prolonged exposure to UV radiation also suppresses the immune system. For instance, infection-fighting phagocytes in the epidermis combat viruses and bacteria. Sunburns interfere with their functioning. This may be why sunburns can trigger the painful blisters called "cold sores" that announce the appearance of *herpes simplex*. Nearly everyone harbors this virus. Usually it becomes localized in a nerve ending near the skin surface, where it remains dormant. Stress factors, including sunburn, can activate the virus and trigger the skin eruptions.

Ultraviolet radiation from sunlight or from the lamps of tanning salons also can activate proto-oncogenes in skin cells (page 426). These bits of DNA can trigger cancer, especially when some factor alters their structure (and hence their function).

Most of the UV radiation to which we are exposed comes from the sun. Until fairly recently, a layer of ozone in the stratosphere intercepted much of the potentially damaging UV radiation (specifically UV-B). Today, however—due in large part to human activities described in Chapter 25—the ozone layer over ever larger regions of the globe is being destroyed faster than natural processes can replace it. The rate of skin cancers now is rapidly increasing. Skin cancers are among the easiest to cure, but only if they are promptly removed surgically. As for the protective ozone layer, experts estimate that even if all ozone-depleting substances were banned tomorrow, it would take about 100 years for the planet to recover.

The Internal Environment

To stay alive, your cells must be continuously bathed in fluid that supplies them with nutrients and carries away metabolic wastes. In this they are no different from an amoeba or any other free-living, single-celled organism. However, trillions of cells are crowded together in your body, and all of them must draw nutrients from and dump wastes into the same 15 liters of fluid. That is less than 4 gallons.

The fluid outside cells is called **extracellular fluid**. Much of it is *interstitial*, meaning it occupies spaces between cells and tissues. The rest is blood plasma. Interstitial fluid exchanges substances with blood and with the cells it bathes.

Functionally, extracellular fluid is closely associated with the fluid inside cells. That's why drastic changes in its composition and volume have drastic effects on cell activities. Its ion concentrations are especially important. They must be maintained at levels that are compatible with cell survival. Otherwise, the body as a whole cannot survive.

Look back at the simple diagram on page 5, where you first read about the concept of homeostasis. The point of that diagram is that *the component parts of the body work together to maintain the stable fluid environment required by its living cells.* This concept has three basic elements:

1. Each body cell engages in metabolic activities that ensure its own survival.

2. At the same time, the cells of a given tissue perform one or more activities that contribute to the survival of the whole organism.

3. The combined contributions of individual cells, tissues, organs, and organ systems help maintain the stable internal environment—the extracellular fluid—required for individual cell survival.

Mechanisms of Homeostasis

Homeostasis literally means "staying the same." It refers to stable operating conditions in the body's internal environment. Three components, called sensory receptors, integrators, and effectors, interact to maintain this state. **Sensory receptors** are cells or cell parts that can detect a **stimulus**, which is a specific change in the environment. For example, when someone kisses you, there is a change in pressure on your lips. Receptors in the skin of your lips translate the stimulus into a signal, which can be sent to the brain. Your brain is an **integrator**, a control point where different bits of information are pulled together in the selection of a response. It can send signals to your muscles or glands (or both). Muscles and glands are

effectors—they carry out the response. In this case, the response might include kissing the person back. Of course, you cannot engage in a kiss indefinitely, because eventually you must eat and perform other tasks that maintain your body's operating conditions.

How does the brain reverse the physiological changes induced by the kiss? Receptors provide it with information only about how things *are* operating. The brain also maintains information about how things *should be* operating—that is, information from "set points." When physical or chemical conditions deviate significantly from a set point, the brain functions to bring them back within an effective operating range. It does this by way of signals that cause specific muscles and glands to increase or decrease their activity. Set points are key elements of many physiological mechanisms, including those that influence eating behavior, breathing, thirst, and urination, to name a few.

Feedback mechanisms are among the controls that help keep physical and chemical aspects of the body within tolerable ranges. In a **negative feedback mechanism** (Figure 3.17), an activity alters a condition in the internal environment, and this triggers a response that reverses the altered condition. To grasp how the mechanism operates, think of a furnace with a thermostat. The thermostat senses the air temperature and mechanically "compares" it to a preset point on a thermometer built into the furnace control system. When the temperature falls below the preset point, the thermostat signals a switching mechanism that turns on the heating unit. When the air becomes heated enough to match the prescribed level, the thermostat signals the switching mechanism, which shuts off the heating unit.

Similarly, feedback mechanisms help keep the body temperature of humans (and many other animals) within a normal range, even during extremely hot or cold weather or under other conditions that alter temperature (Figure 3.18). For example, when the body senses that its

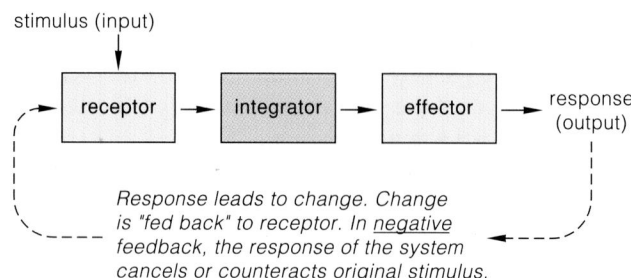

Figure 3.17 Components necessary for negative feedback at the organ level.

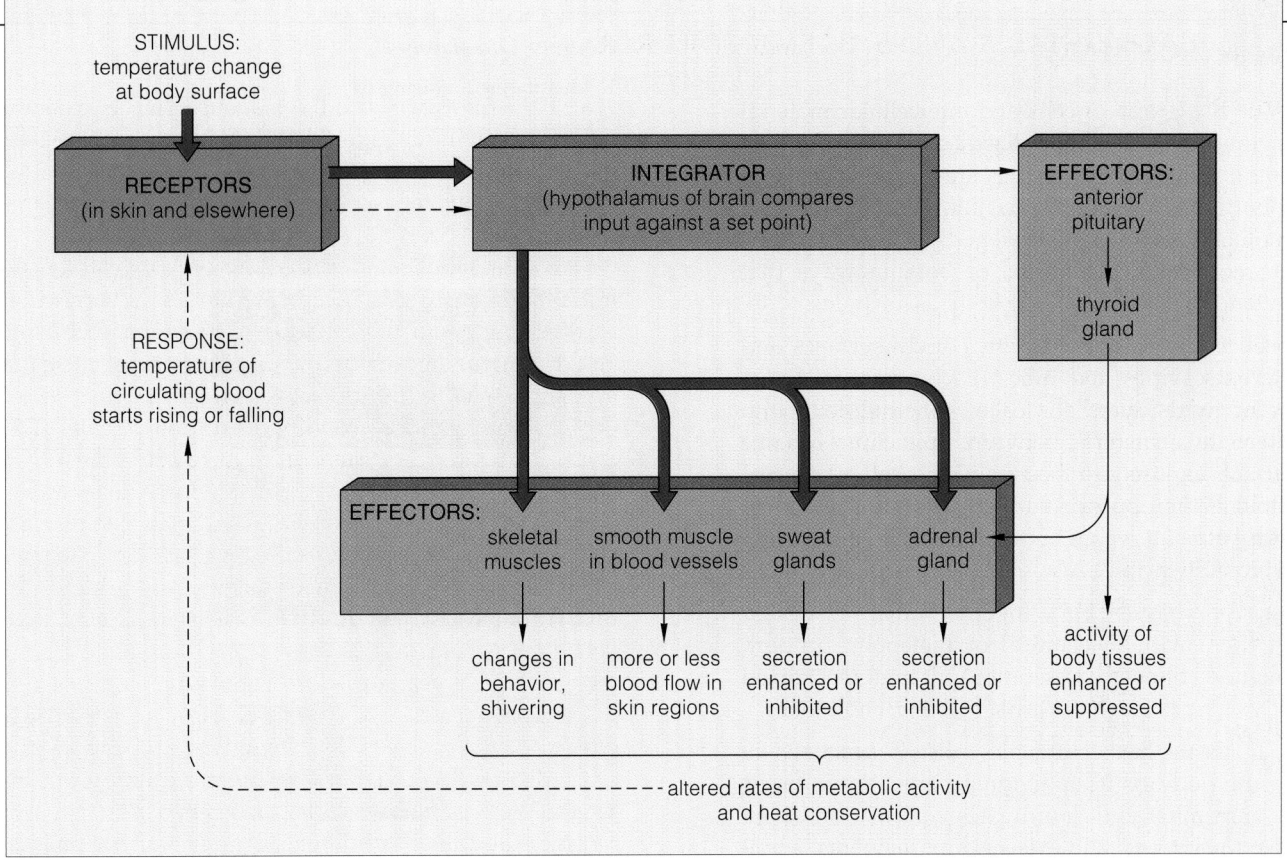

Figure 3.18 Homeostatic controls over the internal temperature of the human body. The dashed line shows how the feedback loop is completed. The *blue* arrows indicate the main control pathways.

skin is getting too hot while you work outside in the summer sun, mechanisms are set in motion that slow down metabolic activity and overall body activity. You move more slowly and may seek the shelter of a shade tree. At the same time, blood flow to the skin increases and your sweat glands are prodded to secrete larger quantities of sweat. As water in sweat evaporates, more heat is lost from the body. These and other mechanisms counter overheating by curbing the body's heat-generating activities and giving up excess heat to the surroundings.

Sometimes **positive feedback mechanisms** operate. These mechanisms set in motion a chain of events that *intensify* a change from an original condition—and after a limited time, the intensification reverses the change. Positive feedback is associated with instability in a system. For example, during childbirth, pressure of the fetus on the walls of the uterus stimulates production and secretion of the hormone oxytocin. Oxytocin causes muscles in the walls to contract and exert pressure on the fetus, which increases pressure on the uterine wall, and so on until the fetus is expelled.

What we have been describing here is a general pattern of monitoring and responding to a constant flow

of information about the external and internal environments. During this activity, organ systems operate together in coordinated fashion.

Throughout this unit we will be asking the following questions about organ systems:

1. What physical or chemical aspect of the internal environment are organ systems working to maintain as conditions change?

2. How are organ systems kept informed of the various changes?

3. How do they process the incoming information?

4. What mechanisms are set in motion in response?

As you will see in chapters to follow, the operation of all organ systems is under precise neural and endocrine control.

Homeostatic control mechanisms maintain physical and chemical aspects of the internal environment within ranges that are most favorable for cell activities.

SUMMARY

1. A tissue is a group of cells and intercellular material that together perform a specialized activity. An organ is a structural unit in which tissues are combined in definite proportions and patterns that allow them to perform a common task. In an organ system, two or more organs interact chemically, physically, or both in ways that contribute to the survival of the body as a whole.

2. Epithelial tissues cover external body surfaces and line internal cavities and tubes. Each epithelial tissue has one or more layers of closely adhering cells with little intercellular material between. Epithelium has one free surface exposed to body fluids or the external environment; the opposite surface rests on a basement membrane that intervenes between it and an underlying connective tissue.

3. Connective tissues bind together other tissues or provide them with mechanical or metabolic support. They include connective tissue proper and specialized connective tissues (such as cartilage, bone, and blood).

4. Muscle tissue has contractile cells. It functions in moving the body or its individual parts. Nerve tissue detects, transmits, and coordinates information about change in the internal and external environments, and it controls responses to change.

5. Tissues, organs, and organ systems work together in ways that help maintain a stable internal environment (the extracellular fluid) required for survival. At homeostasis, conditions in the internal environment are most favorable for cell activities.

6. An example of an organ is skin (the integumentary system), which protects the rest of the body from abrasion, bacterial attack, ultraviolet radiation, and dehydration. It helps control internal temperature, and it serves as a blood reservoir for the rest of the body. Its receptors are essential in detecting environmental stimuli.

7. Feedback controls help maintain internal conditions. In negative feedback (the most common mechanism), a change in some condition triggers a response that reverses the change. In positive feedback, a response reverses a change in some condition by intensifying the change for a limited time.

8. Homeostasis depends on integration of the functioning of the body's organ systems by the nervous system and endocrine system. Specialized receptors detect stimuli, which are specific changes in the environment. Integrating centers such as the brain process the information and direct muscles and glands—the body's effectors—to carry out responses.

Review Questions

1. Identify the following tissues:

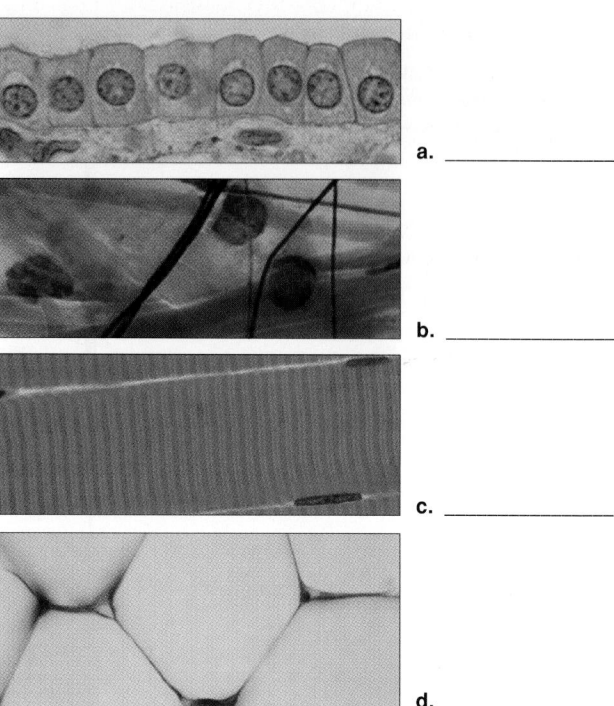

a. _____

b. _____

c. _____

d. _____

2. What is a tissue? An organ? An organ system? List the major organ systems of the human body, along with their functions. *67, 78–79*

3. Define extracellular fluid and interstitial fluid. *82*

4. Epithelial tissue and connective tissue differ from each other in overall structure and function. Describe how. *72*

5. State the overall functions of (a) muscle tissue and (b) nerve tissue. Which type of tissue is shown in this micrograph? *76, 77*

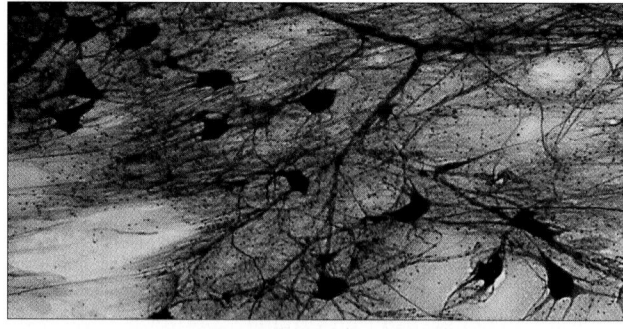

6. What are some of the functions of skin? *80*

7. A major concept in human physiology relates the functioning of cells, organs, and organ systems to the internal environment. Can you state the three main points of this concept? *82*

8. Define homeostasis. What are the three components necessary for homeostatic control over the internal environment? *82*

9. What are the differences between negative feedback and positive feedback mechanisms? *82–83*

Critical Thinking: You Decide (Key in Appendix V)

1. A third-degree burn destroys both the dermis and epidermis layers of skin. What kinds of hazards to homeostasis are posed by a serious burn over a large percentage of the body?

2. The disease called scurvy results from a deficiency of vitamin C, which the body requires for collagen synthesis. Explain why (among other symptoms) scurvy sufferers tend to lose teeth, and why any wounds heal much more slowly than normal, if at all.

Self-Quiz (Answers in Appendix IV)

1. The four main types of tissues in the human body are _____, _____, _____, and _____.

2. The human body is structurally and functionally adapted for which of the following activities:
   a. maintenance of the internal environment
   b. nutrient acquisition, processing, and distribution
   c. self-protection against injury or attack
   d. reproduction
   e. waste disposal
   f. all of the above

3. _____ tissues cover external body surfaces, line internal cavities and tubes, and may form the secretory portions of glands.
   a. Muscle   c. Connective
   b. Nerve   d. Epithelial

4. Most _____ tissues bind or mechanically support other tissues, but one type provides physiological support of other tissues.
   a. muscle   c. connective
   b. nerve   d. epithelial

5. Which of the following is *not* a function of the integumentary system?
   a. to protect the body from abrasion
   b. to protect the body from dehydration
   c. to detect environmental stimuli
   d. to bring about body movements
   e. to serve as a blood reservoir for the rest of the body

6. _____ tissues detect and coordinate information about environmental changes and control responses to those changes.
   a. Muscle   c. Connective
   b. Nerve   d. Epithelial

7. _____ tissues contract and make possible internal body movements as well as movements through the external environment.
   a. Muscle   c. Connective
   b. Nerve   d. Epithelial

8. Cells in the human body do all of the following *except* _____.
   a. engage in metabolic activities that ensure their survival
   b. perform activities that contribute to the survival of the body as a whole
   c. contribute to maintaining the extracellular fluid
   d. digest large food particles

9. In a state of _____, physical and chemical aspects of the body are being kept within tolerable ranges by controlling mechanisms.
   a. positive feedback   c. homeostasis
   b. negative feedback   d. metastasis

10. In negative feedback mechanisms, _____.
    a. a detected change brings about a response that tends to return internal operating conditions to the original state
    b. a detected change suppresses internal operating conditions to levels below the set point
    c. a detected change raises internal operating conditions to levels above the set point
    d. a detected change causes fewer solutes to be fed back to the affected cells

11. Fill in the blanks: _____ detect specific environmental changes, an _____ pulls different bits of information together in the selection of a response, and _____ carry out the response.

12. Match the concepts:
    _____ muscles and glands   a. integrating center
    _____ positive feedback   b. response that reverses an
    _____ sites of body receptors   altered condition
    _____ negative feedback   c. eyes and ears
    _____ brain   d. effectors
                                   e. chain of events that intensifies the original condition

Selected Key Terms

adipose tissue 75	extracellular fluid 82
basement membrane 68	gland 68
blood 75	homeostasis 82
bone 74	integrator 82
cartilage 74	integument 80
compact bone 74	muscle tissue 76
connective tissue 73	negative feedback mechanism 82
dermis 80	nerve tissue 77
effector 82	organ 67
endocrine gland 68	organ system 67
epidermis 80	positive feedback mechanism 83
epithelium 68	sensory receptor 82
exocrine gland 68	tissue 67

Readings

Green, H. November 1991. "Cultured Cells for the Treatment of Disease." *Scientific American*.

Langer, R., and J. P. Vacanti. September 1995. "Artificial Organs." *Scientific American*.

Lipken, R. March 25, 1995. "Bone Fractures: Treatment and Risks." *Science News*.

Nesse, R. M., and G. C. Williams. 1994. *Why We Get Sick: The New Science of Darwinian Medicine*. New York: Times Books. This thought-provoking book suggests some new ways to think about health and disease.

Nucci, M., and A. Abuchowski. February 1998. "The Search for Blood Substitutes." *Scientific American*.

 For additional readings, go to InfoTrac College Edition, your online library at:

http://www.infotrac-college.com/wadsworth

MUSCULOSKELETAL SYSTEM: SUPPORT AND MOVEMENT

Row, Row, Row Your Boat

Betsy and Mary McCagg are twins—and two of the best competitive rowers in the world (Figure 4.1). Beginning in high school, they built a reputation as elite athletes. Among other honors, they were members of the United States national rowing team in the 1992 and 1996 Olympic Games, and in 1995 they were part of an eight-woman team that won the world championship for the distance of 2,000 meters in women's rowing.

The McCagg sisters have genetically determined characteristics that contribute to their prowess in their chosen sport. Both are tall (188 centimeters, or 6 feet 2 inches) with long thighbones in their legs. This feature is important, because the longer a rower's thighbones (femurs), the more power the rower can generate with each stroke. In addition, however, Betsy and Mary put in grueling hours of work with exercise physiologists and other sport scientists to hone their bodies to peak condition. Like other world-class athletes, their main goal is to maximize the potential of the systems that support and move their bodies.

Long before each race, the McCagg twins follow a strict training regimen. The physical workouts increase the capacity of their musculoskeletal system to support and help move their bodies. How? To begin with,

rigorous exercise stresses bones, and the bone mass increases in response. Muscle cells respond to training by becoming larger or loaded with mitochondria, the organelles that produce ATP energy during aerobic respiration. Also, more blood vessels develop, increasing blood flow for the transport of nutrients and wastes. These and other training-related changes can yield amazing increases in strength and stamina—qualities that are essential for competitive success in rowing, cycling, and other endurance sports.

In this chapter we turn our attention to the human body's systems for structural support—its skeleton—and for movement—its skeletal muscles (Figure 4.2). The functions of bones and skeletal muscles are so closely entwined that they sometimes are termed simply the "musculoskeletal system." Some of the functional links are obvious. For example, we take for granted that our bodies' softer parts have a supporting framework—one service, among several, that your hard, bony skeleton provides. The bones of the skeleton also serve as rigid parts against which muscles can work and so translate the pull of contracting tissues into body movement. And while the long bones of your legs, for instance, are quite sturdy, much of the structural material inside them is surprisingly lightweight. This adaptation minimizes the energy "cost" of movement. Muscles are organized to work with one another as well as with the skeleton. Guided and controlled by the nervous system, this fine-tuning enables a person to execute the movements and positional changes required for the full range of human activities—from rowing a racing scull in the Olympic Games to turning the pages of a textbook.

Figure 4.1 Betsy and Mary McCagg illustrating their bodies' systems of support and movement.

KEY CONCEPTS

1. Bones interact with muscles to bring about movement. They protect and support soft organs and store minerals. Blood cells are produced in tissues of some bones.

2. Joints are sites of contact or near-contact between bones. Joints at the articulating ends of bones allow skeletal movements.

3. Skeletal muscle helps move limbs and other structures. Smooth muscle is found in the walls of organs and blood vessels; cardiac muscle is the primary tissue of the heart.

4. Skeletal muscle contracts in response to stimulation by the nervous system. Such contractions generally are under voluntary control. The contractions of smooth muscle and cardiac muscle are involuntary.

CHAPTER AT A GLANCE

4.1   Bone

4.2   How the Skeleton Grows and Is Maintained

4.3   The Axial Skeleton

4.4   The Appendicular Skeleton

4.5   Joints

4.6   *The Vulnerable Knee*

4.7   The Musculoskeletal System

4.8   Muscle Structure and Function

4.9   Controlling the Contraction of a Skeletal Muscle

4.10   Properties of Whole Muscles

4.11   *Muscle Matters*

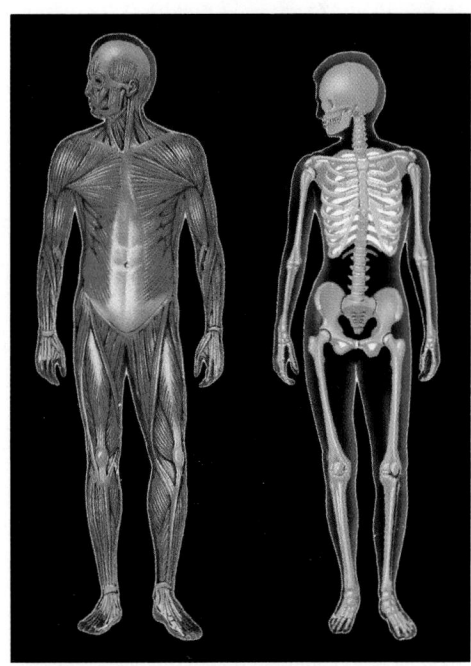

Figure 4.2 Overview of the human body's muscle system and skeletal system.

Basic Bone Functions and Structure

Just as your skin is more than a baglike covering, so is your skeletal system more than a frame to hang muscles on. Its major parts, **bones**, are complex organs that serve these functions:

1. *Movement.* By interacting with skeletal muscles, bones maintain or change the position of body parts.

2. *Protection.* Bones are hard compartments that enclose and protect the brain, lungs, and other organs.

3. *Support.* Bones support and anchor muscles.

4. *Mineral storage.* Bone tissue is a bank for depositing and withdrawing mineral ions, including calcium and phosphate, and so helps maintain body fluids and support metabolic activities.

5. *Blood cell formation.* Some bones contain regions in which blood cells are produced.

Bones of the human skeleton range in size from tiny ear bones to massive thighbones. In shape, bones are long, short or cubelike, flat, or irregular. Like other organs, bones incorporate various tissues. These include bone tissue, cartilage at the sites where one bone articulates with another, nerve tissue, and epithelium in the walls of blood vessels that service bones. Bone surfaces and the cavities within bones are lined with connective tissue.

The main component of bones is *bone tissue*. This connective tissue has both living and nonliving elements. The living cells are **osteocytes**. As a bone develops, precursors of osteocytes called osteoblasts secrete collagen fibers, a few elastin fibers, and a ground substance of proteins and carbohydrates. With time, the osteocytes nestle in spaces called *lacunae* in the ground substance, and the ground substance hardens (becomes mineralized) as calcium salts are deposited. Figure 4.3 gives you an internal view of a typical long bone, the thighbone (femur).

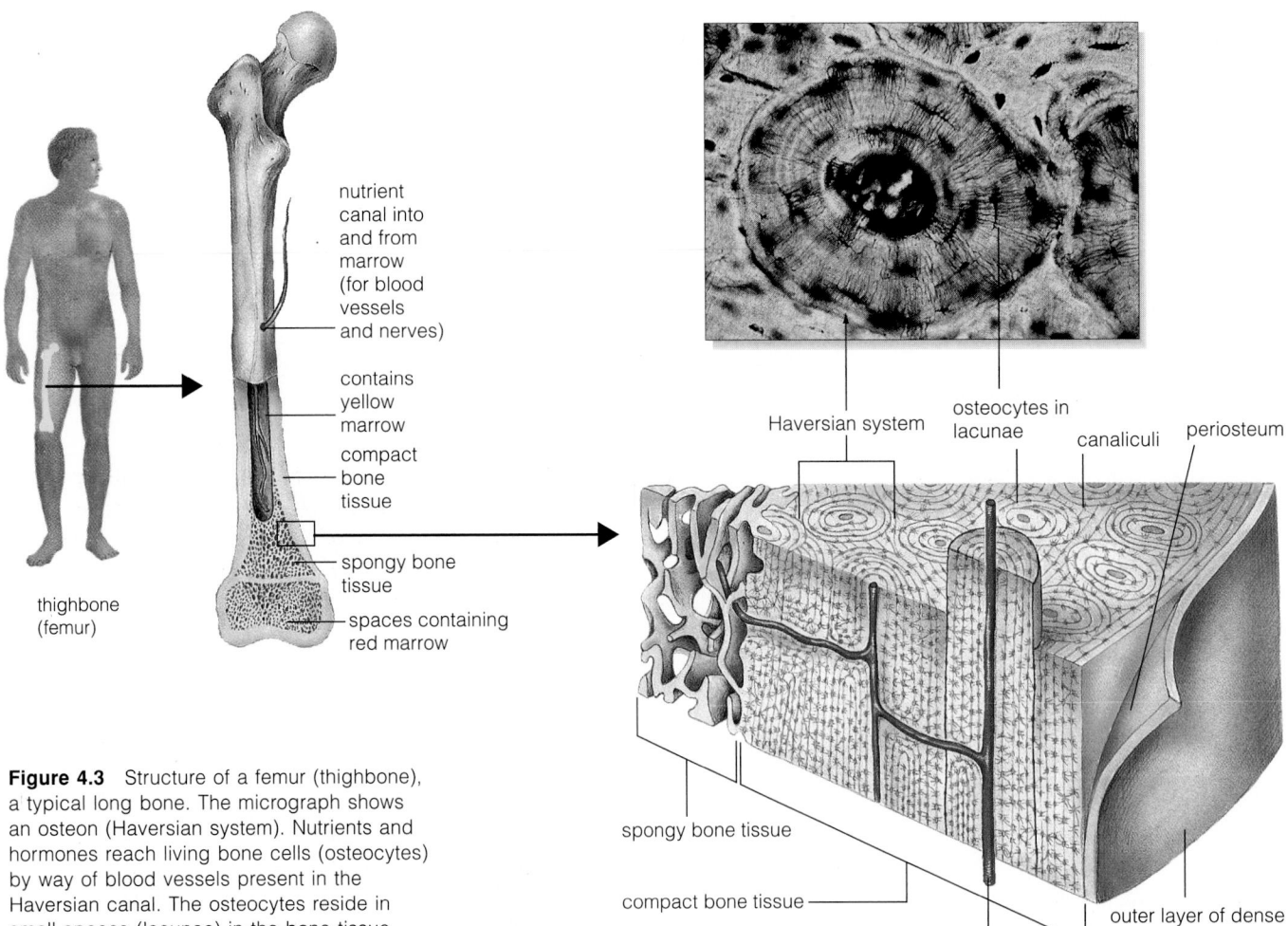

nutrient canal into and from marrow (for blood vessels and nerves)

contains yellow marrow

compact bone tissue

spongy bone tissue

spaces containing red marrow

thighbone (femur)

Haversian system

osteocytes in lacunae

canaliculi

periosteum

spongy bone tissue

compact bone tissue

blood vessel

outer layer of dense connective tissue

Figure 4.3 Structure of a femur (thighbone), a typical long bone. The micrograph shows an osteon (Haversian system). Nutrients and hormones reach living bone cells (osteocytes) by way of blood vessels present in the Haversian canal. The osteocytes reside in small spaces (lacunae) in the bone tissue, and small tunnels (canaliculi) connect neighboring spaces.

COMPACT BONE TISSUE   The tissue that forms the bone's shaft and the outer portion of its two ends is *compact bone*. Such dense tissue forms the shaft of every long bone. The collagen fibers within compact bone give long bones the strength to withstand mechanical stresses. An outer membrane, the periosteum, covers the bone shaft. Compact bone tissue is organized as thin, concentric layers around small canals. Each set of circular layers is an **osteon** (also called a *Haversian system*), and the canals—known as *Haversian canals*—are interconnected channels for blood vessels and nerves that service the osteocytes in compact bone tissue.

Osteocytes extend slender cell processes into narrow channels between lacunae, called *canaliculi* ("little canals"). By way of canaliculi, nutrients can move through the hard ground substance from osteocyte to osteocyte. Wastes can exit by the same route.

SPONGY BONE TISSUE   The bone tissue inside a long bone's shaft and at its ends looks like a sponge. Tiny, flattened struts are fused together in a latticework to make up this *spongy bone tissue*, which actually is quite firm and strong. In some bones, **red marrow** fills the spaces between the struts. Red marrow is a major site for the formation of blood cells. In adults, most red marrow—and, normally, most blood cell production—occurs in irregular bones such as the hipbone and in flat bones such as the sternum (breastbone). This is because, as a person grows into adulthood, the red marrow in the shafts of most long bones is replaced by yellow marrow. **Yellow marrow** is mostly fat. If the need arises, it can convert to red marrow and produce red blood cells.

How Do Bones Develop?

In a growing embryo, a cartilage "model" serves as the template for each long bone (Figure 4.4). Once the outer covering (the periosteum) is in place on the model, it gives rise to osteoblasts, the bone-forming cells. A supportive bony "collar" forms around the cartilage shaft; then the cartilage within the shaft calcifies. An artery, a vein, and other elements (including some osteoblasts) then infiltrate the region of calcified cartilage, and the marrow cavity forms. The osteoblasts continue to secrete matrix materials, which gradually become mineralized. Eventually the osteoblasts are surrounded by the matrix they have secreted. They are now osteocytes (see Figure 4.3), mature living bone cells that maintain mature bones.

A long bone has flaring ends. Each end is called an *epiphysis* (plural: epiphyses). In growing children and young adults epiphyses are separated from the bone shaft by an **epiphyseal plate** of cartilage. The plate develops and is maintained under the influence of growth hormone

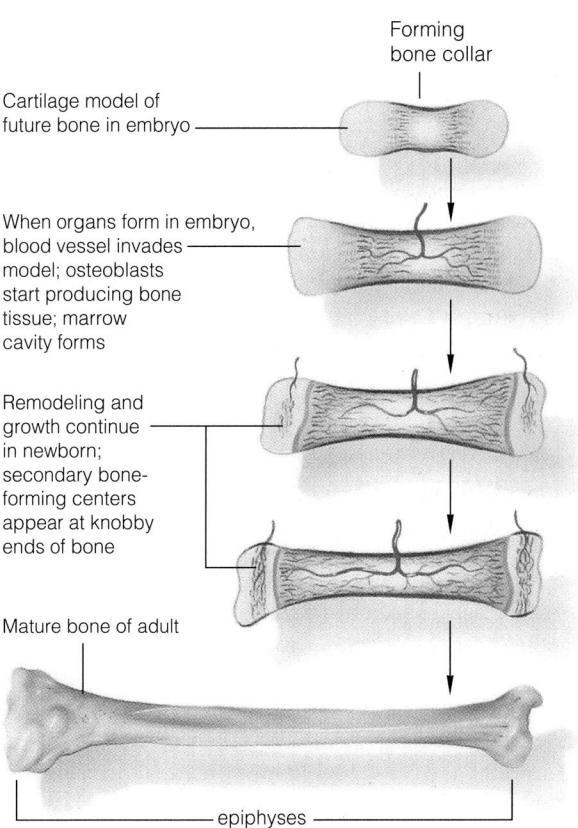

Cartilage model of future bone in embryo

Forming bone collar

When organs form in embryo, blood vessel invades model; osteoblasts start producing bone tissue; marrow cavity forms

Remodeling and growth continue in newborn; secondary bone-forming centers appear at knobby ends of bone

Mature bone of adult

epiphyses

Figure 4.4   Long bone formation, starting with osteoblast activity in a cartilage model (here, already formed in the embryo). Bone-forming cells are active first in the shaft region, then at the knobby ends. In time, cartilage is left only in the epiphyseal plates at the ends of the shaft.

(GH). This hormone is produced by the pituitary gland (located in the brain), which secretes GH in response to other hormones produced by the thyroid gland. As long as the epiphyseal plate is present, the bone can lengthen as a person grows. When growth stops in late adolescence, the cartilage plates are replaced by bone.

Bones are complex organs composed of living cells, a nonliving mineralized matrix of collagen fibers and ground substance, and an outer covering of dense connective tissue.

Bones function in body support, movement, mineral storage, and, in some cases, blood cell production.

Two kinds of marrow fill the spaces in different types of bone. Red marrow is a major site of blood cell formation. Yellow marrow, which in adults replaces red marrow in the shafts of most long bones, is mostly fat. It can be converted to red marrow in time of need.

HOW THE SKELETON GROWS AND IS MAINTAINED

An early human embryo has a skeleton that consists only of cartilage and membranes. After about two months of life in the womb, however, this pliable framework becomes transformed into a bony skeleton as bone cells develop and the process of mineralization gets under way. From childhood onward, two key minerals—calcium and phosphate—are constantly deposited in and withdrawn from bone tissue. This dynamic turnover of minerals in bone tissue is called "remodeling." It occurs as osteoblasts deposit bone and cells called *osteoclasts* break it down.

How Bone Remodeling Works

As you will read later in this book, the mineral calcium is essential for nervous system functioning, for muscle contraction, and for many other physiological processes. When the level of calcium in the blood falls below a set point, the hormone PTH (parathyroid hormone) stimulates osteoclasts to secrete enzymes that break down bone tissue. In adults, osteocytes also respond to PTH and they, too, withdraw calcium from bone tissue. As the component minerals dissolve, the released calcium enters interstitial fluid; from there, it is taken up by the blood. If the amount of calcium in blood is greater than required, another hormone, calcitonin, stimulates osteoblasts to take up calcium from the interstitial fluid and use it to produce new bone. Chapter 12 discusses these and other hormonal controls.

Bone deposits increase relative to withdrawals when mature bone is subjected to increased mechanical stress, partly by way of muscles that pull on the bone during body movements. Brisk walking, jogging, "pumping iron," and similar activities cause affected bones to become denser—and stronger—as more bone tissue is created. Remodeling also occurs after breaks or other bone injuries.

Bone Remodeling over Time

Before adulthood, bones grow by way of remodeling. During this time the body requires ample calcium to meet the combined demands of bone growth and other metabolic needs. Turnover is also especially important. For example, the diameter of the thighbone increases as osteoblasts deposit calcium phosphate at the surface of the shaft. At the same time, osteoclasts destroy a small amount of bone tissue *inside* the shaft. Thus the thighbone becomes thicker and stronger, but not too heavy.

As a person ages, bone may break down faster than it is renewed, especially in older women. The backbone, pelvis (hip bones), and other bones decrease in mass. This progressive bone deterioration is called *osteoporosis* (Figure 4.5). The spine can collapse and curve so much

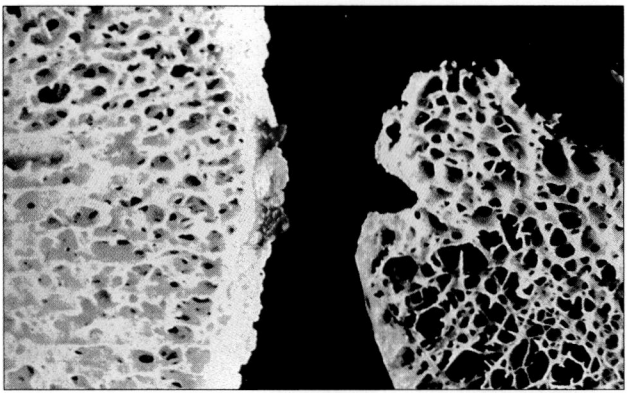

Figure 4.5 Osteoporosis. In normal tissue (**left**), mineral deposits continually replenish mineral withdrawals. After the onset of osteoporosis (**right**), replacements can't keep pace with withdrawals. The tissue gradually erodes, and bones become progressively hollow and brittle.

that the rib cage lowers (a condition called *lordosis*), crowding internal organs. Research has suggested that deficiencies of calcium and sex hormones, eating too much protein, smoking, and a sedentary lifestyle can contribute to osteoporosis. On the other hand, lifelong habits of getting plenty of exercise (to stimulate deposits of bone tissue) and consuming enough calcium in the diet may help minimize bone loss later in life.

The Skeleton: A Preview

A fully formed human skeleton has 206 bones, which grow by way of remodeling for about the first 20 years of life. The bones are organized into two divisions, an **axial skeleton** and an **appendicular skeleton**, as you can see in Figure 4.6. Straps of dense, regular connective tissue called **ligaments** connect the bones at joints. As befits this function, the connective tissue of ligaments contains a large proportion of elastic fibers and thus is extensible and resilient. **Tendons** attach muscles to bones or sometimes to other muscles. In contrast to ligaments, tendons are cords or straps of dense, regular connective tissue that derives its strength from plentiful collagen.

In the next two sections we will examine the functions and structural characteristics of the two skeletal divisions, beginning with the skull and other elements of the axial skeleton.

In childhood and adolescence, bones grow by way of remodeling. This is a process of "turnover" in which osteoblasts deposit bone and osteoclasts break it down.

The fully formed human skeleton consists of 206 bones, organized into axial and appendicular divisions.

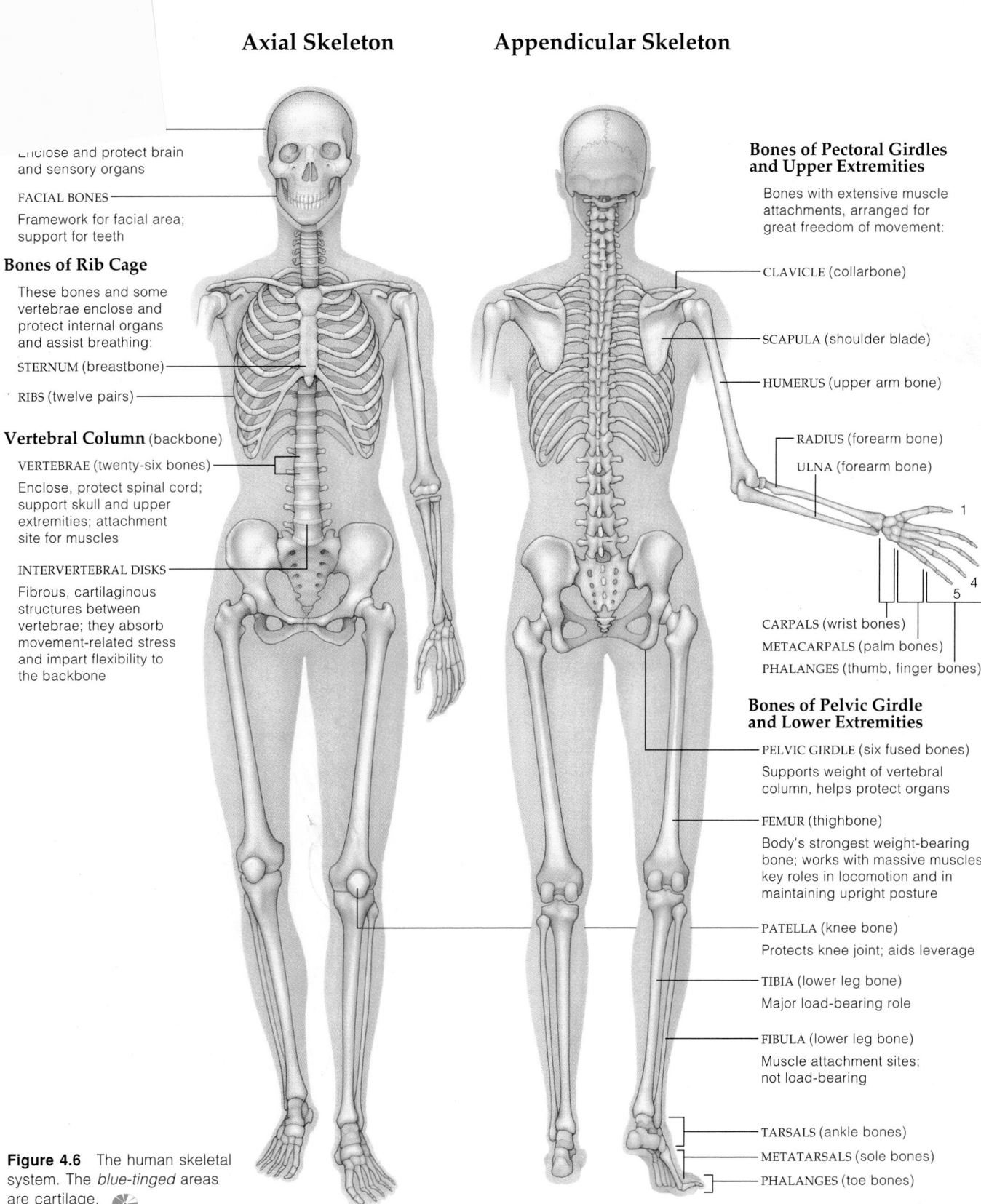

Axial Skeleton

Enclose and protect brain and sensory organs

FACIAL BONES

Framework for facial area; support for teeth

Bones of Rib Cage

These bones and some vertebrae enclose and protect internal organs and assist breathing:

STERNUM (breastbone)

RIBS (twelve pairs)

Vertebral Column (backbone)

VERTEBRAE (twenty-six bones)

Enclose, protect spinal cord; support skull and upper extremities; attachment site for muscles

INTERVERTEBRAL DISKS

Fibrous, cartilaginous structures between vertebrae; they absorb movement-related stress and impart flexibility to the backbone

Appendicular Skeleton

Bones of Pectoral Girdles and Upper Extremities

Bones with extensive muscle attachments, arranged for great freedom of movement:

CLAVICLE (collarbone)

SCAPULA (shoulder blade)

HUMERUS (upper arm bone)

RADIUS (forearm bone)

ULNA (forearm bone)

1
2
3
4
5

CARPALS (wrist bones)

METACARPALS (palm bones)

PHALANGES (thumb, finger bones)

Bones of Pelvic Girdle and Lower Extremities

PELVIC GIRDLE (six fused bones)

Supports weight of vertebral column, helps protect organs

FEMUR (thighbone)

Body's strongest weight-bearing bone; works with massive muscles; key roles in locomotion and in maintaining upright posture

PATELLA (knee bone)

Protects knee joint; aids leverage

TIBIA (lower leg bone)

Major load-bearing role

FIBULA (lower leg bone)

Muscle attachment sites; not load-bearing

TARSALS (ankle bones)

METATARSALS (sole bones)

PHALANGES (toe bones)

Figure 4.6 The human skeletal system. The *blue-tinged* areas are cartilage.

The axial skeleton is made up of bones that roughly form the body's vertical axis. These include the skull, vertebral column (backbone), ribs, and sternum (the breastbone).

The Skull

We begin with the **skull**, which consists of more than two dozen bones that are divided into several groupings. Although many bones are traditionally called by complex-sounding names derived from Latin, their roles are much simpler to grasp. For example, one grouping, the "cranial vault," or **brain case**, includes eight bones that together surround and protect the brain. As Figure 4.7a shows, the *frontal bone* makes up the forehead and upper ridges of the eye sockets. It contains air spaces called **sinuses**, which lighten the skull and are lined with mucous membrane. Sinuses drain into the upper respiratory tract, a frequent source of misery for anyone who has a head cold or pollen allergies. Bacterial infections in the nasal passages can spread to the sinuses, causing *sinusitis*. Figure 4.6c shows the locations of sinuses in cranial and facial bones.

Temporal bones form the lower sides of the cranium and surround the ear canals. Each canal is a tunnel-like space that leads to the middle and inner ear and opens to the outside; within the middle ear are the tiny bones that function in hearing (Chapter 11). A *sphenoid bone* lies just in front of each temporal bone and extends inward to form part of the inner eye socket. The *ethmoid bone* also contributes to the inner socket and helps support the nose.

A pair of *parietal bones* above and behind the temporal bones form a large part of the skull; they sweep upward and meet at the top of the head. An *occipital bone* forms the back and base of the skull; it encloses a large opening, the *foramen magnum* ("great hole"). Here, the spinal cord emerges from the base of the brain and enters the spinal column (Figure 4.7b). Quite a few passageways run through and between various skull bones for nerves and blood vessels, especially at the base of the skull. For instance, the jugular veins, which carry blood leaving the brain, pass through openings between the occipital bone and each temporal bone.

Facial Bones

Figure 4.7 also shows bones of the face. You can easily feel many facial bones with your fingers. The largest is the lower jaw or **mandible**. The upper jaw consists of two *maxillary bones*. Two *zygomatic bones* form the middle portion of the protuberances we call "cheekbones" and the outer parts of the eye sockets. A small, flattened *lacrimal bone* fills out the inner eye socket. Tear ducts pass between this bone and the maxillary bones and drain into the nasal cavity. The upper and lower jaws also contain the teeth in tooth sockets.

Palatine bones make up part of the floor and side wall of the nasal cavity. (Extensions of these bones, together with the maxillary bones, form the back of the hard palate, the "roof" of the mouth.) A *vomer bone* forms part of the nasal septum, a thin "wall" that divides the nasal cavity into two sections.

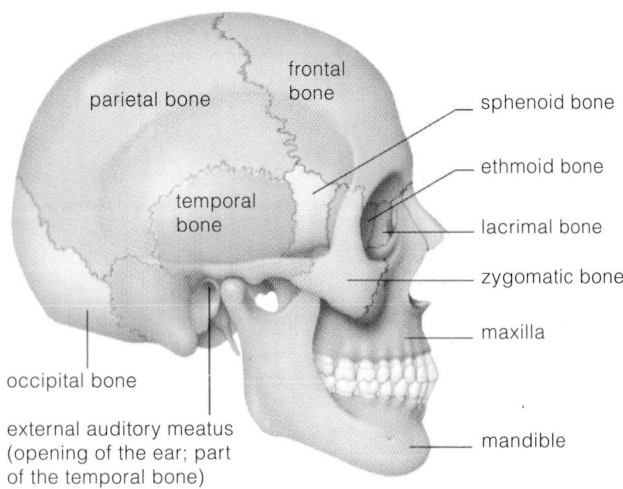

a

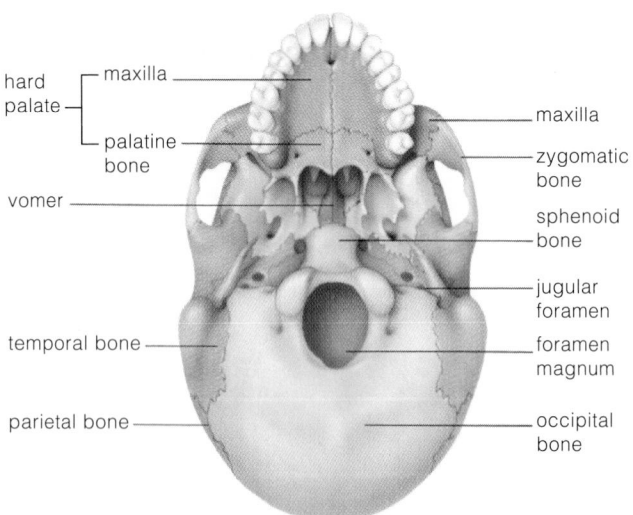

b

Figure 4.7 (**a**) Side view of the skull. The irregular junctions between different bones are called sutures. (**b**) The skull from underneath (inferior view). The large foramen magnum is situated atop the uppermost cervical vertebra. (**c**) Sinuses in bones associated with the nasal cavity.

Vertebral Column: The Backbone

The flexible, curved backbone extends from the base of the skull to the hipbones (pelvic girdle), where it transmits the weight of the torso to the lower limbs. The delicate spinal cord threads through a cavity formed by the vertebrae, which are arranged one above the other. As Figure 4.8 shows, humans have 33 vertebrae, including 7 in the neck region (cervical), 12 in the chest area (thoracic), and 5 in the lower back (lumbar). In the course of human evolution, 5 vertebrae have become fused to form the sacrum, and another 4 vertebrae have become fused to form the coccyx (the "tailbone").

About a quarter of the spine's length is due to the presence of **intervertebral disks**, which occur between the vertebrae and which contain cartilage. The disks serve as shock absorbers and flex points and are thickest between cervical vertebrae and between lumbar vertebrae. However, severe or rapid shocks may cause the core of a disk to *herniate* (sometimes called a "slipped disk"). If the disk ruptures, the jellylike core material may squeeze out. If neighboring nerves or the spinal cord is compressed, excruciating pain can result. Treatment can range from bed rest and use of painkilling drugs to surgery.

The Ribs and Sternum

In addition to protecting the spinal cord, absorbing shocks, and providing flexibility, the vertebral column also serves as an attachment point for 12 pairs of **ribs**,

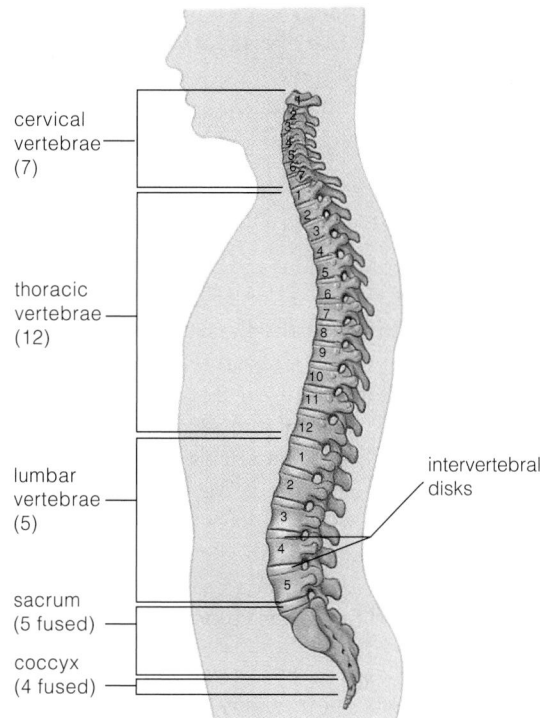

cervical vertebrae (7)

thoracic vertebrae (12)

lumbar vertebrae (5)

intervertebral disks

sacrum (5 fused)

coccyx (4 fused)

Figure 4.8 Side view of the vertebral column or backbone. The cranium balances on the column's uppermost vertebra.

which in turn function as a scaffold for the body cavity of the upper torso. The upper ribs also attach to the **sternum** (refer to Figure 4.6). As you will see in later chapters, this "rib cage" helps protect the lungs, heart, and other internal organs and is vitally important in breathing.

While the axial skeleton provides basic body support and helps protect internal organs, many movements depend on interactions of skeletal muscles with the bones of the next skeletal component we consider, the appendicular skeleton.

Bones of the axial skeleton make up the body's vertical axis. They include the skull (and facial bones), the vertebral column, and the ribs and sternum.

Intervertebral disks between the vertebrae absorb shocks and serve as flex points.

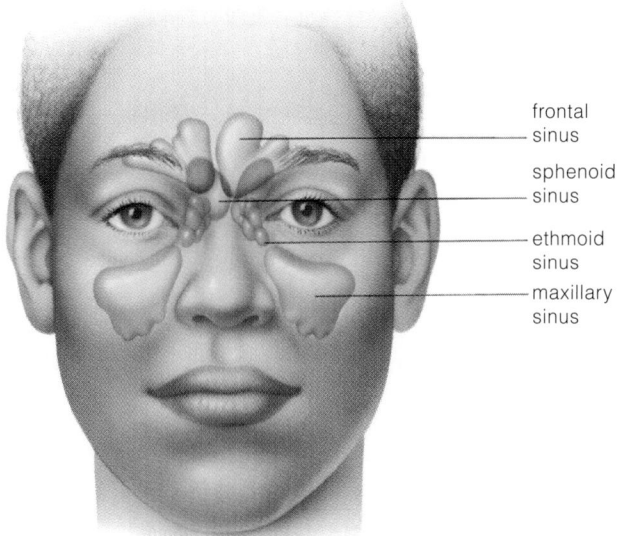

frontal sinus

sphenoid sinus

ethmoid sinus

maxillary sinus

c

THE APPENDICULAR SKELETON

"Append" means to hang, and the appendicular skeleton includes the bones of body parts that we sometimes think of as dangling from the main body frame: arms, hands, legs, and feet. It also includes a pectoral girdle at each shoulder and the pelvic girdle at the hips.

The Pectoral Girdle and Upper Limbs

Each **pectoral girdle** (Figure 4.9) has a large, flat shoulder blade—a **scapula**—and a long, slender collarbone or **clavicle** that connects to the breastbone (sternum). The rounded shoulder end of the **humerus**, the long bone of the upper arm, fits into an open socket in the scapula. The human arm is capable of remarkably versatile movements; it can swing in wide circles and back and forth, lift objects, or tug on a rope. We owe this freedom of movement to the fact that the pectoral girdles and upper limbs are loosely attached to the rest of the body by muscles. Although this arrangement is sturdy enough under normal conditions, it is vulnerable to strong blows.

Fall on an outstretched arm and you might fracture your clavicle or dislocate your shoulder. The collarbone is the bone most frequently broken.

Each of your upper limbs includes some 30 separate bones. The humerus connects with two bones of the forearm, the **radius** (on the thumb side) and the **ulna** (on the "pinky finger" side). The upper end of the ulna joins the lower end of the humerus to form the elbow joint. The prominence sometimes (mistakenly) called the "wristbone" is the lower end of the ulna.

The radius and ulna together join the hand at the wrist joint, where they meet eight small, curved *carpal* bones. Ligaments attach them to the long bones. Blood vessels, nerves, and tendons pass in sheaths over the wrist; when a blow, constant pressure, or repetitive movement (such as prolonged typing) damages these tendons, the result can be a painful disorder called *carpal tunnel syndrome*. The bones of the hand, the five *metacarpals*, end at the knuckles. *Phalanges* are the bones of the fingers.

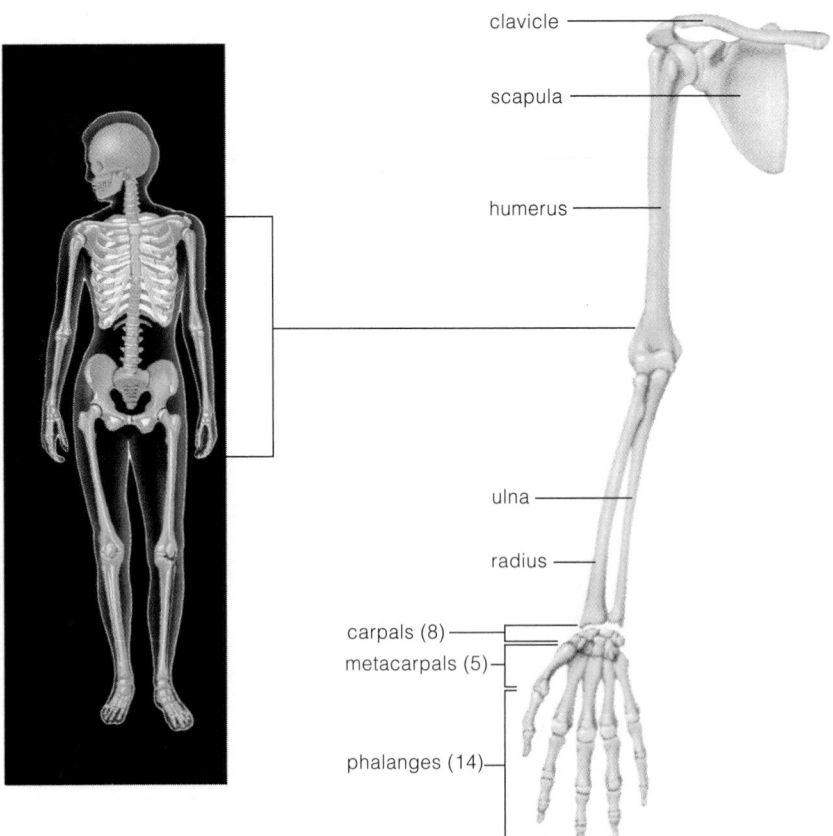

Figure 4.9 Bones of the pectoral girdle, the arm, and the hand.

clavicle

scapula

humerus

ulna

radius

carpals (8)

metacarpals (5)

phalanges (14)

The Pelvic Girdle and Lower Limbs

For most of us, our shoulders and arms are much more flexible than are our hips and legs. Why? Although there are similarities in the basic "design" of both girdles, this lower part of the appendicular skeleton is adapted to bear the body's entire weight when we are standing. The **pelvic girdle** (Figure 4.10) is much more massive than the pectoral girdles, and it is attached to the axial skeleton by exceptionally strong ligaments. It forms an open basin: A pair of *coxal bones* attach to the lower spine (sacrum) in back, then curve forward and meet at the *pubic arch*. ("Hipbones" are actually the upper *iliac* regions of the coxal bones.) This combined structure is the *pelvis*. In females the pelvis is broader than in males, and it shows other structural differences that are evolutionary adaptations for the function of childbearing. A forensic scientist or paleontologist examining skeletal remains can easily establish the sex of the individual if a pelvis is present.

The legs contain the body's largest bones. In terms of length, the thigh bone or **femur** ranks number one. It is also extremely strong. When you run or jump, your femurs routinely withstand stresses of several tons per square inch (aided by contracting leg muscles). The femur's ball-like upper end fits snugly into a deep socket in the coxal (hip) bone. The other end connects with one of the bones of the lower leg, the thick, load-bearing *tibia* on the inner (big toe) side. A slender *fibula* parallels the tibia on the outer (little toe) side. The tibia is your shinbone. A triangular kneecap, the patella, helps protect the knee joint. In spite of this protection, however, knees are among the joints most often damaged by athletes, both amateur and professional.

Bones of the ankle and foot correspond closely with those of the wrist and hand. *Tarsal* bones make up the ankle and heel, and the foot contains five long bones, the *metatarsals*. The largest metatarsal, leading to the big toe, supports a great deal of body weight and is thicker and stronger than the others. Like fingers, the toes contain phalanges.

The appendicular skeleton includes bones of the limbs, a pectoral girdle at the shoulders, and a pelvic girdle at the hips.

The thighbone (femur) is the largest bone in the human body and also one of the strongest. The wrists and hands and ankles and feet have closely corresponding sets of bones known respectively as carpals and metacarpals and tarsals and metatarsals.

Figure 4.10 Bones of the pelvic girdle, the leg, and the foot.

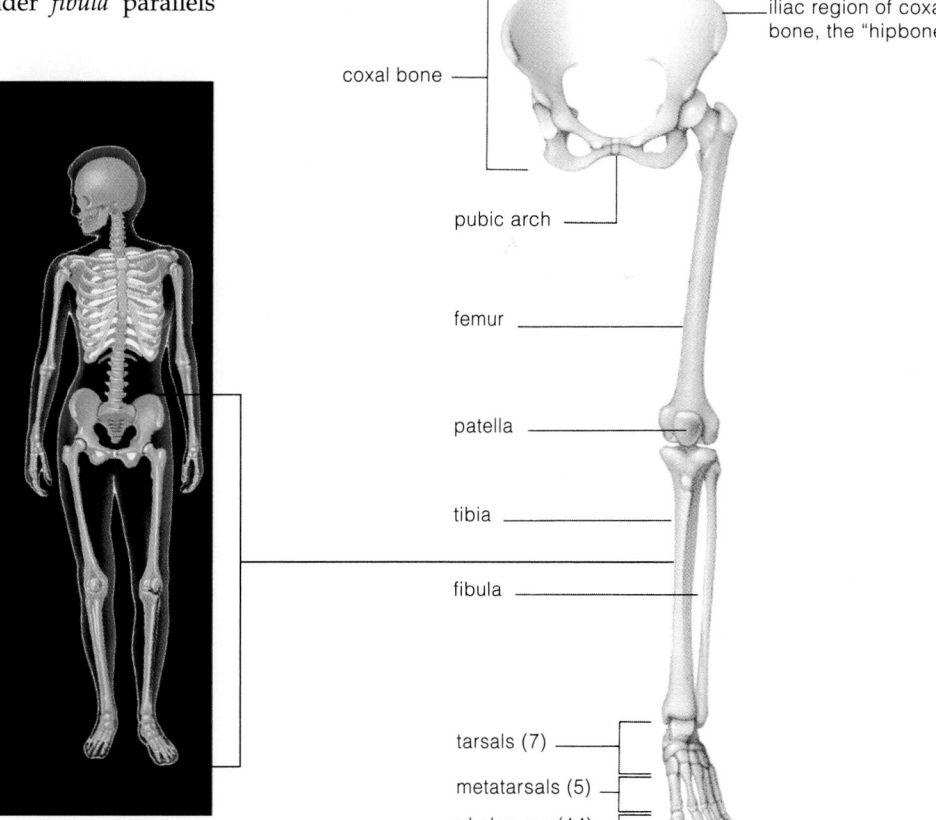

coxal bone

iliac region of coxal bone, the "hipbone"

pubic arch

femur

patella

tibia

fibula

tarsals (7)

metatarsals (5)

phalanges (14)

"Joints" are areas of contact or near-contact between bones. In the most common type, the **synovial joint**, adjoining bones are separated by a cavity. Such joints are freely movable. The articulating ends of the bones are covered with a cushioning layer of cartilage and are stabilized in part by straplike ligaments. A capsule of dense connective tissue surrounds the bones of a synovial joint. Cells that line the interior of the capsule secrete a lubricating *synovial fluid* into the joint cavity.

Synovial joints include the *ball-and-socket* joints at the hips. These types of joints are capable of a wide range of movements, including rotation and movements in different planes (up-down, side-to-side, and so on). *Hingelike* synovial joints such as the knee and elbow move in one plane only; that is, they are limited to simple flexing and extending (straightening), like a door hinge. Figure 4.11 shows various ways body parts can move at joints.

As a person ages, the cartilage covering the bone ends of freely movable joints may degenerate into a condition called *osteoarthritis*. Often, the arthritic joint becomes painfully inflamed. Another condition, *rheumatoid arthritis*, is a degenerative disorder that results when a person's immune system malfunctions and mounts an attack against tissues in the affected joint. The synovial membrane becomes inflamed and thickened, cartilage is eroded away, and the bones fall out of proper alignment. The bone ends may eventually fuse together. As described in the *Science Comes to Life* on the next page, injuries can also disrupt the functioning of a freely movable joint. Surgeons can now use artificial joints (Figure 4.12) to replace certain joints when they become seriously damaged.

In **cartilaginous joints**, cartilage fills the space between bones and permits only slight movement. Such joints occur between vertebrae and between the breastbone and some of the ribs.

In **fibrous joints**, fibrous connective tissue unites the bones, and no cavity is present. Fibrous joints loosely connect the flat skull bones of a fetus. During childbirth, the loose connections allow the bones to slide over each

Figure 4.11   Examples of ways body parts can move at joints. The synovial joint at the shoulder permits the greatest range of movement.

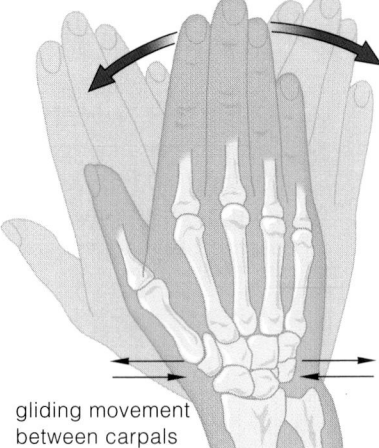

gliding movement between carpals

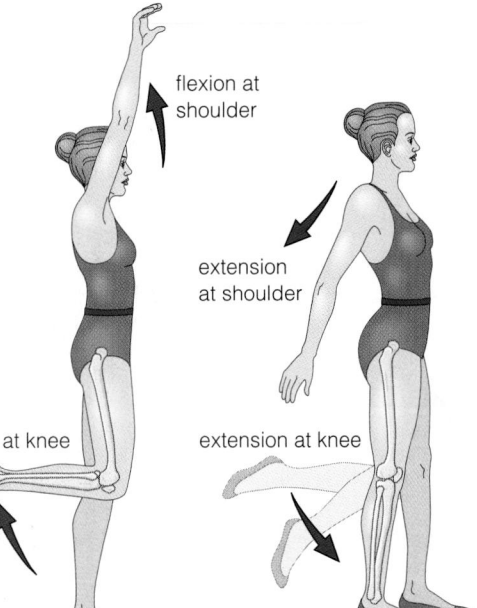

flexion at shoulder

extension at shoulder

at knee   extension at knee

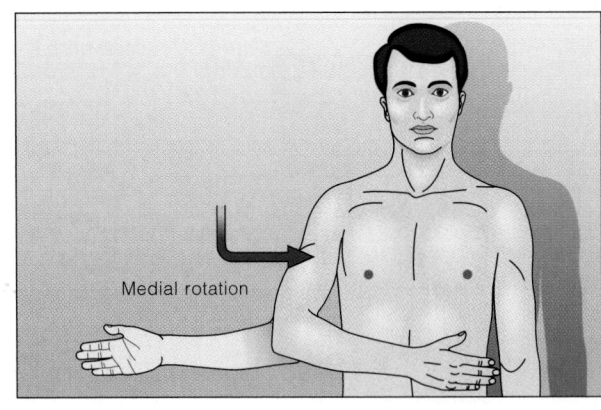

Medial rotation

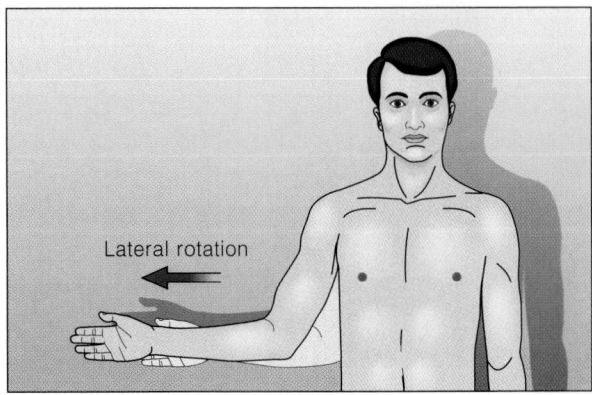

Lateral rotation

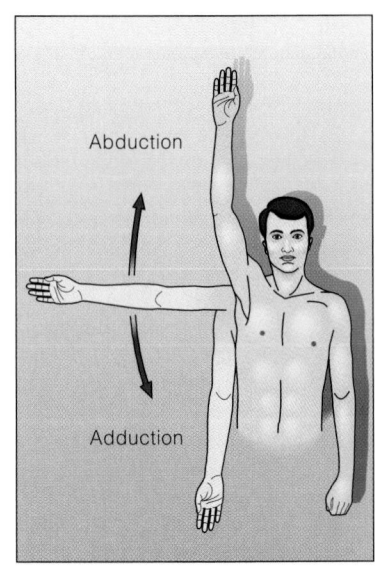

Abduction

Adduction

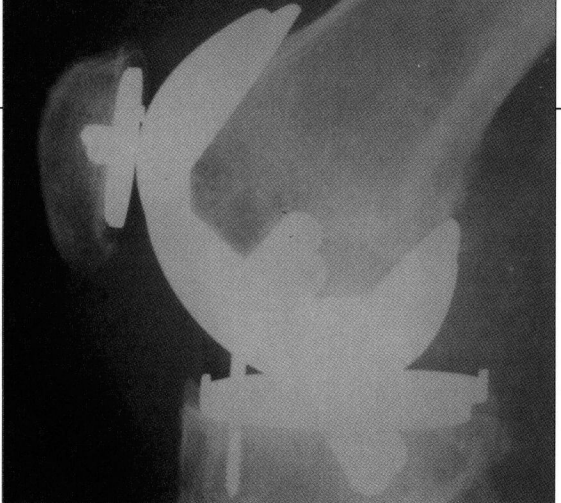

Figure 4.12 An artificial knee joint. Most artificial joints consist of metal or a combination of metal and plastic. The patient's bone can grow into tiny pits in the replacement joint, making the bond between joint and bone stronger than if adhesives are used.

other and so prevent skull fractures. A newborn baby's skull still has fibrous joints and membranous areas called *fontanels* ("soft spots"). During childhood, the joints harden into *sutures*. Much later in life the skull bones may become completely fused.

Joints are areas of contact or near-contact between bones.

In the human body the major types of joints include freely movable synovial joints, such as the ball-and-socket joints at the hips, cartilaginous joints, and fibrous joints.

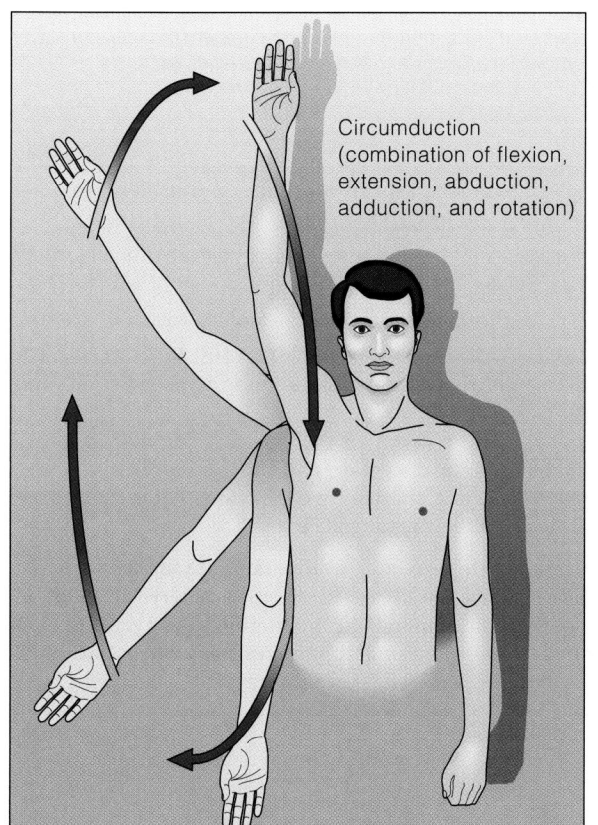

Circumduction
(combination of flexion,
extension, abduction,
adduction, and rotation)

Joggers, skiers, rowers, and players of other sports know that mobility—even an athlete's career—can hinge on the health of the knees.

The knee joint links two long bones, the femur and the tibia. Those bones are separated by a cavity and are held together by ligaments and tendons that form a capsule around the joint (Figure 4.13). A membrane lining the capsule produces fluid that lubricates the joint; where the bone ends meet, they are capped with a cushioning layer of cartilage. Wedges of fibrocartilage between the femur and the tibia called menisci (singular: meniscus) add stability. Fluid-filled sacs (bursae) reduce friction.

When the knee is hit hard or twisted too much, its ligaments can tear. If the damage is slight, the resulting *sprain* is painful but easily treatable with an ice pack, rest, and wrapping the injured knee with a flexible bandage. A severed ligament must be surgically repaired within about 10 days of the injury. If not, scavenger cells in the synovial fluid will destroy the injured tissue.

Torn fibrocartilage also requires medical intervention. If most or all of the damaged tissue is not removed, it can cause arthritis. A common procedure today is arthroscopy, an operation in which the surgeon makes a small slit and inserts an instrument that carries a miniaturized light source and lens hooked up to a television camera. The apparatus enables the physician to see and correct the injury while minimizing scarring and other side effects of major surgery.

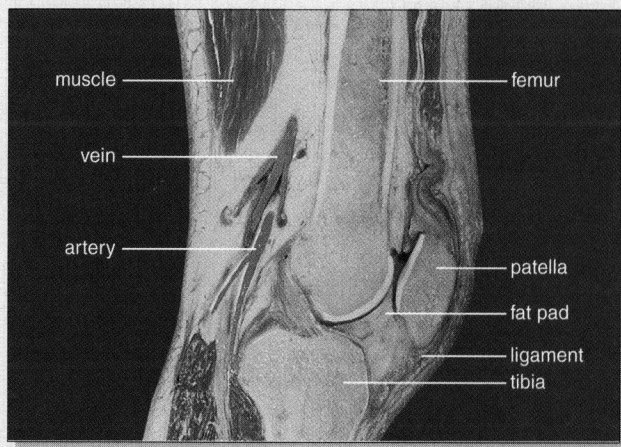

Figure 4.13 Longitudinal section through the knee joint. In this view, the lateral and medial menisci are not readily visible.

You may recall from Chapter 3 that there are three types of muscle tissue: skeletal, cardiac, and smooth. Although skeletal and cardiac muscle are quite different in appearance from smooth muscle, all muscle cells are specialized to generate force or shorten by contracting. Muscles also can relax and lengthen after contracting. Later sections examine the ATP-requiring mechanisms by which muscle contraction and relaxation occur.

Skeletal muscle contracts in response to stimulation by the nervous system. Cardiac muscle can respond to nerve stimulation, but it beats (contracts) spontaneously—that is, without outside stimulation, by way of a mechanism that is a property of the muscle itself. Smooth muscle responds to a variety of control systems, including nerves, hormones, and spontaneous mechanisms. Smooth muscle occurs mostly in the walls of internal organs, where it helps maintain organ shape. It also functions in movement; for example, contractions of smooth muscle in the stomach and in the walls of the intestine help propel substances through your digestive tract. Cardiac muscle occurs only in the heart. Chapter 6 describes its action.

How Skeletal Muscles and Bones Interact

We turn now to skeletal muscle, the functional partner of bone. **Skeletal muscle** is the only type of muscle tissue that interacts with the skeleton to bring about movement. The human body has more than 600 skeletal muscles, often arranged as pairs or groups. Some of them work together (*synergistically*) to promote the same movement. Others work in opposition (*antagonistically*) so that the action of one opposes or reverses the action of the other.

When a skeletal muscle contracts, it pulls on the bones to which it is attached. In general, one end of a muscle, the **origin**, is attached to a bone that stays relatively motionless during a movement. The other end of the muscle, called the **insertion**, is attached to the bone that moves most. Together, the skeleton and the muscles attached to it are like a system of levers in which bones (rigid rods) move about joints (fixed points). Most joints have nearby muscle attachments. This means that a muscle must contract only a short distance to produce a large movement of some body part—and to do so rapidly.

Figure 4.14 shows an antagonistic pair of muscles, the biceps and triceps of a human arm. When the biceps contracts, the elbow joint flexes (bends). As it relaxes and its partner (the triceps) contracts, the limb extends and straightens. Such coordinated action results partly from *reciprocal innervation* by nerves from the spinal cord. By this mechanism, when one muscle group is stimulated, no signals are sent to the opposing group and it does not

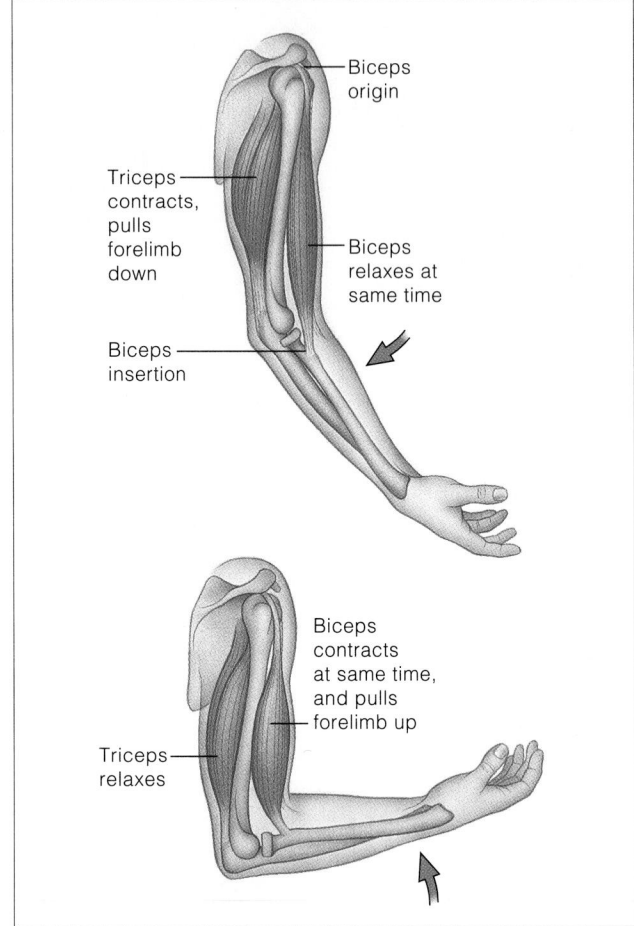

Figure 4.14 Opposing muscle groups in a human arm. When a triceps contracts, the forearm can extend straight out. When the triceps relaxes and its opposing partner (biceps) contracts, the elbow joint flexes and the forearm bends up.

contract. In addition, the nervous system uses signals from stretch receptors in tendons or muscles to coordinate the contractions.

Figure 4.15 shows the major skeletal muscles of the human body. Each one is composed of a few hundred to many thousands of muscle cells ("muscle fibers"). Fibrous connective tissue encloses the muscle cells and blends with the tendons that attach the muscle to bone.

The cells of skeletal, cardiac, and smooth muscle tissue all generate force and shorten by contracting. After contracting they return to their resting state.

Skeletal muscle responds to stimulation by the nervous system. Cardiac muscle contracts without nerve stimulation. Smooth muscle responds to stimulation by nerves and hormones, but some smooth muscles also can contract on their own.

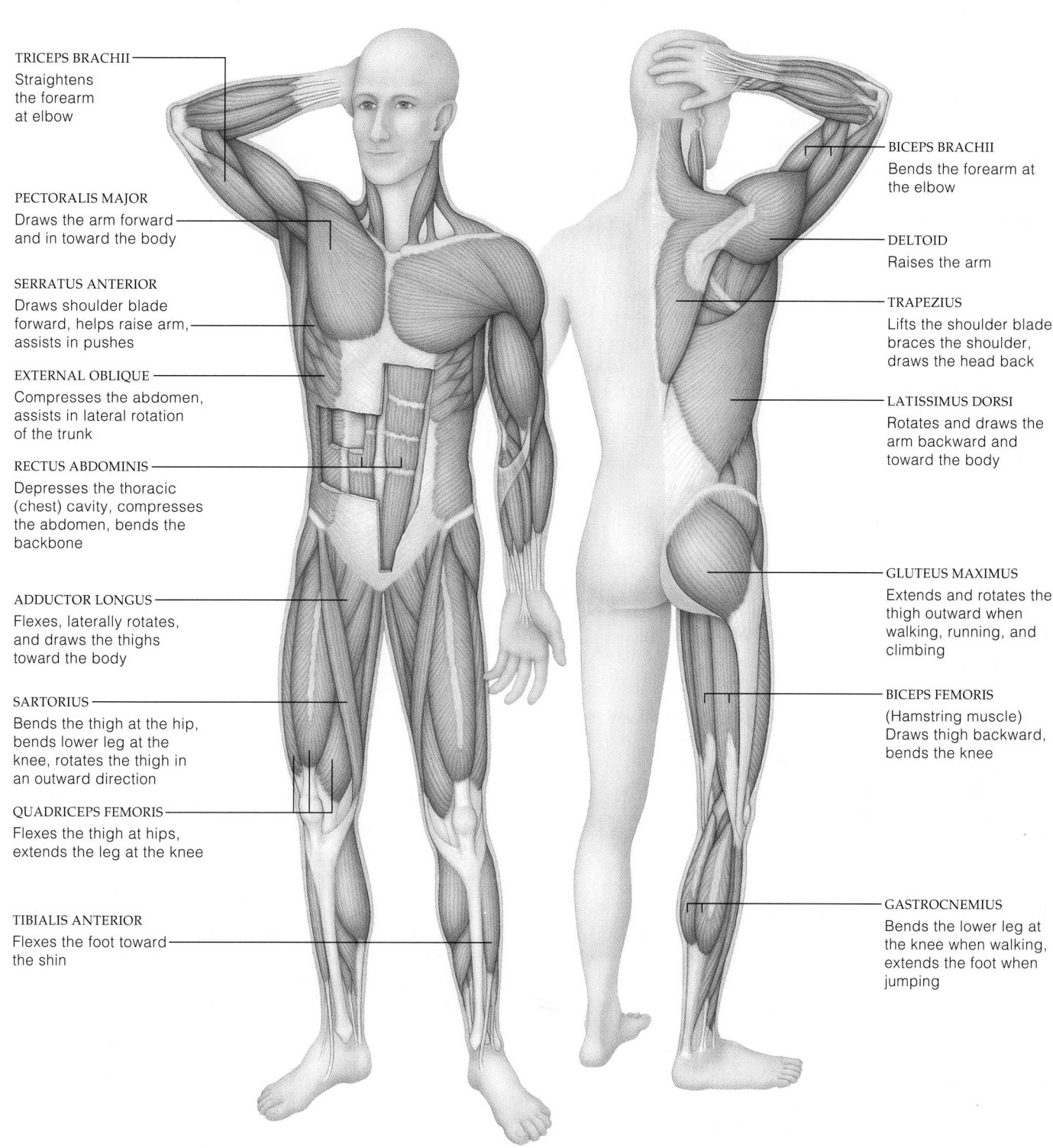

TRICEPS BRACHII
Straightens the forearm at elbow

PECTORALIS MAJOR
Draws the arm forward and in toward the body

SERRATUS ANTERIOR
Draws shoulder blade forward, helps raise arm, assists in pushes

EXTERNAL OBLIQUE
Compresses the abdomen, assists in lateral rotation of the trunk

RECTUS ABDOMINIS
Depresses the thoracic (chest) cavity, compresses the abdomen, bends the backbone

ADDUCTOR LONGUS
Flexes, laterally rotates, and draws the thighs toward the body

SARTORIUS
Bends the thigh at the hip, bends lower leg at the knee, rotates the thigh in an outward direction

QUADRICEPS FEMORIS
Flexes the thigh at hips, extends the leg at the knee

TIBIALIS ANTERIOR
Flexes the foot toward the shin

BICEPS BRACHII
Bends the forearm at the elbow

DELTOID
Raises the arm

TRAPEZIUS
Lifts the shoulder blade, braces the shoulder, draws the head back

LATISSIMUS DORSI
Rotates and draws the arm backward and toward the body

GLUTEUS MAXIMUS
Extends and rotates the thigh outward when walking, running, and climbing

BICEPS FEMORIS
(Hamstring muscle) Draws thigh backward, bends the knee

GASTROCNEMIUS
Bends the lower leg at the knee when walking, extends the foot when jumping

Figure 4.15 Some of the major skeletal muscles of the human muscular system.

MUSCLE STRUCTURE AND FUNCTION

How Skeletal Muscle Is Organized

Chapter 3 described some structural features of skeletal muscle cells. You may remember that each muscle cell contains threadlike structures packed together in a parallel array. You can see these **myofibrils** in Figure 4.16. Each threadlike myofibril is divided along its length into sarcomeres, which occur one after another. A **sarcomere** is the functional unit of muscle contraction.

As Figure 4.16c shows, a sarcomere contains many filaments oriented parallel with its long axis. Certain differences in their length and positioning give rise to the striped ("striated") appearance of skeletal muscle (and cardiac muscle). Some of the filaments are thin, others are thick. Each thin filament is like two strands of pearls twisted together. The "pearls" are molecules of **actin**, a globular protein with contractile functions:

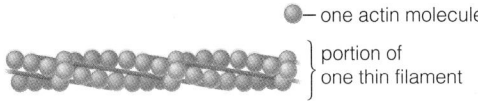

– one actin molecule

} portion of one thin filament

Other proteins (coded blue) are near actin's surface grooves. Each thick filament is made of molecules of **myosin**, another contractile protein. Myosin has a long tail and a double head that projects from the filament's surface:

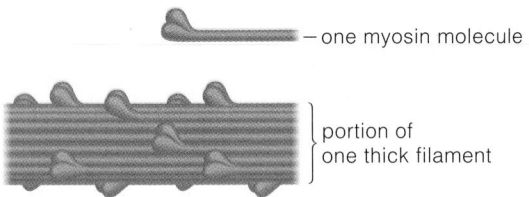

– one myosin molecule

} portion of one thick filament

Thus the myofibrils, muscle cells, and muscle bundles of a skeletal muscle all run in the same direction. What function does this consistent, parallel orientation serve? It focuses the force of muscle contraction onto a bone in a particular direction.

Sliding-Filament Model of Contraction

Skeletal muscles never push on bones; they always pull. Why? The only way that skeletal muscles can move the body parts to which they are attached is to shorten. When a skeletal muscle shortens, its cells are shortening. And when a muscle cell shortens, its component sarcomeres are shortening. The combined shortening of all the sarcomeres results in contraction of the whole muscle.

How does a sarcomere contract? Figure 4.17 shows the arrangement of actin and myosin in sarcomeres. Myosin filaments physically bind onto the actin filaments

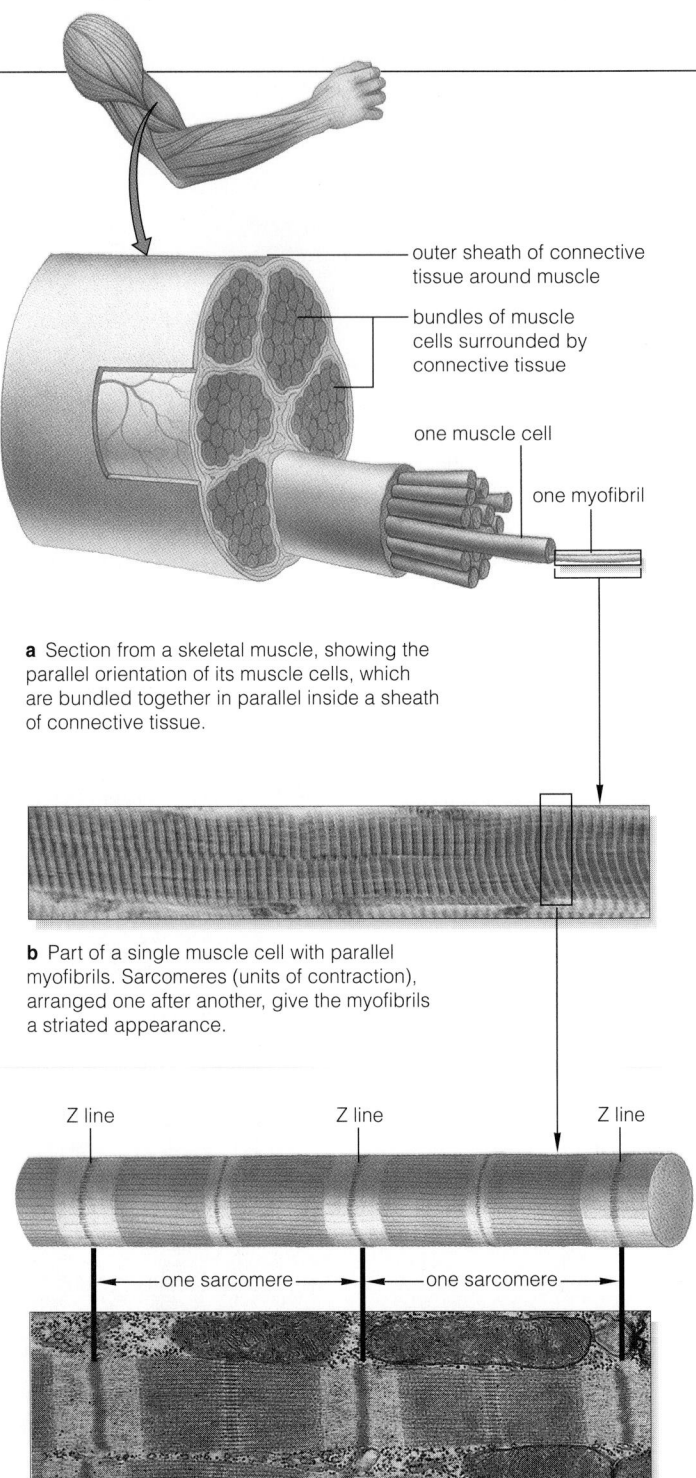

a Section from a skeletal muscle, showing the parallel orientation of its muscle cells, which are bundled together in parallel inside a sheath of connective tissue.

b Part of a single muscle cell with parallel myofibrils. Sarcomeres (units of contraction), arranged one after another, give the myofibrils a striated appearance.

c Diagram and transmission electron micrograph of two of the sarcomeres from a myofibril. This closer view shows how dark "bands," called Z lines, define the two ends of each sarcomere. Mitochondria (the oval-shaped organelles) adjacent to the sarcomeres provide ATP energy for muscle action.

Figure 4.16 Components of a skeletal muscle. (**a**) Each muscle cell contains myofibrils. (**b**) Light micrograph of skeletal muscle cells. (**c**) Each myofibril is functionally divided into sarcomeres, the basic units of contraction. Dark "bands" (Z lines) define the two ends of each sarcomere.

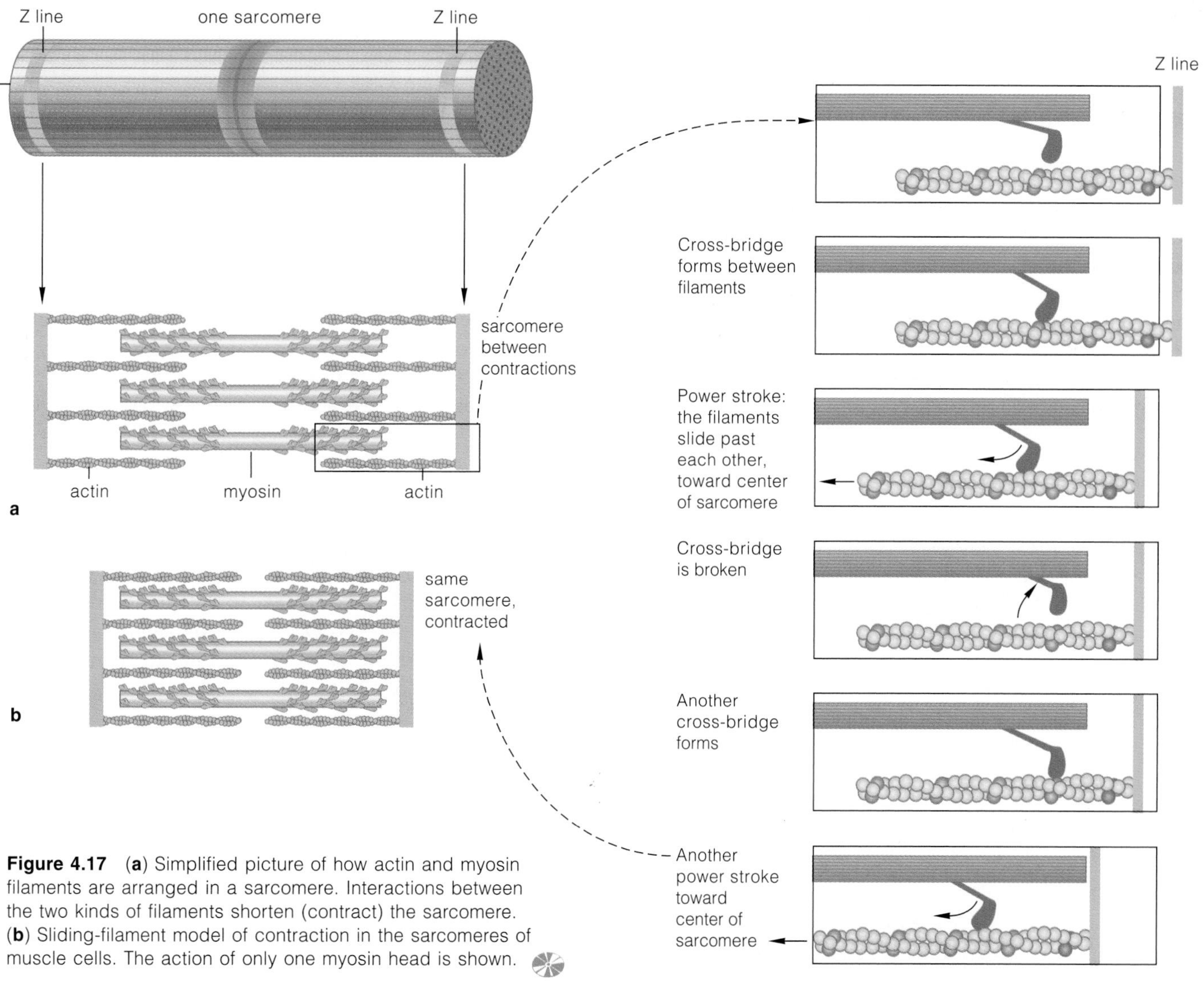

Z line   one sarcomere   Z line

sarcomere
between
contractions

actin   myosin   actin

a

b

same
sarcomere,
contracted

Z line

Cross-bridge
forms between
filaments

Power stroke:
the filaments
slide past
each other,
toward center
of sarcomere

Cross-bridge
is broken

Another
cross-bridge
forms

Another
power stroke
toward
center
of sarcomere

Figure 4.17 (**a**) Simplified picture of how actin and myosin filaments are arranged in a sarcomere. Interactions between the two kinds of filaments shorten (contract) the sarcomere. (**b**) Sliding-filament model of contraction in the sarcomeres of muscle cells. The action of only one myosin head is shown.

and pull toward the center of a sarcomere during contraction. The thick and thin filaments slide over each other, and so the mechanism is known as the **sliding-filament model** of muscle contraction.

Interactions between myosin and actin filaments are cyclic. In each cycle, a bridgelike link forms between a myosin head and a binding site on actin; appropriately, these links are called **cross-bridges.** The linked myosin and actin filaments move; then the myosin heads detach from the actin (Figure 4.17a, right). We can analyze the cycle beginning at a point when enzyme action splits an ATP molecule into ADP and phosphate, briefly creating a "high-energy" form of myosin. In this state, the myosin binds actin. (As you will read, the actual attachment of myosin to actin relies on calcium ions, which are released in a muscle cell when it is stimulated by the nervous system.) After myosin binds actin, the energy released from ATP splitting drives a short "power stroke" in which the

myosin head tilts toward the center of the sarcomere. This stroke also pulls the attached actin filaments toward the center of the sarcomere, which shortens. Then the cycle repeats. Energy from ATP causes each myosin head to detach, the attachment steps occur with the next actin binding site in line, and the actin filaments move a bit more. A single contraction takes a series of power strokes.

In the absence of ATP, myosin cross-bridges cannot detach. When a person dies, ATP production stops, cross-bridges stay locked in place, and skeletal muscles become rigid. This *rigor mortis* lasts up to 60 hours after death.

A skeletal muscle shortens through the combined decreases in length of its sarcomeres, the basic units of contraction.

Energy-driven interactions between myosin and actin filaments shorten the many sarcomeres of a muscle cell and collectively account for its contraction.

CONTROLLING THE CONTRACTION OF A SKELETAL MUSCLE

Skeletal muscle cells contract under commands from the nervous system. *Motor neurons* of the nervous system deliver signals that stimulate contraction. We'll look closely at the nature of these signals in Chapter 10. For the time being, it is enough to understand what happens when commands from the nervous system stimulate a muscle cell. Figure 4.18 summarizes the pathway.

From the point of stimulation, signals spread and rapidly reach small, tubular extensions of the plasma membrane. The small tubes, called *T tubules*, connect with a system of membrane-bound chambers that lace around the cell's myofibrils. That system, called the **sarcoplasmic reticulum** (SR), is a modification of ER, the endoplasmic reticulum (see Section 2.8), and it takes up, stores, and releases calcium ions in controlled ways. The arrival of stimulation from the nervous system triggers an outward flow of calcium ions from the SR. The ions diffuse into myofibrils and reach actin filaments. Molecules of two proteins, *troponin* and *tropomyosin*, are associated with actin filaments (Figure 4.19a and b).

When incoming calcium binds to troponin, the binding site on the actin filament is uncovered (Figure 4.19c). This allows myosin cross-bridges to attach, and the cycle continues. When nervous system stimulation shuts off, the SR membranes actively take up calcium, a process that is powered by ATP. Now the binding site on actin once more is covered up, myosin can't bind to actin, and the muscle cell relaxes.

Neuromuscular Junctions

Acetylcholine (ACh) is a *neurotransmitter*, a substance released by neurons that can have excitatory or inhibitory effects on the cells of muscles and glands throughout the

section from spinal cord

motor neuron

a Signals from the nervous system travel along spinal cord, down motor neuron.

b Endings of motor neuron terminate next to a muscle cell.

section from a skeletal muscle

part of one muscle cell

c Signals travel along muscle cell's plasma membrane to sarcoplasmic reticulum around cell's myofibrils.

Figure 4.18 Pathway for signals from the nervous system that stimulate contraction of skeletal muscle. The plasma membrane of each muscle cell surrounds myofibrils and connects with inward-threading tubes (T tubules). The membrane-bound tubes are close to the sarcoplasmic reticulum, a calcium-storing system that functions in the control of contraction.

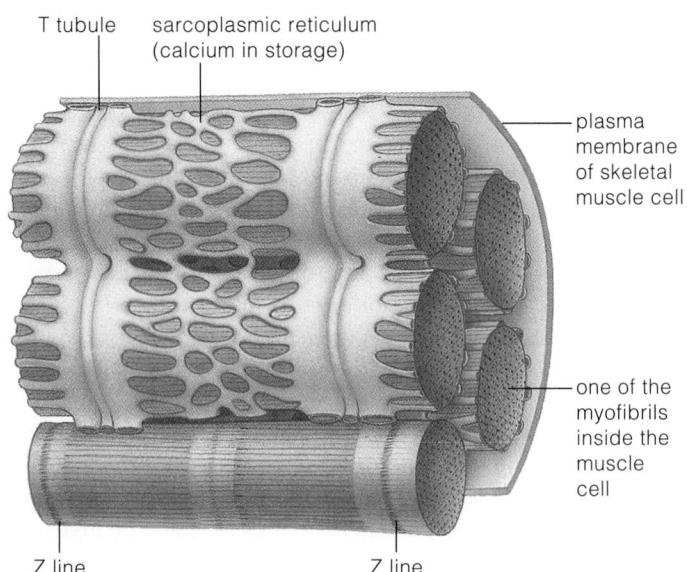

T tubule   sarcoplasmic reticulum (calcium in storage)

plasma membrane of skeletal muscle cell

one of the myofibrils inside the muscle cell

Z line   Z line

d Signals trigger the release of calcium ions from sarcoplasmic reticulum threading among the myofibrils. The arrival of calcium allows actin and myosin filaments in the myofibrils to interact and bring about contraction.

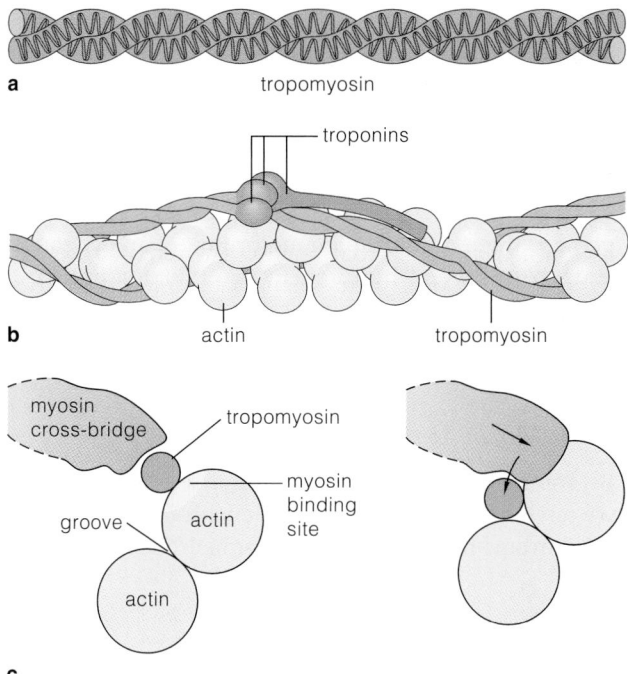

Figure 4.19 Arrangement of troponin, tropomyosin, and actin filaments in skeletal muscle cells. When calcium binds with a troponin, tropomyosin moves away from the actin and so exposes the cross-bridge binding sites.

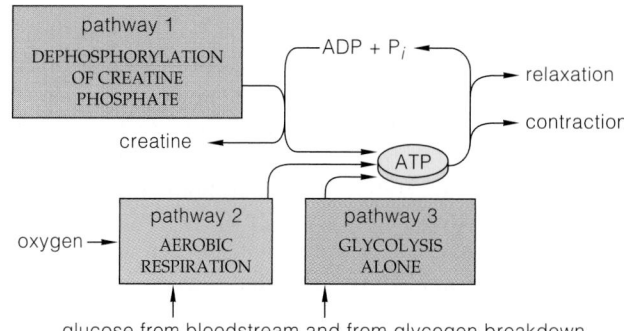

Figure 4.20 Three possible metabolic pathways by which ATP can form in muscles in response to physical exercise.

body. ACh also acts on certain cells in the brain and spinal cord. At **neuromuscular junctions**, the branched endings of certain extensions of a motor neuron (axons) are positioned on the muscle cell membranes, as in Figure 4.18b. (The narrow gap between the neuron ending and the plasma membrane of the other cell is a type of *synapse*.) When the neuron is stimulated, calcium channels open in the plasma membrane of the endings, and calcium ions (Ca^{++}) from the extracellular fluid flow inward. This causes vesicles in the axon ending to release ACh. When ACh binds to receptors on the muscle cell membrane, it has an excitatory effect that may set in motion the events that cause the muscle cell to contract. The poison curare, which is extracted from a South American shrub, blocks the binding of ACh by muscle cells. In so doing, it prevents muscles—including those required for breathing—from contracting.

ATP Formation and Levels of Exercise

A resting muscle cell has a small reservoir of ATP. When the cell is called upon to contract, phosphate donations from ATP must proceed 20 to 100 times faster. At such times, the cell forms ATP by a rapid pathway—an enzyme simply transfers phosphate from creatine phosphate, an organic compound. Muscle cells have about five times as much creatine phosphate as ATP, and this reaction makes enough ATP to power contraction until another, slower ATP-producing pathway begins to operate (Figure 4.20).

During prolonged moderate exercise, the oxygen-requiring reactions of aerobic respiration provide most of the ATP required for contraction. For the first five to ten minutes of muscle activity, a muscle cell taps its store of glycogen for glucose (the starting substrate). For the next half hour or so of sustained activity, it depends on glucose and fatty acid deliveries from the bloodstream. For contractile activity longer than this, fatty acids are the main fuel source (see Figure 2.25).

What happens if exercise is so intense that it exceeds the capacity of the respiratory and circulatory systems to deliver oxygen for the aerobic pathway? Glycolysis and lactate fermentation will now contribute more of the total ATP being formed. By this anaerobic pathway, recall, glucose breakdown ends with lactate formation. The net ATP yield from each glucose molecule is small, but the pathway can proceed as long as glucose from the blood and from glycogen stores is available.

After intense exercise, deep, rapid breathing helps repay the body's **oxygen debt**, incurred when ATP use by muscles exceeded the aerobic pathway's deliveries.

Commands from the nervous system initiate action potentials in muscle cells. These action potentials are the signals for cross-bridge formation—hence for contraction.

During exercise, the availability of ATP in muscle cells affects whether and for how long contraction will proceed.

PROPERTIES OF WHOLE MUSCLES

Muscle Tension

Whether a muscle actually shortens during cross-bridge formation depends on the external forces acting on it. Collectively, the cross-bridges exert **muscle tension**. By definition, this is a mechanical force that a contracting muscle exerts on an object, such as a bone. Opposing it is a load, either the weight of an object or gravity's pull on the muscle. Only when muscle tension exceeds the load does a stimulated muscle shorten.

An *isometrically* contracting muscle develops tension but does not shorten. It supports a load in a constant position, as when you hold a glass of lemonade in front of you. An *isotonically* contracting muscle shortens at a constant load. With *lengthening* contraction, though, an external load is greater than the muscle tension, so the muscle lengthens during the period of contraction. This happens to leg muscles when you walk downstairs.

A muscle's tension relates to the formation of cross-bridges in its cells and the number of cells recruited into action. A skeletal muscle contains a large number of cells, but not all of them contract at the same time. Together, a motor neuron and the muscle cells under its control are called a **motor unit** (Figure 4.21).

The number of cells in motor units varies from perhaps three or four in the motor unit of an eye muscle and up to several hundred in some motor units of a leg muscle. Small motor units provide the "fine-tuning" required for precise control of your muscles.

How Motor Units Function

By stimulating a motor unit with an electrical impulse, we can induce the sort of isometric contraction that would result from stimulation by a motor neuron. It takes a few

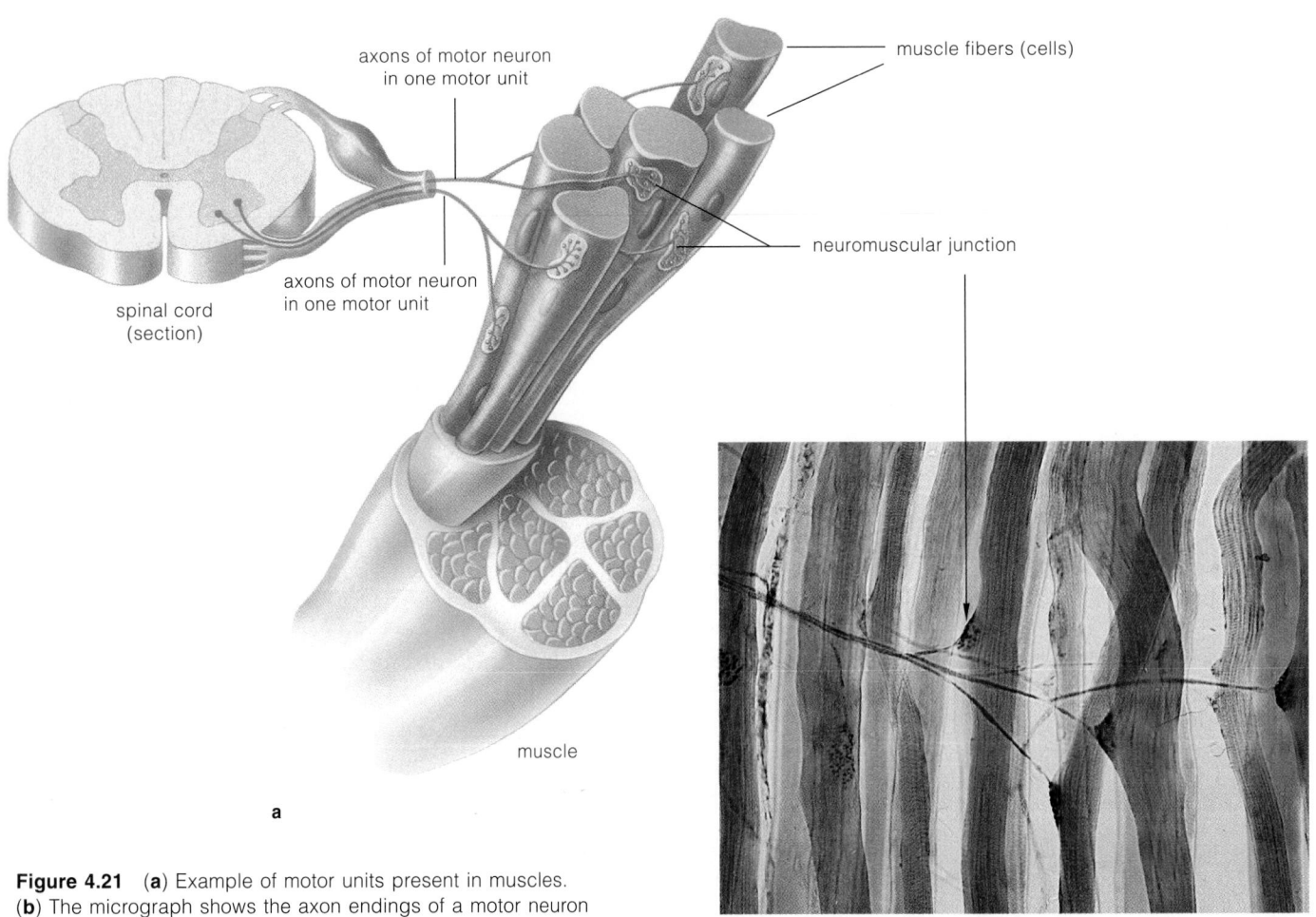

Figure 4.21 (**a**) Example of motor units present in muscles. (**b**) The micrograph shows the axon endings of a motor neuron that acts on individual muscle cells in the muscle.

seconds for tension to increase, then it peaks and declines. This response is a **muscle twitch** (Figure 4.22). The duration of a twitch depends on the load and cell type. For example, fast-acting muscle cells rely on glycolysis (not efficient but fast) and use up their supply of ATP energy faster than slow-acting cells do. When a motor unit is stimulated again before a twitch response is completed, it twitches again. The strength of the contraction depends on how far the twitch response has proceeded by the time the second signal arrives. The effect of the new contraction is added to that of the contraction already under way, a phenomenon called **temporal summation** (Figure 4.22c). As a result, with additional stimuli the strength of contraction increases.

Our muscles normally operate near or at maximum temporal summation, a condition called **tetany**. (In the disease *tetanus*, muscles remain contracted, possibly fatally, due to the effects of a bacterial toxin.) In order to postpone *muscle fatigue*, in which hardworking muscle cells outstrip their supply of ATP and temporarily become unable to respond to stimulation, sets of motor units "take turns" sustaining the contraction of a muscle.

Individual cells in a motor unit always contract according to an **all-or-none principle**: They either contract fully in response to stimulation, or they do not respond at all. If a muscle is contracting only weakly—as might happen in your forearm muscles when you pick up a pencil—it is because the nervous system is activating only a few motor units. In a stronger contraction—when you heft a stack of books, say—a larger number of motor units are activated. Even when muscles are relaxed, some of their motor units are contracted. This steady, low-level contracted state, called *muscle tone*, helps stabilize joints and maintain general muscle health.

"Fast" and "Slow" Muscle

Humans have two general types of skeletal muscle. One type, called "slow" or "red" muscle, appears crimson because its cells are packed with red, iron-rich, oxygen-binding proteins and are served by larger numbers of blood capillaries. It contracts relatively slowly, but because its cells are well equipped to generate lots of ATP aerobically, those contractions can be sustained for a long period. For example, some muscles of the back and legs—called postural muscles because they aid body support—must contract for long periods when a person is standing. These muscles have a high proportion of red muscle cells. By contrast, muscles of your hand have fewer capillaries and relatively more "fast" or "white" muscle cells, which contain fewer mitochondria and less myoglobin. Fast muscle cannot sustain contractions, but it can contract rapidly and powerfully for short periods.

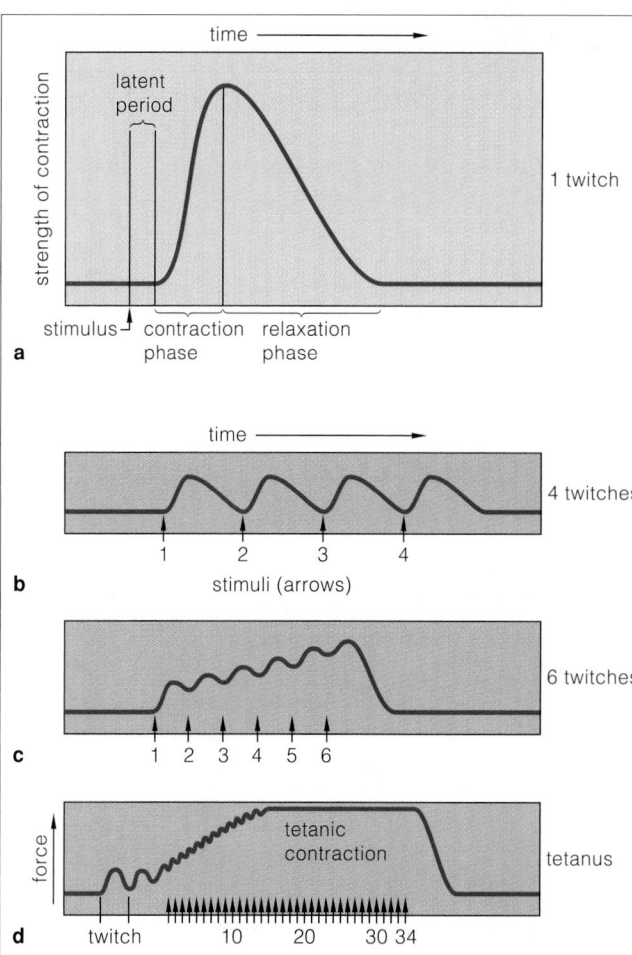

Figure 4.22 Recordings of twitches in artificially stimulated muscles. (**a**) A single twitch. (**b**) Two stimulations per second cause a series of twitches. (**c**) Six per second cause a summation of twitches. (**d**) About 20 per second cause a tetanic contraction.

When athletes train rigorously, one goal is to increase the relative size and contractile strength of fast or slow fibers in their muscles. A sprinter will benefit from larger, stronger fast muscle fibers in the thighs, while a distance swimmer will follow a regimen designed to increase the number of mitochondria in shoulder muscle cells. Some athletes resort to the synthetic hormones called anabolic steroids to rapidly build muscle mass, a practice this chapter's *Focus on Your Health* essay examines.

A motor neuron and the muscle cells under its control are called a motor unit.

A muscle twitch is a brief muscle contraction that occurs when a single, brief stimulus of a given strength activates a certain number of motor units.

A motor unit and its cells always contract according to an all-or-none principle.

Muscles normally operate at or near tetany, a state near or at maximum temporal summation.

4.11   MUSCLE MATTERS

Muscle makes up more than 40 percent of the human body by weight, and most of that is skeletal muscle. Muscle cells adapt to the activity demanded of them. When severe nerve damage or prolonged bed rest prevents a muscle from being used at all, the muscle will rapidly begin to waste away, or *atrophy*. Over time, affected muscles can lose up to three-fourths of their mass, with a corresponding loss of strength. More commonly, the skeletal muscles of a sedentary person stay basically healthy but cannot respond to physical demands in the same way that well-worked muscles can.

Exercise: Making the Most of Your Muscles

One of the best ways to maintain or improve the work capacity of your muscles is to **exercise** them—that is, to increase the level of contractile activity. To increase muscle endurance, nothing beats regular *aerobic exercise*—activities such as walking, biking, organized aerobics classes, jogging, and swimming (Figure 4.23). Aerobic exercise works muscles at a rate at which the body can keep them supplied with oxygen, and it has the following effects on muscle cells:

1. There is an increase in the number and the size of mitochondria, the organelles that make ATP.

2. The number of blood capillaries supplying muscle tissue increases. This increased blood supply brings more oxygen and nutrients to the muscle tissue and removes metabolic wastes more efficiently.

3. There is more of the oxygen-binding pigment myoglobin in muscle tissues.

Together, these changes produce muscles that are more efficient metabolically and can work longer without becoming fatigued. By contrast, *strength training* involves intense, short-duration exercise, such as weight lifting. It affects fast muscle cells, which form more myofibrils and more enzymes of glycolysis. These changes translate into whole muscles that are larger and stronger (Figure 4.24), but such bulging muscles don't have much endurance. They fatigue rapidly.

Starting at about age 30, a person's muscle tension gradually begins to decrease. As a practical matter, this means that, once you enter your third decade of life, you may exercise just as long and intensely as a younger person, but your muscles cannot adapt (change) in response to the same extent. Even so, some adaptation is possible and it is highly beneficial. Aerobic exercise improves your blood circulation and endurance, and even modest strength training slows the loss of muscle tissue that is an inevitable part of the aging process.

Figure 4.23 Swimming—an excellent form of aerobic exercise.

Uses—and Abuses—of Anabolic Steroids

Aerobic exercise may replace fat with muscle, but it does not build significantly larger skeletal muscles. The only way to do that is through dedicated *strength training*, such as weight lifting. Some people, especially competitive athletes, desire significantly larger (and stronger) muscles, and they want them *now*. In spite of legal sanctions and persistent warnings from health professionals, they choose to use anabolic steroids.

Anabolic steroids are synthetic hormones that were developed in the 1930s as therapeutic drugs that could mimic a sex hormone, testosterone. Synthetic testosterone has numerous legitimate medical uses. For example, it is used to help treat impotence in men and some menopausal symptoms in women. Boys who are deficient in growth hormone may receive testosterone as part of hormone-replacement therapy.

Researchers also have looked at the potential use of synthetic testosterone supplements to counteract aging-related losses of lean muscle mass and bone. The results of these preliminary studies have been encouraging, and the National Institutes of Health may fund additional research into possible benefits of synthetic testosterone for preventing or minimizing age-related ailments that arise from changes in sex hormone levels.

As with other potent drugs, only prescription use is legal. The roughly 20 varieties of anabolic steroids stimulate the synthesis of protein molecules, including muscle proteins. Using anabolic steroids while engaged in a weight-training exercise program can lead to rapid gains in lean muscle mass and strength. For this reason, anabolic steroids are popular among weight lifters, football players, and other athletes who specialize in events that call for "brute power" or explosive muscle responses. Users commonly combine daily oral doses with a single hefty injection each month.

A DARKER SIDE OF STEROID USE On the other hand, physicians, researchers, and athletes themselves report a long list of minor and major side effects of steroid use. In men, acne, baldness, shrinking testes, and infertility are the first signs of toxicity. The drugs may be linked to early onset of a cardiovascular disease, atherosclerosis. There is some evidence that consistent use contributes to kidney damage and to various types of cancer—including cancer of the testicles.

In women, anabolic steroids trigger the development of a deep voice, pronounced facial hair, and irregular menstrual periods. A woman's breasts may shrink and her clitoris may become grossly enlarged.

'ROID RAGE Not all steroid users develop such severe physical side effects. In fact, more common are mental difficulties, called *'roid rage* or *bodybuilder's psychosis*. Some steroid-using men experience irritability and increased aggressiveness. Many competitive athletes look upon the added aggressiveness as a plus. Other men, however, experience uncontrollable aggression, delusions, and wildly manic behavior. For example, one steroid-using athlete purposely revved his car up to high speed, and then drove it into a tree.

With all of the suspected dangers associated with anabolic steroids, you might wonder why anyone would place his or her body and future in such jeopardy. Not everyone may be convinced that the drugs do enough damage to outweigh the "edge" they give in competition. What should a competitor do in a world that accords winning athletes wealth and the status of hero while relegating others to the pile of also-rans? What would *you* do?

Figure 4.24 The oversized muscles of a bodybuilder.

SUMMARY

1. Bones are the structural elements of the human skeleton. They function in movement by interacting with skeletal muscles to which they are attached. Bones also function in protection and support of other body parts, mineral storage, and blood cell formation.

2. Bones are organs, and as such they include more than one type of tissue. Bone tissue is a connective tissue having both living and nonliving components. The living cells are osteocytes. In addition to bone tissue, bones incorporate other types of connective tissue, nerve tissue, and epithelium (in the walls of associated blood vessels).

3. A bone develops as precursors of osteocytes, called osteoblasts, secrete collagen fibers and a ground substance of protein and carbohydrate. The secretions eventually surround each osteoblast; the ground substance becomes hardened (mineralized) as calcium salts are deposited within it, and the living cells reside within spaces (lacunae) in the bone tissue.

4. In spongy bone tissue, tiny needlelike struts are fused together in a latticework. In some bones, red marrow, a major site of blood cell formation, fills the spaces between struts. In the long bones of adults, most of the red marrow is replaced by yellow marrow.

5. Dense, compact bone forms the shaft and outer portion of the ends of long bones. This type of bone tissue is organized as thin, concentric layers (osteons) around small canals (canaliculi). The canals are channels for nerves and blood vessels that serve the bone tissue.

6. Bones develop following a cartilage model. In growing humans, the growth of long bones occurs at the bone ends, called epiphyses. Until long bone growth stops in late adolescence or early adulthood, each epiphysis is separated from the bone shaft by an epiphyseal plate of cartilage. The plates are replaced by bone when growth terminates.

7. Bone tissue constantly "turns over" as minerals are deposited and withdrawn from it. Osteoblasts deposit bone and osteoclasts break it down. This process, called "remodeling," is largely controlled by hormones and is a key element in maintaining calcium homeostasis in the body. In mature bones mechanical stresses, such as the pull of muscles during body movements, increase deposits relative to withdrawals. Before adulthood, bones grow by way of remodeling.

8. The human skeleton has 206 bones. It has an axial portion (skull, backbone, ribs, and breastbone) and an appendicular portion (limb bones, pelvic girdle, and pectoral girdles). The axial skeleton forms the body's vertical axis and is a central support structure. The appendicular skeleton provides support for upright posture and interacts with skeletal muscles during most movements. Intervertebral disks are shock pads and flex points in the backbone. Table 4.1 reviews skeletal structure and functions of bones.

9. Together with skeletal muscles, the skeleton works like a system of levers in which rigid rods (bones) move about at fixed points (joints). In a synovial joint, a fluid-filled cavity separates adjoining bones. Such joints are freely movable. In cartilaginous joints, cartilage fills the space between bones and only slight movement is possible. In fibrous joints, there is no cavity, and fibrous connective tissue unites the bones. A limb can be moved and rotated around a joint because of the way pairs or groups of muscles are arranged relative to joints.

10. The body has more than 600 muscles, arranged as pairs or as muscle groups. Smooth, cardiac, and skeletal muscle tissue all contract (shorten) when stimulated. Only skeletal muscle interacts with the skeleton to bring about movement of the body or its hard parts. The origin end of a skeletal muscle is attached to a bone that stays relatively motionless during a movement. The insertion end is attached to the muscle that moves most. Some muscles work together (synergistically) to promote the same movement. Others work antagonistically; the action of one opposes or reverses the action of the other.

11. Skeletal and cardiac muscle cells contain many thread-like myofibrils, which contain actin (thin) and myosin (thick) filaments. The filaments are organized in orderly arrays in sarcomeres, which are the basic units of muscle contraction.

12. Sarcomeres contract when nerve stimulation triggers the release of calcium ions from a membrane system (sarcoplasmic reticulum) in the muscle cell. Calcium binding alters the actin filaments so that the heads of adjacent myosin molecules (embedded in thick filaments) can bind to them. ATP provides the energy to drive the cross-bridge power strokes that cause actin filaments to slide past the myosin filaments and so shorten the sarcomere. This is the sliding-filament model of muscle contraction.

13. A motor neuron and the muscle cells under its control make up a motor unit. The fewer cells in the motor units in a muscle, the more finely the muscle's contractions can be controlled.

14. Neuromuscular junctions are synapses between a motor neuron and muscle cells. Neural stimulation triggers the release of a neurotransmitter (ACh) that sets in motion the biochemical events that lead to contraction. Individual cells in a motor unit always contract in an all-or-none fashion.

Table 4.1 Review of the Skeleton

FUNCTIONS OF BONE:

1. *Movement.* Interact with skeletal muscles to maintain or change the position of body parts.

2. *Support.* Support and anchor muscles.

3. *Protection.* Many bones form hard compartments that enclose and protect soft internal organs.

4. *Mineral storage.* Reservoir for mineral ions, which are deposited or withdrawn and so help maintain ion concentrations in body fluids.

5. *Blood cell formation.* May contain regions where blood cells are produced.

BONES OF THE SKELETON:

Appendicular portion
Pectoral girdles: clavicle and scapula
Arm: humerus, radius, ulna
Wrist and hand: carpals, metacarpals, phalanges (of fingers)
Pelvic girdle (six fused bones at the hip)
Leg: femur (thighbone), patella, tibia, fibula
Ankle and foot: tarsals, metatarsals, phalanges (of toes)

Axial portion
Skull: cranial bones and facial bones
Rib cage: sternum (breastbone) and ribs (12 pairs)
Vertebral column: vertebrae (26)

Review Questions

1. What are some of the functions of bone tissue? *88*

2. Discuss the properties of the three types of muscle tissue. *98*

3. Sketch and label the fine structure of a muscle, down to one of its individual myofibrils. What is the basic unit of contraction in a myofibril? *100*

4. How do actin and myosin interact in a sarcomere to bring about muscle contraction? What roles do ATP and calcium play? *100–101*

5. What is a motor unit? Why does a rapid series of muscle twitches yield a stronger overall contraction than a single twitch? *104–105*

Critical Thinking: You Decide *(Key in Appendix V)*

1. You are training young athletes for the 100-meter dash. They need muscles specialized for speed and strength, *not* endurance. What muscle characteristics would your training regimen aim to develop? How would you alter it to train marathoners?

2. Growth hormone, or GH, is used clinically to spur growth in children who are unusually short because they have a GH deficiency. However, it is useless for a short but otherwise normal 25-year-old to request GH treatment from a physician. Why?

3. If some bleached human bones found lying in the desert were carefully examined, which of the following would not be present? Haversian canals, a marrow cavity, osteocytes, calcium.

Self-Quiz *(Answers in Appendix IV)*

1. _____ and _____ systems work together to move the body and specific body parts.

2. The three types of muscle tissue are _____, _____, and _____.

3. Which of the following serve as shock pads and flex points in the human backbone?
   a. vertebrae   c. lumbar bones
   b. cervical bones   d. intervertebral disks

4. Haversian canals are characteristic of what tissue?
   a. adipose   d. epithelial
   b. bone   e. muscle
   c. cartilage

5. The smallest unit of contraction in skeletal muscle cells is the
_____.
   a. myofibril   c. muscle fiber
   b. sarcomere   d. myosin filament

6. Muscle contraction will not occur _____.
   a. in the absence of calcium ions   c. both a and b
   b. in the absence of ATP   d. neither a nor b

7. Match the following terms concerning muscle structure and function.
   ____ myofibril
   ____ sarcoplasmic reticulum
   ____ sarcomere
   ____ muscle cell
   ____ ATP

   a. contains many myofibrils
   b. the contractile unit of muscle
   c. indirectly provides energy to drive the power stroke that slides actin filaments past myosin filaments
   d. composed of actin and myosin filaments
   e. calcium ion storage site in a muscle

Selected Key Terms

actin *100*	origin *98*
all-or-none principle *105*	osteocyte *88*
appendicular skeleton *90*	osteon *89*
axial skeleton *90*	pectoral girdle *94*
bone *88*	pelvic girdle *95*
brain case *92*	radius *94*
cartilaginous joint *96*	red marrow *89*
clavicle *94*	rib *93*
cross-bridge *101*	sarcomere *100*
epiphyseal plate *89*	sarcoplasmic reticulum *102*
femur *95*	scapula *94*
fibrous joint *96*	sinus *92*
humerus *94*	skeletal muscle *98*
insertion *98*	skull *92*
intervertebral disk *93*	sliding-filament model *100–101*
ligament *90*	sternum *93*
mandible *92*	synovial joint *96*
motor unit *104*	temporal summation *105*
muscle twitch *105*	tendon *90*
myofibril *100*	tetany *105*
myosin *100*	ulna *94*
neuromuscular junction *103*	yellow marrow *89*

Readings

Hoberman, J. M., and C. E. Yesalis. February 1995. "The History of Synthetic Testosterone." *Scientific American.*

Kearney, J. T. June 1996. "Training the Olympic Athlete." *Scientific American.* This article by a renowned sports physiologist describes the specialized training demands for elite athletes in several sports.

Sherwood, L. 1997. *Human Physiology.* Belmont, California: Wadsworth.

 For additional readings, go to InfoTrac College Edition, your online library at:

http://www.infotrac-college.com/wadsworth

5 DIGESTION AND NUTRITION

Tales of a Tube

Surrounded by snow and ice, an Aleut hunter slices slabs of raw whale blubber for dinner. In a village in Nepal, a grandmother eats only a bowl of rice. Ten thousand miles away, an American college student downs a bacon cheeseburger, a salad, and coffee (Figure 5.1). People in various cultures feast on snake meat, insects, hot chilies, salted fish, and cassava tubers.

Our digestive systems receive whatever food we choose to swallow and break it down—*digest* it—into nutrient molecules that the body can absorb. These molecules include simple sugars, fatty acids, and amino acids—the subunits of the various carbohydrates, lipids, and proteins contained in the food we eat. Food also supplies essential vitamins and minerals.

The body uses nutrients for energy and to build, repair, and maintain its tissues. In making nutrients available, the digestive system thus plays a key role in survival. The system consists of a food-processing tube, the gastrointestinal tract, and some accessory organs.

Figure 5.1 Taking in food, which will be digested into nutrients the body requires to survive.

Many Americans have poor eating habits. They may eat large quantities of snack foods, skip meals, or eat too much and too fast. Worse yet, the typical American diet tends to be high in fat, cholesterol, and salt. The world distribution of some illnesses suggests that the body may pay a high price for certain dietary habits. For example, in the United States a fat-laden diet is a major contributor to health-threatening obesity and heart disease. Obesity also increases an adult's risk of developing diabetes and contributes to other health problems. In regions where salt-cured and smoke-cured foods are staples, the rates of stomach and esophageal cancer are disturbingly high. Nongenetic colon cancer is much more prevalent among people whose diets lack sufficient fiber, as is common in the United States. Most people of rural Africa and India cannot afford to eat much more than whole grains, which are high in fiber content, and they rarely suffer colon cancer.

In this chapter we'll look at how the components of the digestive system carry out their life-supporting functions and at the roles various nutrients play in human health. As you read, keep in mind the digestive system's overall role in homeostasis (Figure 5.2). When the digestive system has made nutrients available, other essential activities—including circulation, respiration, and excretion—come into play to help maintain your body in the living state.

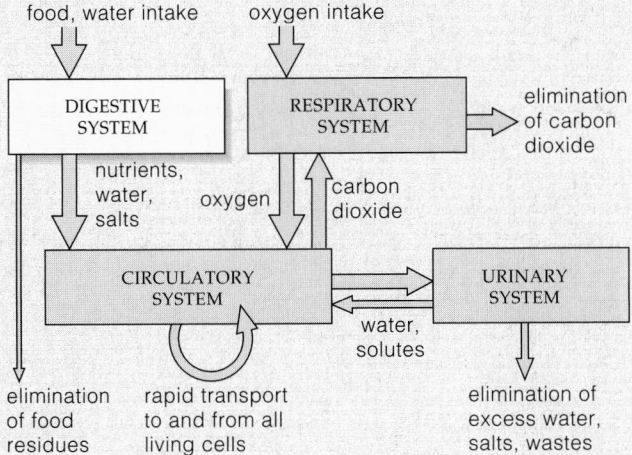

Figure 5.2 Functional links between the digestive system and other organ systems that work together to supply cells with raw materials and eliminate wastes.

KEY CONCEPTS

1. Digestion and absorption of food are an essential aspect of homeostasis. Interactions among the digestive, circulatory, respiratory, and urinary systems supply the body's cells with raw materials, dispose of wastes, and maintain the volume and composition of extracellular fluid.

2. The digestive system has specialized parts for food transport, processing, and storage. These include the mouth, stomach, and small and large intestines. As food moves through these parts, it is mechanically broken apart and chemically broken down, nutrients are absorbed, and unabsorbed residues are eliminated. Accessory organs produce enzymes and other substances that aid in the digestive process.

3. To maintain an acceptable body weight and overall health, energy intake must balance energy output (by way of metabolic activity and physical exertion). Complex carbohydrates provide most of the glucose, which typically is the body's main source of immediately usable energy.

4. Nutrition also involves the intake of vitamins, minerals, and certain amino acids and fatty acids that the body itself cannot produce.

CHAPTER AT A GLANCE

5.1 Overview of the Digestive System

5.2 Chewing and Swallowing

5.3 Digestion in the Stomach and Small Intestine

5.4 Absorption in the Small Intestine

5.5 Liver Functions in Digestion

5.6 The Large Intestine

5.7 Digestion Controls and Nutrient Turnover

5.8 Nutrition

5.9 Vitamins and Minerals

5.10 Food Energy and Body Weight

5.11 *Malnutrition and Undernutrition*

5.1 OVERVIEW OF THE DIGESTIVE SYSTEM

The **digestive system** functions to mechanically break apart and chemically break down food into nutrient molecules, which can then be absorbed into the bloodstream and so enter the internal environment. The system is a tube with many specialized regions. It extends from the mouth to the anus and is also called the **gastrointestinal (GI) tract**. The space inside the tube is the GI tract *lumen*. Stretched out, the gastrointestinal tract would be 6.5 to 9 meters (21 to 30 feet) long in an adult. Its main regions are the mouth (oral cavity), pharynx, esophagus, stomach, small intestine, and large intestine. The large intestine terminates in the rectum, anal canal, and anus. Mucous epithelium lines the entire surface exposed to the lumen. Figure 5.3 shows the digestive system of an adult and summarizes the functions of its component parts.

Figure 5.3 Major components of the human digestive system and their functions. Organs with accessory roles in digestion are also labeled.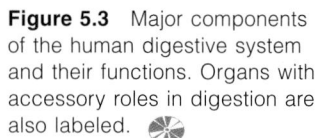

Major Components:

MOUTH (ORAL CAVITY)

Entrance to system; food is moistened and chewed; polysaccharide digestion starts.

PHARYNX

Entrance to tubular part of system (and to respiratory system); moves food forward by contracting sequentially.

ESOPHAGUS

Muscular, saliva-moistened tube that moves food from pharynx to stomach.

STOMACH

Muscular sac; stretches to store food taken in. Gastric fluid mixes with food and kills many pathogens; protein digestion starts.

SMALL INTESTINE

First part (duodenum, C-shaped, about 10 inches long) receives secretions from liver, gallbladder, and pancreas.

In second part (jejunum, about 3 feet long), most nutrients are digested and absorbed.

Third part (ileum, 6–7 feet long) absorbs some nutrients; delivers unabsorbed material to large intestine.

LARGE INTESTINE (COLON)

Concentrates and stores undigested matter by absorbing mineral ions, water; about 5 feet long: divided into ascending, transverse, and descending portions.

RECTUM

Distension stimulates expulsion of feces.

ANUS

End of system; terminal opening through which feces are expelled.

Accessory Organs:

SALIVARY GLANDS

Glands (three main pairs, many minor ones) that secrete saliva, a fluid with polysaccharide-digesting enzymes, buffers, and mucus (which moistens and lubricates food).

LIVER

Secretes bile (for emulsifying fat); roles in carbohydrate, fat, and protein metabolism.

GALLBLADDER

Stores and concentrates bile that the liver secretes.

PANCREAS

Secretes enzymes that break down all major food molecules; secretes buffers against HCl from the stomach.

The various parts of the digestive system work in coordination and carry out these overall tasks:

1. **Mechanical processing and motility**. Movements that break up, mix, and propel food material.

2. **Secretion.** Release of digestive enzymes and other substances into the digestive tube lumen.

3. **Digestion**. Chemical breakdown of food into particles, then into nutrient molecules small enough to be absorbed.

4. **Absorption.** Passage of digested nutrients and fluid across the tube wall and into blood or lymph.

5. **Elimination.** Expulsion of undigested and unabsorbed residues from the end of the GI tract.

Various accessory organs secrete enzymes and other substances that have essential roles in different aspects of digestion and absorption. These "organs" include glands in the wall of the GI tract, the salivary glands, and the liver, gallbladder, and pancreas.

Structure of the Digestive Tube

From the esophagus onward, the wall of the digestive tube consists of four basic layers (Figure 5.4). The *mucosa* (the innermost layer of epithelium) faces the lumen, through which food passes. The mucosa is surrounded by the *submucosa,* a connective tissue layer with blood and lymph vessels and local networks of nerve cells. The next layer is *smooth muscle*—usually two sublayers, one in a circular orientation and the other oriented lengthwise. An outer layer of connective tissue, the *serosa,* is almost as thin as kitchen plastic wrap. Circular arrays of smooth muscle in sections of the gastrointestinal tract are *sphincters.* Contractions of the sphincter muscles can close off a passageway. In the stomach, they help control the forward movement of food and prevent backflow.

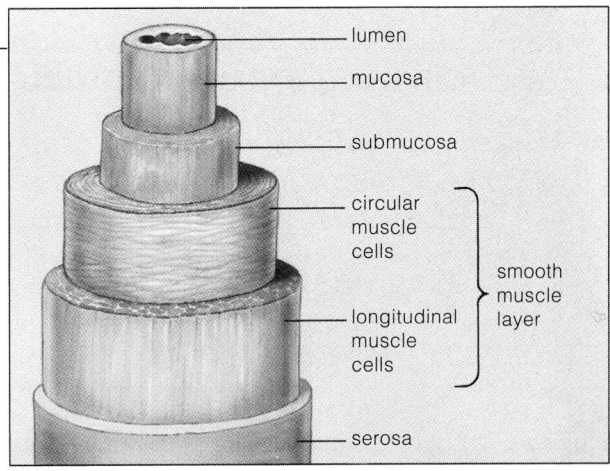

Figure 5.4 The four-layered wall of the gastrointestinal tract. The layers are not drawn to scale. Compare Figure 5.5a.

Motility

The muscle layers of the digestive tube mix the tube's contents and propel them from one region to the next by wavelike contractions called **peristalsis** (Figure 5.5a). In peristalsis, rings of circular smooth muscle contract behind food and relax in front of it. The food distends the tube wall, peristaltic movement forces the food onward and expands the next wall region, and so on. In the intestines, **segmentation** also occurs (Figure 5.5b). Rings of smooth muscle in the wall repeatedly contract and relax, creating an oscillating (back-and-forth) movement. This movement constantly mixes the contents of the lumen and forces the material against the wall's absorptive surface.

The digestive tube extends from the mouth to the anus. For most of the length the tube wall consists of four layers, including smooth muscle.

Muscular movements move swallowed food from one tube region to the next, mix the lumen contents, and increase their contact with the tube's absorptive surface.

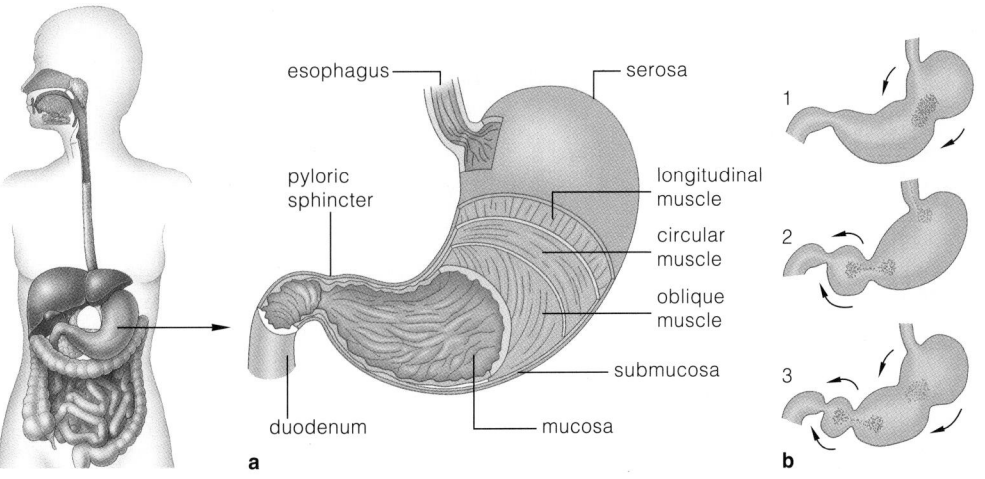

Figure 5.5 (**a**) Peristaltic wave down the stomach, produced by alternating contraction and relaxation of muscles in the stomach wall. (**b**) Segmentation, or oscillating movement, in the intestines.

Mouth, Teeth, and Salivary Glands

In the **oral cavity** or mouth, food begins to be broken apart by chewing, and enzymes begin the chemical digestion of polysaccharides (starch). Adults normally have 32 teeth (Figure 5.6a). Young children have just 20 "primary teeth." Figure 5.5b shows a tooth's main regions, the *crown* and the *root*. The crown is coated with hardened calcium deposits, the tooth enamel, which is the hardest substance in the body. It covers a thick bonelike layer of living material called dentin. Dentin and an inner pulp extend into the root. The pulp cavity contains nerves and blood vessels.

Teeth are engineering marvels, able to withstand years of chemical insults and mechanical stress. The chisel-shaped incisors bite off chunks of food, the cone-shaped cuspids (canines) tear it, and the flat-topped molars and premolars grind it. Bacteria that cause tooth decay *(caries)* abound in the mouth, living on food residues. Such microbes thrive on the sugar sucrose, among other carbohydrates. Daily flossing and gentle brushing are essential to maintaining healthy teeth and gums. Bacterial infection in the gums can lead to an inflammation called *gingivitis*, which can spread to the *periodontal membrane* that helps anchor a tooth in the jaw. In the ensuing *periodontal disease*, bacteria gradually destroy the bone around a tooth, the tooth loosens, and other complications can arise.

Chewing mixes food with a fluid called saliva, which is secreted from several strategically located **salivary glands** (Figure 5.6c). A large *parotid gland* nestles just in front of each ear. *Submandibular glands* lie just below the lower jaw in the floor of the mouth, and *sublingual glands* are under the tongue. Saliva flows through ducts that lead from the glands to the free surface of the mouth's lining.

Although saliva is mostly water, it includes a starch-degrading enzyme, called **salivary amylase**, a buffer (bicarbonate or HCO_3^-), and mucins. The buffering action of bicarbonate keeps the pH of your mouth between 6.5 and 7.5, a range within which salivary amylase can function, even when you eat acidic foods. Mucins are modified proteins that help bind bits of food into a softened, lubricated ball called a **bolus**. Once food is swallowed, starch digestion continues in the stomach until acidic secretions penetrate the bolus and inactivate the salivary amylase.

The roof of the mouth, a bone-reinforced section of the **palate**, serves as a hard surface against which the tongue can press food as it mixes it with saliva. Contractions of the tongue muscle force the food bolus into the **pharynx** (throat). As Figure 5.7 shows, this passageway connects with both the windpipe, or *trachea*, which leads to the lungs, and the **esophagus**, a muscular tube that leads to the stomach. The membrane lining the pharynx and esophagus secretes mucus, which lubricates the food and aids in moving the bolus to the stomach.

molars
(12)

premolars
(8)

canines (4)

incisors
(8)

lower jaw

upper jaw

a

enamel

dentin

pulp cavity
(contains
nerves and
blood vessels)

root canal

periodontal
membrane

bone

crown

gingiva
(gum)

root

b

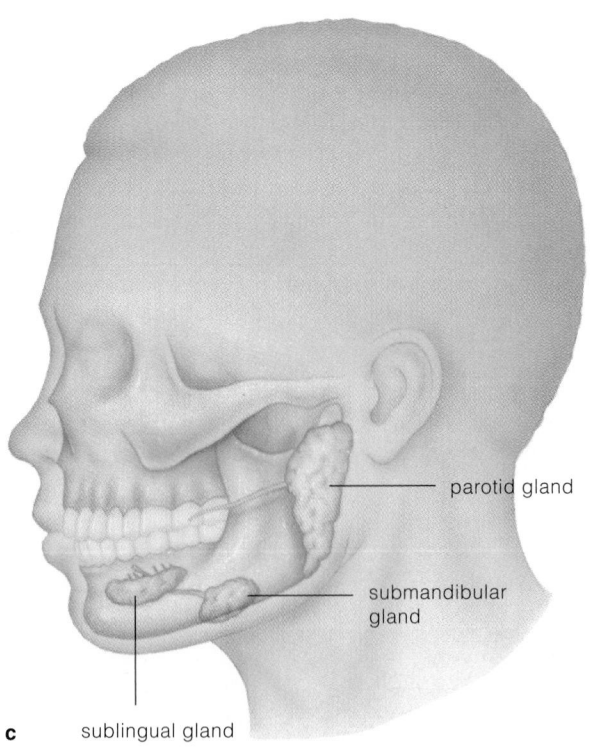

parotid gland

submandibular
gland

c   sublingual gland

Figure 5.6 (**a**) Locations of the different types of teeth. (**b**) Anatomy of a human tooth. (**c**) Locations of the salivary glands.

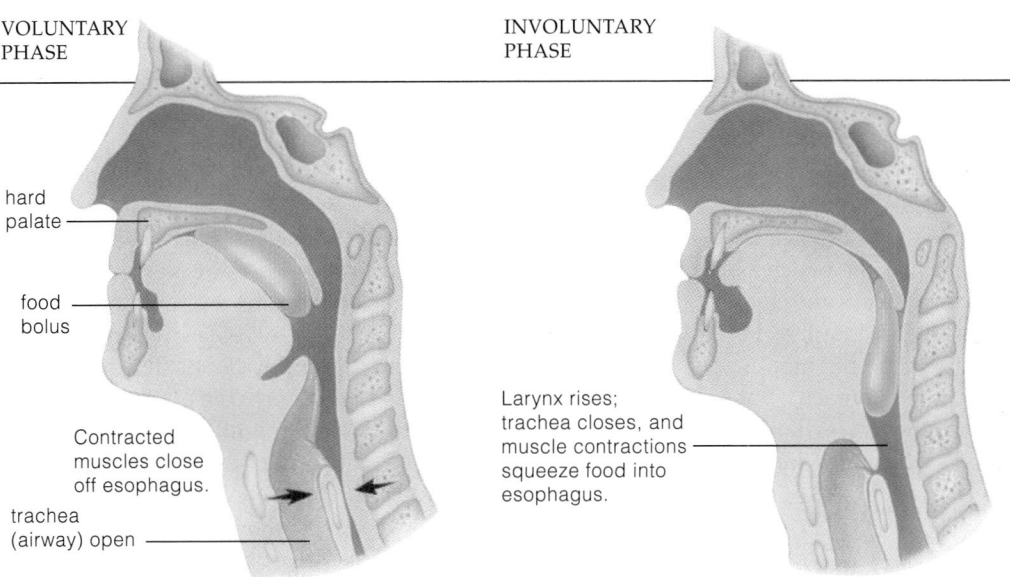

VOLUNTARY PHASE

INVOLUNTARY PHASE

hard palate

food bolus

Contracted muscles close off esophagus.

trachea (airway) open

a

Figure 5.7 Swallowing and peristalsis. (**a**) Contractions of the tongue push the food bolus into the pharynx. Next, the vocal cords seal off the larynx and the epiglottis bends downward, helping to keep the trachea closed. Contractions of throat muscles then squeeze the food bolus into the esophagus. (**b,c**) Finally, peristalsis in the esophagus moves the bolus through a sphincter, and food enters the stomach.

Larynx rises; trachea closes, and muscle contractions squeeze food into esophagus.

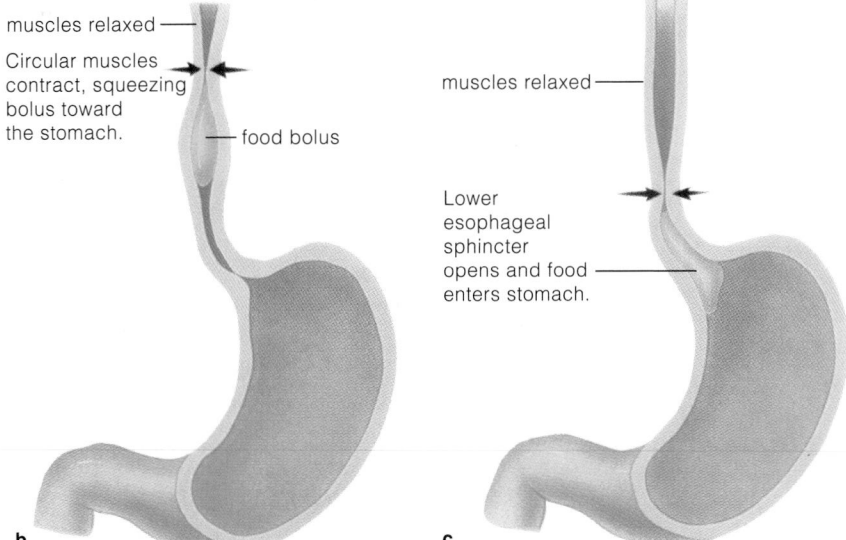

muscles relaxed

Circular muscles contract, squeezing bolus toward the stomach.

food bolus

b

muscles relaxed

Lower esophageal sphincter opens and food enters stomach.

c

Swallowing

Swallowing a mouthful of food might seem straightforward, but it actually involves an elaborate sequence of events. Swallowing begins when voluntary movements push a bolus into the pharynx, stimulating sensory receptors in the pharynx wall. The receptors trigger the "swallowing reflex," in which involuntary muscle contractions prevent food from moving up into the nose and help keep it from entering the trachea. First, the vocal cords are stretched tightly across the entrance to the larynx ("voicebox"). Then, a flaplike valve, the *epiglottis*, is pressed down over the vocal cords as a secondary seal. For a moment, breathing is prevented as food moves into the esophagus—hence, you normally don't choke when you swallow. When swallowed food reaches the lower end of the esophagus (Figure 5.7b), it enters the stomach through a sphincter (Figure 5.7c).

When swallowed food enters the trachea instead of the esophagus, a person can choke to death in as little as four minutes. The Heimlich maneuver (Figure 5.8), which is *an emergency procedure only*, can dislodge food from the trachea by forcibly elevating the diaphragm muscle, which separates the chest cavity and abdominal cavity. Upward movement of the diaphragm reduces the volume of the chest cavity, forcing air up the trachea—sometimes with enough force to eject an obstruction.

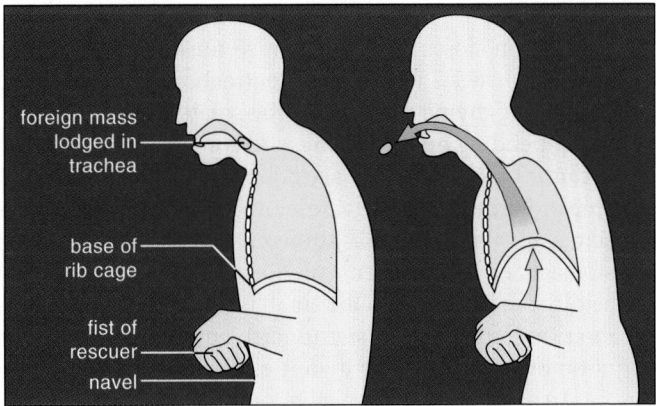

foreign mass lodged in trachea

base of rib cage

fist of rescuer

navel

Figure 5.8 The Heimlich maneuver. To perform the maneuver, stand behind the victim, make a fist, then position the fist, thumb side in, against the victim's abdomen. The fist must be slightly above the navel and well below the rib cage. Next, press the fist into the abdomen with a sudden upward thrust. Repeat the thrust several times if needed. Once the blockage is dislodged, a physician must examine the person at once.

The teeth and tongue begin the mechanical breakup of food. Enzymes in saliva begin the chemical digestion of food.

Swallowing has both voluntary and involuntary phases. Peristalsis moves a food bolus down the esophagus toward the lower esophageal sphincter, the gateway to the stomach.

The stomach and small intestine are the premier food-processing organs. Digestive enzymes and other secretions enter the lumen of both (Table 5.1). Together with secretions of accessory organs such as the pancreas and liver, these substances chemically break nutrients into fragments, then into molecules small enough to be absorbed. Carbohydrate breakdown *starts* in the mouth, and protein breakdown *starts* in the stomach. But digestion of nearly all of the carbohydrates, lipids, proteins, and nucleic acids in food is *completed* in the small intestine.

The Stomach

The **stomach** is a muscular, expandable sac (Figure 5.9). In an adult it can hold up to several liters of food. When it is empty, its walls crumple into folds called *rugae.* The stomach has three main functions. First, it stores and mixes food. Second, its secretions help dissolve food and begin chemical breakdown of food particles. Third, it helps control passage of food into the small intestine.

Glandular epithelium lines the stomach wall facing the lumen. Each day, glandular cells in the lining secrete about two liters of substances, including hydrochloric acid (HCl), mucus, and pepsinogens—precursors of digestive enzymes called **pepsins**. Along with water, these make up the stomach's acidic **gastric juice**. Stomach acidity helps dissolve bits of food to form a liquid mixture called **chyme**. The acidity also kills most microbes in food and causes "heartburn" when gastric juice backs up into the esophagus. Gland cells in the stomach lining secrete *intrinsic factor*, a protein required for absorption of vitamin B_{12} in the small intestine.

When you see, smell, or taste food, your brain signals secretory cells in the salivary glands and stomach lining to step up their activity. As food enters the stomach, it distends the wall and activates sensory receptors in the lining. This also leads to increased secretory activity. Secretory cells also are stimulated directly by substances such as partially digested proteins and caffeine.

The high acidity produced by HCl secretion denatures proteins and exposes the peptide bonds. It also converts inactive pepsinogens to active pepsins, which break the bonds. Accumulating protein fragments trigger secretion of *gastrin*, a hormone that stimulates even more secretion of pepsinogen and HCl.

Why don't HCl and pepsin break down the stomach lining? Control mechanisms usually assure that enough mucus and bicarbonate are secreted to protect the lining. However, these protective mechanisms, collectively called the "gastric mucosal barrier," can be disrupted. Consider infection by a bacterium (*Helicobacter pylori*). *H. pylori* secretes toxins that cause inflammation, a defensive counterattack (Chapter 7) that, among other effects, breaks down tight junctions between cells of the stomach lining that normally prevent HCl from passing between cells. Now, hydrogen ions and pepsins diffuse into the lining, triggering a sequence of events that further damage tissues. The resulting open sore, a *peptic ulcer,* may bleed, leading to anemia (from blood loss) or secondary infections. Most people who have intestinal ulcers or stomach ulcers also have an *H. pylori* infection, but not everyone who is infected develops peptic ulcers. Some people may be genetically predisposed to the disorder. Chronic stress, smoking, and excessive use of aspirin and alcohol also can be contributing factors.

The stomach empties by waves of peristalsis. The waves mix chyme and build up force as they approach the powerful pyloric sphincter between the stomach and the opening of the small intestine. The arrival of a strong contraction closes the sphincter, so most of the chyme is squeezed back. With each contraction, only a small amount of chyme moves into the small intestine.

Depending on the volume and composition of chyme, it can take from two to six hours for a full stomach to empty. Large meals activate more receptors in the stomach wall; these call for stronger contractions, and the stomach empties faster. On the other hand, increased acidity or fat content in the small intestine stimulates hormone secretions that *slow* stomach emptying—so food is not moved along faster than it can be processed.

Alcohol and a few other substances begin to be absorbed across the stomach wall. Liquids imbibed on an empty stomach pass rapidly from the stomach to the small intestine, where further absorption occurs, but the rate of gastric emptying slows when food is also in the stomach. That is why a person feels the effects of alcohol less quickly when drinking accompanies a meal.

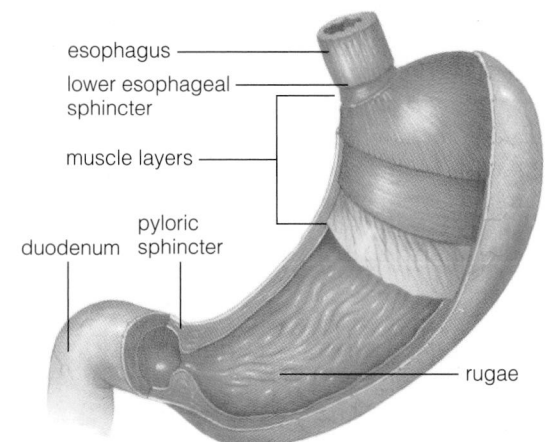

esophagus
lower esophageal sphincter
muscle layers
pyloric sphincter
duodenum
rugae

Figure 5.9 Cutaway view of the stomach.

Table 5.1   Major Digestive Enzymes and Their Breakdown Products

Enzyme	Source	Where Active	Substrate	Main Breakdown Products
CARBOHYDRATE DIGESTION:				
Salivary amylase	Salivary glands	Mouth, stomach	Polysaccharides	Disaccharides, oligosaccharides
Pancreatic amylase	Pancreas	Small intestine	Polysaccharides	Disaccharides, monosaccharides
Disaccharidases	Intestinal lining	Small intestine	Disaccharides	MONOSACCHARIDES* (e.g., glucose)
PROTEIN DIGESTION:				
Pepsins	Stomach lining	Stomach	Proteins	Protein fragments
Trypsin and chymotrypsin	Pancreas	Small intestine	Proteins	Protein fragments
Carboxypeptidase	Pancreas	Small intestine	Peptides	AMINO ACIDS*
Aminopeptidase	Intestinal lining	Small intestine	Peptides	AMINO ACIDS*
FAT DIGESTION:				
Lipase	Pancreas	Small intestine	Triglycerides	FREE FATTY ACIDS, MONOGLYCERIDES*
NUCLEIC ACID DIGESTION:				
Pancreatic nucleases	Pancreas	Small intestine	DNA, RNA	NUCLEOTIDES*
Intestinal nucleases	Intestinal lining	Small intestine	Nucleotides	NUCLEOTIDE BASES, MONOSACCHARIDES*

*Breakdown products small enough to be absorbed into the internal environment.

The Small Intestine

Your small intestine is about an inch and a half in diameter and 6 meters long—some 20 feet! Its three sections are the duodenum, the jejunum, and the ileum. The following sketch diagrams the small intestine wall:

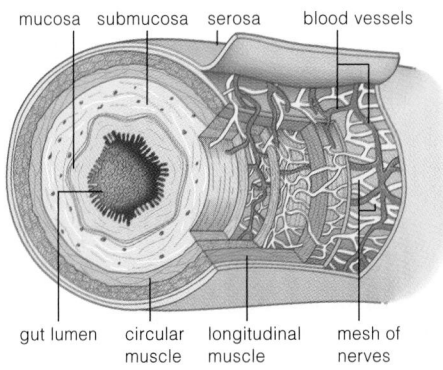

mucosa   submucosa   serosa   blood vessels

gut lumen   circular muscle   longitudinal muscle   mesh of nerves

Digestion in the small intestine depends on substances secreted by three accessory organs: the pancreas, the liver, and the gallbladder. Chyme entering the duodenum triggers hormonal signals that stimulate a brief flood of digestive enzymes from the **pancreas**, an elongated organ just above and behind the opening to the small intestine. As part of "pancreatic juice," these enzymes act on carbohydrates, fats, proteins, and nucleic acids. For example, like pepsin in the stomach, the pancreatic enzymes trypsin and chymotrypsin digest the polypeptide chains of proteins into peptide fragments. The fragments are then broken down to amino acids by *carboxypeptidase* from the pancreas and by *aminopeptidase* (present on the surface of the intestinal mucosa). The pancreas also secretes bicarbonate, which buffers HCl from the stomach and so maintains a chemical environment in which pancreatic

enzymes can function. Two pancreatic hormones, insulin and glucagon, are released into the bloodstream. They do not function in digestion but do have roles in nutrition.

Fat digestion also depends on *bile*, which is secreted continually by the liver. Bile contains bile salts, bile pigments, cholesterol, and lecithin, which is a phospholipid. When food is not moving through the GI tract, a sphincter closes off the main bile duct and bile backs up into the saclike **gallbladder**. As food is digested, the gallbladder contracts and empties bile into the small intestine.

By a process called emulsification, bile salts speed up fat digestion. Most fats in the human diet are triglycerides, which are insoluble in water and tend to clump into fat globules in the chyme. When intestinal wall movements mix chyme, fat globules break up into droplets that become coated with bile salts. Bile salts carry negative charges, so the coated droplets repel each other and form an emulsion—that is, they stay separated. Emulsion droplets give fat-digesting enzymes a much greater surface area upon which to act, so triglycerides can be broken down more rapidly to fatty acids and monoglycerides.

Liver cells use cholesterol to synthesize bile salts, and they also secrete cholesterol into bile. This is how the body excretes cholesterol. When there is chronically more cholesterol than available bile salts can dissolve, the excess may precipitate. *Gallstones*, which are mostly hard clumps of cholesterol, can develop in the gallbladder and cause severe pain if they become lodged in bile ducts.

Carbohydrate breakdown starts in the mouth, and protein breakdown starts in the stomach.

In the small intestine, most large organic molecules are digested to molecules small enough to be absorbed into the internal environment.

ABSORPTION IN THE SMALL INTESTINE

Each day about 9 liters of fluid from the stomach, liver, pancreas, and intestinal glandular cells enter your small intestine, and 95 percent of that fluid—and nutrients it contains—is absorbed across the intestinal lining. Unlike the stomach, which can only absorb alcohol and a few other substances, the small intestine is where the vast majority of nutrients are absorbed. Why is this portion of the GI tract so adept at absorption?

The answer lies with the structure of the wall of the small intestine (Figure 5.10). The wall lining is densely folded into structures called **villi** (singular: villus), which are tiny, fingerlike extensions of the mucosa. Although each villus is only about a millimeter long, the mucosa has millions of them. Their very density gives the mucosa a velvety appearance. Inside each villus are small blood vessels (an arteriole and a venule) and a lymph vessel, all of which function in moving substances to and from the general circulation (Figure 5.10e).

The cells making up the epithelial covering of each villus bear a brushlike crown of microvilli (singular: microvillus). Each **microvillus** is an ultrafine, threadlike projection of the epithelial cell's plasma membrane. The microvilli are exposed to the intestinal lumen, and collectively they are termed the intestinal "brush border." Those near the tip of the villus produce some digestive enzymes, as listed in Table 5.1. Microvilli tremendously increase the intestinal surface area that is exposed to chyme—and that is available for absorption.

Mechanisms of Absorption

As you read in Section 5.1, absorption is the passage of nutrients, water, salts, and vitamins into the internal environment. The vast absorptive surface area of the intestinal wall greatly enhances the process. And so does segmentation, the action of smooth muscle in the intestinal wall. Remember from Section 5.1 that segmentation helps mix the contents of the lumen and force it against the wall's lush surface "carpet" of villi and microvilli.

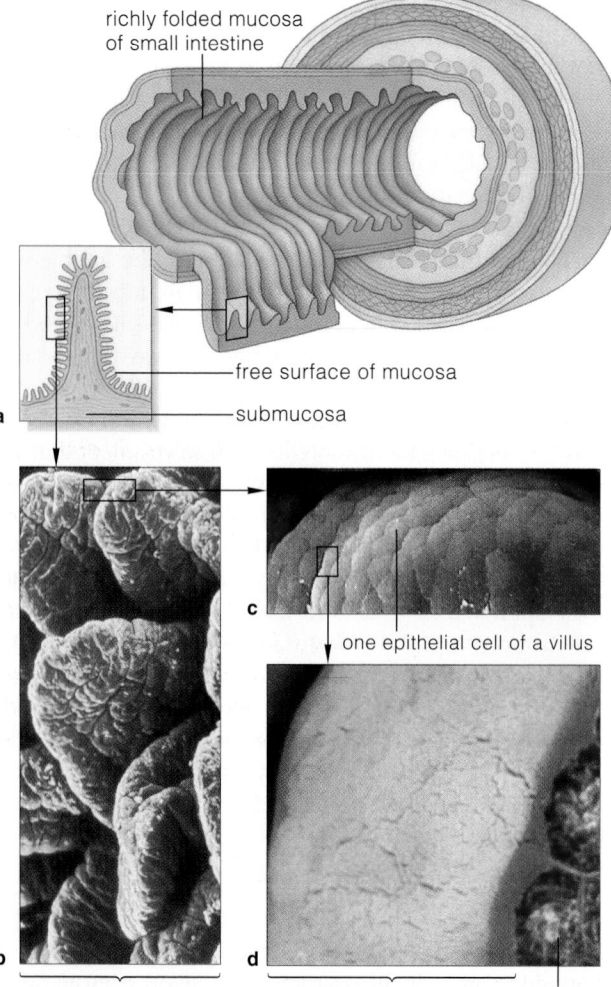

richly folded mucosa of small intestine

free surface of mucosa

submucosa

a

one epithelial cell of a villus

b   profusion of villi at free surface of the intestinal mucosa

c   d   profusion of microvilli at the surface of one epithelial cell

part of the cytoplasm

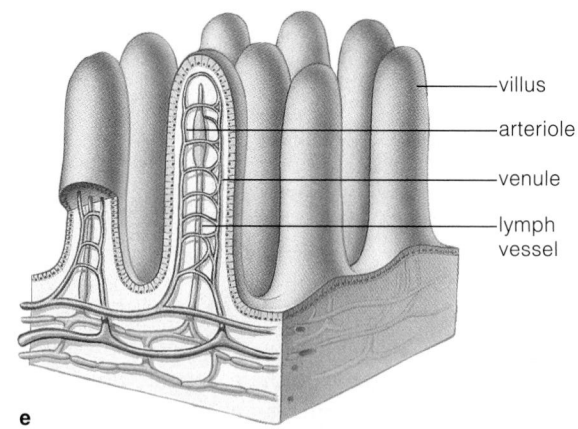

villus

arteriole

venule

lymph vessel

e

Figure 5.10 Surface structures of the small intestine at increasing magnifications.

(**a,b**) Surface view of the deep, permanent folds of the intestinal mucosa. Each fold bears a profusion of fingerlike villi, which are absorptive structures. (**c,d**) Individual epithelial cells at the free surface of a villus. Each has a dense crown of microvilli facing the intestinal lumen. (**e**) Blood and lymph vessels in intestinal villi. Monosaccharides and most amino acids that cross the intestinal lining enter blood vessels. Fats enter lymph vessels.

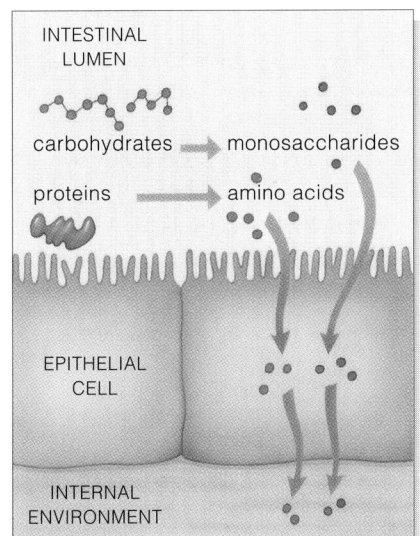

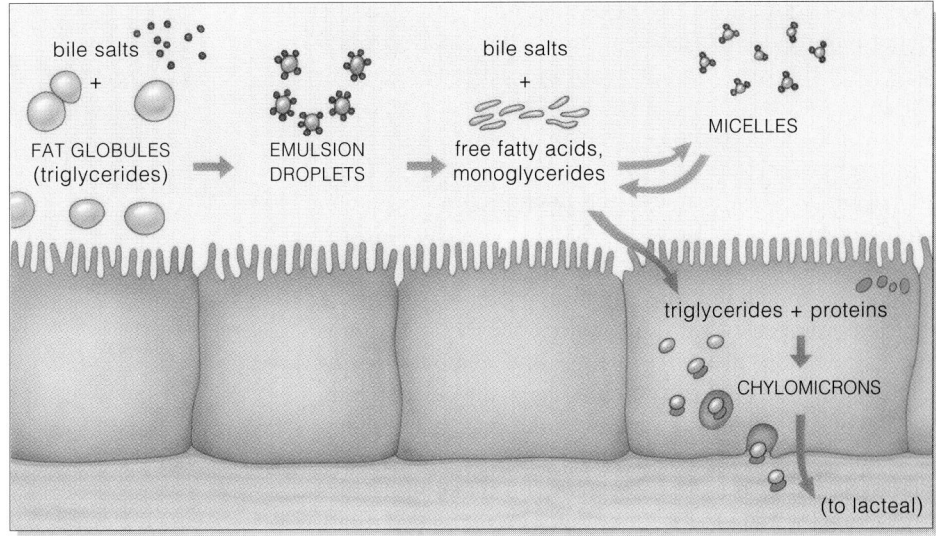

1 Digestion of carbohydrates to monosaccharides, and proteins to amino acids, by pancreatic enzymes.

2 Active transport of monosaccharides and amino acids across plasma membrane of cells; then nutrients move by facilitated diffusion into blood.

1 Emulsification. Wall movements break up fat globules into small droplets. Bile salts keep globules from re-forming. Pancreatic enzymes digest droplets to fatty acids and monoglycerides.

2 Formation of micelles as bile salts combine with digested products and phospholipids. Products readily slip into and out of the micelles.

3 Concentration of monosaccharides and fatty acids in micelles enhances gradients that lead to their diffusion across lipid bilayer of cells' plasma membranes.

4 Reassembly of products into triglycerides inside cells. These join with proteins, then are expelled (by exocytosis) into the internal environment.

Figure 5.11 Summary of digestion and absorption processes in the small intestine.

By the time food is halfway through your small intestine, most of it has been broken down and absorbed. Plasma membrane proteins of the brush border actively transport monosaccharides and amino acids directly into the epithelial cells of villi. Some water also enters by osmosis. The plasma membrane proteins also actively transport mineral ions across the intestinal lining. These nutrients move through the cells, on through extracellular fluid, and into the small blood vessels inside the villus.

Bile salts assist with the absorption of fatty acids and monoglycerides that result when lipids are digested. Such molecules have hydrophobic regions and so they do not dissolve in watery chyme. In fact, left alone they would float like a film of oil on a puddle and not come near the surface of absorptive cells. Instead, however, the molecules clump together with bile salts, lecithin, cholesterol, and other substances, and form tiny droplets called *micelles*.

If concentration gradients are favorable when a micelle comes into close contact with the absorptive cell surface, nutrient molecules diffuse out of it, into the solution bathing the microvilli, and into epithelial cells. Inside an epithelial cell, fatty acids and monoglycerides recombine into triglycerides. Then the triglycerides combine with proteins into particles that leave the cells by exocytosis and enter the internal environment. Once absorbed in a villus, the particles (called *chylomicrons*) enter a lymph vessel called a **lacteal**. Lacteals drain into other lymphatic vessels, which drain into blood vessels. Absorbed glucose, amino acids, and other water-soluble molecules also enter blood vessels. Figure 5.11 summarizes the different absorption routes.

With its richly folded intestinal mucosa, millions of villi, and hundreds of millions of microvilli, the small intestine offers a vast surface area for absorbing nutrients.

Substances pass through the brush border cells that line the free surface of each villus by active transport, osmosis, and diffusion across the lipid bilayer of plasma membranes.

LIVER FUNCTIONS IN DIGESTION

Nutrient-laden blood in intestinal villi flows to the **hepatic portal vein**. This vein carries the blood through vessels in the **liver**, one of the body's largest organs. In the liver, excess glucose is taken up before a *hepatic vein* returns the blood to the general circulation (Figure 5.12). The liver converts much of this glucose to a storage form, glycogen. As already noted, the liver's role in digestion is to secrete bile—as much as 1,500 ml every day—which is stored in the gallbladder and released when chyme enters the small intestine. It also processes incoming nutrient molecules into substances the body requires (such as blood plasma proteins) and removes toxins ingested in food or already circulating in the bloodstream.

Beyond its digestive functions, the liver helps maintain the concentrations of blood's organic substances (Table 5.2). It also inactivates many hormone molecules and sends them on to the kidneys for excretion (in urine). Ammonia (NH_3) is produced when cells break down amino acids, and it can be toxic to cells. The circulatory system carries ammonia to the liver, where it is converted to a much less toxic waste product, urea, which also is excreted in urine.

Cells of the liver can be injured by viruses that cause *hepatitis*. Without treatment, hepatitis can permanently harm liver function. Type A hepatitis is transmitted via contaminated food or water. Symptoms include nausea, abdominal discomfort, loss of appetite, and jaundice (yellowing of the skin). Type B hepatitis can be transmitted sexually (Chapter 15). It can cause serious liver

Table 5.2	Some Activities That Depend on Liver Functioning

1. Carbohydrate metabolism

2. Partial control of synthesis of proteins in blood; assembly and disassembly of certain other proteins

3. Urea formation from nitrogen-containing wastes

4. Assembly and storage of some fats; fat digestion (bile is formed by the liver)

5. Inactivation of many chemicals (such as hormones and some drugs)

6. Detoxification of many poisons

7. Degradation of worn-out red blood cells

8. Immune response (removal of some foreign particles)

9. Red blood cell formation (liver absorbs, stores factors needed for red blood cell maturation)

scarring, called *cirrhosis*. Long-term, heavy alcohol use is another common cause of cirrhosis. Liver transplantation is becoming more successful, but people who suffer from advanced cirrhosis still often die of liver failure.

Simple sugars and amino acids directly enter the epithelial cells of villi.

Both bile produced by the liver and pancreatic enzymes help in the digestion of fats.

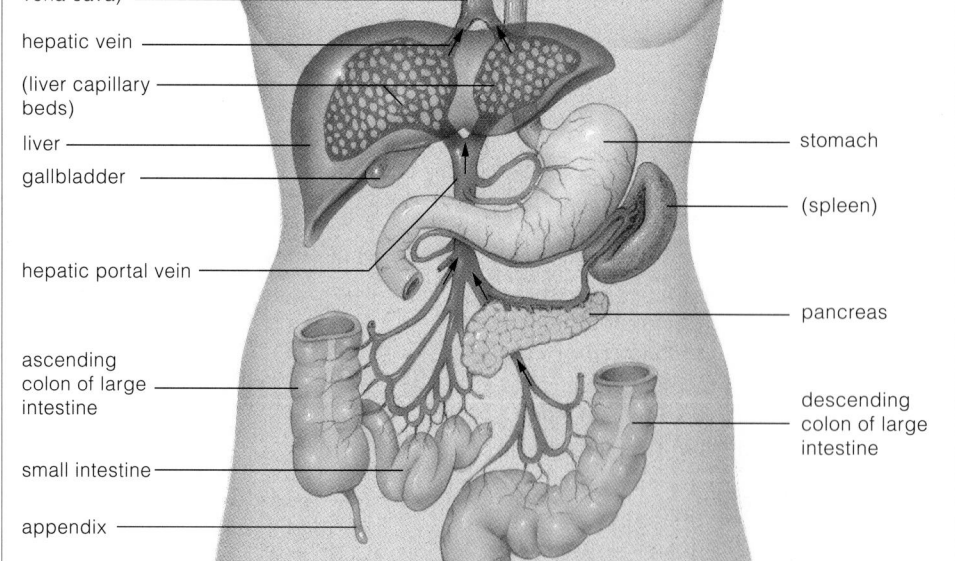

Figure 5.12 Hepatic portal system. Arrows show the direction of blood flow.

(inferior vena cava)

hepatic vein

(liver capillary beds)

liver

gallbladder

hepatic portal vein

ascending colon of large intestine

small intestine

appendix

stomach

(spleen)

pancreas

descending colon of large intestine

rectum

Anything not absorbed in the small intestine moves into the **large intestine**, where "leftovers" are concentrated as water and salts are reabsorbed. Cells in the lining actively transport sodium ions out of the lumen; and as the ion concentration in the lumen drops, water moves out of the lumen by osmosis. The concentrated material is stored and then eliminated as *feces,* a mixture of undigested and unabsorbed food, water, and bacteria. The typical brown color of feces comes mainly from bile pigments in it.

About 30 percent of the dry weight of feces consists of bacteria. Such microorganisms, including *Escherichia coli,* normally inhabit the intestines and are nourished by the food residues there. Their metabolic activity produces useful fatty acids and some vitamins (such as vitamin K), which also are absorbed into the bloodstream. However, feces of humans and other animals also can contain disease-causing organisms. Health authorities use the presence of *E. coli* ("coliform bacteria") in water and food supplies as a measure of fecal contamination.

Relative to the small intestine, the large intestine is short—about 1.2 meters (5 feet) long. It begins as a blind pouch, the *cecum* (Figure 5.13). The *appendix,* a slender projection from the pouch, has no known digestive function, but it may have roles in defense against infections. *Appendicitis* is a serious inflammation of the appendix caused by bacterial infection. Symptoms include pain near the navel or right side of the abdomen, nausea, and vomiting. A suspected case of appendicitis requires urgent medical attention. If the infected appendix bursts, bacteria are spewed into the abdominal (peritoneal) cavity, and the victim can rapidly develop *peritonitis,* a potentially life-threatening infection.

The cecum merges with the **colon**, which is divided into four regions in an inverted U-shape. The *ascending colon* travels up the right side of the abdominal cavity, the *transverse colon* continues across to the left side, and the *descending colon* then turns downward. The *sigmoid colon* makes an S-shaped curve and connects with the **rectum**. Feces in the rectum eventually move into the *anal canal.* When feces distend the rectal wall, the stretching triggers defecation—expulsion of feces from the body. Defecation is controlled by the nervous system, which can stimulate or inhibit contractions of sphincter muscles at the **anus**, the terminal opening of the GI tract.

Colon problems can range from annoying to extremely serious. *Diarrhea* can occur when an irritant (such as a bacterial toxin) causes the mucosal lining of the small intestine to secrete more water and salts than the large intestine can absorb. It can also arise when infections, stress, or other factors speed up peristalsis in the small intestine, so that there is not time for sufficient water to

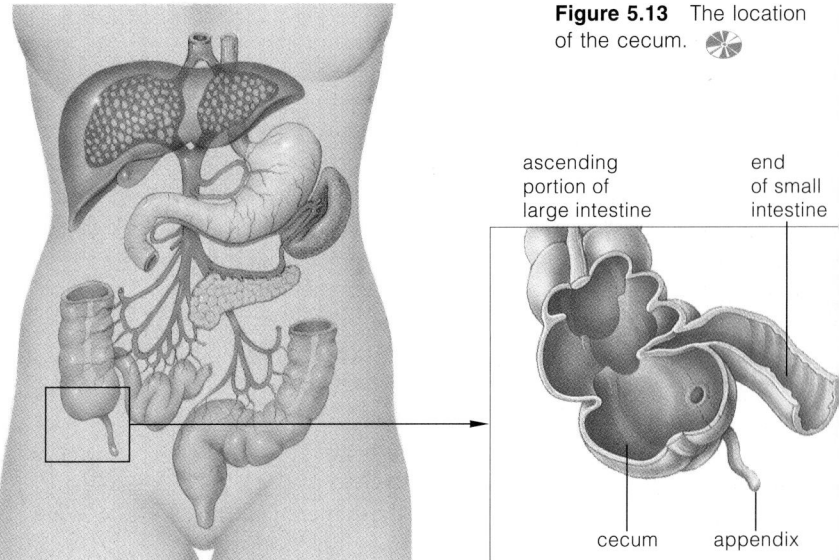

Figure 5.13 The location of the cecum.

ascending portion of large intestine

end of small intestine

cecum

appendix

be absorbed. (The normal feces-concentrating mechanism involves the active transport of sodium ions across the colon lining; water follows passively as a result.) Diarrheal diseases are dangerous largely because they deplete the body of the water and salts needed for proper functioning of nerve and muscle cells, among other things. In *constipation,* food residues remain in the colon for too long, too much water is reabsorbed, the feces become dry and hard, and defecation can be difficult. Stress, lack of exercise, and other factors can cause constipation; a major cause is a lack of "bulk" in the diet. *Bulk* is the volume of fiber (mainly cellulose from plants) and other undigested food material that cannot be decreased by absorption in the colon.

Soluble fiber consists of plant carbohydrates (such as fruit pectins) that swell or dissolve in water or are nutrient sources for beneficial intestinal bacteria. *Insoluble fiber* such as cellulose and other plant structural compounds does not easily dissolve in water and cannot be digested in the human GI tract. Wheat bran is just one example of insoluble fiber, but virtually all plant foods contain both fiber types. People whose diets lack sufficient fiber tend to produce a low volume of feces, and as a result of the low volume, feces move more slowly through the colon. Studies suggest that prolonged contact with feces not only irritates the colon, but over time may also trigger changes that lead to colon cancer. Constipation is a common cause of the enlarged rectal blood vessels called *hemorrhoids.*

In the large intestine, water and salts are reabsorbed from food residues entering from the small intestine.

The remaining concentrated residues are stored and later eliminated as feces.

DIGESTION CONTROLS AND NUTRIENT TURNOVER

Controls over Digestion

Recall from Chapter 3 that homeostatic feedback loops operate when physical and chemical conditions change in the internal environment. By contrast, controls over the digestive system act *before* food is absorbed into the internal environment. They respond to the volume and composition of material in the GI tract lumen. The nervous system, local nerve networks in the digestive tube wall, and the endocrine system interact to exert control (Figure 5.14).

For instance, in the first few hours after a meal, food distends the walls of the stomach and later those of the small intestine. This distension triggers signals from sensory receptors in the tube wall, which can lead to muscle contractions in the wall or secretion of digestive enzymes and other substances into the lumen. Various parts of the brain monitor such activities and coordinate them with other events, such as blood flow to the small intestine, where nutrients are being absorbed.

There are a number of gastrointestinal hormones. *Gastrin* is secreted into the bloodstream by endocrine cells in the stomach's lining when amino acids and peptides are in the stomach. It mainly stimulates the secretion of acid into the stomach. Acid secretion can be inhibited by *somatostatin*, which is produced by another type of endocrine cell in the stomach lining. Endocrine cells in the lining of the small intestine secrete other hormones. As acid arrives in the small intestine, *secretin*, a peptide hormone, stimulates the pancreas to secrete bicarbonate. *Cholecystokinin* (CCK) is released in response to fat in the small intestine. It enhances the actions of secretin, stimulates pancreatic enzyme secretion, and stimulates gallbladder contractions. Secretin and CCK also slow the rate of stomach emptying. *Glucose insulinotropic peptide* (GIP) is released in response to the presence of glucose and fat in the small intestine. It slows both the secretions of gastric glands and the rate of gastric emptying. GIP also stimulates the release of insulin, which is necessary for glucose uptake by cells or storage (as glycogen) by the liver.

Nutrient Turnover and Organic Metabolism

Once nutrient molecules have entered the body, they are shuffled and reshuffled according to body needs at any given time. The nervous system and endocrine system interact to control these and other aspects of organic metabolism; Figure 5.15 summarizes the main pathways. Most carbohydrates, lipids, and proteins are broken down continually, with their component parts picked up and reused as building blocks or as energy sources.

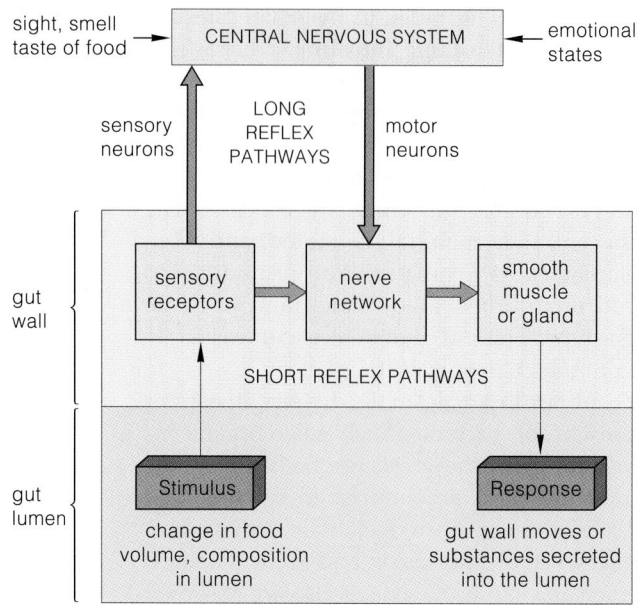

Figure 5.14 Local and long-distance reflex pathways called into action when food is in the digestive tract.

When you eat, your body builds up pools of organic compounds. Excess carbohydrates and other dietary molecules are transformed mostly into fats, which are stored in adipose tissue. Some carbohydrates are converted to glycogen in the liver and in muscle tissue. When organic molecules are being absorbed and stored, most cells use glucose as their main energy source.

As already described, during a meal and for several hours thereafter, glucose moves into cells, where it will be used for energy. However, in a fasting individual or during rigorous physical activity, glucose supplies become depleted. Then the body taps into its fat stores for energy. Fat stored mainly in adipose tissue is broken down to glycerol and fatty acids, which are released into blood. Glycerol is converted to glucose in the liver. Cells can take up the circulating fatty acids and use them for ATP production.

Disorders That Disrupt Absorption

A variety of mild to serious disorders result from maladies that disrupt the production of needed enzymes or harm the absorptive intestinal lining. Their ill effects can range from flatulence (gas) and diarrhea to life-threatening malnutrition. Anything that interferes with the uptake of nutrients across the lining of the small intestine can lead to a **malabsorption disorder**. For example, *lactose intolerance* results from a deficiency in the enzyme lactase. The condition occurs most commonly in adults.

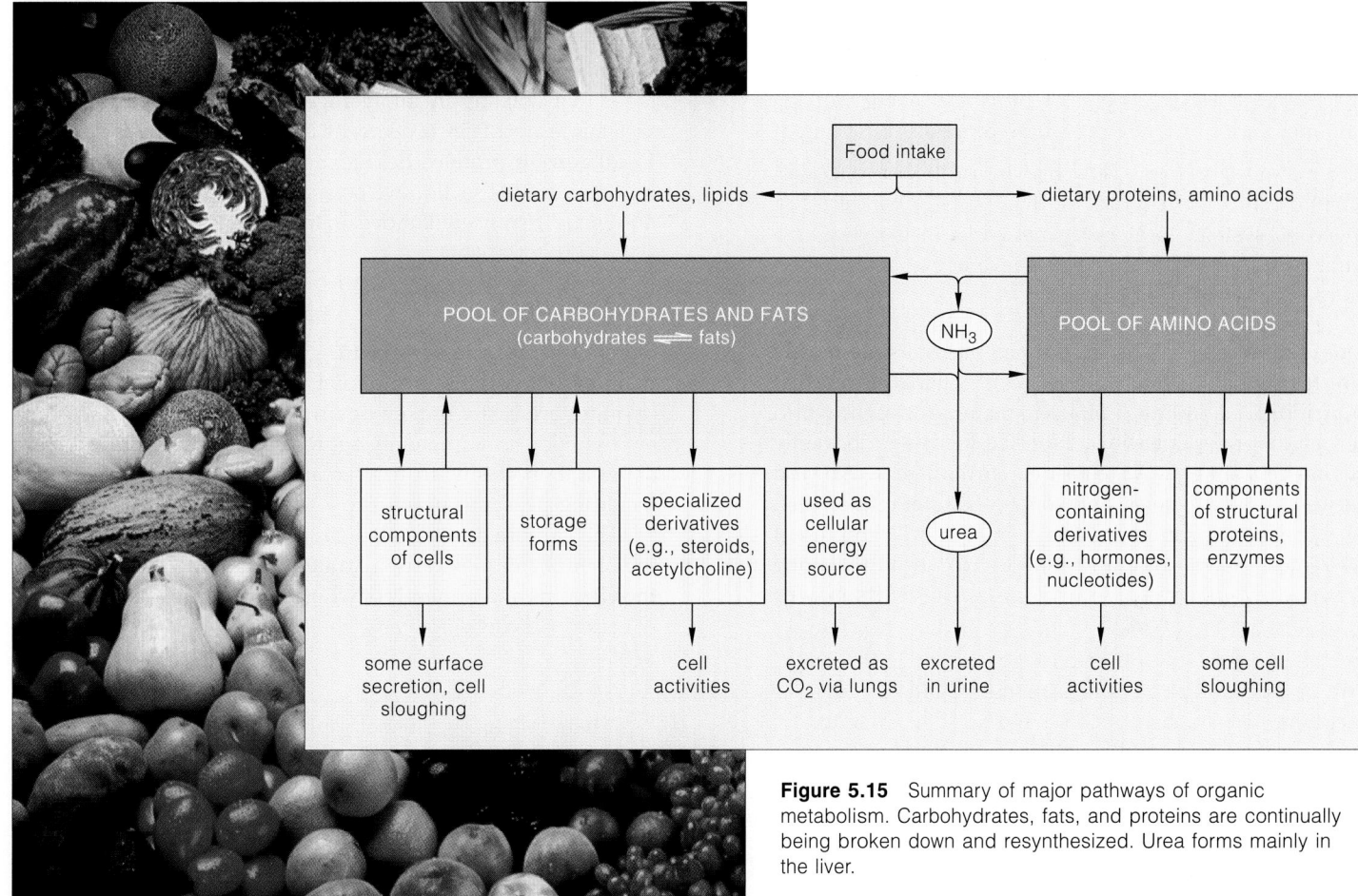

Figure 5.15 Summary of major pathways of organic metabolism. Carbohydrates, fats, and proteins are continually being broken down and resynthesized. Urea forms mainly in the liver.

It prevents normal breakdown and absorption of lactose, the sugar found in milk, ice cream, and many other milk products. People who have certain diseases affecting the pancreas, including the genetic disorder *cystic fibrosis*, do not produce pancreatic enzymes necessary for normal digestion and absorption of fats and other nutrients. Cystic fibrosis also has devastating effects on the lungs, which we consider in later chapters.

Disorders such as *Crohn's disease* and intolerance to the gluten in wheat can so severely damage the intestinal lining that large portions of the intestine must be removed. In some cases a patient must be fed artificially, receiving "meals" of a nutrient-rich liquid through a feeding tube.

Food allergies also disrupt digestion and absorption, but for a different reason. Like other allergies (described in Chapter 7), they are skewed responses of the immune system in which a particular food is interpreted as an "invader." The most common food culprits are shellfish, eggs, and wheat. Reactions can be immediate or occur over a period of hours. Depending on the person and the food involved, symptoms typically include diarrhea, vomiting, and sometimes swelling or tingling of mucous membranes. A food allergy or any other disorder that produces prolonged, severe diarrhea or vomiting can lead to electrolyte imbalances that disrupt nervous system functions and threaten the body's ability to maintain homeostasis in other ways as well.

The nervous system and the endocrine system exert control over conditions in the digestive system in response to the volume and composition of material in the digestive tract.

When glucose supplies run low, the body metabolizes fat molecules as the main energy source.

You grow and maintain yourself by eating certain foods that are suitable sources of energy and raw materials. Nutritionists measure the energy in **kilocalories** (kcal). Each kilocalorie is 1,000 calories of heat energy, or the amount needed to raise the temperature of 1 kilogram of water by 1°C. We'll look carefully at our bodies' energy requirements in Section 5.10; for the moment let's focus on the raw materials.

A Food Pyramid

A million years ago, our prehuman ancestors ate mostly fresh fruits and other fibrous plant material. From that nutritional beginning, humans in many parts of the world came to prefer low-fiber, high-fat foods—a decidedly unhealthy shift, if we consider current trends toward heart disease, diabetes, and some other diet-related ills.

The food pyramid and charts in Figure 5.16 represent the most recent recommendations for healthy eating devised by government nutritionists. The charts may be revised from time to time as additional research results become available.

The current food pyramid shows the daily portions of foods that will supply an average-size adult male with a diet made up of 58–60 percent complex carbohydrates, 15–20 percent proteins (less for females), and 20–25 percent fats. In this section we consider these basic types of nutrients. Then we will move on to vitamins and minerals.

Carbohydrates

Starch and, to a lesser extent, glycogen should be the main carbohydrates in our diet. These complex carbohydrates are easily broken down to glucose units, which are the body's main energy source (Chapter 2). Starch is abundant in fleshy fruits, cereal grains, and legumes, including peas and beans.

The carbohydrates called simple sugars do not have the fiber of complex carbohydrates. Neither do they have the vitamins and minerals of whole foods. Each week, the

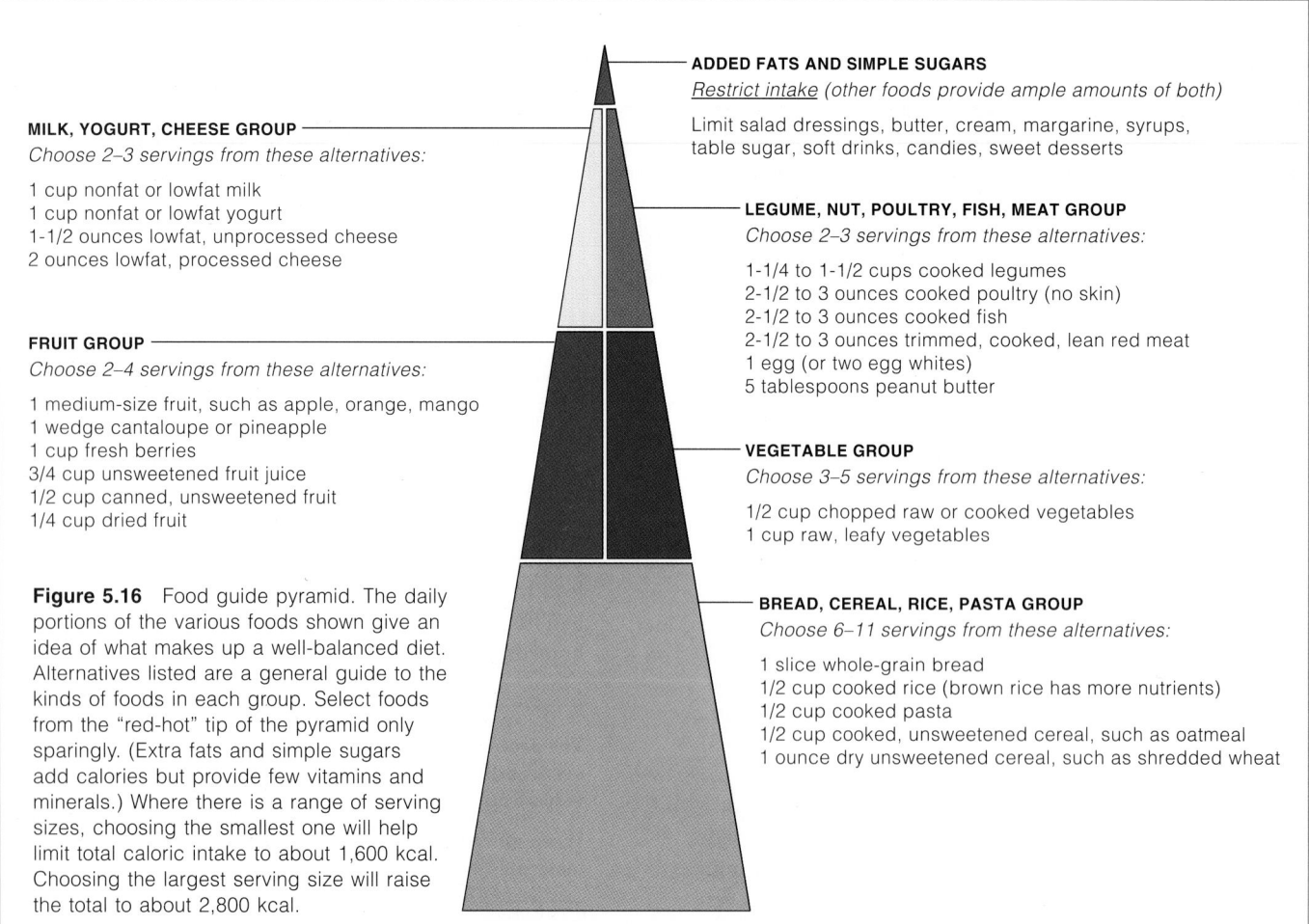

MILK, YOGURT, CHEESE GROUP
Choose 2–3 servings from these alternatives:

1 cup nonfat or lowfat milk
1 cup nonfat or lowfat yogurt
1-1/2 ounces lowfat, unprocessed cheese
2 ounces lowfat, processed cheese

FRUIT GROUP
Choose 2–4 servings from these alternatives:

1 medium-size fruit, such as apple, orange, mango
1 wedge cantaloupe or pineapple
1 cup fresh berries
3/4 cup unsweetened fruit juice
1/2 cup canned, unsweetened fruit
1/4 cup dried fruit

Figure 5.16 Food guide pyramid. The daily portions of the various foods shown give an idea of what makes up a well-balanced diet. Alternatives listed are a general guide to the kinds of foods in each group. Select foods from the "red-hot" tip of the pyramid only sparingly. (Extra fats and simple sugars add calories but provide few vitamins and minerals.) Where there is a range of serving sizes, choosing the smallest one will help limit total caloric intake to about 1,600 kcal. Choosing the largest serving size will raise the total to about 2,800 kcal.

ADDED FATS AND SIMPLE SUGARS
<u>Restrict intake</u> *(other foods provide ample amounts of both)*

Limit salad dressings, butter, cream, margarine, syrups, table sugar, soft drinks, candies, sweet desserts

LEGUME, NUT, POULTRY, FISH, MEAT GROUP
Choose 2–3 servings from these alternatives:

1-1/4 to 1-1/2 cups cooked legumes
2-1/2 to 3 ounces cooked poultry (no skin)
2-1/2 to 3 ounces cooked fish
2-1/2 to 3 ounces trimmed, cooked, lean red meat
1 egg (or two egg whites)
5 tablespoons peanut butter

VEGETABLE GROUP
Choose 3–5 servings from these alternatives:

1/2 cup chopped raw or cooked vegetables
1 cup raw, leafy vegetables

BREAD, CEREAL, RICE, PASTA GROUP
Choose 6–11 servings from these alternatives:

1 slice whole-grain bread
1/2 cup cooked rice (brown rice has more nutrients)
1/2 cup cooked pasta
1/2 cup cooked, unsweetened cereal, such as oatmeal
1 ounce dry unsweetened cereal, such as shredded wheat

No Limiting Amino Acid	Low in Lysine	Low in Methionine, Other Sulfur-Containing Amino Acids	Low in Tryptophan
legumes: soybean tofu soy milk cereal grains: wheat germ nuts milk cheeses (except cream cheese) yogurt eggs meats	legumes: peanuts cereal grains: barley buckwheat corn meal oats rice rye wheat nuts, seeds: almonds cashews coconut English walnuts hazelnuts pecans pumpkin seeds sunflower seeds	legumes: beans (dried) black-eyed peas garbanzos lentils lima beans mung beans peanuts nuts: hazelnuts fresh vegetables: asparagus broccoli green peas mushrooms parsley potatoes soybeans Swiss chard	legumes: beans (dried) garbanzos lima beans mung beans peanuts cereal grains: corn meal nuts: almonds English walnuts fresh vegetables: corn green peas mushrooms Swiss chard

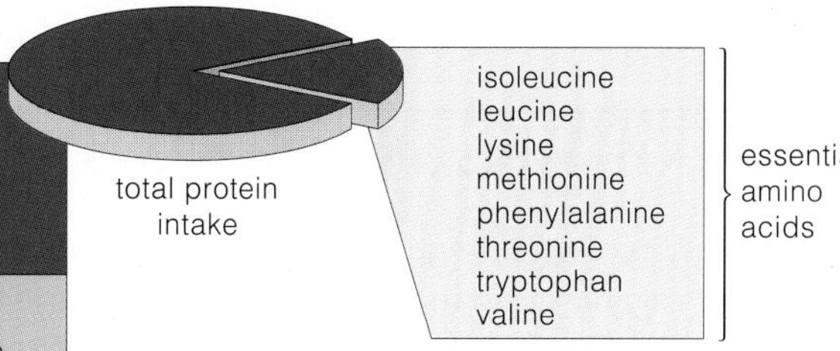

total protein intake

isoleucine
leucine
lysine
methionine
phenylalanine
threonine
tryptophan
valine

essential amino acids

Figure 5.17 Essential amino acids—a small portion of the total protein intake. All eight must be available at the same time, in certain amounts, if cells are to build their own proteins. Milk and eggs have high amounts of all eight in proportions that humans require; they are among the complete proteins.

Nearly all plant proteins are incomplete, so vegetarians should construct their diet carefully to avoid protein deficiency. For example, they can combine different foods from the three columns of incomplete proteins shown to the left. Also, vegetarians who avoid dairy products and eggs should take vitamin B_{12} and B_2 (riboflavin) supplements. Animal protein is a luxury in most societies. Yet their cuisines include good combinations of plant proteins, including rice/beans, chili/cornbread, tofu/rice, and lentils/wheat bread.

average American eats as much as two pounds of refined sugar (sucrose). In the ingredients lists on packaged foods, these sugars often are "hidden" as corn syrup, corn sweeteners, dextrose, and so on.

Lipids

The body can't function without fats and other lipids. For instance, the phospholipid lecithin is a required component of cell membranes. Fats serve as energy reserves, they cushion many organs (such as the eyes and kidneys), and they provide insulation beneath the skin. Adipose tissues also help the body store fat-soluble vitamins.

The liver can synthesize most of the fats the body needs, including cholesterol, from protein and carbohydrates. The ones it cannot produce are called **essential fatty acids**. Linoleic acid is an example. One teaspoon a day of corn oil, olive oil, or some other polyunsaturated fat in food provides enough of it.

Today, 40 percent of the kilocalories in the average diet in the United States comes from fats. Most of the medical community agrees the proportion *should be* less than 30 percent. Aside from questions of weight control, studies show that a diet high in fats increases the risk of colon cancer and heart disease—to name a few disorders. Butter and other animal fats are largely saturated fats, which tend to raise the level of cholesterol in the blood. The body needs cholesterol for the synthesis of bile salts, steroid hormones, and cell membrane, but for many people too much cholesterol may damage the circulatory system. Foods containing "fake fats"—edible but nondigestible oils, such as sucrose polyester (olestra)—can be useful for people who have trouble cutting back on their fat intake. On the down side, though, fat substitutes often produce intestinal gas and other unwanted side effects.

Proteins

When proteins are digested and absorbed, their amino acids become available for the body's own protein synthesis. Of the twenty common amino acids, eight are **essential amino acids**. Our cells cannot synthesize them, so we must obtain them from food. The eight are isoleucine, leucine, lysine, methionine, phenylalanine, threonine, tryptophan, and valine (Figure 5.17).

Most animal proteins are "complete," meaning their ratios of amino acids match human nutritional needs. Plant proteins are "incomplete," meaning they lack one or more of the essential amino acids. To get adequate amounts of needed amino acids, vegetarians must eat certain combinations of different plants.

Because enzymes and other proteins are vital for the body's structure and function, a protein-deficient diet may have severe consequences. Protein deficiency is most damaging among the young, for the brain grows and develops rapidly during early life. Even mild protein starvation during the mother's pregnancy or for some months after a child is born can retard the child's growth and affect mental and physical performance. The final section in this chapter describes some of the severe impacts of undernutrition, which is a critical problem in the poorest nations.

In terms of total caloric intake, the bulk of a well-balanced diet consists of complex carbohydrates, which are the body's preferred energy sources.

Lipids—which are essential components of cell membranes, serve as energy reserves, and have other roles—should make up no more than about 30 percent of calories.

Proteins, which provide the body's essential amino acids, should make up the remainder of dietary calories.

VITAMINS AND MINERALS

Metabolic activity depends on small amounts of more than a dozen organic substances called **vitamins**. In the course of evolution, animal cells have lost the ability to synthesize these substances, so we must obtain vitamins from food.

Human cells need at least the 13 vitamins listed in Table 5.3. Many reactions require several vitamins, so the absence of one affects the functions of others. Healthy metabolism also depends on certain inorganic substances called **minerals** (Table 5.4). For example, most of your cells use calcium and magnesium in many reactions. All cells require iron in electron transport chains. Red blood cells can't function without iron in hemoglobin, the oxygen-carrying pigment in blood. Nerve cells can't function without sodium and potassium.

Table 5.3   Vitamins: Sources, Functions, and Effects of Deficiencies or Excesses*

Vitamin	Common Sources	Main Functions	Signs of Severe Long-Term Deficiency	Signs of Extreme Excess
FAT-SOLUBLE VITAMINS:				
A	Its precursor comes from beta-carotene in yellow fruits, yellow or green leafy vegetables; also in fortified milk, egg yolk, fish liver	Used in synthesis of visual pigments, bone, teeth; maintains epithelia	Dry, scaly skin; lowered resistance to infections; night blindness; permanent blindness	Malformed fetuses; hair loss; changes in skin; liver and bone damage; bone pain
D	D_3 formed in skin and in fish liver oils, egg yolk, fortified milk; converted to active form elsewhere	Promotes bone growth and mineralization; enhances calcium absorption	Bone deformities (rickets) in children; bone softening in adults	Retarded growth; kidney damage; calcium deposits in soft tissues
E	Whole grains, dark green vegetables, vegetable oils	Possibly inhibits effects of free radicals; helps maintain cell membranes; blocks breakdown of vitamins A and C in gut	Lysis of red blood cells; nerve damage	Muscle weakness, fatigue, headaches, nausea
K	Colon bacteria form most of it; also in green leafy vegetables, cabbage	Blood clotting; ATP formation via electron transport	Abnormal blood clotting; severe bleeding (hemorrhaging)	Anemia; liver damage and jaundice
WATER-SOLUBLE VITAMINS:				
B_1 (thiamine)	Whole grains, green leafy vegetables, legumes, lean meats, eggs	Connective tissue formation; folate utilization; coenzyme action	Water retention in tissues; tingling sensations; heart changes; poor coordination	None reported from food; possible shock reaction from repeated injections
B_2 (riboflavin)	Whole grains, poultry, fish, egg white, milk	Coenzyme action	Skin lesions	None reported
Niacin	Green leafy vegetables, potatoes, peanuts, poultry, fish, pork, beef	Coenzyme action	Contributes to pellagra (damage to skin, gut, nervous system, etc.)	Skin flushing; possible liver damage
B_6	Spinach, tomatoes, potatoes, meats	Coenzyme in amino acid metabolism	Skin, muscle, and nerve damage; anemia	Impaired coordination; numbness in feet
Pantothenic acid	In many foods (meats, yeast, egg yolk especially)	Coenzyme in glucose metabolism, fatty acid and steroid synthesis	Fatigue, tingling in hands, headaches, nausea	None reported; may cause diarrhea occasionally
Folate (folic acid)	Dark green vegetables, whole grains, yeast, lean meats; colon bacteria produce some folate	Coenzyme in nucleic acid and amino acid metabolism	A type of anemia; inflamed tongue; diarrhea; impaired growth; mental disorders	Masks vitamin B_{12} deficiency
B_{12}	Poultry, fish, red meat, dairy foods (not butter)	Coenzyme in nucleic acid metabolism	A type of anemia; impaired nerve function	None reported
Biotin	Legumes, egg yolk; colon bacteria produce some	Coenzyme in fat, glycogen formation, and amino acid metabolism	Scaly skin (dermatitis), sore tongue, depression, anemia	None reported
C (ascorbic acid)	Fruits and vegetables, especially citrus, berries, cantaloupe, cabbage, broccoli, green pepper	Collagen synthesis; possibly inhibits effects of free radicals; structural role in bone, cartilage, and teeth; role in carbohydrate metabolism	Scurvy, poor wound healing, impaired immunity	Diarrhea, other digestive upsets; may alter results of some diagnostic tests

*The guidelines for appropriate daily intakes are being worked out by the Food and Drug Administration.

Table 5.4 Major Minerals: Sources, Functions, and Effects of Deficiencies or Excesses*

Mineral	Common Sources	Main Functions	Signs of Severe Long-Term Deficiency	Signs of Extreme Excess
Calcium	Dairy products, dark green vegetables, dried legumes	Bone, tooth formation; blood clotting; neural and muscle action	Stunted growth; possibly diminished bone mass (osteoporosis)	Impaired absorption of other minerals; kidney stones in susceptible people
Chloride	Table salt (usually too much in diet)	HCl formation in stomach; contributes to body's acid-base balance; neural action	Muscle cramps; impaired growth; poor appetite	Contributes to high blood pressure in susceptible people
Copper	Nuts, legumes, seafood, drinking water	Used in synthesis of melanin, hemoglobin, and some electron transport chain components	Anemia, changes in bone and blood vessels	Nausea, liver damage
Fluorine	Fluoridated water, tea, seafood	Bone, tooth maintenance	Tooth decay	Digestive upsets; mottled teeth and deformed skeleton in chronic cases
Iodine	Marine fish, shellfish, iodized salt, dairy products	Thyroid hormone formation	Enlarged thyroid (goiter), with metabolic disorders	Goiter
Iron	Whole grains, green leafy vegetables, legumes, nuts, eggs, lean meat, molasses, dried fruit, shellfish	Formation of hemoglobin and cytochrome (electron transport chain component)	Iron-deficiency anemia, impaired immune function	Liver damage, shock, heart failure
Magnesium	Whole grains, legumes, nuts, dairy products	Coenzyme role in ATP-ADP cycle; roles in muscle, nerve function	Weak, sore muscles; impaired neural function	Impaired neural function
Phosphorus	Whole grains, poultry, red meat	Component of bone, teeth, nucleic acids, ATP, phospholipids	Muscular weakness; loss of minerals from bone	Impaired absorption of minerals into bone
Potassium	Diet provides ample amounts	Muscle and neural function; roles in protein synthesis and body's acid-base balance	Muscular weakness	Muscular weakness, paralysis, heart failure
Sodium	Table salt; diet provides ample to excessive amounts	Key role in body's acid-base balance; roles in muscle and neural function	Muscle cramps	High blood pressure in susceptible people
Sulfur	Proteins in diet	Component of body proteins	None reported	None likely
Zinc	Whole grains, legumes, nuts, meats, seafood	Component of digestive enzymes; roles in normal growth, wound healing, sperm formation, and taste and smell	Impaired growth, scaly skin, impaired immune function	Nausea, vomiting, diarrhea; impaired immune function and anemia

*The guidelines for appropriate daily intakes are being worked out by the Food and Drug Administration.

People who are in good health get all the vitamins and minerals they need from a balanced diet of whole foods. Generally speaking, specific vitamin and mineral supplements are necessary only for strict vegetarians, the elderly, and individuals suffering from a chronic illness or taking medication that affects the body's use of specific nutrients. For example, in one comprehensive study, supplements of vitamin K helped older women retain calcium and so diminish bone loss (osteoporosis) by 30 percent. There is strong evidence that vitamins E and C and some of the pigments known as carotenes serve as antioxidants. You may recall from Chapter 2 that antioxidants combine with the highly reactive molecules called free radicals and counteract their destructive effects on cell structures.

No one should take massive doses of any vitamin or mineral supplement except under medical supervision.

As Tables 5.3 and 5.4 indicate, excess amounts of many vitamins and minerals actually can damage the body. For example, very large doses of the fat-soluble vitamins A and D can accumulate in tissues, especially in the liver, and interfere with normal metabolic functions. Similarly, sodium is present in plant and animal tissues and is a component of table salt. Sodium has roles in the body's salt-water balance, muscle activity, and nerve function. However, over the long term, excessive intake of sodium may contribute to high blood pressure.

Severe shortages or self-prescribed, massive excesses of vitamins and minerals can disturb the delicate balances in body function that promote health.

FOOD ENERGY AND BODY WEIGHT

To maintain an acceptable weight and keep the body functioning normally, caloric intake must be balanced with energy output. Physiologists define basal metabolic rate (BMR) as the amount of energy it takes to sustain the body when a person is resting, awake, and has not eaten for 12–18 hours. BMR varies from person to person; it is generally higher in males. Other factors that influence a person's energy output include differences in physical activity, age, hormone activity, and emotional state.

How many kilocalories should you take in each day to maintain "acceptable" body weight? To estimate your body's energy requirements, first multiply your desired weight (in pounds) by 10 if you are not very active physically, by 15 if you are moderately active, and by 20 if you are quite active. Then, depending on your age, subtract the following amount from the value obtained from the first step:

Age	Subtract
20–34	0
35–44	100
45–54	200
55–64	300
Over 65	400

For example, if you want to weigh 120 pounds and are quite active, 120 × 20 = 2,400 kilocalories. If you are 35 years old and moderately active, then you should take in a total of 1,800 − 100, or 1,700 kcal a day. Along with this rough estimate, factors such as height and gender also must be considered. Males tend to have more muscle and so burn more calories; hence an active woman needs fewer kilocalories than an active man of the same height and weight. Nor does she need as many as another active woman who weighs the same but is several inches taller.

Assessing Body Weight

Traditionally, insurance companies have developed "ideal weight" charts mainly so they can identify people who might be insurance risks. The medical profession has used the same charts. By a standard used in some countries, **obesity** is the condition of being 30 percent heavier than one's ideal weight. It can result from a variety of interacting factors, as described shortly. Cultural values can affect views of obesity. In some African nations, for example, a person who is "fat" in the United States or Europe might be considered highly desirable.

Cultural variables aside, clear medical evidence links obesity with conditions such as diabetes and heart disease. And by 1995, some people were questioning the insurance charts. For example, after a 14-year study of nearly 116,000 women, researchers at Harvard University found a strong link between the risk of heart attack and gaining even 11 to 18 pounds in adult life. Women who gained weight after age 18 had by far the greatest risk.

Figure 5.18 How to estimate the "ideal" weight for an adult. The values given are consistent with a long-term Harvard study into the link between excessive weight and increased risk of heart disorders. Depending on certain factors, such as having a small, medium, or large frame, the "ideal" may vary by plus or minus 10 percent.

Weight Guidelines for Women

Starting with an ideal weight of 100 pounds for a woman who is 5 feet tall, add five additional pounds for each additional inch of height. Examples:

Height (feet)	Weight (pounds)
5' 2"	110
5' 3"	115
5' 4"	120
5' 5"	125
5' 6"	130
5' 7"	135
5' 8"	140
5' 9"	145
5' 10"	150
5' 11"	155
6'	160

Weight Guidelines for Men

Starting with an ideal weight of 106 pounds for a man who is 5 feet tall, add six additional pounds for each additional inch of height. Examples:

Height (feet)	Weight (pounds)
5' 2"	118
5' 3"	124
5' 4"	130
5' 5"	136
5' 6"	142
5' 7"	148
5' 8"	154
5' 9"	160
5' 10"	166
5' 11"	172
6'	178

Table 5.5	Calories Expended in Some Common Activities			
		Hours needed to lose 1 lb. fat		
Activity	Kcal/hour per pound of body weight	120 lbs	155 lbs	185 lbs
Basketball	3.78	7.7	6.0	5.0
Cycling (9 mph)	2.70	10.8	8.4	7.0
Hiking	2.52	11.6	8.9	7.5
Jogging	4.15	7.0	5.4	4.5
Mowing lawn (push mower)	3.06	9.5	7.4	6.2
Racquetball	3.90	7.5	5.8	4.8
Running (9-minute mile)	5.28	5.5	4.3	3.6
Snow skiing (cross-country)	4.43	6.6	5.1	4.3
Swimming (slow crawl)	3.48	8.4	6.5	5.4
Tennis	3.00	9.7	7.5	6.3
Walking (moderate pace)	2.16	13.5	10.4	8.7

To calculate these values for your own body weight, first multiply your weight by the kcal/hour expended for an activity to determine total kcal you use during one hour of the activity. Then divide that number into 3,500 (kcal in a pound of fat) to obtain the number of hours you must perform the activity to burn a pound of body fat.

In making their correlation, the Harvard researchers took into account age, smoking habits, family history of heart disorders, and other risk factors. *They also found conclusive evidence that exercise is the most crucial factor in controlling weight gain and avoiding heart attacks.*

Generally, thin people simply live longer. You can use the chart in Figure 5.18 to get an idea of how a 1995 definition of "ideal weight" applies to you.

My Genes Made Me Do It

In most adults, energy input balances output, so body weight remains stable over long periods. It is as if the body has a set point for its weight and works to counteract deviations from the set point. As you have probably noticed, some people have far more trouble keeping off excess weight than others do. To some extent, a person's age and emotional state have something to do with it. More important, *genes* have a lot to do with it.

In 1995, researchers working with mice isolated a gene that operates only in adipose tissue—fat-storing cells. Named the *ob* gene, this bit of DNA's biochemical information translates into a hormone, which its discoverer called leptin. An almost identical hormone was isolated in humans. Hormones can be released into the bloodstream, and thereby may hang the tale of obesity.

Leptin is only one of several factors that influence the brain's commands to stimulate or suppress appetite, but it's a crucial one. By assessing the blood concentrations of incoming hormonal signals, an appetite control center in the brain "decides" whether the body has taken in enough food to provide enough fat for the day. If so, commands go out to increase metabolic rates—and to stop eating. However, if a faulty *ob* gene disrupts the normal feedback loop, then proper controls won't operate. Investigators theorize that some obese people may inherit such malfunctioning controls. Eventually it may become possible to treat obesity with hormone-based therapy.

Dieting and Exercise

Many overweight people attempt to shed excess fat by dieting, but most eventually regain what they lose. Why? A person's weight set point contributes. Another reason is that the fat-storing cells of adipose tissue are an adaptation for survival—an energy reserve for times when food isn't available. Research suggests that the body interprets dieting as "starvation," triggering metabolic changes. With these changes, food is used more conservatively so that fewer calories are burned. Meanwhile, the dieter's appetite surges, and the "starved" fat cells quickly refill when a diet ends.

As already noted, exercise is essential to long-term weight control. Losing a pound of fat requires expending about 3,500 kilocalories. For example, a 120-pound woman can burn a pound of fat by playing about 10 hours of tennis (Table 5.5). Permanent weight loss requires combining a moderate reduction in caloric intake with increased physical activity. This strategy seems to minimize the "starvation" response and increase the rate at which the body burns calories. Exercise also increases muscle mass, and even at rest muscle burns more calories than other types of tissues.

To maintain an acceptable body weight, energy input (caloric intake) must be balanced with energy output in the form of metabolic activity and exercise.

A person's energy output is influenced by basal metabolic rate, physical activity, age, hormone activity, and emotional state.

5.11 MALNUTRITION AND UNDERNUTRITION

If you are reading this book, you almost certainly enjoy the luxury of a varied diet and adequate food supplies. Like millions of affluent people, you may carefully tailor your meals, picking and choosing foods that optimize your intake of certain vitamins, antioxidants, proteins, minerals, fats, and other substances. You may even be in the enviable position of needing to limit your food intake and exercise more because you have access to much more food energy than your body needs.

Many of your fellow humans are less fortunate. Over a lifetime, an average person in a country such as the United States, Italy, or Australia will consume 20 or 30 times as much food as an individual born in poor, undeveloped and developing nations of Africa, Asia, and South America. Even in affluent countries, people who lack the resources or knowledge to obtain a healthy diet can suffer from malnutrition.

Malnutrition is a state in which body functions or development suffers due to inadequate or unbalanced food intake. A malnourished person may suffer from *undernutrition*—that is, the individual lacks food and thus does not obtain sufficient kilocalories or nutrients to sustain proper growth, body functioning, and development. Another word for undernutrition is *starvation*, and globally it is the most common form of malnutrition. Less often, but probably more common in affluent countries, a malnourished person consumes plenty of food from the standpoint of kilocalories, but has an unbalanced diet that lacks essential vitamins or minerals.

Global Undernutrition

Nutrition researchers estimate that at least half a billion people in the world go hungry every day. Virtually all of them are too poor to obtain adequate food. Over the long term, lack of food energy or essential nutrients disrupts homeostasis and weakens the body's immune defenses, among other effects. Every year up to 20 million people, three-fourths of them children, die of starvation or from diseases that easily ravage a chronically undernourished body. Many other poorly nourished people, including elderly persons, are chronically weak, lethargic, prone to illness, or mentally impaired. Let's consider some of the most serious effects of severe nutritional deficiencies.

PROTEIN DEFICIENCY Many undernourished people take in not only too few kilocalories but especially too little protein. The result, *protein-energy malnutrition*, causes weakness, weight loss, impaired immunity, and other symptoms in adults, but it is especially devastating to infants and children. The child in Figure 5.19 shows the classic signs of *kwashiorkor*, a condition in which a child who may consume near-normal amounts of calories fails to grow and gain weight and shows other evidence of chronic protein deficiency. The swollen abdomen is due to edema (fluid retention) because there are too few proteins in the child's blood plasma. The osmotic imbalance causes water to leave the bloodstream and collect in tissue spaces. Proteins are also essential for proper functioning of the immune system, so affected children are sickly and susceptible to infection. Impaired brain development often leads to permanent mental retardation. Kwashiorkor typically develops when a child is weaned and its diet shifts from protein-rich breast milk to low-protein starchy foods.

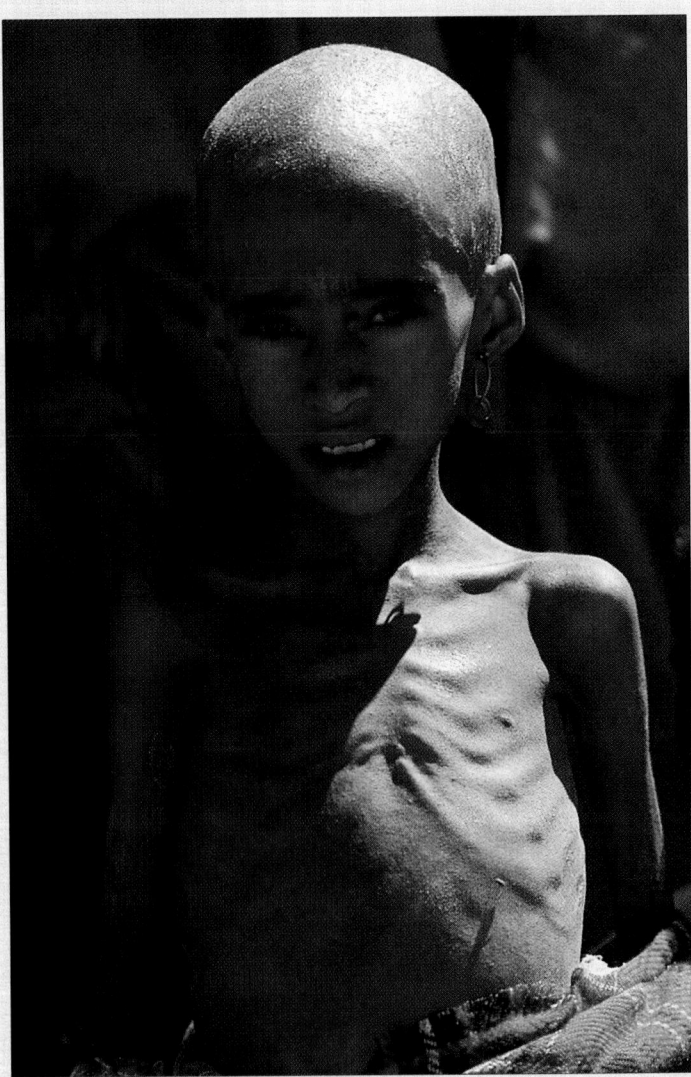

Figure 5.19 A child with kwashiorkor.

Many of the gruesome photographs of skeletal infants in Africa and elsewhere show the body-wasting disease *marasmus*. Usually marasmus affects very young children who lack both protein and food calories. It is common among bottle-fed infants of poor families in which parents dilute prepared formula, essentially "nourishing" their children mainly with water. Symptoms include low weight and muscle wasting, dry hair and skin, and retarded growth and mental development.

Some effects of kwashiorkor and marasmus can be halted or reversed if the child's diet is improved before the damage is too severe. As a practical matter, however, relief efforts are often too little or too late to save many affected children from death. Those that do survive are likely to function at abnormally low levels both physiologically and intellectually.

OTHER NUTRIENT DEFICIENCIES People in many under-developed countries are malnourished because their food sources are extremely limited. A starchy food, such as rice, corn, millet, or cassava, may make up the lion's share of most meals. This lack of variety translates into deficiencies of key vitamin and mineral nutrients. For example, nutritionists estimate that one-sixth of all humans, almost 1 billion people, are anemic because their diet lacks iron. Iron is a component of hemoglobin, the oxygen-transporting protein in blood. Major sources include meat and green, leafy vegetables. Iron deficiency impairs oxygen delivery to cells, and it also limits the ability of cells to extract energy from carbohydrates. In addition, it impairs the functioning of the immune system. Authorities cite a lack of dietary iron as a major factor contributing to high death rates from intestinal and respiratory infections in many countries. It also is the most common nutritional deficiency of children in the United States.

Lack of vitamin A is responsible for *xeropthalmia,* a form of blindness that annually afflicts a quarter of a million children. Globally, xeropthalmia is the leading cause of preventable blindness, and it is irreversible.

Children are also the chief victims of iodide deficiency. This mineral, an ionic form of iodine, is a key component of thyroid hormones, which help govern normal growth and metabolism. In adults, low iodide causes goiter—overgrowth of the thyroid gland, which may be noticeable as a lump in the front or side of the neck. In children an iodide deficiency can irreversibly retard physical and mental development. Other common ailments associated with undernutrition in poor nations include scurvy (vitamin C deficiency), beriberi (vitamin B_1 deficiency), and rickets (vitamin D deficiency).

Groups at Highest Risk

As you surely have gathered by now, children are the main victims of malnutrition. The damage can begin during fetal development when a pregnant mother does not have access to sufficient protein and other nutrients. In general, babies born to undernourished mothers may be one to three weeks premature; they may have low birth weight and an underdeveloped respiratory system, as well as other potentially life-threatening complications. Elderly individuals also are at greater risk of undernutrition. The causes may be economic, social, or psychological (including clinical depression), and over time the predictable effects are declining immunity and impairment of other physiological functions.

Social and Political Factors in Global Malnutrition

The United Nations and relief organizations estimate that current global food supplies are adequate to meet the minimum nutritional needs of all or most of the earth's human population. Why then do people starve in places like Somalia, Ethiopia, and North Korea? Although the situation is complex, part of the answer is that world food supplies and human populations are not equally distributed. Most of the world's grain and other basic foods are produced in relatively wealthy countries, while 75 percent of the human population lives in developing regions such as Africa. Also, people in developed nations have more access to food because they tend to have more money to pay for it. Efforts to increase food production in poor countries are complicated by the high costs of fuel, fertilizers, and other inputs to agriculture and by the fact that arable land is being converted to other uses as the human population expands (see Chapter 25).

Politics also can get in the way of food distribution. For example, in recent years in Sudan, Ethiopia, Somalia, and the Czech Republic, food earmarked for the general public has been stolen by warring factions and then sold to raise funds or has rotted in storage because warlords wish to ensure that it isn't used to feed their enemies.

Within the next three decades the human population is projected to rise from its current 6 billion to more than 11 billion. Given the recurring food crises we now face, it is difficult to imagine how we can adequately feed so many people. Clearly, it is crucial that the human population stabilize at a number that the planet's resources can sustain over the long term. We will return to this very troubling and challenging issue in Chapter 25.

SUMMARY

1. Digestion and nutrition include all the processes by which the body takes in, digests, absorbs, and uses food.

2. Four main activities proceed in the digestive system:

a. Mechanical processing and motility (movements that break up, mix, and propel ingested food through the gastrointestinal tract).

b. Secretion (release of digestive enzymes and other substances from the salivary glands, pancreas, liver, and glandular epithelium into the lumen of the digestive tube).

c. Digestion (chemical breakdown of food into particles, then into nutrient molecules small enough to be absorbed).

d. Absorption (the passage of digested organic compounds, fluid, and ions into the internal environment).

Undigested and unabsorbed residues are expelled at the end of the system.

3. The human digestive system (the gastrointestinal tract) includes the mouth, pharynx, esophagus, stomach, small intestine, and large intestine (colon). Accessory organs associated with digestion include the salivary glands, liver, gallbladder, and pancreas (see Table 5.6).

4. The GI tract is lined with mucous membrane, and from the esophagus onward the tube wall consists of four basic layers. The innermost mucosa is surrounded by the submucosa, a connective tissue layer. The next layer is smooth muscle, which is surrounded by a thin layer of connective tissue called the serosa. Thickened muscular sphincters at either end of the stomach and at other locations help control the forward movement of ingested material and prevent backflow.

5. Starch digestion begins in the mouth, and protein digestion begins in the stomach. Digestion is completed and most nutrients are absorbed in the small intestine. Following absorption, monosaccharides and most amino acids go directly to the liver. Fatty acids and triglycerides enter lymph vessels, then the general circulation.

6. Digestion in the small intestine depends on enzymes and other substances secreted by the pancreas, the liver, and the gallbladder. Pancreatic juice includes enzymes that act on all four main categories of organic molecules in food—carbohydrates, fats, proteins, and nucleic acids. Bile, which is secreted by the liver and then stored and released into the small intestine by the gallbladder, contains bile salts that speed up fat digestion via the process of emulsification.

7. Most nutrient absorption occurs across the lining of the small intestine, which is densely folded into villi (projections of the mucosa). Individual epithelial cells of villi bear microvilli on their surface. Microvilli greatly increase the surface area available for absorption.

Table 5.6   Summary of the Digestive System

MOUTH (oral cavity)	Start of digestive system where food is chewed and moistened and where polysaccharide digestion begins
PHARYNX	Passageway both to the tubular part of the digestive system and to the respiratory system; moves food forward by contracting sequentially
ESOPHAGUS	Muscular tube, moistened by saliva, that moves food from pharynx to stomach
STOMACH	Sac where food mixes with gastric fluid and where protein digestion begins; can stretch to store food taken in faster than can be processed; gastric fluid also destroys many microbes
SMALL INTESTINE	The first part (duodenum, C-shaped and about 10 inches long) receives secretions from the liver, gallbladder, pancreas
	In the second part (jejunum, 3 feet long), most nutrients are digested and absorbed
	Some nutrients are absorbed in the last part (ileum, 6–7 feet long), which delivers unabsorbed material to the colon
LARGE INTESTINE (colon)	Concentrates and stores undigested matter (by absorbing mineral ions and water). It is about 5 feet long and is divided into ascending, transverse, and descending portions.
RECTUM	Passageway for fecal material

Accessory Organs:

SALIVARY GLANDS	Glands (three main pairs and many minor ones) that secrete saliva, a fluid with polysaccharide-digesting enzymes, buffers, and mucus (moistens food)
PANCREAS	Secretes enzymes able to break down all major food molecules; secretes buffers against HCl from the stomach
LIVER	Secretes bile (used in fat emulsification); roles in carbohydrate, fat, and protein metabolism
GALLBLADDER	Stores and concentrates bile from the liver

8. Nervous and endocrine controls over the digestive system operate in response to the volume and composition of food passing through. The response can be a change in muscle activity, in the secretion rate of hormones or enzymes, or in all of these.

9. To maintain acceptable weight and overall health, caloric intake must balance energy output. Complex carbohydrates are the body's preferred energy source. The body converts excess carbohydrates and proteins into fat and stores it in fat cells. The diet must provide eight essential amino acids, a few essential fatty acids, vitamins, and minerals.

Review Questions

1. What are the main functions of the stomach? The small intestine? The large intestine? *112*

2. Name four kinds of molecules that are small enough to be absorbed across the intestinal lining and into the internal environment. *118*

3. A glass of whole milk contains lactose, protein, triglycerides (in butterfat), vitamins, and minerals. Explain what happens to each component when it passes through your digestive tract. *114–123*

4. Give reasons why each of the following is nutritionally important: carbohydrates, proteins, fats, vitamins, and minerals. *126–129*

Critical Thinking: You Decide *(Key in Appendix V)*

1. Some nutritionists claim that the secret to long life is to be slightly underweight as an adult. If a person's weight is related partly to diet, partly to activity level, and partly to genetics, what underlying factors could be at work to generate statistics that support this claim?

2. As a person ages, the number of body cells steadily decreases and their energy needs decline. If you were planning an older person's diet, what kind(s) of nutrients would you emphasize, and why? Which ones would you recommend less of?

3. Along the lines of question 2, formulate a healthy diet for an actively growing seven-year-old.

4. Raw poultry (Figure 5.20) can carry *Salmonella* or *campylobacter* bacteria, both of which produce toxins that can cause serious diarrhea, among other symptoms. Aside from the discomfort, why does such an infection require immediate medical attention?

Self-Quiz *(Answers in Appendix IV)*

1. The _____, _____, _____, and _____ systems interact in supplying body cells with raw materials, disposing of wastes, and maintaining the volume and composition of extracellular fluid.

2. Various specialized regions of the digestive system function in _____ and _____ food and in _____ unabsorbed food residues.

3. Maintaining good health and normal body weight requires that _____ intake be balanced by _____ output.

4. The preferred energy sources for the body are complex _____.

5. The human body cannot produce its own vitamins or minerals, and it also cannot produce certain _____ and _____.

6. Which of the following structures is *not* associated with digestion?
   a. salivary glands   d. gallbladder
   b. thymus gland   e. pancreas
   c. liver

7. Digestion is completed and breakdown products are absorbed in the _____.
   a. mouth   c. small intestine
   b. stomach   d. large intestine

8. After absorption, fatty acids and monoglycerides move into the _____.
   a. bloodstream   c. liver
   b. intestinal cells d. lacteals

9. Excess carbohydrates and proteins are converted to _____ for storage.
   a. amino acids   c. fats
   b. starches   d. monosaccharides

Figure 5.20 Chicken, a meat notoriously susceptible to *Salmonella* infection.

10. Match each of the following digestive system components with its description.
    ____ liver
    ____ small intestine
    ____ human digestive system
    ____ stomach
    ____ large intestine

   a. begins at mouth, ends at anus
   b. where protein digestion begins
   c. where water is reabsorbed
   d. where most digestion is completed
   e. receives monosaccharides and amino acids

11. Body weight is controlled by
    a. caloric intake   d. age and sex
    b. level of metabolism   e. all of the above
    c. energy utilization

12. Four of the five answers listed below are conditions caused by vitamin deficiency. Select the exception.
    a. scurvy   c. rickets   e. goiter
    b. pellagra   d. beriberi

Selected Key Terms

absorption *113*
anus *121*
basal metabolic rate *128*
digestion *113*
digestive system *112*
elimination *113*
esophagus *114*
essential amino acid *125*
essential fatty acid *125*
gallbladder *117*
gastrointestinal (GI) tract *112*
hepatic portal vein *120*
kilocalorie *128*
lacteal *119*
large intestine *121*
malnutrition *130*

mineral *126*
motility *113*
obesity *128*
palate *114*
pancreas *117*
pepsin *116*
peristalsis *113*
pharynx *114*
rectum *121*
salivary amylase *114*
salivary gland *114*
secretion *113*
small intestine *118*
stomach *116*
vitamin *126*

Readings

Jaret, P. January–February 1995. "The Way to Lose Weight." *Health.*

Wardlaw, G. M., P. M. Insel, and M. F. Seyler. 1992. *Contemporary Nutrition: Issues and Insights.* St. Louis: Mosby YearBook.

 For additional readings, go to InfoTrac College Edition, your online library at:

http://www.infotrac-college.com/wadsworth

6

BLOOD AND CIRCULATION

Heartworks

For Dr. Augustus Waller, Jimmie the bulldog was no ordinary pooch. Connected to wires and soaked to his ankles in buckets of salty water, Jimmie was a four-footed window into the workings of the heart (Figure 6.1a). Press your fingers to your chest a few inches left of the center, between the fifth and sixth ribs, and feel the repetitive thumpings of your heart. The same rhythms intrigued Waller and other nineteenth-century physiologists (Figure 6.1b). They wondered: Does each heartbeat produce a pattern of electrical currents? Could they find out by devising a painless way to record such currents at the body surface?

That's where Jimmie and the buckets of salty water came in. Saltwater happens to be an efficient conductor of electricity—so efficient that it carried faint signals from Jimmie's beating heart, through the skin of his legs, to a crude monitoring device. With this device, in the late 1880s Waller made one of the world's first graphic recordings of a beating heart—an *electrocardiogram*, more commonly known as an ECG (Figure 6.1c).

A graph of the normal electrical activity of your own heart would look much the same. The pattern emerged when you were an embryo inside your mother. Patches of newly formed cardiac tissue started to contract. Eventually, one patch took the lead and has been the pacemaker ever since, and if all goes well, it will continue

a

b

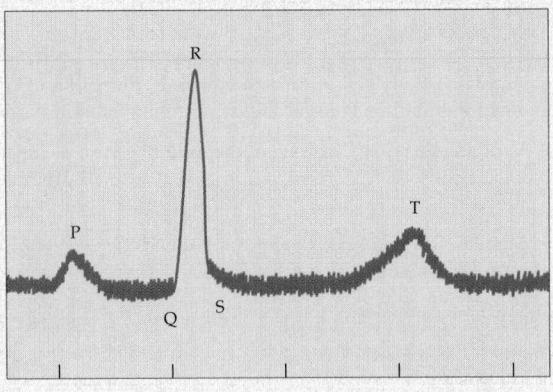

c

Figure 6.1 A bit of history in the making. (**a**) Jimmie the bulldog, taking part in a painless experiment. (**b**) Augustus Waller and his beloved pet bulldog sharing a quiet moment in Waller's study. (**c**) One of the world's first electrocardiograms.

to do so until the day you die. That natural pacemaker sets, adjusts, and resets the rate at which blood is pumped from your heart, through a vast network of blood vessels, then back to the heart.

This service of the pacemaker is vital, in part because blood performs such a remarkable array of functions. It brings oxygen and nutrients to cells, transports cell secretions such as hormones, and carries away metabolic wastes. Blood helps stabilize internal pH, and it serves as a highway for phagocytic cells that scavenge tissue debris and fight infections. It also helps equalize body temperature by carrying excess heat from regions of high metabolic activity (such as skeletal muscles) to the skin, where heat can be dissipated.

When your body is at rest, the demands on blood as a transport medium are at their lowest. Consequently, at rest your heart rate will be moderate, somewhere around 70 beats a minute. However, play a fast game of volleyball and the demands by your muscles for blood-borne oxygen and glucose will escalate. Then, your heart may start pounding 150 times a minute or more to deliver sufficient blood to muscles.

We have come a long way from Jimmie in monitoring the heart. Sensors now detect the faintest signals of an impending heart attack. Computers are used to analyze a patient's beating heart and build images of it on a video screen. Surgeons substitute battery-powered pacemakers for malfunctioning natural ones.

With this chapter we turn to the cardiovascular system, which moves substances rapidly to and from all living cells. As you probably already suspect, the system is absolutely central to the body's ability to maintain stable operating conditions in the internal environment.

KEY CONCEPTS

1. Cells survive by exchanging substances with their surroundings. The cardiovascular system allows rapid movement of substances to and from all living cells. The system consists of the heart, blood, and blood vessels, which are supplemented by a lymphatic system.

2. Blood, a fluid connective tissue confined within the heart and blood vessels, is the transport medium of the cardiovascular system.

3. Blood pumped by the heart flows in two circuits. In the pulmonary circuit, the heart pumps deoxygenated blood to the lungs, where it releases carbon dioxide and picks up oxygen. Then blood flows back to the heart. In the systemic circuit, the heart pumps oxygenated blood to all body regions. After giving up oxygen and picking up carbon dioxide in those regions, blood flows back to the heart.

4. In addition to oxygen and carbon dioxide, blood also transports plasma proteins, vitamins, hormones, lipids, and other solutes.

5. The blood vessels called arteries and veins are large-diameter transport tubes. Capillaries and, to some extent, venules are narrow-diameter tubes for diffusion. Venules also collect blood from capillaries and channel it to veins. Arterioles have adjustable diameters. They are control points for the distribution of different volumes of blood flow to different regions.

CHAPTER AT A GLANCE

6.1 Blood

6.2 Oxygen Transport in Blood

6.3 Life Cycle of Red Blood Cells

6.4 Blood Typing

6.5 Overview of the Cardiovascular System

6.6 The Heart: A Durable Pump

6.7 *Can We Make Artificial Blood?*

6.8 Blood Circulation

6.9 How the Heart Contracts

6.10 Blood Pressure and Velocity in the Cardiovascular System

6.11 How Vessel Structure Affects Blood Pressure

6.12 Exchanges of Fluid and Solutes at Capillaries

6.13 Hemostasis and Blood Clotting

6.14 Cardiovascular Disorders

6.15 *Atherosclerosis, Hypertension, and Arrhythmias*

6.16 The Lymphatic System

BLOOD

The volume of blood in the body depends on body size and on the concentrations of water and solutes. For an average-size adult, blood volume is generally about 4–5 liters, which is about 6–8 percent of body weight. As in the old saying, blood *is* thicker than water, and it flows more slowly. This is because **blood** is a rather sticky fluid consisting of plasma and a cellular portion that includes red blood cells, white blood cells, and platelets (Figure 6.2). Let's consider each of these components in turn.

Plasma

The straw-colored **plasma** is mostly water, and it makes up about 55 percent of whole blood. Plasma serves as a transport medium for blood cells and platelets. The water in plasma is a solvent for ions and molecules, including hundreds of different plasma proteins. The concentrations of these proteins influence the osmotic movement of water between blood and interstitial fluid—and hence blood's fluid volume.

Albumin is important in this water-balancing act, for it represents nearly two-thirds of all plasma proteins. Albumin also carries a variety of chemicals through the system, from metabolic wastes to therapeutic drugs. Some alpha and beta globulins transport lipids (including cholesterol) and fat-soluble vitamins. As you will see in this chapter and Chapter 7, gamma globulins function in immune responses, and fibrinogen serves in blood clotting.

Plasma also contains ions, glucose and other simple sugars, lipids, amino acids, vitamins, hormones and other communication molecules, and dissolved gases (mostly oxygen, carbon dioxide, and nitrogen). The ions (such as Na^+, Cl^-, H^+, and K^+) help maintain extracellular pH and fluid volume. The lipids include fats, phospholipids, and cholesterol. Lipids that are being transported from the liver to different regions typically are bound with proteins, forming lipoproteins (introduced in Chapter 1).

Red and White Blood Cells

About 45 percent of whole blood consists of erythrocytes, or **red blood cells**. Each erythrocyte is a biconcave disk, rather like a thick pancake with an indentation on each side (Figure 6.3a). The cell's red color comes from the iron-containing protein *hemoglobin*, which transports oxygen used in aerobic respiration. Erythrocytes also carry away some carbon dioxide wastes. You will read about other features of red blood cells in upcoming sections.

Leukocytes, or **white blood cells**, make up a tiny fraction of whole blood, but they have vital functions in day-to-day housekeeping and defense. Some leukocytes

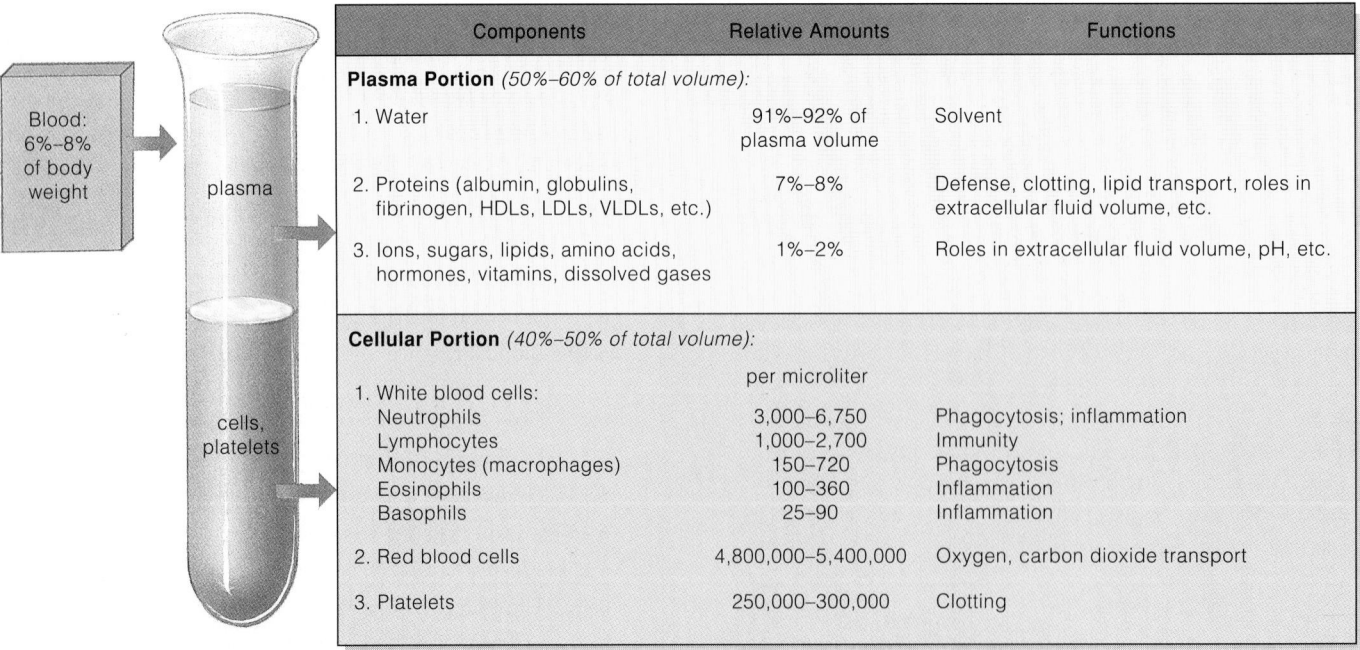

Components	Relative Amounts	Functions
Plasma Portion (50%–60% of total volume):		
1. Water	91%–92% of plasma volume	Solvent
2. Proteins (albumin, globulins, fibrinogen, HDLs, LDLs, VLDLs, etc.)	7%–8%	Defense, clotting, lipid transport, roles in extracellular fluid volume, etc.
3. Ions, sugars, lipids, amino acids, hormones, vitamins, dissolved gases	1%–2%	Roles in extracellular fluid volume, pH, etc.
Cellular Portion (40%–50% of total volume):	per microliter	
1. White blood cells:		
Neutrophils	3,000–6,750	Phagocytosis; inflammation
Lymphocytes	1,000–2,700	Immunity
Monocytes (macrophages)	150–720	Phagocytosis
Eosinophils	100–360	Inflammation
Basophils	25–90	Inflammation
2. Red blood cells	4,800,000–5,400,000	Oxygen, carbon dioxide transport
3. Platelets	250,000–300,000	Clotting

Blood: 6%–8% of body weight

plasma

cells, platelets

Figure 6.2 Components of blood. If a blood sample placed in a test tube is prevented from clotting, it will separate into a layer of straw-colored liquid, the plasma, that floats over the darker-colored cellular portion of blood.

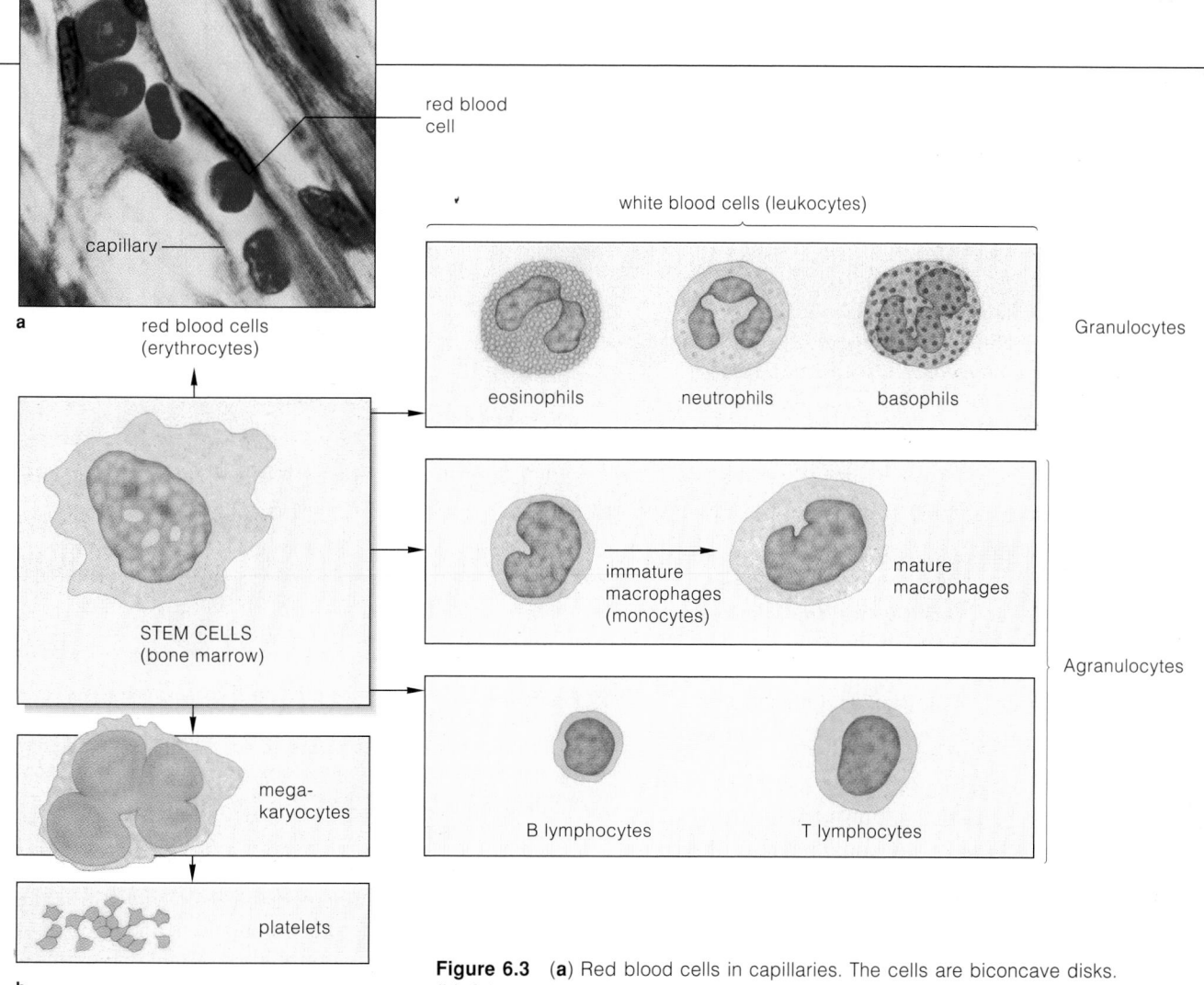

a

red blood
cell

capillary

red blood cells
(erythrocytes)

STEM CELLS
(bone marrow)

mega-
karyocytes

platelets

b

white blood cells (leukocytes)

eosinophils   neutrophils   basophils

Granulocytes

immature
macrophages
(monocytes)

mature
macrophages

B lymphocytes   T lymphocytes

Agranulocytes

Figure 6.3 (**a**) Red blood cells in capillaries. The cells are biconcave disks.
(**b**) Other cellular components of blood.

scavenge dead or worn-out cells, as well as any material that is identified as foreign to the body. Others target or destroy specific bacteria, viruses, or other disease agents. Leukocytes arise from stem cells in bone marrow. They circulate in blood, but most go to work after they squeeze out of blood capillaries and enter tissues. The number of white blood cells varies, depending on whether a person is sedentary or highly active, healthy or battling infection.

There are five types of white blood cells, divided into two major classes (Figure 6.3b). The **granulocytes** include *neutrophils*, *eosinophils*, and *basophils*. All have a lobed nucleus. When the cell is stained, various types of granules are visible in the cytoplasm. About two-thirds of all leukocytes are neutrophils, "search and destroy" cells that follow chemical trails to infected, inflamed, or damaged tissues. Eosinophils take part in allergic responses and attacks on invading parasites. Basophils appear to have roles in inflammation and allergic responses.

The leukocytes known as **agranulocytes** do not have visible granules in the cytoplasm. One type, *monocytes*, differentiate into phagocytic macrophages that engulf infectious microbes and cellular debris in tissues. Another

type, *lymphocytes* (B cells and T cells), have key roles in specific immune responses. Most types of white cells live only for a few days or, during a major infection, a few hours. Others may live for years.

Platelets

Some stem cells in bone marrow develop into "giant" cells called megakaryocytes. These cells shed fragments of their cytoplasm, which become enclosed in a bit of plasma membrane. The membrane-bound fragments, known as **platelets** (or thrombocytes), are disks about 2–4 micrometers across. Each lasts only 5 to 9 days, but millions are always circulating in blood. Substances released from platelets initiate blood clotting.

Blood consists of plasma, in which proteins and other substances are dissolved; red blood cells; white blood cells; and platelets.

Red blood cells transport oxygen. White blood cells have roles in defense, and platelets function in blood clotting.

Hemoglobin, the Oxygen Carrier

If you were to analyze a liter of blood drawn from an artery, you would find only a small amount of oxygen dissolved in the plasma—just 3 milliliters. Yet, like all large, active, warm-blooded animals, humans require a great deal of oxygen to maintain the metabolic activity of their cells. Hemoglobin (Hb) meets this need. In addition to 3 milliliters of dissolved oxygen, a liter of arterial blood generally carries roughly 194 milliliters of oxygen bound to heme groups of hemoglobin molecules.

Factors That Affect Oxygen Binding

As conditions vary in different tissues and organs, so does the tendency of hemoglobin to bind with and hold onto oxygen. Several factors influence this process, which is vital in helping to maintain homeostasis. The most important factor is the amount of oxygen present relative to the amount of carbon dioxide. Other factors are the temperature and the acidity of tissues. Hemoglobin binds oxygen most readily where blood plasma contains a relatively large amount of oxygen, where the temperature is relatively cool, and where the pH is roughly neutral. This is exactly the environment in the lungs, where the body's blood supply must become oxygenated. By contrast, metabolic activity requires oxygen, and it increases both the temperature and the acidity (lowers the pH) of tissues. Under those conditions, the oxyhemoglobin of red blood cells arriving in tissue capillaries tends to give up oxygen, which then becomes available to metabolically active cells. We can summarize these events this way:

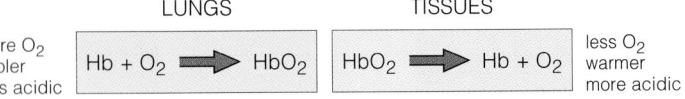

	LUNGS		TISSUES	
ore O_2 oler ss acidic	$Hb + O_2 \Rightarrow HbO_2$		$HbO_2 \Rightarrow Hb + O_2$	less O_2 warmer more acidic

Hemoglobin also transports some of the carbon dioxide wastes of aerobic metabolism and carries hydrogen ions that help control the pH of body fluids. We will return to hemoglobin's functions in Chapter 8, which considers the array of interacting elements that enable the respiratory system to efficiently transport gases to and from body cells.

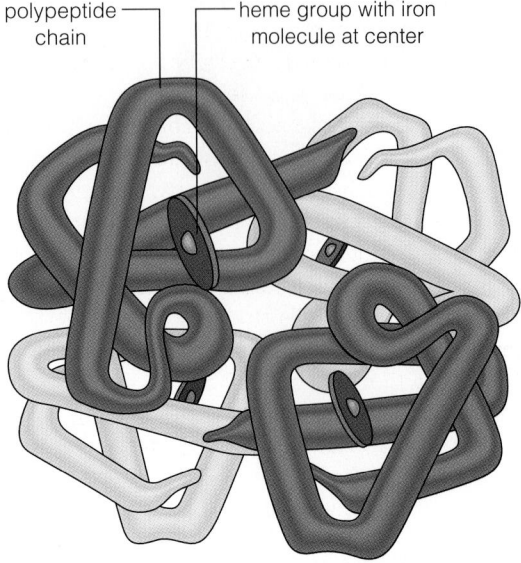

polypeptide chain — heme group with iron molecule at center

Figure 6.4 The structure of hemoglobin. Recall from Chapter 1 (Figure 1.24) that hemoglobin is a globular protein consisting of four polypeptide chains and four iron-containing heme groups. Oxygen binds to the iron in heme groups, which is one reason why humans require iron as a mineral nutrient.

Figure 6.4 shows a hemoglobin molecule's structure. It has two components: the protein *globin* and nonprotein *heme groups* that contain iron. Globin is built of four linked polypeptide chains, and a heme group is associated with each chain. It is the iron molecule at the center of each heme group that binds oxygen.

Oxygen in the lungs diffuses into the blood plasma, then into individual red blood cells, where it binds with the iron in hemoglobin. This oxygenated hemoglobin, or *oxyhemoglobin*, is bright red. Hemoglobin depleted of oxygen appears bluer, especially when it is observed through blood vessel walls and skin.

Hemoglobin in red blood cells transports oxygen, which becomes bound to iron molecules in heme groups in each hemoglobin molecule.

Factors that influence hemoglobin binding—and the amount of oxygen available to tissues—include the relative amounts of oxygen and carbon dioxide present in blood and the temperature and the acidity of tissues.

LIFE CYCLE OF RED BLOOD CELLS

As Figure 6.3 shows, red blood cells are derived from stem cells in red bone marrow, which in adults occurs in certain bones of the skull, the vertebrae, and the sternum (breastbone). (In general, a **stem cell** is unspecialized. It can give rise to descendants that differentiate into specialized cells.) The kidneys produce a hormone called *erythropoietin*, which stimulates certain stem cells to produce red blood cells. Roughly 3 million new red blood cells enter your bloodstream each second! As they mature (Figure 6.5), they lose their nucleus, ribosomes, and other structures. Red blood cells do not divide or synthesize new proteins, but they have enough enzymes and other proteins to function for about 120 days.

In the procedure known as "blood doping," some of an athlete's blood is withdrawn and stored. In response to the loss of red blood cells, erythropoietin stimulate the production of replacements. The stored blood is then reinjected several days prior to an athletic event, so that the athlete has a greater-than-normal number of red blood cells to carry oxygen to muscles—and a temporary, if unethical, competitive advantage.

As red blood cells near the end of their useful life, die, or become damaged or abnormal, phagocytes called *macrophages* ("big eaters") remove them from the blood. Much of this "cleanup" occurs in the spleen, which is located in the upper left abdomen. As a macrophage dismantles a hemoglobin molecule, amino acids from its proteins are returned to the bloodstream, and the iron in its heme groups is returned to the red bone marrow, where it may be recycled in new red blood cells. The rest of the heme group is converted to the orangish pigment bilirubin, which travels to the liver and is mixed with bile secreted during digestion.

Ongoing replacements from stem cells keep the red blood cell count fairly constant. A *cell count* tallies the number of cells in a microliter of blood. The red blood cell count averages 5.4 million in males, 4.8 million in females.

Normally, feedback mechanisms help stabilize the red blood cell count. For example, if you take a ski trip in the Rockies, your body must work harder to obtain sufficient oxygen because air contains less oxygen per unit volume at high altitudes. First your kidneys secrete erythropoietin, which stimulates stem cells in bone marrow to step up production of red blood cells. New red blood cells enter the bloodstream, increasing the oxygen-carrying capacity of the blood. As the oxygen level rises in the blood and in tissues, this information feeds back to the kidneys, erythropoietin production falls, and red blood cell production drops.

Anemia develops when the blood's capacity to carry oxygen is reduced. Anemic people tend to feel tired and cold and may look pale and be short of breath. In *iron-deficiency anemia*, red blood cells contain less hemoglobin than normal. This disorder generally results from an iron-poor diet and is treatable with a mineral supplement. *Pernicious anemia* is caused by a deficiency of vitamin B_{12}, without which red blood cells do not mature properly. It usually reflects an underlying disorder in which the intestines are unable to absorb that vitamin from food. Treatment involves regular injections of vitamin B_{12}. Other types of anemia develop when a person has too few red blood cells, as when a hemorrhage causes severe blood loss or when disease or radiation therapy destroys stem cells in bone marrow.

Red blood cells arise from stem cells in red bone marrow and live about 120 days. Ongoing replacements from stem cells keep the red blood cell count fairly constant.

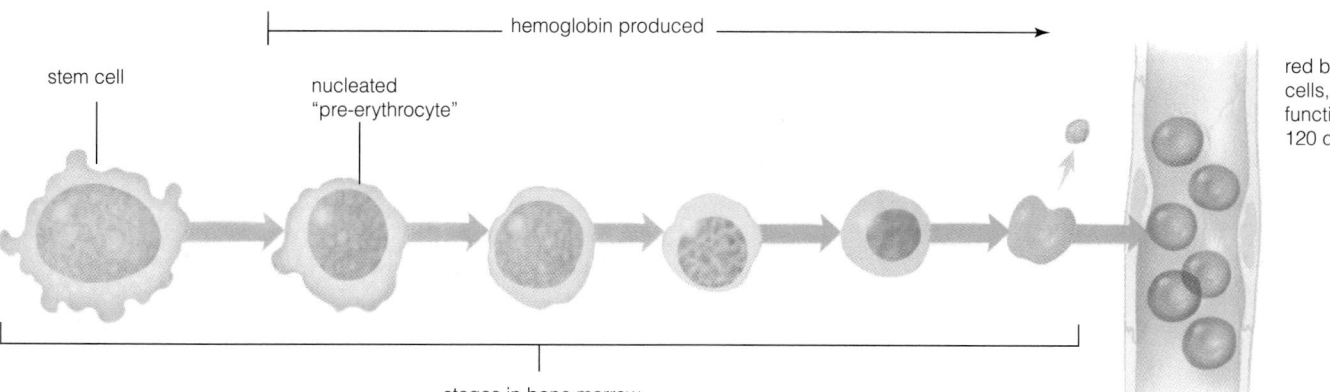

Figure 6.5 Life cycle of a red blood cell. A stem cell in red bone marrow gives rise to a nucleated "pre-erythrocyte." Soon this cell begins making hemoglobin. Later, when the cell's cytoplasm contains a large amount of hemoglobin, its machinery for protein synthesis shuts down. The nucleus is expelled, the center of the cell caves in, and the cell, now an erythrocyte, is released from the bone marrow into the bloodstream.

Genetically determined surface proteins mark each of your body cells, including red blood cells, as "self" (see Figure 2.3). You probably know that there are different human blood types; the differences are due to variations in the surface markers on red blood cells. Such variations are medically important because immune system proteins called **antibodies** recognize and organize an attack on most entities, including viruses and bacteria, that aren't one of the body's own cells. This happens precisely because the foreign material has "nonself" markers. Any protein marker that prompts such a defensive attack is called an **antigen**. We will examine antigen-antibody interactions in Chapter 7. For the time being, we are interested in what happens when the blood of two people mixes, as it does during a transfusion.

ABO Blood Typing

One kind of marker on human red blood cells has variant forms known as A, B, and O. This group is of particular interest because severe immune responses result when incompatible types are mixed. In type A blood, red blood cells bear A markers. Type B blood has B markers, type AB has both A and B, and type O has neither. If you are type A, your body does not have antibodies against A markers but does have them against B markers. If you are type B, you lack antibodies against B markers, but do have antibodies against A markers. If you are type AB, you lack antibodies against either form of the marker, so you can tolerate donations of type A, B, AB, or O blood. However, if you are type O, you have antibodies against both forms of the marker, so you are limited to type O donations.

Figure 6.6 shows what happens when blood from incompatible donors and recipients intermingles. In a defensive response called **agglutination**, antibodies act against the foreign cells and cause them to clump. The clumps can clog small blood vessels, causing severe tissue damage and even death.

Rh Blood Typing

Other kinds of surface markers on red blood cells also may cause agglutination responses. **Rh blood typing** is based on the presence or absence of an Rh marker (so named because it was first identified in the blood of *Rh*esus monkeys). If you are type Rh$^+$, your blood cells bear this marker. If you are type Rh$^-$, they don't. Ordinarily, people do not have antibodies against Rh markers. But a recipient of an Rh$^+$ blood transfusion will produce antibodies against the marker, and these will continue circulating in the blood.

If an Rh$^-$ woman becomes pregnant by an Rh$^+$ man, there is a chance the fetus will be Rh$^+$. During pregnancy or childbirth, some of the fetal red blood cells may leak into the mother's bloodstream. If they do, her body will

Figure 6.6 (**a**) Agglutination responses in blood types O, A, B, and AB when mixed with blood samples of the same and different types. (**b**) Micrographs showing the absence of agglutination in a mixture of two different but compatible types (**top**) and agglutination in a mixture of incompatible blood types (**bottom**).

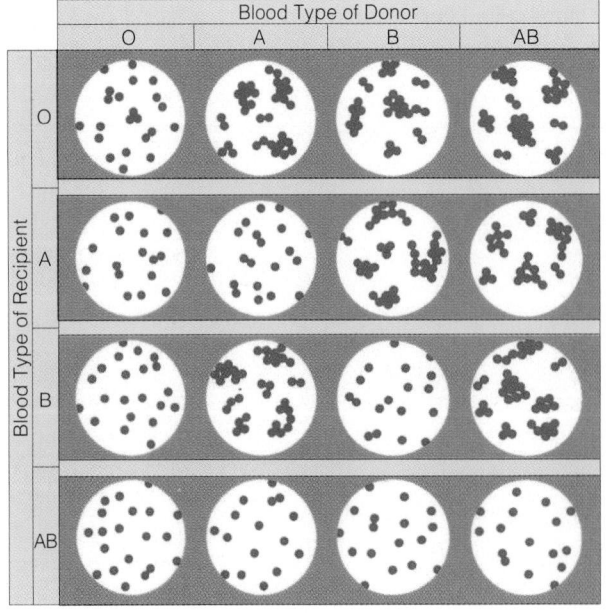

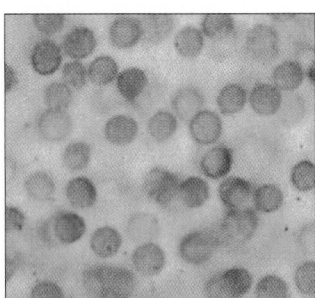

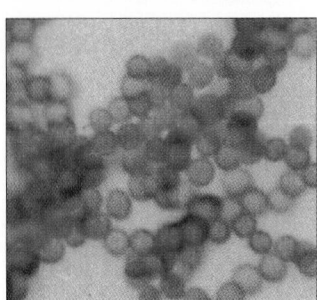

a

b

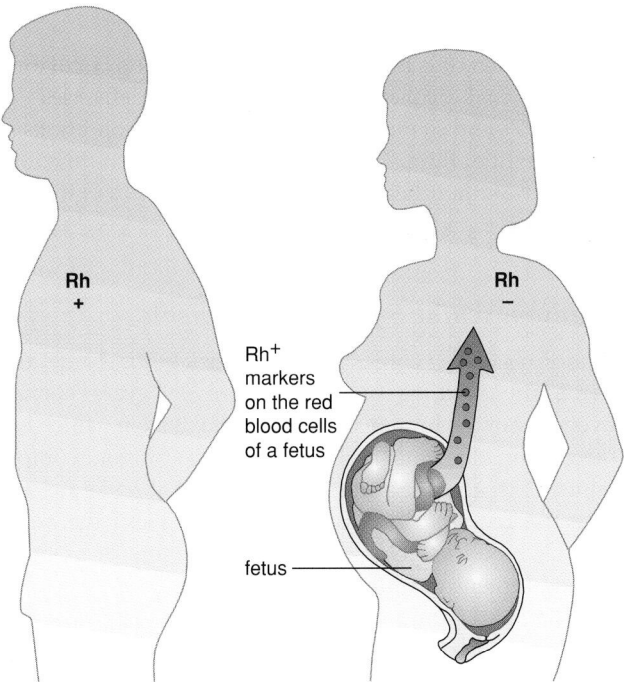

a A forthcoming child of an Rh⁻ woman and Rh⁺ man inherits the genetic instructions for the Rh⁺ marker. Some blood cells from the growing fetus can leak into the mother's bloodstream, passing through the placenta (a complex tissue that forms during pregnancy).

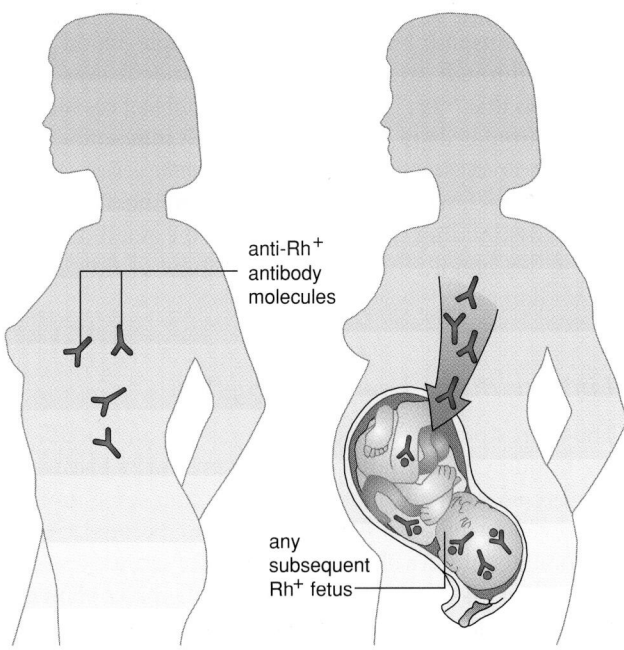

b The foreign markers in the mother's body stimulate production of antibody molecules. Suppose the woman becomes pregnant again. If the new fetus (or any other one) inherits instructions for the Rh⁺ marker, the anti-Rh⁺ antibodies will act against them.

Figure 6.7 Development of antibodies in response to Rh⁺ blood.

produce antibodies against Rh (Figure 6.7). If she becomes pregnant *again*, Rh antibodies will enter the bloodstream of this new fetus. If its blood is type Rh⁺, maternal antibodies will cause fetal red blood cells to swell and burst.

In extreme cases of this disorder, called *hemolytic disease of the newborn*, too many red blood cells are destroyed and the fetus dies. If the condition is diagnosed before or at a live birth, the baby can survive by having its blood replaced with transfusions free of Rh antibodies.

Currently, a known Rh⁻ woman can be treated after her first pregnancy with an anti-Rh gamma globulin (RhoGam) that will protect her next fetus. The drug will inactivate Rh⁺ fetal blood cells circulating in the mother's bloodstream before she can become sensitized and begin producing anti-Rh antibodies. In non-maternity cases, an Rh⁻ person who receives a transfusion of Rh⁺ blood also can have a severe negative reaction if he or she has previously been exposed to the Rh marker.

Besides the ABO and Rh blood markers, hundreds of others are now known to exist. These markers are a bit like needles in a haystack, for they are widely scattered within the human population and typically do not cause problems in transfusions. Reactions do occur, however; and except in cases of extreme emergency, hospitals use a technique called *cross-matching* to exclude the remote possibility that blood to be transfused and that of a patient might be incompatible due to the presence of a rare blood cell marker outside the ABO and Rh groups.

Other Applications of Blood Typing

ABO markers occur not only in blood, but also in semen and saliva. Because blood groups are determined by genes, they are a useful repository of information about a person's genetic heritage. For example, it is now common for criminal investigations of murders, rapes, and sometimes other crimes to focus in part on comparisons of the blood groups of victims and any possible perpetrators. Beyond this, blood samples provide DNA for testing, a modern application of biotechnology that we consider in Chapter 21. The same techniques are used to help establish the identity of a child's father in cases of disputed paternity.

Like all cells, red blood cells bear genetically determined proteins on their surface. These proteins serve as "self" markers and determine a person's ABO (and Rh) blood type.

When incompatible blood types mix, an agglutination response occurs in which antibodies cause potentially fatal clumping of red blood cells.

OVERVIEW OF THE CARDIOVASCULAR SYSTEM

Like the freeways and streets crisscrossing a metropolis, the **cardiovascular system** is the body's internal rapid-transport system. It is the highway by which cells receive such essential substances as oxygen, nutrients extracted from food, and secretions (like hormones) from other cells. It is also the disposal route for waste products of metabolism. As you will see, blood serves as the "carrier" for these various substances, among its other roles.

The cardiovascular system's basic role is to circulate blood to the immediate neighborhood of every living cell in the body. This task is vital because the various types of cells in the human body have only a few mechanisms for adjusting to drastic changes in the composition, volume, and temperature of interstitial fluid. As Chapter 3 described, interstitial fluid is the "external environment"

within which cells live. Hence cells depend on circulating blood to make constant "pickups" and "deliveries" of substances and so maintain stable operating conditions in the interstitial fluid. The cardiovascular system functions together with other organ systems, as illustrated in Figure 6.8.

Cardiovascular System Components

"Cardiovascular" comes from the Greek *kardia* (heart) and the Latin *vasculum* (vessel). Figure 6.9 gives you an overview of the human cardiovascular system, which has three main elements: (1) blood, with its different components; (2) the heart, a muscular pump that generates the pressure required to move blood throughout the body; and (3) blood vessels, which you can think of as tubes having lumens of different diameters.

The heart pumps blood into large-diameter **arteries**. From there blood flows into smaller, muscular **arterioles**, which branch into even smaller-diameter **capillaries**. Blood flows from capillaries into small **venules**, then into large-diameter **veins** that return blood to the heart. The heart pumps constantly, so the volume of flow through the entire system each minute is equal to the volume of blood returned to the heart each minute.

As you will read later, the rate and volume of blood flow through the cardiovascular system can be adjusted as conditions in the body vary. For example, blood flows rapidly through arteries, but in capillaries it must flow slowly so that there is enough time for substances moving to and from cells to be exchanged with interstitial fluid. The required slowdown occurs in *capillary beds*, where blood flow fans out through vast numbers of capillaries. By dividing up the blood flow, the small-diameter capillaries handle the same total volume of flow as the large-diameter vessels, but at a slower pace.

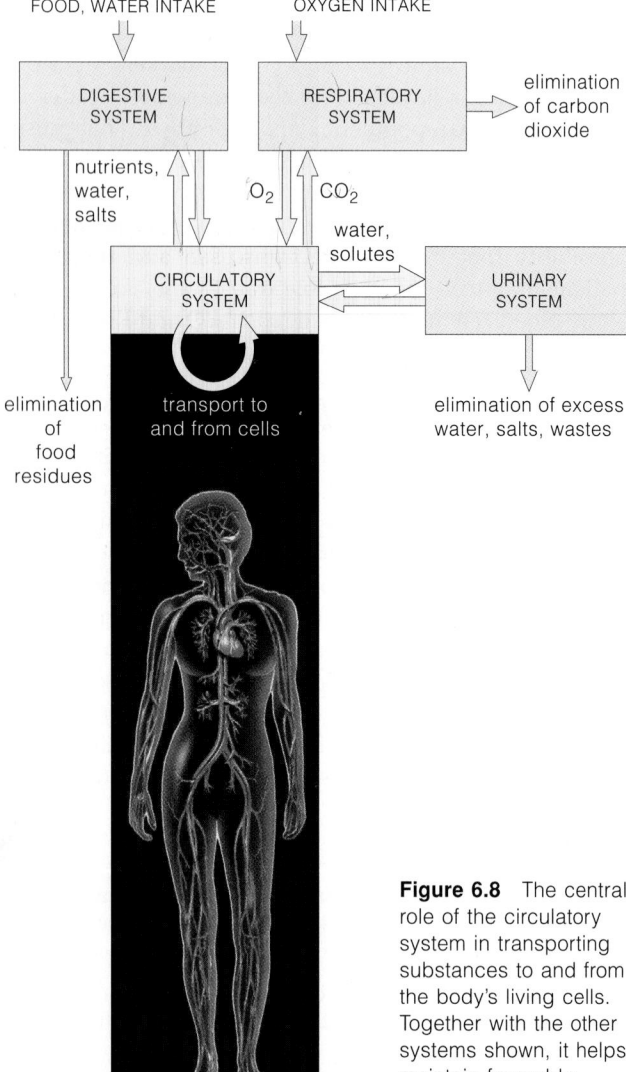

Figure 6.8 The central role of the circulatory system in transporting substances to and from the body's living cells. Together with the other systems shown, it helps maintain favorable operating conditions in the internal environment.

Links with the Lymphatic System

The heart's pumping action puts pressure on blood flowing through the cardiovascular system. Partly because of this pressure, small amounts of water and some proteins dissolved in blood are forced out and become part of interstitial fluid. An elaborate network of drainage vessels picks up excess interstitial fluid and reclaimable solutes—including water, proteins, fatty acids, and glycerol—and returns them to the cardiovascular system. This network is part of the lymphatic system, about which you will learn more later in this chapter.

The cardiovascular system transports substances to and from the interstitial fluid that bathes all living cells.

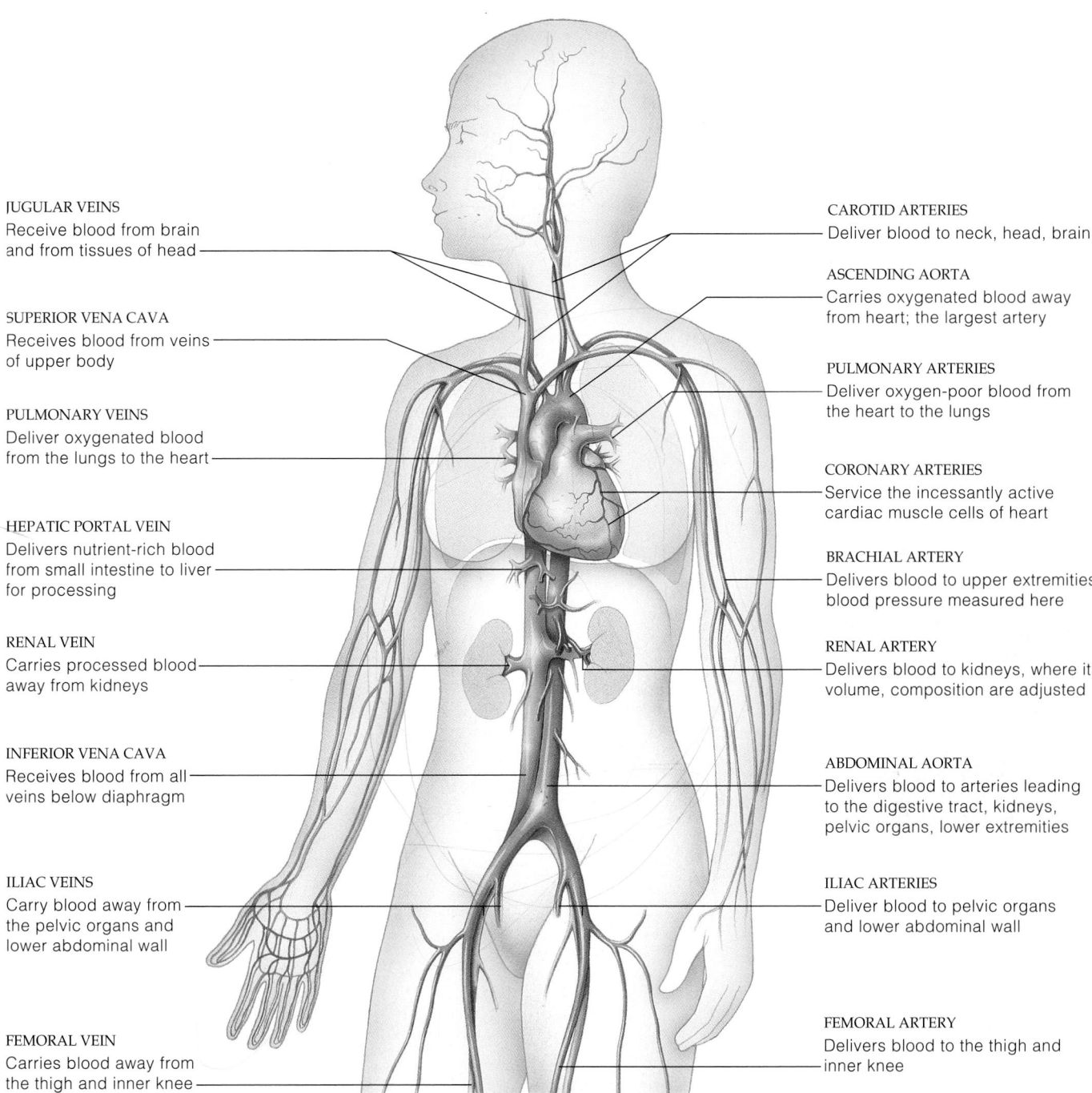

JUGULAR VEINS
Receive blood from brain and from tissues of head

SUPERIOR VENA CAVA
Receives blood from veins of upper body

PULMONARY VEINS
Deliver oxygenated blood from the lungs to the heart

HEPATIC PORTAL VEIN
Delivers nutrient-rich blood from small intestine to liver for processing

RENAL VEIN
Carries processed blood away from kidneys

INFERIOR VENA CAVA
Receives blood from all veins below diaphragm

ILIAC VEINS
Carry blood away from the pelvic organs and lower abdominal wall

FEMORAL VEIN
Carries blood away from the thigh and inner knee

CAROTID ARTERIES
Deliver blood to neck, head, brain

ASCENDING AORTA
Carries oxygenated blood away from heart; the largest artery

PULMONARY ARTERIES
Deliver oxygen-poor blood from the heart to the lungs

CORONARY ARTERIES
Service the incessantly active cardiac muscle cells of heart

BRACHIAL ARTERY
Delivers blood to upper extremities; blood pressure measured here

RENAL ARTERY
Delivers blood to kidneys, where its volume, composition are adjusted

ABDOMINAL AORTA
Delivers blood to arteries leading to the digestive tract, kidneys, pelvic organs, lower extremities

ILIAC ARTERIES
Deliver blood to pelvic organs and lower abdominal wall

FEMORAL ARTERY
Delivers blood to the thigh and inner knee

Figure 6.9 Human cardiovascular system. Arteries, which carry oxygenated blood to tissues, are shaded *red*. Veins, which carry deoxygenated blood away from tissues, are shaded *blue*. Notice, however, that for the pulmonary arteries and veins the roles are reversed.

THE HEART: A DURABLE PUMP

In a lifetime of 70 years, the human **heart** beats some 2.5 billion times. The heart's structure, shown in Figure 6.10, reflects this marvelous organ's role as a durable pump. The heart is mostly cardiac muscle tissue, the **myocardium**. A tough, fibrous sac, the *pericardium*, surrounds, protects, and lubricates the heart. The heart's inner chambers have a smooth lining (*endocardium*) composed of connective tissue and a layer of epithelial cells. The epithelial cell layer, known as *endothelium*, also lines the inside of blood vessels.

A thick wall, the **septum**, divides the heart into two halves, right and left. Each half has two chambers: an **atrium** (plural: atria) located above a **ventricle**. Flaps of membrane separate the two chambers and serve as a one-way **atrioventricular valve** (AV valve) between them. The AV valve in the right half of the heart is termed a *tricuspid valve* because its three flaps come together in pointed cusps (Figure 6.11). In the heart's left half the AV

valve consists of just two flaps; it is called the *bicuspid valve* or *mitral valve*. Tough, collagen-reinforced strands (*chordae tendineae,* or "heartstrings") connect the AV valve flaps to cone-shaped *papillary muscles* that extend out from the ventricle wall. When a blood-filled ventricle contracts, this arrangement prevents the flaps from opening backward into the atrium. Each half of the heart also has a **semilunar valve** between the ventricle and the arteries leading away from it. During a heartbeat, it opens and closes in ways that keep blood moving in one direction through the body.

The heart has its own "coronary circulation." Two **coronary arteries** lead into a capillary bed that services most of its cardiac muscle cells (Figure 6.12). They branch off the **aorta**, the major artery carrying oxygenated blood away from the heart. Coronary arteries become dangerously clogged in some cardiovascular disorders, as described in Sections 6.14 and 6.15.

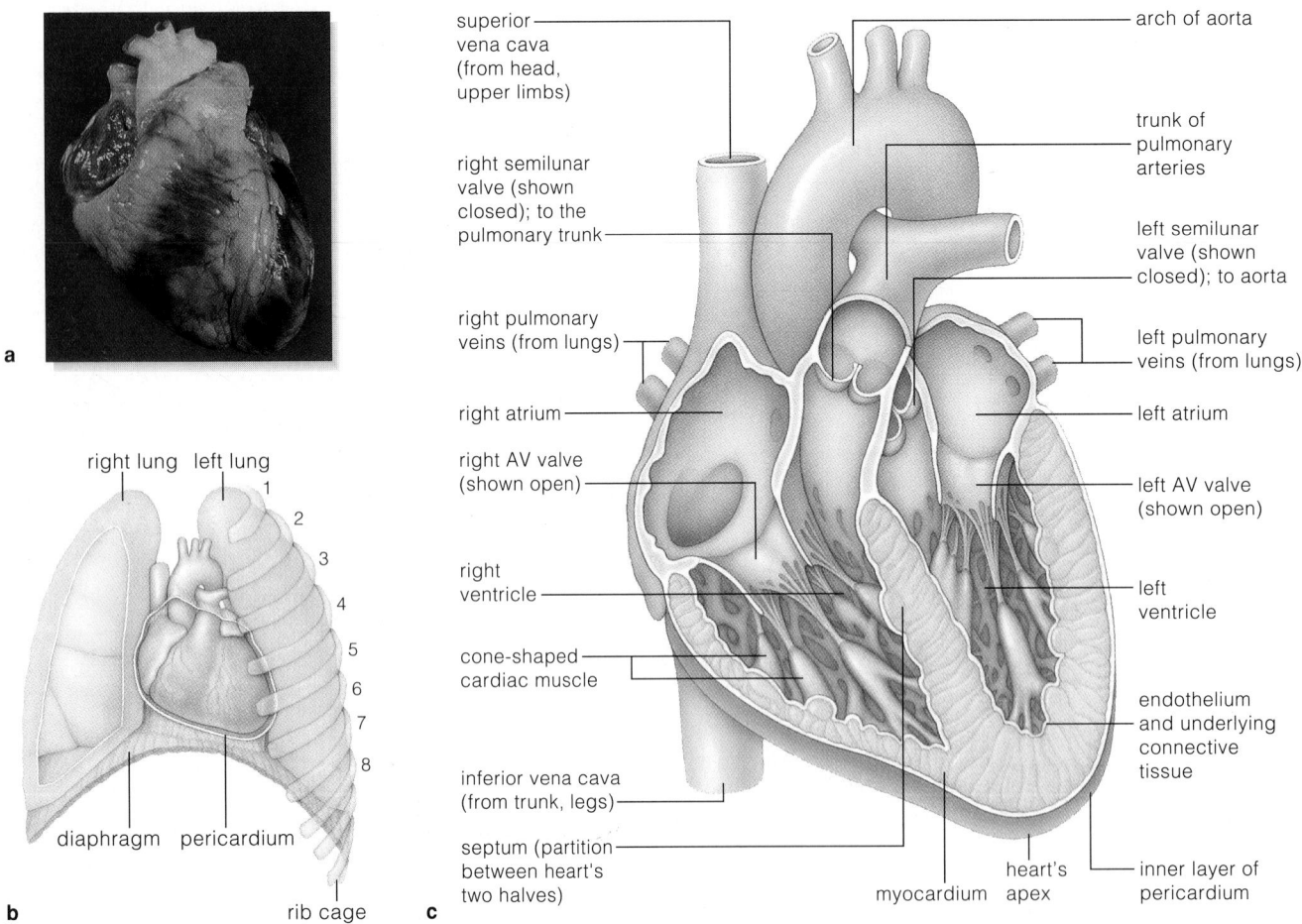

Figure 6.10 (**a**) The human heart, and (**b**) its location. (**c**) Cutaway view showing the heart's internal organization.

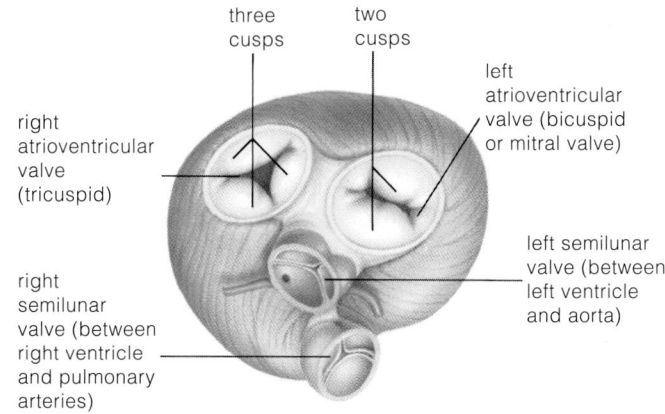

Figure 6.11 The valves of the human heart. In this drawing, you are looking down at the heart, and the atria have been removed so that the atrioventricular and semilunar valves are visible.

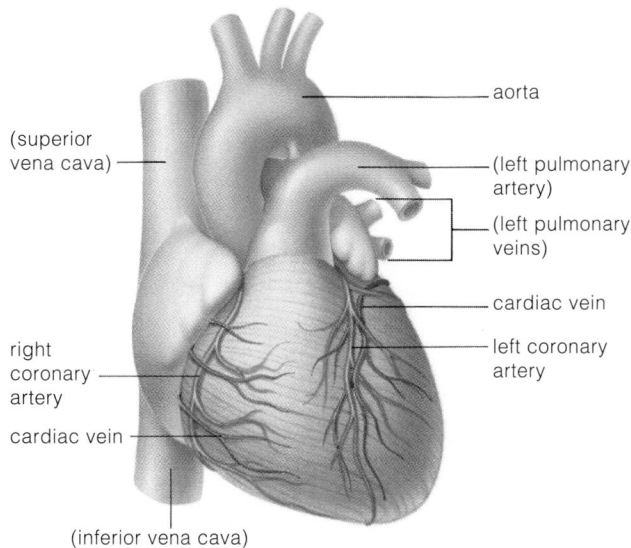

Figure 6.12 Coronary arteries and veins.

With this overview of heart structure in mind, we now can look more closely at the vascular routes blood travels each time the heart pumps. Research aimed at developing effective blood substitutes has advanced tremendously in recent years, and before turning the page you may be interested to learn more about some pioneering efforts that are the subject of this chapter's *Science Comes to Life.*

The heart consists of two halves, each divided into an atrium and a ventricle. Valves in each half of the heart help control the direction of blood flow.

CAN WE MAKE ARTIFICIAL BLOOD?

Suppose you are rushed to the hospital because you are suffering from severe internal bleeding. You may need a blood transfusion—a serious emergency, even more serious if you happen to have a rare blood type. Or suppose you have a cancer for which the most effective treatment involves transfusing you with platelets—the cell fragments in blood that have roles in blood clotting and are now becoming weapons in various advanced cancer therapies. Your life and health might depend precariously on whether a hospital or blood bank can quickly obtain a quantity of blood cells, platelets, or fresh whole blood of your blood type. Each year, hospital patients in the United States require more than 20 million units of red blood cell products and many more units of other blood products.

Whole blood and its elements are highly perishable. Once obtained from a donor, fresh blood products must be used within a few days. Blood and blood elements can be frozen, but at a biological price. As water molecules form ice, the resulting crystals can rupture the plasma membrane and other membranes within a cell. Using traditional methods, freezing platelets generally destroys at least half of them. Red blood cells must be stored in costly specialized freezers and thawed by way of complex procedures. These and other problems—such as worries about contamination by disease agents such as HIV and the hepatitis C virus—often create life-threatening shortages of blood products for patients who need them.

A logical solution (literally!) is to create a safe, efffective, and cheap blood substitute. As it turns out, however, that is a tall order, because blood performs so many physiological functions—not the least of which are oxygen transport and the ability to clot. A substitute must not trigger defense responses to "foreign" blood described in Section 6.4. Also, transfused blood usually is required in large amounts, so availability and a reasonably long shelf life are concerns.

Recent advances have dramatically improved methods for storing donated platelets. In addition, several candidate blood substitutes now are being tested in humans. At least two experimental products incorporate hemoglobin purified from donated blood that has grown too stale to be used safely. Some others are based on hemoglobin extracted from cow's blood. Still, all these substances have potentially dangerous side effects, ranging from kidney damage to high blood pressure (hypertension). Efforts to create a totally synthetic blood substitute have focused on perfluorocarbons (PFCs), compounds that can dissolve large amounts of oxygen and carbon dioxide and can be stored for up to two years. PFC-containing solutions would be excellent candidates for blood substitutes, except that they can actually trigger toxic reactions from *too much* oxygen in tissues, along with other side effects. As the search for blood substitutes continues, it underscores the high regard in which we should hold the remarkable fluid tissue that has aptly been called the "elixir of life."

BLOOD CIRCULATION

Each half of the heart (atrium and ventricle) pumps blood. The two side-by-side pumps are the basis of two cardiovascular circuits through the body, the pulmonary and systemic circuits (Figure 6.13). Each circuit has its own set of arteries, arterioles, capillaries, venules, and veins.

The Two Circuits

The **pulmonary circuit** receives blood from tissues and circulates it through the lungs for gas exchange. It begins as blood from tissues enters the right atrium, then moves through the AV valve into the right ventricle. As the ventricle fills, the atrium contracts. Blood arriving in the right ventricle is fairly low in oxygen and high in carbon dioxide. When the ventricle contracts, the blood moves through the right semilunar valve into the *main* pulmonary artery, then into the *right* and *left* pulmonary arteries. These arteries carry the blood to the two lungs, where (in capillaries) it picks up oxygen and gives up carbon dioxide that will be exhaled. The freshly oxygenated blood returns through two sets of pulmonary veins to the heart's left atrium, completing the circuit.

The **systemic circuit** carries blood to and from tissues. Oxygenated blood pumped by the left half of the heart moves through the body and returns to the right atrium. The left atrium receives blood from pulmonary veins, and this blood moves through an AV (bicuspid) valve to the left ventricle. This chamber contracts forcefully to send blood coursing through a semilunar valve into the aorta.

As the aorta descends into the torso (see Figure 6.9), major arteries branch off it, funneling blood to organs and tissues where O_2 is used and CO_2 is produced.

Deoxygenated blood returns to the right half of the heart, where it enters the pulmonary circuit. Notice that in both the pulmonary and systemic circuits, blood travels through arteries, arterioles, capillaries, and venules, and finally returns to the heart in veins. Blood from the head, arms, and chest arrives through the *superior vena cava*, and the *inferior vena cava* collects blood from the lower body.

Arteries branch off the aorta, carrying blood directly to capillary beds in specific tissues and organs. For example, in a resting person, each minute roughly a quarter of the blood pumped into the systemic circulation enters the kidneys via *renal arteries*. Chapter 9 discusses kidney functions, which include removing metabolic wastes.

Substances actually move between blood and tissues in **capillary beds**. A few capillary beds have two types of vessels: "true capillaries" where exchanges between blood and tissues take place, and "thoroughfare channels"

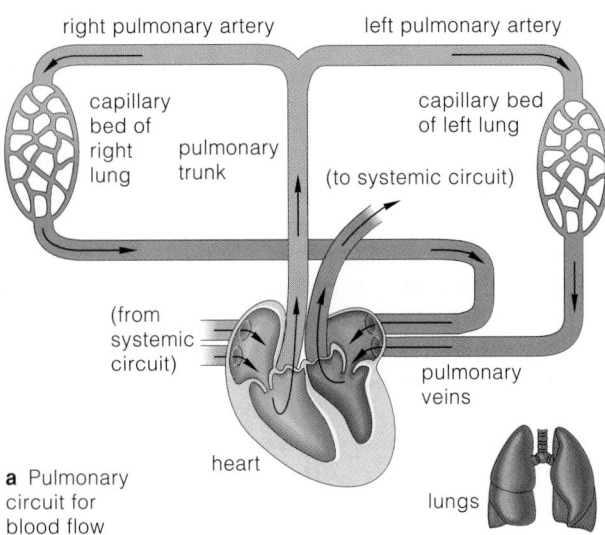

a Pulmonary circuit for blood flow

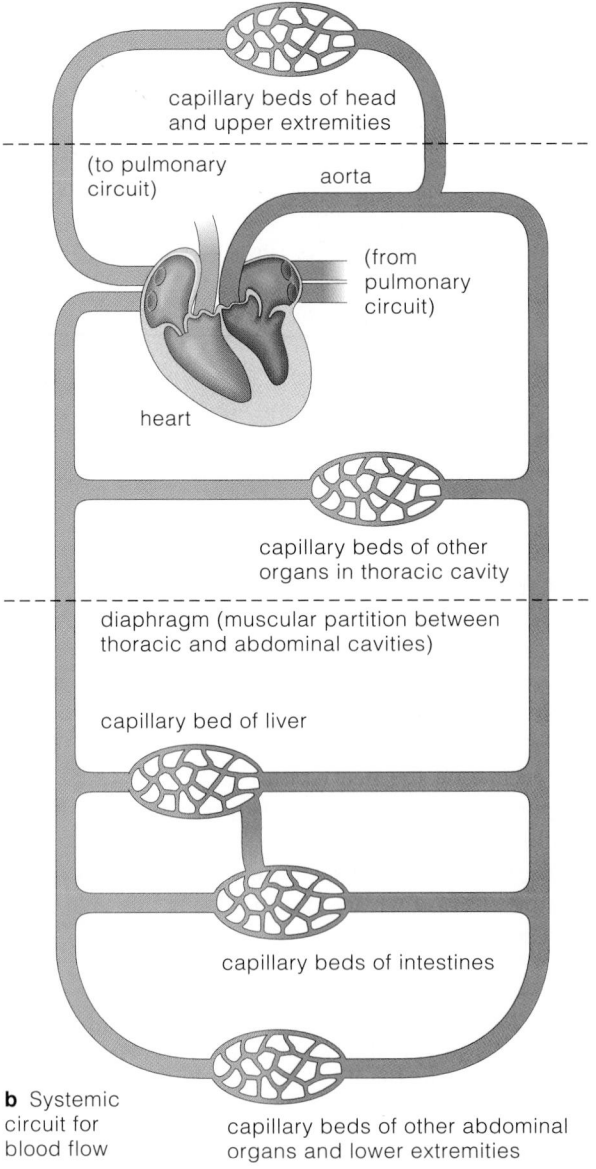

b Systemic circuit for blood flow

Figure 6.13 The (**a**) pulmonary and (**b**) systemic circuits for blood flow in the cardiovascular system. (**c**) Distribution of the the heart's output in a person at rest.

CARDIAC CYCLE

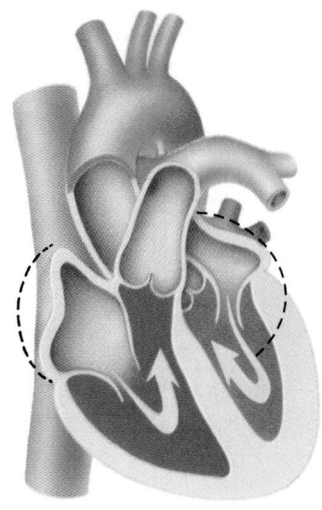

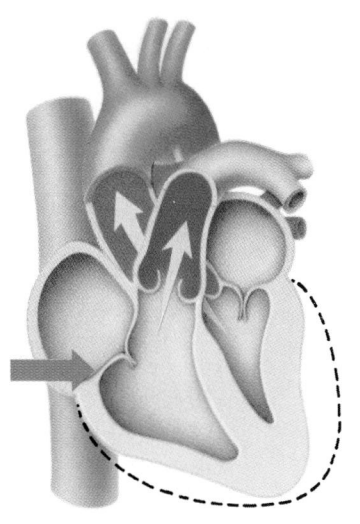

superior vena cava

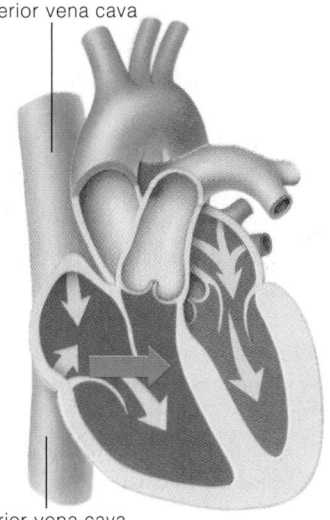

inferior vena cava

Atria contract (systole) and blood enters ventricles through AV valves. Semilunar valves are closed. Ventricles are in diastole.

Atria relax (diastole) and ventricles contract (systole). Blood from right ventricle moves via right semilunar valve to pulmonary artery. Blood from left ventricle moves via left semilunar valve to aorta.

Semilunar valves close. Atria and ventricles are briefly in diastole. Blood is entering right heart chambers through the venae cavae, and filling begins again.

Figure 6.14 Path of blood flow through the heart during part of a cardiac cycle.

---

that connect arterioles and venules. Blood flow into true capillaries is controlled by collars of smooth muscle cells where the capillaries branch from thoroughfare channels.

After a meal the blood passing through capillary beds in the GI tract (taking up nutrients) detours through the *hepatic portal vein* to another capillary bed in the liver. As blood seeps through this second bed, the liver can remove impurities and process absorbed substances. Blood leaving the bed enters the general circulation through a *hepatic vein*. (The liver also receives its own supply of oxygenated blood via the *hepatic artery*.)

Heartbeat: The Cardiac Cycle

Blood is pumped each time the heart beats. A "heartbeat" is one sequence of contraction and relaxation of the heart chambers. The sequence occurs almost simultaneously in both sides of the heart. The contraction phase is called **systole** (SISS-toe-lee), and the relaxation phase is called **diastole** (dye-ASS-toe-lee). This sequence of muscle contraction and relaxation is a **cardiac cycle** (Figure 6.14). Each heartbeat lasts about eight-tenths of a second.

During the cycle, the ventricles relax before the atria contract, and the ventricles contract when the atria relax. When the relaxed atria are filling with blood, fluid pressure inside them rises and the AV valves open. Blood flows into the ventricles, which are 80 percent filled by the time the atria contract. As the filled ventricles begin to contract, fluid pressure inside *them* increases, forcing the AV valves shut. Their ongoing contraction boosts ventricular pressure above that in blood vessels leading away from the heart. The pressure forces the semilunar valves open, and blood flows out of the heart and into the aorta and pulmonary artery. After blood has been ejected, the ventricles relax, and the semilunar valves close. For about half a second the atria and ventricles are all in diastole; then the blood-filled atria contract, and the cycle repeats.

The blood and heart movements during the cardiac cycle generate an audible "lub-dup" sound made by the forceful closing of the heart's one-way valves. At each "lub," the AV valves are closing as the two ventricles contract. At each "dup" the semilunar valves are closing as the ventricles relax.

A short pulmonary circuit carries blood through the lungs for gas exchange. A long systemic circuit transports blood to and from tissues.

During a cardiac cycle, contraction of the atria helps fill the ventricles. Contraction of the left ventricle pumps blood into the systemic circuit.

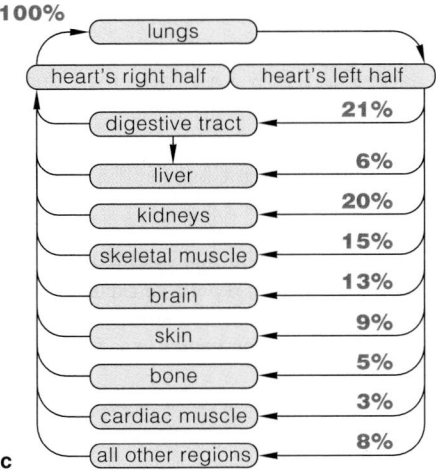

c

The Cardiac Conduction System

The heart contains subsets of specialized cells that make up the **cardiac conduction system**. As this chapter's opening vignette described, the system includes self-excitatory pacemaker cells. These cells spontaneously produce and conduct the electrical impulses (action potentials) that stimulate contractions in the heart's contractile cells. Even if all nerves leading to the heart are severed, the heart will keep on beating.

In skeletal muscle tissue, the ends of the fiberlike cells are attached to bones or tendons. In cardiac muscle tissue, the ends of cardiac muscle cells branch, then connect with one another at their endings (Figure 6.15). Communication junctions (called *intercalated discs*) bridge the plasma membranes on the ends of abutting cells. With each heartbeat, signals calling for contraction spread so rapidly across the junctions that cardiac muscle cells contract together, almost as if they were a single unit.

Normally, excitation begins with a small mass of cells in the upper wall of the right atrium. This **sinoatrial (SA) node** is the **cardiac pacemaker**. The SA node generates one wave of excitation after another, usually 70 or 80 times a minute. Each wave spreads over both atria, causes them to contract, then rapidly reaches the **atrioventricular node** (AV node) in the septum dividing the two atria. Notice in Figure 6.16 that bundles of conducting fibers extend from the AV node to each ventricle. At intervals along each bundle, conducting cells called *Purkinje fibers* branch off and make contact with contractile muscle cells in the ventricles. When a stimulus reaches the AV node, it slows, almost pausing, before it quickly passes along

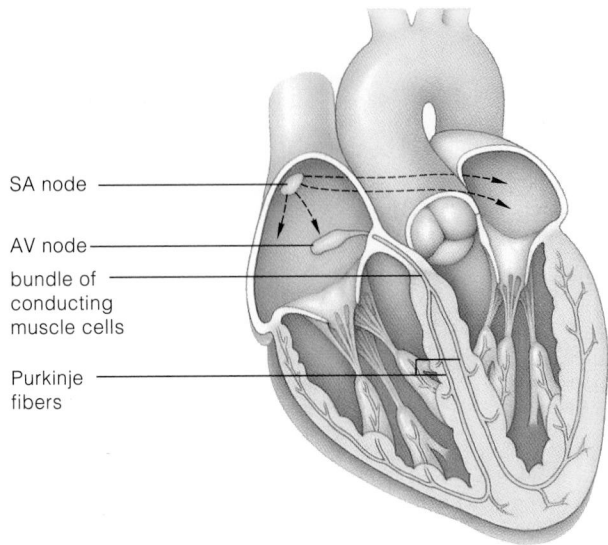

Figure 6.16 Location of specialized cardiac muscle cells that conduct signals for contraction through the heart.

the bundles to Purkinje fibers and on to contractile muscle fibers in each ventricle. The slow conduction in the AV node gives the atria time to finish contracting before the wave of excitation spreads to the ventricles.

Of all cells of the cardiac conduction system, the SA node fires off impulses at the highest frequency and is the first region to respond in each cardiac cycle. Its rhythmic firing is the basis for the normal rate of heartbeat. The ECG in Figure 6.26 traces a normal heartbeat sequence. People whose SA node chronically malfunctions may have an artificial pacemaker implanted to provide a regular stimulus for heart contraction.

Neural Controls over Heart Rate

Whereas the nervous system triggers the contraction of skeletal muscle, it can only *adjust* the rate and strength of cardiac muscle contraction. Stimulation by sympathetic nerves of the autonomic nervous system increases the force and rate of heart contractions, while conversely parasympathetic nerves can slow heart activity. Centers for neural control of heart functions lie in the spinal cord, the brain's medulla oblongata, and other brain regions about which you will read more in Chapter 10.

The SA node is the cardiac pacemaker that establishes a regular heartbeat. Its spontaneous, repetitive excitation spreads along a system of muscle cells that stimulate contractile tissue in the atria, then the ventricles, in a rhythmic cycle.

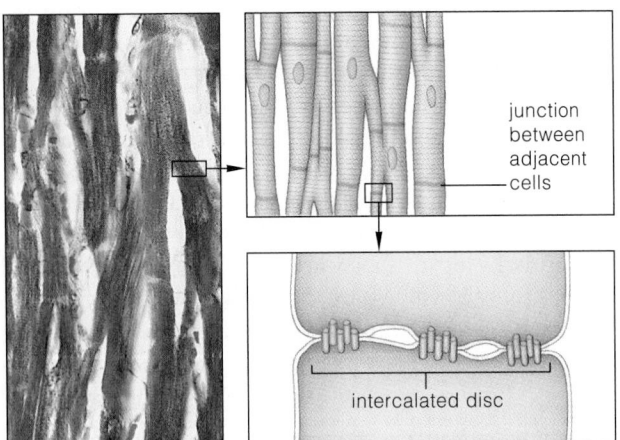

Figure 6.15 Intercalated discs containing communication junctions at the ends of adjacent cardiac muscle cells. Signals travel rapidly across the junctions and cause cells to contract nearly in unison.

Blood Pressure

Heart contractions generate **blood pressure**, the fluid pressure blood exerts against vessel walls. Pressure is highest in the aorta, then drops along the systemic circuit. Blood pressure typically is measured when a person is at rest (Figure 6.17). For an adult, 120/80 is in the normal range. The first number is *systolic pressure*, the pressure peak in the aorta while the left ventricle contracts and pushes blood into the aorta. The second number, *diastolic pressure*, measures the lowest blood pressure in the aorta, when the heart is relaxed and blood is flowing out of the aorta.

Values for systolic and diastolic pressure provide vital health information. Elevated blood pressure can be associated with various ailments, such as atherosclerosis (Section 6.15) and kidney disease.

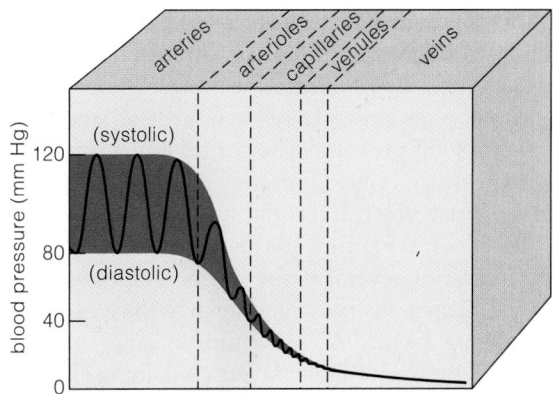

Figure 6.18 Blood pressure. This diagram plots measurements of the drop in fluid pressure for a given volume of blood moving through the systemic circuit.

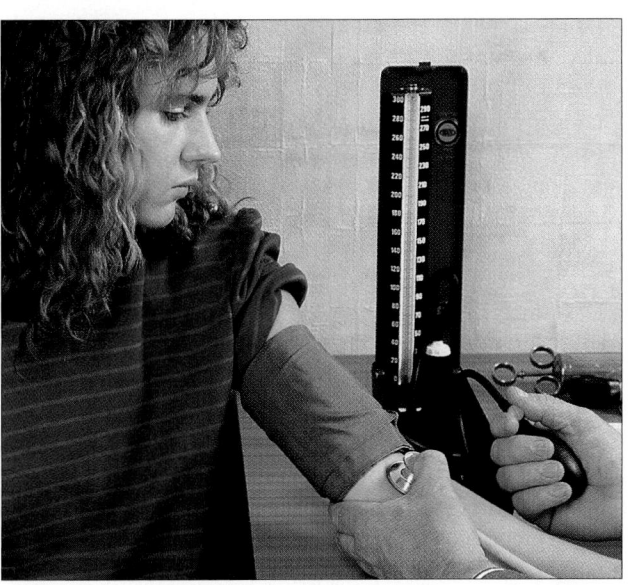

Figure 6.17 Measuring blood pressure with a device called a sphygmomanometer. A hollow cuff attached to a pressure gauge is wrapped around the upper arm. Then it is inflated with air to a pressure above the highest pressure of the cardiac cycle (at systole, when the ventricles contract). Above the systolic pressure, no sounds are heard through a stethoscope positioned below the cuff (because no blood is flowing through the vessel).

Air in the cuff is slowly released, so some blood flows into the artery. The turbulent flow causes soft tapping sounds, and when this first occurs, the value on the gauge is the systolic pressure—about 120 mm mercury (Hg) in young adults at rest. (This means the measured pressure would force mercury to move upward 120 millimeters in a narrow glass column.)

More air is released from the cuff. Just after the sounds become dull and muffled, blood flows continuously, and so the turbulence and tapping sounds stop. The silence corresponds to the diastolic pressure at the end of a cardiac cycle, just before the heart pumps out blood. Generally the reading is about 80 mm Hg. In this example, the *pulse* pressure (the difference between the highest and lowest pressure readings) is 120 − 80, or 40 mm Hg.

Friction and other factors combine to create resistance to the movement of blood in vessels. When friction occurs, energy is lost (as heat). In the systemic circulation, resistance increases dramatically as flowing blood moves from arteries to arterioles. The increasing resistance causes circulating blood to lose energy, and so blood pressure drops (Figure 6.18).

Resting blood pressure depends mainly on centers in the medulla oblongata of the brain (Chapter 10). The centers integrate information from sensory receptors in cardiac muscle tissue and in certain arteries, such as the aorta and the **carotid arteries** in the neck. They use this information to coordinate the rate and strength of heartbeats with changes in the diameter of arterioles and, to some extent, of veins. You will learn more about how the controls regulate vessel diameter in the next section.

Blood Velocity

Pumped by the muscular left ventricle, blood entering the systemic circulation is moving rather rapidly when it leaves the heart in the aorta. Just as a river's flow slows when it becomes divided into many small channels, the flow of arterial blood slows as the total cross-sectional area of vessels increases. Its velocity is greatest in the aorta, decreases in the more numerous arterioles, and slows to a relative crawl in countless narrow capillaries. Hence, there is sufficient time for materials to diffuse between capillaries and tissues. Velocity increases again as blood moves into veins for the return trip to the heart.

Blood pressure is highest in the aorta and drops along the rest of the systemic circuit. The velocity of blood flow varies with the total cross-sectional area of the vascular network.

HOW VESSEL STRUCTURE AFFECTS BLOOD PRESSURE

The aorta and other systemic arteries receive oxygenated blood from the heart. Pulmonary arteries receive deoxygenated blood destined for the lungs. Either way, the arteries serve as pressure reservoirs that "smooth out" surges in blood pressure. The surges are generated during each cardiac cycle. In arteries near the body surface, as in the wrist, you can feel the surges as a **pulse**.

Take a look at Figure 6.19a and you will see that artery walls consist of several tissue layers. The outer layer is mainly collagen fibers, which anchor the vessel to the surrounding tissue. A thick middle layer of smooth muscle lies between thinner layers containing elastin. The thick, muscular, and elastic wall of a large artery bulges slightly under the pressure surge caused by a ventricular contraction.

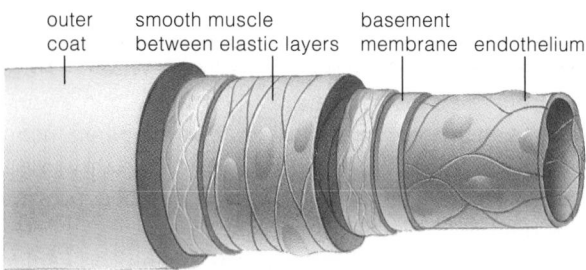

a ARTERY

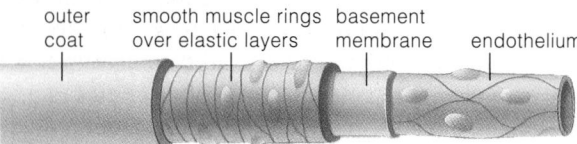

b ARTERIOLE

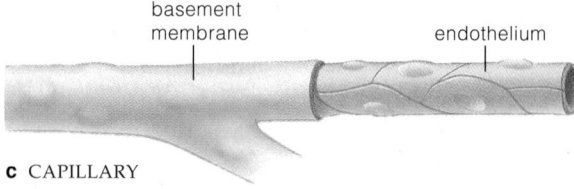

c CAPILLARY

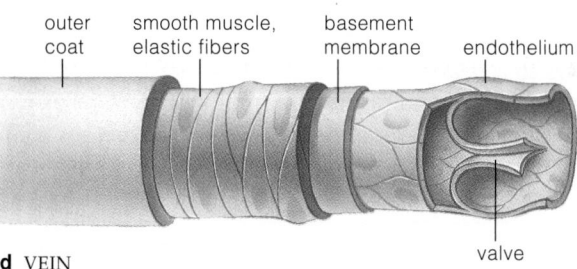

d VEIN

Figure 6.19 Structure of blood vessels. The basement membrane around the endothelium contains a form of collagen especially rich in proteins and polysaccharides.

The bulging of artery walls helps keep blood flowing on through the system. Some of the blood volume pumped during the systole phase of each cardiac cycle is momentarily stored in the "bulge"; the elastic recoil of the artery then forces that stored blood onward during diastole, when heart chambers are relaxed. Arteries also have large diameters. They present little resistance to blood flow, so pressure does not drop much in the large arteries of the systemic and pulmonary circuits.

Resistance at Arterioles

Arteries branch into arterioles, which have a smaller diameter. The wall of an arteriole has rings of smooth muscle over a single layer of elastic fibers (Figure 6.19b). This enables arterioles to dilate (enlarge in diameter) when the smooth muscle relaxes or to constrict (shrink in diameter) when the smooth muscle contracts. As indicated in Figure 6.18, arterioles offer more resistance to blood flow than other vessels do. As it slows, the blood flow can be controlled in ways that divert greater or lesser portions of the total volume to different regions. As you will see shortly, in response to signals from the nervous system and endocrine system, even in response to a change in local chemical conditions, smooth muscle cells contract or relax—and so increase or decrease the arteriole diameter.

Capillaries

Capillary beds are diffusion zones for exchanges of gases, nutrients, and wastes between blood and interstitial fluid. Of all blood vessels, capillaries have the thinnest wall—a single layer of flat endothelial cells, separated from one another by narrow clefts (Figure 6.19c). Most capillaries have such a small lumen that red blood cells must squeeze through them single file. Thus a single capillary presents high resistance to blood flow. Yet so many of them are present in a capillary bed that their combined diameters are greater than the combined diameters of arterioles leading into them (Figure 6.20). Thus a capillary bed presents less total resistance to flow than do the arterioles leading into it, and the total drop in blood pressure is less steep in this region.

Venules and Veins

Capillaries merge into venules. In functional terms, venules are a bit like capillaries. Some solutes diffuse across their wall, which is only a little thicker than that of a capillary. Venules merge into veins, the large-diameter transport tubes leading back to the heart.

Veins are blood volume reservoirs. Collectively they contain 50–60 percent of the total blood volume. A vein

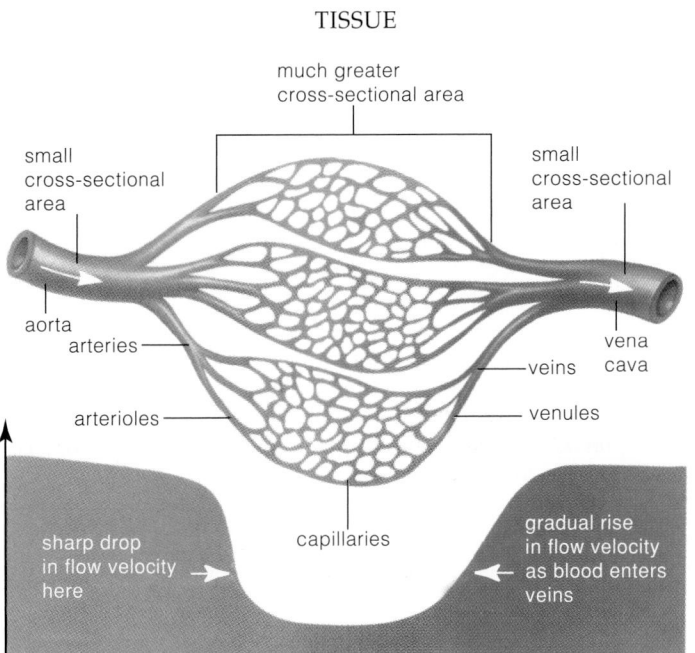

TISSUE

much greater cross-sectional area

small cross-sectional area

small cross-sectional area

aorta
arteries

vena cava

veins

arterioles

venules

capillaries

sharp drop in flow velocity here

gradual rise in flow velocity as blood enters veins

Figure 6.20 Comparison of cross-sectional areas represented by arteries, arterioles, and capillaries in the circulatory system.

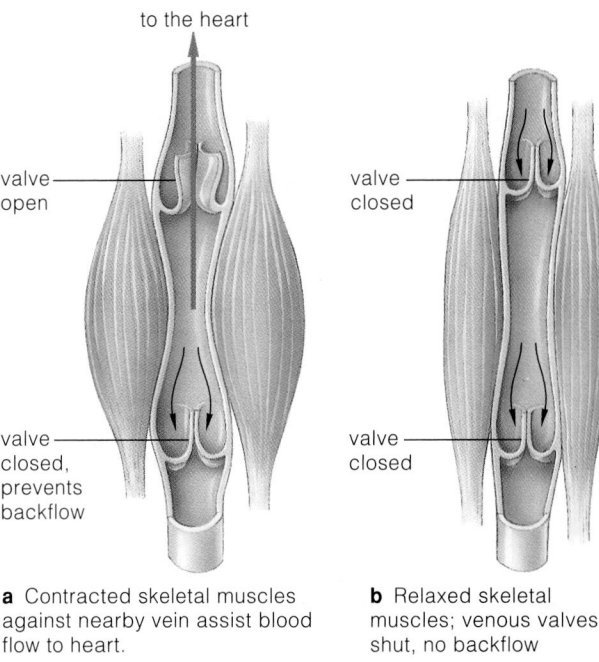

to the heart

valve open

valve closed

valve closed, prevents backflow

valve closed

a Contracted skeletal muscles against nearby vein assist blood flow to heart.

b Relaxed skeletal muscles; venous valves shut, no backflow

Figure 6.21 How contracting skeletal muscles can increase fluid pressure in a vein. Notice the structure of the vein valve.

wall is thin enough to bulge under pressure, more so than an arterial wall. The wall also contains some smooth muscle (Figure 6.19d). When blood must circulate faster (as during exercise) the smooth muscle in veins contracts. The wall stiffens, the vein bulges less, and venous pressure rises and drives more blood to the heart. Venous pressure also rises when contracting skeletal muscle—especially in the legs and abdomen—bulges against adjacent veins (Figure 6.21). This muscle activity is an important factor in returning blood through the venous system.

Some veins, mainly those in the limbs, have valves. When blood starts moving backward due to gravity, it pushes the valves closed, preventing backflow. Inherited defects, obesity, pregnancy, and other factors can result in weakened venous valves. The walls of a *varicose vein* have become overstretched because, over time, weak valves have allowed blood to pool there.

How Vessels Help Control Blood Pressure

Certain arteries, all arterioles, and even veins have key roles in homeostatic mechanisms that work to maintain adequate blood pressure over time. Recall that centers in the brain's medulla control resting blood pressure. When an abnormal *increase* in blood pressure is detected, the brain centers command the heart to beat more slowly and contract less forcefully. They also order smooth muscle

cells in the walls of arterioles to relax. This results in **vasodilation**—an enlargement (dilation) of the vessel diameter. When an abnormal *decrease* in blood pressure is detected, the heart is made to beat faster and contract more forcefully. Smooth muscle cells of arterioles are made to contract and so bring about a decrease in vessel diameter, or **vasoconstriction**.

Hormones help maintain blood pressure. Arterioles in specified regions have receptors for hormones that can trigger vasoconstriction or vasodilation. The nervous system and endocrine system also control the allocation of more or less blood to different body regions at different times. For instance, after you eat a large meal, more blood is diverted to your digestive system.

Local controls also divert blood flow. For example, when you run, the tissue level of oxygen in hardworking skeletal muscles falls, and levels of carbon dioxide, H^+, potassium ions, and other substances rise. The changes in local chemistry cause the smooth muscle in arterioles to relax. With the vasodilation, more blood flows past the active muscles, delivering more raw materials and carrying away cell wastes. At the same time, arterioles in the digestive tract and kidneys constrict.

Overall, blood pressure is an outcome of controls over the heart's output and over the total resistance to blood flow through the vascular system, as exerted mainly at arterioles.

EXCHANGES OF FLUID AND SOLUTES AT CAPILLARIES

Capillaries thread to within 0.01 millimeter of almost every living cell in the body. Most solutes, including oxygen and carbon dioxide, diffuse across the capillary wall. Certain proteins enter or leave by endocytosis or exocytosis, and certain ions probably pass through pores in capillary walls and spaces between endothelial cells.

Fluid also enters and leaves capillaries in response to various types of pressure (Figure 6.22). You may recall from Chapter 1 that the force a fluid exerts against a surface is called hydrostatic pressure. In capillaries, this is the same as blood pressure, and it forces some water out of capillaries, especially at the arterial end. Normally, water also moves *into* capillaries in response to osmotic pressure. That is, the water follows its concentration gradient into capillaries as the concentration of solutes there rises. Hydrostatic pressure exerted by interstitial fluid and the concentration of solutes in that fluid also have some effect on water movements. On balance, more water tends to leave capillaries than to enter them. As water moves in either direction, some solutes follow. Such fluid and solute movements help maintain the proper fluid balance between blood and the surrounding tissues. They are vital to maintaining the blood volume needed for adequate blood pressure.

Capillary beds are diffusion zones for exchanges between the blood and interstitial fluid. Such exchanges not only help maintain the blood volume required for proper blood pressure, but also help maintain the proper fluid balance between blood and tissues.

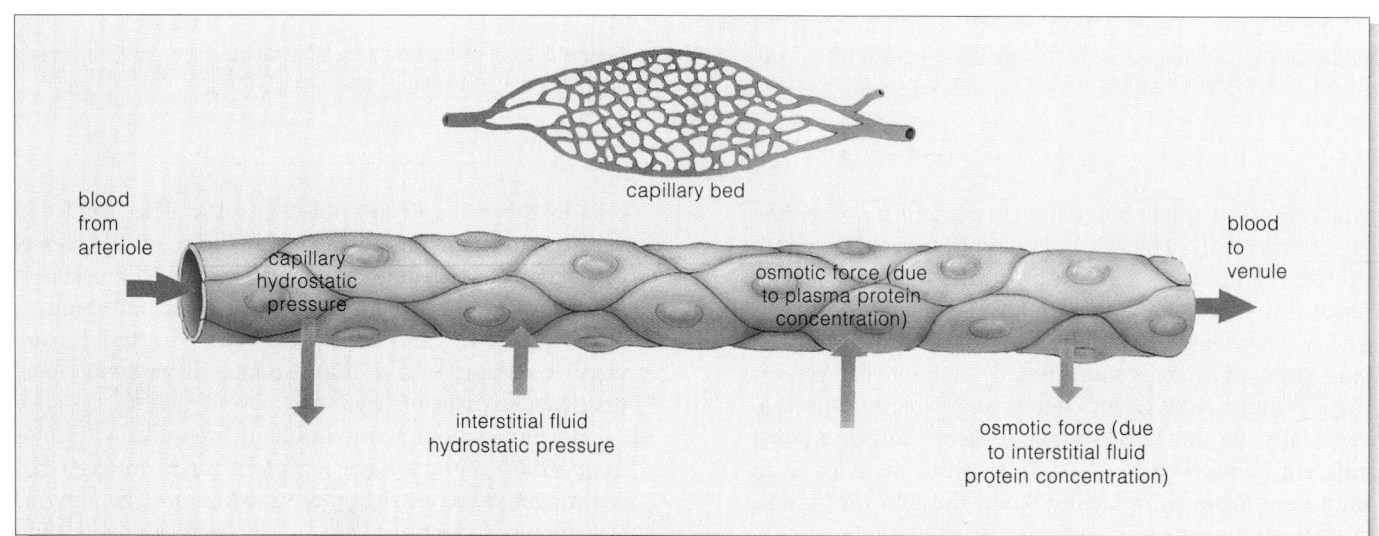

Figure 6.22 Fluid movements in a capillary bed. The movements help maintain the distribution of extracellular fluid between the bloodstream and interstitial fluid.

At the arteriole end of a capillary, the difference between capillary blood pressure and interstitial fluid pressure causes some water to leave the capillary. Part of this water will be picked up by the lymphatic system and eventually return to the bloodstream.

Water enters a capillary by osmosis, following its concentration gradient. (Plasma has a greater solute concentration and therefore a lower water concentration.) Fluid loss at the arteriole end of a capillary bed tends to be balanced by fluid intake at the venule end.

Edema is a condition in which excess fluid accumulates in interstitial spaces. This happens to some extent during exercise. As arterioles dilate in local tissues, capillary blood pressure increases and more fluid is forced out of capillaries into tissues. Edema also can result from an obstructed vein or from heart failure.

HEMOSTASIS AND BLOOD CLOTTING

Overview of Hemostasis

Small blood vessels are vulnerable to ruptures, cuts, or other damage. **Hemostasis** stops the bleeding and so prevents excessive blood loss. This process includes blood vessel spasm, platelet plug formation, and the coagulation, or clotting, of blood.

When a blood vessel first is ruptured, smooth muscle in a damaged vessel wall contracts in an automatic response called a spasm. The blood vessel constricts, so blood flow through it slows or stops. This response can last for up to half an hour and is extremely important in stemming the immediate loss of blood. Second, while vessel spasms reduce blood flow, platelets clump together as a temporary plug in the damaged wall. They also release the hormone serotonin and other chemicals that help prolong the spasm and attract more platelets. Third, blood *coagulates* (converts to a gel) and forms a clot. This happens through one of two responses, which we'll now consider.

Clotting Mechanisms

In the *intrinsic clotting mechanism*, a plasma protein becomes activated and triggers reactions that lead to the formation of an enzyme (thrombin), which acts on a large, rod-shaped plasma protein (fibrinogen). The rods adhere to one another, forming long, insoluble threads (fibrin) that stick to one another. The result is a net in which blood cells and platelets become entangled (Figure 6.23). The entire mass is a blood clot. The clot retracts into a compact mass, drawing the walls of the vessel together and sealing the breach.

Blood also can coagulate through an *extrinsic clotting mechanism*. "Extrinsic" means that the series of reactions leading to blood clotting is triggered by the release of enzymes and other substances *outside* the blood itself. These substances come from damaged blood vessels or from the surrounding tissues. The substances lead to thrombin formation, and the remaining steps parallel those shown in Figure 6.23. Aspirin can suppress the extrinsic pathway, and it also affects platelet clumping. Hence, people who overuse this nonprescription drug may be prone to bleeding.

Hemostasis relies on mechanisms that slow or stop the flow of blood from a ruptured vessel. The mechanisms include constriction of blood vessel walls, formation of platelet plugs, and blood clotting.

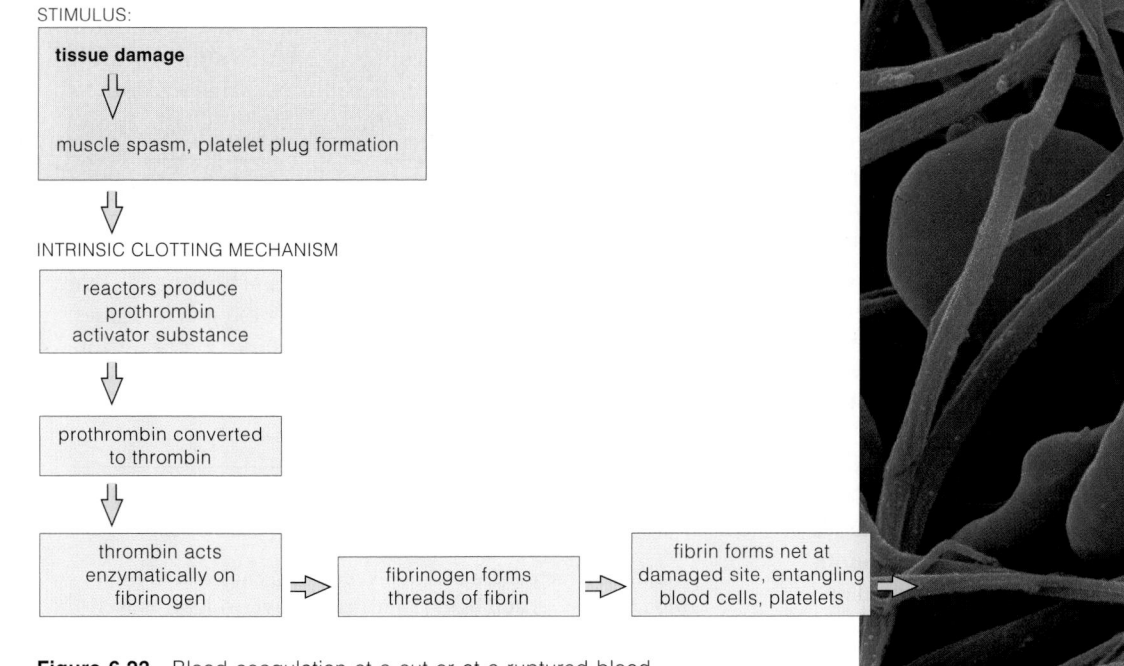

Figure 6.23 Blood coagulation at a cut or at a ruptured blood vessel. The micrograph shows red blood cells trapped in a fibrin net.

STIMULUS:

tissue damage

muscle spasm, platelet plug formation

INTRINSIC CLOTTING MECHANISM

reactors produce prothrombin activator substance

prothrombin converted to thrombin

thrombin acts enzymatically on fibrinogen

fibrinogen forms threads of fibrin

fibrin forms net at damaged site, entangling blood cells, platelets

clot formation

CARDIOVASCULAR DISORDERS

In the United States alone, more than 40 million people have cardiovascular disorders, which claim about 750,000 lives every year. The most common ones are *hypertension* (sustained high blood pressure) and *atherosclerosis*, which is a progressive narrowing of the arterial lumen. Atherosclerosis is the major cause of most *heart attacks*—damage or death of heart muscle due to an interruption of its blood supply. It also can cause *stroke,* damage to the brain due to an interruption of blood circulation to it.

Most heart attacks bring a "crushing" pain behind the breastbone that lasts a half hour or more. Frequently, the pain radiates into the left arm, shoulder, or neck. The pain can be mild but usually it is excruciating. Often it is accompanied by sweating, nausea, vomiting, and dizziness or loss of consciousness.

Cardiovascular disorders are the leading cause of death in the United States. Many risk factors associated with those disorders have been identified. They include cholesterol deposits in arteries and a range of other factors. Table 6.1 on the facing page lists some diseases and disorders that can affect cardiovascular functioning.

Cholesterol and Cardiovascular Disease

Usually, the liver produces enough cholesterol to satisfy body needs. Together with the liver's output, dietary cholesterol also circulates in the blood. If you habitually eat cholesterol-rich food, you may end up with a high blood level of cholesterol. If you have a genetic disorder called *familial hypercholesterolemia,* the same thing might happen no matter what kinds of food you eat.

When cholesterol is transported through the bloodstream, it is bound to protein carrier molecules called lipoproteins. These include *high-density lipoproteins* (HDLs), *low-density lipoproteins* (LDLs), and *very low density lipoproteins* (VLDLs). High levels of LDL are related to a tendency toward heart trouble. LDLs carry cholesterol and triglycerides, and tend to infiltrate arterial walls. HDLs carry only cholesterol. They transport their cargo of cholesterol to the liver, where it can be metabolized. In addition, it appears that unsaturated fats, including olive oil and fish oil, can reduce the level of LDLs in the blood. In some people, not enough LDL is removed, possibly because blood cells may not have enough LDL receptors.

As the blood level of LDL increases, so does the risk of atherosclerosis. LDLs with their bound cholesterol can enter arterial walls. Abnormal cells and cell products multiply at the entry sites. Then calcium salts and a fibrous net form over the mass, creating an atherosclerotic plaque, or "hardening of the arteries" (Figure 6.24). Blood clots may form at plaques and narrow or block the arteries. A heart attack may follow.

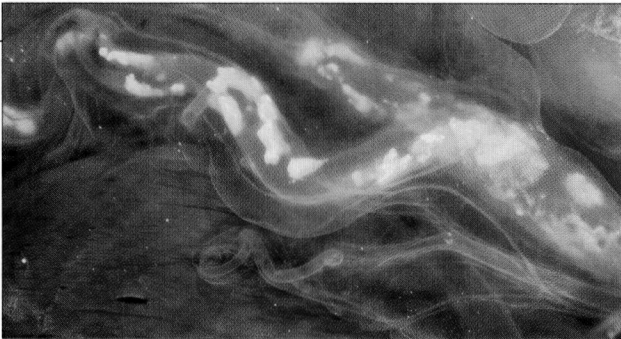

Figure 6.24 Plaques in blood vessels that service the heart. Such plaques form when fats and cholesterol are deposited in the blood vessel wall; then calcium salts and a fibrous net form over the abnormal mass. Blood clots may form here and narrow or block the arteries, leading to a heart attack.

Atherosclerosis is rare in rats, which have mostly HDLs. It is common in humans, who in general have mostly LDLs. In the next section, *Focus on Your Health* considers this serious health threat in some detail.

People who want to reduce LDL levels are usually advised to restrict their intake of cholesterol and saturated fat. Some research suggests that HDL levels are higher in people who exercise regularly. They also appear to be higher in people who do not use tobacco and who drink moderate amounts of alcohol.

Other "Healthy Heart" Issues

A variety of other factors also directly affect the heart's health. The more body fat you carry, the more blood capillaries must develop to service the increased number of cells, and the harder the heart has to work to pump blood through the increasingly divided vascular circuit. Among smokers, the nicotine in tobacco stimulates the adrenal glands to secrete epinephrine, which constricts blood vessels and triggers an accelerated heartbeat and a rise in blood pressure. Cancer-causing substances (carcinogens) in cigarette smoke may also contribute to plaque development. The carbon monoxide present in cigarette smoke has a greater affinity for binding sites on hemoglobin than does oxygen, and its action means that the heart has to pump harder to deliver oxygen to tissues. Other heart disease risk factors include hypertension, lack of exercise, diabetes mellitus, a family history of heart disorders, and increasing age.

Cardiovascular disorders are a leading cause of disability and death in some countries, including the United States.

Major risk factors are a high level of LDL cholesterol, high blood pressure, and being significantly overweight, as well as lifestyle elements such as smoking and lack of exercise.

Table 6.1 The Cardiovascular System under Attack

Cause/Risk Factors	Major Effects	Symptoms
HEART/VASCULAR SYSTEM		
Coronary heart disease Blockage of coronary arteries, usually by fatty deposits and/or blood clots. Risk factors include heredity, hypertension, smoking, obesity, lack of exercise, and elevated blood cholesterol.	Reduced supply of blood and hence oxygen to the heart muscle.	Heart attack; pain in the chest, arm, or neck (angina pectoris).
Angina pectoris Pain caused by reduced blood supply to the heart due to atherosclerosis, anemia, arrhythmia, or narrowing of the aortic valve.	Lack of oxygen for heart muscle cells.	Pain in the chest, arms, or jaw.
Myocardial infarction (heart attack) Unhealthy diet, smoking, hypertension, stress, diabetes mellitus, age, heredity.	Sudden death of part of the heart muscle; may lead to heart failure (reduced pumping efficiency).	Chest pain, shortness of breath, clammy skin, nausea, fainting.
Myocarditis and endocarditis Usually triggered by viral or bacterial infection.	Inflammation of the heart muscle and heart valves, respectively.	Irregular heartbeat, breathlessness, chest pain, heart failure.
Circulatory shock Reduced blood volume and/or blood pressure arising from heart attack, blood or fluid loss due to injury, illness, burns, poisoning, or spinal injury.	Severe reduction of blood flow throughout body tissues.	Rapid, shallow breathing and rapid, weak pulse; dizziness; clammy skin, fainting.
BLOOD		
Anemia Iron deficiency, failure of bone marrow to produce stem cells, destruction of red blood cells, vitamin deficiencies that result in RBCs with reduced capacity to transport oxygen.	Reduced concentration of hemoglobin in the blood; hence reduced oxygen transport.	Tiredness, lethargy, headache, dizziness, pale skin; in severe cases, angina pectoris.
Leukemias Cancers in which stem cells in bone marrow overproduce white blood cells. May be chronic or acute.	Brain, liver, and other organs begin to fail due to invasion of cancerous white blood cells. Bone marrow becomes choked with the abnormal cells, with resulting failure to produce normal blood cells of all types.	Enlarged lymph nodes, liver, and spleen; headache; bruising; fever; night sweats; repeated infections; anemia.
Infectious mononucleosis Epstein-Barr virus or cytomegalovirus.	Viral infection of lymphocytes; most common in teenagers and young adults.	Fever, headache, severe sore throat; swollen lymph glands in armpits, neck, and groin.

6.15

ATHEROSCLEROSIS, HYPERTENSION, AND ARRHYTHMIAS

In *arteriosclerosis,* arteries thicken and lose elasticity. In **atherosclerosis**, this condition worsens as cholesterol and other lipids build up in the arterial wall and cause the lumen to narrow.

Section 6.14 described how an atherosclerotic plaque can develop in an artery wall and protrude into the lumen of the artery. Figure 6.25a shows a normal artery; Figure 6.25b shows what a plaque might look like in cross section. As you have read, the development of plaques is related to cholesterol intake, but other factors also contribute. Smoking and a diet high in saturated fats appear to aggravate plaque formation, and researchers are probing the possible roles of certain bacteria and viruses in some cases.

Sometimes platelets become caught on the rough edges of a plaque and are stimulated into secreting some of their chemicals. When they do, a blood clot forms. As the clot and plaque grow, the artery narrows. Blood flow to the tissue that the artery supplies diminishes or may be blocked entirely. A blood clot that stays in place is called a *thrombus.* If it becomes dislodged and travels the bloodstream, it is called an *embolus.*

With their narrow diameters, coronary arteries and their branches are highly susceptible to clogging through plaque formation or occlusion by a clot. When such an artery becomes narrowed to one-quarter of its former diameter, the resulting symptoms can range from mild chest pain (*angina pectoris*) to a full-scale heart attack.

Atherosclerosis involving coronary arteries can be diagnosed through a stress electrocardiogram. This is a recording of the electrical activity of the cardiac cycle while a person is exercising on a treadmill. It can also be diagnosed by *angiography.* This procedure involves injection of a dye that causes plaques to show up as contrasting masses on X rays. A severe blockage may require surgery. In *coronary bypass surgery*, a section of a large vessel taken from the arm or leg is stitched to the aorta and to the coronary artery below the affected region (Figure 6.25c). In *laser angioplasty*, laser beams vaporize the plaques. In *balloon angioplasty*, a small balloon is inflated within a blocked artery to flatten a plaque and so increase the arterial diameter. Such procedures can reduce the immediate threat of a heart attack, but they only buy time; they do not cure the underlying problem.

Implicated in many cases of artery damage is the amino acid *homocysteine*, which forms when the body breaks down another amino acid, methionine (present in meats and dairy products). Although methionine is an important nutrient, a chronically high level of homocysteine in the blood can encourage the formation of atherosclerotic plaques. Research suggests that B vitamins (including folic acid) help mop up homocysteine before it can trigger damaging changes in arteries.

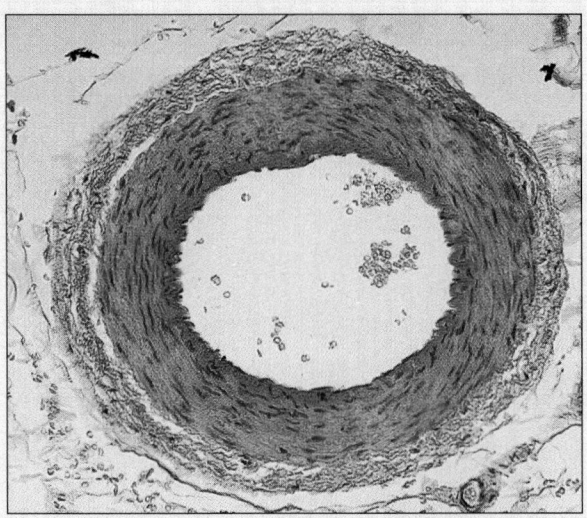

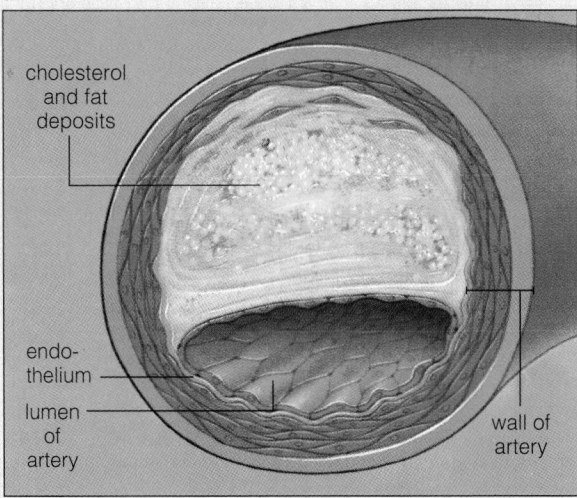

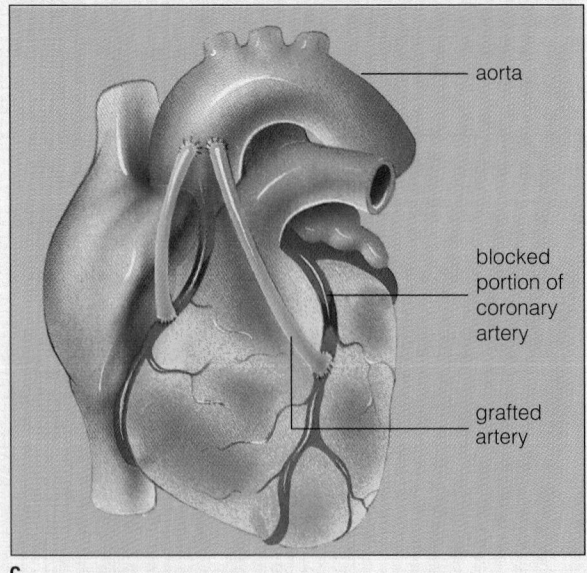

Figure 6.25 (**a**) Cross section of a normal artery. (**b**) Diagram of an atherosclerotic plaque. (**c**) Two coronary bypasses (*green*).

Women under age 55 generally are at less risk of heart disease, probably because the female hormone estrogen helps protect against the formation of atherosclerotic plaques. However, when a woman reaches menopause, usually in her early 50s, her estrogen level plummets and her risk of heart disease eventually rises to equal that of a man.

HYPERTENSION Hypertension results from gradual increases in the resistance to blood flow through small arteries. In time, blood pressure stays elevated, even when a person is resting. Heredity may be a factor; the disorder tends to run in families. Diet also is a factor. For instance, high salt intake can raise blood pressure in susceptible people.

In general, high blood pressure increases the heart's workload. Eventually the heart may enlarge in a way that hampers blood pumping and can lead to heart failure. High blood pressure also can cause arterial walls to "harden" and so influence the delivery of oxygen to the brain, heart, kidneys, and other vital organs. It is a significant cause of stroke.

Hypertension has been called the silent killer because affected people may have no outward symptoms. Even when they know their blood pressure is high, some tend to resist medication, changes in diet, and regular exercise. Of the estimated 23 million Americans who are hypertensive, most do not undergo treatment. About 180,000 die each year.

ARRYTHMIAS ECGs reveal **arrhythmias**, irregular heart rhythms (Figure 6.26). Not all arrhythmias are abnormal. For example, endurance athletes may have a below-average resting cardiac rate, a condition called *bradycardia*. In an adaptation to ongoing strenuous exercise, their nervous system has adjusted the cardiac pacemaker's rate of contraction downward. There is more time for the ventricles to fill, so each contraction pumps blood more efficiently.

A cardiac rate above 100 beats per minute, called *tachycardia*, occurs normally during exercise or stressful situations. Serious tachycardia can be triggered by drugs (including caffeine, nicotine, alcohol, and cocaine), excessive hormone output by the thyroid gland (hyperthyroidism), and other factors.

Coronary artery disease or some other disorders may cause abnormal rhythms that can degenerate rapidly into a dangerous condition called *ventricular fibrillation*. This is an extreme medical emergency in which cardiac muscle in portions of the ventricles contracts haphazardly, and blood pumping suffers. Within seconds, the person loses consciousness, and this may signify impending death. If the patient is lucky, a strong electrical shock to the heart, or the administration of defibrillating drugs, may restore a normal rhythm before the damage becomes too serious.

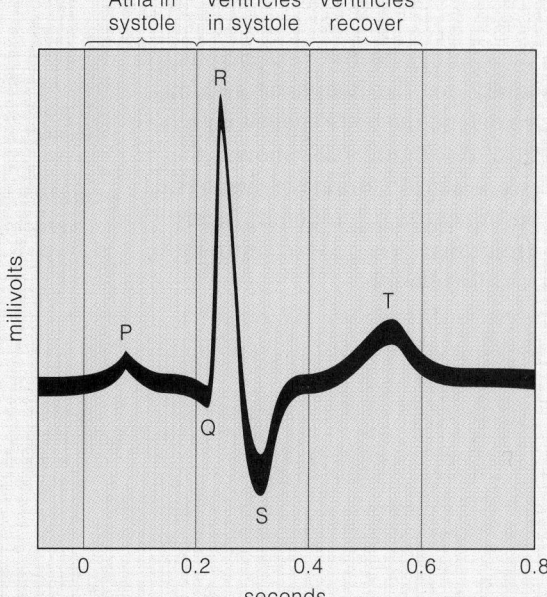

a ECG of a single, normal heartbeat

Bradycardia (here, 46 beats per minute):

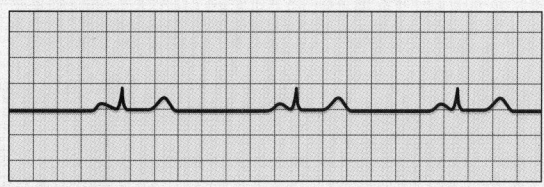

Tachycardia (here, 136 beats per minute):

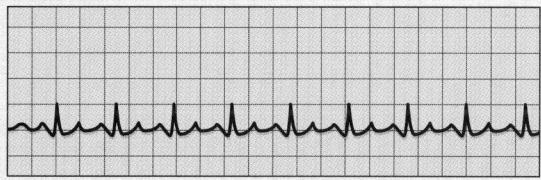

Ventricular fibrillation:

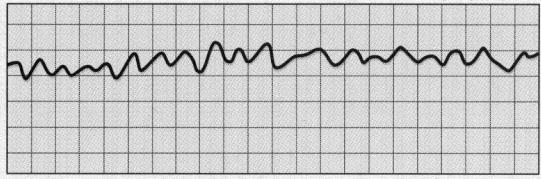

b Examples of ECG readings

Figure 6.26 Examples of ECG readings. (**a**) A single, normal heartbeat. The P wave is generated by electrical signals from the SA node that stimulate contraction of the atria. As the stimulus moves over the cardiac muscle of the ventricles by way of Purkinje fibers, it is recorded as the QRS wave complex. After the ventricles contract they go through a brief period of recovery. Electrical activity during this period is marked by the T wave. (There is also an atrial recovery period "hidden" in the QRS complex.) (**b**) ECG readings for bradycardia, tachycardia, and ventricular fibrillation.

THE LYMPHATIC SYSTEM

We conclude this chapter with a brief look at how the lymphatic system supplements blood circulation. But think of this section as a bridge to the next chapter, on immunity, for the lymphatic system also helps defend the body against injury and attack. As Figure 6.27 shows, the system consists of drainage vessels, lymphoid organs, and lymphoid tissues. Tissue fluid that has moved into the vessels is called **lymph**.

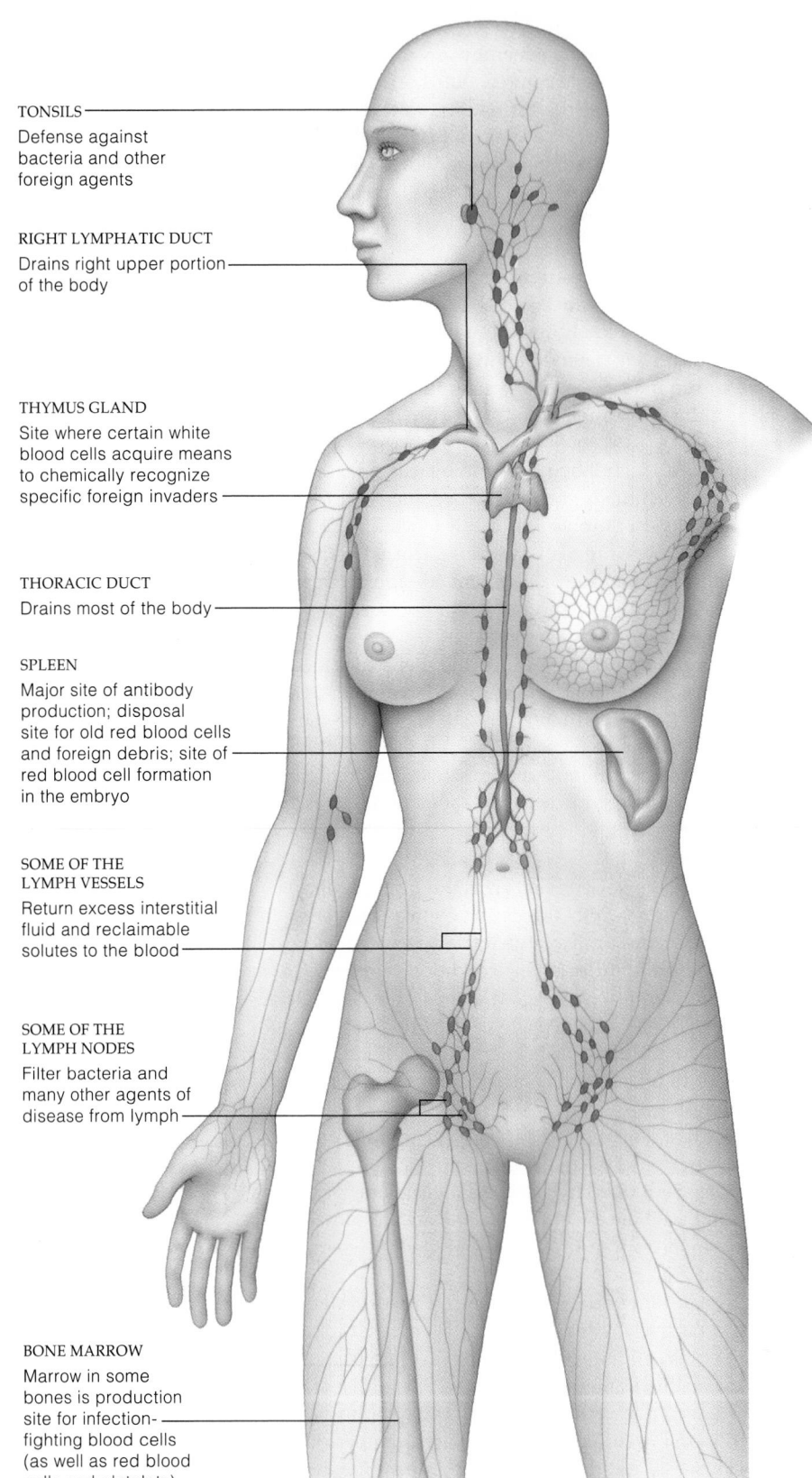

TONSILS
Defense against bacteria and other foreign agents

RIGHT LYMPHATIC DUCT
Drains right upper portion of the body

THYMUS GLAND
Site where certain white blood cells acquire means to chemically recognize specific foreign invaders

THORACIC DUCT
Drains most of the body

SPLEEN
Major site of antibody production; disposal site for old red blood cells and foreign debris; site of red blood cell formation in the embryo

SOME OF THE LYMPH VESSELS
Return excess interstitial fluid and reclaimable solutes to the blood

SOME OF THE LYMPH NODES
Filter bacteria and many other agents of disease from lymph

BONE MARROW
Marrow in some bones is production site for infection-fighting blood cells (as well as red blood cells and platelets)

Figure 6.27 Components of the human lymphatic system and their functions.

The *green* dots show some of the major lymph nodes. Patches of lymphoid tissue in the small intestine and in the appendix also are part of the lymphatic system.

The Lymph Vascular System

The vascular portion of the lymphatic system, which includes **lymph capillaries** and **lymph vessels**, has three functions. First, it takes up water and plasma proteins that have leaked out at capillary beds and returns them to the blood. Second, it transports fats absorbed from the small intestine. Third, it transports foreign particles and cellular debris from tissue spaces to lymph nodes, where such substances are broken down.

The lymph vascular system starts at capillary beds. Tissue fluid enters lymph capillaries. These have no obvious entrance. Instead, their tips contain regions of overlapping endothelial cells (Figure 6.28a). Water and solutes move inward at these flaplike "valves."

Lymph capillaries merge with lymph vessels. These have a larger diameter, smooth muscle in their wall, and flaplike valves that prevent backflow. As you can see in Figure 6.27, the lymph vessels eventually converge into collecting ducts, which drain into veins in the lower neck. In this way, the lymph fluid is returned to the blood circulation. Movements of skeletal muscle and of the rib cage (during breathing) help move fluid through lymph vessels, just as they do for veins.

Lymphoid Organs and Tissues

The defense portion of the lymphatic system includes the **lymph nodes**, **spleen**, and **thymus**. It also includes the tonsils and patches of tissue in the small intestine, appendix, and airways leading to the lungs (the bronchi).

The lymph nodes are located at intervals along lymph vessels (Figure 6.28b). Lymph trickles through at least one of these nodes before entering the bloodstream. A lymph node has several inner chambers. The chambers are sites where many white blood cells take up residence after they have been produced in bone marrow. During the course of an infection, lymph nodes become battlegrounds where great armies of lymphocytes form and where foreign agents are destroyed. Macrophages in the nodes help clear the lymph of bacteria, cellular debris, and other substances.

The largest lymphoid organ, the **spleen**, is a filtering station for blood and a holding station for lymphocytes. The spleen has inner chambers filled with red and white "pulp." The red pulp is a large reservoir of red blood cells and macrophages. In a developing embryo, the spleen produces red blood cells.

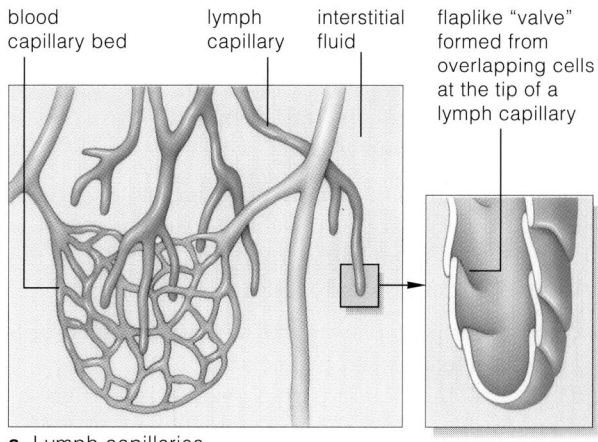

a Lymph capillaries

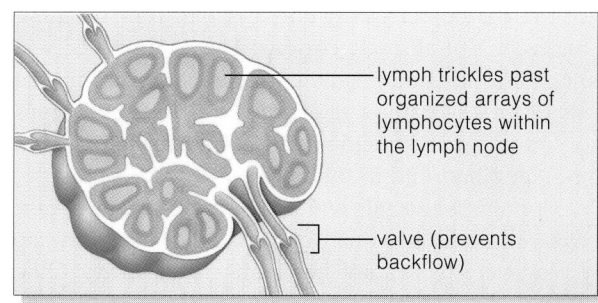

b A lymph node, cross section

Figure 6.28 (**a**) Some of the lymph capillaries at the start of the drainage network called the lymph vascular system. (**b**) Cutaway diagram of a lymph node. Its inner chambers are packed with highly organized arrays of infection-fighting white blood cells.

In the thymus, lymphocytes multiply, differentiate, and mature into fighters of specific disease agents. The thymus produces hormones that influence these events. It is central to immunity, the focus of Chapter 7.

The lymph vascular system returns tissue fluid to the blood, transports fats, and carries debris and foreign material to lymph nodes.

Lymph nodes and other lymphoid organs function in the body's systems of defense against infection and disease.

SUMMARY

1. The human cardiovascular system consists of the heart, a muscular pump; blood vessels, including arteries, arterioles, capillaries, venules, and veins; and blood. The system's function is the rapid internal transport of substances to and from cells.

2. Blood, a fluid connective tissue, helps maintain favorable conditions for cells. It delivers oxygen and other substances to the interstitial fluid around cells. It also picks up cell products and wastes from that fluid.

3. Blood consists of plasma, red and white blood cells, and cell fragments called platelets.

   a. Plasma, the liquid part of blood, is a transport medium for blood cells and platelets. Plasma water is a solvent for plasma proteins, simple sugars, amino acids, mineral ions, vitamins, hormones, and several gases.

   b. Red blood cells transport oxygen from the lungs to the tissues of all body regions. They are packed with hemoglobin, an iron-containing pigment molecule that binds reversibly with oxygen. Red blood cells also transport some carbon dioxide (also bound to hemoglobin) from interstitial fluid to the lungs.

   c. Certain phagocytic white blood cells scavenge dead or worn-out cells and other debris and cleanse tissues of anything detected as not belonging to the body. Other white blood cells (lymphocytes) form armies that destroy specific bacteria, viruses, and other disease agents.

   d. Platelets function in blood clotting.

4. An internal partition divides the human heart into two halves, each with two chambers (an atrium and a ventricle). The partition separates the blood flow into two circuits, one pulmonary and the other systemic.

   a. In the pulmonary circuit, deoxygenated blood in the heart's *right* half is pumped to capillary beds in the lungs. The blood picks up oxygen, then flows to the heart's left atrium.

   b. In the systemic circuit, the *left* half of the heart pumps oxygenated blood to all body tissues. There, cells take up oxygen and give up carbon dioxide. The blood, now deoxygenated, flows to the heart's right atrium.

5. Ventricular contraction drives blood through both circuits. Blood pressure is highest in contracting ventricles, then progressively drops in arteries, arterioles, capillaries, then veins. It is lowest in relaxed atria.

   a. Arteries are an elastic pressure reservoir. They smooth out pressure changes resulting from heartbeats and so smooth out blood flow through capillaries.

   b. Arterioles are control points for distributing different volumes of blood to different regions.

   c. Capillary beds are diffusion zones where blood and interstitial fluid exchange substances.

   d. Venules overlap capillaries and veins somewhat in function.

   e. Veins are a blood-volume reservoir that can be tapped to adjust the volume of flow back to the heart.

6. The vascular portion of the lymphatic system takes up water and plasma proteins that seep out of blood capillaries, then returns them to the blood circulation. It also transports absorbed fats. Some lymphatic system components have major roles in general body defense and in immune responses.

Review Questions

1. What are the functions of blood? *136*

2. What is blood plasma? What are the cellular components of blood? *136–139*

3. Contrast the functions of the cardiovascular system and the lymphatic system. *142–143; 158–159*

4. Distinguish between the systemic and pulmonary circuits. *146*

5. Label the heart's components:

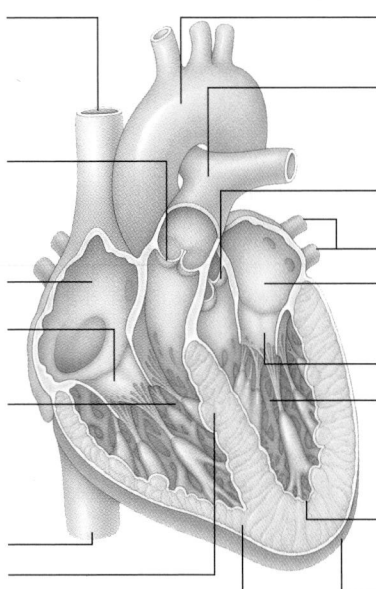

6. Explain how the medulla oblongata of the brain helps regulate blood flow to different body regions. *148, 149*

7. State the main function of blood capillaries. What drives water and solutes out of and into capillaries in capillary beds? *150, 152*

8. State the main functions of venules and veins. What forces work together in returning venous blood to the heart? *150–151*

Critical Thinking: You Decide (Key in Appendix V)

1. A patient suffering from hypertension may receive drugs that decrease the heart's output, dilate arterioles, or increase urine production. In each case, how would the drug treatment help relieve hypertension?

2. Aplastic anemia develops when certain drugs or radiation destroys red bone marrow, including the stem cells that give rise to red and white blood cells and platelets. Predict some symptoms a person with aplastic anemia would be likely to develop. Include at least one symptom related to each type of formed element in blood.

Self-Quiz (Answers in Appendix IV)

1. A _____ system functions in the rapid exchange of substances to and from all living cells, and it is supplemented by a _____ system.

2. _____ and _____ are large-diameter blood vessels for fluid transport; _____ and _____ are fine-diameter, thin-walled blood vessels for diffusion; and _____ serve as control points over the variable distribution of blood to different body regions.

3. Which of the following are *not* components of blood?
   a. red and white blood cells
   b. platelets and plasma
   c. assorted solutes and dissolved gases
   d. all of the above are components of blood

4. Red blood cells are produced in the _____ and function in transporting _____ and some _____.
   a. liver; oxygen; mineral ions
   b. liver; oxygen; carbon dioxide
   c. bone marrow; oxygen; hormones
   d. bone marrow; oxygen; carbon dioxide

5. White blood cells are produced in the _____ and function in both _____ and _____.
   a. liver; oxygen transport; defense
   b. lymph glands; oxygen transport; pH stabilization
   c. bone marrow; day-to-day housekeeping; defense
   d. bone marrow; pH stabilization; defense

6. In the pulmonary circuit, the _____ half of the heart pumps _____ blood to capillary beds inside the lungs, then _____ blood flows back to the heart.
   a. left; deoxygenated; oxygenated
   b. right; deoxygenated; oxygenated
   c. left; oxygenated; deoxygenated
   d. right; oxygenated; deoxygenated

7. In the systemic circuit, the _____ half of the heart pumps _____ blood to all body regions, then _____ blood flows back to the heart.
   a. left; deoxygenated; oxygenated
   b. right; deoxygenated; oxygenated
   c. left; oxygenated; deoxygenated
   d. right; oxygenated; deoxygenated

8. Fluid pressure in the circulatory system is _____ at the beginning of a circuit, then _____ in arteries, arterioles, capillaries, and then veins. It is _____ in the relaxed atria.
   a. low; rises; highest   c. low; drops; lowest
   b. high; drops; lowest   d. high; rises; highest

9. Match the type of blood vessel with its major function.
   _____ arteries   a. diffusion
   _____ arterioles   b. control of blood distribution
   _____ capillaries   c. transport, blood volume reservoirs
   _____ veins   d. blood transport and pressure regulators

10. Match the following circulation components with their descriptions.
   _____ capillary beds   a. two atria, two ventricles
   _____ lymph vascular system   b. driving force for blood
   _____ heart chambers   c. zones of diffusion
   _____ heart contractions   d. return of interstitial fluid to blood

Selected Key Terms

agglutination 140
agranulocyte 137
aorta 144
arrhythmia 157
arteriole 142
artery 142
atrioventricular node 148
atrioventricular valve 144
atrium 144
blood 136
blood pressure 149
capillary 142
capillary bed 146
cardiac conduction system 148
cardiac cycle 147
cardiovascular system 142
coronary artery 144
diastole 147
granulocyte 137
heart 144
hemostasis 153
lymph 158
lymphatic system 158

lymph node 159
lymphoid organ 159
lymph vascular system 159
myocardium 144
platelet 137
pulmonary circuit 146
pulse 150
red blood cell (erythrocyte) 136
semilunar valve 144
septum 144
sinoatrial node 148
spleen 159
stem cell 139
systemic circuit 146
systole 147
thymus 159
vasoconstriction 151
vasodilation 151
vein 142
ventricle 144
venule 142
white blood cell (leukocyte) 136

Readings

"Heart Disease: Women at Risk." May 1993. *Consumer Reports.*

Nucci, M., and A. Abuchowski. February 1998. "The Search for Blood Substitutes." *Scientific American.*

Radetsky, P. March 1995. "The Mother of All Blood Cells." *Discover.*

Vogel, S. 1992. *Vital Circuits: On Pumps, Pipes, and the Workings of the Circulatory System.* New York: Oxford University Press.

 For additional readings, go to InfoTrac College Edition, your online library at:

http://www.infotrac-college.com/wadsworth

7 IMMUNITY

Russian Roulette, Immunological Style

Until about a century ago, smallpox epidemics swept repeatedly through the world's cities. Some outbreaks were so severe that only half of those who were stricken survived. The survivors had permanent scars on the face, neck, shoulders, and arms—but they seldom contracted the disease again. They were "immune" to smallpox.

No one knew what caused smallpox. But the idea of acquiring immunity was dreadfully appealing. In 12th-century China, healthy people were gambling with deliberate infections. They sought out survivors of mild cases of smallpox (who were only mildly scarred), then removed crusts from the scars, ground them up, and inhaled the powder. By the late 17th century, Mary Montagu, wife of the English ambassador to Turkey, was championing inoculation. She even went so far as to inject bits of smallpox scabs into her children. So did the Prince of Wales. Others soaked threads in the fluid from the sores, then poked the threads into the skin.

Individuals who survived these practices acquired immunity to smallpox—but many developed raging infections. As if the odds were not dangerous enough, the crude inoculation procedures also invited the acquisition of several other infectious diseases.

While this immunological Russian roulette was going on, Edward Jenner was growing up in the English countryside. At the time, it was known that people who contracted cowpox never got smallpox. (Cowpox is a mild disease that can be transmitted from cattle to humans.) In 1796, Jenner, by now a physician, injected material from a cowpox sore into the arm of an uninfected boy. Six weeks later, after the reaction subsided, Jenner injected fluid from smallpox sores into the boy (Figure 7.1). He hypothesized that the earlier injection might provoke immunity to smallpox, and he was right. The boy remained free of smallpox.

The French mocked Jenner's procedure, calling it "vaccination" (which translates as "encowment"). Much later a French chemist, Louis Pasteur, devised similar procedures for other diseases. Pasteur also called his procedures vaccinations, and only then did the term become respectable.

By Pasteur's time, improved microscopes were revealing diverse bacteria, fungal spores, and other previously invisible forms of life. As Pasteur himself discovered, microorganisms abound in ordinary air.

Did some cause diseases? Probably. Could they settle into food or drink and cause it to spoil? He proved that they did. Pasteur also found a way to kill most of the

Figure 7.1a Statue honoring Edward Jenner's development of an immunization procedure against smallpox, one of the most dreaded diseases in human history.

suspect disease agents in food or beverages. He and others knew that boiling killed these agents. Being a wine connoisseur, he also knew you cannot boil wine—or beer or milk, for that matter—and end up with the same beverage. He devised a way to heat the beverages at a temperature low enough not to ruin them but high enough to kill most of the microorganisms. We still depend on his pioneering methods, which were named *pasteurization* in his honor.

In the late 1870s Robert Koch, a German physician, linked a specific microorganism to a specific disease—namely, anthrax. In one experiment, Koch injected blood from infected animals into uninfected ones. The recipients of the injection ended up with blood that teemed with cells of a bacterium (*Bacillus anthracis*)—and they developed anthrax. Even more convincing, injections of bacterial cells cultured outside the body also caused the disease!

Thus, by the beginning of the 20th century, the promise of understanding the basis of infectious disease and immunity loomed large—and the battles against those diseases were about to begin in earnest. Since that time, advances in microscopy, biochemistry, and

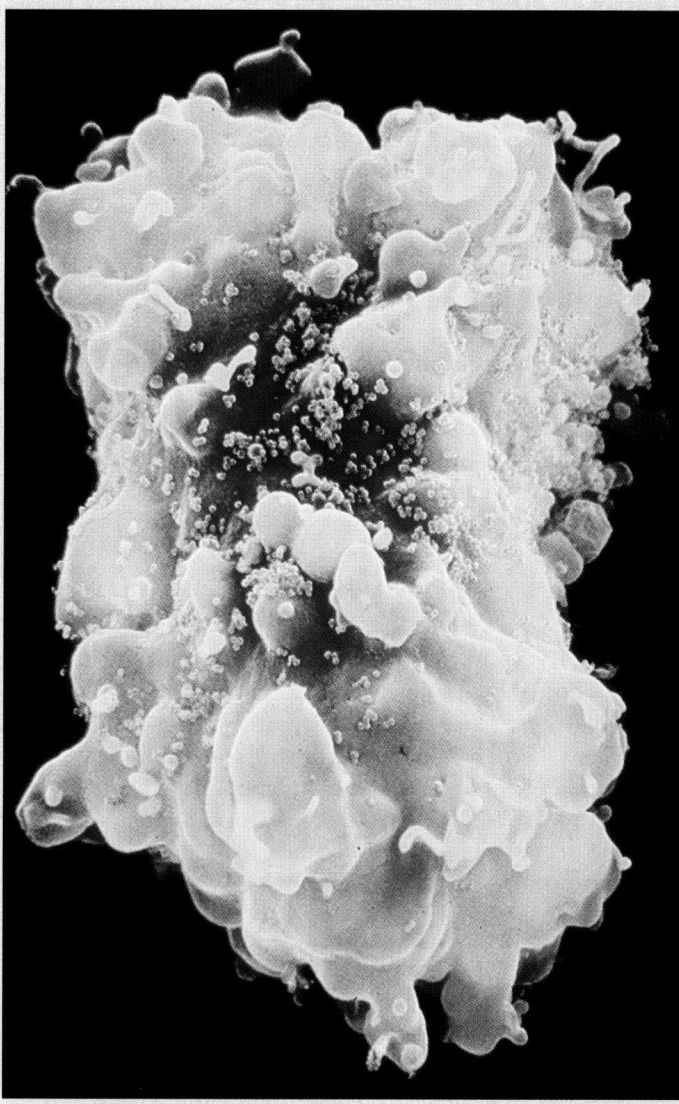

Figure 7.1b Micrograph of a white blood cell being attacked by the virus (*blue* particles) that causes AIDS. Immunologists are working to develop weapons against this modern-day scourge.

molecular biology have enormously increased our understanding of the body's defenses. We now have greater insights into its responses to tissue damage in general—and into its immune responses to specific pathogens or cancerous body cells. These responses are the focus of this chapter.

KEY CONCEPTS

1. The human body has physical, chemical, and cellular defenses against harmful microorganisms, cancer cells, and other agents that can destroy tissues and sometimes even cause death.

2. During early stages of tissue invasion and damage, white blood cells and plasma proteins take part in a rapid, general counterattack. This is an inflammatory response. Phagocytic white blood cells ingest the invaders. The plasma proteins promote phagocytosis, and some also destroy invaders directly.

3. If the invasion persists, certain white blood cells make immune responses. These cells can chemically recognize configurations on molecules that are abnormal or foreign to the body, such as those on bacteria and viruses. If the foreign or abnormal molecule triggers an immune response, it is called an antigen.

4. In one type of immune response, some of the white blood cells produce enormous quantities of antibodies. Antibodies are proteins that bind to a specific antigen and tag it for destruction.

5. In another type of immune response, executioner cells directly destroy body cells that have become abnormal, as by infection or a tumor-producing process.

CHAPTER AT A GLANCE

7.1   Three Lines of Defense

7.2   Complement Proteins

7.3   Inflammation

7.4   The Immune System

7.5   Lymphocyte Battlefields

7.6   Cell-Mediated Responses

7.7   Antibody-Mediated Responses

7.8   *Immune Responses and Organ Transplantation*

7.9   Immune Specificity and Memory

7.10 Immunization and Other Practical Applications of Immunology

7.11 Abnormal or Deficient Immune Responses

7.12 Immunity Defied: Infectious Diseases

Throughout life we encounter a huge assortment of pathogens. In the modern world, new disease threats are emerging even as ancient ones such as smallpox are being conquered. We use the term **pathogen** for viruses, bacteria, fungi, protozoa, and parasitic worms that cause disease. The body surface bars most pathogens. If pathogens do breach the barriers, cells and chemical weapons mount specific and nonspecific attacks against them. Table 7.1 lists these three lines of defense.

Surface Barriers to Invasion

Usually, pathogens can't get past skin or the linings of other body surfaces. Skin, for instance, is relatively dry, with a thick layer of dead cells at the surface—conditions that harmless bacteria can tolerate. Few pathogens can compete with these dense populations of harmless bacteria normally found on the surface unless conditions change. For example, the skin between the toes of sweaty feet encased in sneakers is moist and warm. These conditions favor the growth of locker room fungi that cause *athlete's foot*.

Similarly, normally "friendly" bacteria in the mucosal lining of the GI tract and the vagina help keep pathogens in check. For example, lactate produced by *Lactobacillus*

bacteria in the vagina helps maintain a low vaginal pH that most bacteria and fungi cannot tolerate. In females, some antibiotics routinely prescribed to cure bacterial infections can trigger a vaginal yeast infection because the drug also kills *Lactobacillus*.

Patches of tissue in mucous membranes also contain white blood cells (lymphocytes) that function in specific immune responses, as Section 7.4 describes.

The inner walls of the branching, tubular respiratory airways leading to the lungs are coated with a sticky mucus. In that mucus are protective substances such as **lysozyme**, an enzyme that chemically attacks and helps destroy many bacteria. Broomlike cilia in the airways sweep out the pathogens.

Lysozyme and some other chemicals in tears, saliva, and gastric fluid offer more protection (as when tears give eyes a sterile washing). Urine's low pH and flushing action help bar pathogens from moving into the urinary tract. In adults, mild diarrhea can rid the lower GI tract of irritating pathogens; blocking it can prolong infection. In children, however, diarrhea must be controlled to prevent dangerous dehydration.

Nonspecific and Specific Responses

Specialized types of white blood cells and plasma proteins serve as another line of defense if a pathogen breaches the surface barriers. When a tissue becomes damaged, these defenders take part in a *nonspecific response*—they react to tissue damage in general, not to specific pathogens. You will read about a major nonspecific response, the sequence of events we call *inflammation*, in Section 7.3. The inflammatory response includes activity by complement proteins, the subject of Section 7.2.

The white blood cells known as lymphocytes may mount a *specific* response against a particular pathogen. This phenomenon, called an **immune response**, occurs when lymphocytes recognize a unique molecular configuration on the invading pathogen. An immune response runs its course regardless of whether tissues are damaged. The immune response is the body's third line of defense against invasion and tissue damage.

Table 7.1	The Human Body's Three Lines of Defense against Pathogens

BARRIERS AT BODY SURFACES (*NONSPECIFIC* TARGETS):

1. Intact skin; mucous membranes at other body surfaces

2. Infection-fighting substances in tears, saliva, and so on

3. Normally harmless bacterial inhabitants of body surfaces that outcompete pathogenic visitors

4. Flushing effect of urination and diarrhea; sneezing, coughing, tears

NONSPECIFIC RESPONSES (*NONSPECIFIC* TARGETS):

1. Inflammation
   a. Fast-acting white blood cells (neutrophils, eosinophils, basophils)
   b. Macrophages (also take part in immune responses)
   c. Complement proteins, blood-clotting proteins, other infection-fighting substances

2. Phagocytic functions of macrophages in lymph nodes, spleen, liver, other organs

IMMUNE RESPONSES (*SPECIFIC* TARGETS):

1. White blood cells (macrophages, T cells, B cells)

2. Communication signals (e.g., interleukins) and chemical weapons (e.g., antibodies, complement proteins)

Intact skin, mucous membranes, antimicrobial secretions, and other barriers at the body surface are the body's first line of defense against invasion and tissue damage.

Inflammation and other internal, generalized responses to invasion are the second line of defense.

Immune responses against specific invaders, executed by armies of lymphocytes and their chemical weapons, are the third line of defense.

COMPLEMENT PROTEINS

A set of plasma proteins has roles in both nonspecific and specific defenses. Collectively, these proteins are called the **complement system**. About 20 kinds of complement proteins circulate in blood in inactive form. The system can be activated in two ways. A complement protein called C1 can bind to a "complex" that consists of a defender protein (an antibody) that is already bound to part of an invader (an antigen). Alternatively, a different complement protein can interact with carbohydrate molecules that are present on the surfaces of some microorganisms. If even a few molecules of one complement protein are activated, they trigger a cascade of reactions that become amplified as they progress. Molecules of one complement protein activate huge numbers of another kind of complement molecule at the following reaction step, and so on until the response deploys vast numbers of molecules.

Some complement proteins unite to form **membrane attack complexes**. These structures have an interior channel (Figure 7.2). They are inserted into the plasma membrane of many pathogens, forming pores that induce lysis (Figure 7.3). **Lysis**—cell disintegration—occurs when the plasma membrane's structure is severely disrupted, and the affected cell dies. Membrane attack complexes also

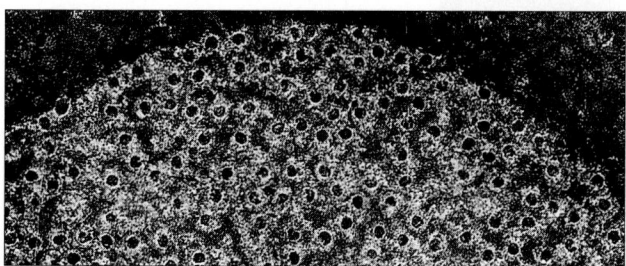

| ACTIVATION | CASCADE REACTIONS | FORMATION OF ATTACK COMPLEXES | LYSIS OF TARGET |

antibodies

activated complement

pathogen

During certain immune responses, complement proteins become activated when they bind to proteins called antibodies (the Y-shaped molecules). The antibodies have already become bound to a pathogen.

activated complement

bacterium

Complement proteins also are activated when they bind directly to bacterial surfaces.

activated complement

Cascading reactions yield huge numbers of different types of complement proteins. These are assembled into molecules that form many membrane attack complexes.

cutaway view of a membrane attack complex

lipid bilayer of pathogen

Attack complexes become inserted into the plasma membrane or lipid envelope of the target and form large pores.

The pores induce lysis; the pathogen dies.

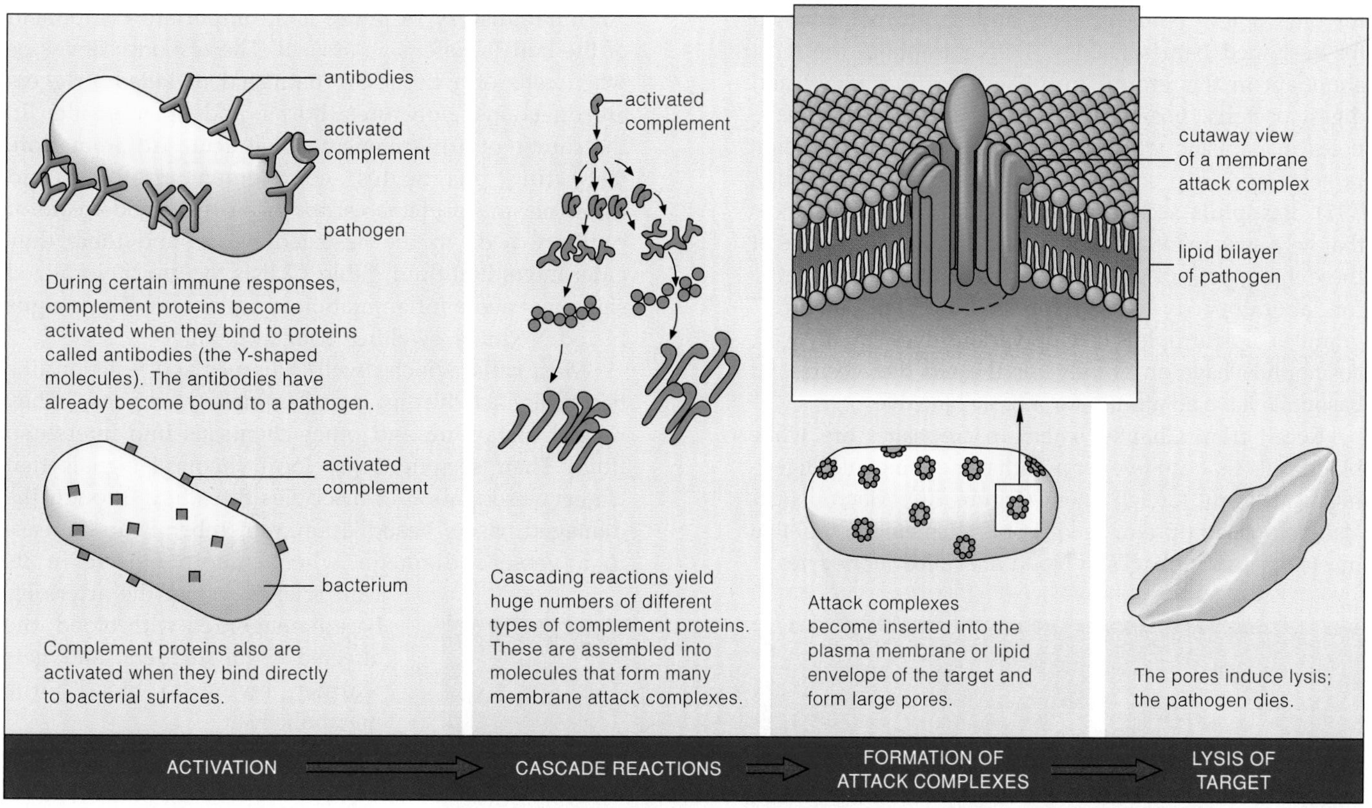

Figure 7.2 Formation of membrane attack complexes by complement proteins. One reaction pathway starts when complement binds to bacterial surfaces; another operates during immune responses to specific invaders. Both produce membrane attack complexes that induce lysis in the target pathogen.

are inserted into the lipid coats of some bacteria. Lysozyme molecules diffuse through the resulting pores and digest a key structural element of the bacterial cell.

Some activated complement proteins promote the process of inflammation, as you'll read shortly. Their cascades of reactions create concentration gradients that attract phagocytes to an irritated or damaged tissue. Also, activated complement proteins can bind to many invaders. Phagocytes, in turn, have receptors that bind to the complement proteins. The "complement-coated" invader sticks to the phagocyte and is ingested and killed.

The complement system operates in both nonspecific and specific defenses. Its plasma proteins circulate in blood and take part in cascades of reactions that help defend against many bacteria, some parasites, and enveloped viruses.

Figure 7.3 Micrograph of a cell surface, showing pores formed by membrane attack complexes.

The Roles of Phagocytes and Their Kin

Certain white blood cells take part in an initial response to tissue damage. White blood cells, remember, arise from stem cells in bone marrow. Many of them circulate in blood and lymph, then enter damaged tissues by squeezing between the endothelial cells of capillary walls. Many others take up stations in the lymph nodes and spleen (Figure 6.27), as well as in the liver, kidneys, lungs, and brain. Still others spend their lives in connective tissue beneath skin and in mucous membranes.

Three kinds of white blood cells, all granulocytes (see Section 6.1), react swiftly to danger in general, but live for only a few hours or days and so are not adapted for sustained battles (Table 7.1). **Neutrophils**, the most abundant in this group, are phagocytes that ingest and digest bacteria. **Eosinophils** secrete enzymes that make holes in parasitic worms; they also phagocytize foreign proteins and help control allergic responses (see Section 7.11). **Basophils** secrete histamine and other substances that help sustain inflammation after it starts. The names of these three types of granulocytes come from their affinity for laboratory dyes used for staining. The "neutral" granules of neutrophils take up various dyes equally well, eosinophils have an affinity for the red dye eosin, and basophils have an affinity for a basic blue dye.

Recall from Chapter 6 that macrophages are white blood cells that can live for months, engulfing and digesting nearly any foreign agent (Figure 7.4). Macrophages also help clean up damaged tissue. (Remember, too, that immature macrophages in blood are called monocytes.)

Table 7.2   Local Signs and Causes of Inflammation

Redness	Vasodilation, increased blood flow to site
Warmth	Vasodilation, greater flow of blood carrying more metabolic heat to site
Swelling	Capillaries made more permeable, plasma and plasma proteins move into interstitial spaces
Pain	Increased fluid pressure and local chemical signals stimulate pain receptors

The Inflammatory Response

An inflammatory response is an important component of the body's defensive arsenal. The response develops when cells of a tissue are damaged or killed. Triggers are infections, punctures, burns, and other insults. In the course of a mechanism called **acute inflammation**, fast-acting phagocytes, complement proteins, and other plasma proteins escape from the bloodstream at capillary beds in the besieged tissue, and there they enter interstitial fluid. Table 7.2 lists the typical localized signs that acute inflammation is under way. These signs include redness, swelling, heat, and pain.

Mast cells, which dwell in tissues and function like basophils, act during an inflammatory response. They release **histamine** and other chemicals into interstitial fluid. Their secretions are local chemical signals that trigger vasodilation of arterioles threading through the damaged tissue. Vasodilation, remember, is an increase in a vessel's diameter when smooth muscle in its wall relaxes. When the arterioles become engorged with blood, the affected tissue reddens and gets warmer, owing to blood-borne metabolic heat.

Released histamine also increases the permeability of the thin-walled capillaries in the tissue. It induces the endothelial cells making up the capillary wall to pull apart farther at the narrow clefts between them. Thus the

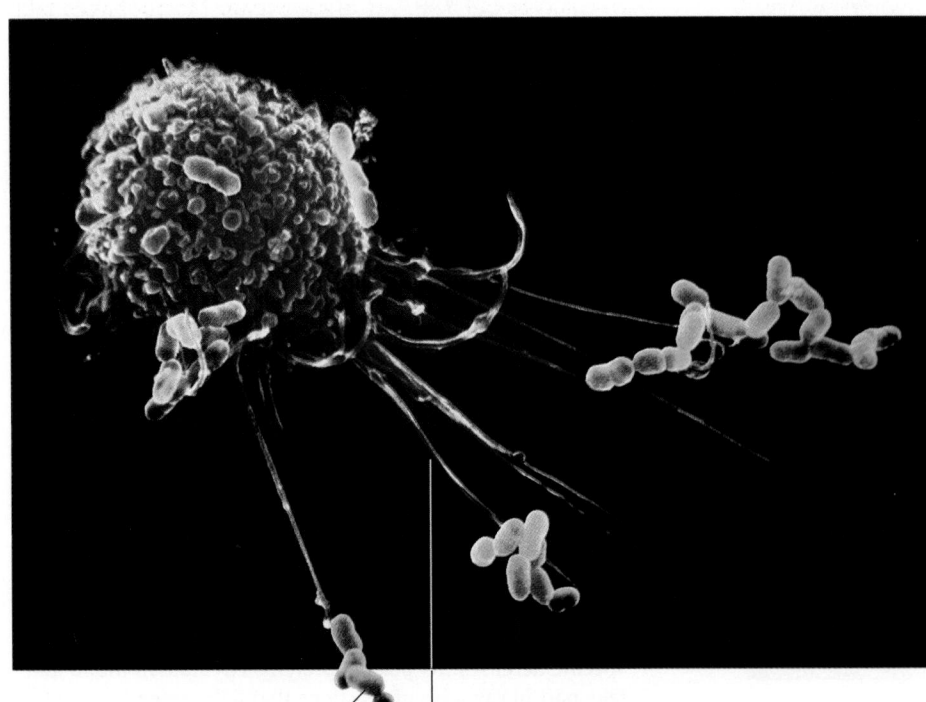

bacterial cells ——— cytoplasmic extension of macrophage

Figure 7.4 False-color scanning electron micrograph of a macrophage (*red*) with some cytoplasmic extensions that made contact with bacteria (*green*) in its surroundings Bacteria are engulfed by this type of phagocyte.

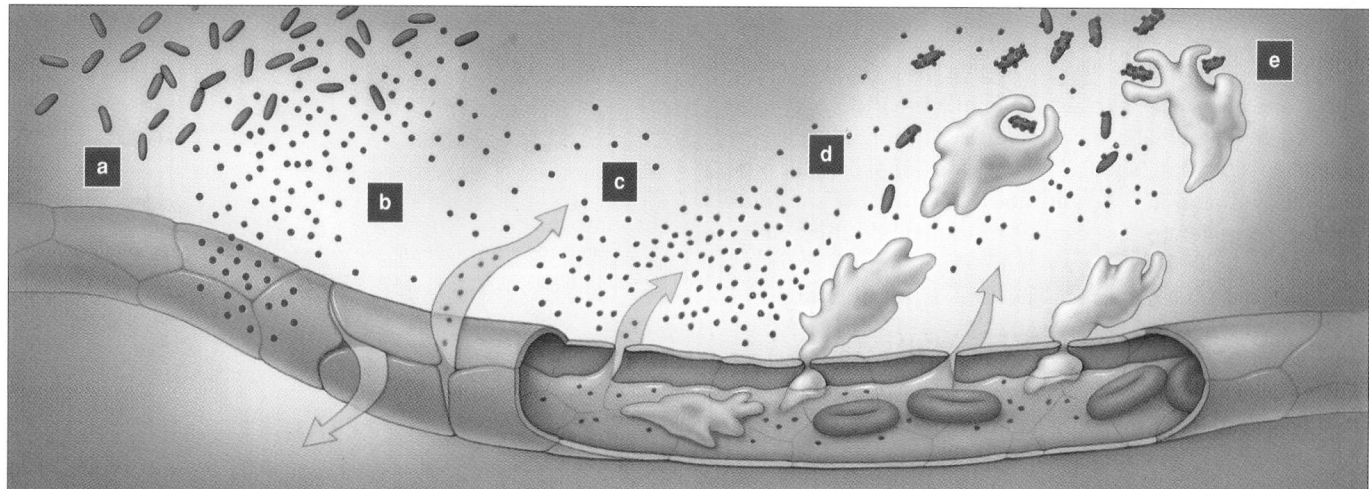

a Bacteria invade a tissue and directly kill cells or release metabolic products that damage tissue.

b Mast cells in tissue release histamine, which then triggers arteriolar vasodilation (hence redness and warmth) as well as increased capillary permeability.

c Fluid and plasma proteins leak out of capillaries; localized edema (tissue swelling) and pain result.

d Plasma proteins attack bacteria. Clotting factors wall off inflamed area.

e Neutrophils, macrophages, and other phagocytes engulf invaders and debris. Activated complement attracts phagocytes and directly kills invaders.

Figure 7.5 Acute inflammation in response to a bacterial invasion. The response involves delivering phagocytes and plasma proteins to the tissue. Together, these components of blood inactivate, destroy, or isolate the invaders, remove chemicals and cellular debris, and prepare the tissue for subsequent repair. These are their functions in all inflammatory responses.

capillaries become "leaky" to water and solutes (plasma) that normally do not leave the blood. When some proteins leak out, osmotic pressure increases in the surrounding interstitial fluid. Ultrafiltration increases and reabsorption decreases across the capillary wall. Localized edema is the outcome of the shift in the fluid balance across the capillary wall. (You can review these events by referring back to Section 6.4.) The tissue swells with fluid, and the swelling and inflammatory chemicals cause pain. Typically, the individual avoids voluntary movements that might aggravate the pain, and this behavior change promotes tissue repair.

Within hours of the first physiological responses to tissue damage, neutrophils are squeezing across capillary walls. They swiftly go to work. Monocytes arrive later, differentiate into macrophages, and engage in sustained action (Figure 7.5). While macrophages are engulfing pathogens, they secrete several substances that act as chemical mediators—communication signals. Mediators called *chemotaxins* attract more phagocytes. Proteins called **interleukins** carry signals between B and T lymphocytes, as you'll soon read. *Lactoferrin* directly kills bacteria. Endogenous pyrogen might trigger the release of substances called prostaglandins (discussed in Chapter 11), which in turn trigger an increase in the "set point" on the body's thermostat. A **fever** is a body temperature that has climbed to the higher set point.

A fever of about 39°C (100°F) is actually a helpful mechanism that promotes body defense activities. It also increases a person's body temperature to a level that is too hot for the functioning of most pathogens. During the fever, *interleukin-1* induces drowsiness, which reduces the body's demands for energy, so more energy can be devoted to defense and tissue repair. Macrophages take part in the cleanup and repair operations.

Among the plasma proteins that leak into the tissue are complement proteins and clotting factors (Section 6.13). The resulting blood clots wall off the inflamed area and typically prevent or delay the spread of pathogens and their toxic secretions into the surrounding tissues. After the inflammation subsides, anticlotting factors that had also escaped from the capillaries dissolve the clots.

A local inflammatory response develops when cells are damaged or killed. Arterioles in the tissue dilate, and capillary walls become more permeable to water and solutes (plasma). The altered fluid balance leads to edema and pain.

Neutrophils, eosinophils, basophils, and then macrophages leave the bloodstream. They defend the damaged tissue and help repair it.

Histamine, complement, chemical mediators, blood-clotting proteins, and other substances take part in the inflammatory response or in the cleanup and repair operations.

THE IMMUNE SYSTEM

Defining Features

Sometimes physical barriers and inflammation are not enough to overwhelm an invader, so an infection may become well established. Then, armies of white blood cells known as **B** and **T lymphocytes** form and join the battle. Lymphocytes are central to the body's third line of defense—the **immune system**. The immune system has two defining features. The first is immunological *specificity*, whereby lymphocytes zero in on specific pathogens and eliminate them. The second feature is immunological *memory*, whereby some lymphocytes that form during a first-time confrontation are set aside for a future battle with the same pathogen.

The operating principle for the system is this: *Each kind of cell, virus, or substance bears unique molecular configurations that give it a unique identity.* The unique configurations on an individual's own cells serve as *self* markers. Lymphocytes recognize self markers and will normally ignore them. They also can recognize *nonself* molecular configurations, which are unique to specific foreign agents. When they do, B and T lymphocytes that recognize a specific nonself marker are stimulated to divide repeatedly—and huge populations of the lymphocytes form.

As the divisions proceed, subpopulations of the new cells become specialized to respond to the foreign agent in different ways. Some consist of **effector cells**—fully differentiated cells that engage and destroy the enemy. Other subpopulations consist of **memory cells**, which enter a resting phase. Instead of attacking the specific agent that triggered the initial response, memory cells "remember" it. And they will undertake a larger, more rapid response if it shows up again.

Any molecular configuration that triggers formation of lymphocyte armies and is their target is an **antigen**. The most important antigens are certain proteins at the surface of pathogens or tumor cells and others that are free but toxic. As you will see, lymphocytes produce receptor molecules that bind to these configurations. That is how they are able to recognize nonself.

In short, immunological specificity and memory involve three events: *recognition* of antigen, *repeated cell divisions* that form huge populations of lymphocytes, and *differentiation* into subpopulations of effector and memory cells with receptors for one kind of antigen.

Antigen-Presenting Cells— The Triggers for Immune Responses

The plasma membrane of every nucleated cell in every human individual incorporates a variety of proteins. Among these proteins are **MHC markers**, named after the genes that encode the instructions for making them.

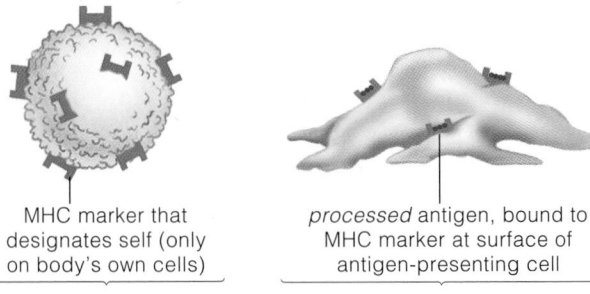

MHC marker that designates self (only on body's own cells)

T and B cells ignore this

processed antigen, bound to MHC marker at surface of antigen-presenting cell

T cells start an immune response

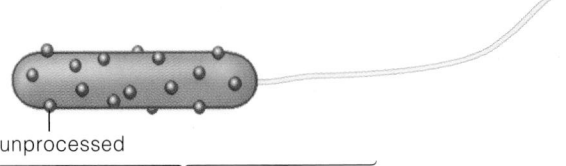

unprocessed

B cells start an immune response

Figure 7.6 Molecular cues that stimulate lymphocytes to make immune responses.

Certain MHC markers are common at the surface of every nucleated cell in the body. Others are found only on the body's macrophages and lymphocytes.

Suppose bacteria enter a cut on your finger. Lymph vessels pick up some of the inflamed tissue's interstitial fluid and deliver it, along with some invading cells, to the lymph nodes in the vicinity. There, macrophages join the fray. Foreign cells are engulfed and enclosed in vesicles with digestive enzymes that break apart the antigen molecules into fragments. The fragments bind to MHC molecules and form **antigen-MHC complexes.** When the vesicles move to the plasma membrane and fuse with it, the complexes are automatically displayed at the macrophage surface.

Any cell that processes and displays an antigen that is bound with a suitable MHC molecule is an **antigen-presenting cell**. When lymphocytes do encounter such a cell, they respond (Figure 7.6). *This response is the antigen recognition that promotes the cell divisions by which great armies of lymphocytes form.*

Key Players in Immune Responses

The same kinds of white blood cells are called into action during each immune response. Figure 7.7 is an overview of how the cells interact. In brief, recognition of antigen-MHC complexes activates **helper T cells**. These T lymphocytes produce and secrete substances that induce any responsive T or B lymphocyte to divide and give rise to large populations of effector cells and memory cells. Recognition also activates **cytotoxic T cells**. These can eliminate infected body cells or tumor cells by "touch-killing." When they touch a target, they

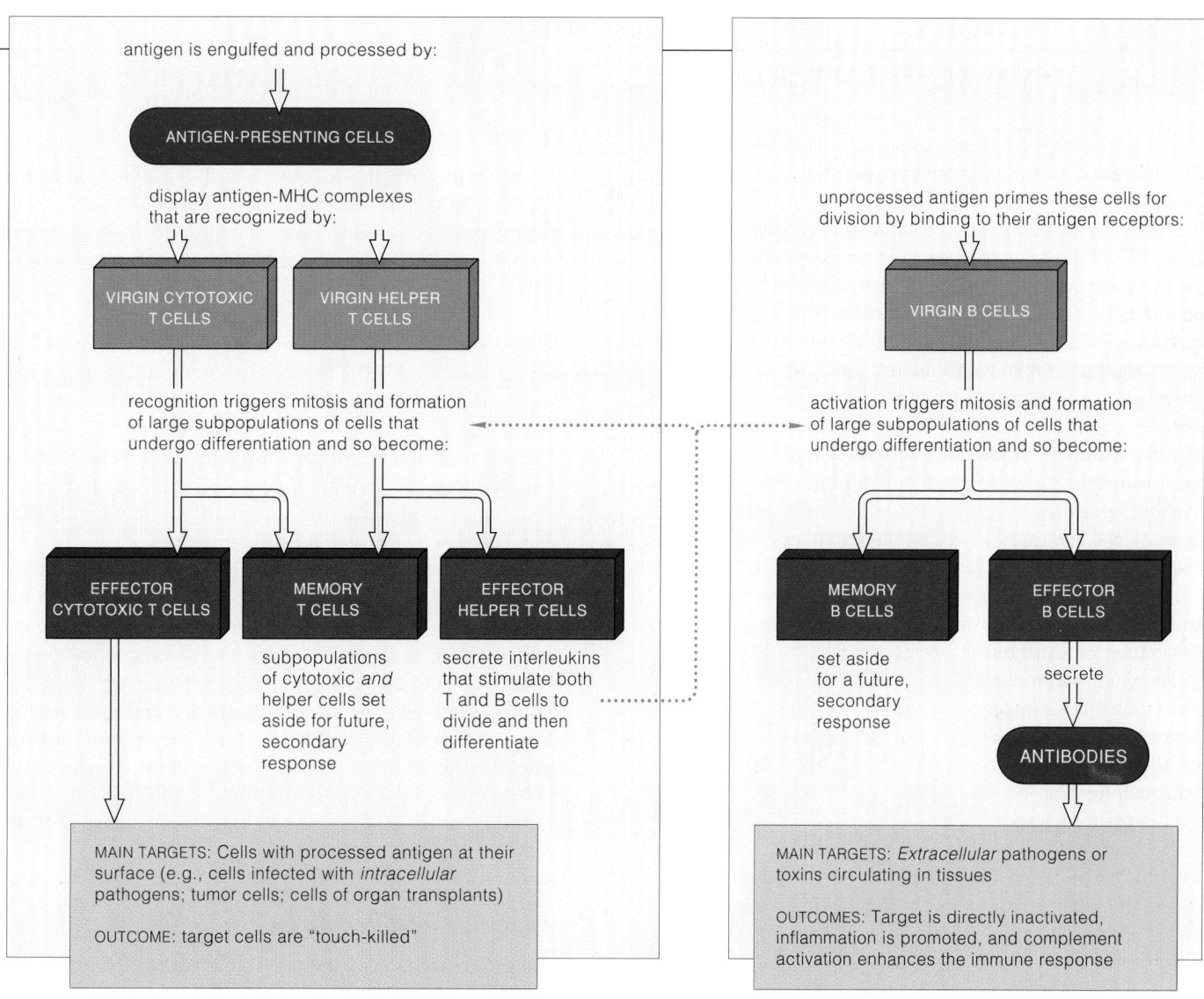

antigen is engulfed and processed by:

ANTIGEN-PRESENTING CELLS

display antigen-MHC complexes that are recognized by:

VIRGIN CYTOTOXIC T CELLS

VIRGIN HELPER T CELLS

recognition triggers mitosis and formation of large subpopulations of cells that undergo differentiation and so become:

EFFECTOR CYTOTOXIC T CELLS

MEMORY T CELLS

EFFECTOR HELPER T CELLS

subpopulations of cytotoxic *and* helper cells set aside for future, secondary response

secrete interleukins that stimulate both T and B cells to divide and then differentiate

MAIN TARGETS: Cells with processed antigen at their surface (e.g., cells infected with *intracellular* pathogens; tumor cells; cells of organ transplants)

OUTCOME: target cells are "touch-killed"

CELL-MEDIATED IMMUNE RESPONSES

unprocessed antigen primes these cells for division by binding to their antigen receptors:

VIRGIN B CELLS

activation triggers mitosis and formation of large subpopulations of cells that undergo differentiation and so become:

MEMORY B CELLS

EFFECTOR B CELLS

set aside for a future, secondary response

secrete

ANTIBODIES

MAIN TARGETS: *Extracellular* pathogens or toxins circulating in tissues

OUTCOMES: Target is directly inactivated, inflammation is promoted, and complement activation enhances the immune response

ANTIBODY-MEDIATED IMMUNE RESPONSES

Figure 7.7 Overview of key interactions among B and T lymphocytes during an immune response. Usually, both types of lymphocytes are activated when an antigen has been detected. A first-time encounter with an antigen elicits a *primary* response. A subsequent encounter with the same type of antigen quickly elicits a *secondary* response, which is larger and more rapid. Memory cells that formed but were not used during the first response can immediately engage in the second one.

deliver cell-killing chemicals into it. By contrast, **B cell**s produce antigen-binding receptor molecules called **antibodies**. When an immune response is under way, effector B cells secrete huge numbers of antibody molecules. B cells thus are the basis of *antibody-mediated* responses.

Control of Immune Responses

An antigen provokes an immune response, and removal of the antigen stops it. For example, by the time the tide of battle turns, the effector cells and their secretions have already destroyed most antigen-bearing agents in the body. With fewer antigen molecules present to stimulate the cells, the response declines, then stops. As a final example, inhibitory signals from cells with suppressor roles help shut down immune responses.

Antigens are nonself molecular configurations that, when recognized by certain lymphocytes, trigger immune responses. Helper T cells, cytotoxic T cells, and B cells and their secretions execute these responses.

Following antigen recognition, repeated cell divisions produce huge armies of T and B cells. These differentiate into subpopulations of effector cells and memory cells, all of which are sensitized to the triggering antigen.

The antigen-presenting cells and lymphocytes we have just introduced interact within **lymphoid organs** that promote immune responses (Figure 7.8).

For example, consider the location of tonsils and most other lymph nodules just beneath the mucous membranes of the respiratory, digestive, and reproductive systems. Just after invaders penetrate surface barriers, antigen-presenting cells and lymphocytes housed here can intercept them. Or think about antigen in tissue fluid that is entering the lymph vascular system. Because lymph vessels eventually drain into the bloodstream, antigen could become distributed to every body region. However, before antigen can reach the blood, it must trickle through lymph nodes—which are packed with defending cells. Even in the few cases where antigen manages to enter the blood, defending cells in the spleen intercept it.

In lymph nodes, cells are organized for maximum effectiveness. Antigen-presenting cells make up the front line, engulfing invaders. They process and display an antigen, thus calling lymphocyte comrades into action.

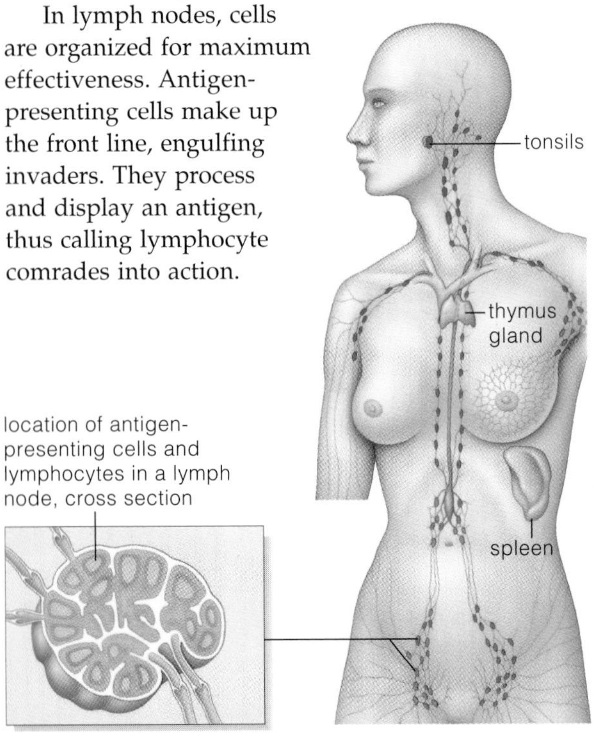

location of antigen-presenting cells and lymphocytes in a lymph node, cross section

Figure 7.8 Organized arrays of antigen-presenting cells and lymphocytes in lymph nodes.

The cell divisions that produce subpopulations of effector and memory cells take place in lymph nodes. As lymph drains through, it moves effector activities to the back of the organ and beyond. And all the while, virgin and memory cells circulate through the lymph node, reconnoitering at the front line.

Antigen-presenting cells and lymphocytes intercept and battle pathogens in organized ways within lymphoid organs and tissues, especially the lymph nodes.

T Cell Formation and Activation

Let's begin with the functions of T lymphocytes in an immune response. Recall from Chapter 6 that T cells arise from stem cells in bone marrow. However, they do not finish developing until they travel to the thymus gland, where they become differentiated into helper T cells and cytotoxic T cells. Each T cell acquires a large number of identical T cell receptor molecules, or TCRs, on its surface. A TCR can bind to a specific antigen. At least a billion different TCRs are made in the thymus; hence there are enough T cells to specifically react with billions of different antigens. Bristling with TCRs, T cells exit the thymus, circulate in lymph and blood, and move into lymph nodes and the spleen as "virgin" T cells that have not been stimulated by an antigen.

The TCRs of virgin T cells ignore unadorned MHC markers on the body's own cells. They also ignore unbound antigens. *But they recognize and bind with antigen-MHC complexes at the surface of antigen-presenting cells.* The binding, and chemicals released by the APC, stimulates T cells to divide repeatedly and give rise to large clones (Figure 7.9). (A clone is a population of genetically identical cells.) These clonal descendants differentiate into subpopulations of effector cells and memory cells—*and each descendant has the same TCR for one kind of antigen-MHC complex.*

Functions of Effector T Cells

Subgroups of effector helper T cells secrete communication signals—interleukins. Interleukins fan repeated cell divisions and differentiation of responsive virgin T cells and virgin B cells, as described shortly. The effector cytotoxic T cells respond to antigen-MHC complexes on body cells that have been infected by intracellular pathogens (such as viruses) and on tumor cells. The complex serves as a "double signal" to destroy the cells that bear it.

Effector cytotoxic T cells destroy infected cells with a touch-kill mechanism. They secrete **perforins**, protein molecules that form doughnut-shaped pores in a target cell's plasma membrane, similar to the ones shown in Figure 7.2. These effectors also secrete chemicals that induce cell death. By a process called **apoptosis**, a target cell commits suicide, so to speak. Its cytoplasm is expelled, its organelles break down, and its DNA becomes fragmented. Having made its lethal hit, the cytotoxic T cell disengages and moves on (Figure 7.9a).

Cytotoxic T cells also contribute to the rejection of tissue and organ transplants (see *Focus on Your Health*). Parts of MHC markers on donor cells are different enough from the recipient's to be recognized as antigens—but other parts are similar enough to complete the double signal. MHC typing and matching donors to recipients minimize the risk.

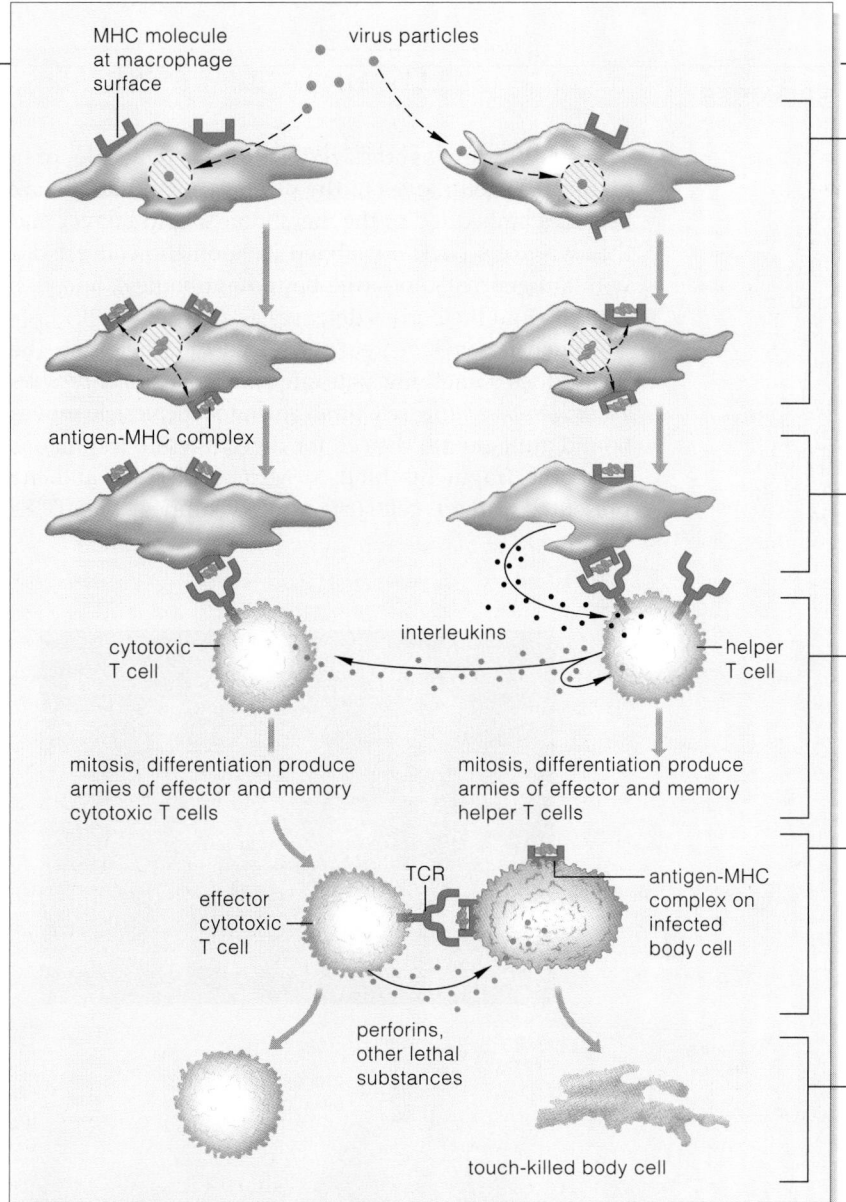

a A virus penetrates a macrophage and pirates its metabolic machinery. The host cell churns out viral proteins, which are antigenic. These are processed into fragments, bound to MHC molecules, and displayed at the cell surface. And *TCRs of cytotoxic T cells recognize antigens-MHC complexes that have been synthesized inside a cell.*

Another macrophage engulfs a particle of the same virus and encloses it in an endocytic vesicle. Digestive enzymes cleave the particle's antigen into fragments. These bind to MHC molecules and are presented at the macrophage surface.

b A responsive T cell binds with the antigen-MHC complex. Binding stimulates the macrophage to secrete interleukins (*black* dots).

c These communication signals stimulate the helper T cell to secrete different kinds of interleukins (*blue* dots). These new signals stimulate cell divisions and differentiations by which large populations of effector T cells and memory T cells form.

d The same virus penetrated a cell in the lining of the respiratory tract, and antigen that was produced in the cytoplasm has been processed. Antigen is bound to MHC molecules and displayed at the cell's surface.

An effector cytotoxic T cell encounters the target. This type of effector specializes in touch-killing. It releases perforins and toxic substances (*green* dots) onto its target and so programs it for death.

e The effector disengages from the doomed cell and reconnoiters for more targets. Meanwhile, perforins make holes in the cell's plasma membrane. Other chemicals can move into the cell, disrupt its organelles, and make the DNA disassemble. The infected cell dies.

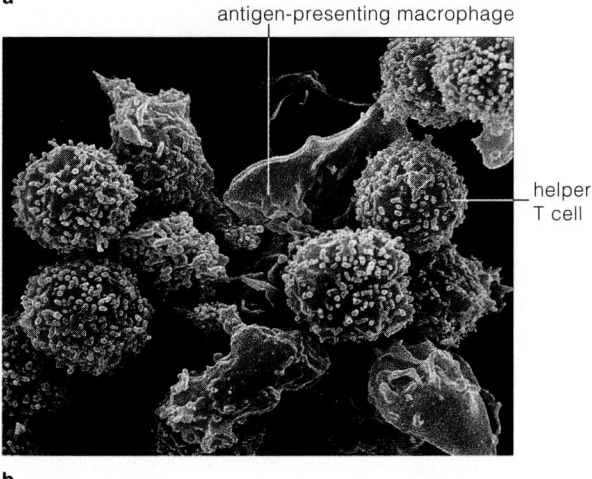

Figure 7.9 (**a**) Example of a T-cell-mediated immune response. In this case, the response involves an antigen-MHC complex that activates T cells. (**b**) Scanning electron micrograph of helper T cells associating with an antigen-presenting macrophage.

Natural Killer Cells

Other cytotoxic cells, including **natural killer cells** (NK cells), also arise from stem cells in bone marrow. They appear to be lymphocytes, but not T or B cells. Their arousal does not depend on a double signal (antigen-MHC complex). NK cells reconnoiter for tumor cells and virus-infected cells, then touch-kill them. They possibly recognize odd molecular configurations at the surface of their targets.

T cells arise in bone marrow. Later on, in the thymus, they acquire TCRs (receptors for self markers and bound antigen).

Effector helper T cells secrete interleukins that trigger the cell divisions and differentiation into armies against specific antigens. Effector cytotoxic cells touch-kill infected cells or tumor cells, even foreign cells of transplants.

ANTIBODY-MEDIATED RESPONSES

B Cells and the Targets of Antibodies

Like T cells, the B cells also arise from stem cells in bone marrow and start down a pathway that will culminate in full differentiation. Along *their* pathway, however, B cells start synthesizing numerous copies of a single kind of antibody molecule.

Although all antibodies are proteins, each kind has binding sites that match up only to a particular antigen. They are more or less Y-shaped, with a tail and two arms that bear identical antigen receptors. Section 7.9 provides a closer look at these molecules. For now, we can simply think of them as Y-shaped structures, as shown in Figure 7.10.

Each freshly synthesized antibody molecule of a maturing B cell moves to the plasma membrane. Its tail becomes embedded in the membrane's lipid bilayer and the two arms stick out above it. Soon the cell bristles with antigen receptors (the bound antibodies), and it is ready to join the body's defenses as a virgin B cell.

When its antigen receptors lock onto a target, the B cell does something you might not expect. *It becomes an antigen-presenting cell.* First, an endocytic vesicle moves bound antigen into the cell for digestion into fragments. Next, the fragments bind to MHC molecules and are presented at the B cell surface. Now suppose that TCRs

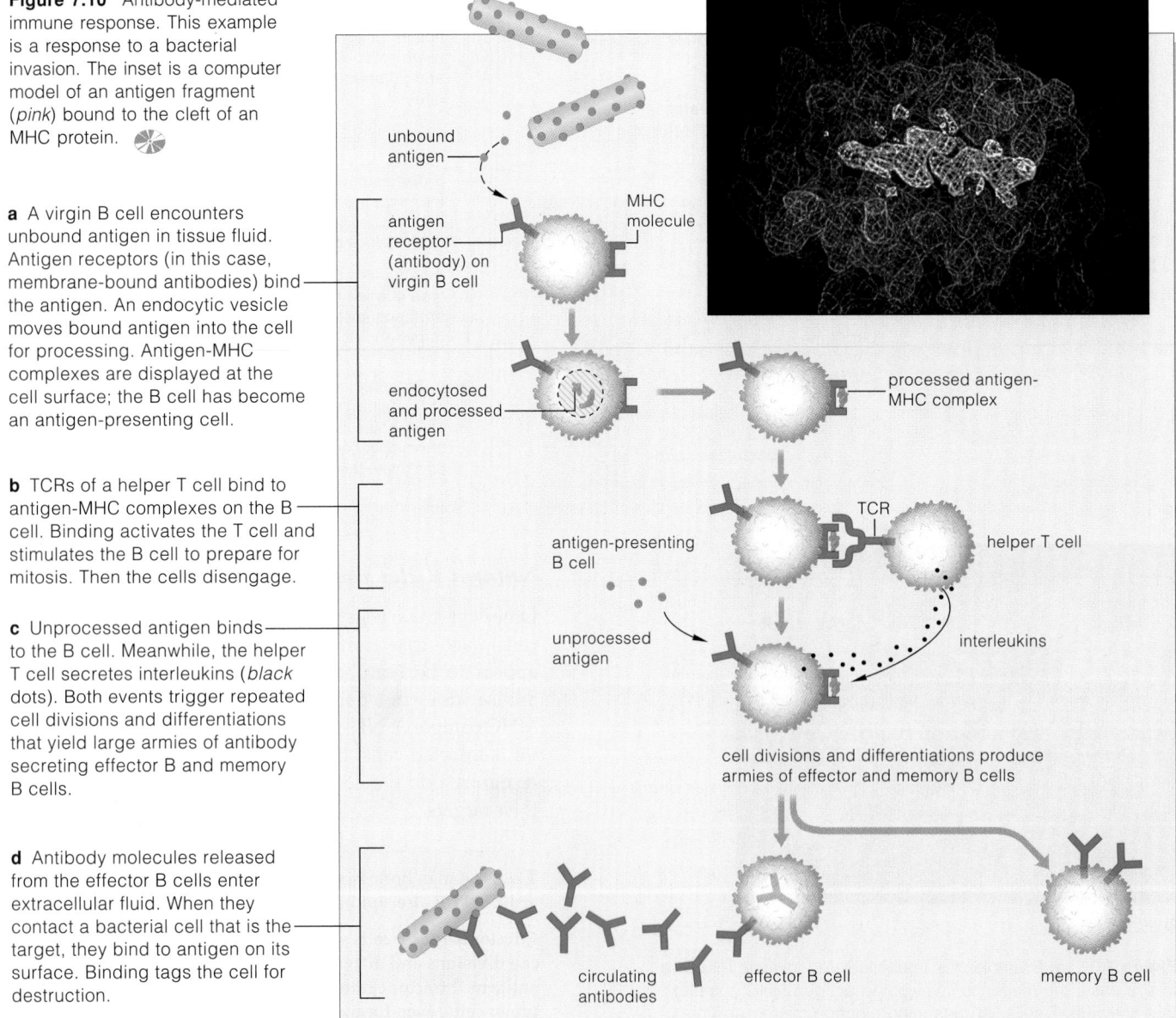

Figure 7.10 Antibody-mediated immune response. This example is a response to a bacterial invasion. The inset is a computer model of an antigen fragment (*pink*) bound to the cleft of an MHC protein.

a A virgin B cell encounters unbound antigen in tissue fluid. Antigen receptors (in this case, membrane-bound antibodies) bind the antigen. An endocytic vesicle moves bound antigen into the cell for processing. Antigen-MHC complexes are displayed at the cell surface; the B cell has become an antigen-presenting cell.

b TCRs of a helper T cell bind to antigen-MHC complexes on the B cell. Binding activates the T cell and stimulates the B cell to prepare for mitosis. Then the cells disengage.

c Unprocessed antigen binds to the B cell. Meanwhile, the helper T cell secretes interleukins (*black* dots). Both events trigger repeated cell divisions and differentiations that yield large armies of antibody secreting effector B and memory B cells.

d Antibody molecules released from the effector B cells enter extracellular fluid. When they contact a bacterial cell that is the target, they bind to antigen on its surface. Binding tags the cell for destruction.

unbound antigen

MHC molecule

antigen receptor (antibody) on virgin B cell

endocytosed and processed antigen

processed antigen-MHC complex

antigen-presenting B cell

TCR

helper T cell

unprocessed antigen

interleukins

cell divisions and differentiations produce armies of effector and memory B cells

circulating antibodies

effector B cell

memory B cell

of a responsive helper T cell bind to the antigen–MHC complex and that signals are transferred between the T and B cell. The cells soon disengage. When the B cell encounters unprocessed antigen, its surface antibodies bind the antigen. The binding, in combination with interleukins secreted from nearby helper T cells, drives the B cell to divide. Its clonal descendants differentiate into effector and memory B cells. The effectors (also called plasma cells) produce and secrete huge numbers of antibody molecules. When circulating antibody molecules bind an antigen, they tag an invader for destruction, as by phagocytes and the activation of complement.

The main targets of antibody-mediated responses are extracellular pathogens and toxins, which are freely circulating in tissues or body fluids. *Antibodies cannot bind to pathogens or toxins hidden in a host cell.*

The Immunoglobulins

During immune responses, B cells produce four classes of antibody molecules in abundance and lesser quantities of another. Collectively the five classes of antibodies are called **immunoglobulins,** or Igs. They are the protein products of genetic events that proceed while B cells mature and while an immune response is under way. The molecules in each class have antigen-binding sites, as well as other sites with specialized functions.

IgM antibodies are the first to be secreted during immune responses. They trigger complement cascades, and they bind targets together in clumps—which are more readily eliminated by phagocytes. *IgD* antibodies associate with IgM on virgin B cells, but their function is not yet understood.

IgG antibodies activate complement proteins and neutralize many toxins. The IgGs are long-lasting. They are the only antibodies to cross the placenta during pregnancy and protect a fetus with the mother's acquired immunities. IgGs also are secreted into the early milk produced by mammary glands, then are absorbed into the suckling newborn's bloodstream.

IgA antibodies enter mucous-coated surfaces, such as those in the respiratory, digestive, and reproductive tracts, where they neutralize infectious agents. Mother's milk delivers them to the mucous lining of a newborn's GI tract.

IgE triggers inflammation after attacks by parasitic worms and other pathogens. As described later, it also figures in allergies. The tails of IgE antibodies bind to basophils and mast cells, and the antigen receptors face outward. Antigen-binding induces basophils and mast cells to release substances that promote inflammation.

Antibodies that are secreted by B cells bind to antigens of extracellular pathogens or toxins and tag them for disposal, as by phagocytes and complement proteins.

IMMUNE RESPONSES AND ORGAN TRANSPLANTATION

Every year, thousands of people with severe heart, lung, kidney, and liver disease receive donated replacement organs. Organ transplants are risky, in part because, except in the case of identical twins, the organ recipient's immune system will perceive donated tissues as foreign and attempt to reject them from the body.

STRATEGIES FOR PREVENTING REJECTION To help prevent rejection, before an organ is to be transplanted the MHC markers of a potential donor are analyzed to determine how closely they match those of the patient. Because such tissue grafts generally succeed only when the donor and recipient share at least 75 percent of their MHC markers, the favored donor is a close relative of the recipient, such as a parent or sibling, who is likely to have a similar genetic makeup. More commonly, the donated organ comes from a fresh cadaver. In addition to having well-matched MHC markers, donor and recipient also must have compatible blood types, as discussed in Section 6.4.

After surgery, the organ recipient receives drugs that suppress the immune system, as well as other therapies (such as radiation) that may be needed to prevent an attack by activated B and T cells. These therapies all have serious side effects, and suppression of the immune system means that the patient must take large doses of antibiotics to control infections. In spite of the difficulties, many organ recipients survive an average of one to five years beyond the surgery, and some live much longer.

There are active efforts to genetically alter pigs to create varieties with organs bearing common human MHC markers. Why choose pigs for this research? A major reason is that, anatomically and physiologically, pig organs are quite similar to those of humans. In theory, such transgenic animals could be sources of readily transplantable organs. Transplantation of organs from one species of animal to another is called *xenotransplantation*. It is a fascinating topic about which you will learn more in Chapter 21.

IMMUNE-PRIVILEGED TISSUES There are some intriguing exceptions to the "rule" that transplanted foreign tissues provoke a recipient's immune defenses. Two examples are tissues of the eye, and the testicles. In simple terms, the plasma membrane of cells of these organs apparently bears receptor proteins that can detect activated lymphocytes in the surroundings. Before such a lymphocyte can launch an attack, the protein signals the soon-to-be-besieged cell to secrete a chemical that triggers apoptosis ("cell suicide") in the approaching lymphocytes—averting an attack. Our ability to readily transplant the cornea—the outermost layer of the eye that is vital to clear vision—is a practical benefit of this mechanism.

IMMUNE SPECIFICITY AND MEMORY

Formation of Antigen-Specific Receptors

A mind-boggling variety of pathogens, each with a unique antigen, occur in food, water, air, soil, people, and almost everything else in your surroundings. Collectively, however, antigen receptors of all T and B cell populations in your body show staggering diversity—by one estimate, enough to recognize about a billion different antigens. How does the diversity arise?

For the answer, start with the knowledge that all antigen receptors of a single B or T cell are identical—and all are proteins. For example, a Y-shaped antibody molecule consists of four polypeptide chains bonded together (Figure 7.11). Certain parts of each chain, the *variable* regions, fold in ways that produce grooves and bumps with a certain charge distribution. Only antigen that has complementary grooves, bumps, and charge distribution will be able to bind with them.

All cells, including T and B cells, carry genetic instructions in the form of DNA. As you will read in later chapters, a gene consists of a particular region of DNA. Within each B cell's DNA are extensive regions that code for antigen receptors. Each region has multiple segments that differ slightly in their chemical makeup.

As a B cell completes its development in bone marrow, a single gene segment from each DNA region is "chosen." The same is true of each T cell as it matures in the thymus. *Which* segment is tapped from each DNA region is a matter of chance. Next, the segments from the different regions are put together in a special version of a process called *genetic recombination*. Following some additional steps, the outcome is a randomly produced gene sequence (a linear arrangement of gene segments). In each T or B cell, that sequence codes for one of at least billions of possible antigen receptors.

The same kinds of DNA rearrangements also help produce the variable regions of the TCR molecules of maturing T cells.

Some time ago, Macfarlane Burnet developed a **clonal selection hypothesis** that helped point the way to our current view of receptor diversity. He proposed that antigen "chooses" (binds to) one lymphocyte from all the various types in the body, because that lymphocyte has the receptor specific for it. Repeated cell divisions then give rise to a clone of cells that carry out the response (Figure 7.12).

antigen
binding site

antigen
binding site

variable region
of heavy chain

variable
region of
light chain

hinge
region
(flexible)

constant region
of light chain

constant region
of heavy chain

a

Figure 7.11 Antibody structure. (**a**) An antibody molecule has four polypeptide chains, often joined in a Y shape. (**b**) Some regions are almost the same in all antibodies. But the molecular configuration in one region is unique for each kind of antibody; it is the binding site for one kind of antigen only. The antigen fits into the site's grooves and onto its protrusions, as shown in Figure 7.9.

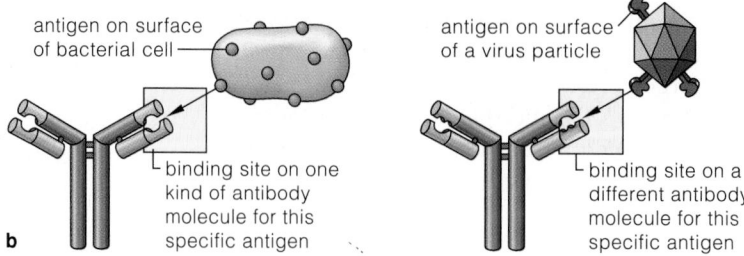

antigen on surface
of bacterial cell

antigen on surface
of a virus particle

binding site on one
kind of antibody
molecule for this
specific antigen

binding site on a
different antibody
molecule for this
specific antigen

b

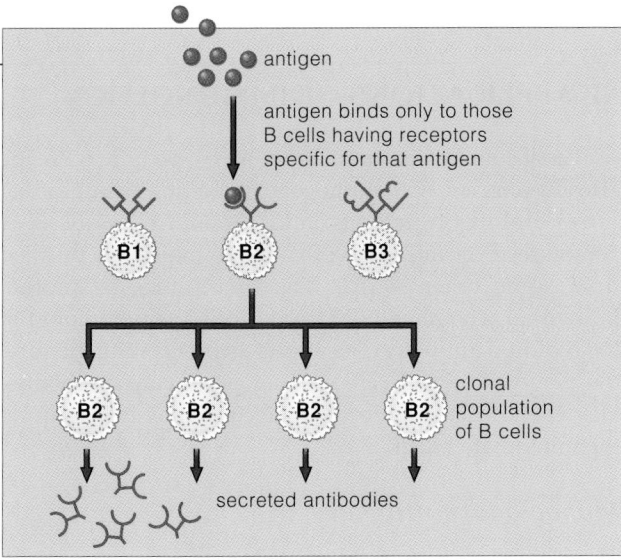

Figure 7.12 Clonal selection of a B cell that produced the specific antibody that can combine with a specific antigen. Only antigen-selected B cells (and T cells) are activated and give rise to a clonal population of immunologically identical cells.

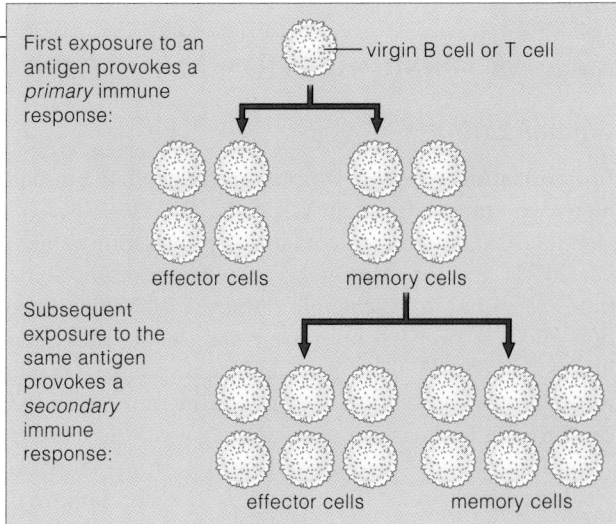

Figure 7.13 Immunological memory. Not all B and T cells are used in a primary immune response to an antigen. A large number continue to circulate as memory cells, which become activated during a secondary immune response.

Immunological Memory

The clonal selection theory explains how an individual can have "immunological memory" of a first encounter with antigen. The term refers to the body's capacity to make a *secondary* immune response to any subsequent encounter with the same type of antigen that provoked the primary response (Figure 7.13).

Memory cells that form during a primary immune response do not engage in battle. They circulate for years or for decades. Compared to the virgin cells that initiate a primary response, these patrolling battalions have far more cells, so they intercept antigen far sooner. Effector cells form sooner, in greater numbers, so the infection is terminated before the host (you) gets sick. Even greater numbers of memory T and B cells form during a secondary response (Figure 7.14). These advance preparations

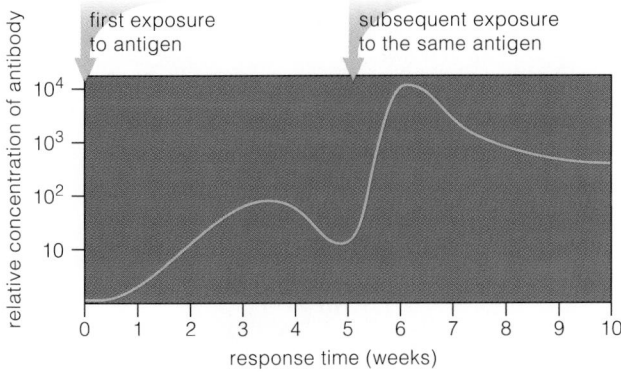

Figure 7.14 Differences in magnitude and duration between a primary and a secondary immune response to the same antigen. (In this example, a secondary response starts at week 5.)

against subsequent encounters with a pathogen bestow a significant survival advantage on the individual.

In people who have a positive reaction to the skin test for tuberculosis, a hard, red swelling develops at the test site, usually within 48 hours of the test. Even in someone who has no medical history of this disease, this response is visible evidence of immunological memory. It indicates the presence of antibodies against the tuberculosis bacterium, which the person's immune system must have encountered at some time in the past.

As this example suggests, over time, the composition of memory T and B cell populations is shaped by the particular kinds of antigens to which a person has been exposed. For instance, if you grew up in North America in the late 20th century, you almost certainly have hosts of memory cells that can fight off pathogens common in your environment. You may *not* have ready contingents of memory cells primed to battle, say, bacteria present in the water supply of a different geographical region. That is one reason why wise tourists travel prepared to cope, at least temporarily, with intestinal problems caused by exposure to unfamiliar waterborne bacteria.

Recombination of segments drawn at random from receptor-encoding regions of DNA gives each new T or B cell a gene sequence for one of billions of possible antigen receptors.

Specificity **means an antigen-selected cell will give rise to a clone of cells that will react only with the selecting antigen.**

Memory **means the individual has the capacity to make a secondary immune response to the pathogen that caused the primary response. A secondary response is greater in magnitude and occurs more rapidly.**

IMMUNIZATION AND OTHER PRACTICAL APPLICATIONS OF IMMUNOLOGY

Immunization

Immunization refers to various processes that promote increased immunity against specific diseases. In *active immunization*, an antigen-containing preparation called a **vaccine** is injected into the body or taken orally, sometimes according to a schedule (Figure 7.15). A first injection elicits a primary immune response. A subsequent injection (booster) elicits a secondary response, with the formation of more effector cells and memory cells that can provide long-lasting protection against the disease.

Many vaccines are manufactured from killed or weakened pathogens. For example, weakened poliovirus particles are used for the Sabin polio vaccine. Other vaccines are based on inactivated forms of natural toxins, such as the bacterial toxin that causes tetanus. Still others are made with harmless genetically engineered viruses (Chapter 21). These incorporate genes from three or more different viruses in their genetic material. After a person is vaccinated with an engineered virus, the incorporated genes are expressed, antigens are produced, and immunity is established.

Passive immunization often helps people who are already infected with pathogens, including the ones that cause diphtheria, tetanus, measles, and hepatitis B. A person receives injections of antibody molecules purified from some other source. The best source is another individual who already has produced a large amount of the required antibody. The effects are not lasting, because the person's own B cells are not producing antibodies. However, injected antibody molecules may help counter the immediate attack.

Monoclonal Antibodies

The antibodies produced by B cells make such a powerful contribution to immune responses because of their remarkable specificity. Commercially prepared **monoclonal antibodies** harness this power for medical and research uses. They are obtained by laboratory techniques that yield large quantities of a desired antibody derived from B cells (Figure 7.16). Commonly, a mouse is immunized with a specific antigen, and then B cells are extracted from its spleen. These are fused with cancerous cells

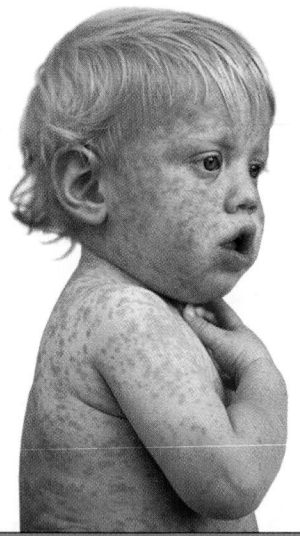

Figure 7.15 A common immunization schedule for children in the United States. Pediatricians routinely immunize infants and children during office visits. Low-cost or free vaccinations also are available at many community clinics.

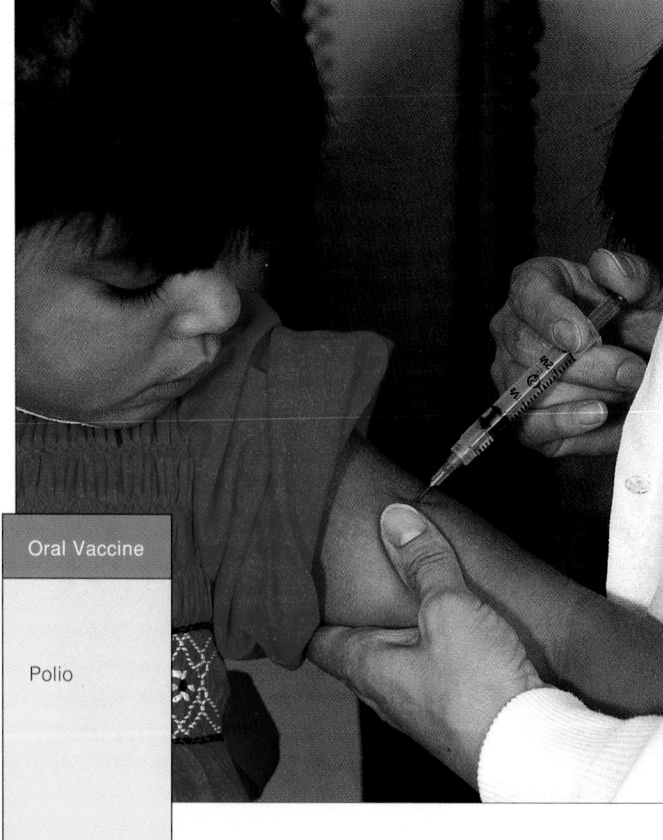

Age	Combined Injected Vaccines	Oral Vaccine
1–2 months	Hepatitis B (two injections, one shortly after birth, another before two months old)	
2 months	Tetramune (diphtheria, whooping cough, tetanus), *Hemophilus influenzae*	Polio
4 months	Tetramune, *H. influenzae*	
6 months	Tetramune, *H. influenzae*	
12–18 months	Measles, mumps, rubella (German measles)	
15 months	Tetramune, hepatitis B, *H. influenzae*	
11–12 years	Measles, mumps, rubella	

known as *myeloma cells*. The new hybrid is called a *hybridoma cell*. Like cancer cells, the hybridoma cells multiply rapidly, and some of them produce the antibody of interest. In a laboratory setting, a clonal population of an antibody-producing hybridoma cell can be maintained indefinitely. The term "monoclonal antibody" refers to the fact that the antibody molecules are produced by a population of cells cloned from just one antibody-producing cell.

Monoclonal antibodies now are put to various uses, including home pregnancy tests and screening for prostate cancer and some sexually transmitted diseases. They are also used for passive immunization against several diseases. In addition, they have a range of potential uses in cancer treatment, including acting as highly specific "magic bullets" to deliver lethal drugs to cancer cells only, sparing the body's healthy tissues.

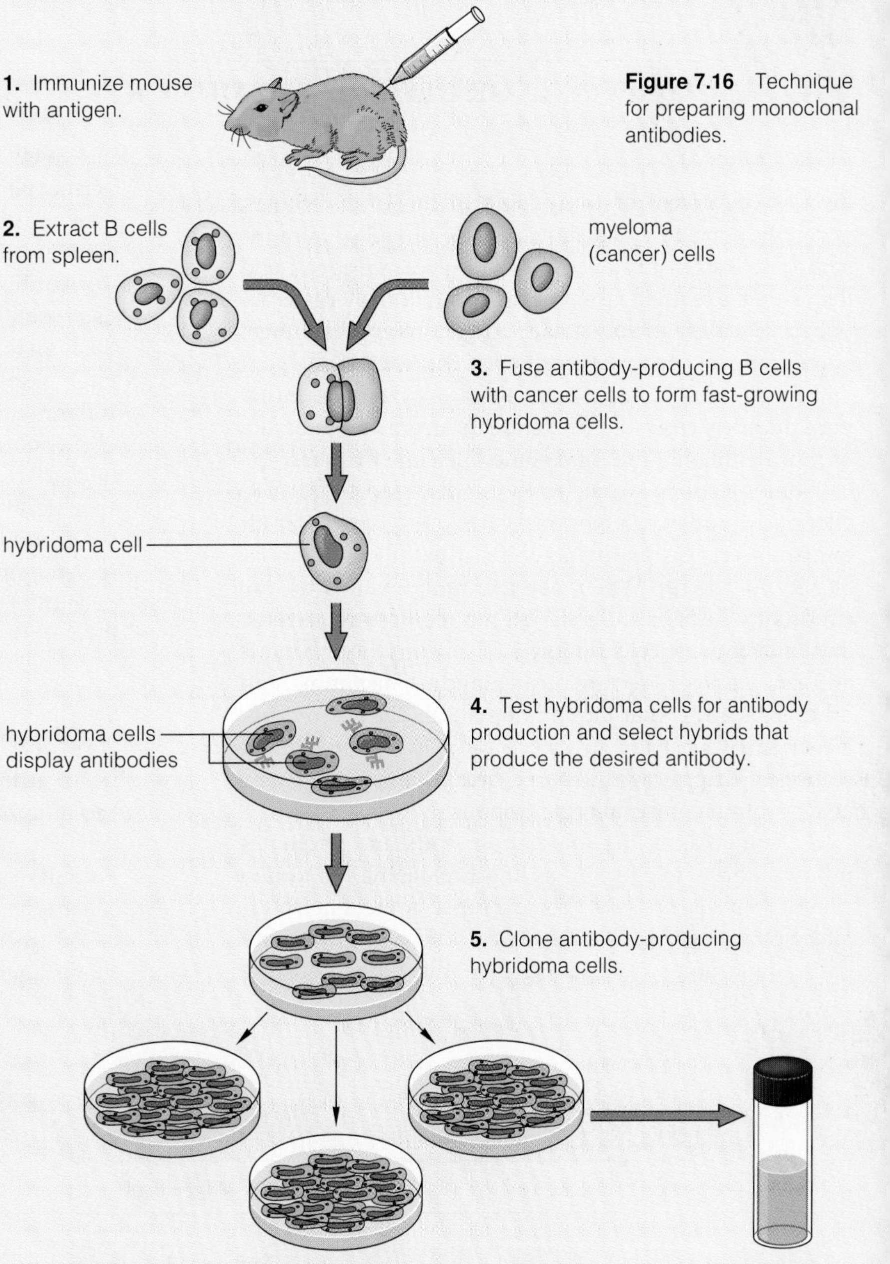

Cytokines

Signaling molecules such as the interleukins produced by lymphocytes are known as **cytokines**. Like monoclonal antibodies, cytokines are rapidly becoming important medical tools. For example, experiments with interleukin-2 have shown it to be effective in treating kidney cancer. The cytokine called *tumor necrosis factor* (TNF), which is secreted by cytotoxic T cells, also can kill cancerous cells. In an experimental treatment of patients with the skin cancer malignant melanoma, researchers removed cytotoxic T cells from patients' tumors and, using recombinant DNA techniques, altered the T cells genetically so that they produced much more TNF than normal. Patients then received doses of the altered cells, along with an interleukin to further stimulate T cell activity. It is possible that eventually a therapy based on this research may be able to kill cancer cells with a level of success that can save lives.

Various types of body cells, including ones infected by a virus, can produce and secrete **interferons**. When they reach an uninfected cell, these defensive proteins trigger a chemical attack that inhibits viral replication. *Gamma* interferon, a cytokine produced by T cells, has other functions. It calls natural killer cells into action and also enhances the activity of macrophages. Genetically engineered gamma interferon now is being used to treat hepatitis C, a chronic, serious viral disease. Some other kinds of cells (not lymphocytes) produce *beta* interferon. This protein has recently been approved for the treatment of one type of *multiple sclerosis*, a disease in which the immune system attacks portions of the nervous system. We consider such abnormal immune responses next.

Immunization promotes enhanced immunity to specific diseases by stimulating the production of both effector and memory lymphocytes.

Monoclonal antibodies and cytokines (such as interleukins and gamma interferon) have become important tools in medical research, testing, and the treatment of disease.

Figure 7.16 Technique for preparing monoclonal antibodies.

1. Immunize mouse with antigen.

2. Extract B cells from spleen.

myeloma (cancer) cells

3. Fuse antibody-producing B cells with cancer cells to form fast-growing hybridoma cells.

hybridoma cell

4. Test hybridoma cells for antibody production and select hybrids that produce the desired antibody.

hybridoma cells display antibodies

5. Clone antibody-producing hybridoma cells.

6. Grow large numbers of antibody-producing cells in culture.

7. Extract pure monoclonal antibodies

Allergies

In 8 to 10 percent of the people in the United States, normally harmless substances can provoke immune responses. Such substances are known as *allergens*, and the response to them is an **allergy**. Common allergens are pollen, a variety of foods and drugs, dust mites, fungal spores, insect venom, and cosmetics. Some responses start within minutes of exposure; others are delayed. Either way, the allergens cause moderate to severe inflammation of mucous membranes and sometimes other tissues.

Some people are genetically predisposed to develop allergies. Infections, emotional stress, or changes in air temperature also may trigger reactions that otherwise might not occur. When allergic individuals first are exposed to certain antigens, IgE antibodies are secreted and bind to mast cells (Figure 7.17). When the IgE binds antigen, mast cells secrete prostaglandins, histamine, and other substances that fan inflammation. They also cause airways to constrict. In *asthma* and *hay fever*, stuffed sinuses, a drippy nose, labored breathing, and sneezing are symptoms of the allergic response.

In some cases, the allergen spreads and promotes inflammation, triggering a life-threatening condition called *anaphylactic shock*. For example, a person who is allergic to wasp or bee venom can die within minutes of a single sting. Air passages to the lungs constrict massively. Fluid escapes swiftly from grossly dilated, permeable (and hence extremely leaky) blood vessels all over the body. Blood pressure plummets and can lead to circulatory collapse.

Antihistamines (anti-inflammatory drugs) are often used to relieve the short-term symptoms of allergies. Over the long term a person may try a desensitization program. First, skin tests are used to identify offending allergens. Inflammatory responses to some of them can be blocked if the patient's body can be stimulated to make IgG instead of IgE. Gradually, larger and larger doses of specific allergens are administered. Each time, the person's body produces more circulating IgG molecules and IgG memory cells. The IgG will bind with allergen it encounters and block its attachment to IgE—and thereby block inflammation.

Autoimmune Disorders

In an **autoimmune response**, the powerful weapons of the immune system are unleashed against normal body cells or proteins. An example is *rheumatoid arthritis*. Patients are genetically predisposed to this disorder. Macrophages, T cells, and B cells get activated by antigens associated with the joints. Immune responses are made against the body's own collagen molecules and

---

Table 7.3   The Immune System under Attack

Cause/Risk Factors	Major Effects	Symptoms
AIDS Human immunodeficiency virus (HIV), transmitted in body fluids such as blood, semen, breast milk	Destroys CD4 (helper T) cells, thereby disrupting immune responses	Unexplained weight loss, fatigue, enlarged lymph glands, fever, diarrhea, infections and tumors that are uncommon in the general population
Immunodeficiency (non-HIV) Genetic disorder; radiation exposure; various drugs used during cancer therapy, organ transplants, and other medical procedures; protein-deficient diet	Destruction of lymphoid tissues (radiation) or impairment of growth and division of lymphocytes; immunosuppressive effects of stress are strongly suspected but not confirmed	Extreme vulnerability to infection and development of tumors
Lymphoma Any of a group of cancers of lymphocytes; sometimes associated with immune system suppression by drugs administered during an organ transplant. Also may be associated with HIV infection; *Burkitt's lymphoma* is caused by the Epstein-Barr virus.	Unchecked growth of lymphocytes in the spleen, lymph nodes, bone marrow, and many other locations, leading to severe impairment or destruction of those tissues	Enlargement of lymph nodes in the spleen and in any other organ where lymphocyte growth is unchecked

Allergen (antigen)
enters the body

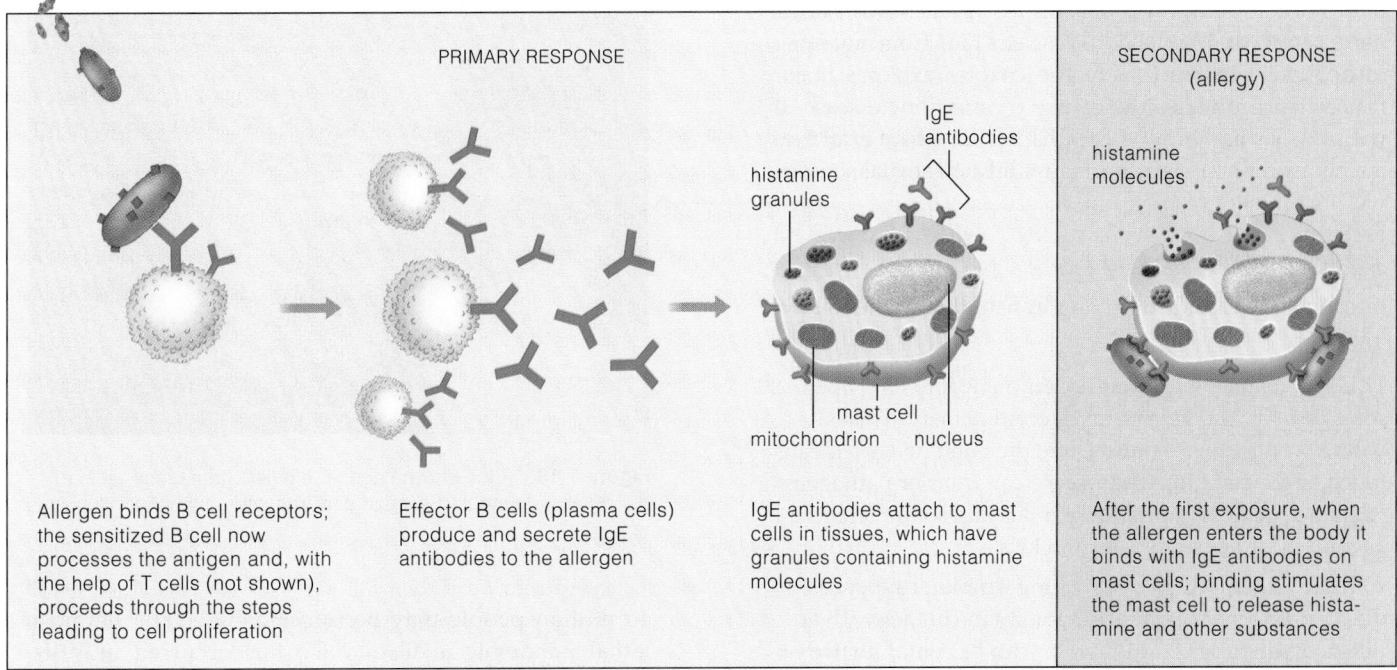

Figure 7.17 The basic steps leading to an allergic response.

---

apparently against antibody that has bound to an as-yet unknown antigen. Complement activation and inflammation cause more damage in tissues of the joints. So do skewed repair mechanisms. Eventually the joints fill with synovial membrane cells and become immobilized.

The autoimmune disease *systemic lupus erythematosus* (SLE) primarily affects younger women, but males can develop it as well. Patients develop antibodies to their own DNA and other self-components. Antigen-antibody complexes accumulate in joints, blood vessel walls, the skin, and the kidneys. Symptoms include fatigue, a rash, painful arthritis, and in some cases major kidney dysfunction. Various drug treatments can help relieve many symptoms, but there is currently no cure for SLE. Table 7.3 lists some other immune system disorders.

Deficient Immune Responses

When the body has inadequate numbers of functioning lymphocytes, immune responses are not effective. Such *severe combined immune deficiencies* (SCIDs) can be inherited or can result from various assaults on the body by outside agents. Either way, deficient or nonexistent immune responses make the person highly vulnerable to infections that are not life-threatening to the general population.

Infection by the human immunodeficiency virus (HIV) causes **AIDS** (acquired immunodeficiency syndrome). HIV is transmitted when bodily fluids of an infected person enter another person's tissues. The virus cripples the immune system by attacking helper T cells and macrophages. This leaves the body dangerously susceptible to opportunistic infections and to some otherwise rare forms of cancer. Chapter 15 on sexually transmitted diseases describes various aspects of HIV infection, including how the virus replicates inside a human host and prospects for treating HIV infections.

An allergic reaction is an immune response to a generally harmless substance.

Autoimmune responses are attacks by lymphocytes on normal body cells or proteins.

An immunodeficiency is a weakened or nonexistent capacity to mount an immune response.

Whenever the body cannot mobilize its defenses quickly enough to prevent a pathogen's activities from interfering with normal body functions, *disease* results. Worldwide, more people die of infectious diseases than from any other cause. This final section of the chapter explores major factors that influence the course of infectious disease. It will serve as background for Chapter 15, which examines infectious diseases transmitted mainly via sexual activity.

Modes of Transmission

Infectious diseases are usually transmitted in the following ways:

1. Direct contact with a pathogen, as by touching open sores or body fluids from an infected person. (This is where "contagious" comes from; the Latin *contagio* means touch or contact.) Infected people can transfer pathogens from their hands, mouth, or genitals and so transmit the agent through a handshake or a kiss.

2. Indirect contact, as by touching doorknobs, pencils, diapers, or other objects previously in contact with an infected person. Food and water can be contaminated by pathogens, including ones responsible for amoebic dysentery, bacterial dysentery (*Salmonella* and *Shigella*), typhoid fever, and hepatitis A.

3. Inhaling pathogens that have been ejected into the air by coughs and sneezes, the most common mode of transmission (Figure 7.18).

4. Encounters with *vectors*. Examples include mosquitoes, flies, fleas, and ticks. Vectors transport pathogens from infected people or contaminated material to new hosts. In some cases, part of the pathogen's life cycle must occur inside the vector, which serves as an *intermediate host*. Mosquitoes serve this function for the parasite *Plasmodium falciparum*, which causes malaria.

Patterns of Occurrence

Infectious diseases often are described in terms of their patterns of occurrence. *Sporadic* diseases, such as whooping cough, break out irregularly and affect few people. *Endemic* diseases occur more or less continuously. Ringworm and the common cold are examples. Impetigo, a contagious bacterial infection, often is endemic to day-care centers.

During an *epidemic*, a disease rate increases to a level above the predicted rate. When cholera broke out all through Peru in 1991, this was an epidemic. The bubonic plague epidemic in 14th-century Europe killed 25 million people. When epidemics break out in several countries around the world in a given time span, they collectively are called a *pandemic*. AIDS is pandemic; as many as

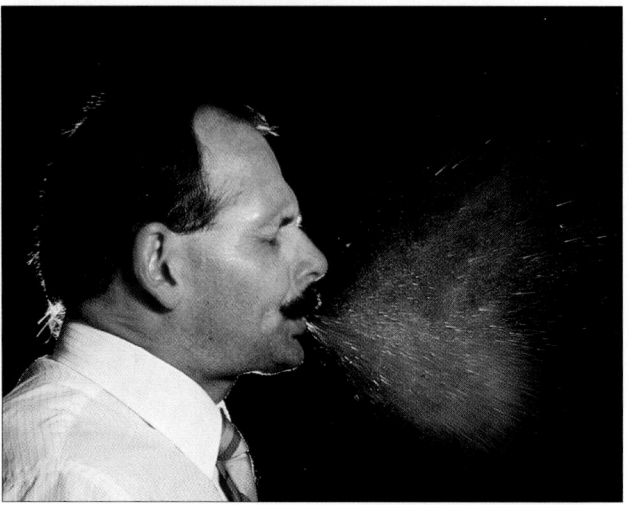

Figure 7.18 A full-blown sneeze. Inhaled pathogens ejected during an unprotected sneeze can transmit colds and influenza.

40 million people may be infected by 2000. The most lethal pandemic in history (so far) occurred in 1918, when viral influenza killed 21 million people around the globe.

Virulence

Virulence refers to the relative ability of a pathogen to cause serious disease. It depends on how fast the pathogen can invade tissues, how severe is the damage it causes, and which tissues it targets. Viruses that can cause pneumonia are more virulent than viruses causing the common cold. Rabies viruses are highly virulent because they target the brain.

What Antibiotics Can and Cannot Do

Today there are more than 160 prescribed **antibiotics**—substances that destroy or inhibit the growth of bacteria and certain other microorganisms. They are produced mainly by bacteria and fungi. Antibiotics such as penicillins, tetracyclines, and streptomycin kill microorganisms by interfering with different metabolic functions. Hence, streptomycin blocks protein synthesis, and penicillins disrupt the bonds necessary to hold molecules together in the cell walls of bacteria.

Antibiotics can have potent side effects. For example, penicillins, tetracyclines, and other drugs inhibit or destroy normal intestinal bacteria as well as pathogens, so they can cause digestive upsets. Some are allergenic and others can even reduce the effectiveness of oral contraceptives.

Table 7.4 Infectious Diseases: Global Health Threats*

Disease	Type of Pathogen	Estimated Deaths per Year
Tuberculosis	Bacterium	3.3 million
Malaria	Protozoan	1–2 million
Hepatitis (includes A, B, C, D, E)	Virus	1–2 million
Various respiratory infections (pneumonia, viral influenza, diphtheria, strep infections)	Virus, bacterium	6.9 million
Diarrheas (includes amoebic dysentery, cryptosporidiosis)	Protozoa, virus, and bacterium	4.2 million
Measles	Virus	220,000
Schistosomiasis	Worm	200,000
Whooping cough	Bacterium	100,000
Hookworm	Worm	50,000+

* Does not include AIDS-related deaths. Recent statistics on worldwide HIV infections and deaths due to AIDS are presented in Chapter 15.

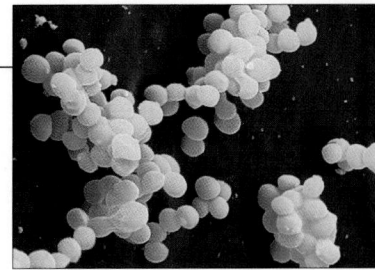

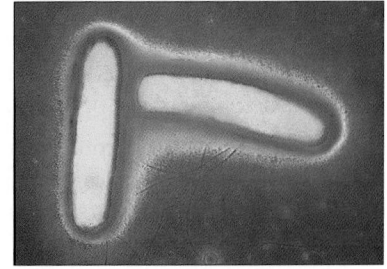

Figure 7.19 False-color micrographs of two microbes that defy most antibiotics. (**a**) *Staphylococcus aureus*, the "flesh-eating" bacterium, and (**b**) *Mycobacterium tuberculosis*.

Antibiotics are not effective against viruses, but there are some *antiviral drugs*. One, called acyclovir, is used to treat cold sores and genital herpes. AZT and a few other antiviral drugs often are part of the treatment plan for some AIDS patients.

Antibiotic Resistance

Bacterial resistance to antibiotics is rapidly becoming a human health crisis. The problem began in the 1950s, when antibiotic use increased dramatically. In many countries antibiotics are available as over-the-counter drugs, so laypeople can "self-prescribe" them whenever they don't feel well. Many people also fail to take the full recommended dose of an antibiotic. This kills off only the most susceptible bacteria, leaving the naturally more antibiotic-resistant microbes to flourish. Physicians may compound the problem when they needlessly prescribe antibiotics or order a broad-spectrum antibiotic (usually effective against many types of microorganisms) when a more specific one would do.

Today, the list of drug-resistant microbes includes strains responsible for certain cases of tuberculosis, gonorrhea, malaria, urinary tract infections, bacterial dysentery, pneumonias, and many surgical-wound infections. One of these, *Staphylococcus aureus* or simply staph A (Figure 7.19a), is a virulent microbe that can cause pneumonia and some rapidly progressing wound infections. Infectious disease specialists have proposed a worldwide surveillance system to identify new resistant strains before they can become established.

Case Study: The Resurgence of Tuberculosis

You may be surprised to learn that, globally, tuberculosis kills more humans than any other single infectious disease (Table 7.4). At least a third of the world's people are infected with *Mycobacterium tuberculosis* (Figure 7.19b), which causes TB. The bacteria are transmitted in airborne droplets produced by coughing or sneezing. In most cases, the immune system kills invading bacilli. If small lesions (tubercles) form in the lungs, their healing will leave a scar visible on a chest X ray. Although TB once was a major health problem in the United States, improved hygiene, nutrition, and medical care—including effective antibiotics—caused a steady decline in new cases for most of the 20th century. Since 1985, however, TB has begun to make a comeback.

Many recent cases have developed in people infected with HIV, whose immune systems are weakened. Intravenous drug abusers also account for a growing number of reported cases of active TB. Many of these cases occur in correctional facilities and public shelters, where the close quarters encourage transmission. In addition, drug-resistant strains of the TB bacterium have turned up in about 25 states, and could slowly spread to the general population. The only recourse may be to reinstitute widespread TB screening programs and hope that researchers can develop new anti-TB drugs.

Infectious diseases are transmitted by direct or indirect contact with a pathogen and may have different patterns of occurrence.

Antibiotics are effective against bacteria and certain microorganisms; they are not effective against viruses.

SUMMARY

1. The body protects itself from many pathogens (infection-causing agents) with physical and chemical barriers at body surfaces. It also is protected by the nonspecific and specific responses of the white blood cells listed in Table 7.5. Nonspecific responses to tissue irritation or damage include inflammation and involve organs with phagocytic functions, such as the liver, spleen, and lymph nodes. Immune responses are made against specific pathogens, foreign cells, or abnormal body cells.

2. Intact skin and mucous membranes lining various body surfaces are physical barriers to infection. Glandular secretions (as in tears, saliva, and gastric juice) are examples of chemical barriers. So are the metabolic products of resident bacteria on body surfaces.

3. Tissue redness, warmth, swelling, and pain are signs of inflammation. The inflammatory response begins with changes in blood flow to a damaged tissue. Pathogens and dead or damaged body cells release substances that make blood capillaries leaky. White blood cells enter the tissue and destroy invaders. Plasma proteins also enter the tissue. Complement proteins bind and induce lysis of pathogens, and they attract phagocytes. Blood-clotting proteins help repair damaged blood vessels.

4. Each immune response is triggered by antigen, a unique molecular configuration that lymphocytes recognize as foreign (nonself). Each response also has specificity, meaning it is directed against one antigen alone. Each response also shows memory. A subsequent encounter with the same antigen will trigger a more rapid, secondary response of greater magnitude. An immune response normally is not made against the body's own self marker proteins.

5. Macrophages and other antigen-presenting cells process and display fragments of antigen attached to self-MHC markers. T lymphocytes have receptors that can bind to antigen-MHC complexes. Binding is the start signal for an immune response.

6. An immune response starts with recognition of antigen. It proceeds through repeated cell divisions that form clones of B and T lymphocytes, which differentiate into subpopulations of effector and memory cells. Interleukins and other signals among white blood cells drive the responses. Effector helper T and cytotoxic T cells, as well as effector B cells and antibodies, act at once. Memory cells are set aside for secondary responses.

7. T cells arise in bone marrow but continue to develop in the thymus, where they acquire TCRs. These T-cell receptors will be able to recognize and bind antigen-MHC complexes at the surface of antigen-presenting cells. B cells arise in bone marrow. While they mature, they start to synthesize antigen receptors (antibodies) that become positioned at their surface.

8. Effector cytotoxic T cells directly destroy virus-infected cells, tumor cells, and cells of tissue or organ transplants. Effector B cells (plasma cells) produce and secrete great numbers of antibodies that freely circulate.

9. Antibodies are protein molecules, often Y-shaped, and each has binding sites for one kind of antigen. Only B cells produce them. When antibody binds to antigen, toxins may be neutralized, pathogens may be tagged for destruction, or the attachment of pathogens to body cells may be prevented.

10. In active immunization, vaccines provoke an immune response, with production of effector and memory cells. In passive immunization, injections of purified antibodies help patients combat an infection.

11. An allergic reaction is an immune response to a generally harmless substance. An autoimmune response is an attack by lymphocytes on normal body cells. An immune deficiency is a weakened or nonexistent capacity to mount an immune response.

Table 7.5	Summary of Major White Blood Cells and Their Roles in Defense
Cell Type	**Main Characteristics**
MACROPHAGE	Phagocyte; has roles in nonspecific defense responses; presents antigen to T cells; cleans up and helps repair tissue damage
NEUTROPHIL	Fast-acting phagocyte; takes part in inflammation, not in sustained responses; most effective against bacteria
EOSINOPHIL	Secretes enzymes that attack certain parasitic worms
BASOPHIL AND MAST CELL	Secrete histamines and other substances that act on small blood vessels, thereby producing inflammation; also contribute to allergic reactions
LYMPHOCYTES	(All take part in most immune responses; following antigen recognition, all form clonal populations of effector cells and memory cells.)
B cell	Effectors secrete five classes of antibodies (IgA, IgD, IgE, IgG, and IgM) that protect the host in specialized ways
Helper T cell	Effectors secrete interleukins that stimulate rapid divisions and differentiation of both B cells and T cells
Cytotoxic T cell	Effectors kill infected cells, tumor cells, and foreign cells by a touch-kill mechanism
NATURAL KILLER (NK) CELLS	Cytotoxic cell of undetermined affiliation; kills virus-infected cells and tumor cells

Review Questions

1. While jogging barefoot in the surf, your toes accidentally land on a jellyfish. Soon the bottoms of your toes are swollen, red, and warm to the touch. Describe the events that result in these signs of inflammation. *166–167*

2. The immune system is characterized by specificity and memory. Describe what these terms mean. *168*

3. Distinguish between:
   a. neutrophil and macrophage
   b. cytotoxic T cell and natural killer cell
   c. effector cell and memory cell
   d. antigen and antibody

4. Are phagocytes deployed during nonspecific defense responses, immune responses, or both? *166, 168*

5. Describe antigen processing. *168*

6. What is the difference between an allergy and an autoimmune response? AIDS is neither of these; what is it? *178–179*

Critical Thinking: You Decide *(Key in Appendix V)*

1. HIV wreaks havoc with the immune system by attacking certain T cells and macrophages. Would the impact be altered if the virus attacked macrophages only? Explain.

2. Given what you now know about how foreign invaders trigger immune responses, explain why mutated forms of viruses, which have altered surface proteins, pose a monitoring problem for a person's memory cells.

Self-Quiz *(Answers in Appendix IV)*

1. _____ are barriers to pathogens at body surfaces.
   a. Intact skin and mucous membranes   c. Resident bacteria
   b. Tears, saliva, and gastric fluid   d. all are correct

2. Macrophages are derived from white blood cell precursors called _____.
   a. lymphocytes   d. monocytes
   b. basophils   e. neutrophils
   c. eosinophils

3. Activated complement functions in defense by _____.
   a. neutralizing toxins   c. promoting inflammation
   b. enhancing resident bacteria   d. forming holes in memory lymphocyte membranes

4. _____ are large molecules that lymphocytes recognize as foreign and that elicit an immune response.
   a. Interleukins   c. Immunoglobulins   e. Histamines
   b. Antibodies   d. Antigens

5. Immunoglobulins designated _____ increase antimicrobial activity in mucus.
   a. IgA   c. IgG   e. IgD
   b. IgE   d. IgM

6. Antibody-mediated responses work best against _____.
   a. intracellular pathogens   d. b and c are correct
   b. extracellular pathogens   e. all are correct
   c. toxins

7. Antigens (nonself molecular markers) are _____.
   a. membrane lipids   c. steroids   e. nucleotides
   b. triglycerides   d. proteins

8. _____ would be a target of an effector cytotoxic T cell.
   a. Extracellular virus particles in blood
   b. Cervical tumor cells
   c. Parasitic flukes in the liver
   d. Bacterial cells in pus
   e. Pollen grains in nasal mucus

9. Development of a secondary immune response is based on populations of _____
   a. memory cells   d. effector cytotoxic T cells
   b. circulating antibodies   e. mast cells
   c. effector B cells

10. Match the following immunity concepts:
   _____inflammation   a. neutrophil
   _____antibody secretion   b. effector B cell
   _____phagocyte   c. nonspecific response
   _____immunological memory   d. deliberately provoking memory cell production
   _____vaccination   e. basis of secondary immune response
   _____allergy   f. nonprotective immune response

Selected Key Terms

acute inflammation *166*   immunization *176*
AIDS *179*   immunoglobulin *173*
allergy *178*   interferon *177*
antibody *169*   interleukin *167*
antigen *168*   lymphoid organ *170*
antigen-MHC complex *168*   lysis *165*
antigen-presenting cell *168*   lysozyme *164*
apoptosis *170*   macrophage *166*
autoimmune response *178*   memory cell *168*
B lymphocyte (B cell) *168*   MHC marker *168*
clonal selection hypothesis *174*   monoclonal antibodies *176*
complement system *165*   natural killer cell *171*
cytotoxic T cell *168*   pathogen *164*
effector cell *168*   perforin *170*
fever *167*   TCR *170*
helper T cell *168*   T lymphocyte (T cell) *168*
immune system *168*   vaccine *176*

Readings

Duke, R., D. Ojcius, and J. Young. December 1996. "Cell Suicide in Health and Disease." *Scientific American.*

Engelhard, V. August 1994. "How Cells Process Antigens." *Scientific American.* Includes an excellent summary of how the immune system functions.

Johnson, H., F. Bazer, B. Szente, and M. Jarpe. May 1994. "How Interferons Fight Disease." *Scientific American.*

Levy, S. March 1998. "The Challenge of Antibiotic Resistance." *Scientific American.*

"Life, Death, and the Immune System." September 1993. *Scientific American.* Special issue devoted to immunity, including articles on AIDS, allergies, cancer, and immune therapies.

 For additional readings, go to InfoTrac College Edition, your online library at:

http://www.infotrac-college.com/wadsworth

THE RESPIRATORY SYSTEM

Conquering Chomolungma

To experienced climbers, Chomolungma may be the ultimate challenge (Figure 8.1). The summit of this Himalayan mountain, also known as Everest, is 9,700 meters (29,108 feet) above sea level. It is the highest place on Earth.

Chomolungma's challenge is not merely iced-over vertical rock, driving winds, blinding blizzards, and heart-stopping avalanches. It is the extreme danger that oxygen-poor air poses to the brain. In 1996, nine climbers, including two of the world's best professional mountaineers, perished in one episode near the mountain's summit. Severe weather—and bodily effects of extreme altitude—largely caused the tragedy.

Most of us live at low elevations. Of the air we breathe, one molecule in five is oxygen. In mountains higher than 3,300 meters (10,000 feet), the breathing game changes. The Earth's gravitational pull is not as great, and gas molecules spread out—so breathing the way we do at sea level will not deliver enough oxygen to our lungs. The oxygen deficit can cause headaches, shortness of breath, heart palpitations, even loss of appetite, nausea, and vomiting. These are the telling symptoms of "altitude sickness."

At Chomolungma's base camp, climbers are 6,300 meters (19,000 feet) above sea level. More than half of the atmospheric oxygen is below them. At 7,000 meters

(23,000 feet), oxygen and other gaseous molecules are extremely diffuse. The exceedingly low air pressure and scarcity of oxygen combine to make small blood vessels leak. Plasma fluid trickles out through gaps between the endothelial cells that line the blood vessels. In a climber's brain and lungs, tissues swell with plasma fluid, a condition called *edema*. Unless the edema is reversed, a climber may become comatose and die.

Given the risks, experienced climbers keep their bodies in top physical condition and live for a time at high elevations before their assault on the summit. They know that breathing "thinner" air containing less oxygen than air at sea level triggers some important physiological responses—including the formation of billions of additional red blood cells and more blood capillaries. Cells also make more mitochondria—the organelles in which *aerobic cellular respiration* occurs (Section 2.12). As you know from your reading in Chapter 2, aerobic cellular respiration uses oxygen and produces carbon dioxide wastes that must be removed from the body. The human body's adaptations for meeting these requirements include structures (such as lungs) and functions (such as breathing) that facilitate gas exchange. The body's organs and mechanisms for gas exchange are what we mean in this chapter when we talk about "respiration" and the respiratory system.

Few of us will ever find ourselves near the peak of Chomolungma, pushing to the limit our ability to obtain oxygen. We won't need the bottled oxygen that elite climbers use. (Although at extreme altitude extra oxygen is no guarantee; most of the climbers who died on Everest in 1996 were using it.) Here in the lowlands, however, disease, smoking, and other environmental insults may affect gas exchange in more ordinary ways. And the risks can be just as great, as you will clearly understand by this chapter's end.

KEY CONCEPTS

1. Humans are highly active animals. Energy to power their metabolism comes mainly from *aerobic cellular respiration* in mitochondria. Aerobic cellular respiration in turn depends on the efficient exchange of oxygen and carbon dioxide gas between the internal environment and the external world.

2. Structures and functions of the body's respiratory system are adapations for gas exchange, bringing oxygen into the body and removing waste carbon dioxide. The steps of this process often are called simply "respiration."

3. Oxygen diffuses into the body as a result of a pressure gradient. Its pressure is higher in air than it is in metabolically active tissues, where cells rapidly use oxygen. Carbon dioxide follows a gradient in the reverse direction. Its pressure is higher in tissues (where it is a by-product of aerobic cellular respiration) than in air.

4. The respiratory system includes a pair of lungs, which provide an efficient respiratory surface—an extensive, thin, moist layer of epithelium. Blood flowing through the body's circulatory system picks up oxygen and gives up carbon dioxide at this respiratory surface.

5. Normally, controls over breathing operate primarily in response to changing carbon dioxide levels in the blood.

CHAPTER AT A GLANCE

8.1   Overview of the Respiratory System

8.2   Factors That Influence Gas Exchange

8.3   *Breathing as a Health Hazard*

8.4   Breathing—Cyclic Reversals in Air Pressure Gradients

8.5   Gas Exchange and Transport

8.6   Controls over Gas Exchange

8.7   *Tobacco and Other Threats to Respiratory Health*

Figure 8.1   A climber approaching the summit of Chomolungma, where oxygen is dangerously scarce.

OVERVIEW OF THE RESPIRATORY SYSTEM

Airways to the Lungs

The **respiratory system** (Figure 8.2) brings in oxygen (in air) that each body cell requires for aerobic respiration. It also releases carbon dioxide wastes to the environment. **Respiration** is what we call the process by which oxygen moves inward and CO_2 is released.

During quiet breathing, air typically enters (and leaves) the respiratory system by way of the nose. Hairs at the entrance to the *nasal cavity* and in the cavity's ciliated epithelial lining filter out dust and other large particles. Incoming air also is warmed in the nose and picks up moisture from mucus. The nasal cavity's two chambers are separated by a *septum* of cartilage and bone. *Paranasal sinuses* above and behind the nasal cavity are linked with it by channels. (This is why nasal sprays prescribed for colds or allergies can help unclog mucus-filled sinuses.) Tear glands constantly produce moisture that drains into the nasal cavity. When you cry, the flow increases and you get "sniffles." From the nasal cavity air moves on into the **pharynx**. This is the entrance to both the **larynx** (an airway) and the esophagus (the tube leading to the stomach). Nine pieces of cartilage form the larynx.

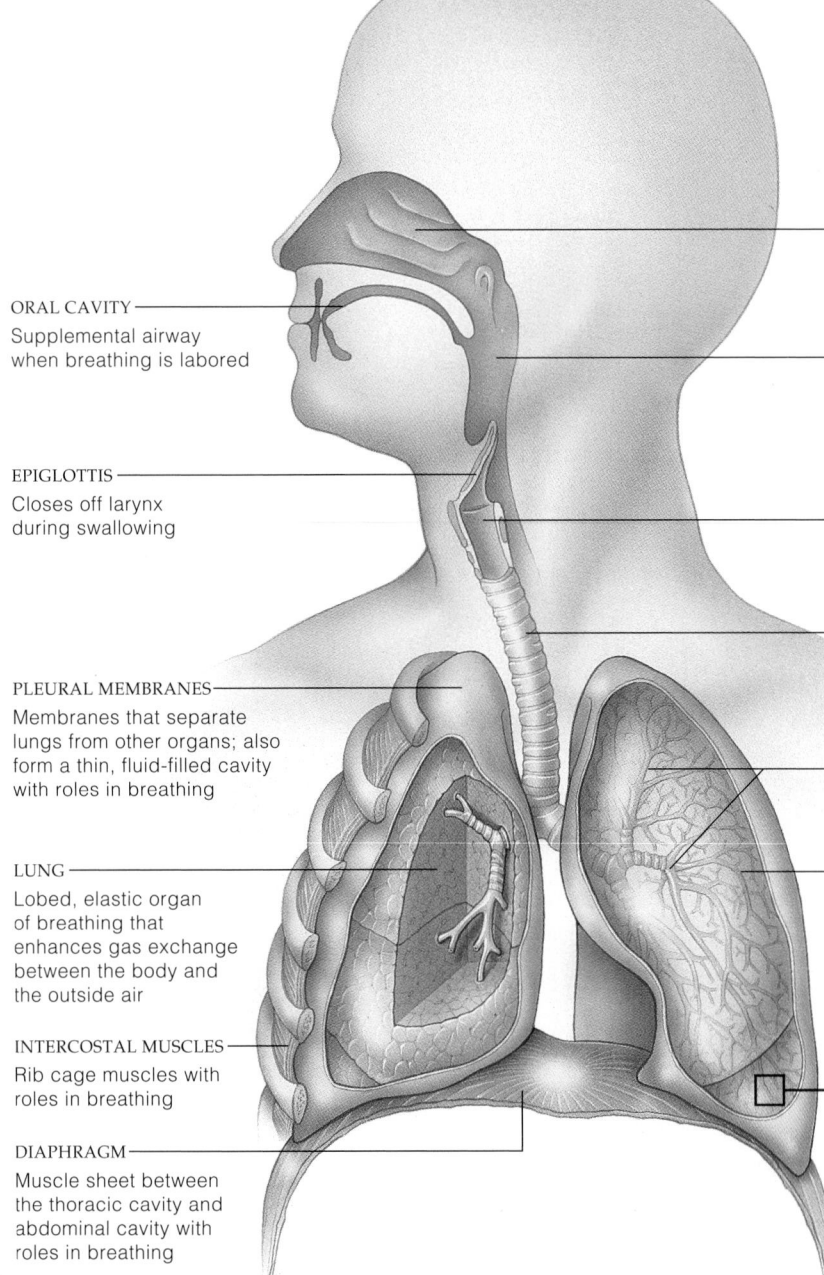

ORAL CAVITY
Supplemental airway
when breathing is labored

EPIGLOTTIS
Closes off larynx
during swallowing

PLEURAL MEMBRANES
Membranes that separate
lungs from other organs; also
form a thin, fluid-filled cavity
with roles in breathing

LUNG
Lobed, elastic organ
of breathing that
enhances gas exchange
between the body and
the outside air

INTERCOSTAL MUSCLES
Rib cage muscles with
roles in breathing

DIAPHRAGM
Muscle sheet between
the thoracic cavity and
abdominal cavity with
roles in breathing

NASAL CAVITY
Chamber in which air is warmed,
moistened, and initially filtered;
and in which sounds resonate

PHARYNX (throat)
Airway that connects the nose and mouth
with the larynx; enhances speech sounds;
also connects with the esophagus, which
leads to the stomach

LARYNX (voice box)
Airway where breathing is blocked while
swallowing and where sound is produced

TRACHEA (windpipe)
Airway that connects the larynx
with the bronchial tree

BRONCHIAL TREE
Increasingly branched airways
between the trachea and alveoli

ALVEOLI
Thin-walled air sacs where oxygen
diffuses into the internal environment
and carbon dioxide diffuses out

a

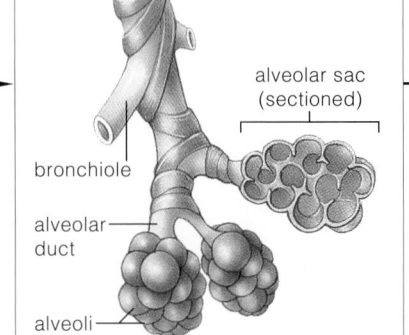

alveolar sac
(sectioned)

bronchiole

alveolar
duct

alveoli

b

Figure 8.2 Components of the human respiratory system and their functions. Also shown are the diaphragm and other structures with secondary roles in respiration.

Viewed from above, it resembles a triangular doughnut. One point of the triangle, the thyroid cartilage, is the "Adam's apple."

Not surprisingly, airways are nearly always open. The flaplike *epiglottis*, attached to the larynx, points up during breathing. However, recall from Chapter 5 that when you swallow, the larynx moves upward so that the epiglottis partly covers the opening of the larynx. This helps prevent food from entering the respiratory tract and causing choking.

From the larynx, air moves into the **trachea** or windpipe. Press gently at the lower front of your neck, and you can feel some of the bands of cartilage that ring the tube, adding strength and helping to keep it open. The trachea branches into two airways, one leading to each lung. Each airway is a **bronchus** (plural: bronchi). Its epithelial lining has abundant cilia and mucus-secreting cells (Figure 8.3). Bacteria and airborne particles stick in the mucus; then the upward-beating cilia sweep the debris-laden mucus up toward the mouth.

Sites of Gas Exchange in the Lungs

The **lungs** are elastic, cone-shaped organs separated from each other by the heart. The left lung has two lobes, the right lung three. The lungs are located within the *rib cage* above the **diaphragm**, a sheet of muscle between the thoracic (chest) and abdominal cavities. The lungs are soft and spongy and are not attached directly to the wall of the chest cavity. Instead, each is positioned within a pair of thin membranes called **pleurae**.

Imagine pushing a closed fist into a fluid-filled balloon (Figure 8.4). A lung occupies the same kind of position as your fist, and the pleural membrane folds back on itself (as does the balloon) to form a closed *pleural sac*. An extremely narrow *intrapleural space* separates the membrane's two facing surfaces. A thin film of lubricating *intrapleural fluid* fills the space and decreases chafing between the membranes. Pneumonia and some other ailments can cause one or both pleurae to become inflamed and swollen. The membranes then rub against each other, making breathing painful—a condition called *pleurisy*.

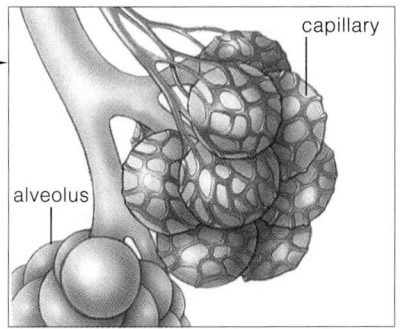

c

capillary

alveolus

Location of alveoli relative to the lung capillaries

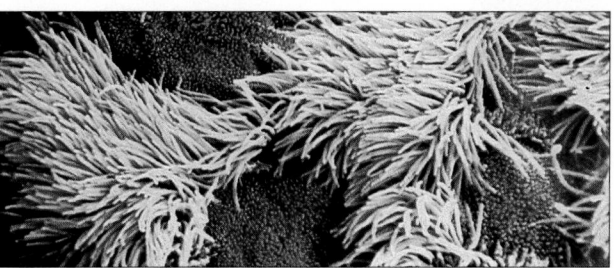

Figure 8.3 Color-enhanced scanning electron micrograph of cilia (*gold*) and mucus-secreting cells (*rust-colored*) in the respiratory tract.

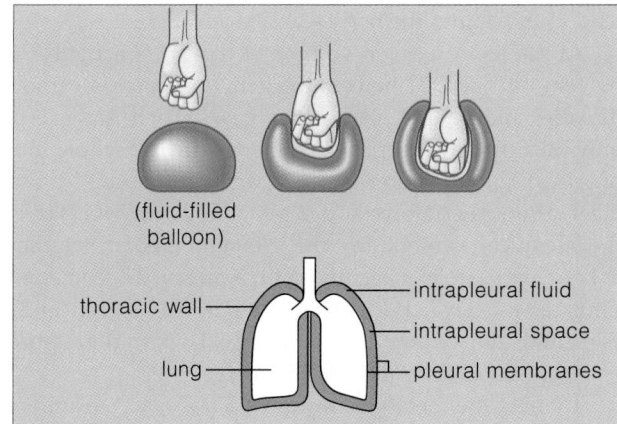

(fluid-filled balloon)

thoracic wall

lung

intrapleural fluid

intrapleural space

pleural membranes

Figure 8.4 Position of the lungs and pleural sac relative to the chest cavity. A lung is analogous to the fist; the pleural sac is analogous to the balloon. For clarity, the intrapleural fluid volume is exaggerated.

Inside each lung, the bronchi narrow as they branch repeatedly, forming "bronchial trees." These narrowing airways are **bronchioles**. Their endings, the **respiratory bronchioles**, have outpouchings from their walls. Each cup-shaped outpouching is an **alveolus** (plural: alveoli), and each lung has about 150 million of them. Most often, alveoli are clustered as a larger pouch, an **alveolar sac**. The alveoli are the sites of gas exchange with lung capillaries (Figure 8.2b and c). In *asthma*, bronchioles can constrict or even close off completely, preventing air from entering or leaving (see Section 8.3).

Collectively the alveoli provide a tremendous surface area for exchanging gases with blood. If they were stretched out as a single layer, they would cover the body several times over—or the floor of a racquetball court! In the next two sections we explore some factors that influence gas exchange and then look closely at the mechanism by which air moves into and out of the lungs.

Airways of the respiratory system are the routes by which oxygen enters the lungs *from* the external environment and carbon dioxide is released *to* the external environment.

Deep within the lungs, at outpouchings (alveoli) at the ends of respiratory bronchioles, gases are exchanged with lung capillaries.

FACTORS THAT INFLUENCE GAS EXCHANGE

Pressure Gradients and Transport Pigments

Gas exchange in the body relies on the tendency of oxygen and carbon dioxide to diffuse down their respective concentration gradients—or, as we say for gases, their *pressure gradients*. Said another way, when molecules of either gas are more concentrated outside the body, they tend to move into the body—and vice versa.

At sea level, a given volume of dry air is roughly 78 percent nitrogen, 21 percent oxygen, 0.04 percent carbon dioxide, and 0.96 percent other gases. Each gas exerts only part of the total pressure exerted by the whole mix of gases. That is, each exerts a "partial pressure." At sea level, atmospheric pressure is about 760 mm Hg. (Hg is the chemical symbol for the element mercury; atmospheric pressure is measured with a mercury barometer.) Thus, at sea level the partial pressure of oxygen is 21 percent of 760, or about 160 mm Hg. The partial pressure of carbon dioxide is about 0.3 mm Hg.

To meet the metabolic needs of a large, active animal such as a human, gas exchange must be efficient. Various factors influence the process. To begin with, gases enter and leave the body by crossing a thin **respiratory surface** of epithelium. The respiratory surface must be moist, because gases cannot diffuse across it unless they are dissolved in fluid. Also, the number of gas molecules moving across the respiratory surface in a given amount of time depends on the surface area and the partial pressure gradient across it. The larger the surface area and the steeper the partial pressure gradient, the faster diffusion occurs. In healthy lungs, millions of thin-walled alveoli provide a huge surface area for gas exchange.

Rates of gas exchange also get a boost by the binding of gases by respiratory pigments, mainly the hemoglobin in red blood cells. Each hemoglobin molecule binds loosely with up to four oxygen molecules in the lungs, where the oxygen concentration is high. When circulating blood carries red blood cells into tissues where the oxygen concentration is low, hemoglobin *releases* oxygen. Thus, by transporting oxygen away from the respiratory surface, hemoglobin helps maintain the pressure gradient that helps draw oxygen into the lungs—and into blood in lung capillaries. Gas exchange is enhanced further by ventilation, the mechanism of breathing (Section 8.4). *Focus on Our Environment* on the facing page examines some external factors that can disrupt normal breathing.

Gas Exchange in Unusual Environments

At high altitude or deep under water the "rules" of gas exchange change. For instance, recall from this chapter's

Figure 8.5 A swimmer responds to neural commands to breathe.

introduction that the partial pressure of oxygen decreases as altitude increases. People who aren't acclimatized to the thinner air at high altitudes can become *hypoxic*—their tissues are chronically short of oxygen. Higher than about 2,400 meters (about 8,000 feet) above sea level, brain respiratory centers work to compensate for the oxygen deficiency by triggering hyperventilation (breathing faster and more deeply than normal).

Underwater, pressure greatly increases with depth. To prevent their lungs from collapsing under the increased pressure, a diver may inhale pressurized air (from tanks), which increases the total pressure of gases in the diver's lungs. As a diver ascends, however, the pressure of the surrounding water *decreases* and gaseous nitrogen, N_2, tends to move from tissues into the bloodstream much more rapidly than it would at sea level. If the ascent is too rapid, the N_2 comes out of solution faster than the diver can exhale it—and bubbles of N_2 form. Too many bubbles cause pain in joints and elsewhere. You may have heard the expression *the bends* for what is otherwise called *decompression sickness*. Bubbles that block blood flow to the brain can damage hearing, vision, and even lead to paralysis. Deeper than about 150 meters, the high partial pressure of N_2 can produce extreme euphoria (*nitrogen narcosis*). Affected divers have reportedly offered their airtank mouthpieces to fish! For deep descents knowledgeable divers use a nitrogen-free gas mixture.

Skin divers and swimmers sometimes purposely hyperventilate. Doing so doesn't increase the oxygen available to tissues. It does increase pH, and it decreases the blood level of carbon dioxide. Both effects dangerously prolong the time before neural controls prod the swimmer to surface and take a breath (Figure 8.5).

Hypoxia also occurs in *carbon monoxide poisoning*. Carbon monoxide, a colorless, odorless gas, is present in automobile exhaust fumes and in smoke from burning tobacco, coal, charcoal, or wood. It binds to hemoglobin at least 200 times more tightly than oxygen does. Even tiny amounts can tie up half of the body's hemoglobin and seriously disrupt oxygen delivery to tissues.

Gas exchange depends on steep partial pressure gradients between the outside and inside of the body. The greater the area of the respiratory surface and the larger the partial pressure gradient, the faster diffusion will occur.

Focus on Our Environment

BREATHING AS A HEALTH HAZARD

In cities, in certain occupations, and anyplace near a smoker, airborne particles and irritating gases put extra workloads on the lungs. Breathing polluted air over a long period of time—and, for some people, even breathing "clean" air that contains allergens—can result in any of several disorders.

Bronchitis can be brought on when air pollution increases mucus secretions and interferes with ciliary action in the lungs. Ciliated epithelium in the bronchioles is especially sensitive to cigarette smoke. Mucus and the particles it traps—including bacteria—accumulate in airways, coughing starts, and the bronchial walls become inflamed. Bacteria or chemical agents start destroying the wall tissue. Cilia are lost from the lining, and mucus-secreting cells multiply as the body attempts to get rid of the accumulating debris. Eventually scar tissue forms and can block parts of the respiratory tract.

In an otherwise healthy person, even acute bronchitis is easily treated. When inflammation continues, however, scar tissue builds up and the bronchi become clogged with more and more mucus. Also, the walls of alveoli break down, and stiffer fibrous tissue comes to surround them. Remaining alveoli enlarge, and the balance between air flow and blood flow is skewed. The result is *emphysema*, a disorder in which the lungs are so distended and inelastic that gases cannot be exchanged efficiently (see Figure 8.6). Running, walking, even exhaling can be difficult. About 1.3 million people in the United States have emphysema.

Smoking, chronic colds, and other respiratory ailments sometimes make a person susceptible to emphysema later in life. "Secondhand smoke" inhaled by a nonsmoker also may contribute to emphysema and other ills, including lung cancer. Many emphysema sufferers lack a functional gene coding for antitrypsin, a protein that inhibits tissue-destroying enzymes produced by bacteria. Emphysema

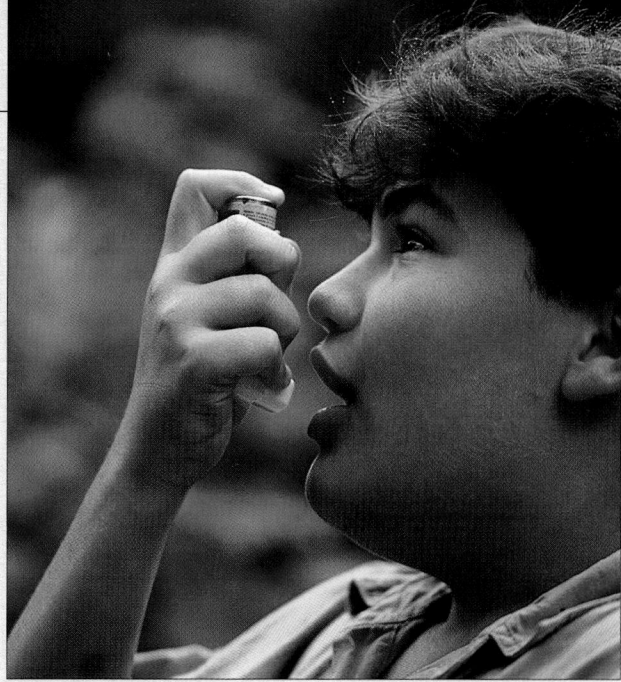

Figure 8.7   An asthma sufferer using an aerosol inhaler.

can develop over 20 or 30 years. By the time the disease is detected, lung tissue is permanently damaged.

Millions of people suffer from *asthma*, a chronic, incurable respiratory condition in which breathing can become so difficult so quickly that the victim literally feels in imminent danger of suffocating. Among other triggers, allergens can set off an asthma attack. Pollen, dairy products, shellfish, flavorings and other foods, pet hairs or dandruff—even the dung of tiny mites in house dust—all can be culprits. In susceptible people, attacks also can result from noxious fumes, stress, strenuous exercise, or from a respiratory infection.

Asthma's scary symptoms are related to a basic feature of respiratory anatomy. Stiff rings and plates of cartilage in the walls of bronchi hold the tubes open (as they do the trachea). However, this cartilage gives way to smooth muscle as the bronchi branch into bronchioles in the lungs. In an asthma attack the smooth muscle contracts in strong spasms that cause the bronchioles to suddenly narrow. Bronchioles even may close off completely.

At the same time, mucus pours from bronchial epithelium, clogging the constricted passages even more.

Various strategies and medications can help asthma sufferers. One tactic is to identify the sources of allergens and limit exposure to those substances. Preventive drugs also can help. Aerosol inhalants squirt a fine mist into the airways (Figure 8.7). A drug in the mist dilates bronchial passages and helps restore free breathing. Aerosol inhalants must be used only with medical supervision. ✺

Figure 8.6   (**a**) Normal human lungs (this lung tissue looks darker than normal because it has been chemically preserved). (**b**) Lungs from a person with emphysema.

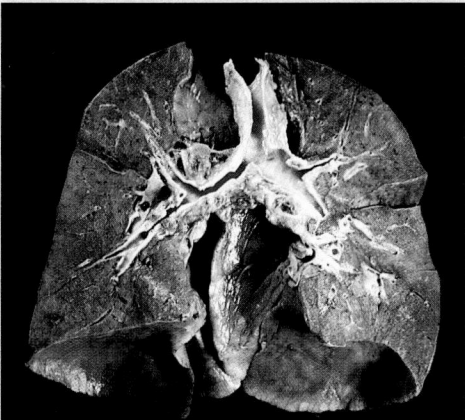

a

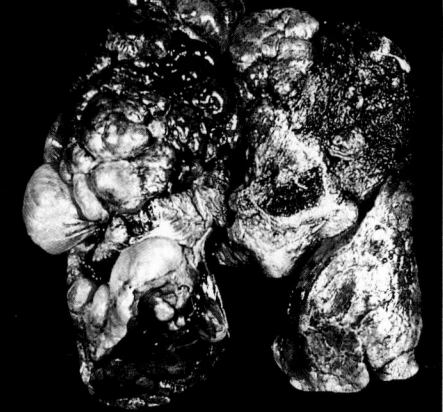

b

The Respiratory Cycle

At 12 breaths per minute, you will take about 500 million breaths by age 75—even more if you consider that young children breathe more rapidly than adults. If breathing stops for as little as five minutes, the person can suffer irreversible brain damage.

Breathing occurs in a cyclic pattern—known as a **respiratory cycle**—that ventilates the respiratory surfaces of your lungs. Ventilation consists of **inspiration** (a single breath of air drawn into the airways) and **expiration** (a single breath out). Inspiration always is an active, energy-requiring action. When someone is breathing quietly, inspiration begins with the contraction of the diaphragm and, to a lesser extent, the external intercostal muscles. These muscle movements increase the volume of the thoracic cavity. If you take a deep breath, the volume increases further because neck muscles contract and raise the sternum and the first two ribs.

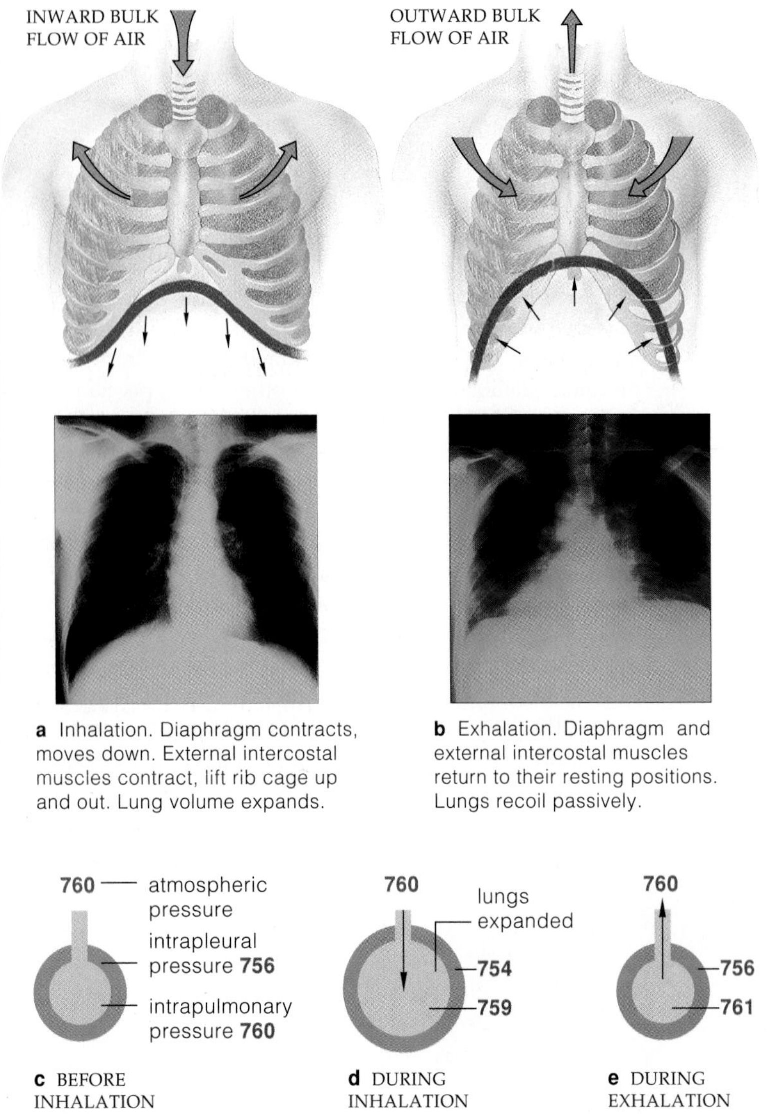

During each respiratory cycle, the volume of the thoracic cavity increases, then decreases. *When the volume changes, pressure gradients between the lungs and the air outside the body are reversed.* To understand how this reversal affects breathing, begin by remembering that the atmospheric pressure—760 mm Hg at sea level—is exerted by the combined weight of all the atmospheric gases on all the airways. Before inspiration, the pressure inside all alveoli (called *intrapulmonary pressure*) is also 760 mm Hg (Figure 8.8c).

Another pressure gradient helps keep the lungs close to the thoracic cavity wall all during the respiratory cycle, even during expiration, when lungs have a far smaller volume than the thoracic cavity (Figure 8.8b). When the cavity expands, so do the lungs, owing to a *negative pressure gradient* across the lung wall. Here is how the mechanism works in a resting person. To start with, pressure inside the pleural sac is about 756 mm Hg—4 mm Hg *less* than atmospheric pressure. And pleural sac pressure is exerted within the thoracic cavity but *outside* the lungs (see Figure 8.4). As pleural sac pressure pushes in on the lung's wall, the pressure inside the lungs pushes outward. The pressure difference—a bit lower in the pleural sac than inside the lungs—is enough to make the lungs stretch, filling the expanded thoracic cavity.

The cohesiveness of water molecules in the fluid inside the pleural sac also helps keep lungs close to the thoracic wall, in much the same way that two wet panes of glass resist being pulled apart. The panes easily slide back and forth, however. Similarly, intrapleural fluid "glues" the lungs to the thoracic wall. So, when the thoracic cavity expands as you inhale, so do your lungs.

Figure 8.8a shows what happens as you start to inhale. The dome-shaped diaphragm flattens, and the rib cage is lifted up and outward. As the thoracic cavity expands, the lungs expand with it. At that time, the air pressure in alveolar sacs is lower than the atmospheric pressure. Fresh air follows the gradient and flows down the airways, almost to the respiratory bronchioles.

During normal "quiet" breathing, the second action of the respiratory cycle is passive. The

a Inhalation. Diaphragm contracts, moves down. External intercostal muscles contract, lift rib cage up and out. Lung volume expands.

b Exhalation. Diaphragm and external intercostal muscles return to their resting positions. Lungs recoil passively.

760 —— atmospheric pressure

intrapleural pressure 756

intrapulmonary pressure 760

c BEFORE INHALATION

760 — lungs expanded — 754 — 759

d DURING INHALATION

760 — 756 — 761

e DURING EXHALATION

Figure 8.8 **(a,b)** Changes in the size of the thoracic cavity during a respiratory cycle. The *blue* line represents the diaphragm. The x-ray images show how the maximum possible inhalation changes the thoracic cavity volume. **(c–e)** Changes in lung volume and pressure during a respiratory cycle.

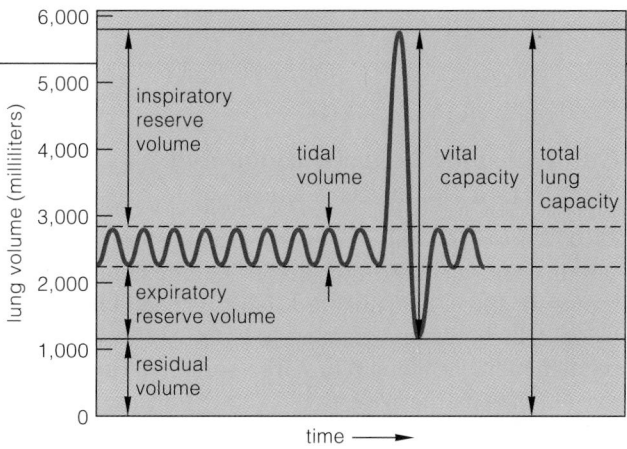

Figure 8.9 Lung volume. During quiet breathing, a tidal volume of air enters and leaves the lungs. Forced inspiration delivers more air to them; forced expiration releases some air that normally stays in the lungs. A residual volume is trapped in partially filled alveoli even during the strongest expiration.

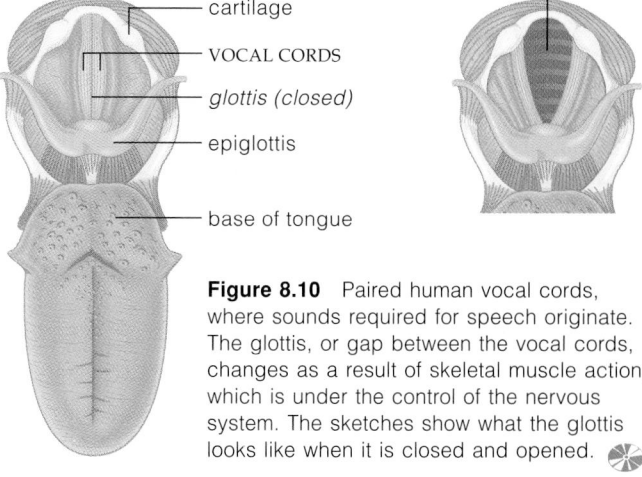

Figure 8.10 Paired human vocal cords, where sounds required for speech originate. The glottis, or gap between the vocal cords, changes as a result of skeletal muscle action, which is under the control of the nervous system. The sketches show what the glottis looks like when it is closed and opened.

muscles that brought about inspiration relax, and the lungs passively recoil, without any further expenditure of energy. The ensuing decrease in lung volume compresses the air in the alveolar sacs, so pressure in the sacs *exceeds* atmospheric pressure. Air follows the gradient, out of the lungs (Figure 8.8b). Expiration becomes an active, energy-requiring action only when more air must be expelled rapidly, as during exercise. In active expiration, muscles in the abdominal wall contract, increasing the pressure in the abdomen and pushing the diaphragm upward. Now the thoracic cavity volume decreases. Internal intercostal muscles contract and pull the thoracic wall down and inward. The chest wall flattens, further reducing the thoracic cavity's dimensions. Lung volume falls as well when the elastic tissue of the lungs recoils.

Lung Volumes

A **tidal volume** of about 500 milliliters (two cupfuls) of air enters or leaves the lungs in a normal breath. You can increase the volume of air you inhale or exhale, however. In addition to air taken in as part of the tidal volume, a person can forcibly inhale roughly 3,100 milliliters of air, called the *inspiratory reserve volume*. Forcibly exhaling, you can expel an additional *expiratory reserve volume* of about 1,200 milliliters. **Vital capacity**—the maximum volume of air that can move out of the lungs after a person inhales as deeply as possible—is about 4,800 milliliters for a healthy young man and about 3,800 milliliters for a healthy young woman. People rarely take in more than half their vital capacity, even when they breathe deeply during strenuous exercise. At the end of your deepest exhal;ation, your lungs still are not completely emptied of air; roughly another 1,200 milliliters of *residual volume* remain (Figure 8.9).

How much of the 500 milliliters of inspired air is actually available for gas exchange? About 150 milliliters of exhaled "dead" air remain in the air-conducting tubes between breaths and never reach the alveoli. Thus only 350 (500 − 150) milliliters of fresh air reach the alveoli with each inhalation. An adult typically breathes at least 12 times per minute. This rate of ventilation supplies the alveoli with (350 × 12) or 4,200 milliliters of fresh air—a little more than the volume of soda pop in four 1-liter bottles—every 60 seconds.

Breathing and Sound Production

Near the entrance to the larynx, part of a mucous membrane forms two pairs of horizontal folds. The lower pair form the **vocal cords** (Figure 8.10). With each exhalation, air is forced through the **glottis**, the gap between the vocal cords. This causes the cords to vibrate. By controlling the vibrations, we can produce different kinds of sounds.

Within the folds are bands of elastic ligaments, connected to various cartilage tissues. When muscles of the larynx contract and relax, the ligaments tighten or slacken—and so change the extent to which the folds are stretched. Under commands from the nervous system, coordinated muscle action can narrow or widen the glottis. For example, by increasing muscle tension in the vocal cords, you decrease the gap between them and make high-pitched sounds or squeaks.

Sometimes an infection causes the mucous lining of the vocal cords to become irritated and inflamed. The tissues become swollen, interfering with the capacity of the vocal cords to vibrate. If hoarseness follows, this condition is called *laryngitis*.

In the respiratory cycle, the air movements of breathing occur because of cyclic increases and decreases in the volume of the chest cavity and the corresponding changes in pressure gradients between the lungs and outside air.

Ventilation draws oxygen into your lungs and moves carbon dioxide out. But ventilation is not the same as respiration. *Respiration,* remember, is the process by which the body as a whole acquires oxygen for aerobic cellular respiration and disposes of carbon dioxide. In the phase called **external respiration**, oxygen moves from alveoli into the blood, and carbon dioxide moves in the opposite direction. In **internal respiration**, oxygen moves from blood into tissues, and carbon dioxide moves from tissues into the blood.

Gas Exchange in Alveoli

Each alveolus is one layer of epithelial cells surrounded by a thin basement membrane. A film of interstitial fluid separates the walls of alveoli from the walls of lung capillaries. This fluid-filled space is narrow (Figure 8.11), and gases can diffuse rapidly across it. Figure 8.12 shows the partial pressure gradients for oxygen and carbon dioxide throughout the respiratory system of a resting person. Oxygen diffuses across the respiratory surface and into the bloodstream, and carbon dioxide diffuses outward.

Certain cells in the epithelium of alveoli secrete *pulmonary surfactant.* This substance reduces the surface tension of the watery fluid film between alveoli. Without it, the force of surface tension can collapse the delicate alveoli. This happens to premature babies whose incompletely developed lungs do not yet have working surfactant-secreting cells. The result is a potentially lethal disorder called *infant respiratory distress syndrome.*

Gas Transport between Blood and Metabolically Active Tissues

Blood can carry only so much dissolved oxygen and carbon dioxide. To meet the body's requirements, the gas transport must be enhanced. The hemoglobin in red blood cells binds and transports both O_2 and CO_2. This pigment enables blood to carry some 70 times more oxygen than it otherwise would and to transport 17 times more carbon dioxide away from tissues.

OXYGEN TRANSPORT Inhaled air that reaches the alveoli has plenty of oxygen and relatively little carbon dioxide. The opposite is true of blood entering the lung capillaries. Thus, in the lungs, oxygen diffuses into the blood plasma, then into red blood cells, where up to four oxygen molecules rapidly form a weak, reversible bond with each hemoglobin molecule. Hemoglobin with oxygen bound to it is called **oxyhemoglobin**, or HbO_2.

The amount of HbO_2 that forms depends on the interplay of several factors. One is the partial pressure of oxygen. In general, the higher its partial pressure, the more oxygen will be picked up, until all hemoglobin binding sites have oxygen attached to them. HbO_2 holds onto oxygen rather weakly and will give it up in tissues where the partial pressure of oxygen is lower than in the blood.

Certain conditions reduce hemoglobin's affinity for binding oxygen. They occur in tissues with high metabolic activity—and with correspondingly greater demands for oxygen. For example, the binding of oxygen weakens as

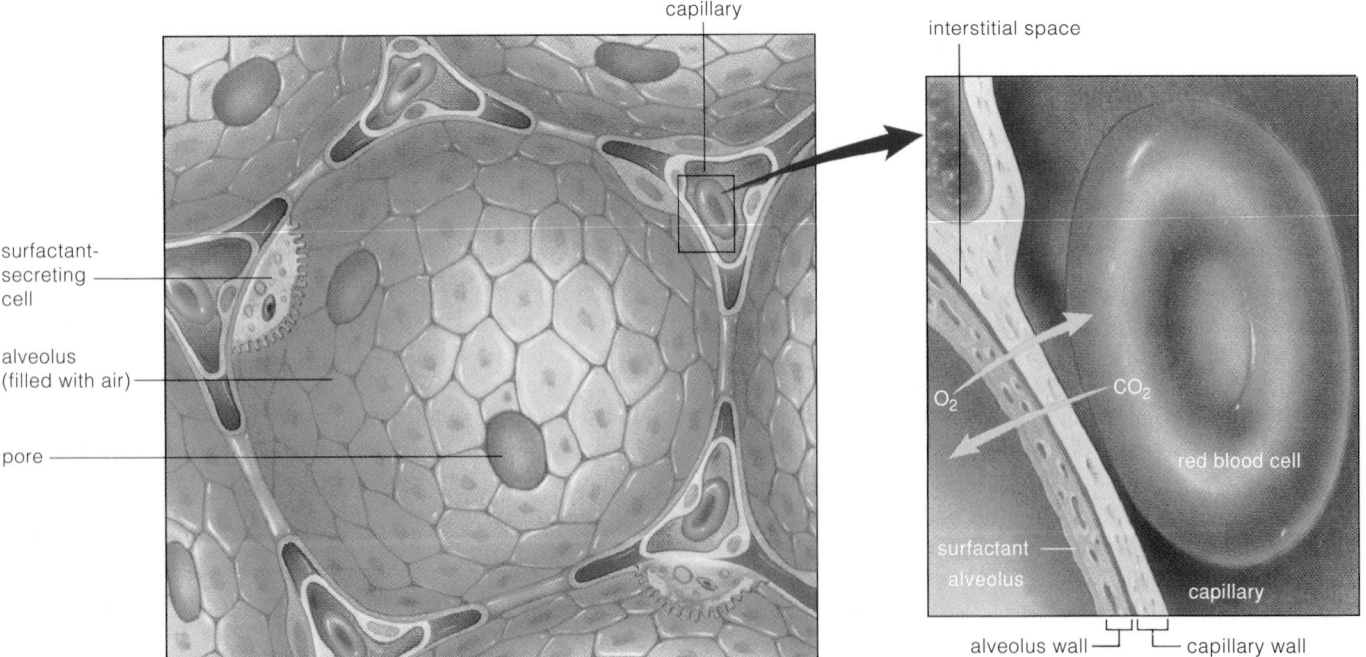

Figure 8.11 Diagram of a section through an alveolus and an adjacent blood capillary. Notice that the total diffusion distance across the capillary wall, the interstitial fluid, and the alveolar wall is small.

temperature rises or as pH declines. Several events contribute to a falling pH: the reaction that forms HbO_2 releases hydrogen ions, and as the level of H^+ increases, the blood becomes more acidic. Blood pH also declines as the level of carbon dioxide given off by active cells increases.

When tissues chronically receive less oxygen than proper functioning requires, red blood cells increase their production of a compound called 2,3-diphosphoglycerate (DPG), which reversibly binds hemoglobin. The more DPG bound to hemoglobin, the *less* affinity hemoglobin has for binding oxygen and thus the more oxygen is available to tissues.

CARBON DIOXIDE TRANSPORT Aerobic cellular respiration in cells produces carbon dioxide as a waste. For this reason, there is relatively more carbon dioxide in metabolically active tissues than in blood flowing through the adjacent capillaries. Carbon dioxide diffuses into these capillaries, which transport it toward the lungs in one of three ways. About 7 percent of the carbon dioxide remains dissolved in plasma. About 23 percent binds with hemoglobin, forming the compound **carbaminohemoglobin** ($HbCO_2$). But most of it—approximately 70 percent—is transported in plasma in the form of bicarbonate (HCO_3^-). The bicarbonate forms after carbon dioxide combines with water in plasma. The resulting carbonic acid (H_2CO_3) dissociates (separates) into bicarbonate and hydrogen ions:

$$CO_2 + H_2O \rightleftharpoons \underset{\text{carbonic acid}}{H_2CO_3} \rightleftharpoons \underset{\text{bicarbonate}}{HCO_3^-} + H^+$$

The reactions just described aren't very significant in plasma, where only 1 in every 1,000 molecules of carbon dioxide is converted to bicarbonate. However, much of the carbon dioxide diffuses into red blood cells, which contain the enzyme **carbonic anhydrase**. With this enzyme, the reaction rate increases by 250 times. In red blood cells, most of the carbon dioxide that is not bound to hemoglobin is converted to carbonic acid and its dissociation products. The enzyme-mediated conversion makes the blood level of carbon dioxide drop rapidly. This helps maintain the gradient that keeps carbon dioxide diffusing from interstitial fluid into the bloodstream.

The bicarbonate ions formed during the reaction tend to diffuse out of the red blood cells and into blood plasma. Hemoglobin binds some of the hydrogen ions and thus acts as a buffer (Chapter 1). Certain plasma proteins also bind H^+. Such buffering mechanisms help prevent an abnormal decline in blood pH.

The reactions are reversed in the alveoli, where the partial pressure of carbon dioxide is lower than it is in

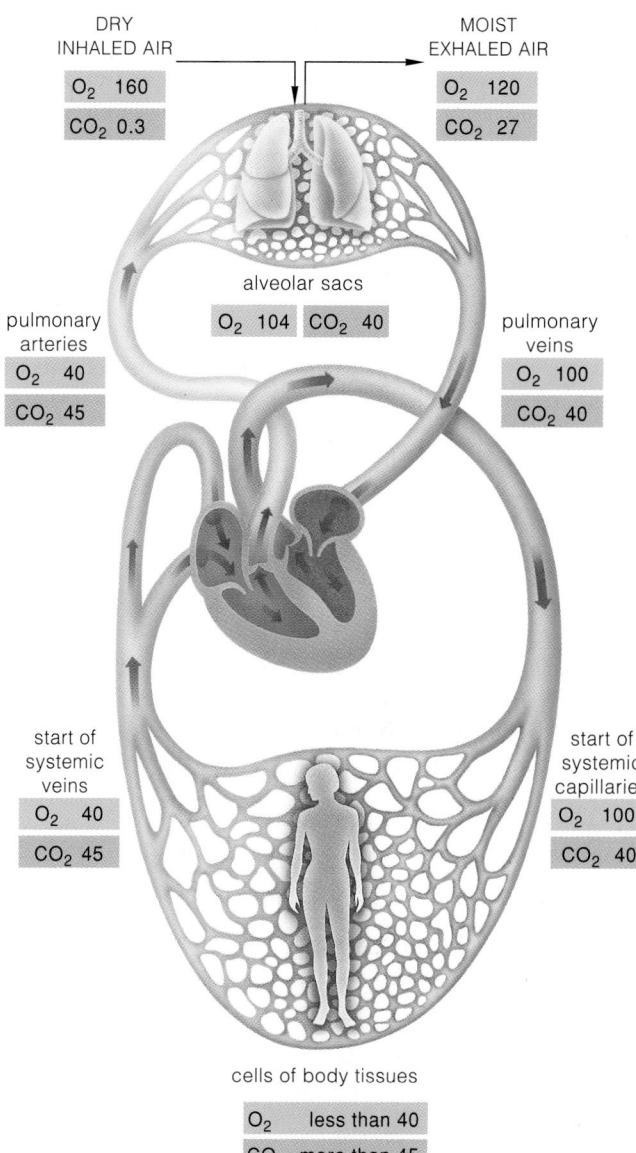

Figure 8.12 Partial pressure gradients for oxygen and carbon dioxide through the respiratory tract. Remember that *each gas moves from regions of higher to lower partial pressure.*

surrounding capillaries. The carbon dioxide that forms as a result of the reactions diffuses into the alveolar sacs. From there the carbon dioxide is exhaled.

Driven by its partial pressure gradient, oxygen diffuses from alveoli, through interstitial fluid, and into lung capillaries. Carbon dioxide diffuses in the opposite direction, driven by its partial pressure gradient.

Hemoglobin in red blood cells greatly enhances the oxygen-carrying capacity of the blood.

Most carbon dioxide is transported in plasma in the form of bicarbonate.

CONTROLS OVER GAS EXCHANGE

To maintain homeostasis in the body, gas exchange in the respiratory system must be efficient enough to deliver a steady stream of oxygen to respiring cells and to carry away carbon dioxide wastes. At least two levels of controls accomplish this all-important regulation.

Matching Air Flow with Blood Flow

Gas exchange is most efficient when a person's breathing is regulated so that the rate of air flow is matched with the rate of blood flow. The nervous system acts to balance them by controlling the breathing *rhythm* and also its *magnitude*—the rate and depth of breathing.

The nervous system governs both inhalation and exhalation, by way of controls that respond to the levels of both oxygen and carbon dioxide in arterial blood. Respiratory centers in the brain play a central role in these events.

BREATHING RHYTHM   Contraction of the diaphragm and the muscles that move the rib cage are under the control of neurons in a system of nerve cells (neurons) running through the brain stem. This system is called the reticular formation, and we will consider its functions more fully in Chapter 10. At this point we are concerned with two small clusters of cells in the reticular formation (in the medulla region of the brain stem). One tiny group of cells coordinates signals calling for inspiration; the other coordinates the signals calling for expiration. The resulting rhythmic contractions are fine-tuned by centers

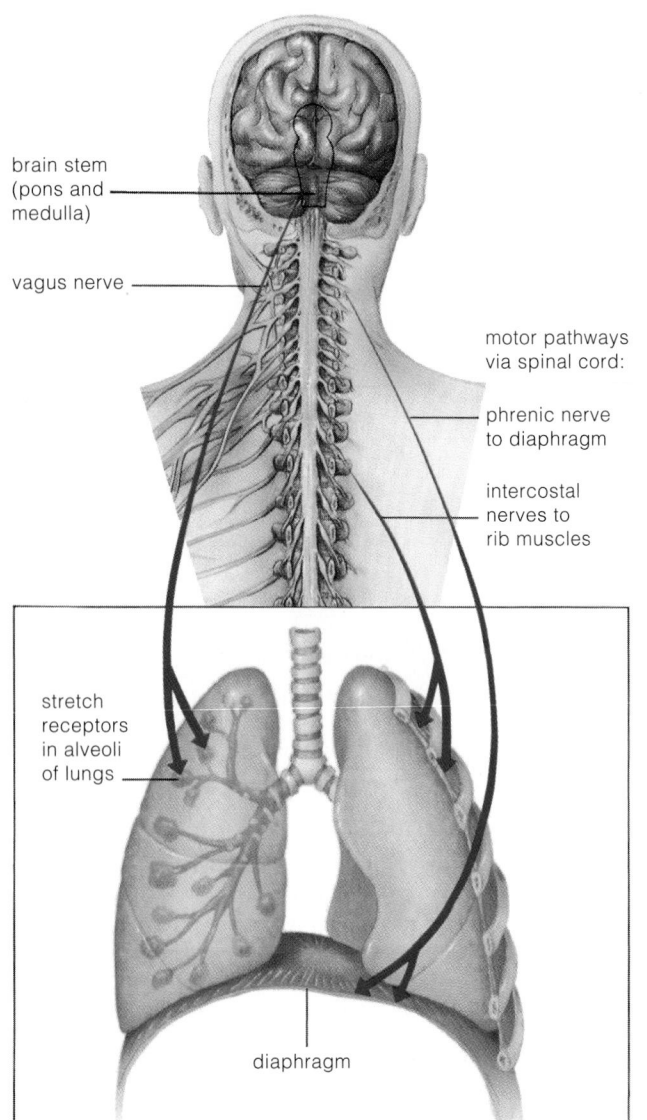

brain stem
(pons and
medulla)

vagus nerve

motor pathways
via spinal cord:

phrenic nerve
to diaphragm

intercostal
nerves to
rib muscles

stretch
receptors
in alveoli
of lungs

diaphragm

a

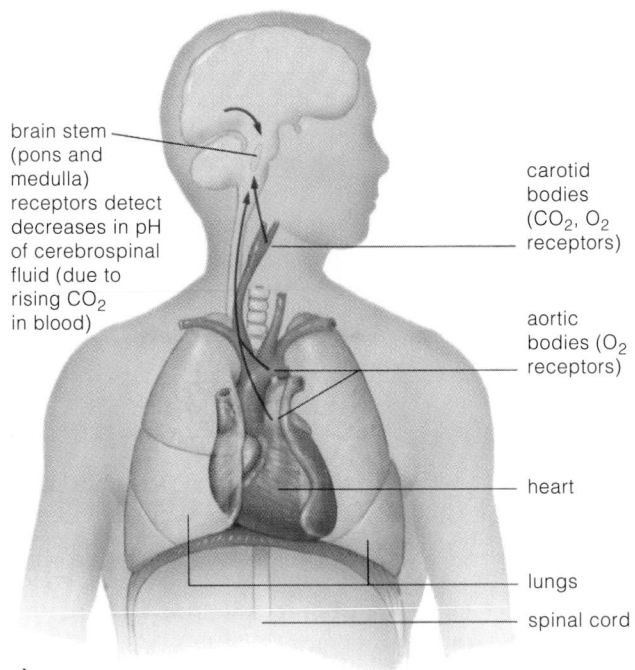

brain stem
(pons and
medulla)
receptors detect
decreases in pH
of cerebrospinal
fluid (due to
rising CO_2
in blood)

carotid
bodies
(CO_2, O_2
receptors)

aortic
bodies (O_2
receptors)

heart

lungs

spinal cord

b

Figure 8.13   Controls over breathing. (**a**) In normal quiet breathing, cell clusters (reticular formation) in the brain stem coordinate signals to the diaphragm and muscles that move the rib cage, triggering inhalation. When a person breathes deeply or rapidly, another cell cluster receives signals from stretch receptors in the lungs and coordinates signals for exhalation. (**b**) Different sensory receptors detect changes in the concentrations of carbon dioxide and oxygen in the blood.

in another part of the brain (the pons) that can stimulate or inhibit both cell clusters.

As Figure 8.13a suggests, when you inhale, signals from the reticular formation travel nerve pathways to the diaphragm and chest. These signals stimulate the rib muscles and diaphragm to contract. As described in Section 8.4, this causes the rib cage to expand, and air moves into the lungs. When the diaphragm and chest muscles relax, elastic recoil returns the rib cage to its unexpanded state, and air in the lungs moves outward. When you breathe rapidly or deeply, stretch receptors in airways send signals to brain control centers, which respond by inhibiting contraction of the diaphragm and rib cage muscles—so you exhale.

BREATHING MAGNITUDE You might assume that nervous system controls over breathing mainly involve monitoring the level of oxygen in blood. However, the nervous system is more sensitive to changes in the level of carbon dioxide. Both gases are monitored in blood flowing through arteries. When the conditions warrant, nervous system signals adjust contractions of the diaphragm and muscles in the chest wall and so adjust the rate and depth of breathing.

Sensory receptors in the medulla of the brain can detect rising carbon dioxide levels. How? The mechanism is indirect, but (fortunately!) extremely sensitive. Through a sequence of chemical reactions, the increased partial pressure of CO_2 in blood lowers the pH of cerebrospinal fluid, which bathes the medulla. In cerebrospinal fluid, a downward pH shift stimulates receptors that signal the change to the brain's respiratory centers (Figure 8.13b). In response, both the rate and the depth of breathing increase—and the blood level of CO_2 falls. This is another example of a negative feedback loop that helps maintain homeostasis.

The brain also receives input from other sensory receptors, including **carotid bodies**, where the carotid arteries branch to the brain, and **aortic bodies** in arterial walls near the heart. Both types of receptors detect changes in levels of CO_2 and of oxygen in arterial blood and also changes in blood pH. The brain responds by increasing the ventilation rate, so more oxygen can be delivered to tissues.

Local Chemical Controls

Some local controls over air flow operate in the lungs, at alveoli. For example, if your heart begins pumping hard and fast but your lungs aren't ventilating at a corresponding pace, blood flows too fast and air moves too sluggishly for efficient disposal of carbon dioxide. An increase in the blood level of carbon dioxide affects smooth muscle in the bronchiole walls. Bronchioles dilate, enhancing the air flow. Similarly, a decrease in carbon dioxide levels causes the bronchiole walls to constrict, so the air flow decreases.

Local controls also work on lung capillaries. When air flow is too great relative to blood flow, oxygen levels rise in some parts of the lungs. The increase affects smooth muscle in blood vessel walls. The vessels dilate, so blood flow to the region increases. Conversely, if the volume of air flow relative to blood flow is too small, the vessels constrict and blood flow decreases.

Disrupted Controls over Gas Exchange

You can voluntarily hold your breath, but not for long. As CO_2 builds up in your blood, neural "orders" force you to take a breath. The mechanisms by which the nervous system regulates the respiratory cycle normally operate under involuntary control, as you've just read. However, in some situations, a person can fail to breathe when the arterial CO_2 level falls below a set point. Breathing that stops briefly and then resumes spontaneously is called *apnea*. During certain times in the normal sleep cycle (see Section 10.10), breathing may stop for one or two seconds or even minutes—in a few cases, as often as 500 times a night!

Sudden infant death syndrome (SIDS) occurs when an infant's normal respiratory cycle cannot be restored after an apneic episode. Some research suggests that this situation may occur because the baby's respiratory control centers or the necessary receptors (especially the carotid bodies) are not yet fully developed. Infants who sleep on their back or sides are at less risk of SIDS than those positioned on their abdomen. The risk is significantly higher among the babies of women who smoked cigarettes or were exposed to secondhand smoke during pregnancy.

When the respiratory system is healthy, gas exchange mechanisms in the body are like parts of a fine Swiss watch—intricately coordinated and smoothly operating. The *Focus* essay in Section 8.7 outlines some ailments in which this seamless functioning falters, including those closely linked to tobacco use.

Through neural controls and local chemical controls (such as those in the lungs), the rates of air flow and blood flow are adjusted in ways that enhance gas exchange.

Respiratory centers in the brain stem control the rhythmic pattern of breathing and the rate and depth of breathing. These controls contribute to homeostasis by helping to maintain appropriate levels of carbon dioxide, oxygen, and hydrogen ions in arterial blood.

8.7 TOBACCO AND OTHER THREATS TO RESPIRATORY HEALTH

When humans start smoking tobacco, they begin wreaking havoc with their lungs. Smoke from just one cigarette can keep cilia in bronchioles from beating for hours. Noxious particles it contains can stimulate mucus secretions and kill the infection-fighting phagocytes that normally patrol the respiratory epithelium. The ensuing "smoker's cough" can pave the way for bronchitis and emphysema. Research suggests that smoking dramatically increases the risk of developing cataracts, a disorder in which the lens of the eye clouds over.

Today we know that cigarette smoke—including "secondhand smoke" inhaled by a nonsmoker—causes lung cancer and contributes to other ills. In the body, some compounds in coal tar and cigarette smoke are converted to highly reactive forms. These are the real carcinogens; they provoke genetic damage leading to lung cancer. Susceptibility to lung cancer is related to the number of cigarettes smoked per day and how often and how deeply the smoke is inhaled (Figure 8.14). Cigarette smoking causes at least 80 percent of all lung cancer deaths. The chart below lists the health risks known to be associated with tobacco smoking, as well as the benefits of quitting. Table 8.1 summarizes some health problems, from lung cancer and pneumonia to various injuries, that can impair respiratory system functioning.

Figure 8.14 Cigarette smoke swirling through the human windpipe and down the two bronchial tubes to the lungs.

Risks Associated with Smoking	Benefits of Quitting
SHORTENED LIFE EXPECTANCY: Nonsmokers live 8.3 years longer on average than those who smoke two packs daily from the mid-20s on	Cumulative risk reduction; after 10 to 15 years, life expectancy of ex-smokers approaches that of nonsmokers
CHRONIC BRONCHITIS, EMPHYSEMA: Smokers have 4–25 times more risk of dying from these diseases than do nonsmokers	Greater chance of improving lung function and slowing down rate of deterioration
LUNG CANCER: Cigarette smoking is the major cause of lung cancer	After 10 to 15 years, risk approaches that of nonsmokers
CANCER OF MOUTH: 3–10 times greater risk among smokers	After 10 to 15 years, risk is reduced to that of nonsmokers
CANCER OF LARYNX: 2.9–17.7 times more frequent among smokers	After 10 years, risk is reduced to that of nonsmokers
CANCER OF ESOPHAGUS: 2–9 times greater risk of dying from this	Risk proportional to amount smoked; quitting should reduce it
CANCER OF PANCREAS: 2–5 times greater risk of dying from this	Risk proportional to amount smoked; quitting should reduce it
CANCER OF BLADDER: 7–10 times greater risk for smokers	Risk decreases gradually over 7 years to that of nonsmokers
CORONARY HEART DISEASE: Cigarette smoking is a major contributing factor	Risk drops sharply after a year; after 10 years, risk reduced to that of nonsmokers
EFFECTS ON OFFSPRING: Women who smoke during pregnancy have more stillbirths, and weight of liveborns averages less (hence, babies are more vulnerable to disease, death)	When smoking stops before fourth month of pregnancy, risk of stillbirth and lower birth weight eliminated
IMPAIRED IMMUNE SYSTEM FUNCTION: Increase in allergic responses, destruction of macrophages in respiratory tract	Avoidable by not smoking
BONE HEALING: Evidence suggests that surgically cut or broken bones require up to 30 percent longer to heal in smokers, possibly because smoking depletes the body of vitamin C and reduces the amount of oxygen reaching body tissues. Reduced vitamin C and reduced oxygen interfere with production of collagen fibers, a key component of bone. Research in this area is continuing.	Avoidable by not smoking

Information provided by the American Cancer Society.

Table 8.1 Respiratory System under Attack

Cause/Risk Factors	Major Effects	Symptoms
LUNG CANCER Most cases associated with tobacco smoking or regular exposure to smoke (*passive smoking*). Other risk factors include inhaling asbestos fibers, exposure to radioactivity, and living where air pollution is high	Tumors develop in lung tissue, with resulting loss of respiratory surface for gas exchange. Tumor spread may trigger pneumonia, lung collapse, or accumulation of fluid between the pleural membranes	Persistent cough or chronic bronchitis, chest pain, shortness of breath, coughing up blood. Weight loss, especially if cancer has spread to other tissues
INFECTIONS *Bronchitis* Often associated with colds or flu. Can be *acute* or *chronic;* may be a symptom of lung cancer	Increased mucus secretion in bronchi and trachea, which leads to coughing. Chronic form can result in inflammation of bronchial walls and scarring	Cough, breathlessness, chest pain. Yellow or greenish phlegm; sometimes, fever
Pneumonia Infection by viruses or bacteria; inhaling toxic fumes	Inflammation of lung tissue; fluid buildup in alveoli	Dry cough, fever; shortness of breath. Symptoms sometimes include chills, blood in phlegm, chest pain, blue tinge to skin
Histoplasmosis Fungal infection, most common in southeastern United States	Inflammation and increased mucus secretion in respiratory tract; can spread to retina of the eye and cause blindness	Flulike symptoms, including cough and fever
Influenza Viral infection, usually spread from person to person by contact with infectious particles (coughs, sneezes, discarded tissues)	Initial infection of mucous membranes in nose and throat, spreading into the trachea and lungs. May be followed by bacterial infection, which leads to bronchitis or pneumonia	Chills, fever, headache, muscle pain, sore throat, usually followed by cough, chest pain, and runny nose due to increased mucus secretion
Tuberculosis Infection by the bacterium *Mycobacterium tuberculosis.* TB had become rare in the United States until the advent of HIV and AIDS. Now this disease is becoming common once again among AIDS sufferers	Destruction of patches of lung tissue; possible spread of infection to other body regions	Flulike illness; if infection progresses, mild fever, night sweats, fatigue, cough, and chest pain. Phlegm containing blood and/or pus. Can be fatal if not treated
Allergic rhinitis (hay fever) Airborne pollen, animal hair, feathers, or other irritants	Release of histamine, which triggers inflammation and fluid production in nasal passages, sinuses, eyelids, and surface layer of eyes	Sneezing, runny nose, watery, itchy eyes; dry throat, wheezing
INJURIES *Pneumothorax Injury,* such as knife or bullet wound or broken rib, that permits air to enter pleural space	Air enters space between pleural membranes, eliminating pressure difference between lungs and external environment	Lung collapses
Asbestosis and *silicosis* Inhalation of particles or fibers of asbestos and silica, respectively	Irritation of lung tissue followed by development of scar tissue, so lungs become stiff and inflexible. Occasionally, lung cancer	Breathlessness; chronic cough with heavy phlegm. Increased susceptibility to other lung disorders
EMPHYSEMA Inflammation resulting from chronic bronchitis or severe asthma	Enzymatic destruction of walls of alveoli and deterioration of gas exchange	Difficulty breathing, usually becoming progressively worse. Can lead to heart failure and death
PULMONARY EMBOLISM Blood clot in artery leading to the lungs	Blockage of blood flow to lungs	Heart failure; general collapse of the circulatory system
RESPIRATORY DISTRESS SYNDROME Lack of *surfactant* in the lungs of a premature newborn. Normally, surfactant acts to keep lung alveoli open so gas exchange can proceed	Alveoli begin to close up within hours of birth	Labored, rapid breathing. Can cause death in premature infants. Also affects infants born to diabetic mothers

SUMMARY

1. Body cells rely mainly on aerobic cellular respiration, a metabolic pathway that requires oxygen and produces carbon dioxide. By way of gas exchange, a function of the respiratory system, the human body as a whole acquires oxygen and disposes of carbon dioxide.

2. Air is a mixture of oxygen, carbon dioxide, and other gases, each exerting a partial pressure. Each gas tends to move from areas of higher to lower partial pressure. The respiratory system makes use of this tendency.

3. In the lungs, oxygen and carbon dioxide diffuse across a moist, thin layer of epithelium (the respiratory surface). Airways carry gases to and from one side of the respiratory surface, and blood vessels carry gases to and away from the other side.

4. The airways of the human respiratory system include the nasal cavity, pharynx, larynx, trachea, bronchi, and bronchioles. Alveoli, located at the end of the terminal bronchioles, are the sites of gas exchange.

5. During inhalation, the chest cavity expands, pressure in the lungs falls below atmospheric pressure, and air flows into the lungs. During normal exhalation, these events are reversed.

6. Driven by its partial pressure gradient, oxygen in the lungs diffuses from alveolar air spaces into the pulmonary capillaries. Then it diffuses into red blood cells and binds weakly with hemoglobin. In tissues where cells are metabolically active, hemoglobin gives up oxygen, which diffuses out of the capillaries, across interstitial fluid, and into cells.

7. Hemoglobin combines with or releases oxygen in response to shifts in oxygen levels, carbon dioxide levels, pH, and temperature.

8. Driven by its partial pressure gradient, carbon dioxide diffuses from cells, across interstitial fluid, and into the bloodstream. Most reacts with water to form bicarbonate; the reactions are reversed in the lungs. There, carbon dioxide diffuses from lung capillaries into the air spaces of the alveoli, then is exhaled.

9. Gas exchange is governed both by local chemical controls in the lungs and by neural controls. The nervous system governs oxygen and carbon dioxide levels, monitoring those gases in arterial blood by way of sensory receptors. These include carotid bodies (at branches of carotid arteries leading to the brain), aortic bodies (in an arterial wall near the heart), and receptors in the medulla of the brain. Blood levels of carbon dioxide are most important in triggering neural responses that regulate breathing. The responses are generated by respiratory centers in the brain, which coordinate neural signals that regulate inspiration and exhalation.

Review Questions

1. Label the components of the human respiratory system and the structures that enclose some of its parts:

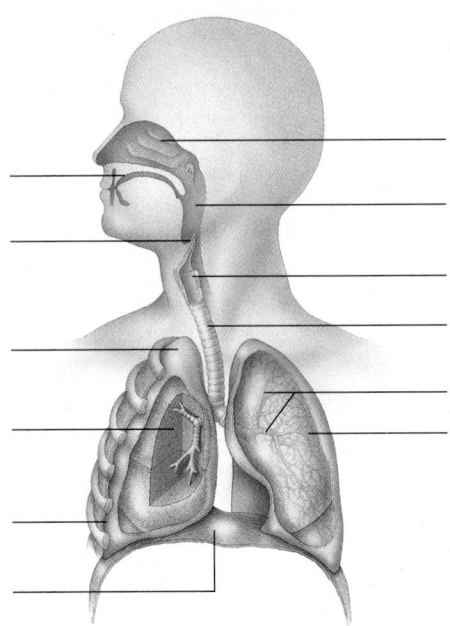

2. What role do partial pressure gradients play in gas exchange? *188*

3. What is oxyhemoglobin? Where does it form? *192*

4. What drives oxygen from alveolar air spaces, through interstitial fluid, and across capillary epithelium? What drives carbon dioxide in the reverse direction? *190, 192–193*

5. How does hemoglobin help maintain the oxygen partial pressure gradient during gas transport in the body? What reactions enhance the transport of carbon dioxide throughout the body? *192–193*

6. Gas exchange is most efficient when the rates of air flow and blood flow are balanced. Give an example of a local control that comes into play in the lungs when the two rates are imbalanced. Do the same for a neural control. *194*

7. Why do your breathing and heart rates increase as you exercise? *194–195*

Critical Thinking: You Decide *(Key in Appendix V)*

1. Some cigarette manufacturers have conducted public relations campaigns urging their customers to "smoke responsibly." What are some social and biological issues in this controversy? How do these issues apply to the nonsmoking spouse or children of a smoker? To nonsmoking patrons in a restaurant? To the unborn child of a pregnant smoker? In your opinion, what behavior would constitute "responsible smoking"?

2. People occasionally poison themselves with carbon monoxide by building a charcoal fire in an enclosed area. Assuming help arrives in time, what would be the *most* effective treatment: placing the victim outdoors in fresh air or administering pure oxygen? Why?

3. When you sneeze or cough, abdominal muscles contract suddenly, pushing your diaphragm upward. After reviewing the discussion of the respiratory cycle in Section 8.4, explain why this change expels air out your nose and mouth with explosive force.

Self-Quiz *(Answers in Appendix IV)*

1. A partial pressure gradient of oxygen exists between _____.
   a. air and lungs
   b. lungs and metabolically active tissues
   c. air at sea level and air at high altitudes
   d. all of the above

2. The _____ is an airway that connects the nose and mouth with the _____.
   a. oral cavity; larynx
   b. pharynx; trachea
   c. trachea; pharynx
   d. pharynx; larynx

3. Oxygen in the air must diffuse across _____ as it follows its partial pressure gradient into the blood.
   a. pleural sacs   c. a moist respiratory surface
   b. alveolar sacs   d. both b and c

4. Each lung encloses a _____.
   a. diaphragm   c. pleural sac
   b. bronchial tree   d. both b and c

5. Gas exchange occurs at the _____.
   a. two bronchi   c. alveolar sacs
   b. pleural sacs   d. both b and c

6. Breathing _____.
   a. ventilates the lungs
   b. draws air into airways
   c. expels air from airways
   d. causes reversals in pressure gradients
   e. all of the above

7. After oxygen diffuses into lung capillaries it also diffuses into _____ and binds with _____.
   a. interstitial fluid; red blood cells
   b. interstitial fluid; carbon dioxide
   c. red blood cells; hemoglobin
   d. red blood cells; carbon dioxide

8. Due to its partial pressure gradient, carbon dioxide diffuses from cells, into interstitial fluid, and into the _____; in the lungs, carbon dioxide diffuses into the _____.
   a. alveoli; bronchioles
   b. bloodstream; bronchioles
   c. alveoli; bloodstream
   d. bloodstream; alveoli

9. Hemoglobin performs which of the following respiratory functions:
   a. transports oxygen
   b. transports some carbon dioxide
   c. acts as a buffer to help maintain blood pH
   d. all of the above

10. Most carbon dioxide in the blood is in the form of _____.
    a. carbon dioxide   c. carbonic acid
    b. carbon monoxide   d. bicarbonate

11. Match the following respiratory components with their descriptions.
    ____ bronchus   a. site of gas exchange
    ____ alveolus   b. fine bronchial tree branching
    ____ hemoglobin   c. throat
    ____ trachea   d. air-conducting tube; windpipe
    ____ bronchiole   e. respiratory pigment
    ____ pharynx   f. airway leading into a lung

Selected Key Terms

alveolar sac 187	larynx 186
alveolus 187	lung 187
aortic body 195	oxyhemoglobin 192
bronchiole 187	pharynx 186
bronchus 187	pleura 187
carbaminohemoglobin 193	respiration 186
carbonic anhydrase 193	respiratory bronchiole 187
carotid body 195	respiratory surface 188
diaphragm 187	tidal volume 191
expiration 190	trachea 187
external respiration 192	ventilation 190
inspiration 190	vital capacity 191
internal respiration 192	vocal cords 191

Readings

American Cancer Society. *Dangers of Smoking; Benefits of Quitting and Relative Risks of Reduced Exposure*, rev. ed. New York: American Cancer Society.

Mortality from Smoking in Developed Countries 1950–2000. A 1994 publication by scientists of Britain's Imperial Cancer Research Fund, the World Health Organization, and the American Cancer Research Fund. Research finds that worldwide, smoking now kills 3 million people every year. If current patterns do not change, by the time today's young smokers reach middle age, 10 million may be dying annually because of their habit. That is one person every three seconds.

Gorman, C. 22 June 1992. "Asthma: Deadly but Treatable." *Time.*

Sherwood, L. 1997. *Human Physiology.* 3d. ed. Belmont, California: Wadsworth.

 For additional readings, go to InfoTrac College Edition, your online library at:

http://www.infotrac-college.com/wadsworth

9

THE INTERNAL ENVIRONMENT

Survival at Stovepipe Wells

In Death Valley, the sand dunes at Stovepipe Wells cover some 14 square miles. Just south of the dunes are 200 square miles of waterless salt flats. The whole parched region has claimed more than a few human lives.

Without adequate water, a person can't survive for long anywhere, let alone in Death Valley (Figure 9.1). Homeostasis in the internal environment is a major theme in our study of human anatomy and physiology, and maintaining water balance is crucial to this internal stability. Among other considerations, living cells are mostly water, and water is required for many critical chemical reactions within them. Water in blood and interstitial fluid acts as a solvent for important salts and other molecules. And most of the potentially toxic nitrogen-containing wastes of protein metabolism are moved out of the body as part of the watery fluid called urine.

Although individuals differ, an adult routinely loses about 2,400 milliliters (two and a half quarts) of water daily, mainly in urine, feces, vapor in exhaled air, and sweat. In the sweltering Death Valley summer, however, sweating increases dramatically as a mechanism to help dissipate heat. Whereas a person might normally lose about 100 milliliters of water in sweat a day, in extreme heat (or during heavy exercise) the loss can rise to 900 milliliters or more *per hour*. The body cannot long tolerate such rapid water loss. So, within the body of a hiker at Stovepipe Wells, a remarkable homeostatic juggling act goes on. In response to changing internal conditions, the brain issues commands you experience as thirst—telling you, in effect, to drink water. Water-conserving mechanisms are also set in motion while other mechanisms operate simultaneously to cool the

body, dispose of metabolic wastes, and make other required exchanges with the external environment.

We also may take in too much water, eat foods that contain large amounts of sodium, or do something else that upsets the optimal balance of water and solutes in blood and body fluids. Fortunately, two fist-sized organs, the kidneys, come to the rescue. Water and solutes from the blood flow continuously through the kidneys, where adjustments occur in how much water and which solutes are reabsorbed into body fluids or disposed of as urine. Those adjustments help the body cope with everything from water imbalances to antibiotic drugs, and they will be our primary focus in this chapter. We will also examine homeostatic mechanisms that operate to maintain body temperature within limits that support life.

KEY CONCEPTS

1. The body continually gains and loses water and dissolved substances (solutes). It continually produces metabolic wastes. Even with all the inputs and outputs, the overall volume and composition of the body's extracellular fluid remain relatively constant.

2. The urinary system is crucial to balancing the intake and output of water and solutes. The urinary system eliminates excess water and solutes and conserves water when necessary.

3. Kidneys are blood-filtering organs, and the urinary system has a pair of them. Packed inside each kidney are a great number of tubelike structures called nephrons.

4. Water and solutes from blood enter kidney nephrons, where selective filtering takes place. After this processing, most of the water and necessary solutes are returned to the bloodstream.

5. Water and solutes not returned to the blood leave the body as urine. At any given time, control mechanisms influence whether the urine is concentrated or dilute. Two hormones play key roles in these adjustments.

6. Internal (core) body temperature must be maintained in a range suitable for metabolism. Overall, core temperature depends on the balance between heat produced through metabolism, heat absorbed from the environment, and heat lost to the environment.

7. Internal body temperature is maintained within a favorable range through controls over a person's metabolic activity, as well as through adaptations in body structure, physiology, and behavior.

CHAPTER AT A GLANCE

9.1   The Challenge: Shifts in Extracellular Fluid

9.2   *Is Your Drinking Water Polluted?*

9.3   The Urinary System

9.4   The Formation of Urine

9.5   *Avoiding Trouble in the Urinary Tract*

9.6   Reabsorption of Water and Sodium

9.7   Hormonal Adjustments of Reabsorption

9.8   Acid–Base Balance

9.9   *Kidney Disorders*

9.10 Maintaining the Body's Core Temperature

Figure 9.1 Hikers in Death Valley National Monument must carry ample water to replenish the body's supply. Otherwise, they risk a potentially serious disruption of homeostasis in the body's finely tuned balance of water and solutes.

THE CHALLENGE: SHIFTS IN EXTRACELLULAR FLUID

Recall that *interstitial* fluid fills the spaces between living cells and other components of the body's tissues. Another fluid, blood, circulates inside blood vessels. Taken together, interstitial fluid and blood plasma are the body's **extracellular fluid** (ECF).

The fluid within cells is *intracellular fluid*. From previous chapters you know that there is a constant exchange of gases and other substances between intracellular and extracellular fluid. Those exchanges are crucial for keeping cells functioning smoothly, and they cannot occur properly unless the volume and composition of the ECF remain stable.

Yet moment to moment, the ECF changes as gases, cell secretions, ions, and other materials enter or leave it. To maintain stable conditions in the ECF, especially the concentrations of water and vital ions, there must be mechanisms that remove substances as they enter the extracellular fluid or add them as they leave it. In humans (and other mammals) a **urinary system** performs this task. Figure 9.2 gives a general picture of the links between the urinary system and other organ systems.

Water Gains and Losses

Ordinarily, each day you take in about as much water as your body loses (Table 9.1). The body *gains* water by two processes:

1. Absorption from ingested liquids and solid foods

2. Metabolism—specifically, the breakdown of organic molecules in reactions that yield water as a by-product

Thirst influences water gain. When a water deficit occurs, the brain compels us to seek out a water fountain or take a cold drink out of the refrigerator. A later section describes the mechanisms involved.

The human body *loses* water by way of the following processes:

1. Urinary excretion

2. Evaporation from the lungs and through the skin

3. Sweating

4. Elimination (in feces)

Urinary excretion affords the greatest control over water loss. This process eliminates excess water and excess or harmful solutes in the form of urine. Some water also evaporates from the respiratory surfaces of the lungs and from skin. These are called "insensible water losses" because a person is not always consciously aware that they are taking place. As noted in Chapter 5, normally very little water that enters the gastrointestinal tract is lost; most is absorbed, not eliminated in feces.

Solute Gains and Losses

The body's extracellular fluid *gains* solutes mainly as a result of four processes:

1. Absorption from ingested liquids and solid food

2. Secretion from cells

3. Respiration

4. Metabolism

Eating provides the body with a variety of nutrients (including glucose) and mineral ions (such as potassium and sodium ions). These are absorbed from the GI tract, as are drugs and food additives. Throughout the body, living cells secrete substances into interstitial fluid and blood. The respiratory system puts oxygen into the blood, and respiring cells put carbon dioxide into it.

Extracellular fluid *loses* mineral ions and metabolic wastes in these ways:

1. Urinary excretion

2. Respiration

3. Sweating

Carbon dioxide, the most abundant metabolic waste, is exhaled from the lungs. Other major wastes leave in urine. One of them, *uric acid*, is formed in nucleic acid breakdown. If allowed to accumulate, it can crystallize and collect in the joints, causing *gout*.

Other major metabolic wastes include several by-products of protein metabolism. One of these is *ammonia*, which is formed in "deamination" reactions whereby nitrogen-containing amino groups are stripped from amino acids. Ammonia can be highly toxic if it is allowed

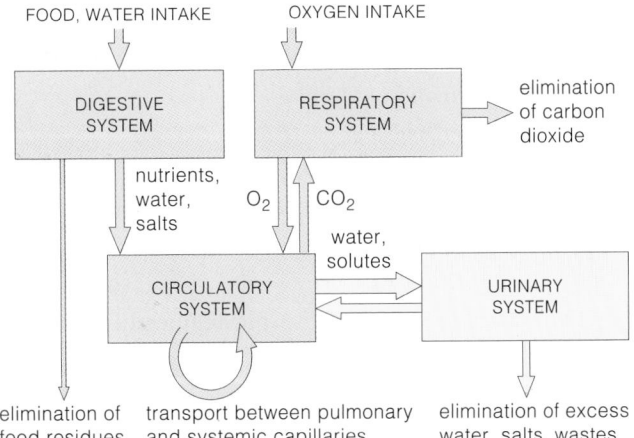

Figure 9.2 Links between the urinary system and other organ systems that maintain stable operating conditions in the internal environment.

Table 9.1	Normal Daily Balance between Water Gain and Water Loss in Adult Humans		
Water Gain (milliliters)		**Water Loss (milliliters)**	
Ingested in solids:	850	Urine:	1,500
Ingested as liquids:	1,400	Feces:	200
Metabolically derived:	350	Evaporation:	900
	2,600		2,600

to accumulate in the body. *Urea* is produced in the liver when two ammonia molecules are combined with carbon dioxide. It is the main waste product of protein breakdown. About 40–60 percent of the urea filtered from blood in the kidneys is reabsorbed, and the rest is excreted. Protein breakdown also produces phosphoric acid, sulfuric acid, and small amounts of other, nitrogen-containing compounds, some of which are toxic. These too are excreted, along with creatinine, a by-product of reactions during muscle activity. Sweat carries away a small percentage of nitrogen-containing wastes (urea and uric acid).

The remarkable kidneys remove many of the waste products we have been discussing. They perform this service as they continuously filter water, mineral ions, organic wastes, and other substances from the blood. As the *Focus* essay in Section 9.2 describes, the increasingly serious problem of polluted drinking water only adds to the kidneys' burden, sometimes in ways that disrupt their functioning or that have other harmful effects on the body.

When your kidneys are healthy, they regulate the volume and solute concentrations of the body's extracellular fluid with great precision. Normally only a tiny portion of the water and solutes entering the kidneys leaves as urine. In fact, except when you take in a great deal of fluid (without exercise), all but about 1 percent of the water is returned to the blood. The composition of the fluid that is returned has been adjusted in vital ways, however. Just how this happens will be our focus as we turn our attention to the urinary system, examining kidney structure and function in some detail.

Each day the body gains water through ingestion of liquids and solid foods and from metabolism. It loses an approximately equal amount of water through urinary excretion, evaporation, sweating, and elimination in feces.

The body gains solutes by way of ingestion, secretion, metabolism, and respiration. Solutes exit the body via urinary excretion, respiration, and sweating.

By adjusting blood's volume and composition, the kidneys help maintain tolerable conditions in the extracellular fluid.

Focus on Our Environment

9.2

IS YOUR DRINKING WATER POLLUTED?

G. Tyler Miller, Jr.

In many parts of the world, the quality of drinking water is being degraded. Aquifers used as sources of drinking water in many countries, including the United States, are becoming contaminated with pesticides, fertilizers, and hazardous organic chemicals (Figure 9.3). In China, for example, 41 large cities get their drinking water from polluted groundwater. In the United States, hundreds of potentially dangerous chemicals have been detected in groundwater supplies.

Some of the more than 700 synthetic organic chemicals found in trace amounts in drinking-water supplies in the United States can cause kidney disorders, birth defects, and various types of cancer in laboratory animals. At high levels, water-soluble inorganic chemicals—including acids, salts, and compounds of toxic metals such as mercury and lead—can also make water unfit to drink, harm aquatic life, and have other negative effects. Excessive nitrates (from agricultural and industrial chemicals) in drinking water can reduce the oxygen-carrying capacity of the blood and even cause stillbirths.

In the United States, up to 25 percent (by volume) of the usable groundwater is contaminated. In some areas, up to 75 percent is contaminated—including every major aquifer in New Jersey. In California, at least 1 million people drink water contaminated with pesticides. Some environmentalists believe that long-lasting groundwater contamination will soon emerge as one of our most serious water resource problems. So: Do you know where your drinking water comes from—and what is in it?

Figure 9.3 Leaking barrels in a hazardous waste dump near Washington, D.C. Such dumps are now illegal.

THE URINARY SYSTEM

The kidneys are the central components of the urinary system (Figure 9.4). Each **kidney** is a bean-shaped organ about the size of a large dinner roll. It is divided internally into several lobes. An outer *cortex* wraps around a central region, the *medulla* (Figure 9.5). The whole kidney is enclosed within a tough coat of connective tissue, the *renal capsule* (from the Latin *renes*, meaning "kidneys").

Each kidney lobe contains blood vessels and slender tubes called **nephrons**. Nephrons filter water and solutes from blood. Most of the filtrate is reabsorbed from them. The rest continues on through tubelike *collecting ducts,* where more reabsorption occurs, and into the kidney's central cavity, the *renal pelvis*.

In addition to the two kidneys, the urinary system includes several other structures. Once urine has formed, it flows from each kidney into a **ureter**, a tubular channel, then into a storage organ, the **urinary bladder**. It leaves the bladder through the **urethra**, a muscular tube that opens at the body surface.

More than a million nephrons are packed inside each kidney. The nephron wall consists of a single layer of epithelial cells, but the cells and junctions between them vary in different regions of the tube. Some regions are highly permeable to water and solutes. Others prevent

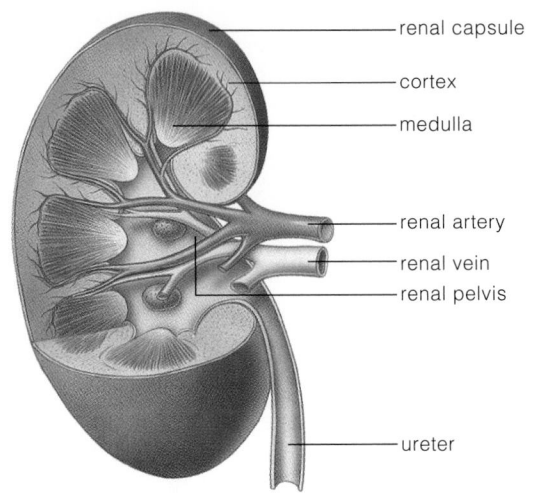

Figure 9.5 Internal structure of the kidney and the major blood vessels leading into and out of it.

Figure 9.4 (**right**) Components of the human urinary system and their functions. (**inset**) The two kidneys, two ureters, and urinary bladder are located outside the abdominal cavity's lining, the peritoneum.

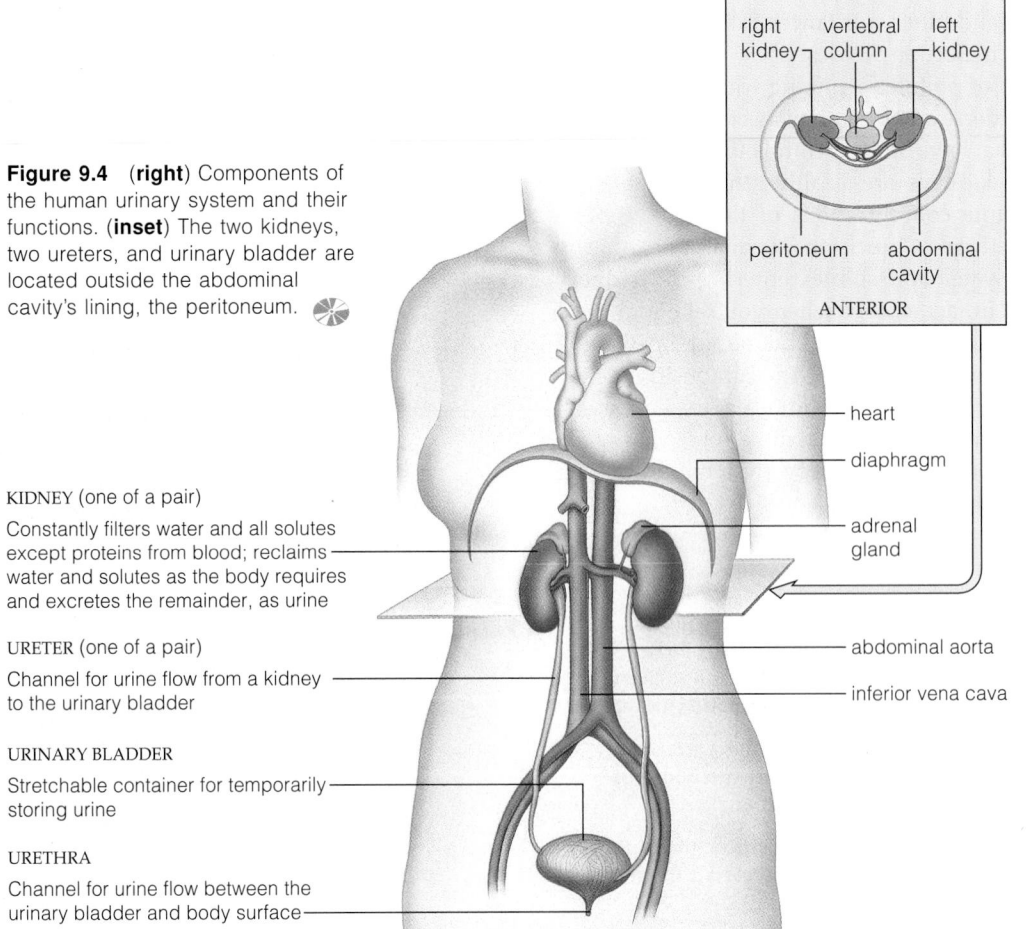

KIDNEY (one of a pair)

Constantly filters water and all solutes except proteins from blood; reclaims water and solutes as the body requires and excretes the remainder, as urine

URETER (one of a pair)

Channel for urine flow from a kidney to the urinary bladder

URINARY BLADDER

Stretchable container for temporarily storing urine

URETHRA

Channel for urine flow between the urinary bladder and body surface

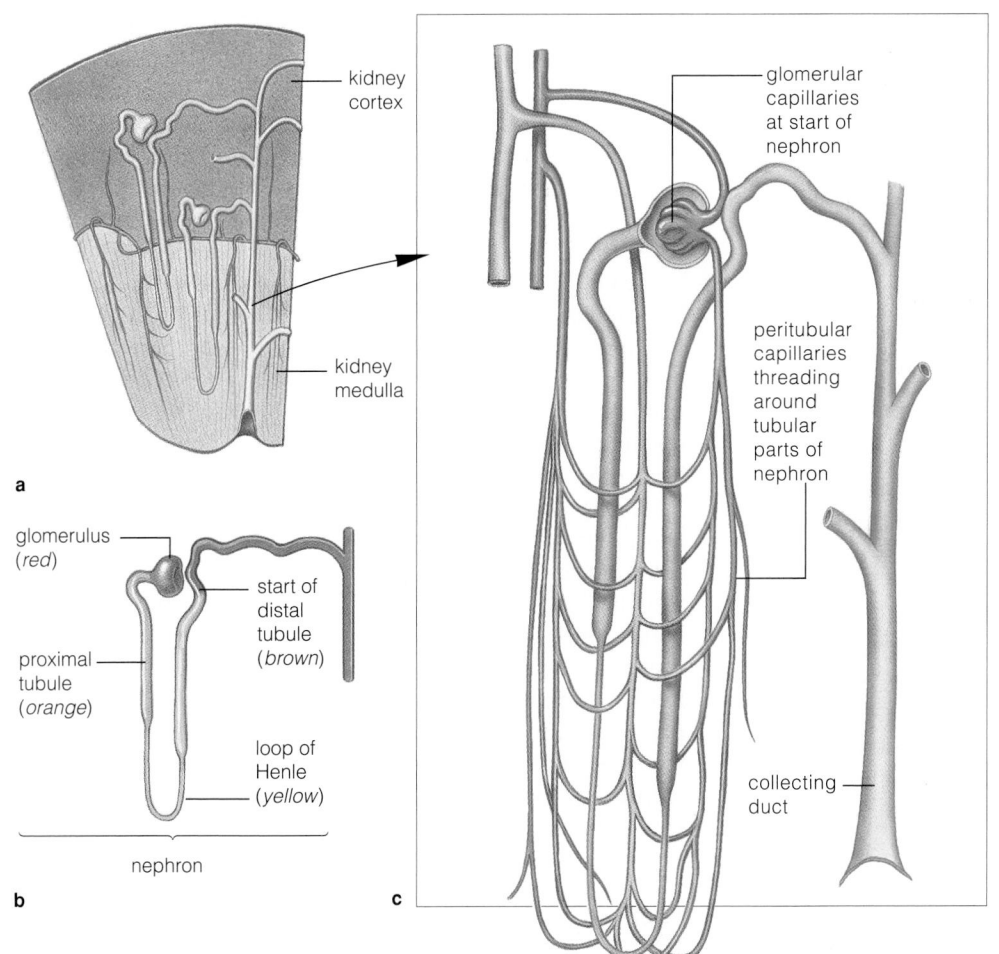

kidney cortex

glomerular capillaries at start of nephron

kidney medulla

peritubular capillaries threading around tubular parts of nephron

collecting duct

a

glomerulus (*red*)

proximal tubule (*orange*)

start of distal tubule (*brown*)

loop of Henle (*yellow*)

nephron

b

c

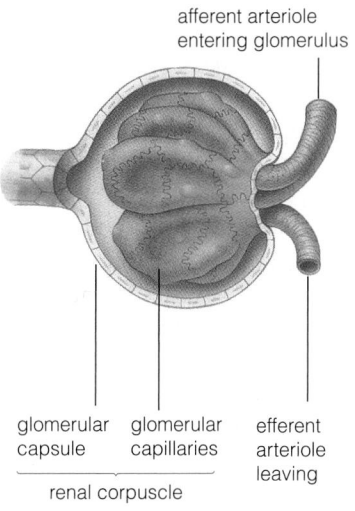

afferent arteriole entering glomerulus

glomerular capsule | glomerular capillaries | efferent arteriole leaving

renal corpuscle

d

Figure 9.6 Diagrams of a nephron and its association with two sets of blood capillaries.

(**a**) Orientation of a nephron relative to the kidney cortex and medulla. (**b**) Functional regions of a nephron. (**c**) The two sets of blood capillaries associated with the nephron. In this sketch, the second set, peritubular capillaries, is highly simplified for clarity; these capillaries actually thread around all tubular parts of the nephron. (**d**) A closer look at the glomerulus, the nephron's blood-filtering unit.

the passage of solutes except via active transport systems built into the plasma membrane (Chapter 2). As you will see, such differences influence the movement of water and solutes across different parts of the nephron wall.

An *afferent arteriole* delivers blood to each nephron. Filtration starts at the **renal corpuscle**, where the nephron wall balloons around a tiny cluster of blood capillaries called the **glomerulus** (Figure 9.6). The ballooned, cuplike wall region, the **glomerular (Bowman's) capsule**, receives water and solutes filtered from blood. The rest of the nephron is tubular. Filtrate flows from the cup into the proximal tubule (closest to the glomerular capsule), then through a hairpin-shaped **loop of Henle** and into the **distal tubule** (most distant from the glomerular capsule). This distal tubule empties into a collecting duct.

Unlike capillaries in most other parts of the body, the glomeruli (capillaries inside a glomerular capsule) do not link arterioles and venules and so do not channel blood immediately back to the general circulation. Instead, they

converge to form an *efferent arteriole* that branches into *another* set of **peritubular capillaries** (Figure 9.6c). The peritubular capillaries weave around the nephron's tubular parts; water and essential solutes reabsorbed from the tubule enter them and so return to the bloodstream. Peritubular capillaries merge into venules, which carry filtered blood out of the kidneys.

The urinary system consists of two kidneys, two ureters, the urinary bladder, and the urethra.

Kidney nephrons receive water and solutes filtered from blood. A tiny cluster of capillaries called a glomerulus is the nephron's blood-filtering unit.

From the glomerulus, filtrate enters the nephron's tubular portion, which ultimately delivers the fluid that enters it to a collecting duct. Peritubular capillaries surrounding tubular nephron regions recapture water and essential solutes.

THE FORMATION OF URINE

Filtration, Reabsorption, and Secretion

Urine is a fluid that rids the body of water and solutes that are in excess of the amounts needed to maintain the extracellular fluid (Table 9.2). Urine forms through a sequence of three processes, called filtration, reabsorption, and secretion.

Filtration starts and ends in the glomerulus. Blood pressure, generated by heart contractions, drives the process. It forces water and various small solutes (such as glucose, sodium, and urea) out of the blood in glomerular capillaries and into the cupped region inside the glomerular capsule. Blood cells, proteins, and other large solutes are left behind. The filtrate moves from the glomerular capsule into the proximal tubule.

Reabsorption proceeds along the nephron's tubular parts. Most of the filtrate's water and solutes—including sodium ions, vitamins, amino acids, and glucose—move *out* of the nephron (by diffusion or active transport) across the tubule wall, then into adjacent capillaries.

Secretion also occurs across tubule wall regions, but in the *opposite* direction. Substances diffuse *out* of the capillaries and into cells of the wall. They then move across the cells, which secrete them into the forming urine. Among other things, this highly controlled process rids the body of excess hydrogen ions (H^+) and potassium ions. It also prevents some metabolic wastes (such as uric acid and some breakdown products of hemoglobin) and water-soluble foreign substances (such as penicillin and certain pesticides) from accumulating in blood. Drug testing of athletes and employees in certain professions relies on the use of urinalysis to detect drug residues secreted into the urine. The *Focus* essay in Section 9.5 describes how urinalysis can help evaluate a person's

health. Table 9.3 lists average daily values for reabsorption of some substances from urine.

Urination, or urine flow from the body (also called "micturition"), is a reflex response. As the urinary bladder fills, tension increases in its strong, smooth-muscled walls. Where the bladder joins the urethra, smooth muscle acts as an *internal urethral sphincter* that helps prevent urine from flowing into the urethra. As bladder tension increases, the sphincter relaxes; at the same time the bladder walls contract and force fluid through the urethra. The internal sphincter cannot be controlled voluntarily, but a person can exert control over an *external urethral sphincter* formed by skeletal muscle closer to the urethral opening. Learning to control this external sphincter is the basis of urinary "toilet training" in young children.

Kidney stones are deposits of uric acid, calcium salts, and other substances that have settled out of urine and collected in the renal pelvis. Smaller kidney stones usually pass naturally from the body during urination. Larger ones can become lodged in the renal pelvis or ureter or, on rare occasions, in the bladder or urethra. The blockage can interfere with urine flow and cause intense pain and kidney damage. Large kidney stones must be removed by medical or surgical procedures. In *lithotripsy*, high-energy sound waves reduce the stone to fragments small enough to pass out in the urine.

Factors That Influence Blood Filtration

Each day, more blood flows through the kidneys than through any other organ except the lungs. Each minute, about 1.5 quarts of blood course through them! That is nearly one-fourth of the cardiac output (at rest). How can the kidneys handle blood flowing through on such a massive scale? There are two mechanisms.

To begin with, the afferent arterioles delivering blood to a glomerulus have a wider diameter—and less resistance to flow—than do most arterioles. The hydrostatic pressure caused by heart contractions therefore does not drop much when blood enters them. By contrast, the efferent arteriole that receives blood from glomeruli offers *high* resistance to blood flow. Because of the properties of the afferent and efferent arterioles, pressure in the glomerular capillaries is higher than in other capillaries.

Second, glomerular capillaries are highly permeable. They do not allow blood cells or protein molecules to escape, but compared to other capillaries they are 10–100 times more permeable to water and small solutes.

Table 9.2 Typical Kinds and Daily Amounts of Solutes in Normal Urine	
Solute	Amount of Solute
Urea	20.0–35.0 grams
Sodium	4.0–6.0 grams
Chloride	6.0–9.0 grams
Potassium	2.5–3.5 grams
Creatinine	1.0–1.5 grams
Calcium	0.01–0.30 grams

Table 9.3	Average Daily Reabsorption Values for a Few Substances		
	Amount Filtered	Amount Excreted	Proportion Reabsorbed
Water	180 liters	1.8 liters	99%
Glucose	180 grams	None, normally	100%
Sodium ions	630 grams	3.2 grams	99.5%
Urea	54 grams	30 grams	44%

Because of the high hydrostatic pressure and greater capillary permeability, the kidneys can filter an average of 45 gallons (180 liters) per day.

At any time, the flow volume to the kidneys affects the filtration rate. Neural, endocrine, and local controls keep that volume fairly constant even when blood pressure changes. For example, when you run a race or dance until dawn, the nervous system diverts an above-normal volume of blood away from the kidneys toward your skeletal muscles. The neural signals direct arterioles in different parts of the body to constrict or dilate in coordinated ways, so that less of the flow volume reaches the kidneys.

As another example, cells in the walls of arterioles leading to glomeruli respond to arterial blood pressure. When the pressure decreases, they secrete chemicals that induce vasodilation, so more blood flows in. When it rises, they constrict, so less blood flows in.

Finally, the filtration rate depends on how fast the kidney tubules are reabsorbing water. As you will see, reabsorption is partly under hormonal control.

Urine forms through the sequential processes of filtration, reabsorption, and secretion. It includes water and solutes not needed to maintain the extracellular fluid, as well as water-soluble wastes.

Two mechanisms permit the kidneys to filter a large amount of blood at a rapid rate. First, blood pressure in glomerular capillaries is higher than in other capillaries. Second, glomerular capillaries are highly permeable.

Neural, endocrine, and local controls ensure that the volume of blood flowing to kidney glomeruli remains relatively constant, even when blood pressure varies.

9.5 AVOIDING TROUBLE IN THE URINARY TRACT

Urinary tract infections routinely plague millions of people. Women especially are susceptible to bladder infections because of their urinary anatomy: The female urethra is short, just a little over an inch long. (In males, the urethra is about 9 inches long.) Hence it is relatively easy for bacteria from outside the body to make their way to a female's bladder and trigger the inflammation known as *cystitis*—or even all the way to the kidneys to cause *pyelonephritis*. An increasingly large number of urinary tract infections in both sexes result from sexually transmitted microbes. As Chapter 15 describes, the organisms that cause *gonorrhea* and *chlamydia* are major culprits.

In males, the prostate gland wraps around the urethra. As a man ages, the prostate may swell either occasionally or chronically, narrowing the urethra and preventing urine from draining effectively. When this happens, bacterial growth can trigger an infection. Urinary problems can also be an early warning sign of prostate cancer.

URINALYSIS The procedure called *urinalysis* analyzes the composition of urine and is used to help diagnose illness. For example, the presence of glucose in urine may be a sign of diabetes, whereas white blood cells (pus) frequently indicate a urinary tract infection. Red blood cells can reveal bleeding due to infection, kidney stones, cancer, or an injury. High levels of albumin and other proteins in urine may indicate severe hypertension, kidney disease, and some other disorders. Bile pigments enter the urine when liver functions are impaired by cirrhosis and hepatitis.

There are simple measures everyone can take to help keep their urinary tract healthy. Drink plenty of fluids, and practice careful hygiene to minimize the opportunity for bacteria to enter the urethra. People who are susceptible to bladder infections may also want to limit their intake of alcohol, caffeine, and spicy foods, all of which can irritate the bladder.

REABSORPTION OF WATER AND SODIUM

Your kidneys precisely adjust how much water and sodium your body excretes or conserves. It makes no difference whether you drink too much or too little water at lunch, eat a bag of salty potato chips, or follow a low-sodium diet—your kidneys will make the necessary adjustments to keep the water and salt content of your body relatively constant. In this section we consider how this regulation is accomplished. Figure 9.7 will help you track the sequence of events.

Role of the Proximal Tubule

Of all the water and sodium filtered in the kidneys, about two-thirds is promptly reabsorbed at the **proximal tubule**—the part of the nephron closest to the glomerulus

(Figure 9.7c). As Figure 9.8 shows, epithelial cells of the proximal tubule wall have transport proteins at their outer surface. Nearly all cells have such proteins, which function as sodium "pumps." In this case, the proteins actively transport sodium ions from the filtrate inside the tubule into the interstitial fluid. Sodium ions (Na^+) are positively charged; negatively charged ions, including chloride (Cl^-), follow the sodium. Glucose and amino acids are reabsorbed by other transport mechanisms that also are linked to sodium reabsorption.

This outward movement reduces the solute concentration inside the tubule and increases it in the interstitial fluid outside the tubule. The wall of the proximal tubule is quite permeable to water, so water follows the osmotic gradient and moves passively out of the tubule.

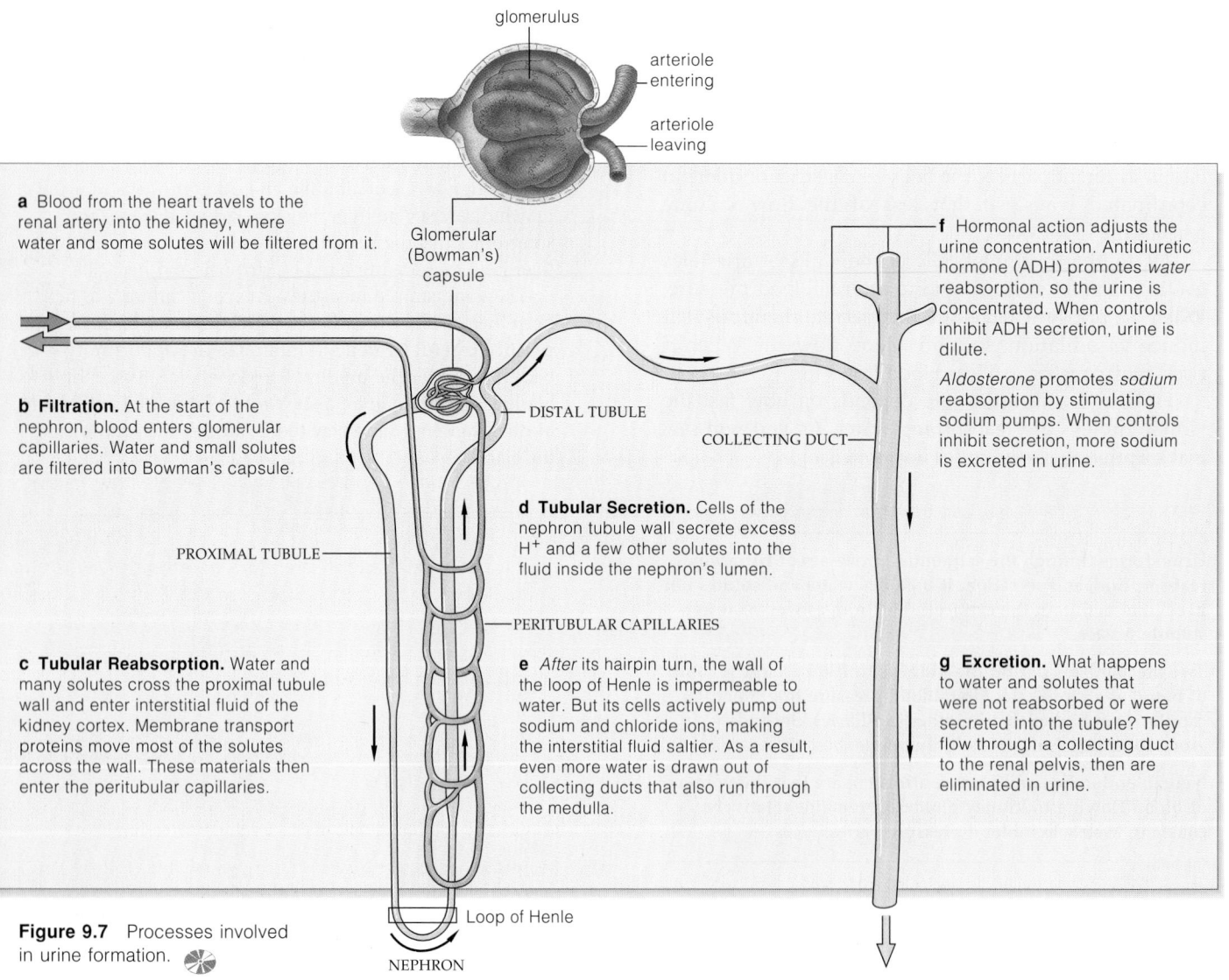

Figure 9.7 Processes involved in urine formation.

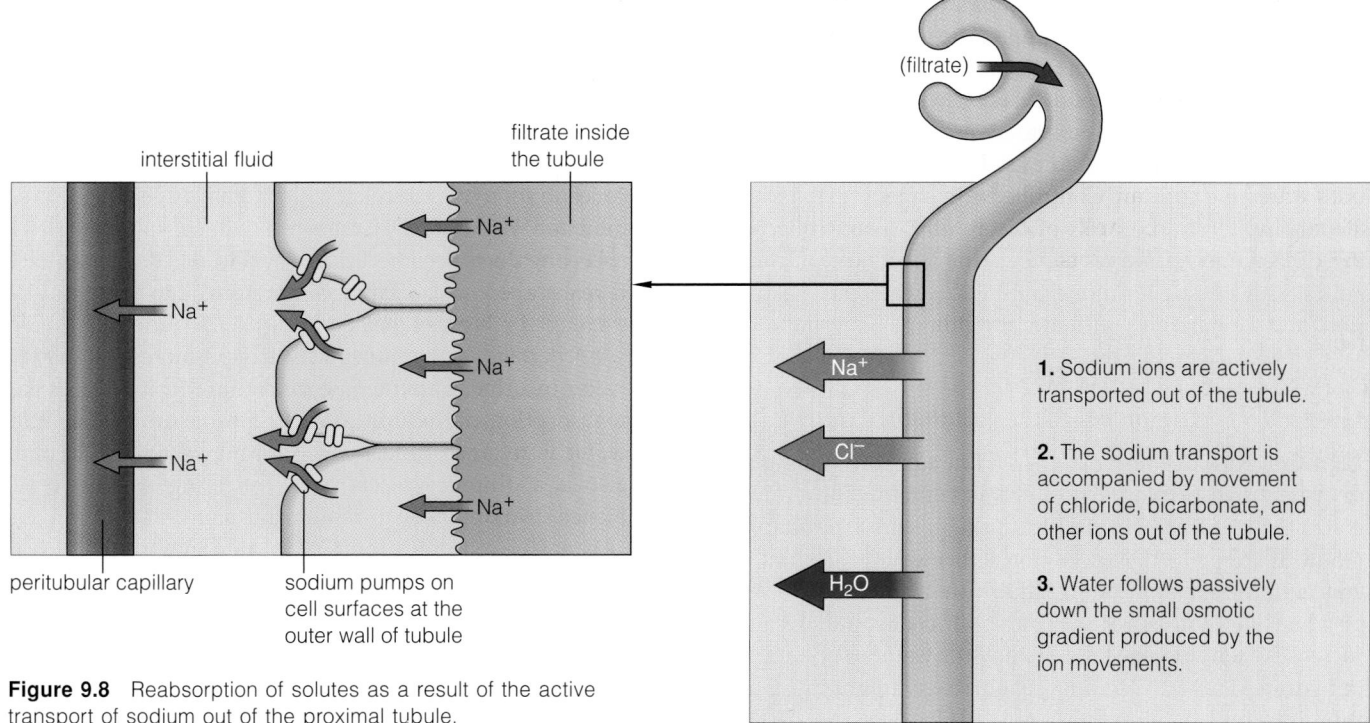

Figure 9.8 Reabsorption of solutes as a result of the active transport of sodium out of the proximal tubule.

1. Sodium ions are actively transported out of the tubule.

2. The sodium transport is accompanied by movement of chloride, bicarbonate, and other ions out of the tubule.

3. Water follows passively down the small osmotic gradient produced by the ion movements.

In this fashion, the volume of fluid remaining within the proximal tubule decreases greatly, but the total concentration of solutes—sodium, especially—changes only a little.

Urine Concentration and Dilution

The situation changes after filtrate moves on through the proximal tubule and enters the loop of Henle. This hairpin-shaped structure descends into the kidney medulla. In the interstitial fluid surrounding the loop, the solute concentration increases progressively with depth within the medulla.

The descending limb of the loop is permeable to water. Water moves out of the descending limb by osmosis, and the solute concentration in the fluid remaining inside increases until it matches that in the interstitial fluid. In the ascending limb, water cannot cross the tubule wall, but sodium is actively transported out. As sodium (and chloride) ions move out of the filtrate, the solute concentration rises outside the tubule and falls inside it. This increase in solute concentration outside the tubule favors the movement of water out of the descending limb.

As solutes leave the tubule, a very high solute concentration develops in the deeper portions of the medulla. Urea contributes to this steep gradient. As water is reabsorbed, the urea left behind in the filtrate becomes concentrated. Some will be excreted in urine, but as filtrate moves into the final portion of the collecting duct, some urea diffuses out. A portion of it enters the interstitial fluid in the inner medulla, further increasing the concentration of solutes there. Solute concentration is highest in the very deepest parts of the inner medulla.

With so many solutes—but no water—having left the fluid in the ascending limb of the tubule, solutes are not very concentrated there. Hence, the filtrate that finally reaches the distal tubule in the kidney cortex is quite dilute, with a low sodium concentration. As you will see in the next section, the stage is set for the excretion of urine that is either highly dilute or highly concentrated—or anywhere in between.

A reabsorption mechanism operates in the tubular portion of the nephron. By way of this mechanism, which is adjustable by hormones, water and solutes can be retained or excreted as required to maintain the extracellular fluid.

Most water is reabsorbed across the permeable walls of the proximal tubule and the descending limb of the loop of Henle. Sodium is pumped outward in the ascending limb. Hormone-induced adjustments of reabsorption will come into play in the distal tubule and collecting duct.

Because so much water and sodium are reabsorbed from the nephron's proximal tubule and loop of Henle, the volume of dilute urine reaching the start of the distal tubule (farthest from the glomerulus) has been greatly reduced. Yet if even that reduced volume were excreted without adjustments, the body would rapidly become depleted of both water and sodium. Controlled adjustments are made at cells located in the walls of distal tubules and collecting ducts. Two hormones serve as the agents of control. **ADH** (antidiuretic hormone) influences water reabsorption, and **aldosterone** influences sodium reabsorption.

How ADH Influences Water Reabsorption

Let's first consider the role of ADH in adjusting the rate of water reabsorption. The hypothalamus in the brain controls the release of ADH from the posterior pituitary gland. It triggers ADH secretion either when the solute concentration of extracellular fluid rises above a set point or when blood pressure falls. The solute concentration can increase when a person takes in too little water or becomes dehydrated; severe bleeding (hemorrhage) might cause a rapid decline in blood pressure. ADH acts on distal tubules and collecting ducts, making their walls more permeable to water (Figure 9.9). Thus, in the kidney cortex, water is reabsorbed from the dilute filtrate inside the distal tubules. The volume of fluid inside is now reduced somewhat, and it passes down through the collecting ducts, which plunge down into the medulla. Remember that solute concentrations in the surrounding interstitial fluid are high in the inner medulla. This encourages the reabsorption of even more water, so only a small volume of very concentrated urine is excreted.

Conversely, when water intake is excessive, the solute concentration in extracellular fluid falls. ADH secretion is inhibited. Without ADH, the walls of the distal tubules and collecting ducts become less permeable to water. Less water is reabsorbed, and a large volume of dilute urine can be excreted. In this way the body rids itself of the excess water.

A *diuretic* is any substance that promotes the loss of water in urine. Caffeine is a mild diuretic; it reduces the reabsorption of sodium along nephron tubules, so less water is retained (and hence more is excreted). Alcohol also is a diuretic; it acts by suppressing ADH release. Hence drinking beer to replenish body fluids after exercise may be self-defeating. Cold water does the job a great deal more effectively.

How Aldosterone Influences Sodium Reabsorption

Let's now examine how aldosterone helps adjust the rate of sodium reabsorption. When the body loses more sodium than it takes in, the volume of extracellular fluid falls. This is because, as you've just read, where sodium goes, water follows. Sensory receptors in the walls of

Figure 9.9 Permeability characteristics of different parts of the nephron. (Both the distal tubule and the collecting duct have very limited permeability to solutes; most of the solute movements in these regions are related directly or indirectly to active transport mechanisms.)

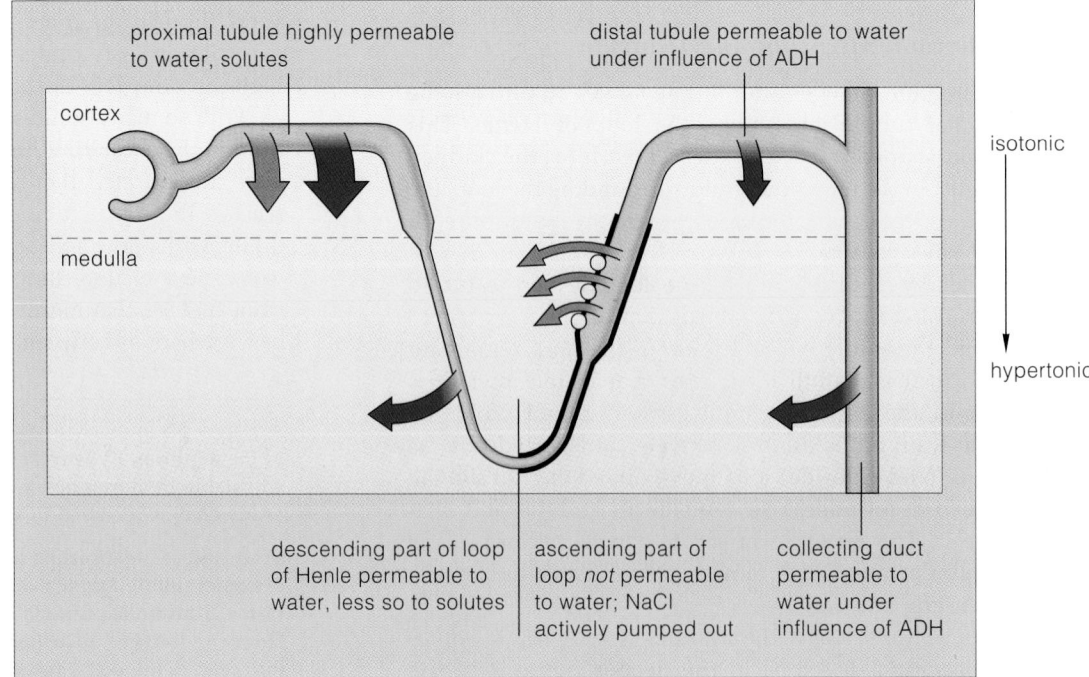

proximal tubule highly permeable to water, solutes

distal tubule permeable to water under influence of ADH

cortex

medulla

isotonic

hypertonic

descending part of loop of Henle permeable to water, less so to solutes

ascending part of loop *not* permeable to water; NaCl actively pumped out

collecting duct permeable to water under influence of ADH

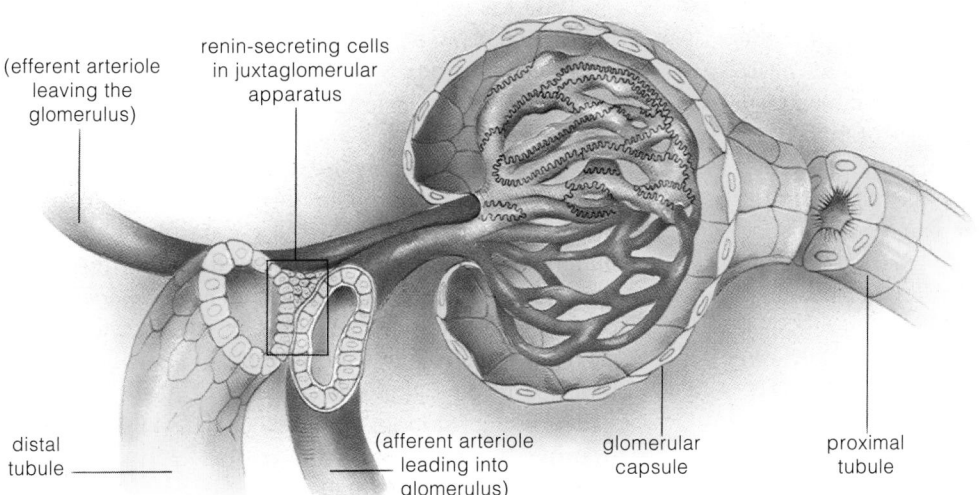

Figure 9.10 Location of juxtaglomerular apparatus and renin-secreting cells that play a role in sodium reabsorption.

(efferent arteriole leaving the glomerulus)

renin-secreting cells in juxtaglomerular apparatus

distal tubule

(afferent arteriole leading into glomerulus)

glomerular capsule

proximal tubule

blood vessels in the kidneys, heart, and elsewhere detect the decrease, and renin-secreting cells are called into action. Those cells, mainly in the afferent arteriole, are part of the **juxtaglomerular apparatus** (Figure 9.10). The name refers to a region of contact between the arterioles of the glomerulus and the distal tubule of the nephron.

Renin acts on molecules of an inactive protein (angiotensinogen) that circulates in the bloodstream. In effect, enzyme action removes part of the molecule, leaving angiotensin I. A subsequent reaction converts angiotensin I to angiotensin II. Among other effects, angiotensin II stimulates cells of the adrenal cortex, the outer portion of a gland perched on top of each kidney, to secrete aldosterone. Aldosterone causes cells of the distal tubules and collecting ducts to reabsorb sodium faster (by stimulating active transport of Na^+), and less sodium is excreted in the urine. Conversely, when the extracellular fluid contains too much sodium, aldosterone secretion is inhibited. Less sodium is reabsorbed, and more is excreted.

In both instances, water "follows the salt." When less sodium is excreted, so is less water, and when more sodium is excreted, more water leaves the body as well.

For various reasons, sodium-regulating mechanisms do not operate properly in some people, and the body cannot fully rid itself of excess sodium. Inevitably, tissues retain excess water, which leads to a rise in blood pressure. Abnormally high blood pressure (hypertension) can adversely affect the kidneys as well as the vascular system and brain (Chapter 6). Some hypertensive people can help control their blood pressure by restricting their intake of sodium chloride. This means limiting their intake of table salt, which is virtually 100 percent NaCl, and salt-laden processed foods.

Salt–Water Balance and Thirst

The body does not rely entirely on events in the kidneys when it needs water. The same stimuli that lead to ADH secretion and increased reabsorption of water in the kidneys also stimulate thirst. Suppose you eat a box of salty popcorn at the movies. Soon the salt is absorbed into your bloodstream. The solute concentration in your extracellular fluid rises, and the hypothalamus detects the increase and prompts ADH secretion. In addition, the **thirst center** is stimulated. Signals from this cluster of nerve cells in the hypothalamus can inhibit saliva production. Your brain interprets the resulting sensation of mouth dryness as "thirst" and leads you to seek fluids.

In fact, a "cottony mouth" is one of the early signs that the body is becoming dehydrated. Dehydration commonly results after severe blood loss, burns, or diarrhea. It also results after profuse sweating, after the body has been deprived of water for a long time, or as a side effect of some medications. Under such circumstances, thirst becomes exceptionally intense.

ADH enhances water reabsorption at distal tubules and collecting ducts when the body must conserve water. As a result, the urine is concentrated.

When excess water must be excreted, ADH secretion is inhibited. Less water is reabsorbed, and dilute urine is produced.

Aldosterone enhances sodium reabsorption at distal tubules and collecting ducts when the body must conserve sodium.

A nerve cell cluster in the hypothalamus responds to the same stimuli that prompt increased water reabsorption. Signals from the center trigger the sensation of thirst.

In addition to maintaining the volume and composition of extracellular fluid, the kidneys also help keep the extracellular fluid from becoming too acidic or too basic (alkaline). This overall **acid–base balance** is maintained through control over the concentrations of hydrogen ions (H^+) and other dissolved ions. Buffer systems, respiration, and urinary excretion provide the control.

Normal extracellular pH in the human body must be maintained between 7.37 and 7.43. As you know, acids lower the pH and bases raise it. A variety of acidic and basic substances enter the blood by absorption from the gut and as a result of normal metabolism. Typically, cell activities produce excess acids, which dissociate into H^+ and other fragments. This lowers the pH. The effect is minimized when excess hydrogen ions react with buffer molecules. One example is the **bicarbonate–carbon dioxide buffer system**:

$$H^+ + \underset{\text{bicarbonate}}{HCO_3^-} \rightleftharpoons \underset{\text{carbonic acid}}{H_2CO_3} \rightleftharpoons H_2O + CO_2$$

In this case, the excess H^+ is neutralized, and the carbon dioxide that forms during the reactions is exhaled from the lungs.

The urinary system contributes to acid–base balance in two major ways. It not only eliminates excess H^+, but also *restores the supply of bicarbonate*. How? The reactions just described for the lungs proceed in reverse in the cells of the nephron tubule's walls, as diagrammed in Figure 9.11. HCO_3^- produced by the reverse reactions moves into interstitial fluid, then into the peritubular capillaries. From there, it moves into the general circulation and buffers excess acid. What happens to the H^+ formed in the cells? It is secreted into the nephron. There, it may combine with bicarbonate ions in the filtrate (which can't cross the tubule wall) to form CO_2—which can be returned to the blood and exhaled. Or it may combine with phosphate ions or ammonia (NH_3), then be excreted in urine. In such ways, hydrogen ions are permanently removed from extracellular fluid.

Adjustments in acid–base balance and other kidney functions are essential to maintaining homeostasis. This chapter's *Science Comes to Life* describes some debilitating health problems that develop when the kidneys cannot effectively add or remove solutes from blood.

Along with buffering systems and the respiratory system, kidneys help keep the extracellular fluid from becoming too acidic or too basic (alkaline).

Buffering systems neutralize acids. The urinary system eliminates excess hydrogen ions and also restores bicarbonate used up in buffering reactions.

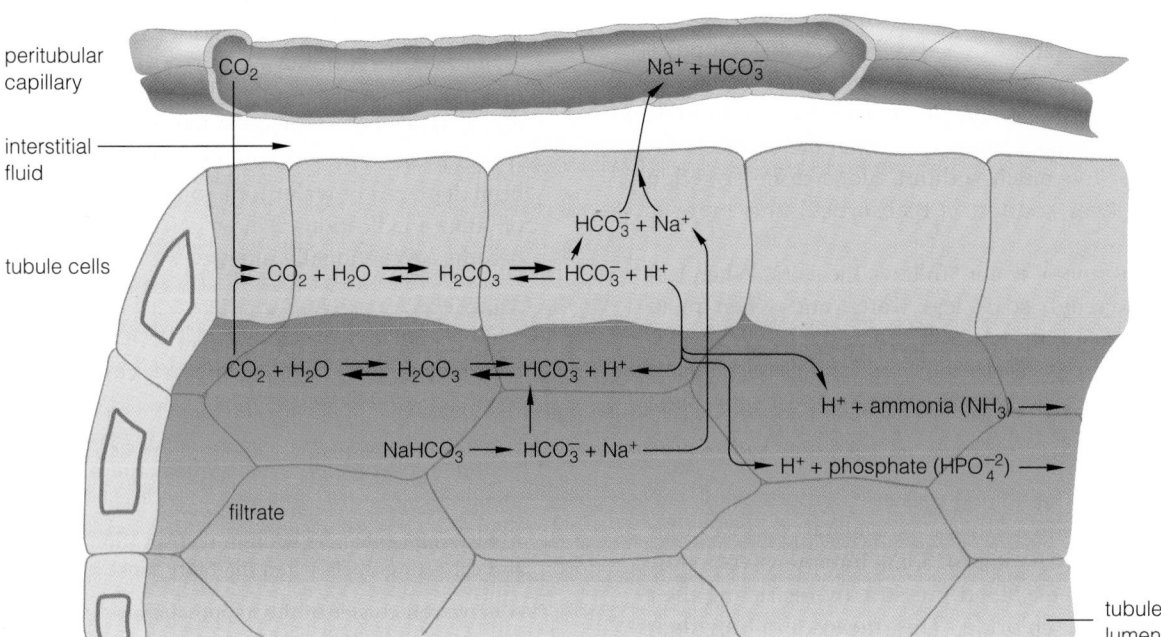

Figure 9.11 How the bicarbonate–carbon dioxide buffer system in the kidneys helps regulate pH.

9.9 KIDNEY DISORDERS

From the preceding sections, you can sense that good health depends absolutely on normal kidney function. Disorders or injuries that interfere with it can have mild to severe effects. As you've already read, for example, uric acid, calcium salts, and other wastes can settle out of urine and collect in the renal pelvis as kidney stones.

Nephritis is an inflammation of the kidneys. It can be caused by a range of factors, including bacterial infections. As you may recall from Chapter 7, an inflamed tissue tends to swell as fluid accumulates within it. However, because a kidney is "trapped" inside the tough renal capsule, it cannot increase in size. As a result, hydrostatic pressure builds up in or around glomerular capillaries, blocking them and preventing the passage of blood.

Glomerulonephritis is an umbrella term for a large number of disorders (often involving faulty immune responses) that can severely damage the kidneys. Hypertension and diabetes can disrupt blood circulation to and within the kidneys, sometimes virtually blocking the flow of blood through the glomeruli. An estimated 13 million people in the United States have kidneys in which the nephrons have been so damaged that the filtering of blood *and* formation of urine are seriously impaired. Control of the volume and composition of the extracellular fluid is disturbed, and toxic by-products of protein breakdown can accumulate in the bloodstream. Patients can suffer nausea, fatigue, and memory loss. In advanced cases, death may result. A kidney dialysis machine can restore the proper solute balances (Figure 9.12). Like the kidney itself, the machine helps maintain extracellular fluid by selectively removing and adding solutes to the bloodstream.

"Dialysis" refers to the exchange of substances across a membrane between solutions of differing compositions. In *hemodialysis*, the dialysis machine is connected to an artery or a vein, and then blood is pumped through tubes made of a material similar to cellophane. The tubes are submerged in a warm-water bath. The precise mix of salts, glucose, and other substances in the bath sets up the correct gradients with the blood. Dialyzed blood is returned to the body.

Hemodialysis generally takes about four hours; blood must circulate repeatedly before solute concentrations in the body are improved. The procedure must be performed

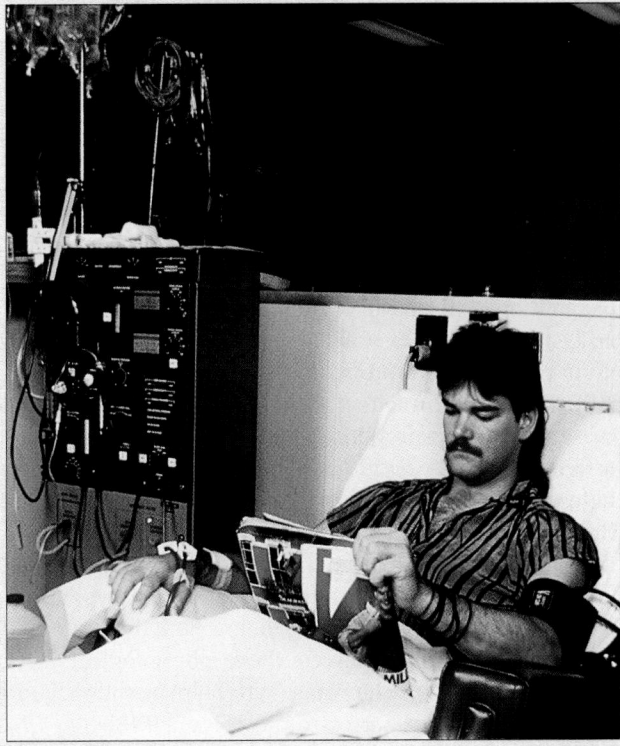

Figure 9.12 Patient undergoing kidney dialysis.

three times a week and is used as a temporary measure in patients with reversible kidney disorders. In chronic cases, the procedure must be used for the rest of the patient's life or until a functional kidney is transplanted.

As an alternative to hemodialysis, in *peritoneal dialysis* exchanges are made several times every day. Fluid of the proper composition is put into the abdominal cavity, left in place for a period of time, and then drained out. Here, the lining of the cavity (the peritoneum) serves as the dialysis membrane.

Although chronic kidney disease can impose some inconveniences, with proper treatment and a controlled diet, many people are able to pursue a surprisingly active, close-to-normal lifestyle.

Temperatures in the environment just outside the body often change quickly, and exercise can send metabolic rates soaring. Such changes trigger slight increases or decreases in the body's normal **core temperature**. "Core" refers to the body's innermost tissues, as opposed to the tissues near its surface. Normal human core temperature is about 37°C (98.6°F).

Heat is an inevitable by-product of metabolic activity. (Even as you sit reading this book, you are producing roughly one kilocalorie of heat per hour per kilogram of body weight.) If that heat were to accumulate internally, your body temperature would steadily rise. Above 41°C (105.8°F), some protein enzymes become denatured and cease to function properly. Likewise, the rate of enzyme activity generally *decreases* by at least half when body temperature drops by 10°F. As body core temperature drops below 35°C (95°F), reduced enzyme functioning causes the heart rate to fall, and heat-generating mechanisms such as shivering stop. Breathing slows, and a person may lose consciousness. Below 80°F the heart may stop beating entirely. Given these physiological facts, humans require mechanisms that help maintain body temperature within narrow limits (Figure 9.13).

You and all other humans are **endotherms**, which means "heat from within." Our body temperature is controlled mainly by (1) metabolic activity and (2) controls over heat conservation and dissipation. We also can make behavioral adjustments (such as building a fire or changing clothes) that supplement the physiological controls.

Responses to Cold Stress

Table 9.4 summarizes the major responses to cold stress, which are governed by the hypothalamus. **Peripheral vasoconstriction** follows a drop in temperature. In this response, thermoreceptors at the body surface detect the falling temperature. When their signals reach the hypothalamus, a hormonal message goes out to smooth muscle in the walls of arterioles in the skin, calling for the muscles to contract. The resulting vasoconstriction reduces blood flow to capillaries near the body surface, so body heat is retained. This response to cold stress is remarkably effective. For example, when your fingers or toes get cold, up to 99 percent of the blood that would otherwise flow to their skin is diverted.

In the **pilomotor response** to a drop in outside temperature, smooth muscle controlling the erection of body hair is stimulated to contract. This creates a layer of still air that reduces heat losses from the body. (Of course, this response is much more effective in mammals with more body hair than humans.) Heat loss can be restricted even more by behaviors that reduce the amount of body surface exposed for heat exchange—as when you put on a sweater or hold both arms tightly against your body.

When other responses can't counter cold stress, the hypothalamus calls for increased skeletal muscle activity that leads to **shivering**. Rhythmic tremors begin as the muscles contract about ten to twenty times per second, and heat production throughout the body increases several times over.

Prolonged or severe cold exposure can lead to a hormonal response that elevates the rate of metabolism. This **nonshivering heat production** is especially notable in brown adipose tissue (found in the neck, armpits, and near the kidneys), where heat is generated as the lipid molecules are broken down. Human babies have this tissue; adults have little unless they are cold-adapted.

When defenses against cold are not adequate, the result is *hypothermia*, a condition in which the core temperature falls below normal. A drop of only a few degrees affects brain function and leads to confusion; further cooling can lead to coma and death. Some victims of extreme hypothermia, mainly children, have survived prolonged immersion in ice-cold water. One reason is that mammals, including humans, have a **dive reflex**. When a mammal is submerged, the heart rate slows and blood is shunted to the brain and other vital organs.

Tissue destruction through localized freezing is called *frostbite*. Cells that become frozen may be destroyed unless thawing is precisely controlled, as sometimes can be done in a hospital.

Responses to Heat Stress

Table 9.4 also summarizes the main responses to heat stress. When core temperature rises above a set point, the hypothalamus once again plays a central role in ordering responses. In **peripheral vasodilation**, hypothalamic signals cause blood vessels in the skin to dilate. More blood

Table 9.4   Summary of Human Responses to Cold Stress and to Heat Stress

Environmental Stimulus	Main Responses	Outcome
Drop in temperature	Vasoconstriction of blood vessels in skin; pilomotor response; behavior changes (e.g., putting on a sweater)	Heat is conserved
	Increased muscle activity; shivering; nonshivering heat production	More heat produced
Rise in temperature	Vasodilation of blood vessels in skin; sweating; changes in behavior; heavy breathing	Heat is dissipated from body
	Reduced muscle activity	Less heat produced

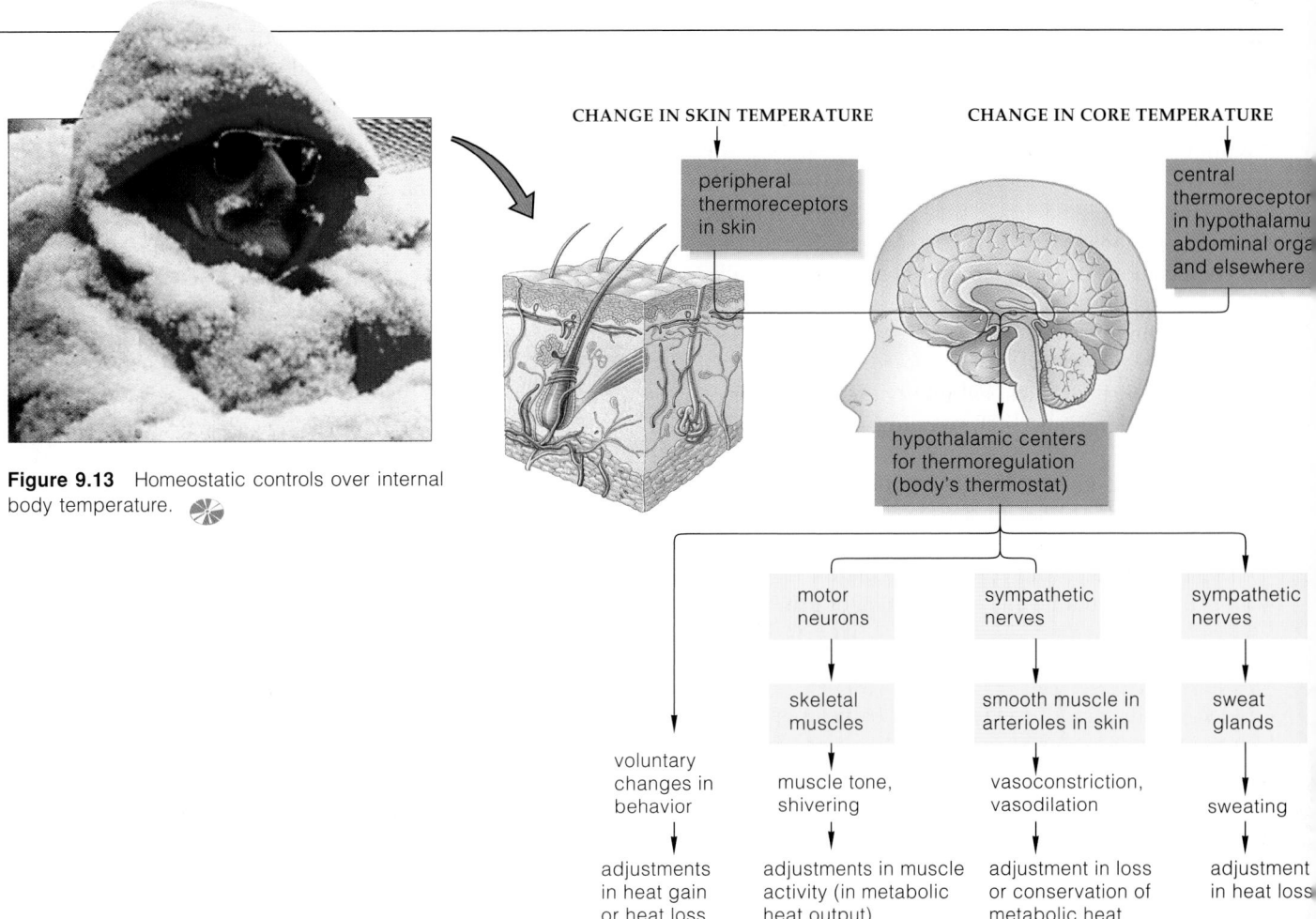

Figure 9.13 Homeostatic controls over internal body temperature.

CHANGE IN SKIN TEMPERATURE

peripheral thermoreceptors in skin

CHANGE IN CORE TEMPERATURE

central thermoreceptor in hypothalamu abdominal orga and elsewhere

hypothalamic centers for thermoregulation (body's thermostat)

motor neurons → skeletal muscles → muscle tone, shivering → adjustments in muscle activity (in metabolic heat output)

sympathetic nerves → smooth muscle in arterioles in skin → vasoconstriction, vasodilation → adjustment in loss or conservation of metabolic heat

sympathetic nerves → sweat glands → sweating → adjustment in heat loss

voluntary changes in behavior → adjustments in heat gain or heat loss

---

flows to the skin, where the excess heat the blood carries is dissipated.

Evaporative heat loss is another response that can be influenced by the hypothalamus, which can activate sweat glands. Your skin has 2.5 million or more sweat glands, and considerable heat is dissipated when the water they give up to the skin surface evaporates. With prolonged heavy sweating the body loses important salts—especially sodium chloride—as well as a great deal of water. Such losses may so alter the character of the internal environment that the affected person may faint.

Sometimes peripheral blood flow and evaporative heat loss can't adequately counter heat stress. The result is *hyperthermia*, a condition in which the core temperature increases above normal. If the increase isn't too great, a person can suffer *heat exhaustion*, in which vasodilation and water losses from heavy sweating cause a drop in blood pressure. The skin feels cold and clammy, and the affected person may collapse.

When heat stress is great enough to completely break down the body's temperature controls, *heat stroke* occurs. Sweating stops, the skin becomes dry, and body temperature rapidly increases to a level that can be lethal.

When someone has a fever, the hypothalamus has reset the "thermostat" that dictates what the body's core temperature will be. The normal response mechanisms are brought into play, but they are carried out to maintain a higher temperature.

At the onset of fever, heat loss decreases, heat production increases, and the individual feels chilled. When a fever "breaks," peripheral vasodilation and sweating increase as the body attempts to restore the normal core temperature; then, the person feels warm. The controlled increase in body temperature during a fever seems to enhance the body's immune response. For that reason, using fever-reducing drugs such as aspirin or ibuprofen may actually interfere with fever's beneficial effects. A severe fever can be quite dangerous, however, and in such cases the drugs are essential.

Physiological responses to cold stress include peripheral vasoconstriction, the pilomotor response, shivering, and sometimes nonshivering heat production.

Responses to heat stress include peripheral vasodilation and evaporative heat loss. The hypothalamus governs temperature regulation.

SUMMARY

Control of Extracellular Fluid

1. Within the body, the cellular environment consists of certain types and amounts of substances dissolved in water. This extracellular fluid fills tissue spaces and (in the form of blood plasma) blood vessels. Its volume and composition are maintained only when the daily intake of water and solutes are in balance. The following processes maintain the balance:

a. Water is gained by absorption (from the GI tract) and by metabolism. It is lost by urinary excretion, evaporation from the lungs and skin, sweating, and elimination in feces.

b. Solutes are gained by absorption from the gut, secretion, respiration, and metabolism. They are lost by excretion, respiration, and sweating.

c. Losses of water and solutes are controlled mainly by adjusting the volume and composition of urine.

2. The human urinary system consists of two kidneys, two ureters, a urinary bladder, and a urethra.

3. Blood is filtered and urine forms in kidney nephrons. Each nephron interacts closely with two sets of blood capillaries (glomerular and peritubular).

a. A nephron has a cup-shaped beginning (glomerular capsule), then three tubelike regions (proximal tubule, loop of Henle, and distal tubule, which empties into a collecting duct).

b. The glomerular (Bowman's) capsule surrounds a set of highly permeable capillaries. Together, they are a blood-filtering unit, the glomerulus.

c. Blood pressure forces water and small solutes out of the capillaries, into the cup. Most of the filtrate is reabsorbed by the tubules and returned to the blood. A portion is excreted as urine.

4. Urine forms in the nephron by three processes:

a. Filtration of blood at the glomerulus of a nephron, which puts water and small solutes into the nephron.

b. Reabsorption. Water and solutes to be retained leave the nephron's tubular parts and enter the peritubular capillaries that thread around them. A small volume of water and solutes remains in the nephron.

c. Secretion. Some ions and a few other substances leave the peritubular capillaries and enter the nephron, for disposal in urine.

5. Reabsorption of many solutes may occur passively, following concentration gradients. In other instances, active transport is required. Sodium reabsorption occurs by active transport. Water reabsorption is always passive, occurring along its osmotic gradient.

6. Urine becomes more or less concentrated by the action of two hormones, ADH and aldosterone. These act on cells of distal tubules and collecting ducts as follows:

a. ADH is secreted when the body must conserve water; it enhances reabsorption from the distal nephron tubule and collecting ducts. Inhibition of ADH allows more water to be excreted.

b. Aldosterone conserves sodium by enhancing its reabsorption in the distal tubule. Inhibition of aldosterone allows more sodium to be excreted. Because "water follows salt," aldosterone indirectly influences water reabsorption.

7. Together with the respiratory system and other mechanisms, the kidneys also help maintain the body's overall acid–base balance.

Control of Body Temperature

1. Body temperature is determined by the balance between metabolically produced heat and the heat absorbed from and lost to the environment.

2. Humans are endotherms. Their body temperature is controlled largely by metabolic activity and by precise controls over heat produced and heat lost.

Review Questions

1. Label the regions of the nephron, and identify where filtration, reabsorption, and secretion occur: *206*

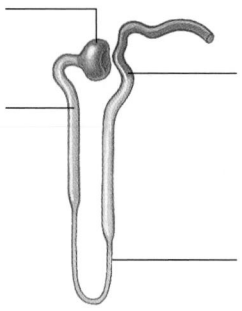

2. How does urine formation help maintain the body's internal environment? *202–203*

3. Label the kidney's component parts: *204*

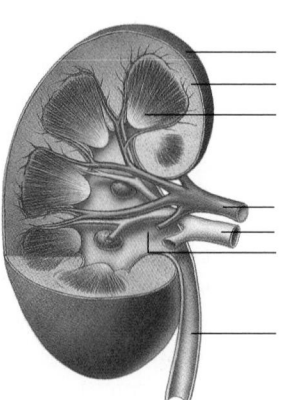

4. Which hormone influences water reabsorption and conservation? Sodium reabsorption and conservation? *210*

5. Fatty tissue holds the kidneys in place. Extremely rapid weight loss may cause this tissue to shrink so that the kidneys slip from their normal position. On rare occasions, the slippage may put a kink in one or both ureters and block urine flow. Speculate on what might then happen to the kidneys. *203, 213*

Critical Thinking: You Decide (Key in Appendix V)

1. A urinalysis reveals that the patient's urine contains glucose, hemoglobin, and white blood cells (pus). Are any of these substances abnormal in urine? Explain.

2. As a person ages, nephron tubules lose some of their ability to concentrate urine. What is the effect of this change?

3. What homeostatic mechanism is visibly operating in the photograph below? What are its physiological effects?

Self-Quiz (Answers in Appendix IV)

1. The body gains water by _____.
   a. gastrointestinal absorption   c. both a and b
   b. metabolism   d. neither a nor b

2. The body loses water by _____.
   a. evaporation from lungs and skin
   b. elimination in feces
   c. urinary excretion
   d. all of the above

3. Each human kidney contains about _____ nephrons.
   a. 1,000   c. 100,000
   b. 10,000   d. 1,000,000

4. Water and small solutes return to the blood during_____.
   a. filtration   c. secretion
   b. reabsorption   d. both a and b

5. Water and small solutes leave blood during _____.
   a. filtration   c. secretion
   b. reabsorption   d. both a and c

6. A few substances move out of the peritubular capillaries and are moved into the nephron during _____.
   a. filtration   c. secretion
   b. reabsorption   d. both a and c

7. A nephron's reabsorption mechanism depends on _____.
   a. osmosis across the nephron wall
   b. active transport of sodium across the nephron wall
   c. a steep solute concentration gradient starting at the kidney cortex and descending into the medulla
   d. all of the above

8. Water is conserved and urine becomes more concentrated by the action of _____.
   a. ADH   c. aldosterone
   b. renin   d. both b and c

9. Sodium is conserved and urine becomes more concentrated by the action of _____.
   a. ADH   c. aldosterone
   b. renin   d. both b and c

10. Match the following salt–water balance concepts:
   _____ aldosterone   a. blood filter of a nephron
   _____ nephron   b. controls sodium reabsorption
   _____ thirst mechanism c. occurs at nephron tubules
   _____ reabsorption   d. site of urine formation
   _____ glomerulus   e. controls water gain

Selected Key Terms

ADH *210*
aldosterone *210*
core temperature *214*
distal tubule *205*
endotherm *214*
extracellular fluid *202*
filtration *206*
glomerular (Bowman's) capsule *205*
glomerulus *205*
juxtaglomerular apparatus *211*
kidney *204*
loop of Henle *205*
nephron *204*
peripheral vasoconstriction *214*

peripheral vasodilation *214*
peritubular capillary *205*
pilomotor response *214*
proximal tubule *208*
reabsorption *206*
renal corpuscle *205*
secretion *206*
thirst center *211*
ureter *204*
urethra *204*
urinary bladder *204*
urinary excretion *202*
urinary system *202*
urine *206*

Readings

Flieger, Ken. March 1990. "Kidney Disease: When Those Fabulous Filters Are Foiled." *FDA Consumer*. Excellent, nontechnical survey of kidney functions and treatments for kidney disease.

Sherwood, L. 1997. *Human Physiology*. 2d. ed. Belmont, California: Wadsworth.

Smith, H. 1961. *From Fish to Philosopher*. New York: Doubleday. Paperback. This entertaining, classic book traces the evolutionary path that produced the human kidney.

Vander, A., J. Sherman, and D. Luciano. 1994. "The Kidneys and Regulation of Water and Inorganic Ions," in *Human Physiology*, 6th ed. New York: McGraw-Hill, Chapter 15.

For additional readings, go to InfoTrac College Edition, your online library at:

http://www.infotrac-college.com/wadsworth

10

NERVOUS SYSTEM

Why Crack the System?

Suppose your biology instructor asks you to volunteer for an experiment. A microchip will have to be implanted in your brain, and it will make you feel *really* good. But it may harm your health, take years off your life, and perhaps destroy part of your brain. Your behavior will change for the worse, so you may have trouble completing school, getting or keeping a job, even having a normal family life. The longer the chip is implanted, the less you will want to give it up. You won't get paid to participate in this experiment—*you* pay the experimenter, first at bargain rates, then a little more each week. The chip is illegal. If you get caught using it, you and the experimenter will go to jail.

Sometimes Jim Kalat, a professor at North Carolina State University, proposes this experiment (which, of course, is hypothetical). Few students volunteer. Then he substitutes *drug* for "microchip" and *dealer* for "experimenter"—and an amazing number of students come forward! Like 30 million other Americans, the "volunteers" seem ready to engage in self-destructive uses of drugs that alter emotional and behavioral states.

The destruction shows up in unexpected places. Each year, for instance, about 300,000 infants are born addicted to crack cocaine smoked or injected by their mothers. Crack relentlessly stimulates a region deep within the brain. In response, the brain cells produce chemicals that dampen normal urges to eat and sleep and that cause blood pressure to rise. Elation and sexual desire intensify. In time, brain cells that produce the stimulatory chemicals can't keep up with the incessant demands. The chemical vacuum makes crack users frantic, then profoundly depressed. Only crack makes them "feel good" again.

Many legal drugs, including alcohol, can also disrupt normal functioning of the human nervous system. For example, chronic heavy drinkers typically have problems with short-term memory. A pregnant woman who drinks, especially if she drinks heavily, puts her fetus at risk of developing fetal alcohol syndrome. Along with this devastating disorder of the nervous system come severe learning disabilities and behavioral problems (Section 14.9).

Directly or indirectly, the nervous system is responsible for our ability to perceive joy and pain, to read a book, and to remember what we've read. It is also responsible for nearly every body movement (Figure 10.1). Constant, coordinated signals that travel rapidly through its communication lines—nerves composed of cells called neurons—control the muscles concerned with facial expressions, food digestion, breathing, and other vital functions. In this regard it may be useful to keep in mind that substance abuse—whether the substance is cocaine or a legal drug such as alcohol—can distort or even eliminate these remarkable possibilities.

The nervous system has two divisions. The *central nervous system* (CNS) consists of the brain and spinal cord. It receives and integrates signals from the external world and from elsewhere in the body and coordinates responses. Cordlike communication lines called *nerves* make up the *peripheral nervous system*, which carries messages between the central nervous system and other body regions. Two general types of cells make up both divisions of the nervous system. One type is specialized for communication, and the other type provides structural and functional support. The remainder of this chapter will flesh out this general picture. Let's begin by examining the structure and functioning of nervous system cells.

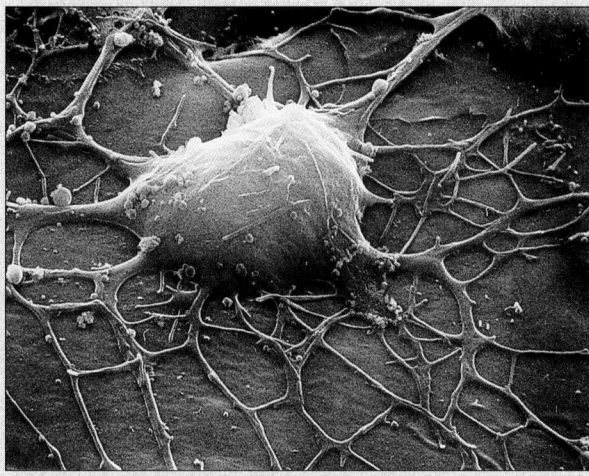

Figure 10.1 **(left)** Owners of an evolutionary treasure—a complex nervous system that is the foundation for our memory, for reasoning, for coordinated movement, and for our future. **(above)** A motor neuron, one of the communication cells of the nervous system.

KEY CONCEPTS

1. Neurons are the basic units of communication of the nervous system. Collectively, they detect and integrate information about external and internal conditions, then select or control muscles and glands in ways that produce suitable responses.

2. The inside of a neuron is negatively charged relative to the outside. When a neuron is stimulated, this polarity of charge across the membrane may briefly and abruptly reverse. Such reversals—called action potentials—are the means by which messages are sent through the nervous system.

3. Action potentials travel along a neuron, but they cannot cross the small gaps *between* neurons or between neurons and some other kinds of cells. Chemical signals bridge the gaps and stimulate or inhibit the adjoining neuron, muscle cell, or gland cell.

4. The flow of information through the nervous system depends on the moment-to-moment integration of excitatory and inhibitory signals that act upon neurons.

5. The nervous system includes the brain, spinal cord, and many nerves. The brain and spinal cord make up the central nervous system; paired nerves that thread through the rest of the body make up the peripheral nervous system.

6. Nerves of the peripheral nervous system are further divided into somatic nerves, which service skeletal muscles, and autonomic nerves, which service the heart, lungs, and other internal organs.

7. The brain is complex in both structure and function. It has three main regions: the forebrain, the midbrain, and the hindbrain. Within these regions are centers for receiving, integrating, storing, and responding to information.

CHAPTER AT A GLANCE

10.1	Neurons—The Communication Specialists
10.2	Action Potentials
10.3	High-Speed Signals Along Sheathed Axons
10.4	Chemical Synapses
10.5	The Nervous System: An Overview
10.6	The Peripheral Nervous System
10.7	The Central Nervous System
10.8	Other Aspects of CNS Structure
10.9	*An Environmental Assault on the Nervous System*
10.10	A Closer Look at the Cerebral Cortex
10.11	*Sperry's Split-Brain Experiments*
10.12	Memory
10.13	How Psychoactive Drugs Affect the Central Nervous System
10.14	Commonly Abused Drugs

NEURONS—THE COMMUNICATION SPECIALISTS

Neurons are the communication cells of the brain, spinal cord, and nerves, and they are divided into three classes. A **sensory neuron** is adapted to respond to a specific type of stimulus (light, pressure, or another form of energy). A sensory neuron relays information about the stimulus to the spinal cord and brain. In the spinal cord and brain, **interneurons** receive sensory input, integrate it with other information, then influence the activity of other neurons. **Motor neurons** relay information away from the brain and spinal cord to muscles or glands—the body's effectors, which carry out responses.

Neurons make up less than half the volume of the human nervous system, and no more than 20 percent of its cells. The rest are **neuroglial cells**, which physically support and protect neurons and help maintain proper concentrations of important ions in the fluid around neurons. Star-shaped neuroglia called *astrocytes* give structure to the brain, as connective tissues do elsewhere in the body. Other neuroglia separate groups of neurons; still others, called *Schwann cells*, insulate the long extensions (axons) of sensory and motor neurons and enable

them to conduct impulses very rapidly. In the CNS *microglia* are macrophages specialized to engulf and dispose of dead cells and microorganisms.

Functional Zones of a Neuron

Neurons have a nucleated cell body and cytoplasmic extensions that can differ in number and length (Figure 10.2). Typically, the cell body and slender extensions called **dendrites** are *input* zones, regions where a neuron receives information. A slender, often longer extension called an **axon** is a *conducting* zone. Axons rapidly propagate signals that arise at a neuron's trigger zone. Except in sensory neurons, the *trigger* zone is a patch of plasma membrane at the junction between the cell body and an axon. In motor neurons and interneurons this junction, called the axon hillock ("little hill"), is where electrical signals are initiated. The branched endings of an axon are *output* zones, where messages are sent to other cells.

A Neuron at Rest, Then Moved to Action

Different forms of signals arise in the nervous system. Let's start with one that begins and ends on the same neuron and is not transferred to another cell. When a neuron is resting, a difference in electric charge (a voltage difference) is being maintained across its plasma membrane. The cytoplasmic fluid next to the membrane remains negatively charged, compared to the interstitial fluid right outside. We measure these charges in units called millivolts. The steady voltage difference across the membrane—for many neurons, about –70 millivolts—is called the **resting membrane potential**.

Suppose a stimulus causes a voltage change on a patch of membrane in a neuron's input zone. The voltage difference across the patch may change only slightly, if at all. By contrast, a strong signal might trigger an **action potential**—an abrupt, short-lived reversal in the voltage difference across the plasma membrane. For a fraction of a second, the inside becomes positive with respect to the outside. The reversal triggers another action potential at the adjoining patch of membrane, this triggers another at the next patch, and so on away from the initiation point. In short, *stimulation of a neuron disturbs the distribution of electric charge across its plasma membrane.*

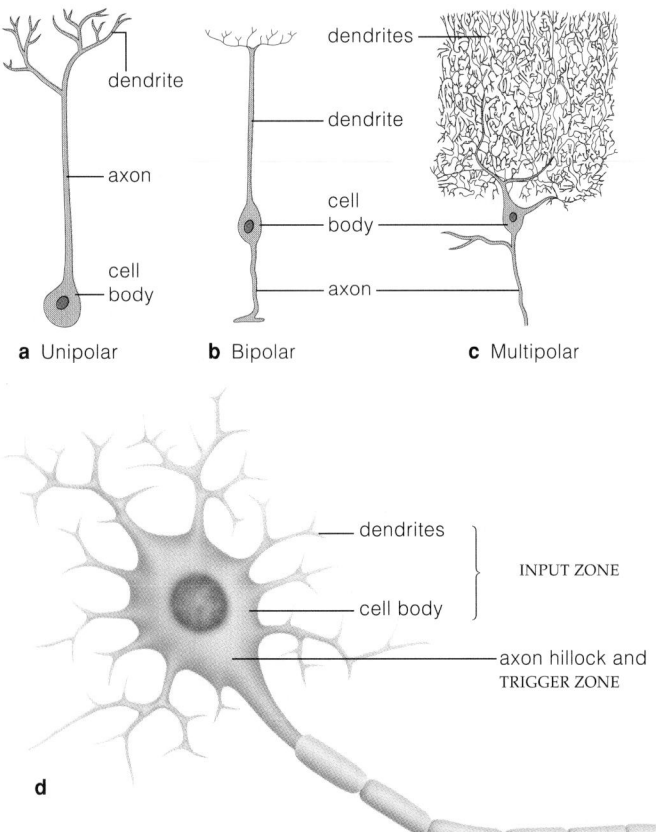

a Unipolar   **b** Bipolar   **c** Multipolar

dendrites
cell body
} INPUT ZONE

axon hillock and
TRIGGER ZONE

d

CONDUCTING ZONE

axon

axon endings

OUTPUT ZONE

Figure 10.2 (**a–c**) Classification of neurons based on the number of cytoplasmic extensions of the cell body. Unipolar cells have one extension, an axon, with dendritic branchings. Bipolar cells have one dendrite and one axon. Multipolar cells, such as the motor neuron pictured on page 219, have one axon and many dendrites. (**d**) Functional zones of a motor neuron.

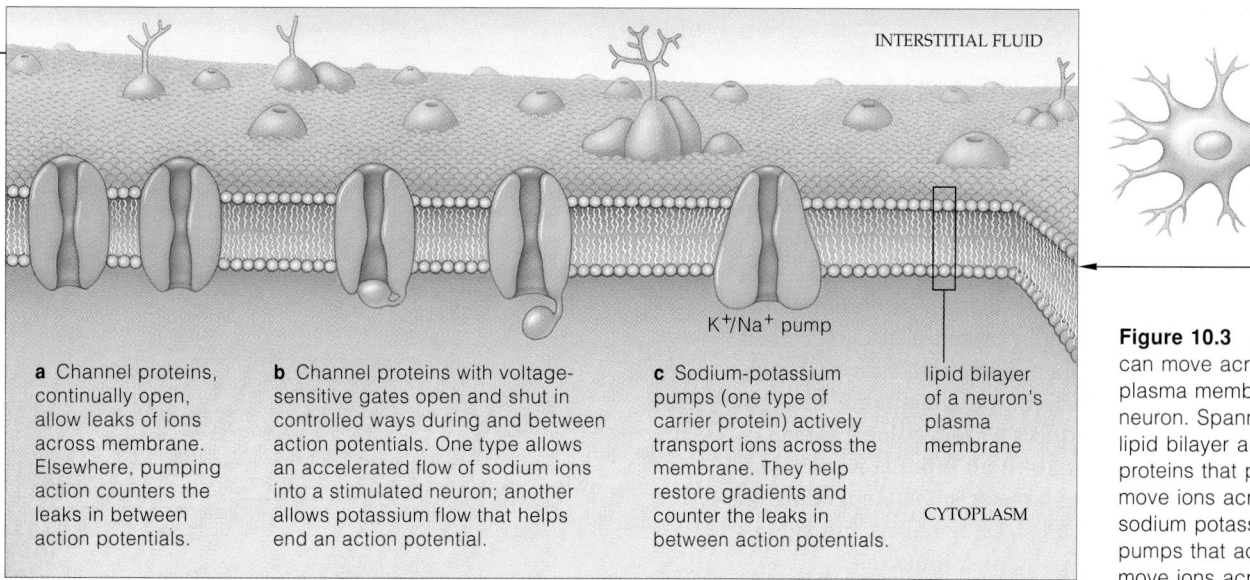

a Channel proteins, continually open, allow leaks of ions across membrane. Elsewhere, pumping action counters the leaks in between action potentials.

b Channel proteins with voltage-sensitive gates open and shut in controlled ways during and between action potentials. One type allows an accelerated flow of sodium ions into a stimulated neuron; another allows potassium flow that helps end an action potential.

c Sodium-potassium pumps (one type of carrier protein) actively transport ions across the membrane. They help restore gradients and counter the leaks in between action potentials.

lipid bilayer of a neuron's plasma membrane

CYTOPLASM

K^+/Na^+ pump

INTERSTITIAL FLUID

trigger zone

Figure 10.3 How ions can move across the plasma membrane of a neuron. Spanning the lipid bilayer are channel proteins that passively move ions across and sodium potassium pumps that actively move ions across.

Restoring and Maintaining Readiness

How does a neuron restore and maintain the voltage difference across a patch of plasma membrane in between action potentials? *First*, the membrane's lipid bilayer bars the passage of charged substances—including potassium ions (K^+) and sodium ions (Na^+). Thus the *differences* in ion concentrations can build up across the membrane. *Second*, ions can flow from one side to the other in controllable ways, through the interior of proteins that span the bilayer (Figure 10.3). Some channels are permanently open, so that ions can diffuse ("leak") through them all the time. Other channels have regions that serve as molecular gates, opening only when the neuron is adequately stimulated.

In a resting motor neuron the gated sodium channels are shut. The neuron's plasma membrane also is much more permeable to K^+ than it is to Na^+. Hence, each ion has a concentration gradient across the membrane:

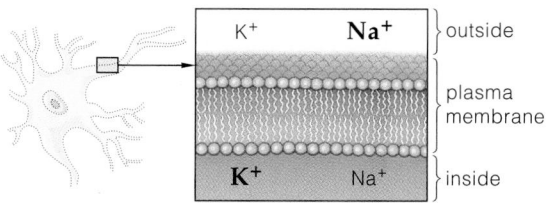

The concentration gradients determine the direction in which Na^+ and K^+ tend to diffuse across the membrane (from the larger letter to the smaller letter) through the interior of channel proteins. Sodium tends to move inward, and potassium tends to move outward. There also is an *electric* gradient across the neuron's plasma membrane. The inside of the cell is slightly negative with respect to the outside, partly because the cytoplasm contains many negatively charged proteins, and partly because K^+ can so readily move down its concentration gradient out of the neuron. Together, these factors mean that in a resting motor neuron, some Na^+ is constantly

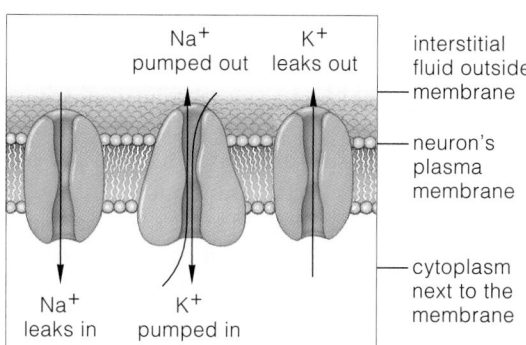

Na^+ pumped out K^+ leaks out — interstitial fluid outside membrane

— neuron's plasma membrane

— cytoplasm next to the membrane

Na^+ leaks in K^+ pumped in

Figure 10.4 Pumping and leaking processes that maintain the distribution of sodium and potassium ions across a resting neuron's plasma membrane. Notice how the total inward and outward movements for each kind of ion are balanced.

leaking *into* the cell, down its electrochemical gradient, and K^+ is leaking *out* down its concentration gradient.

The concentration and electric gradients across the neuron's plasma membrane are what enable the neuron to respond to stimulation. Yet, as you've just read, the Na^+ and K^+ leaks never stop—and one might think that the net amount of K^+ in the cell would continue to fall while the amount of Na^+ would slowly and surely increase. This potentially disastrous imbalance doesn't develop because a resting neuron expends energy on an active transport mechanism that maintains the gradients (Figure 10.4). Spanning the membrane are carrier proteins called **sodium-potassium pumps**. With energy from the cell's supply of ATP, they actively transport potassium *in* and sodium *out* at the same time.

An undisturbed neuron is maintaining a resting membrane potential—a difference in electric charge across the plasma membrane. An action potential is an abrupt, short-lived reversal in that difference. Ion pumps restore and maintain the resting membrane potential.

ACTION POTENTIALS

When a neuron's input zone is sufficiently stimulated, the "resting" ion balance across the plasma membrane is disturbed. For instance, suppose you stroke your arm and put a bit of pressure on its skin. In tissues beneath the skin surface are receptor endings—input zones of sensory neurons. Patches of plasma membrane at the receptor endings deform under pressure, causing certain types of ion channels to open. Some ions flow across, so the voltage difference across the membrane changes slightly. The pressure produces a graded, local signal.

Graded means that signals at an input zone can vary in magnitude. They can be large or small, depending on the intensity of the stimulus. *Local* means the signals don't spread far from the point of stimulation.

When a stimulus is intense enough or long-lasting, graded signals can spread out of the neuron's input zone and into an adjacent trigger zone. At this zone, a certain minimum amount of change in the voltage difference across the plasma membrane can trigger an action potential. That amount, the **threshold level**, can be reached at any patch of membrane that has voltage-sensitive, gated channels for sodium ions. Although you may barely touch your arm, the fact that you can feel it means that action potentials must have been triggered.

As Figure 10.5 shows, an appropriate stimulus causes sodium ions to flow into the neuron. With the influx of these positively charged ions, the cytoplasm side

of the membrane becomes less negative. In response, more gates open, more sodium enters, and so on. The "snowballing" inward flow of sodium is an example of positive feedback, in which an event intensifies as a result of its own occurrence:

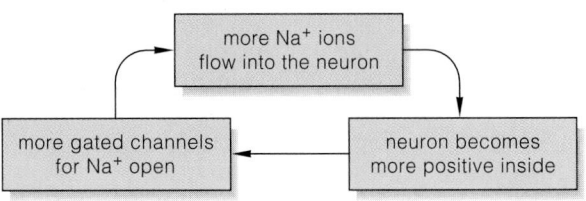

At threshold, the opening of more sodium gates no longer depends on the strength of the original stimulus. Because the positive-feedback cycle is now under way, the inward-rushing sodium itself is enough to cause more sodium gates to open.

An All-or-Nothing Spike

Figure 10.6 is a recording of the voltage difference across the neuron plasma membrane before and during an action potential. Notice how the membrane potential rapidly changes to an inside positive value once threshold is reached. Each rapid deflection is called a "spike." Action potentials in a neuron always reach the same height; that is, they are *all-or-nothing* events. Once a positive-feedback cycle starts, it continues to completion. However, if threshold is *not* reached, the membrane disturbance will subside when the stimulus is removed.

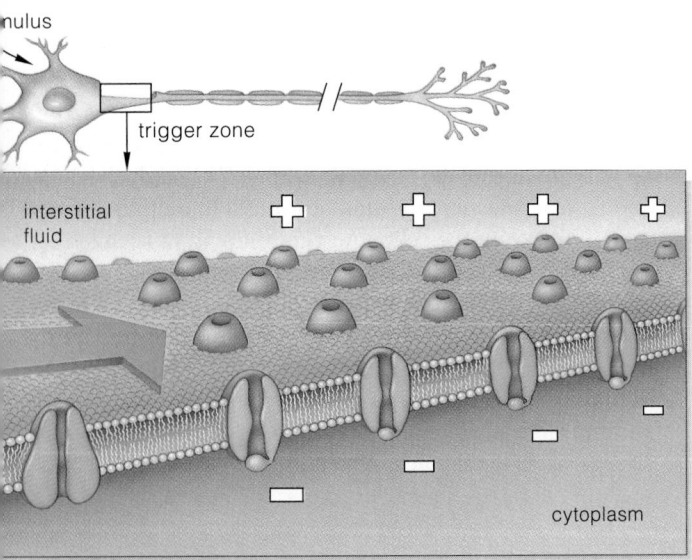

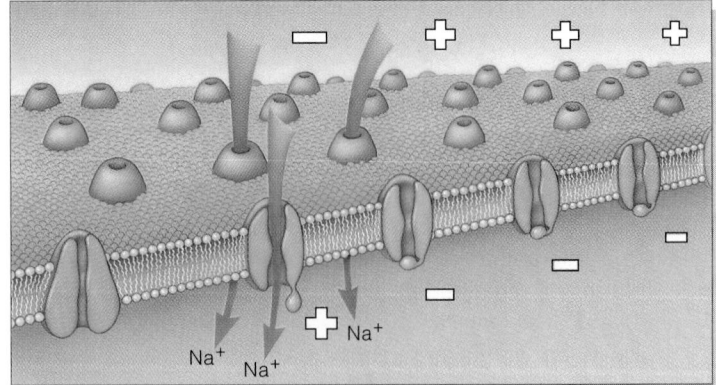

Membrane at rest (inside negative with respect to the outside). An electrical disturbance (*red* arrow) spreads from an input zone to an adjacent trigger region of the membrane, which has many gated sodium channels.

b A strong disturbance initiates an action potential. Sodium gates open, the sodium inflow decreases the negativity inside; this causes more gates to open, and so on, until threshold is reached and the voltage difference across the membrane reverses.

Figure 10.5 Propagation of an action potential along the axon of a motor neuron.

Each spike lasts for about a millisecond. At the membrane site of the charge reversal, the gated sodium channels close and shut off the sodium inflow. About halfway through the reversal, potassium channels open, so potassium ions flow out and restore the original voltage difference across the membrane. Later, sodium-potassium pumps restore the ion gradients. After the resting membrane potential has been restored, most potassium gates close and sodium gates are in their initial state, ready to be opened with the arrival of a suitable disturbance.

Propagation of Action Potentials

The membrane disturbances leading up to an action potential are self-propagating—that is, they spread by themselves—and they don't diminish in magnitude. When they spread to an adjacent membrane patch, the opening of an equivalent number of gated channels is repeated. The disturbance causes gated channels to open in the next patch, and so on. For a brief period after the disturbance ends, each membrane patch remains insensitive to stimulation because its sodium gates cannot open. That is why action potentials do not spread back into the trigger zone, but propagate *away* from it.

Cytoplasm next to the plasma membrane of a resting neuron is more negative than the interstitial fluid just outside the membrane.

During an action potential, the inside of a disturbed patch of membrane is more positive than the outside.

After an action potential, resting conditions are restored at the membrane patch.

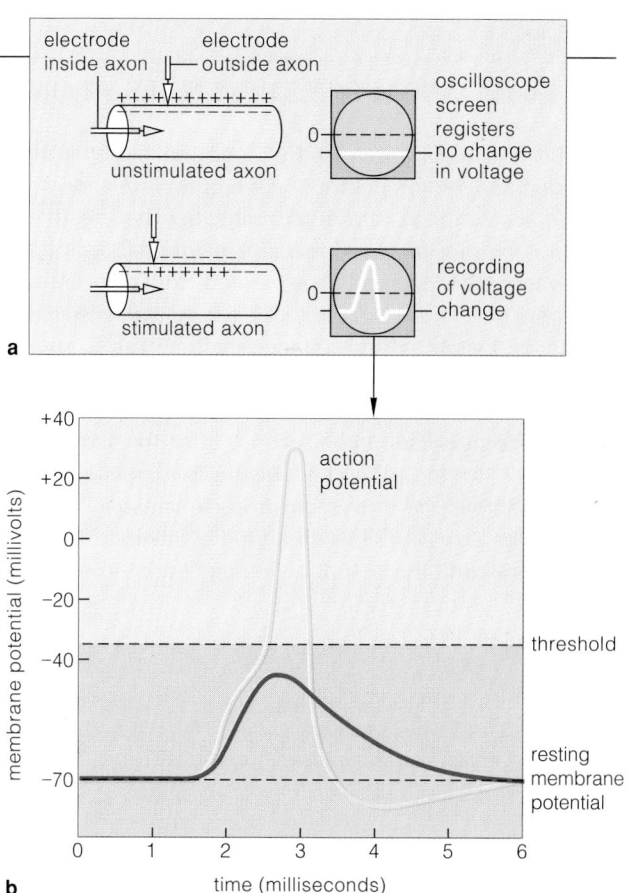

a

b

Figure 10.6 Action potentials. (**a**) Electrodes positioned inside and outside an axon can be used to measure voltage across the membrane and changes when the axon is stimulated. Changes show up as deflections in a beam of light on an oscilloscope screen. The solid *white* line is a recording of an action potential. (**b**) The *yellow* line is a typical waveform for an action potential. The *red* line represents a local signal that did not reach the threshold value, so no spiking occurred.

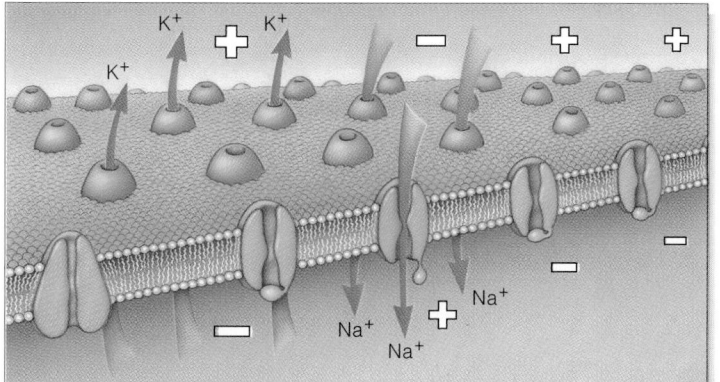

c The reversal causes sodium gates to shut and potassium gates to open (at *pink* arrows). Potassium follows its gradient (out of the neuron). Voltage is restored. The disturbance produced by the action potential triggers an action potential at the adjacent membrane site, and so on, away from the point of stimulation.

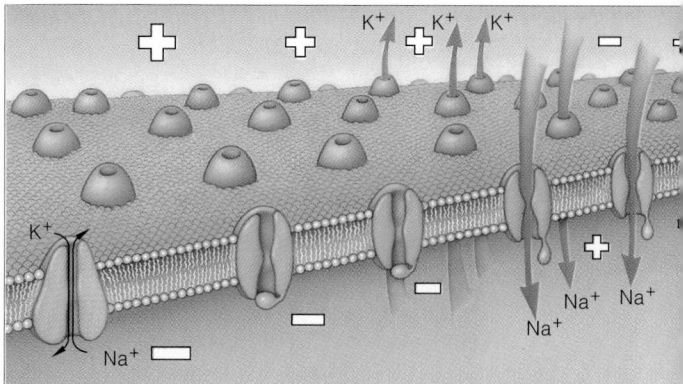

d The inside of the membrane becomes negative again following each action potential, but the sodium and potassium concentration gradients are not yet fully restored. Active transport at sodium-potassium pumps restores the gradients.

HIGH-SPEED SIGNALS ALONG SHEATHED AXONS

Action potentials arise rapidly along the axons of many sensory and motor neurons. As Figure 10.7 shows, the axons are wrapped in a **myelin sheath**. The sheath consists of the plasma membranes of neuroglial cells called **Schwann cells.** Each cell is separated from adjacent ones by a small, exposed gap (node) where the axon membrane is loaded with voltage-sensitive gated sodium channels. In a manner of speaking, the action potentials jump from node to node. (Hence biologists sometimes use the term *saltatory* conduction, after the Latin word meaning "to jump.") The sheathed regions between nodes hamper the movement of ions across the plasma membrane, so disturbances tend to flow along the plasma membrane until the next node in line—where the flow of

ions can produce a new action potential. In the largest sheathed axons, action potentials propagate themselves at a remarkable 120 meters per second. There aren't any Schwann cells in the central nervous system. There, in the brain and spinal cord, processes from other glial cells (called *oligodendrocytes*) form the sheaths of myelinated axons. Figure 10.8 gives a general idea of the differences being described.

In *multiple sclerosis*, patches of the myelin sheath around axons in the brain and spinal cord are slowly destroyed. Depending on where the destruction occurs, typical symptoms include numbness, muscle weakness or paralysis, and inability to control urination and elimination. Table 10.1 lists some other serious attacks on the nervous system.

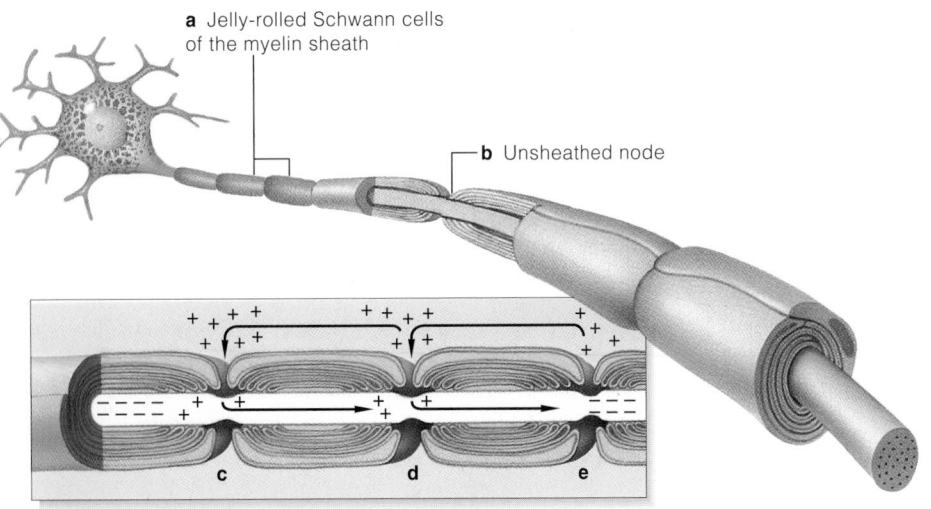

a Jelly-rolled Schwann cells of the myelin sheath

b Unsheathed node

Figure 10.7 Propagation of an action potential along a motor neuron having a myelin sheath.

(**a**) The sheath is a series of Schwann cells, each wrapped like a jelly roll around the axon. Each "jelly roll" blocks ion movements across the membrane. But ions can cross it at nodes between Schwann cells. (**b**) The unsheathed nodes have dense arrays of gated sodium channels. (**c,d**) An action potential propagates down the axon. When the potential charge reaches a node, sodium gates open, Na+ rushes in, and another action potential results. (**e**) This new disturbance spreads rapidly to the next node and triggers another action potential, and so on down the line. (**f**) An axon wrapped in a myelin sheath (cross section).

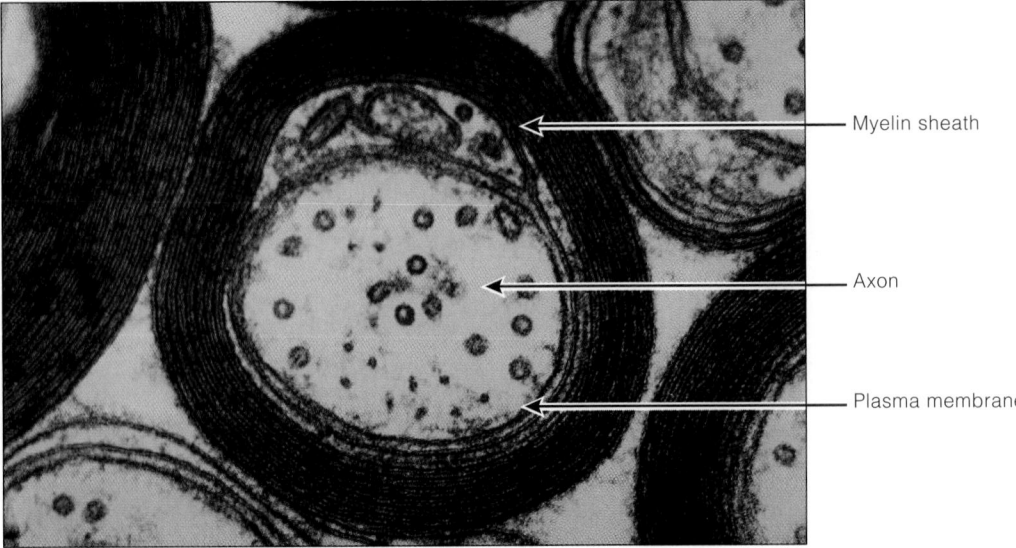

Myelin sheath

Axon

Plasma membrane

f

Although speeding the travel of action potentials is a prime function of myelin sheathing, this fatty insulation has some other benefits. For one, myelination saves the neuron energy, because a sheathed axon's ion pumps need only operate at the exposed nodes. Another vital benefit involves the repair of neuron injuries. In the peripheral nervous system, myelination helps damaged axons regenerate—for example, after an injury or surgery that severs the connections between muscles and nerves.

With the tubelike Schwann cells as a guide, axons can reestablish their links with a particular region of skeletal muscle—probably by following chemical cues to their destination.

In the central nervous system, where there are no Schwann cells, severed axons can't readily regenerate; this is why, given the current state of our knowledge, it is virtually impossible to repair damage to neurons cut in spinal injuries. However, advances in research on special communication molecules called nerve growth factors are offering significant promise in this arena—and new hope for people who suffer spinal cord injuries. You will read more about this exciting topic in Chapter 12.

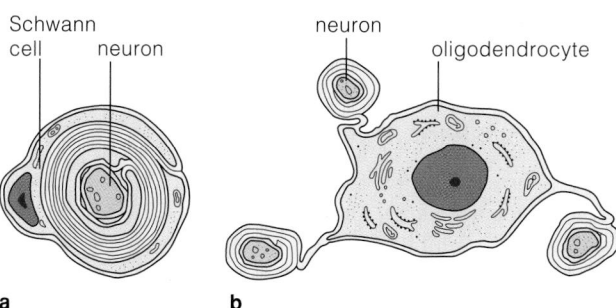

a   b

Figure 10.8 Organization of a sheathed axon of a motor neuron (**a**) in the peripheral nervous system and (**b**) in the central nervous system.

Myelin sheathing provided by Schwann cells insulates axons in the peripheral nervous system, saving cellular energy and greatly increasing the speed at which action potentials can travel along axons.

In the central nervous system, axons are sheathed not by Schwann cells but by the processes of oligodendrocytes. This difference is a key reason why severed CNS axons cannot regenerate, whereas PNS axons can.

Table 10.1 **The Nervous System under Attack**		
Cause/Risk Factors	Major Effects	Symptoms
Meningitis Viral or bacterial infection of meninges; often fatal	Inflammation of the membranes covering the brain and/or spinal cord	Severe headache, fever, stiff neck, nausea, vomiting
Encephalitis Usually, infection by a virus, such as herpes simplex; also, HIV infection	Inflammation of the brain	Fever, headache, mental confusion, memory loss, seizures, gradual loss of consciousness
Epilepsy Brain injury or infection, drug overdose, metabolic imbalance. Often, no specific cause can be identified	Abnormal electrical activity that temporarily alters one or more brain functions. Attacks are sometimes triggered by fatigue, stress, or flashing lights	Seizures of various types. *Grand mal* seizures involve loss of consciousness, jerking body movements. *Petit mal* attacks involve only momentary loss of consciousness
Multiple sclerosis Cause unknown. Possibly an autoimmune disease with viral origins. Some people may have genetic susceptibility	Progressive destruction of myelin sheaths of neurons in brain and spinal cord	Numbness, muscle weakness, fatigue, incontinence; vision, gait, and speech may also be affected. Some victims become severely disabled and bedridden
Alzheimer's disease Cause unknown. Victims have lower than normal levels of acetylcholine in brain tissues; some families show a genetic predisposition to the disorder	Progressive degeneration of neurons and shrinking of the brain. Abnormal buildup of masses of amyloid protein	Mild memory loss followed by severe short-term memory loss and disorientation; sudden mood changes; severe confusion, paranoid delusions, extreme antisocial or childlike behavior
Concussion Violent blow to the head or neck	Disruption of electrical activity of brain neurons	Brief unconsciousness, followed by mental confusion, blurred vision, vomiting
Neuralgia Physical injury or viral infection; specific cause often unknown	Irritation of or damage to a nerve	Usually, intermittent bouts of severe pain; often, a shooting pain along the affected nerve. Nerves of the face and head are commonly affected, as in *migraine headache*

CHEMICAL SYNAPSES

When action potentials reach a neuron's output zone, they usually do not go any farther. But their arrival may induce the neuron to release one or more of the signaling molecules called **neurotransmitters**. These are substances that diffuse across junctions called **chemical synapses**. The junctions are narrow gaps between one neuron's output zone and the input zone of a neighboring cell (Figure 10.9). Some chemical synapses occur between two neurons, others between a neuron and a muscle cell or gland cell.

At a chemical synapse, *one* of the two cells stores neurotransmitter molecules in synaptic vesicles in its cytoplasm. This is the *pre*synaptic cell. Gated channels for calcium ions span the cell's plasma membrane, and they open when an action potential arrives. There are more calcium ions outside the cell, and when they flow in (down their gradient), synaptic vesicles fuse with the plasma membrane. Now, neurotransmitter molecules are released into the synaptic cleft and diffuse across it. They bind with specific receptor proteins on the membrane of the *post*synaptic cell. Binding changes the shape

of these proteins, so that a channel opens up through their interior. Ions now cross the plasma membrane by diffusing through the channels (Figure 10.9d).

A postsynaptic cell's response depends on the type and concentration of neurotransmitter molecules in the cleft, what kinds of receptors the cell bears, and the number and responsiveness of gated channels in its membrane. Such factors help determine whether a neurotransmitter will have an *excitatory* effect and help drive the postsynaptic cell's membrane toward the threshold of an action potential. Or they may influence whether it has an *inhibitory* effect, driving the membrane away from the threshold.

A Smorgasbord of Signals

The neurotransmitter **acetylcholine** (ACh) is a case in point. Depending on the circumstances, ACh can have excitatory *or* inhibitory effects on cells in muscles, glands, the brain, and the spinal cord. Figure 10.9d shows a chemical synapse between a motor neuron and a muscle cell. ACh released from the motor neuron diffuses across the cleft and binds to receptors on the muscle cell membrane. It has excitatory effects on this kind of cell, triggering action potentials that initiate muscle contraction. ACh receptors at some neuromuscular junctions are destroyed in the disease *myasthenia gravis*, causing drooping eyelids, muscle weakness, and fatigue.

Serotonin is another neurotransmitter. It acts on brain cells that govern sleeping, sensory perception, regulation

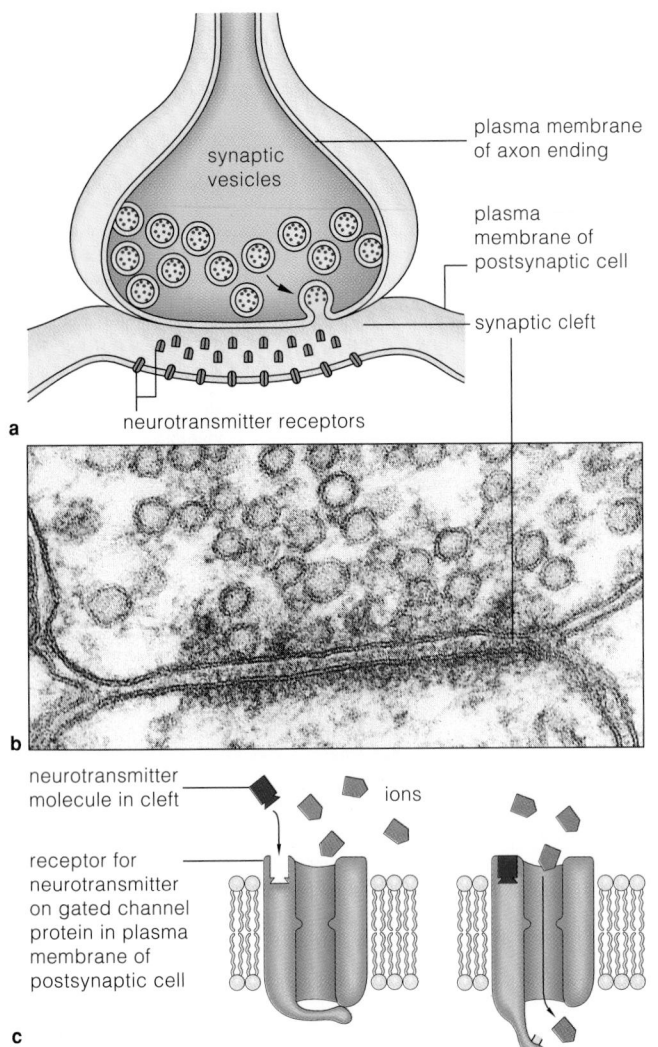

a neurotransmitter receptors

- synaptic vesicles
- plasma membrane of axon ending
- plasma membrane of postsynaptic cell
- synaptic cleft

b

c
- neurotransmitter molecule in cleft
- ions
- receptor for neurotransmitter on gated channel protein in plasma membrane of postsynaptic cell

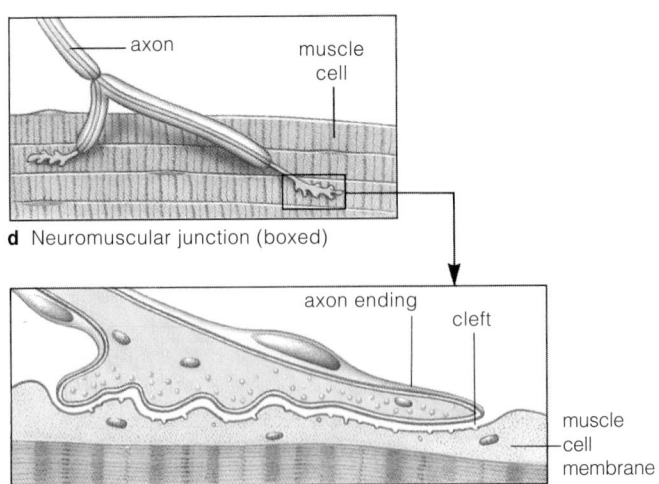

d Neuromuscular junction (boxed)
- axon
- muscle cell

e Motor end plate (troughs in muscle cell membrane)
- axon ending
- cleft
- muscle cell membrane

Figure 10.9 Chemical synapses. (**a**) A chemical synapse between two neurons. (**b,c**) A neurotransmitter carries signals from the presynaptic cell to the postsynaptic cell. (**d**) One type of chemical synapse, the neuromuscular junction, occurs. The axon's myelin sheath stops at the junction, so the membranes of the two cells are exposed. (**e**) Troughs in the muscle cell membrane where the axon endings are positioned.

of body temperature, and emotional states. *Norepinephrine* affects brain regions concerned with emotional states, dreaming, and awaking. Other neurotransmitters that act on different parts of the brain are *dopamine* and *GABA* (gamma aminobutyric acid). Appropriately stimulated sensory neurons release *Substance P,* a neurotransmitter that induces the perception of pain.

Two debilitating degenerative diseases underscore how important neurotransmitters are to normal life. *Parkinson's disease,* which involves a progressive, severe loss of voluntary muscle control, results from a lack of dopamine in certain parts of the brain. A loss or degeneration of neurons that manufacture and release acetylcholine may play a key role in *Alzheimer's disease,* with its attacks on personality and intellect.

Signaling molecules known as **neuromodulators** can magnify or reduce the effects of a neurotransmitter on neighboring or distant neurons. Neuromodulators include *endorphins*—natural painkillers that inhibit the release of Substance P from nerves. Athletes who push themselves beyond normal fatigue can experience a euphoric "high" as endorphin release increases. Endorphins also may have roles in memory, learning, and some mental disorders, and in control of body temperature and sexual behavior.

Synaptic Integration

Between 1,000 and 10,000 communication lines form synapses with a typical neuron in your brain. And your brain contains at least *100 billion* neurons, humming with messages about doing what it takes to be a human.

At any moment, many excitatory and inhibitory signals are washing over the input zones of a postsynaptic cell. Some signals drive its membrane voltage closer to threshold; others maintain the resting level or drive it away from threshold. *The signals are competing for control of the membrane potential at the trigger zone.*

All synaptic signals are graded potentials. The ones called EPSPs (for excitatory postsynaptic potentials) have a *depolarizing* effect. This simply means they bring the membrane closer to threshold. IPSPs (inhibitory postsynaptic potentials) may have a *hyperpolarizing* effect (driving the membrane away from threshold) or may help maintain the membrane at its resting level.

In the process of **synaptic integration**, competing signals that reach an input zone of a neuron at the same time are summed. Summation is the means by which signals arriving at a neuron are reinforced, suppressed, or sent onward to other cells in the body. Figure 10.10 shows what recordings of an EPSP, an IPSP, and their summation might look like.

Integration involves both spatial and temporal summation. *Spatial summation* occurs when neurotransmitter

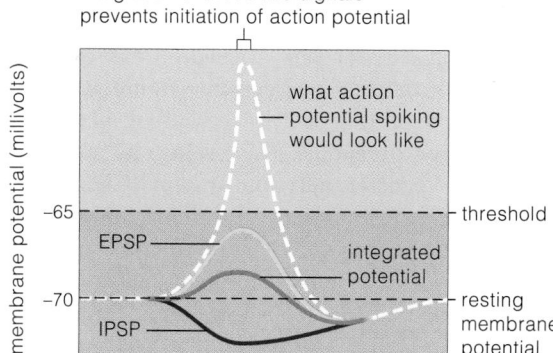

Figure 10.10 Synaptic integration. The *yellow* line shows how an EPSP of a certain magnitude would register on an oscilloscope screen *if it were acting alone.* The *purple* line shows the effect for a single IPSP. The *red* line shows their effect on a postsynaptic cell membrane when they arrive at the same time. When these two signals are integrated (the *red* line), threshold is not reached, so no action potential is initiated in the target cell.

molecules from several presynaptic cells reach the neuron's input zone at the same time. On the other hand, if the cell is excited by a rapid series of action potentials so that a neurotransmitter is released repeatedly, over a short period, the response is a *temporal summation.*

Removing Neurotransmitter from the Synaptic Cleft

The flow of signals through the nervous system depends on the rapid, controlled removal of neurotransmitter molecules from synaptic clefts. Some neurotransmitter molecules diffuse out of the cleft. Enzymes cleave others right in the cleft, as when acetylcholinesterase breaks down ACh. Also, membrane transport proteins actively pump the neurotransmitter molecules back into presynaptic cells or into neighboring neuroglia.

Some drugs block the reuptake of certain neurotransmitters. For example, cocaine blocks the uptake of dopamine, which then lingers in synaptic clefts and keeps on stimulating target cells—with the disastrous effects described in this chapter's introduction. Some antidepressant drugs (such as Prozac) alter a person's mood by blocking the reuptake of serotonin.

Neurotransmitters cross the synapse between two neurons or between a neuron and a muscle cell or gland cell. These signaling molecules may promote or inhibit activity of different kinds of receiving cells.

In synaptic integration, excitatory and inhibitory signals acting on a postsynaptic cell are combined (summed). In this way messages traveling through the nervous system can be reinforced or downplayed, sent onward or suppressed.

THE NERVOUS SYSTEM: AN OVERVIEW

Thus far our discussion has focused on the signals that travel through the nervous system. Figure 10.11a now gives you an overview of how the system is organized into two major divisions. The brain and spinal cord make up the **central nervous system** (CNS); all the nerves that carry signals to and from the brain and spinal cord make up the **peripheral nervous system** (PNS). A **nerve** is a cablelike bundle of neuron axons, as you can see in Figure 10.12. Most axons are bundled in parallel within connective tissue, and each axon has a

myelin sheath, which enhances the rate at which action potentials propagate.

The peripheral nervous system consists of 31 pairs of spinal nerves that connect with the spinal cord, and 12 pairs of cranial nerves that connect directly with the brain (Figure 10.11b). At some sites in the peripheral nervous system, cell bodies of several neurons occur in clusters called *ganglia* (singular: ganglion). Both the CNS and the PNS also have neuroglial cells, such as the oligodendrocytes (CNS) and Schwann cells (PNS) previously described.

Figure 10.11 (**a**) Divisions of the human nervous system. (**b**) View of the nervous system showing the brain, spinal cord, and some major peripheral nerves. Twelve pairs of cranial nerves extend from different regions of the brain stem.

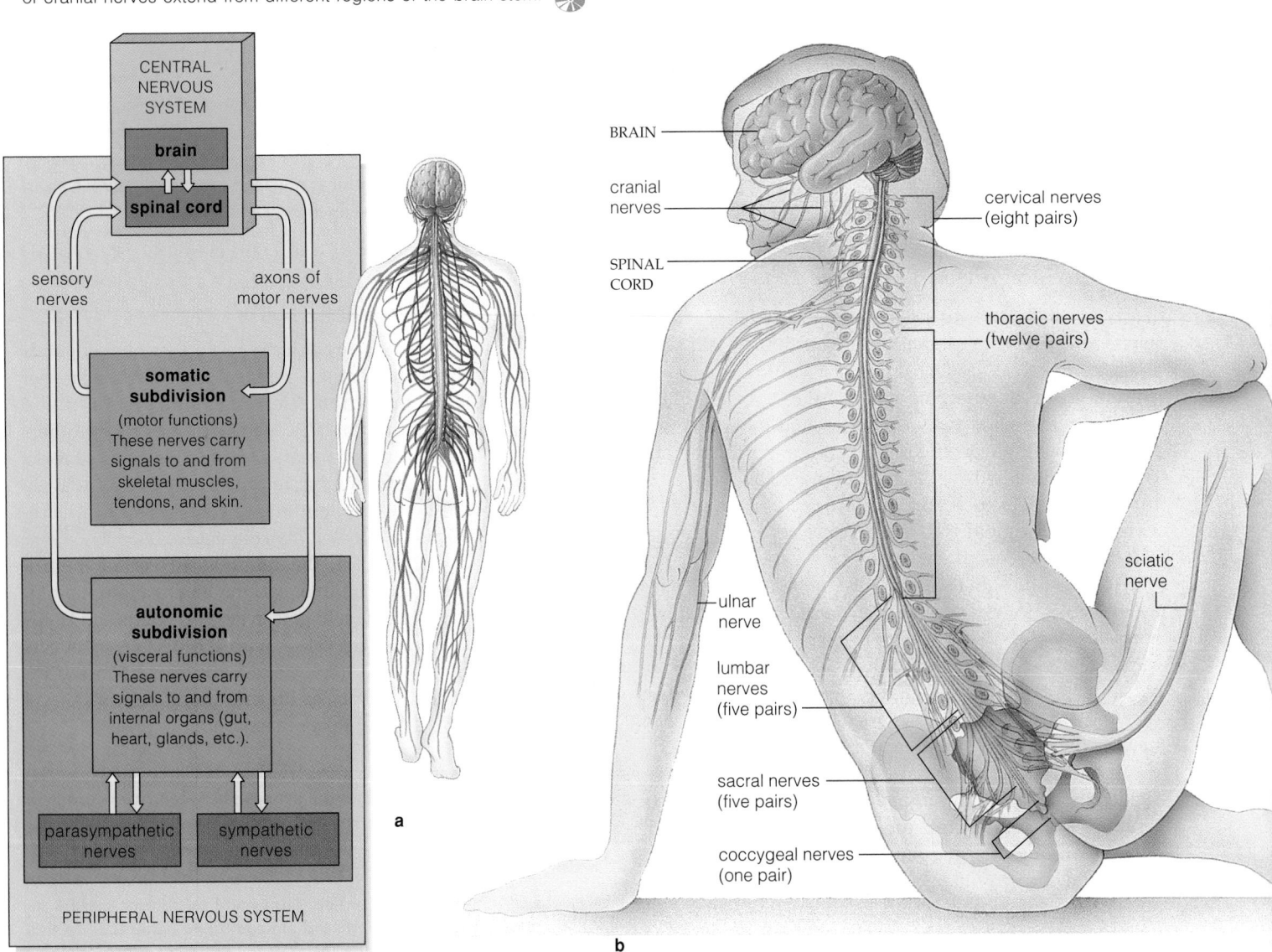

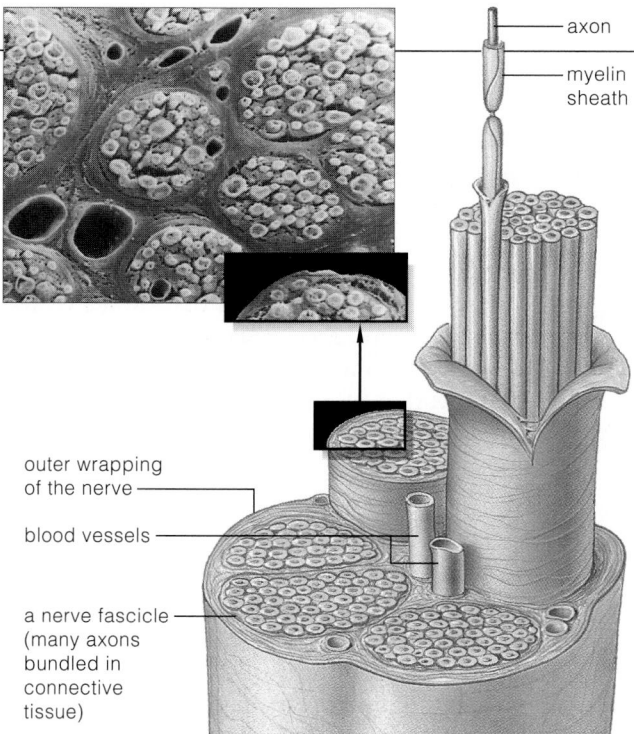

axon

myelin sheath

outer wrapping of the nerve

blood vessels

a nerve fascicle (many axons bundled in connective tissue)

Figure 10.12 Structure of a nerve. Axons in the nerve are bundled together inside wrappings of connective tissue.

General Paths of Information Flow

Through synaptic integration, signals arriving at any neuron in the body can be reinforced or dampened, sent on or suppressed. The direction in which a given signal will travel depends on the organization of neurons in different body regions (Figure 10.13).

For example, your brain deals with many of its 100 billion neurons in a manner analogous to block parties.

Regional blocks of hundreds or thousands of neurons receive excitatory and inhibitory signals. They integrate the signals entering the block, then send out signals in response. In some regions the neurons are organized as *divergent* circuits, as when their processes fan out from one block and form connections with many others. In different regions, neurons form *convergent* circuits, with signals from many sent on to just a few. In still other regions, neurons synapse back on themselves, repeating signals among themselves over and over, in *reverberating* circuits. Such neurons include those that make your eye muscles rhythmically twitch as you sleep.

In nerves, long axons of many sensory neurons, motor neurons, or both permit long-distance communication between the brain, the spinal cord, and the rest of the body. A *nerve tract* is a bundle of axons of interneurons in the spinal cord and brain.

The sensory and motor neurons of many nerves take part in **reflexes**—simple, stereotyped movements in response to specific sensory information. In the simplest **reflex arc**, sensory neurons synapse directly on motor neurons. They are common in the peripheral nervous system, our topic in Section 10.6.

The nervous system is subdivided into the central nervous system and the peripheral nervous system. The CNS consists of the brain and spinal cord, and the PNS consists of nerves that thread through the rest of the body and carry signals into and out of the central region.

Figure 10.13 Information flow through the nervous system. Sensory nerves relay information *into* the spinal cord and brain, where their neurons synapse with interneurons in the spinal cord and brain, which integrate the signals. Interneurons also relay information to motor neurons, which carry signals *away* from the spinal cord and brain.

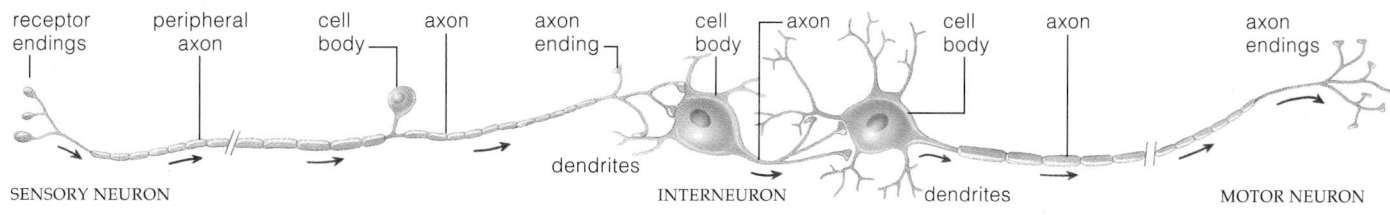

receptor endings — peripheral axon — cell body — axon — axon ending — cell body — axon — cell body — axon — axon endings

dendrites

SENSORY NEURON — INTERNEURON — dendrites — MOTOR NEURON

THE PERIPHERAL NERVOUS SYSTEM

Somatic and Autonomic Subdivisions

As Figure 10.14 indicates, cranial and spinal nerves can be further classified according to function. The ones that help move your head, trunk, and limbs are **somatic nerves**. The sensory axons of these nerves deliver information from receptors in the skin, skeletal muscles, and tendons into the central nervous system. Their motor axons deliver commands from the brain and spinal cord to the body's skeletal muscles.

Sometimes the nerves carrying sensory input to the central nervous system are said to be "afferent," a word meaning to bring to. Nerves carrying motor output away from the central nervous system to muscles and glands are "efferent," meaning to carry outward. The reflex pathway shown in Figure 10.15 is a simple example of how somatic nerves work.

By contrast, spinal and cranial nerves dealing with smooth muscle, cardiac (heart) muscle, and gland cells are **autonomic nerves**. They deal with internal organs and structures.

Unlike somatic neurons, single autonomic neurons don't extend the entire distance between muscles or glands and the central nervous system. Instead, *preganglionic* ("before a ganglion") *neurons* have cell bodies within the spinal cord or brain stem, and their axons travel through nerves to autonomic system ganglia outside the central nervous system. There, the axons synapse with *postganglionic* ("after a ganglion") *neurons*. Axons of this second set of neurons make the actual connection with the body's muscles and glands (effectors).

Sympathetic and Parasympathetic Nerves

There are two categories of autonomic nerves, called parasympathetic and sympathetic. They normally work antagonistically, with the signals from one opposing those of the other. Both types carry excitatory and inhibitory signals to internal organs. Often their signals arrive at the same time at muscle cells or gland cells and so compete for control over them. In such cases,

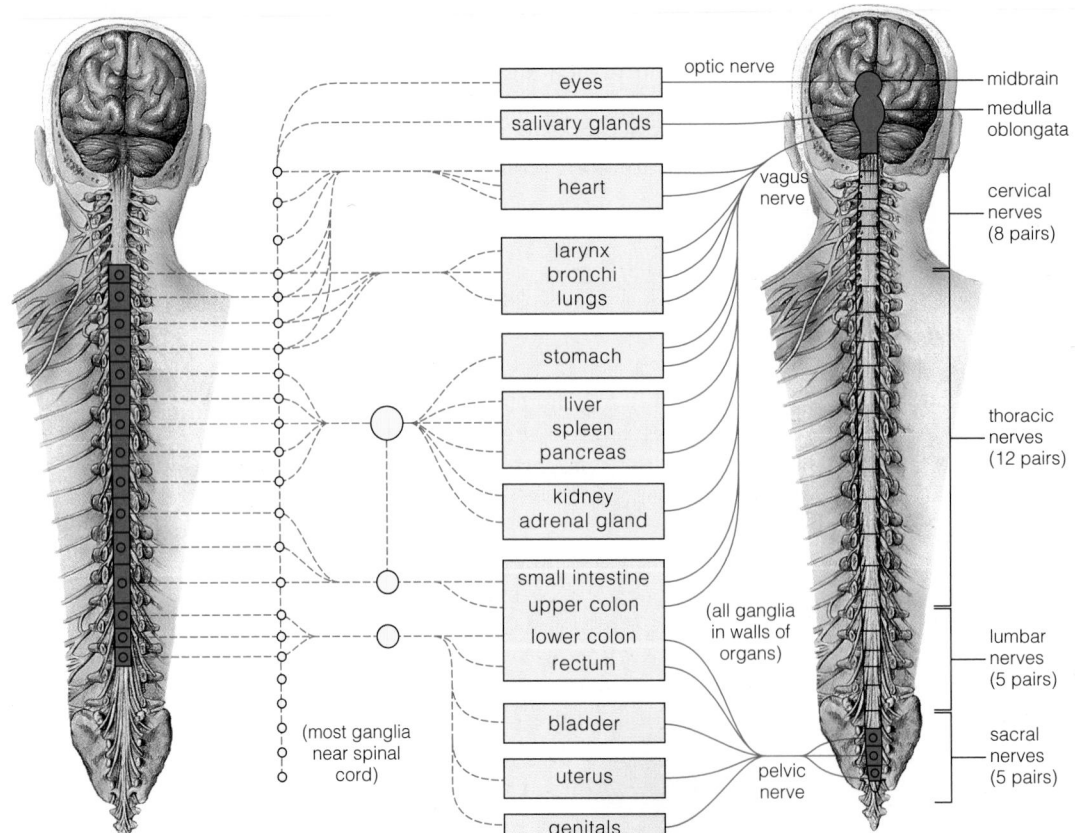

SYMPATHETIC OUTFLOW:

Examples of effects:
Dilation of pupil in eye
Inhibits bronchial glandular secretion
Stimulates thick salivary secretions
Inhibits stomach, intestinal movements
Contracts sphincters

PARASYMPATHETIC OUTFLOW:

Examples of effects:
Contraction of pupillary muscles
Stimulates bronchial glandular secretion
Stimulates watery salivary secretions
Stimulates stomach, intestinal movements
Relaxes sphincters

Figure 10.14 Autonomic nervous system. This is a diagram of the major sympathetic and parasympathetic nerves leading out from the central nervous system to some major organs. Remember, there are *pairs* of both kinds of nerves, servicing the right and left halves of the body. The ganglia are simply clusters of cell bodies of the neurons that are bundled together in nerves.

eyes
salivary glands
heart
larynx bronchi lungs
stomach
liver spleen pancreas
kidney adrenal gland
small intestine upper colon
lower colon rectum
bladder
uterus
genitals

optic nerve
vagus nerve
(most ganglia near spinal cord)
(all ganglia in walls of organs)
pelvic nerve

midbrain
medulla oblongata
cervical nerves (8 pairs)
thoracic nerves (12 pairs)
lumbar nerves (5 pairs)
sacral nerves (5 pairs)

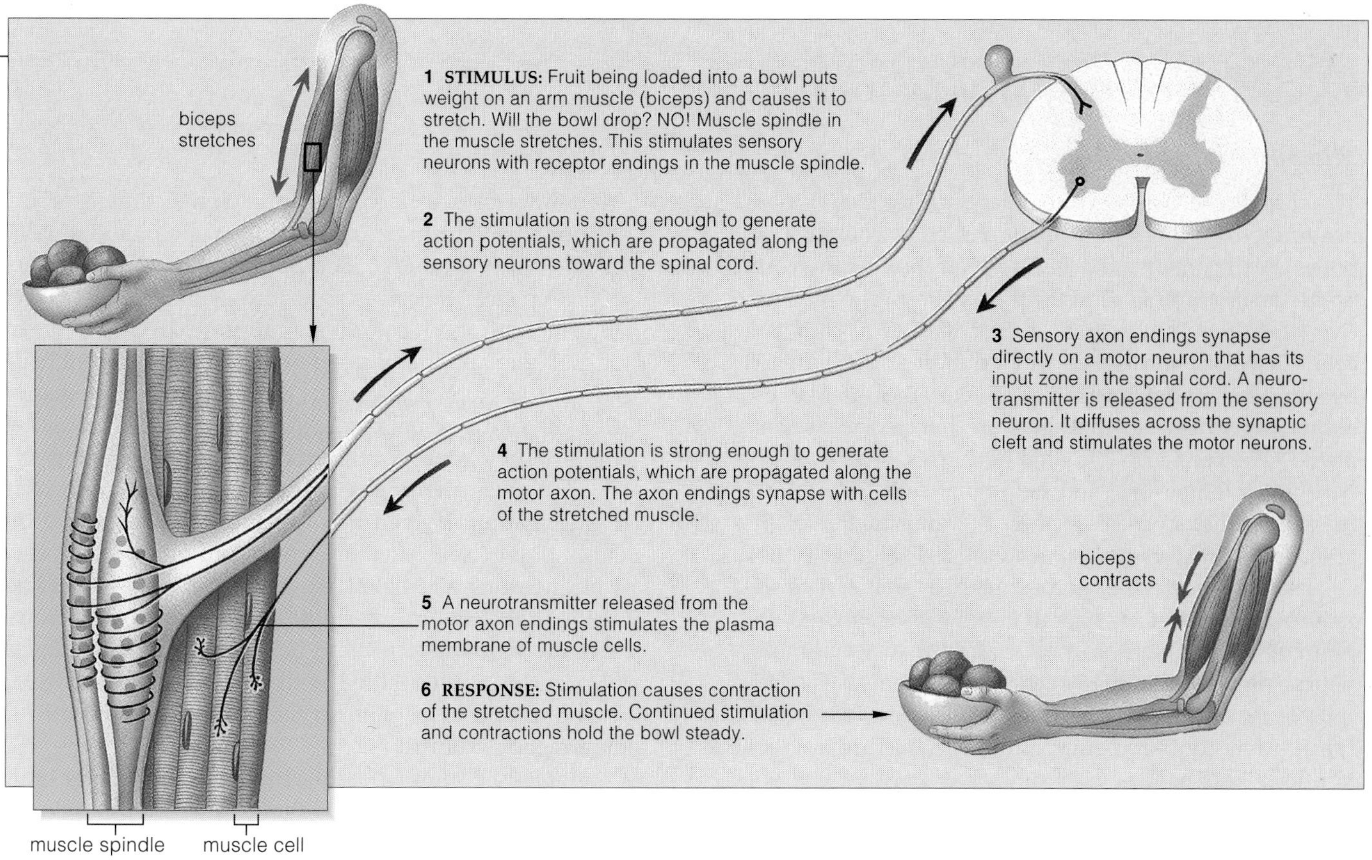

1 STIMULUS: Fruit being loaded into a bowl puts weight on an arm muscle (biceps) and causes it to stretch. Will the bowl drop? NO! Muscle spindle in the muscle stretches. This stimulates sensory neurons with receptor endings in the muscle spindle.

biceps stretches

2 The stimulation is strong enough to generate action potentials, which are propagated along the sensory neurons toward the spinal cord.

3 Sensory axon endings synapse directly on a motor neuron that has its input zone in the spinal cord. A neurotransmitter is released from the sensory neuron. It diffuses across the synaptic cleft and stimulates the motor neurons.

4 The stimulation is strong enough to generate action potentials, which are propagated along the motor axon. The axon endings synapse with cells of the stretched muscle.

5 A neurotransmitter released from the motor axon endings stimulates the plasma membrane of muscle cells.

6 RESPONSE: Stimulation causes contraction of the stretched muscle. Continued stimulation and contractions hold the bowl steady.

biceps contracts

muscle spindle   muscle cell

Figure 10.15 **(1–6)** Organization of nerves in a reflex arc dealing with muscle stretching. Inside a skeletal muscle, stretch-sensitive receptors of a sensory neuron are located in muscle spindles. These are sensory organs in which small, specialized cells are enclosed in a sheath that runs parallel with the muscle. The receptor endings are input zones of sensory neurons the axons of which synapse with motor neurons in the spinal cord. Axons of the motor neurons lead back to the stretched muscle, and action potentials that reach the axon endings trigger the release of ACh, which initiates contraction. Action potentials are generated when stretching—here, in a biceps muscle, caused by fruit being loaded into a bowl held in the hand—stimulates the receptors. Continued receptor activity excites the motor neurons further, allowing them to maintain the hand's position.

synaptic integration at the cellular level leads to minor adjustments in the organ level of activity.

Parasympathetic nerves dominate when the body is not receiving much outside stimulation. They tend to slow down the body overall and divert energy to basic "housekeeping" tasks, such as digestion.

Sympathetic nerves dominate during heightened awareness, excitement, or danger. They tend to shelve housekeeping tasks. At the same time, they prepare the animal to fight or escape (when threatened) or to frolic (as in play and sexual behavior).

As you read this, sympathetic nerves are commanding your heart to beat a bit faster, and parasympathetic nerves are commanding it to beat a bit slower. Integration of these opposing signals influences the heart rate. If something scares or excites you, parasympathetic input to your heart drops. Sympathetic nerves cause the release of norepinephrine, a signaling molecule that makes your heart beat faster and makes you breathe

faster and sweat. In this state of intense arousal, you are primed to fight (or play) hard or to get away fast. Hence the term **fight-flight response**.

When the stimulus for the fight-flight response stops, sympathetic activity may decrease and parasympathetic activity may rise. You might see this "rebound effect" after someone has been instantly mobilized to rush onto a highway to save a child from an oncoming car. The person may well collapse as soon as the child has been swept out of danger.

The peripheral nervous system consists of the communication lines to and from the brain and spinal cord.

Its somatic nerves deal with skeletal muscle movements. Its autonomic nerves deal with the functions of internal organs, such as the heart and glands.

The Spinal Cord

The **spinal cord** (Figure 10.16a) lies within a closed channel formed by the bones of the vertebral column. The bones and ligaments attached to them protect the cord. So do the *meninges*, a series of three coverings of connective tissue that are layered around the spinal cord and brain. The cord consists partly of bundles of myelinated axons called **nerve tracts**. The myelin sheaths of these axons are white, and the tracts are referred to as *white matter*. The cord also contains dendrites, cell bodies of neurons, interneurons, and neuroglial cells. These form its *gray matter*. In cross section, the gray matter of the spinal cord looks vaguely like a butterfly (Figure 10.16b).

The spinal cord is a vital expressway that carries signals between the peripheral nervous system and the brain. It is also a reflex control center. Sensory and motor neurons involved in many reflex movements of skeletal muscle make direct connections in the spinal cord; such *spinal reflexes* do not require input from the brain. When you jerk your hand away from a hot stove burner, you are experiencing a spinal reflex in action. Higher brain centers also receive information about the sensory stimulus, and so you become aware of "hot burner!" even as your hand is moving away from it. The spinal cord also contributes to some *autonomic reflexes*, which deal with internal organ functions (such as bladder emptying).

The Brain

Your impressively large, intricate **brain** contains about 100 billion neurons and probably weighs about 1,300 grams (three pounds). The brain is the body's master control panel. It receives, integrates, stores, and retrieves information, and it coordinates appropriate responses by stimulating and inhibiting the activities of different body parts. The brain starts out as a continuation of the anterior end of the spinal cord. And, like the spinal cord, it is protected by bones (of the cranium) and by meninges.

The three **meninges** are membranes of connective tissue that are layered between the skull bones and the brain tissue itself. Meninges cover and protect the fragile CNS neurons and blood vessels that service the tissue. Folds in the tough, outermost one (the *dura mater*) separate the right and left brain hemispheres. The meninges also enclose spaces filled with *cerebrospinal fluid*, which cushions and helps nourish the brain. We will return to its functions shortly.

Table 10.2 summarizes the functions of the three brain divisions: the hindbrain, midbrain, and forebrain. In the hindbrain and midbrain, the most ancient nervous tissue still contains centers that control basic reflexes, and we call it the **brain stem**.

The **hindbrain** consists of the medulla oblongata, the cerebellum, and the pons. The **medulla oblongata** houses

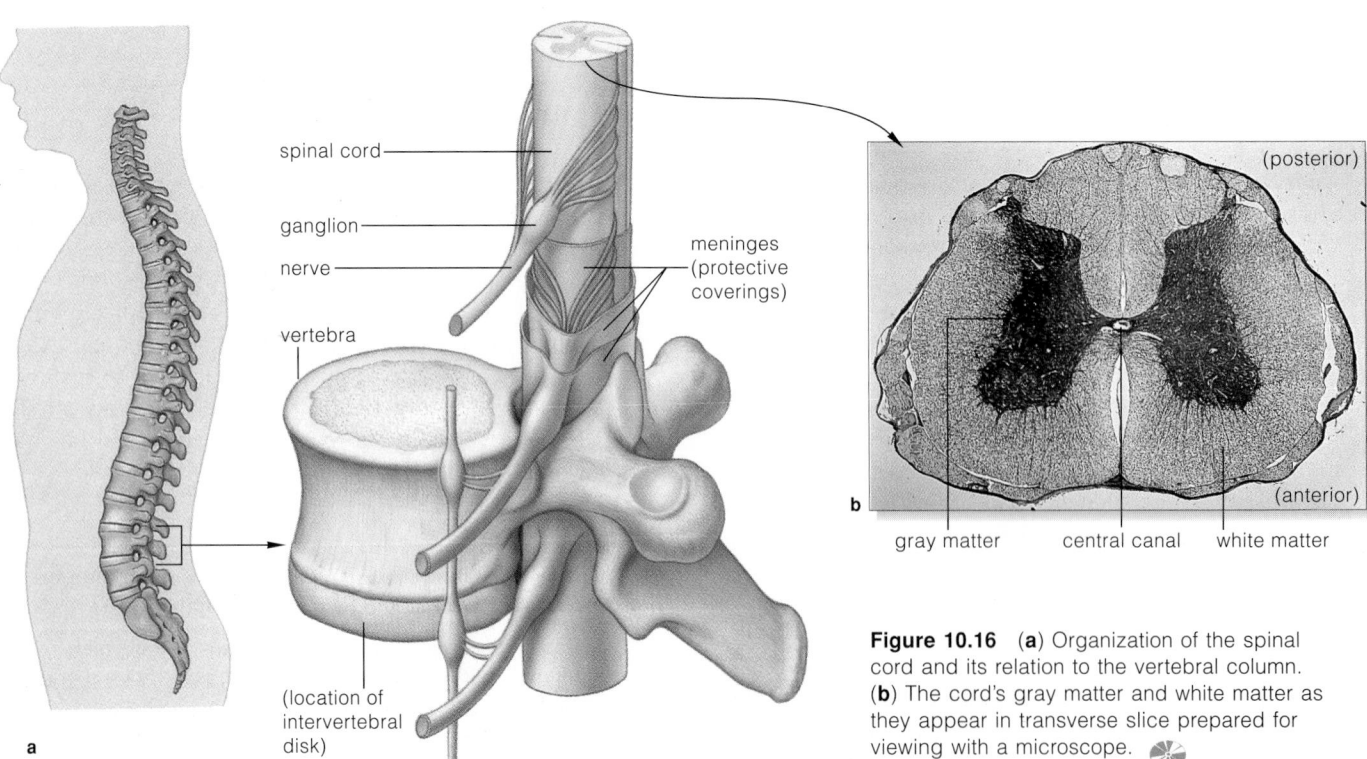

Figure 10.16   (a) Organization of the spinal cord and its relation to the vertebral column. (b) The cord's gray matter and white matter as they appear in transverse slice prepared for viewing with a microscope.

Table 10.2 Regions of the Brain

Division	Main Parts	Typical Functions
FOREBRAIN	Cerebrum	In two cerebral hemispheres, centers for coordinating sensory and motor functions, for memory, and for abstract thought. Most complex coordinating center; intersensory association, memory circuits
	Olfactory lobes	Relaying of sensory input from the nose to olfactory structures of cerebrum
	Thalamus	Major coordinating center for sensory signals; relay station for most sensory impulses to cerebrum
	Hypothalamus	Neural-endocrine coordination of visceral activities (e.g., water-solute balance, temperature control, carbohydrate metabolism)
	Limbic system	Via interacting scattered brain centers (including hypothalamus), coordination of skeletal muscle and internal organ activity underlying emotional expression
	Pituitary gland	"Master" endocrine gland (controlled by hypothalamus). Control of growth, metabolism, etc.
	Pineal gland	Control of some daily (circadian) rhythms
MIDBRAIN	Tectum	Largely reflex coordination of visual, tactile, auditory input; contains nerve tracts ascending to thalamus, descending from cerebrum
HINDBRAIN	Pons	"Bridge" of nerve tracts from cerebrum to both sides of cerebellum. Also contains longitudinal tracts connecting forebrain and spinal cord
	Cerebellum	Unconscious coordination of motor activity underlying limb movements, maintaining posture, spatial orientation
	Medulla oblongata	Contains tracts extending between pons and spinal cord; reflex centers involved in respiration, cardiovascular function, etc.

reflex centers for respiration, circulation, and other vital tasks. It coordinates movements and complex reflexes (such as coughing). Also, signals for voluntary muscle movements travel along nerve tracts that pass from higher brain centers via the medulla to the spinal cord. The medulla also influences brain centers that help govern waking and sleeping.

The **cerebellum** integrates signals from the eyes, inner ears, and muscle spindles with motor commands from the forebrain to coordinate movement and balance. It keeps higher brain centers informed about how the body trunk and limbs are positioned, how much different muscles are contracted or relaxed, and in which direction the body or limbs are moving. Activities of the cerebellum may also be crucial in human language and other forms of mental dexterity.

Bands of axons extend from the cerebellum into the **pons** ("bridge"). Although lodged in the brain stem, the pons directs the signal traffic between the cerebellum and the forebrain's higher integrating centers.

The **midbrain** coordinates reflex responses to sights and sounds. It has a roof of gray matter, the *tectum*, where visual and auditory sensory input converges before being sent on to higher brain centers.

By now, you may be thinking that the brain is tidily subdivided into three main regions. This is not the case. An evolutionarily ancient mesh of interneurons still extends from the uppermost part of the spinal cord, on through the brain stem, and on into higher integrative centers of the cerebral cortex. This major network of interneurons is the **reticular formation**. It operates as a low-level pathway to motor centers of the medulla oblongata and spinal cord. Through these links the reticular formation helps govern muscle activity associated with maintaining balance, posture, and muscle tone. It also can activate centers in the cerebral cortex and thereby help govern activities of the whole nervous system.

The **forebrain** is the most highly developed region of the human brain. Its parts include the **cerebrum**, where information processing occurs and sensory input and motor responses are integrated; a pair of olfactory lobes that deal with the sense of smell; and the **thalamus** and **hypothalamus**. The thalamus mainly serves as a relay switchboard for sensory information. Within it, signals arriving via sensory nerve tracts are "projected" onto clusters of neuron cell bodies (called *nuclei*) and are relayed onward to the cerebrum or elsewhere. The nuclei also process some outgoing motor information. The hypothalamus monitors internal organs and influences behavior related to their activities, such as thirst, hunger, and sex.

The spinal cord is a vital expressway for signals between the peripheral nerves and the brain. Some of its interneurons also exert direct control over certain reflex pathways.

The brain is divided into a hindbrain, midbrain, and forebrain. The forebrain includes the cerebrum, where information processing occurs and sensory inputs are integrated with motor responses. Neuron groups in the medulla oblongata of the hindbrain are associated with reflexes governing basic functions such as respiration and heart rate. The midbrain functions mainly in coordination of sensory inputs.

OTHER ASPECTS OF CNS STRUCTURE

The Cerebral Hemispheres

The human cerebrum vaguely resembles the much-folded nut inside a walnut shell. A deep fissure divides the human cerebrum into two parts, the left and right *cerebral hemispheres* (Figure 10.17a). Other fissures and folds in each hemisphere follow certain patterns and divide it into four main lobes whose roles we examine in Section 10.10.

Much of the gray matter of the cerebral hemispheres is arranged as a thin surface layer, the **cerebral cortex**. The cerebral cortex weighs about a pound, and if you were to stretch it flat, it would cover a surface area of two and a half square feet. The white matter consists of major nerve tracts that keep the hemispheres in communication with each other and with the rest of the body. Some of the tracts originate in the brain stem, where they are aligned close together, then fan out in the cerebral hemispheres. Each hemisphere has its own set of tracts that serve as communication lines among its different regions. A prominent tract, the corpus callosum (Figure 10.17b), runs between the two hemispheres and keeps them in communication with each other.

Many experiments have focused on the functioning of the cerebral hemispheres. We know, for example, that each hemisphere can function separately. However, the left cerebral hemisphere responds primarily to signals from the right side of the body. The opposite is true for the right cerebral hemisphere. Signals that travel by way of the corpus callosum coordinate the functioning of the two hemispheres.

The main regions responsible for spoken language skills generally reside in the left hemisphere. The main regions responsible for nonverbal skills such as music, mathematics, and other abstract abilities generally reside in the right hemisphere.

Brain Cavities and Canals

To get an idea of how delicate nervous tissue is, try holding a jiggling blob of Jell-O in your hands. The brain and spinal cord would be highly vulnerable to damage if they were not protected by bones and meninges. In addition, both are surrounded by **cerebrospinal fluid.** This clear extracellular fluid helps cushion the brain and spinal cord from sudden, jarring movements. It is secreted from specialized capillaries within four ventricles, which are part of a continuous system of fluid-filled cavities and canals. The ventricles connect with one another and with the central canal of the spinal cord. Cerebrospinal fluid fills the ventricles, and it also fills the space between the innermost layer of the meninges and the brain itself (Figure 10.18).

The bloodstream exchanges substances with the extracellular fluid, which in turn exchanges substances with neurons. However, the structure of brain capillaries makes them relatively impermeable to many substances.

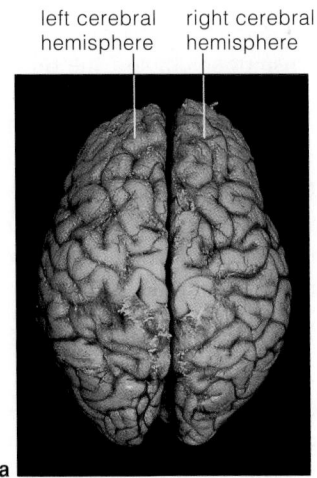

left cerebral hemisphere · right cerebral hemisphere

a

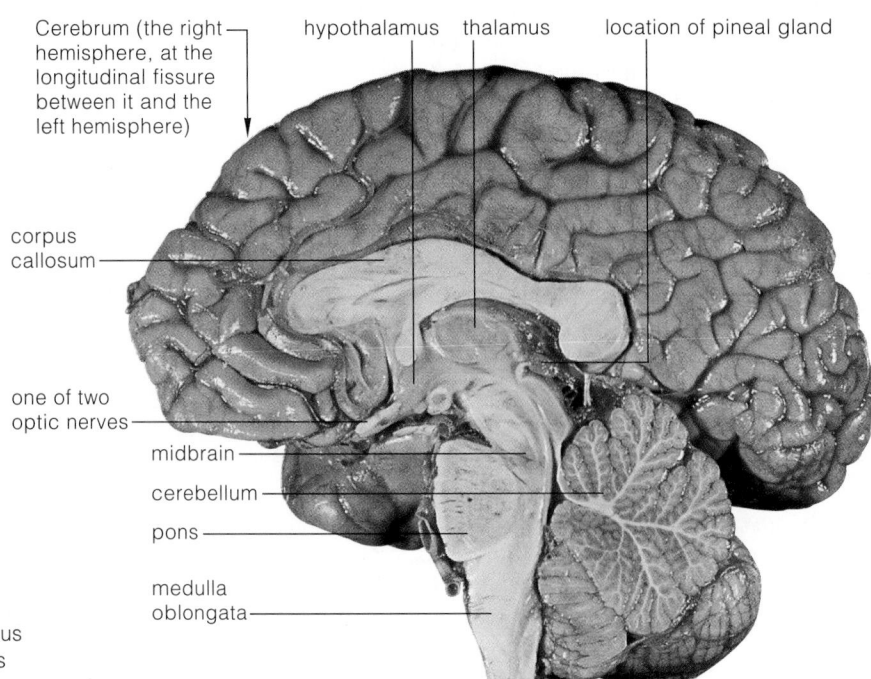

Cerebrum (the right hemisphere, at the longitudinal fissure between it and the left hemisphere)

hypothalamus · thalamus · location of pineal gland

corpus callosum

one of two optic nerves

midbrain

cerebellum

pons

medulla oblongata

b

Figure 10.17 Human brain: (**a**) Looking down on the longitudinal fissure between its two cerebral hemispheres, and (**b**) in sagittal section. The corpus callosum, a major transverse nerve tract, connects the two hemispheres.

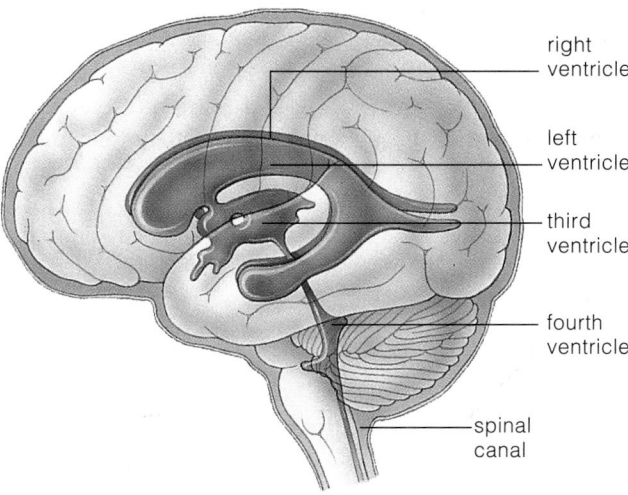

Figure 10.18 Location of the cerebrospinal fluid in the human brain. This extracellular fluid surrounds and cushions the brain and spinal cord. It also fills the four interconnected cavities (cerebral ventricles) within the brain and the central canal of the spinal cord.

This **blood-brain barrier** helps control *which* blood-borne substances are allowed to reach the neurons. Endothelial cells making up the walls of the capillaries are fused together by continuous tight junctions (refer to Figure 3.4). This means that substances must pass *through* the cells, rather than between them, to reach the brain. Transport proteins embedded in the plasma membrane of those cells selectively transport glucose and other water-soluble substances across the barrier. In addition, lipid-soluble substances quickly diffuse through the lipid bilayer of the plasma membrane. This "loophole" is one reason why caffeine, nicotine, alcohol, barbiturates, heroin, and other lipid-soluble drugs have such rapid effects on brain function. Certain environmental pollutants, such as the heavy metal mercury, are lipid-soluble and so can cross the blood-brain barrier. As the *Focus* essay explains, this fact has sobering implications for human health.

The brain's highest integrating centers reside in the gray matter of the right and left cerebral hemispheres. Typically, each hemisphere responds to signals from the opposite side of the body.

Cerebrospinal fluid surrounds and helps cushion the brain and spinal cord. The blood-brain barrier helps control which substances enter cerebrospinal fluid from the blood, and hence which blood-borne substances come into contact with neurons.

AN ENVIRONMENTAL ASSAULT ON THE NERVOUS SYSTEM

Have you ever heard someone called "mad as a hatter"? The phrase originated a century ago in Great Britain, when hat makers used the heavy metal mercury to process felt. Traces of that substance entered the bloodstream, crossed the blood-brain barrier, and caused irreparable neurological damage. In an adult, symptoms of mercury poisoning include muscle tremors, impaired vision, difficulty walking and talking, and emotional problems. Victims were often believed to be mentally deranged.

Industrial mercury compounds may be dumped or washed from the atmosphere into waterways, lakes, and marshlands. There, metabolic processes in bacteria convert such compounds to methyl mercury, which dissolves in fatty tissues in fish and other animals. Consumed in food (especially fish) by a pregnant woman, methyl mercury becomes concentrated in adipose (fat) cells and breast milk—and also wreaks havoc on the developing fetus. In Minimata, Japan, in the 1960s, more than 200 women living near a mercury-discharging factory gave birth to mentally retarded, physically deformed babies. In the late 1980s, authorities in Michigan, Minnesota, and Wisconsin warned nursing mothers and women who were even *considering* pregnancy not to eat *any* fish from many lakes in the region. Those advisories are still in force. In fact, the U.S. Environmental Protection Agency estimates that mercury contamination in fish is rising by 3–5 percent per year. Ocean fish have also been found to carry large amounts of mercury in their tissues.

From a health standpoint, consumers are well advised to limit their intake of fish taken from waters where mercury pollution may be a problem. Pregnant women or those considering pregnancy should be especially careful. Citizens also can support action on the part of government, industry, and individuals to minimize the entry of mercury-containing pollutants into the environment. Recent studies show that emissions from garbage incinerators and coal-fired power plants may now be even more serious sources of mercury than are wastewater discharges. In Sweden, where stringent controls over waste-burning were instituted in the early 1980s, the amount of mercury released into the air by human activities has declined by 90 percent.

A CLOSER LOOK AT THE CEREBRAL CORTEX

The **cerebral cortex** is a thin layer of gray matter—that is, the cell bodies of interneurons at or near the cerebrum's surface, above the information highways of axons. Figure 10.19a shows some of its centers for receiving, encoding, and processing information. These centers commonly are localized to the four lobes of each cerebral hemisphere—the occipital, temporal, parietal, and frontal lobes.

The rear portion, the *occipital* lobe, has centers for vision. Blows to it may cause loss of vision in some portion of the **visual field** (the outside world that the individual can actually see). The *temporal* lobe, near each temple, is a processing center for hearing and for complex associations related to vision. Damage here won't leave you blind, but it can impair your ability to recognize complex visual patterns (say, human faces). Centers in the temporal lobe also influence emotions.

The *parietal* lobe is centrally located at the upper rear of each cerebral hemisphere. It encompasses the somatosensory cortex—the main receiving area for signals from the skin and joints. The *frontal* lobe includes the motor cortex, which generates instructions for motor responses. Thumb, finger, and tongue muscles get much of the brain's attention, which gives you an idea of how much control is required for hand movements and verbal expression (Figure 10.20).

The prefrontal cortex is located in front of the motor cortex. It is crucial for planning and integrating information about movements, for inhibition of unsuitable behaviors, and for aspects of memory, as described shortly. PET scans suggest that this cortical region and part of the cerebellum (especially a structure called the dentate nucleus) may interact to govern motor abilities underlying language and some thought processes. *Science Comes to Life* on page 238 describes some classic experiments from which we have learned a great deal about some of these brain operations.

States of Consciousness

Throughout the spectrum of consciousness, which includes sleeping and aroused states, neural chattering shows up as wavelike patterns in electroencephalograms (EEGs). EEGs are electrical recordings of the frequency and strength of membrane potentials at the brain surface (Figure 10.21). PET scans, of the sort shown in Figure 10.19b, also provide information on the precise location of brain activity as it takes place.

Certain neurons of the reticular formation make up a **reticular activating system** (RAS) that controls the changing levels of consciousness. The RAS sends signals to the spinal cord, cerebellum, and cerebrum as well as back to itself. Inhibitory or excitatory chemical changes accompanying the flow of RAS signals affect whether you stay awake or fall asleep. One RAS "sleep center" releases serotonin. This neurotransmitter inhibits other neurons that arouse the brain and maintain wakefulness. Thus, high serotonin levels are linked to drowsiness and sleep. Substances released from another brain center inhibit serotonin's effects and bring about wakefulness.

Damage to some RAS circuits can lead to unconsciousness and coma. Some drugs used in general anesthesia induce unconsciousness by suppressing the RAS.

primary motor cortex

primary somato-sensory cortex

parietal lobe (body sensations)

frontal lobe (planning movements; some aspects of memory; inhibition of unsuitable behavior)

temporal lobe (hearing, advanced visual processing)

occipital lobe (vision)

a

Figure 10.19 (**a**) Primary receiving and integrating centers for the human cerebral cortex. Primary cortical areas receive signals from receptors on the body's periphery. Association areas coordinate and process sensory input from different receptors. The PET scans (**b**) show which brain regions were active when a person performed three specific tasks: speaking, generating words, and observing words.

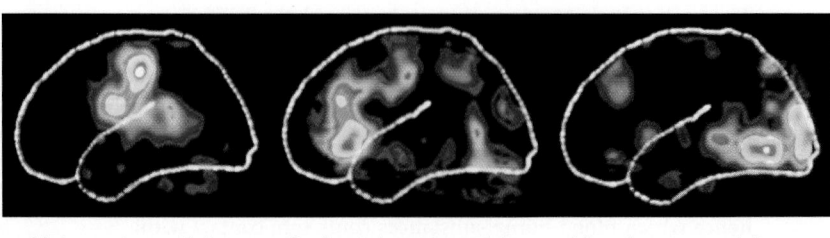

Motor cortex activity when speaking

Prefrontal cortex activity when generating words

Visual cortex activity when observing words

b

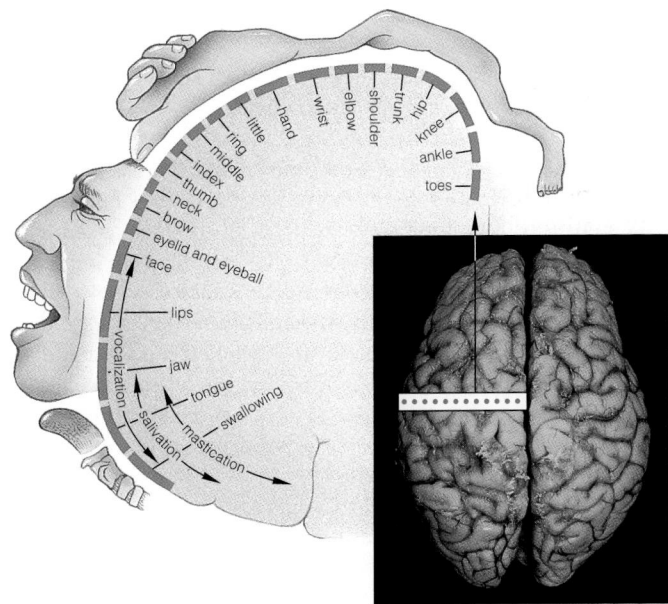

Figure 10.20 Body parts controlled by the primary motor cortex. This diagram is a slice through the motor cortex of the right hemisphere of someone facing you. It controls muscles on the body's left side. The distortions to the body draped over it indicate which body parts receive the most precise control.

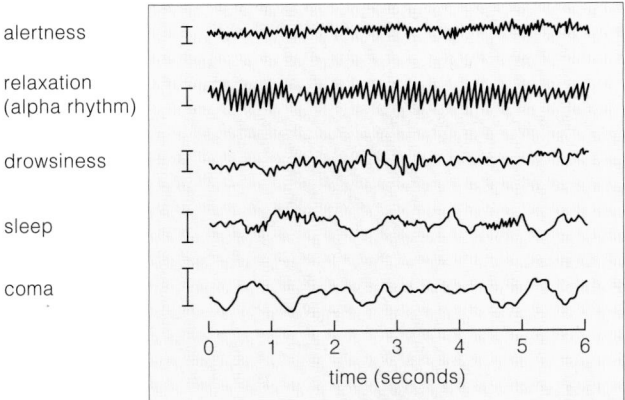

Figure 10.21 EEG patterns. Vertical bars indicate a range of 50 microvolts of electrical activity. The horizontal graph lines indicate the electrical response over time.

The prominent wave pattern for someone who is relaxed, with eyes closed, is an *alpha rhythm*. The EEG waves are recorded in "trains" of one after the other, about ten per second. With a transition to sleep, wave trains gradually become larger, slower, and more erratic. This *slow-wave sleep* pattern shows up in about 80 percent of the total sleeping time for adults. It occurs when sensory input is low. Subjects awakened from slow-wave sleep usually report that they were not dreaming. Often they seemed to be mulling over recent, ordinary events.

Slow-wave sleep is punctuated by brief spells of *REM sleep*. *Rapid Eye Movements* accompany this pattern (the eyes jerk beneath closed lids), as do irregular breathing, faster heartbeat, and twitching fingers. Most people awakened from REM sleep report they were experiencing vivid dreams. The transition from sleep or deep relaxation into alert wakefulness is marked by a shift to low-amplitude, higher frequency wave trains. The transition, *EEG arousal,* occurs when a person makes a conscious effort to focus on external stimuli or even thoughts.

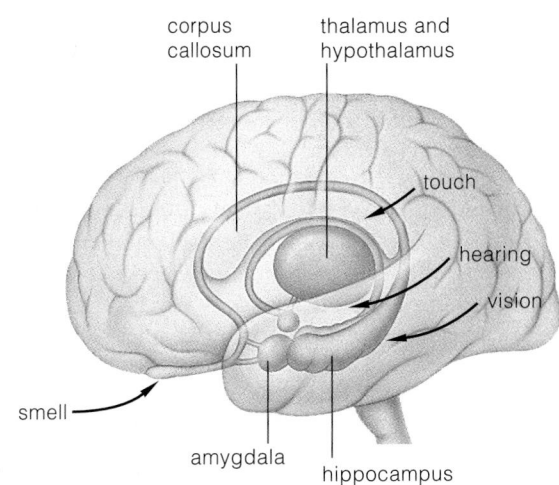

Figure 10.22 The limbic system, our "emotional brain." It includes brain regions called the thalamus, hypothalamus, amygdala, and hippocampus. The limbic system also plays key roles in memory.

Connections with the Limbic System

The **limbic system**, located in the middle of the cerebral hemispheres, governs our emotions and has roles in memory (Figure 10.22). It is distantly related to olfactory lobes and still deals with the sense of smell. That's one reason why you may feel warm and fuzzy when your brain recalls the cologne of a special person who wore it. Connections from the cerebral cortex and other brain centers pass through the limbic system, which includes the amygdala, hippocampus, hypothalamus, and parts of the thalamus. The connections enable us to correlate organ activities with self-gratifying behavior, such as eating and sex. Neural signals based on reasoning in the cerebral cortex often can override or dampen rage, hatred, or other "gut" behaviors. Hypothalamic functions are disrupted by psychoactive drugs, as you will read in the final sections of this chapter.

In each cerebral hemisphere, different centers for receiving and processing information reside in the four lobes of the cerebral cortex.

The left hemisphere affords most of the control over speech, mathematics, and analytical skills. The right hemisphere affords most of the control over spatial abilities, music, and other abstract, nonverbal skills.

Certain neurons of the reticular formation make up a reticular activating system that controls the changing levels of consciousness, including sleep and wakefulness.

The cerebral cortex makes functional connections with the limbic system, which deals with emotions and memory.

SPERRY'S SPLIT-BRAIN EXPERIMENTS

Roger Sperry and his coworkers demonstrated intriguing differences in perception between the two halves of the cerebrum of epileptics. Severe *epilepsy* is characterized by seizures, sometimes as often as every half hour. The seizures are analogous to an electrical storm in the brain. Sperry's patients were so debilitated by their condition that it was impossible for them to lead normal lives. Conventional treatments had been ineffective, so Sperry asked: Would *cutting* the corpus callosum of epileptics confine the electrical storm to one hemisphere, leaving at least the other hemisphere to function normally? Earlier studies of laboratory animals and of humans whose corpus callosum had been damaged suggested this might be so.

He performed the surgery, and the electrical storms did subside in frequency and intensity. Cutting the neural bridge ended what must have been positive feedback of ever intensifying electrical disturbances between the two hemispheres. The "split-brain" patients were able to lead what seemed, on the surface, entirely normal lives. But then Sperry devised some elegant experiments to test whether their conscious experience was indeed "normal." Given that the corpus callosum contains 200 million axons, surely *something* was different. Something was. "The surgery," he later reported, "left these people with two separate minds, that is, two spheres of consciousness. What is experienced in the right hemisphere seems to be entirely outside the realm of awareness of the left."

Sperry presented the two hemispheres of split-brain patients with two different portions of the same visual stimulus. It was known at the time that visual connections to and from one hemisphere are mainly concerned with the opposite half of the visual field (Figure 10.23a). Sperry

projected words—say, COWBOY—onto a screen so that COW fell in the left half of the visual field, and BOY fell in the right (Figure 10.23b).

The subjects of this experiment reported *seeing* the word BOY. The left hemisphere, which controls language, perceived only the letters BOY. However, when asked to write the perceived word with the left hand—a hand that was deliberately blocked from a subject's view—the subject wrote COW. The right hemisphere "knew" the other half of the word (COW) and had directed the left hand's motor response. But it couldn't tell the left hemisphere what was going on because of the severed corpus callosum. The subject knew that a word was being written, but could not say what it was!

Thus Sperry showed that signals across the corpus callosum coordinate the functioning of the two cerebral hemispheres, each of which had responded to visual signals from the opposite side of the body.

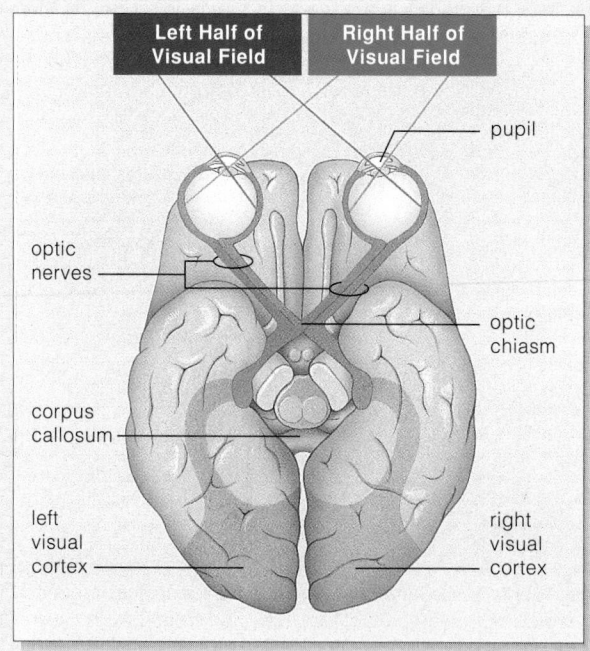

Figure 10.23 **(a)** The human eye gathers visual information at the retina, a layer of densely packed light receptors. Light from the *left* half of the visual field strikes receptors on the right side of both retinas. Parts of two optic nerves carry signals from the receptors to the right cerebral hemisphere. Light from the *right* half of the visual field strikes receptors on the left side of both retinas. Parts of the optic nerves carry signals from them to the left hemisphere.

(b) Response of a split-brain patient to different portions of a visual stimulus.

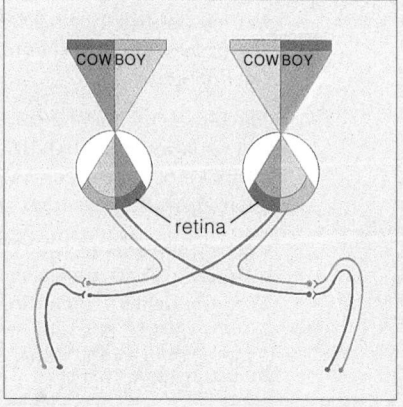

10.12 MEMORY

Even before you were born, your brain began to build your **memory**—to store and retrieve information about an individual's unique experiences. Learning and adaptive modifications of our behavior would be impossible without it. Information is stored in stages. *Short-term* storage is a stage of neural excitation that lasts a few seconds to a few hours. It is limited to bits of sensory information—numbers, words of a sentence, and so on. In *long-term* storage, seemingly unlimited amounts of information get tucked away more or less permanently, as shown in Figure 10.24.

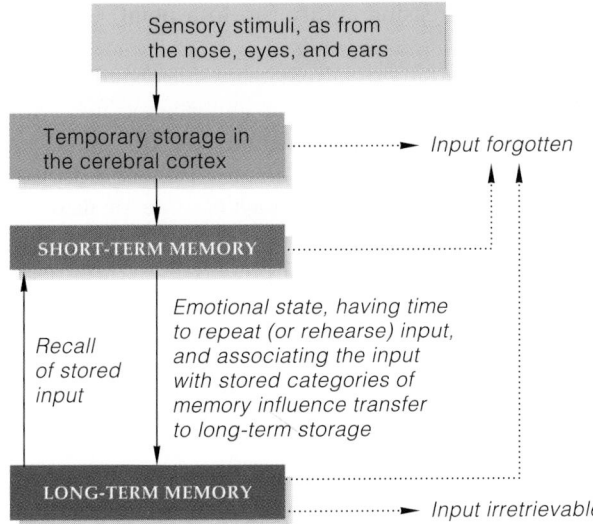

Figure 10.24 Stages of memory processing, starting with the temporary storage of sensory inputs in the cerebral cortex.

Not all of the sensory input bombarding the cerebral cortex ends up in memory storage. Only some is selected for transfer to brain structures involved in short-term memory. Information in these holding bins is processed for relevance, so to speak. If irrelevant, it is forgotten; otherwise it is consolidated with the banks of information in long-term storage structures.

The human brain processes facts separately from skills. Dates, names, faces, words, odors, and other bits of explicit information are *facts*, soon forgotten or filed away in long-term storage, along with the circumstance in which they were learned. Hence you might associate, say, the smell of baking bread with your grandmother's kitchen. By contrast, *skills* are gained by practicing specific motor activities. A skill such as slam-dunking a basketball or playing a piano concerto is best recalled by actually performing it, rather than by recalling the circumstances in which the skill was first learned.

Separate memory circuits handle different kinds of input. A circuit leading to fact memory (Figure 10.25a)

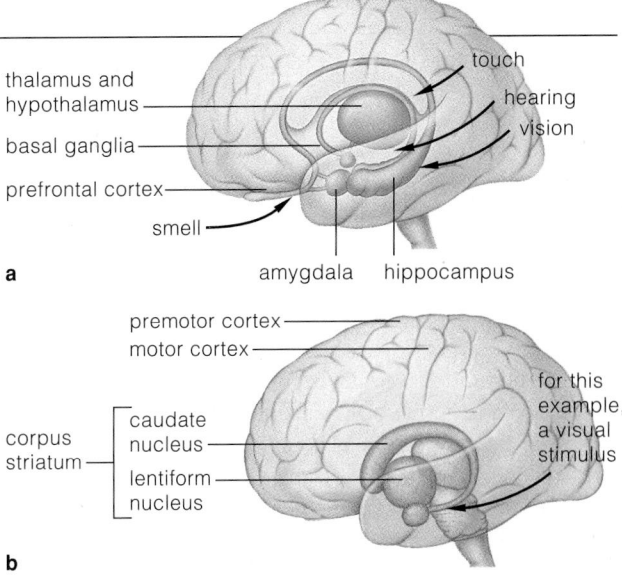

Figure 10.25 Possible circuits involved in (**a**) fact memory and (**b**) skill memory.

starts with inputs at the sensory cortex that flow to the amygdala and hippocampus in the limbic system. The amygdala is the gatekeeper, connecting the sensory cortex with parts of the thalamus and parts of the hippocampus that govern emotional states. Information flows on to the prefrontal cortex, where multiple banks of fact memories are retrieved and used to stimulate or inhibit other parts of the brain. The new input also flows to basal ganglia, which send it back to the cortex in a feedback loop that reinforces the input until it can be consolidated in long-term storage.

Skill memory also starts at the sensory cortex, but this circuit routes sensory input to the corpus striatum ("layered body"), which promotes motor responses (Figure 10.25b). Motor skills entail muscle conditioning, and as you might suspect, the circuit extends to the cerebellum, the brain region that coordinates motor activity.

Amnesia is a loss of memory, the severity of which depends on whether the hippocampus, amygdala, or both are damaged, as by a severe head blow. Amnesia does not affect capacity to learn new skills. By contrast, basal ganglia are destroyed and learning ability is lost during Parkinson's disease, yet skill memory is retained. Alzheimer's disease, which usually has its onset in later life, is linked to structural changes in the cerebral cortex and hippocampus. Affected people often can remember long-standing information, such as their Social Security number, but they have trouble remembering what just happened to them. In time they grow confused, depressed, and unable to complete a train of thought.

Memory, the storage and retrieval of sensory information, results from circuits between the cerebral cortex and parts of the limbic system, thalamus, and hypothalamus. Sensory input is processed through short-term and long-term storage.

A drug is any substance introduced into the body to provoke a specific physiological response. For thousands of years people in nearly every culture have used drugs as medicines, to alter mental states as part of religious or social rituals, or simply for individual effects. **Drug abuse** can be defined as use of a drug in a way that harms health or interferes with a person's ability to function in society.

In general, drug abuse involves **psychoactive drugs**, which act on the central nervous system by binding to receptors in the neuron plasma membrane. Such receptors normally bind neurotransmitter molecules, which transmit chemical messages among neurons (page 226). When drug molecules bind such receptors, the result is a change in the chemical messages neurons send or receive—and in associated mental and physical states.

Drug Effects

Psychoactive drugs exert their effects on brain regions that govern states of consciousness and behavior. Various drugs also influence physiological events, such as heart rate, respiration, sensory processing, and muscle coordination. Many affect a pleasure center in the hypothalamus and artificially fan the sense of pleasure we associate with eating, sexual activity, or other behaviors.

Often the body develops *tolerance* to a drug, meaning that it takes larger or more frequent doses of the drug to produce the same effect. Tolerance reflects *physical drug dependence*. As you may recall from Chapter 5, the liver produces enzymes that detoxify drugs circulating in the bloodstream. Tolerance develops when the level of detoxifying liver enzymes increases in response to the consistent presence of the drug in the blood. In effect, a user must increase his or her intake to keep one step ahead of the liver's increasing ability (up to a point) to break down the drug.

Psychological drug dependence, or *habituation,* develops when a user begins to crave the feelings associated with using a particular drug. The inability to "feel good" or function normally without a regular supply of a drug—whether it is caffeine, alcohol, crack cocaine, or nicotine (Figure 10.26)—is a clear indication of habituation. Table 10.3 lists some warning signs of potentially serious drug

Figure 10.26 Two examples of psychoactive drugs. People who regularly take in the nicotine in cigarettes develop psychological dependence on the drug, as well as physical dependence. Regular alcohol use can have a similar effect.

Table 10.3 Warning Signs of Drug Dependence

There may be cause for concern about drug dependency if a person consistently or often shows three or more of the following characteristics with respect to a substance:

1. Noticeable tolerance to a substance, marked by significant increase in the amount needed to produce the desired effect

2. Using a substance regularly for an extended period of time

3. Difficulty stopping or cutting down, even when a person consistently wishes to do so

4. Use of a substance to avoid withdrawal symptoms and feel "normal"

5. Concealing substance use from others

6. Taking extreme or dangerous measures to get the drug or to ensure its supply. (Examples: stealing money to buy cocaine; drinking on the job in spite of severe penalties for doing so; going from doctor to doctor until finding one who will prescribe a particular tranquilizer)

7. Having intoxication or withdrawal symptoms that interfere with performance at work or school and/or cause problems in personal relationships

8. Extreme anger and defensiveness if someone suggests there may be a problem

9. Reducing, avoiding, or giving up customary activities because they have begun to interfere with substance use

dependence. Table 10.4 will give you an idea of the use level of some common "recreational drugs."

Habituation and tolerance are evidence of **addiction** to a particular substance. Although all drugs can be misused and abused, the psychoactive drugs described here include substances that can injure health and in some cases can lead to widespread harm in other ways as well.

Table 10.4 Most Commonly Used Recreational Drugs in the United States	

This table tallies recreational drug use only. Drugs are ranked by estimated number of users. In interpreting the chart, keep in mind that a large number of people regularly use or have used more than one drug.

Drug	Number of Users
Alcohol	102,919,000
Nicotine	60,744,000
Marijuana, hashish	10,206,000
Cocaine (non-crack)	1,601,000
Analgesics (painkillers)	1,536,000
Inhalants	1,188,000
Stimulants (smoked or injected methamphetamine, amphetamine)	957,000
Tranquilizers (Valium, etc.)	568,000
Sedatives (Quaalude, etc.)	568,000
Hallucinogens	553,000
Crack cocaine	494,000
Heroin	48,000

Estimates are based on information published by the National Institute on Drug Abuse (1991) for the population of the United States age 12 and over.

Drug Action and Interactions

Not surprisingly, a drug's effects depend on the dose, and typically the effects of a higher dose are more intense than those of a smaller dose. For example, smoking just a little marijuana may pleasurably intensify some sensory perceptions (such as listening to music), but a higher dose may provoke wildly exaggerated, and sometimes even paranoid, perceptions of events.

When different psychoactive drugs are used simultaneously, they can interact, sometimes in ways that are extremely dangerous. In a *synergistic* interaction, two drugs used together have a much more powerful effect than they would have if used separately. Alcohol and barbiturates (such as Seconal and Nembutal) both depress the central nervous system, for example; and when they are used at the same time they can lethally depress respiratory centers in the brain. Some drug combinations are *antagonistic*—one drug blocks the effects of another. Others are *potentiating,* in that one drug enhances the effects of another. For instance, you may have noticed that labels on allergy medications often warn against drinking alcohol while using such products. The warnings are necessary because even a little alcohol may deepen the drowsiness antihistamines cause in certain people and seriously impair a person's ability to function normally.

Psychoactive drugs act on the nervous system, exerting their effects on brain regions that govern states of consciousness and behavior.

Certain psychoactive drugs also influence physiological events, such as heart rate, respiration, sensory processing, and muscle coordination.

Habituation and tolerance to a drug are evidence of addiction.

Stimulants

Stimulants include caffeine, nicotine, amphetamines, and cocaine. First a stimulant increases alertness and body activity, then it leads to depression.

Coffee, tea, chocolate, and many soft drinks contain **caffeine**, one of the most widely used stimulants. Low doses stimulate neurons of the cerebral cortex first, and cause increased alertness and restlessness. Higher doses act at the medulla oblongata to disrupt motor coordination and transmission of signals from other brain regions governing intellectual functions.

Nicotine, a component of tobacco, mimics the neurotransmitter acetylcholine and can directly stimulate various sensory receptors. In less than 10 seconds, about 15 percent of inhaled nicotine has entered the bloodstream and reached the brain, where it triggers secretion of adrenal hormones that increase a smoker's heart rate and blood pressure. These and other metabolic changes lead to effects a smoker comes to crave. Other short-term effects can include water retention, irritability, and gastric upsets. An addicted smoker can quickly learn to regulate nicotine intake to maintain the blood level necessary for the nicotine "lift" while avoiding many of these unpleasant side effects. Because people who smoke or chew tobacco also take in tars and other substances that cause lung and oral cancers, the long-term impacts of nicotine addiction can be devastating.

Amphetamines (including "speed") are synthetic chemicals that resemble dopamine and norepinephrine, natural signaling molecules that stimulate the pleasure center. Small doses tend to make a person feel alert and interested in surrounding activities. Some people come to rely on them to relieve fatigue or boredom or curb the appetite. Over time, however, the brain responds to amphetamines by producing less and less of its own signaling molecules. It comes to depend on artificial stimulation. As tolerance develops, the user requires increasing doses to achieve the desired effect. Those effects can also wear off quickly, causing the person to suddenly "crash" in exhaustion. Chronic users may become seriously malnourished and develop circulatory problems. They may also exhibit hostile, violent behaviors.

Unlike amphetamines, **cocaine** stimulates the pleasure center by *blocking* the reabsorption of dopamine, norepinephrine, and other signaling molecules that are normally released at synapses. Depending on whether it is smoked or injected, cocaine can act on the brain within 10 to 20 seconds of entering the bloodstream. Receptors are incessantly stimulated over an extended period. Heart rate and blood pressure rise; sexual appetite increases. Then the effects change. The signaling molecules that have accumulated in synaptic clefts diffuse away, but the cells that produce them cannot make up for the loss. The sense of pleasure evaporates as the receptor cells (which are now hypersensitive) demand stimulation. After prolonged, heavy use of cocaine, it becomes impossible to experience "pleasure." The addict loses weight and cannot sleep restfully. The immune system becomes compromised, and heart abnormalities develop. Otherwise-healthy people have been known to die of heart failure after a single cocaine "experiment."

Granular cocaine is inhaled ("snorted") or injected in solution. Crack cocaine, a cheaper but more potent form, is burned (Figure 10.27) and the smoke inhaled. Many authorities rank the social ills associated with crack use as the most serious of all illegal drug problems.

Drugs That Reduce Brain Activity

Drugs that act as sedatives, hypnotics, or antianxiety agents lower the activity in nerves and parts of the brain, so they reduce overall body activity. Some act at synapses in the reticular formation and in the thalamus. Depending on the dose, most can produce responses ranging from emotional relief, sedation, sleep (hypnosis), anesthesia and coma, to death. At low doses the person at first feels excited or euphoric. Higher doses suppress excitatory synapses in the CNS, leading to depression.

Unlike other drugs in this group, **alcohol** acts directly on the plasma membrane to alter cell function. It is water-soluble and is absorbed across the mucosal lining of the stomach and small intestine. Alcohol's effects can vary according to circumstances. In general, it is absorbed more slowly when there is food in the stomach. Due to gender-related differences in enzyme secretions, females tend to become intoxicated more rapidly than do males.

Some people think of alcohol as a harmless substance that helps them relax or eases tension. Evidence suggests that in moderate amounts (up to one or two drinks per day) it may not do serious damage in many people. Even so, alcohol is one of the most powerful psychoactive drugs known, and alcohol abuse is a major cause of illness, death, and disruption of personal relationships. As Chapter 14 will describe, there is growing evidence that even small amounts of alcohol can damage a developing fetus. Among adults, under certain circumstances, as little as an ounce or two can produce disorientation, uncoordinated motor functions, and diminished judgment. Long-term addiction destroys nerve cells and causes permanent damage to the brain, heart, and liver. It is the most common cause of the liver scarring known as cirrhosis (Chapter 5) and is a major factor in more than 50 percent of all fatal traffic accidents.

Blood alcohol concentration (BAC) measures the percentage of alcohol in the blood. In most states, someone with a BAC of 0.08 is considered legally drunk. When the

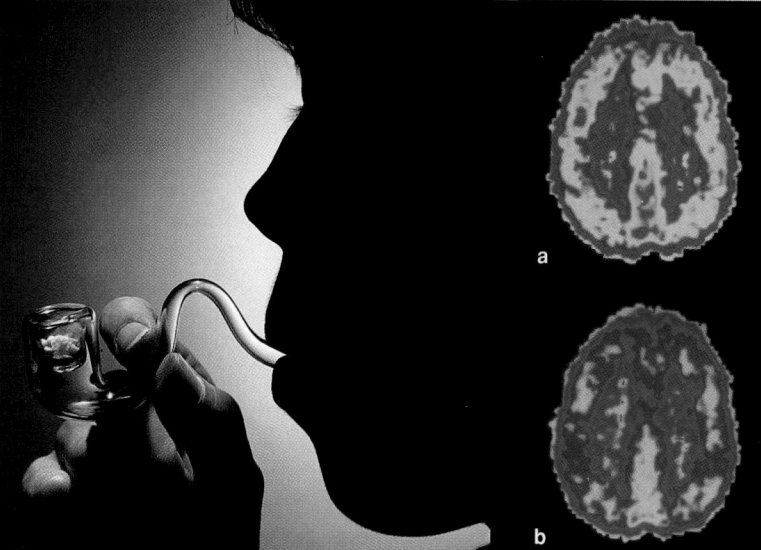

Figure 10.27 Smoking crack puts cocaine in the brain in less than 8 seconds. (**a**) PET scan of a horizontal section of the brain, showing normal activity. (**b**) PET scan of a comparable section showing cocaine's effect. *Red* indicates greatest activity; *yellow*, *green*, and *blue* indicate successively inhibited activity.

BAC reaches 0.15 to 0.4, a person becomes obviously intoxicated and cannot function normally physically or mentally. A BAC greater than 0.4 can be lethal.

Barbiturates include drugs such as phenobarbitol and secobarbitol. Prescribed medically, various forms help control epileptic seizures and induce relaxation. Used abusively as "downers," they can cause extreme drowsiness and markedly impaired judgment. Barbiturates may be prescribed medically to combat extreme anxiety, severe insomnia, and panic attacks. Used long-term or at high dosages, they can lead to dependence, in part because withdrawal symptoms (including headaches, anxiety, nausea, vomiting, and even seizures) can be severe.

Methaqualone, sometimes sold as Quaalude, has legitimate medical uses as a sedative, but it can be habit-forming. At levels that constitute abuse, it causes mental confusion, impaired coordination and judgment, and abusive behavior toward others.

Analgesics

When stress leads to physical or emotional pain, the brain produces its own analgesics, or natural pain relievers. Endorphins and enkephalins are examples. These substances seem to inhibit activity in many parts of the nervous system, including central nervous system centers concerned with emotions and perception of pain.

Narcotic analgesics such as codeine, morphine, and heroin sedate the body and relieve pain. Such substances are *opioids* derived from opium. They have medical uses but are habit-forming; some can be seriously addictive. Deprivation following massive doses of heroin leads to

fever, chills, hyperactivity and anxiety, violent vomiting, cramping, and diarrhea. A synthetic form of heroin known as 3-methylfentanyl or China White is one type of "designer drug"—a laboratory-made chemical that differs from the natural compound but mimics its effects.

Psychedelics and Hallucinogens

These so-called mind-expanding drugs alter sensory perception and other aspects of consciousness. The potent drug LSD (lysergic acid diethylamide) affects the activity of serotonin, a brain hormone. It causes physical changes, such as increases in heart rate and blood pressure, as well as vivid hallucinations. Some people become psychotic while under the drug's influence.

Marijuana, made from the crushed leaves, flowers, and stems of the plant *Cannabis sativa*, is another hallucinogen. Its main psychoactive ingredient is tetrahydrocannibinol (THC), which enters the bloodstream when marijuana is smoked or eaten in food. In low doses THC acts like a depressant, slowing motor activity and eliciting mild euphoria. However, it can produce disorientation, extreme anxiety, and paranoid delusions. Chronic marijuana smoking irritates the lungs and throat. Like alcohol, marijuana can affect one's ability to perform complex tasks, such as driving or studying.

Deliriants

Certain substances alter brain function in ways that lead to *toxic psychosis* or delirium. One type is PCP (phencyclidine). PCP not only skews perceptions. It also acts as an anesthetic, impairing proprioception (awareness of the position of body parts). PCP may trigger permanent psychosis or temporary but dangerous outbursts of violent behavior. Large doses can produce symptoms ranging from dizziness to seizures and heart failure.

Other deliriants are **inhalants**, which produce psychoactive vapors that a user inhales. Commonly abused ones include cleaning fluids, butane, model airplane glue, "poppers" (amyl nitrate, a heart medication), and gases in certain aerosol products. Some inhalants work directly on brain neurons, while others interfere with gas exchange and only indirectly affect the brain. All can be highly toxic to the liver, kidneys, respiratory system, and the heart. "Sniffers" may develop hepatitis, arrhythmias (irregular heartbeat), and even permanent brain damage.

Abuse of psychoactive drugs may trigger destructive behaviors and permanently damage the brain and other organs.

SUMMARY

1. The nervous system detects, interprets, and responds directly to stimuli.

2. A neuron can receive and respond to stimuli because of its membrane properties. An unstimulated neuron shows a steady voltage difference across its plasma membrane (the inside is more negative than the outside). The resting neuron maintains concentration gradients of potassium ions, sodium ions, and other ions across the membrane.

3. With adequate stimulation, the voltage difference across the membrane changes, exceeding a certain threshold level. Then, gated sodium channels across the membrane open and close rapidly and suddenly reverse the voltage difference, which recording devices register as a spike (action potential).

4. Action potentials propagate themselves along the neuron membrane until they reach an output zone, where axon endings form a chemical synapse with another neuron or a muscle or gland cell. The presynaptic cell releases a neurotransmitter into the synaptic cleft. The neurotransmitter excites or inhibits the postsynaptic cell. Integration is the moment-by-moment combining of all signals—excitatory and inhibitory—acting on a neuron.

5. The direction of information flow through the body depends on the organization of neurons into circuits and pathways. Local circuits are sets of interacting neurons confined to a single region in the brain or spinal cord. Nerve pathways extend from neurons in one body region to neurons or effectors in different regions. Reflex arcs, in which sensory neurons directly signal motor neurons that act on muscle cells, are the simplest pathways. In more complex reflexes, interneurons coordinate and refine the responses.

6. The central nervous system consists of the brain and spinal cord. The peripheral nervous system consists of nerves and ganglia in other body regions. Nerves are pathways for signals between the central nervous system and the peripheral nervous system.

7. The peripheral nervous system's somatic subdivision deals with skeletal muscles concerned with voluntary body movements and sensations arising from skin, muscles, and joints. Its autonomic subdivision deals with the functions of the heart, lungs, glands, and other internal organs.

8. The spinal cord has nerve tracts that carry signals between the brain and the peripheral nervous system. It also is a center for some direct reflex connections that underlie limb movements and internal organ activity.

9. The brain has three divisions:

 a. The hindbrain includes the medulla oblongata, pons, and cerebellum and contains reflex centers for vital functions and muscle coordination.

 b. Midbrain centers coordinate and relay visual and auditory information. The midbrain, medulla oblongata, and pons make up the brain stem. The reticular formation, a network of interneurons, extends the length of the brain stem. It helps govern activities of the entire nervous system.

 c. The forebrain includes the cerebrum, thalamus, hypothalamus, and limbic system. The thalamus relays sensory information and helps coordinate motor responses. The hypothalamus monitors internal organs and influences behaviors related to their functioning (such as thirst and sexual activity). The limbic system, which has roles in learning, memory, and emotional behavior, includes pathways that link parts of the thalamus and hypothalamus with other forebrain regions.

10. The cerebral cortex has regions devoted to specific functions, such as receiving and integrating information from sense organs and coordinating motor responses. States of consciousness vary between total alertness and deep coma. The levels are governed by the reticular activating system.

11. Memory occurs in short-term and long-term stages. Long-term storage depends on chemical or structural changes in the brain.

Review Questions

1. Define sensory neuron, interneuron, and motor neuron. *220*

2. Distinguish between a local signal at the input zone of a neuron and an action potential. *222*

3. What are the functional zones of a motor neuron? *224*

4. What is a synapse? Explain the difference between an excitatory and an inhibitory synapse. Define synaptic integration. *226–227*

5. What is a reflex? Describe the events in a stretch reflex. *229, 231*

6. What constitutes the central nervous system? The peripheral nervous system? *228–230*

7. Label the parts of the brain: *232*

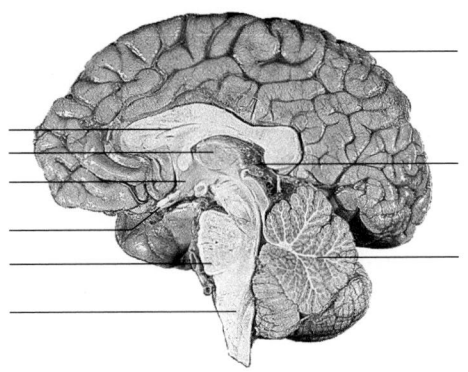

8. Distinguish between the following:
   a. neurons and nerves *220, 228*
   b. somatic system and autonomic system *230*
   c. parasympathetic and sympathetic nerves *231*

Critical Thinking: You Decide *(Key in Appendix V)*

1. In some cases of ADD (attention deficit disorder) the impulsive, erratic behavior typical of so-called hyperactive youngsters can be normalized with drugs that *stimulate* the central nervous system. Explain this finding in terms of neurotransmitter activity in the brain.

2. Meningitis is an inflammation of the meninges that cover the brain and spinal cord. Diagnosis involves making a "spinal tap" (lumbar puncture) and analyzing a sample of cerebrospinal fluid for signs of infection. Why analyze this fluid and not blood?

Self-Quiz *(Answers in Appendix IV)*

1. The nervous system senses, interprets, and issues commands for responses to _____. Its communication lines are organized gridworks of nerve cells, or _____.

2. A neuron responds to adequate stimulation with _____, a type of self-propagating signal.

3. When action potentials arrive at a synapse between a neuron and another cell, they stimulate the release of molecules of a _____ that diffuse over to that cell.

4. The moment-by-moment combining of all signals acting on all the different synapses on a neuron is called _____.

5. Interactions among neurons in your body _____.
   a. involve neurotransmitters
   b. are mediated by memory
   c. are mediated by learning and reasoning
   d. all of the above are correct

6. In the simplest kind of reflex, _____ directly signal _____, which act on muscle cells.
   a. sensory neurons; interneurons
   b. interneurons; motor neurons
   c. sensory neurons; motor neurons
   d. motor neurons; sensory neurons

7. The accelerating flow of _____ ions through gated channels across the membrane is the actual trigger for an action potential.
   a. potassium
   b. sodium
   c. hydrogen
   d. a and b are correct

8. _____ nerves slow down the body overall and divert energy to basic housekeeping tasks; _____ nerves slow down housekeeping tasks and increase overall activity during times of heightened awareness, excitement, or danger.
   a. Autonomic; somatic
   b. Sympathetic; parasympathetic
   c. Parasympathetic; sympathetic
   d. Peripheral; central

9. Match each of the following central nervous system regions with some of its functions.
   ____ spinal cord
   ____ medulla oblongata
   ____ hypothalamus
   ____ limbic system
   ____ cerebral cortex

   a. receives sensory input, integrates it with stored information, coordinates motor responses
   b. monitors internal organs and related behavior (e.g., thirst, hunger, sex)
   c. governs emotions
   d. coordinates reflexes (e.g., for respiration, blood circulation)
   e. makes reflex connections for limb movements, internal organ activity

---

Selected Key Terms

action potential *220*	nerve *228*
autonomic nerve *230*	nerve tract *232*
axon *220*	neuroglial cell *220*
blood-brain barrier *235*	neuromodulator *227*
brain *232*	neuron *220*
central nervous system *228*	neurotransmitter *226*
cerebral cortex *234*	parasympathetic nerve *231*
cerebrospinal fluid *234*	peripheral nervous system *228*
cerebrum *233*	reflex *229*
chemical synapse *226*	reflex arc *229*
dendrite *220*	resting membrane potential *220*
forebrain *233*	reticular activating system *236*
hindbrain *232*	reticular formation *233*
hypothalamus *233*	Schwann cell *224*
interneuron *220*	sensory neuron *220*
limbic system *237*	sodium-potassium pump *221*
memory *239*	somatic nerve *230*
meninges *232*	spinal cord *232*
midbrain *233*	sympathetic nerve *231*
motor neuron *220*	synaptic integration *227*
myelin sheath *224*	thalamus *233*

Readings

Changeux, J. P. November 1993. "Chemical Signaling in the Brain." *Scientific American.*

Fischbach, G. September 1992. "Mind and Brain." *Scientific American.* This and the following reading are from a special issue devoted to the brain.

Hinton, G. September 1992. "How Neural Networks Learn from Experience." *Scientific American.*

Streit, W. J., and C. A. Kincaid-Colton. November 1995. "The Brain's Immune System." *Scientific American.* Fascinating article on the role of microglia in immunity—and how they may wreak serious damage in the brain when their functioning goes awry.

 For additional readings, go to InfoTrac College Edition, your online library at:

http://www.infotrac-college.com/wadsworth

SENSORY RECEPTION

Nosing Around

That charming protuberance called your nose knows more than you might think. In the upper reaches of the nasal cavity, patches of yellowish epithelium are crammed with some 10 million specialized neurons. The neurons sport hairlike cilia. And at the surfaces of the cilia are protein receptors for odor molecules. Molecules arriving in inhaled air are sensory calling cards of a rose's perfume, garlic's pungency, and the human body's secretions, just to mention a few of the 10,000 different scents the average person can distinguish (Figure 11.1). Sensory receptors are gateways into the nervous system, as we will see in the following pages.

A recent exciting discovery indicates that there are several hundred *different* types of protein receptors in the olfactory epithelium, each one constructed according to a genetic blueprint in a particular segment of DNA (a gene). There may even be as many as 1,000 different receptor types coded by a large family of "smell genes." Each receptor can latch onto odor molecules of a specific size and shape. Most likely, many "smells" result from stimulation of a combination of receptors.

By contrast, your taste buds distinguish only four basic types of flavors (sweet, sour, salty, and bitter),

Figure 11.1 The sense of smell—an ancient sensory capacity, and just one of the ways humans gain information about the world.

and your ability to visually distinguish colors relies on just three general types of photoreceptors (for red, green, or blue light wavelengths). In short, your ability to sniff the difference between pizza pie and apple pie may be the product of one of the more complex elements of the human nervous system.

In this chapter we turn to the means by which humans receive signals from the external and internal environments—and how we then decode those signals in ways that give rise to awareness of sounds, sights, odors, pain, and other sensations. Sensory neurons, nerve pathways, and brain regions are required for these tasks. Together they represent the portions of the nervous system that are called *sensory systems*.

KEY CONCEPTS

1. Sensory systems are part of the nervous system. Each one consists of specific types of sensory receptors, nerve pathways from those receptors to the brain, and brain regions that deal with sensory information.

2. A stimulus is a form of energy that activates a specific type of sensory receptor, which is either a sensory neuron or a specialized cell adjacent to it. Photoreceptors detect light energy, thermoreceptors detect heat energy, and so on.

3. A sensation is a conscious awareness of change in external or internal conditions. It begins when sensory receptors detect a specific stimulus. The energy of the stimulus is converted to graded, local signals that may help initiate an action potential.

4. Information about the stimulus becomes encoded in the number and frequency of action potentials sent to the brain along particular nerve pathways. Then specific brain regions translate the information into a sensation.

5. The somatic sensations include touch, pressure, temperature, pain, and muscle sense.

6. The special senses include taste, smell, hearing, and vision.

CHAPTER AT A GLANCE

11.1   Sensory Pathways and Receptors

11.2   Somatic Sensations

11.3   Taste and Smell

11.4   *A Taste of Some Sensory Science*

11.5   Hearing

11.6   Balance

11.7   *Noise Pollution: An Attack on the Ears*

11.8   Vision: An Overview

11.9   *Disorders of the Eye*

11.10 From Neuron Signaling to Visual Perception

Sensory systems, the front doors of the nervous system, receive and notify the spinal cord and brain of specific changes outside and inside the body. Such a change is a **stimulus**. The systems all have sensory receptors, nerve pathways from the receptors to the brain, and brain regions where sensory information is processed and translated into a **sensation**—conscious awareness of a stimulus. It is not the same as **perception** (understanding what the sensation means). Compound sensations arise when information about different stimuli is integrated at the same time. For example, "wetness" is not a single stimulus; our perception of it arises from simultaneous inputs concerning pressure, touch, and temperature.

Types of Sensory Receptors

Sensory receptors are the receptionists at the front door of the nervous system. As listed in Table 11.1, there are six major categories, based on the type of stimulus energy that each type detects:

Mechanoreceptors detect forms of mechanical energy (changes in pressure, position, or acceleration).

Thermoreceptors detect infrared energy (heat).

Nociceptors (pain receptors) detect tissue damage.

Chemoreceptors detect chemical energy of specific substances dissolved in the fluid surrounding them.

Osmoreceptors detect changes in water volume (solute concentration) in the surrounding fluid.

Photoreceptors detect visible light.

Sensory Pathways

Recall, from Chapter 10, that sensory axons carry signals from receptors to the brain. Before they can do so, the stimulus energy must be converted to action potentials, the basis of neural messages. Briefly, when a stimulus disturbs the plasma membrane of a receptor ending, some ions flow through protein channels across a local patch of the membrane. In other words, the disturbance has triggered a local, graded potential. Such signals don't spread far from the point of stimulation, and they vary in magnitude. However, when a stimulus is intense or repeated fast enough for summation of those local signals, action potentials may result. Action potentials self-propagate from receptor endings to the axon endings of sensory neurons (Figure 11.2). There, the release of a neurotransmitter stimulates or inhibits the activity of interneurons (or motor neurons) adjacent to it. Action potentials may be generated in the interneurons, which may be part of nerve tracts leading to the brain.

The action potentials being propagated along sensory neurons are not like a wailing ambulance siren. *They do not vary in amplitude.* How, then, does the brain assess the nature of a given stimulus? It depends on which sensory area receives signals from nerve pathways, the *frequency* of signals that are traveling along each axon of the pathway, and the *number* of axons that are recruited into action.

First, specific sensory areas of the brain can interpret action potentials only in certain ways. That is why you

Table 11.1 Major Categories of Sensory Receptors		
Category	Examples	Stimulus
MECHANORECEPTORS		
Touch, pressure	Certain free nerve endings and Pacinian corpuscles in skin	Mechanical pressure against body surface
Baroreceptor	Carotid sinus	Pressure changes in fluid that bathes them
Stretch	Muscle spindle in skeletal muscle	Stretching
Auditory	Hair cells in organ inside ear	Vibrations (sound or ultrasound waves)
Balance	Hair cells in organ inside ear	Fluid movement
THERMORECEPTORS	Certain free nerve endings	Change in temperature (heating, cooling)
NOCICEPTORS (PAIN RECEPTORS)*	Certain free nerve endings	Tissue damage (e.g., distortions, burns)
CHEMORECEPTORS		
Internal chemical sense	Carotid bodies in blood vessel wall	Substances (O_2, CO_2, etc.) dissolved in extracellular fluid
Taste	Taste receptors of tongue	Substances dissolved in saliva, etc.
Smell	Olfactory receptors of nose	Odors in air, water
OSMORECEPTORS	Hypothalamic osmoreceptors	Change in water volume (solute concentration) of fluid that bathes them
PHOTORECEPTORS		
Visual	Rods, cones of eye	Wavelengths of light

*Extremely intense stimulation of any sensory receptor also may be perceived as pain.

"see stars" when your eye is poked, even in the dark. The mechanical pressure on its photoreceptors generated signals that an optic nerve carried to a specific sensory area. That area interprets incoming signals as "light." If the signals were somehow rerouted to the sensory area for sound, you would "hear" the poke in the eye.

Second, when a stimulus is stronger, receptors fire action potentials more frequently. The same receptor detects the sounds of a throaty whisper and a wild screech. The brain senses the difference through frequency variations in the signals that the receptor sends to it (Figure 11.3).

Third, strong stimulation recruits more receptors. Tap a spot of skin on your arm and you activate some receptors. Press hard on the same area and you activate many more in a larger area. The increased disturbance sets off action potentials in many sensory axons at the same time. The brain interprets the combined activity as an increase in stimulus intensity.

Sometimes the frequency of action potentials decreases or stops even when a stimulus is being maintained at a constant strength. This phenomenon is called **sensory adaptation**. After you have put on clothing, for example, you no longer are aware of its pressure against your skin. Some mechanoreceptors in your skin are a rapidly adapting type; they fire only at the onset of a stimulus. By contrast, if you pick up a full sack of sugar, slowly adapting stretch receptors in your arm's skeletal muscle will fire off action potentials and continue to do so for as long as you hold the sack.

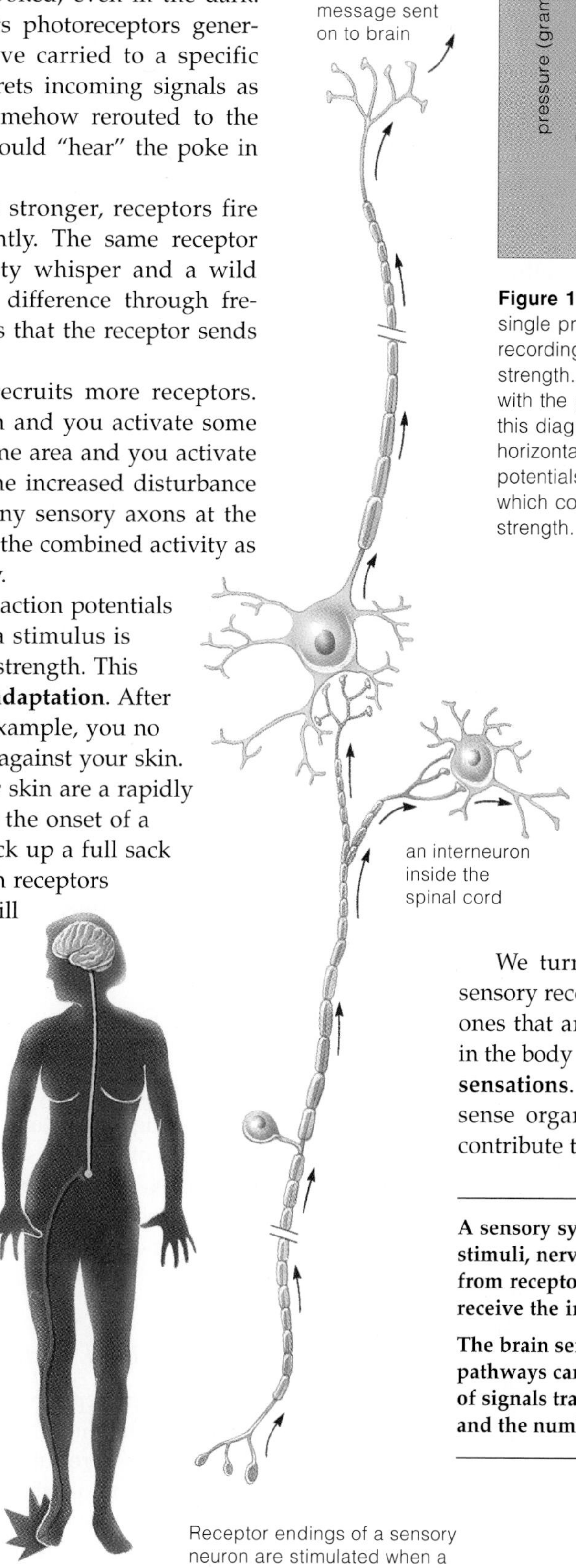

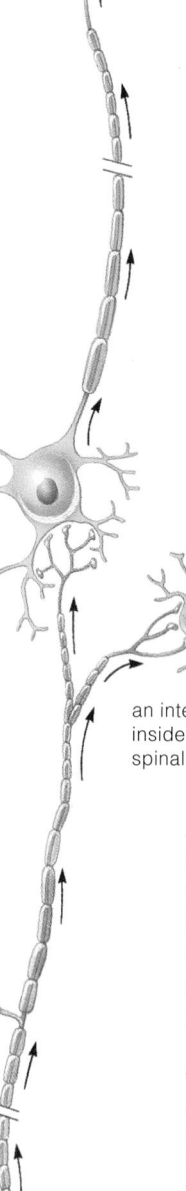

message sent on to brain

an interneuron inside the spinal cord

Receptor endings of a sensory neuron are stimulated when a bare foot lands on a tack.

Figure 11.2 Example of a sensory nerve pathway leading from a sensory receptor to the brain. The sensory neuron is coded *red*; interneurons are coded *yellow*.

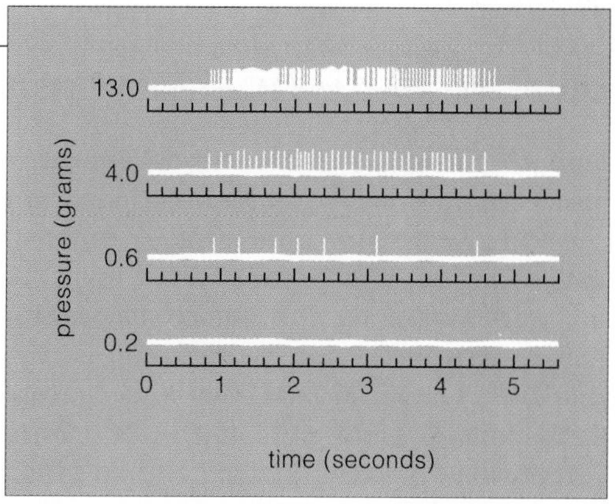

pressure (grams)

13.0

4.0

0.6

0.2

0   1   2   3   4   5

time (seconds)

Figure 11.3 Action potentials recorded from a single pressure receptor of the human hand. The recordings correspond to variations in stimulus strength. A thin rod was pressed against the skin with the pressure indicated on the vertical axis of this diagram. Vertical bars above each thick horizontal line represent individual action potentials. Notice the increases in frequency, which correspond to increases in stimulus strength.

We turn now to specific examples of the sensory receptors we have been discussing. The ones that are present at more than one location in the body contribute to **somatic** ("of the body") **sensations**. Other receptors are restricted to sense organs, such as the eyes or ears. They contribute to the **special senses**.

A sensory system has sensory receptors for specific stimuli, nerve pathways that conduct information from receptors to the brain, and brain regions that receive the information.

The brain senses a stimulus based on which nerve pathways carry the incoming signals, the frequency of signals traveling along each axon of that pathway, and the number of axons that have been recruited.

Somatic sensations begin with receptors in the body's surface tissues, in skeletal muscles, and in the walls of internal organs. Information travels into the spinal cord and to the **somatosensory cortex**, part of the surface layer of gray matter of the cerebral cortex. Interneurons of this part of the brain are organized in a way that maps the body's surface. The largest areas of the map correspond to body parts where the density of sensory receptors is greatest. Such body parts have the most sensory acuity and require the most intricate control. They include the fingers, thumbs, and lips (Figure 11.4).

Receptors Near the Body Surface

You can discern sensations of touch, pressure, cold, warmth, and pain near the body surface. Regions with the greatest number of sensory receptors, such as the fingertips and the tip of the tongue, are most sensitive to stimulation. Other regions, such as the back of the hand and neck, do not have nearly as many receptors per unit of area and are far less sensitive.

Free nerve endings are the simplest receptors. They are thinly myelinated or unmyelinated (naked) branched endings of sensory neurons in the epidermis or in the underlying dermis of skin. Different types function as mechanoreceptors, thermoreceptors, and nociceptors; all adapt very slowly to stimulation. One subpopulation

of free nerve endings can give rise to a sensation of prickling pain—like the "jab" you feel when you stick your finger with a pin. Another subpopulation contributes to the sensations of itching or warming that are elicited by certain chemicals, including histamine. Two thermoreceptive types of free nerve endings have peak sensitivities that are higher or lower than the normal body temperature, respectively. One mechanoreceptive type coils around hair follicles and detects the movement of the hair inside (Figure 11.5).

Encapsulated receptors, which are surrounded by a capsule of epithelial or connective tissue, also are common near the body surface. A Meissner's corpuscle (see Figure 11.5) adapts slowly to low-frequency vibrations. These receptors are abundant in the lips, fingertips, eyelids, nipples, and genitals. The bulb of Krause is an encapsulated thermoreceptor that is activated by temperatures below 20°C (68°F). Below 10°C it gives rise to a painful sensation of freezing. Ruffini endings are other slowly adapting encapsulated receptors. They are sensitive to steady touching and pressure, and to temperatures greater than 45°C (113°F).

The Pacinian corpuscle is an encapsulated receptor that is widely distributed in dermis, where it may have a role in perception of fine textures. Pacinian corpuscles also are present deeper in the body, as in the membrane near freely movable joints and in some internal organs. Concentric, onionlike layers of connective tissue alternating with fluid-filled spaces surround this sensory ending. This structural arrangement enhances the detection of rapid pressure changes associated with touch and vibrations.

Muscle Sense

Sensing limb motions and the body's position in space requires mechanoreceptors in skeletal muscles, joints, tendons, ligaments, and skin. Examples include stretch receptors of muscle spindles. As you may recall from Chapter 4, muscle spindles are embedded in skeletal muscle tissue and run parallel with the muscle cells. Their responses to stimulation depend partly on how much and how fast the muscle stretches.

A Closer Look at the Sensation of Pain

Pain is the perception of injury to some body region. The most important nociceptors (from the Latin word *nocere*, "to do harm") are subpopulations of free nerve endings, several million of which are distributed throughout the skin and internal tissues. They trigger awareness of pain in the thalamus, but the type and intensity of pain are interpreted in the cerebral cortex.

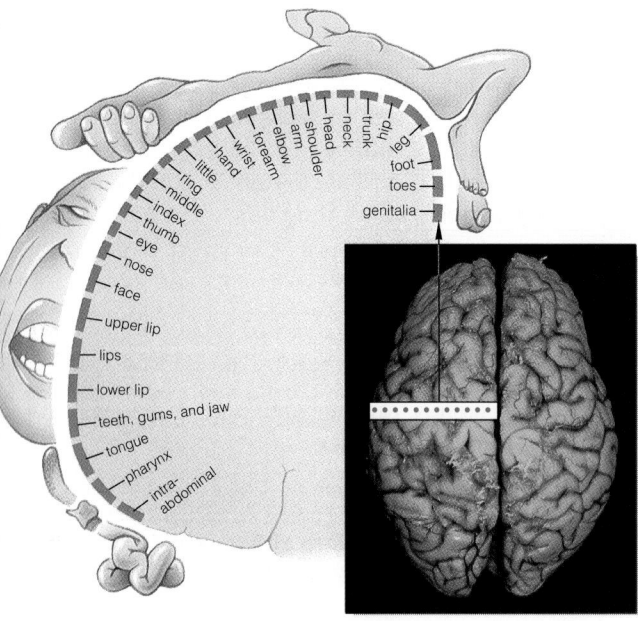

Figure 11.4 Body regions represented in the somatosensory cortex. This region is a strip a little more than an inch (2.5 centimeters) wide, running from the top of the head to just above the ear at the surface of each cerebral hemisphere.

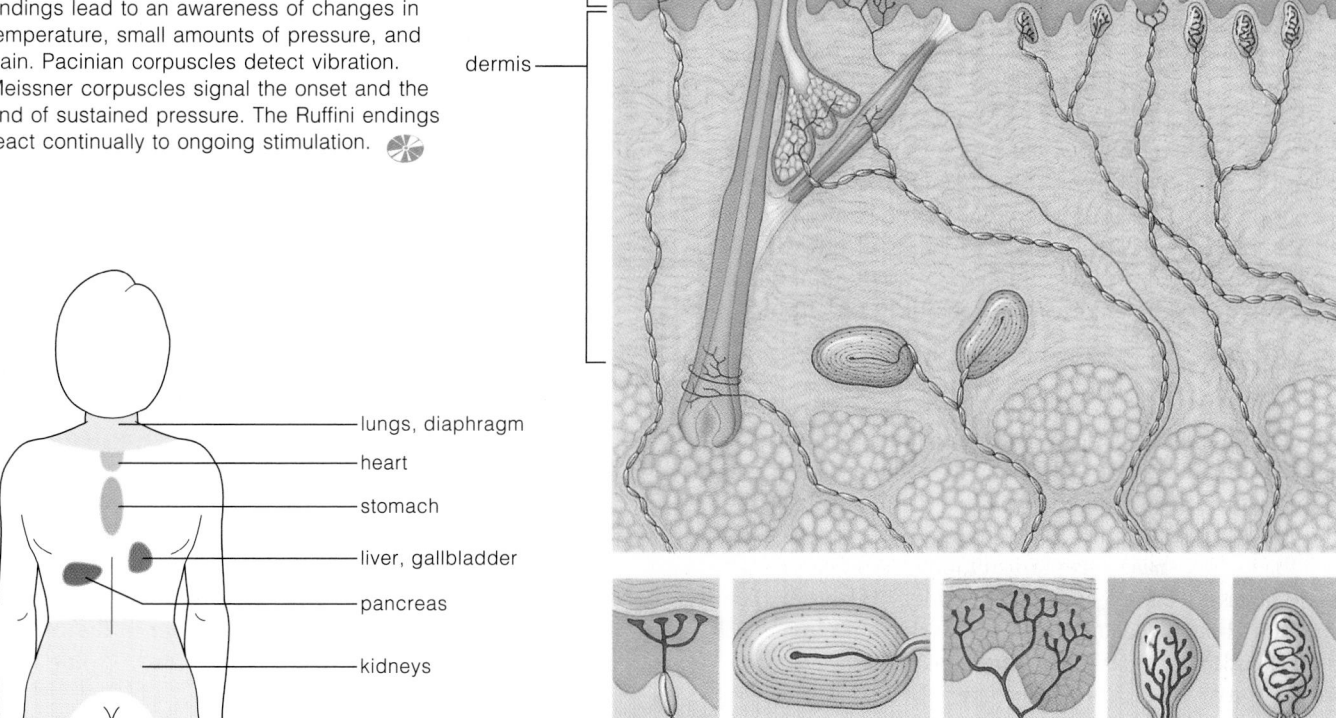

Figure 11.5 Tactile receptors in human skin. Signals from the types called free nerve endings lead to an awareness of changes in temperature, small amounts of pressure, and pain. Pacinian corpuscles detect vibration. Meissner corpuscles signal the onset and the end of sustained pressure. The Ruffini endings react continually to ongoing stimulation.

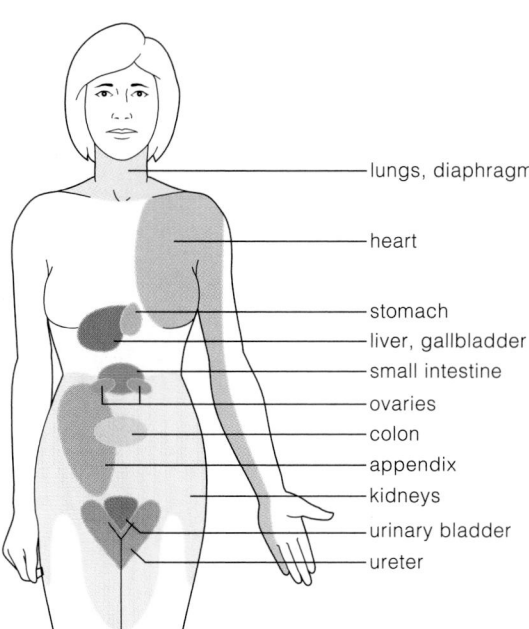

Figure 11.6 Referred pain. In certain disorders, pain associated with the affected internal organ is localized to the skin areas indicated.

Sensations of *somatic pain* start with nociceptors in skin, skeletal muscles, joints, and tendons. Sensations of *visceral* pain, which is associated with the internal organs, are related to excessive chemical stimulation, muscle spasms, muscle fatigue, inadequate blood flow to organs, and other abnormal conditions.

A person's perception of pain often depends on the brain's ability to identify the affected tissue. Get hit in the face with a snowball and you "feel" the contact on facial skin. However, sensations of pain from some internal organs may be wrongly projected to part of the skin surface. This response is called "referred pain," and it is related to the way the nervous system is constructed. Possibly, sensory inputs from the skin and from certain internal organs enter the spinal cord along common nerve pathways, so the brain can't accurately identify their source. For example, a heart attack can be felt as pain in skin above the heart and along the left shoulder and arm (Figure 11.6).

Diverse mechanoreceptors detect touch, pressure, heat and cold, pain, limb motions, and changes in the body's position in space. Responses to signals from these receptors give rise to somatic sensations.

We turn now to examples of the special senses, starting with those of taste and smell. Both are *chemical* senses; their sensory pathways start at chemoreceptors, which are activated when they bind a chemical substance that is dissolved in the fluid bathing them. The receptors themselves regularly wear out, and new ones replace them. In both pathways, sensory information travels from the receptors through the thalamus and on to the cerebral cortex, where perceptions of the stimulus take shape and undergo fine-tuning. The input also travels to the limbic system, which can integrate it with emotional states and stored memories (Sections 10.10 and 10.12).

Gustation: The Sense of Taste

In the human body, *taste receptors* are part of sense organs called taste buds (Figure 11.7). You have about 10,000 taste buds, which are scattered over the tongue, roof of the mouth (palate), and throat.

Taste buds bear receptors for one or more of four general taste classes: sweet, sour, salty, and bitter. The receptors bind molecules that are dissolved in saliva. The stimulated receptor in turn stimulates a sensory neuron, which conveys the message to centers in the brain, where the stimulus is interpreted.

Strictly speaking, the flavors of most foods are some combination of the four basic "tastes," plus sensory input from olfactory receptors in the nose. Simple as this sounds, researchers are beginning to discover that our taste sense actually encompasses a richly complex set of molecular mechanisms. *Science Comes to Life* examines the findings of some of these intriguing studies.

The olfactory element of taste is extremely important. In addition to odor molecules in inhaled air, molecules of volatile chemicals are released as you chew food. These waft up into the nasal passages. There, as described next, the "smell" inputs contribute to the perception of a smorgasbord of complex flavors. This is why anything that dulls your sense of smell—such as a head cold—also seems to diminish food's flavor.

Olfaction: The Sense of Smell

Olfactory receptors (Figure 11.8) detect water-soluble or volatile (easily vaporized) substances. Recent studies suggest that when odor molecules bind to receptors on olfactory neurons in cells of the nose's olfactory epithelium, the resulting action potential travels directly to olfactory bulbs in the frontal area of the brain. There, other neurons forward the message to a center in the cerebral cortex, which interprets it as "fresh bread," "pine tree," or some other substance.

From an evolutionary perspective, olfaction is one of the most ancient senses—and for good reason. Food, potential mates, and predators give off chemical substances that can diffuse through air (or water) and so

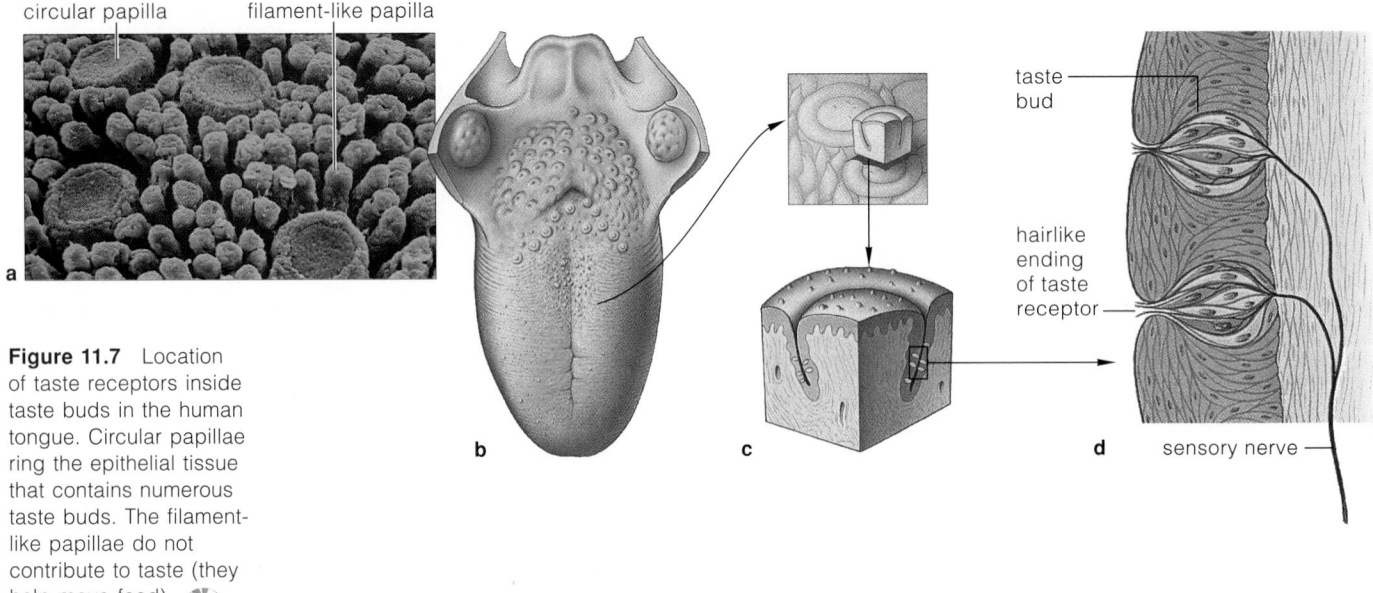

Figure 11.7 Location of taste receptors inside taste buds in the human tongue. Circular papillae ring the epithelial tissue that contains numerous taste buds. The filament-like papillae do not contribute to taste (they help move food).

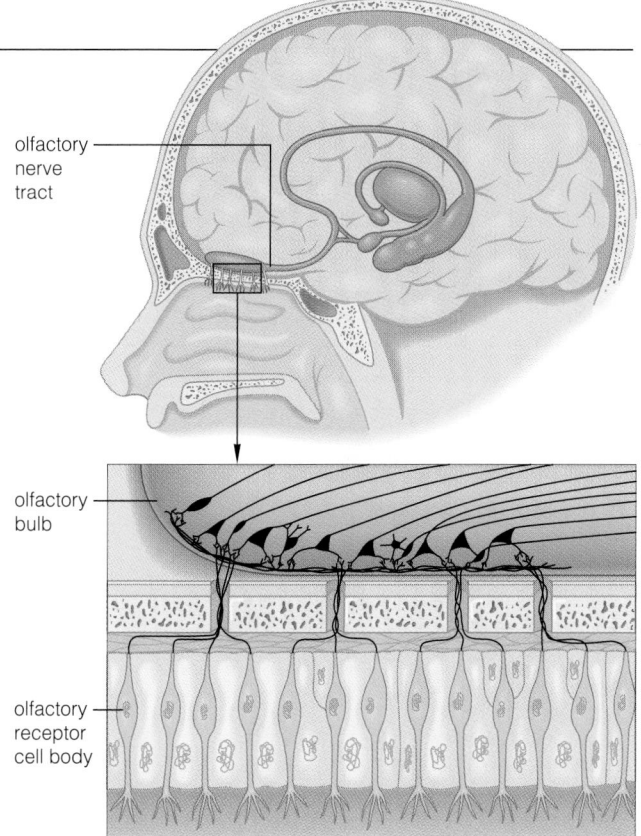

olfactory nerve tract

olfactory bulb

olfactory receptor cell body

Figure 11.8 Sensory nerve pathway leading from olfactory receptors in the nose to primary receiving centers in the brain.

give clues or advance warning of their whereabouts. Even humans, with their rather insensitive sense of smell, have about 10 million olfactory receptors in patches of olfactory epithelium in the upper nasal passages. (The nose of a bloodhound has more than 200 million.)

Like some other animals, humans have a vomeronasal organ, or "sexual nose." Its receptors detect pheromones, which are signaling molecules with roles in animals' social interactions. These exocrine gland secretions affect the behavior—and sometimes, perhaps, the physiology—of other individuals. For instance, there is evidence that one or more pheromones in the sweat of females accounts for the common observation that women of reproductive age who are in regular close contact with one another often come to have their menstrual periods on a similar schedule.

Taste (gustation) relies on receptors in sensory organs called taste buds. The receptors bind molecules dissolved in saliva. Associated sensory neurons relay the message to the brain. The four basic taste classes are sweet, sour, salty, and bitter.

Olfaction (smell) relies on olfactory receptors in patches of epithelium in the upper nasal passages. Neural signals along olfactory neurons travel directly to the olfactory bulbs in the brain.

A TASTE OF SOME SENSORY SCIENCE

Without your taste buds, every meal would be an essential but dull exercise in food intake. With these sensory organs, eating is not only a means of securing nutrients, but also one of life's pleasures. In ways that have been poorly understood until the last decade, the sensory receptors in taste buds distinguish sweet, sour, salty, and bitter—which then become mingled (together with olfactory inputs) into our conscious perceptions of countless flavors. Thanks to methods of molecular biology, we have begun to understand the cellular and molecular details underlying this powerful chemical sense.

Essentially, a taste bud is an organized array of 50 to 150 taste receptor cells. Each receptor cell functions for at most two weeks before being replaced. In that time, it can respond to "tastants" (molecules we can perceive as having a taste) in at least two—and in some cases as many as all four—of the taste classes.

Different kinds of molecules and mechanisms are responsible for each of the four basic tastes. Which taste class (or combination of classes) is ultimately perceived depends on the chemical nature of the "signal" and also on how it is processed by the receptor. In each case, some event causes the receptor cell to release a neurotransmitter (Chapter 10), which triggers action potentials in a nearby sensory neuron.

Many chemical salts, including sodium ions, are perceived as salty. The receptor cell's response is due to the influx of Na^+ through sodium ion channels in its plasma membrane. A receptor responds with a "sour" message when tastant molecules release hydrogen ions that block membrane potassium ion channels. Caffeine and quinine taste bitter, while many "sweet" tastants are sugars (of course!), alcohols, and amino acids. Experiments indicate that both the bitter and sweet taste messages sent on to the nervous system are mediated by the activity of specific proteins within the receptor cell.

There also are differences in taste receptor sensitivity to various tastants. The receptors tend to be exquisitely sensitive to bitter tastants, which can be detected in extremely low concentrations. Some researchers speculate that this may be an evolutionary adaptation to the fact that many naturally occurring substances that are poisonous to mammals (strychnine is one) are bitter-tasting. Taste receptors must be exposed to somewhat higher concentrations of sour tastants before the stimulus registers; they are least sensitive to (and hence require the highest concentrations of) sweet and salty substances. Relatively small quantities of artificial sweeteners can significantly sweeten foods because their molecular characteristics make them 150 times (aspartame) to more than 600 times (saccharin) as potent as plain sucrose.

The sense of hearing starts with vibration-sensitive mechanoreceptors deep within the ear. A vibration is a wavelike form of mechanical energy. For example, clapping produces waves of compressed air. Each time hands clap together, molecules are forced outward, so a low-pressure state is created in the region they vacated. The pressure variations can be depicted as a wave form, and the *amplitude* of its peaks corresponds to loudness. The *frequency* of a sound is the number of wave cycles per second. Each cycle extends from the start of one wave to the start of the next.

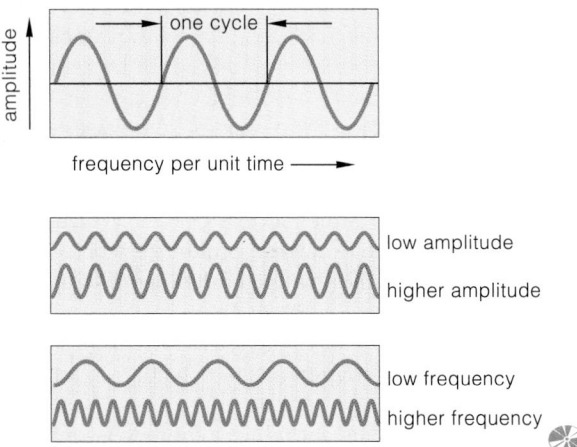

When sound waves travel into the ear's auditory canal, they eventually reach a membrane and make it vibrate. The vibrations then cause a fluid inside the ear to move, the way water in an inflatable swimming pool sloshes if you push back and forth on the pool's side wall. In your ear, the movement of fluid bends the tips of hairs on mechanoreceptors. With enough bending, the end result will be action potentials that ultimately reach the brain, where they are interpreted as a "sound."

The Ear

A human ear consists of three regions (Figure 11.9a), each with its own role in hearing. The *outer ear* provides a pathway by which sound waves can enter the ear, setting up vibrations. As described shortly, the vibrations are amplified in the *middle ear*. Within the *inner ear*, vibrations of different sound frequencies are "sorted out" as they stimulate different patches of receptors. The inner ear houses several structures, including *semicircular canals* which are involved in balance—the topic of Section 11.6. It also contains the coiled **cochlea,** where key events in hearing take place. As you'll now read, a coordinated sequence of events in the ear's various regions provides the brain with the auditory input it can interpret to give us a hearing sense.

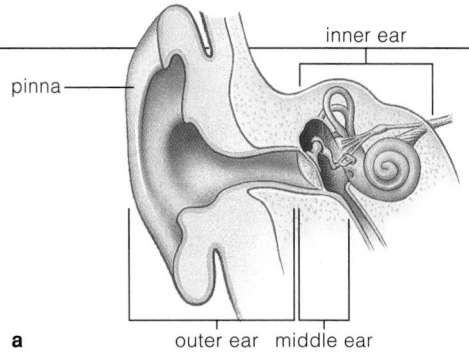

a

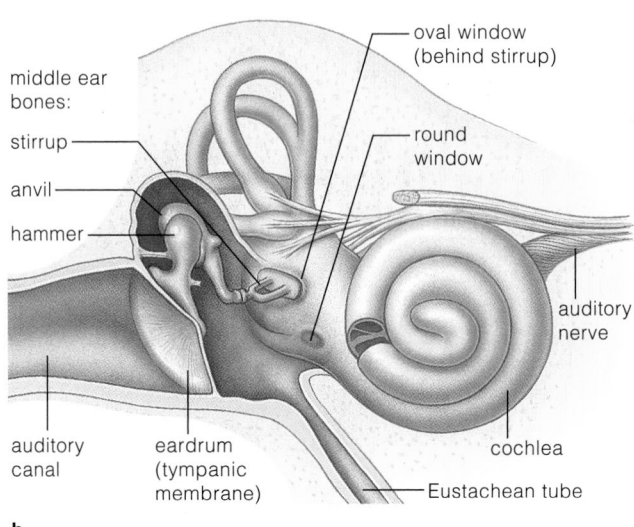

b

The "How" of Hearing

Hearing begins when the external flaps of the outer ear collect sound waves and channel them inward through the auditory canal to the **tympanic membrane** (the eardrum). Sound waves cause the tympanic membrane to vibrate. This vibration in turn causes vibrations in a leverlike array of three tiny bones of the middle ear: the *malleus* ("hammer"), *incus* ("anvil"), and stirrup-shaped *stapes.* The vibrating bones transmit their motion to the *oval window,* an elastic membrane over the entrance to the cochlea. The oval window is much smaller than the tympanic membrane, and as the middle-ear bones vibrate against its small surface with the full energy that struck the tympanic membrane, the force of the original vibrations becomes amplified.

Now the action, so to speak, shifts to the cochlea. If we could uncoil the cochlea, we would see that a fluid-filled chamber folds around an inner *cochlear duct* (Figure 11.9c). Each long "arm" of the outer chamber functions as a separate compartment (the *scala vestibuli* and *scala tympani,* respectively). The amplified vibrations of the oval window are strong enough to produce pressure waves in the fluid within the chambers. In turn, these waves are transmitted to the fluid in the cochlear duct. On the floor

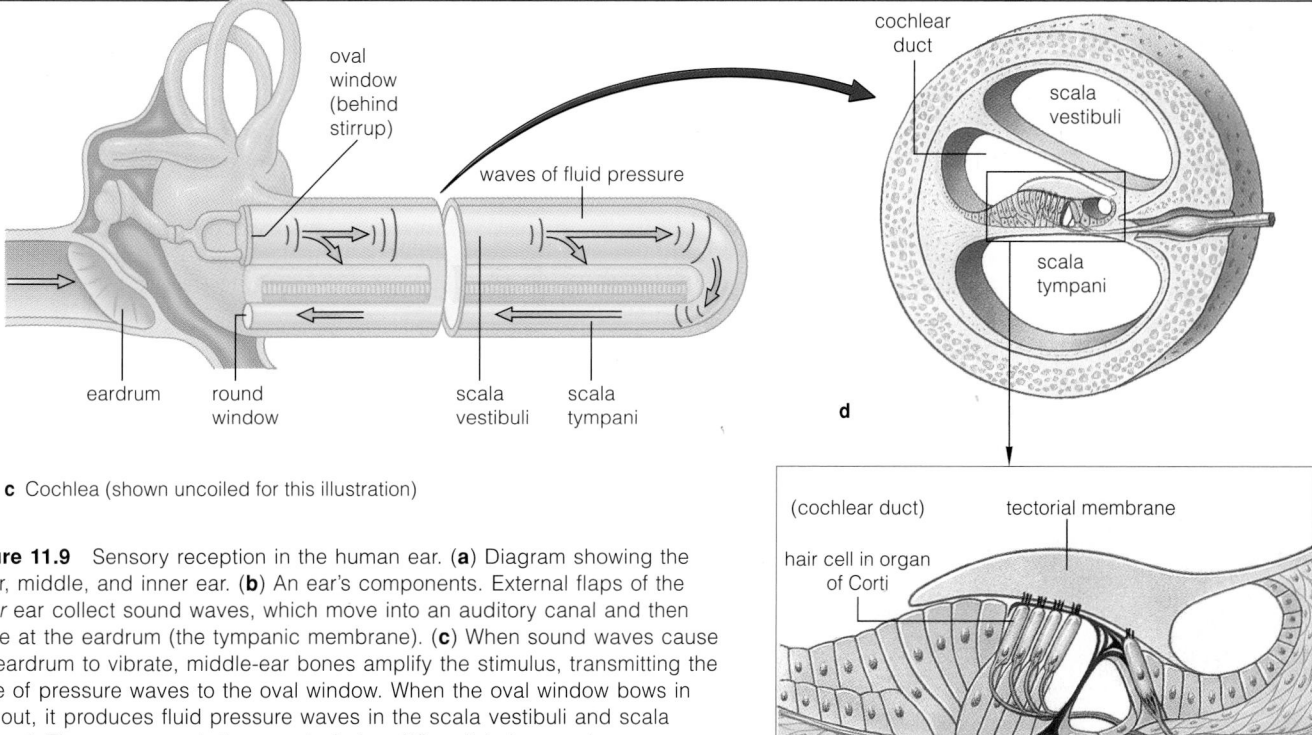

oval window (behind stirrup)

waves of fluid pressure

cochlear duct

scala vestibuli

scala tympani

eardrum

round window

scala vestibuli

scala tympani

d

c Cochlea (shown uncoiled for this illustration)

(cochlear duct)

tectorial membrane

hair cell in organ of Corti

scala tympani

basilar membrane

to auditory nerve

e

Figure 11.9 Sensory reception in the human ear. (**a**) Diagram showing the outer, middle, and inner ear. (**b**) An ear's components. External flaps of the *outer* ear collect sound waves, which move into an auditory canal and then arrive at the eardrum (the tympanic membrane). (**c**) When sound waves cause the eardrum to vibrate, middle-ear bones amplify the stimulus, transmitting the force of pressure waves to the oval window. When the oval window bows in and out, it produces fluid pressure waves in the scala vestibuli and scala tympani. The waves reach the round window. When it bulges under pressure, fluid moves back and forth in the inner ear. (**d**) Pressure waves are "sorted" at the cochlear duct. At different points, its basilar membrane vibrates more strongly to sounds of different frequencies. (**e**) When the basilar membrane and tectorial membrane vibrate, hair cells sandwiched between them bend. With enough bending, they fire off action potentials that travel via the auditory nerve to the brain.

of the cochlear duct is a *basilar membrane*, and resting on the basilar membrane is a specialized **organ of Corti**, which includes sensory **hair cells**. Slender projections at the tips of the cells abut against an overhanging **tectorial** ("rooflike") **membrane**, which is not a membrane at all but a firm, jellylike noncellular structure. Pressure waves in the cochlear fluid vibrate the basilar membrane, and its movements can press hair cell projections against the tectorial membrane so that the projections bend. Affected hair cells release a neurotransmitter, which triggers action potentials in neurons of the auditory nerve—the route by which the action potentials reach the brain.

Different sound frequencies cause different parts of the basilar membrane to vibrate—and, accordingly, to bend different patches of hair cells. Pressure waves that vibrate at a higher frequency (more waves per second) set up vibrations in regions of the membrane nearer the entrance to the coil, and waves of lower frequency cause vibrations nearer the tip of the coil. Apparently, the total number of hair cells that are stimulated in a given region determines the loudness of a sound. The perceived tone or "pitch" of a sound depends on the frequency of the vibrations that excite different groups of hair cells—the higher the frequency, the higher the pitch.

Eventually, pressure waves en route through the cochlea push against the *round window*, a membrane at the far end of the cochlea. As the round window bulges outward toward the air-filled middle ear, it serves as a "release valve" for the force of the waves. Air also moves through an opening in the middle ear into the *Eustachian tube*. This tube runs from the middle ear to the throat (pharynx) and permits air pressure in the middle ear to be equalized with the pressure of outside air. When you change altitude (say, riding in aircraft), this equalizing process makes your ears "pop."

Sounds such as amplified music and the thundering of jet engines are so intense that prolonged exposure to them can cause permanent damage to the inner ear (see *Focus on Our Environment*, page 257). Sounds produced by these modern technologies exceed the functional range of the evolutionarily ancient hair cells in the ear.

The sense of hearing relies on mechanoreceptors called hair cells, which are attached to membranes within the cochlea of the inner ear. Pressure waves generated by sound cause membrane vibrations that bend hair cells. The bending results in action potentials in neurons of the auditory nerve.

BALANCE

Like most other animals, humans must have a sense of the "natural" position for the body (and its parts), given the predictable way they return to it after being tilted or turned upside down. The baseline against which the brain assesses displacement from the natural position is called the "equilibrium position."

The sense of balance relies partly on input from receptors in the eyes, skin, and joints. It also relies on the **vestibular apparatus**, a closed system of fluid-filled canals and sacs in the inner ear (Figure 11.10a). The three **semicircular canals** are positioned at right angles to one another, corresponding to the three planes of space. Within them, sensory receptors detect rotational head movement in all directions, as well as acceleration and deceleration.

vestibular apparatus, a system of fluid-filled sacs and canals inside the ear

A vestibular apparatus (part of each inner ear) consists of a utricle, a saccule, and the three canals labeled here.

superior canal

posterior canal

utricle

saccule

nerve

horizontal canal

fluid pressure (*blue* arrows)

a gelatinous cupula that stimulates hair cells inside when it bends under pressure

hair cells, which synapse with sensory endings leading to nerve

a

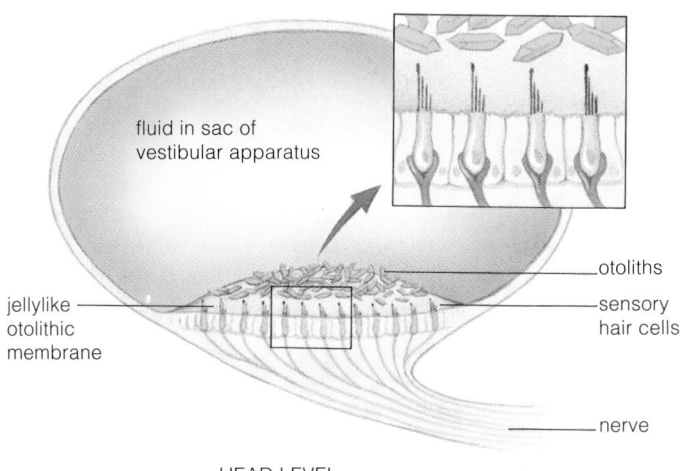

fluid in sac of vestibular apparatus

jellylike otolithic membrane

otoliths

sensory hair cells

nerve

HEAD LEVEL

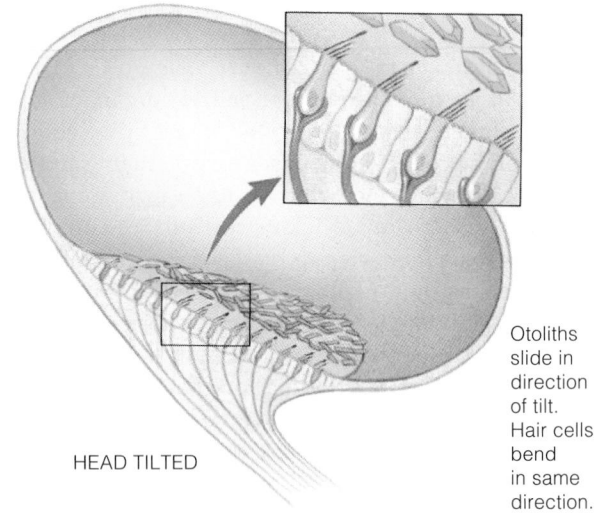

HEAD TILTED

b

Otoliths slide in direction of tilt. Hair cells bend in same direction.

Figure 11.10 (**a**) Vestibular apparatus, an organ of equilibrium. Its three semicircular canals, positioned at angles corresponding to the three planes of space, detect changes in angular (rotational) movements. (**b**) Moving in response to gravity, the otoliths in otolith organs bend projections of hair cells and so stimulate the endings of sensory neurons that are part of a nerve.

At the base of each semicircular canal is a jellylike mass (the *cupula*); this mass rests atop a specialized region of epithelium that contains hair cells just like those in the cochlea. When fluid in the canals moves and presses on the soft mass, projections of the hair cells bend. As in the cochlea, this bending is the first step in a sequence of events leading to action potentials that travel to the brain—in this case, along the vestibular nerve.

The vestibular apparatus also holds two fluid-filled sacs (the *utricle* and *saccule*), each of which contains an *otolith organ* with hair cells embedded in a jellylike "membrane." This material also contains hard bits of calcium carbonate called *otoliths* ("ear stones"). Movements of the membrane and otoliths signal changes in the head's orientation relative to gravity, as well as straight-line acceleration and deceleration. For example, if you tilt your head to one side, the otoliths slide in that direction, the membrane mass shifts, and projections of the hair cells are bent (Figure 11.10b). The otoliths also press on hair cells if your head accelerates, such as happens if you start running down a street, set off in an automobile, or begin a downward plunge on a roller-coaster.

Action potentials from the different parts of the vestibular apparatus travel to reflex centers in the brain stem. The brain integrates the incoming signals with information from the eyes and muscles and then orders compensatory movements that help you keep your balance when you stand, walk, or move your body in other ways.

Motion sickness can occur when extreme or continuous linear, angular, or vertical motion overstimulates hair cells in the vestibular apparatus. Visual sensations often contribute to the sickness; fear and anxiety can also play roles. Action potentials triggered by the sensory input reach a brain center that governs the vomiting reflex. Nausea and vomiting are the chief symptoms in individuals who get carsick, airsick, or seasick. Medications that relieve motion sickness act on brain centers that govern these responses.

Balance involves a sense of the natural position for the body or its parts.

The sense of balance relies in large measure on input from the vestibular apparatus, a closed system of fluid-filled canals and sacs in the inner ear.

The semicircular canals are positioned at angles corresponding to the three planes of space. Sensory receptors within them detect rotational head movement, acceleration, and deceleration.

The fluid-filled utricles and saccules contain otolith organs, in which sensory hair cells are embedded in a jellylike membrane. Movements of the membrane and otoliths signal changes in the head's orientation relative to gravity, as well as straight-line acceleration and deceleration.

NOISE POLLUTION: AN ATTACK ON THE EARS

In the United States, roughly 10 million people, including a growing number of musicians, have lost some hearing due to damage caused by loud noise. The National Institutes of Health has estimated that high noise levels in the home, on the job, or in recreational pursuits put another 20 million Americans at risk of hearing loss. One-third of adults in the United States will suffer significant damage to their hearing by the time they are 65. Researchers believe that most cases are due to the long-term effects of living in a noisy world.

The loudness of a sound is measured in *decibels*. A quiet conversation occurs at about 50 decibels, whereas rustling papers make noise at a mere 20 decibels. The delicate sensory hair cells in the inner ear (Figure 11.11) begin to be damaged when a person is exposed to sounds louder than about 75–85 decibels over long periods. Unfortunately, our society is permeated with sounds above that threshold.

For example, a snowmobile or "boom box" stereo system cranks out sound at well over 100 decibels, as can a personal "Walkman"-type stereo if the volume is turned up too high. At 130 decibels—typical of a rock concert or shotgun blast—permanent damage can occur much more quickly. Many health professionals recommend protective earwear for anyone who regularly operates a power mower or other noisy lawn-care equipment. Such protection is (or should be) mandatory for workers in some noisy professions, such as airline service personnel or construction workers who operate jackhammers or chain saws.

For many people it is difficult to avoid noise pollution completely. However, individuals *can* use ear protection when it seems prudent and avoid frequent, prolonged exposure to very loud noise. Following these simple guidelines can help ensure that your hearing lasts as long as you do.

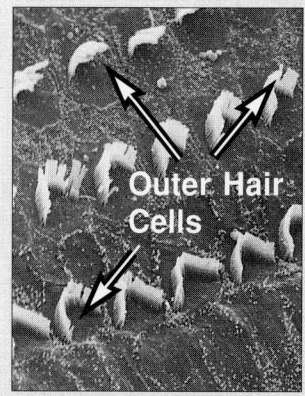
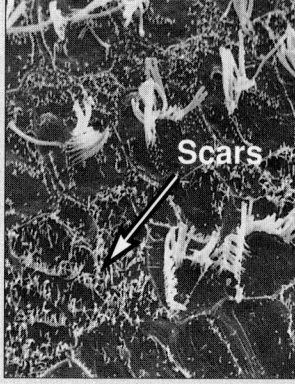

a   b

Figure 11.11 (a) Healthy sensory hair cells of the inner ear. (b) Hair cells damaged by exposure to loud noise.

VISION: AN OVERVIEW

All organisms are sensitive to light. **Vision**, however, requires (1) a system of photoreceptors and (2) brain centers that can receive and interpret the patterns of action potentials from different parts of the photoreceptor system. The sense of vision is an awareness of the position, shape, brightness, distance, and movement of visual stimuli. **Eyes** are photoreceptor organs that contribute to image formation.

Eye Structure

The eye has three layers (Table 11.2), sometimes called "tunics." The outer layer consists of a sclera and transparent **cornea**. The middle layer consists mainly of a choroid, ciliary body, and iris. The key feature of the inner layer is the retina (Figure 11.12).

The *sclera*—the dense, fibrous "white" of the eye—protects most of the eyeball, except for a "front" region formed by the cornea. Moving inward, the thin, dark-pigmented *choroid* underlies the sclera. It prevents light from scattering inside the eyeball and contains most of the eye's blood vessels. Behind the transparent cornea is a circular, pigmented **iris** (after *irid*, which means "colored circle") with a "hole" in its center. This *pupil* is the entrance for light. When bright light hits the eye, circular muscles in the iris contract and shrink the pupil. In dim light, radial muscles contract and enlarge the pupil. Behind the iris is a saucer-shaped **lens**, with onionlike layers of transparent proteins. Ligaments attach the lens to smooth muscle of the *ciliary body*; this muscle functions in focusing light, as we will see shortly. The lens focuses incoming light onto a dense layer of photoreceptor cells behind it, in the retina. A clear fluid (*aqueous humor*) bathes both sides of the lens, and a jellylike substance (*vitreous humor*) fills the chamber behind the lens.

The **retina** is a thin layer of neural tissue at the back of the eyeball. It has a basement layer—a pigmented epithelium that covers the choroid. Resting on the basement layer are densely packed photoreceptors that are functionally linked with a variety of neurons. Axons from some of these neurons converge to form the optic nerve at the back of the eyeball. The optic nerve is the trunk line to the thalamus, which sends information on to the **visual cortex**. The site where the optic nerve exits the eye is a "blind spot" because no photoreceptors are present there.

Table 11.2 Components of the Eye

THREE LAYERS FORMING THE WALL OF THE EYEBALL

Fibrous Tunic	Sclera. *Protects eyeball*
	Cornea. *Focuses light*
Vascular Tunic	Choroid. *Blood vessels nutritionally support wall cells; pigments prevent light scattering*
	Ciliary body. *Its muscles control lens shape; its fine fibers hold lens in upright position*
	Iris. *Adjustments here control incoming light*
	Pupil. *Serves as entrance for light*
Sensory Tunic	Retina. *Absorbs and transduces light energy*
	Fovea. *Increases visual acuity*
	Start of optic nerve. *Carries signals to brain*

INTERIOR OF THE EYEBALL

Lens	*Focuses light on photoreceptors*
Aqueous humor	*Transmits light, maintains pressure*
Vitreous body	*Transmits light, supports lens and eyeball*

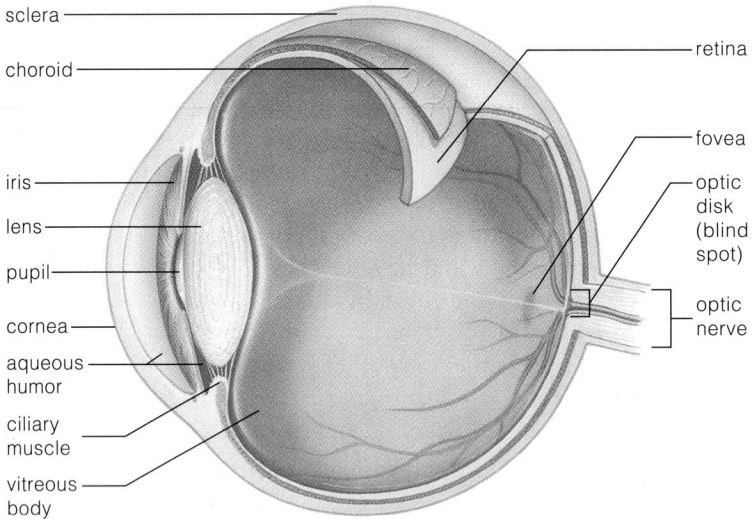

Figure 11.12 Structure of the human eye.

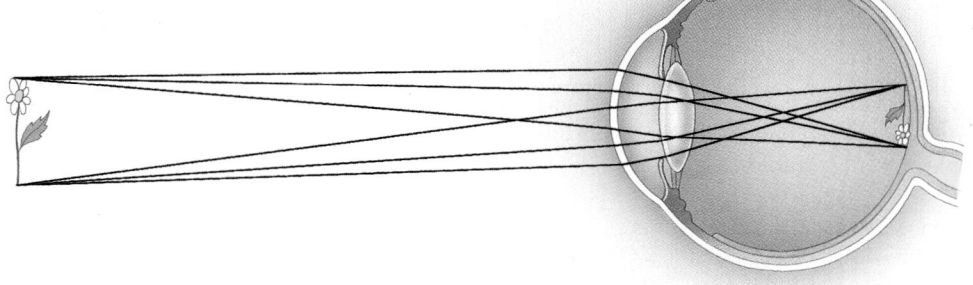

Figure 11.13 Bending of light rays by the curvature of the cornea.

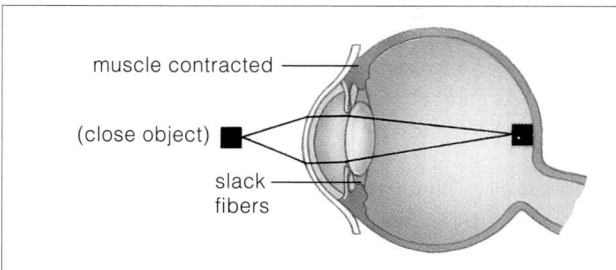

a Accommodation for close objects (lens bulges)

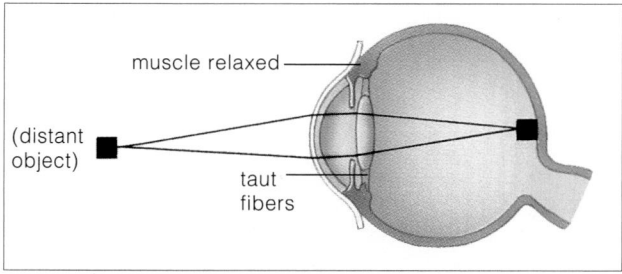

b Accommodation for distant objects (lens flattens)

Figure 11.14 Focusing light on the retina by adjusting the lens (visual accommodation). A muscle encircling the lens attaches to it by fiberlike ligaments. (**a**) Close objects are brought into focus when the muscle contracts and makes the lens bulge, so the focal point moves closer. (**b**) Distant objects are brought into focus when the muscle relaxes and makes the lens flatten, so the focal point moves farther back.

Focusing Mechanisms

The surface of the cornea is curved. Incoming light rays hit it at different angles, and as they pass through the cornea their trajectories (paths) bend. The rays converge at the back of the eyeball. There, because of the way the rays were bent at the curved cornea, they form a pattern of stimulation that is upside-down and reversed left to right relative to the original light source (Figure 11.13).

This "upside-down and backwards" orientation is corrected in the brain.

Light rays from sources at different distances from the eye strike the cornea at different angles and will be focused at different distances behind it. Therefore, adjustments must be made so that the light will be focused precisely on the retina. Normally, the lens can be adjusted so that the focal point coincides exactly with the retina. Ciliary muscle adjusts the shape of the lens. The muscle encircles the lens and attaches to it by fiberlike ligaments (Figure 11.12). When the muscle contracts, the lens bulges, so the focal point moves closer. When the muscle relaxes, the lens flattens, so the focal point moves farther back (Figure 11.14). Such adjustments are called **accommodation**. If they are not made, rays from very distant objects will be in focus at a point slightly in front of the retina, and rays from very close objects will be focused behind the retina.

Sometimes the lens cannot be adjusted enough to place the focal point on the retina. Sometimes also, the eyeball is not shaped quite right. The lens is too close or too far away from the retina, so accommodation alone cannot bring about a precise match. Eyeglasses can correct these problems, which are called farsightedness and nearsightedness, respectively. *Focus on Your Health*, page 260, examines these and some more serious eye disorders.

Eyes are sensory organs specialized for photoreception.

In the outer eye layer, the sclera protects the eyeball and the cornea focuses light.

In the middle layer, the choroid prevents light scattering, the iris controls incoming light, and the ciliary body and lens aid in focusing light on photoreceptors.

The inner layer of the retina absorbs light. Photoreceptors in this thin layer of neural tissue are functionally linked to neurons by which action potentials travel to the brain.

11.9 DISORDERS OF THE EYE

Two-thirds of all the sensory receptors that your body possesses are in your eyes. Those photoreceptors do more than detect light. They also allow you to see the world in a rainbow of colors. Your eyes are the single most important source of information about the outside world.

Injuries, disease, inherited abnormalities, and advancing age can disrupt functions of the eyes. The consequences range from relatively harmless conditions, such as nearsightedness, to total blindness. Each year, many millions of people must deal with such consequences.

COLOR BLINDNESS Consider a common heritable abnormality, *red-green color blindness*. It is an X-linked, recessive trait that shows up most often in males. The retina lacks some or all of the cone cells with pigments that normally respond to light of red or green wavelengths. Most of the time, color-blind persons merely have trouble distinguishing red from green in dim light. However, some cannot distinguish between the two even in bright light. The rare few who are totally color-blind have only one of three kinds of pigments that selectively respond to red, green, or blue wavelengths. They see the world only in shades of gray.

FOCUSING PROBLEMS Other heritable abnormalities arise from misshapen features of the eye that affect the focusing of light. *Astigmatism*, for example, results from corneas with an uneven curvature; they cannot bend incoming light rays to the same focal point.

Nearsightedness (myopia) commonly occurs when the horizontal axis of the eyeball is longer than the vertical axis. It also occurs when the ciliary muscle responsible for adjustments in the lens contracts too strongly. The outcome is that images of distant objects are focused in front of the retina instead of on it (Figure 11.15a). *Farsightedness* (hyperopia) is the opposite problem. The vertical axis of the eyeball is longer than the horizontal axis (or the lens is "lazy"), so close images are focused behind the retina (Figure 11.15b).

Even a normal lens loses some of its natural flexibility as a person grows older. That is why people over 40 years old often must start wearing eyeglasses.

EYE DISEASES The eye and its functions are vulnerable to infection and disease. In some parts of the United States, for example, a fungal infection of the lungs (*histoplasmosis*) can lead to retinal damage when the pathogen moves to the eye. This complication can cause partial or total loss of vision. As another example, *herpes simplex*, a virus that causes skin sores, also can infect the cornea and cause it to ulcerate.

Trachoma is a highly contagious disease that has blinded millions, mostly in North Africa and the Middle East. The culprit is a bacterium that also is responsible for the

a Focal point in nearsighted vision. The example shows flamingos in Tanzania, East Africa. In a nearsighted person, the birds in flight in the background would be out of focus.

Figure 11.15 Examples of nearsighted and farsighted vision.

sexually transmitted disease chlamydia (Chapter 15). The eyeball and the lining of the eyelids (conjunctiva) become damaged. The damaged tissues are entry points for bacteria that can cause secondary infections. In time the cornea can become so scarred that blindness follows.

AGE-RELATED PROBLEMS *Cataracts*, the gradual clouding of the lens, are a problem associated with aging, although they also may arise through injury or diabetes. Possibly the condition arises when the transparent proteins making up the lens undergo structural changes. The clouding may skew the trajectory of incoming light rays. If the lens becomes totally opaque, light cannot enter the eye at all.

Glaucoma results when excess aqueous humor accumulates inside the eyeball. Blood vessels that service the retina collapse under the increased fluid pressure. Vision deteriorates as neurons of the retina and optic nerve die off. Although chronic glaucoma often is associated with advanced age, the problem actually starts in middle age. If detected early, the fluid pressure can be relieved by drugs or surgery before the damage becomes severe.

EYE INJURIES *Retinal detachment* is the eye injury we read about most often. It may follow a physical blow to the head or an illness that tears the retina. As the jellylike vitreous body oozes through the torn region, the retina becomes

lifted from the underlying choroid. In time it may peel away entirely, leaving its blood supply behind. Early symptoms of the damage include blurred vision, flashes of light that occur in the absence of outside stimulation, and loss of peripheral vision. Without medical intervention, the person may become totally blind in the damaged eye.

NEW TECHNOLOGIES Today many different procedures are used to correct eye disorders. In *corneal transplant surgery*, the defective cornea is removed; then an artificial cornea (made of clear plastic) or a natural cornea from a donor is stitched in place. Within a year, the patient is fitted with eyeglasses or contact lenses. Similarly, cataracts can be surgically corrected by removing the lens and replacing it with an artificial one, although the operation is not always successful.

Severely nearsighted people may opt for *radial keratotomy*, a still-controversial surgical procedure in which tiny, spokelike incisions are made around the edge of the cornea to flatten it more. When all goes well, the adjustment eliminates the need for corrective lenses. Sometimes, however, the result is overcorrected or undercorrected vision.

Retinal detachment may be treatable with *laser coagulation*, a painless technique in which a laser beam seals off leaky blood vessels and "spot welds" the retina to the underlying choroid.

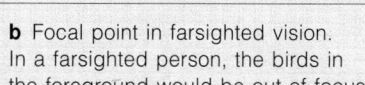

b Focal point in farsighted vision. In a farsighted person, the birds in the foreground would be out of focus.

The pathway from the retina into the brain stands as one of the best examples of neuronal architecture. Along this sensory pathway, raw visual information is received, transmitted, and combined in ways that lead to awareness of light and shadows, of colors, and of near and distant objects in the outside world.

Organization of the Retina

The flow of information begins as light reaches the retina, at the back of the eyeball. The retina's basement layer, a pigmented epithelium, covers the choroid. Resting on the epithelium are densely packed photoreceptors called **rod cells** and **cone cells** (Figure 11.16). Rod cells are much more sensitive to light than cone cells. At night, they contribute to perception of movement by detecting changes in light intensity across the field of vision. Cone cells detect bright light. They contribute to daytime vision and color perception.

Neurons in the eye are organized in distinct layers above the rods and cones. As Figure 11.17 shows, information flows from these photoreceptors to the *bipolar* interneurons, then to the interneurons called *ganglion* cells, the axons of which form the two optic nerves to the visual cortex.

Before visual signals depart from the retina, they converge dramatically. Input from 125 million photoreceptors converges on a mere 1 million ganglion cells. Signals also flow laterally among *horizontal* cells and *amacrine* cells. Both types of neurons act in concert to dampen or enhance the signals before they reach ganglion cells. Thus, *a great deal of synaptic integration and processing goes on even before visual information is sent to the brain.*

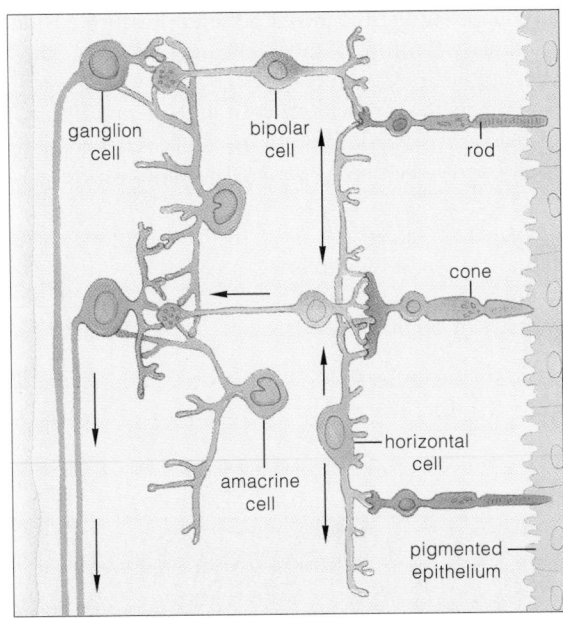

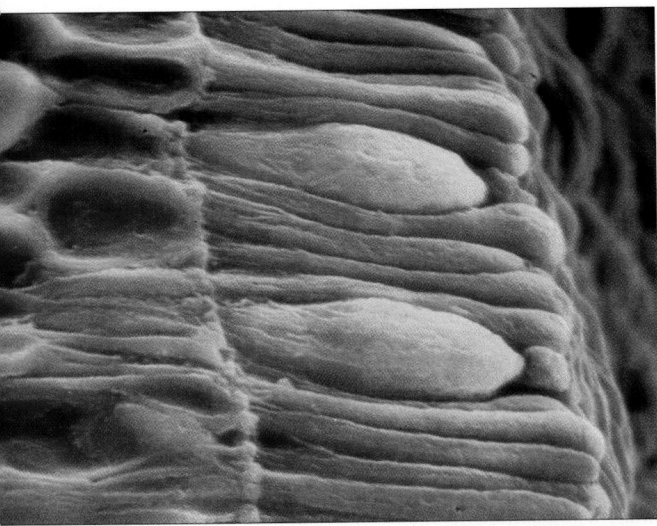

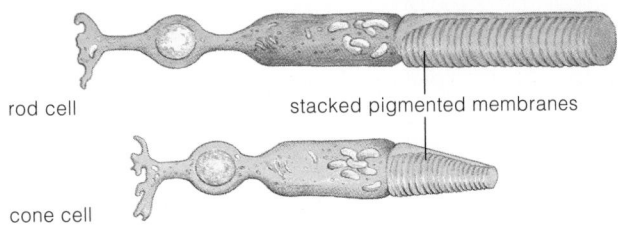

Figure 11.16   Photoreceptors: rods and cones.

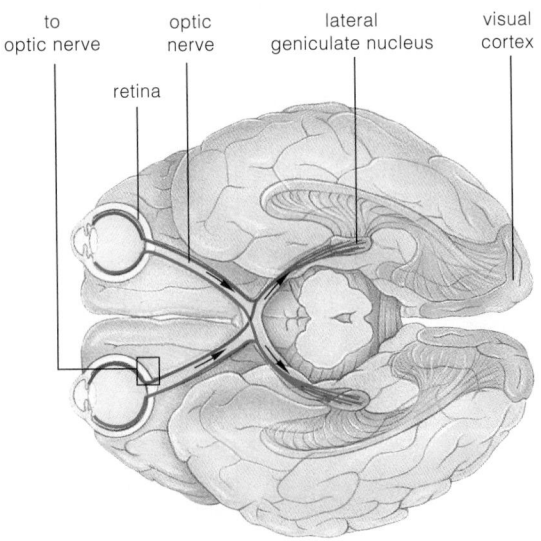

Figure 11.17   Sensory pathway from the retina to the brain.

Neuronal Responses to Light

ROD CELLS A rod cell's outer segment consists of several hundred membranous disks, each peppered with 108 molecules of **rhodopsin** (a visual pigment). The membrane stacking and the sheer density of pigments increase the chance of photon interception. Signals resulting from even a few photons can lead to conscious awareness of incoming light.

Each rhodopsin molecule consists of a protein (opsin) to which a signal molecule, *cis*-retinal, is bound. The retinal is derived from vitamin A, which is one reason why lack of vitamin A in the diet can impair vision, especially at night. When the cis-retinal absorbs light energy, it is temporarily converted to a slightly different form, called *trans*-retinal:

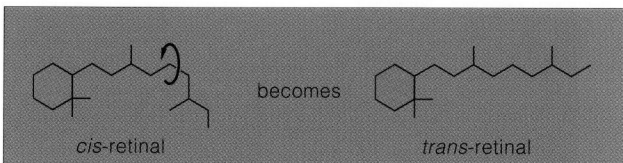

cis-retinal   becomes   *trans*-retinal

In this altered form, the retinal initiates a series of chemical reactions within the photoreceptor. The reactions lead to a voltage change across the photoreceptor's plasma membrane. This local, graded potential causes the rod or cone to *reduce* its release of a neurotransmitter. That chemical change alters the activity of neighboring neurons.

CONE CELLS Rod cells respond to dim light, and they are most effective in absorbing light wavelengths in the green part of the spectrum. Color and daytime vision depend on red, green, and blue cone cells, each with a different kind of visual pigment. Here again, the absorption of photons reduces the release of a neurotransmitter that otherwise inhibits signal formation in the adjacent neurons.

Near the center of the retina is a funnel-shaped depression called the **fovea**. Here, photoreceptors are arrayed in the greatest density, and the visual acuity (ability to discriminate) also is greatest. The fovea's dense array of cone cells enables you to discriminate between adjacent points in space with precision.

RECEPTIVE FIELDS The retina's surface is organized into "receptive fields," which are restricted areas that influence the activity of specific neurons. For example, a given ganglion cell will respond best only when light stimulates a particular (typically circular) region of the retina. Cells in the visual cortex respond to motion, to a rapid change in light intensity, or to some other specific stimulus. In one experiment, cells generated signals about a suitably oriented bar (Figure 11.18). Such cells do not respond to diffuse, uniform illumination, which could produce a confusing array of signals to the brain.

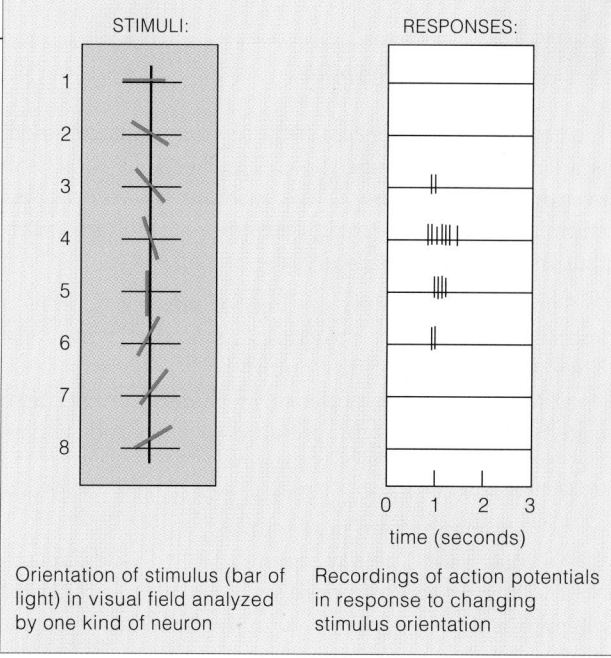

STIMULI: RESPONSES:

time (seconds)

Orientation of stimulus (bar of light) in visual field analyzed by one kind of neuron

Recordings of action potentials in response to changing stimulus orientation

Figure 11.18 Example of experiments into the nature of receptive fields for visual stimuli. David Hubel and Torsten Wiesel implanted an electrode in the brain of an anesthetized cat. They positioned the cat in front of a small screen upon which different patterns of light were projected—in this case, a hard-edged bar. Light or shadow falling on a restricted portion of the screen excited or inhibited signals by a single neuron in the visual cortex.

Tilting the bar at different angles produced changes in the neuron's activity. A nearly vertical bar image produced the strongest signal (*numbered 4 in the sketch*). When the bar image tilted slightly, signals were less frequent. When the image tilted past a certain angle, signals stopped.

ON TO THE VISUAL CORTEX The part of the outside world that a person actually sees is the individual's "visual field." The right side of both retinas intercepts light from the left half of the visual field; the left side intercepts light from the right half. The optic nerve leading out of each eye delivers the signals about a stimulus from the left visual field to the right cerebral hemisphere, and signals from the right to the left hemisphere (Figure 11.17).

Axons of the optic nerves end in a layered brain region, the lateral geniculate nucleus. Each layer has a map corresponding to receptive fields of the retina. Its interneurons deal with one aspect of a visual stimulus—form, movement, depth, color, texture, and so on. After initial processing all the visual signals travel rapidly, at the same time, to different parts of the visual cortex. There, final integration produces organized electrical activity—and the sensation of sight.

Each eye is an outpost of the brain, collecting and analyzing visual information, then sending it on for further processing in the brain through a highly organized sensory pathway.

SUMMARY

1. A stimulus is a specific form of energy that the body is able to detect by specific receptor regions of sensory neurons. A sensation is a conscious awareness that stimulation has occurred. Perception is understanding that the sensation has been experienced.

2. Sensory receptors are endings of sensory neurons or specialized accessory structures associated with them. They respond to specific stimuli.

   a. Mechanoreceptors detect mechanical energy associated with changes in pressure (e.g., sound waves), changes in position, or acceleration.

   b. Thermoreceptors detect the presence of or changes in radiant energy from heat sources.

   c. Nociceptors (pain receptors) detect tissue damage.

   d. Chemoreceptors, such as taste receptors, detect chemical substances dissolved in the body fluids that are bathing them.

   e. Osmoreceptors detect changes in water volume (hence solute concentrations) in the surrounding fluid.

   f. Photoreceptors, such as rods and cones of the retina, detect light.

3. A sensory system has sensory receptors for specific stimuli and nerve pathways from those receptors to receiving and processing centers in the brain. The brain assesses a particular stimulus based on which nerve pathway is delivering the signals, the frequency of signals traveling along each axon of the pathway, and the number of axons that were recruited into action.

4. Somatic sensations include touch, pressure, pain, temperature, and muscle sense. The receptors associated with these sensations are not localized in a single organ or tissue.

5. The special senses include taste, smell, hearing, balance, and vision. The receptors associated with these senses typically reside in sense organs or another specific body region.

6. The senses of taste and smell involve sensory pathways from chemoreceptors to processing regions in the cerebral cortex.

7. The sense of hearing requires components of the outer, middle, and inner ear that respectively collect, amplify, and respond to sound waves. Organs of equilibrium in the inner ear detect gravity, velocity, acceleration, and other factors that affect body positions and movements.

8. Vision requires eyes (sensory organs with a dense array of photoreceptors) and a capacity for image formation in the brain, based upon incoming patterns of visual stimulation.

Review Questions

1. What is a stimulus? When a receptor cell detects a specific kind of stimulus, what happens to the stimulus energy? *248*

2. Name six categories of sensory receptors and the type of stimulus energy that each type detects. *248*

3. What is the difference between a sensory receptor and a sense organ? *249*

4. How do somatic sensations differ from special senses? *249*

5. What is pain? Describe one type of pain receptor. *250–251*

6. What is sound? How is vibration frequency related to sound? *254–255*

7. In the ear, sound waves cause the tympanic membrane to vibrate. What happens next in the middle ear? In the inner ear? *254-255*

8. Label the component parts of the eye: *258*

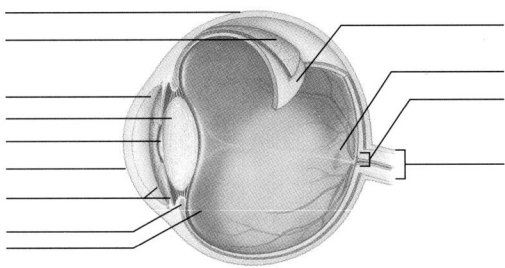

9. How does vision differ from photoreception? What sensory apparatus does vision require? *262–263*

10. How does the eye focus the light rays of an image? What do nearsighted and farsighted mean? *260*

Critical Thinking: You Decide *(Key in Appendix V)*

1. Astronauts living in a space vehicle eat much of their food directly from sealed pouches. They have often complained that such food tastes bland. Given what you know about the senses of taste and smell and the link between them, can you explain why?

2. Juanita consulted her doctor because she was experiencing recurring episodes of dizziness. Her doctor immediately asked her whether "dizziness" meant she had sensations of lightheadedness, as if she were going to faint, or whether it meant she had sensations of *vertigo*—that is, a feeling that she herself or objects near her were spinning around. Why did her doctor consider this clarification important early in his evaluation of her condition?

3. Michael, a 3-year old, experiences chronic *middle-ear infection*, which is becoming quite common among youngsters, in part due to an increase in antibiotic-resistant bacteria. This year, despite antibiotic treatment, an infection became so advanced that he had trouble hearing. Then his left eardrum ruptured and a jellylike substance dribbled out. The pediatrician told Michael's parents not to worry, that if the eardrum had not ruptured on its own she would have had to insert a drainage tube into it. Speculate on why the pediatrician concluded that this would have been a necessary procedure.

Self-Quiz (Answers in Appendix IV)

1. A _____ is a specific form of energy that is capable of eliciting a response from a sensory receptor.

2. Awareness of a stimulus is called a _____ .

3. _____ is understanding what particular sensations mean.

4. Each sensory system is composed of _____ .
   a. nerve pathways from specific receptors to the brain
   b. sensory receptors
   c. brain regions that deal with sensory information
   d. all of the above

5. _____ detect mechanical energy associated with changes in pressure, in position, or acceleration.
   a. Chemoreceptors   c. Photoreceptors
   b. Mechanoreceptors   d. Thermoreceptors

6. Detecting chemical substances present in the body fluids that bathe them is the function of _____ .
   a. thermoreceptors   c. mechanoreceptors
   b. photoreceptors   d. chemoreceptors

7. Which of the special senses is based on the following events: Membrane vibrations cause fluid movements, which lead to bending of mechanoreceptors and initiation of action potentials.
   a. taste   c. hearing
   b. smell   d. vision

8. Detecting light energy is the function of _____.
   a. chemoreceptors   c. photoreceptors
   b. mechanoreceptors   d. thermoreceptors

9. Vision requires _____.
   a. a tissue with dense arrays of photoreceptors
   b. eyes
   c. image-forming centers in the brain
   d. all of the above
   e. b and c only

10. The outer layer of the eye includes the _____ .
    a. lens and choroid   c. retina
    b. sclera and cornea   d. both a and c are correct

11. The inner layer of the eye includes the _____ .
    a. lens and choroid   c. retina
    b. sclera and cornea   d. start of optic nerve

12. Match each of the following terms with the appropriate description.

   ____ somatic senses   a. produced by strong
         (general senses)   stimulation
   ____ special senses   b. endings of sensory
   ____ variations in   neurons or specialized
         stimulus intensity   cells next to them
   ____ action potential   c. taste, smell, hearing,
   ____ sensory receptor   balance, and vision
                                d. frequency and number of
                                action potentials
                                e. touch, pressure,
                                temperature, pain, and
                                muscle sense

Selected Key Terms

accommodation 259
chemoreceptor 248
cochlea 254
cone cell 262
cornea 258
eye 258
fovea 263
hair cell 254
iris 258
lens 258
mechanoreceptor 248
nociceptor 248
organ of Corti 254
perception 248
photoreceptor 248
retina 258

rhodopsin 263
rod cell 262
semicircular canal 256
sensation 248
sensory receptor 248
sensory system 248
somatosensory cortex 250
somatic sensation 249
stimulus 248
tectorial membrane 255
thermoreceptor 248
tympanic membrane 254
vestibular apparatus 256
vision 258
visual cortex 258

Readings

Discover, June 1993. Entire issue devoted to articles on current understanding of the special senses.

McLaughlin, S., and R. Margolskee. November–December 1994. "The Sense of Taste." *American Scientist*.

Sherwood, L. 1997. *Human Physiology*. 3d ed. Belmont, Calif.: Wadsworth.

Wright, K. April 1994. "The Sniff of Legend." *Discover*. Speculation on the existence of a sensory pathway activated by human pheromones.

Zeki, S. September 1992. "The Visual Image in Mind and Brain." *Scientific American*.

For additional readings, go to InfoTrac College Edition, your online library at:

http://www.infotrac-college.com/wadsworth

INTEGRATION AND CONTROL: ENDOCRINE SYSTEM

Rhythms and Blues

The pineal gland, a lump of tissue in your brain, secretes the hormone melatonin. When the brain receives sensory signals from your eyes about the waning light at sunset, the pineal gland steps up its melatonin secretion. Molecules of melatonin are picked up by the bloodstream and transported to target cells—in this case, to certain brain neurons. Those neurons are involved in sleep behavior, a lowering of body temperature, and possibly other physiological events. At sunrise, when the eye detects the light of a new day, melatonin production slows down. Your body temperature increases, and you wake up and become active.

The cycle of sleep and arousal is evidence of an internal *biological clock* that seems to tick in synchrony with day length. The clock apparently is influenced by melatonin, and it can be disturbed by a change in circumstances. Jet lag is an example. A traveler from the United States to Paris starts her vacation with four days of disorientation. Two hours past midnight she is sitting up in bed, wondering where the coffee and croissants are. Two hours past noon she is ready for bed. Her body will gradually shift to a new routine as melatonin secretion becomes adjusted so that the hormone signals begin arriving at their target neurons on Paris time.

Figure 12.1 Too little light and too much melatonin, a hormone, may trigger seasonal affective disorder (SAD), or "winter blues," in some people.

Some people are affected by severe *winter blues*, or seasonal affective disorder (SAD). Symptoms typically include depression and an overwhelming desire to sleep (Figure 12.1). Their discomfort may result from a biological clock that is out of synchrony with the changes in day length during winter (days are shorter and nights longer). Their symptoms worsen when they are given doses of melatonin. And they improve dramatically when they are exposed to intense light, which shuts down pineal activity.

Hormones, our main topic in this chapter, are part and parcel of life's tempos. Some hormones are crucial to basic biological events, such as sexual development and reproduction. They and other signaling molecules are secreted by various organs and tissues and, as we will see, hormones have a range of targets throughout the human body. Indeed, they ultimately control many body functions and influence many aspects of behavior.

KEY CONCEPTS

1. Hormones and other signaling molecules have a central role in integrating the activities of individual cells in ways that benefit the whole body.

2. Only the cells with molecular receptors for a specific hormone are its targets. Hormones operate by serving as signals for change in the activity of target cells.

3. Many types of hormones influence the activity of a target cell's genes and protein-synthesizing machinery. Other types call for alterations in existing cell molecules and cell structures. Some hormones exert their effects by altering membrane characteristics—for example, changing the permeability of the plasma membrane to a particular solute.

4. Some hormones help the body adjust to short-term changes in diet and activity levels. Other hormones help induce long-term adjustments in cell activities that bring about body growth, development, and reproduction.

5. The hypothalamus and pituitary gland interact in ways that coordinate the activities of a number of endocrine glands. Together, they exert control over many body functions.

6. In addition to hormonal signals, neural signals, changes in local chemical conditions, and environmental cues are the triggers for hormone secretion.

CHAPTER AT A GLANCE

12.1   The Endocrine System

12.2   Signaling Mechanisms

12.3   The Hypothalamus and Pituitary Gland

12.4   Examples of Abnormal Pituitary Output

12.5   Sources and Effects of Other Hormones

12.6   Pancreatic Islets

12.7   The Adrenal Glands

12.8   *Are "Endocrine Disrupters" Harming Human Health?*

12.9   Hormones from the Thyroid, Parathyroid, and Some Other Sources

12.10   Some Final Examples of Integration and Control

12.11   *Communication Molecules Provide Hope for Healing*

THE ENDOCRINE SYSTEM

Hormones and Other Signaling Molecules

Throughout their lives, cells must respond to changing conditions by taking up and releasing various chemical substances. In humans (as in other vertebrates), the responses of millions to many billions of cells must be integrated in ways that benefit the whole body.

Integration of cell activities depends upon signaling molecules. These include hormones, neurotransmitters, local signaling molecules, and pheromones. Each type of signaling molecule acts on target cells, which are any cells that have receptors for the molecule and that may alter their activities in response to it. A target may or may not be adjacent to the cell that sends the signal.

Hormones are secretions of endocrine glands, endocrine cells, and some neurons, and they travel the bloodstream to distant target cells. They are this chapter's focus.

By contrast, **neurotransmitters**, released from axon endings of neurons, then act swiftly on abutting target cells by diffusing across the tiny gap (synapse) that separates the cells. Section 10.3 describes their sources and their action. Also, **local signaling molecules**, released by many types of body cells, alter conditions within localized regions of tissues. For instance, targets of prostaglandins include smooth muscle cells in the walls of bronchioles, which then constrict or dilate and so alter air flow in lungs. **Pheromones**, the nearly odorless secretions of certain exocrine glands, diffuse through water or air to targets outside the body. Pheromones act on cells of other animals of the same species and so help integrate social behavior, such as behaviors related to reproduction. In humans, the vomeronasal organ described in Section 11.3 is a pheromone detector. Do phermomes act at the subconscious level to trigger inexplicable impressions, such as spontaneous good or bad "feelings" about someone you just met? We do not know the answer, but studies are under way.

Discovery of Hormones and Their Sources

The word hormone dates back to the early 1900s. W. Bayliss and E. Starling were trying to find out what triggers the secretion of pancreatic juices when food travels through the canine gut. As they knew already, acids mix with food in the stomach, and the pancreas secretes an alkaline solution after the acidic mixture moves into the small intestine. Was the nervous system or something else stimulating the pancreatic response?

To find the answer, Bayliss and Starling blocked nerves—but not blood vessels—leading to a laboratory animal's upper small intestine. Later, when acidic food entered the intestine, the pancreas was still able to respond. Extracts of cells from the intestinal lining—

a glandular epithelium—also induced the response. Glandular cells had to be the source of the pancreas-stimulating substance.

The substance came to be called secretin. Proof of its existence and mode of action confirmed a centuries-old idea: *Internal secretions that are picked up by the bloodstream influence the activities of the body's organs.* Starling coined the word "hormone" for such internal glandular secretions (after the Greek *hormon*, "to set in motion").

Later, researchers identified other hormones and their sources. Figure 12.2 shows the locations of the major sources of hormones in the human body.

Collectively, these sources of hormones came to be viewed as the **endocrine system**. The name implies that there is a separate control system for the body, apart from the nervous system. (*Endon* means within; *krinein* means separate.) However, biochemical studies and electron microscopy have revealed that endocrine sources and the nervous system function in intricately connected ways, as you will see shortly.

Today we know that hormones also often interact with one another. Three kinds of hormone interactions are common:

1. *Opposing interaction.* The effect of one hormone opposes the effect of another. Insulin, for example, promotes a decrease in the glucose level in the blood, and glucagon promotes an increase.

2. *Synergistic interaction.* The sum total of the actions of two or more hormones is necessary to produce the required effect on target cells. For instance, women cannot produce and secrete milk without the synergistic interaction of the hormones prolactin, oxytocin, and estrogen.

3. *Permissive interaction.* One hormone exerts its effect only when a target cell has become "primed" to respond in an enhanced way to that hormone. The priming is accomplished by previous exposure to another hormone. Pregnancy, for example, depends on the lining of the uterus being exposed first to estrogens, then to progesterone.

Hormones are signaling molecules secreted by endocrine glands, endocrine cells, and some neurons. The bloodstream distributes hormones to distant target cells.

Together, the various glands and cells that secrete hormones make up the endocrine system.

Hormones may interact in opposition, as when the effect of one blocks the effect of another; synergistically, as when the actions of two or more hormones are necessary to produce the effect on target cells; or permissively, as when a target cell must be primed by exposure to one hormone before it can respond to a second one.

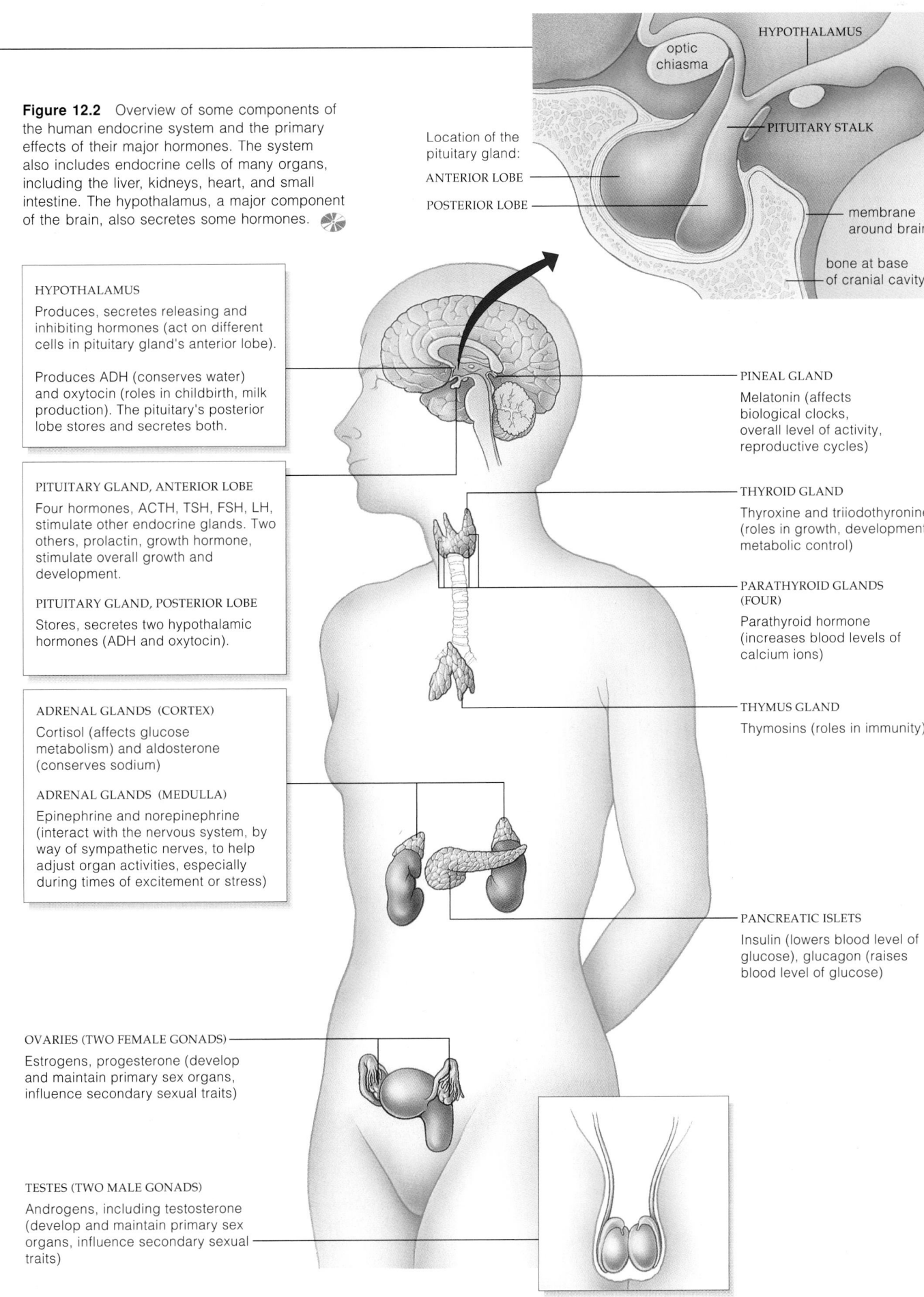

Figure 12.2 Overview of some components of the human endocrine system and the primary effects of their major hormones. The system also includes endocrine cells of many organs, including the liver, kidneys, heart, and small intestine. The hypothalamus, a major component of the brain, also secretes some hormones.

HYPOTHALAMUS

optic chiasma

PITUITARY STALK

Location of the pituitary gland:

ANTERIOR LOBE

POSTERIOR LOBE

membrane around brain

bone at base of cranial cavity

HYPOTHALAMUS

Produces, secretes releasing and inhibiting hormones (act on different cells in pituitary gland's anterior lobe).

Produces ADH (conserves water) and oxytocin (roles in childbirth, milk production). The pituitary's posterior lobe stores and secretes both.

PITUITARY GLAND, ANTERIOR LOBE

Four hormones, ACTH, TSH, FSH, LH, stimulate other endocrine glands. Two others, prolactin, growth hormone, stimulate overall growth and development.

PITUITARY GLAND, POSTERIOR LOBE

Stores, secretes two hypothalamic hormones (ADH and oxytocin).

ADRENAL GLANDS (CORTEX)

Cortisol (affects glucose metabolism) and aldosterone (conserves sodium)

ADRENAL GLANDS (MEDULLA)

Epinephrine and norepinephrine (interact with the nervous system, by way of sympathetic nerves, to help adjust organ activities, especially during times of excitement or stress)

OVARIES (TWO FEMALE GONADS)

Estrogens, progesterone (develop and maintain primary sex organs, influence secondary sexual traits)

TESTES (TWO MALE GONADS)

Androgens, including testosterone (develop and maintain primary sex organs, influence secondary sexual traits)

PINEAL GLAND

Melatonin (affects biological clocks, overall level of activity, reproductive cycles)

THYROID GLAND

Thyroxine and triiodothyronine (roles in growth, development, metabolic control)

PARATHYROID GLANDS (FOUR)

Parathyroid hormone (increases blood levels of calcium ions)

THYMUS GLAND

Thymosins (roles in immunity)

PANCREATIC ISLETS

Insulin (lowers blood level of glucose), glucagon (raises blood level of glucose)

SIGNALING MECHANISMS

Hormones and other signaling molecules have diverse effects. Some kinds induce their target cells to take up glucose or some other substance from the surroundings. Other kinds stimulate or inhibit their target cells in ways that lead to altered rates of protein synthesis, to the modification of existing proteins or of structures in the cell cytoplasm, even to changes in the cell's shape.

Two factors exert major influence over the responses to hormonal signals. First, different hormones activate different cellular mechanisms. Second, not all cells can respond to a given signal. Many kinds of cells have receptors for cortisol, for instance, so that hormone's effects are widespread. If only a few cell types have receptors for a hormone, its effects in the body are highly specific.

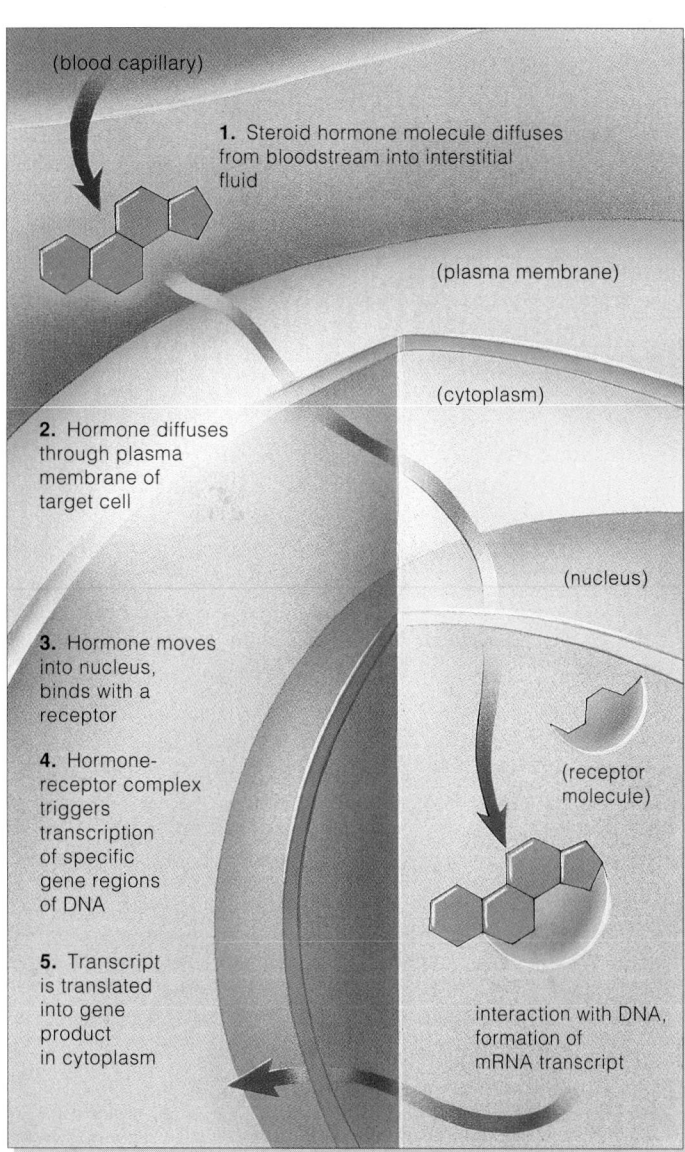

Figure 12.3 Example of a mechanism by which a steroid hormone initiates changes in a target cell's activities.

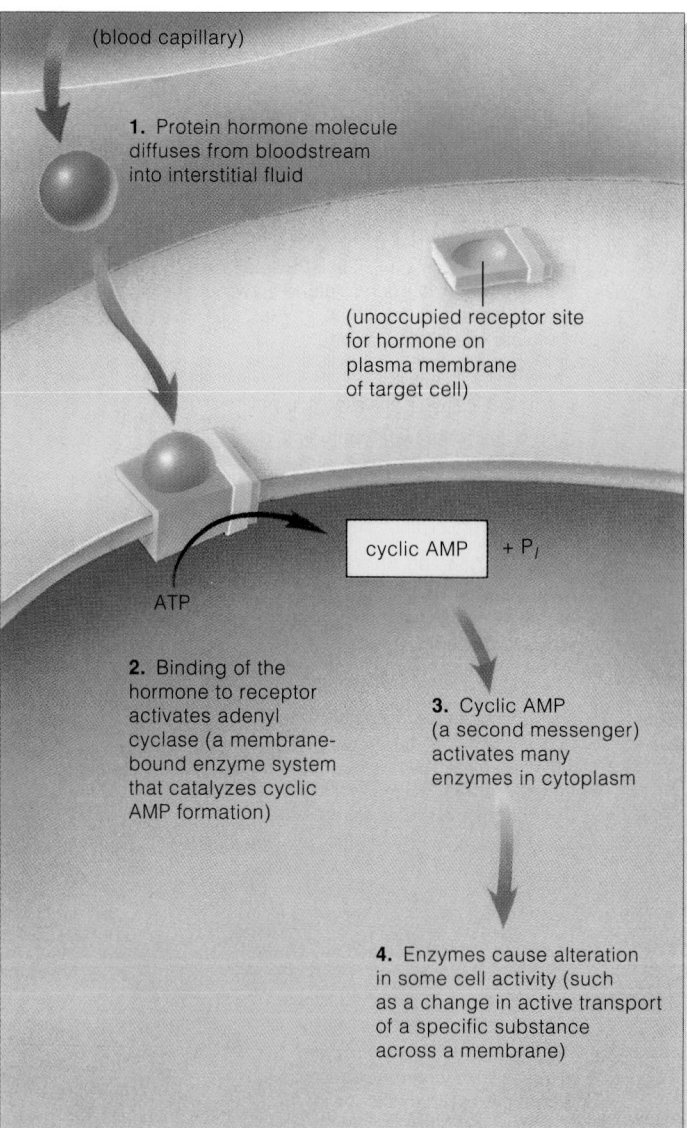

Figure 12.4 Example of a mechanism by which a peptide hormone initiates changes in the activity of a target cell. When the hormone glucagon binds at a receptor, it initiates reactions inside the cell. In this case, cyclic AMP, which is one type of second messenger, relays the signal into the cell interior.

Table 12.1	Main Categories of Hormones
STEROIDS	Estrogens, testosterone, progesterone, aldosterone, cortisol
NONSTEROIDS:	
Amines	Norepinephrine, epinephrine
Peptides	ADH, oxytocin, TRH
Proteins	Insulin, growth hormone, prolactin
Glycoproteins	FSH, LH, TSH

We assign hormones to chemical families, depending on their "parent" molecule. As indicated in Table 12.1, steroid hormones are derived from cholesterol. A second group, the amines, are synthesized from the amino acid tyrosine. A third group, peptide hormones, encompass nonsteroid hormones that are derived from amino acids *other than* tyrosine.

Steroid Hormones

Steroid hormones, synthesized from cholesterol, are lipid-soluble. So, while some cells have membrane receptors for them, typically steroid hormones diffuse directly across the lipid bilayer of a target cell's plasma membrane (Figure 12.3). Once inside the cytoplasm, a steroid hormone molecule usually moves into the nucleus, where it binds to some type of receptor. In some cases, the molecule binds to a receptor molecule that is located in the cytoplasm; then the hormone-receptor complex moves into the nucleus.

The molecular configuration of the complex allows it to interact with a specific region of the cell's DNA. Regions of DNA are the genes that contain instructions for making proteins. The complex stimulates (or inhibits) protein synthesis by switching certain genes on (or off). The resulting proteins carry out the cellular response to the hormonal signal. (As you'll read shortly, thyroid hormones, which also are lipid-soluble, exert their effects this way, too.)

Testosterone is an example of a steroid hormone. The development of male sexual traits depends on it, and proceeds normally when target cells have functional receptors for testosterone. If target cells can't respond to the hormone, this aspect of development goes awry. Consider *testicular feminization syndrome*, in which the target cell receptors are defective. Genetically, an affected person is male (XY), and has functional testes that secrete testosterone. However, because target cells cannot respond to the hormone, the person's secondary sexual traits are like those of females.

Amines and Peptide Hormones

Amine hormones and **peptide hormones** are derived from amino acids. The peptides include several polypeptides, as well as some glycoproteins. Because of their chemical makeup, amine and peptide hormones are soluble in water, but not in lipid. Hence, like other water-soluble signaling molecules, these hormones cannot cross the lipid-rich plasma membrane of a target cell. Instead, they require the assistance of membrane proteins in order to exert their effect on target cells.

For example, polypeptide hormones bind to receptors at the plasma membrane. In certain cases, the hormone-receptor complex moves into the cytoplasm by way of endocytosis; then further action takes place in the cell. In most cases, however, a hormone binds to receptors at the membrane surface. The hormone-receptor complex may activate transport proteins. Alternatively, it may trigger the opening of channel proteins across the membrane. Either way, specific ions or other substances move into the cell, and their cytoplasmic concentrations change. The changes influence specific cell activities.

Often hormones in this group activate what are called **second messengers**. These molecules form within a target cell and act as a go-between, eliciting the cell's response. *The hormone molecule itself only makes contact with a membrane protein on target cells.*

Cyclic AMP (cyclic adenosine monophosphate) is an example of a second messenger. Suppose, for instance, that a polypeptide hormone binds to a receptor on a target cell, as diagrammed in Figure 12.4. The binding triggers activity at a membrane-bound enzyme system. Now the enzyme adenyl cyclase speeds the conversion of ATP to cyclic AMP. The hormone-receptor complex activates many molecules of the enzyme, in a cascade of reactions. Each molecule speeds the conversion of many ATP molecules to cyclic AMP. Each cyclic AMP molecule switches on many other enzyme molecules. These then convert many substrate molecules into activated enzymes, and so on. In short order, an enormous cadre of molecules represents the target cell's final response to the hormonal signal.

Steroid hormones stimulate or inhibit protein synthesis by entering the nucleus of target cells and switching certain genes on or off.

Amine and peptide hormones enter cells either by receptor-mediated endocytosis or by activating membrane proteins. They bind to receptors and so activate second messengers in the cell.

Second messengers have various functions. Often they trigger an amplified response to the hormone.

THE HYPOTHALAMUS AND PITUITARY GLAND

Recall from Chapter 10 that the **hypothalamus**, located deep in the forebrain, tracks internal organs and states (such as fluid balance) related to their functioning. The hypothalamus also secretes hormones. Suspended from its base by a slender stalk is a lobed gland, about the size of a pea. This **pituitary gland** and the hypothalamus interact as a major neural-endocrine control center.

The pituitary's *posterior* lobe stores and secretes two hormones that are synthesized in the hypothalamus. The *anterior* lobe produces and secretes other hormones, most of which govern the release of hormones from other endocrine glands (Table 12.2).

Posterior Lobe Secretions

Figure 12.5 shows the cell bodies of certain neurons in the hypothalamus. Notice that their axons extend down into the posterior lobe, then terminate next to a capillary bed. The neurons produce antidiuretic hormone (ADH) and oxytocin, then store them in the axon endings. When either hormone is released, it diffuses through interstitial fluid and enters capillaries, then travels the bloodstream to its targets. ADH acts on cells of nephrons and collecting ducts in the kidneys. Kidneys, remember, filter blood and rid the body of excess water and salts (in urine). ADH promotes water reabsorption when the body must conserve water.

Through the gamut of events that can cause blood pressure to drop—from the loss of body water through sweating in a warm classroom to severe blood loss due to injury—the hypothalamus keeps tabs on the fluctuations,

so to speak, and releases ADH into the bloodstream when blood pressure falls below a set point. ADH causes the arterioles in some tissues to constrict, and so systemic blood pressure rises. (For this reason, ADH is sometimes called *vasopressin*.)

Oxytocin has roles in reproduction in both males and females. For example, it triggers muscle contractions in the uterus during labor and causes milk to be released when a mother nurses her infant.

Anterior Lobe Secretions

Inside the pituitary stalk, a capillary bed picks up other hormones secreted by the hypothalamus and delivers them to a different capillary bed in the anterior lobe (Figure 12.6). These hormones leave the blood and act on target cells, which variously secrete the following six hormones:

Corticotropin	ACTH
Thyrotropin	TSH
Follicle-stimulating hormone	FSH
Luteinizing hormone	LH
Prolactin	PRL
Growth hormone (or somatotropin)	GH (or STH)

ACTH acts on the outer portion (cortex) of adrenal glands, and TSH acts on the thyroid gland (Section 12.9). FSH and LH have major roles in reproduction, the main topic of Chapter 13. Prolactin has general effects, but it is best known for stimulating and sustaining milk production in mammary glands, after other hormones have primed the tissues.

Table 12.2 Hormones Released from the Pituitary Gland

Pituitary Lobe	Secretions	Designation	Main Targets	Primary Actions
POSTERIOR Nervous tissue (extension of hypothalamus)	Antidiuretic hormone	ADH	Kidneys	Induces water conservation required in control of extracellular fluid volume (and, indirectly, solute concentrations)
	Oxytocin		Mammary glands	Induces milk movement into secretory ducts
			Uterus	Induces uterine contractions
ANTERIOR Mostly glandular tissue	Corticotropin	ACTH	Adrenal cortex	Stimulates release of adrenal steroid hormones
	Thyrotropin	TSH	Thyroid gland	Stimulates release of thyroid hormones
	Gonadotropins: Follicle-stimulating hormone	FSH	Ovaries, testes	In females, stimulates egg formation; in males, helps stimulate sperm formation
	Luteinizing hormone	LH	Ovaries, testes	In females, stimulates ovulation, corpus luteum formation; in males, promotes testosterone secretion, sperm release
	Prolactin	PRL	Mammary glands	Stimulates and sustains milk production
	Growth hormone (also called somatotropin)	GH (STH)	Most cells	Promotes growth in young; induces protein synthesis, cell division; roles in glucose, protein metabolism in adults

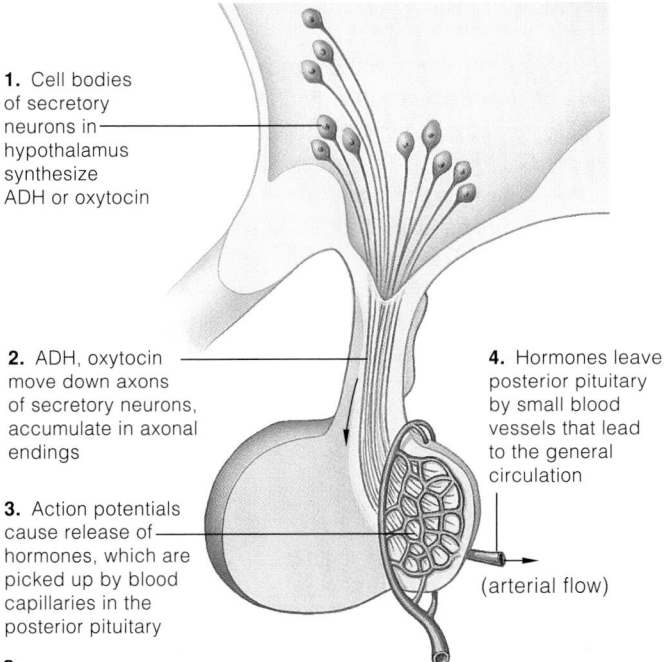

1. Cell bodies of secretory neurons in hypothalamus synthesize ADH or oxytocin

2. ADH, oxytocin move down axons of secretory neurons, accumulate in axonal endings

3. Action potentials cause release of hormones, which are picked up by blood capillaries in the posterior pituitary

4. Hormones leave posterior pituitary by small blood vessels that lead to the general circulation

(arterial flow)

a

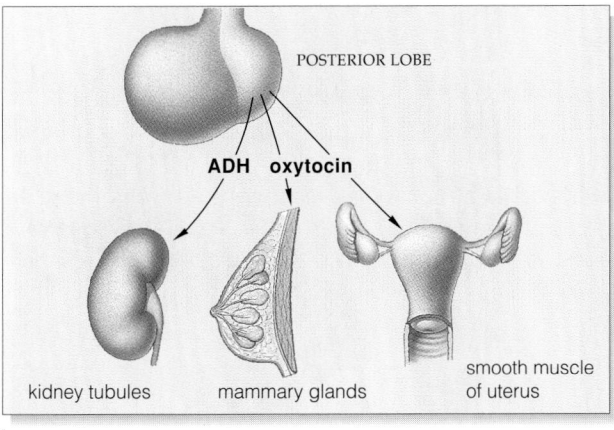

POSTERIOR LOBE

ADH   oxytocin

kidney tubules   mammary glands   smooth muscle of uterus

b

Figure 12.5 (**a**) Functional links between the hypothalamus and the posterior lobe of the pituitary. (**b**) Main targets of the posterior lobe secretions.

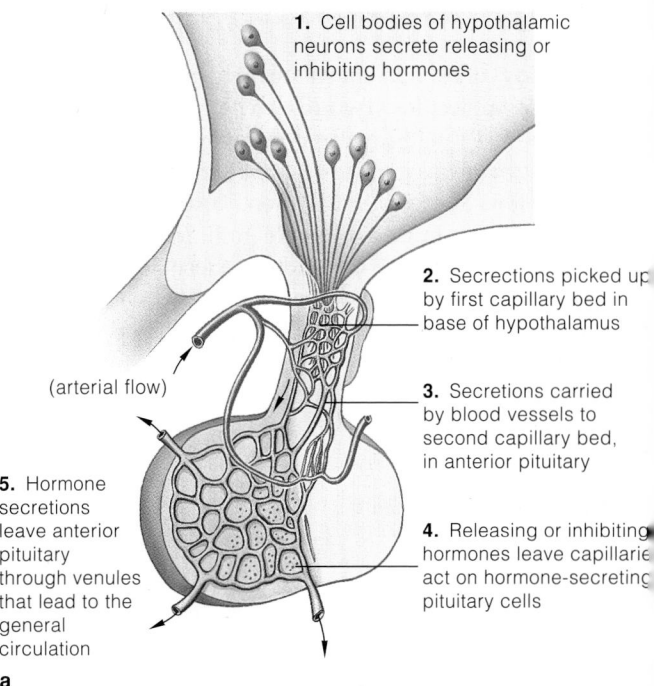

1. Cell bodies of hypothalamic neurons secrete releasing or inhibiting hormones

2. Secretions picked up by first capillary bed in base of hypothalamus

3. Secretions carried by blood vessels to second capillary bed, in anterior pituitary

4. Releasing or inhibiting hormones leave capillaries act on hormone-secreting pituitary cells

5. Hormone secretions leave anterior pituitary through venules that lead to the general circulation

(arterial flow)

a

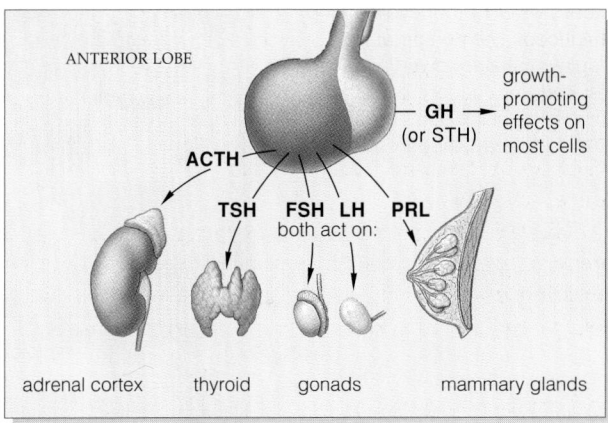

ANTERIOR LOBE

ACTH

TSH   FSH LH   PRL
both act on:

GH (or STH)   growth-promoting effects on most cells

adrenal cortex   thyroid   gonads   mammary glands

b

Figure 12.6 (**a**) Functional links between the hypothalamus and the anterior lobe of the pituitary. (**b**) Main targets of the anterior lobe secretions.

Growth hormone (GH) affects most body tissues. It stimulates protein synthesis and cell division in target cells, and it profoundly influences growth, especially of cartilage and bone. GH is equally important as a "metabolic hormone." Throughout life it stimulates cells to take up amino acids and promotes the breakdown and release of fat stored in adipose tissues, thereby increasing the amount of fatty acids available to cells. GH also moderates the rate at which cells take up glucose, which helps maintain appropriate blood levels of that cellular fuel.

Most hypothalamic hormones acting in the anterior lobe of the pituitary are **releasers** that stimulate secretion from target cells. For example, GnRH (short for gonadotropin-releasing hormone) brings about secretion of FSH and LH—both of which are classified as gonado-

tropins. Similarly, TRH stimulates the secretion of TSH. Other hypothalamic hormones are **inhibitors** of secretion from targets in the anterior pituitary. An example is somatostatin, which inhibits the secretion of growth hormone and thyrotropin.

The hypothalamus and pituitary gland produce eight hormones and interact to control their secretion. The pituitary's posterior lobe stores and secretes ADH and oxytocin, both of which target specific cell types. The anterior lobe of the pituitary produces and secretes six hormones—ACTH, TSH, FSH, LH, PRL, and GH. These trigger the release of other hormones from other endocrine glands, with wide-ranging effects throughout the body.

EXAMPLES OF ABNORMAL PITUITARY OUTPUT

The body does not churn out enormous numbers of hormone molecules. Two researchers, Roger Guilleman and Andrew Schally, realized this when they isolated the first known releasing hormone. After 4 years of dissecting 500 tons of sheep brains, then 7 tons of hypothalamic tissue, they extracted a single milligram of TSH. Yet normal body function depends on those tiny amounts.

For example, when anterior pituitary cells produce too much growth hormone during childhood, *gigantism* results. Proportionally, affected adults are similar to a normal person but much larger (Figure 12.7a). When not enough growth hormone is produced during childhood, *pituitary dwarfism* results. Affected adults are proportionally similar to a normal person but much smaller (Figure 12.7b).

What happens if growth hormone output becomes excessive during adulthood, when long bones no longer can lengthen? Then, *acromegaly* will be the result. Bone, cartilage, and other connective tissues in the hands, feet, and jaws will thicken abnormally. So will epithelia of the skin, nose, eyelids, lips, and tongue. Figure 12.8 shows an example.

Figure 12.7 (**a**) Manute Bol, an NBA center, is 7 feet 6-3/4 inches tall owing to excessive GH production during childhood. (**b**) Effect of growth hormone (GH) on overall body growth. The person at the center is affected by gigantism, which resulted from excessive GH production during childhood. The person at right displays pituitary dwarfism, which resulted from underproduction of GH during childhood. The person at the left is average in size.

a
b

Figure 12.8 Acromegaly, which resulted from excessive production of GH during adulthood. Before this female reached maturity, she was symptom-free.

age nine   sixteen   thirty-three   fifty-two

SOURCES AND EFFECTS OF OTHER HORMONES

Table 12.3 lists hormones from endocrine sources other than the pituitary. The remainder of this chapter will provide you with a few examples of their effects and of the controls over their output. The examples will make more sense if you keep the following points in mind.

First, hormones often interact with one another. Second, negative feedback mechanisms usually control the secretions. When a hormone's concentration increases or decreases in some body region, the change triggers events that respectively dampen or stimulate further secretion. Third, a target cell may react differently to a hormone at different times. Its response depends on the hormone's concentration and on the functional state of the cell's receptors. Fourth, environmental cues, such as light, may be important mediators of hormone secretion.

The secretion of a hormone and its effects are influenced by hormone interactions, feedback mechanisms, variations in the state of target cells, and sometimes environmental cues.

Table 12.3 Hormone Sources Other Than the Hypothalamus and Pituitary

Source	Secretion(s)	Main Targets	Primary Actions
PANCREATIC ISLETS	Insulin	Muscle, adipose tissue	Lowers blood sugar level
	Glucagon	Liver	Raises blood sugar level
	Somatostatin	Insulin-secreting cells	Influences carbohydrate metabolism
ADRENAL CORTEX	Glucocorticoids (including cortisol)	Most cells	Promote protein breakdown and conversion to glucose
	Mineralocorticoids (including aldosterone)	Kidney	Promote sodium reabsorption; control salt–water balance
ADRENAL MEDULLA	Epinephrine (adrenalin)	Liver, muscle, adipose tissue	Raises blood level of sugar, fatty acids; increases heart rate, force of contraction
	Norepinephrine	Smooth muscle of blood vessels	Promotes constriction or dilation of blood vessel diameter
THYROID	Triiodothyronine, thyroxine	Most cells	Regulate metabolism; have roles in growth, development
	Calcitonin	Bone	Lowers calcium levels in blood
PARATHYROIDS	Parathyroid hormone	Bone, kidney	Elevates levels of calcium and phosphate ions in blood
THYMUS	Thymosins, etc.	Lymphocytes	Have roles in immune responses
GONADS:			
Testes (in males)	Androgens (including testosterone)	General	Required in sperm formation, development of genitals, maintenance of sexual traits; influence growth, development
Ovaries (in females)	Estrogens	General	Required in egg maturation and release; prepare uterine lining for pregnancy; required in development of genitals, maintenance of sexual traits; influence growth, development
	Progesterone	Uterus, breasts	Prepares, maintains uterine lining for pregnancy; stimulates breast development
PINEAL	Melatonin	Hypothalamus	Influences daily biorhythms
ENDOCRINE CELLS OF STOMACH, GUT	Gastrin, secretin, etc.	Stomach, pancreas, gallbladder	Stimulate activity of stomach, pancreas, liver, gallbladder
LIVER	IGFs (Insulin-like growth factors)	Most cells	Stimulate cell growth and development
KIDNEYS	Erythropoietin	Bone marrow	Stimulates red blood cell production
	Angiotensin*	Adrenal cortex, arterioles	Helps control blood pressure, aldosterone secretion
	Vitamin D_3*	Bone, gut	Enhances calcium resorption and uptake
HEART	Atrial natriuretic hormone	Kidney, blood vessels	Increases sodium excretion; lowers blood pressure

*These hormones are not produced in the kidneys but are formed when enzymes produced in kidneys activate specific substances in the blood.

PANCREATIC ISLETS

The pancreas is a gland with both exocrine and endocrine functions. Its exocrine cells secrete digestive enzymes and sodium bicarbonate. (Recall from Chapter 3 that exocrine gland products are not secreted into the bloodstream; they are secreted through a duct onto a free epithelial surface.) Many of its endocrine cells, which we now consider, do not respond primarily to signals from other hormones or nerves. Instead, they make a homeostatic response to chemical change in their immediate surroundings.

Secretions from Pancreatic Islets

The endocrine cells of the pancreas are located in about 2 million scattered clusters. Each cluster, a **pancreatic islet**, is permeated with blood capillaries and contains three types of hormone-secreting cells:

1. *Alpha cells* secrete the hormone **glucagon**. Between meals, cells use the glucose delivered to them by the bloodstream. The blood glucose level decreases, at which time glucagon secretions cause glycogen (a storage polysaccharide) and amino acids to be converted to glucose in the liver and muscle. In such ways, *glucagon raises the glucose level in the blood.*

2. *Beta cells* secrete the hormone **insulin**. After meals, when the blood glucose level is high, insulin stimulates uptake of glucose by muscle and adipose cells. It also promotes synthesis of fats, glycogen, and to a lesser extent, proteins, and inhibits protein conversion to glucose. Thus *insulin lowers the glucose level in the blood.*

3. *Delta cells* secrete *somatostatin*. This hormone acts on beta cells and alpha cells to inhibit secretion of insulin and glucagon, respectively. Somatostatin is part of several hormonal control systems. For example, it is released from the hypothalamus to block secretions of growth hormone; it is also secreted by cells of the GI tract, where it acts locally to inhibit secretion of various substances involved in digestion.

Diabetes—Faulty Regulation of Blood Glucose

Figure 12.9 shows how interplay among the pancreatic hormones helps keep blood glucose levels fairly constant despite great variation in when—and how much—we eat. When the body cannot produce enough insulin or when insulin's target cells cannot respond to it, the body does not store glucose in a normal fashion, and disorders in carbohydrate, protein, and fat metabolism occur.

Insulin deficiency can lead to *diabetes mellitus*. Because target cells cannot take up glucose from blood, that sugar accumulates in blood and is lost in the urine. (*Mellitus* means honey in Greek, and early physicians reportedly often tasted their patients' urine to confirm the diagnosis.) As the kidneys shunt excess glucose into the urine, water is lost as well—skewing the body's water–solute balance. Affected people become dehydrated and excessively thirsty. Their insulin-deprived (glucose-starved) cells start breaking down protein and fats for energy, and this leads to weight loss. Ketones—normal acidic products of fat breakdown—build up in the blood and urine. One result is excess water loss, which can lead to dangerously low blood pressure and heart failure. Another outcome is *metabolic acidosis*, a lower than optimal pH in blood that can seriously disrupt brain functioning.

In type 1 or "insulin-dependent" diabetes, the body mounts an autoimmune response against its beta cells and destroys them. Only about 10 percent of diabetics are type 1, which is the more immediately dangerous of the two types of diabetes. It may be caused by a combination of genetic susceptibility and viral infection. Because symptoms usually appear during childhood and adolescence, type 1 diabetes also is called "juvenile-onset diabetes." Patients survive with insulin injections, but their life span may be shortened by associated cardiovascular problems.

In type 2 diabetes, insulin levels are close to or above normal, but for various reasons target cells cannot respond properly to the hormone. As affected people grow older, their beta cells deteriorate and they produce less and less insulin. Type 2 diabetes usually occurs in middle age and is sometimes called "maturity-onset diabetes." Obesity increases a person's risk of developing the disorder. Affected people can lead a normal life by controlling their diet and weight and sometimes by taking drugs that improve insulin action or secretion.

Three types of endocrine cells in the pancreatic islets secrete hormones.

Alpha cells secrete glucagon, which acts to raise the blood glucose level when it falls below a set point. Beta cells secrete insulin, which stimulates glucose uptake from the blood and acts in other ways to lower blood glucose levels.

Delta cells secrete somatostatin, which acts on alpha and beta cells to inhibit secretion of glucagon and insulin, respectively.

The interplay of pancreatic hormones stabilizes blood glucose levels despite variations in food intake.

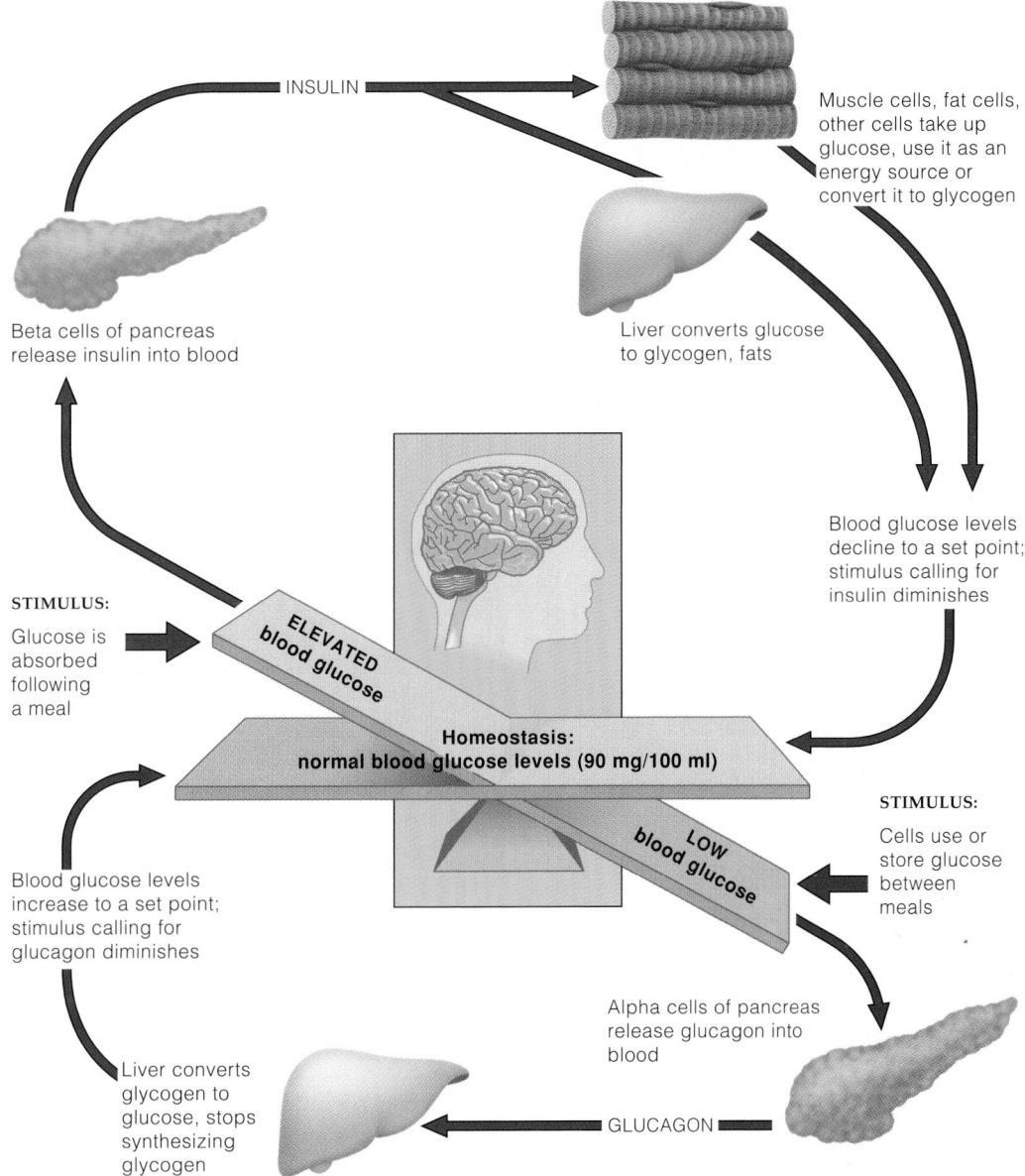

Figure 12.9 A simplified picture of some homeostatic controls over glucose metabolism.

Following a meal, glucose enters the bloodstream faster than cells can use it. The blood glucose level rises, and pancreatic beta cells are stimulated to secrete insulin. Insulin's targets (mainly liver, fat, and muscle cells) use glucose or store excess amounts as glycogen, or use it to synthesize lipids.

Between meals, the blood glucose level decreases. Pancreatic alpha cells are stimulated to secrete glucagon. This hormone's target cells convert glycogen back to glucose, which enters the blood. Also, the hypothalamus prods the pituitary gland to secrete other hormones that slow down conversion of glucose to glycogen in liver, fat, and muscle cells.

THE ADRENAL GLANDS

Hormones of the Adrenal Cortex

Humans have a pair of adrenal glands, one perched above each kidney. The outer portion of each gland is the **adrenal cortex**. Here, two major types of hormones, glucocorticoids and mineralocorticoids, are secreted.

Glucocorticoids influence metabolism in ways that help raise the blood level of glucose when it falls below a set point. In humans, *cortisol* is the primary glucocorticoid. Among other effects, it promotes protein breakdown in muscle and stimulates the liver to take up amino acids, from which liver cells synthesize glucose in a process called *gluconeogenesis*. Cortisol also dampens the uptake of glucose from the blood and its use for cellular fuel by tissues such as skeletal muscle. This effect is sometimes called "glucose sparing." Lastly, cortisol promotes fat breakdown and use of the resulting fatty acids for energy. Glucose sparing is an extremely important mechanism of homeostasis: It helps ensure that the concentration of glucose in blood will be adequate to meet the needs of the brain, which generally cannot use other molecules for fuel.

Under routine conditions, a negative feedback loop to the neuroendocrine control center operates to control cortisol secretion. When the blood level of the hormone rises above a set point, the hypothalamus begins to secrete less of the releasing hormone CRH (Figure 12.10), the anterior pituitary responds by secreting less ACTH, and the adrenal cortex slows its secretion of cortisol. Daily cortisol secretion is highest in a healthy person when the blood glucose level is lowest, generally in the early morning. Chronic severe *hypoglycemia*, a persistent low concentration of glucose in the blood, can develop when a person has a cortisol deficiency. Then, the mechanisms that spare glucose and generate new supplies in the liver do not operate properly.

Glucocorticoids suppress the inflammatory response to tissue injury or infection. In fact, their secretion increases during any abnormal physiological stress. A painful injury, severe illness, or allergic reaction may trigger shock, tissue inflammation, or both. Then, increased secretion of cortisol and other signaling molecules is essential to recovery. That is why cortisol-like drugs, such as cortisone, are administered following an asthma attack and episodes of serious inflammatory disorders. (Prolonged use of large doses of glucocorticoids has serious side effects, including suppression of the immune system.) Cortisone is the active ingredient in many over-the-counter creams and lotions for treating skin irritations.

Mineralocorticoids mainly regulate the concentrations of mineral salts, such as potassium and sodium, in extracellular fluid. The most abundant mineralocorticoid is *aldosterone*. Recall from Section 9.7 that aldosterone acts

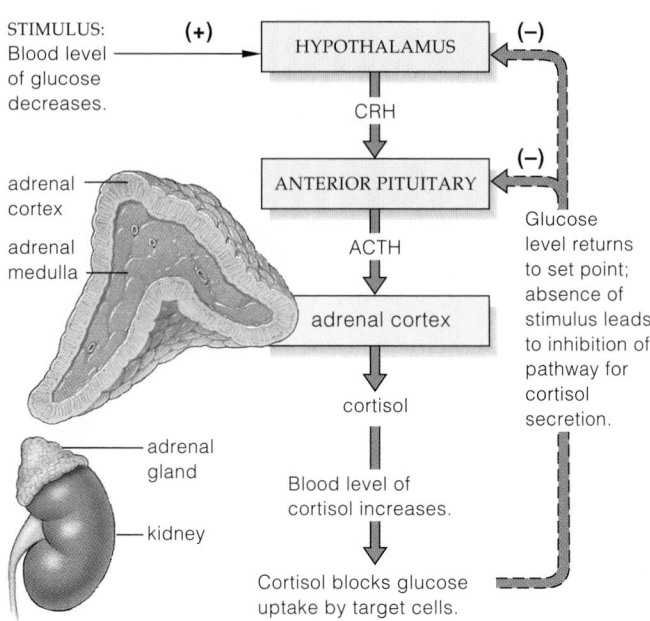

Figure 12.10 Location of the adrenal glands. The diagram shows a negative feedback loop that governs secretion of cortisol from the adrenal cortex.

on the distal tubules of kidney nephrons, stimulating the reabsorption of sodium ions and the excretion of potassium ions. Sodium reabsorption in turn promotes reabsorption of water from the tubules as urine is forming. Various circumstances, such as falling blood pressure or falling blood levels of sodium (hence, falling blood volume as "water follows salt"), can trigger aldosterone secretion.

A condition called *Addison's disease* arises when the adrenal cortex fails to secrete sufficient mineralocorticoids and glucocorticoids. Sufferers lose weight, and blood levels of glucose and sodium drop while blood potassium rises. The large sodium losses limit the kidney's ability to conserve water. One result can be dangerously low blood pressure and dehydration as the body loses large amounts of water. Figure 12.11 shows a famous Addison's disease patient. The condition can be easily treated with drugs that replace the missing hormones.

In a developing fetus and early in puberty, the adrenal cortex also secretes notable amounts of sex hormones. The main ones are male sex hormones (androgens), although some female sex hormones (estrogens and progesterone) are also produced. However, in adults most sex hormones are generated by the reproductive organs (Chapter 13). Adrenal androgens may be responsible for the female sex drive.

Some researchers are concerned that industrial chemicals released into the environment may have hor-

Figure 12.11 President John F. Kennedy reportedly suffered from Addison's disease but kept the ailment a secret for political reasons.

monelike effects on human tissues. The *Focus* essay on this page explores the scientific debate about what have come to be dubbed "endocrine disrupters."

Adrenal Medulla Hormones

The **adrenal medulla** is the inner region of the adrenal gland (Figure 12.10). It contains modified neurons that secrete epinephrine and norepinephrine (once commonly called adrenalin and noradrenalin). These substances are considered hormones when they are released into the bloodstream, and neurotransmitters when they are released (in tiny amounts) by neurons. Sympathetic nerves carry stimulatory signals to the adrenal medulla from the hypothalamus and other brain regions.

Both substances help regulate blood circulation and carbohydrate metabolism when the body is excited or stressed. For example, they increase heart rate, dilate arterioles in some body regions and constrict them in others, and dilate bronchioles. Thus the heart beats faster and harder, more blood volume is shunted to heart and muscle cells from other regions, and more oxygen flows to energy-demanding cells throughout the body. These are features of the "fight-flight" response described in Chapter 10.

Glucocorticoids released by the adrenal cortex help raise blood levels of glucose when they fall below a set point. Mineralocorticoids mainly regulate blood concentrations of mineral salts.

Epinephrine and norepinephrine released from the adrenal medulla both help regulate blood circulation and carbohydrate metabolism when the body is stressed.

ARE "ENDOCRINE DISRUPTERS" HARMING HUMAN HEALTH?

A steady stream of hormone-modulating chemicals trickles into the air we breathe, the water we drink, and the soil in which we grow food. Such compounds may be naturally occurring or the products of modern chemistry. They are called "hormone-modulators" because they can bind to hormone receptors of body cells and either stimulate or inhibit the response of tissues to particular hormones, including estrogen. Dubbed "endocrine disrupters" by some researchers, the compounds are at the center of an ongoing scientific controversy.

Endocrinologists generally agree that many synthetic and natural estrogenlike compounds can disrupt normal functioning of the animal endocrine system. There, the consensus ends. In one camp are researchers who cite altered functions linked to environmental estrogens such as the pesticide DDT, dioxins, PCBs, and their breakdown products. Much evidence comes from pollution and industrial incidents that have harmed humans, and from experiments on wildlife species and lab animals. For example, scientists have documented physical deformities, "feminization" of males, and increased incidence of embryo death among bald eagles and other bird species in the Great Lakes region, once heavily polluted with estrogen-mimicking industrial compounds. Experiments with rodents have linked exposure to DDT and many other pesticides to the development of mammary tumors in females and testicular cancer in males. In 1993, an article in the British journal *Lancet* even speculated that an overall decline in human male sperm counts since the 1940s might be due to increasing exposure to estrogenlike pollutants. The findings are being debated, as research continues.

Critics fault the conclusions of some investigators—for instance, worldwide, human sperm-count studies indeed show a significant decline in sperm counts from 1938 to 1970, but the data from 1970 to 1990 indicate a slight rebound. This finding has prompted calls for continued study of the relationship (if any) between sperm counts and pollutants. Some scientists and policymakers are skeptical of conclusions based on studies of nonhuman species. Also, laboratory experiments may use extremely high doses of the chemical under study, possibly far greater than a person would be exposed to by way of background pollution. To counter this objection, many respected studies have exposed the test subjects to doses that are known to occur in the real world.

More than ever, scientists and citizens now are wondering about the cumulative effects on human health of a lifetime of exposure to endocrine disrupters. At this writing, Congress has called on the Environmental Protection Agency to develop practical means for testing chemicals for their potential to do endocrine harm.

The Thyroid and Parathyroids—and the Importance of Feedback Control

The **thyroid gland** is located at the base of the neck, in front of the trachea (Figure 12.12). The main secretions of this gland, *thyroxine* (T$_4$) and *triiodothyronine* (T$_3$), influence overall metabolic rate, growth, and development. The thyroid also makes *calcitonin*, a hormone that helps lower the level of calcium (and phosphate) in blood. Proper feedback control is essential to the functioning of these hormones, a point that is brought into sharp focus by cases of abnormal thyroid output.

Consider how thyroid hormone synthesis requires iodine, a trace element in food. In the absence of iodine, thyroid hormone levels in the blood decrease. The anterior pituitary responds to the decrease by secreting the thyroid-stimulating hormone thyrotropin (TSH). Too much TSH overstimulates the thyroid gland, the gland enlarges, and the result is a form of *goiter* (Figure 12.13). Goiter caused by iodine deficiency is now rare in countries where iodized table salt is used. Although goiter itself is not usually a serious health threat, it can be a symptom of a serious disorder.

Insufficient thyroid output is called *hypothyroidism*. Hypothyroid adults have a slowed heart rate; they are sluggish, dry-skinned, intolerant of cold, and sometimes feel confused and depressed. Affected women often have menstrual disturbances. *Cretinism* is a severe hypothyroid condition in children, and it can arise from a genetic disorder that affects the thyroid gland in the fetus. If affected children do not receive treatment, their growth is stunted and they are mentally retarded. However, these effects can be prevented if the disorder is detected soon after birth. In the United States, newborns commonly are tested for hypothyroidism.

Excessive thyroid output can lead to *hyperthyroidism*. The most common hyperthyroid disorder is the curious condition known as *Graves' disease*. Affected people have increased metabolic rates, heart rate, and blood flow, and they lose weight even when they take in increased amounts of food. They may be excessively nervous and agitated and typically have trouble sleeping. They are intolerant of heat and sweat profusely. Apparently, Graves' disease is an autoimmune disorder in which a thyroid-stimulating antibody binds to thyroid cells and causes overproduction of thyroid hormones. Often, affected people are genetically predisposed to Graves' disease, which may be triggered by some environmental event. For unknown reasons, females are affected more often than males. Although surgery and drug treatment are options, today treatment typically involves administering radioactive iodine, which destroys thyroid gland function. The patient then takes controlled doses of thyroid hormones for life to maintain normal functioning.

Like the pancreatic islets described in Section 12.6, the four **parathyroid glands** respond to chemical change in their immediate surroundings in ways that help maintain homeostatic balance. The parathyroids are located at the back of the thyroid (Figure 12.12b). They secrete parathyroid hormone (PTH) when the concentration of calcium ions in their surroundings falls. Their action affects how

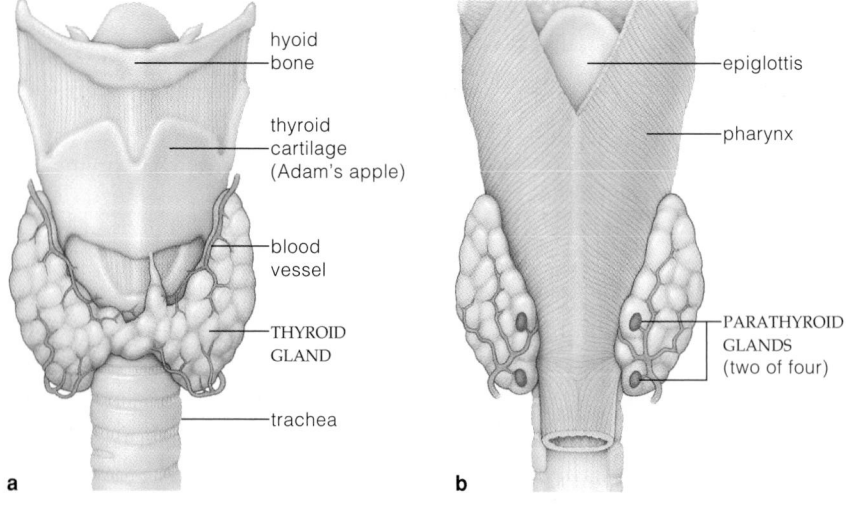

hyoid bone
thyroid cartilage (Adam's apple)
blood vessel
THYROID GLAND
trachea

epiglottis
pharynx
PARATHYROID GLANDS (two of four)

a   **b**

Figure 12.12   **(a)** Human thyroid gland, anterior view. **(b)** Four parathyroid glands, next to the back of the thyroid.

goiter

Figure 12.13   A mild case of goiter, as displayed by Maria de Medici in 1625. During the late Renaissance, a rounded neck was a sign of beauty. It occurred regularly in parts of the world where iodine supplies in food were insufficient for normal thyroid function.

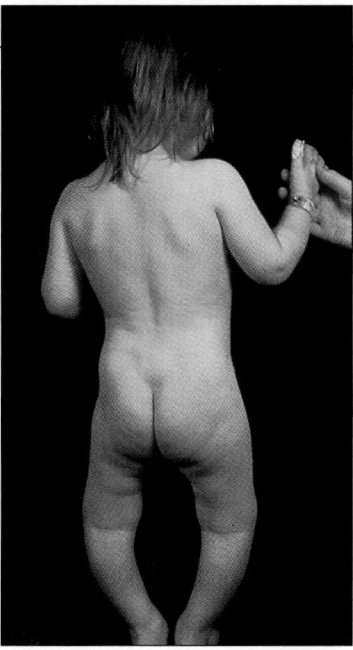

Figure 12.14 A child with rickets. Bowed legs are typical of the disorder.

much calcium is available to activate enzymes, and for muscle contraction, blood clotting, and many other tasks.

PTH prompts bone cells to release calcium and phosphate ions and stimulates the kidneys to conserve them. PTH also helps activate vitamin D. The activated form, a hormone, enhances calcium absorption from food. In vitamin D deficiency, too little calcium and phosphorus are absorbed, so bones develop improperly. Figure 12.14 shows the evidence of this condition, called *rickets*.

As you might guess, disorders related to parathyroid functioning disrupt calcium homeostasis in the body. Calcium has so many key roles in the body that such disorders can be extremely serious. For example, excess PTH (hyperparathyroidism) causes so much calcium to be withdrawn from bones that the bone tissue becomes dangerously weakened. Muscle function deteriorates, and the excess circulating calcium may lead to the development of kidney stones. Operations of the central nervous system may be so disrupted that the affected person becomes comatose and eventually dies.

Thymus Hormones

The lobed **thymus** is located behind the breastbone (the sternum), between the lungs. As described in Chapter 7, T lymphocytes multiply, differentiate, and mature in this gland under the influence of hormones collectively called *thymosins*. You may recall that different categories of mature T cells, such as helper T cells and cytotoxic T cells, are key elements of immune system functioning (Section 7.4). The thymus is large in children but shrinks to a relatively small size as a person matures into adulthood.

The Pineal Gland: Melatonin

So far, we have seen how endocrine glands and endocrine cells respond to other hormones, to signals from the nervous system, and to chemical changes in their surroundings. In humans and other animals, certain hormones are secreted or inhibited in response to cues from the external environment.

Until about 240 million years ago, vertebrates commonly had a light-sensitive "third eye" on top of the head. Some, such as lampreys (a type of jawless, parasitic fish), still have one, beneath the skin. A modified form of this photosensitive organ, the **pineal gland** described at the beginning of this chapter, persists in nearly all vertebrates. The pineal gland secretes the hormone melatonin into the blood and cerebrospinal fluid. Melatonin functions in the development of gonads and in reproductive cycles and sleep/wake cycles.

As this chapter's opening vignette explained, melatonin is secreted in the absence of light. This means that melatonin levels vary from day to night. The levels also change with the seasons, for winter days are shorter than summer days. In humans, decreased melatonin secretion might help trigger the onset of *puberty,* the age at which reproductive organs and structures start to mature. If disease destroys the pineal gland, puberty can begin prematurely.

Hormones from the Heart, GI Tract, and Elsewhere

Beyond the endocrine glands we have just considered, hormones are also manufactured and secreted by specialized cells in the heart, gastrointestinal tract, and elsewhere. For example, the heart atria secrete *atrial natriuretic peptide*, or ANP, a hormone with various effects that include helping to regulate blood pressure. When blood pressure rises, ANP acts to inhibit the reabsorption of sodium ions—and hence water—in the kidneys. As a result, more water is excreted, blood volume decreases, and blood pressure falls.

The thyroid gland secretes two hormones, thyroxine (T_4) and triiodothyronine (T_3), that influence overall metabolic rate, growth, and development. A third hormone produced by the thyroid is calcitonin. It helps regulate calcium homeostasis in blood. The parathyroids secrete PTH, which has key roles in calcium homeostasis.

The normal functioning of the thyroid and parathyroid glands relies on proper feedback control of their hormone secretions.

Other sources of hormones in the body include the thymus gland (which produces thymosins active in lymphocyte maturation), the pineal gland (the source of melatonin), and endocrine cells in the heart and gastrointestinal tract.

Local Signaling Molecules

In humans (and mammals generally), many cells can detect changes in the surrounding chemical environment and alter their activity, often in ways that either counteract or amplify the change. The cells secrete **local signaling molecules,** which act only on the secreting cell itself or in the immediate vicinity of change. Most of the signaling molecules are taken up so rapidly that not many are left to enter the general circulation. Prostaglandins and growth factors are examples of such secretions.

Prostaglandins

More than 16 different kinds of the fatty acids called **prostaglandins** have been identified in tissues throughout the body. In fact, the plasma membranes of most cells contain the precursors of prostaglandins. They are released continually, but the rate of synthesis often increases in response to local chemical changes. The stepped-up secretion can influence neighboring cells as well as the prostaglandin-releasing cells themselves.

At least two prostaglandins help adjust blood flow through local tissues. When their secretion is stimulated by epinephrine and norepinephrine, they cause smooth muscle in the walls of blood vessels to constrict or dilate. Prostaglandins have similar effects on smooth muscle of airways in the lungs. Tissue inflammation and allergic responses to airborne dust and pollen may be aggravated by prostaglandins. Cells in the kidneys also produce prostaglandins that act as vasodilators or as vasoconstrictors, but exactly how they affect overall nephron function is not well understood.

Prostaglandins have major effects on some reproductive events. Many women experience painful cramping and excessive bleeding when they menstruate, and both effects have been traced to prostaglandin action on smooth muscle of the uterus. (Aspirin and other antiprostaglandin drugs block synthesis of this local signaling molecule and alleviate the discomfort.)

Prostaglandins also influence the *corpus luteum,* a glandular structure that develops from cells that earlier surrounded a developing ovum (egg) in the ovary. When pregnancy does not follow ovulation (the release of an ovum from the ovary), a corpus luteum produces copious amounts of prostaglandins. In response, capillaries that service the corpus luteum constrict, shutting off the blood supply—and the corpus luteum dies. Along with oxytocin, prostaglandins also have roles in stimulating uterine contractions during labor. Prostaglandins in semen may stimulate uterine contractions that help move sperm deeper into the female reproductive tract. We will look much more closely at some of these events in Chapter 14.

Growth Factors

Signaling molecules called **growth factors** influence growth by regulating the rate at which certain cells divide. *Epidermal growth factor* (EGF), discovered by Stanley Cohen, influences the growth of many cell types. So does insulin-like growth factor (IGF) secreted by the liver. *Nerve growth factor* (NGF) is another example. NGF, discovered by Rita Levi-Montalcini, promotes survival and growth of neurons in the developing embryo. One experiment demonstrated that certain immature neurons survive indefinitely in tissue culture when NGF is present but die within a few days if it is not. NGF also may define the direction of growth for these embryonic neurons, laying down a chemical path that leads the elongating processes to target cells.

In the years since NGF was discovered, several other growth factors with effects on neural cells have been identified. *Science Comes to Life* on the facing page will give you an inkling of how such growth factors may one day be important tools for treating victims of heart attack, paralyzing spinal cord injuries, and perhaps other conditions as well.

Pheromones Revisited

Chapter 11 touched on possible human pheromones in its discussion of the sense of smell (Section 11.3). It's not too surprising that we humans may produce and respond to pheromones; after all, some of our close primate relatives (including rhesus monkeys), and bears, coyotes, dogs, various insects, and many other animals produce pheromones that serve as sex attractants, territory markers, and other communication signals to members of their own species. Pheromones are released outside of the individual and then pass through the air or water and so reach another individual.

As noted previously, hypotheses about pheromone activity in humans are still quite tentative, but research on this fascinating subject is continuing.

Unlike hormones, local signaling molecules act on the secreting cell itself or in the immediate vicinity.

Prostaglandins are a family of local signaling molecules that have varied effects throughout the body. Several serve as vasodilators or vasoconstrictors and so help adjust blood flow in affected regions. Other prostaglandins cause smooth muscle (as in airways or the uterus) to contract.

Growth factors help regulate the rate at which certain cells divide. Researchers are exploring their potential clinical use to stimulate regeneration of damaged neurons.

COMMUNICATION MOLECULES PROVIDE HOPE FOR HEALING

In the not too distant future, a doctor treating an automobile accident victim who has sustained severe injuries, including cuts, broken bones, and major damage to the spinal cord, could have a powerful new chemical arsenal at her disposal. Researchers are unraveling how an array of communication molecules, including growth factors, operate in body tissues. They are also seeking ways such knowledge can be transformed into treatments for specific kinds of tissue damage.

The list of known growth factors is growing rapidly. They include the epidermal growth factor (EGF), insulin-like growth factor (IGF), and nerve growth factor (NGF) described in the text, as well as fibroblast growth factor (FGF), platelet-derived growth factor (PDGF), transforming growth factors (TGFs), and tumor angiogenesis factors (TAFs).

Studies on rodents have shown that EGF and one type of TGF enhance the healing of some types of wounds. In one experiment, an antibiotic cream that contained TGF-*alpha* accelerated the rate at which burned skin regenerated. TGF-*beta* helps regulate the formation of bone and could possibly be marshaled to speed the healing of broken bones. TAFs that stimulate the development of new blood

vessels in tumors could conceivably be harnessed to perform that function in cardiac muscle damaged by a heart attack. PDGF, which is released by platelets in blood plasma at the site of a wound, seems to promote the regeneration of smooth muscle cells in torn vessel walls—and could perhaps do the same for damaged smooth muscle elsewhere in the body.

Some of the most tragic injuries involve spinal cord damage that quickly results in the death of neurons—and in paralysis (Figure 12.15). Experiments have already demonstrated that NGF can help severed nerves regrow; there is also evidence that FGF serves to help maintain healthy neurons and to help repair damaged tissues. (Fibroblasts are cells in loose connective tissue that are thought to give rise to various fiberlike molecules with structural roles.) In experimental trials, a substance called GM_1 appears to derail continued deterioration of damaged spinal cord neurons. GM_1 is a ganglioside, a normal component of nerve cell membranes that seems to serve as a communication link between cells. It is already being used experimentally on spinal injury patients, with encouraging results.

Figure 12.15 One goal of research on nerve growth factor (NGF) is to develop a treatment that could help repair spinal cord injuries that have paralyzed an estimated 200,000 Americans. Even with such nerve damage, people can live active, enjoyable lives—as this photograph attests.

SUMMARY

1. Cells continually take up and release substances. In complex animals, these constant withdrawals and secretions must be integrated in ways that ensure cell survival through the whole body.

2. Integration requires the stimulatory or inhibitory effects of signaling molecules.

   a. Signaling molecules are chemical secretions by one cell that adjust the behavior of other, target cells.

   b. Any cell with receptors for the signal is the target. It may or may not be next to the signaling cell.

   c. Hormones, neurotransmitters, local signaling molecules, and pheromones are different kinds of signaling molecules.

3. Steroid and nonsteroid hormones exert their effects on target cells by different mechanisms.

   a. Steroid (and thyroid) hormones have receptors inside their target cells. The hormone-receptor complex binds to the DNA. Binding triggers gene activation and protein synthesis.

   b. Amines and the peptide hormones (peptides, polypeptides, and glycoproteins) interact with receptors on the plasma membrane of target cells. Responses to them are often mediated by a second messenger, such as cyclic AMP, inside the cell.

   c. Most nonsteroid (amine and peptide) hormones alter the activity of existing proteins in target cells. The ensuing target cell responses help maintain the internal environment or contribute to normal development or reproductive functioning.

4. The hypothalamus and pituitary gland interact to integrate many body activities.

   a. ADH and oxytocin, two hypothalamic hormones, are stored in and released from the posterior lobe of the pituitary. ADH influences extracellular fluid volume. Oxytocin has roles in reproduction.

   b. Six additional hypothalamic hormones, called releasing and inhibiting hormones, control the secretions by different cells of the pituitary's anterior lobe.

   c. Of the six hormones produced in the anterior lobe, two (prolactin and growth hormone) have general effects on body tissues. Four (ACTH, TSH, FSH, and LH) act on specific endocrine glands.

5. Responses to hormones may be influenced by hormone interactions and homeostatic feedback loops to the hypothalamus and pituitary. They may be influenced by variations in hormone concentrations. They also may be influenced by the number and kind of receptors on a target cell.

6. Fast-acting hormones such as parathyroid hormone or insulin generally come into play when the extracellular concentration of a substance must be controlled by homeostatic feedback loops.

7. Hormones such as growth hormone have more prolonged, gradual, and often irreversible effects, such as those on development.

Review Questions

1. Which secretions of the posterior and anterior lobes of the pituitary gland have the targets indicated? (*Fill in the blanks; see pages 272–273.*)

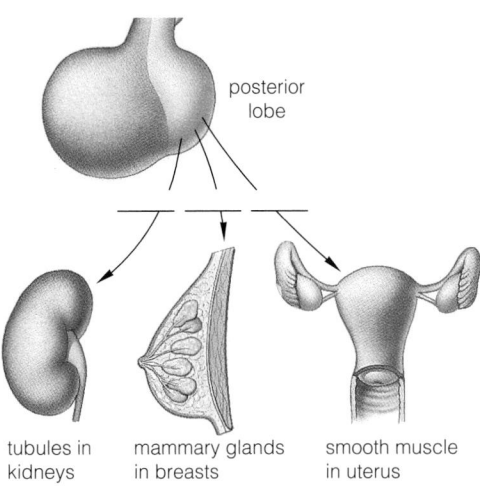

posterior lobe

tubules in kidneys   mammary glands in breasts   smooth muscle in uterus

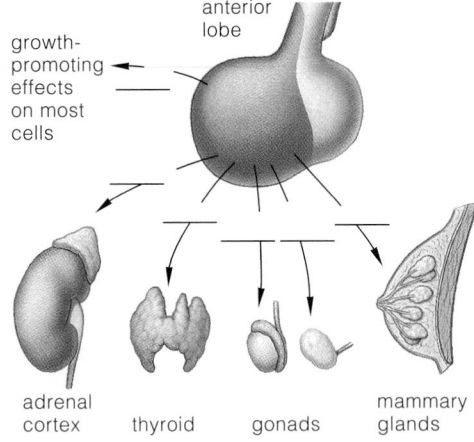

growth-promoting effects on most cells

anterior lobe

adrenal cortex   thyroid   gonads   mammary glands

2. Name the main endocrine glands and state where each is located in the human body. *269*

3. Distinguish among hormones, neurotransmitters, local signaling molecules, and pheromones. *268*

4. A hormone molecule binds to a receptor on a cell membrane. It doesn't enter the cell; rather, the binding activates a second messenger inside the cell that triggers an amplified response to the hormonal signal. Is the signaling molecule a steroid or a peptide hormone? *270–271*

Critical Thinking: You Decide *(Key in Appendix V)*

1. A 20-year-old woman with a malignant brain tumor has her pineal gland removed. What might be some side effects of the loss of the gland?

2. A physician sees a patient whose symptoms include sluggishness, depression, and intolerance to cold. After eliminating several other possible causes, the doctor diagnoses an endocrine disorder. What endocrine disorder fits the symptoms? Why does the doctor suspect that the underlying cause is a malfunction of the anterior pituitary gland?

3. Marianne is affected by type 1 insulin-dependent diabetes. One day, after injecting herself with too much insulin, she starts to shake and feels confused. Following her doctor's suggestion, she drinks a glass of orange juice—a ready source of glucose—and soon her symptoms subside. What caused her symptoms? How would a glucose-rich snack help?

Self-Quiz *(Answers in Appendix IV)*

1. _____ are molecules released from a signaling cell that have effects on target cells.
   a. Hormones
   b. Neurotransmitters
   c. Local signaling molecules
   d. Pheromones
   e. a and b
   f. All of the above

2. Hormones are products of _____ .
   a. endocrine glands and cells
   b. some neurons
   c. exocrine cells
   d. a and b
   e. a and c
   f. a, b, and c

3. ADH and oxytocin are hypothalamic hormones secreted from the pituitary's _____ lobe.
   a. anterior
   b. posterior
   c. primary
   d. secondary

4. _____ has effects on body tissues in general.
   a. ACTH
   b. TSH
   c. LH
   d. Growth hormone

5. Which of the following stimulate hormone secretions?
   a. neural signals
   b. local chemical changes
   c. hormonal signals
   d. environment cues
   e. all of the above can stimulate hormone secretion

6. _____ lowers blood sugar levels; _____ raises it.
   a. Glucagon; insulin
   b. Insulin; glucagon
   c. Gastrin; insulin
   d. Gastrin; glucagon

7. The pituitary detects a rising hormone concentration in blood and inhibits the gland secreting the hormone. This is a _____ feedback loop.
   a. positive
   b. negative

8. Second messengers assist _____ .
   a. steroid hormones
   b. nonsteroid hormones
   c. only thyroid hormones
   d. both a and b

9. Match the hormone source with the closest description.
   ____ adrenal cortex
   ____ adrenal medulla
   ____ thyroid gland
   ____ parathyroids
   ____ pancreatic islets
   ____ pineal gland
   ____ thymus
   a. affected by day length
   b. cortisol source
   c. roles in immunity
   d. adjust(s) blood calcium level
   e. epinephrine source
   f. insulin, glucagon
   g. hormones require iodine

10. Match the endocrine control concepts.
    ____ oxytocin
    ____ ACTH
    ____ ADH
    ____ estrogen
    ____ growth hormone
    a. released by the anterior pituitary amd affects the adrenal gland
    b. influences extracellular fluid volume
    c. has general effects on growth
    d. triggers uterine contractions
    e. a steroid hormone

Selected Key Terms

adrenal cortex *278*
adrenal medulla *279*
amine hormone *271*
endocrine system *268*
hormone *268*
hypothalamus *272*
local signaling molecule *282*
pancreatic islet *276*

parathyroid gland *280*
peptide hormone *271*
pineal gland *281*
pituitary gland *272*
second messenger *271*
steroid hormone *271*
thymus *281*
thyroid gland *280*

Readings

Bower, B. 25 July 1992. "Here Comes the Sun." *Science News.* Current research on seasonal affective disorder.

Colburn, T. 1996. *Our Stolen Future.* New York: NAL/Dutton. A provocative discussion of endocrine disrupters.

Diamond, J. February 1992. "Sweet Death." *Natural History.* High rates of type 2 diabetes in westernized societies may reflect the fact that humans lack the genetic resources to cope with a high-sugar diet.

Hadley, M. 1995. *Endocrinology.* 4th ed. Englewood Cliffs, N.J.: Prentice-Hall.

 For additional readings, go to InfoTrac College Edition, your online library at:

http://www.infotrac-college.com/wadsworth

13

REPRODUCTIVE SYSTEMS

Girls and Boys

An old joke says that you can tell the sex of a baby by the booties it wears—pink or blue. Actually, the way a person's sex is determined is a bit more interesting than that. When you were conceived, a sperm cell from your father fertilized an egg cell in your mother's body. Among its total of 23 chromosomes, the egg carried an X chromosome, and among *its* 23 chromosomes, the sperm carried either an X or a Y chromosome. Normally, from the time of conception, the embryo that develops from the fertilized egg carries a pair of each of the 23 chromosomes, including a combination of the parental sex chromosomes—either XX or XY. But for the first month or so of development, anatomically *you were neither male nor female.*

A gene on the Y chromosome governs the fork in the developmental road that determines a person's gender. In XY embryos, testes—the primary reproductive organs of males—begin to develop about 6 weeks following conception (Figure 13.1). If the embryo is XX, however, it lacks the Y chromosome gene and develops ovaries—the primary female reproductive organs. When there is no Y chromosome present, female body structures begin to develop automatically. In humans, the Y chromosome literally makes all the difference.

Once their development is under way, the testes start producing sex hormones, including testosterone. These hormones influence the development of the

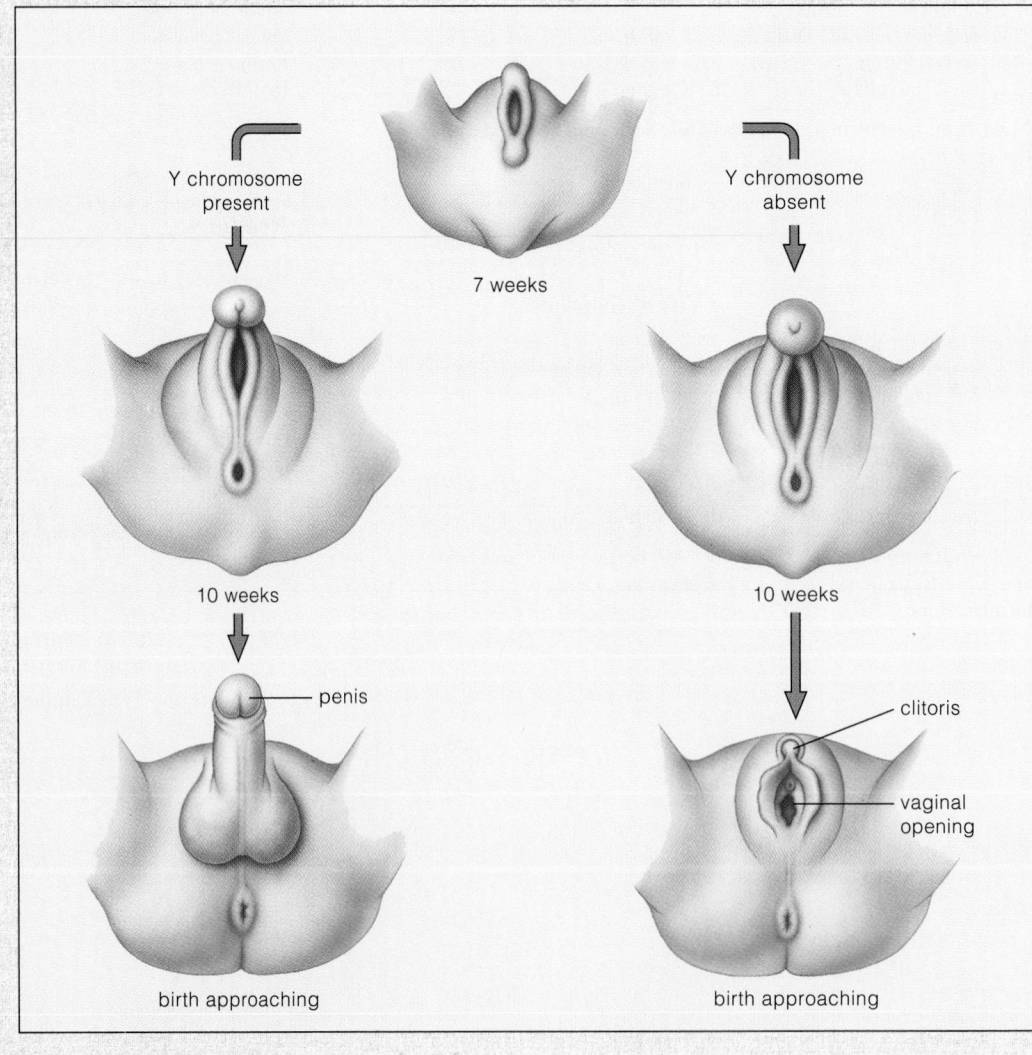

different organs that make up the male reproductive system.

Researchers have identified a particular region of the Y chromosome that appears to be the "master gene" for male sex determination. It is called *Sry* (for *sex-determining region* of the Y chromosome). So far, the same gene has been identified in the DNA of human males and in male chimpanzees, rabbits, pigs, horses, cattle, and tigers. In all females tested, the gene was absent because females lack the Y chromosome. Other tests with mice indicate that the gene becomes active about the time that testes start developing. This chapter focuses on the male and female reproductive systems that come into being by way of these fascinating events.

For both men and women, the reproductive system consists of a pair of primary reproductive organs (gonads), accessory glands, and ducts. Male gonads are **testes** (singular: testis), and female gonads are **ovaries**. Testes produce sperm; ovaries produce eggs. Both secrete sex hormones that influence reproductive functions and the development of **secondary sexual traits**. Such traits are distinctly associated with maleness and femaleness, although they do not play a direct role in reproduction. Examples are the amount and distribution of body fat, hair, and skeletal muscle.

Gonads look the same in all early human embryos. After 7 weeks of development, activation of genes on the sex chromosomes and hormone secretions trigger their development into testes *or* ovaries. The gonads and accessory organs are already formed at birth, but they will not reach full size and become functional until 12 to 16 years later. In the chapter that follows, the story continues as we trace the step-by-step unfolding of human development from the fertilized egg to ripe old age.

KEY CONCEPTS

1. The human reproductive system consists of a pair of primary reproductive organs (testes in males, ovaries in females), accessory glands, and ducts. Testes produce sperm; ovaries produce eggs. Both release sex hormones in response to signals from the hypothalamus and pituitary gland.

2. Human males continually produce sperm from puberty onward. The hormones testosterone, LH, and FSH control male reproductive functions.

3. Human females are fertile on a cyclic basis. Each month during their reproductive years, an egg is released from an ovary, and the lining of the uterus is prepared for pregnancy. The hormones estrogen, progesterone, FSH, and LH control this cyclic activity.

4. Sexual intercourse is the natural mechanism that brings together male and female gametes (sperm and eggs), permitting fertilization and the conception of a new individual.

CHAPTER AT A GLANCE

13.1 The Male Reproductive System
13.2 Male Reproductive Function
13.3 The Female Reproductive System
13.4 Female Reproductive Function
13.5 Visual Summary of the Menstrual Cycle
13.6 Sexual Intercourse and Fertilization
13.7 Control of Fertility
13.8 Coping with Infertility
13.9 *Dilemmas of Fertility Control*

Figure 13.1 Photograph: The child this woman is expecting has inherited chromosomes from its parents that will determine its sex—and the reproductive structures it will have. **Diagram:** External appearance of developing reproductive organs.

THE MALE REPRODUCTIVE SYSTEM

Where Sperm Form

Figure 13.2 shows the organs of the reproductive system in an adult male, and Table 13.1 lists their functions. In an embryo that is genetically destined to become male, a pair of testes form on the abdominal cavity wall. Before birth, the testes descend into the scrotum, an outpouching of skin below the pelvic region. By the time of birth, testes are fully formed miniatures of the adult organs.

Each testis is only about 5 centimeters long, yet 125 meters of **seminiferous tubules** are packed into it! Within the tubules, sperm develop in a process (spermatogenesis) we will explore more fully in the next section and in Chapter 16.

As many as 300 wedge-shaped lobes, of the sort shown in Figure 13.3a, partition the interior of each testis. Two or three seminiferous tubules are coiled inside each lobe. Sperm develop properly when the temperature in the interior of the scrotum remains a few degrees cooler than the rest of the body's normal core temperature. A mechanism that controls the contractions of smooth muscles in the scrotum helps assure that the internal temperature does not stray far from 95°F. When it is cold just outside the body, contractions draw the scrotum closer to the (warmer) body. When it is warm outside, the muscles relax and lower the scrotum.

Where Semen Forms

Human sperm are not quite mature when they leave the testes. First they enter a pair of long, coiled ducts, the epididymides (singular: epididymis). Secretions from cells in the wall of these ducts trigger the finishing touches on

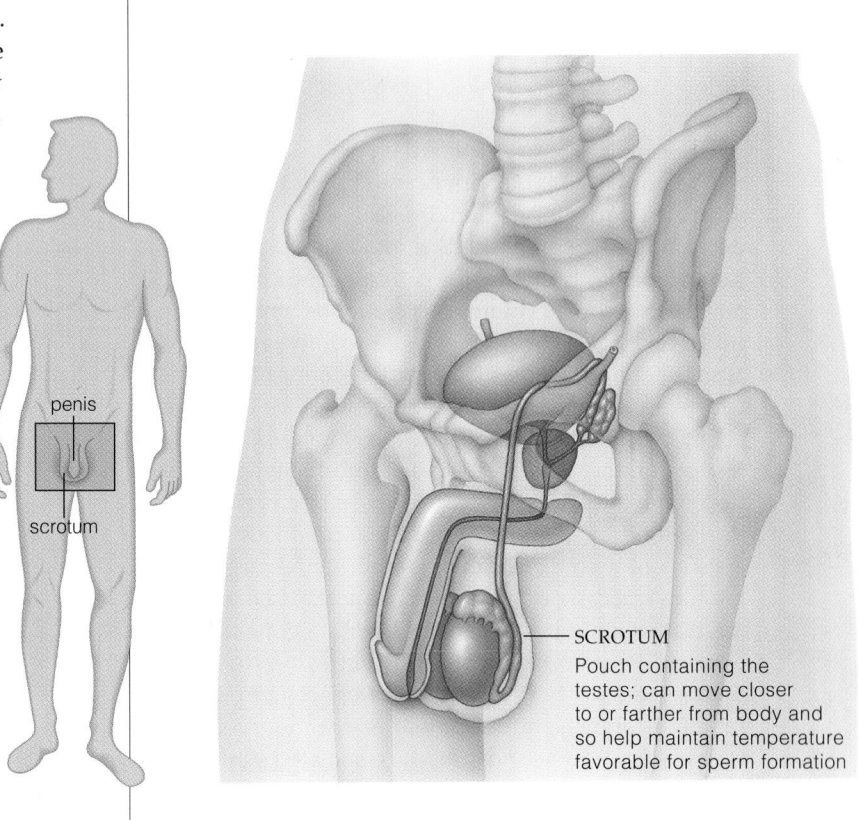

penis

scrotum

SCROTUM
Pouch containing the testes; can move closer to or farther from body and so help maintain temperature favorable for sperm formation

Figure 13.2 Above and facing page: Components of the male reproductive system and their functions.

Table 13.1	Organs and Accessory Glands of the Male Reproductive Tract
REPRODUCTIVE ORGANS:	
Testis (2)	Production of sperm, sex hormones
Epididymis (2)	Sperm maturation site and sperm storage
Vas deferens (2)	Rapid transport of sperm
Ejaculatory duct (2)	Conduction of sperm to penis
Penis	Organ of sexual intercourse
ACCESSORY GLANDS:	
Seminal vesicle	Secretion of large part of semen
Prostate gland	Secretion of part of semen
Bulbourethral gland (2)	Production of lubricating mucus

sperm. Until the time that sperm depart from the body, they are stored in the last stretch of each epididymis.

When a male becomes sexually aroused, muscle contractions in the walls of reproductive organs propel sperm into and through a pair of thick-walled tubes, the **vas deferentia** (singular: vas deferens; also called the ductus deferens). From there, contractions propel sperm through a pair of ejaculatory ducts and then through the urethra. This last tube passes through the penis, the male sex organ, and opens at its tip. The urethra, recall, also functions in urine excretion.

During the trip to the urethra, glandular secretions mix with sperm. The result is **semen**, a thick fluid that is eventually expelled from the penis during sexual activity. Early in the formation of semen, a pair of seminal vesicles secrete the sugar fructose, which provides

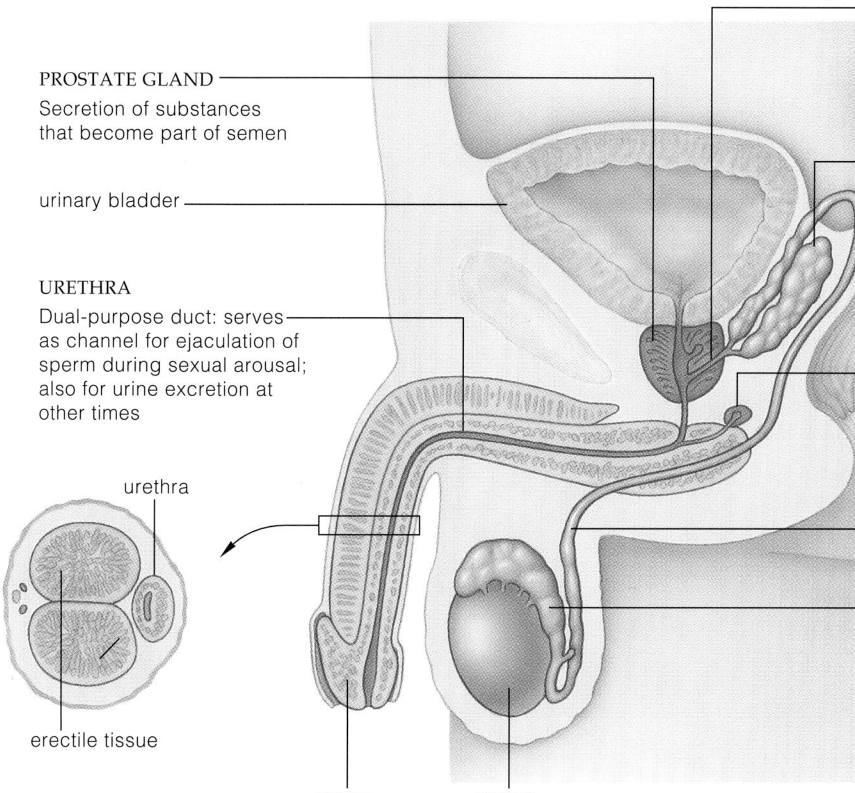

PROSTATE GLAND
Secretion of substances that become part of semen

urinary bladder

URETHRA
Dual-purpose duct: serves as channel for ejaculation of sperm during sexual arousal; also for urine excretion at other times

urethra

erectile tissue

EJACULATORY DUCT
One of a pair of sperm-conducting ducts

SEMINAL VESICLE
One of a pair of glands that secrete fructose and prostaglandins, which become part of semen

BULBOURETHRAL GLAND
One of a pair of glands that secrete a lubricating mucus

anus

VAS DEFERENS
One of a pair of ducts for rapid transport of sperm

EPIDIDYMIS
One of a pair of ducts in which sperm complete maturation; the portion farthest from testis stores mature sperm

PENIS
Organ of sexual intercourse

TESTIS
One of a pair of primary reproductive organs; packed with sperm-producing tubules and cells that secrete testosterone, other hormones

energy for sperm cells. Seminal vesicles also secrete certain prostaglandins. These signaling molecules can induce muscle contractions. (As noted in Chapter 12, it might be that the prostaglandins trigger contractions in the female reproductive tract and assist sperm movement through it.) Secretions from the **prostate gland** probably help buffer the acidic environment that sperm encounter in the vagina. Vaginal pH is about 3.5 to 4.0, but sperm motility improves at pH 6.

A pair of **bulbourethral glands** secrete some mucus-rich fluid into the urethra when the male is sexually aroused. This fluid neutralizes acids in any traces of urine in the urethra, creating a more hospitable environment for the 150 to 350 million sperm that pass through the channel in a typical ejaculation.

The testes and prostate gland both are sites where cancer can develop in males. At least 5,000 cases of testicular cancer are diagnosed each year in the United States, mostly among young men, and the cancer kills about half of its victims. Prostate cancer, which is more common among men over 50, kills 40,000 older men annually—almost the same mortality rate recorded for breast cancer in women. As with other cancers, early detection is the key to survival. The causes and treatments of cancers are the subject matter of Chapter 20.

A pair of testes are the primary reproductive organs (gonads) of an adult male. The male reproductive system also includes accessory glands and ducts.

Sperm, the male gametes, develop mainly in the seminiferous tubules of the testes. When sperm are nearly mature, they leave each testis and enter the long, coiled epididymis, where they remain until ejaculated.

Secretions from the seminal vesicles and the prostate gland mix with sperm to form semen.

Sperm Formation

Just inside the walls of seminiferous tubules are undifferentiated cells called *spermatogonia*. The cells undergo divisions, including a type called *mitosis* and a type called *meiosis*. Chapter 16 gives the details of mitosis and meiosis; here it is important to keep in mind that meiosis results in the specialized reproductive cells called *gametes* (sperm and eggs). As the introduction noted, human gametes have 23 chromosomes, including one sex chromosome. This is termed a "haploid" number of chromosomes because it is one-half the normal chromosome number (a "diploid" number of 46) of other human body cells. When two haploid gametes unite at fertilization, the diploid chromosome number is restored.

Spermatogonia develop into *primary spermatocytes*, which become *secondary spermatocytes* after a first round of meiotic division (meiosis I). A second round of division (meiosis II) results in *spermatids*. The spermatids gradually develop into *spermatozoa*, or simply **sperm**—the male gametes. The "tail" (flagellum) of each sperm arises at the end of the process, which takes 9 to 10 weeks. All the while, the developing cells receive nourishment and chemical signals from adjacent **Sertoli cells**, which are the cells that line the seminiferous tubule.

From puberty onward, sperm are produced continuously. Millions are in different stages of development on any given day. A mature sperm has a tail, a midpiece, and a head (Figure 13.3d). Within the head, a nucleus contains DNA organized into chromosomes. An enzyme-containing cap, the **acrosome**, covers most of the head. Its enzymes help the sperm penetrate the extracellular material around an egg at fertilization. In the midpiece, mitochondria supply energy for the tail's whiplike movements.

Hormonal Controls

Male reproductive function depends on testosterone, LH, and FSH (Figure 13.4). **Leydig cells** (also called interstitial cells), located in tissue between the seminiferous tubules in testes, secrete **testosterone**. This hormone governs the growth, form, and functions of the male reproductive tract. It stimulates sexual behavior, and promotes development of secondary sexual traits, including facial hair growth and deepening of the voice at puberty.

LH and **FSH** are secreted by the anterior lobe of the pituitary gland. They were initially named for their effects in females. (As indicated in Chapter 12, the two abbreviations stand for luteinizing hormone and follicle-stimulating hormone.) However, their molecular structure is identical in males and females.

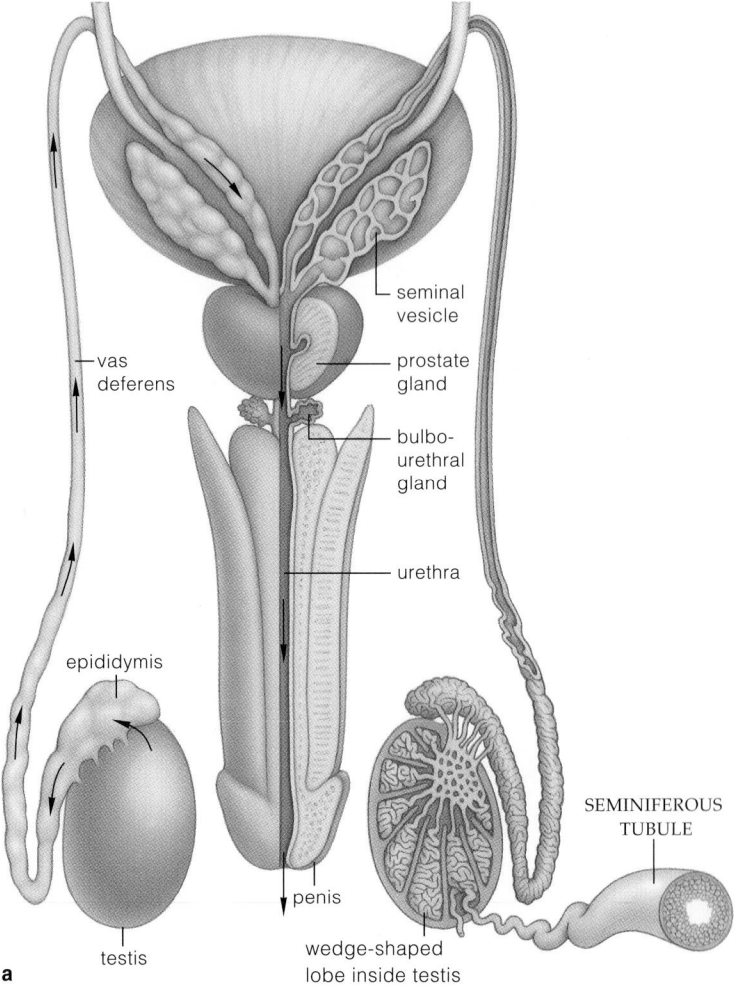

a

vas deferens

seminal vesicle

prostate gland

bulbo-urethral gland

urethra

epididymis

testis

penis

wedge-shaped lobe inside testis

SEMINIFEROUS TUBULE

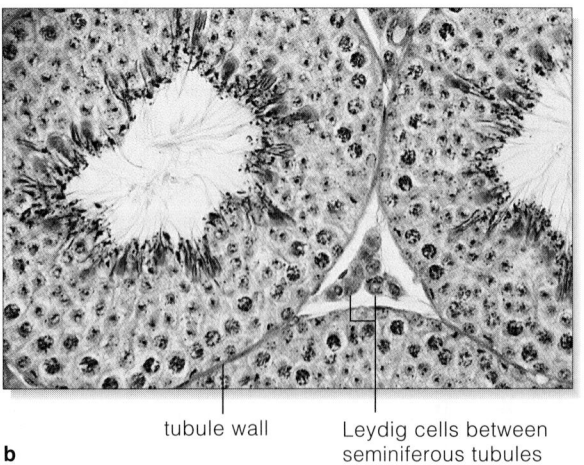

b

tubule wall

Leydig cells between seminiferous tubules

Figure 13.3 Above and facing page: (**a**) Male reproductive tract, posterior view. Arrows show the route that sperm take before ejaculation from a sexually aroused male. (**b**) Micrograph of cells in seminiferous tubules.

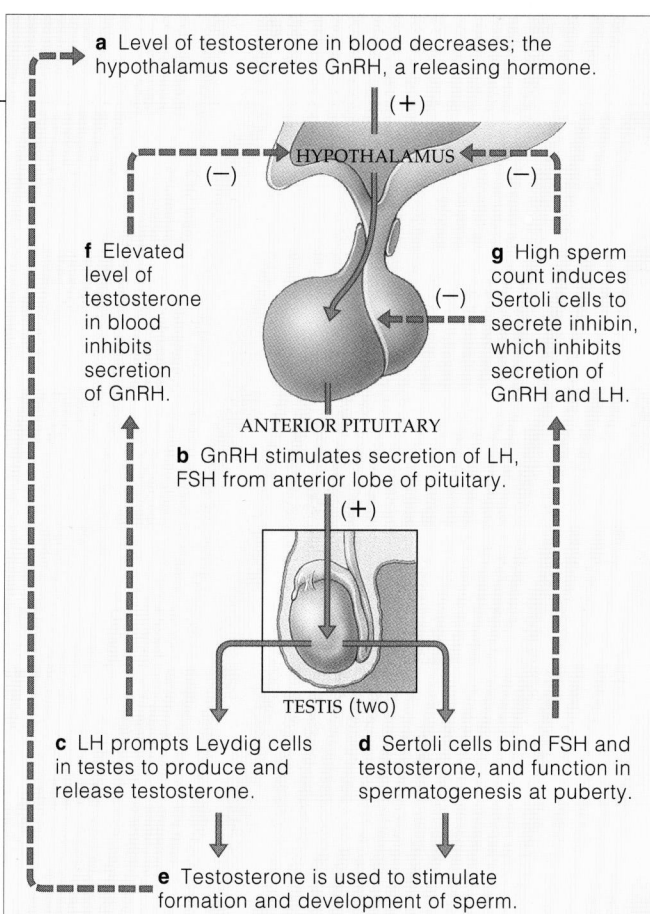

a Level of testosterone in blood decreases; the hypothalamus secretes GnRH, a releasing hormone.

(+)

HYPOTHALAMUS

(−)   (−)

f Elevated level of testosterone in blood inhibits secretion of GnRH.

(−)

g High sperm count induces Sertoli cells to secrete inhibin, which inhibits secretion of GnRH and LH.

ANTERIOR PITUITARY

b GnRH stimulates secretion of LH, FSH from anterior lobe of pituitary.

(+)

TESTIS (two)

c LH prompts Leydig cells in testes to produce and release testosterone.

d Sertoli cells bind FSH and testosterone, and function in spermatogenesis at puberty.

e Testosterone is used to stimulate formation and development of sperm.

Figure 13.4 Negative feedback loops to the hypothalamus and pituitary gland from the testes. Through these loops, excess testosterone production shuts off the mechanisms leading to its production. This helps maintain the testosterone level in amounts required for sperm formation.

The hypothalamus controls secretions of LH, FSH, and testosterone—and thus controls sperm formation (Figure 13.4). When the testosterone level in blood decreases past a set point, the hypothalamus secretes GnRH. This releasing hormone prompts the pituitary's anterior lobe to release LH and FSH, which have targets in the testes. LH stimulates Leydig cells to secrete testosterone, which stimulates diploid germ cells to become sperm. Sertoli cells have FSH receptors. FSH is crucial to establishing spermatogenesis at the time of puberty, but researchers do not know whether this hormone is essential for the normal functioning of mature testes in males.

A high testosterone level in blood has an inhibitory effect on GnRH release. Also, when the sperm count is high, Sertoli cells release inhibin, a protein hormone that acts on the hypothalamus and pituitary to inhibit the release of GnRH and FSH. Hence, feedback loops to the hypothalamus begin to operate—with a resulting decrease in testosterone secretion and sperm formation.

Sperm formation depends on the hormones testosterone, LH, and FSH. Feedback loops from the testes to the hypothalamus and pituitary gland control their secretion.

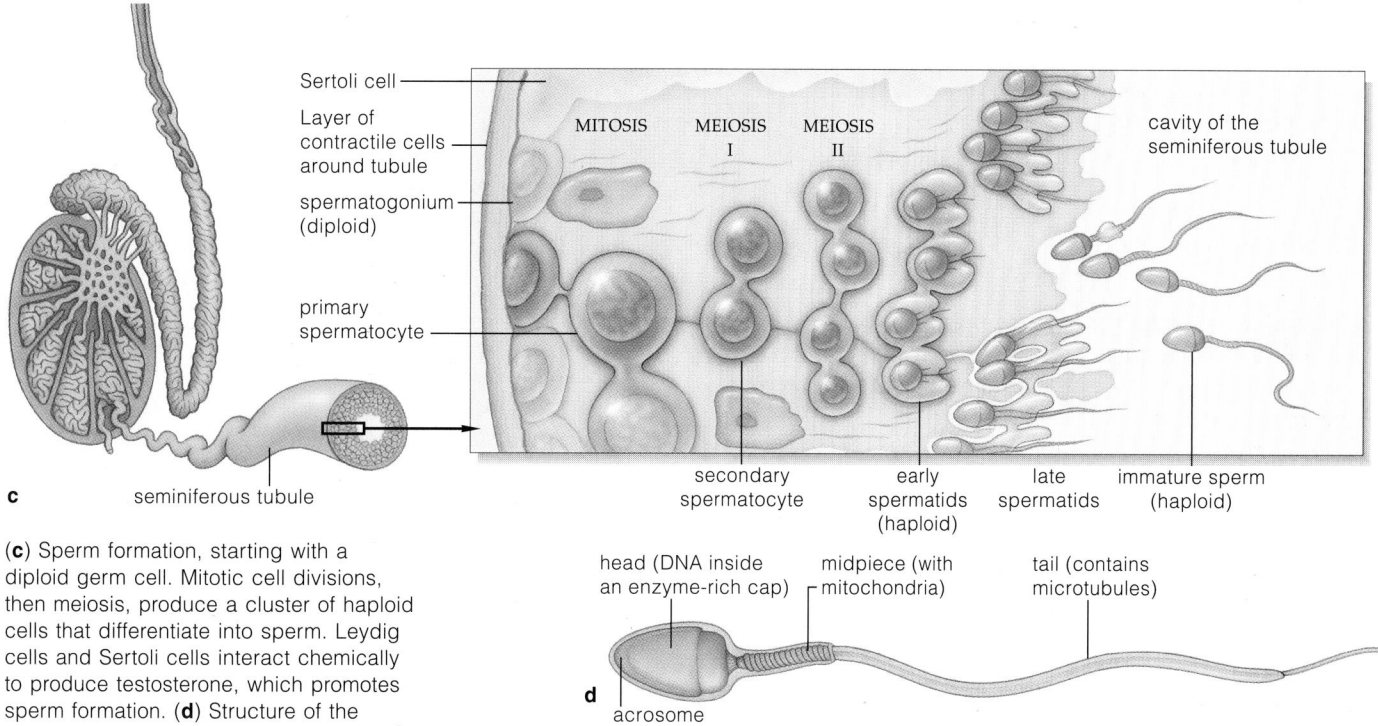

Sertoli cell

Layer of contractile cells around tubule

spermatogonium (diploid)

primary spermatocyte

MITOSIS   MEIOSIS I   MEIOSIS II

cavity of the seminiferous tubule

secondary spermatocyte

early spermatids (haploid)

late spermatids

immature sperm (haploid)

c   seminiferous tubule

(**c**) Sperm formation, starting with a diploid germ cell. Mitotic cell divisions, then meiosis, produce a cluster of haploid cells that differentiate into sperm. Leydig cells and Sertoli cells interact chemically to produce testosterone, which promotes sperm formation. (**d**) Structure of the mature human sperm.

head (DNA inside an enzyme-rich cap)

midpiece (with mitochondria)

tail (contains microtubules)

d   acrosome

The Reproductive Organs

Figure 13.5 shows the main components of the female reproductive system and lists their functions (see also Table 13.2). About 8 weeks after a female embryo is conceived, the two ovaries (like the testes) begin to develop from buds in the abdominal wall. After further development, they eventually reside in the pelvic cavity. Ovaries secrete sex hormones, including **estrogens** and **progesterone**, and during a woman's reproductive years they release eggs. Ovarian hormones influence the development of female secondary sexual traits, including the "filling out" of breasts, hips, and buttocks as fat deposits accumulate in those areas.

An immature egg, or **oocyte**, released from an ovary moves into an **oviduct** (sometimes called a *Fallopian tube*), which is where fertilization may occur. Fertilized or not, the egg then continues down the oviduct into the hollow, pear-shaped **uterus**. In this organ, a new individual can grow and develop. The wall of the uterus consists of a

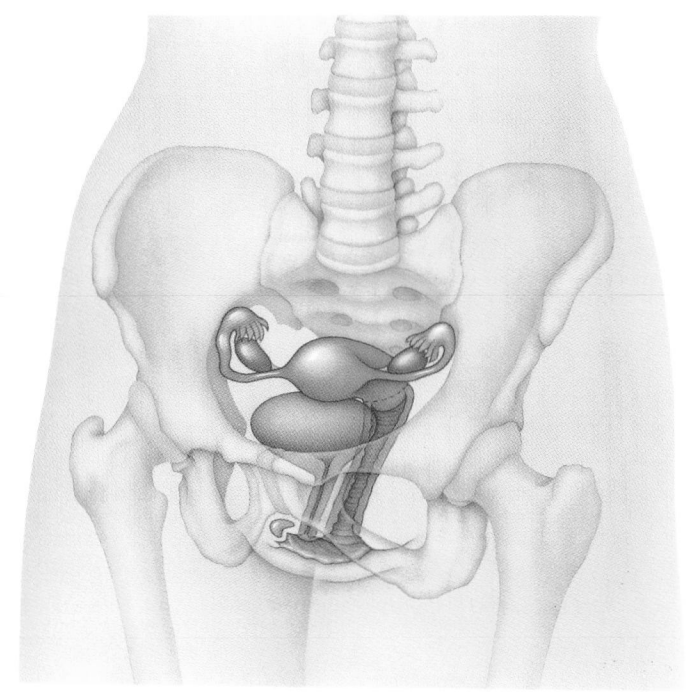

Figure 13.5   **Above and facing page:** Components of the human female reproductive system and their functions.

Table 13.2   **Female Reproductive Organs**

Ovaries	Oocyte production, sex hormone production
Oviducts	Conduction of oocyte from ovary to uterus
Uterus	Chamber in which new individual develops
Cervix	Secretion of mucus that enhances sperm movement into uterus and (after fertilization) reduces the embryo's risk of bacterial infection
Vagina	Organ of sexual intercourse; birth canal

Table 13.3   **Events of the Menstrual Cycle**

Phase	Events	Days of the Cycle*
Follicular phase	Menstruation; endometrium breaks down	1–5
	Follicle matures in ovary; endometrium rebuilds	6–13
Ovulation	Secondary oocyte released from ovary	14
Luteal phase	Corpus luteum forms; endometrium thickens and develops	15–28

*Assumes a 28-day cycle.

thick layer of smooth muscle (the myometrium) and an interior lining, the **endometrium**, which includes epithelial tissue, connective tissue, glands, and blood vessels. The lower portion of the uterus is the *cervix*. A muscular tube, the *vagina*, extends from the cervix to the body surface. This tube receives the penis and sperm and functions as part of the birth canal.

At the body surface are external genitals, collectively called the *vulva*, which include organs for sexual stimulation. Outermost are a pair of fat-padded skin folds, the *labia majora*. They enclose a smaller pair of skin folds, the *labia minora*, that are highly vascularized but have no fatty tissue. The labia minora partly enclose the **clitoris**, a small organ sensitive to stimulation that is developmentally analogous to the penis (see Figure 13.1). The urethra's opening is about midway between the clitoris and the vaginal opening. Notice that while in males the urethra carries both urine and sperm, in females the urethra is separate and is not involved in reproduction.

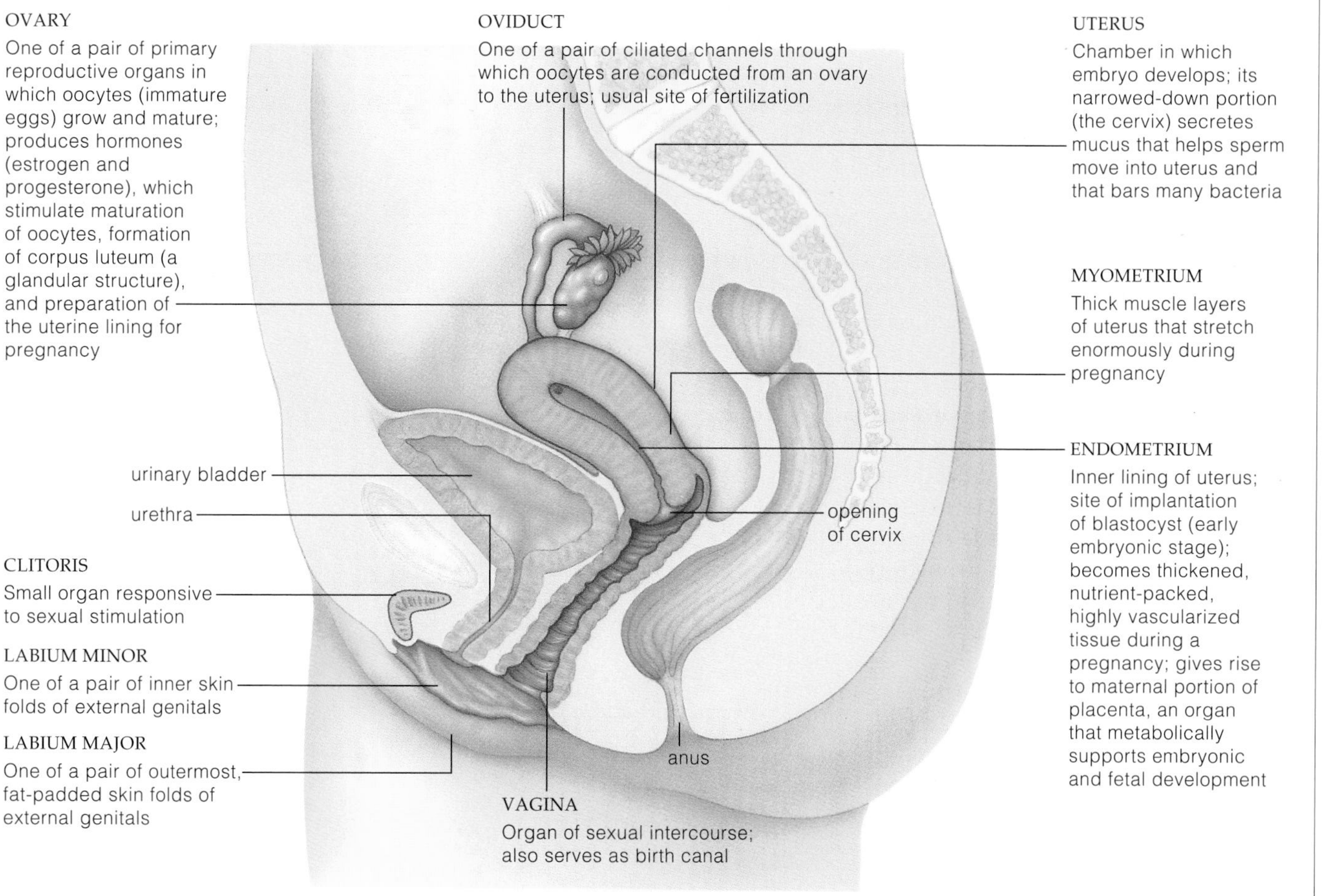

OVARY

One of a pair of primary reproductive organs in which oocytes (immature eggs) grow and mature; produces hormones (estrogen and progesterone), which stimulate maturation of oocytes, formation of corpus luteum (a glandular structure), and preparation of the uterine lining for pregnancy

OVIDUCT

One of a pair of ciliated channels through which oocytes are conducted from an ovary to the uterus; usual site of fertilization

UTERUS

Chamber in which embryo develops; its narrowed-down portion (the cervix) secretes mucus that helps sperm move into uterus and that bars many bacteria

MYOMETRIUM

Thick muscle layers of uterus that stretch enormously during pregnancy

ENDOMETRIUM

Inner lining of uterus; site of implantation of blastocyst (early embryonic stage); becomes thickened, nutrient-packed, highly vascularized tissue during a pregnancy; gives rise to maternal portion of placenta, an organ that metabolically supports embryonic and fetal development

urinary bladder

urethra

opening of cervix

CLITORIS

Small organ responsive to sexual stimulation

LABIUM MINOR

One of a pair of inner skin folds of external genitals

LABIUM MAJOR

One of a pair of outermost, fat-padded skin folds of external genitals

anus

VAGINA

Organ of sexual intercourse; also serves as birth canal

Overview of the Menstrual Cycle

Like other female primates, female humans follow a **menstrual cycle**. As Table 13.3 indicates, it takes about 28 days to complete one cycle, although this can vary from month to month and from woman to woman. On the first day of the cycle, blood-rich fluid, about 4 to 6 tablespoons total, starts flowing out through the vaginal canal. While the cycle progresses, an oocyte matures and is released from an ovary. As you will see, hormones are also priming the endometrium to receive and nourish an embryo. If fertilization occurs and an embryo becomes implanted in the endometrium, menstrual cycling will stop for the duration of the pregnancy. If the egg is not fertilized, the endometrium will break down, and a new cycle will begin with the menstrual flow.

The first menstruation, or *menarche*, usually occurs between the ages of 10 and 16. Menstrual cycles continue until *menopause*, in the late 40s or early 50s. By then, a woman's hormone secretions slow down, and her sensitivity to pituitary reproductive hormones diminishes.

Decreasing estrogen levels in a menopausal woman may trigger various temporary symptoms, including moodiness and "hot flashes." These sudden bouts of sweating and feeling uncomfortably warm result from widespread vasodilation of the blood vessels in skin. Section 14.14, which discusses aging in various body systems, describes other physiological changes associated with menopause. Eventually, a woman's menstrual cycles stop, and fertility is over.

Ovaries are the female gonads. Together with the oviducts, uterus, cervix, and vagina, they make up the female reproductive system.

The reproductive period of a female's life is characterized by a menstrual cycle of about 28 days. In the course of each cycle, an oocyte (immature egg) matures and is released.

The first menstruation (menarche) usually occurs between the ages of 10 and 16. Menstrual cycles continue until menopause.

FEMALE REPRODUCTIVE FUNCTION

Cyclic Changes in the Ovary

At birth, a normal baby girl has about 2 million primary oocytes in her ovaries. By the time she is 7 years old, about 300,000 remain (her body has resorbed the rest). These oocytes entered meiosis I during fetal life, but then the division process was arrested. Meiosis will resume in one oocyte at a time, starting with the first menstrual cycle. Only about 400 or 500 will be released during her reproductive years.

Figure 13.6 shows a *primary* oocyte located near the surface of an ovary. A layer of **granulosa cells** surrounds and nourishes the primary oocyte. Together, the primary oocyte and the cell layer are a **follicle**. At the start of a menstrual cycle, the hypothalamus is secreting GnRH, which stimulates the anterior pituitary to release FSH and LH (Figure 13.7). These hormones stimulate the follicle to grow. FSH means "follicle-stimulating hormone." The oocyte starts to increase in size, and more cell layers form around it. Glycoprotein deposits build up between the oocyte and the cell layers, widening the space between them. In time they form a noncellular coating, the **zona pellucida**, around the oocyte.

In response to FSH and LH, cells outside the zona pellucida secrete several estrogens. Estrogen-containing fluid starts to accumulate inside the follicle (now called a secondary or Graafian follicle), and estrogen levels in the blood start to rise. About 8 to 10 hours before being released from the ovary, the oocyte completes the meiotic cell division (called meiosis I) that was arrested years before. Now, two cells are present: a large, **secondary oocyte** and a much smaller *polar body*. The allocation of chromosomes to each cell assures that the secondary oocyte will be haploid—the chromosome number required for gametes.

About midway through the cycle, the pituitary gland detects the rising estrogen level and responds with a brief outpouring of LH. This triggers cellular contractions that make the fluid-filled follicle balloon outward, then rupture. The fluid escapes, carrying the secondary oocyte and polar body with it (Figure 13.6). *The midcycle surge of LH has triggered ovulation—the release of a secondary oocyte from the ovary.*

A secondary oocyte released from an ovary enters an oviduct. Fingerlike projections from the oviduct (called *fimbriae*) extend over part of the ovary. Each "finger" bears beating cilia, and movements of the projections and their cilia sweep the oocyte into the channel. If fertilization

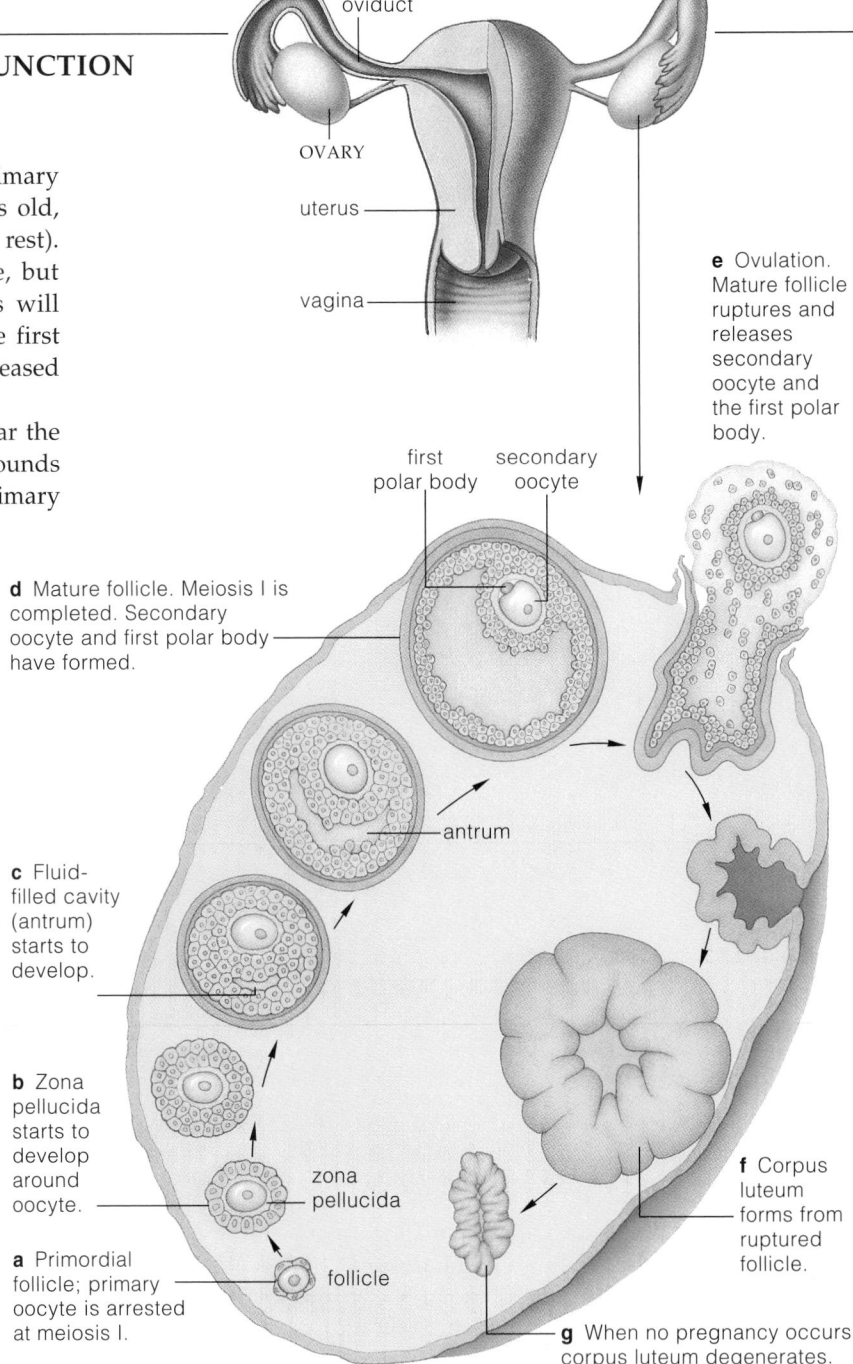

e Ovulation. Mature follicle ruptures and releases secondary oocyte and the first polar body.

d Mature follicle. Meiosis I is completed. Secondary oocyte and first polar body have formed.

first polar body   secondary oocyte

antrum

c Fluid-filled cavity (antrum) starts to develop.

b Zona pellucida starts to develop around oocyte.

zona pellucida

a Primordial follicle; primary oocyte is arrested at meiosis I.

follicle

f Corpus luteum forms from ruptured follicle.

g When no pregnancy occurs, corpus luteum degenerates.

Figure 13.6 Section through a human ovary. The events proceed at the same location; the sketch merely shows the order in which they occur. In the first phase of the menstrual cycle, a follicle grows and matures. At ovulation, the mature follicle ruptures, releasing a secondary oocyte. In the second phase, a corpus luteum forms from the follicle's remnants. It self-destructs if pregnancy does not occur.

takes place, it typically occurs while the oocyte is in the oviduct. At fertilization, meiosis II is completed, giving rise to a mature ovum, the egg. Chapter 14 looks at the events following fertilization in detail.

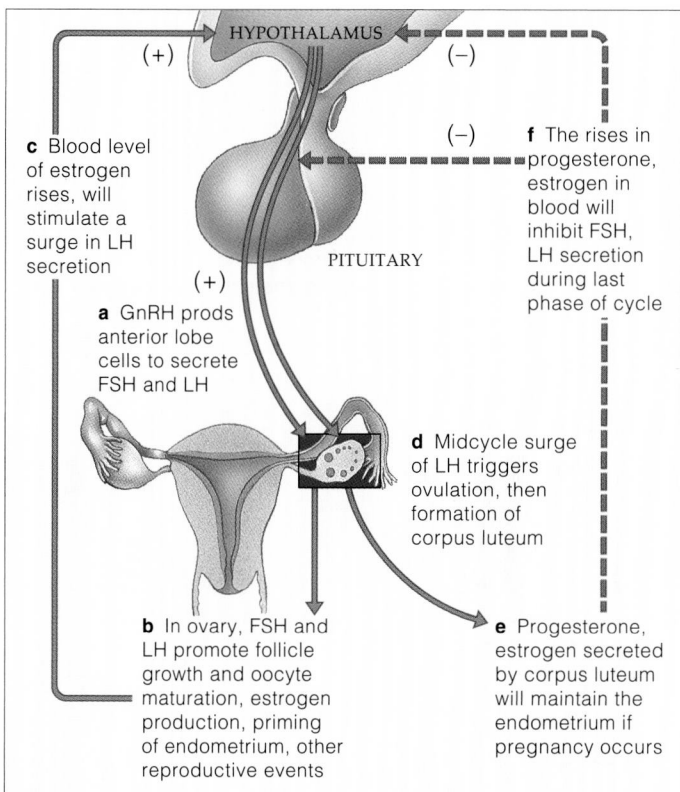

secondary oocyte

surface of ovary

Figure 13.7 Feedback control of hormonal secretion during a menstrual cycle. A positive feedback loop from an ovary to the hypothalamus causes a surge in LH secretion. This surge triggers ovulation. The micrograph above shows a secondary oocyte being released from an ovary at this time. Afterward, negative feedback loops to the hypothalamus and pituitary inhibit FSH secretion. They prevent another follicle from maturing until the cycle is completed.

Cyclic Changes in the Uterus

The estrogens released during the early phase of the menstrual cycle also help pave the way for pregnancy. They stimulate growth of the endometrium and its glands. Then, just before the midcycle LH surge, cells of the follicle wall start secreting progesterone as well as estrogens. Blood vessels grow rapidly in the thickened endometrium. At ovulation, estrogens act on tissue around the cervical canal, the narrowed portion of the uterus that leads to the vagina. The cervix now secretes large amounts of a thin, clear mucus—an ideal medium for sperm travel.

After ovulation, another structure dominates events. Granulosa cells left behind in the follicle differentiate into a yellowish glandular structure, the **corpus luteum** (meaning "yellow body"). Formation of the corpus

luteum results from the midcycle surge of LH. Hence its name, luteinizing hormone.

The corpus luteum secretes progesterone and some estrogen. Progesterone prepares the reproductive tract for the arrival of an embryo. For example, it causes mucus in the cervix to become thick and sticky. The mucus may prevent normal vaginal bacteria from entering the uterus. Progesterone also maintains the endometrium during a pregnancy.

A corpus luteum persists for about 12 days. During that time, the hypothalamus signals for a decrease in FSH secretion, which prevents any other follicles from developing. If a developing embryo does not arrive and burrow into the endometrium (a process called *implantation*), the corpus luteum self-destructs during the last days of the menstrual cycle. It does so by secreting prostaglandins, which apparently disrupt the corpus luteum's own functioning.

After the corpus luteum breaks down, progesterone and estrogen levels fall rapidly, so the endometrium also starts to break down. Deprived of oxygen and nutrients, its blood vessels constrict, and its tissues die. Blood escapes from the ruptured walls of weakened capillaries. The blood and sloughed endometrial tissues make up the menstrual flow, which continues for 3 to 6 days. Then the cycle begins anew, and rising estrogen levels stimulate the repair and growth of the endometrium.

Coordinated secretions of estrogen, progesterone, LH, and FSH bring about changes in the ovary and uterus during the menstrual cycle.

A midcycle surge of LH triggers ovulation, the release of the secondary oocyte and the polar body from the ovary.

The hormones released during various phases of the menstrual cycle help pave the way for fertilization and prepare the endometrium and other parts of a female's reproductive tract for pregnancy.

VISUAL SUMMARY OF THE MENSTRUAL CYCLE

By now you may have come to the conclusion that the menstrual cycle is not a simple tune on a biological banjo—it is a full-blown hormonal symphony!

Before continuing your reading, take a moment to review Figure 13.8. It may leave you with a better understanding of the coordinated changes in hormone levels during each menstrual cycle, as correlated with the cyclic changes that they bring about in the ovary and uterus.

Figure 13.8 Changes in the ovary and uterus, correlated with changing hormone levels during the menstrual cycle. *Green* arrows indicate which hormones dominate the cycle's first phase (when the follicle matures), then the second phase (when the corpus luteum forms). (**a,b**) FSH and LH cause changes in ovarian structure and function. (**c,d**) Estrogen and progesterone from the ovary cause changes in the endometrium.

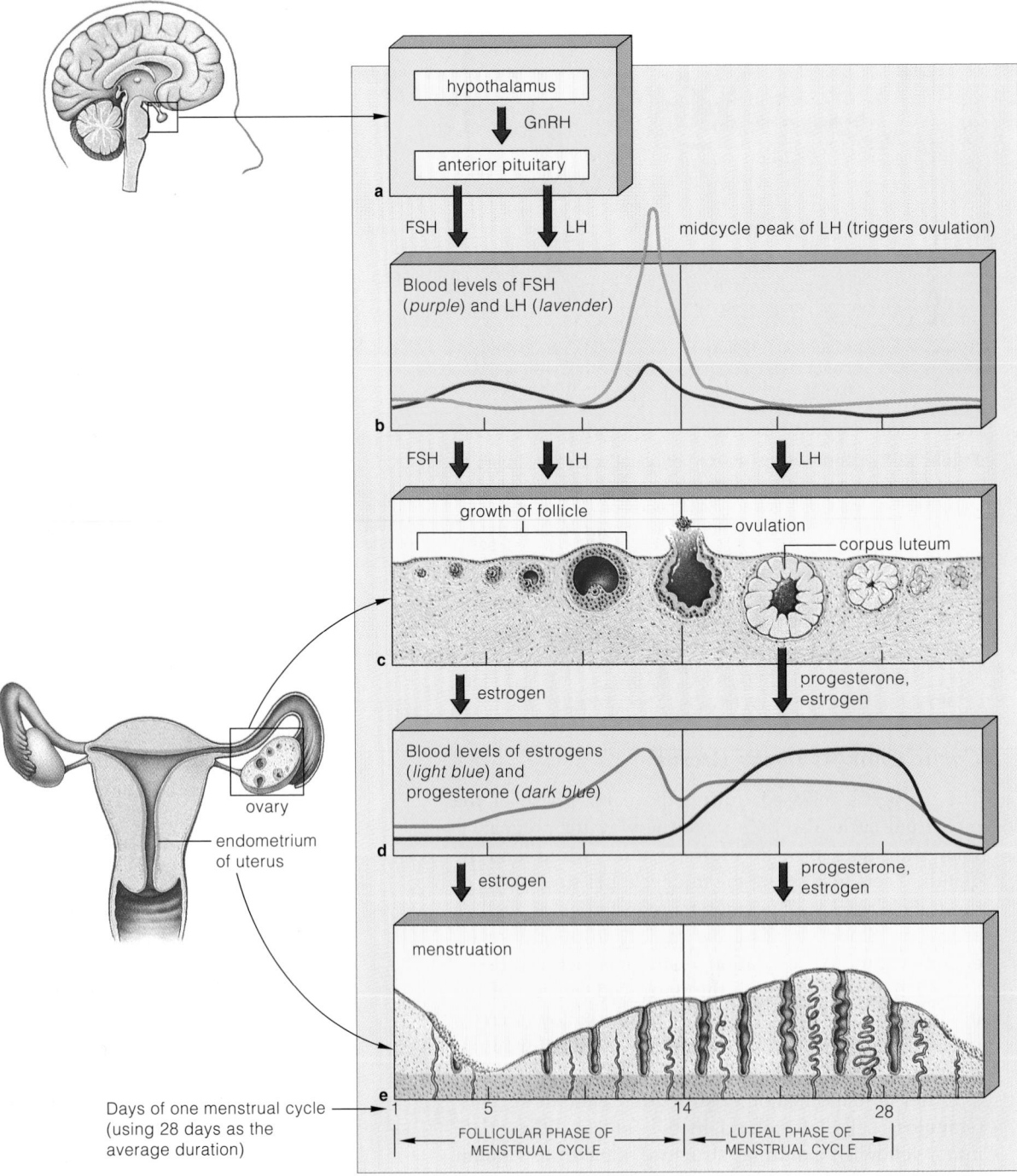

SEXUAL INTERCOURSE AND FERTILIZATION

Sexual Intercourse

Suppose a secondary oocyte is on its way down an oviduct when a female and male are engaged in sexual intercourse, or **coitus**. The male sex act typically requires *erection*, in which the limp penis stiffens and lengthens, and ejaculation, a forceful expulsion of semen into the urethra and out from the penis. As Figure 13.2 shows, the penis contains cylinders of spongy tissue. One cylinder has a mushroom-shaped tip (the glans penis). Within it are densely arrayed sensory receptors that are activated by friction. In sexually unaroused males, the large blood vessels leading into the cylinders are vasoconstricted. In aroused males, these blood vessels vasodilate, so blood flows into the cylinders faster than it flows out. Blood collects in the spongy tissue, and the engorgement stiffens and lengthens the organ—a mechanism that helps the penis penetrate into the female's vagina.

In females, arousal includes vasodilation of blood vessels in the genital area. This causes vulvar tissues to engorge with blood and swell. Secretions flow from the cervix, lubricating the vagina.

During coitus, pelvic thrusts stimulate the penis as well as the female's clitoris and vaginal wall. The mechanical stimulation triggers rhythmic, involuntary contractions in smooth muscle in the male reproductive tract, especially the vas deferens and the prostate. The contractions rapidly force sperm out of each epididymis. They force the contents of seminal vesicles and the prostate gland into the urethra. The resulting mixture, semen, is ejaculated into the vagina. During ejaculation, a sphincter closes off the neck of the male's bladder and prevents urine from being excreted. Ejaculation is a reflex response; once it begins, it cannot be halted.

Emotional intensity, hard breathing, and heart pounding, as well as generalized skeletal muscle contractions, accompany the rhythmic throbbing of the pelvic muscles. At **orgasm**, the culmination of the sex act, strong sensations of release, warmth, and relaxation dominate. During female orgasm, similar events occur, including intense vaginal awareness, involuntary uterine and vaginal contractions, and sensations of relaxation and warmth.

Some people mistakenly believe that unless a woman experiences orgasm, she cannot become pregnant. Don't believe it. A female can become pregnant from intercourse *regardless* of whether orgasm occurs, and even if she is not sexually aroused. All that is required is that a sperm encounter a secondary oocyte that is traveling down an oviduct. Chapter 14 considers the events of conception in some detail. However, it will help round out this chapter's portrait of reproduction to briefly preview the biological sequel when sexual intercourse brings a sperm and an oocyte together.

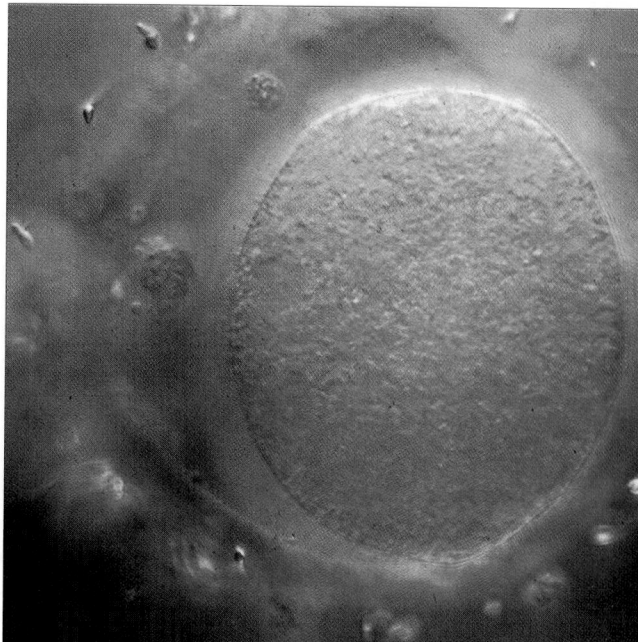

Figure 13.9 A secondary oocyte surrounded by sperm. If fertilization ensues, it will set the stage for the development of a new individual, continuing the human life cycle.

Fertilization: A Preview

Now, sperm are in the vagina. If they arrive a few days before or after ovulation or anytime between, fertilization may be the outcome. Within thirty minutes after ejaculation, muscle contractions move the sperm deeper into the female reproductive tract. Only a few hundred sperm will actually reach the upper portion of the oviduct, which is where fertilization usually takes place. The remarkable micrograph in Figure 13.9 shows living sperm around a secondary oocyte.

As you will read more fully in Chapter 14, the meeting of sperm and secondary oocyte is only the first of several intricately orchestrated events that culminate in actual **fertilization**—the fusion of the sperm nucleus with the nucleus of a mature egg.

Sexual intercourse (coitus) typically involves a sequence of physiological changes in both partners.

During arousal, dilation of blood vessels causes increased blood flow to the penis (males) and vulva (females). Orgasm involves muscular contractions (including those leading to ejaculation of semen into the vagina) and sensations of release, warmth, and relaxation.

A female may become pregnant through intercourse even if she is not sexually aroused or does not experience orgasm.

Fertilization can occur when a sperm encounters a secondary oocyte, usually in the oviduct.

Many sexually active people choose to exercise control over whether their activity will produce a child. The *Choices* feature (page 301) explores some social and ethical dilemmas that control of fertility can entail. Here we consider the biological bases of different forms of birth control.

Birth Control Options

The most effective method of birth control is complete *abstinence*—no sexual intercourse whatsoever. However, the motivation to engage in sex has been evolving for more than 500 million years, and so it is probably unrealistic to expect many people to practice abstinence for long periods.

A modified form of abstinence is the *rhythm method*, also called the "fertility awareness" or *sympto-thermal method*. The idea is to avoid intercourse during the woman's fertile period, beginning a few days before ovulation and ending a few days after. Her fertile period is identified and tracked by keeping records of the length of her menstrual cycles and sometimes by examining her cervical secretions and taking her temperature each morning when she wakes up. (Core body temperature rises by one-half to one degree just after ovulation.) But ovulation can be irregular, and it can be easy to miscalculate. Also, sperm deposited in the vaginal tract a few days before ovulation may survive until ovulation. The method is inexpensive (a woman need only buy a thermometer) and does not require fittings and periodic checkups by a physician. But its practitioners do run a very high risk of pregnancy (Figure 13.10).

Withdrawal, or removing the penis from the vagina before ejaculation, dates back at least to biblical times. It is not very effective. Not only does withdrawal require strong willpower, but, in addition, fluid released from the penis before ejaculation may contain sperm.

Douching, or rinsing out the vagina with a chemical right after intercourse, is next to useless. Sperm can move past the cervix and out of reach of the douche within 90 seconds after ejaculation.

Other methods involve physical or chemical barriers to prevent sperm from entering the uterus and moving to the oviducts. *Spermicidal foam* and *spermicidal jelly* are toxic to sperm. They are packaged in an applicator and placed in the vagina just before intercourse. These products are not reliable unless used with another device, such as a diaphragm or condom.

A *diaphragm* is a flexible, dome-shaped device that is inserted into the vagina and positioned over the cervix before intercourse. A diaphragm is fairly effective when fitted initially by a doctor, used with foam or jelly before each sexual contact, and inserted correctly with each use. A variation on the diaphragm, the *cervical cap*, is smaller and can be left in place for as long as 3 days with just

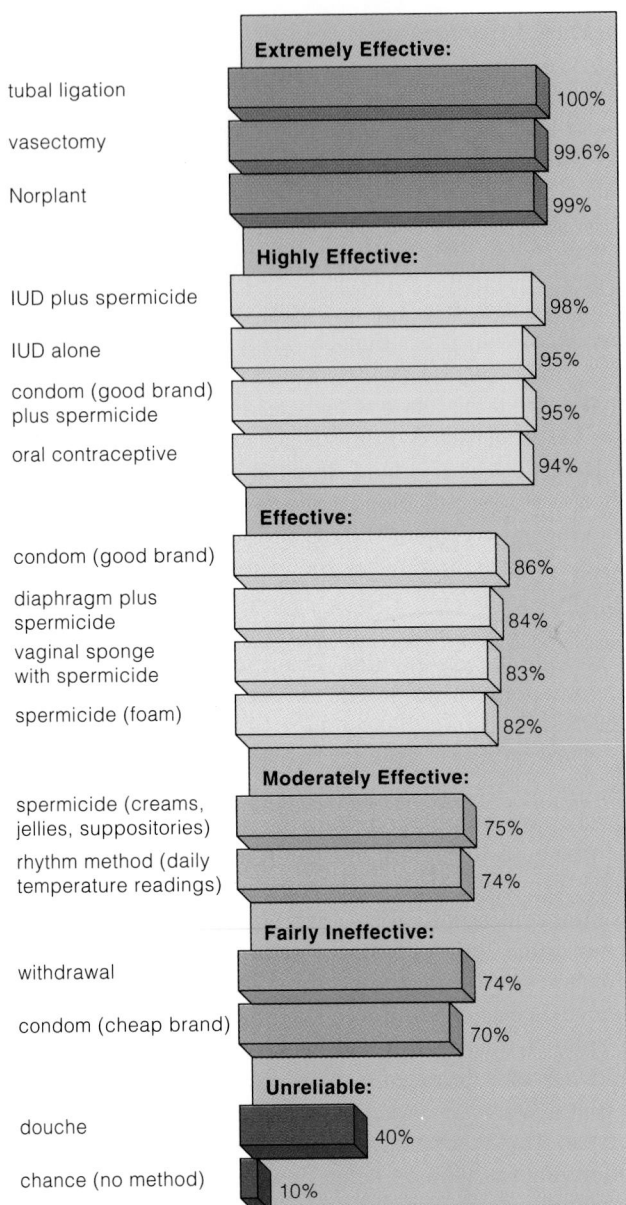

Figure 13.10 Comparison of the effectiveness of some contraceptive methods in the United States. Percentages shown are based on the number of unplanned pregnancies per 100 couples who used the method as the only form of birth control for one year. For example, "94% effectiveness" for oral contraceptives (the Pill) means that, on average, 6 of every 100 women using them will become pregnant.

a single dose of spermicide. It can be more difficult to fit properly, however. The *contraceptive sponge* is a soft, disposable disk that contains a spermicide and covers the cervix. It does not require a prescription or special fitting; it is simply wetted and inserted up to 24 hours before intercourse. Because the sponge can slip and become displaced, the method is about 84 percent reliable.

The *intrauterine device,* or IUD, is a small plastic or metal device that is placed into the uterus for up to 2 years at a time. It interferes with implantation of a fertilized egg into the uterine wall. Available by prescription, IUDs have been associated with severe menstrual cramping, increased risk of infection or perforation of the uterus, and other problems. "Improved" types may possibly address such concerns, but any woman considering an IUD should discuss the matter fully with her physician.

Condoms are thin, tight-fitting sheaths of latex or animal skin worn over the penis during intercourse. They are about 85 to 93 percent reliable, and latex condoms help prevent the spread of sexually transmitted diseases. However, condoms can tear and leak, which renders them useless. A pouchlike latex "female condom" that is inserted into the vagina has also been developed.

The most widely used method of fertility control is the oral contraceptive ("the Pill")—any of a number of formulations of synthetic estrogens and progesterones. An oral contraceptive suppresses the normal release of anterior pituitary hormones (LH and FSH) required for eggs to mature and be released at ovulation. The Pill is a prescription drug. Formulations vary and are selected to match each patient's needs. That is why it is not wise for a woman to borrow oral contraceptives from someone else.

When a woman unfailingly remembers to take her daily dosage, this is one of the most reliable methods of controlling fertility. It does not interrupt sexual intercourse, and the method is easy to follow. Often, taking an oral contraceptive corrects erratic menstrual cycles and decreases cramping, and studies suggest that using the Pill reduces a woman's risk of ovarian cancer. However, a few users experience (usually temporary) side effects, including nausea, weight gain, tissue swelling, and minor headaches. Continued use may lead to blood clotting in the veins of women predisposed to this disorder. (3 out of 10,000). There may be other side effects as well. Complications are much more likely to develop in women who smoke, and as a rule a physician will not prescribe an oral contraceptive for a smoker.

Norplant, an implant inserted under the skin of a woman's upper arm, is currently considered to be one of the most effective reversible methods of fertility control. The most common version of Norplant consists of six matchsticklike capsules that contain levonorgestrel, a hormone that prevents implantation of a fertilized egg. The capsules release the hormone over several years. There can be side effects, including heavier menstrual periods. In addition, the implant procedure can cost hundreds of dollars.

In 1993, a progesterone drug called *Depo-Provera* was approved for contraceptive use in the United States. (The drug was previously approved for treating endometriosis and some other female reproductive disorders.)

A single injection provides protection against pregnancy for 3 months by suppressing ovulation and is more than 99 percent effective. This drug's side effects include weight gain, fatigue, abdominal pain, and gradual bone loss. Some studies suggest a link between Depo-Provera and development of breast cancer in younger women. For these and other reasons, its use is controversial.

In *vasectomy,* a tiny incision is made in a man's scrotum, and each vas deferens is severed and tied off. The procedure can be performed in 20 minutes in a doctor's office, using only a local anesthetic. After vasectomy, sperm cannot leave the testes and so will not be present in semen. There is no firm evidence that vasectomy disrupts the male hormone system, and there seems to be no noticeable difference in sexual activity. Technically, a vasectomy can be reversed, but residual scar tissue can block the tubes, preventing the passage of sperm.

In *tubal ligation,* a woman's oviducts are cauterized or cut and tied off. Tubal ligation is usually performed in a hospital. Afterward, some women suffer recurring pain and inflammation of tissues where the surgery was performed. The operation can be reversed, although major surgery is required and is not always successful.

A pregnancy test doesn't register positive until after implantation. According to one view, a woman isn't pregnant until that time. From this viewpoint, RU-486, the *morning-after pill*, intercepts pregnancy. It interferes with hormonal signals that control the events between ovulation and implantation. Three pills, taken within 72 hours after intercourse, block fertilization or prevent implantation. RU-486 sometimes has side effects, mainly nausea, vomiting, and breast tenderness. However, because RU-486 disrupts complex hormonal interactions, it should be taken only with a doctor's supervision. There is evidence that RU-486 may be helpful in treating endometriosis as well as some types of cancer. At this writing, RU-486 is available in Europe. In the United States, its use is still a subject of controversy.

Future Options for Fertility Control

Researchers are working to develop new and better methods of fertility control. Examples include biodegradable implants that don't require surgical removal, a 2-year "pregnancy vaccine," a contraceptive for men that reduces sperm count, chemicals for nonsurgical sterilization, and male and female sterilization procedures that can be reversed more easily.

There are many options for controlling fertility, some safer or more effective than others. Under development are methods designed to offer long-term, highly effective control with few side effects.

COPING WITH INFERTILITY

Many couples cherish the prospect of becoming parents (Figure 13.11). Hence, controls over fertility also extend in the other direction—to help childless couples who wish to conceive a child but have problems doing so. In the United States, about 15 percent of all couples cannot conceive a child because of sterility or infertility. For example, hormonal imbalances may prevent ovulation, or the oviducts may be blocked by effects of disease. Alternatively, the male's sperm count may be low, or his sperm may be defective in some way that reduces the likelihood of successful fertilization.

In Vitro Fertilization

With *in vitro fertilization*—literally "fertilization in glass"—conception can occur externally, provided that sperm and oocytes obtained from the couple are normal. First, FSH is administered to prepare the ovaries for ovulation. Then, using an instrument called a laparascope, a physician locates and removes preovulatory oocytes with a suction device. Before the oocytes are removed, a sample of the male's sperm is placed in a glass laboratory dish in a solution that simulates the fluid in oviducts. When suctioned oocytes are placed with the sperm, fertilization may occur. About 12 hours later, zygotes (fertilized eggs in their earliest stages of development) are transferred to a solution that will support further development. Two to four days after that, one or more embryos are transferred to the woman's uterus. An embryo implants in about 20 percent of cases, and each attempt costs several thousand dollars.

In vitro fertilization can produce several viable embryos at one time, and those not used in a given procedure can be frozen and stored for long periods. The fate of unused embryos has prompted ethical debates and even bitter court battles between divorcing couples.

Recently, doctors in Belgium pioneered a variation on in vitro fertilization in which a single sperm is injected into an egg using a tiny glass needle. The method appears to have a higher success rate than traditional in vitro fertilization, and so far it has produced healthy babies for several hundred couples.

Intrafallopian Transfers

In a technique called GIFT, for *gamete intrafallopian transfer*, a couple's sperm and oocytes are collected and then placed into an oviduct (fallopian tube). About 20 percent of the time, the oocyte becomes fertilized and a normal pregnancy ensues. An alternative procedure is ZIFT (*zygote intrafallopian transfer*). Here, oocytes and sperm are brought together in a laboratory dish, where fertilization can give rise to a zygote. The zygote is then placed in one of the woman's oviducts. GIFT and ZIFT are about as successful as in vitro fertilization.

Figure 13.11 An image of parental devotion. For both men and women, the urge to rear children can be extremely strong, and infertility can be emotionally devastating.

A variation on these methods is *embryo transfer*, in which a fertile female "donor" is inseminated with sperm from a male whose female partner is infertile. If the donor becomes pregnant, the developing embryo is transferred to the infertile woman's uterus. Embryo transfer is technically difficult and can have legal complications. It is not yet a common solution to infertility.

Artificial Insemination

Artificial insemination is the placing of semen into the vagina or uterus by artificial means, usually a syringe, around the time of ovulation. The semen may come from a woman's partner, especially if the man has a low sperm count, because his sperm can be concentrated prior to the procedure. In *artificial insemination by donor* (AID), an anonymous donor can sell his sperm to a "sperm bank," which then charges a fee for insemination. AID produces about 20,000 babies in the United States every year.

In in vitro fertilization, sperm and oocytes are brought together in a laboratory dish, where conception may occur. Other techniques for overcoming infertility include intrafallopian transfers and artificial insemination.

13.9 DILEMMAS OF FERTILITY CONTROL

About 3 percent of American women of childbearing age have had an abortion; more than 1,500,000 abortions are performed in the United States each year. At one time, abortions were generally forbidden by law in the United States unless the pregnancy endangered the mother's life. In 1973 the Supreme Court ruled (in *Roe v. Wade*) that the government does not have the right to forbid abortions during the early stages of pregnancy (typically up to 5 months). Before this ruling, there were dangerous, traumatic, and often fatal attempts to abort embryos, either by pregnant women themselves or by quacks. During the first trimester (12 weeks) abortions performed by competent medical personnel are relatively rapid, painless, and free of complications. Even with judicial approval, however, abortions in the second and third trimesters are highly controversial unless the mother's life is threatened.

Recent court action has reaffirmed the legality of abortion in the United States while permitting states to impose restrictions. Yet court rulings have not done much to quell debate over the issue of legalized abortion. Currently, legal and social conflict rage on. Both sides protest, hold marches, and lobby lawmakers (Figure 13.12). Fights have broken out between the factions, clinics have been bombed and burned, and clinic staff have been harassed. Several doctors have been shot and killed. Abortion foes charge that the physicians themselves are committing murder.

There are few easy solutions to the ethical issues associated with fertility control. For some people, a crucial question is, *When does life begin?* In formulating your own answer, keep in mind that during her lifetime, a human female can produce as many as 500 eggs, all of which are alive. During one ejaculation, a human male can release up to half a billion sperm, which also are living cells. Does "life begin" only when a sperm and an egg fuse? Is there a difference between life in general and *a particular life*? Does "meaningful" life begin only when a fetus attains a certain age in the womb? Does a woman have a right to control her reproductive life, even if doing so means terminating a pregnancy? Does the government have the right to intervene in such a decision? These are only a sampling of the moral, ethical, and philosophical questions that underlie the public debate on fertility control.

A GLOBAL PERSPECTIVE ON FERTILITY CONTROL Globally, fertility control is equally controversial. Some cultures and religions strongly discourage efforts to limit births. Meanwhile, the human population growth rate spirals out of control (Chapter 25). For example, in the time it takes you to read this chapter, about 10,700 babies will be born. By the time you go to bed tonight, there will be 257,000 more people on earth than there were last night at that hour. Within a week, the number will reach 1,800,000— about the population of Massachusetts. Worldwide population growth has so outstripped resources that each year millions face the horrors of starvation.

Figure 13.12 Demonstration symbolizing the abortion debate in the United States. It is an understatement to say that emotions run high on both sides of the issue.

Is massive, global birth control a viable solution? Where do individual rights fit into the picture? In China, where the population is expected to top 1.5 billion by 2025 and famine is an ongoing concern, the government has established the most extensive family planning program in the world. Couples who limit themselves to one child receive benefits such as extra food, better housing, free medical care, and salary bonuses. Those who exceed the one-child limit are severely penalized. In this environment, abortion is officially encouraged as an alternative to child-bearing.

In some countries, especially poorer ones with rapidly growing populations, some factions are promoting the large-scale use of Norplant and the contraceptive injection Depo-Provera (Section 13.7). Critics of these proposals fear that the drugs' side effects have not been well studied. Moreover, they worry that in some circumstances (with poor or uneducated women or those having a history of child abuse or drug addiction) authorities might coerce the women into receiving injections or implants.

As diverse societies now move into the 21st century, with burgeoning populations and their associated economic, social, and environmental pressures, one thing seems certain: These and other dilemmas of fertility control will not be easily resolved.

SUMMARY

1. Humans have a pair of primary reproductive organs (sperm-producing testes in males, egg-producing ovaries in females), accessory ducts, and glands. Testes and ovaries also produce hormones that influence reproductive functions and secondary sexual traits.

2. The hormones testosterone, LH, and FSH control sperm formation. They are part of feedback loops among the hypothalamus, anterior pituitary, and testes.

3. The hormones estrogen, progesterone, FSH, and LH control egg maturation and release, as well as changes in the lining of the uterus (endometrium). They are part of feedback loops involving the hypothalamus, anterior pituitary, and ovaries.

4. In both males and females, GnRH from the hypothalamus stimulates the anterior pituitary to release LH and FSH.

5. The following events occur during a menstrual cycle:
   a. The cycle begins with the onset of menstrual flow, as the endometrial lining of the uterus is shed.
   b. A follicle, which is an oocyte surrounded by a cell layer, matures in an ovary. Under the influence of hormones produced by the follicle, the endometrium of the uterus starts to rebuild.
   c. A midcycle peak of LH triggers the release of a secondary oocyte from the ovary. This event is called ovulation.
   d. A corpus luteum forms from the remainder of the follicle. Its secretions prime the endometrium for fertilization. Secretions from the corpus luteum help maintain the endometrium. When fertilization does not occur, the corpus luteum degenerates and the endometrial lining breaks down and is shed through menstruation.

6. Control of human fertility raises important ethical questions. These questions extend to the physical, chemical, surgical, or behavioral interventions used in the control of unwanted pregnancies and in attempts to help infertile couples.

Review Questions

1. Distinguish between:
   a. Sertoli cell and Leydig cell *290*
   b. sperm and semen *288, 290*
   c. primary oocyte and ovum *294, 295*
   d. follicle and corpus luteum *294, 295*
   e. ovulation and implantation *294, 295*

2. Which hormones influence male reproductive function? *290–291*

3. What is the menstrual cycle? Which four hormones influence this cycle? *293–295*

4. List four events that are triggered by the surge of LH at the midpoint of the menstrual cycle. *294–295*

5. What changes occur in the endometrium during the menstrual cycle? *295*

6. Label the components of the human male and female reproductive systems and state their functions. *288–291, 292–295*

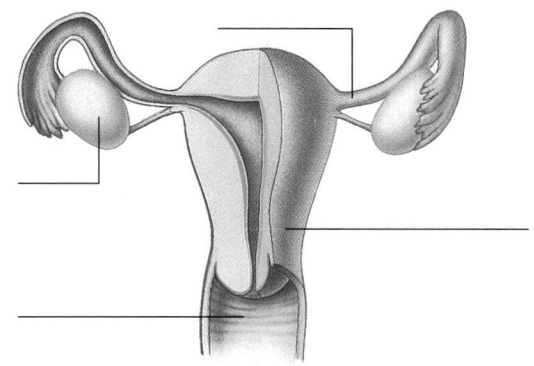

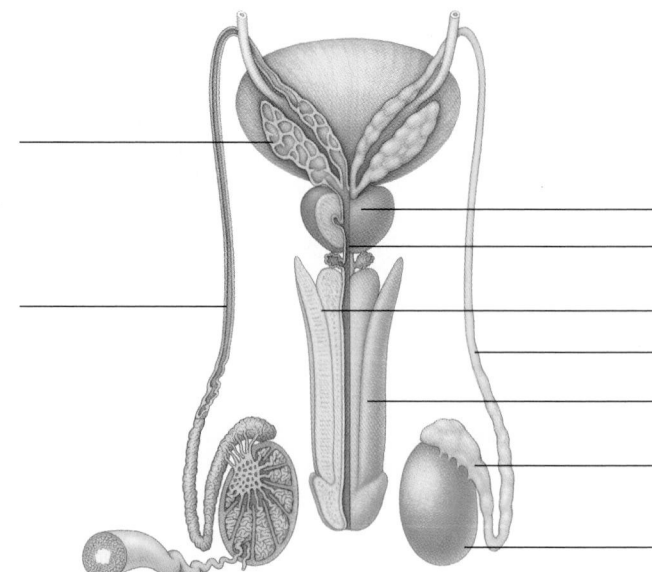

Critical Thinking: You Decide (Key in Appendix V)

1. Counselors sometimes advise a couple who wish to conceive a child to use an alkaline (basic) douche immediately before intercourse. Speculate about what the doctors' reasoning might be.

2. In the "fertility awareness" method of birth control, a woman gauges her fertile period each month by monitoring changes in the consistency of her vaginal mucus. What kind of specific information does such a method provide? How does it relate to the likelihood of getting pregnant?

3. Some women experience *premenstrual syndrome* (PMS), which can include a distressing combination of mood swings, fluid retention (edema), anxiety, backache and joint pain, food cravings, and other symptoms. PMS usually develops after ovulation and lasts until just before or just after menstruation begins. A woman's doctor can recommend strategies for managing PMS, which often include diet changes, regular exercise, and use of diuretics or other drugs. Many women find that doses of vitamin B$_6$ and vitamin E help reduce pain and other symptoms. Although the precise cause of PMS is unknown, it seems clearly related to the cyclic production of ovarian hormones. After reviewing Figure 13.8, suggest which hormonal changes may trigger PMS in affected females.

4. The absence of menstrual periods, or *amenorrhea*, is normal in pregnant and postmenopausal women and in girls who have not yet reached puberty. However, in females of reproductive age amenorrhea can result from tumors of the pituitary or adrenals. Based on discussion in this chapter and Chapter 12, speculate about why such tumors might disrupt cyclic menstruation.

Self-Quiz (Answers in Appendix IV)

1. Besides producing gametes, human male and female gonads also produce sex hormones. The _____ and the pituitary gland control secretion of both.

2. _____ production is continuous from puberty onward in males; _____ production is cyclic and intermittent in females.
   a. Egg; sperm c. Testosterone; sperm
   b. Sperm; egg d. Estrogen; egg

3. Sperm formation is controlled through _____ secretions.
   a. testosterone c. FSH
   b. LH d. all of the above are correct

4. During the menstrual cycle, a midcycle surge of _____ triggers ovulation.
   a. estrogen c. LH
   b. progesterone d. FHS

5. Which is the correct order for one turn of the menstrual cycle?
   a. corpus luteum forms, ovulation, follicle forms
   b. follicle grows, ovulation, corpus luteum forms

Selected Key Terms

acrosome 290	ovary 287
bulbourethral gland 289	oviduct 292
clitoris 292	ovulation 294
coitus 297	progesterone 292
corpus luteum 295	prostate gland 289
endometrium 292	secondary oocyte 294
estrogen 292	secondary sexual trait 287
fertilization 297	semen 288
follicle 294	seminiferous tubule 288
FSH 290	Sertoli cell 290
granulosa cell 294	sperm 290
Leydig cell 290	testis 287
LH 290	testosterone 290
menstrual cycle 293	uterus 292
oocyte 292	vas deferens 288
orgasm 297	zona pellucida 294

Readings

Alexander, N. September 1995. "Future Contraceptives." *Scientific American*.

"The Evolution of Sexes." 17 July 1992. *Science*. Fascinating discussion of why (perhaps) there are two sexes of human beings and many other organisms.

Franklin, D. July/August 1992. "The Birth Control Bind." *Health*.

Hales, D. 1992. *An Invitation to Health*. 5th ed. Redwood City, Calif.: Benjamin Cummings. Chapters 10 and 11 contain a wealth of useful information on human sexuality and reproductive choices.

Sherwood, L. 1997. *Human Physiology*. 3d ed. Belmont, California: Wadsworth.

Smolowe, J. 14 June 1993. "New, Improved, and Ready for Battle." *Time*. Clear presentation of issues surrounding the controversy over the "abortion pill," RU-486.

 For additional readings, go to InfoTrac College Edition, your online library at:

http://www.infotrac-college.com/wadsworth

14 DEVELOPMENT

How Your Journey Began

The early development of the human body is a truly remarkable journey. Insofar as we understand the sequence of events—using your own beginnings as a handy example—in general the first few weeks of early development unfold in the following way.

In the upper end of one of your mother's oviducts, an ovulated oocyte is fertilized. As the fertilized egg, now the embryo, travels down the oviduct toward the uterus (Figure 14.1), it undergoes several rounds of cell division to produce a hollow ball of about 64 tiny cells. At this stage, most of the cells make up the wall of the ball, but a small cluster of cells—perhaps 3 to 6 of them—is situated on the inner wall of the sphere. This cluster of cells, known as the *inner cell mass*, will eventually give rise to all of the embryo proper. The cells of the outer wall will contribute to vital tissues outside the embryo, about which you will learn much more later in this chapter.

As development proceeds, cell divisions continue, adding more and more cells to the inner cell mass. In short order, the structure reorganizes, forming a flattened disk with two tissue layers. Less than a millimeter across, the disk could stretch out across the head of a straight pin with room to spare. But that speck is an embryo.

A third cell layer forms between the original two. Within those three layers, cells follow commands to change shape, to get moving. Part of one layer curves

up and folds over. Other cells are elongating, making their component layer thicker. And still other cells are migrating to new destinations. These changes transform the pancakelike embryo into a pale, translucent crescent with surface bumps, tucks, and hollows.

By the third week after fertilization, the tubelike forerunners of the nervous system and circulatory system appear. By the fourth week, a heart starts beating rhythmically in the thickening wall of one tube. On the embryo's sides, budding regions of tissue mark the beginning of upper and lower limbs.

Five weeks into the journey, the limb buds have paddles—the start of hands and feet. Round dots appear under the transparent skin; the eyes are forming. Around and below them, cells migrate in concert, interacting in ways that sculpt out a nose, cheeks, and a mouth. By now, the embryo is recognizably a human in the making.

This was your beginning. Later, embryonic and fetal events filled in the details, rounded out contours, added flesh and fat and hair and nails to your peanut-sized body. That beginning, and all the subsequent events that unfolded inside your mother's body, is the story of this chapter.

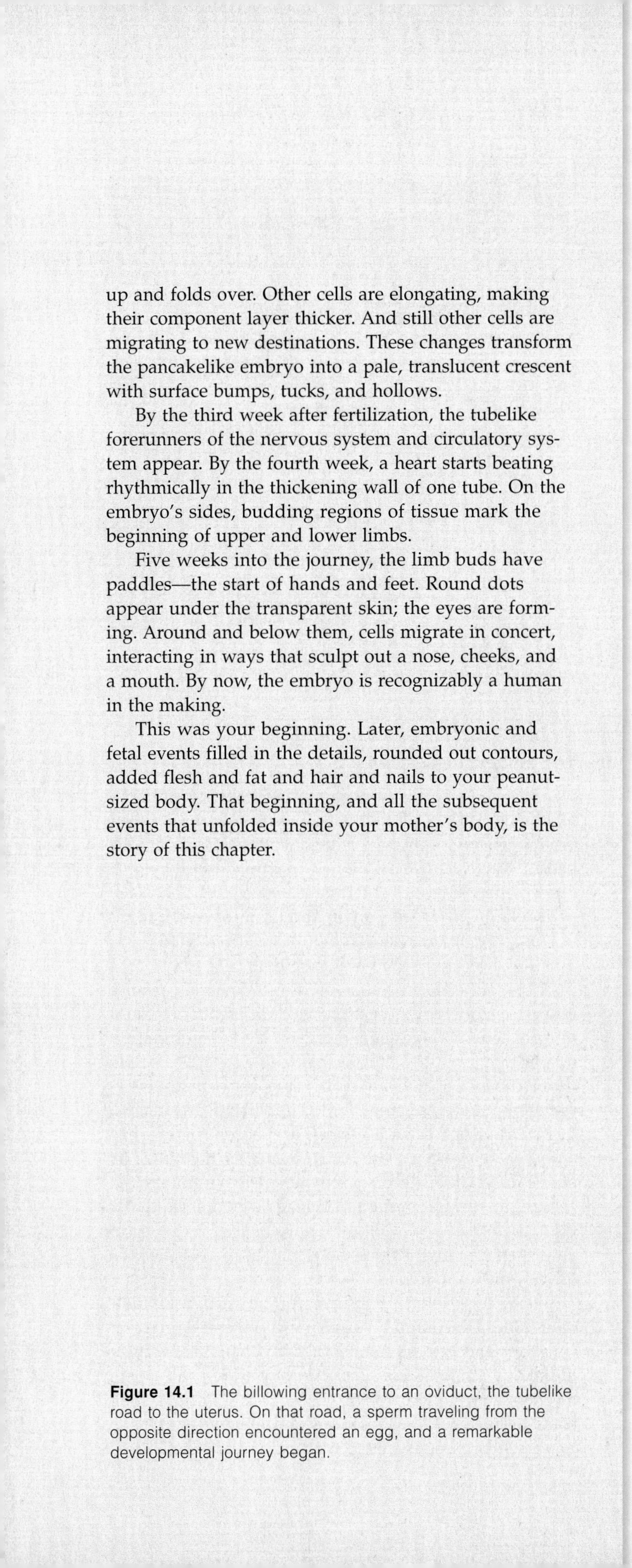

KEY CONCEPTS

1. The human life cycle begins when gametes form in the parents.

2. The life cycle then proceeds through five development stages, called fertilization, cleavage, gastrulation, organ formation, and growth and tissue specialization.

3. Embryonic development is a continuum in which each stage builds on the tissues and structures that formed during the stage that preceded it. Also, each new stage establishes a more complex body than was present at the end of the previous stage.

4. A key event in early embryonic development is the formation of three distinct tissue layers: endoderm, mesoderm, and ectoderm. Every tissue of the body arises from one of these layers. Interactions between the layers stimulate the development of specific organs.

5. Development does not end with the birth of a baby. It continues through childhood and adolescence and culminates as an adult undergoes aging. In a sense, development ends only with death.

CHAPTER AT A GLANCE

14.1 Stages of Development

14.2 The Beginnings of You—Early Events in Development

14.3 Extraembryonic Membranes

14.4 The Making of an Early Embryo—A Closer Look

14.5 Emergence of Distinctly Human Features

14.6 Fetal Development

14.7 From Birth Onward

14.8 *Saving Babies and Making Clones*

14.9 How the Mother's Lifestyle Affects Early Development

14.10 *Prenatal Diagnosis: Detecting Birth Defects*

14.11 Summary of Developmental Stages

14.12 The Body As It Ages

14.13 Aging of Skin, Muscle, the Skeleton, and Internal Transport Systems

14.14 Age-Related Changes in Some Other Body Systems

Figure 14.1 The billowing entrance to an oviduct, the tubelike road to the uterus. On that road, a sperm traveling from the opposite direction encountered an egg, and a remarkable developmental journey began.

Development begins as sperm or eggs—the **gametes**—form and mature within the parents. **Fertilization** starts when the plasma membranes of a sperm and egg fuse. It ends when the sperm nucleus fuses with the egg nucleus and gives rise to the **zygote**, the first cell of the new individual.

Next comes **cleavage**, when cell divisions typically convert the zygote to a ball of cells. This is the point in your existence when you first became a multicellular creature. Figure 14.2 is a simple diagram of the first two cleavage divisions. Notice that the first round of cell division in cleavage divides the zygote into two cells; then, in the second round, the two cells each divide but in different planes—one perpendicular to the other. (This "rotational" pattern of cleavage occurs in the embryos of all mammals.) After the third round of cleavage, there are 16 embryonic cells arranged in a compact ball called a **morula** (from a Latin word for mulberry).

One of the more interesting outcomes of cleavage is that each new cell—called a *blastomere*—ends up with a particular region of the egg's cytoplasm. And which bit of cytoplasm a blastomere receives helps determine the developmental fate of the blastomere's descendants. For instance, its cytoplasm alone may have the molecules of a protein that can activate, say, the gene coding for a certain hormone. Hence, *only* its descendants will make the hormone.

Gastrulation: Primary Tissues Form

Cleavage gives way to **gastrulation**, a stage of major cell rearrangements. The organizational framework for the whole body is laid out as cells become arranged into three primary tissues, called **germ layers**. The three layers are called **endoderm**, **mesoderm**, and **ectoderm**, respectively. As you can see in Table 14.1, they give rise to the human body's various tissues and organs.

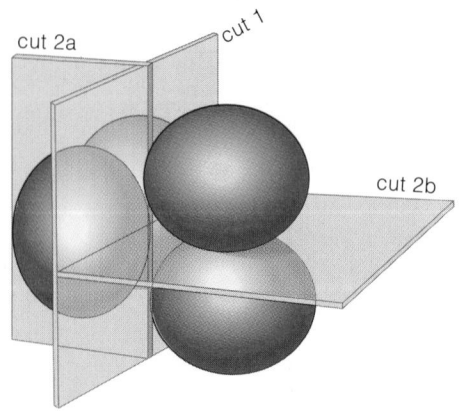

cut 2a

cut 1

cut 2b

Figure 14.2 Early cleavage planes in a human egg.

Table 14.1	Tissues and Organs Derived from the Three Germ Layers in Human Embryos
Germ Layer	Main Derivatives in the Adult
Endoderm	Various epithelia, as in the gut, respiratory tract, urinary bladder and urethra, and parts of the inner ear; also portions of the tonsils, thyroid and parathyroid glands, thymus, liver, and pancreas
Mesoderm	Cartilage, bone, muscle, and various connective tissues; gives rise to cardio-vascular system (including blood), lymphatic system, spleen, adrenal cortex, excretory and reproductive systems
Ectoderm	Central and peripheral nervous systems; sensory epithelia of the eyes, ears, and nose; epidermis and its derivatives (including hair and nails), mammary glands, pituitary gland, subcutaneous glands, tooth enamel, and adrenal medulla

Organogenesis, Growth, and Tissue Specialization

Once the three germ layers form, they split into subpopulations of cells. This marks the onset of a phase called **organogenesis**, or organ formation. Different sets of cells become unique in structure and function, and they give rise to different tissues and organs. During the final stage of development, called **growth and tissue specialization**, organs grow in size and take on specialized properties. This stage continues into adulthood.

Three crucial processes of development are those known as cell determination, cell differentiation, and morphogenesis of tissues and organs. **Cell determination** occurs first, and it establishes which of several possible developmental paths an embryonic cell can follow—rather like students being divided into liberal arts majors, business majors, science majors, and so on. In an early embryo, a given cell's fate (and, eventually, the fate of its descendants) depends on exactly where in the embryo the cell originates (the portion of the egg's cytoplasm that a blastomere receives). Cell fate is influenced by physical and chemical interactions between groups of cells as development progresses.

Next, a gene-guided process of **cell differentiation** takes place. To continue our analogy, think of a group of college science majors, some of whom go on to specialize in biology while others specialize in physics, still others in chemistry, and so forth. As cells differentiate, they come to have specific structures and the ability to make products that are associated with particular functions.

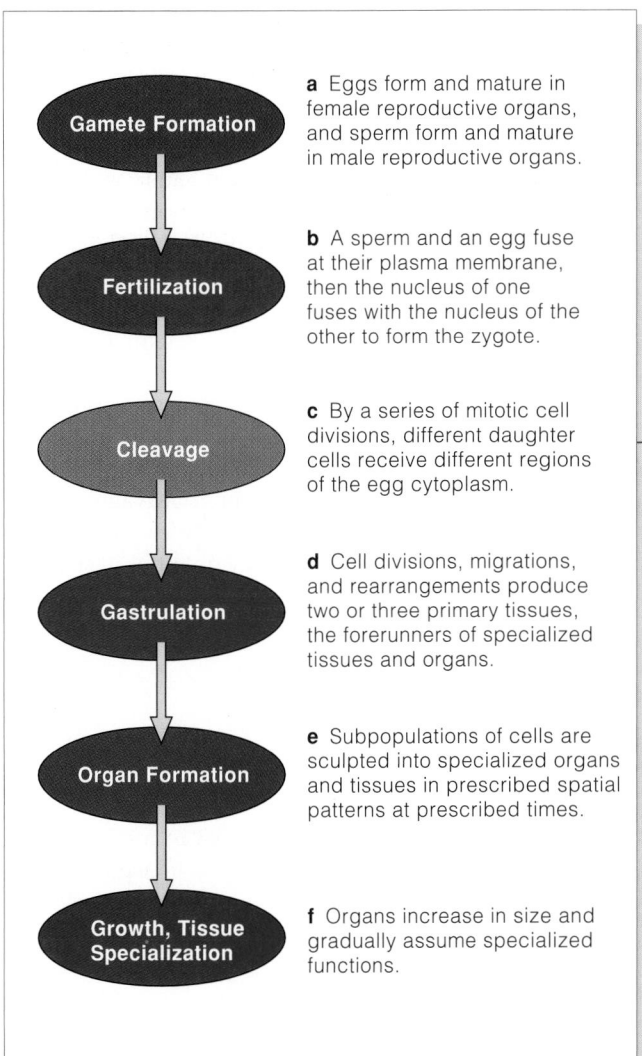

a Eggs form and mature in female reproductive organs, and sperm form and mature in male reproductive organs.

b A sperm and an egg fuse at their plasma membrane, then the nucleus of one fuses with the nucleus of the other to form the zygote.

c By a series of mitotic cell divisions, different daughter cells receive different regions of the egg cytoplasm.

d Cell divisions, migrations, and rearrangements produce two or three primary tissues, the forerunners of specialized tissues and organs.

e Subpopulations of cells are sculpted into specialized organs and tissues in prescribed spatial patterns at prescribed times.

f Organs increase in size and gradually assume specialized functions.

a

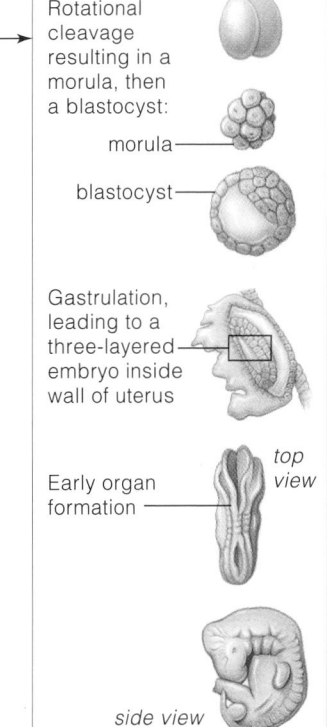

Rotational cleavage resulting in a morula, then a blastocyst:

morula

blastocyst

Gastrulation, leading to a three-layered embryo inside wall of uterus

Early organ formation

top view

side view

b

Figure 14.3 An overview of development. (**a**) The stages of development. (**b**) The developmental journey of a human embryo from fertilization to about 8 weeks. For clarity, the membranes surrounding the embryo are not shown from cleavage onward. Early cleavages result in a morula, which becomes transformed into a blastocyst (Section 14.2). Several stages are shown here in cross section.

Morphogenesis (quite literally, "the beginning of form") produces the shape and structure of particular body regions. It's a remarkable and complicated process, involving localized cell division and growth and movements of differentiated cells and entire tissues from one site to another. It also involves the folding of sheetlike tissues as well as the programmed death (apoptosis) of certain cells. For example, morphogenesis at the ends of limb buds first produced paddle-shaped hands at the ends of your arms; then epithelial cells between lobes in the paddles died on cue, leaving separate fingers.

Figure 14.3a lists the stages of human development and Figure 14.3b diagrams examples of some of the stages. It's worth taking a few moments to study these examples, for they reveal an important concept. By the end of each developmental stage, the embryo has become more complex than it was before. As you will see in sections to follow, the structures that develop during one stage serve as the foundation for the stage that follows it. The successful development of an embryo depends on the formation of these structures according to normal patterns, in a prescribed sequence.

Development begins with the formation of gametes in each parent. It proceeds through a sequence of stages: fertilization followed by cleavage of the zygote, followed by gastrulation, organogenesis, and tissue growth and specialization.

In the gene-guided process of cell differentiation, cells in different body regions become specialized for particular functions. By way of morphogenesis, various body regions come to have a particular shape and overall structure.

Each stage of embryonic development builds on structures that were formed during the stage preceding it. Development cannot proceed properly unless each stage is successfully completed before the next begins.

Fertilization and Cleavage

A female's fertile period begins a few days before she ovulates and lasts until a few days afterward. Recall from Chapter 13 that if sperm enter her vagina at any time during this period, fertilization can occur. Of the millions of sperm deposited in the vagina during one ejaculation, a few hundred reach the upper region of the oviduct, where fertilization most commonly takes place. Within 30 minutes after ejaculation, uterine muscle contractions have helped move sperm toward the oviducts. As sperm swim through the cervix and uterus and into the oviducts, *capacitation* occurs. In this process, changes occur in the region of membrane covering the sperm cell's acrosome. Only a sperm that is capacitated ("made able") can fertilize an oocyte (Figure 14.4).

When a capacitated sperm encounters an oocyte, the now-fragile region of cell membrane enables the sperm to release digestive enzymes from its acrosome. Many sperm can reach and bind to the oocyte, and acrosome enzymes clear a path through the zona pellucida. Usually, however, only one sperm fuses with the oocyte; rapid chemical changes in the oocyte cell membrane block the entry of additional sperm. Fusion with a sperm stimulates the second meiotic division in the oocyte which, recall, was arrested in meiosis I (page 294). There are now three polar bodies and a mature egg, or **ovum** (plural: ova). The sperm nucleus fuses with the nucleus of the ovum, and a new nucleus forms around their mingled chromosomes. Remember that a sperm or oocyte has 23 chromosomes, only half the number of chromosomes present in body cells. When the nuclei of a sperm and an egg fuse during fertilization, these two "halves" become a "whole"—a zygote that has a full complement of 46 chromosomes.

As the ensuing embryo develops, each cell will have all the DNA required to guide proper development and functioning of body parts.

Fertilization is followed by cell divisions. Sometimes the two cells produced by the first cleavage separate from one another and develop into independent embryos. This produces *identical twins*, or two complete, normal individuals having the same genetic makeup. *Fraternal twins* occur when two different eggs are fertilized at roughly the same time by two different sperm. Fraternal twins can have different sexes and do not necessarily resemble each other any more than do any other siblings.

Subsequent cleavages convert the zygote, a single cell, into a multicelled ball (Figure 14.5). (Each new cell receives a full complement of 46 chromosomes.) The embryo doesn't grow larger while the initial cleavages are proceeding. The daughter cells collectively occupy the same volume as did the zygote, but they do differ in size, shape, and activity.

Implantation

For 3 or 4 days after fertilization, the zygote travels down the oviduct. Sustained by nutrients stored in the ovum

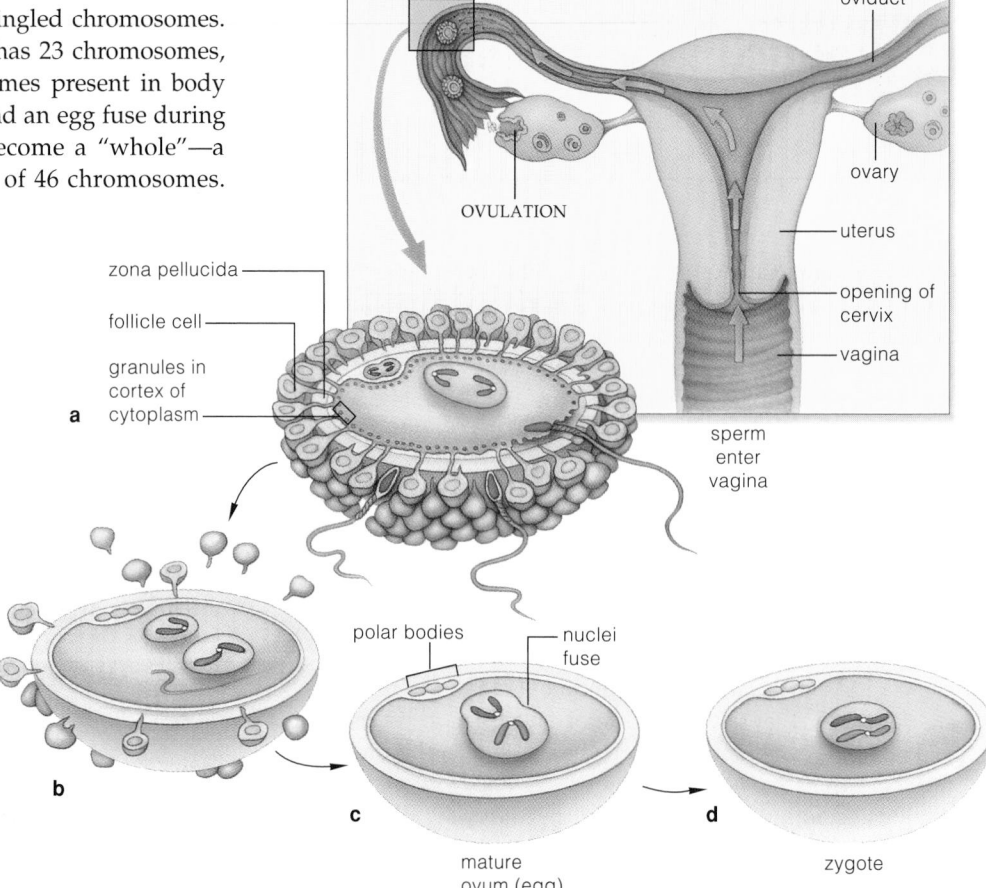

Figure 14.4 Fertilization. (**a**) A number of sperm surround a secondary oocyte. Acrosomal enzymes clear a path through the zona pellucida. (**b**) When a sperm manages to penetrate the secondary oocyte, cortical granules in the egg cytoplasm release substances that make the zona pellucida impenetrable to other sperm. Penetration also stimulates the second meiotic division of the oocyte's nucleus. (**c**) The sperm tail degenerates. The sperm nucleus enlarges and fuses with the egg nucleus. (**d**) With that fusion, fertilization is over. The zygote has formed.

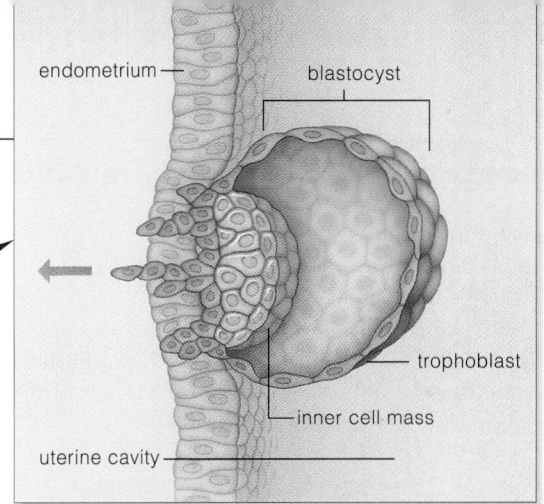

endometrium

blastocyst

trophoblast

inner cell mass

uterine cavity

f DAYS 6–7. The blastocyst attaches to the endometrium and starts burrowing into it. Implantation is under way.

actual size

surface layer of cells (trophoblast)

inner cell mass

e DAY 5. On day 5, a blastocyst has developed from the morula. It consists of a fluid-filled cavity and an inner cell mass.

d DAY 4. By 96 hours, divisions have produced a ball of sixteen to thirty-two cells, the morula. The morula will give rise to the embryo and extraembryonic membranes.

c DAY 3. By 72 hours, divisions have produced a ball of six to twelve cells.

b DAY 2. The second division, which is completed by about 40 hours, produces the four-cell stage.

a DAY 1. Cleavage begins within 24 hours after fertilization. The first cut is along a plane at right angles to the zygote's equator, in line with the polar bodies.

FERTILIZATION

oviduct   uterus

ovary

IMPLANTATION

endometrium

Figure 14.5 Steps from fertilization through implantation.

and present in maternal secretions, the steadily developing embryo undergoes the first three cleavages. By the time the cluster of dividing cells reaches the uterus, it is a morula, a solid ball of cells. Next, a fluid-filled cavity develops, an event that transforms the morula into a **blastocyst**. The blastocyst consists of two tissues: a surface epithelium and a cluster of a few cells—the **inner cell mass** (Figure 14.5e). The embryo, which ultimately will become a baby, develops from the inner cell mass.

Implantation occurs about a week after fertilization. First, the blastocyst breaks out of the zona pellucida, then cells of the epithelium invade across the endometrium and into the underlying connective tissue. This gives the blastocyst a foothold in the uterus, so to speak. It now will sink deep into the connective tissue of the uterus and the endometrium closes over it.

Occasionally a fertilized egg implants not in the uterine wall, but in the oviduct or sometimes even in the external surface of the ovary or in the abdominal wall. This condition is called *ectopic (tubal) pregnancy*. Ectopic pregnancy cannot go to full term and must be terminated by surgery. Sometimes the condition results in permanent infertility.

Implantation is completed 14 days after ovulation. Menstruation, which would begin at this time in the absence of pregnancy, fails to occur because the newly implanted blastocyst secretes HCG (human chorionic gonadotropin). This hormone prods the corpus luteum to keep on secreting estrogen and progesterone, which maintain the uterine lining. By the third week of pregnancy, HCG can be detected in the mother's blood or urine. At-home pregnancy tests use chemicals that change color when HCG is present in urine.

Fertilization of an egg by a sperm produces a zygote, a single cell with a full complement of parental chromosomes. Cleavage and further development produce a multicellular blastocyst that implants in the endometrium of the uterus.

EXTRAEMBRYONIC MEMBRANES

During implantation, the inner cell mass of the blastocyst is transformed into an **embryonic disk** (Figure 14.6a). This disk is the pancakelike structure described on page 304. Some cells of the disk will give rise to the embryo during the week after implantation, when all three germ layers will form. Other cells give rise to extraembryonic membranes, which we now consider.

Yolk Sac, Amnion, Allantois, and Chorion

A few days after implantation, several **extraembryonic membranes** form. One of them, the **yolk sac**, forms below the embryonic disk. It is the source of early blood cells and of germ cells that will become gametes; parts of the yolk sac give rise to the embryo's digestive tube. Early on, however, the yolk sac stops functioning. Three other extraembryonic membranes, which merit a good deal of our attention, are the amnion, allantois, and chorion.

The embryo develops in a fluid-filled sac enclosed by the **amnion** (Figure 14.6c). Amniotic fluid insulates the embryo, keeps it moist, and absorbs shocks. Just outside the amnion is the **allantois**. As this structure develops, mesoderm on its surface gives rise to several blood vessels that will invade the **umbilical cord**. These vessels are the embryo's contribution to circulatory "plumbing" (Figure 14.7) that will link the embryo with the placenta, which we discuss shortly.

The **chorion** is a protective membrane around the embryo and the other membranes (Figure 14.6b and 14.6c). It develops from the trophoblast and continues the secretion of HCG that began at implantation. The HCG maintains the uterine lining for the first three months of pregnancy. Thereafter, the placenta produces enough progesterone and estrogen to maintain the lining.

The Importance of the Placenta

Three weeks after fertilization, almost a fourth of the inner surface of the uterus has become a spongy tissue, the developing **placenta**. This structure is a physiological "broker," mediating the transfer of nutrients and oxygen from the mother to the embryo and waste products from the embryo to its mother's bloodstream.

Although the fully developed placenta is considered an organ, you may find it helpful to think of it as an intimate association of the embryonic chorion and the superficial cells of the mother's endometrial lining where the embryo implanted. The maternal side of the placenta consists of a layer of endometrial tissue that contains arterioles and venules. As the chorion develops (from the trophoblast), the tiny projections sent out from the blastocyst as it implants in the endometrium develop into large numbers of *chorionic villi,* and each villus is endowed with small blood vessels (Figure 14.7).

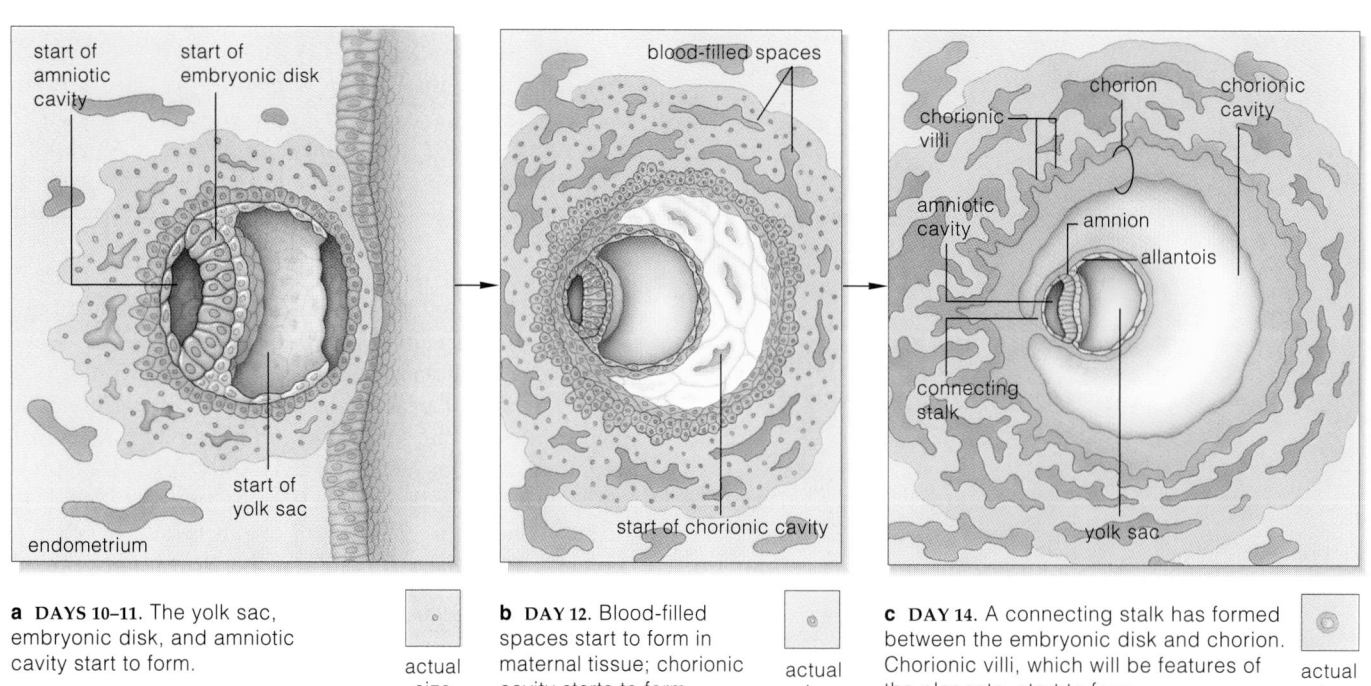

a DAYS 10–11. The yolk sac, embryonic disk, and amniotic cavity start to form. *actual size*

b DAY 12. Blood-filled spaces start to form in maternal tissue; chorionic cavity starts to form. *actual size*

c DAY 14. A connecting stalk has formed between the embryonic disk and chorion. Chorionic villi, which will be features of the placenta, start to form. *actual size*

Figure 14.6 Early formation of extraembryonic membranes (the amnion, chorion, and yolk sac).

While chorionic villi are developing, the erosion of endometrium that began with implantation continues. As capillaries in the endometrium are broken down, spaces in the eroding endometrial tissue fill with maternal blood. The chorionic villi extend into these spaces. As the embryo develops, its blood never mixes with that of its mother. Oxygen and nutrients simply diffuse out of the mother's blood vessels, across the blood-filled spaces in the endometrium, then into the embryo's blood vessels. Carbon dioxide and other wastes diffuse in the opposite direction, leaving the embryo. This chapter's *Science Comes to Life* (page 322) describes some methods for screening for birth defects, including using cells removed from chorionic villi in a procedure called *chorionic villus sampling*.

Besides nutrients and oxygen, many other substances taken in by the mother—including alcohol, caffeine, drugs, pesticide residues, and toxins in cigarette smoke—can cross the placenta, as can the virus that causes AIDS.

Extraembryonic membranes—the yolk sac, amnion, allantois, and chorion—begin to form shortly after implantation. A region of allantois tissue gives rise to blood vessels that become housed within the umbilical cord.

The umbilical cord links the embryo with the placenta, a spongy tissue in which maternal and embryonic blood vessels are closely associated. By way of the placenta, the embryo's bloodstream can take up nutrients and oxygen from the mother and also discharge wastes that her bloodstream will transport away.

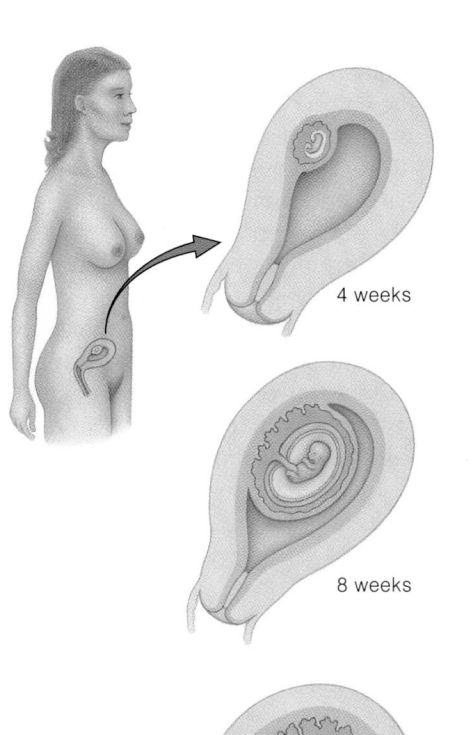

4 weeks

8 weeks

12 weeks

appearance of the placenta at full term

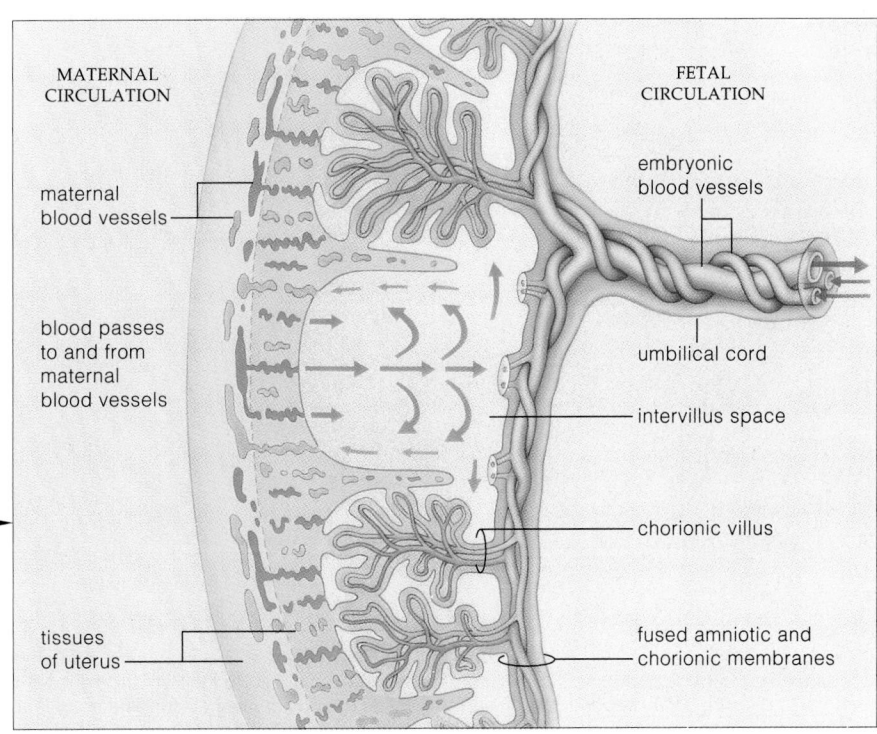

MATERNAL CIRCULATION

FETAL CIRCULATION

maternal blood vessels

embryonic blood vessels

blood passes to and from maternal blood vessels

umbilical cord

intervillus space

chorionic villus

tissues of uterus

fused amniotic and chorionic membranes

Figure 14.7 Relationship between fetal and maternal blood circulation in a full-term placenta. Blood vessels extend from the fetus, through the umbilical cord, and into chorionic villi. Maternal blood spurts into spaces between villi. Oxygen, carbon dioxide, and other small solutes diffuse across the placental membrane surface; there is no gross intermingling of the two bloodstreams.

THE MAKING OF AN EARLY EMBRYO—A CLOSER LOOK

The embryonic period begins shortly after fertilization and lasts for 8 weeks. It makes up most of the "first trimester," or 3 months, of the 9 months of human gestation. This first trimester is a vulnerable period of embryonic development. By about 7 days after fertilization the embryonic disk has formed. Soon, gastrulation begins, leading to the formation of the three germ layers: ectoderm, mesoderm, and endoderm. After the germ layers are established, organogenesis begins and the embryo's organs and organ systems start to develop.

Gastrulation Revisited

By the time a woman has missed her first menstrual period, gastrulation is under way. As the third week of development begins, the two-layered embryonic disk has been produced. Two fluid-filled sacs—the amnion and chorion—surround the disk, except where a stalk connects the disk to the inner wall of the chorionic cavity.

As a result of some of the cell rearrangements that take place during gastrulation, a faint "primitive streak" appears at the midline of the disk (Figure 14.8). Next, ectoderm along the midline thickens to establish the forerunner of a **neural tube**, which will give rise to the brain and spinal cord. Some of its cells also give rise to a flexible rod of cells called a notochord around which the vertebral column will form.

These events establish the body's long axis and its forthcoming bilateral symmetry. In other words, *the embryonic disk is undergoing morphogenetic events that will provide your body with the basic form that is characteristic of all vertebrates.*

Early in the third week the allantois appears on the yolk sac. As described in Section 14.3, it has roles in early blood vessel formation.

Meanwhile, on the embryonic disk surface near the neural tube, the third primary tissue layer—mesoderm—has been forming. Toward the end of the third week, some mesoderm gives rise to **somites**—paired blocks of mesoderm that will give rise to most bones and skeletal muscles of the neck and trunk, as well as to the dermis overlying these regions. Pharyngeal arches start to form; these structures ultimately contribute to development of the face, neck, mouth, and other, associated structures. In other mesodermal tissues spaces open up, and in time these will coalesce to form the cavity (called the *coelom*) between the body wall and the gastrointestinal tract.

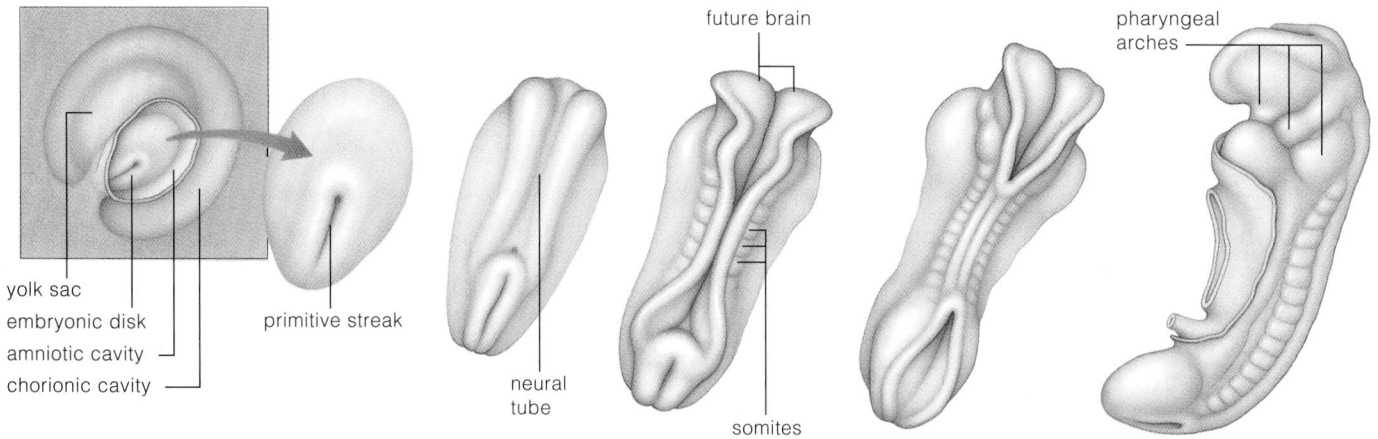

yolk sac
embryonic disk
amniotic cavity
chorionic cavity

primitive streak

neural tube

future brain

somites

pharyngeal arches

DAY 15. A primitive streak appears along the axis of the embryonic disk. This thickened band of cells marks the onset of gastrulation.

DAYS 19–23. Cell migrations, tissue folding, and other morphogenetic events lead to the formation of a hollow neural tube and to somites (bumps of mesoderm). The neural tube gives rise to the brain and spinal cord. Somites give rise to most of the axial skeleton, skeletal muscles, and much of the dermis.

DAYS 24–25. By now, some cells have given rise to pharyngeal arches, which contribute to the face, neck, mouth, nasal cavities, larynx, and pharynx.

Figure 14.8 Hallmarks of the embryonic period of development—the appearance of a primitive streak foreshadowing the brain and spinal cord—and the formation of somites and pharyngeal arches. These are dorsal views (of the embryo's back) except for day 24–25, which is a side view.

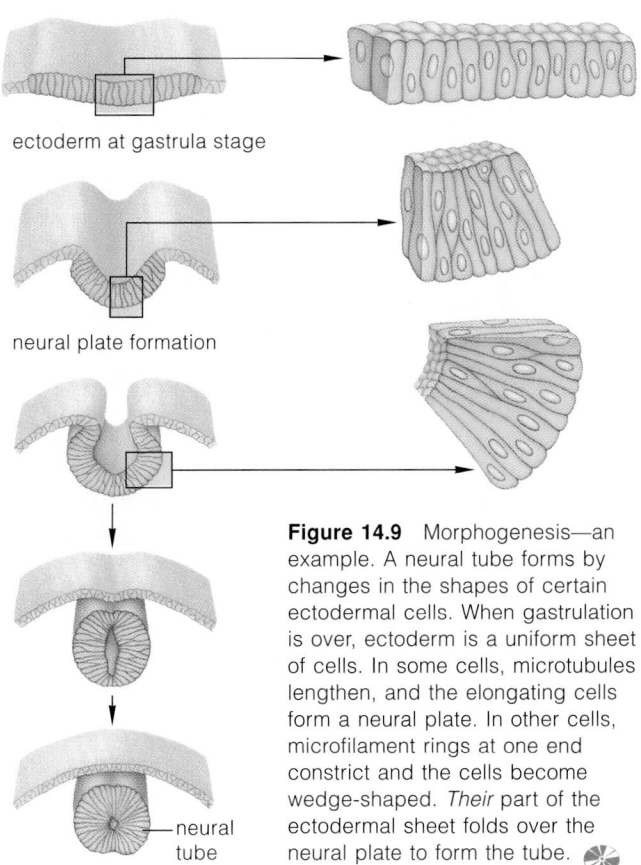

ectoderm at gastrula stage

neural plate formation

neural tube

Figure 14.9 Morphogenesis—an example. A neural tube forms by changes in the shapes of certain ectodermal cells. When gastrulation is over, ectoderm is a uniform sheet of cells. In some cells, microtubules lengthen, and the elongating cells form a neural plate. In other cells, microfilament rings at one end constrict and the cells become wedge-shaped. *Their* part of the ectodermal sheet folds over the neural plate to form the tube.

Morphogenesis Revisited

The end of gastrulation marks the beginning of organ formation and morphogenesis—the process, recall, that produces the shape and structure of internal body regions. Consider neurulation, the all-important first stage in the development of the nervous system. Figure 14.9 shows how ectodermal cells at the midline of the embryo elongate and form a neural plate, the first sign that a region of ectoderm is on its way to developing into nervous tissue. Next, cells near the middle become wedge-shaped. Collectively, the changes in cell shape cause the neural plate to fold over and meet at the embryo's midline to form the neural tube.

One form of birth defect, called *spina bifida* ("split spine"), occurs when the neural tube fails to develop properly and doesn't close and separate from ectoderm. The infant may be born with a portion of its spinal cord exposed within a cyst and have various kinds of neurological problems, including poor bowel and bladder control. Infection is a constant danger. Neural tube defects afflict 2 of every 1,000 babies born in the United States. Some cases may be preventable simply through proper maternal nutrition, a topic discussed in Section 14.9. Some neural tube defects can be cured through surgery after a child is born.

"Marching Orders" for Cells

Active cell migration from one place to another is essential for morphogenesis. Cells send out and use extensions of the cytoplasm called pseudopods (pseudopod means false foot) that move them along prescribed routes. When they reach their destination, they establish contact with cells already there. For example, forerunners of neurons interconnect this way as your nervous system is forming.

How do cells "know" where to go while they are migrating? They respond to adhesive cues, as when migrating Schwann cells stick to adhesion proteins on the surface of axons but not on blood vessels. They also respond to chemical gradients. Their migrations may be coordinated by the synthesis, release, deposition, and removal of specific chemicals in the extracellular matrix; biologists still have a great deal to learn about this process. Adhesive cues also tell the cells when to stop. Cells will migrate to regions where the adhesive cues are strongest, but once there, migration is blocked.

As you can see in Figure 14.9, another aspect of morphogenesis is the folding of sheets of cells—events that depend on the lengthening of microtubules and constriction of rings of microfilaments in cells (refer to Section 2.10). The assembly and disassembly of such components of the cytoskeleton are aspects of control mechanisms that bring about the changes. The size, shape, and proportion of your body parts emerge through such controlled, localized events. We do not fully understand why some embryonic tissues expand more than others. Apparently, selective controls over the activity of certain genes play major roles.

Finally, apoptosis, the mechanism of genetically programmed cell death, helps sculpt body parts. Molecular signals switch on chemical weapons of self-destruction that are stockpiled inside cells destined to die.

During the third week after fertilization (a time when the woman has missed her first menstrual period), the basic vertebrate body plan emerges in the new individual.

Through morphogenesis, orderly changes take place in an embryo's size, shape, and proportions. Morphogenesis involves cell division, active cell migration, tissue growth and foldings, changes in cell size and shape, and programmed cell death by apoptosis.

By the end of the fourth week of the embryonic period, the embryo has grown to 500 times its original size. The placenta has been sustaining this remarkable growth spurt. However, now the pace slows. The embryo embarks on a prescribed, intricate program of cell differentiation and morphogenesis. Limbs form, and fingers and toes are sculpted from paddle-shaped structures. The circulatory system becomes more intricate, and the umbilical cord forms. Much emphasis is placed on the all-important head; its growth now surpasses that of any other body region (Figure 14.10a). The embryonic period ends as the eighth week draws to a close. The embryo is no longer merely "a vertebrate." As you can see from Figure 14.10c, its features now clearly define it as a human fetus.

Gonad Development

Male or female gonads begin to develop by the second half of the first trimester. As described in the introduction to Chapter 13, in an embryo that has inherited both X and Y sex chromosomes, a sex-determining region of the Y chromosome (*Sry*) triggers development of testes at this time. Sex hormones produced by the embryonic testes then influence the development of the entire reproductive system. An embryo with XX sex chromosomes will be female, and ovaries and other female reproductive structures begin to form in her body. Notice that no hormones are required to stimulate development of a female—all that is necessary is the *absence* of testosterone.

After 8 weeks the embryo is just over 1 inch long, its organ systems are formed, and it is designated a **fetus**. As the first trimester ends, a heart monitor can detect the fetal heartbeat. The genitals are well formed, and a physician can often detect the baby's sex using ultrasound.

Miscarriage

Miscarriage, the spontaneous expulsion of the uterine contents, occurs in more than 20 percent of all conceptions and usually during the first trimester. Often a woman may not realize she was ever pregnant. Although many factors can trigger a miscarriage (also called spontaneous abortion), as many as half of all cases occur because the embryo (or the fetus) suffers from one or more genetic disorders that prevent normal development.

Distinctly human features emerge in the first 8 weeks of embryonic development. At the end of this period the developing individual is termed a fetus.

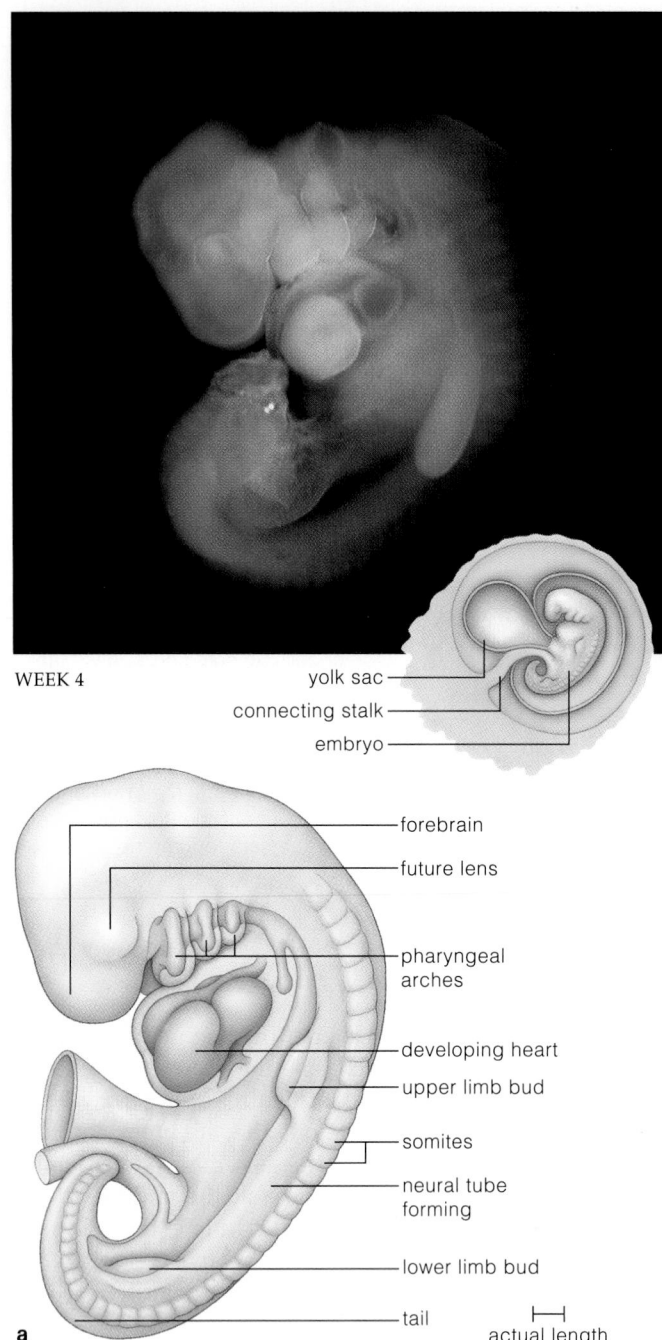

WEEK 4

yolk sac
connecting stalk
embryo

forebrain

future lens

pharyngeal arches

developing heart

upper limb bud

somites

neural tube forming

lower limb bud

tail

actual length

a

Figure 14.10 (**a**) Human embryo at 4 weeks. As is true of all vertebrates, it has a tail and pharyngeal arches. (**b**) The embryo at 5 to 6 weeks after fertilization. (**c**) An embryo poised at the boundary between the embryonic and fetal periods. It now has features that are distinctly human. It is floating in fluid within the amniotic sac. The chorion, which normally covers the amniotic sac, has been opened and pulled aside.

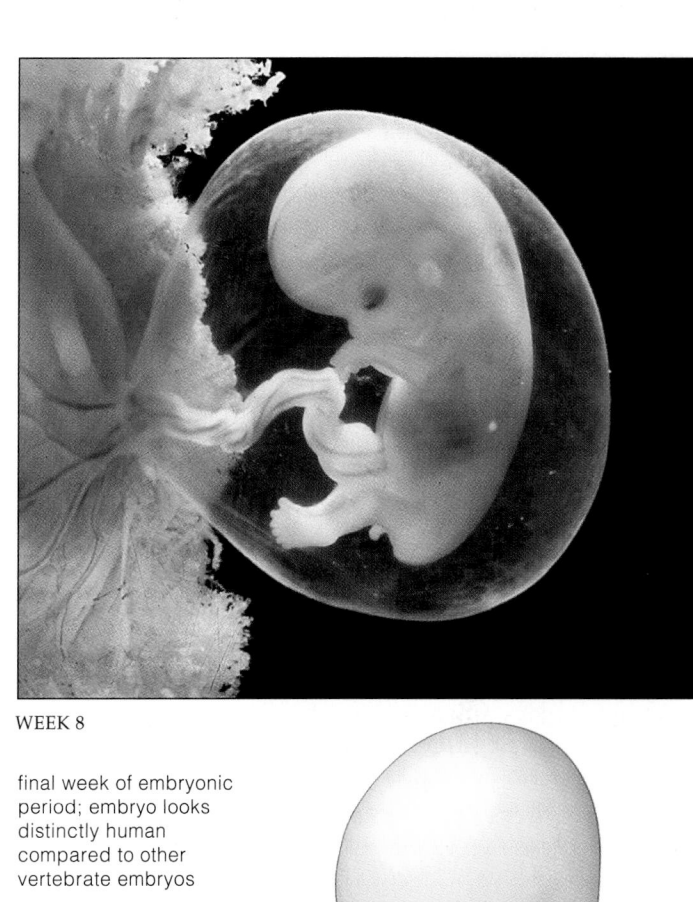

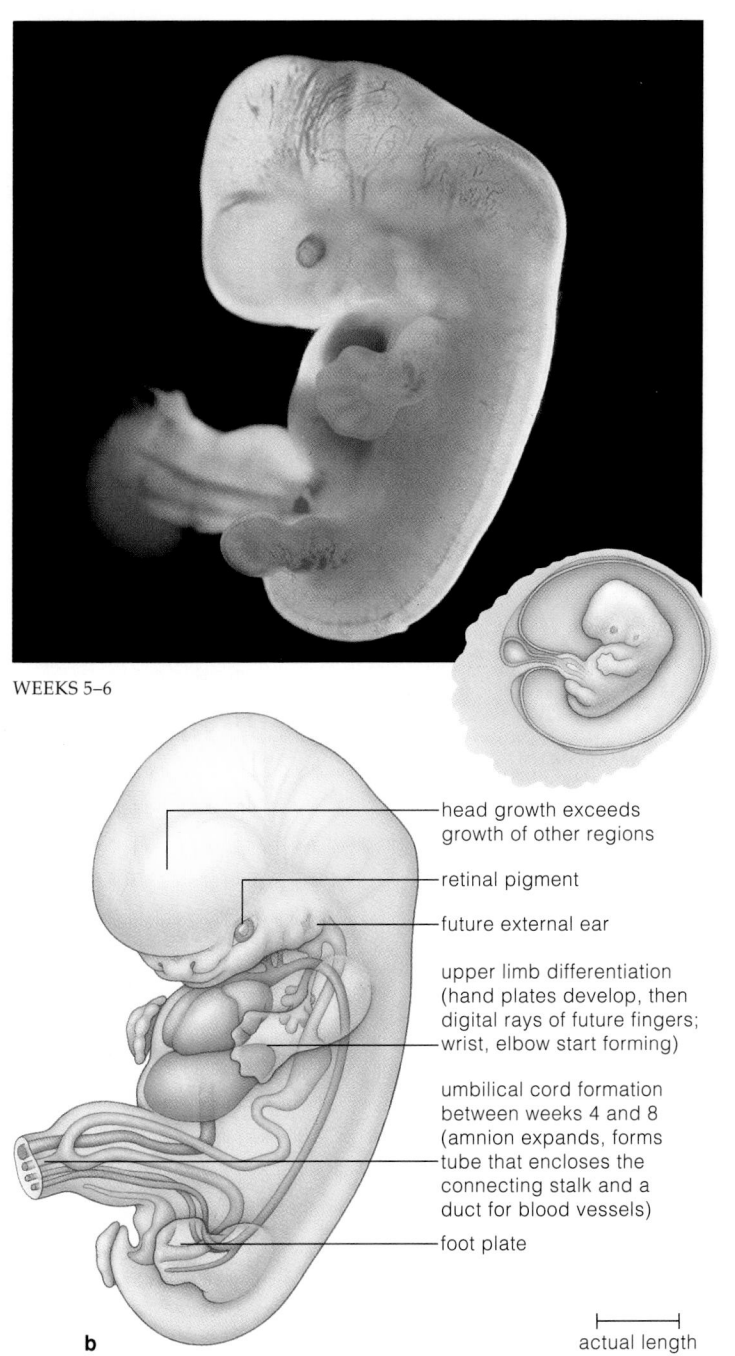

WEEKS 5–6

WEEK 8

head growth exceeds growth of other regions

retinal pigment

future external ear

upper limb differentiation (hand plates develop, then digital rays of future fingers; wrist, elbow start forming)

umbilical cord formation between weeks 4 and 8 (amnion expands, forms tube that encloses the connecting stalk and a duct for blood vessels)

foot plate

actual length

b

final week of embryonic period; embryo looks distinctly human compared to other vertebrate embryos

upper and lower limbs well formed; fingers and then toes have separated

primordial tissues of all internal, external structures now developed

tail has become stubby

actual length

c

FETAL DEVELOPMENT

When the fetus is 3 months old, it is about 4.5 inches long. Soft, fuzzy hair (the *lanugo*) covers its body. Its reddish skin is wrinkled and protected from abrasion by a thick, cheesy coating called the *vernix caseosa*.

The second trimester of development extends from the start of the fourth month to the end of the sixth. Figure 14.11 shows what the fetus looks like at the sixteenth week of development. Movements of facial muscles produce frowns and squints. The sucking reflex also is evident. Before the second trimester draws to a close, the mother can easily sense movements of the fetal arms and legs. During the sixth month, eyelids and eyelashes form.

From the Seventh Month to Birth

The third trimester extends from the seventh month until birth. At 7 months the fetus is about 11 inches long, and during the seventh month its eyes open. Although the fetus is growing much larger and rapidly becoming "babylike," not until the middle of the third trimester will it be able to survive on its own if born prematurely or removed surgically from the uterus. Although development appears to be relatively complete by the seventh month, few fetuses would be able to maintain a normal body temperature or breathe normally, even with the best medical care. However, with intensive medical support, fetuses as young as 23 to 25 weeks have survived early delivery. Babies born before 7 months' gestation often suffer from *respiratory distress syndrome*, as described in Chapter 8, because their lungs lack surfactant and cannot expand adequately. The longer the baby can remain in its mother's uterus, the better its chances. By the ninth month, survival chances increase to about 95 percent.

Fetal Circulation

Over time, maturation of organs and organ systems gradually readies the fetus for life outside its mother's body. In the case of the circulatory system, however, the path toward independence requires a detour. Several temporary bypass vessels form and will function until birth. As Figure 14.12 shows, two *umbilical arteries* within the umbilical cord transport deoxygenated blood and metabolic wastes from the fetus to the placenta. There, the fetal blood gives up wastes, takes on nutrients, and exchanges gases with maternal blood. Fetal hemoglobin binds oxygen more readily than "normal" hemoglobin, which helps ensure that adequate oxygen will reach developing fetal tissues. The oxygenated blood, enriched with nutrients, returns from the placenta to the fetus in the *umbilical vein*.

Other temporary vessels divert blood past the lungs and liver. These organs do not develop as rapidly as some

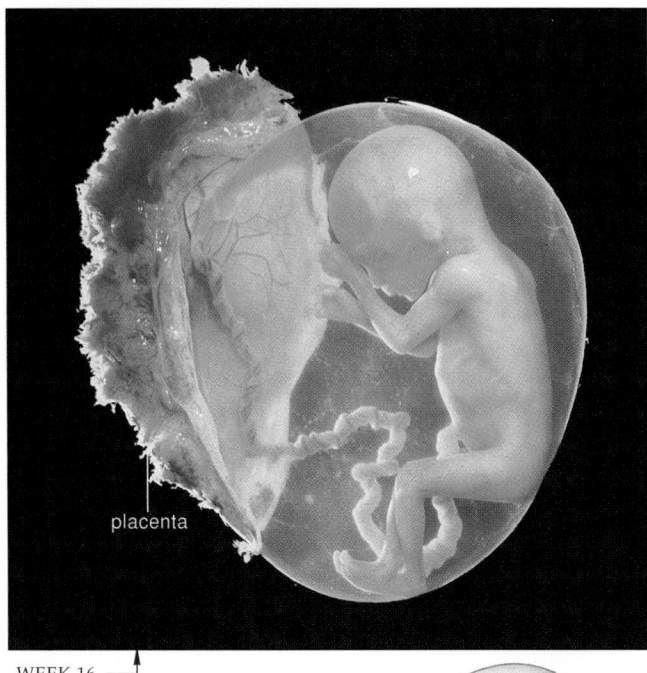

placenta

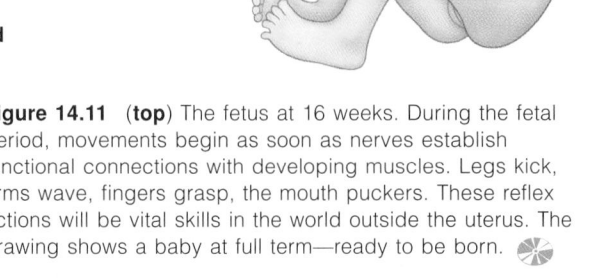

WEEK 16

Length: 16 centimeters
(6.4 inches)
Weight: 200 grams
(7 ounces)

WEEK 29

Length: 27.5 centimeters
(11 inches)
Weight: 1,300 grams
(46 ounces)

WEEK 38 (full term)

Length: 50 centimeters
(20 inches)
Weight: 3,400 grams
(7.5 pounds)

During fetal period, length measurement extends from crown to heel (for embryos, it is the longest measurable dimension, as from crown to rump).

d

Figure 14.11 (**top**) The fetus at 16 weeks. During the fetal period, movements begin as soon as nerves establish functional connections with developing muscles. Legs kick, arms wave, fingers grasp, the mouth puckers. These reflex actions will be vital skills in the world outside the uterus. The drawing shows a baby at full term—ready to be born.

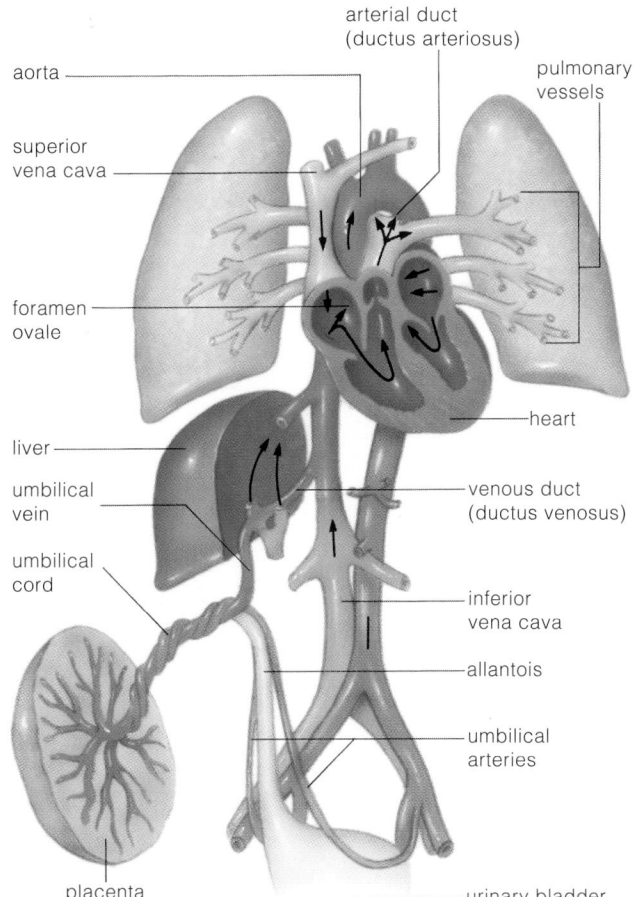

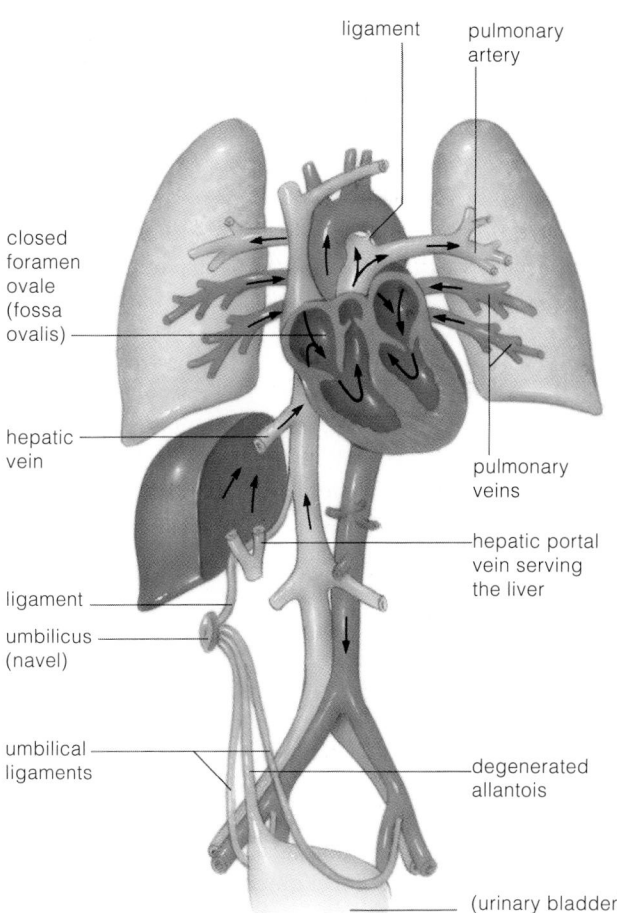

Figure 14.12 The direction of blood flow in fetal circulation (arrows). (**a**) Umbilical arteries carry deoxygenated blood from fetal tissues to the placenta. There blood picks up oxygen and nutrients from the mother's bloodstream and returns to the fetus via the umbilical vein. Blood bypasses the lungs, moving through the foramen ovale and the arterial duct. It bypasses the liver by moving through the venous duct.

(**b**) At birth the foramen ovale closes, and the pulmonary and systemic circuits of blood flow become completely separate. The arterial duct, venous duct, umbilical vein, and portions of the umbilical arteries become ligaments, and the allantois degenerates.

others, because (by way of the placenta) the mother's body can perform their functions. The lungs of a fetus are collapsed and will not become functional for gas exchange until the newborn takes its first breaths outside the womb. Hence, lung tissues receive only enough blood to sustain their development. A little of the blood entering the heart's right atrium flows into the right ventricle and moves on to the lungs. Most of it, however, travels through a gap in the interior heart wall called the *foramen ovale* ("oval opening") or into an arterial duct (*ductus arteriosus*) that bypasses the nonfunctioning lungs entirely.

Likewise, most blood bypasses the fetal liver because the mother's liver performs most liver functions (such as nutrient processing) until birth. Nutrient-laden blood from the placenta travels through a venous duct (*ductus*

venosus) past the liver and on to the heart, which pumps it to body tissues. At birth, blood pressure in the heart's left atrium increases. This causes a valvelike flap of tissue to close off the foramen ovale, which gradually seals. The closure separates the pulmonary and systemic circuits of blood flow, and the arterial duct collapses (Figure 14.12b). The venous duct gradually closes during the first few weeks after birth.

During the second and third trimesters, fetal organs and organ systems gradually mature. Because the fetus exchanges gases and receives nutrients via the mother's bloodstream prior to birth, the fetal circulatory system develops temporary vessels that bypass the lungs and liver. At birth, normal circulatory routes begin functioning.

FROM BIRTH ONWARD

Birth

Birth, or **parturition**, takes place about 39 weeks after fertilization, usually within 2 weeks of the "due date" (280 days from the start of the last menstrual period). The birth process, "labor," begins when smooth muscle in the uterus starts to contract. Evidently, uterine contractions are the indirect outcome of a cascade of hormones from the fetus's hypothalamus, pituitary, and adrenal glands— triggered by an as-yet-unknown signal that says, in effect, it's time to be born. The hormonal flood prompts the placenta to produce more estrogen. Rising estrogen in turn calls for a rush of oxytocin and of prostaglandins (also produced by the placenta), which stimulate contractions of the uterus. For the next 2 to 18 hours, the contractions will become stronger, more painful, and more frequent.

Three Stages of Labor

Labor is divided into three stages, somewhat akin to "before, during, and after." In the first stage, uterine contractions push the fetus against the mother's cervix. Initial contractions are relatively mild and occur about every 15 to 30 minutes. As the cervix gradually dilates to a diameter of about 10 centimeters (4 inches, or "5 fingers"), contractions become more frequent and intense. Usually, the amniotic sac ruptures during this stage, which can last up to 12 hours or more.

The second stage of labor, actual birth of the fetus, typically occurs less than an hour after full dilation of the cervix. This stage is generally brief—under 2 hours. Strong contractions of the uterus and abdominal muscles occur every 2 or 3 minutes, and the mother feels an urge to push. Her efforts and the intense contractions move the soon-to-be newborn through the cervix and out through the vaginal canal, usually head first (Figure 14.13). Complications can develop if the baby begins to emerge in a "bottom-first" (*breech*) position, and the attending physician may use hands or forceps to aid the delivery.

After the baby is expelled, the third stage of labor gets under way. Uterine contractions force fluid, blood, and the placenta (now called the *afterbirth*) from the mother's body. The umbilical cord—the lifeline to the mother—is now severed. Without the placenta to remove wastes, carbon dioxide builds up in the baby's blood. Together with other factors, including handling by medical personnel, this stimulates control centers in the brain, which respond by triggering inhalation—the newborn's crucial first breath.

As the lungs begin to function, the bypass vessels of the fetal circulation begin to close and soon shut completely. The fetal heart's opening, the foramen ovale, normally closes slowly during the first year of life. A lasting reminder of this final stage is the scar we call the navel, the site where the umbilical cord was once attached.

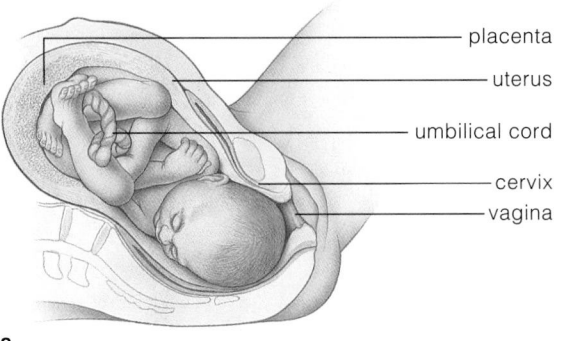

placenta
uterus
umbilical cord
cervix
vagina

a

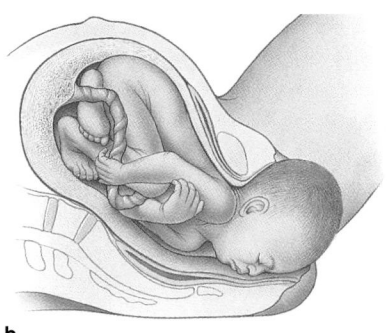

b

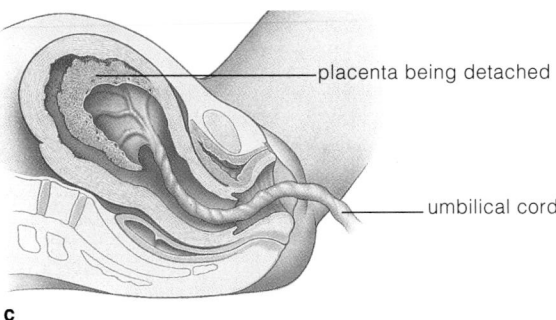

placenta being detached

umbilical cord

c

Figure 14.13 Expulsion of the fetus during birth. The afterbirth— the placenta, fluid, and blood—is expelled shortly afterward.

Most full-term pregnancies end in the birth of a healthy infant. Yet babies born prematurely—especially before about 8 months of uterine life—can suffer complications because their organs have not developed to the point where they can function independently of the mother. In such cases, attempts to sustain the baby's life under hospital conditions that will permit further needed development may require the most advanced medical technology. Even then, the vast majority of extremely premature infants do not survive. The *Choices* essay on

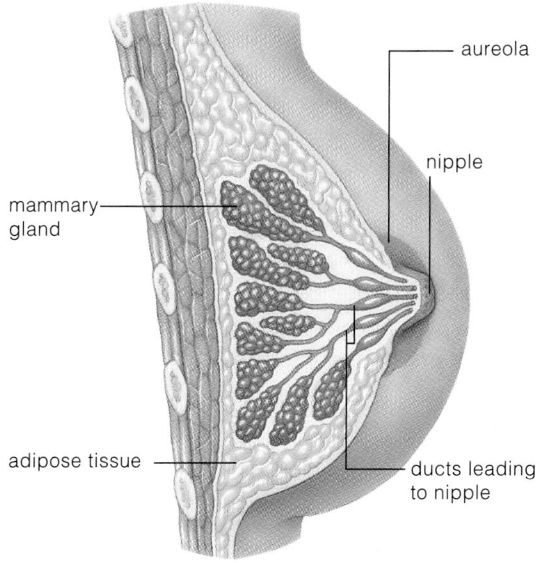

Figure 14.14 Breast of a lactating female. This cutaway view shows the mammary glands and ducts. Chapter 20 describes how to examine breast tissues for cancer.

this page addresses this and other issues related to uses of advanced technology in medical care of desperately ill patients.

Nourishing the Newborn

During pregnancy, estrogen and progesterone stimulate the growth of mammary glands and ducts in the mother's breasts (Figure 14.14). For the first few days after birth, those glands produce *colostrum*, a colorless fluid low in fat but rich in proteins, antibodies, minerals, and vitamin A. Then prolactin secreted by the pituitary stimulates milk production, or **lactation**.

When the newborn suckles, the pituitary also releases oxytocin, which causes breast tissues to contract and so force milk into the ducts. Oxytocin also triggers uterine contractions that "shrink" the uterus back to its normal size.

The mother's cervix dilates during the first stage of labor, and the baby is born during the second stage. In the third stage, uterine contractions expel the placenta.

Lactation (milk production) begins a few days after birth. It is stimulated by the hormone prolactin, which is released from the mother's pituitary and acts on breast (mammary gland) tissues. Oxytocin released from the pituitary acts to force milk into mammary ducts.

SAVING BABIES AND MAKING CLONES

Our ability to devise ever more sophisticated medical technologies raises serious social issues. One concern is medical treatment for gravely ill newborns, and the high cost of care for extremely premature infants or those born with devastating abnormalities. Nearly 85 percent of early infant deaths occur in premature babies—many still only partially developed fetuses. They cannot be saved no matter how elaborate the technology that is brought to bear. Those that do live often have moderate to severe medical problems. Common ones include blindness and cerebral palsy, a neuromuscular disorder that damages or destroys a person's control over voluntary muscle movements.

Today, there is great societal pressure to control the costs of medical care. To reduce the approximately $5 billion spent annually in the United States on caring for "preemies," not to mention the anguish of parents and affected youngsters alike, some researchers are refining methods for identifying and treating causes of prematurity, including uterine infections in the mother. Another approach involves efforts to improve our understanding of biochemical indicators—typically, "markers" present in the mother's blood—that can be used to predict when a woman is about to enter premature labor. With such information, her physician could mount preventive countermeasures early on.

More wrenching are cases in which babies are born at full term, but with birth defects so serious that there is little realistic hope for long-term survival. Sometimes an ill infant's death is a foregone conclusion. This is the case with *anencephaly*, in which most of the brain fails to develop. Even here, some parents understandably want high-tech efforts to prolong the child's life. In other instances, only extreme measures, such as organ transplants and complex heart surgeries, can provide a glimmer of hope. Although the ethical issues can be complex, health insurers and government agencies are increasingly reluctant to pay for expensive or experimental therapies. Should such care be provided only in certain cases? And what if *your* child's life or health were in the balance?

In the late 1990s, improving technology also ushered in a new, and to some people, heart-stopping possibility— the ability to produce a clone (an exact genetic copy) of a mammal from adult cells. In 1997, after 277 attempts, researchers in Scotland introduced the world to Dolly, a sheep cloned from adult mammary cells. Not long afterward, an American scientist announced his intention to pursue the cloning of adult *human* cells. Although the ensuing public outcry quickly led to government action aimed at banning such an attempt, no one can be sure that someone, somewhere is not going forward with such an enterprise. Some reputable researchers even argue in *favor* of human cloning efforts. All of us would do well to think about the possible implications if—or, perhaps, when— the cloning of humans becomes a reality.

HOW THE MOTHER'S LIFESTYLE AFFECTS EARLY DEVELOPMENT

The placenta is a highly selective filter that prevents many noxious substances in the mother's bloodstream from gaining access to the embryo or fetus. Even so, the rapidly developing infant is at the mercy of its mother's diet and health habits.

Maternal Nutrition

During pregnancy, a balanced diet that includes sufficient carbohydrates, amino acids, and fats or oils usually provides most necessary vitamins and minerals. The mother's vitamin needs are definitely increased, and most physicians recommend that pregnant women take vitamin supplements. The developing fetus is more resistant to vitamin and mineral deficiencies because the placenta preferentially absorbs vitamins and minerals from the mother's blood. Even so, recent evidence suggests that folic acid supplements may be important in preventing certain birth defects, including neural tube disorders.

In most cases, a woman should gain from 20 to 35 pounds during pregnancy. If she restricts her food intake too severely, especially during the last trimester, the newborn will be underweight. Significantly underweight infants face more postdelivery complications than do infants of normal weight and represent nearly half of all newborn deaths. As birth approaches, the growing fetus demands more and more nutrients from the mother's body. During this last phase of pregnancy, the mother's diet profoundly influences the course of development. Poor nutrition damages most organs—particularly the brain, which undergoes its greatest growth in the weeks just before birth.

Risk of Infections

Throughout pregnancy, some antibodies cross the placenta and help protect the developing individual from all but the most severe bacterial infections. However, certain

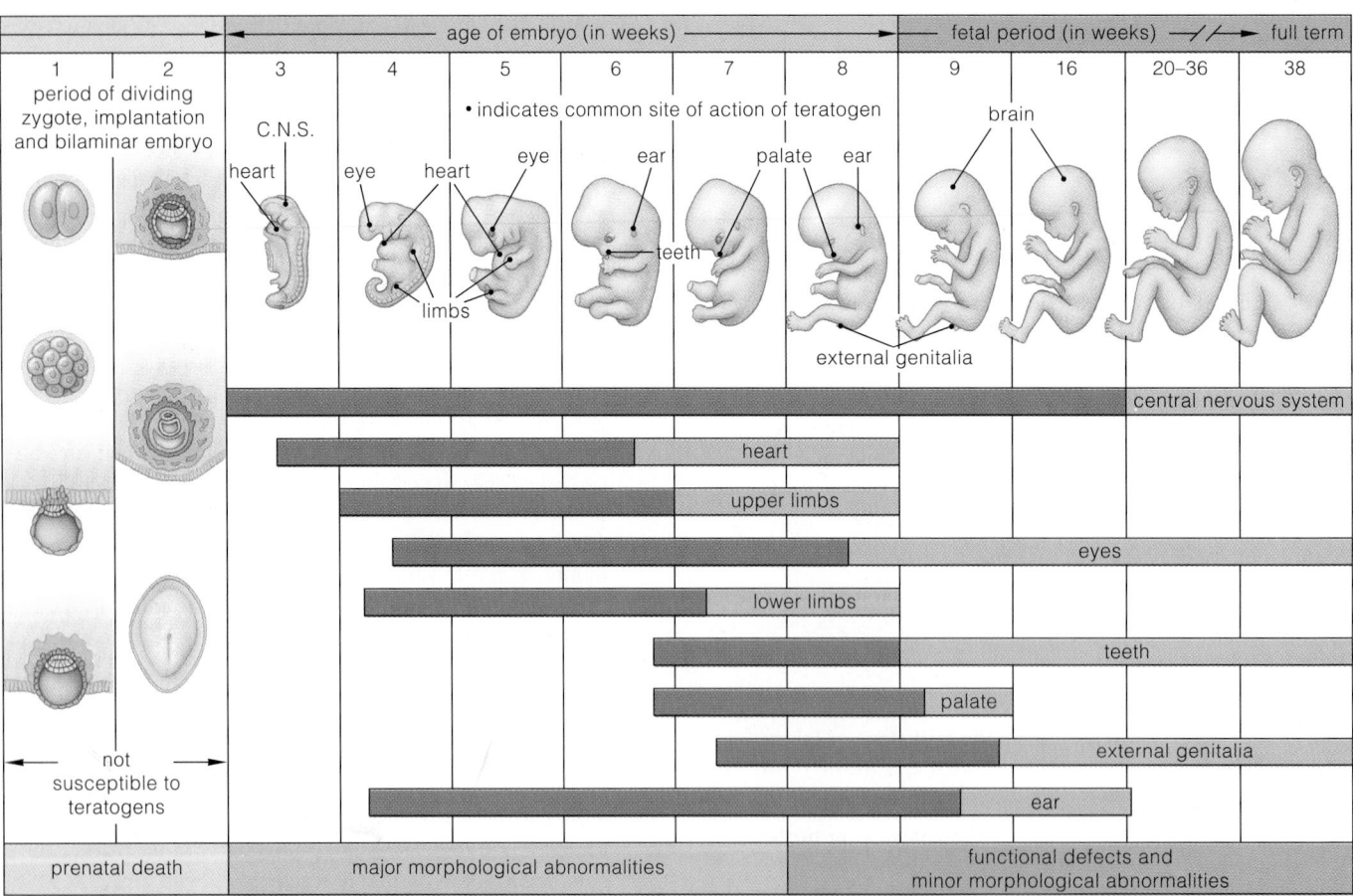

Figure 14.15 Critical periods of embryonic and fetal development. *Light blue* also indicates periods in which organs are most sensitive to damage from alcohol, viral infection, and so on. Numbers signify the week of development.

viral diseases can have damaging effects (Figure 14.15) if they are contracted during the first 6 weeks after fertilization, the critical time of organ formation. For example, if the woman contracts German measles during this period, there is a 50 percent chance that her embryo will become malformed. If she contracts the virus when the embryo's ears are forming, her newborn may be deaf. (German measles can be avoided by vaccination *before* pregnancy.) The risk of damage to the embryo diminishes after the first 6 weeks. After the fourth month, the same infection has no apparent effect on the fetus.

Prescription Drugs, Illegal Drugs, and Alcohol

During the first trimester the embryo is highly sensitive to drugs. In the 1960s many women using the drug thalidomide during the first trimester gave birth to infants with missing or severely deformed arms and legs. Although it wasn't known at the time, thalidomide disrupts the events that induce development of limbs. Thalidomide was withdrawn from the market, but there is evidence that other tranquilizers (and sedatives and barbiturates) might cause similar, although less severe, damage. Use of certain anti-acne drugs, such as retinoic acid (Retin-A), increases the risk of facial and cranial deformities. The antibiotic tetracycline causes yellowed

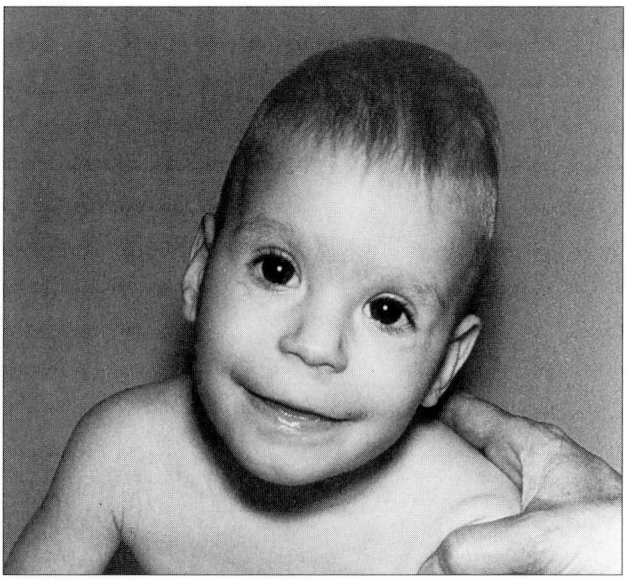

Figure 14.16 An infant affected by fetal alcohol syndrome (FAS). Symptoms include a small head, low and prominent ears, poorly developed cheekbones, and a long, smooth upper lip. The child may be likely to encounter growth problems and abnormalities of the nervous system. About 1 newborn in 750 in the United States is affected by this disorder.

teeth. Streptomycin causes hearing problems and may affect the nervous system.

Cocaine, particularly crack cocaine, disrupts the functioning of the fetal nervous system as well as the mother's (Chapter 10). Cocaine-addicted newborns are chronically irritable. They are also abnormally small because during prenatal development their body tissues were not provided with enough oxygen and nutrients. (Crack causes blood vessels to constrict.) Even though crack babies are abnormally fussy, they do not respond to rocking and other kinds of normal stimulation. It may be a year or more before they recognize their mother. Although authorities disagree about the long-term effects of infant crack addiction, there is some evidence that even with treatment, severely affected babies may develop into children with moderate to severe problems in their emotional and social adjustment.

Like many other drugs, alcohol crosses the placenta and affects the fetus. *Fetal alcohol syndrome* (FAS) is a constellation of defects that can result from alcohol use by a pregnant woman. *FAS is the third most common cause of mental retardation in the United States.* It is characterized by facial deformities, poor coordination, and, sometimes, heart defects (Figure 14.16). Between 60 and 70 percent of alcoholic women give birth to infants with FAS. Some researchers suspect that drinking any alcohol at all during pregnancy may be dangerous for the fetus.

Effects of Cigarette Smoke

Cigarette smoking harms fetal growth and development. Research shows that women who smoke are at greater risk of miscarriage, stillbirth, and premature delivery. Newborns of women who have smoked every day throughout pregnancy have a low birth weight. That is true even when the woman's weight, nutritional status, and all other relevant variables are identical to those of pregnant women who do not smoke. Studies also suggest that babies of smoking mothers may be at greater risk for heart abnormalities and be slower learning to read.

The critical period appears to be the last half of pregnancy. In one study, newborns of women who had stopped smoking by the middle of the second trimester were indistinguishable from those born to women who had never smoked.

A woman who is pregnant or planning to become pregnant can adopt a lifestyle that helps protect her child from harm during early development. Major risks to an embryo and fetus include poor maternal nutrition, certain viral infections, unsupervised use of some prescription drugs, alcohol use, and cigarette smoking.

14.10

PRENATAL DIAGNOSIS: DETECTING BIRTH DEFECTS

A growing number of options now enable us to detect more than 100 genetic disorders, such as cystic fibrosis, before a child is born.

Amniocentesis samples fluid from within the amnion, the sac that contains the fetus (Figure 14.17a). During the fourteenth to sixteenth week of pregnancy, the thin needle of a syringe is inserted through the mother's abdominal wall, into the amnion. The physician must take care that the needle doesn't puncture the fetus and that no infection occurs. Amniotic fluid contains sloughed fetal cells; as the syringe withdraws fluid, some of those cells are included. They are then cultured and tested for genetic abnormalities.

Chorionic villus sampling (CVS) uses tissue from the chorionic villi of the placenta. CVS is tricky. Using ultrasound, the physician must guide a tube through the vagina, past the cervix, and along the uterine wall, then remove a small sample of chorionic villus cells by suction. The method can be used by the eighth week of pregnancy, and results are available within days. However, CVS can damage the placenta—and hence the fetus. Both CVS and amniocentesis involve a small risk of triggering miscarriage.

A relatively new method called FISH (for fluorescent in situ hybridization) analyzes the genetic material of fetal cells that have crossed the placenta and entered the mother's blood. The genetic material is exposed to segments of DNA, called probes, that are tagged with dyes that glow (fluoresce) when exposed to ultraviolet light. The DNA probes attach to chromosome regions that are associated with various genetic birth defects. In experiments, the method has correctly detected abnormalities in hundreds of fetal samples. FISH takes only about 2 days and does not disturb the protected fetal environment in the uterus.

Physicians also have available several methods of *embryo screening*. In preimplantation diagnosis, an embryo conceived by in vitro fertilization is analyzed for genetic defects using recombinant DNA technology. The testing occurs at the eight-cell stage (Figure 14.17b), which might be considered a "prepregnancy" stage because under normal circumstances the tiny ball of cells would not yet have implanted in the mother's uterus. A method called embryoscopy can be used to view older embryos that have already implanted. A physician threads a tiny viewing device through a hollow needle inserted into the mother's uterus. Embryo screening is designed to help parents who are at high risk of having children with a genetic birth defect. Even so, it is controversial, raising questions we consider in the *Choices* essay in Chapter 18.

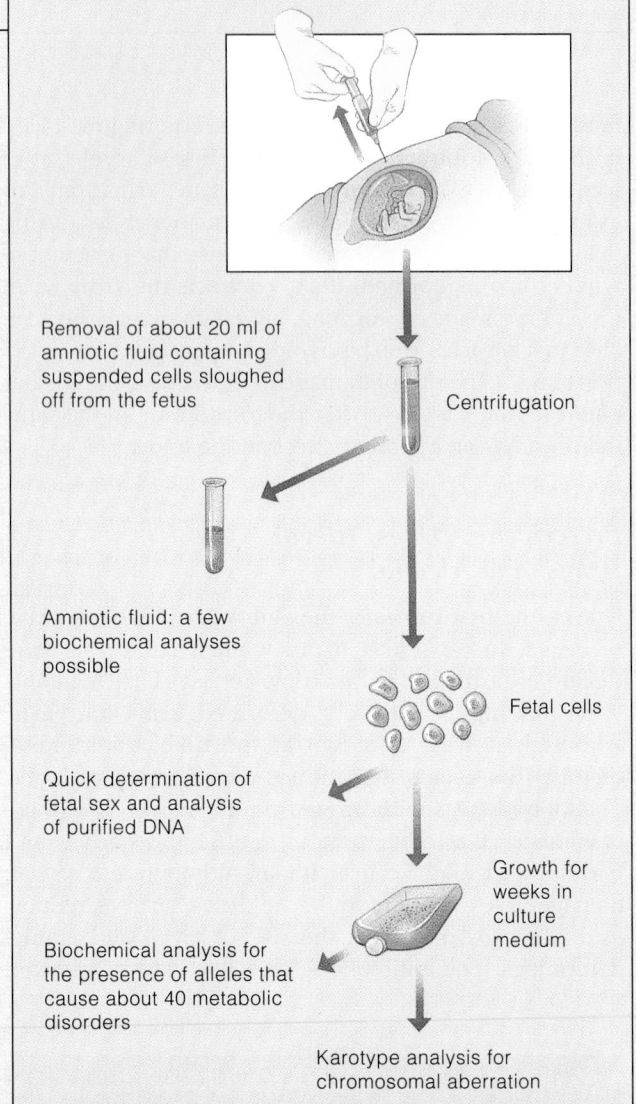

Removal of about 20 ml of amniotic fluid containing suspended cells sloughed off from the fetus

Centrifugation

Amniotic fluid: a few biochemical analyses possible

Fetal cells

Quick determination of fetal sex and analysis of purified DNA

Growth for weeks in culture medium

Biochemical analysis for the presence of alleles that cause about 40 metabolic disorders

Karotype analysis for chromosomal aberration

a

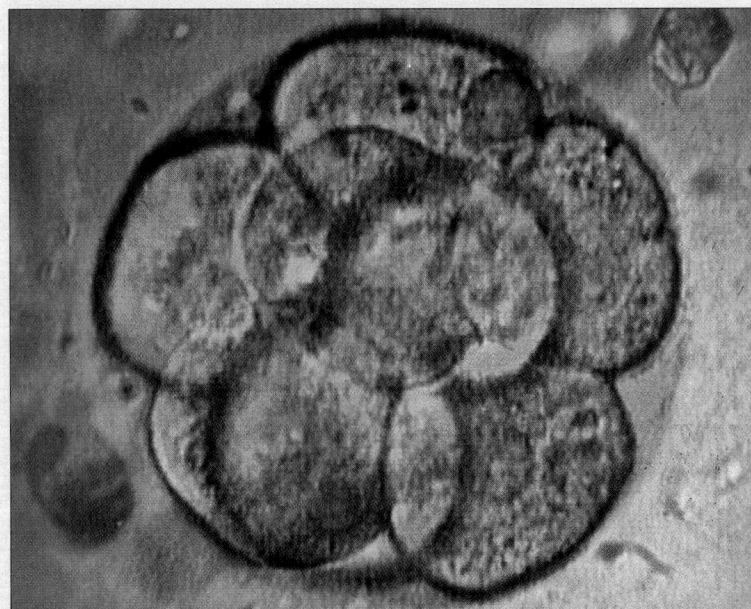

b

Figure 14.17 (**a**) Procedure for amniocentesis. (**b**) An eight-cell-stage human embryo.

SUMMARY OF DEVELOPMENTAL STAGES

Birth to Adulthood

Following birth, a prescribed course of further growth and development leads to adulthood. Table 14.2 summarizes all the prenatal (before birth) and postnatal (after birth) stages of life. A newborn is called a *neonate*. Infancy lasts until about 15 months of age. Although a baby is born with all the brain neurons it will ever have, during this period the nervous and sensory systems mature rapidly, and the body becomes longer through a series of "growth spurts." Figure 14.18 shows how body proportions change as development proceeds through childhood and adolescence. A key feature of adolescence is *puberty*, the arrival of sexual maturity as the reproductive organs begin to function. Sex hormones trigger the appearance of pubic and underarm hair, other secondary sex characteristics, and behavior changes. A combination of hormones triggers another growth spurt at this time. Boys usually grow most rapidly between the ages of 12 and 15,

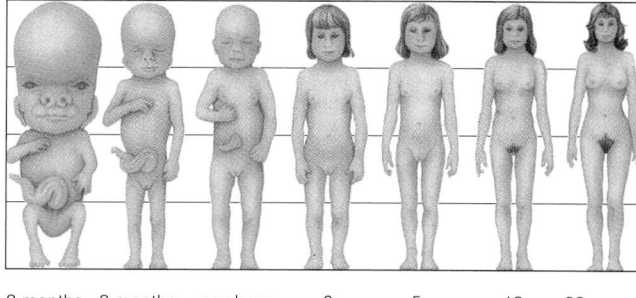

2 months 3 months newborn   2   5   13   22 years

Figure 14.18 Diagram of changes in the proportions of the human body during prenatal and postnatal growth.

whereas girls tend to grow most rapidly between the ages of 10 and 13. After several years, the influence of sex hormones causes the cartilaginous plates near the ends of long bones to harden into bone, and growth stops by the early 20s.

Adult Life

Following growth and differentiation, the cells of all complex animals gradually deteriorate. Humans are no different. Although in the United States the average life expectancy is 72 years for males and 79 years for females, we reach the peak of our physical potential in adolescence and early adulthood. A healthy diet, regular exercise, and other beneficial lifestyle habits can go far in keeping a person vigorous for decades of adult life. Even so, after about age 40, body parts begin to undergo structural changes, and there is a gradual loss of efficiency in bodily functions, as well as increased sensitivity to environmentally induced stress. This progressive cellular and bodily deterioration is built into the life cycle of all organisms in which differentiated cells become highly specialized. The process is called *senescence*, or simply **aging**.

Aging in humans leads to many structural changes in the body. Beginning around age 40, there is a gradual decline in bone and muscle mass, increased skin wrinkling, and more fat deposition. Less obvious are gradual physiological changes. This chapter's concluding sections explore the phenomenon of aging, including some ideas about its causes.

Table 14.2 Stages of Human Development: A Summary	
PRENATAL PERIOD	
1. Zygote	Single cell resulting from fusion of sperm nucleus and egg nucleus at fertilization
2. Morula	Solid ball of cells produced by cleavages
3. Blastocyst	Ball of cells with surface layer and inner cell mass
4. Embryo	All developmental stages from 2 weeks after fertilization until end of eighth week
5. Fetus	All developmental stages from the ninth week until birth (about 39 weeks after fertilization)
POSTNATAL PERIOD	
6. Newborn (neonate)	Individual during the first 2 weeks after birth
7. Infant	Individual from 2 weeks to about 15 months after birth
8. Child	Individual from infancy to about 12 or 13 years
9. Pubescent	Individual at puberty, when secondary sexual traits develop; girls between 10 and 16 years, boys between 13 and 16 years
10. Adolescent	Individual from puberty until about 3 or 4 years later; physical, mental, emotional maturation
11. Adult	Early adulthood (between 18 and 25 years); bone formation and growth completed. Changes proceed very slowly afterward.
12. Old age	Aging culminates in general body deterioration

Following birth, development proceeds through childhood and adolescence, which includes the arrival of sexual maturity at puberty. Puberty is the gateway to the adult phase of life. After about age 30, developmental changes associated with aging become increasingly apparent. The average life span for males currently is about 72 years; it is about 79 years for females.

THE BODY AS IT AGES

Technically, aging begins the day we are born, and it continues until we die. However, most people equate aging with "getting old," a gradual loss of vitality as cells, tissues, and organs function less and less efficiently. Beginning at about age 40, our skin begins to noticeably wrinkle and sag, body fat tends to accumulate, and injuries to muscles and joints occur more readily and take longer to heal. Stamina declines, and we become increasingly susceptible to cardiovascular disorders, cancer, and degenerative ailments such as arthritis and Alzheimer's disease.

Structural alterations in body proteins—including collagen, a key component of various connective tissues—may contribute to many changes we associate with aging (Figure 14.19). A collagen molecule consists of three polypeptide chains twisted into a spiral. This shape is stabilized by molecular bonds that form "cross-links" between segments of the chain. As a person grows older, new cross-links develop, and the protein becomes more and more rigid. As collagen's structure changes, so may the structure and functioning of organs and blood vessels that contain it. Age-related cross-linking is also thought to affect many enzymes and possibly DNA.

What Causes Aging?

No one yet knows for sure what factors cause the degenerative changes we associate with aging. Many hypotheses have been proposed, and there is tantalizing evidence for some of them.

OXIDATIVE DAMAGE Some researchers believe that many of the changes seen in aging tissues could result from damage caused by free radicals of oxygen. As Chapter 1 described, free radicals are by-products of normal cell metabolism. They carry an unpaired electron and can combine with (oxidize) and damage proteins, DNA, lipids, and other molecules. Aging cells in the heart, brain, skin, and other tissues typically contain clumps of lipofuscins, which are oxidized lipids. In tissue samples, the amount of oxidized proteins, including key cell enzymes, also increases with the age of the sample donor. Moreover, studies show that the DNA in cell mitochondria (page 49) is especially susceptible to free radical damage. This discovery suggests why, as the years pass, an organ might function less and less well. As more and more of its cells contain increasing numbers of damaged mitochondria, the cells would undergo a steady decline in the energy available to sustain metabolic activities.

DECLINE IN DNA REPAIR A related hypothesis focuses on a possible decline in the ability of cells to repair damaged DNA, either in mitochondria or in the cell nucleus. DNA makes up genes, the instructions for building cell proteins, including many vital enzymes. Some genes

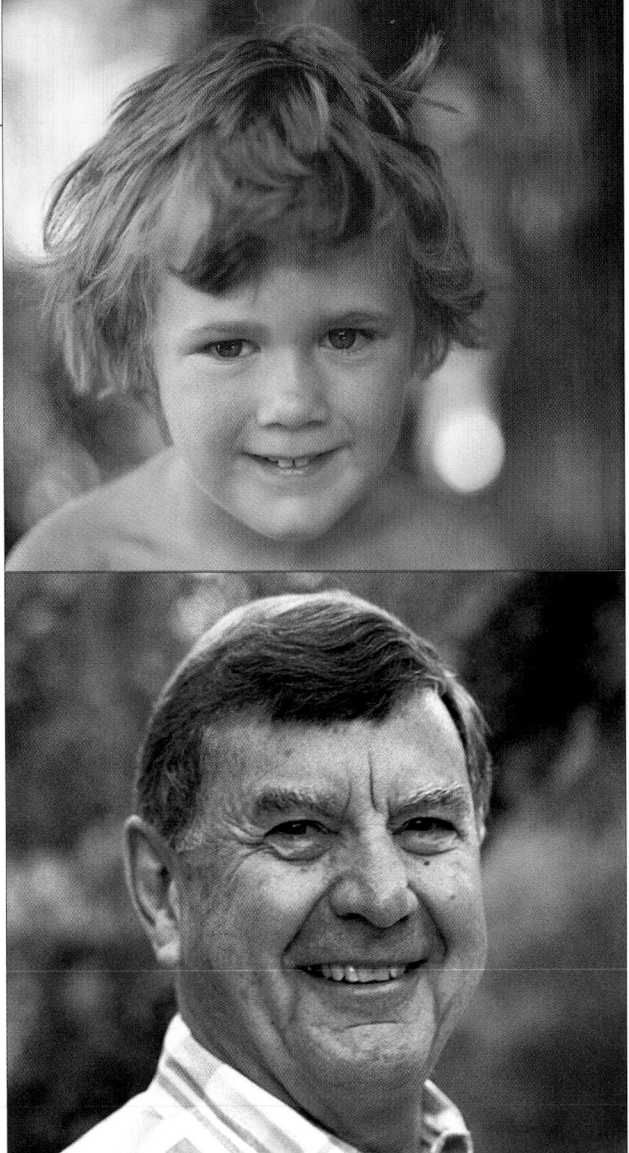

Figure 14.19 A predictable sequence of changes unfolds as we age. To a greater or lesser degree, these changes affect virtually all body tissues and organs.

regulate the activity of others and so indirectly control cell operations. If cells gradually lose the capacity for DNA self-repair, over time gene mutations would accumulate. Needed enzymes or other molecules might be in increasingly short supply, or cell operations might become deficient in other ways. Perhaps genetic changes cause cells to lose their ability to repair free radical damage. We already know that changes in various types of genes underlie most cancers (Chapter 20) and that the incidence of cancer rises sharply in older age groups.

Cell division is the mechanism by which the body grows, maintains certain tissues (such as skin and intestinal epithelium), and repairs tissue damage. Although the reasons are not clear, normal embryonic cells cultured in the laboratory tend to divide ever more slowly as they grow "older," and after about 50 rounds of division the process stops altogether. Researchers are now looking for evidence of how the gradual loss of the ability of body cells to proliferate may contribute to aging.

Skin, Muscles, and the Skeleton

Skin changes often are early obvious signs of aging. The normal replacement of sloughed cells of the epidermis (through cell division) begins to slow. Cells called fibroblasts are a major component of connective tissue, and the number of fibroblasts in the dermis starts to decrease. Also, elastic fibers that give skin its flexibility become replaced with more rigid collagen. As a result of these changes, the skin becomes thinner and less elastic, and sags and wrinkles develop. Wrinkling increases as fat stores decline in the hypodermis, the subcutaneous layer beneath the dermis. The skin becomes dryer as sweat and oil glands begin to break down and are not replaced. The loss of sweat glands and subcutaneous fat is one reason why older people tend to have difficulty regulating their body temperature. Hair follicles die or become less active, so there is a general loss of body hair. As pigment-producing cells die and are not replaced, remaining body hair begins to appear gray or white.

Fibers in skeletal muscle atrophy, in part because of a corresponding loss of motor neurons that synapse with muscle fibers. Muscles lose mass and strength, and lost muscle tends to be replaced by fat and later by collagen, although the extent of this replacement depends on a person's diet and level of exercise. In general, an 80-year-old has about half the muscle strength he or she had at 30. A program of regular physical activity can help retard all these changes (Figure 14.20).

Bones become weaker, more porous, and brittle as a person ages. Over the years bones begin to lose some of their collagen and elastin, but the main cause of bone weakening is loss of calcium and other minerals. Some individuals may inherit a gene for faulty vitamin D receptors, which reduces the efficiency with which calcium is absorbed and used in bone formation. After about age 40, minerals begin to be removed from bone faster than they are replaced (see Section 4.2). Women naturally have less bone calcium than men do, and they begin to lose it earlier. As a result, older women are prone to develop osteoporosis. Without medical intervention, some women lose as much as 50 percent of their bone mass by age 70. In both women and men, the spinal vertebrae become closer together as the intervertebral disks deteriorate. This is why people often get shorter, losing about a centimeter every 10 years from middle age onward.

Nine out of ten people over 40 have some degree of joint breakdown. Cartilage in joints deteriorates over time, the surfaces of the joined bones begin to wear away, and the joints become more difficult to move. About 15 percent of adults develop *osteoarthritis*, a chronic inflammation of cartilage, in one or more joints. Although many factors probably contribute to osteoarthritis, it is most common in older people. This suggests that age-related changes in bone are often involved.

Figure 14.20 Moderate physical activity, consistently maintained as a person ages, slows age-related deterioration of the musculo-skeletal system and promotes cardiovascular health. Throughout life, regular exercise aids in weight control and promotes a positive outlook.

Aging in the Cardiovascular and Respiratory Systems

The heart and lungs function less efficiently with increasing age. In the lungs, walls of alveoli break down, so there is less total respiratory surface available for gas exchange. If a person does not develop an enlarged heart due to cardiovascular disease, the heart muscle becomes slightly smaller, and its strength and ability to pump blood diminish. As a result, less blood and oxygen are delivered to muscles and other tissues. In fact, decreased blood supply may be a factor in age-related changes throughout the body. Blood transport is also affected by structural changes in aging blood vessels. Elastic fibers in blood vessel walls are replaced with connective tissue containing collagen or become hardened with calcium deposits, and so vessels become stiffer. Cholesterol plaques and fatty deposits often further narrow arteries and veins (page 154). Hence older people often have higher blood pressure. However, as with the musculoskeletal system, lifestyle choices such as not smoking, eating a low-fat diet, and regular exercise can help a person maintain a vigorous respiratory and cardiovascular system well past middle age.

The Nervous System and Senses

Neurons stop dividing once they form. You are born with all the neurons you will ever have, and any that are lost will not be replaced. Brain neurons die steadily throughout life, and as they do the brain shrinks slightly, losing about 10 percent of its mass after 80 years. On the other hand, the brain has more neurons than are required for various functions, and there is evidence that when some types of neurons are lost or damaged, other neurons may produce new dendrites and synaptic connections and so take up the slack.

Over time, however, the death of some neurons and structural changes in others apparently do interfere with nervous system functions. For instance, in nearly anyone who lives to old age, tangled clumps of cytoplasmic fibrils develop in the cytoplasm of many neuron cell bodies. These *neurofibrillary tangles* may disrupt normal cell metabolism, although their exact effect is not understood. Clotlike plaques containing protein fragments called *beta amyloid* also develop between neurons.

In people who exhibit the dementia called *Alzheimer's disease* (AD), the brain tissue contains masses of neurofibrillary tangles. It also is riddled with large numbers of beta amyloid plaques (Figure 14.21). For the time being, however, researchers disagree over whether the amyloid plaques are a cause of Alzheimer's or simply one effect of another, unknown disease process. For instance, the brains of Alzheimer's patients also have lower-than-normal amounts of the neurotransmitter acetylcholine,

and some evidence suggests that this shortage may be related to beta amyloid buildup. Other possibilities under investigation include the hypothesis that AD results from chronic inflammation of brain tissue, similar to the inflammation that triggers arthritis. Certain therapeutic drugs can temporarily help alleviate some symptoms of Alzheimer's disease, which include memory loss and disruptive personality changes.

We are beginning to understand the genetic foundations for at least some inherited forms of AD. In families that show a pattern of early-onset (before age 60) Alzheimer's, the disease has been linked to a mutated gene for the amyloid precursor protein on chromosome 21 or to an unknown gene on chromosome 14. In families in which late-onset AD is prevalent, chromosome 19 of affected individuals carries at least one copy of a gene calling for a variant form of a lipid-binding protein. Although we do not yet know the exact role the variant protein plays in AD, individuals who have two copies of the gene that encodes it (the maximum number of copies possible) have a 90 percent chance of developing Alzheimer's disease.

Even otherwise healthy people begin to have problems with short-term memory after about age 60 and may find it takes longer to process new information. Perhaps because aging CNS neurons tend to lose some of their insulating myelin sheath, older neurons do not conduct action potentials as efficiently. In addition, neurotransmitters such as acetylcholine may be released

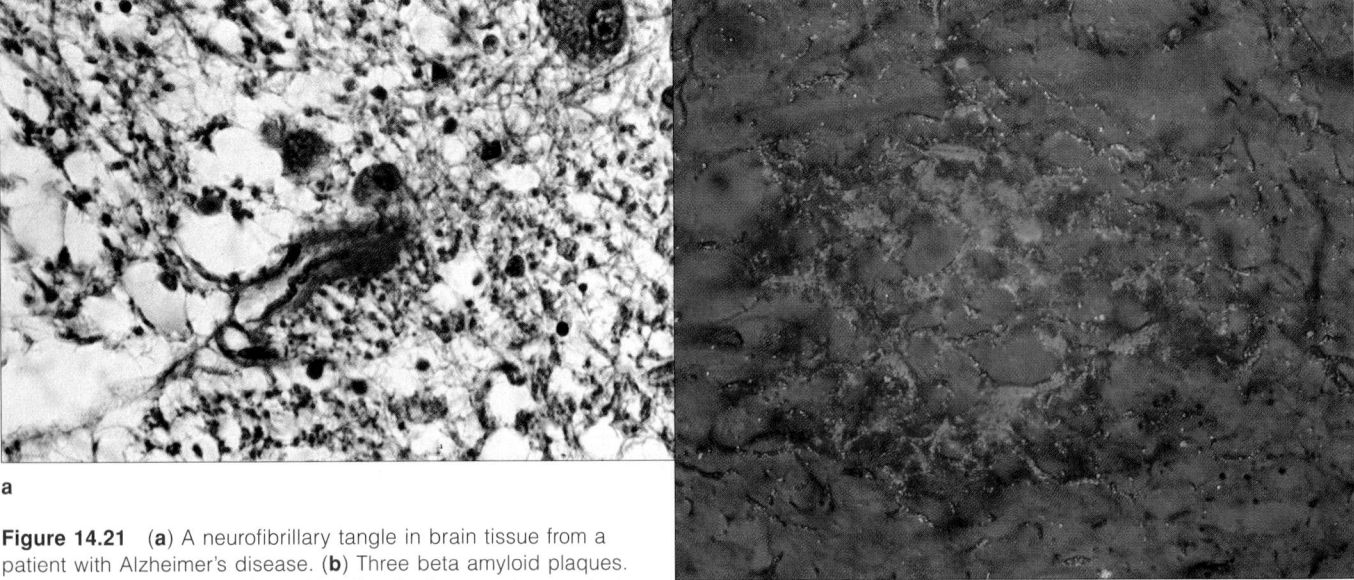

a

Figure 14.21 (a) A neurofibrillary tangle in brain tissue from a patient with Alzheimer's disease. (b) Three beta amyloid plaques. Older people who do not have Alzheimer's disease also develop neurofibrillary tangles and beta amyloid plaques, but not nearly as many.

b

more slowly. This may be due to age-related changes in plasma membrane proteins, which cause the membranes of synaptic vesicles to stiffen. As a result of such changes, movements and reflexes become slower, and some coordination is lost.

Sensory organs become less efficient at detecting or responding to stimuli. For example, people tend to become farsighted as they grow older because the lens of the eye loses its elasticity and is altered in other ways that prevent it from properly flexing to bend incoming light during focusing.

Reproductive Systems and Sexuality

Levels of most hormones remain steady throughout life, but reduced secretion of estrogens and progesterone triggers menopause in women, whereas falling testosterone levels cause reduced fertility in older men. Although menopause brings a woman's reproductive period to an end, men can and have fathered children into their 80s. Males and females both retain their capacity for sexual response well into old age.

Menopause usually begins in a woman's late 40s or early 50s. Over a period of several years, her menstrual periods become irregular, then stop altogether as the ovaries become less and less sensitive to the hormones FSH and LH (Chapter 12) and gradually stop secreting estrogen in response. Many women report relatively few unpleasant symptoms during menopause. However, declining estrogen levels may trigger "hot flashes" (intense sweating and uncomfortable warmth), thinning of the vaginal walls, and some loss of natural lubrication. Postmenopausal women are also at increased risk of osteoporosis and heart disease. Small doses of estrogen (with progesterone) administered as *hormone replacement therapy* (HRT) can counteract severe side effects of estrogen loss and may even help protect against uterine cancer. However, HRT is not without risks and must be carefully discussed with a knowledgeable physician.

After about age 50, men gradually begin to take longer to achieve an erection due to vascular changes that cause the penis to fill with blood more slowly. In the United States, more than half of men over 50 also experience bladder infections, urinary frequency, and other problems caused by age-related enlargement of the prostate gland.

Immunity, Nutrition, and the Urinary System

Other organs and organ systems also change as the years go by. In the immune system, the number of T cells falls, B cells become less active, and the ability to recognize self

Table 14.3 Some Physiological Changes in Aging

Age	Maximum Heart Rate	Lung Capacity	Muscle Strength	Kidney Function
25	100%	100%	100%	100%
45	94%	82%	90%	88%
65	87%	62%	75%	78%
85	81%	50%	55%	69%

Note: Age 25 is the benchmark for maximal efficiency of physiological functions.
Source: "The Search for the Fountain of Youth," *Newsweek*, 5 March 1990.

markers on body cells declines. These and other events help account for the fact that older people are more prone to many illnesses, including autoimmune diseases such as rheumatoid arthritis. In the aging digestive tract, glands in mucous membranes lining the stomach and small and large intestines gradually degenerate, and the pancreas secretes fewer digestive enzymes.

Although it is vital for older people to maintain adequate nutrition, we require fewer food calories as we age. By age 50, the basal metabolic rate (page 128) is only 80 to 85 percent of what it was in childhood and will keep falling about 3 percent every decade. Hence people tend to gain weight as they enter middle age, unless they compensate for a falling BMR by consuming fewer calories, increasing physical activity, or both.

The muscular walls of the large intestine, bladder, and urethra become weaker and less flexible. As the urinary sphincter is affected, many older people experience the urine leakage called *urinary incontinence*. Women who have borne children may have more trouble with urinary incontinence because their pelvic floor muscles are weak. The kidneys may continue to function well throughout life, despite the fact that nephrons gradually break down and lose some of their ability to maintain the balance of water and ions in body fluids (Table 14.3). At birth, a person's kidneys generally contain more than enough nephrons for satisfactory functioning. This may explain why the loss of nephrons in normal aging can often occur without major effects.

Relatively little is known about the exact causes of aging, though researchers currently are pursuing several promising hypotheses.

Ultimately, however, aging is a steady decline in the finely tuned ebb and flow of substances and chemical reactions that maintain homeostasis.

SUMMARY

1. Human development proceeds through six stages:

a. Gamete formation, during which the egg and sperm mature within the reproductive organs of the parents.

b. Fertilization, which begins when a sperm penetrates an egg and is completed when the sperm and egg nuclei fuse.

c. Cleavage, when the fertilized egg (zygote) undergoes cell divisions that form the early multicelled embryo (the blastocyst). The destiny of cell lineages is established in part by the sector of cytoplasm inherited at this time.

d. Gastrulation, when the organizational framework of the whole body is laid out. Endoderm, ectoderm, and mesoderm form; all the tissues of the adult body will develop from these germ layers.

e. Organ formation (organogenesis), when the different organs start developing by a tightly orchestrated program of cell differentiation and morphogenesis. This is a critical time of early development.

f. Growth and tissue specialization, when organs enlarge overall and acquire their specialized chemical and physical properties. The maturation of tissues and organs continues into the fetal stage and beyond.

2. Cell differentiation and morphogenesis are based on cell interactions that begin during organogenesis. Both are guided by genes. During differentiation, cells come to have specific structures and functions; morphogenesis produces the shape and structure of particular body regions.

3. Embryonic development depends on the formation of four extraembryonic organs:

a. Yolk sac: contributes to the embryo's digestive tube and helps form blood cells and germ cells.

b. Allantois: its blood vessels become umbilical arteries and veins and blood vessels of the placenta; they function in oxygen transport and waste excretion.

c. Amnion: a fluid-filled sac that surrounds and protects the embryo from mechanical shocks and keeps it from drying out.

d. Chorion: a protective membrane around the embryo and the other membranes; a primary component of the placenta.

4. The embryo and the mother exchange nutrients, gases, and wastes by way of the placenta (a spongy organ of endometrium and extraembryonic membranes).

5. The placental barrier provides some protection for the embryo/fetus, but may allow transfer of harmful agents (e.g., various drugs, alcohol, and chemicals from tobacco smoke) from mother to fetus.

6. Estrogen and progesterone stimulate growth of the mammary glands. At delivery, contractions of the uterus dilate the cervix and expel the fetus and afterbirth. After delivery, nursing causes the secretion of hormones that stimulate milk production and release.

7. As with all complex animals, the human body gradually shows changes in structure and a decline in efficiency (aging). Although aging is part of the life cycle of all animals having extensively specialized cell types, its precise cause is unknown.

Review Questions

1. Define and describe the main features of the following developmental stages: fertilization, cleavage, gastrulation, and organogenesis. *306–309*

2. Label the following stages of early development: *307*

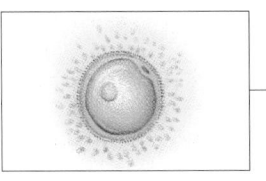

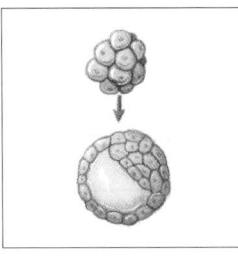

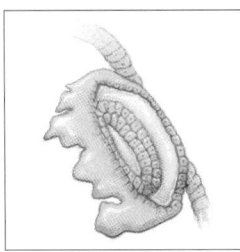

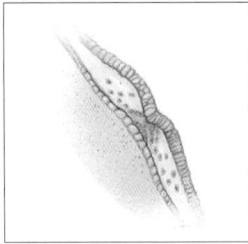

3. Define cell differentiation and morphogenesis, two processes that are critical for development. Which two mechanisms serve as the basic foundation for cell differentiation and morphogenesis? *306*

4. Summarize the development of an embryo and fetus. When are body parts such as the heart, nervous system, and skeleton largely formed? *312–317*

Critical Thinking: You Decide *(Key in Appendix V)*

1. How accurate is the statement "A pregnant woman must do everything for two"? Give some specifics to support your answer.

2. The renowned developmental biologist Lewis Wolpert once observed that birth, death, and marriage are not the most important events in human life—rather, gastrulation is. In what sense was he correct?

Self-Quiz *(Answers in Appendix IV)*

1. Development cannot proceed properly unless each stage is successfully completed before the next begins, starting with
_____.
  a. gamete formation   d. gastrulation
  b. fertilization   e. organ formation
  c. cleavage   f. growth, tissue specialization

2. During cleavage, the _____ becomes converted to a ball of cells, which in turn is transformed into the _____.
  a. zygote; blastocyst   c. ovum; embryonic disk
  b. trophoblast; embryonic disk   d. blastocyst; embryonic disk

3. In the week following implantation, cells of the _____ will give rise to the embryo.
  a. blastocyst   c. embryonic disk
  b. trophoblast   d. zygote

4. The developmental process called _____ produces the shape and structure of particular body regions.

5. _____ is the gene-guided process by which cells in different locations in the embryo become specialized.
  a. Implantation   c. Cell differentiation
  b. Neurulation   d. Morphogenesis

6. In a human zygote, the cell divisions of cleavage produce an embryonic stage known generally as a _____.
  a. zona pellucida   c. blastocyst
  b. gastrula   d. larva

7. Match each developmental stage with its description.
  ____ cleavage   ____ cell differentiation
  ____ gamete formation   ____ gastrulation
  ____ organ formation   ____ fertilization
  a. egg and sperm mature in parents
  b. sperm, egg nuclei fuse
  c. formation of germ layers
  d. zygote becomes a ball of cells called a morula
  e. cells come to have specific structures and functions
  f. starts when germ layers split into subpopulations of cells

8. Parts of the _____, an extraembryonic membrane, give rise to the embryo's digestive tube.
  a. yolk sac   c. amnion
  b. allantois   d. chorion

9. The _____, a fluid-filled sac, surrounds and protects the embryo from mechanical shocks and keeps it from drying out.
  a. yolk sac   c. amnion
  b. allantois   d. chorion

10. Blood vessels of the _____, an extraembryonic membrane, transport oxygen and nutrients to the embryo.
  a. yolk sac   c. amnion
  b. allantois   d. chorion

11. Substances are exchanged between the embryo and mother through the _____, which is composed of maternal and embryonic tissues.
  a. yolk sac   d. amnion
  b. allantois   e. chorion
  c. placenta

Selected Key Terms

aging *323*
allantois *310*
amnion *310*
blastocyst *309*
cell differentiation *306*
chorion *310*
cleavage *306*
ectoderm *306*
embryonic disk *310*
endoderm *306*
extraembryonic membranes *310*
fertilization *306*
fetus *314*
gamete formation *306*
gastrulation *306*
germ layer *306*
implantation *309*
inner cell mass *309*
lactation *319*
mesoderm *306*
morphogenesis *307*
organogenesis *306*
ovum *308*
parturition *318*
placenta *310*
trophoblast *309*
umbilical cord *310*
yolk sac *310*
zygote *306*

Readings

Austad, S. 1997. *Why We Age—What Science Is Discovering about the Body's Journey through Life.* New York: Wiley.

Caldwell, M. November 1992. "How Does a Single Cell Become a Whole Body?" *Discover.*

McGinnis, W., and M. Kuziora. February 1994. "The Molecular Architects of Body Design." *Scientific American.*

Nathanielz, P. November-December, 1996. "The Timing of Birth." *American Scientist.*

Nusslein-Volhard, C. August 1996. "Gradients That Organize Embryonic Development." *Scientific American.*

Sherwood, L. 1997. *Human Physiology.* 3d ed. Belmont, Calif.: Wadsworth.

 For additional readings, go to InfoTrac College Edition, your online library at:
http://www.infotrac-college.com/wadsworth

15

SEXUALLY TRANSMITTED DISEASES

Safer Sex

Every year about 14 million people in the United States, a quarter of them teenagers, confront a harsh reality of life: They discover that they have contracted a **sexually transmitted disease** (STD).

Reported cases of sexually transmitted diseases have reached epidemic proportions, even in countries with high medical standards. The number of unreported cases may be up to 10 times higher. In many cases, the pathogen responsible is a bacterium or virus. Most often, the pathogens are transmitted from infected to uninfected people during sexual contacts. Some STDs result from infection by a **parasite**—an organism that must live on or within a host. At the host's expense, the parasite gains nutrients or some other substance that is essential to its survival and its ability to reproduce.

STDs are more than socially embarrassing. Untreated, various ones can lead to terrible pain, organ damage, prolonged illness, sterility, babies with birth defects, even death. This health problem's economics also are sobering, with yearly treatment costs in the billions of dollars—*not* including the cost of treating AIDS.

AIDS, caused by the human immunodeficiency virus (HIV), has become a leading cause of death among Americans ages 25–44, and its incidence is steadily increasing among certain groups. In many developing countries, this modern plague threatens to overwhelm health care delivery systems and to unravel hard-won economic progress. In the United States, the rate of new HIV infections is greatest among women and teenagers.

Reading the grim statistics—including 130,000 reported new cases each year of syphilis, 450,000 of gonorrhea, and half a million new infections by the virus that causes genital herpes—you might conclude that there is no such thing as "safe sex." However, it's more accurate to say that sexual encounters (Figure 15.1a) can sometimes be a source of infection by viruses, bacteria, fungi, and other disease agents (Figure 15.1b). That is one reason why we include this chapter in a text on human biology. Another is that "safer sex" *is* possible. By learning more about the causes and symptoms of sexually transmitted diseases, you will be equipped not only to adopt behaviors that can reduce your personal risk, but perhaps to help educate others as well.

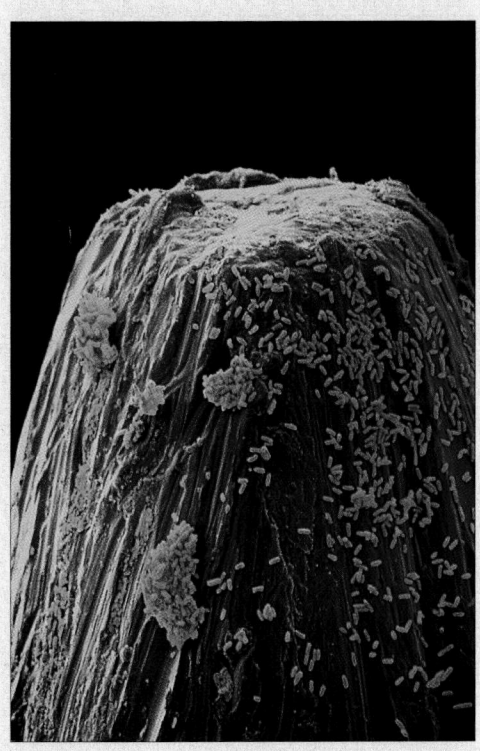

b

Figure 15.1 There is goodness in togetherness (**a**), but for sexually active people there is also a very real threat of sexually transmitted diseases. As with other types of diseases, an ounce of prevention is worth a pound of cure—when a cure exists. (**b**) How small are bacteria? Shown here are *Bacillus* bacteria peppering the tip of a pin.

KEY CONCEPTS

1. Most sexually transmitted diseases are caused by viruses and bacteria. They are usually transmitted from infected to uninfected people during sexual intercourse.

2. Complications of untreated STDs can result in sterility and birth defects, irreversible damage to organs, and death. There are no cures for genital herpes, HIV infections, and others.

3. The chances of contracting STDs can be reduced by adopting behaviors that help prevent or significantly reduce the likelihood of exposure to disease agents.

CHAPTER AT A GLANCE

15.1 Viruses and Bacteria: The Unseen Multitudes

15.2 How Viruses and Bacteria Cause Disease

15.3 HIV and AIDS: A Global Perspective

15.4 A Closer Look at HIV

15.5 HIV Infection

15.6 A Trio of Common STDs

15.7 Other Prevalent STDs

15.8 STDs Caused by Parasites or Fungi

15.9 *Protecting Yourself—and Others—from STDs*

VIRUSES AND BACTERIA: THE UNSEEN MULTITUDES

A **virus** is a noncellular infectious agent. The viral life cycle has two distinct components. One component is the virus particle, which consists of only a nucleic acid core and a protein coat that sometimes is enclosed in a lipid envelope (Figure 15.2). The other component is metabolic. Virus metabolism proceeds only after the virus particle enters a host cell and releases its genetic material, which directs the cellular machinery to synthesize the materials that are necessary to produce new viral particles. The genetic material of a given virus is either DNA or RNA. The viral protein coat, or *capsid*, comes in different forms.

After a virus particle enters a cell, it breaks apart, releasing its DNA or RNA. The genetic material is copied repeatedly, viral proteins are synthesized, and new virus particles are assembled inside a suitable host cell (Figure 15.3). A cell is a potential *host* if a virus can chemically recognize and lock onto certain molecular groups at the cell surface. As a result, each type of virus has specificity. Differences in surface receptors of different cell types are why an influenza virus can attack cells of respiratory epithelium but not liver cells, and a hepatitis virus attacks liver cells but can't infect cells of the airways.

In order for a virus to multiply, it first must recognize and attach to specific molecules at the surface of a host cell. Next, the virus must enter the cell, releasing its genetic material (DNA or RNA) into the cell's cytoplasm or nucleus. Now, information in the viral DNA or RNA directs the host cell to copy viral nucleic acids and synthesize viral enzymes, capsid proteins, and other proteins. Then the viral nucleic acids, capsid

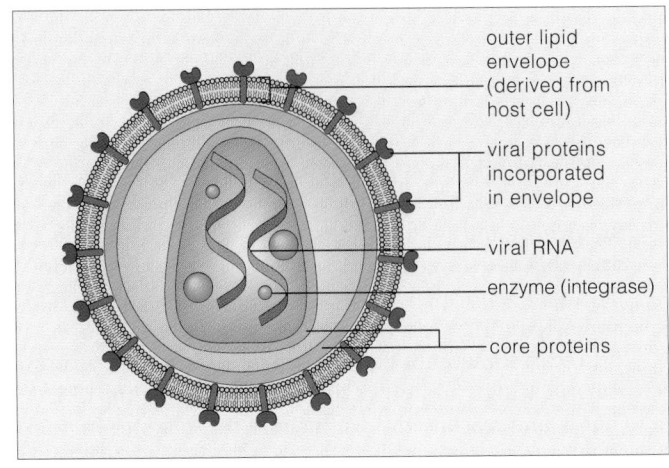

Figure 15.2 Lipid envelope around the capsid of the human immunodeficiency virus (HIV).

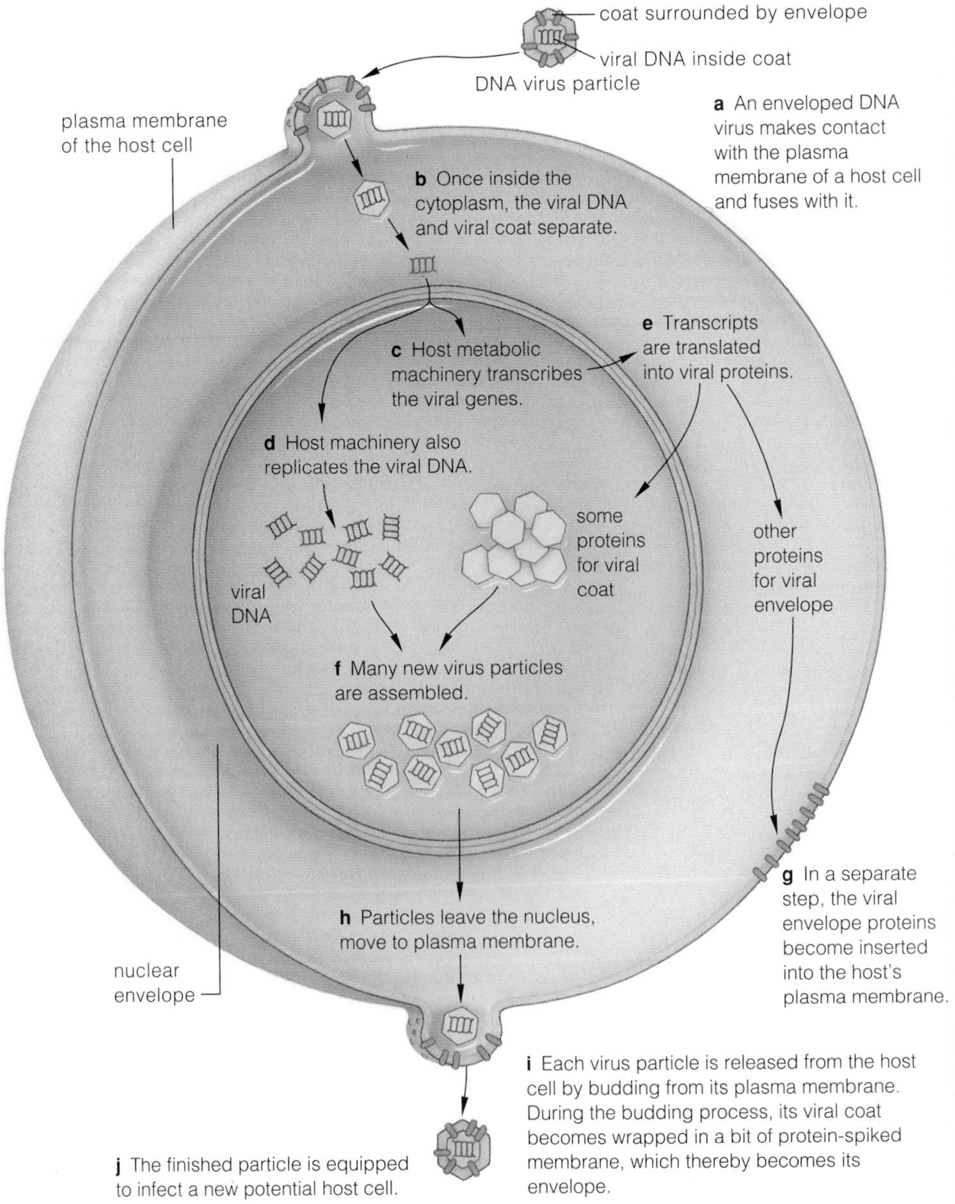

Figure 15.3 Multiplication cycle of one type of animal virus. This diagram shows an enveloped DNA virus infecting an animal cell. When the viral envelope fuses with the plasma membrane of the host cell, its genetic material is dumped automatically into the cytoplasm. The cell's metabolic machinery is subverted into churning out viral DNA and proteins. Some of the viral proteins become embedded in the internal membranes of the host cell; others are used for the assembly of new virus particles.

Before they bud from the cell, the new particles acquire an envelope, spiked with suitable viral proteins, by becoming wrapped in a bit of the cell's nuclear envelope or plasma membrane.

coat surrounded by envelope
viral DNA inside coat
DNA virus particle

a An enveloped DNA virus makes contact with the plasma membrane of a host cell and fuses with it.

plasma membrane of the host cell

b Once inside the cytoplasm, the viral DNA and viral coat separate.

c Host metabolic machinery transcribes the viral genes.

e Transcripts are translated into viral proteins.

d Host machinery also replicates the viral DNA.

viral DNA

some proteins for viral coat

other proteins for viral envelope

f Many new virus particles are assembled.

g In a separate step, the viral envelope proteins become inserted into the host's plasma membrane.

h Particles leave the nucleus, move to plasma membrane.

nuclear envelope

i Each virus particle is released from the host cell by budding from its plasma membrane. During the budding process, its viral coat becomes wrapped in a bit of protein-spiked membrane, which thereby becomes its envelope.

j The finished particle is equipped to infect a new potential host cell.

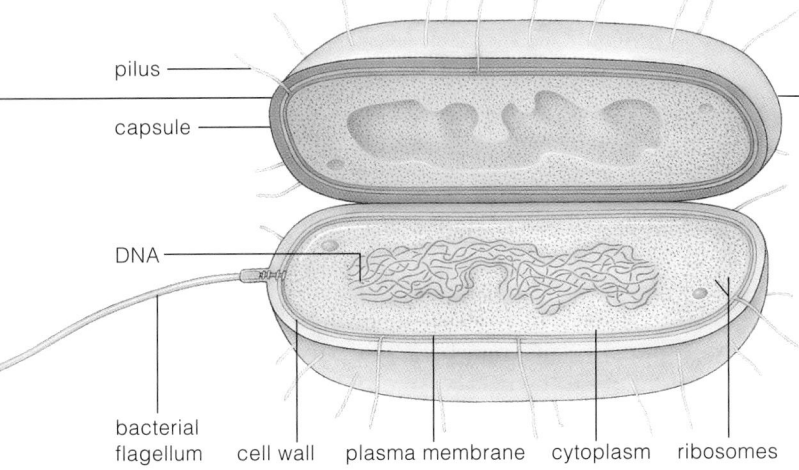

Figure 15.4 Generalized cell structure of a bacterium.

pilus

capsule

DNA

bacterial flagellum　cell wall　plasma membrane　cytoplasm　ribosomes

cross-linked by peptides. Peptidoglycan mesh makes the wall strong and semirigid, and helps maintain the bacterium in one of three common shapes: a ball-shaped coccus, a rodlike bacillus, or a helical spirillum of the type called a spirochete:

coccus　　bacillus　　　　spirillum

proteins, and sometimes enzymes are assembled into new virus particles. Finally, the newly constructed virus particles are released from the infected cell, usually after the cell's death and lysis.

Sometimes the virus does not kill the host cell outright. Instead, the virus enters a period of **latency**, in which viral genes remain inactive inside the host cell.

Retroviruses are RNA viruses that infect animal cells and can become latent. After the viral RNA chromosome enters the cytoplasm, a viral enzyme called reverse transcriptase uses the RNA as a template and synthesizes a DNA molecule. The molecule, called a provirus, integrates into the host's DNA. This is what happens when a person becomes infected with HIV.

Recall from Chapter 2 that bacteria are *prokaryotes*—unlike eukaryotic cells, such as your own, they have no nucleus or other membrane-bound organelles. Their metabolic reactions take place at the plasma membrane.

Most bacteria have a *cell wall* (Figure 15.4). In many types of bacteria the wall is a mesh of *peptidoglycan*—a molecule in which complex polysaccharide strands are

Some kinds of bacteria have threadlike structures anchored to the cell wall and plasma membrane. One type is a *bacterial flagellum*, a stiff protein filament that rotates like a propeller, moving the cell through its fluid surroundings. Some kinds of bacteria have filaments called *pili* that help the cell stick to surfaces. The bacterium that causes gonorrhea uses short pili to attach to mucous membranes.

Bacteria reproduce by *prokaryotic fission* (Figure 15.5). In this kind of cell division, a cell's DNA is copied, then the cell divides into two genetically identical daughter cells. Bacterial cells have only one circular DNA molecule (a chromosome) to copy and parcel out to daughter cells. Prokaryotic fission often is rapid; given optimal circumstances, some bacteria can divide about every 20 minutes.

A virus is a noncellular infectious particle. Viruses multiply by way of a sequence of events in which the host cell's metabolic machinery is usurped and used to make new virus particles.

Bacteria are single-celled prokaryotes. They multiply by way of a cell division mechanism called prokaryotic fission.

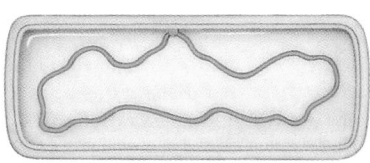

a Bacterium (cutaway view) before its DNA is copied.

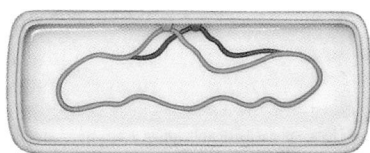

b Replication begins and proceeds in two directions, away from some point on the DNA molecule.

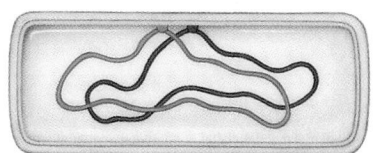

c The DNA copy is attached at a site close to the attachment site of the parent DNA molecule.

d Membrane growth occurs between the two attachment sites and moves the two DNA molecules apart.

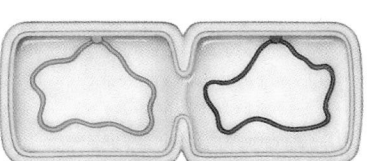

e New membrane and wall material start growing through the cell midsection.

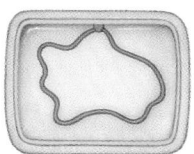

f Membrane and wall material deposited at the cell midsection divide the cytoplasm in two.

Figure 15.5 Bacterial reproduction by prokaryotic fission, a cell division mechanism.

Several Routes to Disease

Numerous types of animal viruses are responsible for diseases as varied as warts, genital herpes, the common cold and influenza, and AIDS as well. Table 15.1 lists a few that are associated with STDs.

Pathogenic viruses cause disease in several ways. A few viruses trigger cancer when they interact with the DNA in a host cell's chromosomes. More commonly, disease symptoms develop when infected cells begin to malfunction or die.

Paradoxically, the immune system's responses to viral infection can also lead to symptoms we associate with disease. Fever and fatigue are just two examples. Another is inflammation. While a controlled inflammatory response helps rid the body of pathogens (see Section 7.3), an inflammatory response gone awry can cause serious damage, especially if a major organ or organ system is affected. For instance, in an example relevant to this chapter's subject matter, inflammation of a male's testes can lead to sterility. This is because, for several reasons, sperm cells normally are shielded from a male's immune system. During testicular inflammation, however, sperm antigens can escape into the bloodstream. Then, the immune system "sees" the sperm antigens as foreign, and proceeds to mount a fertility-destroying immune response against the testes.

As discussed in Section 7.12, antibiotics are useless against viruses. Such drugs act by disrupting living *cells*—they cannot "kill" a virus particle, which has no metabolic pathways to disrupt.

Most disease-causing bacteria secrete toxins that damage cells. If invading bacteria multiply more rapidly than the immune system can respond to them, large quantities of a toxin may enter the bloodstream or other tissues. Bacteria that release toxins into the bloodstream are especially dangerous because the toxin moves rapidly throughout the body. The spreading poison may affect the nervous system and other vital organ systems.

Another Kind of Infectious Particle

Some infectious particles are even smaller—more stripped down, if you will—than bacteria or viruses. In this group are **prions** (pronounced "pree-ons"), which are small, infectious proteins linked to several rare, fatal degenerative diseases of the nervous system.

An example of a prion-linked disease is *Creutzfeldt-Jakob disease* (CJD), in which the victim loses vision and speech, develops spastic paralysis, and experiences rapid mental deterioration. CJD is always fatal. To date, prions have not been associated with any sexually transmitted disease, but they are worth mentioning here because, as HIV has proved, we do well to be prepared, scientifically and personally, for emerging infectious diseases that may be transmitted by agents presently unknown to us.

With the general features of viruses and bacteria as background, we turn now to a survey of the major sexually transmitted diseases. We will focus on a rogue's gallery of the most common STDs, most of which are caused by viruses or bacteria. We begin with the most troubling STD of all, AIDS.

Pathogenic viruses cause disease by killing infected cells or altering their functioning in harmful ways.

Disease-causing bacteria secrete toxins that can enter tissues and damage vital organs.

Prions are infectious particles even smaller than bacteria or viruses. To date, no sexually transmitted disease has been linked to them, but they are implicated in several other degenerative diseases.

Table 15.1	**Some Animal Viruses That Cause Sexually Transmitted Diseases**
Agents	STD
DNA VIRUSES	
Herpesviruses	
Herpes simplex types 1 and 2	Cold sores on genitals, lips, eyes, buttocks
Hepadnaviruses	Hepatitis B
Papovaviruses human papillomavirus	Genital warts
RNA VIRUSES	
Retroviruses HIV	AIDS

AIDS—An International Epidemic

AIDS is a constellation of disorders that follow infection by the **human immunodeficiency virus**, HIV. The World Health Organization (WHO) has estimated that well over 7 million people, mostly adults and teenagers, have died from AIDS. More than 30 million have become infected with HIV since the global epidemic started in 1984. By the turn of the century, the number of infected people worldwide may be 40 to 50 million.

Figure 15.6 gives a recent estimate of the numbers of adults infected with HIV in various parts of the world. The rate of new infections is growing especially rapidly in parts of South and Southeast Asia. In hard-hit African cities, officials estimate that at least one-third of the adult population is infected. These infected adults fill vital roles in society, including those of parents, wage earners, and providers of services. From what we know so far, at least 95 percent will eventually become seriously ill with AIDS and die—with horrible human, social, and economic costs. By 1998, AIDS had claimed roughly 450,000 lives in the United States.

AIDS-type illnesses were first recorded in various countries in the 1970s and early 1980s. HIV itself was finally identified in 1984. Worldwide, most AIDS patients are infected with HIV-1, although HIV-2 is more prevalent in parts of Africa and is also becoming common in India.

Within the two broad HIV classes (1 and 2), there are many genetic strains of the virus. Researchers at the Centers for Disease Control and Prevention in Atlanta have found that the different strains seem to fall into two distinct groups. Some seem best adapted to infecting cells of the immune system, whereas others may be more effective at entering cells in the mucous membranes lining reproductive structures such as the vagina. This second group is most common in Africa and parts of Asia, which could help explain why heterosexuals make up the majority of AIDS cases in those places. At present, in the United States and Europe most AIDS cases are caused by the first type.

AIDS is a constellation of disorders caused by infection with the human immunodeficiency virus (HIV).

HIV infects many millions of people throughout the world. This global epidemic shows little sign of abating; several million new infections are reported each year.

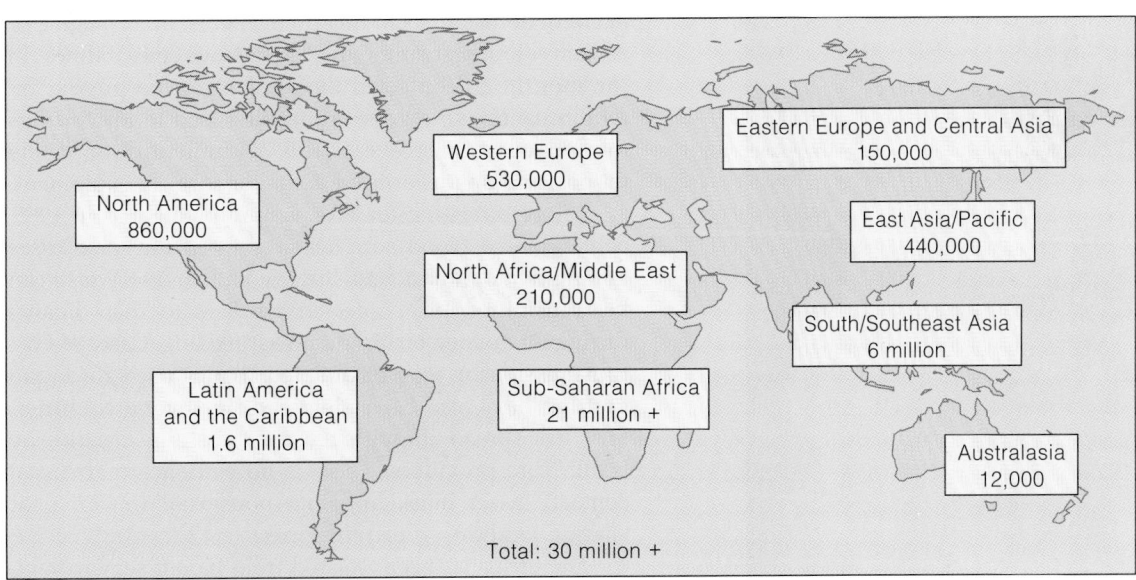

Figure 15.6 Estimates of the world distribution of cumulative HIV infections in adults, as of 1997, according to the World Health Organization. The rate of new infections is increasing most rapidly in Latin America and South/Southeast Asia (including India).

A CLOSER LOOK AT HIV

Figure 15.7 shows HIV particles on the surface of a helper T cell. The virus destroys T cells. In so doing, it cripples the immune system and leaves the body susceptible to infections and some rare forms of cancer. Currently there is no vaccine against HIV, and there is no cure for people already infected. After two decades of intensive research, there is much about HIV and AIDS that we still do not understand. However, as you'll read shortly, researchers are making tremendous progress in explaining such basic questions as how HIV infects cells and how it breaks down the human immune response.

Physicians diagnose AIDS if a patient has a severely depressed immune system, tests positive for HIV, and has one or more of twenty-six "indicator diseases," including types of pneumonia, cancer, recurrent yeast infections, and drug-resistant tuberculosis.

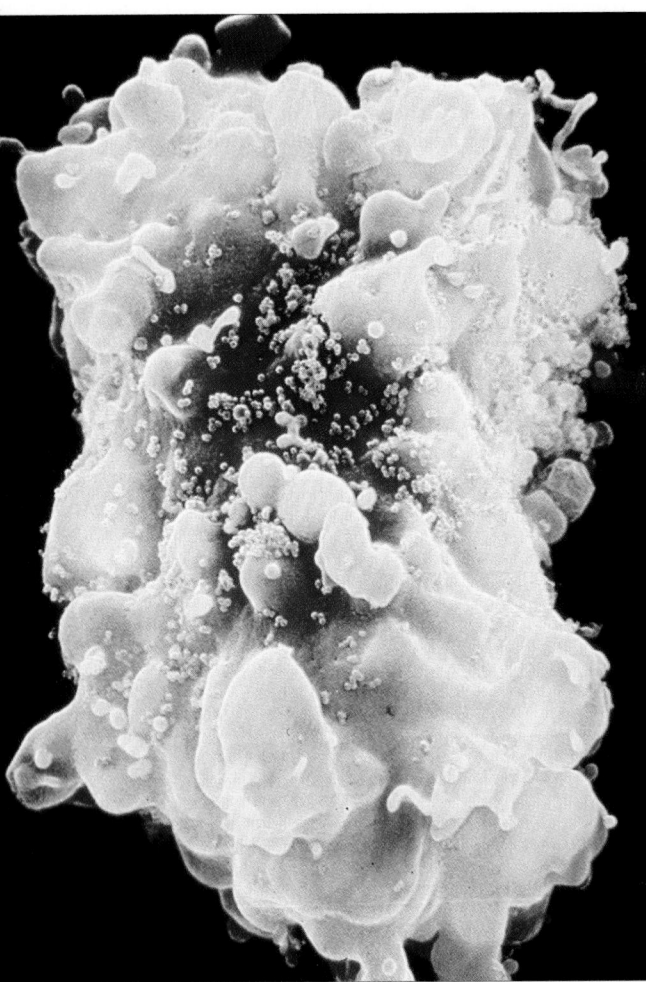

Figure 15.7 In this color-enhanced micrograph, the blue "balls" are spherical HIV particles spread over the surface of a helper T cell. The image is highly magnified. HIV particles are only one ten-thousandth of a millimeter wide—ten thousand of them could line up across the period at the end of this sentence.

HIV Structure and Replication

HIV is a retrovirus—it has genetic material of ribonucleic acid, RNA, rather than of deoxyribonucleic acid, DNA. A protein core surrounds the RNA and several copies of an enzyme called reverse transcriptase. The core itself is wrapped in a lipid envelope derived from the plasma membrane of a host helper T cell. Once inside a host cell, the enzyme uses the viral RNA as a template for making DNA. This DNA, a provirus, then becomes integrated into a host chromosome, the structure into which the host's genetic material is packaged (Chapter 2).

Eventually the provirus's genetic instructions for making new viral particles are read out. A process called *transcription* rewrites the genetic message in DNA as RNA, and these RNA instructions then are "translated" into protein (Chapter 19). Figure 15.8 summarizes how this process generates new HIV particles in an infected cell. We will soon return to this topic to consider efforts to develop drugs that can chemically interfere with HIV replication—and so prevent HIV from reproducing once it enters a host.

How HIV Is Transmitted

HIV is transmitted when bodily fluids, especially blood and semen, of an infected person enter another person's tissues. The virus can enter the body through *any* kind of cut or abrasion, including on the penis, in the vagina or rectum or, less commonly, in mucous membranes in the mouth. HIV-infected blood also can be present on toothbrushes and razors; on needles used to inject drugs intravenously, pierce ears, do acupuncture, or create tattoos; and on unsterilized dental or surgical equipment.

Before screening for HIV was implemented in 1985, contaminated blood supplies accounted for some cases among hemophiliacs and surgery patients. HIV also can be transmitted from infected mothers to their infants during pregnancy, birth, and breast-feeding. Cases of HIV transmission by way of donated tissues used for organ grafting have also been identified. In several countries, HIV has spread through reuse of unsterile needles by health care providers. There is no evidence that casual contact, insect bites, hugging, nonsexual touching, or sharing food can spread the virus.

The virus has been isolated from blood, semen, vaginal secretions, saliva, tears, breast milk, amniotic fluid, cerebrospinal fluid, and even urine. However, apparently only infected blood, semen, vaginal secretions, and breast milk contain HIV in concentrations that are high enough for successful transmission. People who have other sexually transmitted diseases—such as syphilis, herpes, and chancroid (Sections 15.6 and 15.7)—are at greater risk of HIV infection because they may have sores or other

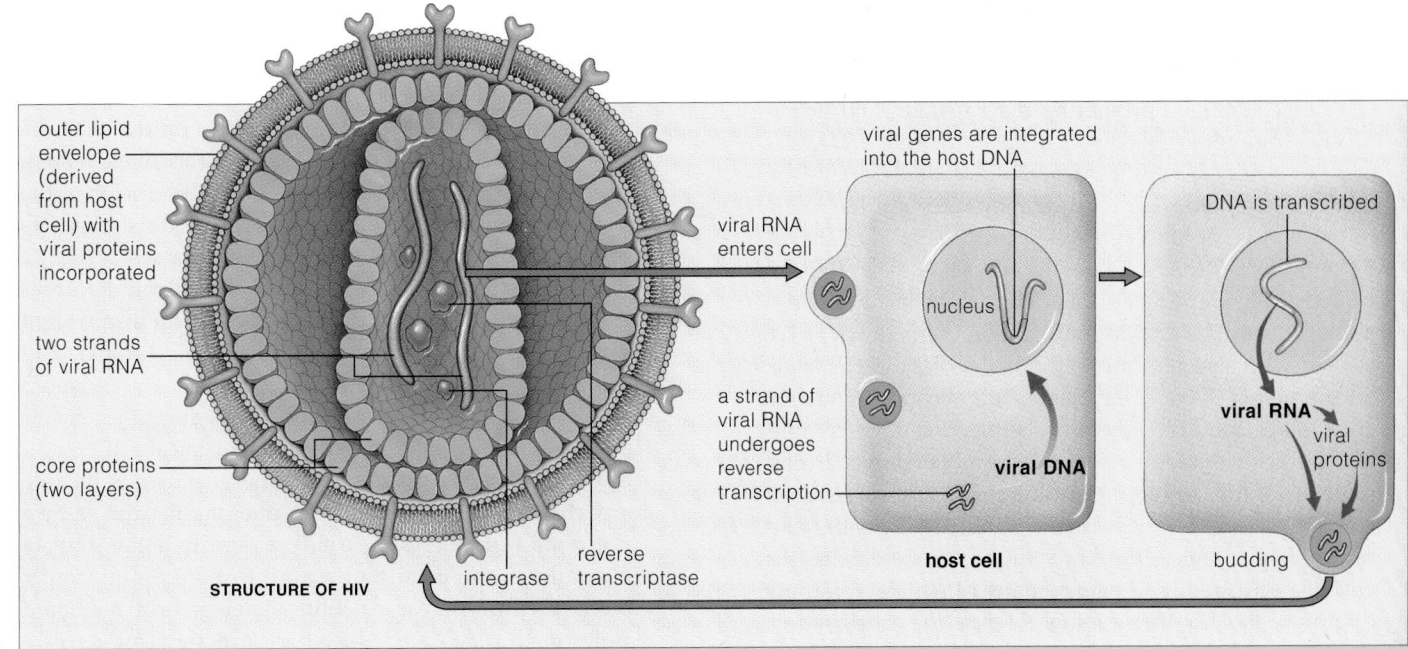

Figure 15.8 Infection cycle of HIV, a retrovirus. For clarity, the drawing shows only a single HIV budding from the host cell. In fact, many particles will be formed and will exit from each infected cell.

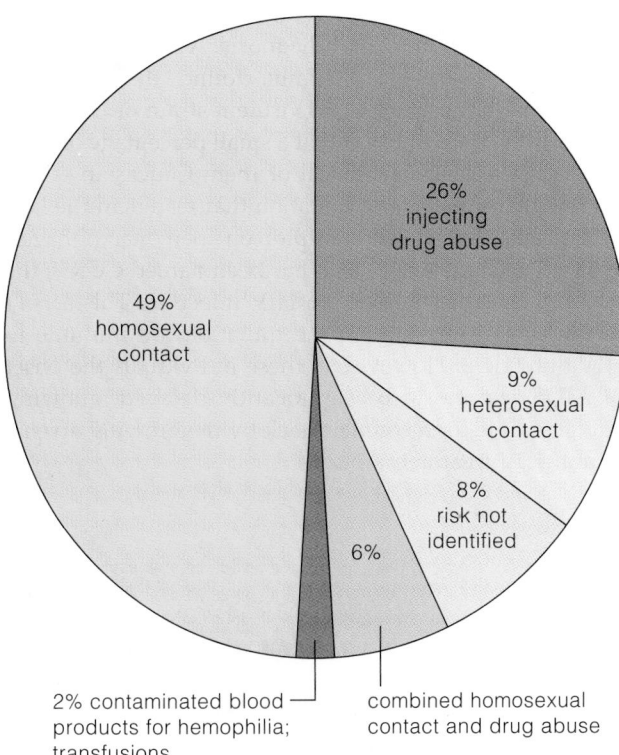

Figure 15.9 Reported AIDS cases in the United States, by mode of transmission, at the end of 1997. These percentages are approximate, and they only cover individuals over the age of 13. Roughly 6,000 young children have been diagnosed with AIDS. Of those, 87 percent were infected when they were exposed to the mother's HIV-infected blood during gestation.

lesions that afford the virus easy entry into the body. HIV generally cannot survive for more than about one or two hours outside an infected cell. Virus particles on needles and other objects are readily destroyed by disinfectants, including household bleach.

In the United States and other developed countries, HIV spread at first within three groups: male homosexuals, hemophiliacs and others who received HIV-tainted blood products, and intravenous drug abusers who shared needles that were contaminated with infected blood. As many as two-thirds of all IV drug abusers may now be infected with the virus. The incidence of HIV infection also is increasing among heterosexual males and females, including teenagers, in the general population (Figure 15.9). Many of those infections were (and are) due to intravenous drug abuse, but others are the result of heterosexual contact. In recent years, more young adults in the United States died from AIDS than from any other single cause. In 1997, women accounted for about 20 percent of new AIDS cases.

HIV is a retrovirus. When it infects a cell, its genetic material becomes integrated into the host cell's DNA, and eventually the host cell begins producing new HIV particles.

HIV is transmitted only when blood, semen, or certain other body fluids of an infected person enter another person's tissues.

HIV INFECTION

The Initial Attack by HIV

HIV infects antigen-presenting macrophages and helper T cells. Immunologists also call helper T cells *CD4 cells,* because helper T cells bear a receptor dubbed CD4 on their surface. And it turns out that this receptor has a central role in what has been called the "secret handshake" by which HIV infects helper T cells. First, a viral glycoprotein called gp120 docks with a CD4 receptor on a helper T cell. The docking interaction then allows gp120 to bind with a second protein, called a *chemokine receptor,* on the helper T cell. Only when this second interaction takes place can HIV enter the helper T cell. Over time, HIV infection depletes a person's supply of helper T cells, until the ability to mount immune responses is severely compromised. The assault on helper T cells and macrophages makes the body highly vulnerable to the entire spectrum of pathogens, many of which would not otherwise be life-threatening. Hence the designation, "opportunistic" infections.

From HIV Infection to AIDS

Infection marks the beginning of a titanic battle between HIV and the host's immune system. Soon after a person is infected, HIV begins to replicate and circulate in the bloodstream. At this stage, many people have a bout of flulike symptoms. In response to HIV antigenic proteins, B cells synthesize antibodies, which are the basis of diagnostic tests for identifying HIV infection. Armies of helper T cells and cytotoxic T cells also form. However, HIV is a formidable opponent. During certain phases of the infection the virus infects an estimated 2 billion helper T cells and produces 100 million to 1 billion new HIV particles each day. They bud from the plasma membrane of the host helper T cell or are released when the membrane ruptures (Figure 15.10).

Every two days the immune system destroys about half of the helper T cells lost in the battle. Huge reservoirs of HIV and masses of infected T cells accumulate in lymph nodes. As the battle proceeds, the number of virus particles in the general circulation rises. Gradually, the balance shifts in favor of HIV. The body produces fewer and fewer helper T cells to replace the ones it lost. Although it may take a decade or more, the erosion of the helper T count inevitably causes the body to lose its capacity to mount effective immune responses.

As the population of functional helper T cells becomes seriously depleted, the person may begin to lose weight and experience joint and muscle pain, fatigue and malaise, nausea, bed-drenching night sweats, enlarged lymph nodes, various minor infections, and other symptoms. Eventually, one or more of the typical AIDS indicator diseases begin to appear. These maladies include opportunistic infections, such as pneumonia caused by a protozoan (*Pneumocystis carinii*); a rapidly lethal form of tuberculosis; and persistent yeast infections of the mouth, throat, rectum, or vagina. Blue-violet or brown-colored spots on the legs and elsewhere are signs of *Kaposi's sarcoma.* This deadly form of cancer strikes males only. It affects blood vessels in the skin and some internal organs and is extremely rare in people who are not HIV-infected.

About 5 percent of HIV-infected people are "nonprogressors" in whom no symptoms of illness develop even 10 years or more after diagnosis. A study of a group of nonprogressors in Australia found that all were infected with a mutated, less virulent strain of HIV. Also intriguing is the discovery that a small percentage of HIV-infected people—virtually all of them Caucasian—were born with a rare genetic mutation that results in the lack of a specific chemokine receptor on their macrophages. Without the receptor (which has been named CCR5), the macrophages are resistant to HIV infection. This *doesn't* mean that people having the mutation are immune to HIV and AIDS. However, in these individuals the onset of AIDS appears to be significantly delayed, allowing valuable time for a counterattack by the growing arsenal of anti-HIV treatment options.

Figure 15.10 An HIV virus particle budding from the host cell's plasma membrane.

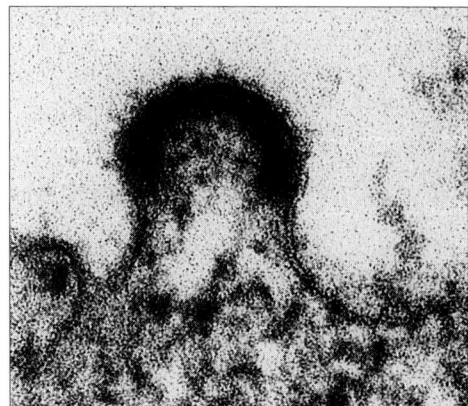
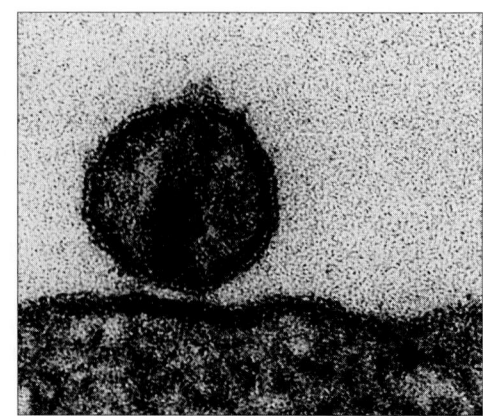

Possibilities for Treatment

Realizing that both HIV infection and replication are multistep processes, researchers have been seeking ways to disrupt one or more vital steps. For example, it might be possible to chemically block the action of HIV protease, an enzyme that has a key role in the assembly of new virus particles. HIV genes also include instructions for proteins that are synthesized early in replication and help regulate subsequent replication steps. The development of protease inhibitors that can knock the proteins out and so halt reproduction of the virus has helped advance the clinical treatments of HIV-positive people, as well as those who have progressed to full-blown AIDS. The search also is proceeding for compounds that might disrupt the ability of HIV to bind to the chemokine receptors by which it gains entry to cells.

Some AIDS patients receive "cocktails" consisting of a protease inhibitor and one or more of the drugs azidothymidine (AZT), ddI (dideoxyinosine), and ddC (zalcitabine or dideoxycytidine). These compounds block the action of reverse transcriptase. The drugs cannot cure infected people, however, because they cannot eliminate the HIV genes that have become incorporated into their DNA. AZT, ddI, and ddC also have problematic side effects.

Vaccines: Problems and Prospects

Modern medicine has developed vaccines to protect human populations from several viral diseases. Yet for a variety of reasons that we can only touch on here, it is much more difficult to develop a vaccine against HIV.

Recall from Chapter 7 that a vaccine confers long-term protection by stimulating the immune system to respond to viral antigens—proteins produced by the invading virus. However, a key problem in developing a vaccine against HIV is that HIV mutates rapidly and can exist in many different genetic forms—and antigenic forms—even in the same person. Vaccine production would have to keep pace with these antigenic variants.

With these and other obstacles in mind, researchers are designing experimental vaccines. Many efforts rely on genetic engineering, which we explore in Chapter 21. One approach is to base vaccines on HIV particles from which key genetic information has been removed. In theory, the altered virus would still be able to stimulate the immune system to produce a protective response but could not replicate or cause disease. Although this approach raises important safety questions, such vaccines may well become candidates for testing because of the rapid spread of HIV. Several dozen different vaccines of various types have already been tested on small groups of HIV-infected people to find out if vaccination can bolster their weakened immune systems. Larger-scale trials are being planned or implemented in countries around the globe.

Such trials are essential to the ultimate development of a safe, effective vaccine but entail many ethical problems. For example, if a vaccine does its job of stimulating anti-HIV antibodies, those who participate in a trial will thereafter test positive for the virus—and could become victims of discrimination. Trial participants carry an identification card that certifies their participation in vaccine studies.

AIDS Prevention

It is crucial that every sexually active person be aware of the danger of HIV infection and adopt behaviors that minimize the risk of exposure. These include limiting the number of sex partners and avoiding any exchange of body fluids if there is a chance either person may be infected. Proper use of high-quality latex condoms, together with the spermicide nonoxynol-9, helps prevent infection. Caressing carries no risk—if there are no body fluids exchanged or no lesions through which the virus can enter the body.

To stop the spread of HIV, any uninfected person who is sexually active and not in a mutually monogamous relationship with an uninfected partner must follow "safer sex" guidelines (see *Focus on Your Health,* page 345).

In addition, public health officials in hard-hit cities are promoting "clean needle" strategies, such as supplying IV-drug abusers with sterile needles. This is not support for drug abuse, but merely recognition that society has a huge stake in halting the spread of HIV. Free or low-cost confidential testing for AIDS is available through public health facilities and many physicians' offices. There are also over-the-counter blood tests for HIV infection, although they may not be as reliable as testing done by medical personnel. As a society we may benefit most by ensuring that people who test positive for the virus have access to nonjudgmental counseling and appropriate treatments.

In the United States and some European countries, the increase in reported new HIV infections has slowed. This trend suggests that people can change their behavior when their lives may hang in the balance. At the same time, in parts of Asia and Latin America, the incidence of new cases has begun to rise. Everywhere, a distressingly large percentage of younger people admit they do not practice safer sex, despite being educated about HIV and AIDS. However, no one is immune to infection. With the mounting medical, social, and economic consequences of AIDS, no one can afford to be complacent. Now and in the future, HIV and AIDS are everyone's problem.

HIV causes AIDS by breaking down helper T cells and other specific cells of the immune system. Because there is no cure and treatment possibilities are limited, prevention is the best defense against HIV infection.

A TRIO OF COMMON STDs

Gonorrhea

Unlike AIDS, **gonorrhea** can be cured by prompt diagnosis and treatment. Gonorrhea is caused by *Neisseria gonorrhoeae* (Figure 15.11). This bacterium (also called gonococcus) can infect epithelial cells of the genital tract, the rectum, eye membranes, and the throat. Each year in the United States there are about 475,000 new cases reported, and there may be up to 10 million unreported cases. Part of the problem is that the initial stages of the disease can be so uneventful that a carrier may be unaware of being infected.

In early stages of infection, males are more likely than females to notice that something is wrong because the symptoms are easier to detect. Within a week, yellow pus begins to ooze from the penis. Urinating is more frequent and may be painful.

In women, the early stages of gonorrhea can be dangerously asymptomatic. For example, a female may or may not experience a burning sensation while urinating. She may or may not have a vaginal discharge; even if there is a slight discharge, she may not perceive it as abnormal. In the absence of worrisome symptoms, a woman's gonorrhea infection may well go untreated—and all the while, the bacteria may spread into her oviducts. Eventually, the woman may experience violent cramps, fever, and vomiting—and even become sterile due to scarring and blocking of her oviducts from pelvic inflammatory disease (Section 15.7).

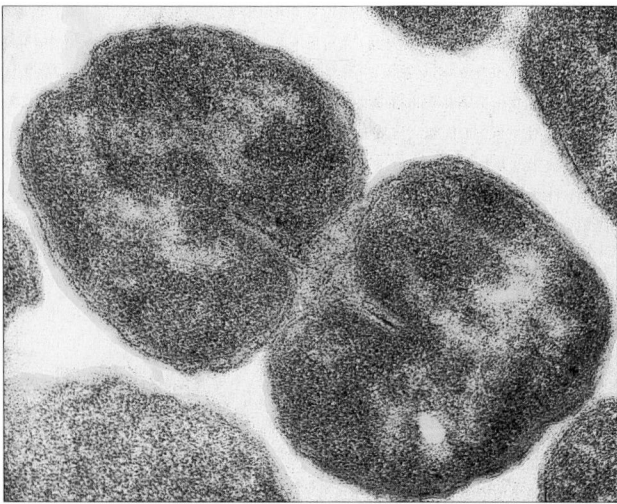

Figure 15.11 *Neisseria gonorrhoeae*, or gonococcus, a bacterium that typically is seen as paired cells, as shown here. Threadlike pili not evident in this electron micrograph help the bacterium attach to its host, upon which it bestows gonorrhea.

A man can become sterile when untreated gonorrhea leads to inflammation of the testicles or scarring of the vas deferens.

Antibiotics can kill the gonococcus and prevent complications of gonorrhea. Penicillin was once the most commonly used drug treatment. Unfortunately, antibiotic-resistant strains of gonococcus have developed in recent years. As a result, many physicians now order testing to determine the strain responsible for a particular patient's illness and then, treat the infection with an appropriate antibiotic.

Many people wrongly believe that once cured of gonorrhea, they are immune to reinfection. On the contrary, reinfections can and do occur. People who have multiple sexual partners can use condoms to help avoid becoming infected. Condom use can also prevent a person who has gonorrhea from infecting others, one of the issues we explore in this chapter's *Focus on Your Health* essay.

Syphilis

Syphilis is caused by a motile, spirochete bacterium, *Treponema pallidum*. Each year in the United States about 30,000 new cases are reported, although the actual incidence may be much higher. Since 1984, the incidence of syphilis has nearly doubled among females between ages 15 and 24. Infected women who become pregnant typically have miscarriages, stillbirths, or sickly and syphilitic infants.

The bacterium is transmitted by sexual contact. After it has penetrated exposed tissues, it produces a type of ulcer called a chancre (pronounced "shanker," Figure 15.12a) that teems with treponeme offspring. Usually the chancre is flat rather than bumpy, and it is not painful. It becomes visible between 1 and 8 weeks following infection and is a symptom of the *primary stage* of syphilis. Using a technique called immunofluorescence, treponemes (Figure 15.12d) can be identified—and syphilis diagnosed—in a cell sample taken from a chancre. By then, however, bacterial cells have already moved into the lymph vascular system and bloodstream.

The *secondary stage* of syphilis begins about 1 to 2 months following the appearance of the chancre. Lesions can develop in mucous membranes, eyes, bones, and the central nervous system. A blotchy rash breaks out over much of the body (Figure 15.12b). Afterward, the infection enters a latent stage, which can last many years. Meanwhile, the disease produces no significant outward symptoms and can be detected only by laboratory tests.

Usually, the *tertiary stage* of syphilis begins from 5 to 20 years after infection. Lesions may develop in the skin (Figure 15.12c) and internal organs, including the liver, bones,

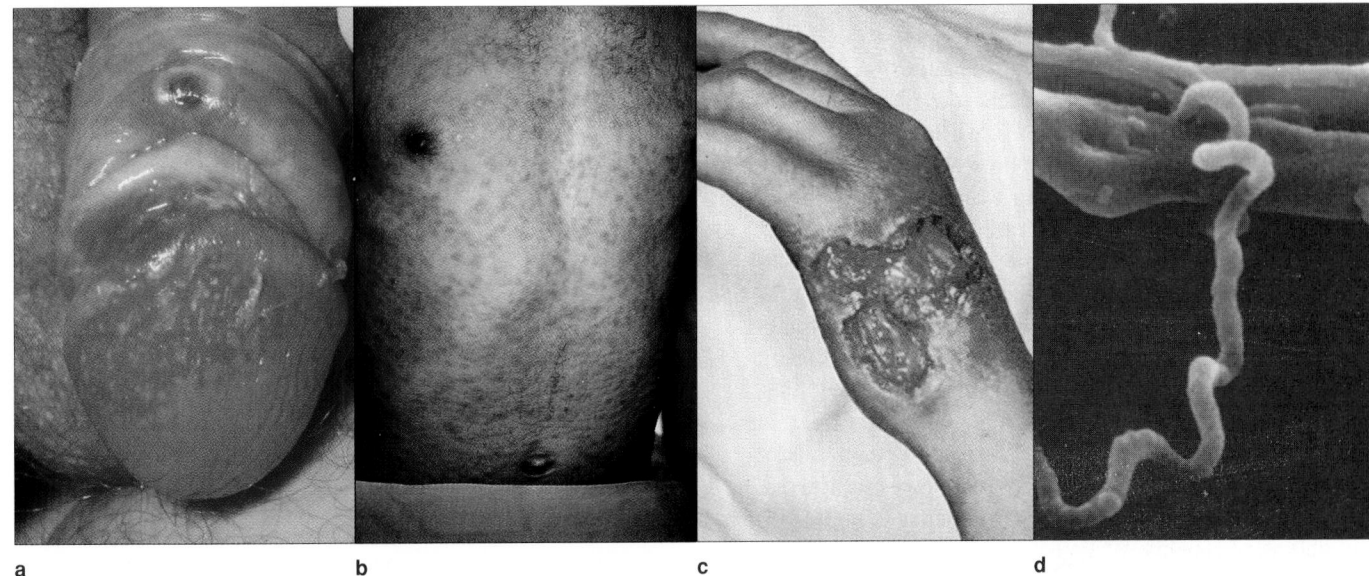

a   b   c   d

Figure 15.12 (**a**) Painless ulcer called a chancre ("shanker"), a sign of the first stage of syphilis. The sore appears anytime from about 9 days to 3 months after infection, on the genitals or near the anus or the mouth. It literally teems with infective bacteria and remains for 6 to 10 weeks if the infection is not treated. (**b**) Rash typical of secondary syphilis. (**c**) Arm tissue showing the white lesions, called *gummas*, that occur during the tertiary stage of syphilis. (**d**) *Treponema pallidum*, the spirochete bacterium that causes syphilis.

and aorta. Scars form; the walls of the aorta can weaken. Treponemes also damage the brain and spinal cord in ways that lead to various forms of insanity and paralysis.

Penicillin effectively cures syphilis during the early stages. Even so, potential health damage is so serious that no one who is sexually active should take the disease lightly.

Chlamydial Infections

The bacterium *Chlamydia trachomatis* is the culprit behind an infection commonly called **chlamydia** (Figure 15.13). Each year, a staggering 3 million to 10 million Americans are affected, a large proportion of them college students. More than 30 percent of newborns who are treated for eye infections and pneumonia developed those disorders after being infected with *C. trachomatis* during birth.

The bacterium infects cells of the genital and urinary tract. Infected men may have a discharge from the penis and a burning sensation when they urinate. Women may have a vaginal discharge as well as burning and itching sensations. Often, however, there is no outward evidence of infection—yet the bacterium can still be passed on to others.

Following infection, the bacteria migrate to lymph nodes, which become enlarged and tender. The enlargement can impair lymph drainage and lead to pronounced swelling of the surrounding tissues. Chlamydial infections can be treated with antibiotics such as tetracycline.

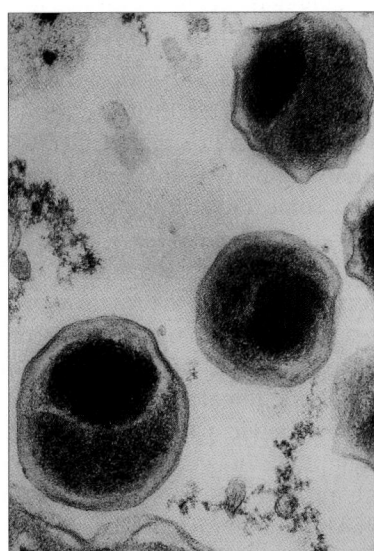

Figure 15.13 The large spheres in this color-enhanced micrograph are *Chlamydia trachomatis*, magnified about 32,000 times.

Gonorrhea, syphilis, and chlamydial infections are three of the most common STDs. All can be cured with antibiotics, but can have severe health consequences if they go untreated.

Pelvic Inflammatory Disease

A condition called **pelvic inflammatory disease** (PID) strikes about 1 million women each year, most often sexually active women under age 25. Millions more suffer the consequences of a past infection—PID is the leading cause of infertility among young women.

PID is one of the serious complications of gonorrhea, chlamydial infections, and other STDs. Physicians estimate that half or more of all cases are sexually transmitted. It also can arise when microorganisms that normally inhabit the vagina ascend into the pelvic region as a result of excessive douching. Most often, the uterus, oviducts, and ovaries are affected. Pain may be so severe that infected women often think they are having an attack of acute appendicitis. The oviducts may become scarred, which in turn can lead to ectopic pregnancy (Chapter 14) as well as to sterility. Affected women can also develop chronic menstrual problems.

As soon as PID is diagnosed, antibiotics such as tetracycline and penicillin can be administered. Severe cases require hospitalization and hysterectomy (removal of the uterus). A woman's partner should also be treated, even if the partner has no symptoms.

Nongonococcal Urethritis (NGU)

Males and females both can develop **nongonococcal** ("not caused by gonococcus") **urethritis**, an inflammation of the urethra that is usually simply called NGU. Doctors are now diagnosing more than a million cases each year. And although some medical experts lump NGU with chlamydia because a percentage of cases are caused by the *Chlamydia* bacterium, in other cases laboratory tests reveal the herpesvirus or fail to turn up any clear evidence of a pathogen. Symptoms in males are discomfort during urination and a puslike discharge from the urethra. As with chlamydia, females may have a slight vaginal discharge, some pelvic pain, and urinary discomfort (similar to symptoms of *cystitis*), or no symptoms at all.

Although an antibiotic such as tetracycline can usually cure NGU, the disorder can be difficult to treat in cases in which no cause can be established. Relapses are common and complications can arise, including inflammation of the prostate or testicles in males and inflammation of the oviducts in women.

Genital Herpes

Infections with herpes simplex viruses, or HSV, are extremely contagious. HSV is transmitted when any part of a person's body comes into direct contact with active viruses or sores that contain them (Figure 15.14). Mucous

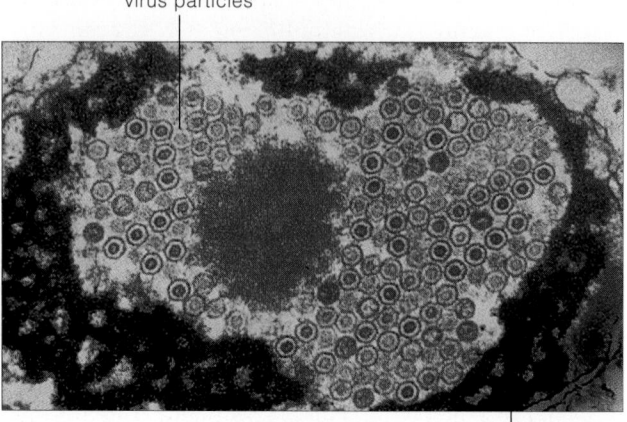

virus particles

cell's plasma membrane

Figure 15.14 Particles of herpesvirus in an infected cell. Herpes infections below the waist may involve either type 1 or type 2 strains, but most genital herpes infections are type 2.

membranes of the mouth or genital area are especially susceptible to invasion, as is broken or damaged skin. Transmission seems to require direct contact; the virus does not survive for long outside the human body.

There are an estimated 30 million or more people with **genital herpes** in the United States alone. From 1965 to 1979, the number of reported cases increased by 830 percent. It probably is no accident that this rise coincided with the advent of widespread use of oral contraceptives and changing attitudes toward sexual activity. About 200,000 to 500,000 new cases of genital herpes are now reported annually in the United States.

The many strains of sexually transmitted HSV are classified as types 1 and 2. Type 1 strains infect mainly the lips, tongue, mouth, and eyes. Type 2 strains cause most (but not all) genital infections. An infected person can have both types. Symptoms most often occur 2 to 10 days after exposure to the virus, although sometimes they are mild or absent. Usually, small, painful blisters appear on the penis, vulva, cervix, urethra, or anal tissues. The sores can also occur on the buttocks, thighs, or back. During the initial flare-up a person may also have a fever and flulike symptoms for several days. Within 3 weeks the sores crust over and heal.

After the first sores disappear, sporadic reactivation of the virus can produce new, painful sores at or near the original site of infection. Recurrences may be triggered by sexual intercourse, emotional stress, menstruation, a rise in body temperature, or other infections.

If a pregnant woman has genital herpes, her infant can become infected at birth during a vaginal delivery. Contact with the mother's active lesions can lead to a form of herpes that is often fatal or can lead to mental retardation; lesions arising in the infant's eyes can cause blindness. If a woman has an active herpes infection near or at the time of delivery, her physician will recommend a cesarean section—surgical removal of the baby through an incision made in the mother's abdomen, avoiding the contaminated cervix, vagina, vulva, and adjacent surfaces.

At present there is no cure for HSV infection. Between flare-ups, the virus remains latent in nervous tissue. Acyclovir (Zovirax), an antiviral drug that is most effective taken in pill form, inhibits the ability of the virus to synthesize DNA and so inhibits viral reproduction. Acyclovir has helped millions of people who are infected with HSV. It seems to reduce the shedding of virus particles from sores, and sores are often less painful and heal more rapidly. Over time, recurrences may also become both less frequent and less severe.

Human Papillomavirus

Genital warts are painless growths caused by infection of epithelium by the **human papillomavirus** (HPV). The warts can develop months or years after exposure to the virus and usually they occur in clusters on the penis, the cervix, or around the anus. There is strong evidence that females who have genital warts on the cervix, or who have had a partner who has genital warts, are at increased risk of developing cervical cancer. In fact, HPV is the most probable cause of cervical cancer. Any woman who has a history of genital warts should inform her physician, who may want to schedule a *Pap smear*, a painless test for abnormal cell growth on the cervix, at least once a year.

Genital warts can be difficult to diagnose because they are not always visible. When detected, the warts are usually removed surgically, by freezing (with liquid nitrogen) or burning, or by application of the drug podophyllin, which destroys the wart tissue.

Type B Viral Hepatitis

The hepatitis B virus (HBV) is a DNA virus that, like HIV, is transmitted in blood or body fluids such as saliva, vaginal secretions, and semen. However, HBV is far more contagious than HIV. **Hepatitis B** is an increasingly common STD, striking about 300,000 people in the United States each year. The virus attacks cells of the liver. A key symptom is jaundice, yellowing of the skin and whites of the eyes as the liver becomes unable to process bilirubin pigments produced by the breakdown of hemoglobin from red blood cells. In about 10 percent of cases the immune system cannot fully eliminate HBV, and the long-term infection becomes chronic. Carriers are individuals without symptoms who can easily spread infection to their intimate contacts. Chronic hepatitis can lead to liver cirrhosis or cancer.

A blood test can reveal HBV or antibodies to it (which indicate that a person has been infected). The only treatment is rest. However, people at known risk for contracting the disease (such as health care workers and anyone who requires repeated blood transfusions) can receive a vaccination that provides immunity against the virus. This HBV vaccination is usually given to infants, along with other routine vaccinations for a range of childhood diseases.

Chancroid

The bacterium *Haemophilus ducreyi* causes an STD called **chancroid**. Chancroid was once mainly a tropical disease, but due to increased international travel it is becoming more common in the United States and Europe. Chancroid produces soft, painful ulcers on the external genitals. Untreated, the infection can spread to lymph nodes in the pelvic area and destroy tissues there. Chancroid (as well as some other STDs) has been implicated in some cases of heterosexual transmission of HIV because the sores provide a site for the virus to enter the body. It can be treated by antibiotics.

Beyond AIDS, gonorrhea, syphilis, and chlamydial infections, the most prevalent sexually transmitted diseases in the United States are genital warts (caused by human papillomavirus), genital herpes, nongonococcal urethritis, type B hepatitis, and chancroid.

Pelvic inflammatory disease is also termed an STD because it appears most often as a complication of chlamydial infection and gonorrhea.

Neither genital herpes, type B viral hepatitis, nor genital warts can be cured with antibiotics, though all can be treated medically.

STDs CAUSED BY PARASITES OR FUNGI

Pubic Lice and Scabies

Two tiny ectoparasites ("ecto" means external) can be transmitted by any kind of close body contact. Both are arthropods, "joint-legged" relatives of crabs and spiders. *Pubic lice*, also called crab lice or simply "crabs" (Figure 15.15), usually turn up in the pubic hair, although they can make their way to any hairy spot on the body. They cling tenaciously to individual hairs and attach their small, whitish eggs ("nits") to the base of the hair shaft. Intense itching and irritation result when the parasites bite into the skin and suck blood.

A mite causes *scabies*. As the parasite burrows in and around the skin and lays eggs, it creates dark, undulating lines in the skin of the pubic area, armpits, around the nipples, or elsewhere. When the eggs hatch, they cause tremendous irritation and extreme itching. Both lice and scabies can be treated by applying antiparasitic drugs to the infested areas.

Vaginitis

The warm, moist environment of the vagina can provide a hospitable environment for a range of organisms, although the vagina's rather acidic pH usually keeps pathogens in check. When certain types of vaginal infections do occur, they can be transmitted to a sex partner during intercourse. Any event that alters the usual chemical balance of the vagina (such as taking an antibiotic)

Figure 15.16 Flagella propel the protozoan parasite *Trichomonas vaginalis* through its fluid environment in the vagina. Structurally, flagella of animal cells such as this one differ from bacterial flagella, but all serve the function of aiding locomotion.

can trigger overgrowth of *Candida albicans*, a type of yeast (a fungus) that can be a common inhabitant of the vagina. Symptoms of a vaginal yeast infection (candidiasis) include a white "cottage cheesy" discharge and itching and irritation of the vulva. An affected male may notice itching, redness, and flaking skin on the penis. Yeast infections are easily treated by over-the-counter and prescription antifungal medications. Health care practitioners often recommend that both partners be treated to prevent reinfection. Genital yeast infections can be an early sign of HIV infection, so they should be brought to the attention of a physician.

Trichomonas vaginalis, a small protozoan parasite (Figure 15.16), can cause a severe inflammation of the vaginal epithelium. Symptoms include a greenish, frothy, foul-smelling vaginal discharge and burning and itching of the vulva. A male who contracts *T. vaginalis* may experience painful urination and a discharge from the penis, the results of an inflamed urethra. When the infection is diagnosed, both partners are treated with the drug metronidazole.

Figure 15.15 Magnified 120 times, this crab louse looks rather vicious. Crab lice are tiny but large enough to be visible on the skin, generally as mobile brownish dots. The mite uses claws at the end of its appendages to cling tenaciously to a hair shaft.

Ectoparasites and fungi are the source of STDs such as pubic lice, scabies, and sexually transmitted vaginal yeast infections. All these conditions are readily curable.

15.9

PROTECTING YOURSELF—AND OTHERS—FROM STDs

The old saying "An ounce of prevention is worth a pound of cure" has never been truer than with sexually transmitted diseases (Figure 15.17). The only people who are not at risk of STDs are those who are celibate (never have sex) or who are in a long-term, mutually monogamous relationship in which no disease is present. Otherwise, health care professionals recommend the following guidelines to help you minimize your risk of acquiring or spreading an STD. These guidelines aren't meant to be "preachy." They do reflect solid *scientific* advice in the realm of STD prevention.

GUIDELINES FOR SAFER SEX

1. Use a latex condom, or make sure that your partner uses one. Doing so during genital *or* oral sex will significantly reduce your risk of exposure to the pathogens that cause HIV, gonorrhea, herpes, and other diseases. With the condom, use a spermicide that contains nonoxynol-9, which may help kill virus particles. Condoms are now available for men and women.

2. Limit yourself to one partner who has sex only with you.

3. Get to know a prospective partner before you have sex. Having a friendly but frank discussion of your sexual histories, including any previous exposure to an STD, is very helpful. The safest policy is to assume you are at risk and to take appropriate precautions.

4. If you decide to become sexually intimate, be alert to the presence of sores, scabs, a discharge, or any other sign of possible trouble in the genital area. With so much at stake, there's nothing embarrassing about looking.

5. Keep yourself and your immune system healthy by getting sufficient rest, eating properly, and learning strategies for coping with stress.

6. If for no other reason than protection from STDs, avoid the abuse of alcohol and drugs. Studies show that alcohol and drug abuse both are correlated with unsafe sex practices.

COPING WITH INFECTION Humans have probably been afflicted with sexually transmitted diseases since prehistoric times. We know for sure that syphilis and gonorrhea have been around for hundreds, if not thousands, of years. And for all those years people have probably felt worried, ashamed, guilty, embarrassed, or angry—or all of the above—about getting an STD. Admitting to a partner that you have an STD—and that *both* of you may require treatment—can be difficult. Yet most people recognize that they have an ethical obligation to inform partners when an STD enters the picture. In addition to simply being honest, dealing realistically with an infection involves the following:

1. Learn about and be alert for symptoms of STDs. If you have reason to think you have been exposed, abstain from sexual contact until a medical checkup rules out any problems. Self-treatment won't help. See a doctor or visit a clinic.

2. Take all prescribed medication. *Do not* share medication with a partner. Unless both of you take a full course of medication, chances of reinfection will be great. Your partner may need to be treated even if he or she does not show symptoms.

3. Avoid sexual activity until medical tests confirm that you are free from infection.

Figure 15.17 Students at a workshop on STD prevention.

SUMMARY

1. Viruses, bacteria, fungi, protozoa, and parasitic arthropods such as pubic lice and mites are all potential pathogens associated with sexually transmitted diseases. They live as parasites on or in their human host, from which they derive nutrients or other substances essential to their survival and reproduction.

2. The most threatening of all STDs is HIV infection and the diminished immune response associated with AIDS (Figure 15.18). HIV cripples the immune system primarily by disabling helper T cells (CD4 cells) and macrophages, leaving the body defenseless against a wide range of pathogens. AIDS appears to be nearly always fatal. Researchers do not yet understand why a small percentage of HIV-infected people do not progress to AIDS within 10 years of infection.

3. Viral STDs also include genital herpes, genital warts (HPV), and type B viral hepatitis. Bacteria are responsible for syphilis, gonorrhea, chlamydial infections, and chancroid.

4. Untreated STDs can lead to serious, even life-threatening, complications. Scarring of the oviducts due to pelvic inflammatory disease is a major cause of female sterility. NGU can result in inflammation of reproductive structures in both males and females; infants born to mothers infected with syphilis or genital herpes may suffer effects ranging from blindness to death. Chronic type B viral hepatitis can lead to liver cancer or cirrhosis.

5. Intimate contact can transmit pubic lice and the tiny mites that cause scabies. Vaginitis caused by overgrowth of the fungus *Candida albicans* also can be transmitted by sexual intercourse.

6. Frequently, sexually transmitted infections are interconnected. Conditions such as genital herpes and syphilis increase the risk of HIV infection during risky sexual encounters.

7. Only people who abstain from sexual contact or who are in an infection-free monogamous relationship can be assured of not being exposed to a sexually transmitted disease. Ethical behavior with respect to STDs requires that an infected person inform his or her partner(s), obtain proper medical treatment, and refrain from activity that might infect others.

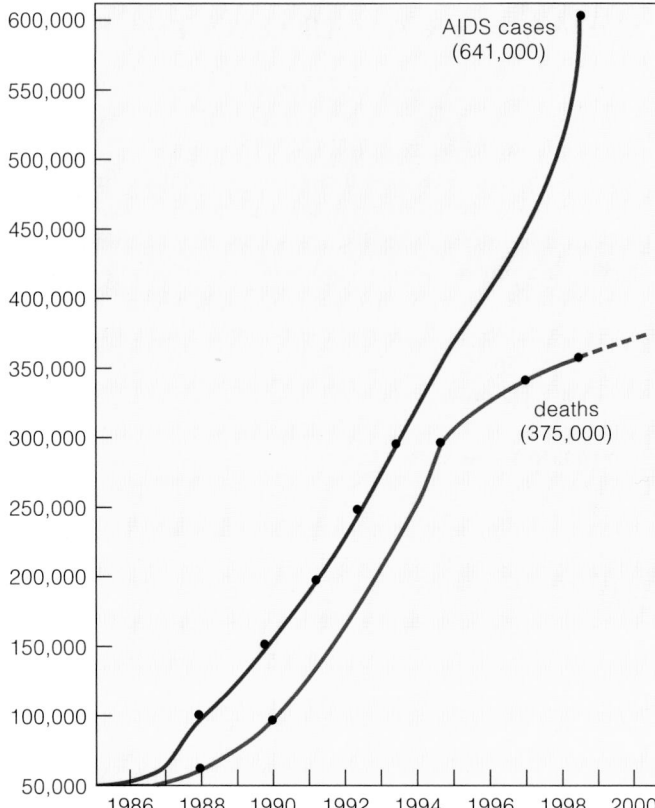

Figure 15.18 Cumulative totals of HIV infections and AIDS cases in the United States as of 1998.

Review Questions

1. How does HIV cause disease? *336*

2. Describe five ways the human immunodeficiency virus (HIV) can be transmitted. *336*

3. List at least four major STDs that are caused by bacteria. What are the major symptoms of each? In addition to AIDS, which STDs are viral in origin? (*throughout chapter*)

4. Pelvic inflammatory disease is a major cause of infertility in young women. Discuss the various STD sources of this disorder. *342*

5. Which STDs pose a danger to a fetus? To an infant during birth? What are those dangers? *336, 340, 341, 343*

6. Compare and contrast the dangers and treatment options for HSV, human papillomavirus, and type B viral hepatitis. *343*

7. Do scabies and pubic lice require the use of antibiotics in order to cure symptoms? Why or why not? *344*

Critical Thinking: You Decide *(Key in Appendix V)*

1. A cleanup worker in the cafe where you work on weekends has been diagnosed as HIV-positive, and a server has come down with type B hepatitis. Some other employees start a petition demanding that both people be required to wear a mask over the mouth and nose, not handle soiled dishes or food, and not use the employee restroom. Both infected employees strenuously object to the plan. You are asked to lead a discussion aimed at resolving the issue, and you decide to prepare a handout giving the scientific basis for making a decision in each case. What does the handout say?

2. You've been dating someone for a while, and now the relationship is getting serious. Everything seems fine, except that your potential partner refuses to consider using a condom if the two of you engage in sex. To allay your fears, the person shows you a lab report indicating a negative HIV blood test. You refuse to forgo condom use, and you terminate the relationship. Why?

Self-Quiz *(Answers in Appendix IV)*

1. The pathogens responsible for sexually transmitted diseases are mostly _____ and _____.
   a. lice; other parasitic animals
   b. viruses; bacteria
   c. bacteria; parasitic animals
   d. bacteria; fungal parasites

2. A _____ is described as a noncellular infectious agent.
   a. virus   c. CD4 cell
   b. bacterium   d. chancre

3. The majority of HIV infections occur through _____. In the vast majority of cases, the virus enters the body of a new host in infected _____, _____, _____, or _____.

4. True or false: Once a person is cured of a gonorrhea infection, that person has immunity to subsequent infection.

5. Of the following, which *cannot* be treated effectively with an antibiotic?
   a. gonorrhea   c. syphilis
   b. chancroid   d. hepatitis B

6. All the following are true of chlamydial infections except:
   a. many victims are young adults
   b. there is often no outward evidence of infection
   c. they often cause sterility in males
   d. they are common culprits in PID

7. A woman with a history of STDs and chronic menstrual problems is having difficulty getting pregnant. Her infertility might well be traceable to _____.
   a. PID   c. human papillomavirus
   b. syphilis   d. HIV infection

8. Type B viral hepatitis resembles HIV in that _____.
   a. it is caused by a retrovirus
   b. the virus is transmitted in blood or body fluids such as semen and vaginal secretions
   c. it strikes about 300,000 people in the United States each year
   d. it causes jaundice

9. Match the following STD concepts:
   _____ NGU
   _____ HPV
   _____ *Trichomonas vaginalis*
   _____ HIV
   _____ genital herpes
   _____ PID
   _____ chancroid

   a. diminished immune response
   b. protozoan parasite
   c. may be caused by either of two types of the responsible virus
   d. may be caused by *Chlamydia* parasite or herpesvirus
   e. causes genital warts
   f. lesion similar to that in primary syphilis
   g. a serious complication of gonorrhea, chlamydial infection, and other STDs

Selected Key Terms

chancroid *343*
chlamydia *341*
genital herpes *342*
gonorrhea *340*
hepatitis B *343*
human immunodeficiency virus (HIV) *335*
human papillomavirus (HPV) *343*
latency *333*
nongonococcal urethritis (NGU) *342*
parasite *330*
pelvic inflammatory disease (PID) *342*
sexually transmitted disease (STD) *330*
syphilis *340*
virus *332*

Readings

Alcamo, I. E. 1993. *AIDS: The Biological Basis.* Dubuque: Wm. C. Brown.

The Battle against Infection. 1992. C. B. Clayman, ed. American Medical Association. Pleasantville, N.Y.: Reader's Digest Association. Clearly written summary of the basic science and medical aspects of sexually transmitted diseases.

Nowak, M., and A. McMichael. August 1995. "How HIV Defeats the Immune System." *Scientific American.*

Rathus, S. A., and S. Boughn. 1993. *AIDS: What Every Student Needs to Know.* Fort Worth: Harcourt Brace.

Richardson, S. May 1995. "The Race Against AIDS." *Discover.*

Additional, current information on STDs, including HIV infections and AIDS, may be obtained by mail or on-line from the Centers for Disease Control and Prevention/National Center for HIV, STD, and TB Prevention, P.O. Box 6003, Rockville, Maryland 20848-6003. Online: info@cdcnac.org

 For additional readings, go to InfoTrac College Edition, your online library at:

http://www.infotrac-college.com/wadsworth

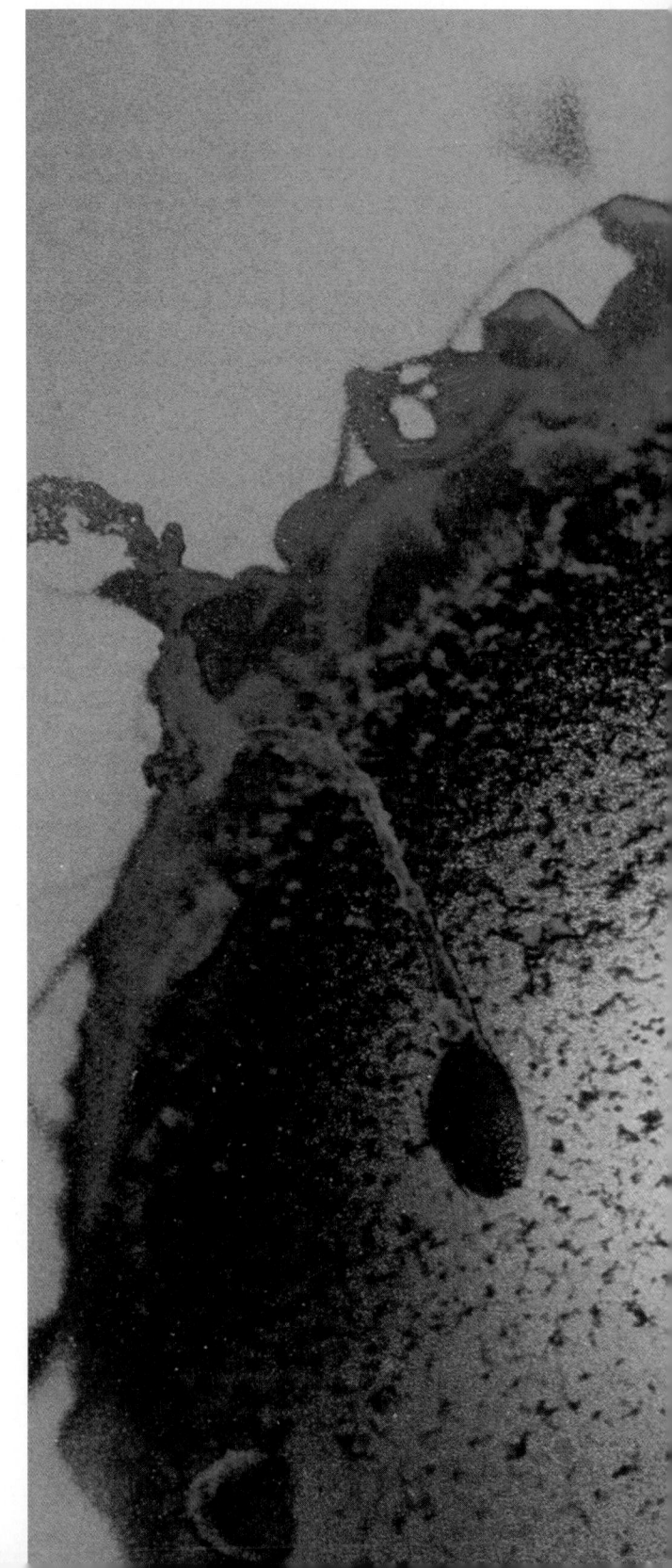

Sperm on the surface of an egg. Only one sperm can successfully fertilize an egg. Their union into a single cell brings together genetic instructions from mother and father, a DNA blueprint that will guide the development and functioning of the new individual.

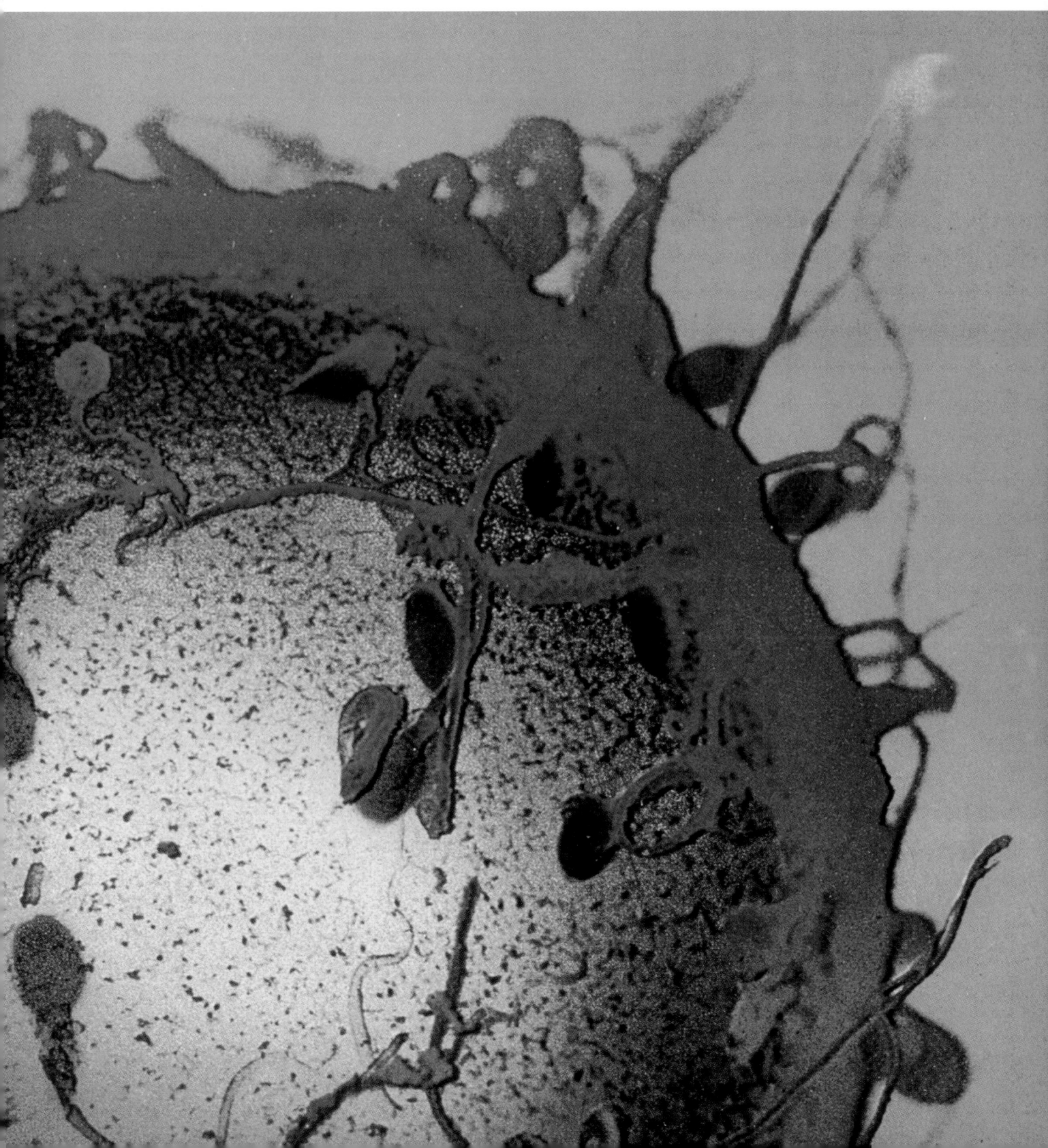

16

CELL REPRODUCTION

Trillions from One

Starting with the fertilized egg in your mother's body, a single cell divided in two, then the two into four, and so on, until billions of cells were growing, developing in specialized ways, and dividing at different times to produce your genetically prescribed body parts (Figure 16.1). Today your body is composed of trillions of cells. Cells still divide within it. Every 5 days, for example, cell divisions replace the entire lining of your small intestine.

Depending on the end result of cell division, its early phase proceeds in one of two ways. One mechanism, mitosis, occurs in all dividing cells except those that give rise to gametes—sperm and eggs. The other mechanism, meiosis, occurs only in the reproductive organs, in gamete-producing stem cells. Mitosis and meiosis clearly resemble each other. But as we'll see and as Table 16.1 suggests, they are fundamentally different in the way they parcel out the bundles

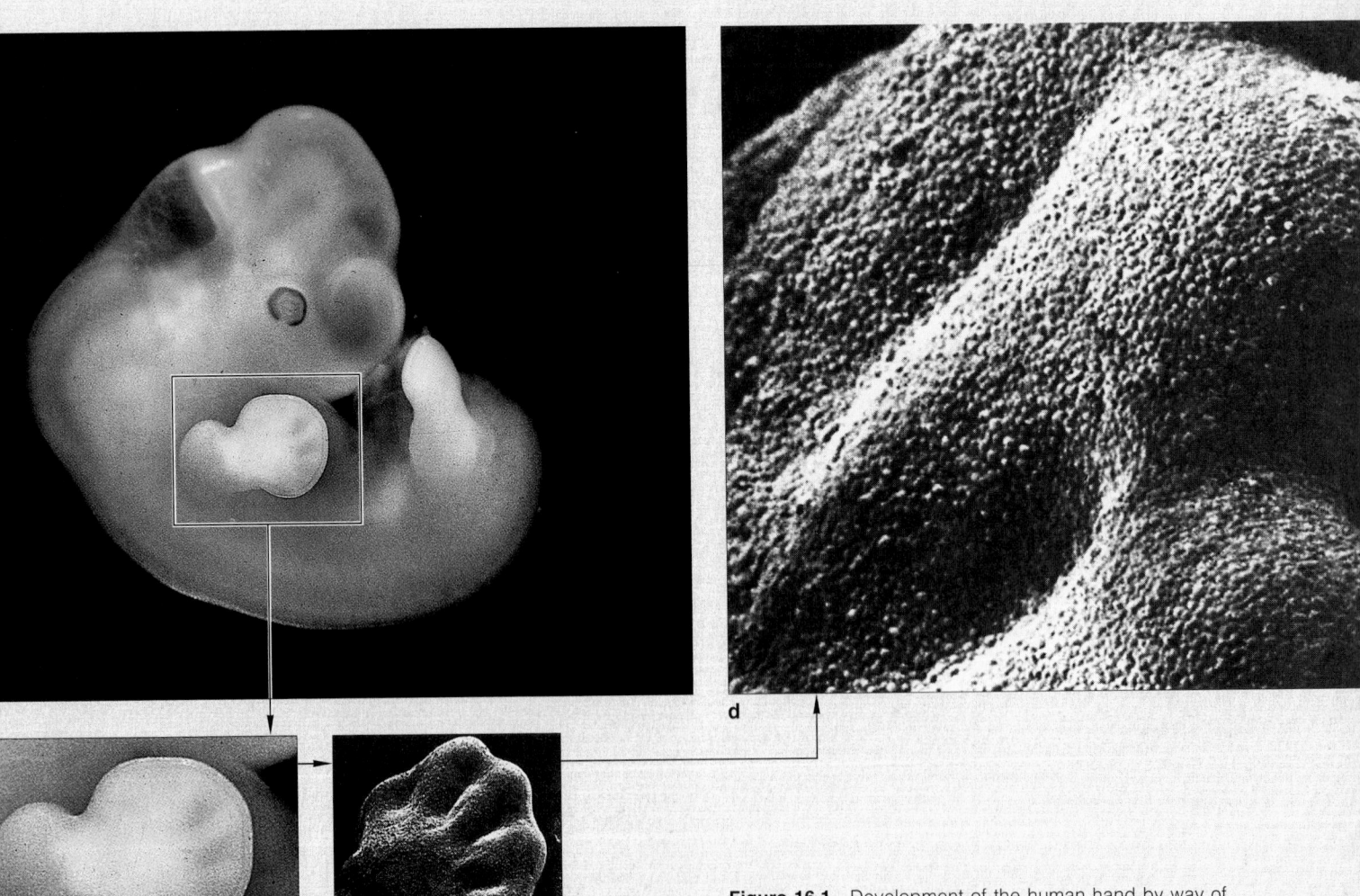

c

d

Figure 16.1 Development of the human hand by way of cell divisions and other processes. Individual cells resulting from the mitotic cell divisions are clearly visible in (**d**). The hand is turned palm upward in (**e**) on the facing page.

of DNA and protein called chromosomes. Thanks to mitosis, the body grows larger, wounds heal, and many tissues are renewed. It is meiosis, followed by the union of sperm and egg at fertilization, that makes sexual reproduction possible.

Microscopic structures called chromosomes carry the instructions—genes—that determine which traits a person inherits. How are chromosomes and their genes distributed into daughter cells? In this chapter and the next, we consider the answers (and best guesses) to questions about cell reproduction and other aspects of human heredity.

Table 16.1 Cell Division Mechanisms

Mechanisms	Functions
Mitosis, cytoplasmic division	Bodily growth and tissue repair
Meiosis, cytoplasmic division	Gamete formation and sexual reproduction

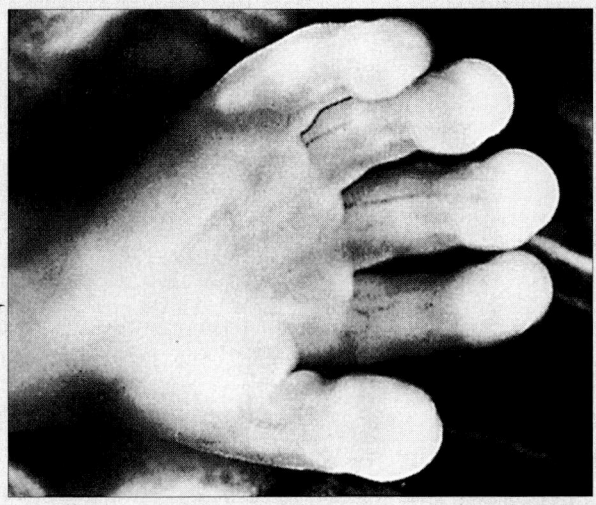

e

KEY CONCEPTS

1. When a cell divides, each of its two daughter cells must receive the same hereditary instructions (DNA) as the original cell had. In humans and other eukaryotes, mitosis sorts the DNA into two new nuclei. Mitosis is the basis of body growth and tissue repair.

2. Daughter cells also must receive some of the parent cell's cytoplasm. After mitosis has produced two new cell nuclei, a mechanism called cytokinesis divides the parent cell's cytoplasm in two.

3. DNA molecules are usually packaged into chromosomes. Except for gametes (sperm and eggs), most cells in the human body contain two of each type of chromosome characteristic of our species. That is, they contain two full sets of hereditary information, one set from each parent.

4. The total number of chromosomes in a body cell is called the *diploid* number, and body cells are thus diploid cells.

5. Mitosis ensures that daughter cells receive the same number of chromosomes as were in the parent cell.

6. Meiosis *reduces* the diploid number of chromosomes by half. A resulting gamete will thus contain only one full set of chromosomes. This single set of chromosomes is called the *haploid* number. The union of two haploid gametes at fertilization restores the diploid number of chromosomes in the new individual.

7. During meiosis, each pair of chromosomes may exchange segments. By doing so, they exchange hereditary instructions about particular traits.

8. Meiosis randomly assigns each chromosome of a pair to a gamete. The random union of gametes at fertilization produces many combinations of parental genes in offspring. All three of these reproductive events lead to variations in the genetic traits of human children. They also ensure that no two people (other than identical twins) will be genetically identical.

CHAPTER AT A GLANCE

16.1   Dividing Cells: The Bridge between Generations
16.2   Mitosis and the Cell Cycle
16.3   *Henrietta's Immortal Cells*
16.4   A Visual Tour of the Stages of Mitosis
16.5   A Closer Look at Mitosis
16.6   Division of the Cytoplasm
16.7   Overview of Meiosis
16.8   A Visual Tour of the Stages of Meiosis
16.9   Key Events during Meiosis I
16.10   Meiosis and the Life Cycle
16.11   *Ionizing Radiation: Invisible Threat to Cells*
16.12   Meiosis and Mitosis Compared

DIVIDING CELLS: THE BRIDGE BETWEEN GENERATIONS

Overview of Division Mechanisms

In biology, reproduction means producing a new generation of individuals *or* a new generation of cells. Reproduction is part of a *life cycle*, a recurring series of events in which individuals grow, develop, maintain themselves, and reproduce according to instructions encoded in DNA, which they inherit from their parents. Reproduction typically begins with the division of single cells. It follows this basic rule: Each cell of a new generation must receive a copy of the parent cell's DNA and enough cytoplasmic machinery to start up its own operation.

DNA contains the genetic instructions for making proteins. Some proteins are structural materials. Many serve as enzymes during the synthesis of carbohydrates, lipids, and other building blocks of the cell or in crucial reactions such as glycolysis and the Krebs cycle. Unless new cells receive the necessary DNA instructions, they will not grow or function properly.

Also, the cytoplasm of the parent cell already has operating machinery—enzymes, organelles, and so forth. When a daughter cell inherits what might look like a blob of cytoplasm, it really is getting "start-up" machinery for its operation until it has time to use its inherited DNA for growing and developing on its own.

The cells of humans (and all other eukaryotic organisms) divide the DNA by **mitosis** or **meiosis**. Both mechanisms sort out and package DNA molecules into new nuclei for forthcoming daughter cells. In other words, mitosis and meiosis partition the chromosomes. A different mechanism divides the cytoplasm and so splits a parent cell into daughter cells.

In contrast to mitosis, meiosis occurs only in **germ cells** set aside for sexual reproduction—the oogonia in ovaries and spermatogonia in testes (Chapter 13). In these cells, recall, meiosis must precede the formation of gametes—that is, eggs and sperm. In order to understand the basic elements of mitosis and meiosis, you must first know more about chromosomes—the structures in the cell nucleus that carry hereditary information.

Some Key Points about Chromosomes

You may recall from Chapter 2 that the DNA in your cells comes packaged in chromosomes. A **chromosome** is one very long DNA molecule combined with a roughly equal amount (by weight) of protein. There are 46 of these chromosomes in the nucleus of a human body cell, and their physical organization differs at different times in a cell's life. When a cell is dividing, each chromosome is tightly coiled, or *condensed*, into a rodlike shape. Between divisions, the chromosomes are not as tightly coiled and are dispersed throughout the nucleus. The DNA and

protein together in a chromosome are called *chromatin.* Each chromosome is still uncondensed and dispersed when it duplicates prior to cell division, and the two duplicate threads of the chromosome remain attached as **sister chromatids**:

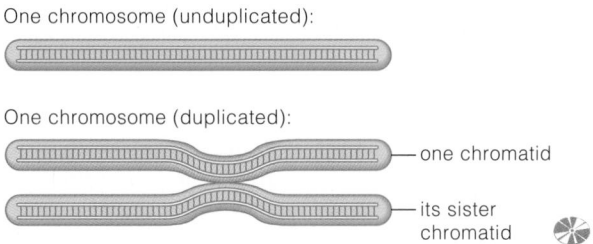

Notice how the two chromatids of a duplicated chromosome are constricted at one small region. This region, the **centromere**, has sites where microtubules will attach and help move the chromosome when the nucleus divides:

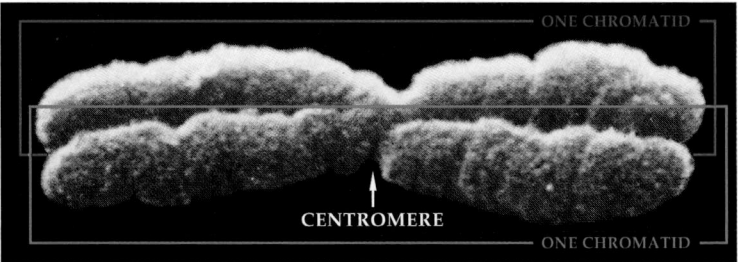

As Figure 16.2 suggests, the centromere's location may vary, depending on the chromosome. Also, the two parallel strands that make up a DNA molecule are twisted together repeatedly like a spiral staircase (a "double helix") and are then condensed further together with proteins. They are also much longer than shown here. You will read more about DNA structure in Chapter 19.

Mitosis and the Chromosome Number

Human cells other than sperm or eggs normally each have two full sets of the chromosomes characteristic of our species—one set from each parent. Your parents each contributed a full set of 23 chromosomes to the fertilized egg that became you, so there are 46 chromosomes in each of your **somatic cells** (somatic means "of the body"). Any cell having two of each type of chromosome is said to be a **diploid** cell. Diploid cells, with two of each type of chromosome, are referred to as 2*n*. The *n* stands for the number of chromosomes in one complete set. The term **chromosome number** refers to the number of each type of chromosome normally present in a cell.

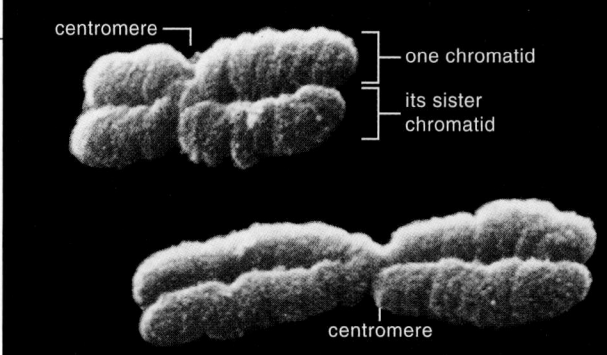

centromere

one chromatid

its sister chromatid

centromere

Figure 16.2 Photomicrograph of two chromosomes, each in the duplicated state. The organization of these chromosomes into tightly coiled, rodlike structures indicates that they are from a dividing cell.

Figure 16.3 Forty-six chromosomes from a human male. Each chromosome is in the duplicated state. There are pairs of chromosomes (two of each type), which tells you that they came from a diploid cell. One member of each pair contains genetic instructions inherited from the father. The other member contains instructions from the mother.

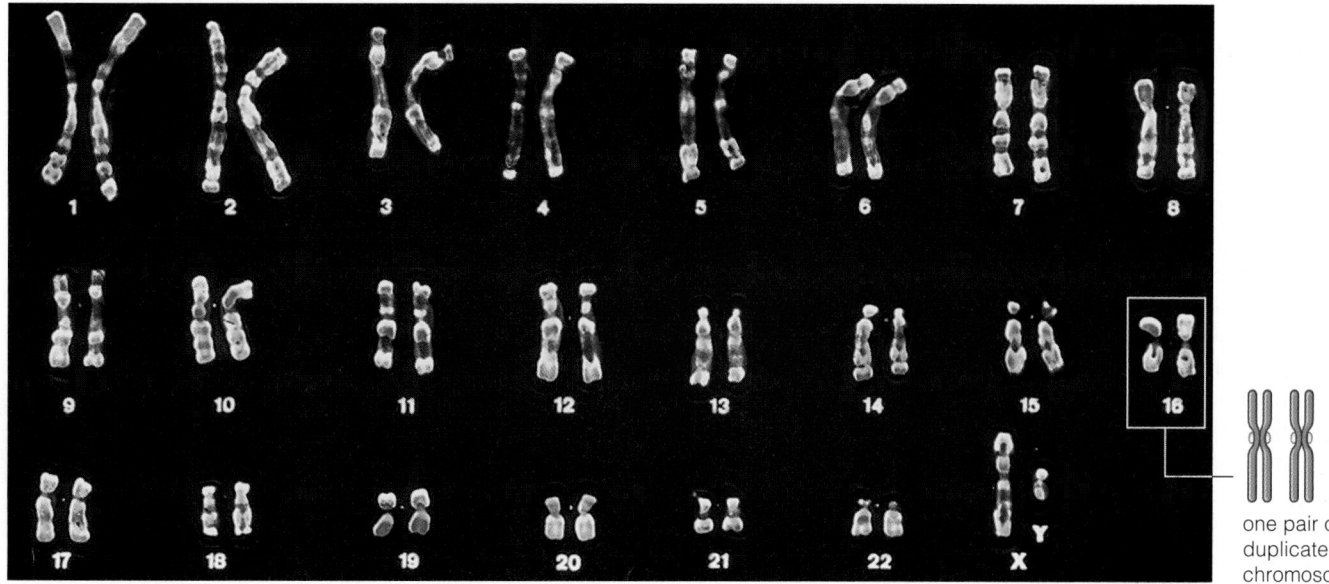

one pair of duplicated chromosomes

With mitosis, a diploid parent cell produces two diploid daughter cells. One member of each chromosome pair is maternal in origin, and the other is paternal in origin. Shortly before a diploid cell divides, the genetic material is duplicated, so that the diploid number of chromosomes doubles (yielding the sister chromatids just described). Mitosis allots half of this doubled genetic material to each new cell, so each daughter cell receives the full diploid number of chromosomes.

Figure 16.3 shows all the chromosomes in a human somatic cell. The chromosomes have been lined up as 23 pairs, an arrangement called a *karyotype* (the Greek root *karyo-* refers to a somatic cell nucleus). Except for the sex chromosomes, X and Y, the two members of each chromosome pair are the same length and shape; both carry hereditary instructions for the same traits. Such corresponding chromosomes, one from each parent, are called **homologous chromosomes**, or simply *homologues* (from a Greek word meaning "to agree"). In general, homologues are the same length, have the same shape, and carry genes for the same traits. Sex chromosomes are exceptions; although they are considered homologues, they differ in size and form and carry different genes. Genes that code for certain products, such as clotting factors in the blood, are X-linked (located on the X chromosome). Only a few traits are influenced by Y-linked genes. Chapter 18 describes such traits more fully.

Mitosis and meiosis sort DNA (in chromosomes) into new nuclei for daughter cells.

In mitosis, the chromosome number remains constant, division after division, from one cell generation to the next. So, if a parent cell is diploid, its two daughter cells will be diploid also. Mitosis takes place in somatic cells. The body grows and tissues are repaired by way of mitosis.

Meiosis takes place only in germ cells set aside for sexual reproduction. It is the mechanism that produces gametes.

16.2 MITOSIS AND THE CELL CYCLE

Mitosis usually proceeds smoothly through four stages—prophase, metaphase, anaphase, and telophase. We will look at these stages more closely in upcoming sections. You can get a preview of them by examining the sequence of micrographs in Figure 16.4.

Mitosis is only one phase of the **cell cycle**. Such cycles begin each time new cells are produced and end when those cells complete their own division. The cycle starts again for each new daughter cell (Figure 16.5).

The Wonder of Interphase

Three other phases occur during interphase, which is usually the longest part of the cell cycle. At this time, a cell increases its mass and the number of cytoplasmic components and duplicates its DNA.

If you could coax the DNA molecules from the nucleus of just one of your somatic cells to stretch in a single line, one after another, they would extend about six feet. The wonder is, enzymes and other proteins selectively scan all of a cell's DNA, switch protein-building instructions on and off, and even produce precise copies of each DNA molecule—all during interphase.

As shown in Figure 16.5, the periods G1, S, and G2 of interphase have distinct patterns of biosynthesis. During G1, most of the carbohydrates, lipids, and proteins for a cell's own use and for export are assembled. During S, the cell copies its DNA and manufactures the proteins that will become organized into structural scaffolding for the condensed versions of chromosomes. Finally, during G2, some of the proteins that will drive mitosis to completion are produced.

Once S begins, events normally proceed at about the same rates in all your dividing somatic cells and continue all the way through mitosis. Given this observation, you might well assume that the cycle has a built-in "brake." Apply the brake that operates in G1, and the cycle stalls. Release the brake, and the cycle runs to completion. Said another way, control mechanisms determine whether or not a cell will divide. In some cells, such as mature neurons of the central nervous system, the process becomes arrested in interphase and the cells never divide again.

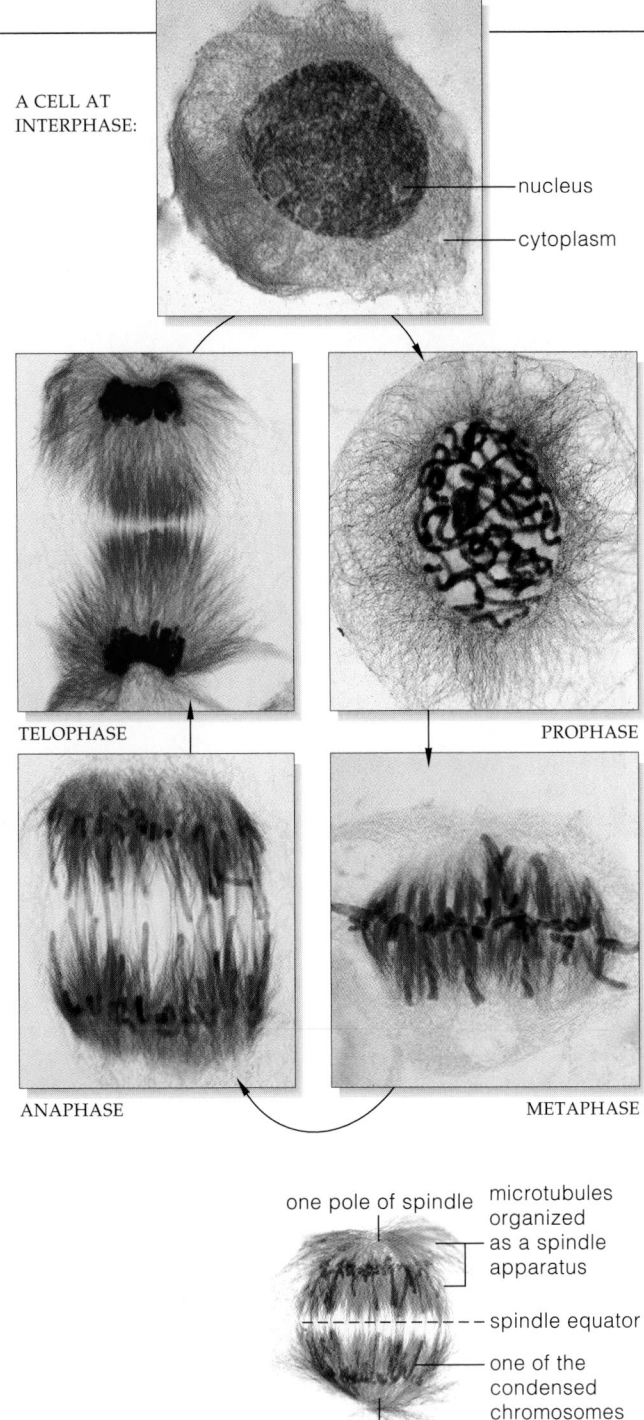

A CELL AT INTERPHASE:

nucleus

cytoplasm

TELOPHASE

PROPHASE

ANAPHASE

METAPHASE

one pole of spindle — microtubules organized as a spindle apparatus

spindle equator

one of the condensed chromosomes

one pole of spindle

Figure 16.4 Light micrographs showing the progress of mitosis in a cell from the African blood lily (*Haemanthus*). The chromosomes are stained *blue*, and microtubules that are moving them about are stained *red*.

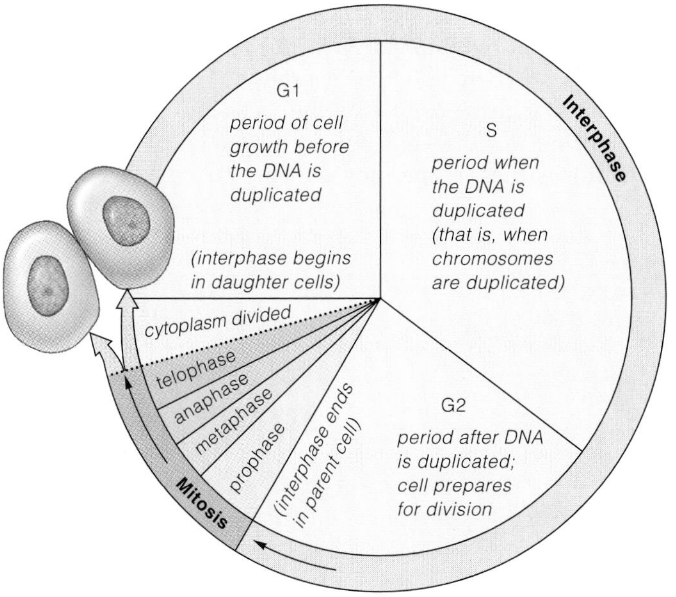

Figure 16.5 Generalized cell cycle. The length of each phase differs among different cell types.

Summary of the Cell Cycle Phases

The following abbreviations signify the four phases of the cell cycle:

M   Mitosis; partitioning of chromosomes, commonly followed by cytoplasmic division

G1 Of interphase, a "Gap" (interval) before the onset of DNA synthesis

S   Of interphase, the time when DNA is synthesized

G2 Of interphase, a second "Gap" between the completion of DNA replication and the onset of mitosis

The duration of the cycle varies, depending on the type of cell. For instance, the cycle lasts 18 hours in bone marrow cells and 25 hours in epithelial cells in your stomach lining.

Good health depends on the successful completion of cell cycles, including DNA replication and chromosome movements. Even the timing and regulation of cell division must be tightly controlled. When normal controls fail, cancer may follow (Chapter 20). *Science Comes to Life* describes a landmark case of rampant cell divisions.

A cell cycle begins at interphase, when a new cell (formed by mitosis and cytoplasmic division) increases its mass and the number of its cytoplasmic components, then duplicates its chromosomes. The cycle ends when the cell divides.

16.3

HENRIETTA'S IMMORTAL CELLS

Each human starts out as a single fertilized egg. At birth a human body has about a trillion cells. Even in an adult, many cells are still dividing. Cells in the stomach's lining divide every day. Liver cells usually don't divide—but if part of the liver becomes injured or diseased, some cells will divide repeatedly and produce new cells until the damaged part is replaced.

In 1951, George and Margaret Gey of Johns Hopkins University were trying to develop a way to keep human cells dividing outside the body. (Researchers could study basic life processes with such cells. They also could study cancer and other diseases, without having to experiment directly on humans.) Local physicians had provided them with normal and diseased human cells from patients. But the Geys just couldn't stop the cell lines from dying out within a few weeks.

Mary Kubicek, one of their assistants, was about to give up after dozens of failed attempts. Still, in 1951, she prepared another sample of cancer cells for culture. The sample was code-named HeLa, for the first two letters of the patient's first and last names.

The cells began to divide. And divide. And divide again. By the fourth day there were so many cells that they had to be subdivided into more cultures. As months passed, the culture continued to thrive. Unfortunately, the cancer cells inside the patient's body were just as vigorous. Six months after the patient was first diagnosed as having cancer, tumor cells had spread through her body. Two months later, Henrietta Lacks, a young woman from Baltimore, was dead.

Although Henrietta was gone, some of her cells lived on in the Geys' laboratory as the first successful human cell culture. HeLa cells were soon being shipped to other researchers, who passed cells on to others, and so HeLa cells came to live in laboratories all over the world. Some even traveled into space aboard the *Discoverer XVII* satellite. Every year, research that is described in hundreds of scientific papers is based on work with HeLa cells.

Henrietta was only 31 years old when runaway cell divisions killed her. Now, more than 40 years later, her legacy is still benefiting humans everywhere, in cells that are still alive and dividing, day after day after day.

A VISUAL TOUR OF THE STAGES OF MITOSIS

After a cell makes the transition from interphase to mitosis, its DNA has been replicated and it stops constructing new cell parts. Within that cell, profound changes now take place smoothly, one after the other, through the four sequential stages described earlier: **prophase**, **metaphase**, **anaphase**, and **telophase**.

Figure 16.6 shows these stages for a dividing animal cell. Compare this illustration with the micrographs in Figure 16.4. When you look closely, it becomes quite clear that the chromosomes within cells move dramatically during mitosis—but not on their own. A **spindle apparatus** harnesses and moves them.

Figure 16.6 Mitosis. This mechanism assures that daughter cells will have the same chromosome number as the parent cell. For clarity, the diagram shows only two pairs of chromosomes from a diploid (2n) animal cell. Most cells have more than two pairs of chromosomes, as indicated by the micrographs of mitosis in a whitefish cell.

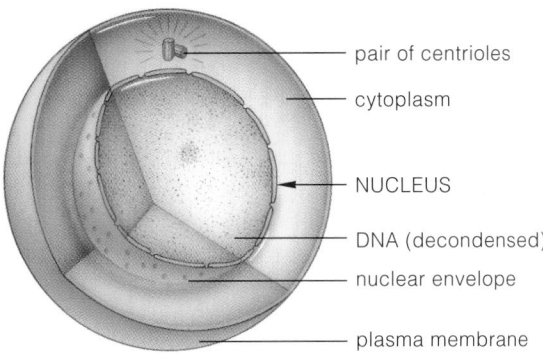

pair of centrioles

cytoplasm

NUCLEUS

DNA (decondensed)

nuclear envelope

plasma membrane

CELL AT INTERPHASE

The DNA is duplicated, then the cell prepares for division.

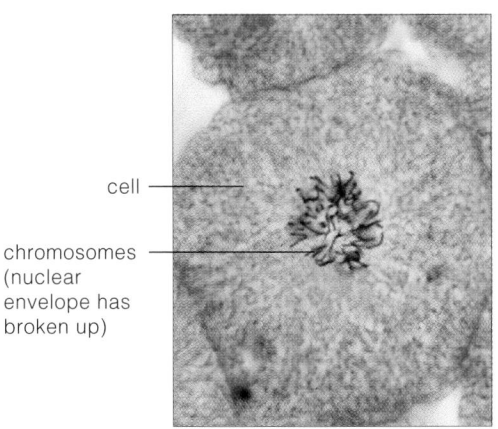

cell

chromosomes (nuclear envelope has broken up)

MITOSIS

microtubules

nuclear envelope

EARLY PROPHASE

The DNA and its associated proteins start to condense. The two chromosomes shaded *purple* were inherited from the male parent. The other two (*blue*) are their counterparts, inherited from the female parent.

LATE PROPHASE

Chromosomes continue to condense. New microtubules are assembled, and they move one of two centriole pairs toward the opposite end of the cell. The nuclear envelope starts to break up.

TRANSITION TO METAPHASE

Microtubules penetrate the nuclear region. Together they form a spindle apparatus. They become attached to the sister chromatids of each chromosome.

A spindle consists of microtubules arranged in two sets. Each set extends from one of the two poles (end points). The two sets overlap each other at the spindle equator, or midway between the two poles. The formation of this bipolar, microtubular spindle establishes the ultimate destinations of chromosomes during mitosis.

Mitosis proceeds through four stages, called prophase, metaphase, anaphase, and telophase.

A bipolar spindle, composed of two sets of microtubules, positions the chromosomes and moves them to specific locations during these stages.

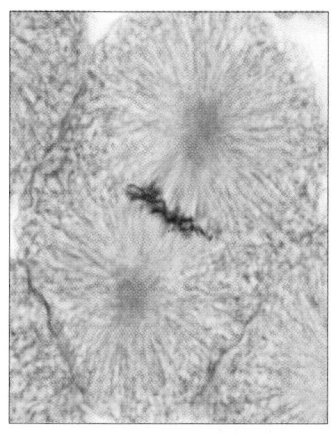 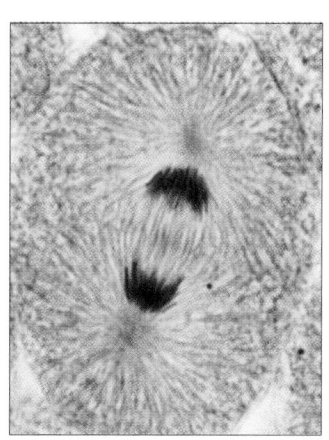 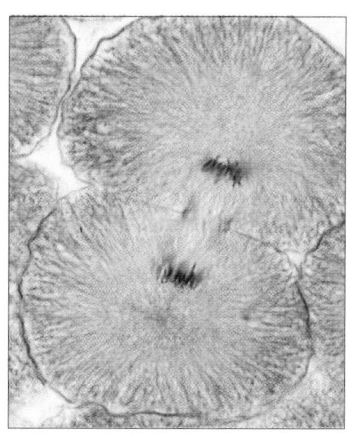

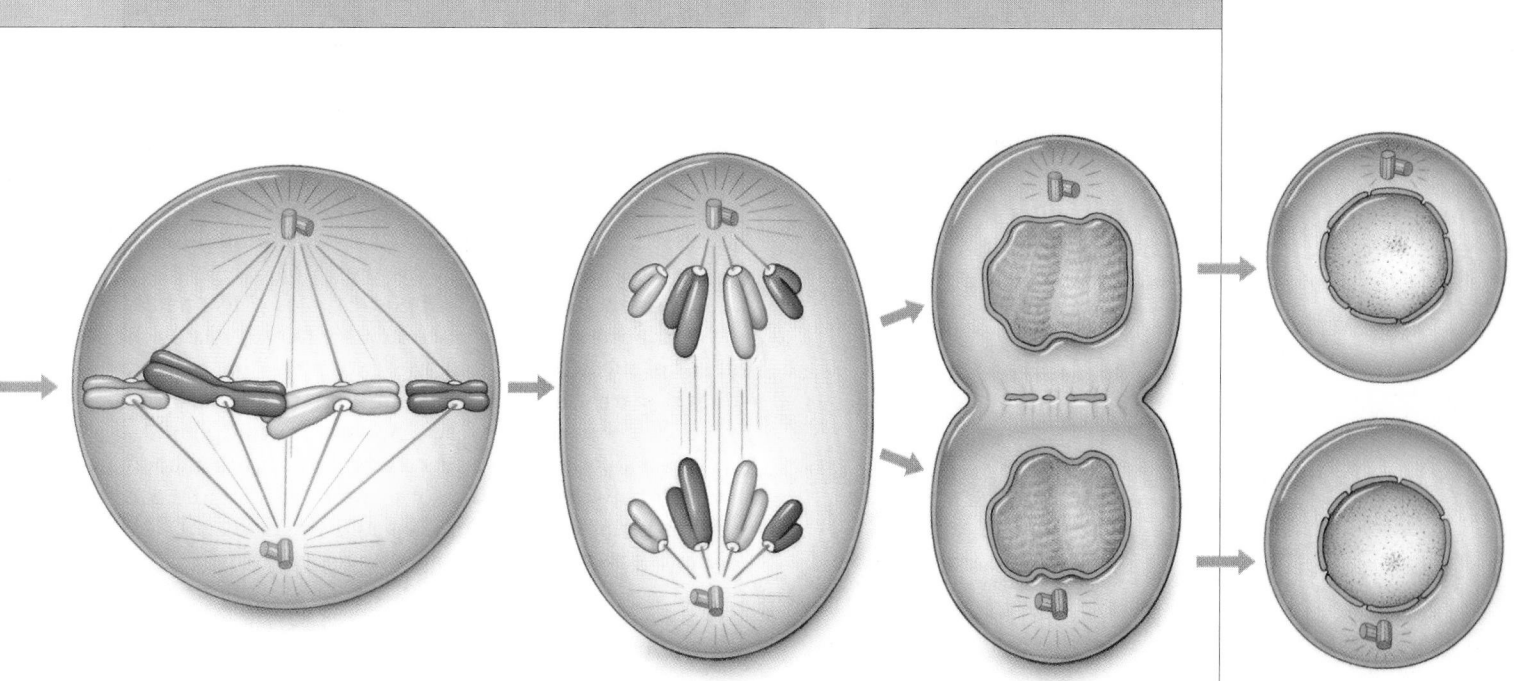

METAPHASE

All chromosomes are lined up at the equator of the spindle. They are now in their most condensed form.

ANAPHASE

The attachment between the two sister chromatids of each chromosome breaks; the two are now chromosomes in their own right. They move to opposite spindle poles.

TELOPHASE

Chromosomes decondense. New patches of membrane join to form nuclear envelopes around them. Most often, cytoplasmic division occurs before telophase is over.

INTERPHASE

Two daughter cells have formed. Each is diploid, with two of each type of chromosome—just like the parent cell.

A CLOSER LOOK AT MITOSIS

Prophase: Mitosis Begins

CHROMOSOMES START CONDENSING   Prophase, the first stage of mitosis, is evident when chromosomes become visible in the light microscope as threadlike forms. ("Mitosis" comes from the Greek *mitos,* meaning thread.) Each was duplicated earlier, during interphase. In other words, each chromosome already consists of two sister chromatids joined at the centromere. Early in prophase, both chromatids twist and fold into a more compact form. By late prophase, all of the chromosomes are condensed into thicker, rod-shaped forms.

Figure 16.7 is an example of a duplicated chromosome in its most condensed form. Notice the centromere. At this constricted region, each sister chromatid of the chromosome has a disk-shaped structure on its surface. This structure, a **kinetochore**, serves as an attachment site for microtubules of the spindle.

THE SPINDLE STARTS FORMING   Meanwhile, in the cytoplasm, most microtubules of the cytoskeleton are breaking apart, into their tubulin subunits (refer to Figure 2.18). The subunits reassemble near the nucleus, as *new* microtubules. Next, groups of the new microtubules move the cell's two centriole pairs toward opposite ends of the cell. This establishes the two spindle poles. Many of the microtubules will extend from one spindle pole or the other to the centromere of a chromosome. The remainder will not interact at all with the chromosomes. Rather, they will extend from the poles and overlap each other.

In many cells, an MTOC (microtubule organizing center) near the nucleus organizes the assembly of microtubules so that they lengthen in specific directions.

THE SPINDLE SEPARATES THE CENTRIOLES   In many cells, an MTOC near the nucleus includes two barrel-shaped **centrioles**. Each centriole started duplicating itself during interphase, and by the time prophase is under way, the cell has two pairs of them. Now microtubules start moving one pair to the opposite pole of the newly forming spindle. You can see this carefully orchestrated movement in Figure 16.6.

Centrioles, recall, also give rise to flagella or cilia. If you observe them in the cells of a particular organism, you can safely bet that flagellated or ciliated cells develop during at least some part of its life cycle. For example, remember that ciliated cells lining airways to your lungs help sweep inhaled bacteria and debris up and out of your body. Ultimately, these and other functions depend on centrioles, so it is no wonder that centrioles are parceled out carefully for the forthcoming daughter cells.

While new microtubules are assembling, the nuclear envelope prevents them from interacting with chromosomes inside the nucleus. However, as prophase ends, the nuclear envelope starts to break up.

Transition to Metaphase

So much happens during the transition from prophase to metaphase that researchers give this transitional period its own name, "prometaphase."

During this time the nuclear envelope breaks up completely. Its two membrane layers have fused together at scattered sites, producing numerous tiny, flattened vesicles. Now the chromosomes are free to interact with microtubules that extend toward them, from both poles of the developing spindle. At first the chromosomes appear to go into a frenzy. This happens when kinetochores are making their first random contacts with the microtubules.

Once a chromosome has both of its kinetochores harnessed, microtubules from *both* poles start pulling on it. The two-way pulling orients the chromosome's two sister chromatids toward opposite poles. Meanwhile, the overlapping spindle microtubules ratchet past each other in such a way that the two poles are pushed farther apart. All of the push–pull forces are balanced when the chromosomes reach the spindle's midpoint.

When all of the duplicated chromosomes are aligned midway between the poles of a completed spindle, we call this metaphase (*meta-* means midway between). The alignment is crucial for the next stage of mitosis.

Anaphase

During anaphase, the two sister chromatids of each chromosome separate and move to opposite poles. Two mechanisms account for the movements. First, the microtubules attached to the kinetochore shorten, pulling each

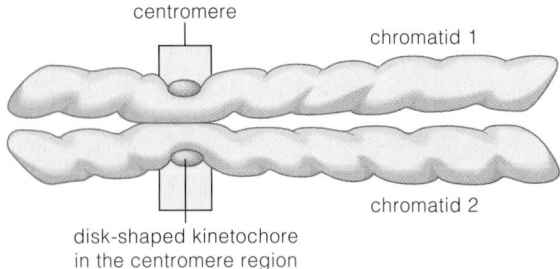

Figure 16.7   Kinetochores of a duplicated chromosome. Spindle microtubules become attached to the kinetochore of each sister chromatid.

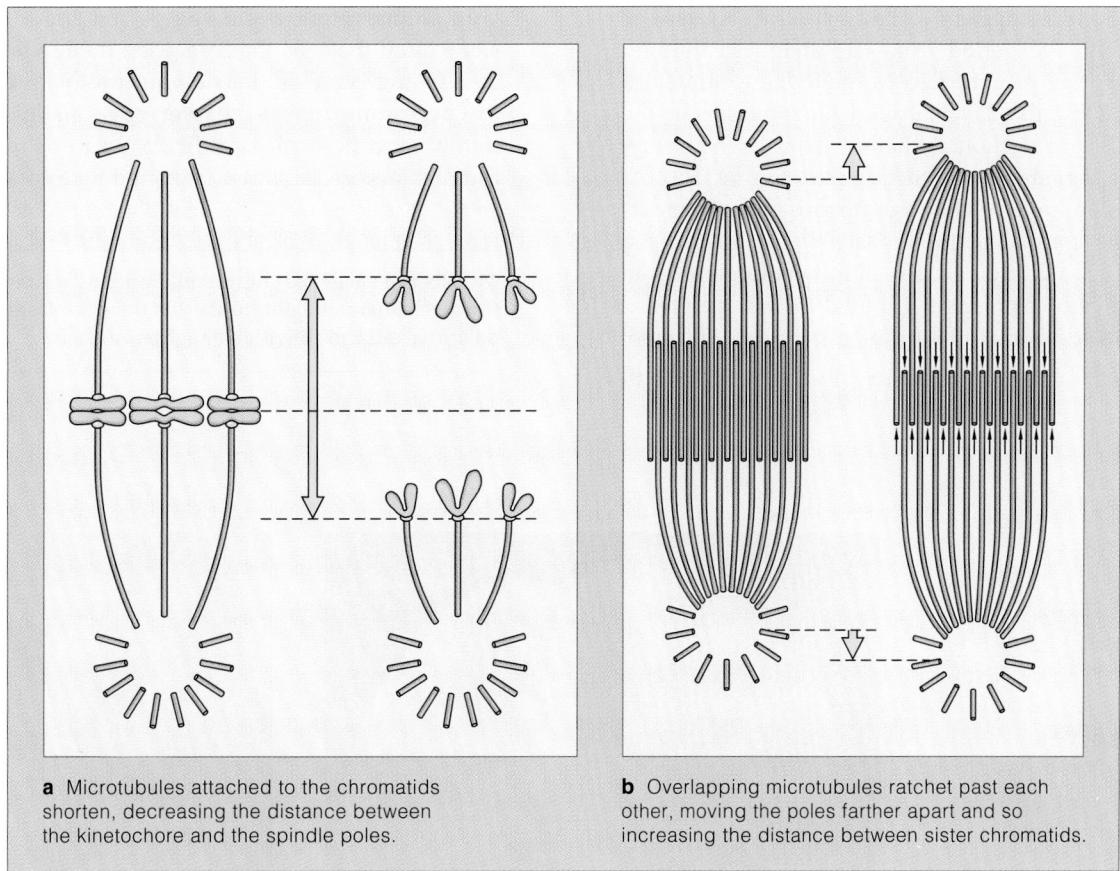

a Microtubules attached to the chromatids shorten, decreasing the distance between the kinetochore and the spindle poles.

b Overlapping microtubules ratchet past each other, moving the poles farther apart and so increasing the distance between sister chromatids.

Figure 16.8 The two mechanisms responsible for separating the sister chromatids of a chromosome at anaphase.

chromatid to a pole (Figure 16.8a). Second, the spindle elongates when overlapping microtubules ratchet past each other and push the spindle poles farther apart (Figure 16.8b). Each sister chromatid is now a chromosome in its own right:

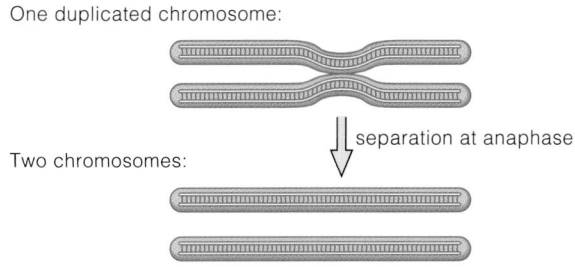

One duplicated chromosome:

Two chromosomes: separation at anaphase

Said another way, every chromosome that was present in the parent cell now has a daughter chromosome at each spindle pole.

Telophase

Telophase begins once the two daughter chromosomes arrive at opposite spindle poles. The chromosomes are no longer harnessed to microtubules, and they return to threadlike form. Vesicles of the old nuclear envelope fuse together to form patches of membrane around the chromosomes. Patch joins with patch, and soon a new nuclear envelope separates each cluster of chromosomes from the cytoplasm. Each cluster contains a pair of each type of chromosome. With mitosis, each new nucleus has the same chromosome number as the parent nucleus. Once the two nuclei form, telophase is over—and so is mitosis.

During mitosis, a microtubular spindle moves sister chromatids of all the chromosomes apart, to opposite spindle poles. A new nuclear envelope forms around the two clusters of chromosomes. Both daughter nuclei thus have the same chromosome number as the parent cell's nucleus.

DIVISION OF THE CYTOPLASM

Division of the cytoplasm, called **cytokinesis**, usually coincides with the period from late anaphase through telophase. For most animal cells, material accumulates and forms a ringlike layer around microtubules at the dividing cell's midsection. A shallow depression appears above the layer, at the cell surface (see Figure 16.9). At this depression, called a **cleavage furrow**, microfilaments made of the contractile protein actin pull the plasma membrane inward and cut the cell in two. Organelles in the divided cytoplasm are distributed to daughter cells.

This concludes our picture of mitotic cell division. Look now at your hands—and think of all the cells in your palms, thumbs, and fingers. Imagine all of the divisions of all the cells that preceded them when you were developing early on, inside your mother. Figure 16.1 at the beginning of this chapter gave an inkling of this unfolding process. It is difficult not to be in awe of the astonishing precision with which it takes place.

Mitosis is a small part of the cell cycle. As it draws to a close, the mechanism of cytokinesis cuts the cytoplasm into two daughter cells, each with a daughter nucleus.

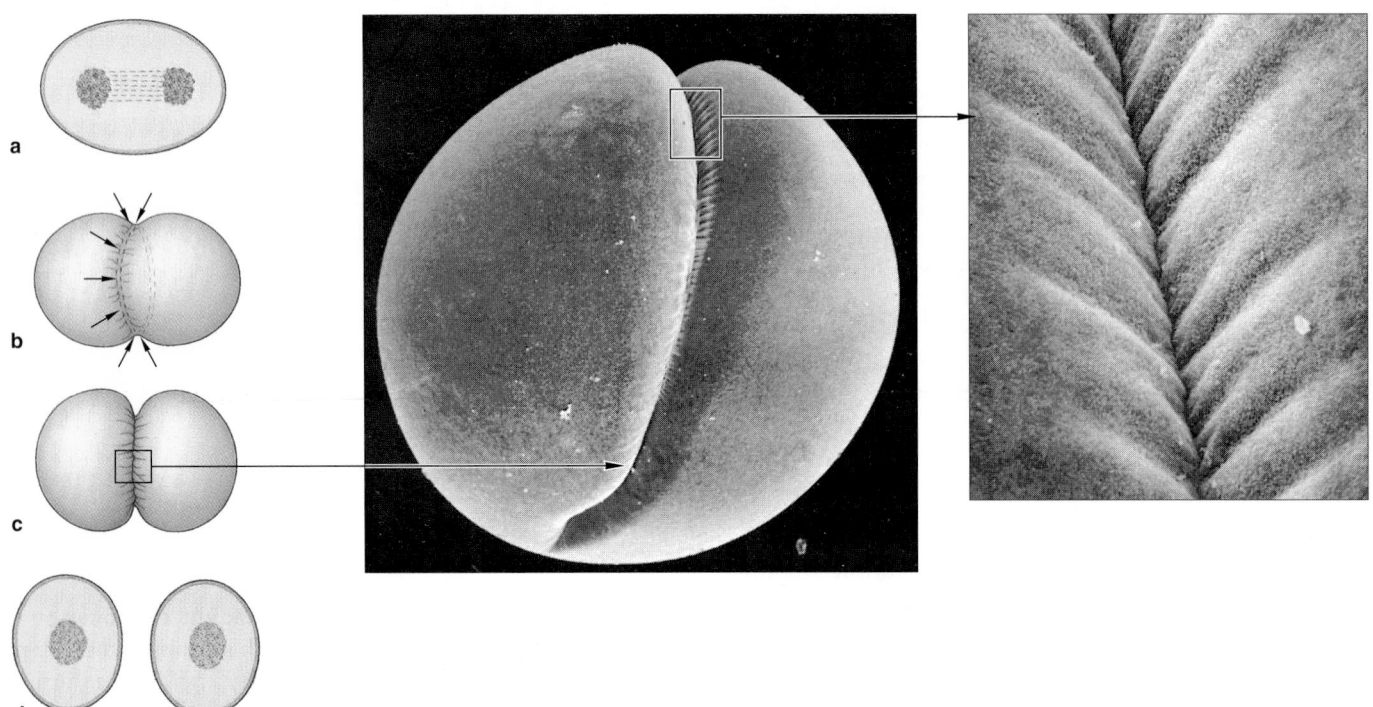

Figure 16.9 Cytokinesis in an animal cell. (**a**) Mitosis is complete and the spindle is disassembling. (**b**) Just beneath the plasma membrane, microfilament rings at the former spindle equator contract, like a purse string closing. (**c,d**) Continuing contractions divide the cell in two. The micrographs show how the plasma membrane sinks inward, defining the plane of cleavage.

OVERVIEW OF MEIOSIS

Making Haploid Gametes

Meiosis occurs in dividing germ cells in ovaries or testes (Figure 16.10). Remember from Chapter 13 that human germ cells are *spermatogonia* in males and *oogonia* in females. Like other body cells, spermatogonia and oogonia are diploid. However, *unlike* other body cells, they give rise to haploid gametes (sperm or eggs) by way of meiosis.

A mature human sperm or oocyte typically contains a single set of 23 chromosomes, rather than pairs of homologous chromosomes. This is because meiosis is a **reductional division** that halves the diploid number of chromosomes (2*n*) to a **haploid** number (*n*). And not just any half: *Each haploid gamete ends up with one partner from each pair of homologous parent chromosomes.*

Two Divisions, Not One

Meiosis resembles mitosis in some respects, even though the outcome is different. Before interphase gives way to meiosis, a germ cell duplicates its DNA. Now each duplicated chromosome consists of two DNA molecules. These remain attached to one another. For as long as the two remain attached, they are sister chromatids:

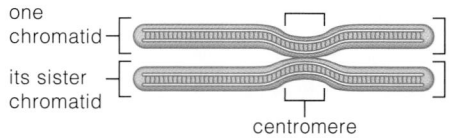

one chromatid —
its sister chromatid —
centromere

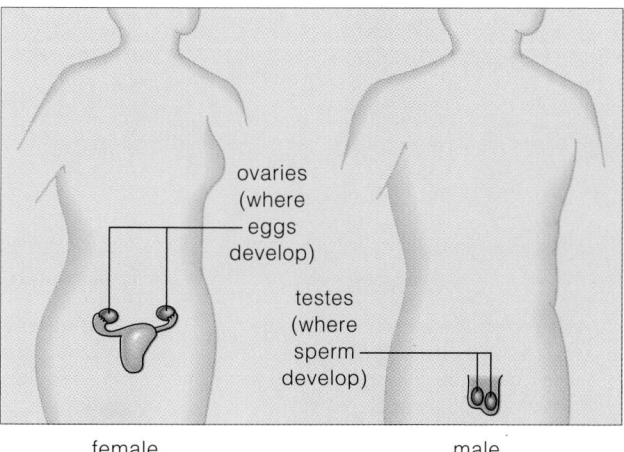

ovaries (where eggs develop)

testes (where sperm develop)

female   male

Figure 16.10 Germ cells in the ovaries and testes give rise to eggs and sperm, respectively. In females, ovaries are the sites where primary oocytes undergo meiosis. In the testes, primary spermatocytes undergo meiosis to give rise to spermatids that differentiate into sperm. Compare Figures 16.14 and 16.15.

As in mitosis, microtubules of a spindle apparatus move the chromosomes in prescribed directions. With meiosis alone, however, *chromosomes proceed through two consecutive divisions, which end with the formation of four haploid nuclei.* The two divisions are called meiosis I and meiosis II:

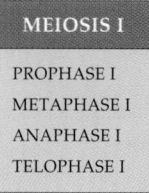

DNA is replicated during interphase

MEIOSIS I
PROPHASE I
METAPHASE I
ANAPHASE I
TELOPHASE I

DNA is *not* replicated between divisions

MEIOSIS II
PROPHASE II
METAPHASE II
ANAPHASE II
TELOPHASE II'

During meiosis I, each duplicated chromosome lines up with its partner, *homologue to homologue*; then the partners separate from each other:

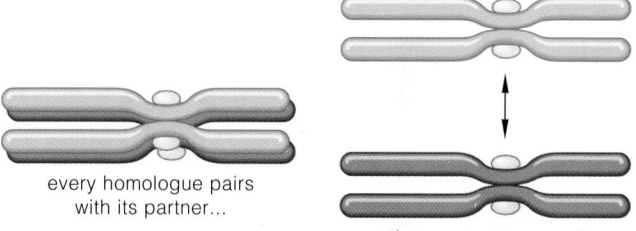

every homologue pairs with its partner...

...then partners separate

The cytoplasm typically divides after completion of meiosos I. The two daughter cells are haploid, with only one of each type of chromosome. But remember, those chromosomes are still duplicated.

During meiosis II, *the two sister chromatids of each chromosome separate from each other:*

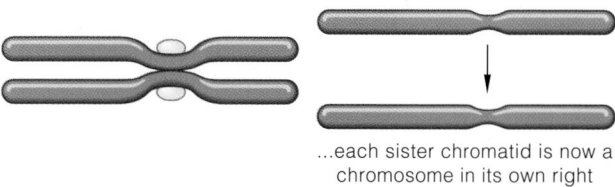

...each sister chromatid is now a chromosome in its own right

After the four nuclei form, the cytoplasm divides once more, the outcome being four haploid cells.

On the next two pages, Figure 16.11 illustrates the key events of meiosis I and II.

Meiosis is a reductional division of a cell's chromosomes. It reduces the parental chromosome number by half—to the haploid number.

Meiosis is the first step leading to the formation of gametes, which are required for sexual reproduction.

A VISUAL TOUR OF THE STAGES OF MEIOSIS

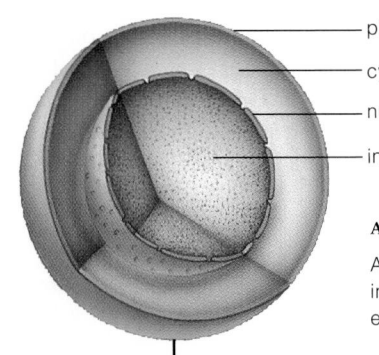

- plasma membrane
- cytoplasm
- nuclear envelope
- interior of nucleus

Figure 16.11 Meiosis: the mechanism by which the parental number of chromosomes is reduced by half (to the haploid number) for forthcoming gametes. Only two pairs of homologous chromosomes are shown. Maternal chromosomes are shaded *purple*, and paternal ones *blue*.

A GERM CELL AT INTERPHASE

A germ cell that happens to have a diploid chromosome number (2n) is about to leave interphase and undergo the first division of meiosis. Its DNA is already duplicated, so each chromosome is in the duplicated state; it consists of two sister chromatids.

MEIOSIS I

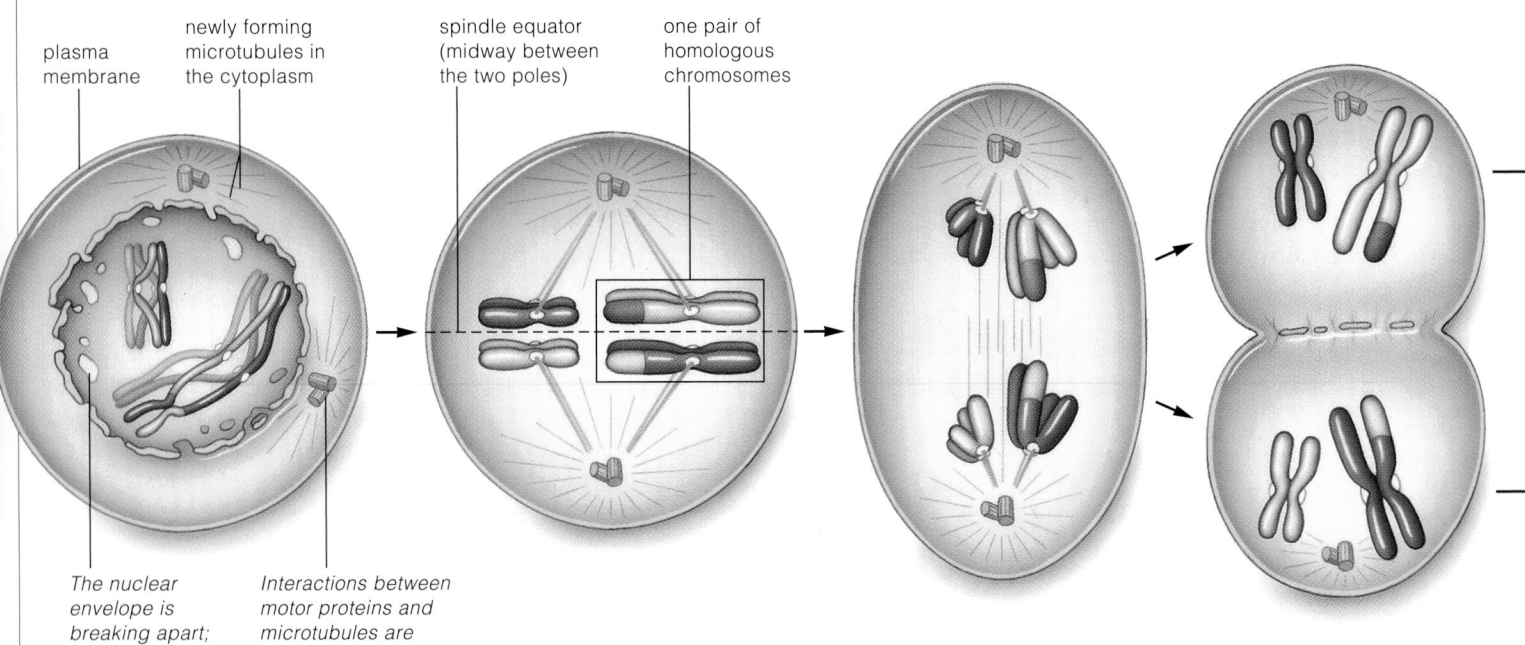

plasma membrane | newly forming microtubules in the cytoplasm | spindle equator (midway between the two poles) | one pair of homologous chromosomes

The nuclear envelope is breaking apart; microtubules will be able to penetrate the nuclear region.

Interactions between motor proteins and microtubules are moving one of two pairs of centrioles toward the opposite spindle pole.

PROPHASE I

Each duplicated chromosome is in threadlike form, but now it starts to twist and fold into more condensed form. It pairs with its homologue, and the two typically swap segments. The swapping, called crossing over, is indicated by the break in color on the pair of larger chromosomes. Each chromosome becomes attached to some microtubules of a newly forming spindle.

METAPHASE I

Motor proteins have been driving the movement of microtubules that became attached to the kinetochores of chromosomes. As a result, the chromosomes have been pushed and pulled into position, midway between the spindle poles. Now the spindle is fully formed, owing to dynamic interactions of motor proteins, microtubules, and the chromosomes themselves.

ANAPHASE I

Microtubules extending from the poles and overlapping at the spindle equator *lengthen* and push the poles apart. At the same time, the microtubules extending from the poles to the kinetochores of chromosomes *shorten,* and each chromosome is thereby pulled away from its homologous partner. These motions move the homologous partners to opposite poles.

TELOPHASE I

At some point, the cytoplasm of the germ cell divides. Two cells, each with a haploid chromosome number (n), result. That is, the cells have one of each type of chromosome that was present in the parent (2n) cell. All chromosomes are still in the duplicated state.

Figure 16.11 traces the stages of meiosis. The stages are similar to the stages of mitosis and are called prophase, metaphase, anaphase, and telophase. They are designated I and II because there are two nuclear divisions in meiosis.

Of the four haploid that vary form by way of meiosis, one or all may develop into gametes.

MEIOSIS II

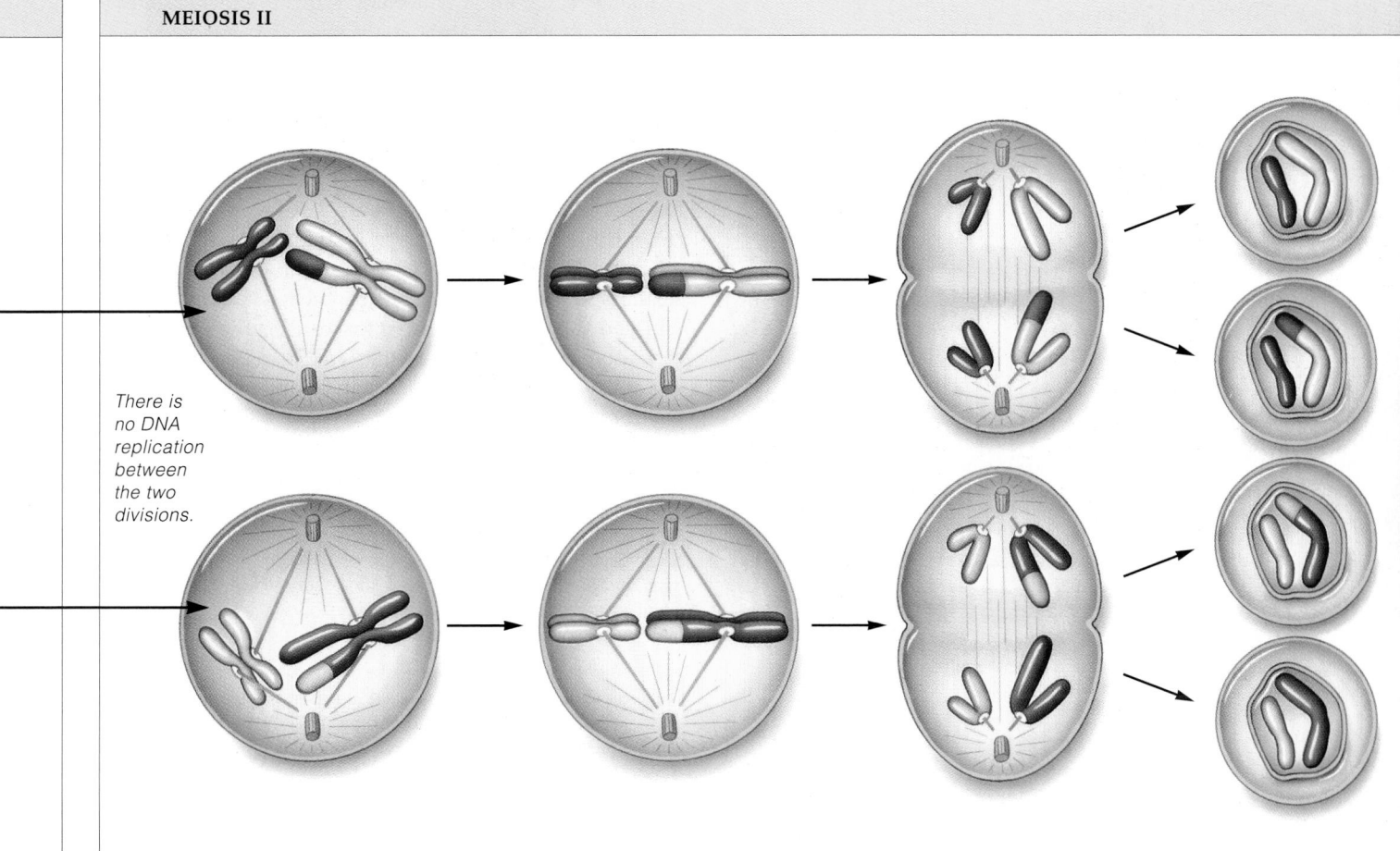

There is no DNA replication between the two divisions.

PROPHASE II

In each of the two daughter cells, microtubules have already moved one member of the centriole pair to the opposite pole of the spindle during the transition to prophase II. Now, at prophase II, microtubules attach to the kinetochores of chromosomes, and motor proteins drive the movement of chromosomes toward the spindle's equator.

METAPHASE II

Now, in each daughter cell, interactions among motor proteins, spindle microtubules, and each duplicated chromosome have moved all the chromosomes so that they are positioned at the spindle equator, midway between the two poles.

ANAPHASE II

The attachment between the two chromatids of each chromosome breaks. The former "sister chromatids" are now chromosomes in their own right. Motor proteins interact with kinetochore microtubules to move the separated chromosomes to opposite poles of the spindle.

TELOPHASE II

By the time telophase II is completed, there will be four daughter nuclei. At the time when the division of the cytoplasm is completed, each daughter cell will have a haploid chromosome number (*n*). All of those chromosomes will be in the unduplicated state.

KEY EVENTS DURING MEIOSIS I

In the first stage, prophase I, a spindle forms and the nuclear envelope disappears. This is also a time of major gene shufflings between homologous chromosomes. Keep in mind that each homologue is in the duplicated state and consists of two sister chromatids joined at a centromere. As prophase I gets under way, homologues begin to pair up in a process called *synapsis* ("bringing together"). It is as if the homologues become stitched point by point along their entire length, with little space between them. (The X and Y chromosomes pair at one end only.) This arrangement favors **crossing over** between *nonsister chromatids* of a pair of homologues. In a crossover, nonsister chromatids exchange corresponding segments at the crossover points. Each segment contains a group of genes. As Figure 16.12 shows, X-shaped configurations called *chiasmata* (singular: chiasma) are evidence of crossovers.

Genes can come in alternative forms called **alleles**. For example, the gene for earlobe shape has two alleles—one calls for attached earlobes, and the other calls for detached earlobes. Frequently, many of the alleles on one chromosome are not exactly identical to the corresponding alleles on its homologue. With each crossover, new combinations of alleles may form.

This exchange of chromosome pieces is called **genetic recombination**, and it leads to variation in the traits of offspring. Such variation can be a major advantage for sexually reproducing organisms. In later chapters we will see that variations in traits can be the source of evolutionary modifications that enable a population of organisms, including humans, to adapt to a changing environment.

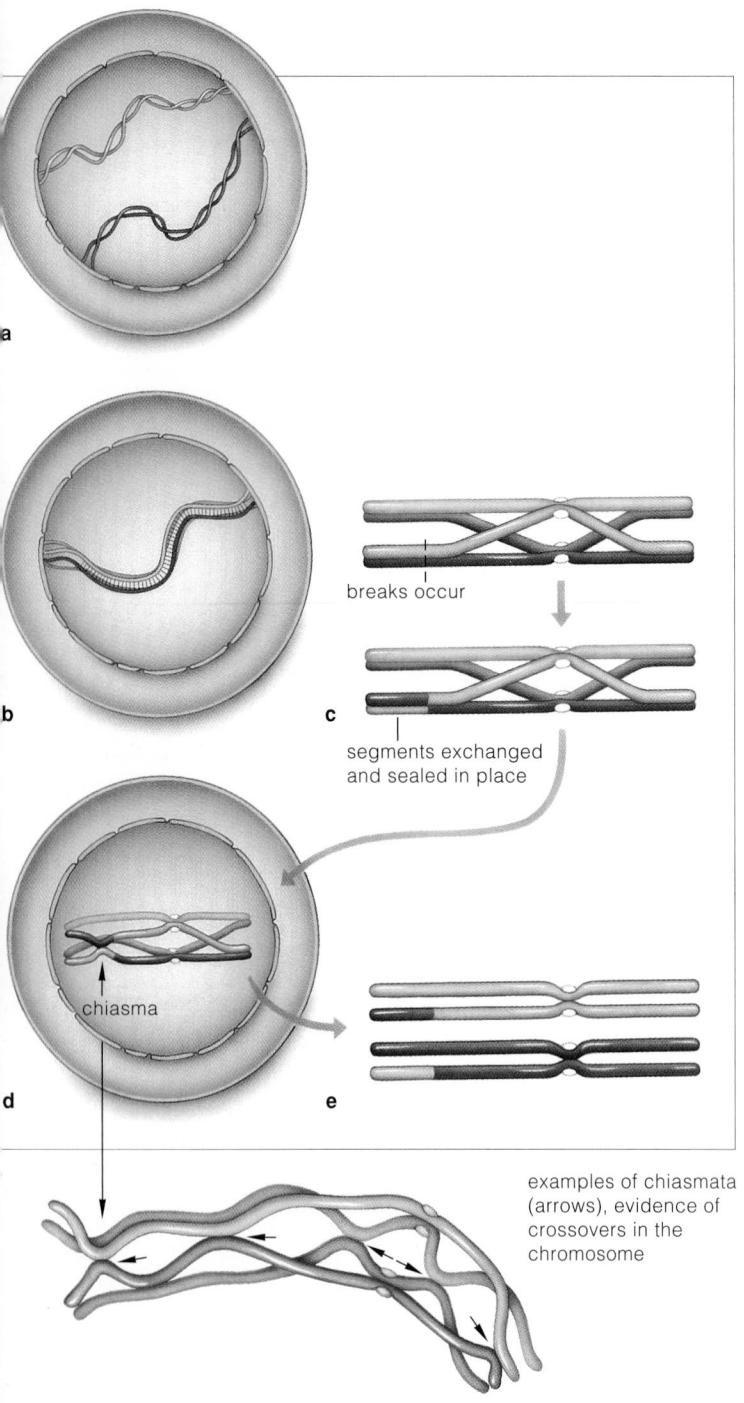

breaks occur

segments exchanged
and sealed in place

a

b

c

chiasma

d

e

examples of chiasmata
(arrows), evidence of
crossovers in the
chromosome

Figure 16.12 Key events during prophase I, the first stage of meiosis. For clarity, only a single pair of homologous chromosomes and only one crossover event are shown. *Blue* signifies the paternal chromosome, and *purple* signifies its maternal homologue.

a Chromosomes become duplicated before meiosis begins. Early in prophase I, each duplicated chromosome is in thread-like form, attached at both ends to the nuclear envelope. Its two sister chromatids are so close together they look like a single thread.

b The two chromosomes become zippered together, so that all four chromatids are intimately aligned. (X and Y chromosomes, which are two forms of sex chromosomes, also become zippered together for a small part of their length.)

c One or more crossovers occur at intervals along the chromosomes. In each crossover, two nonsister chromatids swap segments. (For clarity, the two chromosomes are shown in condensed form and pulled apart. Crossing over may seem more plausible when you realize that it occurs while chromosomes are extended like threads and tightly aligned.)

d As prophase I ends, the chromosomes continue to condense, becoming thicker, rodlike forms. They detach from the nuclear envelope and from each other—except at "chiasmata." Each chiasma is indirect evidence that a crossover occurred at some point in the chromosomes.

e Crossing over breaks up old combinations of alleles and puts new ones together in pairs of homologous chromosomes.

Metaphase I Alignments

During the transition from prophase I to metaphase I of meiosis, whole chromosomes start to be shuffled in major ways. At this point, crossovers have made genetic mosaics out of the chromosomes, but let's put this aside to simplify tracking. Just call the 23 chromosomes from the mother the *maternal* chromosomes; their 23 homologues from the father are the *paternal* chromosomes.

Microtubules have harnessed and oriented one chromosome of each pair toward one spindle pole and its homologue toward the other. Now they are moving all the chromosomes, which soon will be positioned midway between the spindle poles. Are all the maternal chromosomes moved to one pole and all the paternal ones moved to the other? Probably not, because the first contacts between spindle microtubules and chromosomes are random. Hence, *the eventual position of a maternal or paternal chromosome at the spindle equator at metaphase I follows no particular pattern.*

Think of the possibilities when there are merely three pairs of homologues. As Figure 16.13 shows, by metaphase I, these may be arranged in any one of four possible positions. In this case, 2^3 or 8 combinations of maternal and paternal chromosomes are possible for the forthcoming gametes.

Of course, a human germ cell has *twenty-three* pairs of homologous chromosomes. Hence, 2^{23} or *8,388,608 combinations* of maternal and paternal chromosomes are possible every time a germ cell gives rise to a gamete! In each sperm or egg, the genetic instructions from the mother might differ slightly from those from the father. Chapter 17 explains more fully why this is so. But are you beginning to see why striking mixes of traits can show up even in the same family?

During anaphase I, each homologue is separated from its partner, an event called *disjunction*. Each member of a set of sister chromatids moves to one pole of the spindle. So long as the separation takes place, it doesn't matter which sister chromatid moves to which pole. Failure of homologues to separate during meiosis, called nondisjunction, can lead to birth defects (Chapter 18).

Cytokinesis follows telephase I. After a brief resting period (called interkinesis), the second phase of meiosis will get under way. DNA isn't duplicated between the two meiotic divisions. But each chromosome was duplicated during interphase, and it remains in the duplicated form when meiosis II begins.

Separating the Sister Chromatids

Meiosis II is almost identical to mitosis. It occurs in the two daughter cells resulting from meiosis I, and its main function is to separate the two sister chromatids of each

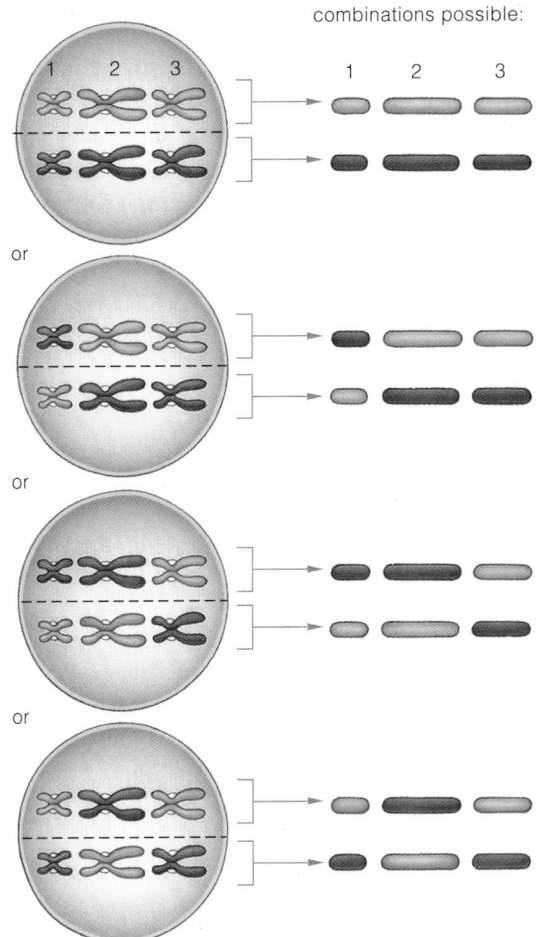

combinations possible:

Figure 16.13 Possible outcomes of the random alignment of three pairs of homologous chromosomes at metaphase I of meiosis. Maternal chromosomes are *purple*; paternal ones *blue*.

chromosome in the daughter cells. Prophase II may be fleeting or may not occur. At metaphase II, each duplicated chromosome moves to the spindle equator. Each duplicated chromosome splits at anaphase II; its former sister chromatids are now chromosomes in their own right.

As Figure 16.11 shows, at the close of anaphase II one of each type of chromosome moves to each spindle pole. During telophase II, new nuclear membranes form around the chromosomes after they have clustered at the two poles. Meiosis is complete. Cytokinesis now pinches each cell in two. The result of the two meiotic divisions is a set of four cells, each cell with a haploid number of unduplicated chromosomes.

Crossing over is an interaction between a pair of homologous chromosomes. It breaks up old combinations of alleles and puts new ones together.

In metaphase I, the random attachment and later positioning of each pair of maternal and paternal chromosomes lead to varied combinations of parental traits in offspring.

MEIOSIS AND THE LIFE CYCLE

As with other multicellular animals, the human life cycle proceeds from meiosis to gamete formation, fertilization, then growth of the new individual by way of mitosis. **Spermatogenesis** is the term for meiosis and gamete formation in males (Figure 16.14). First, a diploid germ cell increases in size. The resulting large, immature cell (a primary spermatocyte) undergoes meiosis. The resulting

four haploid spermatids then change in form, develop tails, and become sperm—mature male gametes.

In females, meiosis and gamete formation are called **oogenesis** (Figure 16.15). Oogenesis differs from spermatogenesis in some important features. Compared to a primary spermatocyte, many more cytoplasmic components accumulate in a primary oocyte, the female germ

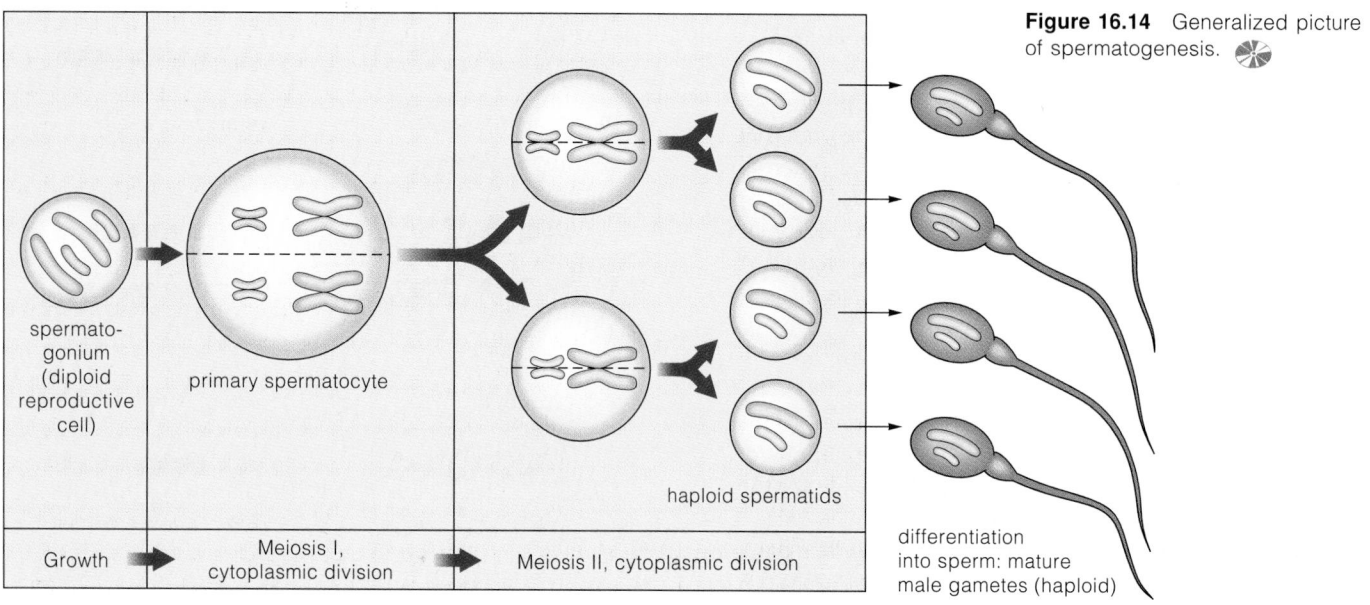

Figure 16.14 Generalized picture of spermatogenesis.

spermato-gonium (diploid reproductive cell)

primary spermatocyte

haploid spermatids

differentiation into sperm: mature male gametes (haploid)

Growth → Meiosis I, cytoplasmic division → Meiosis II, cytoplasmic division

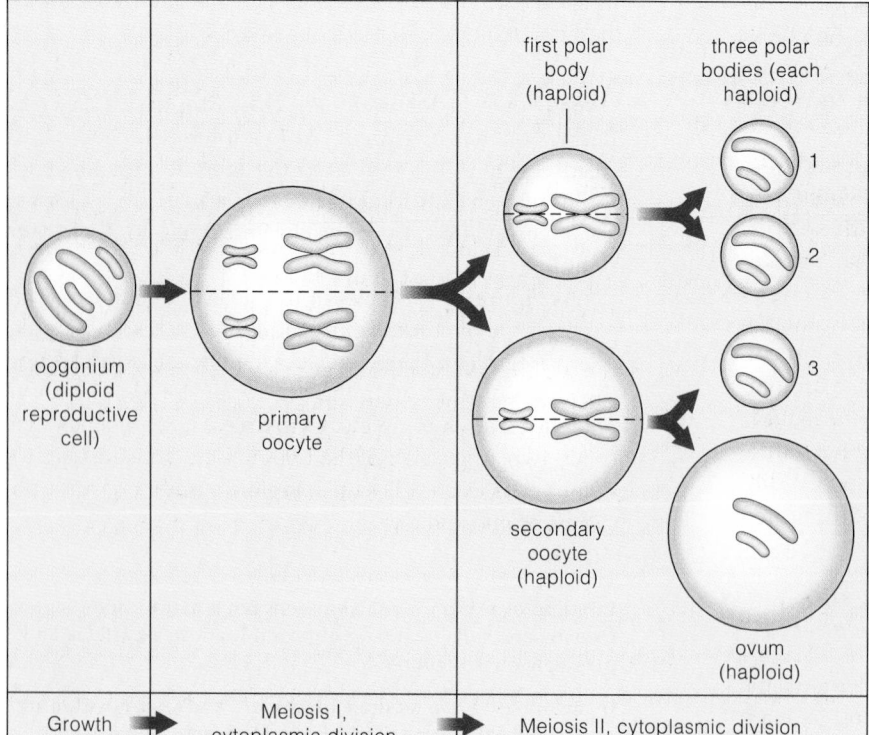

first polar body (haploid)

three polar bodies (each haploid)

1
2
3

oogonium (diploid reproductive cell)

primary oocyte

secondary oocyte (haploid)

ovum (haploid)

Growth → Meiosis I, cytoplasmic division → Meiosis II, cytoplasmic division

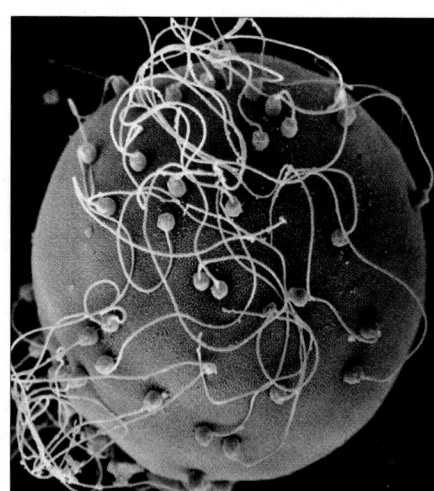

Figure 16.15 Generalized picture of oogenesis. The diagram is not at the same scale as Figure 16.14. An egg becomes *much* larger than a sperm cell, as indicated by the micrograph. Also, the three polar bodies are extremely small compared to a mature egg.

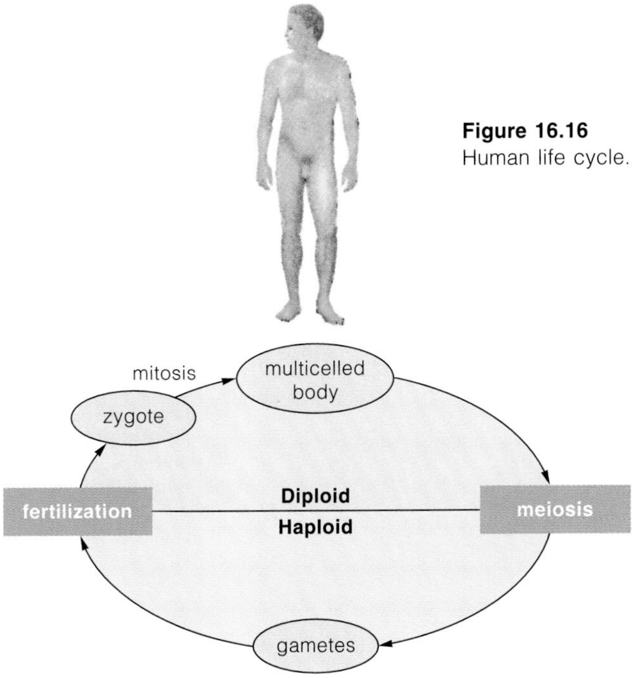

Figure 16.16
Human life cycle.

cell that undergoes meiosis. Also, in females the cells formed after meiosis differ in their size and function.

The early stages of oogenesis take place in a developing female embryo. Until she reaches puberty, however, her primary oocytes remain arrested in prophase I. Then, each month, meiosis resumes in (usually) one oocyte that is ovulated. Following meiosis I, this cell, the secondary oocyte, receives nearly all the cytoplasm; the other, much smaller cell is a polar body (Section 13.5). Both cells enter meiosis II, but the process is arrested again at metaphase II. If fertilization occurs, meiosis II continues, and the outcome is one large cell and three extremely small polar bodies. The polar bodies serve as "dumping grounds" for three sets of chromosomes so that the future egg ends up with the necessary haploid number. The large cell develops into the mature egg (ovum). Its ample supply of cytoplasm contains components that will help guide development of an embryo.

Gametes produced in oogenesis and spermatogenesis are available for fertilization, the next stage in the human life cycle (Figure 16.16). Fertilization restores the diploid number of chromosomes in a forthcoming individual. *Focus on Our Environment* discusses how ionizing radiation can disrupt gamete formation, and other events, with tragic consequences.

Spermatogenesis is the process in males by which meiotic cell divisions followed by further development give rise to sperm, the male's haploid gametes.

Oogenesis is the analogous process in females. It gives rise to ova, the female's haploid gametes.

Focus on Our Environment

16.11

IONIZING RADIATION: INVISIBLE THREAT TO CELLS

G. Tyler Miller, Jr.

Ionizing radiation can take many forms, and all of them can damage human cells. Natural sources include cosmic rays from outer space and radioactive radon gas in rocks and soil. Most nonnatural ionizing radiation comes from medical and dental X rays and from diagnostic tests and treatments using radioactive isotopes. Nuclear power plants are not major sources, *provided* they operate properly. In 1986 a nuclear reactor at Chernobyl in the former Soviet Union "melted down," releasing large amounts of ionizing radiation into the atmosphere. At least 36 people died immediately, and 237 more became seriously ill. Officials estimate that 2 to 4 million others who were exposed to the radiation will eventually develop radiation-related health problems.

Ionizing radiation damages cells by breaking apart chromosomes and altering genes. If the damage occurs in germ cells, the resulting gametes may give rise to infants with genetic defects. If somatic cells are affected, the damage can include burns, miscarriages, eye cataracts, and cancers of the bone, thyroid, breast, skin, and lung. If an affected cell has fragmented chromosomes, the spindle apparatus cannot harness and move the fragments when the cell divides. The cell or its daughters may then die.

As we learned at Chernobyl and in Japan at the end of World War II, when a person receives a large dose of ionizing radiation over a short time, the radiation typically destroys cells of the immune system, epithelial cells of the skin and intestinal lining, red blood cells, and other cell types. The results are raging infections, intestinal hemorrhages, anemia, and wounds that do not heal.

Small doses of ionizing radiation over a long period of time appear to cause less damage than the same total dosage given all at once. This may be due in part to the body's ability to repair some of the damage. Even so, a 1990 study by the U.S. National Academy of Sciences concluded that the likelihood of getting cancer from exposure to a low dose of radiation is three to four times higher than previously thought. A 1991 study of white male workers at the Oak Ridge National Laboratory in Tennessee revealed that those workers exposed to radiation well below established permissible limits had a death rate from leukemia (bone marrow cancer) 63 percent higher than that of white males in the general population.

Many scientists believe that for individuals, the best policy is to avoid unnecessary exposure, having X rays only when they are essential for health or diagnostic purposes.

MEIOSIS AND MITOSIS COMPARED

In this chapter, we have focused on two different mechanisms that divide the DNA in cell nuclei. Mitosis occurs in somatic cells, and it is the basis of body growth and tissue repair. Meiosis occurs only in germ cells; it is the basis of gamete formation and sexual reproduction. Figure 16.17 summarizes the similarities and differences between the two mechanisms.

The two mechanisms also differ in another crucial way. *Mitotic cell division produces genetically identical copies of a parent cell. Meiosis, together with fertilization, promotes variation in traits among offspring. First,* crossing over at prophase I of meiosis puts new combinations of alleles in chromosomes. *Second,* the movement of each member of homologous chromosome pairs to opposite spindle poles after metaphase I puts different mixes of maternal and paternal alleles into gametes. *Third,* different combinations of alleles are brought together by chance at fertilization.

Figure 16.17 Comparison of mitosis and meiosis, using a diploid (2n) animal cell as the example. This diagram is arranged to help you compare the similarities and differences between the two mechanisms. Maternal chromosomes are shaded *purple*, and paternal chromosomes are *blue.*

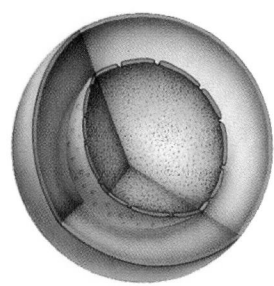

A diploid (2n) *somatic* cell is at interphase. DNA is replicated (all chromosomes are duplicated) before cell division begins.

MEIOSIS I

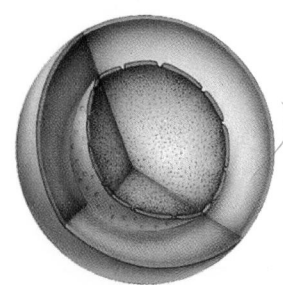

A diploid (2n) *reproductive* cell is at interphase. DNA is replicated (all chromosomes are duplicated) before cell division begins.

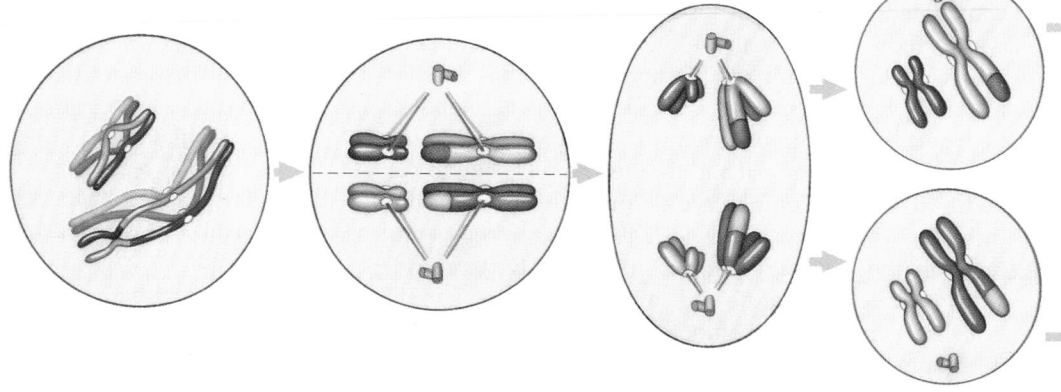

PROPHASE I

Each duplicated chromosome (consisting of two sister chromatids) condenses to threadlike form, then rodlike form. *Crossing over* occurs. Each chromosome unzips from its homologue. Each gets attached to the spindle in transition to metaphase.

METAPHASE I

All chromosomes are now positioned at the spindle's equator.

ANAPHASE I

Each chromosome is separated from its homologue. They are moved to opposite poles of the spindle.

TELOPHASE I

When the cytoplasm divides, there are two cells. Each has a haploid (n) number of chromosomes, but these are still in the duplicated state.

MITOSIS

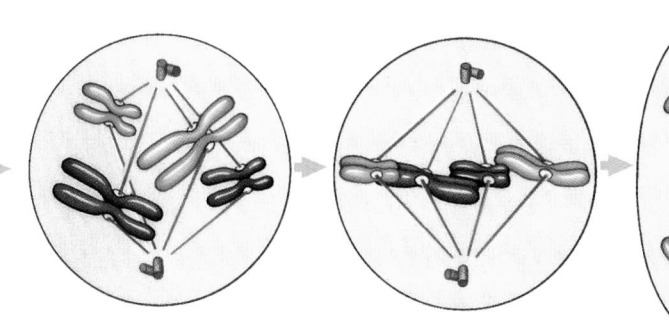

PROPHASE

Each duplicated chromosome (consisting of two sister chromatids) condenses from threadlike form to rodlike form. Each gets attached to the spindle during the transition to metaphase.

METAPHASE

All chromosomes are now positioned at the spindle's equator.

ANAPHASE

Sister chromatids of each chromosome are separated from each other. These new, daughter chromosomes are moved to opposite poles of the spindle.

TELOPHASE

When the cytoplasm divides, there are two cells. Each is diploid (2n)—*it has the same chromosome number as the parent cell.*

MEIOSIS II

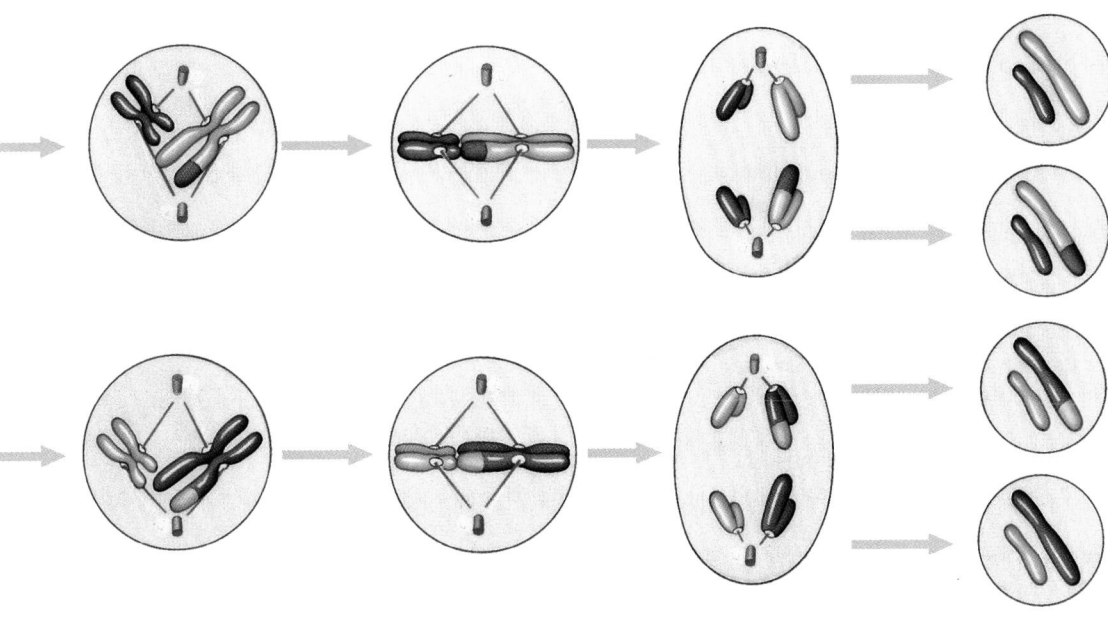

PROPHASE II

Before prophase II, the two centrioles in each new cell were moved apart and a new spindle formed. Now, each chromosome becomes attached to the spindle and starts moving toward its equator.

METAPHASE II

All chromosomes are now positioned at the spindle's equator.

ANAPHASE II

Sister chromatids of each chromosome are separated from each other. These new, daughter chromosomes are moved to opposite poles of the spindle.

TELOPHASE II

Four daughter nuclei form. When the cytoplasm divides, each new cell is haploid (n). *The original chromosome number has been reduced by half.* One or all of these cells may become gametes.

SUMMARY

1. Each cell of a new generation must receive a duplicate of all the parental DNA and enough cytoplasmic machinery to start up its own operation.

2. Somatic (body) cells usually have a diploid number of chromosomes. That is, they have two of each type of chromosome characteristic of the species. Except for sex chromosomes, the pairs of homologous chromosomes are alike in length, shape, and the genes they carry.

3. Chromosomes are duplicated during interphase (between cell divisions). Whereas each was one DNA molecule (and associated proteins), now it consists of two, temporarily attached as sister chromatids.

4. Mitosis maintains the parental number of chromosomes in each of two daughter nuclei. If the parental nucleus is diploid, the nucleus in each daughter cell will also be .diploid. Mitosis is the basis of body growth and tissue repair. Actual division of the cytoplasm, or cytokinesis, occurs toward the end of mitosis or at some point afterward.

5. A cell destined to divide by mitosis spends a significant portion of the cell cycle in interphase, a period between mitotic divisions. Interphase includes a primary growth stage (G1), during which the cell increases in mass and number of cytoplasmic components; a stage (S) in which its chromosomes are replicated; and a final, brief growth stage (G2).

6. Mitosis proceeds through four continuous stages:

a. Prophase. Duplicated, threadlike chromosomes condense into rodlike structures; new microtubules start to assemble in organized arrays near the nucleus; they will form a spindle apparatus. The nuclear envelope disappears.

b. Metaphase. Spindle microtubules harness each chromosome and orient its two sister chromatids toward opposite spindle poles. All chromosomes align at the spindle equator.

c. Anaphase. Sister chromatids of each chromosome separate. Both are now independent chromosomes, and they move to opposite poles.

d. Telophase. Chromosomes decondense and a new nuclear envelope forms around the two clusters of chromosomes. Mitosis is completed. Cytokinesis divides the cytoplasm; the result is two diploid cells.

7. Meiosis is a reductional division mechanism that halves the parental diploid chromosome number— producing the haploid number. It consists of two consecutive rounds of chromosome partitioning that sort out the chromosomes in a germ cell. In meiosis I, each chromosome pairs with and then separates from its homologue. In meiosis II, the sister chromatids of each chromosome separate from each other. In both cases, microtubules of a spindle apparatus move the chromosomes.

8. The following key events occur during meiosis I:

a. At prophase I, homologues pair with each other. Crossing over breaks up old combinations of alleles and puts together new ones in the chromosomes. This genetic recombination leads to variation in traits among offspring.

b. At metaphase I, all pairs of homologous chromosomes are aligned at the spindle equator.

c. At anaphase I, paired homologous chromosomes separate from one another and move to opposite spindle poles.

9. The following key events occur during meiosis II:

a. At metaphase II, all chromosomes are moved to the spindle equator.

b. At anaphase II, the sister chromatids of each chromosome separate and move to opposite poles. Once separated, they are chromosomes in their own right.

10. Following cytokinesis (cytoplasmic division), there are four haploid cells, one or all of which may function as gametes.

Table 16.2 Basics of Mitosis and Meiosis

	Mitosis	Meiosis
Function	Growth, including repair and maintenance	Gamete production (sperm/eggs)
Occurs in	Somatic (body) cells	Germ cells in gonads (testes and ovaries)
Mechanism	One round of chromosome partitioning plus cytokinesis	Two rounds of chromosome partitioning and cytokinesis
Outcome	Maintains diploid chromosome number ($2n \rightarrow 2n$)	Reduces diploid chromosome number ($2n \rightarrow n$)
Effect	Two diploid daughter cells	Four haploid daughter cells

Review Questions

1. Define the two types of chromosome partitioning mechanisms that occur in the human body. What is cytokinesis? *352, 360*

2. Define somatic cell and germ cell. Which type of cell can undergo mitosis? *352*

3. What is a chromosome? What are chromosomes called during interphase, when they are not condensed? *352*

4. What are homologous chromosomes? Relate this concept to the diploid number. *353*

5. Describe the spindle apparatus and its general function in chromosome partitioning. *356, 358–359*

6. Name the four main stages of mitosis, and describe the main features of each stage. *356–360*

7. In a paragraph, summarize the similarities and differences between mitosis and meiosis. *368–369*

8. Explain the significance of fertilization. *367*

Critical Thinking: You Decide *(Key in Appendix V)*

1. Under normal circumstances you can't inherit both copies of a homologous chromosome from the same parent. Why? Assuming that no crossing over has occurred, what is the chance that one of your non-sex chromosomes is an exact copy of the same chromosome possessed by your maternal grandmother?

2. Suppose you have a means of measuring the amount of DNA in a single cell during the cell cycle. You first measure the amount during the G1 phase. At what points during the remainder of the cycle would you predict changes in the amount of DNA per cell?

Self-Quiz *(Answers in Appendix IV)*

1. DNA is distributed to daughter cells by _____ or _____, both of which are mechanisms for allocating chromosomes.

2. Each kind of organism contains a characteristic number of _____ in each cell; each of those structures is composed of a _____ molecule with its associated proteins.

3. A pair of chromosomes that are similar in length, shape, and the traits they govern are called _____.
   a. diploid chromosomes
   b. mitotic chromosomes
   c. homologous chromosomes
   d. germ chromosomes

4. Somatic cells usually have a _____ number of chromosomes.

5. Interphase is the stage when _____.
   a. nothing occurs
   b. a germ cell forms its spindle apparatus
   c. a cell grows and duplicates its DNA
   d. cytokinesis occurs

6. After mitosis, each daughter cell contains genetic instructions that are _____ and _____ chromosome number of the parent cell.
   a. identical to the parent cell's; the same
   b. identical to the parent cell's; one-half the
   c. rearranged; the same
   d. rearranged; one-half the

7. All of the following are stages of mitosis *except* _____.
   a. prophase
   b. interphase
   c. metaphase
   d. anaphase

8. A duplicated chromosome has _____.
   a. one chromatid
   b. two chromatids
   c. three chromatids
   d. four chromatids

9. Crossing over in meiosis _____.
   a. alters the chromosome alignments at metaphase
   b. occurs between sperm DNA and egg DNA at fertilization
   c. leads to genetic recombination
   d. occurs only rarely

10. Because of the _____ alignment of homologous chromosomes at metaphase I, gametes can end up with _____ mixes of maternal and paternal chromosomes.
    a. unvarying; different   c. random; duplicate
    b. unvarying; duplicate   d. random; different

11. Following meiosis and cytokinesis, there are _____ haploid cells, one or all of which may function as _____.
    a. two; body cells   c. four; gametes
    b. two; gametes   d. four; body cells

12. Match each stage of mitosis with the following key events.
    _____ metaphase   a. sister chromatids of each chromosome
    _____ prophase   separate and move to opposite poles
    _____ telophase   b. chromosomes condense and a
    _____ anaphase   microtubular spindle forms
                         c. chromosomes decondense, daughter
                         nuclei re-form
                         d. all chromosomes align at spindle equator

Selected Key Terms

allele *364*   interphase *354*
anaphase *356*   kinetochore *358*
cell cycle *354*   meiosis *352*
centromere *352*   metaphase *356*
chromosome *352*   mitosis *352*
chromosome number *352*   oogenesis *366*
cleavage furrow *360*   prophase *356*
crossing over *364*   reductional division *361*
cytokinesis *360*   sister chromatid *352*
diploid *352*   somatic cell *352*
genetic recombination *364*   spermatogenesis *366*
germ cell *352*   spindle apparatus *356*
haploid *361*   telophase *356*
homologous chromosome *353*

Readings

Glover, D., C. Gonzalez, and J. Raff. June 1993. "The Centrosome." *Scientific American.* New insights about how the microtubule organizing center guides mitosis and affects cell movements and shape.

Gould, S. J. 1980. *The Panda's Thumb.* New York: Norton. The classic essay "Dr. Down's Syndrome" describes one outcome in humans when meiosis falters.

Murray, A., and M. Kirschner. March 1991. "What Controls the Cell Cycle?" *Scientific American.*

 For additional readings, go to InfoTrac College Edition, your online library at:

http://www.infotrac-college.com/wadsworth

OBSERVABLE PATTERNS OF INHERITANCE

A Smorgasbord of Ears and Other Traits

Basketball ace Charles Barkley has them. So does actor Tom Cruise. Actress Joan Chen doesn't, and neither did a monk named Gregor Mendel. To see how *you* fit in with these folks, use a mirror to check your ears. Is the fleshy lobe at the base of each ear attached to the side of your head? If so, you and Barkley and Cruise have something in common. Or is the fleshy lobe unattached, so that you can flap it back and forth? If so, you are like Chen and Mendel (Figure 17.1).

b Charles Barkley

a Tom Cruise

Whether a person is born with detached or attached earlobes depends on a single kind of gene. That gene comes in slightly different molecular forms—alleles. Only one form has information conferring detached lobes. The information is put to use while a human body is developing inside the mother. It calls for a death signal, which is sent to all the cells positioned between the newly forming lobes and the head. Without the signal, the cells don't die, and earlobes don't detach.

We all have genes for thousands of traits, including earlobes, cheeks, lashes, and eyeballs. Most of the traits vary in their details from person to person. Remember, humans inherit pairs of genes, on pairs of chromosomes.

In some pairings, one allele has powerful effects and overwhelms the other's contribution to a trait. The over-whelmed allele is said to be recessive to the dominant one. If you have *detached* earlobes, *dimpled* cheeks, *long* lashes, or *large* eyeballs, you carry at least one and possibly two dominant alleles that influence the trait in a particular way.

When both alleles of a pair are recessive, nothing masks their effect on a trait. You get *attached* earlobes with one pair of recessive alleles (and *flat* feet with another, a *straight* nose with another, and so on).

How did we discover such remarkable things about our genes? It all started with Gregor Mendel. By analyz-ing pea plants generation after generation, Mendel found indirect but *observable* evidence of how parents transmit units of hereditary information—genes—to offspring. This chapter focuses on Mendel's experimental methods and results. They remain a classic example of how a scientific approach can pry open important secrets about the natural world. And to this day, they serve as the foundation for modern genetics.

c Joan Chen

d Gregor Mendel

Figure 17.1 Attached and detached earlobes of representative humans. This sampling provides observable evidence of a trait governed by a single gene, which exists in different molecular forms in the human population. Which version of the trait do you have? It depends on which molecular forms of the gene you inherited from your mother and your father. As Gregor Mendel perceived, such easily observable traits can be used to identify patterns of inheritance that exist from one generation to the next.

KEY CONCEPTS

1. Genes are units of information about heritable traits. Alleles, which are slightly different molecular forms of a gene, specify different versions of the same trait.

2. Each gene has a particular location on a particular chromosome. Humans inherit pairs of genes, at equivalent locations on pairs of homologous chromosomes.

3. When each pair of homologous chromosomes moves apart at meiosis, their paired genes also move apart, and they end up in different gametes. Gregor Mendel found indirect evidence of this gene segregation when he cross-bred plants showing different versions of the same trait (such as purple or white flowers).

4. Each pair of homologous chromosomes (and the genes they carry) is sorted out for distribution into one gamete or another independently of how the other pairs of homologous chromosomes are sorted. Mendel found indirect evidence of this when he tracked plants having observable differences in two traits, such as flower color and height.

5. The contrasting traits that Mendel happened to study were specified by nonidentical alleles. One allele was dominant, in that its effect on a trait masked the effect of a recessive allele paired with it.

6. Not all traits have such clearly dominant or recessive forms. One allele of a pair may be fully or partially dominant over its partner or codominant with it. Also, two or more gene pairs may influence the same trait, and some single genes influence many traits. Environmental conditions may induce further variation in traits.

CHAPTER AT A GLANCE

17.1   Patterns of Inheritance

17.2   Mendel's Theory of Segregation

17.3   Doing Genetic Crosses and Figuring Probabilities

17.4   The Testcross: A Tool for Discovering Genotypes

17.5   Independent Assortment

17.6   A Closer Look at Independent Assortment

17.7   Variations on Mendel's Themes

17.8   Less Predictable Variations in Traits

17.9   *Mom and Pop Genes: Genomic Imprinting*

PATTERNS OF INHERITANCE

The Origins of Genetics

Having read about meiosis in Chapter 16, you already have insight into the mechanisms of sexual reproduction. That is more than Gregor Mendel had. Mendel was born into a farm family in what is now the Czech Republic. By the 1850s he had become a university-educated monk, specializing in studies of both theology and what was then called "natural history." Even so, Mendel did not know about chromosomes, so he could not have known that the chromosome number is reduced by half in gametes, then restored at fertilization. Yet Mendel was interested in discovering how traits were inherited in the offspring of sexually reproducing organisms, and to pursue this question he experimented with the garden pea plant. This plant is self-fertilizing: Its flowers produce sperm (pollen) *and* eggs, and fertilization occurs in the same flower. Based on his research, Mendel hypothesized that fertilization united "factors" from each parent that were the units of heredity. Today we call such factors *genes*.

Some Terms Used in Genetics

The following list expresses some of Mendel's ideas in modern terms (see also Figure 17.2):

1. **Genes** are units of information about specific traits, and they are passed from parents to offspring. Each gene is a segment of DNA and has a specific location (locus) on a chromosome.

2. Diploid cells have a pair of genes for each trait, on pairs of homologous chromosomes.

3. Although both genes of a pair deal with the same trait, they may vary in their information about it. This happens when there are slight molecular differences in the DNA segments that make up each gene. Each version of the gene is called an **allele**. Contrasting alleles account for a great deal of the variation we see in traits. For example, one allele of a gene specifies a straight hairline, and another allele of the same gene specifies a widow's peak.

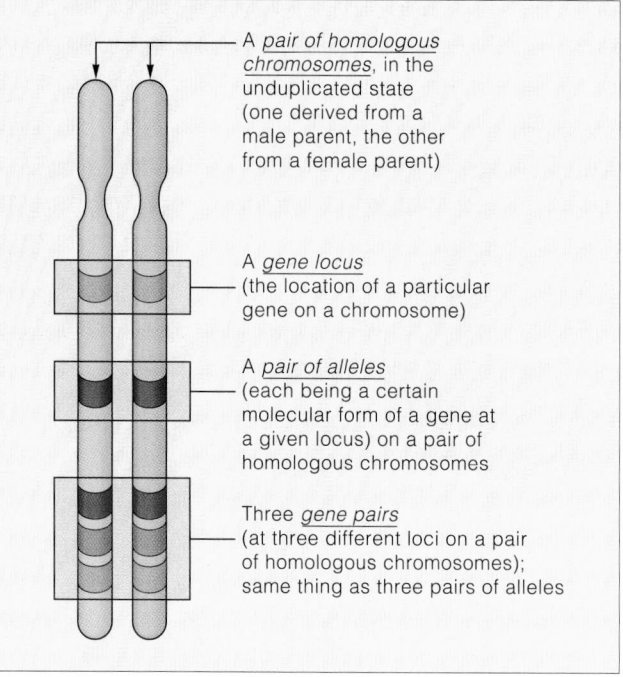

Figure 17.2 A few genetic terms illustrated. Diploid organisms such as humans have pairs of genes on pairs of homologous chromosomes. One chromosome of each pair is inherited from the mother and the other from the father.

The genes themselves may have different molecular forms, called alleles. Different alleles specify slightly different versions of the same trait. An allele at one location on a chromosome may or may not be identical to its partner on the homologous chromosome.

4. If a gene pair consists of two identical alleles, this is a **homozygous** condition (*homo:* same; *zygo:* joined together). If the gene pair consists of different alleles, this is a **heterozygous** condition (*hetero:* different).

5. An allele is **dominant** when its effect on a trait masks that of any **recessive** allele paired with it. An uppercase letter represents a dominant allele, and a lowercase letter represents a recessive allele (for instance, *A* and *a* or *C* and *c*).

6. A *homozygous dominant* individual has a pair of dominant alleles (*AA*) for the trait being studied. A *homozygous recessive* individual has a pair of recessive alleles (*aa*). A *heterozygous* individual has a pair of nonidentical alleles (*Aa*).

7. Two terms help keep the distinction clear between genes and the traits they specify. **Genotype** refers to the alleles present in an individual. **Phenotype** refers to an individual's observable traits.

Table 17.1 summarizes the relationship between genotype and phenotype.

Table 17.1	Genotype and Phenotype Compared	
Genotype	Described as	Phenotype
CC	homozygous dominant	chin fissure
Cc	heterozygous (one of each allele; dominant form of trait observed)	chin fissure
cc	homozygous recessive	smooth chin

MENDEL'S THEORY OF SEGREGATION

Mendel's idea that there are discrete units of inheritance (now called genes) was just the beginning of his search for the basis of heredity. Building on this initial concept, he proposed two hypotheses: (1) each diploid organism inherits two such units for each trait, one from each parent; and (2) in parents, different units assort independently into gametes. For the moment, we'll consider the strategy Mendel pioneered to test the first hypothesis.

To probe the question of how a single trait is passed to offspring, Mendel devised what is now called a **monohybrid cross** (*mono*- means one). In a monohybrid cross, the parents have differing allele pairs for the trait in question. Although we do not carry out controlled crosses between individual humans, monohybrid matings do occur naturally, and they show patterns of inheritance at work.

The following example illustrates a monohybrid cross for the configuration of a person's chin. This trait is governed by a gene that has two allelic forms. One allele, which calls for an indentation called a chin fissure (Figure 17.3), is dominant when it is present; we can represent this allele as *C*. The recessive allele, which codes for a smooth chin, is *c*. In this example, one parent is homozygous for the *C* form of the gene and has a chin fissure, and the other is homozygous for the *c* form and has a smooth chin. The *CC* parent produces only *C* gametes, and the *cc* parent produces only *c* gametes:

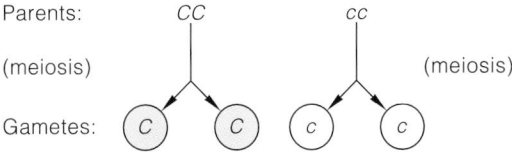

Each gamete receives one allele for the trait because in anaphase I in a germ cell, the homologous chromosomes separate from each other (Figure 16.11, page 362). We know already that the genes on homologues match up into equivalent pairs. So, when homologues separate into different gametes, the two genes of each pair separate also (Figure 17.4). This is known as the principle of **segregation**. Because meiosis also reduces the diploid chromosome number to the haploid number, each gamete contains one member of each chromosome pair.

The two copies of each gene present in a diploid organism segregate from each other during meiosis in germ cells. As a result, each gamete contains only one copy of each gene.

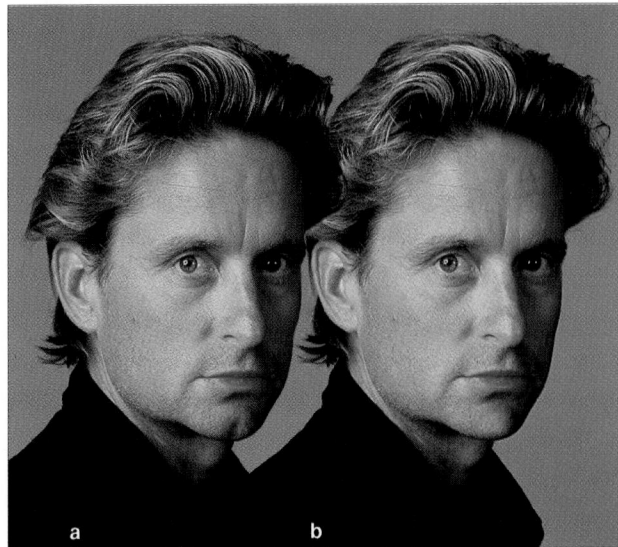

Figure 17.3 (**a**) The chin fissure, a heritable trait arising from a rather uncommon allele of a gene. Actor Michael Douglas received a gene that influences this trait from each of his parents. At least one of those genes was dominant. (**b**) What Mr. Douglas's chin might have looked like if he had inherited two ordinary forms of the gene instead.

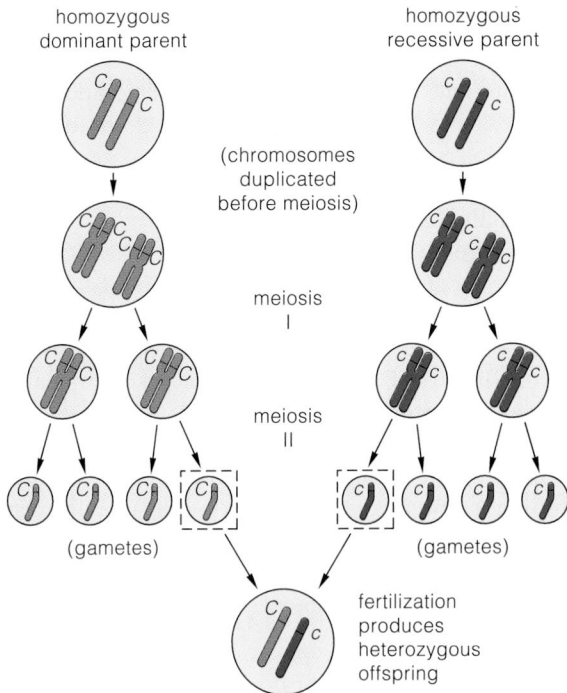

Figure 17.4 A monohybrid cross, showing how one gene of a pair segregates from the other. Two parents that are each homozygous for a different version of a trait give rise to heterozygous offspring only.

DOING GENETIC CROSSES AND FIGURING PROBABILITIES

The parent generation of a cross is designated by a **P**. Offspring of a monohybrid cross are called the **F₁** or *first filial* generation. In the present example, each child will inherit a pair of differing alleles for the trait, one from each homozygous parent. The children will thus each be heterozygous for the chin genotype, or *Cc*. Because *C* is dominant, each child will have a chin fissure.

Suppose now that one of the *Cc* children grows up and marries another *Cc* person. From this second cross we derive the **F₂** (second filial) generation. Because half of each parent's gametes (sperm or eggs) are *C* and half are *c* (due to segregation at meiosis), four outcomes are possible every time a sperm fertilizes an egg:

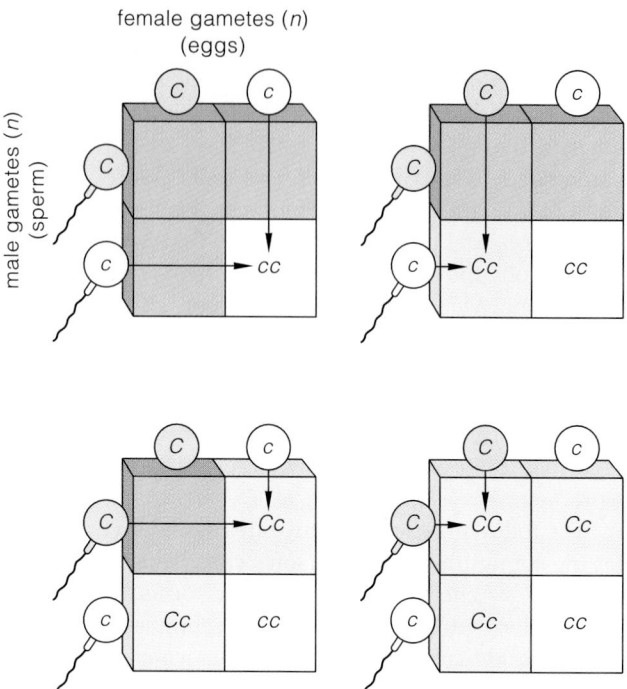

Figure 17.5 Punnett-square method of predicting the probable outcome of a genetic cross. In this example, the cross is between two heterozygous individuals. Circles represent gametes. Letters on gametes represent dominant or recessive alleles. The different squares depict the different genotypes possible among offspring.

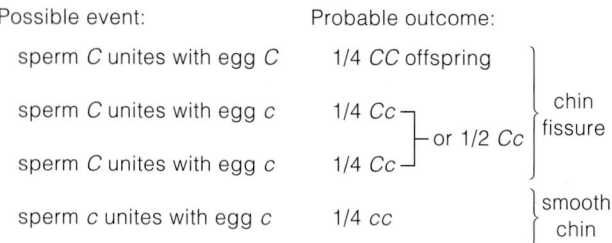

Possible event:	Probable outcome:	
sperm *C* unites with egg *C*	1/4 *CC* offspring	
sperm *C* unites with egg *c*	1/4 *Cc*	chin fissure
sperm *C* unites with egg *c*	1/4 *Cc*	or 1/2 *Cc*
sperm *c* unites with egg *c*	1/4 *cc*	smooth chin

The diagrams in Figure 17.5 show how to construct a **Punnett square**, a convenient tool for determining the probable outcome of genetic crosses. In this case, there is a 75 percent chance that a child from a cross between two *Cc* parents will have the dominant *C* allele and a chin fissure. When a large number of offspring are involved, a ratio of 3:1 is likely (Figure 17.6).

Fertilization is a chance event, which is why rules of probability apply to crosses. **Probability** is a number between zero and one that expresses the likelihood of a particular event. For example, an event that has a probability of one will always occur; an event with a probability of zero will never occur, and an event with a probability of one-half (or 50 percent) is expected to occur in about half of all opportunities (Figure 17.7).

Having a chin fissure doesn't affect a person's health. However, a fair number of human genetic disorders, including cystic fibrosis and sickle-cell anemia, result from single-gene defects and so follow a Mendelian inheritance pattern. Genetic disorders are a major topic of Chapter 18. For the moment, it is important to realize two things:

1. The outcomes (fractions) predicted by probability do not necessarily turn up in a single family. For instance, it's common to see families in which the parents have produced several children, all of the same sex. Fractions predicted by probability turn up consistently only when a large number of events is analyzed. You can test this for yourself by flipping a coin. Probability predicts that heads and tails should each come up about half the time, but you may have to flip the coin a hundred times to consistently achieve a ratio close to 1:1. Likewise, probability predicts that parents with two or more children will have equal numbers of girls and boys, but this does not always happen.

2. In a given genetic situation, *probability is constant*. The likelihood that a certain genotype will occur—say, a baby with the genotype for cystic fibrosis—is the same for every child no matter how many children a couple has. Based on the parents' genotypes, if the probability that a child will inherit a certain genotype is one in four, then each child of those parents has a one-in-four (25 percent) chance of inheriting the genotype. If the parents have three children without the trait, the fourth child still has only a one-in-four chance of inheriting it.

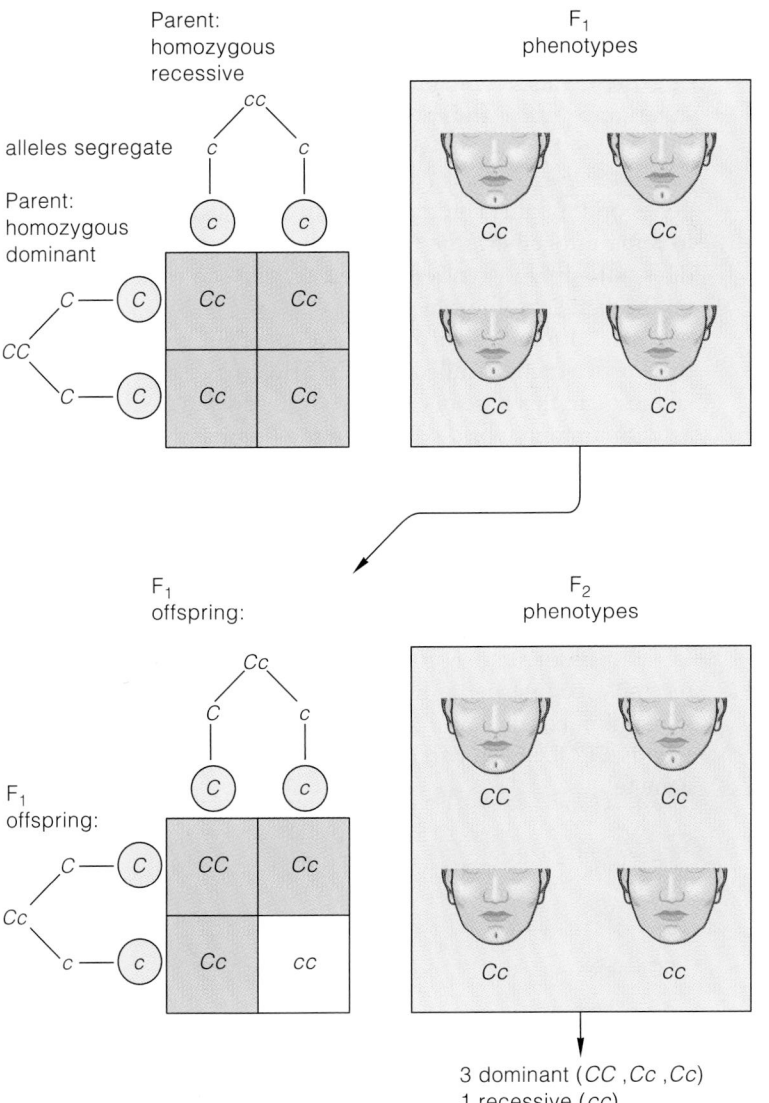

Parent:
homozygous
recessive

alleles segregate

Parent:
homozygous
dominant

F₁
phenotypes

Cc   *Cc*

Cc   *Cc*

F₁
offspring:

F₁
offspring:

F₂
phenotypes

CC   *Cc*

Cc   *cc*

3 dominant (*CC*, *Cc*, *Cc*)
1 recessive (*cc*)

Figure 17.6 Results from a monohybrid cross, in which one parent is homozygous dominant for a trait and the other is homozygous recessive for the same trait. Notice that the dominant-to-recessive ratio is 3:1 for the second generation (F₂) offspring.

How to Calculate Probability

Step 1. Actual genotypes of parental gametes

In the cross *Cc* × *Cc*, gametes have a 50–50 chance of receiving either allele (*C* or *c*) from each parent. Said another way, the probability that a particular sperm or egg will be *C* is 1/2, and the probability that it will be *c* is also 1/2:

probability of *C*:	1/2
probability of *c*:	1/2

Step 2. Probable genotypes of offspring

Offspring receive one allele from each parent. Three different combinations of alleles are possible in this cross. To figure the probability that a child will receive a particular allele combination, simply multiply the probabilities of the individual alleles:

probability of *CC*:	1/2 × 1/2 = 1/4
probability of *Cc*:	1/2 × 1/2 = 1/4
probability of *cC*:	1/2 × 1/2 = 1/4
probability of *cc*:	1/2 × 1/2 = 1/4

} 1/2

Step 3. Probable phenotypes

Chin fissure: (*CC*, *Cc*, *cC*)	1/4 + 1/4 + 1/4 = 3/4
Smooth chin: (*cc*)	1/4

Figure 17.7 Calculating probabilities. Simple multiplication lets you figure the probability that a child will inherit alleles for a particular phenotype.

Rules of probability apply to the inheritance of single gene traits. Thus, if the genotypes of parents are known, it is possible to establish a potential child's chances for inheriting a particular genotype and thus for having a particular phenotype (trait).

THE TESTCROSS: A TOOL FOR DISCOVERING GENOTYPES

In working with nonhuman organisms, researchers often use a simple technique called a **testcross** to ascertain an unknown genotype. One individual in the testcross is known to be homozygous for the recessive allele of the trait being studied. The other has a dominant phenotype that could be the outcome of either a homozygous or a heterozygous condition (Figure 17.8a). The phenotype of the offspring reveal the genotype of the parent with the dominant phenotype.

We can find similar situations in human genetics that shed light on the genotype of a human parent. Suppose, for example, that a woman has smooth cheeks and her husband has dimples. "No dimples" is a recessive trait, so the woman must be *dd*. As Figure 17.8b diagrams, if a child is born with no dimples, then the father must be a heterozygote for this trait, with a genotype of *Dd*; that is the only way he could father a *dd* child. If the child has dimples, the father can be either *DD* or *Dd*. If he is *Dd*, a heterozygote, the probability that he will have a dimpled child is 1/2, a 50–50 chance, every time. If he were *DD*, every child would be dimpled.

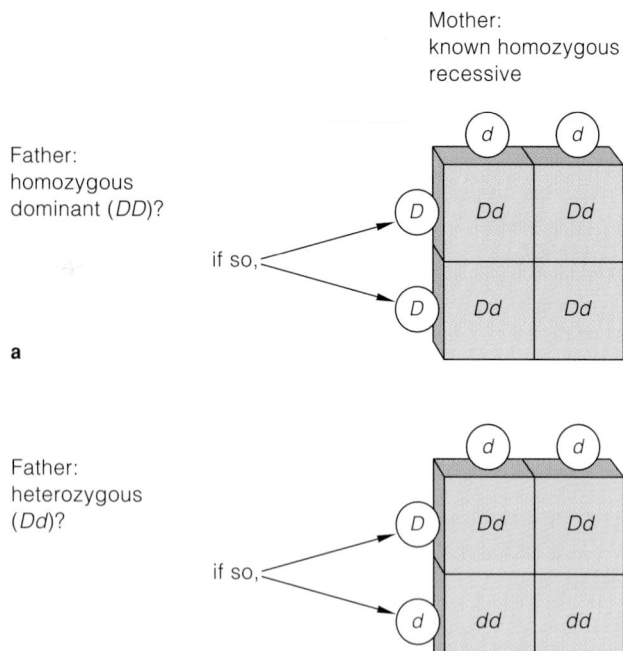

a

b

Figure 17.8 A testcross "works backward" from the phenotypes of offspring to help identify the genotype of a parent. In this example, the father's genotype can be determined if he mates with a woman known to be homozygous recessive for the trait in question *and* if they produce a child who shows the recessive phenotype. Then, the father must be a heterozygote. If the child shows the dominant phenotype, then all we can say for sure is that the father is either a heterozygote or homozygous dominant for the trait.

INDEPENDENT ASSORTMENT

Mendel also addressed the question of how genes for two *different* traits are passed to offspring. Based on the results of experiments, he established another basic principle of inheritance, called **independent assortment**. This general rule states that *alleles for different traits are sorted into gametes independently of one another* (Figure 17.9). Because genes on the same chromosome can be linked (they tend to remain together), the rule of independent assortment does not apply to genes that are very close to each other on the same chromosome.

Mendel found evidence for independent assortment by way of a method called the **dihybrid cross**. As its name suggests, in this type of cross two traits are studied. And as with a monohybrid cross, "simple dominance" exists: There are two contrasting alleles of each gene, one dominant and one recessive. The parents have different combinations of alleles. Independent assortment becomes apparent when we follow two generations of a dihybrid cross.

Let's consider an example using dominant alleles *C* for chin fissure and *D* for dimples and *c* and *d* as their recessive counterparts. In this case, let's say that the P generation consists of a man who is *ccdd* and a woman who is *CCDD*. Each parent can produce just one type of gamete:

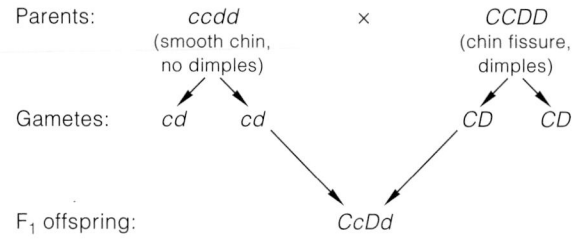

As Mendel would have predicted, all of this couple's children (the F$_1$ generation) are *CcDd* and have a chin fissure and dimples.

What happens in a second generation if two *CcDd* individuals mate? The man and woman can each produce four types of gametes in equal proportions:

1/4 *CD*   1/4 *Cd*   1/4 *cD*   1/4 *cd*

This outcome is possible because gene pairs on different chromosomes assort independently of one another during meiosis.

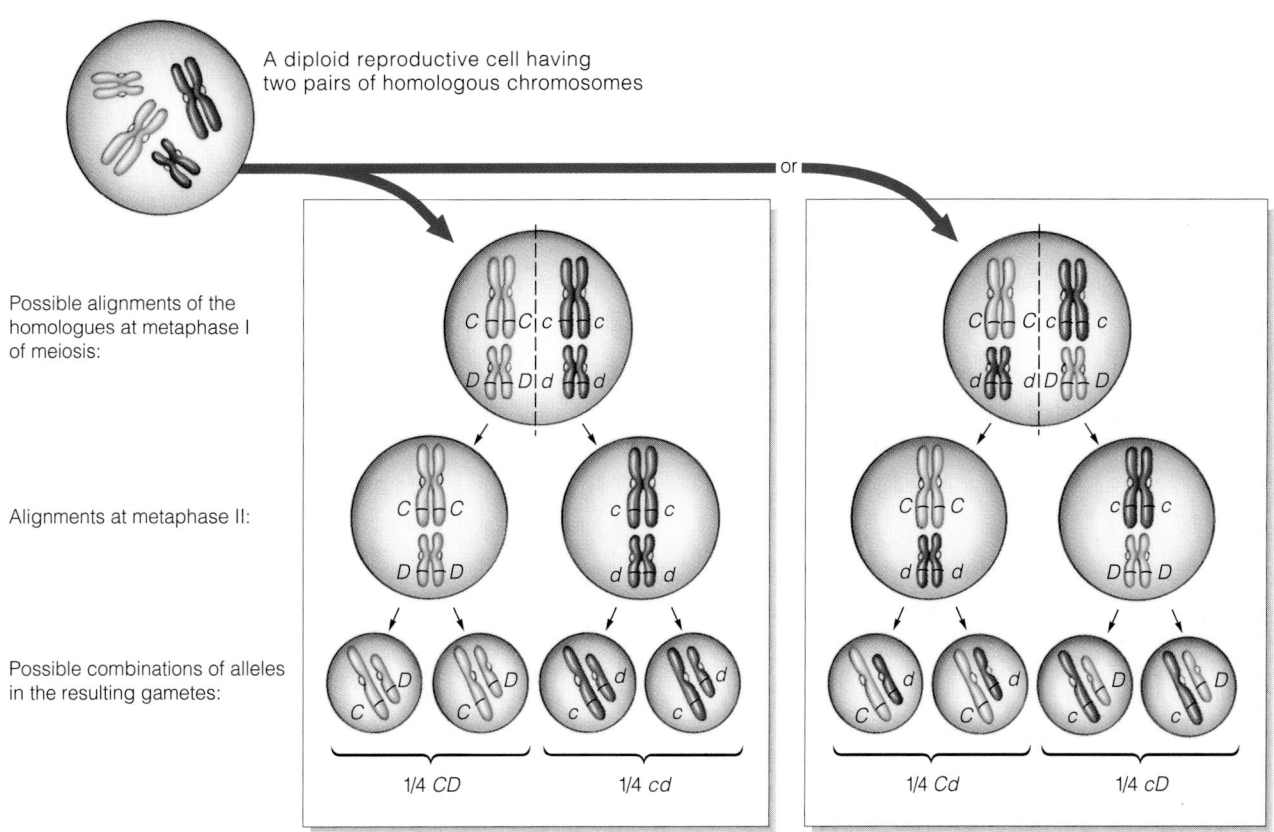

A diploid reproductive cell having two pairs of homologous chromosomes

Possible alignments of the homologues at metaphase I of meiosis:

Alignments at metaphase II:

Possible combinations of alleles in the resulting gametes:

1/4 *CD*   1/4 *cd*   1/4 *Cd*   1/4 *cD*

Figure 17.9 Example of independent assortment, showing just two pairs of homologous chromosomes. The alleles of one chromosome may or may not be identical to those of its homologue. Because either chromosome of the pair can move to one spindle pole or the other during meiosis (page 362), different gametes can end up with different mixes of alleles. Both options shown here are equally probable.

To work out the preceding result yourself, assume that the genes for chin fissure and dimples are located on two different sets of homologous chromosomes. During metaphase I of meiosis, all chromosome pairs become positioned at random at the spindle equator. The chromosome with allele *C* may ultimately move to either spindle pole, and on into any one of four gametes. The same is true for its homologue with the *c* allele. And the same is true for the chromosomes with alleles *D* and *d*. These outcomes will be our starting point in the following section, which takes the dihybrid cross a step further. It explores (as Mendel did with successive generations of pea plants) what kinds of phenotypes are likely in offspring when their parents are both heterozygous for two traits.

Because of events during meiosis, either gene of a pair may end up in a particular gamete.

A dihybrid cross can reveal evidence of independent assortment. This basic principle of inheritance states that a gene pair on one chromosome assorts into gametes independently of other gene pairs of other chromosomes.

A CLOSER LOOK AT INDEPENDENT ASSORTMENT

Simple multiplication (four kinds of sperm times four kinds of eggs) tells us that 16 gamete unions are possible when each parent in a dihybrid cross is heterozygous for the two genes in question. The Punnett square in Figure 17.10 diagrams the possibilities, using the chin fissure and dimple traits. It assumes that each parent is heterozygous at both gene loci. That is, both parents have

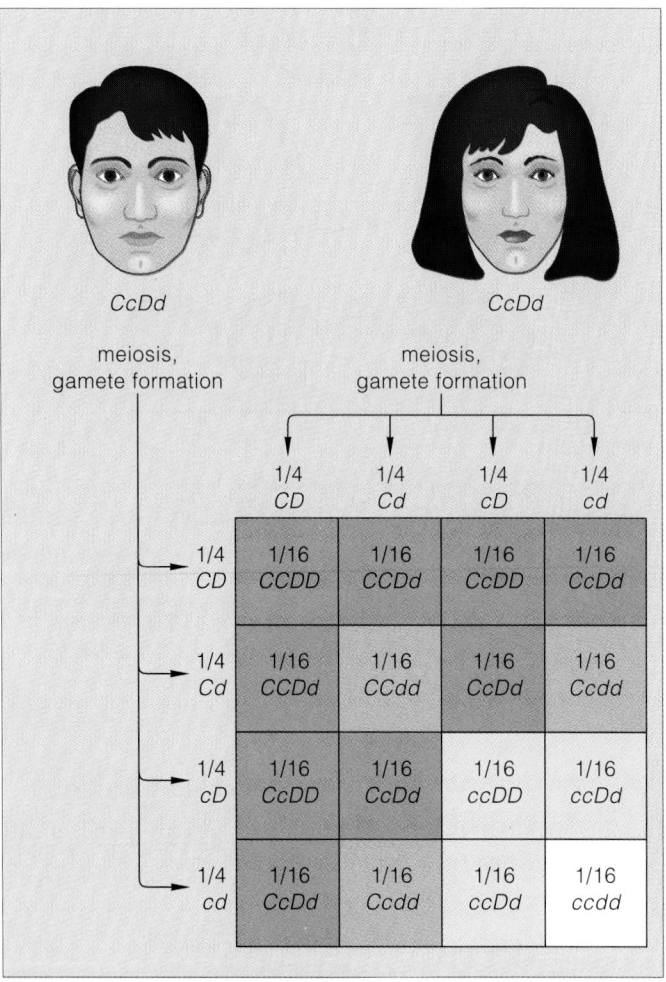

Adding up the combinations possible:

[shaded] 9/16 or 9 chin fissure, dimples

[shaded] 3/16 or 3 chin fissure, no dimples

[shaded] 3/16 or 3 smooth chin, dimples

[white] 1/16 or 1 smooth chin, no dimples

Figure 17.10 Results from a mating in which both parents are heterozygous at both loci. *C* and *c* represent dominant and recessive alleles for chin fissure. *D* and *d* represent dominant and recessive alleles for dimples. Rules of probability predict that certain combinations of phenotypes among offspring of this type of cross occur in a 9:3:3:1 ratio, on average.

Probability in a Mating Where Both Parents Are Heterozygous at Two Loci

A dihybrid cross considers two traits. If both parents are heterozygous for both traits, a dihybrid cross produces the following 9:3:3:1 phenotype ratio (Figure 17.10):

9/16 or 9 chin fissure, dimples

3/16 or 3 chin fissure, no dimples

3/16 or 3 smooth chin, dimples

1/16 or 1 smooth chin, no dimples

Individually, these phenotypes have the following probabilities:

probability of chin fissure (12 of 16) = 3/4
probability of dimples (12 of 16) = 3/4
probability of smooth chin (4 of 16) = 1/4
probability of no dimples (4 of 16) = 1/4

To figure the probability that a child will show a particular *combination* of phenotypes, multiply the probabilities of the individual phenotypes in each possible combination:

Trait combination		Probability
Chin fissure, dimples	3/4 × 3/4	9/16
Chin fissure, no dimples	3/4 × 1/4	3/16
Smooth chin, dimples	1/4 × 3/4	3/16
Smooth chin, no dimples	1/4 × 1/4	1/16

Figure 17.11 Probability applied to independent assortment.

the genotype *CcDd*. Notice that when such individuals mate, there are nine possible ways for gametes to unite that produce a chin fissure and dimples, three for a chin fissure and no dimples, three for a smooth chin and dimples, and one for a smooth chin and no dimples. The probability of any one child having a chin fissure and dimples is 9/16; a chin fissure and no dimples, 3/16; a smooth chin and dimples, 3/16; and a smooth chin and no dimples, 1/16.

Figure 17.11 describes how to calculate the probability that a child will inherit genes for a particular set of two traits on different chromosomes.

If both parents are heterozygous for the two genes in question, 16 genotypes and 4 phenotypes are possible. Following the principles of probability, these phenotypes have probabilities of 9/16, 3/16, 3/16, and 1/16 associated with them.

Many traits do not show up in predicted Mendelian ratios. Such deviations from Mendel's single-gene model, with dominant and recessive alleles, reflect a variety of genetic and environmental influences on the phenotype. Genes are chemical instructions for building proteins. We say that a gene is "expressed" when its instructions are implemented and the protein in question is synthesized.

Dominance Relations

Heterozygotes, recall, have a pair of contrasting alleles for a trait. Sometimes, *both* alleles are expressed. We see a classic example of this **codominance** in people who are heterozygotes for alleles that confer A and B blood types.

If you have type AB blood, for instance, you have a pair of codominant alleles that are both expressed in the stem cells that give rise to your red blood cells. Recall from Chapter 6 that a polysaccharide (a sugar) on the surface of red blood cells has different molecular forms that determine your blood type. ABO blood typing can reveal which form (or forms) of the polysaccharide a person has.

The gene specifying an enzyme that synthesizes this polysaccharide has three alleles, which influence the sugar molecule's form in different ways. Two alleles, *IA* and *IB*, are codominant when paired with each other. A third allele, *i*, is recessive. When paired with either *IA* or *IB*, the effect of *i* is masked. A gene that has three or more alleles is called a **multiple allele system**.

Multiple Effects of Single Genes

In **pleiotropy**, one allele has several phenotypic effects. For instance, a mutant allele in a single gene causes a disorder called *osteogenesis imperfecta*. Infants with this disorder typically develop extremely fragile bones, weak tendons and ligaments, deafness, and blue coloration in the sclera of the eye.

Another example is the disease *sickle-cell anemia*, caused by homozygosity for a recessive allele. The normal allele (*HbA*) encodes functional hemoglobin. When a person is homozygous for the recessive mutant allele, *HbS*, he or she has sickle-cell anemia. Red blood cells, normally biconcave disks, become deformed into a sickle shape when the oxygen content of blood falls below a certain level (Figure 17.12). The sickled cells clump in blood capillaries and may

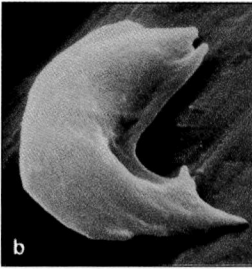

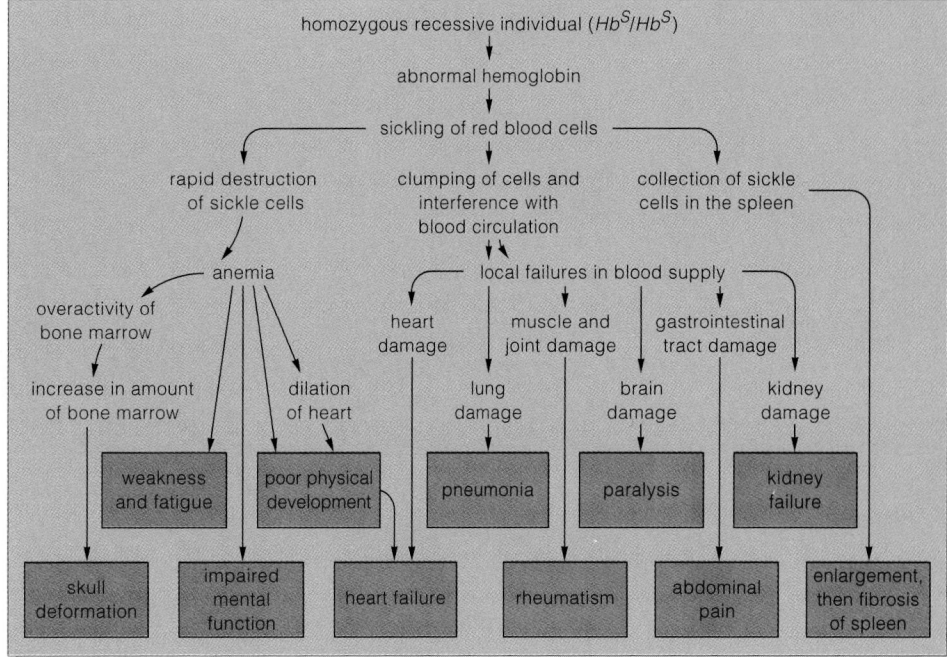

Figure 17.12 Scanning electron micrographs of normal (**a**) and sickled (**b**) red blood cells. (**c**) The range of symptoms characteristic of sickle-cell anemia.

rupture. Blood flow can be so disrupted that oxygen-deprived tissues are severely damaged. Homozygotes (*HbSHbS*) often die early. Heterozygotes (*HbA HbS*), on the other hand, have *sickle-cell trait*. Under most environmental conditions they show few symptoms because the one HbA allele provides enough normal hemoglobin to prevent red blood cells from sickling. During a crisis, sickle-cell anemia patients may receive blood transfusions, oxygen, antibiotics, and painkilling drugs. There is evidence that the food additive butyrate can reactivate "dormant" genes responsible for fetal hemoglobin, an efficient oxygen carrier that normally is produced only before birth. Some states require hospitals to screen newborn infants for sickle-cell anemia so that appropriate action can be taken right away.

The alleles at a single gene location may have positive or negative effects on two or more traits.

Penetrance and Expressivity

A gene can have an all-or-nothing effect on a trait—either you have dimples or you don't. In other cases, gene interactions as well as environmental factors may alter the *degree* to which a gene is actually expressed.

Penetrance refers to the percentage of individuals in which the phenotype associated with a particular genotype is expressed. The recessive allele that produces cystic fibrosis is completely penetrant; 100 percent of children who are homozygous for it develop the disease. The dominant allele for having extra fingers or toes (called *polydactyly*) is incompletely penetrant. Some people who inherit the allele have the normal number of digits, and others have more (Figure 17.13). Some people with *campodactyly* have immobile, bent fingers on both hands. In others, the traits show up on one hand only. When expression of an allele can produce such a range of phenotypes, the allele is said to show "variable expressivity." The allele for campodactyly also is incompletely penetrant. In some people with the allele, the traits don't show up at all.

Studies of some rare genetic disorders have uncovered evidence that a trait may differ from person to person, depending on which parent provided the allele(s) responsible for it. This phenomenon, which is called *genomic imprinting*, is the topic of this chapter's *Science Comes to Life* on the facing page.

Some of the variations we observe in gene expression probably result from gene interactions. With respect to this point, bear in mind that the path from most precursors to their products is actually a series of metabolic steps, each controlled by enzymes, which are the products of genes.

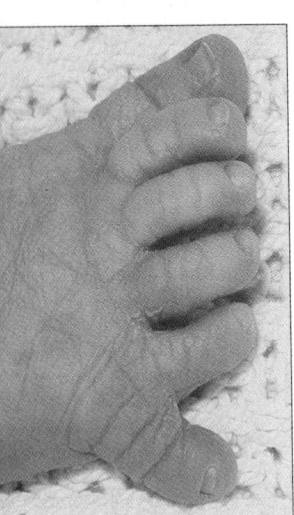

Figure 17.13 An example of polydactyly, a dominant trait.

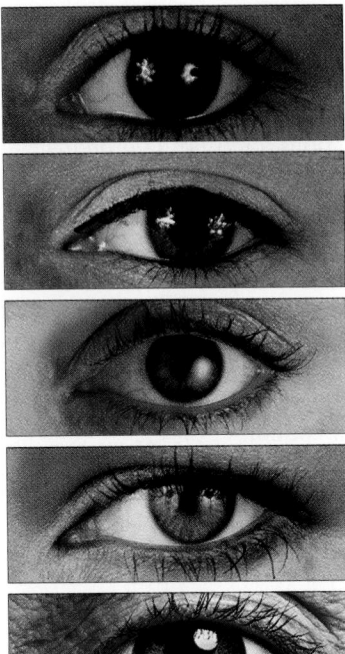

Figure 17.14 Samples from the range of continuous variation in human eye color. Different gene pairs interact to produce and deposit melanin. Among other things, this pigment helps color the eye's iris. Different combinations of alleles result in small differences in eye color. So the frequency distribution for the eye color trait appears to be continuous over the range from black to light blue.

Figure 17.15 Students at Brigham Young University, organized according to height—a vivid example of continuous variation.

1	4	8	10	16	16	15	15	14	13	13	11	9	8	8	5	1	2

Number of individuals

60	61	62	63	64	65	66	67	68	69	70	71	72	73	74	75	76	77

Height (inches)

Genes or their products interact at many steps. Hence, if different people have varying combinations of alleles for those genes, their phenotypes may vary also.

Polygenic Traits

Polygenic traits result from the combined expression of several genes. For example, skin-color and eye-color phenotypes are the cumulative result of many genes involved in the stepwise production and distribution of melanin. Black eyes have abundant melanin deposits in the iris. Dark-brown eyes have less melanin, and light-brown or hazel eyes have still less (Figure 17.14). Green, gray, and blue eyes don't have green, gray, or blue pigments. Instead, they have so little melanin that we readily see blue wavelengths of light being reflected from the iris. Hair color probably results from the interactions of several genes. This explains our real-world observation that there is a tremendous variation in human hair color.

For many polygenic traits a population may show **continuous variation**. That is, its members show a range of continuous, rather than incremental, differences in some trait. Continuous variation is especially evident in traits that are easily measurable, such as height (Figure 17.15).

Do Genes "Program" Behavior?

Identical twins have identical genes and look alike. In addition, they have many behaviors in common. Are these parallels coincidence, clever hoaxes, or evidence that aspects of behavior are inherited characteristics, like hair and eye color? Although the question is intriguing, clear answers have proven very difficult to come by.

There is strong evidence that certain basic human behaviors, such as smiling to indicate pleasure and the crying of an infant when it is hungry, are indeed genetically programmed universals. Recently, scientists have also begun to look for connections between genes and alcoholism, some types of mental illness, violent behavior, and even sexual orientation. Such studies raise controversial social issues, and so far their greatest impact has been to point out how little we know about the biological basis of human behavior. In general, most human behavior is so complex that it is extraordinarily difficult to scientifically test hypotheses about genetic links. For the time being, we can only marvel at this remarkable complexity and ponder the intriguing possibilities of human inheritance.

Gene expression can sometimes vary, so that the resulting trait (phenotype) is unpredictable. Examples include alleles that are incompletely penetrant and polygenic traits that result from the combined expression of two or more genes.

MOM AND POP GENES: GENOMIC IMPRINTING

No doubt Gregor Mendel would have been fascinated by the investigations of modern geneticists. In the case of a phenomenon called *genomic imprinting*, he might have added a footnote to his basic principles. A basic tenet of Mendelian genetics is this: The phenotypic expression of an allele is the same, regardless of the parent from which it is inherited. In the 1980s, however, researchers began to suspect another unusual phenomenon. A "genome" is the sum total of genes a person (or any other organism) inherits from his or her parents. And in some cases, it seems that a given gene can result in *different* traits in offspring, depending on which parent the gene came from. Such genes must somehow be marked, or imprinted, as "from mother" or "from father."

Tantalizing evidence for genomic imprinting has come from odd inheritance patterns for certain disorders. For example, a rare inherited neck tumor (paraganglioma) is caused by a DNA segment on chromosome 11. However, genetic analysis shows that children develop the tumors only if they inherit the tumor-causing gene from their father. Girls who have the gene (and have tumors) do not pass the trait to their children. The same connection between a paternal imprint and disease seems to hold true for several types of childhood cancer, including osteosarcoma, a bone cancer.

Two unusual genetic diseases that cause different types of mental retardation also have shed light on genomic imprinting. The diseases apparently can arise in two ways. In about 40 percent of cases, a chance glitch produces a zygote with two normal copies of chromosome 15, but *both copies are from one parent*. If the chromosomes come from the mother, the result is *Prader-Willi syndrome*, or PWS. If they come from the father, *Angelmann syndrome* (AS) occurs. More often, an affected child inherits a homologous pair of chromosome 15, one from each parent, but one of the chromosomes has a region missing. As you might guess, if the defect is in the chromosome derived from the father, PWS develops, and if the problem is in the maternal chromosome, AS develops.

SUMMARY

1. Mendel's studies with pea plants provided evidence that genes are discrete units passed to offspring.

2. Monohybrid crosses (between parents showing different versions of a single trait) show that a gene can have different alleles, some of which are dominant over other, recessive forms.

3. Homozygous dominant individuals have two dominant alleles (AA) for the trait being studied. Homozygous recessive individuals have two recessive alleles (aa). Heterozygotes have two contrasting alleles (Aa).

4. Matings between two heterozygous individuals (Cc x Cc) produced these combinations of alleles in F_2 offspring:

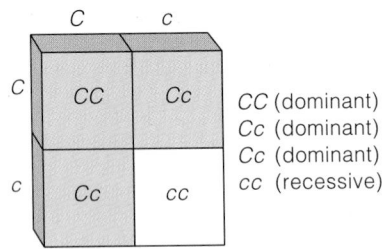

CC (dominant)
Cc (dominant)
Cc (dominant)
cc (recessive)

This results in a probability of 3/4 that any one child will have the dominant phenotype and 1/4 that the child will have the recessive phenotype.

5. Evidence from monohybrid crosses supports the principle of segregation. This principle states that (1) diploid organisms have pairs of genes on pairs of homologous chromosomes, and (2) the two genes of each pair segregate from each other during meiosis, such that each gamete formed ends up with one or the other.

6. A dihybrid cross occurs between two homozygous parents showing different versions of two traits. The result in the F_1 generation is:

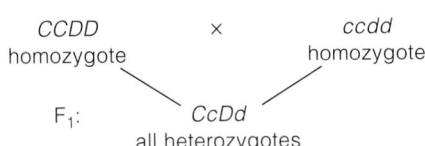

A mating between two heterozygous parents results in the following probable phenotypes:

$CcDd$ x $CcDd$

9 dominant for both traits
3 dominant C, recessive for d
3 dominant D, recessive for c
1 recessive for both traits

7. Dihybrid crosses provide evidence of independent assortment during meiosis. Each pair of homologous chromosomes tends to sort into gametes independently of how other chromosome pairs sort. Thus, the genes on the chromosomes assort independently.

8. Degrees of dominance exist among some alleles of the same gene. Genes can also interact to produce an effect on a trait. In some cases a single gene can influence many seemingly unrelated traits.

Review Questions

1. Define the difference between: (a) gene and allele, (b) dominant allele and recessive allele, (c) homozygote and heterozygote, and (d) genotype and phenotype. *374*

2. State the theory of segregation. Does segregation occur during mitosis or meiosis? *375*

3. Distinguish between monohybrid and dihybrid crosses. What is a testcross, and why is it useful in genetic analysis? *375, 378–379*

4. What is independent assortment? Does independent assortment occur during mitosis or meiosis? *378*

Self-Quiz *(Answers in Appendix IV)*

1. Alleles are _____.
   a. different molecular forms of a gene
   b. different molecular forms of a chromosome
   c. always homozygous
   d. always heterozygous

2. A heterozygote has _____.
   a. only one of the various forms of a gene
   b. a pair of identical alleles
   c. a pair of contrasting alleles
   d. a haploid condition, in genetic terms

3. The observable traits of an organism are its _____.
   a. phenotype   c. genotype
   b. sociobiology   d. pedigree

4. Offspring of a monohybrid cross AA x aa are _____.
   a. all AA d. 1/2 AA and 1/2 aa
   b. all aa e. none of the above
   c. all Aa

5. Second-generation offspring from a cross between two homozygotes are the _____.
   a. F_1 generation   c. hybrid generation
   b. F_2 generation   d. none of the above

6. Assuming complete dominance, offspring of the cross Aa x Aa will show a phenotypic ratio of _____.
   a. 1:2:1   c. 9:1
   b. 1:1:1   d. 3:1

7. Which statement best fits the principle of segregation?
   a. Units of heredity are transmitted to offspring.
   b. Two genes of a pair separate from each other during meiosis.
   c. Members of a population become segregated.
   d. A segregating pair of genes is sorted out into gametes independently of how gene pairs located on other chromosomes are sorted out.

8. Dihybrid crosses of heterozygous individuals ($AaBb$ x $AaBb$) lead to F_2 offspring with phenotypic ratios close to _____.
   a. 1:2:1   c. 3:1
   b. 1:1:1:1   d. 9:3:3:1

9. Match each genetic term appropriately.
   ____ dihybrid cross   a. AA x aa
   ____ monohybrid cross   b. Aa
   ____ homozygous condition   c. $AABB$ x $aabb$
   ____ heterozygous condition d. aa

Critical Thinking: Genetics Problems
(Answers in Appendix III)

1. One gene has alleles *A* and *a*. Another has alleles *B* and *b*. For each genotype listed, what type(s) of gametes can be produced? (Assume independent assortment occurs.)
   - a. *AABB*
   - b. *AaBB*
   - c. *Aabb*
   - d. *AaBb*

2. Still referring to Problem 1, what will be the possible genotypes of offspring from the following matings? With what frequency will each genotype show up?
   - a. *AABB* x *aaBB*
   - b. *AaBB* x *AABb*
   - c. *AaBb* x *aabb*
   - d. *AaBb* x *AaBb*

3. A gene on one chromosome governs a trait involving tongue movement. If you have a dominant allele of that gene, you can curl the sides of your tongue upward (see photo). If you are homozygous for recessive alleles of the gene, you cannot roll your tongue. A gene on a different chromosome controls whether your earlobes are attached or detached. People with detached earlobes have at least one dominant allele of the gene. These two genes are on different chromosomes and so they assort independently. Suppose a tongue-rolling woman with detached earlobes marries a man who has attached earlobes and can't roll his tongue. Their first child has attached earlobes and can't roll its tongue.
   - a. What are the genotypes of the mother, father, and child?
   - b. What is the probability that a second child will have detached earlobes and won't be a tongue-roller?

4. Go back to Problem 1, and assume you now study a third gene having alleles *C* and *c*. For each genotype listed, what type(s) of gametes can be produced?
   - a. *AABBCC*
   - b. *AaBBcc*
   - c. *AaBBCc*
   - d. *AaBbCc*

5. When you decide to breed your purebred Labrador retriever Molly and sell the puppies, you discover that two of Molly's four siblings have developed a hip disorder that is traceable to the action of a single recessive allele. Molly herself shows no sign of the disorder. If you breed Molly to a male Labrador that does not carry the recessive allele, can you assure a purchaser that the puppies will also be free of the condition? Explain your answer.

6. The ABO blood system has been used to settle cases of disputed paternity. Suppose, as a geneticist, you must testify during a case in which the mother has type A blood, the child has type O blood, and the alleged father has type B blood. How would you respond to the following statements:
   - a. *Man's attorney*: "The mother has type A blood, so the child's type O blood must have come from the father. Because my client has type B blood, he could not be the father."
   - b. *Mother's attorney*: "Further tests prove this man is heterozygous, so he must be the father."

7. Soon after a couple marries, tests show that both the man and the woman are heterozygotes for the recessive allele that causes sickling of red blood cells; they are both *HbAHbS*. What is the probability that any of their children will have sickle-cell trait? Sickle-cell anemia?

8. A man is homozygous dominant for 10 different genes that assort independently. How many genotypically different types of sperm could he produce? A woman is homozygous recessive for 8 of these genes and is heterozygous for the other 2. How many genotypically different types of eggs could she produce? What can you conclude about the relationship between the number of different gametes possible and the number of heterozygous and homozygous gene pairs that are present?

9. As is the case with the mutated hemoglobin gene that causes sickle-cell anemia, certain dominant alleles are crucial to normal functioning (or development). Some are so vital that when the mutant recessive alleles are homozygous, the combination is lethal and death results before birth or early in life. However, such recessive alleles can be passed on by heterozygotes (*Ll*). In many cases, these are not phenotypically different from homozygous normals (*LL*). If two heterozygotes mate (*Ll* x *Ll*), what is the probability that any surviving progeny will be heterozygous?

10. Bill and Marie each have flat feet, long eyelashes, and "achoo syndrome" (chronic sneezing). All are dominant traits. The genes for these traits each have two alleles, which we can designate as follows:

	Dominant	Recessive
Foot arch	*A*	*a*
Sneezing	*S*	*s*
Eyelash length	*E*	*e*

Bill is heterozygous for each trait. Marie is homozygous for all of them. What is Bill's genotype? What is Marie's genotype? If they have four children, what is the probability that each child will have the same phenotype as the parents? What is the probability that a child will have short lashes, high arches, and no achoo syndrome?

11. You decide to breed a pair of guinea pigs, one black and one white. In guinea pigs, black fur is caused by a dominant allele (*B*) and white is due to homozygosity for a recessive allele (*b*) at the same locus. Your guinea pigs have 7 offspring, 4 black and 3 white. What are the genotypes of the parents? Why is there a 3:1 ratio in this cross?

Selected Key Terms

allele *374*
codominance *381*
continuous variation *382*
dihybrid cross *378*
dominant (allele or trait) *374*
F₁ (first filial) generation *376*
F₂ (second filial) generation *376*
gene *374*
genotype *374*
heterozygous *374*
homozygous *374*
independent assortment *378*

monohybrid cross *375*
multiple allele system *381*
P (parent generation) *376*
penetrance *382*
phenotype *374*
pleiotropy *381*
probability *376*
Punnett-square method *376*
recessive (allele or trait) *374*
segregation *375*
testcross *378*

For additional readings, go to InfoTrac College Edition, your online library at:

http://www.infotrac-college.com/wadsworth

CHROMOSOME VARIATIONS AND MEDICAL GENETICS

At the Mercy of Our Genes

Several years ago scientists identified the genetic cause of the disease cystic fibrosis. The breakthrough made news because CF is devastating. In the United States roughly one in every 2,000 babies is born with the disease, and many of those who have it are likely to die from it before they reach 30.

Several different mutations in a single gene labeled *CFTR* can lead to the CF phenotype. The gene is carried on human chromosome number 7. When a mutated

form is expressed, the outcome is a defective channel protein in the plasma membrane of cells in exocrine glands. You may remember that such glands secrete a variety of substances, including digestive enzymes (in the pancreas), sweat, and mucus in the lungs and digestive tract. In cystic fibrosis, the genetic flaw affects the flow of certain ions into and out of cells. As a result, exocrine glands secrete an abnormally thick, gluey mucus. Ducts of the pancreas become so clogged with

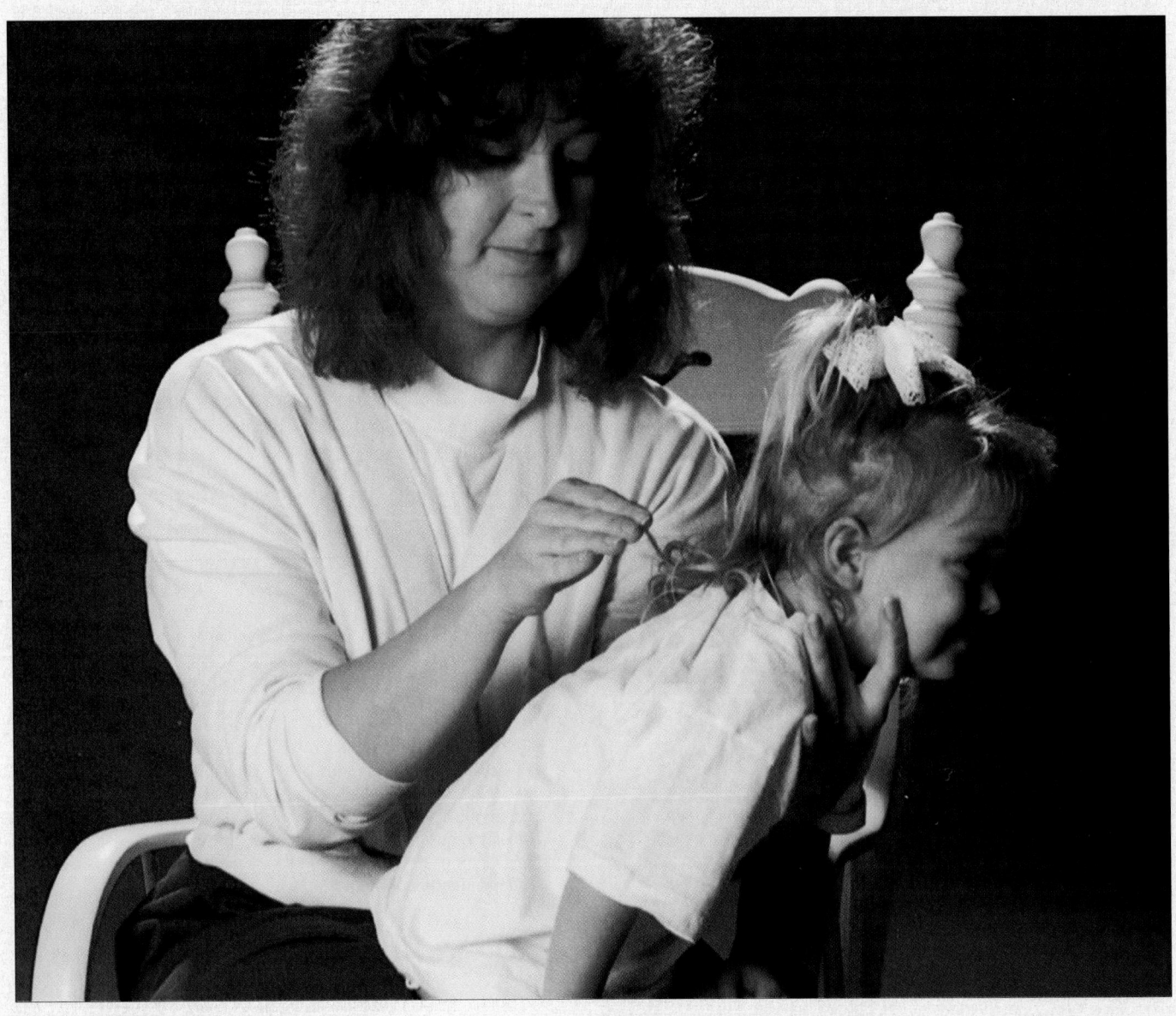

mucus that enzymes required for proper digestion of proteins, lipids, and carbohydrates never make it to the small intestine. Until their disease is diagnosed, affected children often suffer from severe malnutrition.

The most dangerous, frightening effect of CF occurs in the lungs. Cells lining respiratory passages crank out huge quantities of the sticky mucus, overwhelming the normal cleansing action of cilia. Breathing becomes extraordinarily difficult, and the bronchi and lungs become susceptible to infections such as pneumonia. Many CF children begin the day with a regimen of physical therapy designed to literally pound mucus from their lungs (Figure 18.1). Most deaths result from infection and respiratory failure.

In this chapter we delve more deeply into the topic of inheritance, looking especially at chromosomes and the genes they carry. Patterns of inheritance run the gamut from totally unforeseeable changes in genetic instructions to the predictable appearance of a trait through many generations of a family. Cystic fibrosis is only one of several thousand disorders that can arise from errors in genes or chromosomes. Fortunately, as we will see in this chapter and the next, our growing understanding of the basis of genetic disorders holds the promise of more effective treatments—and improved lives.

Figure 18.1 This little girl, born with a gene mutation that causes cystic fibrosis, is receiving percussion therapy as her mother pounds forcefully on her back. The procedure helps clear the child's lungs and bronchi of thick mucus, making breathing easier and reducing the chance of infection.

KEY CONCEPTS

1. Genes are segments of DNA. Genes follow one another along the length of a chromosome, and each gene has its own position in that sequence.

2. Chromosomes are of two types: autosomes and sex chromosomes (X and Y). Diploid human cells contain 22 pairs of autosomes and one pair of sex chromosomes. In normal females the sex chromosome pair consists of two X chromosomes. In normal males the sex chromosome pair consists of an X and a Y.

3. The combination of alleles in a chromosome may not remain intact during meiosis and gamete formation. Through an event called crossing over, some alleles in the sequence exchange places with their partners on the homologous chromosome. The alleles that are exchanged may or may not be identical.

4. The structure of a chromosome may change. For example, a chromosome segment may be deleted, duplicated, inverted, or moved to a new location.

5. The chromosome number may change as a result of improper separation of duplicated chromosomes during meiosis or mitosis.

6. Allele rearrangements, changes in chromosome structure, and changes in chromosome number all contribute to variation in traits. Often they lead to genetic abnormalities or disorders.

CHAPTER AT A GLANCE

18.1   The Chromosomal Basis of Inheritance

18.2   Sex Determination and X Inactivation

18.3   Linkage Groups and Crossing Over

18.4   Human Genetic Analysis

18.5   *Promise—and Problems—of Genetic Screening*

18.6   Patterns of Autosomal Inheritance

18.7   Patterns of X-Linked Inheritance

18.8   Sex-Influenced Inheritance

18.9   Changes in Chromosome Structure

18.10 Changes in Chromosome Number

THE CHROMOSOMAL BASIS OF INHERITANCE

Genes and Their Locations on Chromosomes

From preceding chapters, you already are familiar with some concepts about the structure of chromosomes and how they behave during meiosis. We can now integrate this information with a few concepts that will help explain patterns of human inheritance:

1. Genes are units of information about heritable traits. The genes are distributed among a person's chromosomes, and each gene has its own location (called a *locus*) in one type of chromosome.

2. A cell with a diploid chromosome number (2*n*) has inherited pairs of **homologous chromosomes**. All but one pair are identical in their length, shape, and gene sequence. The exception is the sex chromosomes, which are designated X and Y.

3. In general, each chromosome has the same genes, in the same locations, as its partner. However, the X and Y chromosomes are homologous in one small region only.

4. A pair of homologous chromosomes can carry identical or nonidentical alleles at a particular locus. **Alleles**, which arise through mutation, are slightly different molecular forms of the same gene. Although many different forms may have arisen in a population, a diploid cell can have only two of them.

5. There is no pattern to the way that maternal and paternal chromosomes become attached to the microtubular spindle just before metaphase I of meiosis, so different combinations are possible at anaphase I. As a result, gametes and then new individuals receive a mix of alleles of maternal and paternal chromosomes.

6. Genes that are close to one another on the same chromosome are linked. The farther apart two linked genes are, the more vulnerable they are to **crossing over**, an event by which homologous chromosomes exchange corresponding segments. Crossing over results in **genetic recombination**, or nonparental combinations of linked alleles in gametes, then in offspring.

7. Abnormal events during meiosis or mitosis occasionally change the structure of chromosomes and the parental chromosome number.

8. For any trait, the particular combination of alleles inherited by the new individual may have neutral, beneficial, or harmful effects.

As you will see, these points will help you make sense of the basic patterns of human inheritance.

Autosomes and Sex Chromosomes

In the late 1800s, microscopists discovered a chromosomal difference between the sexes. In 22 of the 23 pairs of chromosomes in a human diploid cell, the maternal and paternal chromosomes are roughly the same size and shape. They also carry genes for the same traits. These are **autosomes**. The remaining pair of chromosomes are **sex chromosomes**, and they differ between males and females. Human females have two sex chromosomes, both designated "X." Males have one X chromosome and another, physically different chromosome designated "Y."

Today, special dyes can be used to delineate human autosomes and sex chromosomes at metaphase, when they are most condensed. At that point, each type has a certain length, banding pattern, and so forth. (The bands appear because some regions condense more than others and absorb dyes differently.) These features are used to create a karyotype, a cut-up, rearranged photograph of the chromosomes of a cell at metaphase. The chromosomes are arranged in order, essentially from largest to smallest, and the sex chromosomes are placed last (Figure 18.2).

Autosomes carry the vast majority of the estimated tens of thousands of human genes. The X chromosome probably carries around 300 genes. Like other chromosomes, it carries some genes associated with sexual traits, such as the distribution of body fat. But most of the genes on the X chromosome are concerned with *nonsexual* traits, such as blood-clotting functions.

A Y chromosome is much smaller than an X chromosome, and it carries very few genes. We know that these include genetic instructions for male sexual development, but geneticists are still sifting through the evidence for other roles of Y chromosome genes.

Genes on sex chromosomes may be called "sex-linked genes." However, researchers now use the more precise designations **X-linked gene** or **Y-linked gene**. As you'll read, X-linkage is a factor in some genetic disorders.

Gene Mutations on Chromosomes

Genes are segments of DNA on chromosomes. As you know, DNA consists of various types of nucleotides linked by chemical bonds (Section 1.13). A *mutation* is a change in one or more of the nucleotides that make up a particular gene. Mutations can arise in various ways, which we consider in Chapter 19. A mutation may have no apparent effect on an individual, it may code for a new,

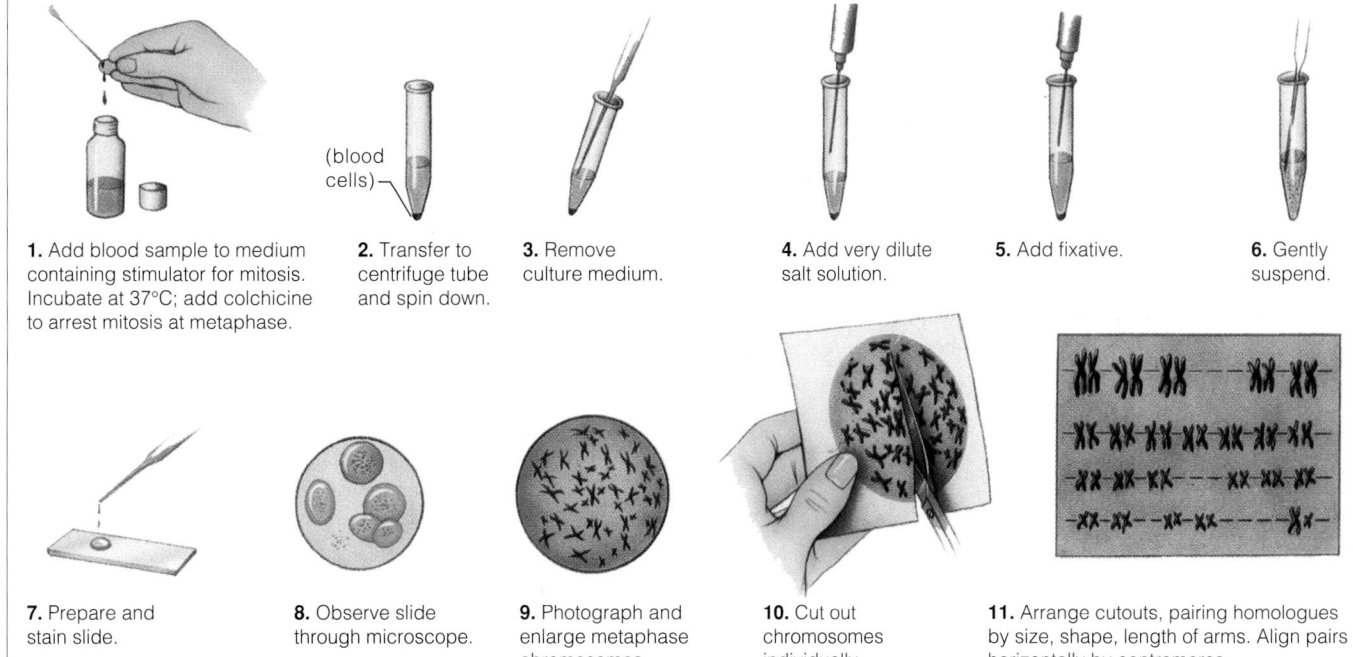

1. Add blood sample to medium containing stimulator for mitosis. Incubate at 37°C; add colchicine to arrest mitosis at metaphase.

2. Transfer to centrifuge tube and spin down.

3. Remove culture medium.

4. Add very dilute salt solution.

5. Add fixative.

6. Gently suspend.

(blood cells)

7. Prepare and stain slide.

8. Observe slide through microscope.

9. Photograph and enlarge metaphase chromosomes.

10. Cut out chromosomes individually.

11. Arrange cutouts, pairing homologues by size, shape, length of arms. Align pairs horizontally by centromeres.

a

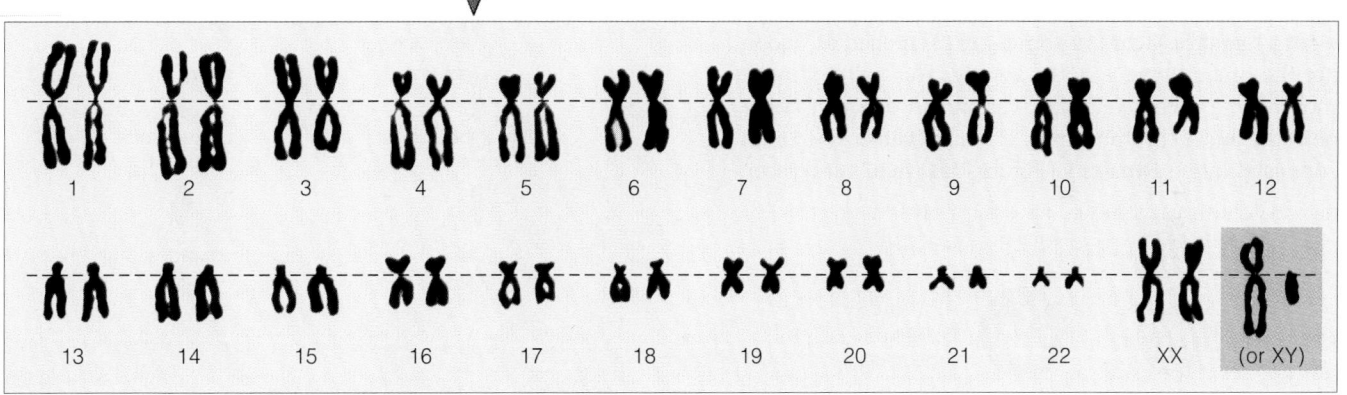

1 2 3 4 5 6 7 8 9 10 11 12

13 14 15 16 17 18 19 20 21 22 XX (or XY)

b

Figure 18.2 Karyotyping. (**a**) Karyotype preparation. (**b**) A human karyotype. Human somatic cells have 22 pairs of autosomes and 1 pair of sex chromosomes (XX or XY). These are metaphase chromosomes; each is in the duplicated state.

beneficial phenotype—or, as in many of the disorders we consider later in this chapter—it may call for a phenotype that seriously disrupts healthy functioning.

Chromosomes also can become rearranged in abnormal ways, as described in Section 18.9. Such changes tend to be quite devastating to the affected individual, with outcomes ranging from mental retardation to death.

Autosomes are the pairs of chromosomes that are the same in males and females. One other pair—the sex chromosomes—govern a new individual's gender.

The two sex chromosomes of females are both X chromosomes. Males have one X chromosome and one Y chromosome.

Genes on the sex chromosomes are sometimes referred to as X-linked or Y-linked, respectively.

Genetic disorders can result from mutations of genes on chromosomes or from an abnormal rearrangement of a chromosome itself.

SEX DETERMINATION AND X INACTIVATION

All gametes produced by a female (XX) carry an X chromosome. Half the gametes produced by a male (XY) carry an X and half carry a Y chromosome. Hence, a new individual's sex is determined by the father's sperm because the mother's oocytes can contain only an X chromosome, while the father's sperm can carry either an X or a Y (Figure 18.3). At fertilization, when both the sperm and the egg carry an X chromosome, the new individual will develop into a female. Conversely, when the sperm carries a Y chromosome, the new individual will be XY and develop into a male.

Recall from Chapter 13 that the few genes on the Y chromosome include a "male-determining gene" (*SRY*). Its product causes a new individual to develop testes. In the absence of that gene product, ovaries will develop. Testes and ovaries both produce hormones that govern the development of particular sexual traits.

Since females have two X chromosomes and males have only one, do females have twice as many X-linked genes as males do and therefore have a double dose of their gene products? Generally the answer is no, because a compensating phenomenon called **X inactivation** occurs in females. Although the mechanism is not fully understood, it appears that most or all of the genes on one of a female's X chromosomes are "switched off" soon after the first cleavages of the zygote. The switch flips at random; in a given cell, inactivation can occur in either one of the two X chromosomes. The inactivated X becomes condensed into a *Barr body* (Figure 18.4) that can be seen

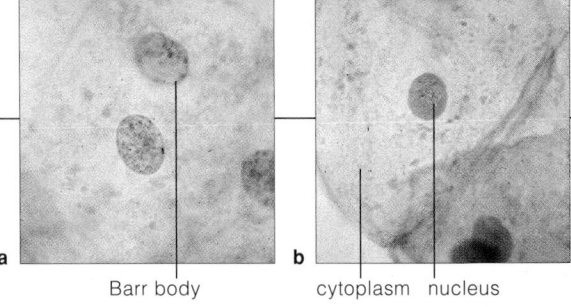

a   Barr body   b   cytoplasm nucleus

Figure 18.4 (**a**) Nucleus from a female cell, showing a Barr body. (**b**) Nucleus from a male cell, which has no Barr body.

under the microscope. Once X inactivation takes place, the embryo continues to develop. Ultimately, a female's body is typically a mosaic of tissues in which patches of cells have one or the other of the original pair of X chromosomes she inherited from her parents.

Human sex is determined by the father's sperm, which can carry either an X chromosome or a Y chromosome. XY individuals develop as males, and XX individuals as females.

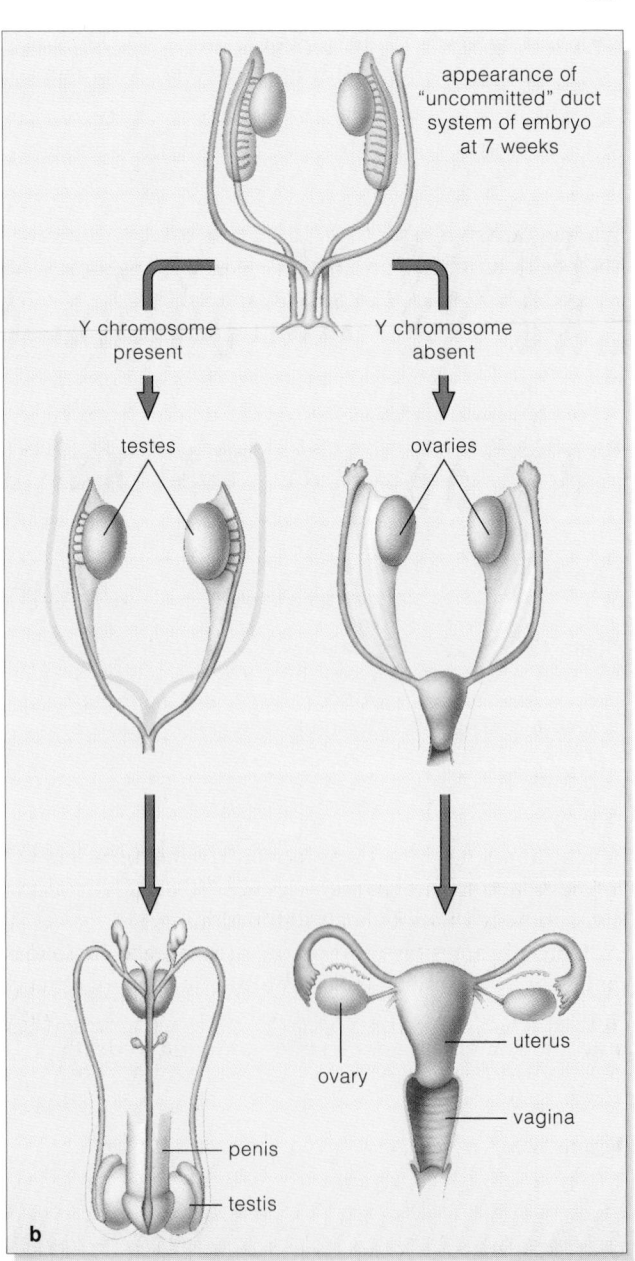

appearance of "uncommitted" duct system of embryo at 7 weeks

Y chromosome present

Y chromosome absent

testes

ovaries

uterus

ovary

vagina

penis

testis

b

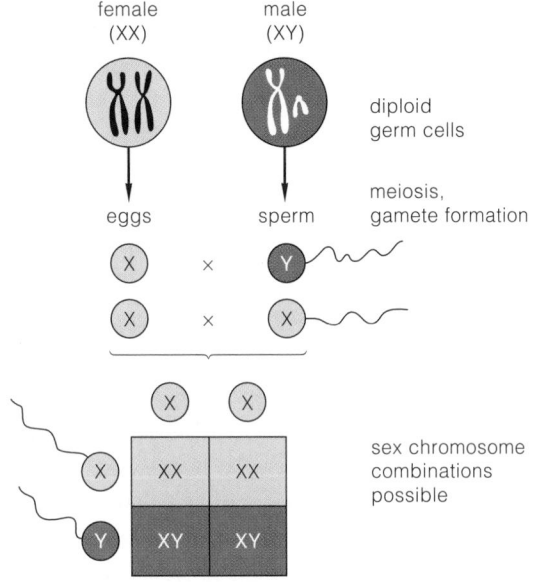

female (XX)   male (XY)

diploid germ cells

eggs   sperm

meiosis, gamete formation

X × Y

X × X

sex chromosome combinations possible

X   X

X | XX | XX

Y | XY | XY

a

Figure 18.3 (**a**) Pattern of sex determination in humans. Males transmit their Y chromosome to sons but not to daughters. Males receive their X chromosome only from their mother. (**b**) Duct system in the early embryo that develops into a male or female reproductive system.

18.3 LINKAGE GROUPS AND CROSSING OVER

Early geneticists, who worked with fruit flies, realized that some traits were inherited *as a group* from one parent or the other. Today, the term **linkage** is used to describe the tendency of genes located on the same chromosome to end up together in the same gamete.

Linkage groups are vulnerable to crossing over (Figure 18.5). In fact, we now know that in most eukaryotic organisms, humans included, meiosis cannot be completed properly unless *every* pair of homologous chromosomes takes part in at least one crossover.

Imagine any two genes at two different locations on the same chromosome. *The probability that a crossover will disrupt their linkage is proportional to the distance that separates them.* If genes *A* and *B* are twice as far apart as two other genes, *C* and *D*, we would expect crossing over to disrupt the linkage between *A* and *B* much more often.

Two genes are very closely linked when the distance between them is small; they nearly always end up in the same gamete. Linked genes are more vulnerable to crossing over when the distance between them is greater. When two genes are very far apart, crossing over disrupts their linkage so often that the genes assort independently of each other into gametes (Figure 18.6).

The patterns in which genes are distributed into gametes are so regular that they can be used to determine the positions of the various genes on a chromosome. This procedure is called **chromosome mapping**.

The farther apart two genes are on a chromosome, the greater will be the frequency of crossing over and recombination between them.

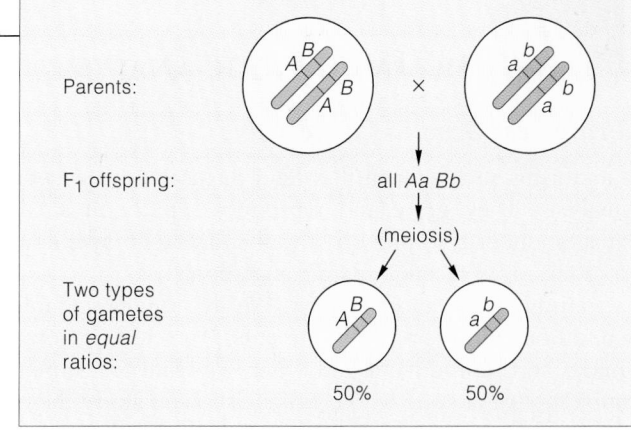

a Complete linkage (no crossing over)

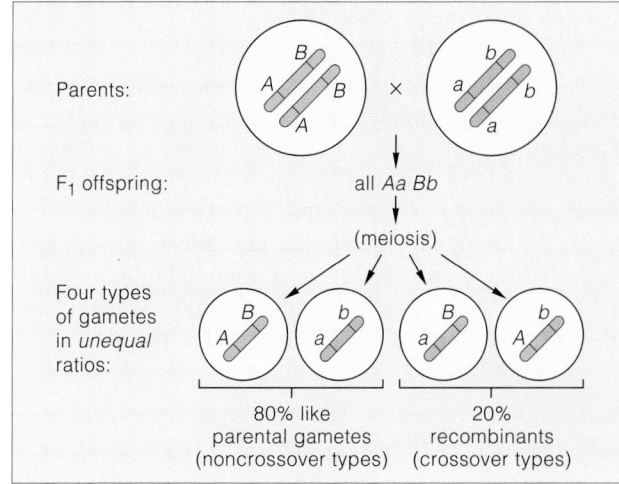

b Incomplete linkage due to crossing over

Figure 18.6 How crossing over can affect gene linkage, using two gene loci as the example.

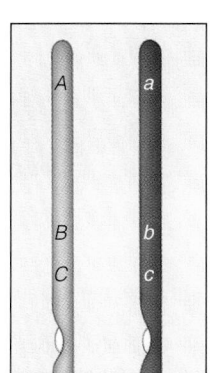 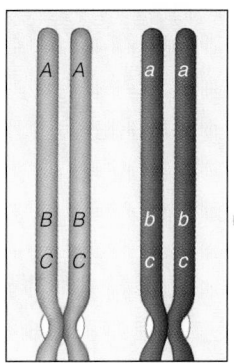 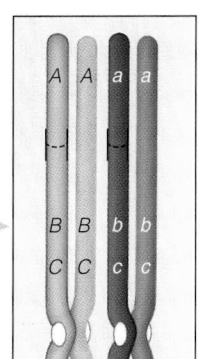 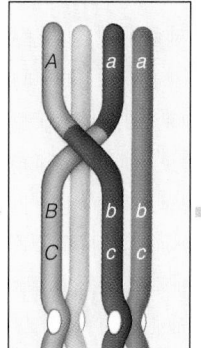 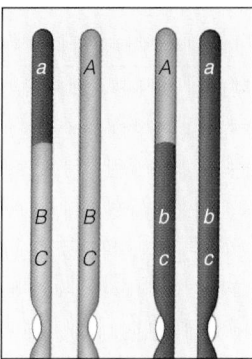

a Diploid (2*n*) cells have pairs of genes at corresponding locations, on pairs of homologous chromosomes. The two genes of each pair may or may not be identical. In this case, the alleles are nonidentical (*A* with *a*, *B* with *b*, and *C* with *c*).

b The same gene regions at interphase, after DNA replication. Both of the homologous chromosomes are in the duplicated state.

c At prophase I of meiosis, two of the nonsister chromatids break while tightly aligned together.

d Nonsister chromatids swap segments, then enzymes seal the broken ends. The breakage and exchange represent one crossover event.

e The outcome of the crossover is genetic recombination between two of four chromatids. (They are shown after meiosis, as unduplicated, separate chromosomes.)

Figure 18.5 Crossing over and genetic recombination.

HUMAN GENETIC ANALYSIS

Pedigrees: Genetic Connections

Organisms such as fruit flies lend themselves to genetic analysis. They grow and reproduce rapidly in small spaces, under controlled conditions. It doesn't take long to track a trait through many generations.

Humans are another story. We live under variable conditions in diverse environments. We select our own mates and reproduce if and when we want to. Human subjects live as long as the geneticists who study them, so tracking traits through generations can be rather tedious. Most human families are so small there aren't enough offspring for easy inferences about inheritance.

To get around some of the problems associated with analyzing human inheritance, geneticists put together **pedigrees**. A pedigree is a chart that is constructed, by standardized methods, to show the genetic connections among relatives. Figure 18.7 is an example, and includes definitions of a few of the standardized symbols.

When analyzing pedigrees, geneticists rely on their knowledge of probability and Mendelian inheritance patterns, which may yield clues to a trait's genetic basis. For instance, they might determine that the responsible allele is dominant or recessive or that it is located on an autosome or a sex chromosome. Gathering many family pedigrees increases the numerical base for analysis. When a trait clearly shows a simple Mendelian inheritance pattern, a geneticist may be justified in predicting the probability of its occurrence among children of prospective parents. We will return to this topic later in the chapter.

A person who is heterozygous for a recessive trait can be designated as a *carrier*. A carrier shows the dominant phenotype (no disease symptoms) but can produce gametes with the recessive allele and potentially pass on that allele to a child. If both parents are carriers for a genetic disorder, a child has a 25 percent chance of being homozygous for the harmful recessive allele.

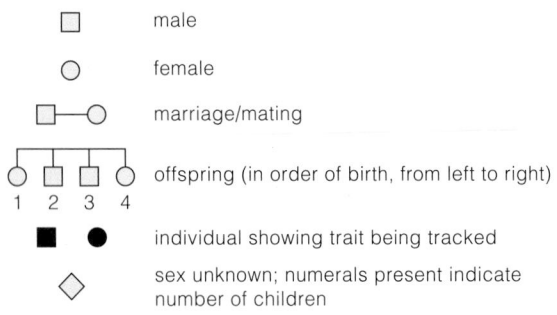

male

female

marriage/mating

offspring (in order of birth, from left to right)
1 2 3 4

individual showing trait being tracked

sex unknown; numerals present indicate number of children

a I, II, III, IV... successive generations

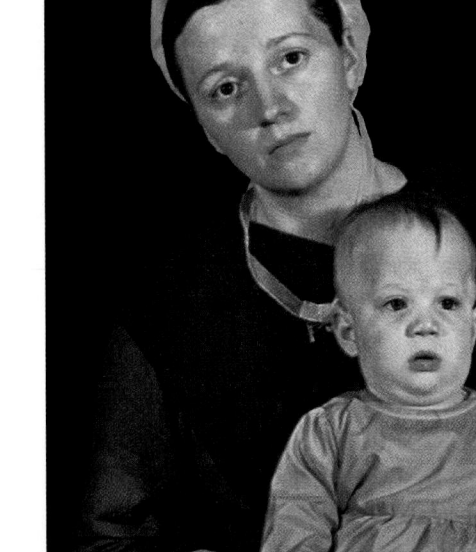

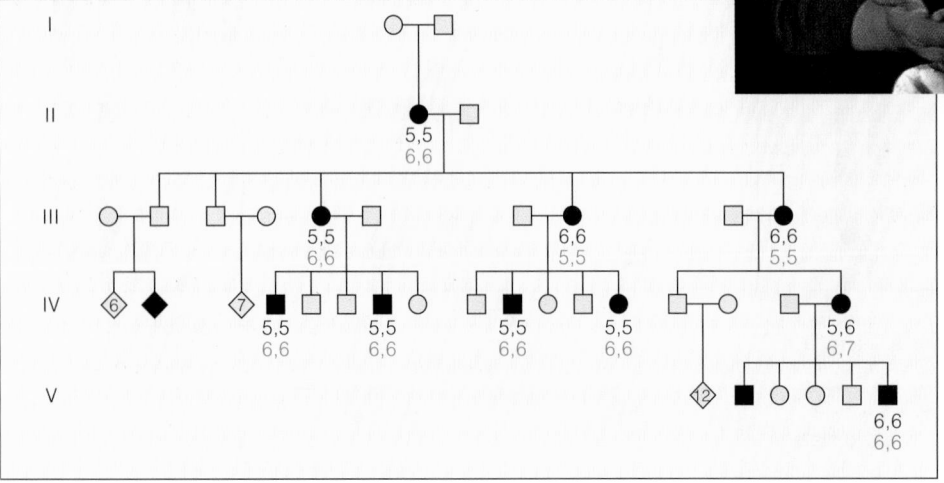

b

Figure 18.7 (**a**) Some symbols used in constructing pedigree diagrams. (**b**) This is an example of a pedigree for *polydactyly*. A person with this condition has extra fingers, extra toes, or both. Expression of the gene governing polydactyly can vary from one individual to the next. Here, *black* numerals designate the number of fingers on each hand. *Blue* ones designate the number of toes on each foot.

Keep in mind that "abnormality" and "disorder" are not necessarily the same thing. *Abnormal* simply means deviation from the average, such as having six toes on each foot instead of five. A *genetic disorder* is an inherited condition that causes mild to severe medical problems.

Every human being carries an average of three to eight harmful recessive alleles. Why don't alleles that cause severe disorders simply disappear from a population? *First,* rare mutations put new copies of the alleles in the population. *Second,* in heterozygotes, a recessive allele is paired with a normal dominant one that prevents phenotypic expression of the recessive allele, but allows it to be transmitted to offspring.

You may hear a genetic disorder called a disease (cystic fibrosis is a case in point), but the terms aren't always interchangeable. For instance, in some cases, a person's genes increase susceptibility or weaken the response to infection by a virus, bacterium, or some other disease agent. Strictly speaking, however, the resulting illness isn't a genetic disease.

Genetic Counseling

Sometimes prospective parents suspect they are very likely to produce a severely afflicted child. Their first child or a close relative may have a genetic disorder, and they wonder if future children also will be affected. In such cases, clinical psychologists, geneticists, social workers, and other consultants may be brought in to provide information and emotional support to parents at risk.

Counseling often begins with a diagnosis of the genotypes of a couple who are contemplating a pregnancy; this may reveal the potential a child will have a specific disorder. Biochemical tests or DNA analysis can be used to detect many genetic disorders, even in an embryo (see the *Choices* essay). Detailed family pedigrees can aid the diagnosis. For disorders showing a simple Mendelian inheritance pattern, it is possible to predict the chances a given child will be affected—but not all follow Mendelian patterns. And those that do can be influenced by other factors, some identifiable, others not. Even when the extent of risk has been determined with some confidence, prospective parents must know that the risk is the same for *each* pregnancy. If a pregnancy has one chance in four of producing a child with a genetic disorder, the same odds apply to every subsequent pregnancy.

For many genes, pedigree analysis may reveal Mendelian inheritance patterns that permit inferences about the probability of their transmission to children.

A genetic abnormality is an uncommon version of an inherited trait. A genetic disorder is an inherited condition that results in mild to severe medical problems.

PROMISE—AND PROBLEMS— OF GENETIC SCREENING

Some years ago in Great Britain, a couple who had a child with cystic fibrosis underwent embryo screening (Chapter 14) to make sure that their next baby would not have the disease. The woman's ovaries were hormonally stimulated to produce several eggs at once, and then the eggs were fertilized in the laboratory with the father's sperm. When the resulting embryos reached the eight-cell stage (photograph), genetic engineering methods were used to test each embryo for the presence of CF alleles. Embryos that tested positive were discarded, while two normal embryos were placed in the mother's uterus. One of them implanted and developed into a baby girl, named Chloe, who today is a healthy teenager.

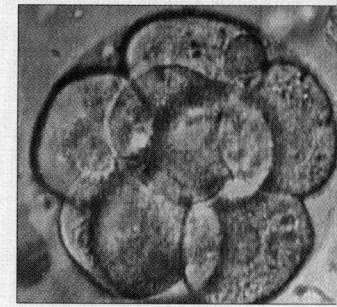

In theory, more than 4,000 genetic disorders can be detected in the early embryo. Researchers already are screening embryos for sickle-cell anemia, Tay-Sachs disease, hemophilia, and both Duchenne and myotonic muscular dystrophy, among other disorders. The procedures are costly. Yet there are couples who are carriers for such disorders and wish desperately to have a child—but only if the child can be guaranteed to be free of the genetic defect.

Procedures that can detect harmful alleles in carriers or in offspring come under the heading of *genetic screening.* Common prenatal screening options include amniocentesis and chorionic villus sampling (Chapter 14). The screening of embryos has fueled controversy over how to handle our growing ability to detect defects in genes or chromosomes.

All of this raises the question: Is it wrong to abort a fetus or discard a developing embryo because it carries a genetic disorder—even one with highly damaging effects? Some people answer with a resounding "Yes!" Others believe, on humanitarian grounds, that parents faced with the certainty of a seriously affected child should have the option of *not* bringing such a child into the world.

In the real world of *embryo screening*, there are many potential moral and ethical dilemmas. Consider the fact that physicians who conduct *in vitro* fertilization for otherwise-infertile couples routinely must discard some healthy embryos, simply because the method often produces more embryos than can successfully be implanted in the mother's uterus. Does this practice raise an ethical red flag for you? Or do you consider it simply an unavoidable result of a remarkable new technology that helps people realize a dream of parenthood?

PATTERNS OF AUTOSOMAL INHERITANCE

Autosomal Recessive Inheritance

For some traits, inheritance patterns reveal two clues that point to a recessive allele on an autosome. *First*, if both parents are heterozygous, any child of theirs will have a 50 percent chance of being heterozygous and a 25 percent chance of being homozygous recessive (Figure 18.8). *Second*, if both parents are homozygous recessive, any child of theirs will be too.

Phenylketonuria (PKU) results from abnormal buildup of the amino acid phenylalanine. Affected people are homozygous for a recessive mutant allele that fails to encode an enzyme that converts phenyl-alanine to tyrosine, so excess phenylalanine accumulates. If excess phenylalanine is diverted into other metabolic pathways, phenylpyruvic acid and other compounds may be produced. At high levels, phenylpyruvic acid can cause mental retardation. Fortunately, a diet that is low in phenylalanine will alleviate PKU symptoms. Many diet soft drinks and other products are artificially sweetened with aspartame, which contains phenylalanine. Such products carry warning labels, so people with phenyl-ketonuria can avoid using them.

Cystic fibrosis (CF) is another autosomal recessive condition. It is one of the most common genetic disorders among Caucasians. Major symptoms are the buildup of huge amounts of sticky mucus in the lungs and ducts of the pancreas.

In recent years, various trials of a gene therapy for CF have been carried out. In some, patients inhale a nasal spray containing genetically engineered virus particles that carry functional genes for the affected protein. The idea is that the virus will infect cells in the patients' airways and thereby shuttle the functional gene into the cells, where it is integrated into the DNA and expressed.

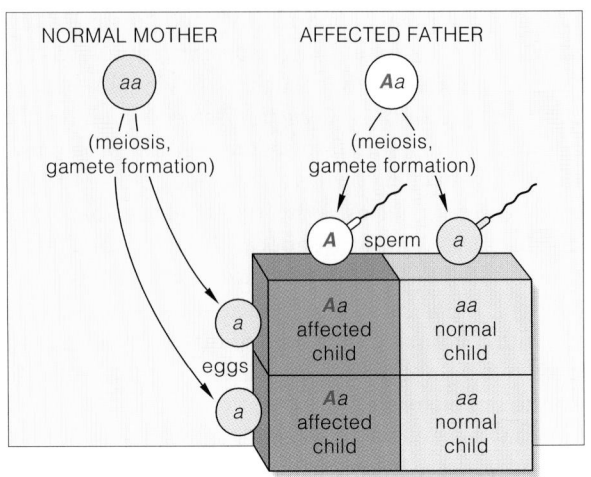

Figure 18.9 One pattern for autosomal dominant inheritance. In this example, the dominant allele (*red*) is fully expressed in the carriers.

Thus far, the trials have achieved moderate success. However, some patients have developed immunity to the genetically engineered viruses. We'll look again at gene therapy in Chapter 21.

In some autosomal recessive disorders, the defective gene product is an enzyme required in lipid metabolism. Infants born with *Tay-Sachs disease* lack hexosaminidase A, which is an enzyme required for the metabolism of sphingolipids, a type of lipid that is especially abundant in the plasma membrane of cells in nerves and the brain. Affected babies seem normal at birth, but they steadily lose motor functions and become deaf, blind, and mentally retarded. Most die in the first three years of life.

Tay-Sachs disease is most common among children of eastern European Jewish descent. Biochemical tests before conception can determine whether either member of a couple carries the recessive allele.

Autosomal Dominant Inheritance

Two clues of a different sort indicate that an autosomal dominant allele is responsible for a trait. First, the trait usually appears in each generation, for the allele is usually expressed even in heterozygotes. Second, if one parent is heterozygous and the other is homozygous for the normal, recessive allele, there is a 50 percent chance that any child of theirs will be heterozygous (Figure 18.9). A few dominant alleles that cause severe genetic disorders persist in populations. Some result from spontaneous mutations. In other cases, expression of a dominant allele may not prevent reproduction, or affected people have children before the disorder's symptoms become severe.

An example is *Huntington disease*, in which the basal nuclei of the brain degenerate (Section 10.10). In about

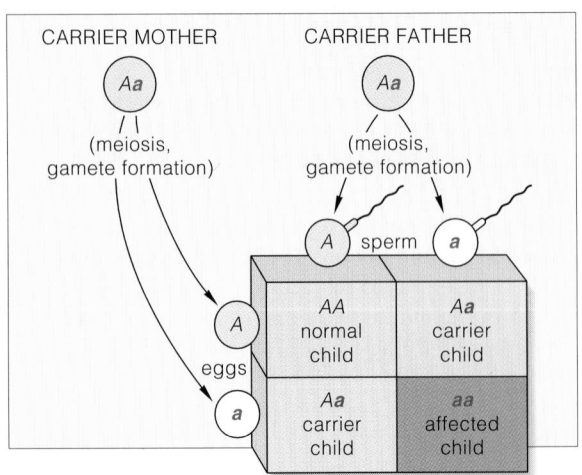

Figure 18.8 Possible phenotypic outcomes for autosomal recessive inheritance when both parents are heterozygous carriers of the recessive allele (shaded *red* here).

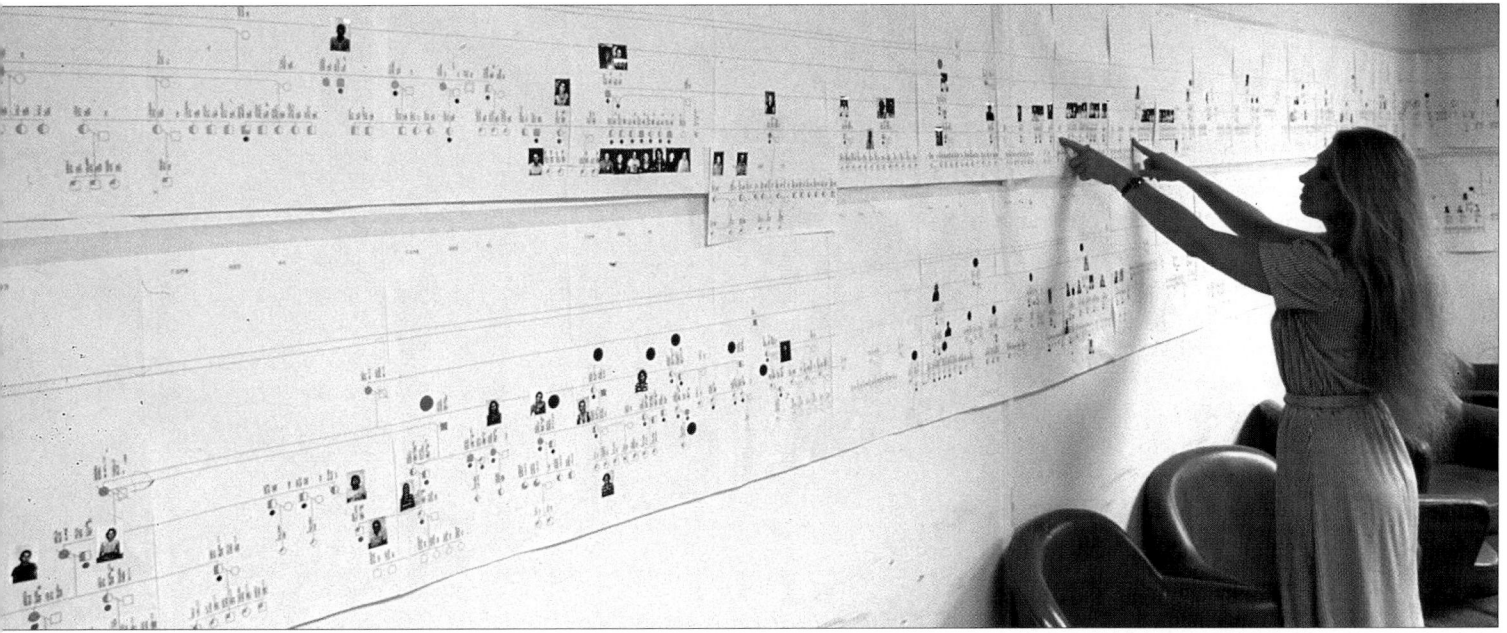

Figure 18.10 From the human genetic researcher Nancy Wexler, a pedigree for Huntington disease, by which the nervous system progressively degenerates. Wexler's team constructed an extended family tree for nearly 10,000 people in Venezuela. Wexler has a special interest in the disorder because she herself has a 50 percent chance of developing it.

Figure 18.11 A painting by Velazquez of Infanta Margarita Teresa of the Spanish court and her maids, including the achondroplasic woman at the far right.

half the cases, symptoms emerge after age 30—when the person may already have had children. Homozygotes for the Huntington allele die as embryos, so affected adults are always heterozygous. A person who has one parent with the disorder thus has a 50 percent chance of carrying the dominant allele. Testing can reveal whether an individual has the disease-causing allele, which causes an abnormal sequence of nucleotides in a gene on chromo-

some 4. Because of the nature of the Huntington defect (described in Section 19.9), there is no cure. Some at-risk people choose not to have the diagnostic test, and many elect to remain childless to avoid passing on the disorder.

Another example, *achondroplasia*, affects about 1 in 10,000 people. Homozygous dominant infants commonly are stillborn. Heterozygotes are able to reproduce. However, while they are young and their limb bones are forming, the cartilage components of those bones cannot form properly. Accordingly, at maturity affected people have abnormally short arms and legs (Figure 18.11). Adults with achondroplasia are less than 4 feet, 4 inches tall. The dominant allele often has no other phenotypic effects.

About 1 person in 500 is heterozygous for a dominant autosomal allele that causes *familial hypercholesterolemia*. People with this allele have dangerously elevated blood cholesterol because the allele fails to encode the normal number of cell receptors for low-density lipoproteins (LDLs). Recall from Chapter 6 that LDLs bind cholesterol in the blood and thereby help remove it from the circulation. Rare individuals who are homozygous for the allele may develop severe cholesterol-related heart disease as children and usually die in early adulthood.

In autosomal recessive inheritance, if both parents are heterozygous carriers of the recessive allele, there is a 25 percent chance that a child will be homozygous for the trait and exhibit the recessive phenotype.

In autosomal dominant inheritance, the trait may appear in each generation, because the allele is expressed even in heterozygotes.

PATTERNS OF X-LINKED INHERITANCE

X-Linked Recessive Inheritance

Distinctive clues often show up when a recessive allele on an X chromosome causes a trait. *First*, males show the recessive phenotype far more often than females do. A recessive allele can be masked in females, who may inherit a dominant allele on their other X chromosome. It cannot be masked in males, who have only one X chromosome (Figure 18.12). *Second*, a son cannot inherit the recessive allele from his father. A daughter can. If a daughter is heterozygous, there is a 50 percent chance that each son of hers will inherit the allele.

Two forms of the bleeding disorder *hemophilia* are inherited as X-linked recessive traits. The most common one is hemophilia A, which is caused by a mutation in the gene that encodes a protein for blood clotting (see Section 6.12), called factor VIII. Males with a recessive allele on their X chromosome are always affected. They risk death from untreated bruises, cuts, or internal bleeding. Blood clotting is normal in heterozygous females, because the nonmutated gene on their normal X chromosome produces enough factor VIII to cover the required function. Symptoms are similar in hemophilia B, but the clotting factor is factor IX.

Hemophilia A affects only about 1 in 7,000 males. However, the frequency of the recessive allele was atypically high among 19th-century European royal families. Queen Victoria of England was a carrier, as were two of

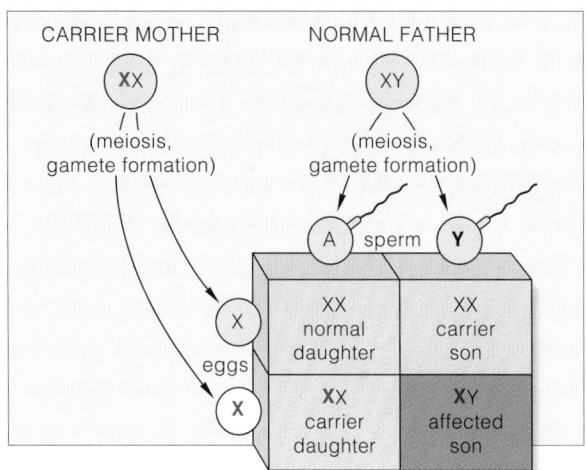

Figure 18.12 One pattern for X-linked inheritance. This example shows the outcomes possible when the mother carries a recessive allele on one of her X chromosomes (shown in *red*).

her daughters. In a pedigree developed some years ago, more than 15 of her 69 descendants at that time were affected males or female carriers (Figure 18.13).

Diseases lumped under "muscular dystrophy" all involve progressive wasting of muscle tissue. One X-linked type, *Duchenne muscular dystrophy* (DMD), affects 1 in 3,500 males, usually in early childhood. As muscles degenerate,

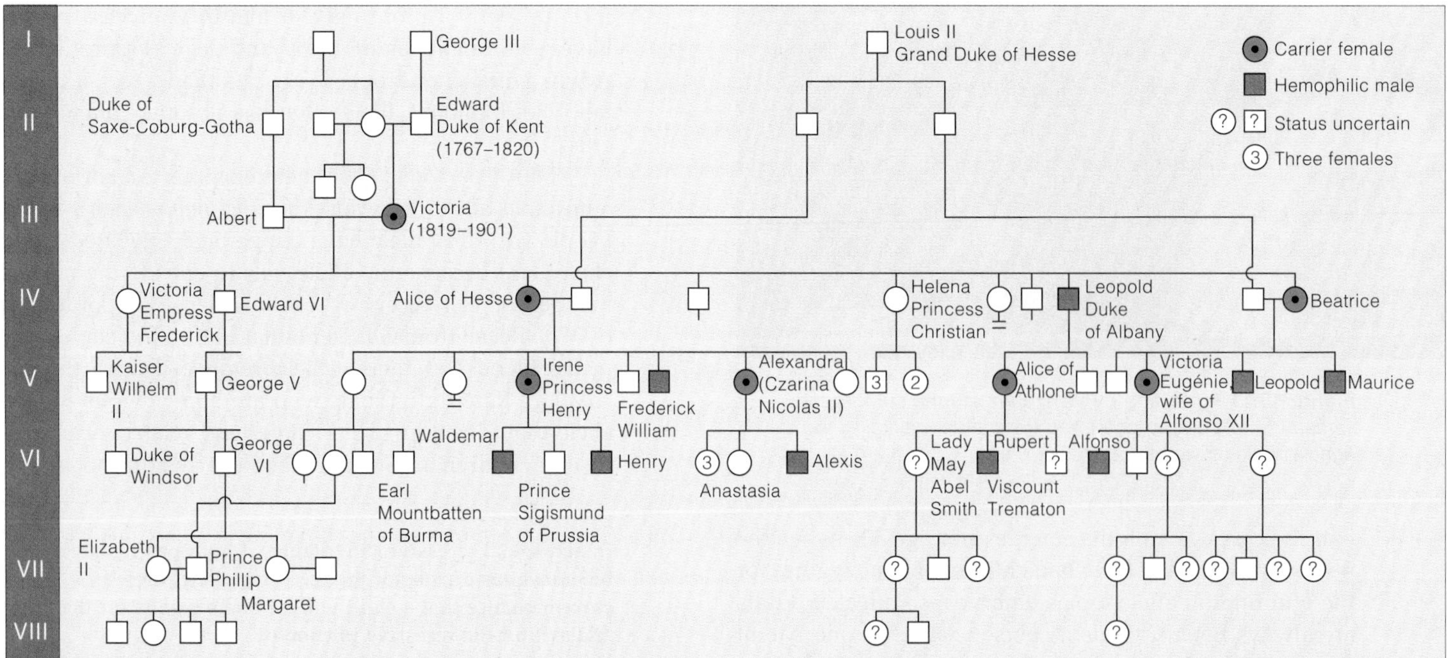

a

Figure 18.13 (a) Partial pedigree of Queen Victoria's descendants, showing carriers and affected males who carried the X-linked allele for hemophilia A. Many individuals of later generations are not included in this pedigree. (**b, facing page**) Photograph of the Russian royal family in the early 1900s.

affected boys become weak and unable to walk. They usually die by age 20 from cardiac or respiratory failure. The normal gene that is mutated in DMD encodes the protein *dystrophin*, which gives structural support to the muscle cell plasma membrane. Lacking dystrophin, muscle fibers cannot withstand the physical stress of contraction and break down. Eventually the whole muscle tissue is destroyed.

Red/green color blindness is an X-linked recessive trait. About 8 percent of males in the United States have this condition, which arises from mutation of an allele that codes for a protein (opsin) that binds visual pigments in cone cells of the retina. Females also can have red/green color blindness, but this occurs rarely because a girl must inherit the recessive allele from both parents.

X-Linked Inheritance of Dominant Mutant Alleles

The *faulty enamel trait* is one of a few known examples of a trait caused by a dominant mutant allele that is X-linked. With this disorder, the hard, thick enamel coating that normally protects teeth fails to develop properly (Figure 18.14). The inheritance pattern resembles that for X-linked recessive alleles, except that the trait is expressed in heterozygous females. An affected father transmits the allele to all of his daughters and none of

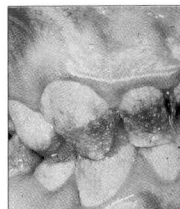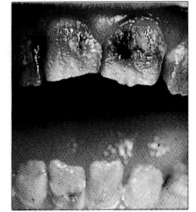

Figure 18.14 The discolored, abnormal tooth enamel of two people affected by the faulty enamel trait.

his sons. If a woman is heterozygous, she will transmit the allele to half her offspring, regardless of their sex.

Another Rare X-Linked Abnormality

Rarely, someone whose sex chromosomes are XY develops as a female. An example is testicular feminizing syndrome, or "androgen insensitivity." In an affected person, a mutation in a gene on the X chromosome results in defective receptors for androgens (male sex hormones), including testosterone and DHT (dihydroxytestosterone). Normally, cells in the testes and other male reproductive organs bind one or more of the hormones and then proceed to develop. With defective receptors, however, they can't bind the hormones and the embryo develops externally as a female. The person is female in every external characteristic, but has no uterus or ovaries.

A Few Qualifications

We include the preceding examples of autosomal and X-linked traits to give you a general idea of the kinds of clues that hold meaning for trained geneticists. Genetic analysis is usually not an easy task. Before diagnosing a case, geneticists often find it necessary to pool together many pedigrees. Typically they make detailed analyses of clinical data and keep abreast of current research. Why? Consider that more than one type of gene may be responsible for a given phenotype. Geneticists already know of dozens of conditions that can arise from a mutated gene on an autosome *or* a mutated gene on the X chromosome. They know of genes on autosomes that show dominance in males and recessiveness in females— so they may initially appear to be due to X-linked recessive inheritance, even though they are not.

b The Russian royal family members. All are believed to have been executed near the end of the Russian Revolution. They were recently exhumed from their hidden graves, but DNA fingerprinting indicated that the remains of Alexis and one daughter, Anastasia, were not among them. At this writing, the search is on for other graves, where the two missing children might have been buried.

A trait that shows up most often in males and that a son can inherit only from his mother most likely is passed on through X-linked recessive inheritance.

A few rare mutant traits are passed to offspring via X-linked dominant inheritance. A heterozygous mother will pass the allele to half her offspring. An affected father will pass the allele only to his daughters.

SEX-INFLUENCED INHERITANCE

Some traits are *sex-influenced*—they appear more frequently in one sex than in the other, or the phenotype differs depending on whether the individual is male or female. The difference may reflect the varying influences of male and female sex hormones on gene expression. Genes for such traits are carried on autosomes, not on sex chromosomes. Pattern baldness, shown in Figure 18.15, is an intriguing example. We can designate the "no baldness" form of the responsible gene b^+ and the "baldness" form b. A man will develop pattern baldness if he is homozygous bb and *also* if he is heterozygous b^+b. Females develop pattern baldness only if they are bb, usually later in life and to a lesser degree than men.

Sex-influenced traits appear more frequently in one sex, *or* the phenotype differs among males and females.

Genotype	Male Phenotype	Female Phenotype
b^+b^+	Hair	Hair
b^+b	Bald	Hair
bb	Bald	Bald

Figure 18.15 Pattern baldness. The bald area on the top of this man's head is a typical instance of pattern baldness. Geneticists usually denote the alleles for a sex-influenced trait with lowercase letters. Here, b^+ is "dominant" and b is recessive. Because of the influence of sex hormones, female heterozygotes have a full head of hair. Male heterozygotes, and all homozygous recessive individuals, show pattern baldness, though females don't show baldness as severely.

CHANGES IN CHROMOSOME STRUCTURE

On rare occasions, a chromosome changes structurally through deletion, duplication, inversion, or translocation of one or more of its segments.

Deletion

A **deletion** is the loss of a chromosome region. It may occur spontaneously or it may be caused by a virus, by irradiation, chemical assaults, or other environmental factors. One or more genes may be lost, as diagrammed in Figure 18.16:

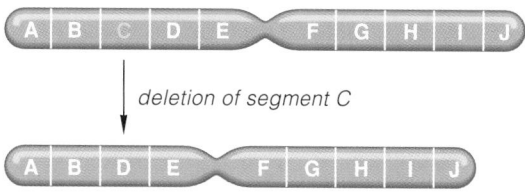

deletion of segment C

Figure 18.16 Simple diagram of a chromosome deletion.

A deletion can occur in any chromosome region. The loss may even occur at one end, which makes the chromosome unstable. Wherever a deletion happens, it permanently excises one or more of the chromosome's genes. Gene loss can mean serious problems. For example, one deletion from human chromosome 5 leads to mental retardation and an abnormally shaped larynx. When affected infants cry, the sounds produced resemble meowing—hence the name of the disorder, *cri-du-chat* (meaning cat cry). Figure 18.17 shows an affected child.

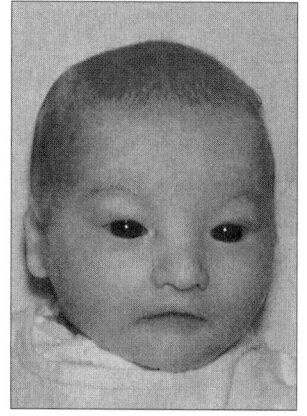
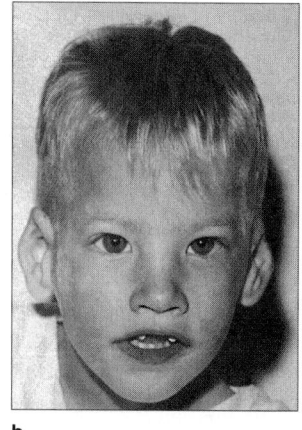

a   b

Figure 18.17 Cri-du-chat syndrome. (**a**) One affected boy just after birth. (**b**) The same boy 4 years later, showing how facial features change. By this age, affected individuals no longer make the mewing sounds typical of the cri-du-chat syndrome.

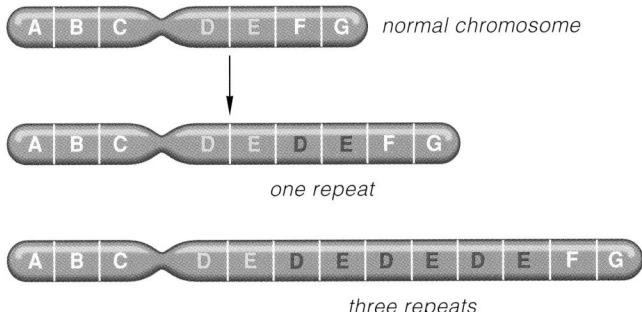

Figure 18.18 Simple diagrams of chromosome duplications.

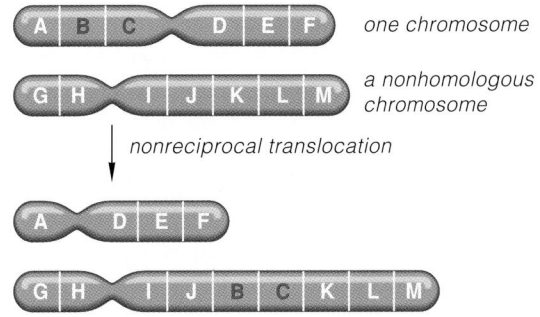

Figure 18.20 Simple diagram of a chromosomal translocation.

You might be wondering whether, given that humans have diploid cells, genes on the affected chromosome's homologue compensate for the loss. In fact, this is often the case if a segment deleted from one chromosome is present—and normal—on the homologous chromosome. However, if the remaining, homologous segment is abnormal or carries a harmful recessive allele, nothing will mask *its* effects.

Duplications

Even normal chromosomes contain the changes called **duplications**, which are gene sequences that are repeated several to many times. Often the same gene sequence has been repeated thousands of times. Such a duplication can take several forms, as you can see in Figure 18.18.

No human genetic disorder has yet been linked to a chromosome duplication, although examples are known from other organisms.

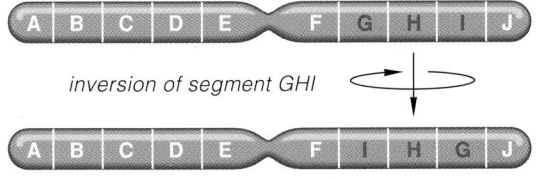

Figure 18.19 Simple diagram of a chromosome inversion.

Inversions and Translocations

An **inversion** is a chromosome segment that separated from the chromosome and then was inserted at the same place—but in reverse orientation. The reversal alters the position and order of the chromosome's genes (Figure 18.19). We know little about the effects of inversions in human genetics, possibly because this type of change in chromosome structure is rarely detected in humans.

Usually, a **translocation** is part of one chromosome that has exchanged places with a corresponding part of another chromosome that is *not* its homologous partner (Figure 18.20). In certain cancers, for example, a region of chromosome 8 has been translocated to chromosome 14. The disease develops because genes in that region are no longer properly regulated, a topic we return to in Chapter 20.

Genetic disorders can arise from changes in the structure of one or more of a person's chromosomes.

Factors such as viral infection and irradiation can delete a chromosome region. Sometimes the genes of a normal segment on the affected chromosome's homologue can compensate for the loss.

In duplication, a gene sequence is repeated in a chromosome.

In inversion, a chromosome segment separates from the chromosome and then is reinserted in the same place but in reverse orientation.

In translocation, a chromosome region is transferred to a nonhomologous chromosome.

CHANGES IN CHROMOSOME NUMBER

Categories and Mechanisms of Change

Various abnormal cellular events can put too many or too few chromosomes into gametes. New individuals end up with the wrong chromosome number. The effects range from minor physical changes to lethal disruption of body function. More often, affected individuals are miscarried, or spontaneously aborted before birth.

With **aneuploidy**, new individuals have too many or too few chromosomes. This condition is a major cause of human reproductive failure, affecting possibly half of all fertilized eggs. Autopsies have shown that many human fetuses that were miscarried (spontaneously aborted before pregnancy reached full term) were aneuploids.

With **polyploidy**, new individuals have three or more of each type of chromosome. This condition is lethal for humans. All but 1 percent of human polyploids die before birth. The rare newborns die soon after birth.

Chromosome numbers can change during mitotic or meiotic cell divisions or at fertilization. Suppose a cell cycle proceeds through DNA duplication and mitosis, then is arrested before the cytoplasm divides. The cell is tetraploid, with four of each type of chromosome. Or suppose one or more pairs of chromosomes fail to separate during mitosis or meiosis. Such events are called **nondisjunction**. Some or all of the resulting cells end up with too many or too few chromosomes (Figure 18.21).

If a gamete with an extra chromosome ($n + 1$) unites with a normal gamete at fertilization, the new individual will be **trisomic**, with three of one type of chromosome

($2n + 1$). If the gamete is missing a chromosome, then the individual will be **monosomic** ($2n - 1$).

Changes in the Number of Autosomes

Most changes in the number of autosomes arise through nondisjunction during gamete formation. Here we consider the most common of the resulting disorders.

A newborn with trisomy 21 has three copies of chromosome 21, and will show the effects of Down syndrome. ("Syndrome" means a set of symptoms characterizing a disorder; typically the symptoms occur together.) Symptoms vary greatly, but most affected individuals show moderate to severe mental impairment. About 40 percent develop heart defects. Because of abnormal skeletal development, older children have shortened body parts, loose joints, and poorly aligned bones of the hips, fingers, and toes. Muscles and muscle reflexes are weaker than normal, and development of speech and other motor functions is quite slow. Yet with special training, people with Down syndrome often take part in normal activities (Figure 18.22). As a group, they are cheerful, affectionate people who typically derive great pleasure from music and dancing.

Down syndrome is one of the genetic disorders that can be detected by prenatal diagnosis (Section 14.10). About one child of every 1,100 live births has the disorder. The probability that a woman will conceive an embryo with Down syndrome rises steeply after age 35.

Figure 18.21 Example of nondisjunction. Here, chromosomes fail to separate during anaphase I of meiosis and so change the chromosome number in resulting gametes.

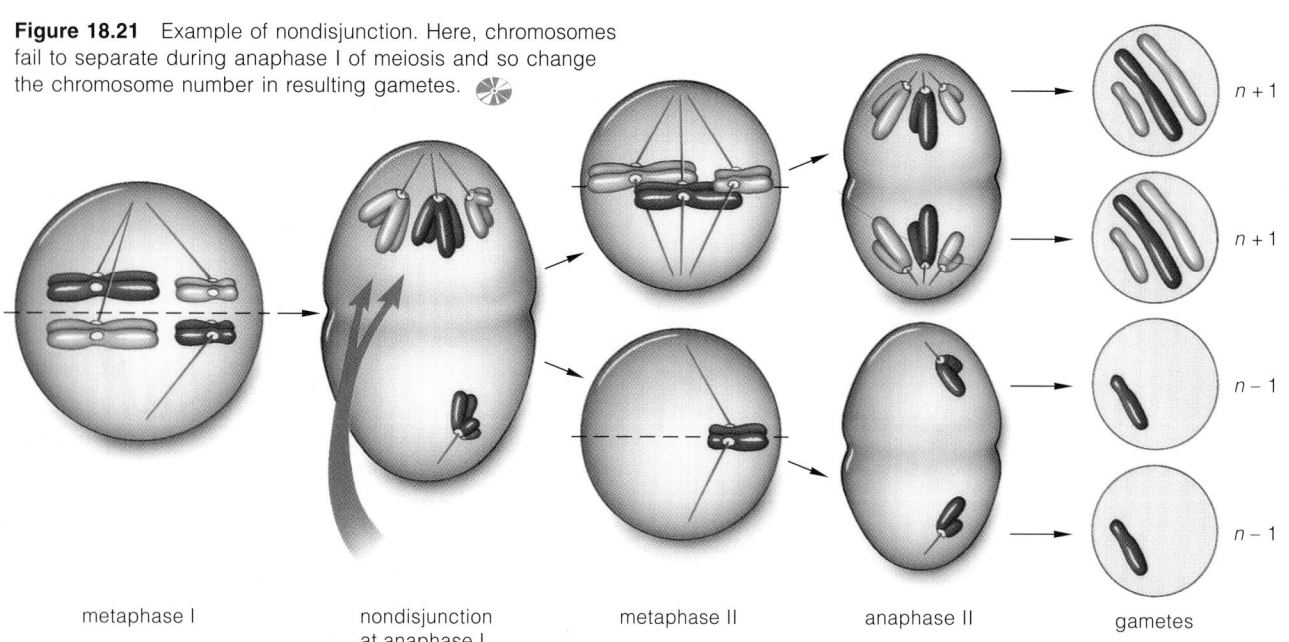

| metaphase I | nondisjunction at anaphase I | metaphase II | anaphase II | gametes |

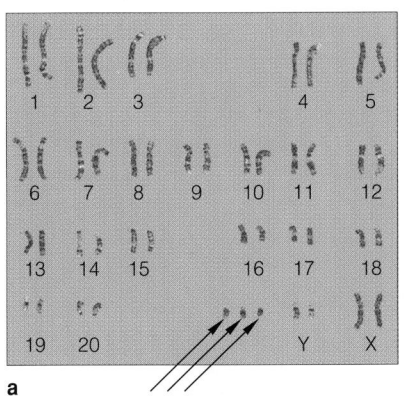

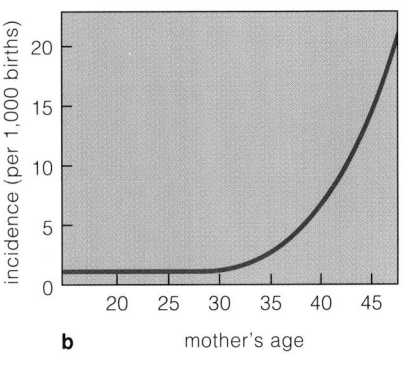

Figure 18.22 (**a**) Karyotype of a girl with Down syndrome; the three arrows identify the trisomy of chromosome 21. (**b**) Relationship between the frequency of Down syndrome and the mother's age. Results are from a study of 1,119 children with the disorder who were born in Victoria, Australia, between 1942 and 1957. (**c**) A child with Down syndrome.

Change in the Number of Sex Chromosomes

Most sex chromosome abnormalities arise as a result of nondisjunction during gamete formation. Let's look at a few phenotypic outcomes.

TURNER SYNDROME AND XXX FEMALES About 1 in every 5,000 newborns is destined to have Turner syndrome. Through a nondisjunction, affected individuals have a chromosome number of 45 instead of 46 (Figure

18.23). They are missing a sex chromosome (and a Barr body), so this is a type of sex chromosome abnormality. It is symbolized as XO. Turner syndrome occurs less often than other sex chromosome abnormalities, probably because most XO embryos are miscarried early in pregnancy. Affected people are female and have a webbed neck and a distorted phenotype. Their ovaries do not function, they are sterile, and secondary sexual traits fail to develop at puberty. Often they age prematurely and have shortened life expectancies.

Roughly 1 in 1,000 females has three X chromosomes. Two of these X chromosomes are condensed to Barr bodies, and most XXX females develop normally.

KLINEFELTER SYNDROME Nondisjunction can give rise to males who show Klinefelter syndrome. In most cases the genotype is XXY (Figure 18.21). This sex chromosome abnormality occurs at a frequency of about 1 in 500 live-born males. XXY males show low fertility and usually some mental retardation. Their testes are much smaller than normal, body hair is sparse, and there may be some breast enlargement. Injections of the hormone testosterone can reverse some aspects of the phenotype but not the infertility or mental retardation.

XYY CONDITION About 1 in every 1,000 males has one X and two Y chromosomes, an XYY condition. XYY males tend to be taller than average, but have a normal male phenotype.

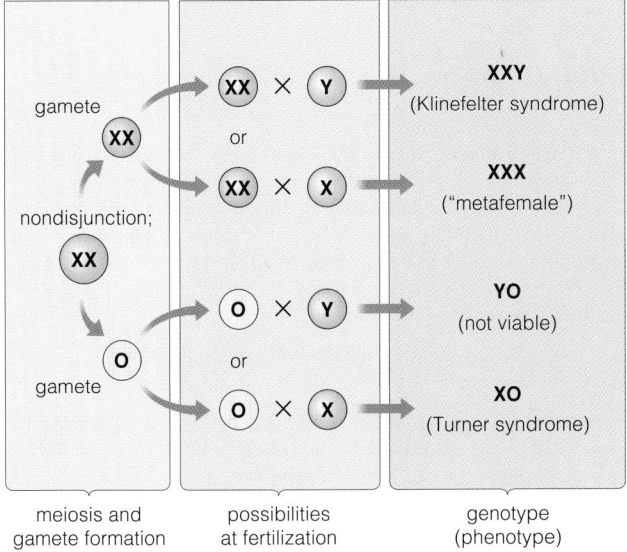

Figure 18.23 Genetic disorders that result from nondisjunction of X chromosomes followed by fertilization involving normal sperm.

Most changes in chromosome number arise as a result of nondisjunction during meiosis and gamete formation.

SUMMARY

1. Genes, the units of instruction for heritable traits, are arranged linearly on a chromosome.

2. Human diploid cells have two chromosomes of each type, one from each parent. The two are homologues and pair at meiosis. Sex chromosomes (X and Y) differ from each other in size, shape, and the genes they carry. By contrast, autosomes are roughly the same in size and shape and carry genes for the same traits.

3. Genes on the same chromosome tend to be linked—they tend to stay together during meiosis and end up in the same gamete. However, the farther apart two genes are on a chromosome, the greater will be the frequency of crossing over and recombination between them.

4. A family pedigree is a chart of genetic relationships. It helps establish inheritance patterns and track genetic abnormalities through several generations. Table 18.1 lists some common genetic disorders.

5. In disorders involving autosomal recessive inheritance, a person who is homozygous for a recessive allele shows the recessive phenotype. Heterozygotes generally have no symptoms.

6. In autosomal dominant inheritance, a dominant allele usually is expressed to some extent.

7. Many genetic disorders are X-linked—the mutated gene occurs on the X chromosome. Males, who inherit only one X chromosome, typically are affected.

8. Chromosome structure can be altered by deletions, duplications, inversions, or translocations. Chromosome number can be altered by nondisjunction (chromosomes don't separate during meiosis or mitosis).

Review Questions

1. How do X and Y chromosomes differ? *388*

2. Why do harmful genes (alleles) persist in a population? *394*

3. What indicates that a trait is coded by a dominant allele on an autosome? *394*

4. How do X-linked traits and sex-influenced traits differ? *396, 398*

Self-Quiz (Answers in Appendix IV)

1. _____ segregate during _____.
   a. Homologues; mitosis
   b. Genes on one chromosome; meiosis
   c. Homologues; meiosis
   d. Genes on one chromosome; mitosis

2. The alleles of a gene on homologous chromosomes end up in separate _____.
   a. body cells
   b. gametes
   c. nonhomologous chromosomes
   d. offspring
   e. both b and d are possible

3. Genes on the same chromosome tend to remain together during _____ and end up in the same _____.
   a. mitosis; body cell
   b. mitosis; gamete
   c. meiosis; body cell
   d. meiosis; gamete
   e. both a and d

4. The probability of a crossover occurring between two genes on the same chromosome is _____.
   a. unrelated to the distance between them
   b. increased if they are closer together on the chromosome
   c. increased if they are farther apart on the chromosome
   d. impossible

5. Chromosome structure can be altered by _____.
   a. deletions
   b. duplications
   c. inversions
   d. translocations
   e. all of the above

6. Nondisjunction can be caused by _____.
   a. crossing over in meiosis
   b. segregation in meiosis
   c. failure of chromosomes to separate during meiosis
   d. multiple independent assortments

7. A gamete affected by nondisjunction could have _____.
   a. a change from the normal chromosome number
   b. one extra or one missing chromosome
   c. the potential for a genetic disorder
   d. all of the above

8. Genetic disorders can be caused by _____.
   a. gene mutations
   b. changes in chromosome structure
   c. changes in chromosome number
   d. all of the above

9. Match the following chromosome terms appropriately.
   _____ crossing over
   _____ deletion
   _____ nondisjunction
   _____ translocation
   _____ gene mutation

   a. a chemical change in DNA that may affect genotype and phenotype
   b. movement of a chromosome segment to a nonhomologous chromosome
   c. disrupts gene linkages during meiosis
   d. causes gametes to have abnormal chromosome numbers
   e. loss of a chromosome segment

Critical Thinking: Genetics Problems
(Answers in Appendix III)

1. A female runner is disqualified from a race because testing shows that this individual is XY. Later, a medical exam reveals "androgen insensitivity," an abnormality in which an embryo's cells lack receptors for the male hormone testosterone. As a result, female sex characteristics develop, and indeed the person's phenotype is clearly female. Do you agree or disagree with the disqualification, and why?

2. If a couple has six boys, what is the probability that a seventh child will be a girl?

3. Human sex chromosomes are XX for females and XY for males.
   a. From which parent does a male inherit his X chromosome?
   b. With respect to an X-linked gene, how many different types of gametes can a male produce?
   c. If a female is homozygous for an X-linked allele, how many different types of gametes can she produce with respect to this allele?
   d. If a female is heterozygous for an X-linked allele, how many different types of gametes can she produce with respect to this allele?

Table 18.1 Examples of Human Genetic Disorders and Genetic Abnormalities

Disorder or Abnormality*	Main Consequences	Disorder or Abnormality*	Main Consequences
AUTOSOMAL RECESSIVE INHERITANCE		**X-LINKED RECESSIVE INHERITANCE**	
Albinism *CT*	Absence of pigmentation	Color blindness *18.7*	Inability to distinguish among all or some colors of the spectrum of visible light
Cystic fibrosis *CI; 18.6*	Excessive glandular secretions leading to tissue, organ damage		
Phenylketonuria (PKU) *18.6*	Mental retardation	Duchenne muscular dystrophy *18.7*	Muscles waste away
Sickle-cell anemia *11.5; 14.1; 18.6*	Adverse pleiotropic effects on organs throughout body	Hemophilia *18.7*	Impaired blood-clotting ability
		Testicular feminizing syndrome *18.7, CT*	XY individual but having some female traits, sterility
AUTOSOMAL DOMINANT INHERITANCE		**SEX-INFLUENCED INHERITANCE**	
Achondroplasia *18.6*	One form of dwarfism	Pattern baldness *18.8*	Loss of hair on the top and upper sides of the head
Achoo syndrome *CT*	Chronic sneezing		
Amyotrophic lateral sclerosis (ALS) *21.8*	Loss of all muscle function	**CHANGES IN CHROMOSOME NUMBER**	
Familial hypercholesterolemia *18.6*	High cholesterol levels in blood; eventually clogged arteries	Down syndrome *18.10*	Mental retardation, heart defects
		Klinefelter syndrome *18.10*	Sterility, retardation
Huntington disease *18.6*	Nervous system degenerates progressively, irreversibly	Turner syndrome *18.10*	Sterility; abnormal ovaries, abnormal sexual traits
Polydactyly *18.4*	Extra fingers, toes, or both	XYY condition *18.10*	Mild retardation or free of symptoms
Tay-Sachs disease *18.6*	Progressive deterioration of the nervous system		
X-LINKED DOMINANT INHERITANCE		**CHANGES IN CHROMOSOME STRUCTURE**	
Faulty enamel trait *18.7*	Problems with teeth	Cri-du-chat syndrome *18.9*	Mental retardation, abnormally formed larynx
		Fragile X syndrome *18.9*	Mental retardation

*Italic numbers indicate sections in which a disorder is described. CI signifies Chapter Introduction. CT signifies an end-of-chapter Critical Thinking question.

4. In cattle, the polled (hornless) phenotype is dominant over the horned phenotype. Suppose you are a rancher, and you are trying to keep a pure polled herd by removing horned cattle before they reproduce. If you keep your herd isolated from other herds (so no interbreeding occurs), why might an occasional horned calf be born?

5. Individuals with Down syndrome have an extra chromosome 21, for a total of 47 chromosomes in body cells. However, in a few cases of Down syndrome, 46 chromosomes are present. This total includes two normal-appearing chromosomes 21, one normal chromosome 14, and a longer-than-normal chromosome 14. Interpret this observation. How can these individuals have 46 chromosomes?

6. Ocular albinism is a recessive X-linked trait that causes a lack of pigment in the eyes. Two first cousins marry, a man with ocular albinism and a woman whose father (her husband's uncle) also had ocular albinism. What is the probability that any one of their children will have ocular albinism? Does the probability differ if the child is male or female?

7. If a trait appears only in males, is this good evidence that the trait is due to a Y-linked allele? Explain why you answered as you did.

8. A woman unaffected by hemophilia A whose father had hemophilia A marries a man who also has hemophilia A. If their first child is a boy, what is the probability he will have the disorder?

9. Among people of European descent, about 4 percent have the allele for the genetic disorder cystic fibrosis. Yet only about 1 in 2,500 people actually have the disorder. What is the most likely reason for this?

10. People with albinism have very pale skin, white hair, and pink eyes. This phenotype, typically caused by a recessive allele, is due to the absence of melanins, which impart color to the skin, hair, and eyes. Suppose a person with albinism marries a person with typical pigmentation and they have one child with albinism and three with typical pigmentation. What is the genotype of the parent with typical pigmentation? Why is the ratio of this couple's offspring 3:1?

Selected Key Terms

aneuploidy *400*
autosome *388*
chromosome mapping *391*
deletion (chromosomal) *398*
duplication *398*
inversion *399*
linkage *391*
nondisjunction *400*
pedigree *392*
polyploidy *400*
sex chromosome *388*
translocation *399*
X inactivation *390*
X-linked gene *388*
Y-linked gene *388*

Readings

Cummings, M. 1994. *Human Heredity: Principles and Issues.* 3d ed. St. Paul, Minn.: West.

"Huntington's Gene Finally Found." April 1993. *Science.*

Travis, J. 30 November 1996. "Clue to Lou Gehrig's Disease Emerges." *Science News.*

 For additional readings, go to InfoTrac College Edition, your online library at:

http://www.infotrac-college.com/wadsworth

19 DNA STRUCTURE AND FUNCTION

Cardboard Atoms and Bent-Wire Bonds

Throughout the early 20th century, biologists pondered one of the most fundamental questions in all science: What molecule serves as the book of genetic information in every living cell?

For decades researchers had assumed that proteins, which could consist of virtually limitless combinations of amino acids, were the genetic material. Heritable traits are awesomely diverse. It made sense for the molecules that held the information for those traits to be structurally diverse. Yet, by the late 1940s, there was something about another cellular substance—DNA—that tugged at more than a few good minds. DNA consists of only four kinds of subunits. For DNA to

be the "master molecule of life," the subunits would have to be arranged in a way that could carry a vast amount of information. And there would have to be a mechanism to copy that structure in a way that would faithfully transfer hereditary information to a new generation. Research published in 1950 suggested strongly that DNA was indeed the genetic material, but still unknown were the crucial structural details that would explain how it functioned.

Figure 19.1 James Watson and Francis Crick posing in 1953 by their newly unveiled model of DNA structure. Behind this photograph is a recent computer-generated model. It is more sophisticated in appearance yet basically the same as the prototype that was built nearly four decades before.

A small cadre of brilliant researchers dreamed in 1950 of winning the race to elucidate the structure of DNA. Biochemist Linus Pauling—who had already discovered that a single gene could cause sickle-cell anemia—came very, very close. In the end, however, the glory went mostly to James Watson, a 25-year-old postdoctoral student from Indiana University, and to Francis Crick, a Cambridge University researcher who was working on his Ph.D. Watson and Crick spent long hours arguing over everything they had read about the size, shape, and bonding requirements of DNA's subunits. They arranged and rearranged cardboard cutouts of the subunits and pestered chemists to identify potential bonds they might have overlooked. Crucial evidence came from the laboratory of Maurice Wilkins, where Rosalind Franklin was analyzing DNA samples using X-ray diffraction, a method that can reveal details of a molecule's structure. Aided by clues in the X-ray images, Watson and Crick painstakingly fashioned models of bits of metal held together with wire "bonds" bent at chemically correct angles. In 1953 they devised a model that fit what was known about DNA and the pertinent biochemical rules (Figure 19.1).

Watson and Crick had found the structure of DNA and ushered in the golden age of molecular genetics. Together with Maurice Wilkins, they won the 1962 Nobel Prize for medicine. Rosalind Franklin died before the prize was awarded.

The remarkable work of these scientists, and of others before and since, is the foundation for the concepts you will study in this chapter. As you read, keep in mind DNA's two main functions. *First*, it stores the vast number of hereditary instructions required for the building and functioning of each and every body cell, tissue, and organ. *Second*, by way of *self-replication*, it copies those hereditary instructions in a way that faithfully conveys them to a new generation. The key to both these remarkable feats lies in DNA's structure, and so that is the point at which we will begin.

KEY CONCEPTS

1. DNA is the cell's storehouse of information about heritable traits.

2. Hereditary information is encoded in the nucleotides that make up the DNA molecule. There are four kinds of nucleotides in DNA, and they differ in one component, a nitrogen-containing base.

3. In a DNA molecule, two strands of nucleotides twist together like a spiral stairway, forming a double helix. Hydrogen bonds connect the bases of one strand to bases of the other.

4. Before a cell divides, its DNA is replicated with the help of enzymes and other proteins. Each double-stranded DNA molecule starts unwinding. A new, complementary strand is assembled on the exposed bases of each parent strand.

5. There is only one DNA molecule in an unduplicated chromosome. Great numbers of proteins are attached to the DNA and function in its structural organization.

6. In each of the body's proteins, the sequence of amino acids corresponds to a gene—which is a sequence of nucleotide bases in DNA.

7. The path leading from genes to proteins has two steps, called transcription and translation. In transcription, the DNA molecule unwinds at a gene region, then an RNA molecule is assembled on the exposed bases of one of the strands. In translation, RNA directs the linkage of one amino acid after another, in the sequence required to produce a specific polypeptide chain.

8. A permanent change in a gene's base sequence is called a mutation. Mutations are the source of genetic variation—they lead to alterations in protein structure, function, or both. These alterations may lead to differences in traits among individuals.

CHAPTER AT A GLANCE

19.1 DNA Structure and Replication

19.2 *DNA Detectives: Discovering the Connection between Genes and Proteins*

19.3 Organization of DNA in Chromosomes

19.4 DNA into RNA—Protein Synthesis Begins

19.5 From mRNA to Proteins

19.6 Translation

19.7 How Mutation Affects Protein Synthesis

19.8 Regulating Gene Action

19.9 *A Tale of Nucleotide Repeats*

A DNA molecule is built from four kinds of **nucleotides**, the building blocks of nucleic acids (Figure 19.2). A DNA nucleotide consists of a five-carbon sugar (deoxyribose), a phosphate group, and one of the following nitrogen-containing bases:

adenine	guanine	thymine	cytosine
(A)	(G)	(T)	(C)

Many researchers contributed knowledge in the race to discover DNA's structure, but Watson and Crick were the first to recognize that DNA consists of two strands of nucleotides twisted together into a double helix. Nucleotides in a strand are linked together, like boxcars in a train, by strong covalent bonds. Weaker hydrogen bonds link the bases of one strand with bases of the other. The two strands are aligned in an antiparallel arrangement—they run in opposite directions, as you can see in Figure 19.3 on the facing page.

The shapes of the bases and their sites available for hydrogen bonding determine which bases can pair up. Adenine pairs with thymine, and guanine pairs with cytosine. Thus, two kinds of **base pairs** typically occur in

Figure 19.2 The nucleotide subunits of DNA. All chromosomes contain DNA. What does DNA contain? Four kinds of nucleotides. A nucleotide has a five-carbon sugar (shaded *red*), which has a phosphate group attached to the fifth carbon atom of its ring structure. It also has one of four kinds of nitrogen-containing bases (shaded *blue*) attached to its first carbon atom. Small numerals on the structural formulas identify the carbon atoms to which these and other parts of the molecule are attached.

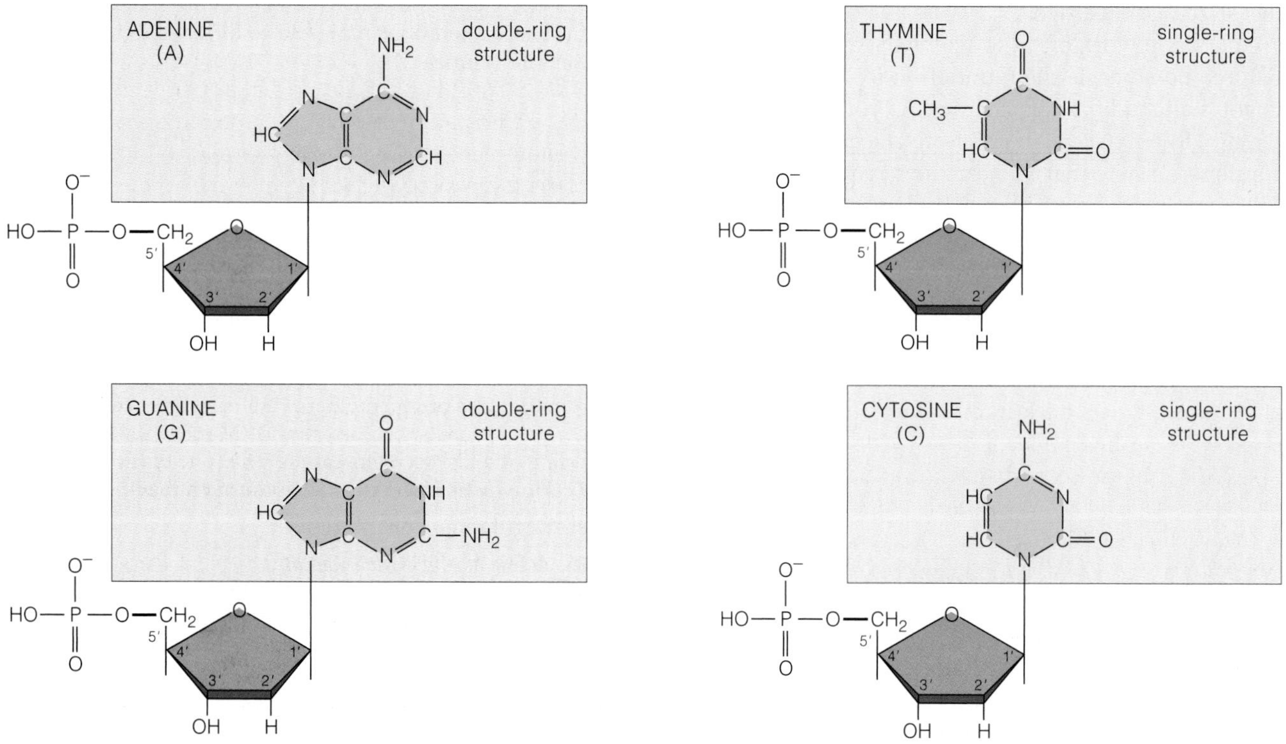

DNA: A—T and G—C. In a double-stranded DNA molecule, the amount of adenine equals the amount of thymine, and the amount of guanine equals the amount of cytosine.

At various times in a cell's life, regions of the double helix unwind so that a cell gains access to specific genes. In general, a **gene** is a region of DNA that codes for a specific polypeptide chain. Such chains, remember, are the basic structural units of proteins. (Later sections describe *how* genes give rise to polypeptide chains.) Some genes code only for certain types of RNA. This chapter's *Science Comes to Life* (page 408) describes some of the classic research that uncovered the basic function of genes.

The two long nucleotide strands of the double helix also unwind and separate from each other when DNA replicates. As described next, the exposed bases of each strand serve as a pattern, or *template*, upon which a new strand is built.

DNA Replication

Before a cell divides, DNA is duplicated in a process called **replication**. Replication yields sister chromatids, the duplicated chromosomes that are parceled out to daughter cells in mitosis and meiosis (Chapter 16).

Hydrogen bonds hold together the two nucleotide strands making up a DNA double helix. Certain enzymes can break those bonds. Through enzyme action, one strand starts to unwind from the other at distinct sites, leaving the bases exposed. Free nucleotides pair with the

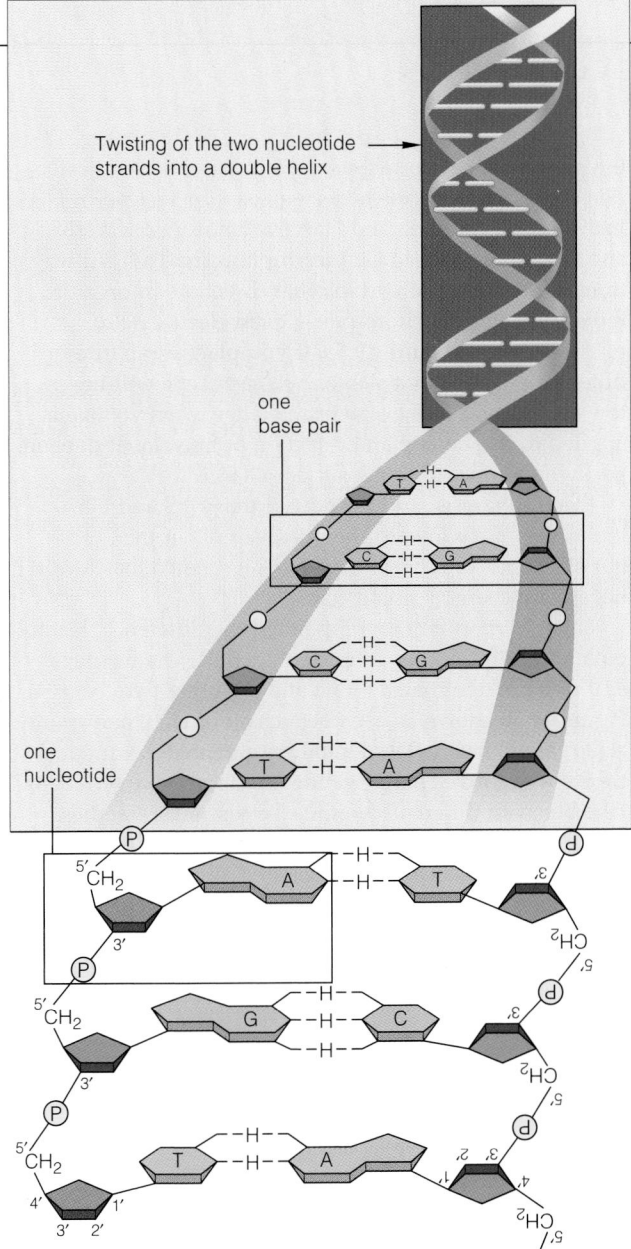

Twisting of the two nucleotide strands into a double helix →

one base pair

one nucleotide

Figure 19.3 Arrangement of bases in the DNA double helix.

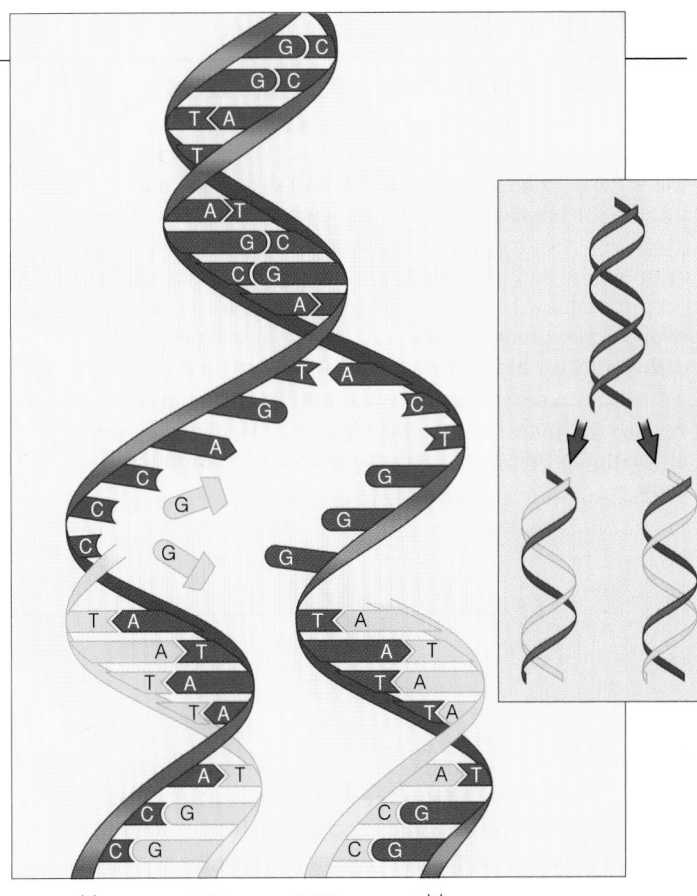

old　　new　　new　　old

Figure 19.4 Semiconservative nature of DNA replication. The original two-stranded DNA molecule is shown in *blue*. A new strand (*yellow*) is assembled on each parent strand.

exposed bases, following the base-pairing rule (A with T and G with C). They are linked by hydrogen bonds. Each parent strand remains intact while a new companion strand is assembled on it, nucleotide by nucleotide.

As replication proceeds, the newly formed double-stranded molecule twists back into a double helix. One strand is from the starting molecule, so that strand is said to be *conserved*. Only the second strand has been freshly synthesized—so each DNA molecule is really half new and half "old." Hence the DNA replication mechanism is called *semiconservative replication* (Figure 19.4).

During replication, enzymes and other proteins unwind the DNA molecule, keep the two strands separated at unwound regions, and assemble a new strand on each one. For example, **DNA polymerases** carry out the assembly of nucleotides on a parent strand.

DNA Repair

DNA polymerases, DNA ligases, and other enzymes also perform DNA repair. If an error occurs during replication, or if a DNA strand is damaged, enzymes may detect and correct the problem, restoring the proper sequence.

When a repair mechanism is impaired (as when mutation alters one or more of the genes that govern it), the impact can be serious. In the genetic disorder *xeroderma pigmentosum*, a faulty DNA repair mechanism doesn't correct damage to the DNA in skin cells exposed to ultraviolet light. In affected people, even brief exposure to sunlight leads to disfiguring skin tumors and skin cancer.

Information for producing heritable traits is encoded in DNA, which consists of two strands of nucleotides. Hydrogen bonds between pairs of the four types of nucleotide bases in DNA hold the strands together in a double helix.

DNA replication is semiconservative. After the double helix unwinds, each parent strand remains intact—it is conserved—and enzymes assemble a new, complementary strand on each one. Normally, certain enzymes repair damage in the DNA molecule.

DNA DETECTIVES: DISCOVERING THE CONNECTION BETWEEN GENES AND PROTEINS

In the early 1900s a physician, Archibald Garrod, was tracking metabolic disorders that seemed to be heritable (they kept recurring in the same families). In one of the disorders he studied, blood or urine samples from affected people contained abnormally high levels of a substance known to be produced at a certain step in one metabolic pathway. Most likely, the enzyme at the *next* step in the pathway was defective and could not use that substance. Because the pathway was blocked from that step onward, unused molecules of the substance accumulated in the body:

$$A \longrightarrow B \overset{C\ C\ C}{\underset{C}{\longrightarrow}} C \times D$$

pathway is blocked

Only one thing distinguished affected people from unaffected ones: The affected people had inherited a metabolic deficiency. Thus, Garrod concluded, specific "units" of inheritance (genes) function through the synthesis of specific enzymes. As later researchers pursued this idea, it became known as the "one gene, one enzyme" hypothesis.

The hypothesis was refined through studies of sickle-cell anemia (page 381). This heritable disorder arises from the presence of abnormal hemoglobin in red blood cells. The abnormal molecule is designated HbS instead of HbA. In 1949 Linus Pauling and Harvey Itano subjected HbS and HbA molecules to **gel electrophoresis**. This is a way to measure how fast and in what direction an organic molecule will move in response to an electric field.

As shown in Figure 19.5a, if you place a mixture of different proteins in a slab of gel, each type will move toward one end of the slab or the other when voltage is applied to it. The rate and direction of movement depend partly on a molecule's net surface charge.

Electrophoresis studies showed that HbS and HbA molecules move toward the positive pole of the field—but HbS does so more slowly. HbS, it seemed, has fewer negatively charged amino acids.

Later, Vernon Ingram pinpointed the difference. Hemoglobin, recall, consists of four polypeptide chains (page 32). Two are designated alpha and the other two, beta. As Figure 19.5b shows, in each beta chain of HbS, one amino acid (valine) has replaced another (glutamate). Glutamate carries a negative charge; valine has no net charge. Thus HbS behaved differently in the electrophoresis studies.

This discovery suggested that two genes code for hemoglobin—one for each kind of polypeptide chain—and that genes code for proteins in general, not just for enzymes.

And so a more precise hypothesis emerged: *One gene codes for the amino acid sequence of one polypeptide chain.*

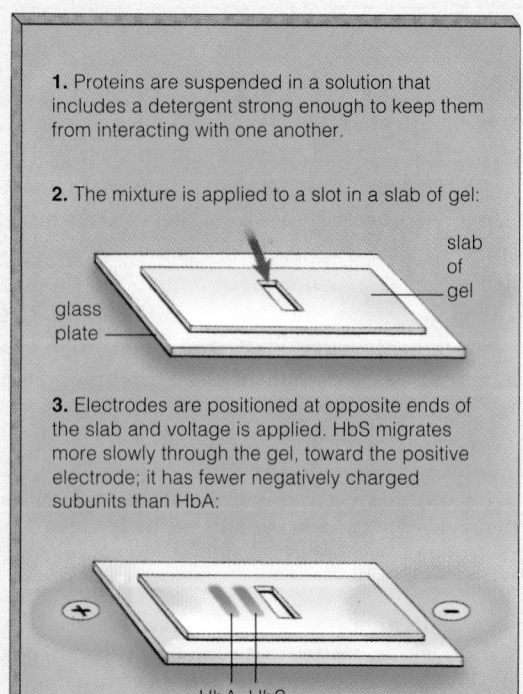

1. Proteins are suspended in a solution that includes a detergent strong enough to keep them from interacting with one another.

2. The mixture is applied to a slot in a slab of gel:

glass plate

slab of gel

3. Electrodes are positioned at opposite ends of the slab and voltage is applied. HbS migrates more slowly through the gel, toward the positive electrode; it has fewer negatively charged subunits than HbA:

HbA   HbS

a

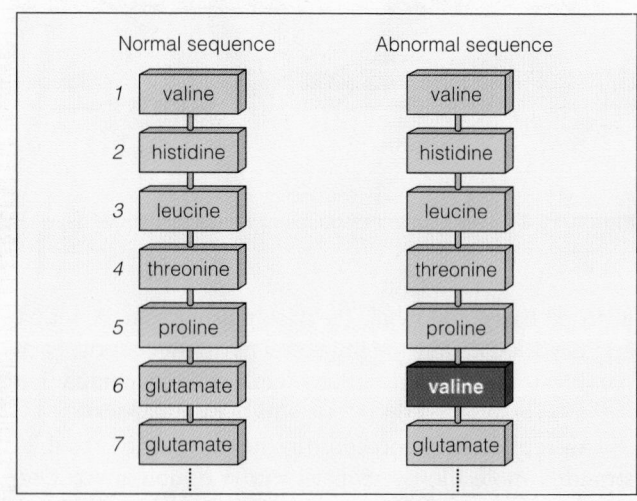

b

Figure 19.5 Gel electrophoresis. (**a**) A simplified diagram of the steps in gel electrophoresis. (**b**) Example of contrasting amino acid sequences of the sort gel electrophoresis can reveal.

	Normal sequence	Abnormal sequence
1	valine	valine
2	histidine	histidine
3	leucine	leucine
4	threonine	threonine
5	proline	proline
6	glutamate	**valine**
7	glutamate	glutamate

ORGANIZATION OF DNA IN CHROMOSOMES

Each chromosome has one DNA molecule. Stretched out and lined up end to end, the DNA in the 46 chromosomes of a human somatic cell is about 2 meters long—a little over 6 feet. All that DNA would become a tangled mess if it were not organized in some way. In fact, the DNA of humans and all other eukaryotes is tightly bound with many proteins, including ones called **histones**. Some histones are like spools for winding up small stretches of DNA. Each histone–DNA spool is a **nucleosome** (Figure 19.6a and b). Another histone stabilizes the spools.

The nucleosome string along the DNA molecule becomes coiled repeatedly through interactions between the histones and DNA. The coiling greatly increases its diameter. Further folding results in a series of loops of various sizes. Proteins other than histones serve as a structural "scaffold" for the loops (Figure 19.6d).

In a chromosome, DNA is tightly bound with histones and other proteins. Interactions between DNA and histones lead to the formation of beadlike nucleosomes, then to coiling and folding of the molecule into loops.

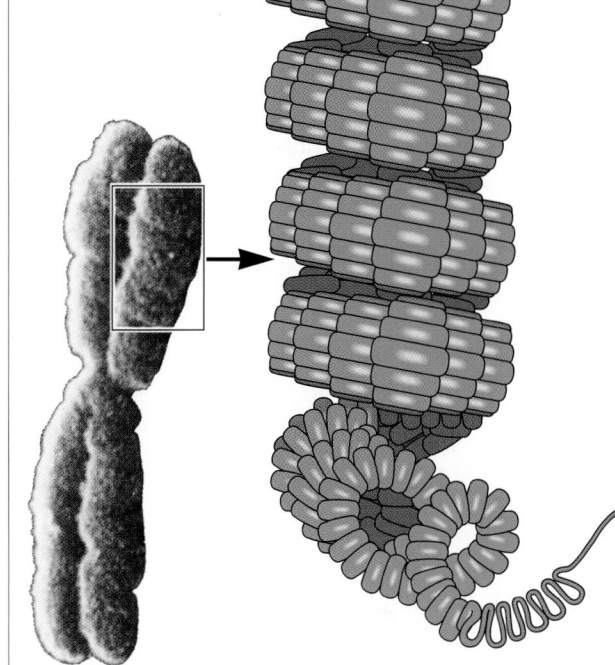

a Human chromosome at metaphase, in its most condensed form. Interactions among certain chromosomal proteins keep loops of DNA tightly packed in a "supercoiled" arrangement.

Figure 19.6 Levels of organization of DNA in a chromosome.

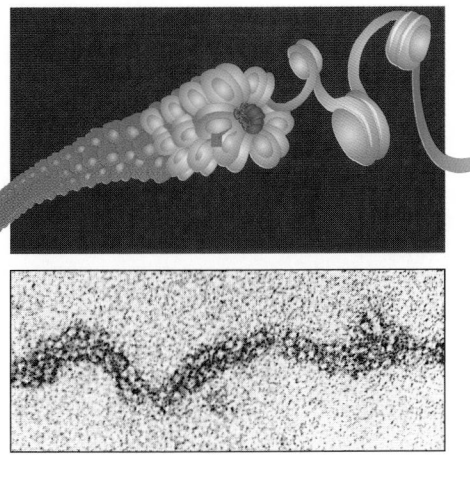

b At a deeper level of organization, the chromosomal proteins and DNA are arranged as a cylindrical fiber (solenoid), 30 nanometers in diameter.

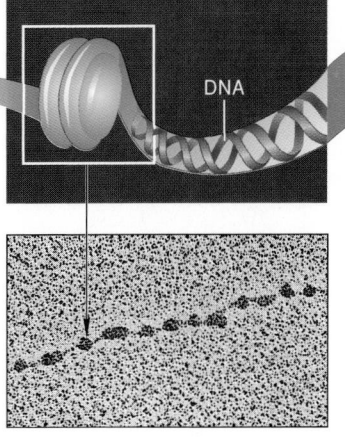

c Immerse a chromosome in a salt solution, and it will loosen up to a beads-on-a-string organization. The "string" is one DNA molecule. Each "bead" is a nucleosome, the smallest organizational unit of the eukaryotic chromosome.

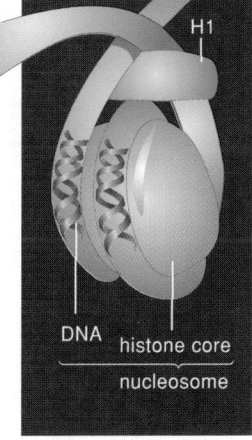

d Each nucleosome consists of a double loop of DNA around a core of eight proteins, of a class called histones. Another histone (H1) stabilizes the arrangement.

A Brief Overview of Protein Synthesis

The path from genes to proteins has two steps, called transcription and translation. Both involve molecules of ribonucleic acid, or **RNA**. Most often, RNA is single-stranded. Structurally, it is much like a strand of DNA. Its nucleotides each consist of a sugar (ribose), a phosphate group, and a nitrogen-containing base. However, its bases are adenine, cytosine, guanine, and **uracil**. We can summarize these differences as follows:

	DNA	**RNA**
Sugar:	deoxyribose	ribose
Bases:	adenine, cytosine, guanine, thymine	adenine, cytosine, guanine, uracil

Like the thymine in DNA, the uracil in RNA base-pairs with adenine.

In **transcription**, molecules of RNA are produced on DNA templates in the nucleus. In **translation**, which you'll read about in Section 19.6, RNA molecules move from the nucleus into the cytoplasm, where they become templates for assembling polypeptide chains. After translation, one or more polypeptide chains become folded into protein molecules. The proteins serve as enzymes in biosynthesis, as membrane receptors and channels, as transport proteins, and so on. They also include most of the enzymes and other proteins that take part in DNA replication, RNA synthesis, and protein synthesis.

The *central dogma of molecular biology* sums up this flow of information in cells. It states that DNA is transcribed into RNA, then RNA is translated into protein.

Genes are transcribed into three types of RNA:

ribosomal RNA (rRNA)	a nucleic acid chain that combines with certain proteins to form a *ribosome*, on which a polypeptide chain is assembled
messenger RNA (mRNA)	a linear sequence of nucleotides that carries protein-building instructions; this "code" is delivered to the ribosome for translation into a polypeptide chain
transfer RNA (tRNA)	another nucleic acid chain that can pick up a specific amino acid *and* pair with an mRNA code word for that amino acid

Notice that only mRNA eventually becomes translated into a protein product. The other two types of RNAs have specific roles during translation.

Transcription

In transcription, an RNA strand is assembled on a DNA template according to the base-pairing rules:

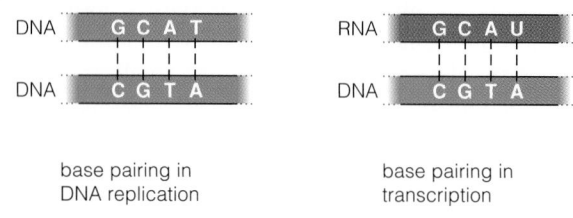

base pairing in DNA replication

base pairing in transcription

Transcription takes place in the cell nucleus, but it differs from DNA replication in key respects. *First*, only the gene segment serves as the template—not the whole DNA strand. *Second*, enzymes called **RNA polymerases** are involved. Third, transcription results in only a single-stranded molecule, not one with two strands.

Transcription starts at a **promoter**, a base sequence that signals the start of a gene. Proteins help position an RNA polymerase on the DNA so that it binds with the promoter. The enzyme moves along the DNA, joining

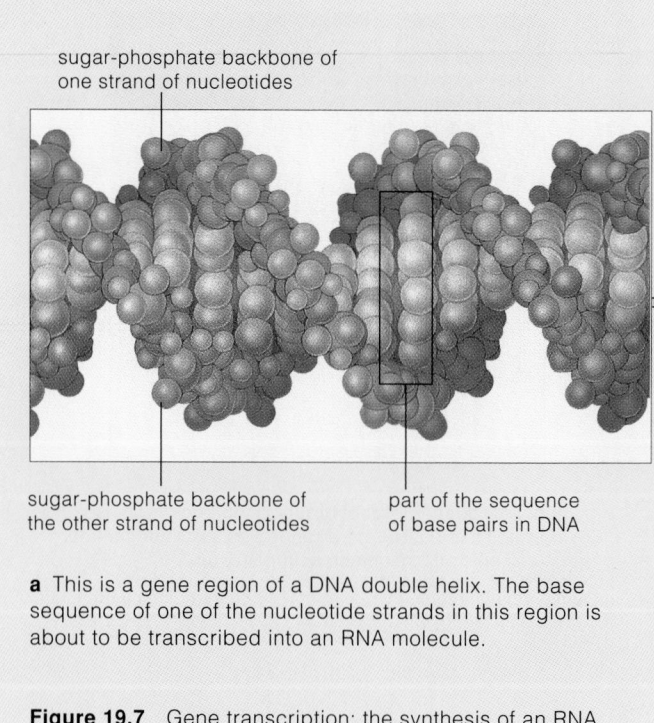

sugar-phosphate backbone of one strand of nucleotides

sugar-phosphate backbone of the other strand of nucleotides

part of the sequence of base pairs in DNA

a This is a gene region of a DNA double helix. The base sequence of one of the nucleotide strands in this region is about to be transcribed into an RNA molecule.

Figure 19.7 Gene transcription: the synthesis of an RNA molecule on a DNA template.

nucleotides together (Figure 19.7). When it reaches a termination sequence that serves as a signal to cut the RNA loose, the RNA is released as a transcript.

Newly formed pre-mRNA is an unfinished molecule that must be modified before its protein-building instructions can be used. Just as a film editor might cut some frames and splice together others to make the final version of a movie, a cell tailors its pre-mRNA. For example, newly formed transcripts contain portions called introns and exons. **Introns** are base sequences that are removed from the pre-mRNA. **Exons** are the regions that remain in the mature mRNA. As Figure 19.8 shows, before the mRNA leaves the nucleus, the introns are snipped out and the exons are spliced together.

Protein synthesis has two steps, transcription and translation.

In transcription, a sequence of bases in one of the two strands of a DNA molecule serves as the template for assembling an RNA strand. The assembly follows base-pairing rules (adenine with uracil, cytosine with guanine).

Before leaving the nucleus, new RNA transcripts undergo modification into final form.

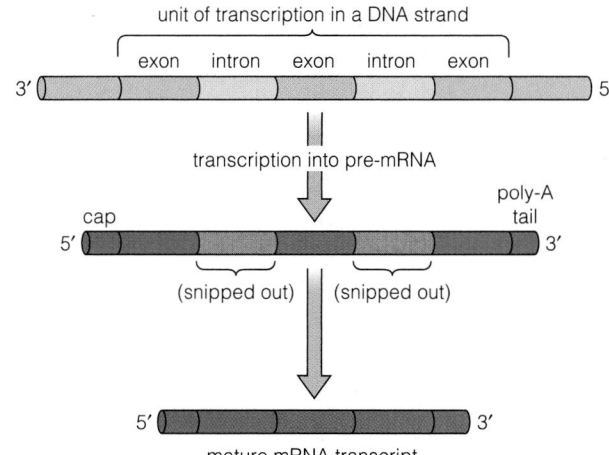

Figure 19.8 Transcription and modification of newly formed mRNA. The cap is simply a nucleotide that serves as a site for ribosome attachment. The tail is a string of adenine nucleotides. The notations 3′ and 5′ indicate noncoding regions of the mRNA.

b An RNA polymerase molecule binds to a promoter in the DNA. It will use the base sequence positioned downstream from that site as a template for linking nucleotides into a strand of RNA.

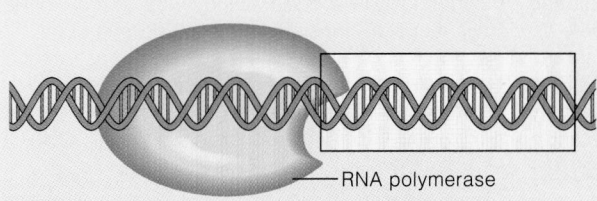

c During transcription, RNA nucleotides are base-paired, one after another, with the exposed bases on the DNA template.

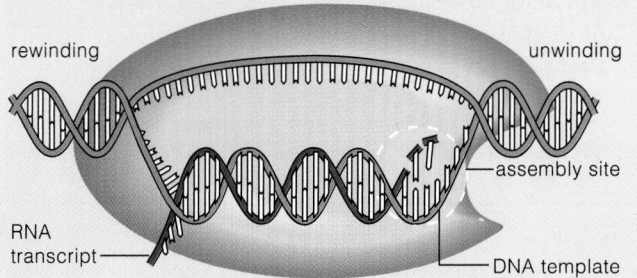

d Throughout transcription, the DNA double helix unwinds just in front of the RNA polymerase. Short lengths of the newly forming RNA strand temporarily wind up with the DNA template strand. Then they unwind from it—and the two strands of DNA wind up together again.

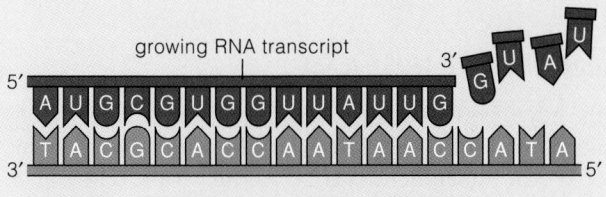

e At the end of the gene region, the last stretch of the RNA is unwound from the DNA template and so is released.

19.5 FROM mRNA TO PROTEINS

The Genetic Code

Like a DNA strand, mRNA is a linear sequence of nucleotides. What are the protein-building "words" encoded in its sequence? We now know that enzymes recognize nucleotide bases *three at a time*, as triplets. In mRNA, the base triplets are called **codons**. Figure 19.9 shows how the order of different codons in an mRNA molecule dictates the order in which particular amino acids are added to a growing polypeptide chain. Count the codons listed in Figure 19.10, and you see there are 64 kinds.

Most of the 20 kinds of amino acids correspond to more than one codon. (Glutamate corresponds to the code words GAA *or* GAG, for example.) The codon AUG also establishes the reading frame for translation. That is, ribosomes start their "three-bases-at-a-time" selections at an AUG that serves as the "start" signal in an mRNA strand. Three codons (UAA, UAG, UGA) are stop signals. They stop ribosomes from adding any more amino acids to a new polypeptide chain.

First Letter	Second Letter				Third Letter
	U	C	A	G	
U	phenylalanine	serine	tyrosine	cysteine	U
	phenylalanine	serine	tyrosine	cysteine	C
	leucine	serine	stop	stop	A
	leucine	serine	stop	tryptophan	G
C	leucine	proline	histidine	arginine	U
	leucine	proline	histidine	arginine	C
	leucine	proline	glutamine	arginine	A
	leucine	proline	glutamine	arginine	G
A	isoleucine	threonine	asparagine	serine	U
	isoleucine	threonine	asparagine	serine	C
	isoleucine	threonine	lysine	arginine	A
	(start) methionine	threonine	lysine	arginine	G
G	valine	alanine	aspartate	glycine	U
	valine	alanine	aspartate	glycine	C
	valine	alanine	glutamate	glycine	A
	valine	alanine	glutamate	glycine	G

Figure 19.10 The genetic code. The codons in mRNA are nucleotide bases, read in blocks of three. Sixty-one of these base triplets correspond to specific amino acids. Three others serve as signals that stop translation. The left column of the diagram shows the first of the three nucleotides in each codon in mRNA. The middle columns show the second nucleotide. The right column shows the third. Reading from left to right, for instance, the triplet UGG corresponds to tryptophan. Both UUU and UUC correspond to phenylalanine.

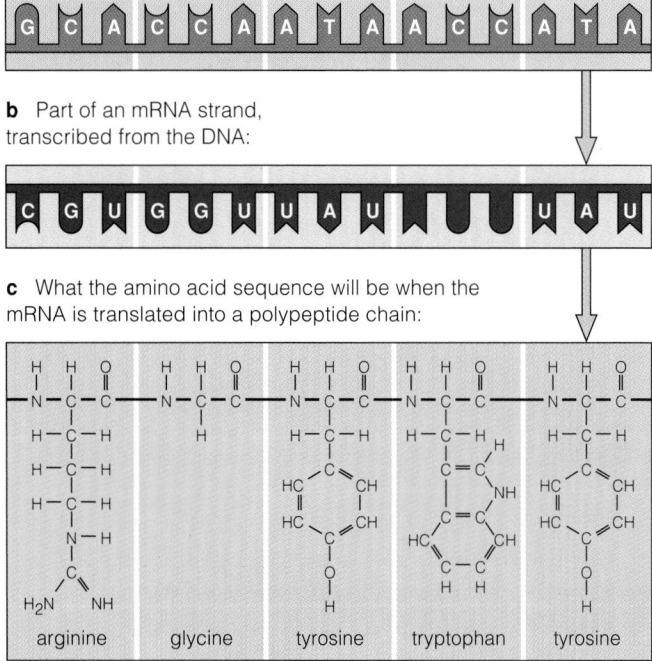

a Base sequence of a gene region in DNA:

G C A C C A A T A A C C A T A

b Part of an mRNA strand, transcribed from the DNA:

C G U G G U U A U U A U

c What the amino acid sequence will be when the mRNA is translated into a polypeptide chain:

arginine glycine tyrosine tryptophan tyrosine

Figure 19.9 The steps from genes to proteins. (**a**) This region of a DNA double helix was unwound during transcription. (**b**) Exposed bases on one strand served as a template for assembling an mRNA strand. In the mRNA, every three nucleotide bases equaled one codon. Each codon called for one amino acid in this polypeptide chain. (**c**) Referring to Figure 19.10, can you fill in the blank codon for tryptophan in the chain?

The set of 64 different codons is the **genetic code**. It is the basis of protein synthesis.

Roles of tRNA and rRNA

Cells have pools of free amino acids and free tRNA molecules in the cytoplasm. tRNAs have a molecular "hook," an attachment site for amino acids. They also have an **anticodon**, a nucleotide triplet that can base-pair with codons (Figure 19.11). When tRNAs bind to codons, they automatically position their attached amino acids in the order specified by mRNA.

A cell has more than 60 kinds of codons but fewer kinds of tRNAs. How do they match up? By the base-pairing rules, adenine pairs with uracil, and cytosine with

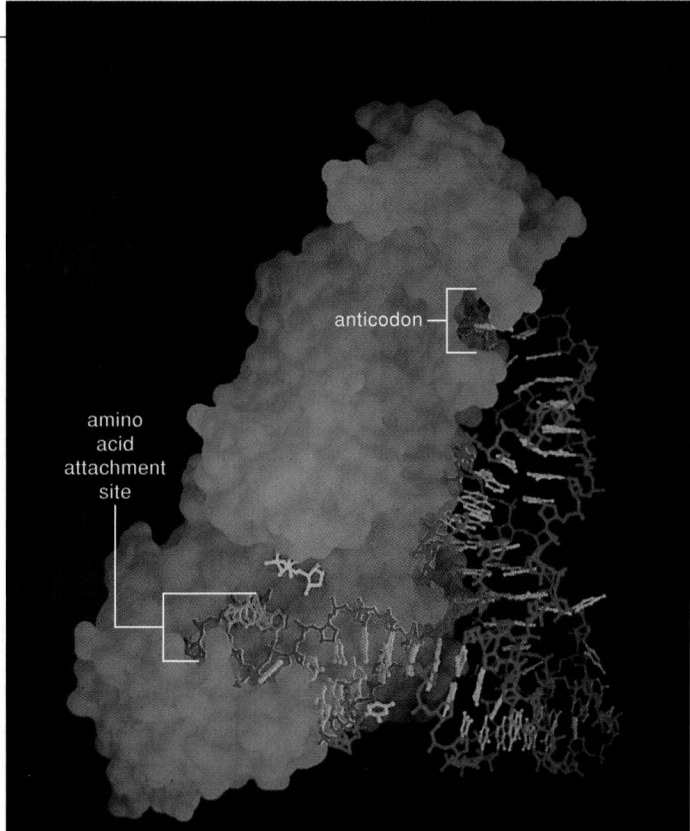

a

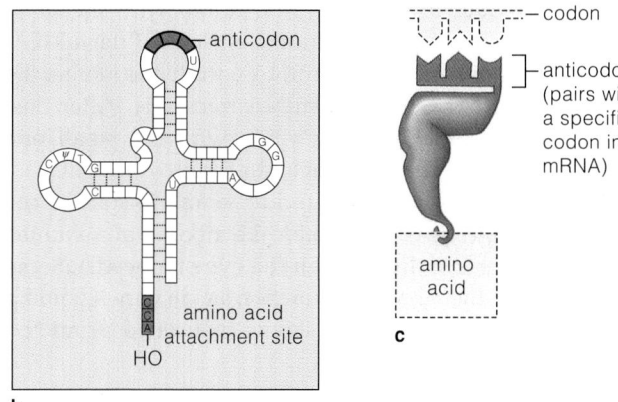

b

codon

anticodon
(pairs with
a specific
codon in
mRNA)

amino
acid

c

Figure 19.11 (**a**) Computer-generated, three-dimensional model of one type of tRNA molecule. The tRNA (*reddish brown*) is shown attached to a bacterial enzyme (*green*), along with an ATP molecule (*gold*). This particular enzyme attaches amino acids to tRNAs. (**b**) Structural features common to all tRNAs. (**c**) Simplified model of tRNA that is used in subsequent illustrations. The "hook" at one end is the site to which a specific amino acid can become attached.

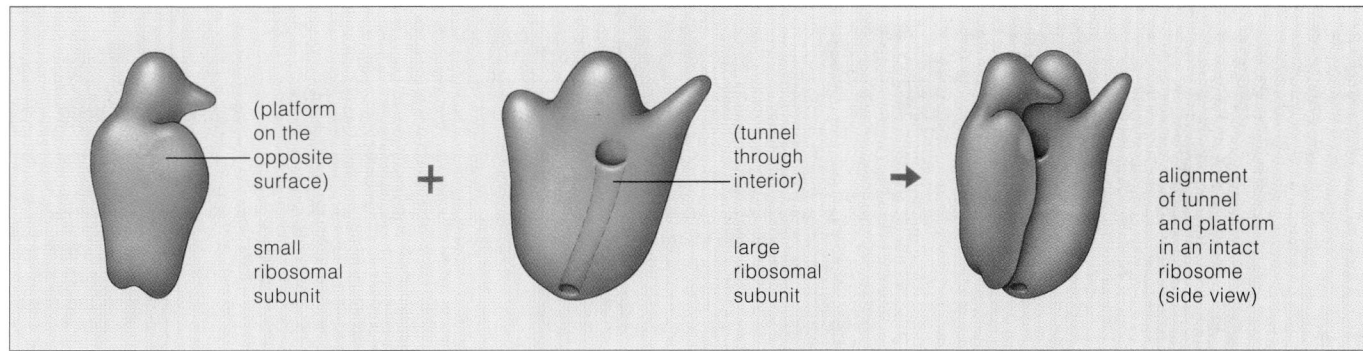

Figure 19.12 Model of eukaryotic ribosomes. Polypeptide chains are assembled on the small subunit's platform. Newly forming chains may move through the large subunit's tunnel.

guanine. However, for codon–anticodon interactions, the rules loosen up for the third base. For example, CCU, CCC, CCA, and CCG all specify proline but require only two tRNAs. Such freedom in codon–anticodon pairing at the third base is called the "wobble effect."

tRNAs interact with mRNA at binding sites on the surface of **ribosomes**. Each ribosome has two subunits (Figure 19.12). These are assembled in the nucleus from rRNA and proteins, then are shipped separately to the cytoplasm. There they will combine as functional units only during translation, as described next.

Protein-building instructions are encoded in the nucleotide sequence of DNA and mRNA. The genetic code is a set of 64 base triplets (nucleotide bases, read in blocks of three). A codon is a base triplet in mRNA.

Different combinations of codons specify the amino acid sequence of different polypeptide chains, start to finish.

mRNAs are the only molecules that carry protein-building instructions from DNA to the cytoplasm.

tRNAs bind to codons and so position their attached amino acids in the order specified by mRNA. Thus they translate the mRNA into a corresponding sequence of amino acids.

rRNAs are components of ribosomes, the sites where amino acids are assembled into polypeptide chains.

TRANSLATION

Translation occurs in the cytoplasm. It consists of three stages, called initiation, elongation, and termination.

In *initiation*, an initiator tRNA binds with the small ribosomal subunit and attaches to one end of the mRNA. This unit moves along the mRNA until it encounters the start codon (AUG) of the mRNA transcript. After this, a large ribosomal subunit binds with the small one (Figure 19.13a). The next stage can begin.

In *elongation*, a polypeptide chain forms as the mRNA strand passes between the ribosomal subunits, like a thread moving through the eye of a needle. Some proteins in the ribosome function as enzymes, joining amino acids together in the sequence dictated by mRNA codons. As Figure 19.13b indicates, they catalyze the formation of a peptide bond (refer to Figure 1.22) between adjacent amino acids.

In *termination*, a stop codon is reached and there is no corresponding anticodon in a tRNA. Now the ribosome interacts with certain proteins (release factors). This causes the ribosome as well as the polypeptide chain to detach from the mRNA (Figure 19.13c). The detached chain may join the pool of free proteins in the cytoplasm. Or it may enter the cytomembrane system, starting with the compartments of rough ER. Many newly formed chains take on their final form in that system before being shipped to their destinations.

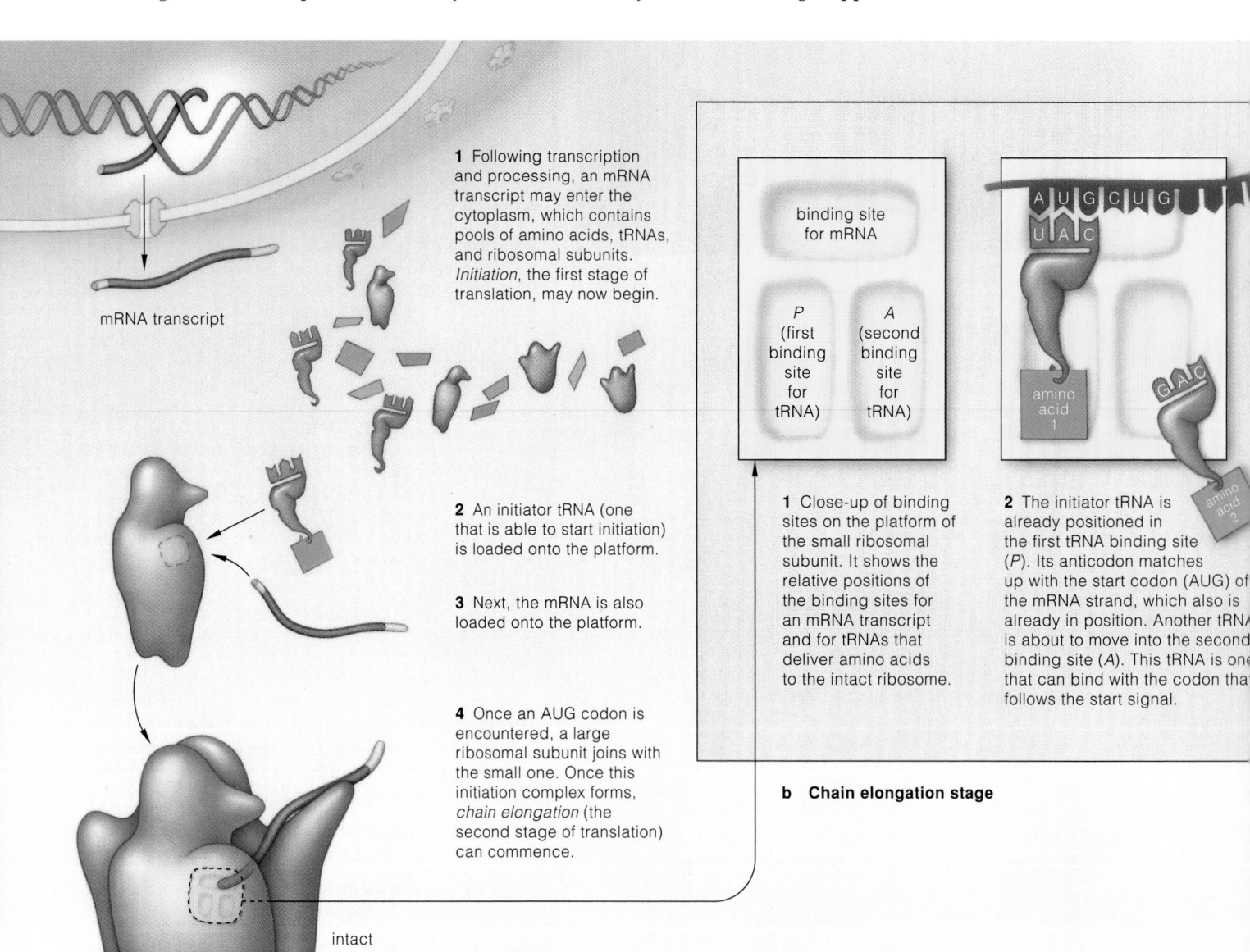

1 Following transcription and processing, an mRNA transcript may enter the cytoplasm, which contains pools of amino acids, tRNAs, and ribosomal subunits. *Initiation*, the first stage of translation, may now begin.

mRNA transcript

2 An initiator tRNA (one that is able to start initiation) is loaded onto the platform.

3 Next, the mRNA is also loaded onto the platform.

4 Once an AUG codon is encountered, a large ribosomal subunit joins with the small one. Once this initiation complex forms, *chain elongation* (the second stage of translation) can commence.

intact ribosome

a   Initiation stage

binding site for mRNA

P (first binding site for tRNA)

A (second binding site for tRNA)

1 Close-up of binding sites on the platform of the small ribosomal subunit. It shows the relative positions of the binding sites for an mRNA transcript and for tRNAs that deliver amino acids to the intact ribosome.

amino acid 1

2 The initiator tRNA is already positioned in the first tRNA binding site (*P*). Its anticodon matches up with the start codon (AUG) of the mRNA strand, which also is already in position. Another tRNA is about to move into the second binding site (*A*). This tRNA is one that can bind with the codon that follows the start signal.

b   Chain elongation stage

Figure 19.13   Translation, the second step of protein synthesis.

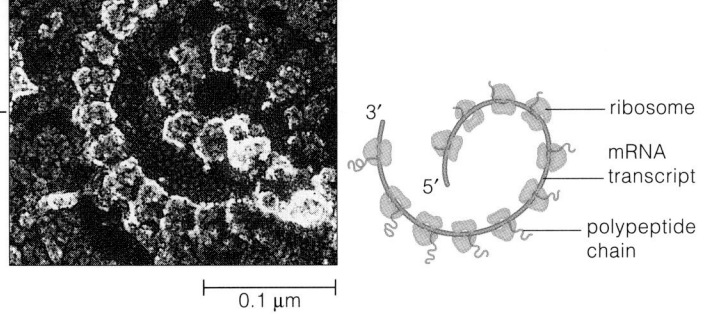

ribosome
mRNA transcript
polypeptide chain

3′
5′

0.1 μm

Figure 19.14 A polysome (many ribosomes simultaneously translating the same mRNA molecule).

Often, enzymes and tRNAs repeatedly translate the same mRNA transcript in a given period. The transcript threads through many ribosomes, which are arranged one after another, close together in assembly-line fashion. Figure 19.14 shows one of these arrangements,

which are called **polysomes**. Their presence indicates that a cell is producing many copies of a polypeptide chain from the same transcript.

Translation is initiated by the convergence of a small ribosomal unit, an initiator tRNA, an mRNA transcript, and then a large ribosomal subunit.

Next, anticodons of tRNAs base-pair with mRNA codons. A polypeptide chain grows through peptide bond formation between every two amino acids delivered to the ribosome.

Translation ends when a stop codon triggers events that cause the chain and the mRNA to detach from the ribosome.

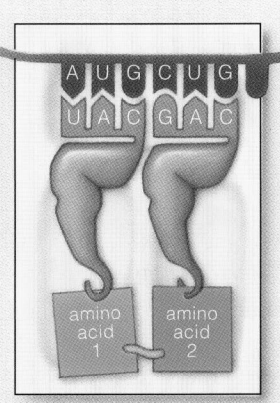

3 Through enzyme action, the bond between the initiator tRNA and the amino acid hooked to it is broken. At the same time, an enzyme catalyzes the formation of a peptide bond between the two amino acids. After these bonding events are over, the initiator tRNA will be released from the ribosome.

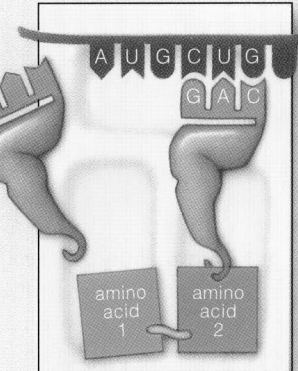

4 Now the first amino acid is attached only to the second one—which is still hooked to the second tRNA. This tRNA will move into the *P* site on the ribosomal platform, sliding the mRNA with it by one codon. When it does so, the third codon will become aligned above the *A* site.

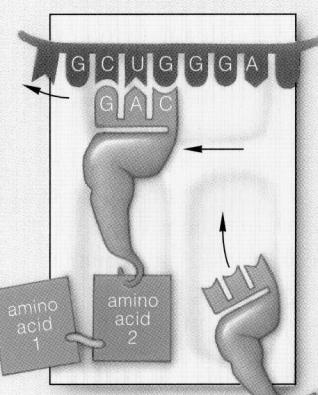

5 A third tRNA is about to move into the *A* site. Its anticodon is capable of base-pairing with the third codon of the mRNA transcript. Next, through enzyme action, a peptide bond will form between amino acids 2 and 3.

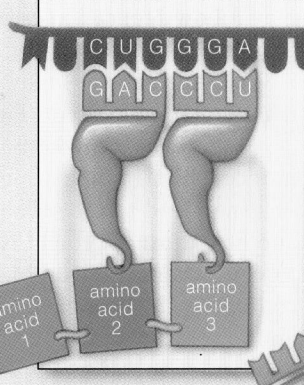

6 Steps 3 through 5 are repeated again and again. The polypeptide chain continues to grow this way until enzymes reach a stop codon in the mRNA transcript. Now *termination*, the last stage of translation, can begin, as shown in (**c**).

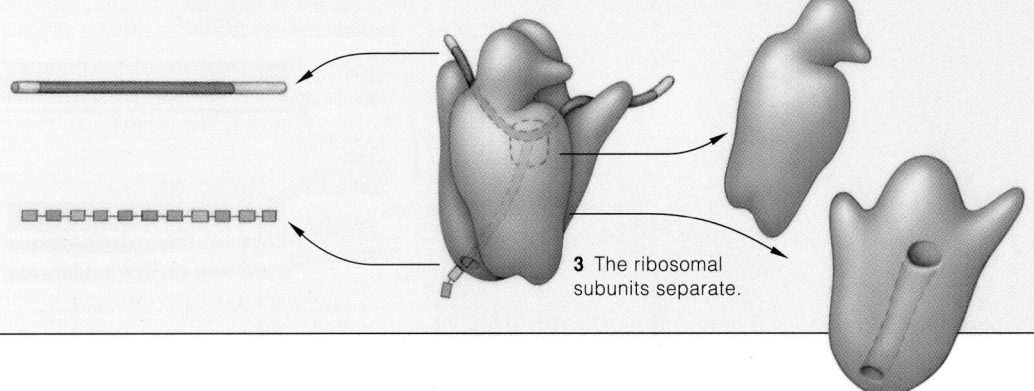

1 Once a stop codon is reached, the mRNA transcript is released from the ribosome.

2 The newly formed polypeptide chain also is released.

c   Chain termination stage

3 The ribosomal subunits separate.

Whenever a cell puts its genetic code into action, it is synthesizing precisely those proteins that it requires for its structure and functioning. If something changes a gene's code words, the resulting protein may also change.

Every so often, genes do change. Maybe one base is substituted for another in the DNA sequence. Or maybe an extra base is inserted or a base is lost. These small-scale changes in the nucleotide sequence of a DNA molecule are **point mutations**. While mutations may occur in any cell, they are inherited only when they occur in the germline, the cells (in gonads) that give rise to gametes.

Some mutations result from exposure to **mutagens**. These are agents that increase the risk of heritable alterations in the structure of DNA. Ultraviolet light, ionizing radiation, and certain substances in tobacco smoke are common mutagens. So are free radicals, the destructive molecules described on page 21.

Even in the absence of mutagens, however, mutations arise in cells. For example, spontaneous mutations may follow replication errors, as when adenine wrongly pairs with a cytosine unit on a DNA template strand. DNA repair enzymes may fail to detect an error, allowing a mutation to occur (Figure 19.15).

Regardless of whether they are spontaneous or induced by a mutagen, base-pair substitutions of this sort may lead to the substitution of one amino acid for another during protein synthesis. Sickle-cell anemia results from a substitution mutation in the DNA strand coding for the beta chain of hemoglobin. In a chain 150 amino acids long, only one amino acid is substituted for another—yet, as we now know, the consequences of the substitution can be severe. (Some other diseases, including some cases of heart disease, are now thought to arise

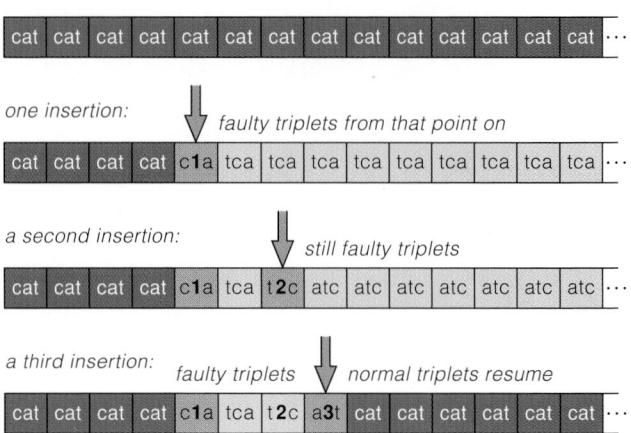

Figure 19.16 Outcomes of three frameshift mutations. In this hypothetical example, three different base insertions occur in the same nucleotide sequence of DNA. The word *cat* represents a code word (base triplet) in mRNA. In the first mutation, an extra nucleotide (arrow) is inserted into a gene. When mRNA is transcribed off the gene, the insertion puts the reading frame out of phase and changes the code words. Thus the wrong amino acids are called up during translation. The second base insertion does not improve matters. The third one restores most of the code words, so only part of the resulting protein will be defective.

from mutations in DNA that resides not in cell nuclei, but in mitochondria; see *Focus on Your Health*.)

Or consider the frameshift mutation, in which one to several base pairs are inserted into a DNA molecule or deleted from it. Remember, polymerases read a nucleotide sequence in blocks of three. As Figure 19.16 shows, an insertion or deletion in a gene region can shift this reading frame. Because the gene is read incorrectly, an abnormal protein is synthesized.

Spontaneous mutations also result when *transposable elements* are on the move. These are segments of DNA that can move from one location to another in the same DNA molecule or in a different one. Often, they inactivate genes into which they are inserted. The DNA of a human diploid cell typically contains hundreds of thousands of copies of a transposable element called *Alu*. A particular insertion of an *Alu* element is known to cause the genetic disorder *neurofibromatosis*.

Whether mutations ultimately prove to be harmful, neutral, or beneficial depends on how the resulting proteins interact with other genes and with the environment. We will revisit these ideas in Chapter 22.

Mutations can be inherited when they occur in the germline. Point mutations are small-scale alterations in the nucleotide sequence of DNA. They may be harmful, beneficial, or neutral, depending on how they alter proteins.

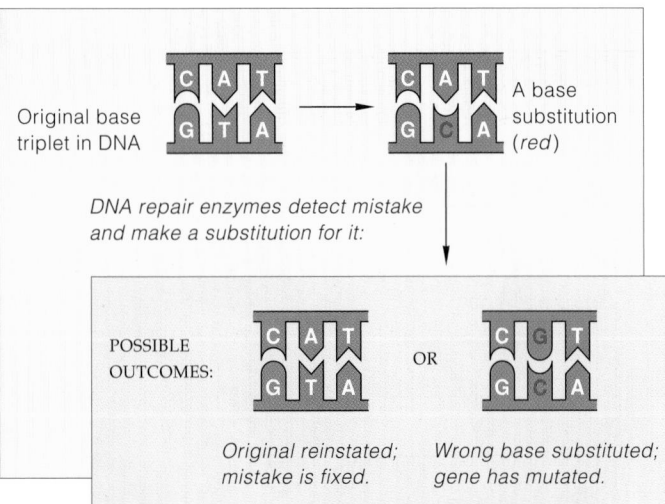

Figure 19.15 Example of a base-pair substitution.

19.8 REGULATING GENE ACTION

Regulatory Proteins

Most of the different cells of your body carry the same genes. Many of those genes carry instructions for synthesizing proteins that are essential to any cell's structure and functioning. Yet each type of cell uses a small subset of genes in specialized ways. For example, every cell carries the genes for hemoglobin, but only the precursors of red blood cells activate those genes. Only the white blood cells known as lymphocytes activate genes for antibodies. *Each cell determines which genes are active and which gene products appear, when, and in what amounts.* Some genes might be switched on and off throughout a person's life. Other genes might be turned on only in certain cells and only at certain times.

Gene regulation is exerted by way of proteins, hormones, and other molecules that interact with DNA, RNA, or gene products. For example, **regulatory proteins** enhance or suppress the rate of transcription.

A gene may be associated with one or more DNA sequences that do *not* code for a protein but instead interact with regulatory proteins. When a regulatory protein binds with one of these noncoding sequences, it may trigger transcription of an adjacent gene into mRNA (or shut it down altogether). For instance, like many other organisms, humans have "heat shock" genes that are activated when a fever raises the body temperature above 37°C (98.6°F). Rising cell temperature first causes a shape change in an inactive regulatory protein called heat shock factor (HSF). The change activates the protein, which binds to a nucleotide sequence "upstream" from a group of heat shock genes. The genes are then transcribed to mRNA, and the mRNA is translated into proteins that help repair damage to other cell proteins.

Hormones as Regulatory Agents

Hormones have widespread effects on gene activity in many cell types. In humans, for instance, the pituitary gland secretes growth hormone (Section 12.3). Most cells have receptors for growth hormone, which helps control synthesis of proteins required for cell division and, ultimately, body growth. On the other hand, some hormones affect only certain cells at certain times. The hormone prolactin, which activates genes in mammary gland cells that have exclusive responsibility for milk production, is a good example. Liver cells and heart cells have the same genes, but they have no means of responding to signals from prolactin.

Although most cells have the same genes, some of those genes are activated or suppressed in different ways to produce differences in cell structure or function.

19.9

A TALE OF NUCLEOTIDE REPEATS

Most genetic disorders have their origin in mutated genes. In 1991, researchers pinpointed the genetic source of a disorder called *fragile X syndrome*—and thereby hangs a curious genetic tale.

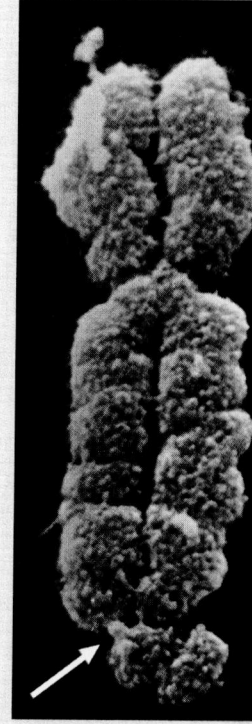

Fragile X syndrome is the second leading cause of mental retardation. It correlates with a peculiar type of gene mutation called a *trinucleotide repeat expansion*. This daunting label refers to the fact that, in such a mutation, multiple copies of a particular nucleotide sequence occur on a given chromosome. In fragile X syndrome, the repeats occur within a gene called *FMR1* (which stands for "fragile mental retardation 1") on the X chromosome. The three repeated nucleotides are CGG (representing an initial cytosine and two following guanines). Associated with the disorder is an abnormal constriction in a segment of the long arm of the X chromosome (arrow in photograph). The protein encoded by the FMR1 gene affects the expression of several other genes that are essential for normal brain development. Here, the plot thickens. Most people have CGG repeats, and apparently there is no harmful effect from having as many as 29 to 50 of them. Someone who has from about 50 to 230 of the repeats is said to have a *premutation*. They rarely have any symptoms, but if the person is female her descendants have an increased risk of developing the disorder. People who exhibit fragile X syndrome may have as many as 700 to more than 1000 CGG repeats. For unknown reasons, chemical changes associated with the repeat region inhibit the transcription of the *FMR1* gene. When that happens, brain cells contain less of the protein encoded by *FMR1*, and the brain's development suffers.

Trinucleotide repeat expansion is at the root of at least nine other genetic disorders. One of them is Huntington disease (Section 18.6). Usually, this degenerative condition results from expansion of a region of CAG repeats (cytosine, adenine, and guanine) in a gene region that encodes a protein that has been named, appropriately, *huntingtin*. As with fragile X syndrome, people who have relatively few of the repeats (10 to 35) don't develop symptoms. But affected persons typically have 36 to 121 CAG repeats—enough to seriously impair the protein's functioning.

Adapted from Fairbanks, D. J., and W. R. Andersen, *Genetics: The Continuity of Life*. 1999: Wadsworth Publishing Company.

SUMMARY

1. A gene is a sequence of nucleotide bases in DNA. Most genes contain protein-building instructions; their nucleotide sequence corresponds to the sequence of amino acids in a polypeptide chain. Proteins are composed of one or more polypeptide chains.

2. Some genes contain instructions for building RNA molecules, which are necessary for protein synthesis. There are three classes of RNA molecules:

a. Messenger RNA (mRNA) carries protein-building instructions from DNA to the cytoplasm. It is the only class of RNA molecules to do this.

b. Transfer RNA (tRNA) is the translator that converts the sequential message of mRNA into a corresponding sequence of amino acids.

c. Ribosomal RNA (rRNA) is a component of ribosomes, the actual sites where amino acids are assembled into polypeptide chains.

3. As summarized in Figure 19.17, the path leading from genes to proteins has two steps. In the first step, called transcription, a DNA template is used to synthesize RNA. In the second step, called translation, three classes of RNA interact in ways that lead to the synthesis of polypeptide chains, which later will become folded into the three-dimensional shape of proteins:

4. Here are the key points concerning transcription:

a. A double-stranded DNA molecule is unwound at a particular gene region; then an RNA molecule is assembled on the exposed bases of one of the strands, which serves as a template.

b. Transcription of DNA into RNA follows the same base-pairing rule that applies to DNA replication. However, uracil takes the place of thymine in an RNA strand. It pairs with adenine:

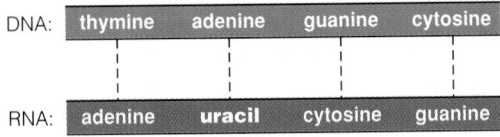

c. New RNA transcripts undergo modification into final form before being shipped from the nucleus. For mRNAs, the noncoding portions (introns) are excised and the coding portions (exons) are spliced together. Only mature mRNA transcripts are translated in the cytoplasm.

5. Here are the key points concerning translation:

a. mRNA interacts with tRNAs and ribosomes in such a way that amino acids become linked one after another, in the sequence required to produce a specific kind of polypeptide chain.

b. Translation is based on the genetic code, a set of 64 base triplets (nucleotide bases that are read in blocks of three).

c. In mRNA, a base triplet is called a codon. An anticodon is a complementary triplet in tRNA. Some combination of codons specifies the amino acid sequence of a polypeptide chain, start to finish.

6. Translation has three stages:

a. Initiation. A small ribosomal subunit binds with an initiator tRNA, then with an mRNA transcript. The small subunit then binds with a large ribosomal subunit to form the initiation complex once the AUG start codon is encountered.

b. Chain elongation. tRNAs deliver amino acids to the ribosome. Each has an anticodon, a base triplet that is complementary to a codon in mRNA and that can base-pair with it. Each amino acid brought to the ribosome by tRNAs is linked (by a peptide bond) to the growing polypeptide chain.

c. Chain termination. A stop codon triggers events that cause the polypeptide chain and mRNA to detach from the ribosome.

7. Point mutations are heritable, small-scale alterations in the nucleotide sequence of DNA. They may be harmful, neutral, or beneficial, depending on how the proteins they specify interact with other genes and with the environment.

8. Mutations can be induced by exposure to mutagens, which are agents that cause heritable alterations in the structure of DNA. They also may arise spontaneously—for example, replication errors and movements of transposable elements.

9. Gene expression is controlled by many interacting elements, including control sites built into DNA molecules, regulatory proteins, and hormones. Their interactions govern which gene products appear, at what times, and in what amounts.

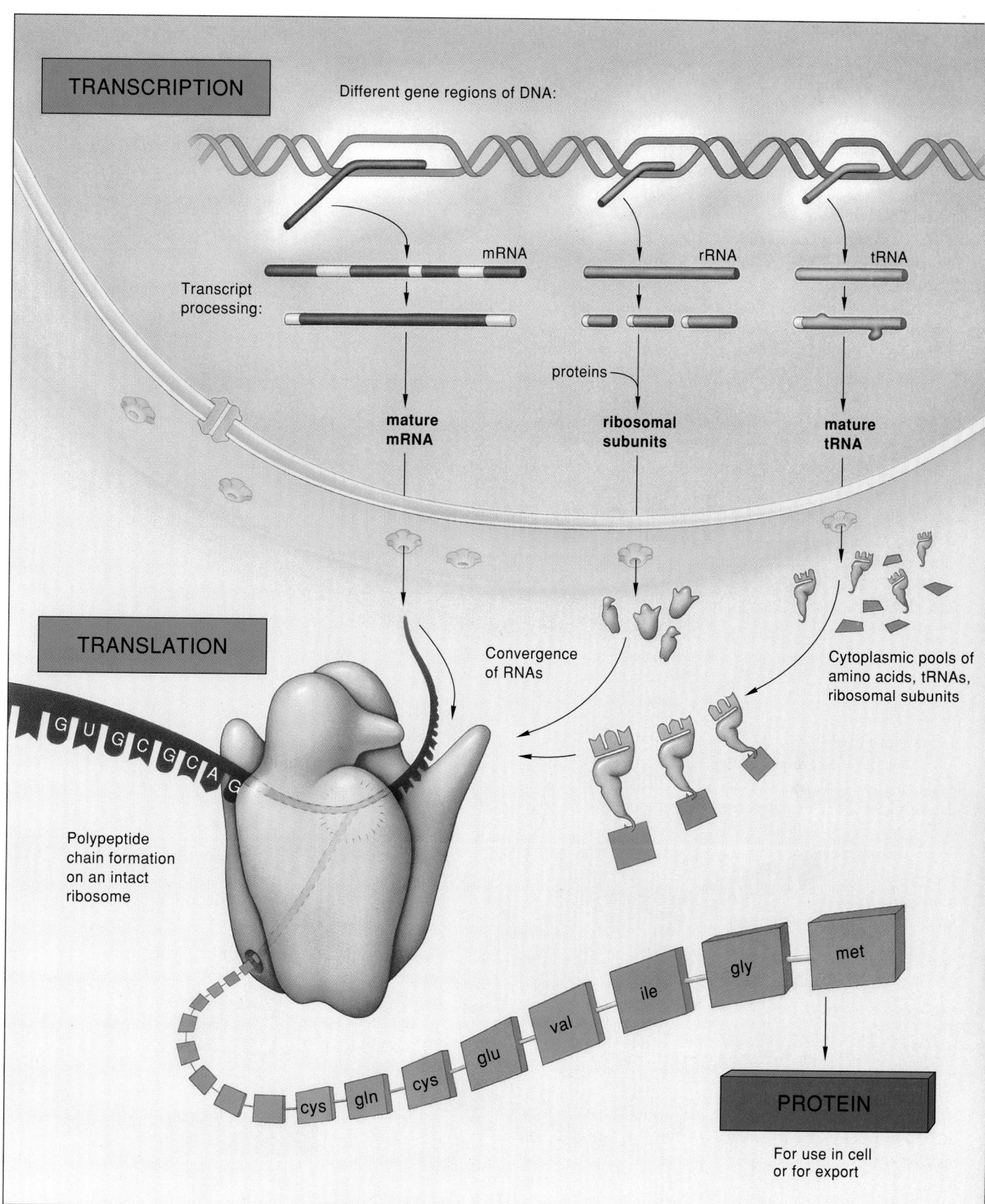

Figure 19.17 Summary of transcription and translation, the two steps leading to protein synthesis.

1. Are the proteins specified by DNA assembled *on* the DNA molecule? If so, state how. If not, tell where they are assembled and on which molecules. *410*

2. Protein synthesis requires the interaction of three classes of RNA molecules. Name the three classes and define the functions of each. *410*

3. In what key respect does an RNA molecule differ from a DNA molecule? *410*

4. Describe the steps of gene transcription. In what respects does this process resemble DNA replication? In what respects does it differ? *410–411*

5. Before an mRNA transcript leaves the nucleus, are its introns or exons snipped out? *411*

6. What is the genetic code? Is the code the basis of protein synthesis in all organisms? *412*

7. Distinguish between a codon and an anticodon. *412–413*

8. If 61 codons in mRNA actually specify amino acids, and if there are only 20 common amino acids, then more than one codon combination must specify some of the amino acids. How do triplets that code for the same thing usually differ? *412*

9. Describe the events that unfold during the three steps of translation. *414–415*

10. Describe one kind of mutation that leads to an altered protein. What determines whether the altered protein will have beneficial, neutral, or harmful effects on the individual? *416*

11. Fill in the blanks on the diagram below. *419*

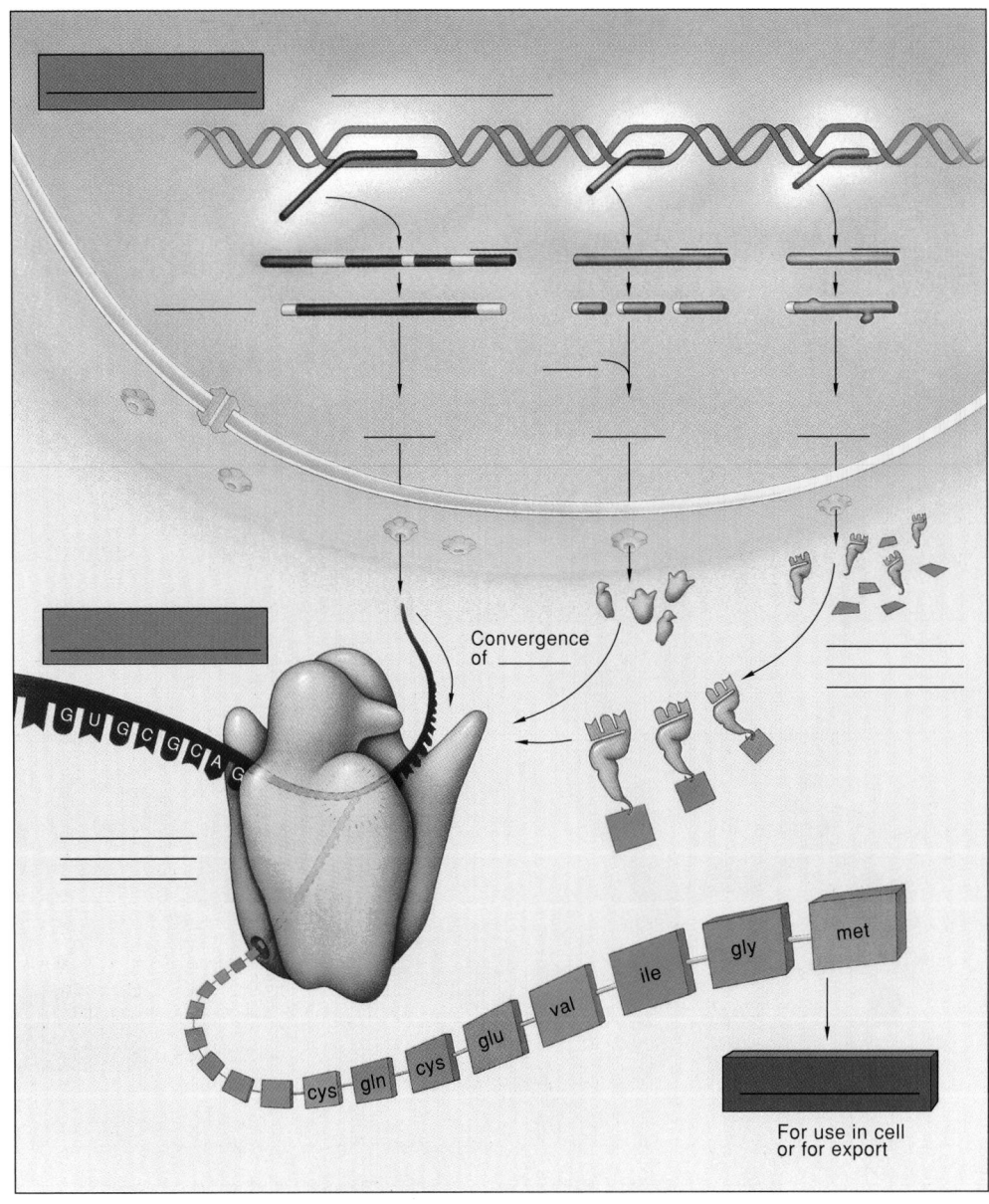

Convergence of _____

For use in cell
or for export

Critical Thinking: You Decide *(Key in Appendix V)*

1. Which mutation would be more harmful: a mutation in DNA or one in mRNA? Explain your answer.

2. A pathogenic strain of *E. coli* has acquired an ability to produce a dangerous toxin that has caused illness and fatalities. This pathogen is especially dangerous for young children who have ingested undercooked, contaminated beef. Based on your reading in this chapter, develop a hypothesis to explain how a harmless strain of *E. coli* can become a pathogen.

3. A DNA polymerase made an error during the replication of an important gene region of DNA. None of the DNA repair enzymes detected or repaired the damage. A portion of the DNA strand with the error is shown here:

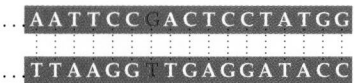

... AATTCC**G**ACTCCTATGG
... TTAAGG**T**TGAGGATACC

After the DNA molecule is replicated and two daughter cells have formed, one cell carries a mutation and the other cell is normal. Develop a hypothesis to explain this observation.

Self-Quiz *(Answers in Appendix IV)*

1. Nucleotide bases, read _____ at a time, serve as the "code words" of genes.

2. DNA contains different genes that are transcribed into _____.
   a. proteins
   b. mRNAs
   c. rRNAs
   d. tRNAs
   e. b, c, and d

3. RNA molecules are _____.
   a. a double helix
   b. usually single-stranded
   c. always double-stranded
   d. usually double-stranded

4. mRNA is produced by _____.
   a. replication
   b. duplication
   c. transcription
   d. translation

5. _____ carries coded instructions for an amino acid sequence to the ribosome.
   a. DNA
   b. rRNA
   c. mRNA
   d. tRNA

6. tRNA _____.
   a. delivers amino acids to ribosomes
   b. picks up genetic messages from rRNA
   c. synthesizes mRNA
   d. all of the above

7. Each codon (except for stop codons) calls for a specific _____.
   a. protein
   b. polypeptide
   c. amino acid
   d. carbohydrate

8. Assuming that the reading frame begins after the start signal AUG, how many amino acids are coded for in this mRNA sequence: AUGUUACACCGUCAC?
   a. three
   b. five
   c. six
   d. seven
   e. more than seven

9. An anticodon pairs with the nitrogen-containing bases of _____.
   a. mRNA codon
   b. DNA codons
   c. tRNA anticodon
   d. amino acids

10. The loading of mRNA onto the small ribosomal subunit occurs during _____.
   a. initiation of transcription
   b. transcript processing
   c. translation
   d. chain elongation

11. Use the genetic code (Figure 19.10) to translate the mRNA sequence AUGCGCACCUCAGGAUGAGAU. (All human reading frames start with AUG.) Which amino acid sequence is being specified?
   a. meth-arg-thr-ser-gly-stop-asp . . .
   b. meth-arg-thr-ser-gly . . .
   c. meth-arg-tyr-ser-gly-stop-asp . . .
   d. none of the above

12. Match the terms related to protein building.

   _____ alters genetic instructions
   _____ codon
   _____ transcription
   _____ translation
   _____ stages of transcription and translation

   a. initiation, elongation, termination
   b. conversion of genetic messages in RNA into polypeptide chains
   c. base triplet for an amino acid
   d. RNA synthesis
   e. mutation

Selected Key Terms

anticodon *412*
codon *412*
exon *411*
gel electrophoresis *408*
genetic code *412*
histone *409*
intron *411*
messenger RNA (mRNA) *410*
mutagen *416*
nucleotide *406*
point mutation *416*
polysome *415*

promoter *410*
regulatory protein *417*
replication *406*
ribosomal RNA (rRNA) *410*
ribosome *413*
RNA *410*
RNA polymerase *410*
transcription *410*
transfer RNA (tRNA) *410*
translation *410*
uracil *410*

Readings

Beardsley, T. August 1991. "Smart Genes." *Scientific American.* A fascinating discussion of how organisms from humans to fruit flies control their genes by regulating transcription.

Grunstein, M. October 1992. "Histones as Regulators of Genes." *American Scientist.*

Murray, A., and M. Kirschner. March 1991. "What Controls the Cell Cycle?" *Scientific American.*

Watson, J. 1978. *The Double Helix.* New York: Atheneum. Highly personal view of scientists and their methods, interwoven into an account of how DNA structure was discovered.

Wolfe, S. 1995. *Introduction to Molecular and Cellular Biology.* Belmont, Calif.: Wadsworth. This text is comprehensive, current, and accessible.

For additional readings, go to InfoTrac College Edition, your online library at:

http://www.infotrac-college.com/wadsworth

20 CANCER: A CASE STUDY OF GENES AND DISEASE

The Body Betrayed

The tiny mass of tissue, only a few million cells strong, may go unnoticed for 10, 20, even 30 years. In fact, not until the mass becomes a clump of perhaps 10 billion cells will it be felt, worried about, and finally—perhaps reluctantly—discussed with a doctor.

Cancer is a grim fact of human life. It strikes one in three people in the United States (Figure 20.1) and kills one in five—about 1,400 cancer deaths every day, just over half a million in a year. On average, cancer strikes more males than females, but the pattern varies

Figure 20.1 One out of three people in the United States will be diagnosed with cancer during their lifetime. These young women are taking steps to limit their risk of lung cancer, the chief cancer killer of both men and women.

depending on the type of cancer involved. Thus the number of new lung cancer cases is increasing faster among women due to a corresponding increase in the number of women who began smoking cigarettes in the last several decades. Adult leukemias and bladder cancers occur more often in men, but males and females are about equally susceptible to colon cancer. Cancer strikes more often in older age groups, although it is the leading cause of death among children ages 3–14. Fortunately, we are rapidly increasing our understanding of and ability to treat many kinds of cancer, including those of the breast, ovary, colon, and skin.

Cancer typically begins with cells in which the controls over cell division are lost. In this sense cancer is a betrayal, a bit of the body that turns *against* the body. More than 100 specific cancer types are known. They have names like retinoblastoma and fibrosarcoma that identify the type of tissue in which the first cancer cells develop. Some cancers arise from genetic changes caused by viral infection; some come from an inherited mutation; others develop due to mutations triggered by environmental factors, including certain chemicals and radiation. Establishing causes is not a simple matter, however. Multiple triggering events almost certainly are the root of many cancers.

Statistics hint at the impact of cancer on human beings, but they don't tell us much about the biological causes and effects of cancers, and they don't reveal recent advances in our understanding of the genetic events that underlie this group of diseases. These are the topics we consider in this chapter, along with cancer treatments and lifestyle choices that can affect personal cancer risk.

KEY CONCEPTS

1. Cancer arises when genetic controls over cell division are lost.

2. Cancer develops through a multistep process. It requires changes in a series of genes, including tumor suppressor genes, that normally operate in cells.

3. Cancers may be triggered by "environmental" factors. These include mutagens such as ultraviolet radiation and chemical carcinogens that switch on oncogenes (cancer-causing genes), turn off tumor suppressor genes, or do both.

4. Lifestyle choices can limit a person's risk of developing many types of cancer. The most important choices appear to involve diet, exposure to sunlight, and tobacco use.

CHAPTER AT A GLANCE

20.1 Cancer Defined
20.2 The Genetic Triggers for Cancer
20.3 *Cancer and Agricultural Chemicals*
20.4 Diagnosing Cancer
20.5 Cancer Treatment and Prevention
20.6 Some Major Types of Cancer
20.7 Cancers of the Breast and Reproductive System
20.8 A Survey of Other Common Cancers

Characteristics of Tumors

As genes switch on and off, they determine what a cell's specialized function will be, when and how rapidly the cell will grow and divide, even when it will *stop* dividing and when it will die. If cells overgrow, the result is a defined mass of tissue called a **tumor**. This mass is a *neoplasm*, which literally means new growth.

A tumor may not be "cancer." As Figure 20.2a shows, the cells of a *benign* tumor are often enclosed by a capsule of connective tissue, and within the capsule they are organized in an orderly array. They also tend to grow slowly and to be well differentiated (structurally specialized), much like normal cells of the same tissue. Benign tumors typically stay put in the body, push aside but don't invade surrounding tissue, and usually can be easily removed by surgery. Benign tumors *can* threaten health, as when they occur in the brain. Nearly everyone has at least several of the benign tumors we call moles. Most of us also have or have had some other type of benign neoplasm, which is usually destroyed by the immune system. Table 20.1 compares the main features of malignant and benign tumors.

Dysplasia ("bad form") is an abnormal change in the sizes, shapes, and organization of cells in a tissue. It is often a precursor to cancer. Under the microscope, the borders of a cancerous tumor often appear ragged (Figure 20.2b), and its cells form a disorganized clump. Cancer cells usually also have characteristics that enable them to behave very differently from normal body cells.

Characteristics of Cancer Cells

STRUCTURAL ABNORMALITIES   A cancer cell generally has an abnormally large nucleus and less cytoplasm than usual. It also is *poorly differentiated*. That is, cancer cells often do not have the clear structural specializations of cells in mature body tissues. The extent of differentiation of cancer cells can be medically important. In general, the less differentiated cancer cells are, the more readily they break away from the primary tumor and spread the disease.

When a normal cell becomes transformed into a cancerous cell, additional changes occur. The cytoskeleton shrinks, becomes disorganized, or does both. Proteins of the plasma membrane are lost or altered, and new, different ones appear. These changes are passed on to the cell's descendants: When a transformed cell divides, its daughter cells are cancerous cells too.

Table 20.1	Comparison of Benign and Malignant Tumors	
	Malignant Tumor	Benign Tumor
Rate of growth	Rapid	Slow
Nature of growth	Invades surrounding tissue	Expands within tissue
Spread	Metastasis by way of bloodstream and lymphatic system	Stays localized
Cell differentiation	Usually poor	Nearly normal

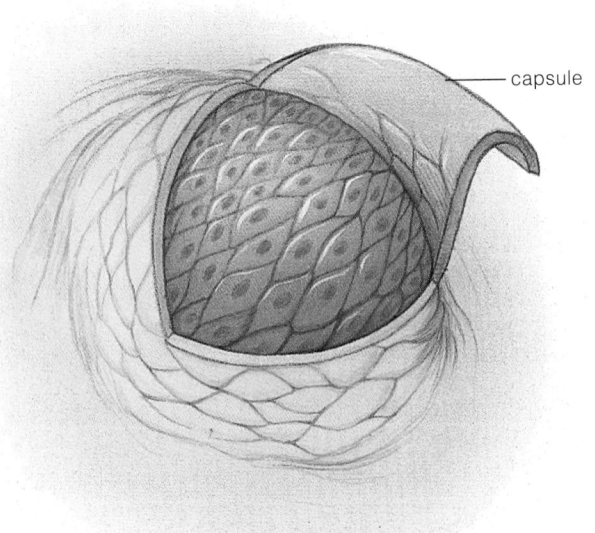

capsule

a   b

Figure 20.2   (**a**) Sketch of a benign tumor. Cells appear nearly normal, and the tumor mass is encapsulated within connective tissue. (**b**) A cancerous neoplasm. Due to the abnormal growth of cancer cells, the tumor is a disorganized heap of cells, some of which break off and invade surrounding tissues (metastasis).

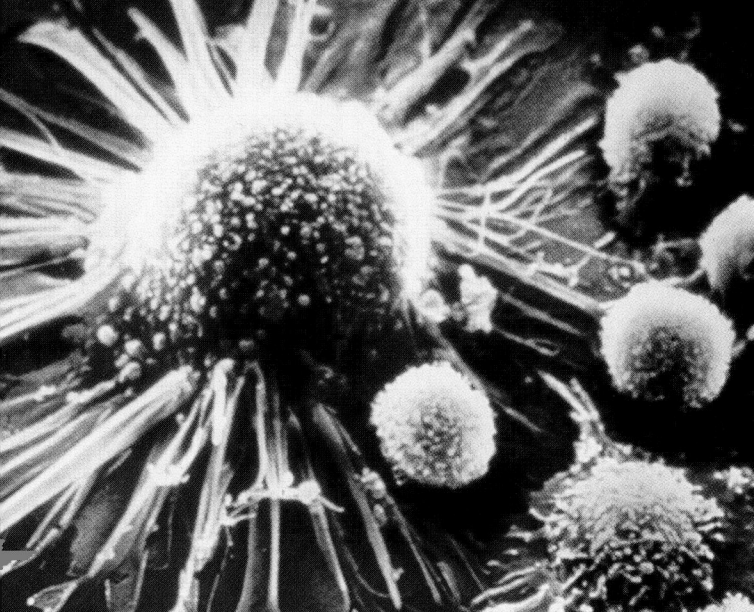

Figure 20.3 Scanning electron micrograph of a cancer cell surrounded by some of the body's white blood cells that may or may not be able to destroy it.

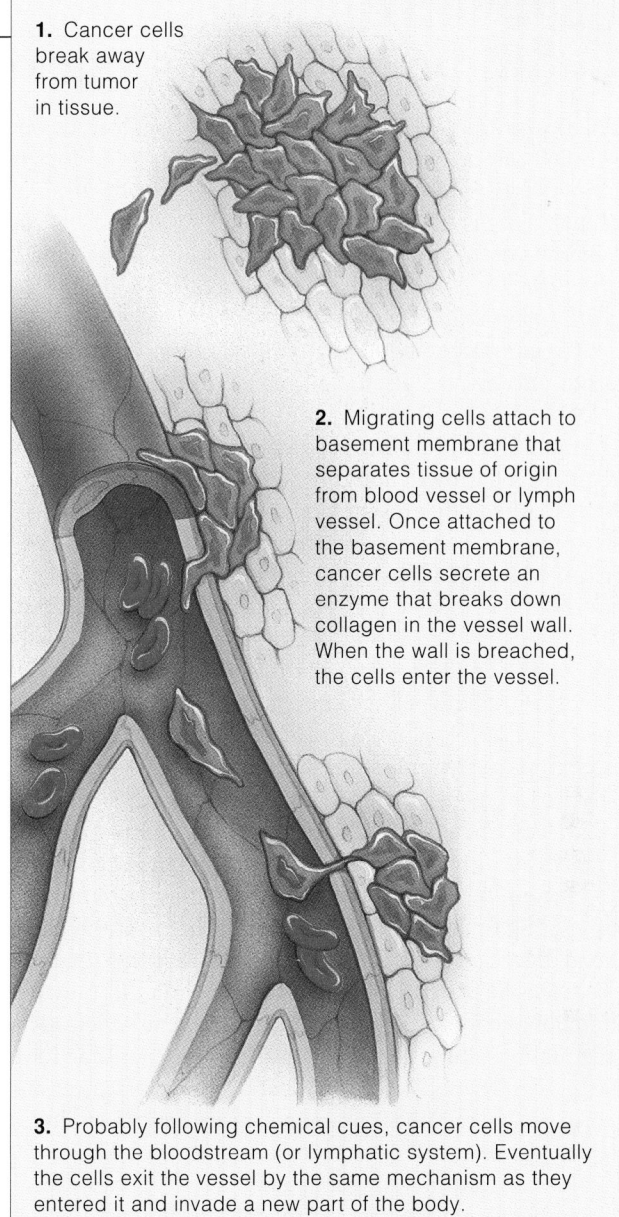

1. Cancer cells break away from tumor in tissue.

2. Migrating cells attach to basement membrane that separates tissue of origin from blood vessel or lymph vessel. Once attached to the basement membrane, cancer cells secrete an enzyme that breaks down collagen in the vessel wall. When the wall is breached, the cells enter the vessel.

3. Probably following chemical cues, cancer cells move through the bloodstream (or lymphatic system). Eventually the cells exit the vessel by the same mechanism as they entered it and invade a new part of the body.

Figure 20.4 Stages in metastasis.

UNCONTROLLED GROWTH Cancer cells lack normal controls over cell division. Contrary to popular belief, cancer cells do not necessarily divide more rapidly than normal cells do, but they do increase in number faster. Normally, the death of cells closely balances the production of new ones through mitosis, so that the cells are arranged in an orderly tissue. In a developing cancerous tumor, however, at any given moment more cells are dividing than are dying. As this unregulated cell division continues, the cancer cells do not respond to crowding—that is, normal *contact inhibition* does not occur. Whereas a normal cell stops dividing once it comes into contact with another cell, a cancerous cell keeps on dividing. Hence, cancer cells accumulate in a disorganized heap—which is why cancerous tumors are often lumpy.

Cancer cells also lack strong cell-to-cell adhesion with neighboring cells, and as Figure 20.3 shows, they may form extensions (pseudopodia) that enable them to move about. This property gives cancer cells the dangerous ability to break away from the parent tumor and invade other tissues (**metastasis**), including the lymphatic system and circulatory system (Figure 20.4). The ability to metastasize is what makes cancers *malignant.*

Many cancer cells produce the hormone HCG (human chorionic gonadotropin). Recall from Chapter 14 that HCG maintains the uterine lining at the onset of pregnancy. It also helps prevent the mother's immune system from attacking the fetus as "foreign." One hypothesis is that HCG secreted by cancerous cells serves in some way to shield the abnormal cells from the immune system.

Some cancer cells produce interleukin-2, a cytokine that stimulates cell division (Chapter 7), and also display receptors for it. Cancer cells also secrete a growth factor, *angiogenin,* that encourages new blood vessels to grow around the tumor. Tumor growth requires a large supply of nutrients and oxygen supplied via the bloodstream. Some of the most exciting current cancer research is focused on developing drug therapies that "starve" tumors to death by blocking angiogenin's effects.

Cancer cells lack normal controls over cell division and organization into tissues. They are also structurally abnormal.

Cancer cells are often poorly differentiated. They can leave a primary tumor, invade surrounding tissue, and metastasize to other parts of the body.

The transformation of a normal cell into a cancerous one is called **carcinogenesis** (Figure 20.5). Cancer develops in many vertebrates, some invertebrates, and even some plants. Fossils show that dinosaurs developed cancer. Wherever it turns up, *cancer almost always develops through a multistep process in which genetic changes alter normal controls over cell division.*

Certain DNA sequences on chromosomes are called **proto-oncogenes**. "Proto" means before. An **oncogene** is a DNA segment that can induce cancer. The distinction is important. Proto-oncogenes are normal genes that regulate cell growth and development. They encode proteins that include growth factors (signals sent by one cell to trigger growth in other cells), regulatory proteins involved in cell adhesion, and the protein signals for cell division. If some event changes their structure or expression, they may become altered into oncogenes that do not respond to controls over cell division.

An oncogene acting alone is not wholly responsible for the onset of malignant cancer. That usually requires mutations in several genes, including at least one **tumor suppressor gene**. Such genes encode proteins that are produced when cells are stressed. They typically halt cell growth and division, preventing cancers from occurring.

Several lines of evidence have shed light on the roles of specific tumor suppressor genes. Studies of the childhood eye cancer *retinoblastoma* have revealed that the disease is likely to develop when a child inherits only one functional copy of a tumor suppressor gene called *Rb*. If a mutation alters a functional allele to a nonfunctional form, then retinoblastoma is likely. Colorectal cancer and perhaps several other common cancers have been linked to damage in a tumor suppressor gene called FHIT.

A tumor suppressor gene called p53 is especially interesting because it appears to prevent cancerous changes in many types of tissue. This gene codes for a regulatory protein that seems to stop cell division when cells are stressed or damaged, as often happens in cancer cells. When p53 cannot be produced due to mutations in the p53 gene, the controls don't operate and cell division may continue unchecked. When p53 mutates, the resulting faulty protein may actually *promote* the development of cancer, possibly by activating an oncogene. Researchers have identified mutated p53 genes in cells from cancers of the breast, bone, colon, skin, lung, bladder, cervix, and brain. The fact that an absent or defective p53 gene correlates with so many cancers has become the basis for diagnostic tests that screen for mutations in the gene.

Oncogene Activation

An oncogene may be present in a cell and never cause trouble because some condition in the cell, such as the presence of a suppressor gene, prevents the oncogene from being expressed. Several kinds of events can change this state of affairs. The oncogene or a related suppressor gene may mutate in a way that triggers expression. A break in a chromosome followed by a translocation (page 399) may move an oncogene away from a regulatory nucleotide sequence that would normally prevent it from being expressed. Or new genetic material may be introduced into a cell (as by viral infection) and disrupt controls.

Other Routes to Carcinogenesis

INHERITED SUSCEPTIBILITY TO CANCER Heredity plays a major role in about 5 percent of cancer cases. If a mutation exists in a germ cell (sperm or egg) and it alters a proto-oncogene or tumor suppressor gene on a chromosome, the defect can be passed on to offspring. The first step toward cancer has now occurred, so that an affected person may be more susceptible to cancer if mutations occur in other proto-oncogenes and tumor suppressor genes. Patterns of familial breast, colon, and lung cancer suggest that several genes are involved, including genes that control aspects of cell metabolism and responses to hormones.

VIRUSES Viruses almost certainly cause some cancers. Sometimes a viral infection can alter a proto-oncogene when the viral DNA becomes inserted at a certain position in the host cell DNA. For example, the viral gene could take the place of a regulatory sequence that normally prevents a proto-oncogene from switching on (or off) at the wrong time. Other viruses simply carry oncogenes as part of their genetic material and insert them into the host's DNA. Most viruses linked to human cancer are DNA viruses, although retroviruses are associated with some types of leukemias.

CHEMICAL CARCINOGENS There are thousands of known chemical **carcinogens**, cancer-causing substances that can cause DNA damage and a subsequent mutation in DNA. A carcinogen that directly causes DNA damage is sometimes called an "initiator." The list includes many compounds that are by-products of the industrialization of human societies, such as asbestos, coal tar, vinyl chloride, and benzene. The list also includes hydrocarbons in cigarette smoke (as well as other types of smoke) and on the charred surfaces of barbecued meats and carcinogenic substances in dyes and pesticides (see this chapter's *Focus on Our Environment*, page 428). One of the first carcinogens to be recognized was chimney soot (more accurately, substances soot contains), which frequently caused cancer of the scrotum in chimney sweeps. Plants and fungi can produce carcinogens; aflatoxin, a metabolic by-product of a fungus that attacks stored grain and other seeds, causes liver cancer. For this reason, some authorities advise against eating "raw" peanut butter. (Commercial peanut butters are safe because processing kills the aflatoxin fungus.)

Some chemicals may be "precarcinogens" that cause gene changes only *after* they have been altered by metabolic activity in the cell. Other substances are cancer "promoters"—alone they are not carcinogenic, but they can make a carcinogen more potent if both the carcinogen and the promoter act on a cell at the same time.

RADIATION As you know from previous chapters, radiation can damage DNA, causing mutations that lead to cancer. Common sources include ultraviolet radiation from sunlight and sunlamps, medical and dental X rays, and some radioactive materials used to diagnose diseases. Other sources are "background" radiation from cosmic rays and radon gas in soil and water, and the gamma rays emitted from nuclear reactors and radioactive wastes. Sun exposure is probably the greatest radiation risk factor for most people. (Skin cancer is the most frequently diagnosed cancer in the Northern Hemisphere.) It is also wise to avoid unnecessary X rays.

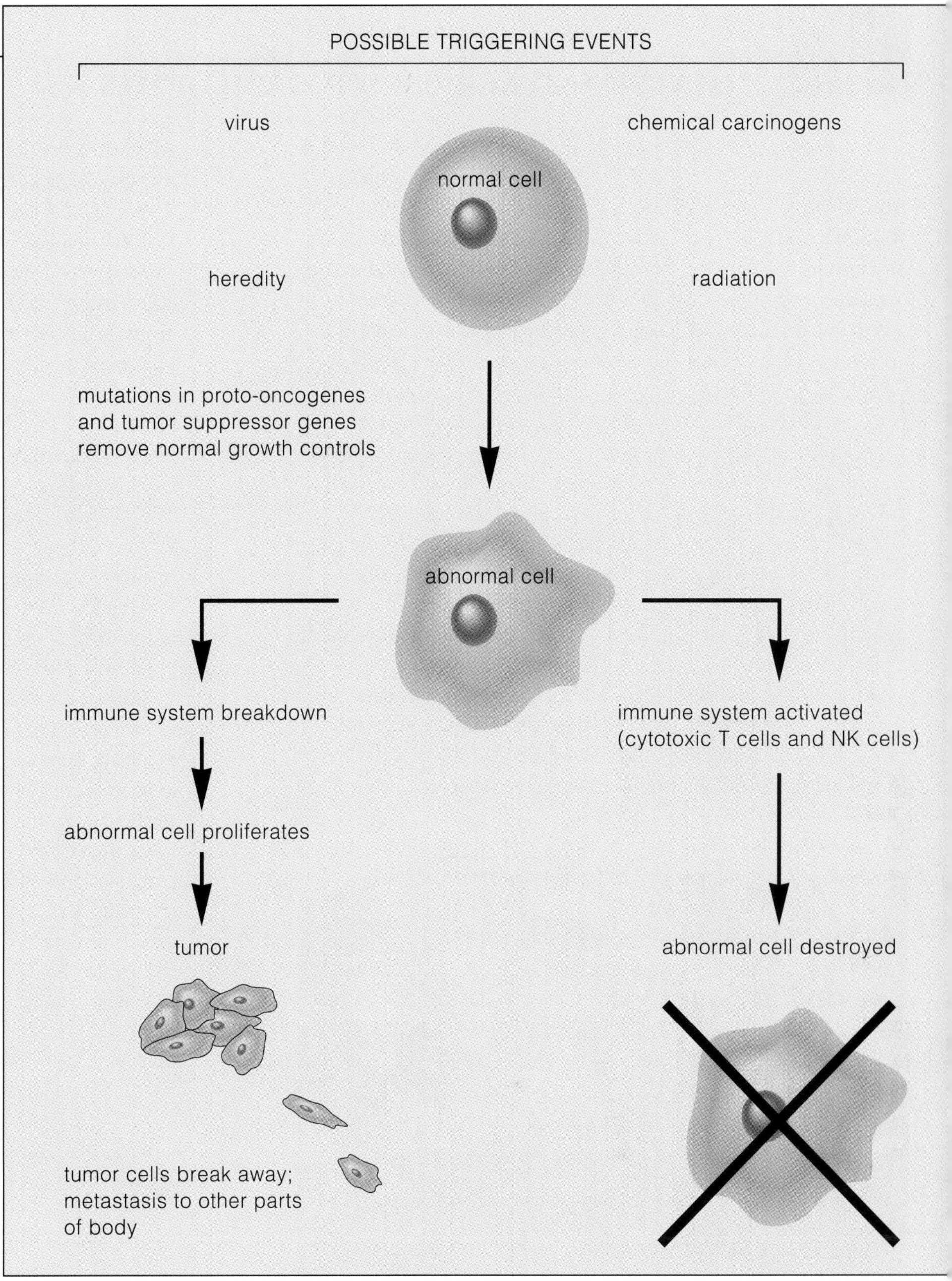

BREAKDOWNS IN IMMUNITY
Typically, when a normal cell becomes cancerous, certain proteins at the cell surface become altered and function like foreign antigens—the "nonself" tags that mark a cell for destruction by cytotoxic T cells and natural killer cells (see Section 7.4). Research has uncovered clear evidence that a healthy immune system regularly detects and destroys some types of cancer cells. This protective function can break down as a person ages, or if the immune system is suppressed by any of a range of factors, including some therapeutic drugs and mental states such as anxiety and severe depression.

The presence of a growing cancer may also suppress the immune system. In addition, cancerous cells that bear tumor antigens may lack other molecules that must bind with receptors on lymphocytes in order to trigger an immune response (Section 7.6). In some cases tumor antigens may be chemically disguised or masked. For whatever reason, the transformed cells are free to divide uncontrollably.

Cancer develops through a multistage process in which gene changes remove normal controls over cell division.

Oncogenes have a major role in inducing cancer in a normal cell. Proto-oncogenes may become oncogenes if some event changes their structure or the way they are expressed.

The development of cancer typically also requires the absence or mutation of at least one tumor suppressor gene.

POSSIBLE TRIGGERING EVENTS

virus   chemical carcinogens

normal cell

heredity   radiation

mutations in proto-oncogenes and tumor suppressor genes remove normal growth controls

abnormal cell

immune system breakdown

immune system activated (cytotoxic T cells and NK cells)

abnormal cell proliferates

abnormal cell destroyed

tumor

tumor cells break away; metastasis to other parts of body

Figure 20.5 Overview of steps in carcinogenesis.

20.3

CANCER AND AGRICULTURAL CHEMICALS

G. Tyler Miller, Jr.

According to the Food and Drug Administration, about 40 percent of the food bought in supermarkets contains detectable residues of one or more of the active ingredients used in pesticides in the United States. Approximately 3 percent of this food has levels of one or more pesticides above the legal limit. The residues are especially likely to be found in tomatoes, grapes, apples, lettuce, oranges, potatoes, beef, and dairy products. Imported produce, such as fruits, vegetables, and coffee beans, can also carry significant pesticide residues—sometimes including pesticides deemed so dangerous that they are banned in the United States. The results of this long-term worldwide experiment, in which the subjects are human consumers, may never be known because it is almost impossible to determine that a certain level of a specific chemical caused a particular cancer or some other harmful effect.

In 1987 the National Academy of Sciences reported that the active ingredients in 90 percent of all fungicides, 60 percent of all herbicides, and 30 percent of all insecticides in use in the United States may cause cancer in humans.

Figure 20.6 A common source of pesticide exposure. Home garden chemicals are just one avenue by which mutagenic or carcinogenic substances can come into contact with the human body.

According to the *worst-case estimate* in this study, exposure to pesticides in food causes 4,000 to 20,000 cases of cancer a year in the United States.

Cancer is only one possible harmful effect of long-term exposure to low levels of pesticides (Figure 20.6), which collectively contain more than 600 different chemicals. Some scientists are becoming increasingly concerned about possible genetic mutations, birth defects, nervous system disorders, and effects on the immune and endocrine systems. In 1993, new questions were raised about the health effects in children of low-level pesticide residues. Levels established as "safe" have traditionally been based on adult body weights and patterns of food consumption, whereas those parameters differ for children. Findings of one study suggest that by the age of one year, an average American child's exposure to some cancer-causing pesticides in and on food will be greater than federal guidelines currently consider safe over an *entire lifetime*.

Although residues of some pesticides can be (and should be) removed from the surfaces of fruits and vegetables by washing before eating, in our modern world avoiding pesticide exposure is not so easy. Many people are exposed to pesticides from community spraying programs to control mosquitoes and other pests; from pesticide-sprayed lawns, gardens, golf courses, and roadsides; and from sprayed croplands, rangelands, and forests near their homes.

Agricultural chemicals are not the only potential threats to human health. Table 20.2 lists some industrial chemicals that have been linked to cancer.

Table 20.2 Selected Industrial Chemicals Linked to Cancer	
Chemical/Substance	Type of Cancer
Benzene	Leukemias
Vinyl chloride	Liver, various connective tissues
Various solvents	Bladder, nasal epithelium
Ether	Lung
Asbestos	Lung, epithelial linings of body cavities
Arsenic	Lung, skin
Radioisotopes	Leukemias
Nickel	Lung, nasal epithelium
Chromium	Lung
Hydrocarbons in soot, tar smoke	Skin, lung

20.4 DIAGNOSING CANCER

Table 20.3 lists seven common cancer warning signs. These can help people spot cancer in its early stages, when treatment is most effective. Routine *cancer screening* becomes important as a person ages; some recommended cancer screening tests are listed in Table 20.4.

Early and accurate diagnosis of cancer is extremely important to maximize the chances that a cancer can be cured. To confirm—or rule out—cancer, various types of tests can refine the diagnosis. Blood tests can detect **tumor markers**, substances produced by specific types of cancer cells or by normal cells in response to the cancer. For example, as we noted earlier, the hormone HCG is a highly specific marker for certain cancers. This is especially true of cancers affecting the reproductive organs in both men and women. Radioactively labeled monoclonal antibodies, which home in on tumor antigens, have become extremely useful for pinpointing the location and sizes of tumors of the colon, brain, bone, and some other tissues. *Medical imaging* of tumors includes methods such as magnetic resonance imaging (MRI; shown in Figure 20.7), X rays, ultrasound, and computerized tomography (CT). The definitive cancer detection tool is **biopsy**. A small piece of suspect tissue is removed from the body through a hollow needle or exploratory surgery. A pathologist then microscopically examines cells of the tissue sample for the characteristic features of cancer cells.

A *DNA probe* is a snippet of radioactively labeled DNA. Such probes can be used to locate gene mutations or alleles associated with some types of inherited cancers.

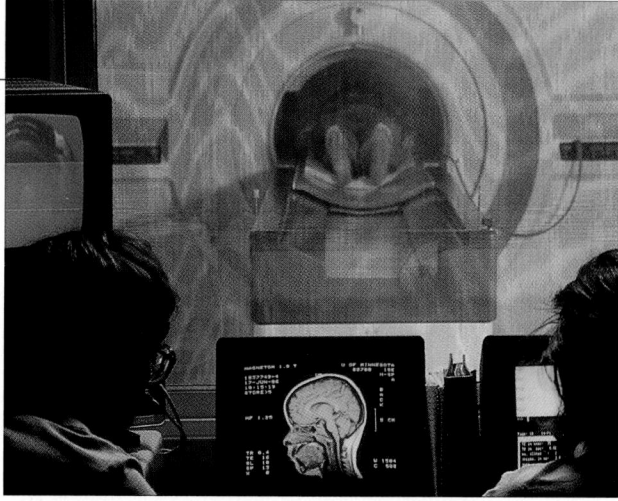

Figure 20.7 MRI scanning. MRI does not use X rays. The patient is placed inside a chamber that is surrounded by a magnet. The machine generates a magnetic field in which nuclei of hydrogen and some other atoms in the body align and absorb energy. A computer analyzes the information and uses it to generate an image of soft tissues.

However, the procedure is expensive, and generally is not covered by health insurance. Some researchers point out, however, that using the technique for widespread screening could allow people with increased genetic susceptibility to make medical and lifestyle choices that could significantly reduce their cancer risk.

Biopsy is a definitive cancer detection tool. Other diagnostic procedures include blood testing for substances produced by cancer cells and medical imaging.

Table 20.3 The Seven Warning Signs of Cancer*
Change in bowel or bladder habits and function
A sore that does not heal
Unusual bleeding or bloody discharge
Thickening or lump
Indigestion or difficulty swallowing
Obvious change in a wart or mole
Nagging cough or hoarseness

* Notice that the first letters of the signs spell the advice CAUTION. *Source:* American Cancer Society.

Table 20.4 Recommended Cancer Screening Tests

Test or Procedure	Cancer	Sex	Age	Frequency
Breast self-examination	Breast	Female	20+	Monthly
Mammogram	Breast	Female	40–49 50+	Every 1–2 years Yearly
Testicle self-examination	Testicle	Male	18+	Monthly
Sigmoidoscopy	Colon	Male, Female	50+	Every 3–5 years
Fecal occult blood test	Colon	Male, Female	50+	Yearly
Digital rectal examination	Prostate, colorectal	Male, Female	40+	Yearly
Pap test	Uterus, cervix	Female	18+ and all sexually active women	Every other year until age 35; yearly thereafter
Pelvic examination	Uterus, ovaries, cervix	Female	18–39 40+	Every 1–3 years w/Pap Yearly
Endometrial tissue sample	Endometrial	Female	At menopause for women at high risk	Once
General checkup		Male, Female	20–39 40+	Every 3 years Yearly

Surgery, drugs, and irradiation of tumors have long been the major weapons against cancer. Surgery may be a cure when a tumor is fully accessible and has not spread.

Chemotherapy uses drugs to kill cancer cells. Most anticancer drugs are designed to kill dividing cells. They disrupt DNA replication during the S phase of the cell cycle or disrupt mitosis by interfering with formation of the mitotic spindle. The drugs fluorouracil and vincristine (derived from the rosy periwinkle) are among the compounds that work this way. Unfortunately, such drugs are also toxic to rapidly dividing healthy cells such as hair cells, stem cells in bone marrow, lymphocytes of the immune system, and epithelial cells of the gut lining. Hence chemotherapy patients can suffer side effects such as nausea and vomiting, hair loss, anemia, and reduced immune responses. Radiation likewise kills cancer cells *and* healthy cells in the irradiated area. *Adjuvant therapy* (*adjuvant* means help) combines surgery and a less toxic dose of chemotherapy. For example, a patient might receive enough chemotherapy to shrink a tumor, then have surgery to remove what remains.

Researchers are also using monoclonal antibodies as "magic bullets" to deliver lethal doses of radiation or anticancer drugs to tumor cells while sparing healthy cells. Experiments with radioactive monoclonal antibodies have resulted in partial or complete remission in a few patients with a type of adult leukemia. Another goal is to link tumor-specific monoclonal antibodies with lethal doses of cytotoxic drugs. So far, results have been mixed, but this is a very active area of cancer research.

Another promising prospect for cancer treatment is **immunotherapy**—giving patients substances that will trigger a strong immune response against cancer cells. Recall from Chapter 7 that many cells can produce and release interferons, which in turn can activate cytotoxic T cells and natural killer cells. These can recognize and kill various types of cancer cells. So far, therapies that involve administering interferon have been useful only against some rare forms of cancer. Another approach is to develop cancer "vaccines" that stimulate cytotoxic T cells to recognize and destroy body cells bearing an abnormal surface protein. In some patients with malignant melanoma, doses of an experimental vaccine have caused tumors to regress, at least temporarily. Still another promising approach is the refinement of anticancer drugs that trigger apoptosis (cell suicide) in cancer cells but not in healthy body cells.

Interleukins—signaling molecules produced by the immune system's lymphocytes—are also potential anticancer weapons. Cytotoxic T cells grown in culture with IL-2 become activated. In clinical trials, several hundred patients with advanced kidney cancer have received doses of such activated cells, combined with large

Figure 20.8 A high-fiber, low-fat diet that includes cruciferous vegetables such as broccoli may help limit personal cancer risk.

amounts of more IL-2. Although the treatment can have serious side effects, enough patients have responded favorably that regulators have approved it for limited use.

None of us can control factors in our heredity or biology that might lead one day to cancer, but each of us can make lifestyle decisions that promote health. The American Cancer Society recommends the following strategies for limiting your cancer risk:

1. Avoid tobacco in any form, including "secondary smoke" from others.

2. Maintain a desirable weight. Being more than 40 percent overweight increases the risk of cancers of the colon, breast, prostate, gallbladder, ovary, and uterus.

3. Eat a low-fat diet that includes plenty of vegetables, fruits, and fiber (Figure 20.8). A low-fiber, high-fat diet is associated with cancer of the colon, prostate, and possibly other tissues. As noted in Chapter 5, antioxidants such as beta carotene and certain vitamins may have roles in cancer prevention.

4. Drink alcohol in moderation. Heavy alcohol use, especially in combination with smoking, increases risk for cancers of the mouth, larynx, esophagus, and liver.

5. If you are a woman entering menopause, consult carefully with your doctor about using estrogen, which *may* increase the risk of breast and endometrial cancer.

6. Learn whether your job or residence exposes you to such industrial agents as nickel, chromate, vinyl chloride, benzene, asbestos, and agricultural pesticides, which are associated with various cancers.

7. Protect your skin from excessive sunlight.

SOME MAJOR TYPES OF CANCER

In general, a cancer is named according to the type of tissue in which it first forms. For instance, cancers of connective tissues such as muscle and bone are *sarcomas*. Various types of *carcinomas* arise from epithelium, including cells of the skin and epithelial linings of internal organs. When a cancer begins in a gland or its ducts, it is called an *adenocarcinoma*. *Lymphomas* are cancers of lymphoid tissues in organs such as lymph nodes, and cancers arising in blood-forming regions—mainly stem cells in bone marrow—are *leukemias*.

The remainder of this chapter surveys some common cancers in the United States. Figure 20.9 summarizes the most recent data for males and females.

A cancer is categorized according to the tissue in which it arises. Examples include sarcomas (connective tissue), carcinomas (epithelium), adenocarcinomas (glandular tissue), lymphomas (lymphoid tissues), and leukemias (blood-forming regions such as bone marrow).

Cancer Incidence by Site and Sex*		Cancer Deaths by Site and Sex	
MALE	**FEMALE**	**MALE**	**FEMALE**
prostate 165,000	breast 182,000	lung 93,000	lung 56,000
lung 100,000	colon and rectum 75,000	prostate 35,000	breast 46,000
colon and rectum 77,000	lung 70,000	colon and rectum 28,800	colon and rectum 28,200
bladder 39,000	uterus 44,500	pancreas 12,000	ovary 13,300
lymphoma 28,500	lymphoma 22,400	lymphoma 11,500	pancreas 13,000
oral 20,300	ovary 22,000	leukemia 10,100	lymphoma 10,500
melanoma of the skin 17,000	melanoma of the skin 15,000	stomach 8,200	uterus 10,100
kidney 16,800	pancreas 14,200	esophagus 7,600	leukemia 8,500
leukemia 16,700	bladder 13,300	liver 6,800	liver 5,800
stomach 14,800	leukemia 12,600	brain 6,600	brain 5,500
pancreas 13,500	kidney 10,400	kidney 6,500	stomach 5,400
larynx 10,000	oral 9,500	bladder 6,500	multiple myeloma 4,600
all sites 600,000	all sites 570,000	all sites 277,000	all sites 249,000

* Excluding basal and squamous cell skin cancer and carcinoma in situ.

Figure 20.9 Summary of annual incidence of and deaths from common cancers, by site and sex. Data are estimates for the United States, 1993. Courtesy American Cancer Society.

Breast Cancer

In the United States, about one woman in nine develops breast cancer. Of all cancers in women, breast cancer currently ranks second only to lung cancer as a cause of death. Breast cancer also occurs in males, but much more rarely. In women, obesity, late childbearing, early puberty, late menopause, and excessive levels of estrogen (and perhaps certain other hormones) seem to play roles in breast cancer development. A family history of breast cancer increases the risk; it may indicate the presence of one of several mutated genes thought to be responsible for many cases of familial breast cancer. Although 80 percent of breast lumps are *not* cancer, a woman should seek medical advice about any breast lump, thickening, dimpling, breast pain, or discharge.

Chances for cure are excellent if breast cancer is detected early and treated promptly. Hence a woman should examine her breasts every month, about a week after her menstrual period. Figure 20.10 shows the steps of a self-exam, as recommended by the American Cancer Society. Low-dose mammography (breast X ray) is the most effective method for detecting small breast cancers. It is 80 percent reliable. The American Cancer Society recommends an annual mammogram for women over 50 and for younger women at high risk.

Treatment depends mainly on the extent of the disease. In *modified radical mastectomy*, the affected breast tissue, overlying skin, and nearby lymph nodes are removed, but muscles of the chest wall are left intact. If a breast tumor is small, the preferred treatment may be *lumpectomy*,

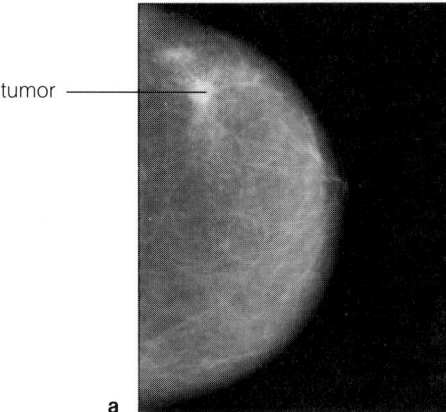

tumor

Figure 20.10 (a, right) A mammogram showing a tumor, which a biopsy indicated to be cancer. The white patches at the front of the breast are milk ducts and fibrous tissue. **(b, below)** How to perform a breast self-examination.

a

1. Lie down and put a folded towel under your left shoulder, then put your left hand behind your head. With the right hand (fingers flat), begin the examination of your left breast by following the outer circle of arrows shown. Gently press the fingers in small, circular motions to check for any lump, hard knot, or thickening. Next, follow the inner circle of arrows. Continue doing this for at least three more circles, one of which should include the nipple. Then repeat the procedure for the right breast. For a complete examination, repeat the procedure while standing in a shower. Hands glide more easily over wet skin.

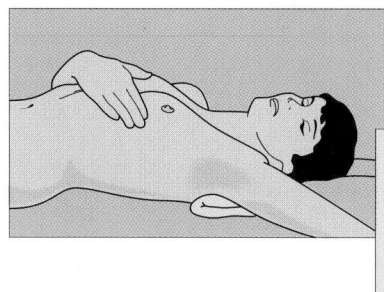

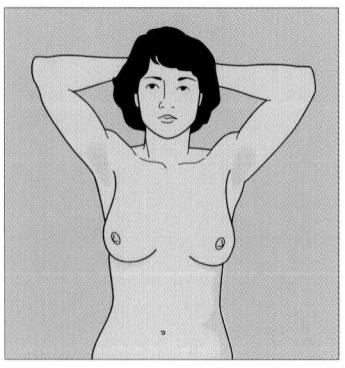

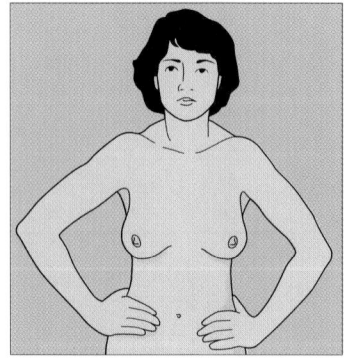

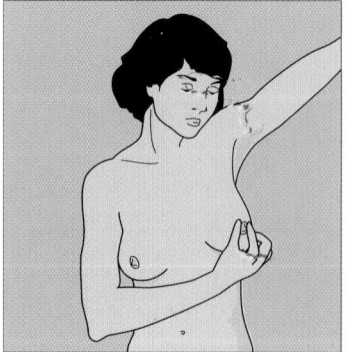

2. Stand before a mirror, lift your arms over your head, and look for any unusual changes in the contour of your breasts, such as a swelling, dimpling, or retraction (inward sinking) of the nipple. Also check for any unusual discharge from the nipple.

If you discover a lump or any other change during a breast self-examination, it's important to see a physician at once. Most changes are not cancerous, but let the doctor make the diagnosis.

b

which is less disfiguring than mastectomy because it leaves some breast tissue in place. In both cases, removed lymph nodes are examined to determine the need for further treatment and to predict the prospects of a cure.

Cancers of the Reproductive System

UTERINE AND OVARIAN CANCER   Cancers of the uterus most often affect the endometrium (uterine lining) and the cervix. Various types are treated by surgery, radiation, or both. The incidence of uterine cancers is falling, in part because precancerous phases of cervical cancer can be easily detected by the *Pap smear* that is part of a routine gynecological examination. Risk factors for cervical cancer include having multiple sex partners, early age of first intercourse, cigarette smoking, and genital warts (Section 15.7). Endometrial cancer is more common during and after menopause; women who take supplemental estrogen during menopause are at higher risk.

Ovarian cancer is often lethal because symptoms, mainly an enlarged abdomen, do not appear until the cancer is advanced and has already metastasized. In a few women the first sign is abnormal vaginal bleeding or vague abdominal discomfort. Risk factors include family history of the disease, childlessness, and a history of breast cancer. New hope for victims of ovarian (and breast) cancers has come from the discovery of taxol, a compound derived from the Pacific yew tree. In 20–30 percent of advanced cases, taxol shrinks tumors significantly. How? It makes a tumor cell's microtubules highly stable, so they can't disassemble and reassemble—key operations, recall, in cell division. Hence, the cancer cells die just as they are about to divide. To meet the demand for taxol, researchers have developed ways of synthesizing taxol-like compounds (called *taxoids*) in quantity.

CANCER OF THE TESTIS AND PROSTATE   About 5,000 cases of cancer of the testis are diagnosed annually in the United States. In its early stages, testicular cancer is painless. However, it can spread to lymph nodes in the abdomen, chest, neck, and, eventually, the lungs. Once a month from high school onward, men should examine each testis separately after a warm bath or shower (when the scrotum is relaxed). The testis should be rolled gently between the thumb and forefinger to check for any type of lump, enlargement, or hardening (Figure 20.11). Such changes may or may not cause discomfort, but only a physician can rule out the possibility of disease. Surgery is the usual treatment, and the success rate is high when the cancer is caught before it can spread.

Prostate cancer is the second leading cause of cancer deaths in men (after lung cancer). There are no confirmed risk factors. Symptoms include various kinds of urinary problems, although these can also signal nothing more

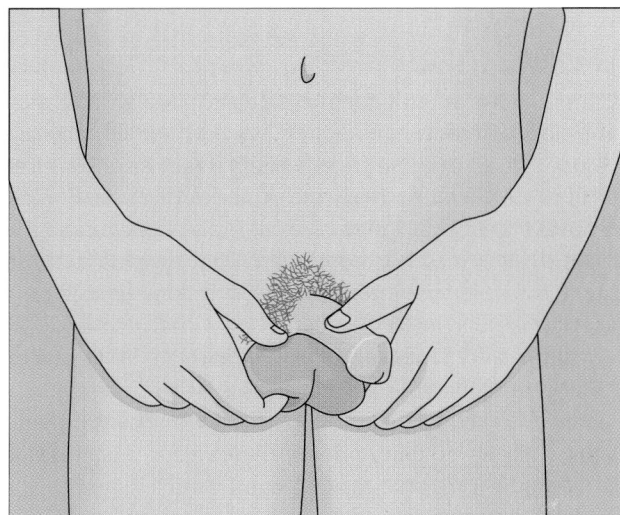

Figure 20.11   Method for testicular self-examination, as recommended by the American Cancer Society.

Perform the exam when the scrotum is relaxed, as it is after a warm bath or shower. Simply roll each testicle between the thumb and forefinger, feeling for any lumps or thickening. By performing the exam regularly, you can detect changes early on and discuss them with your physician. As with breast lumps, most such changes are *not* cancer, but only a doctor can make the diagnosis.

than a noncancerous enlarged prostate. The American Cancer Society recommends that men over 40 have an annual digital rectal examination, which enables a physician to feel the prostate and detect any unusual lumps. A blood test called PSA can also screen for larger than normal amounts of a substance called prostate-specific antigen. If a physician still suspects cancer after these two tests have been performed, the next step is a tissue biopsy. The cure rate for prostate tumors detected early is over 90 percent.

Many prostate cancers grow quite slowly and appear to cause few problems for the patient. In such cases a man's physician may recommend simply monitoring the tumor, taking no other action until the threat to health is well established. Like all medical decisions, however, this one can be made only after carefully weighing the risks and benefits.

Breast cancer affects about one woman in nine. Rarely, it also occurs in males.

The most common reproductive cancers in women develop in the ovaries and uterus. In males, cancers of the testis and prostate gland are the most common reproductive cancers.

Chances for a cure are best when cancer is detected early. Monthly self-examination is a crucial tool for early detection of breast cancer and testicular cancer.

A SURVEY OF OTHER COMMON CANCERS

Oral and Lung Cancers

Cancers of the mouth, tongue, salivary glands, and throat are most common among smokers and people who use smokeless tobacco (snuff), especially if they are also heavy alcohol drinkers. In 1993, the American Cancer Society reported over 29,000 new cases of oral cancer and nearly 8,000 deaths, mostly among men. That trend is continuing. Of people who are diagnosed with oral cancer, only about half survive for 5 years, even with treatment.

Lung cancer kills more people than any other cancer. Long-term tobacco smoking is the overwhelming risk factor, followed by exposure to asbestos, industrial chemicals such as arsenic, and radiation. For a smoker, a combination of these factors significantly increases the odds of developing cancer. For nonsmokers, especially spouses and children of smokers, inhaling "secondhand" tobacco smoke also poses a significant cancer risk. The Environmental Protection Agency estimates that lung cancer resulting from long-term breathing of secondhand smoke kills 3,000 people in the United States each year.

In recent years, the incidence of lung cancer has decreased in men. However, a rise in the relative number of female smokers in the last several decades is now reflected in the fact that lung cancer has surpassed breast cancer as the leading cancer killer of women. Warning signals include a nagging cough, shortness of breath, chest pain, bloody sputum (coughed-up phlegm), unexplained weight loss, and frequent respiratory infections or pneumonia.

Four types of lung cancer account for 90 percent of cases. About one-third of lung cancers are *squamous cell carcinomas*, affecting squamous epithelium in the bronchi. Another 48 percent are either *adenocarcinomas* or *large-cell carcinomas* (Figure 20.12). A fourth type, *small-cell carcinoma*, spreads rapidly and kills most of its victims within 5 years of diagnosis.

Cancers of the Digestive System and Related Organs

Cancers of the stomach and pancreas are usually adenocarcinomas of duct cells. Often they are not detected until they have spread to other organs. Cigarette smoking is a risk factor for pancreatic cancer, while stomach cancer may be associated with heavy alcohol consumption and a diet rich in smoked, pickled, and salted foods. Liver cancer is uncommon in the United States, but growing evidence suggests that hepatitis B infection can trigger it.

Most colon cancers are adenocarcinomas. Warning signs include a change in bowel habits, rectal bleeding, and blood in the feces. Family history of colon cancer

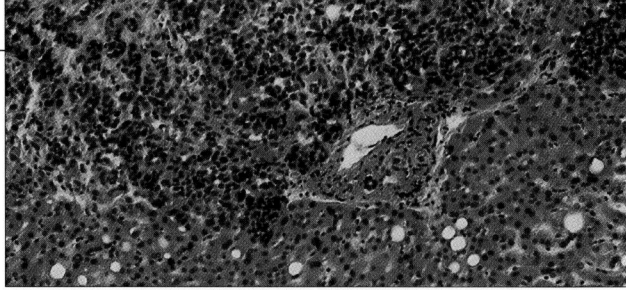

a

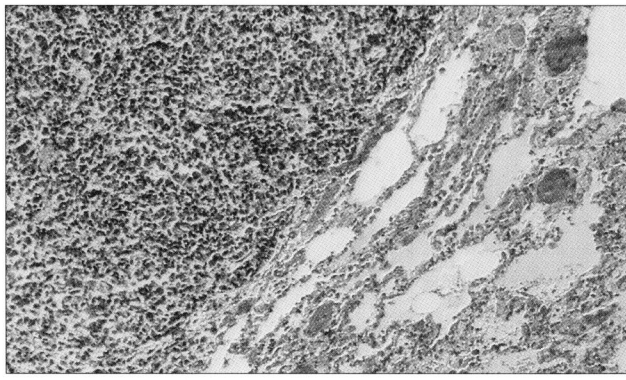

b

Figure 20.12 Photomicrographs of (**a**) adenocarcinoma and (**b**) large-cell carcinoma of the lung.

(indicating a genetic component) or inflammatory bowel disease is a major risk factor, as is a lack of dietary fiber.

Urinary System Cancers

Carcinomas of the bladder and kidney account for about 70,000 new cancer cases each year. The incidence is higher in males, and smoking and exposure to certain industrial chemicals are major risk factors for both. Kidney cancer easily metastasizes via the bloodstream to the lungs, bone, and liver. An inherited type, *Wilms tumor*, is one of the most common of all childhood cancers.

Cancers of the Blood and Lymphatic System

Lymphomas develop in lymphoid tissues in organs such as the lymph nodes, spleen, and thymus. They include the diseases known as *Hodgkin disease, non-Hodgkin lymphoma*, and *Burkitt lymphoma*. Risk seems to increase along with infections—such as HIV—that impair immune system functioning. Burkitt lymphoma is most common in parts of Africa, where it seems to develop especially in children who become infected with both the Epstein-Barr virus (which also causes mononucleosis) and malaria. Lymphoma symptoms include enlarged lymph nodes, rashes, weight loss, and fever. Intense itching and night sweats also are typical of Hodgkin disease. Chemotherapy and radiation are standard treatments, and new

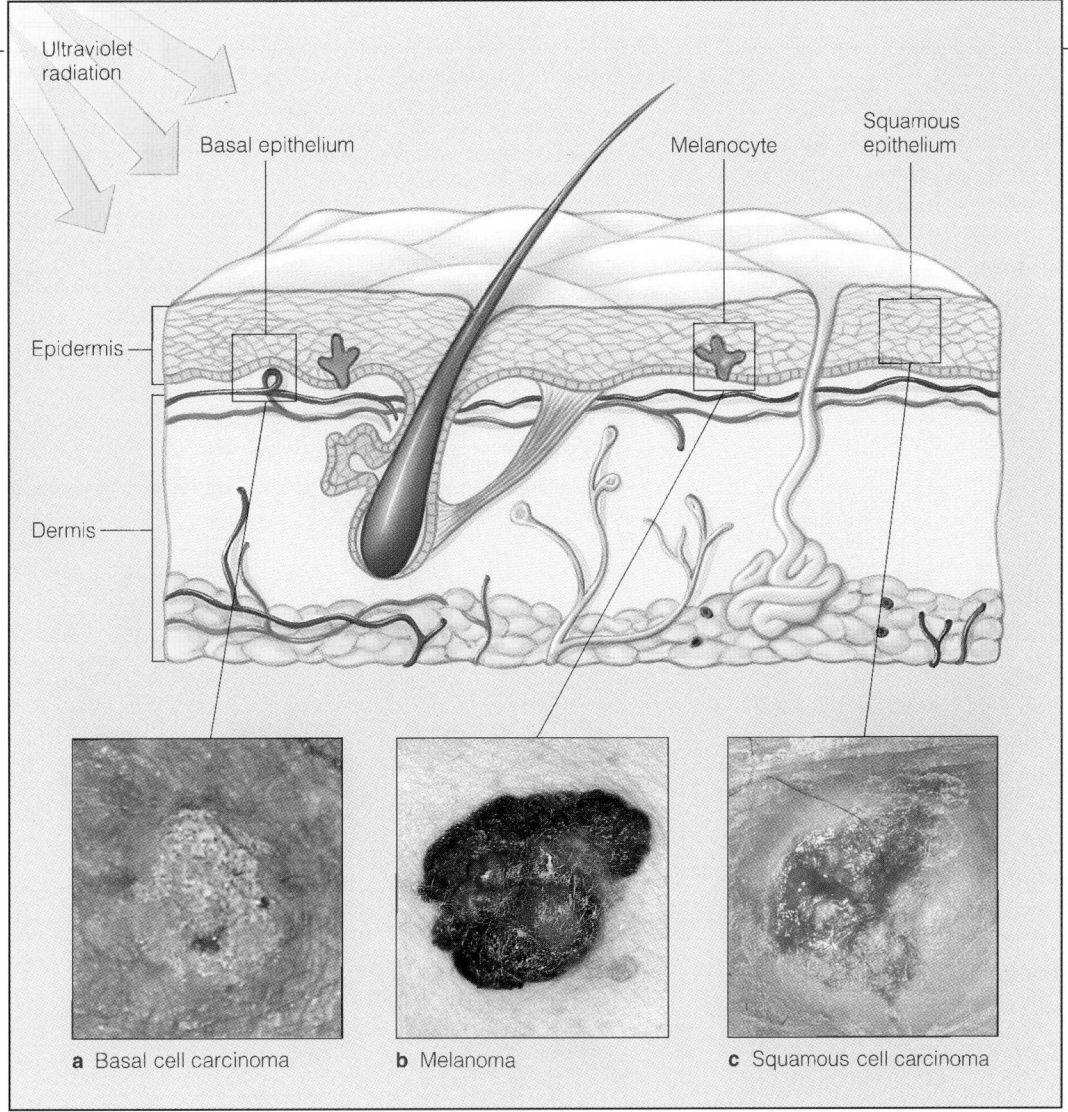

Figure 20.13 (**a**) Basal cell carcinoma, (**b**) malignant melanoma, and (**c**) squamous cell carcinoma. Each type of skin cancer has a distinctive appearance. Malignant melanoma can be deadly because it metastasizes aggressively.

a Basal cell carcinoma   **b** Melanoma   **c** Squamous cell carcinoma

treatments using targeted monoclonal antibodies are being developed.

Leukemias are cancers in which stem cells in bone marrow overproduce white blood cells. Some types are the most common childhood cancers, but other types are diagnosed in more than 70,000 adults each year. Risk factors include Down syndrome, exposure to chemicals such as benzene, and radiation exposure. Many leukemias can be effectively treated with chemotherapy using two compounds—vincristine and vinblastine—derived from species of periwinkle plants that grow only on the African island of Madagascar. Environmentalists often cite these drugs and other plant-derived compounds such as taxol as reasons why humans should attempt to preserve natural areas where as-yet-undiscovered medicinal plants may be present.

Skin Cancer

Skin cancers are the most common of all cancers. For all types, fair skin and exposure to ultraviolet (UV) radiation in sunlight (or tanning salons) are risk factors. UV dam-

age to the DNA in melanocytes, the melanin-producing cells in the skin's epidermis, is the most dangerous kind, called malignant melanoma (see Section 3.11). Malignant melanoma is such a threat because in its later stages it metastasizes aggressively.

Squamous cell (epidermal) carcinomas start out as scaly, reddened bumps (Figure 20.13c). They grow rapidly and can spread to adjacent lymph nodes unless they are surgically removed. Basal cell carcinomas begin as small, shiny bumps and slowly grow into ulcers with beaded margins. Basal cell carcinoma and squamous cell carcinoma are much more common than malignant melanoma, and usually they are easily treated by minor surgery in a doctor's office.

Common cancers include oral and lung cancers and those of the digestive system (especially colon cancer), the urinary system, the blood and lymphatic systems (leukemias and lymphoma), and the skin. Each year, these collectively account for more than 750,000 cases of cancer in the United States alone.

SUMMARY

1. Cancer arises when genetic controls over cell division are lost. Cancer cells have specific characteristics that set them apart from normal cells. These include structural abnormalities, including (typically) poor differentiation and altered surface proteins; uncontrolled growth with absence of contact inhibition; invasion of surrounding tissues; and the ability to metastasize to other body regions.

2. Cancer develops through a multistep process involving several genetic changes. Initially, mutation or some other event may alter a proto-oncogene into a cancer-causing oncogene. Infection by a virus can also insert an oncogene into a cell's DNA or disrupt normal controls over a proto-oncogene. In addition, one or more tumor suppressor genes probably must be missing or become mutated before a normal cell can be transformed into a cancerous one.

3. A predisposition to some cancers is inherited. Causes of carcinogenesis are viral infection, chemical carcinogens, radiation, faulty immune system functioning, and possibly breakdown in normal DNA repair mechanisms.

4. Common methods for cancer diagnosis include blood testing for the presence of tumor markers, which are substances produced either by specific types of cancer cells or by normal cells in response to the cancer. Medical imaging (as by magnetic resonance imaging) also can aid diagnosis. The definitive diagnostic procedure is biopsy.

5. In general, a cancer is named according to the type of tissue in which it arises. Common types include sarcomas (connective tissues such as muscle and bone), carcinomas (epithelium), adenocarcinoma (glands or their ducts), lymphomas (lymphoid tissues), and leukemias (blood-forming regions).

6. Standard cancer treatments include surgery, chemotherapy, and tumor irradiation. Target-specific monoclonal antibodies and immune therapy using interferons and interleukins are other treatment options currently under development.

7. Lifestyle choices such as the decision not to use tobacco, to maintain a low-fat, high-fiber diet, and to avoid overexposure to direct sunlight and chemical carcinogens can help limit personal cancer risk.

8. The incidence of various cancers varies among the sexes. Among adult females, the most prevalent sites are the breasts, colon and rectum, lungs, and uterus. Among adult males the most prevalent sites are the prostate gland, lung, colon and rectum, and bladder.

Review Questions

1. How are cancer cells structurally different from normal cells of the same tissue? What is the relevance of altered surface proteins to uncontrolled growth? *424–425*

2. What are differences between a benign tumor and a cancerous one? *424*

3. Write a short paragraph that summarizes the roles of proto-oncogenes, oncogenes, and tumor suppressor genes in carcinogenesis. *426*

4. List the four main categories of cancer tumors. *431*

5. What are the seven warning signs of cancer? (Remember the American Cancer Society's clue word, CAUTION.) *429*

6. Using the diagram below as a guide, indicate the major steps in cancer metastasis. *425*

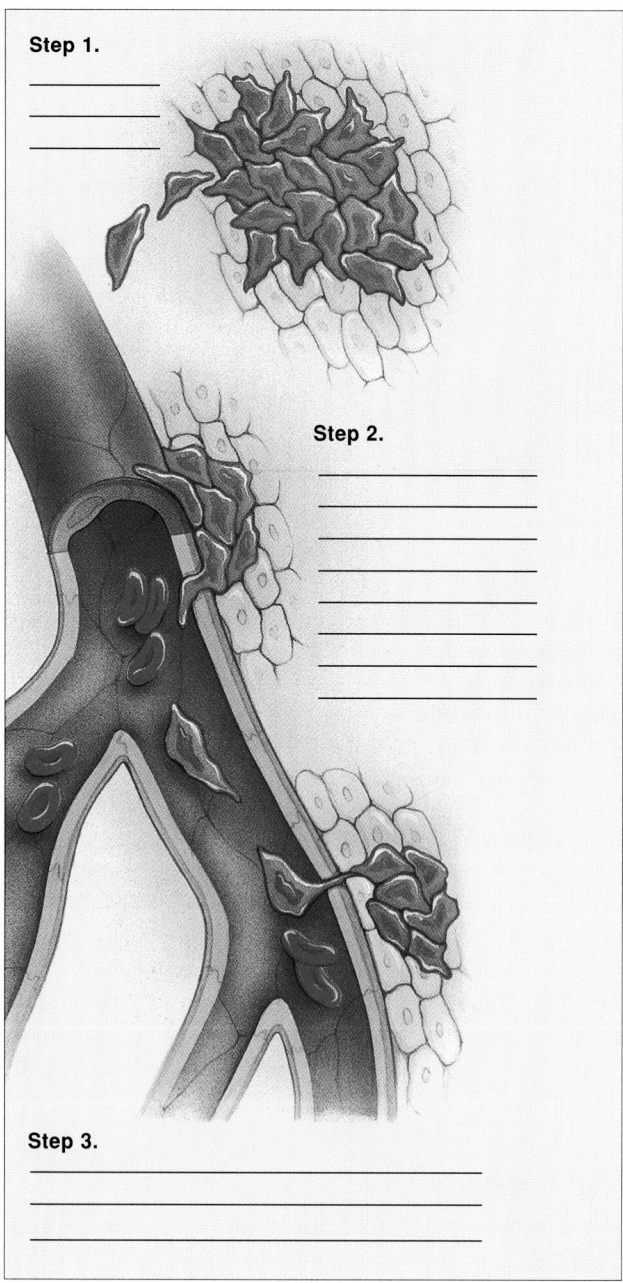

Step 1.

Step 2.

Step 3.

Critical Thinking: You Decide *(Key in Appendix V)*

1. Some people are concerned that exposure to electromagnetic fields (EMFs) emanating from power lines and small appliances may trigger cancerous changes in cells. What are some specific kinds of biological events you would expect researchers to probe in order to establish an EMF–cancer link?

2. A textbook on cancer contains the following statement: "Fundamentally, cancer is a failure of the immune system." Why do you think the author wrote this?

3. Ultimately, cancer kills because it spreads and disturbs homeostasis. Consider, for example, a kidney cancer that metastasizes to the lungs and liver. What are some specific homeostatic mechanisms that the spreading disease could disrupt?

Self-Quiz *(Answers in Appendix IV)*

1. A tumor is _____.
   a. malignant by definition
   b. always enclosed by connective tissue
   c. a mass of tissue that may be benign *or* malignant
   d. always slow-growing

2. Cancer cells _____.
   a. lack normal controls over cell division
   b. secrete the growth factor angiogenin
   c. display altered surface proteins
   d. do not respond to contact inhibition
   e. all of the above

3. The onset of cancer seems to require the activity of an oncogene plus the absence or mutation of at least one _____.

4. Chemical carcinogens _____.
   a. include viral oncogenes
   b. can damage DNA and cause a mutation
   c. must be ingested in food
   d. are not found in foods

5. So far as we know, carcinogenesis is *not* triggered by _____.
   a. breakdowns in DNA repair
   b. a breakdown in immunity
   c. radiation
   d. protein deficiency
   e. inherited gene defects

6. Tumor suppressor genes _____.
   a. occur normally in cells
   b. promote metastasis
   c. enter cells by way of viral infection
   d. probably affect the development of cancer only in rare cases

7. _____ is the definitive method for detecting cancer.
   a. Blood testing
   b. Physician examination
   c. Biopsy
   d. Medical imaging

8. The most common therapeutic approaches to treating cancer include all of the following except:
   a. chemotherapy
   b. irradiation of tumors
   c. surgery to remove cancerous tissue
   d. administering doses of vitamins

9. The goal of immune therapy is to _____.
   a. cause defective T cells in the thymus to disintegrate
   b. activate cytotoxic T cells
   c. dramatically increase the numbers of circulating macrophages
   d. promote the secretion of monoclonal antibodies

10. Currently, _____ cancer is the leading cause of death among adult females; _____ cancer is the leading cause of cancer death among adult males.
    a. lung; prostate
    b. breast; colon
    c. lung; lung
    d. breast; lung

Selected Key Terms

biopsy *429*	metastasis *425*
carcinogen *426*	oncogene *426*
carcinogenesis *426*	proto-oncogene *426*
chemotherapy *430*	tumor *424*
dysplasia *424*	tumor marker *429*
immunotherapy *430*	tumor suppressor gene *426*

Readings

Cancer Facts and Figures. American Cancer Society. Free pamphlet published annually; an excellent summary of the latest information on cancer, risk factors, and treatments.

Cavenee, W. K, and R. L. White. March 1995. "The Genetic Basis of Cancer." *Scientific American.*

Nicolaou, K., R. Guy, and P. Potier. June 1996. "Taxoids: New Weapons against Cancer." *Scientific American.*

Radetsky, P. March 1993. "Magic Missiles." *Discover.* Fascinating nontechnical description of research on monoclonal antibodies as anticancer weapons.

Waldholz, M. 1997. *Curing Cancer: Solving One of the Greatest Medical Mysteries of Our Time.* New York: Simon & Schuster. This popular book's strength lies in its engrossing behind-the-scenes descriptions of how modern cancer research is carried out.

Weber, B. January/February 1996. "Genetic Testing for Breast Cancer." *Scientific American.* SCIENCE & MEDICINE.

 For additional readings, go to InfoTrac College Edition, your online library at:

http://www.infotrac-college.com/wadsworth

RECOMBINANT DNA AND GENETIC ENGINEERING

Ingenious Genes

In the 1960s a world-shaking revolution began quietly in test tubes and laboratory culture dishes. The revolutionaries were scientists who discovered how they could use enzymes to snip DNA molecules into specific pieces and then join the pieces in new arrangements. They learned, too, how to use bacteria and viruses as messengers that could cross the boundaries between cells of one species and cells of another, carrying a cargo of genetic information.

As you read this, some investigators are using such methods to insert human genes into mouse embryos as part of the search for treatments for cancer, autoimmune diseases, and infections. Others are attempting to alter genes in HIV, the AIDS virus, as part of the effort to develop effective vaccines. Various laboratories have refined methods for bioengineering livestock. The result? Sheep, pigs, and cows born with human genes that enable them to produce milk containing human

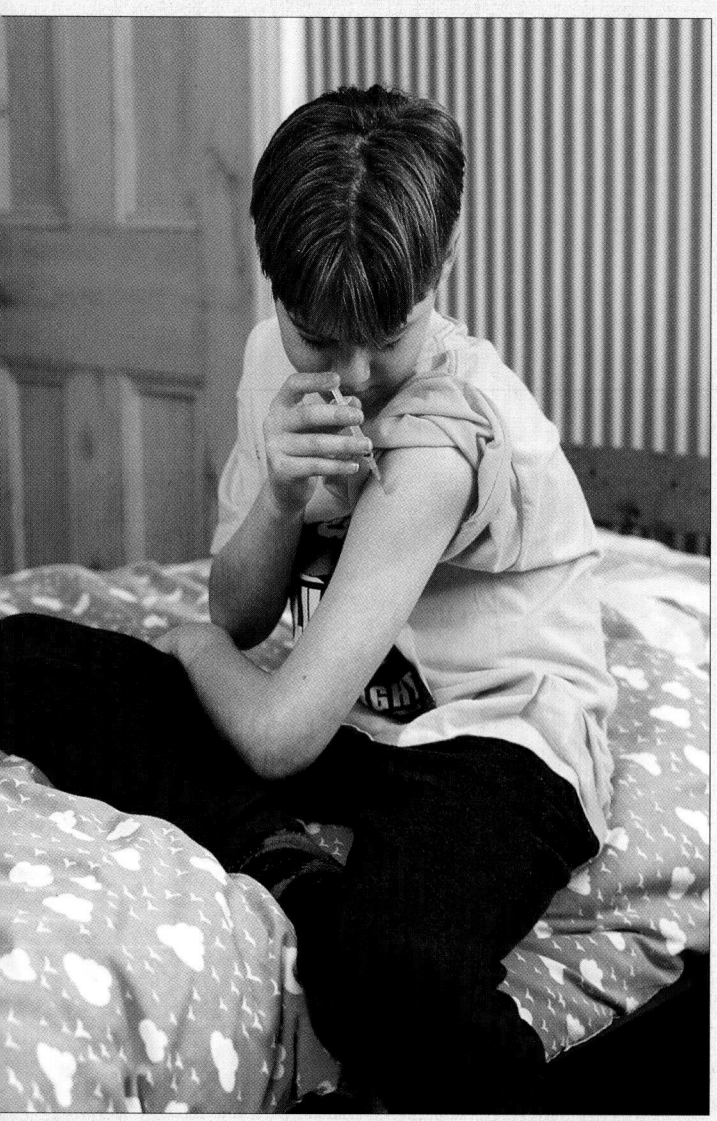

a

b

Figure 21.1 (**a**) Diabetic boy injecting insulin from genetically engineered bacteria. (**b**) Commercial production of interferon.

proteins, such as hemoglobin and blood-clotting factors. More than likely, your local drugstore carries at least some bioengineered products, including the hormone insulin required by people with type 1 diabetes—insulin that is produced by vats of genetically altered bacteria (Figure 21.1).

The biotechnological revolution has staggering potential not only for medicine, but for agriculture and industry as well. By 1998 numerous patents had been granted on bioengineered organisms and products, and scores more were pending government approval. For example, tomatoes from genetically altered plants are already on the market, carrying in their cells "foreign" genes that retard spoilage. At Texas A & M University, plant bioengineers have succeeded in creating genetically altered carrots and onions whose tissues are unusually high in antioxidant compounds. Other plant biotechnology projects are making major strides in developing crops resistant to drought or pests or that contain high levels of essential amino acids.

Around the world, at a mind-bending pace, researchers are solving genetic puzzles and learning how to mix and match genes of various species. At the same time, other people are sounding notes of caution, even fear, about what this "brave new world" could bring. With so many actual and potential applications to human health and well-being, we are well advised to understand the basics of biotechnology. In this chapter we'll also consider some of the ecological, social, and ethical questions such methods raise.

KEY CONCEPTS

1. Genetic experiments have been occurring in nature for billions of years through gene mutations, crossing over and recombination, and other events.

2. Humans are now purposefully bringing about genetic changes by way of recombinant DNA technology. Such enterprises are called genetic engineering.

3. With this technology, researchers isolate, cut, and splice together gene regions from different species, then greatly amplify the number of copies of the genes that interest them. The genes and, in some cases, the proteins they specify are produced in quantities large enough for research and for practical applications.

4. Three activities are at the heart of recombinant DNA technology. First, procedures based on specific enzymes are used to cut DNA molecules into fragments. Second, the fragments are inserted into cloning tools, such as plasmids. Third, fragments containing the genes of interest are identified, then copied rapidly and repeatedly.

5. Genetic engineering involves isolating, modifying, and inserting genes back into the same organism or into a different one. The goal is to beneficially modify traits that the genes influence. Human gene therapy, which focuses on controlling or curing genetic disorders, is an example.

6. The new technology raises social, legal, ecological, and ethical questions regarding its benefits and risks.

CHAPTER AT A GLANCE

21.1 Recombination in Nature—and in the Laboratory
21.2 Working with DNA Fragments
21.3 *DNA Fingerprints*
21.4 Modified Host Cells
21.5 Bacteria, Plants, and the New Technology
21.6 *Bacteria That Clean Up Pollutants*
21.7 Genetic Engineering of Animals
21.8 Methods and Prospects for Gene Therapy
21.9 *Issues for a Biotechnological Society*

RECOMBINATION IN NATURE—AND IN THE LABORATORY

For at least 3 billion years, nature has been conducting countless genetic experiments, through mutation, crossing over, and other events that introduce changes in genetic messages. This is the source of life's diversity.

For many thousands of years, we humans have been changing numerous genetically based traits of species. Through artificial selection, we produced modern crop plants and new breeds of cattle, birds, dogs, and cats from wild ancestral stocks. We developed meatier turkeys and sweeter oranges, larger corn, seedless watermelons, flamboyant ornamental roses, and other useful plants. We produced hybrids such as the tangelo (tangerine × grapefruit) and mule (donkey × horse).

Researchers now use **recombinant DNA technology** to analyze genetic changes. With this technology, they cut and splice DNA from different species, then insert the modified molecules into bacteria or other types of cells that can rapidly replicate genetic material and divide. The cells copy the foreign DNA along with their own. In short order, huge populations produce useful quantities of recombinant DNA molecules. The new technology also is the basis of **genetic engineering**, in which genes are isolated, modified, and inserted back into the same organism or into a different one.

Plasmids, Restriction Enzymes, and the New Technology

The remarkable technology we have been describing is made possible by the genetic workings of bacteria. As you may recall from Chapter 15, bacterial cells have a single chromosome, a circular DNA molecule with all the genes needed for growth and reproduction. Many bacteria also have **plasmids**, which are small, circular molecules of "extra" DNA that contain a few genes (Figure 21.2).

Usually, plasmids are not essential for survival, but some of the genes they carry may benefit the bacterium. For instance, some plasmid genes confer resistance to antibiotics. The bacterium's replication enzymes can copy and reproduce plasmid DNA, just as they do with the bacterium's chromosomal DNA.

Viruses as well as bacteria take part in gene transfer and recombination. And so do most eukaryotic species. As you might imagine, viral infection does a bacterium no good. Over time, bacteria developed an arsenal against

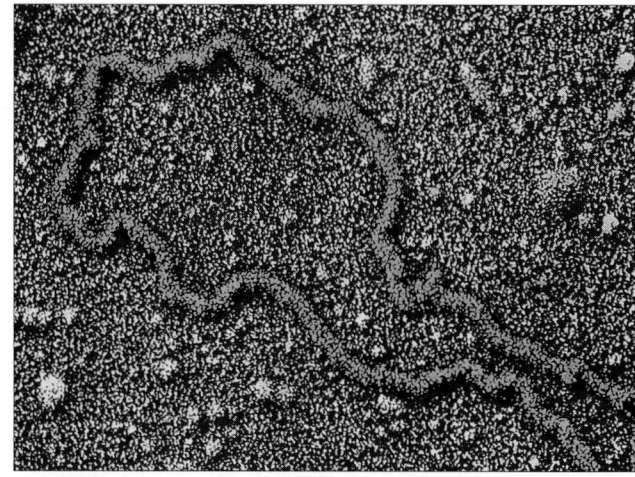

a

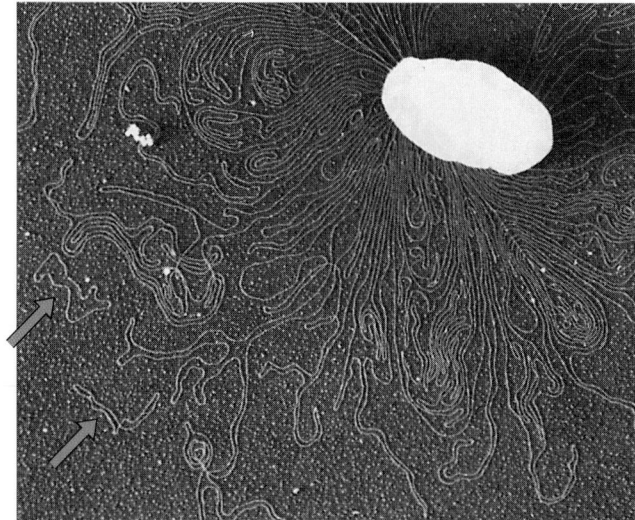

b

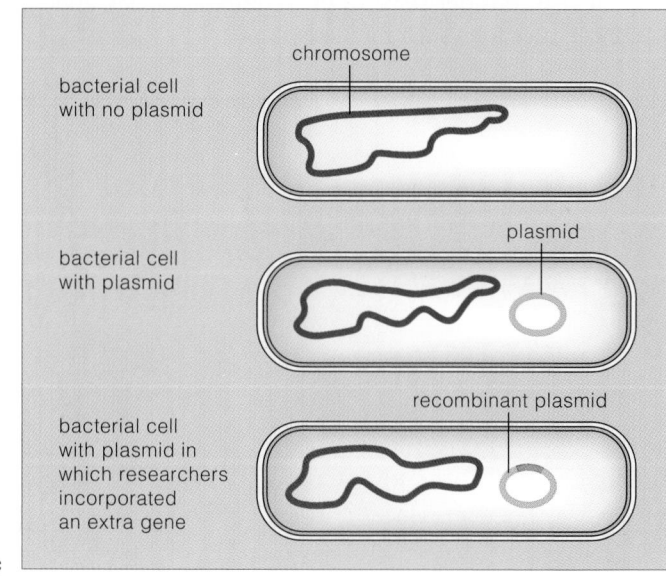

c

chromosome

bacterial cell with no plasmid

plasmid

bacterial cell with plasmid

recombinant plasmid

bacterial cell with plasmid in which researchers incorporated an extra gene

Figure 21.2 (**a**) A plasmid at high magnification. (**b**) Plasmids (*blue* arrows) released from a ruptured *Escherichia coli* cell. (**c**) Naturally occurring and modified plasmids.

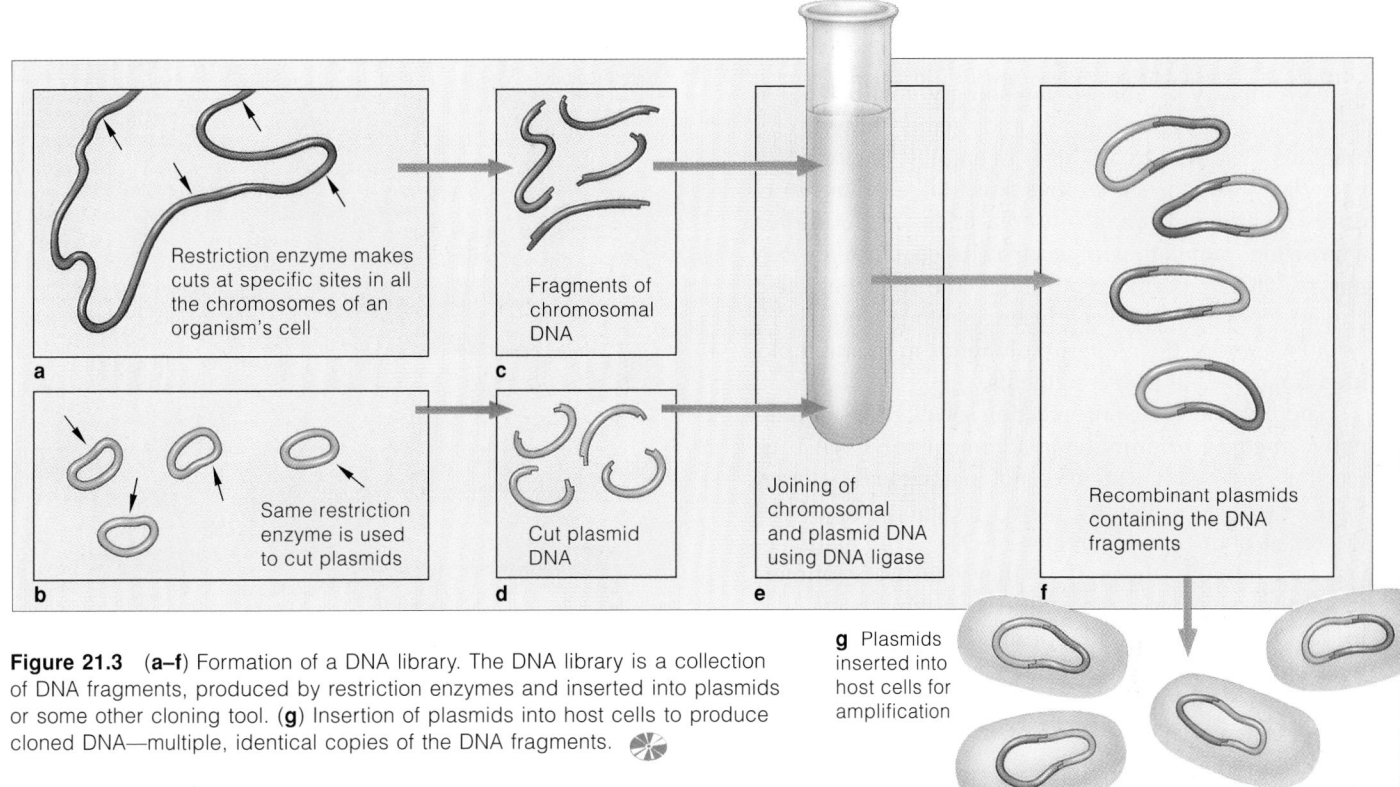

Figure 21.3 (**a–f**) Formation of a DNA library. The DNA library is a collection of DNA fragments, produced by restriction enzymes and inserted into plasmids or some other cloning tool. (**g**) Insertion of plasmids into host cells to produce cloned DNA—multiple, identical copies of the DNA fragments.

g Plasmids inserted into host cells for amplification

invasion by harmful genes. They became equipped with many types of **restriction enzymes**, which are able to recognize and cut apart foreign DNA that may enter a cell. Eventually, researchers learned how to use plasmids and restriction enzymes for genetic recombination in the laboratory.

Producing Restriction Fragments

Each type of restriction enzyme makes a cut wherever it recognizes a specific, very short nucleotide sequence in the DNA. Cuts at two identical sequences in the same DNA molecule produce a fragment. Because some types of enzymes make staggered cuts, some fragments have single-stranded portions at both ends. Sometimes these are called "sticky" ends:

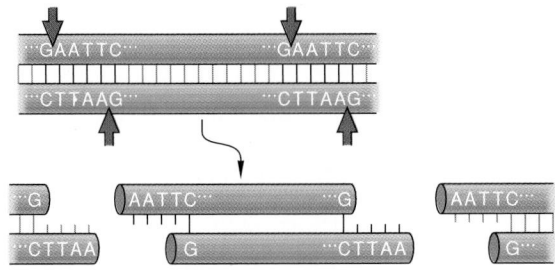

liberated DNA fragment with sticky ends

By "sticky," we mean that the short, single-stranded ends of a DNA fragment have the chemical ability to base-pair with any other DNA molecule that also has been cut by the same restriction enzyme.

For example, suppose you use the same restriction enzyme to cut plasmids *and* DNA molecules that you have isolated from a human cell. When you mix the cut plasmids and human molecules together, they base-pair at the cut sites. Then you add **DNA ligase** to the mixture. This enzyme seals DNA's sugar-phosphate backbone at the cut sites, just as it does during DNA replication. In this way, you create "recombinant plasmids," which have pieces of DNA from another organism inserted into them (Figure 21.2c).

You now have a **DNA library**—that is, a collection of DNA fragments, produced by restriction enzymes, that have been incorporated into plasmids, as illustrated in Figure 21.3.

With recombinant DNA technology, DNA from different species is cut and spliced together; then the recombinant molecules are amplified (by way of copying mechanisms) to produce useful quantities of the genes of interest.

Recombination is made possible by the use of restriction enzymes that make specific cuts in DNA molecules and of DNA ligases that seal the cut ends.

WORKING WITH DNA FRAGMENTS

Amplification Procedures

A DNA library is almost vanishingly small. It must be amplified—copied again and again—into useful amounts. One way to do this is to employ "factories" of bacteria, yeasts, or some other kind of cell that reproduces rapidly and can take up plasmids. In short order, a growing population of such cells contains a DNA library. Their repeated replications and divisions yield **cloned DNA** that has been inserted into plasmids. The "cloned" part of the name simply refers to the multiple, identical copies of DNA fragments.

The **polymerase chain reaction**, or **PCR**, is an alternative method of amplifying fragments of DNA. The reactions proceed in test tubes, not in microbial factories. First, researchers identify short nucleotide sequences located at both ends of a region of DNA from an organism that interests them. Then they synthesize primers (nucleotide sequences that will base-pair with the ones in the DNA). They mix the primers with enzymes, free nucleotides, and the DNA from some of the organism's cells. Then they subject the mixture to precise cycles of temperature variation.

During the cycles, the two strands of the DNA molecules unwind from each other. Then, as Figure 21.4 indicates, the short sequences become positioned at the specific target sites, in accordance with the exposed, complementary bases. *Taq* **DNA polymerase**, a replication enzyme, recognizes the synthesized primers as signals to initiate replication. This particular DNA polymerase comes from a bacterium that thrives in hot springs and even in hot water heaters. The enzyme is not destroyed by the elevated temperatures required to unwind the DNA.

With each round of reactions, the number of DNA molecules defined by the primer binding sites doubles. For example, if there are 10 such molecules in the test tube, there soon will be 20, then 40, 80, 160, 320, and so on. Very quickly, a target region from a single DNA molecule can be amplified to *billions* of molecules.

Thus PCR can amplify samples with very little DNA. Such tiny samples can be obtained from long-buried remains, mummies, or even from a single hair left at the scene of a crime.

Sorting Out DNA Fragments

When restriction enzymes cut DNA, the resulting fragments are not all the same length. Researchers can employ gel electrophoresis to separate the fragments from one another. This laboratory procedure uses an electric field to move molecules through a viscous gel and separate them according to certain properties. For DNA, the sizes of the molecules affect how far they move.

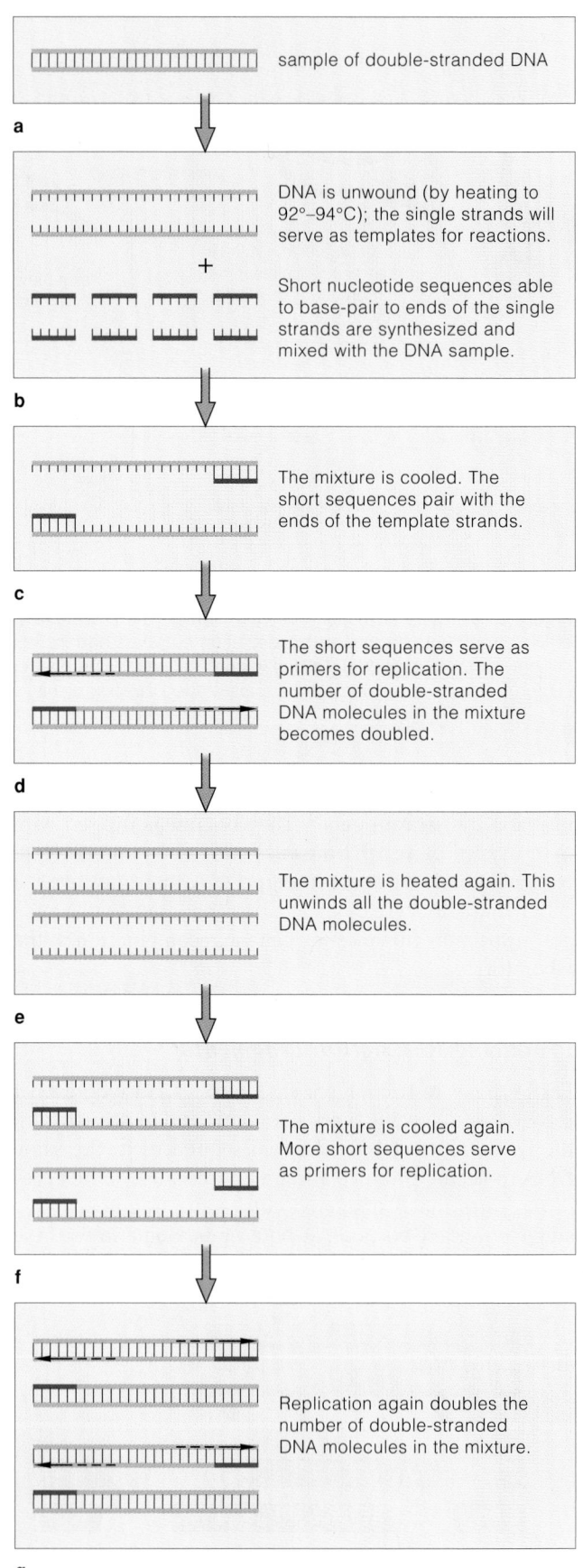

a sample of double-stranded DNA

b DNA is unwound (by heating to 92°–94°C); the single strands will serve as templates for reactions.

+

Short nucleotide sequences able to base-pair to ends of the single strands are synthesized and mixed with the DNA sample.

c The mixture is cooled. The short sequences pair with the ends of the template strands.

d The short sequences serve as primers for replication. The number of double-stranded DNA molecules in the mixture becomes doubled.

e The mixture is heated again. This unwinds all the double-stranded DNA molecules.

f The mixture is cooled again. More short sequences serve as primers for replication.

g Replication again doubles the number of double-stranded DNA molecules in the mixture.

Figure 21.4 The polymerase chain reaction (PCR).

Figure 21.5 A method used for sequencing DNA, as first developed by Frederick Sanger. The method can determine the nucleotide sequence of specific DNA fragments, as this example illustrates. Nowadays, much of the procedure is automated.

a Single-stranded DNA fragments are added to a solution, shown here in four different test tubes:

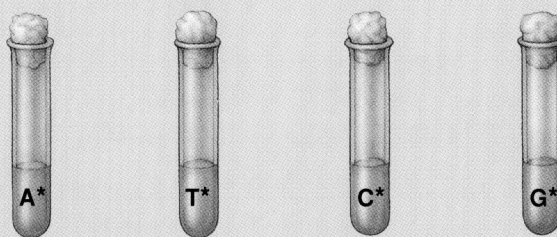

All four tubes contain DNA polymerases, short nucleotide sequences that can serve as primers for replication, and the nucleotide subunits of DNA (A, T, C, and G). Each tube contains a modified, labeled version of only one of the four kinds of nucleotides. We can show these as A*, T*, C*, and G*. The labeled form is present in low concentration, along with a generous supply of the unmodified form of the same nucleotide. Let's follow what happens in the tube with the A* subunits.

b As expected, the DNA polymerase recognizes a primer that has become attached to a fragment, which it uses as a template strand. The enzyme assembles a complementary strand according to base-pairing rules (A only to T, and C only to G). Sooner or later, the enzyme picks up an A* subunit for pairing with a T on the template strand. The modified nucleotide is a chemical roadblock—it prevents the enzyme from adding more nucleotides to the growing complementary strand. In time, the tube contains labeled strands of different lengths, as dictated by the location of each A* in the sequence:

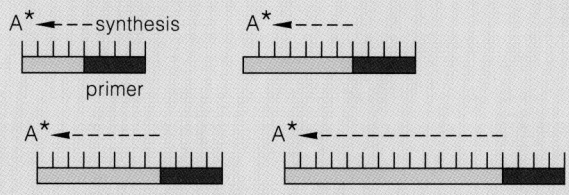

The same thing happens in the other three tubes, with strand lengths dictated by the location of T*, C*, and G*.

c DNA from each of the four tubes is placed in four parallel lanes in the same gel. Then the DNA can be subjected to electrophoresis. The resulting nucleotide sequence can be read off the resulting bands in the gel. Look at the numbers running down the side of this diagram:

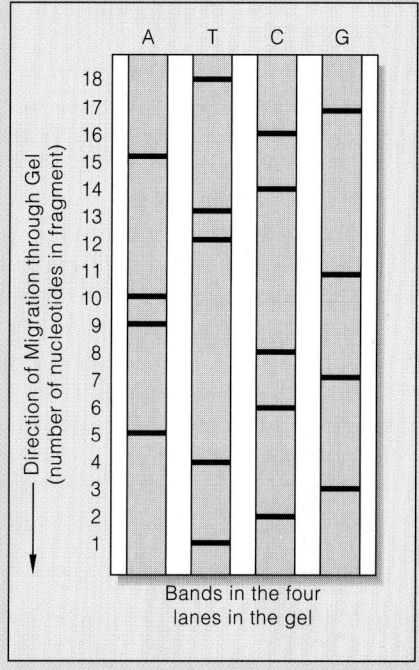

Bands in the four lanes in the gel

Start with "1" and read across the four lanes (A, T, C, G). As you can see, T is the closest to the start of the nucleotide sequence; it has migrated farthest through the gel. At "2," the next nucleotide is C, and so on. The entire sequence, read from the first nucleotide to the last is

T C G T A C G C A A G T T C A C G T

And now, by applying the rules of base-pairing, you can deduce the sequence of the DNA fragment that served as the template.

A gel that contains DNA is immersed in a buffered solution, and electrodes connect the solution to a power source. Apply voltage, and the DNA fragments migrate through the gel toward the positively charged electrode (because of their negatively charged phosphate groups). They separate from one another in the gel. Again, how far they actually migrate toward the positively charged electrode depends only on their size—the larger ones cannot move as fast through the gel. Following a predetermined period of electrophoresis, fragments of different length can be identified by staining the gel.

Once DNA fragments from a sample have been sorted out according to length, researchers can work out the nucleotide sequence of each type. The Sanger method of DNA sequencing is one of the ways this can be done (Figure 21.5). Automated DNA sequencers can determine sequences of hundreds of nucleotides in a day, at a cost of less than twenty dollars.

Restriction fragments can be rapidly amplified by PCR and by large populations of bacterial or yeast cells.

Restriction fragments can be sorted out according to length, and their nucleotide sequences can be determined.

21.3 DNA FINGERPRINTS

Suppose a researcher has isolated a good-size sample of your DNA molecules. She cuts them up with restriction enzymes, then subjects the restriction fragments to gel electrophoresis (refer to Section 19.2). Large ones cannot move as fast as small ones, so the fragments separate from one another in the gel. Now the researcher uses the *Southern blot method*. She transfers the electrophoretically separated fragments from the gel to a membrane filter and then immerses the filter in a solution containing a labeled probe, which can bind to the fragment that interests her. Afterward, she applies the same method to DNA samples from several other people.

When the researcher compares the banding patterns, she finds small variations among them. Why? Although the DNA molecules of any two people are alike in some regions, they differ significantly in others. *The slight molecular differences alter the sizes of some fragments.* As a result, some fragments from corresponding regions of DNA from different individuals differ in length. The differences in DNA electrophoresis patterns among different people are called RFLPs (*riff*-lips). The term stands for **restriction fragment length polymorphism**.

As you know, the set of fingerprints from each human is unique, a marker of his or her identity. Each person also has a **DNA fingerprint**—a unique array of RFLPs, inherited from parents in a Mendelian pattern.

RFLP analysis is one of several important procedures for basic research. It also has major applications in society at large. For instance, evolutionary biologists are analyzing RFLPs to decipher DNA from mummies, from fossilized animals and plants, and from mammoths and humans that were preserved many thousands of years ago in glacial ice. Investigators even used RFLP analysis to confirm that the bones exhumed from a shallow pit in Siberia belonged to five members of the Russian imperial family, all shot to death in secrecy in 1918.

A somewhat different sort of DNA fingerprint consists of characteristic patterns of repeated nucleotides. Short "repeats" are called *microsatellites*; longer ones are *minisatellites*. Such fingerprints often are generated by PCR (Section 21.2).

Analysis of microsatellite and minisatellite patterns and of RFLPs has spectacular potential for medicine. Researchers have already pinpointed unique nucleotide repeats or restriction sites in several mutated genes, including those responsible for sickle-cell anemia, cystic fibrosis, Huntington disease, fragile X syndrome, and other conditions.

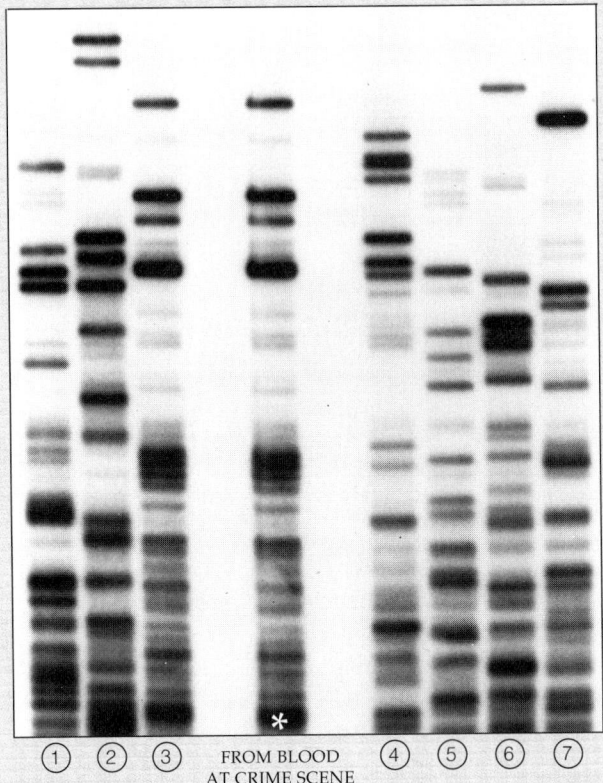

① ② ③   FROM BLOOD   ④ ⑤ ⑥ ⑦
AT CRIME SCENE

Figure 21.6 DNA fingerprint from a crime-scene bloodstain, with DNA fingerprints from blood of seven suspects (circled numbers). Which one of the seven was a match?

DNA fingerprinting has become a factor in various social issues. Think of all the paternity and maternity cases involving celebrities and ordinary folks alike. They can now be resolved by comparing the child's DNA fingerprints with those of the disputed parent. Or think about forensic medicine. A few drops of blood, semen, or even cells from a hair follicle at a crime scene or on a suspect's clothing often yield enough DNA to identify the perpetrator (Figure 21.6).

Britain, which started the first DNA database in the world, is at the forefront of solving crimes with DNA fingerprinting. British police can lawfully collect samples of hair or saliva from crime suspects for analysis. In one infamous case, detectives in Cardiff, Wales, collected blood samples from roughly 5,000 "donors" searching for the man who raped and killed a teenager. Mass DNA screening has already helped solve other murders in Britain and Germany.

MODIFIED HOST CELLS

Blue-White Screening

When you mix DNA with living cells, how can you find out which ones take up the DNA and contain a gene of interest? One common way to do this is to use a method called **blue-white screening**.

In blue-white screening, the test organisms are modified bacteria that normally carry a gene for the enzyme beta-galactosidase. In the experimental bacteria, however, part of the enzyme-coding gene sequence is missing (due to a mutation), so when the gene sequence is expressed, the result is only a non-functioning fragment of the enzyme. Next, the bacteria are mixed with two versions of a plasmid. Both versions carry DNA that codes for the missing beta-galactosidase fragment—the fragment the bacteria cannot make. However, only *some* of the plasmids are recombinant and so carry the gene of interest.

When a bacterial cell incorporates a nonrecombinant plasmid and the two "partner" beta-galactosidase gene fragments are expressed, the two resulting protein fragments associate, forming a functional beta-galactosidase molecule. With the functional enzyme, the bacterial colonies turn bright blue. However, something different happens with bacteria that incorporate the recombinant plasmids. *Those* plasmids carry the gene of interest, and the presence of this DNA alters expression of one of the enzyme-coding gene sequences. The resulting protein fragments can't associate and form a functional beta-galactosidase molecule. Hence, bacterial colonies that have taken up the recombinant plasmid remain white (Figure 21.7). To determine which bacteria have taken up the gene of interest, the experimenter needs only look at the color of colonies in a laboratory dish.

Use of cDNA

Researchers study genes to find out about the protein products and how they are put to use. However, even if a host bacterial cell takes up a gene, the cell may not be able to manufacture the protein. For example, recall that new mRNA transcripts of human genes contain noncoding sequences (introns). They cannot be translated until the introns are snipped out and coding regions (exons) are spliced together into mature form. Bacterial enzymes don't recognize the splice signals, so bacterial host cells cannot always properly translate human DNA.

Sometimes researchers get around the problem by using **cDNA**, which has been "copied" from a *mature* mRNA transcript for the gene. By a backwards process called **reverse transcription**, a complementary DNA strand is assembled on mRNA. The outcome is a hybrid mRNA-cDNA molecule (Figure 21.8). Enzymes remove the RNA, and then synthesize a complementary DNA strand, thus making double-stranded DNA.

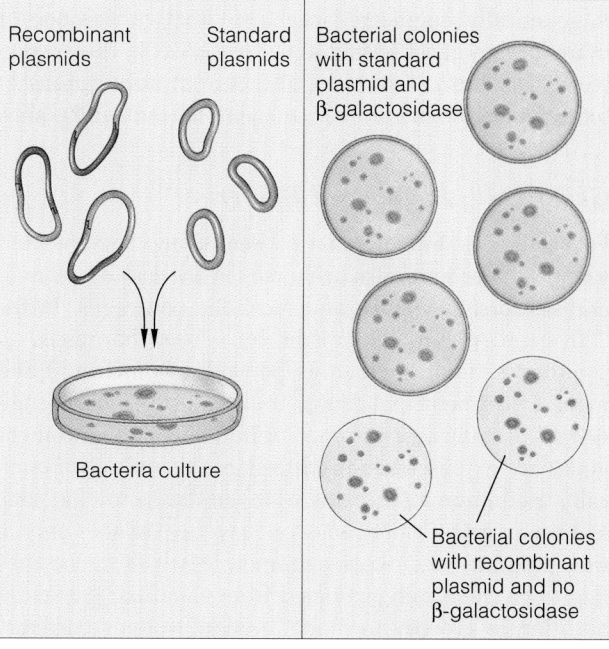

Figure 21.7 Blue-white screening—a convenient way to identify colonies of bacterial cells with recombinant plasmids.

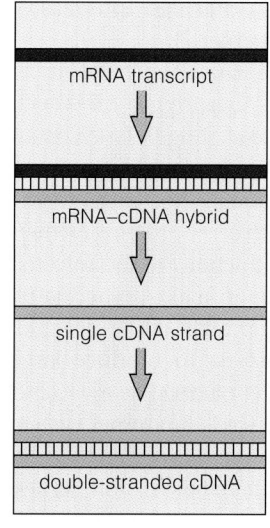

a An mRNA transcript of a desired gene is used as a template for assembling a DNA strand. An enzyme (reverse transcriptase) does the assembling.

b An mRNA-DNA hybrid molecule results.

c Enzyme action removes the mRNA and assembles a second strand of DNA on the remaining DNA strand.

d The result is double-stranded cDNA, "copied" from an mRNA template.

Figure 21.8 Formation of cDNA from an mRNA transcript.

Double-stranded cDNA can be further modified by attaching signals for transcription and translation. The modified cDNA can then be inserted into a plasmid for expression of the gene in bacterial cells.

Host cells that take up modified genes may be identified by procedures such as blue-white screening.

Modified human genes may not be expressed in host bacterial cells unless first introduced into the cells as cDNA.

A genetically engineered organism that carries one or more foreign genes is said to be *transgenic*. In this section we take a look at some of the current applications of biotechnology that involve transgenic bacteria and plants.

Genetically Engineered Bacteria

For centuries humans have been using bacteria and yeasts as biochemical workhorses in the manufacture of cheeses, bread, yogurt, and alcoholic beverages. In the 20th century, microorganisms have been harnessed to produce antibiotics (such as penicillin from fungi) and many other medicinal drugs. Today, plasmids in bioengineered bacteria carry a range of human genes, and those genes are expressed to produce large quantities of clinically useful human proteins. Many of these proteins, such as human growth hormone, once were available only in tiny amounts and were extremely costly because they had to be chemically extracted from glandular tissues. In addition to growth hormone, human proteins currently produced by bacteria include insulin and interferons (Table 21.1). Insulin, once available only from the pancreas glands of slaughtered pigs and cattle, helps regulate glucose metabolism (Section 12.6), and it is essential to the survival of people with type 1 diabetes. Interferons, recall, have roles in immunity (Chapter 7).

Bacteria may also be used as miniature factories to manufacture vaccines, substances that trigger the development of antibodies to a particular antigen (such as an influenza virus) and so provide immunity to that antigen (Section 7.10). Vaccines have traditionally been made using weakened or killed microbes, which carry proteins on their surface that trigger antibody production by the infected individual. However, it is possible to make a vaccine by inserting the gene for the antigen protein into a plasmid, then growing the transgenic bacteria to produce large quantities of the protein. The protein is used as a vaccine, which can then mobilize a person's immune system against that particular antigen. So far, bioengineered vaccines have been developed for hepatitis B and rabies. Researchers in various laboratories are working to genetically engineer an AIDS vaccine.

Genetically engineered bacteria may also become major weapons in efforts to clean up environmental pollution, a process called *bioremediation*. This chapter's *Focus on Our Environment* describes some examples of this "environmental biotechnology." It involves strategies for exploiting genes that permit some types of bacteria to metabolically break down toxic substances to harmless compounds.

SAFETY CONCERNS Some bacteria used in genetic engineering experiments are harmless to begin with, and in many cases they are also modified (by mutation) so that

Table 21.1 Some Cloned Human Gene Products	
Protein	Condition Treated
Insulin	Diabetes
Growth hormone	Pituitary dwarfism
Erythropoietin	Anemia
Factor VIII	Hemophilia A
Factor IX	Hemophilia B
Interleukin-2	Cancer
Tumor necrosis factor	Cancer
Interferons	Some cancers, viral infections
Monoclonal antibodies	Infectious diseases
Atrial natriuretic factor	High blood pressure
Tissue plasminogen factor (tPA)	Heart attack, stroke

they cannot survive outside the laboratory. Even so, there is legitimate concern about possible risks of introducing genetically engineered bacteria into humans or the environment (see *Choices*, Section 21.9). Government regulations now govern the release of genetically engineered organisms into the environment, and those regulations are periodically reevaluated in the light of new research.

Genetic Modification of Plants

Within the last few years, hundreds of field tests of genetically engineered plants have been initiated in some 20 countries. Some tests are geared toward creating improved varieties of crop plants. Others are exploiting plants that, like bacteria, can serve as factories for producing medically and commercially useful proteins.

For example, researchers can insert DNA fragments into a Ti plasmid, which is extracted from a bacterium (*Agrobacterium tumefaciens*) that infects many flowering plants. Some genes on the plasmid normally cause tumors to form. When the Ti plasmid is used to ferry genes, however, the tumor-inducing genes are first removed and desired ones substituted. Inserted into plant cells, the foreign genes are integrated into the plant's chromosomes and are expressed in the plant tissues.

Similar methods have already produced some practical successes, including cotton plants genetically engineered to resist worm attacks (Figure 21.9)—a potential alternative to heavy applications of insecticides. Efforts are also under way to engineer corn and other food plants so that they will be more tolerant to drought and pesticides or have an improved balance of essential amino

a

b

c

d

Figure 21.9 Examples of genetically engineered plants having commercial importance. (**a**) Some worms attack buds of cotton plants but not the modified plant shown in (**b**). A normal potato plant susceptible to viral attack is shown in (**c**); a genetically modified strain is shown in (**d**).

acids. Studies show that gene modifications can also stimulate plants to produce larger than normal quantities of commercially or nutritionally valuable starches, oils, and other substances.

For some species of plants that cannot be grown easily from seeds, methods of *tissue culture* can be used to multiply the number of plants for field transplanting. Basically, somatic cells of the parent plant are removed and placed on a nutrient medium that promotes tissue growth; small lumps of the tissue are then divided and grown in a new chemical solution that promotes development of a fully differentiated plant. The strawberries you buy in the grocery store are often the fruit of plants propagated through tissue culture.

Recently, researchers have begun to insert human genes for certain proteins into plants. One such protein, serum albumin, is widely used in preparations given to burn patients and other ill people to replace lost body fluids. Researchers have already extracted useful amounts of the protein from genetically altered potato and tobacco plants.

Genetic engineering of selected microorganisms and plants is yielding a range of useful products, including medically valuable proteins and improved varieties of crop plants.

Focus on Our Environment

21.6

BACTERIA THAT CLEAN UP POLLUTANTS

As a group, bacteria have a rather astonishing quality. They are so diverse metabolically that they can gain energy for life processes from breaking down substances that are highly toxic, even carcinogenic. In the process, many noxious substances—including crude oil, toxic heavy metals such as mercury, and pesticides—are taken up by the bacterium or chemically altered to a safer form.

"Oil-eating" bacteria, for example, oxidize the hydrocarbons in petroleum to carbon dioxide (Figure 21.10). Mercury-resistant bacteria metabolically process molecules of the metal (which causes severe neurological damage in humans) into a nontoxic compound.

Often, the genes that code for these metabolic feats are carried on plasmids—and so are ready candidates for bioengineering. A number of "de-tox" genes have been isolated, cloned, and transferred to other bacteria. Recently, for example, plasmids with genes that code for enzymes that break down hydrocarbons were successfully transferred to marine bacteria. The bacteria have since been used to help clean up oil spills off Texas and elsewhere.

Recombinant plasmids also have been put together using genes obtained from bacteria that normally break down various contaminants in toxic waste dumps. The goal is to engineer microbes that can be used to treat industrial wastes and possibly help clean up polluted soils. The U.S. Environmental Protection Agency has identified more than 2,000 seriously contaminated toxic-dump sites in the United States, many of them endangering groundwater supplies. Biotechnology may be an invaluable tool for restoring such places to environmental health.

Figure 21.10 Bioremediation, the removal of oil by natural biodegradation, is accelerated by the application of fertilizers that provide oil-eating bacteria with nitrogen and phosphorus nutrients to augment the oil's natural carbon content. Here a cleanup worker sprays a shoreline with a liquid fertilizer (Inipol™).

Super Mice and Biotech Barnyards

Animal cells do not accept plasmids, but they can be "micro-injected" with foreign DNA. In 1982, researchers introduced the gene for growth hormone from rats into fertilized mouse eggs. The mice grew much larger than their normal littermates and had dramatically higher blood concentrations of the hormone. Why? The rat gene had become integrated into the mouse DNA and was being expressed. When the gene for human growth hormone was successfully introduced and expressed in mice, the result was the "super rodent" shown in Figure 21.11a.

Similar experiments with large domesticated animals have begun to show some success. As the chapter opening vignette noted, transgenic sheep, pigs, and cows may become sources of important pharmaceutical drugs. The goat in Figure 21.11b developed from a fertilized egg that was injected with recombinant DNA consisting of goat gene sequences spliced to human genes for tissue plasminogen activator, or tPA. This protein dissolves blood clots in heart attack and stroke patients, and the goat's milk contains it. Although tPA and other human proteins are currently made by recombinant bacteria, so-called barnyard biotechnology is exciting because in some cases transgenic livestock have the potential to produce large quantities of drugs faster and more cheaply than bacteria cultured in huge industrial vats.

People who have hemophilia A now can obtain the needed blood-clotting factor VIII from a product called Recombinate. This drug is produced by Chinese hamster ovary cells in which genes for human factor VIII have been inserted. Factor VIII produced in this way eliminates the need to obtain it from human blood—and so eliminates the risk of transmitting blood-borne pathogens such as HIV.

In 1996 researchers made a genetic duplicate—a clone—of an adult ewe. They inserted the nucleus from one of her mammary gland cells into an egg (from another ewe) from which the nucleus had been removed. Eventually, a cluster of embryonic cells developed and was implanted into a surrogate mother. The resulting lamb, named Dolly, today is a thriving adult ewe herself. As experiments with transgenic animals continue around the globe, they may well be laying the groundwork for commercial development of animals with custom-made genetic characteristics—including, for instance, tissues and organs for use in treating human diseases.

Applying the New Technology to Humans

By now, you probably know that scientists in hundreds of laboratories around the world are collaborating on a massive Human Genome Project. Already, they have created a physical map of all human chromosomes, identifying the locations of specific genes. Another, more daunting goal is to determine the sequence of nucleotides in the entire complement of human DNA.

Researchers use artificial chromosomes to clone copies of human genes for analysis. The two smallest human

a

Figure 21.11 (**a**) Ten-week-old mouse littermates, the one on the left weighing 29 grams, and the one on the right, 44 grams. The larger mouse grew from a fertilized egg into which the gene for human somatotropin (growth hormone) had been inserted. (**b**) This transgenic goat carries human genes that code for the production of tissue plasminogen activator (tPA), an enzyme that helps break up blood clots.

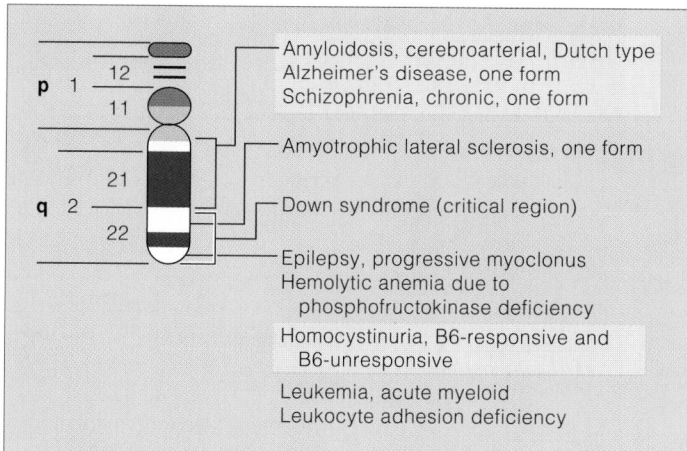

Figure 21.12 Map of human chromosome 21 showing genes correlated with specific diseases. In all, researchers have mapped 13 genes to this chromosome. The upper arm of the chromosome is marked p and the lower arm, q. In this case, each arm consists of a single region, labeled 1 and 2, respectively. In other chromosomes, each arm may have two or more regions. Stains used in chromosome analysis produce a series of bands within each region. The small numbers just to the left of each arm indicate specific bands. A combination of letters and numbers indicates the chromosome region where a given gene is found; for instance, the gene for one form of amyotrophic lateral sclerosis (ALS) is 21q2.22.

chromosomes, the Y chromosome and chromosome 21 (Figure 21.12), were extensively mapped in the early 1990s, and since then tens of thousands of markers have been mapped to other chromosomes. For just a few examples, we know that chromosome 21 carries genes for early-onset Alzheimer's disease, some forms of epilepsy, and one type of ALS (amyotrophic lateral sclerosis), a disease that progressively destroys motor neurons.

Automated DNA sequencers can sequence thousands of DNA nucleotides a day. Even with an estimated 3.2 billion nucleotides in the 23 pairs of human chromosomes, we may have a detailed picture of the human genome sequence within the next few years. Today, anyone with access to the World Wide Web can obtain information about human gene sequences from several huge databases. One of the most useful databases, called EST (for *e*xpressed *s*equence *t*ags), includes sequences of human cDNAs derived from various tissues.

Mapping and sequencing the human genome will undoubtedly increase our understanding of the genetic basis of human traits and diseases. However, it also has raised thorny issues. For example, some people criticize laboratories that attempt to patent the information they uncover—presumably for future commercial gain. On the other hand, various laboratories are making their data freely available to all.

Gene Therapy

Generally speaking, **gene therapy** is the insertion of one or more normal genes into a person's body cells in order to correct a genetic defect. Numerous experiments in human gene therapy are now under way, involving desperately ill victims of skin cancer, cystic fibrosis, and other disorders. We touched on this topic in Chapter 18 in the discussions of various genetic disorders. The next section describes some of the efforts in more detail. Briefly, however, let's examine some other experimental approaches to altering gene activity that are emerging from biotechnological research.

Altering Gene Expression

Several approaches aim to develop drugs that act on DNA or RNA in specific ways. For example, genetic engineering methods are being used to create drugs that can block the synthesis of undesirable proteins by shutting down transcription of DNA to messenger RNA (Section 19.5). One target is the virus that causes herpes infection: Scientists are working to develop a drug that can enter an infected cell and shut down transcription of viral DNA without harming expression of the cell's own genes. Research on such *transcription factors* could also lead to drugs that can turn off oncogenes responsible for various cancers (Section 20.2) or that shut down mutated alleles for various diseases.

Another approach is to block protein synthesis by way of molecules that can prevent mRNA from being translated in the cell cytoplasm. Engineered "antisense" RNA molecules are one option. These are short stretches of nucleic acid that can bind to specific mRNAs and prevent translation. Alternatively, **ribozymes** are bits of RNA that act like enzymes and, when inserted into a cell, cut up specific mRNA sequences before they can be translated. Ribozymes will probably first be used in antiviral drugs.

Applications of genetic engineering include efforts to develop transgenic animals or animal cells capable of producing medically useful substances.

The Human Genome Project is mapping the locations of specific genes on all human chromosomes and determining the nucleotide sequence of each gene. The project holds great promise for scientific and medical research.

Approaches to gene therapy include insertion of one or more normal genes into a person's body cells to correct a genetic defect, and methods geared toward altering the expression of specific genes.

METHODS AND PROSPECTS FOR GENE THERAPY

As you know from previous chapters, diseases such as cystic fibrosis and hemophilia result when a single gene is mutated, producing a nonfunctional protein. Many cancers develop when mutations upset normal gene regulation, and several genes in a cell turn on or off at the wrong time. Gene therapy attempts to replace mutated genes with normal ones, which will encode functional proteins, or to insert genes that restore normal controls over gene activity. An inserted gene might also code for a protein that blocks a cell surface receptor associated with a disorder. For example, it might be possible to treat arthritis by inserting genes coding for proteins that block the binding of substances that cause inflammation. Scientists are actively exploring these and other possibilities (Table 21.2) in clinical trials in the U.S. and elsewhere.

Strategies for Transferring Genes

The size of a gene (the number of its base pairs) helps determine what sort of mechanism can be used to insert it into a host cell. Animal cells do not take up plasmids, so plasmids cannot be directly used as vectors in gene therapy. Smaller genes can be carried into animal cells by viruses; larger ones must enter a host cell some other way.

In *transfection*, cells cultured in the laboratory are directly exposed to DNA segments that contain a gene of interest. Typically, some of the foreign DNA becomes integrated into the host cell's genome. Exposing the host cells to a weak electric current, called *electroporation*, seems to help. Even so, this strategy is inefficient: Sometimes only one cell in 10 million incorporates a new gene.

A small gene can be inserted into a virus, which can then "infect" a host cell and carry the desirable gene with it. Several gene therapy trials are using retroviruses in this way because such RNA viruses easily infect animal cells. Once inside the host cell, the viral RNA is transcribed into DNA (by way of reverse transcriptase), and usually the foreign DNA is integrated into the host cell's DNA. The first step in gene therapy is to remove certain segments of genetic instructions from the virus, so that it cannot replicate and cause disease. In their place, a researcher substitutes the gene to be transferred. Next the virus is permitted to infect target cells. Retroviruses are efficient at integrating into the host DNA, increasing the odds that the new gene will become active. Some gene therapy experiments use other viruses that typically infect the kinds of host cells being manipulated.

There are pitfalls involved with using viruses to introduce genes. Suppose that a virus is engineered to infect cells with a "good" copy of a mutated gene. When such a virus integrates into a host cell's DNA, there is no guarantee that the new gene will be in its normal position—

Table 21.2 Examples of Gene Therapy Trials*

Disease	Inserted Gene Codes for:
ADA deficiency	Adenosine deaminase
Cystic fibrosis	CFTR
Malignant melanoma	Interleukin-2
Hemophilia B	Factor VIII
Advanced cancers	Tumor necrosis factor
Kidney cancer	Interleukin-2
Lung cancer	p53/antisense *ras* RNA
Ovarian cancer, brain cancer	Thymidine kinase (enzyme that promotes attack by the cell-destroying drug gancyclovir)
AIDS	Thymidine kinase

*List compiled as of 1993

and if it is not, the gene may not be transcribed into protein. Sometimes changes to an altered virus render any genes they carry ineffective.

At present, only a few types of human cells can successfully be removed, induced to take up engineered genes, grown in the laboratory, and then reinserted into the body. Thus, this method can be used only with genes that are expressed in those tissues. There are no reliable procedures for making large quantities of certain vectors that are commonly used to carry new DNA into host cells. In addition, many introduced genes turn off within a few days or weeks. In spite of these obstacles, researchers are conducting important clinical trials of various gene therapy methods.

Examples of Gene Therapy in Action

Chapter 18 described gene therapy trials for patients with the inherited disease cystic fibrosis. Gene therapy researchers also are targeting numerous other disorders. We mention just a few of them here to give you an idea of the kinds of efforts that may improve the lives of millions of people in the not too distant future.

ADA In 1990 Ashanthi De Silva (Figure 21.13) became the subject of the first federally approved gene therapy test for humans. Born with a defective gene that codes for the enzyme adenosine deaminase (ADA), Ashanthi suffers from severe combined immune deficiency (SCID). This set of disorders is brought on by a drastic reduction or complete absence of T and B lymphocytes.

Figure 21.13 Ashanthi De Silva when she was a patient in the first approved gene therapy trial. Now in her teens, Ashanthi was a candidate for gene therapy because she suffers from severe combined immune deficiency (SCID) and lacks a functional immune system. At this writing, she is doing well.

Using recombinant DNA methods, researchers introduced copies of the ADA gene into some of the child's T cells, which were then stimulated to divide. Next, about a billion copies of the genetically engineered cells were delivered through a plastic tube into Ashanthi's bloodstream. After a series of monthly infusions, Ashanthi now receives one treatment a year. Tests show that about half of her circulating T cells carry the normal gene. However, in some other patients only about 1 percent of circulating T cells are normal.

Another research team has extracted lymphocyte stem cells (called *lymphoblasts*) from umbilical cord blood of three newborns with ADA deficiency. They then used a recombinant retrovirus as a vector to introduce functional ADA genes into the stem cells. The altered cells were cultured with substances that stimulated growth and division, then reinserted into the newborn patients. So far, only about 10 percent of the children's T cells carry the normal gene. All ADA patients, including Ashanthi De Silva, receive regular doses of a synthetic form of ADA, as "insurance" in case the gene therapy ultimately fails.

RHEUMATOID ARTHRITIS In this crippling autoimmune disease, an excess of the communication molecules called cytokines—especially interleukin-1 (Section 7.10)—accumulate in synovial joints such as the knee. There, they trigger a raging inflammatory response that ultimately destroys the joint's cushioning cartilage. Cells in the membrane that lines synovial joints normally produce a protein called IRAP, which blocks the uptake of IL-1 and so halts inflammation. In rheumatoid arthritis, the cells may make too little IRAP. Working with several patients, a research team is removing synovial membrane cells,

inserting extra copies of the IRAP gene into them (using a retrovirus), then reinserting the altered cells. The experiment is a step toward a therapy in which patients receive transformed cells that take up residence in affected joints and manufacture enough IRAP to slow or even stop progression of the disease.

CANCERS Gene therapy may become a potent anticancer weapon. Early trials have targeted the deadly skin cancer malignant melanoma, leukemia, a fast-growing form of lung cancer, and cancers of the brain, ovaries, and other organs. In some approaches, tumor cells are first removed from a patient and grown in the laboratory. Genes for an interleukin (which helps activate white blood cells of the immune system) are then introduced into the cells, and the cells are returned to the body. In theory, interleukins produced by the tumor cells may act as "suicide tags" that stimulate T cells of the immune system to recognize cancerous cells and attack them. An exciting variation on this theme involves structures called "lipoplexes," laboratory-made packets in which a plasmid is encased in a lipid coat that facilitates its entry into a cell. The plasmid carries a gene encoding a protein marker that can trigger an immune system attack on cancer cells. Thus far, several dozen melanoma patients have been treated with lipoplex therapy, with encouraging results.

Other trials are adding the gene for the powerful anti-tumor agent called tumor necrosis factor (TNF) to tumor-infiltrating lymphocytes, or TILs. TNF helps kill tumors by shutting off their blood supply, but it also can have the same effect on healthy tissues. As their name suggests, TILs home in specifically on tumors. One goal of such experiments is to determine whether TILs carrying the TNF gene can deliver enough TNF to kill a targeted tumor without harming surrounding tissues.

Gene Therapy and Your Future

Gene therapy is currently highly experimental, extremely costly, and available to only a few. Authorities agree that the key to its widespread use is the development of vectors, such as specially engineered viruses and lipoplexes, that can deliver genes to target cells safely and reliably. Without a doubt, as methods of gene therapy are refined, its application will become more widespread, and gene therapy will play a major role in future efforts to treat and cure disease.

In gene therapy, one or more normal genes are inserted into body cells to correct a genetic defect or enhance the activity of specific genes.

21.9 ISSUES FOR A BIOTECHNOLOGICAL SOCIETY

Some people say that no matter what the species of organism, DNA should never be altered. But as we noted earlier, nature alters DNA all the time. The real issue is how to bring about beneficial changes without harming ourselves or the environment.

In the 1980s activists entered a test plot and ripped out strawberry plants that had been sprayed with a bacterium (Figure 21.14) that had been genetically altered to help prevent ice crystals from forming on the plants. Although previous studies had shown that the "ice-minus" bacteria posed no threat to the environment or to human health, many people have voiced similar concerns about releasing recombinant organisms into the environment. Researchers themselves have moved into the vanguard of efforts to develop regulations and guidelines that protect the public interest while permitting beneficial advances. Current concerns include the following:

• Transgenic bacteria or viruses could mutate, possibly becoming new pathogens in the process. However, as described in Section 21.6, most recombinant microbes are altered in ways that will prevent them from reproducing outside of the laboratory.

• Bioengineered plants could escape from test plots and pose an ecological risk by becoming "superweeds" resistant to herbicides and other control measures.

• Crop plants with added insect resistance could set in motion an evolutionary seesaw in which new, even more formidable insect pests arise.

• Transgenic fish that feed voraciously and grow rapidly could displace natural species, wrecking the delicate ecological balance in streams and lakes.

• Major ethical problems could arise if people undertake "eugenic engineering"—using biotechnology to insert genes into normal people (or sperm or eggs) simply for the purpose of creating people that are "more desirable."

• Genetic screening could provide a means of discrimination by insurance companies and others against people who carry a gene predisposing them to a disorder, even if such a person shows no sign of ill health.

Experts point out that some of these examples—mutation of transgenic bacteria or viruses into pathogenic strains or escape of bioengineered plants as "superweeds"—are not

Figure 21.14 Spraying an experimental strawberry patch in California with "ice-minus" bacteria. (Government regulations required that the sprayer use elaborate protective gear.)

very likely. However, others, such as discrimination on the basis of genetic screening, are issues we already face.

These examples only touch on some of the profound social and ethical issues raised by recombinant DNA technology and genetic engineering. With new advances coming every day, it may be time for everyone to think about the benefits *and* risks of living in a biotechnological society.

SUMMARY

1. Genetic "experiments" have been occurring in nature for billions of years. Mutation, crossing over and recombination at meiosis, and other natural events have all contributed to the current diversity that we see among organisms.

2. Humans have been manipulating the genetic character of different species for thousands of years. The emergence of recombinant DNA technology in the past few decades has enormously expanded our capacity to control genetic change. Recombinant DNA technology is founded on procedures by which DNA molecules can be cut into fragments, inserted into plasmids or some other cloning tool, then multiplied in a population of rapidly dividing cells.

3. A DNA clone is a foreign DNA sequence that has been introduced and amplified in dividing cells. DNA sequences also can be amplified in test tubes by the polymerase chain reaction.

4. Recombinant DNA technology and genetic engineering have enormous potential for research and applications in medicine, agriculture, and industry. As with any new technology, potential benefits must be weighed against potential risks, including ecological and social disruptions.

5. Although the new technology has not developed to the extent that human genes can readily be modified, the social, legal, ecological, and ethical questions it raises should be explored in detail before such an application is possible.

Review Questions

1. What is a plasmid? What is a restriction enzyme? Do such enzymes occur naturally in organisms? 440–441

2. Recombinant DNA technology involves producing DNA restriction fragments, amplifying the DNA, and identifying modified host cells. Briefly describe one of the methods used in each of these categories. 441–444, 448

Critical Thinking: You Decide (Key in Appendix V)

1. Having read about examples of genetic engineering in this chapter, can you think of some additional potential benefits of this technology? Can you envision other potential problems?

2. A biotechnology company in New Jersey is attempting to genetically engineer pigs so that their meat, pork, is lower in fat and cholesterol. Their strategy is to make pigs grow faster and produce more muscle and less fat. In a general way, describe the procedures their work might require, including the types of genes they might be trying to alter.

Self-Quiz (Answers in Appendix IV)

1. Gene mutations, crossing over and recombination during meiosis, and other natural events are the basis of the _____ observed in present-day organisms.

2. Causing genetic change by deliberately manipulating DNA is known as _____.

3. _____ are small circles of bacterial DNA that are separate from the bacterial nucleoid.

4. Genetic researchers use plasmids as _____.

5. Rejoined cut DNA fragments from different organisms are best known as _____.
   a. cloning genes
   b. mapping genes
   c. recombinant DNA
   d. conjugating DNA

6. Repeated DNA replications and cell divisions of the host cells yield _____.

7. Using the metabolic machinery of a bacterial cell to produce multiple copies of genes carried on recombinant plasmids is _____.
   a. DNA fingerprinting
   b. bacterial conjugation
   c. mapping a genome
   d. DNA amplification

8. A collection of DNA fragments produced by restriction enzymes, incorporated into plasmids, and introduced into bacterial host cells is a _____.
   a. DNA clone   c. hybridized sequence
   b. DNA library   d. gene map

9. The polymerase chain reaction _____.
   a. is a natural reaction in bacterial DNA
   b. cuts DNA into fragments
   c. amplifies DNA sequences in test tubes
   d. inserts foreign DNA into bacterial DNA

10. Recombination is made possible by _____ that can make cuts in DNA molecules.

Selected Key Terms

cDNA 445	plasmid 440
cloned DNA 442	polymerase chain reaction (PCR) 442
DNA fingerprint 444	recombinant DNA technology 440
DNA library 441	restriction fragment 441
DNA ligase 441	reverse transcription 445
DNA polymerase 442	RFLP 444
gene therapy 449	ribozyme 449

Readings

Aldridge, S. 1997. *The Thread of Life: The Story of Genes and Genetic Engineering.* New York: Cambridge University Press.

Cohen, J. S., and M. E. Hogan. December 1994. "The New Genetic Medicines." *Scientific American.*

Mirsky, S., and J. Rennie. June 1997. "What Cloning Means for Gene Therapy." *Scientific American.*

 For additional readings, go to InfoTrac College Edition, your online library at:

http://www.infotrac-college.com/wadsworth

A smoke cloud over South America's Amazon River basin photographed by astronauts in the space shuttle Discovery *in 1988. The cloud was caused by the clearing and burning of tropical forests, pasture, and croplands, activities that continue today at a rapid pace. If the cloud were placed over North America, it would cover more than one-third of the area of the 48 contiguous United States. Human impacts on Earth's ecosystems are a central topic in Unit IV.*

PRINCIPLES OF EVOLUTION

Floods and Fossils

Nearly 500 years ago, Michelangelo was finishing his painting of the Great Deluge on the ceiling of the Sistine Chapel in Rome (Figure 22.1). Another artist, Leonardo da Vinci, was brooding about seashells that lay embedded in rock layers in mountains of northern Italy. The traditional view was that the shells were strewn about during the Deluge. Yet Leonardo wondered how even a great flood could wash them inland and up the mountainsides, which were hundreds of kilometers from the sea. Leonardo also perceived that the rock layers were arranged in strata, one above the other like cake layers. His journals tell us that he suspected the distinct groups of shells in those layers might be the fossilized remains of ancient marine communities, buried not in a great flood, but gradually over vast stretches of time. By the early 1700s fossil shells, bones, plant parts, and impressions of footprints, animal tracks, and burrows were being accepted as evidence of life in the past.

By the mid-18th century extensive canal building, quarrying, and mining operations were revealing regular patterns in the color and composition of stratified

Figure 22.1 (**left**) From the Sistine Chapel, Michelangelo's painting of the onset of a catastrophic flood, traditionally called the Great Deluge. (**right**) A modern-day photographer captures the intricate structural pattern of fossilized ammonite shells. About 65 million years ago the last ammonites perished, along with many other groups of organisms. That mass extinction is but one piece of the evolutionary puzzle.

rock layers. Soon it became apparent that certain fossils occur in distinctive types of layers, sometimes over considerable distances. For example, cliffs on the English side of the English Channel reveal the same fossil sequences as cliffs on the French side.

Some scientists used the correlation between fossils and strata as a geologic mapping tool. Others began using them to analyze the possible connections between past geologic events and the history of life. As you will see in this chapter, fossils and other sources of information afford convincing evidence of change in populations of organisms through time—in other words, evolution.

KEY CONCEPTS

1. Biological evolution is heritable change in lines of descent (lineages).

2. Members of a population generally have the same number and kinds of genes, which give rise to the same assortment of traits.

3. In a population, each gene may exist in two or more slightly different molecular forms (alleles). Different individuals don't necessarily inherit the same alleles, so they may differ in the details of their traits. Differences in survival and reproduction among individuals that differ in one or more traits are the basis of natural selection.

4. An allele may become more or less common in the population, or it may disappear. Such changes in a population's allele frequencies over time are called *microevolution*.

5. Natural selection is simply the result of a difference in survival and reproduction among individuals who differ in one or more traits.

6. A species consists of one or more populations of individuals that can interbreed under natural conditions and produce fertile offspring, and that are reproductively isolated from other such populations.

7. New species evolve from variant individuals of existing species. Hence all species that have ever lived on Earth are related. *Macroevolution* refers to the patterns, trends, and rates of change among lineages over geologic time.

8. Evidence of macroevolution comes from fossils, the geologic record, and radioactive dating of rocks. Comparing the anatomy and biochemistry of different lineages helps us identify patterns of change through time.

9. Life apparently originated on Earth more than 3.8 billion years ago. Its origin and subsequent evolution have been linked to the physical and chemical evolution of the universe, the stars, and our solar system.

CHAPTER AT A GLANCE

22.1 A Little Evolutionary History

22.2 Some Basic Principles of Evolution

22.3 Processes of Microevolution

22.4 An Introduction to Macroevolution

22.5 Comparisons of Body Form

22.6 Comparing Biochemistry

22.7 Evolutionary Trees and Their Branchings

22.8 *Genes of Endangered Species*

22.9 Evolution and Earth History

22.10 The Origin of Life

A LITTLE EVOLUTIONARY HISTORY

In 1831, confusion reigned in scholarly circles with respect to the origin of the Earth's amazing diversity of life forms. A widely accepted view held that all species had come into existence at the same time in the distant past, at the same center of creation, and had not changed since. Yet, by the mid-19th century, several hundred years of exploration and advances in the sciences of comparative anatomy and geology had raised questions that were difficult to ignore. For example, why were some species found in particular isolated regions, and nowhere else? Why, on the other hand, were similar (but not identical) species found in geographically separate parts of the world? Why did species as different as humans, whales, and bats have certain strikingly similar anatomical structures? What was the significance of geologists' discoveries of similar fossil organisms in similar layers of the Earth's sedimentary rocks—no matter where the layers occurred?

Enter Charles Darwin, who in 1831 was 22 years old, had a freshly minted degree in theology from Cambridge, and was wondering what to do with his life. He was *not* interested in becoming a clergyman. Ever since he was a boy, all he had wanted to do was hunt, fish, collect shells, or simply watch insects and birds. John Henslow, a botanist who befriended Darwin during his university years, perceived the young man's real interests. He arranged for Darwin to function as ship's naturalist aboard the HMS *Beagle*, which was about to embark on a 5-year voyage around the world (Figure 22.2). The *Beagle* sailed first to South America, to complete work on mapping the coastline. During the Atlantic crossing, Darwin collected and examined marine life and studied geology. During stops along the coast and at various islands, he observed diverse species in environments ranging from sandy shores to high mountains. After returning to England in 1836, Darwin began talking with other naturalists about a topic that was on many scholars' minds—the growing evidence that life forms evolve, changing over time.

How could organisms evolve? One clue came from an essay by Thomas Malthus, a clergyman and economist, who proposed that any population tends to outgrow its resources, that its members must compete for what is available. Darwin's own observations suggested that any population has the capacity to produce more individuals than the environment can support. (For instance, a single starfish can release 2,500,000 eggs each year, but because of predation, starvation, and other environmental insults, the oceans are not filled with starfish.) Another clue came from observations by Darwin and others that even members of the same species vary slightly in their traits.

As you will read in the following sections, Darwin's thoughtful melding of his observations of the natural

a

b

Figure 22.2 (**a**) A replica of the HMS Beagle, shown sailing off the coast of South America. (**b**) Charles Darwin and a blue-footed booby, one of the species he encountered during his 5-year voyage around the world.

world with the ideas of other thinkers led him to propose that evolution could occur by way of a process called *natural selection*. Widespread acceptance of this theory would not come until nearly 70 years later, when a new field, genetics, provided insights on the source of variation in traits. Now let's consider some current views of the mechanisms of evolution.

Combining observations of the natural world with ideas about interactions of populations with their environment, Darwin forged his theory of evolution by natural selection.

SOME BASIC PRINCIPLES OF EVOLUTION

The history of life on Earth spans nearly 4 billion years. It is a story of how species originated, persisted or went extinct, and stayed put or radiated into new environments. The overall "plot" of the story is **evolution**, or change in lines of descent over time. **Microevolution** refers to cumulative genetic changes that may give rise to new species. **Macroevolution** applies to the large-scale patterns, trends, and rates of change among *groups* of species. In this chapter we will examine the basic principles of evolutionary change. Chapter 23 looks at what we currently know of the evolutionary pathway leading to modern humans.

Variation in Populations

Evolution occurs when there is change in the genetic makeup of populations of organisms. A **population** is a group of individuals of the same species occupying a given area. There is a great deal of genetic variation within and among populations of the same species.

Overall, the members of a population have similar traits (phenotypes). They have the same general form and appearance (*morphological* traits), their body structures function in the same way (*physiological* traits), and they respond the same way to certain basic stimuli (*behavioral* traits). However, the details of traits vary greatly from one individual to another. For instance, individuals vary in the color of their body hair, as well as in its texture, amount, and distribution over the body. As we know from previous chapters, this example only hints at the immense genetic variation in human populations. Populations of most other species of organisms show the same kind of variation, as Figure 22.3 indicates.

Figure 22.3 Variation in shell color and banding patterns among populations of a snail species found on islands of the Caribbean. Variation in traits arises from different combinations of alleles carried by different members of the populations.

Where Does Variation Come From?

The members of a population generally have inherited the same number and kinds of genes. But remember, a given gene may have slightly different molecular forms, called alleles. Different combinations of alleles are inherited, and this leads to variations in traits among individuals. Whether your hair is black, brown, red, or blond depends on *which* alleles of certain genes you inherited from your mother and father. As described in earlier chapters, five events contribute to that mix of alleles:

1. Gene mutation (produces new alleles)

2. Crossing over at meiosis (leads to new combinations of alleles in chromosomes)

3. Independent assortment at meiosis (leads to mixes of maternal and paternal chromosomes in gametes)

4. Fertilization (puts together combinations of alleles from two parents)

5. Changes in chromosome structure or number (leads to the loss, duplication, or alteration of alleles)

The abundance of each kind of allele in a population is called that allele's *frequency*. The term "microevolution" refers to small-scale changes in allele frequencies that are brought about by mutation, natural selection, and other processes we consider in the following sections.

A population is a group of individuals of the same species occupying a given area.

Although individuals in a population share many general phenotypes, in nearly all populations there is great underlying genetic variation.

Evolution is change in lines of descent over time. It results from changes in the genetic makeup of populations of organisms.

PROCESSES OF MICROEVOLUTION

Mutation: Sole Source of New Alleles

Recall that a mutation is a heritable change in DNA. In evolutionary terms, it is a significant event. *Mutations are the source of alleles that have been accumulating in different lineages for about 4 billion years.*

Mutations occur relatively rarely. Whether a mutation proves to be harmful, neutral, or beneficial depends on how its product interacts with other genes *and* with the environment. Because they are accidental changes in the DNA sequence, most mutations are probably harmful, altering traits in such a way that an individual cannot survive or reproduce as well as other individuals. For example, humans typically live in environments in which small cuts, bruises, or other minor injuries are common. Before effective medical treatments existed, hemophiliacs, whose blood does not clot properly, could die (and often did) at a young age from such minor injuries. Hence, the frequencies of the mutated alleles for various hemophilias historically have been low. Beneficial traits typically *improve* some aspect of an individual's functioning in the environment and so improve the chances of surviving and reproducing. Neutral traits—such as attached earlobes in humans—may be of no great help to an individual's survival, but under prevailing conditions they do not seriously damage it either.

Natural Selection

Natural selection probably accounts for more changes in allele frequencies than does any other microevolutionary process. As you have read, Darwin discovered this major process by correlating his understanding of inheritance with certain features of populations and the environment. In 1859 he published his ideas in a classic book, *On the Origin of Species.* Here we express the main points of Darwin's correlation in light of modern genetics:

1. Individuals of a population vary in form, function, and behavior.

2. Much of the variation in a population is *heritable.* This means that more than one kind of allele exists for genes that give rise to the traits.

3. Some forms of a trait are more "fit" than others, improving chances of surviving and reproducing.

4. **Natural selection** is the *difference in survival and reproduction* that has occurred among individuals that differ in one or more traits. (This differential in survival and reproduction is sometimes informally referred to as "survival of the fittest.")

5. A population is *evolving* when some forms of a trait are becoming more or less common relative to the other forms. The shifts are evidence of changes in the relative abundances of alleles for that trait.

6. Life's diversity is the outcome of changes in allele frequencies in different lines of descent over time. Such changes can result in the splitting of a single species into two related ones.

As natural selection proceeds over time, organisms come to have characteristics that suit them to conditions in a particular environment—a trend we call **adaptation**.

You have probably heard references to the "theory" of evolution. Yet decades of rigorous scientific observations have yielded so much evidence of evolution and natural selection that both are more properly considered fundamental principles of the living world. Later we will consider a few examples of this evidence.

Genetic Drift and Gene Flow

Besides the nonrandom changes in allele frequency caused by natural selection, allele frequencies in a population also can change randomly because of chance events alone. This is called **genetic drift** (Table 22.1). Often the change is most rapid in small populations. In one type of genetic drift, called the *founder effect,* a few individuals leave a population and establish a new one. Compared to the population left behind, the founders usually carry fewer alleles for certain traits. By chance, the allele frequencies in the new population differ from the old. The founder effect probably explains why there are virtually no Native Americans with type B blood. Genes for that blood group must not have been present in the few individuals who originally crossed the Bering Strait from Asia into what is now Alaska at least 10,000 years ago.

Allele frequencies also can change as individuals leave a population (emigration) or new individuals enter it (immigration). This physical movement of alleles, or **gene flow**, helps keep neighboring populations genetically similar. Over time it tends to counter the differences between populations that are brought about through mutation, genetic drift, and natural selection. In this age of international travel, the pace of gene flow among human populations has increased dramatically.

Table 22.1   Major Processes of Microevolution

Mutation	A heritable change in DNA
Genetic drift	Random fluctuation in allele frequencies over time, due to chance occurrences alone
Gene flow	Movement of alleles among populations through migration of individuals
Natural selection	Change or stabilization of allele frequencies due to differences in survival and reproduction among variant members of a population

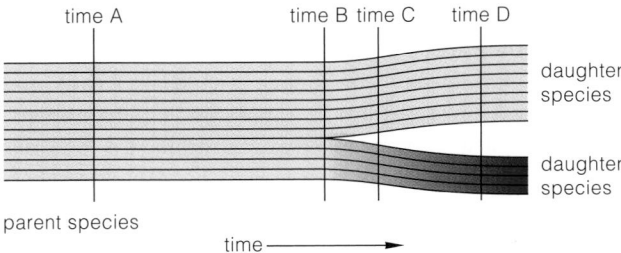

time A   time B time C   time D

daughter species

daughter species

parent species

time ———→

Figure 22.4 Divergence, the first step on the path to speciation. Each horizontal line represents a different population. Because evolution is gradual, we cannot say at any one point in time that there are now two species rather than one. At time A there is only one species. At D there are two. At B and C the split has begun but is far from complete.

Reproductive Isolation and Speciation

For sexually reproducing organisms, a **species** is a unit consisting of one or more populations of organisms that usually closely resemble each other physically and physiologically. Individuals of a species are able to interbreed and produce fertile offspring under natural conditions. No matter how diverse those individuals become, they remain members of the same species so long as they can continue to interbreed successfully and share a common pool of alleles. A woman lawyer in India may never meet up with an Icelandic fisherman, but there is no biological reason why the two could not mate and produce children. Neither individual, however, could mate successfully with a chimpanzee, even though chimps and humans are closely related species and have a large number of genes in common. In evolutionary terminology, we would say that humans and chimpanzees are *reproductively isolated*. There are enough genetic differences between them that they cannot interbreed successfully.

Reproductive isolation develops when gene flow between two populations stops. This can happen in various ways; it often occurs when two populations become separated geographically. For example, geological changes in the eastern United States have frequently split small rivers into separate streams. The newly separated fish populations may then encounter slightly different environments. In those different environments, mutation, natural selection, and genetic drift begin to operate independently in each population. And these processes can change the allele frequencies of each in different ways. Differences among separated populations may lead to **reproductive isolating mechanisms**. These are aspects of structure, function, or behavior that reduce the likelihood of successful interbreeding. The two populations may breed in different seasons or have different mating rituals, or there may be changes in body structure

that physically interfere with mating. Other types of isolating mechanisms prevent zygotes or hybrid offspring from developing properly.

An accumulation of differences in allele frequencies among isolated populations is called *divergence* (Figure 22.4). Divergence may become the first step on the road to **speciation**, the evolutionary process by which species originate. When the genetic differences between isolated populations become great enough, their members will not be able to interbreed successfully under natural conditions. At this point, the populations are separate species.

Rates of Evolutionary Change

We know that factors such as gene mutation and natural selection ultimately trigger speciation; even so, biologists disagree about the pace and timing of microevolution.

According to the traditional model, called *gradualism*, new species emerge through many small changes in form over long spans of time. In other words, microevolution is constantly going on in imperceptibly tiny increments to produce new species. By contrast, according to the model called *punctuated equilibrium*, most evolutionary changes occur in bursts. In this scenario, each species undergoes a spurt of changes in form when it first branches from the parental lineage, then changes little for the rest of its time on Earth.

There is evidence that the driving force behind these rapid changes in species may be dramatic changes in climate or some other aspect of the physical environment. This type of change suddenly (in evolutionary terms) alters the physical conditions to which populations of organisms have become adapted. For example, the onset of an ice age changes the living conditions for many land-dwelling species in affected areas. A major shift in ocean currents (which help determine water temperature) might have the same effect on marine life forms. Punctuated equilibrium could help explain why the fossil record provides little evidence of a continuum of microevolution—one often cannot find forms between closely related species in the fossil record. Both models probably have a place in explaining the history of life.

Mutations are the source of new alleles, hence of heritable variation in traits.

A mutation may be harmful, neutral, or beneficial, depending on how the trait affects an individual's ability to survive and reproduce under prevailing conditions.

A species is a unit of one or more populations of individuals that can interbreed under natural conditions and produce fertile offspring.

Biogeography

Even before Charles Darwin, naturalists were finding indirect support for the idea of evolution in what today is called **biogeography**, the study of the world distribution of plants and animals. Biogeography addresses the basic question of why certain species (and higher groupings) occur where they do. For example, why do Australia, Tasmania, and New Guinea have species of monotremes (egg-laying mammals such as the duckbilled platypus), while such animals are absent from other regions of the world where the living conditions are similar? And why do we find weasels, skunks, and otters on virtually every continent except Australia? The simplest explanation for such biogeographical patterns is that species occur where they do either because they evolved there from ancestral species or because they dispersed there from someplace else.

Modern studies of *plate tectonics*, the movement of plates of the Earth's crust, show that earlier in Earth's history all present-day continents, including Africa and South America, were parts of a massive "supercontinent" called Pangaea (Figure 22.5). By determining the locations of plates at different times in Earth's history, researchers can shed light on possible dispersal routes for some groups of organisms and when (in geological history) such movements occurred. Key support for hypotheses about the evolutionary history of different groups of organisms comes from a range of sources, none of them more important than fossils.

Fossils: Evidence of Ancient Life

In general, **fossils** are recognizable, physical evidence of ancient life. Figure 22.6 shows examples. The most

Figure 22.5 (a) Plate tectonics, the arrangement of the Earth's crust in rigid plates that split apart, move about, and collide with one another. (b) About 240 million years ago (mya) Earth's land masses were joined in a massive supercontinent, Pangea. By about 40 million years ago plate movements had split Pangea into isolated land masses, including Africa, South America, Australia, and Eurasia.

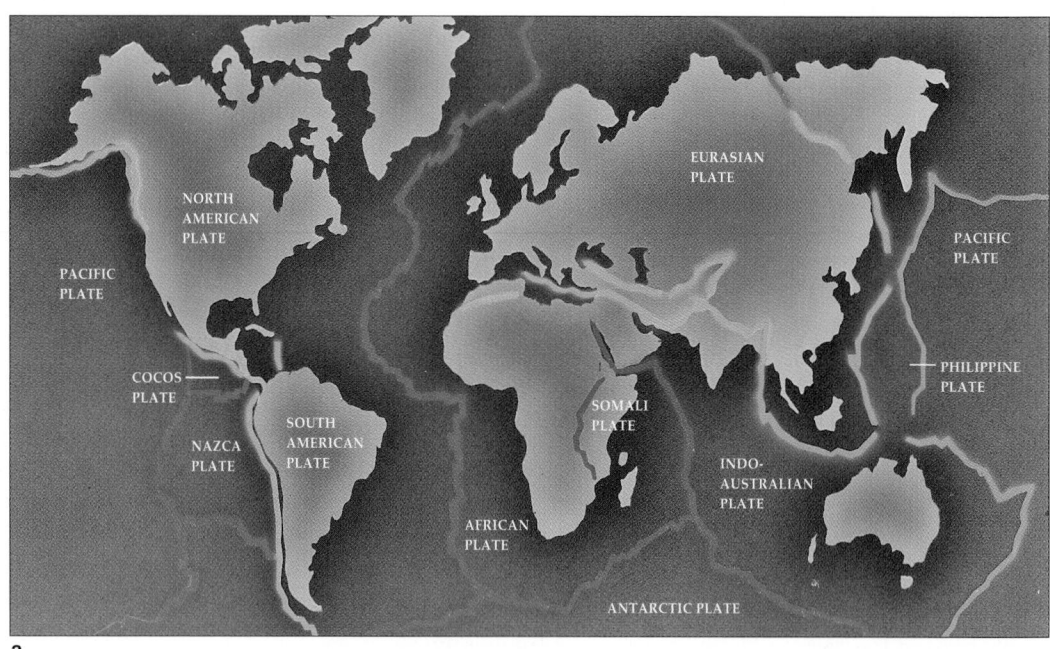

a

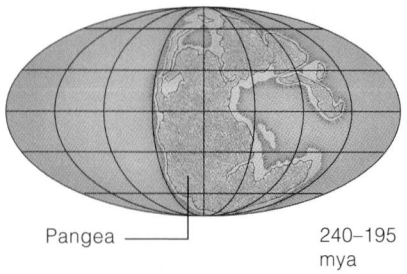

Pangea — 240–195 mya

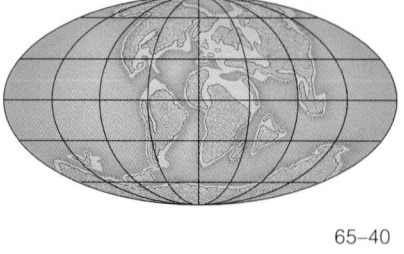

65–40 mya

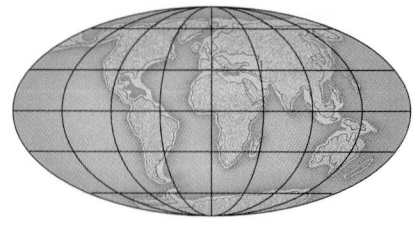

10 mya

b

a

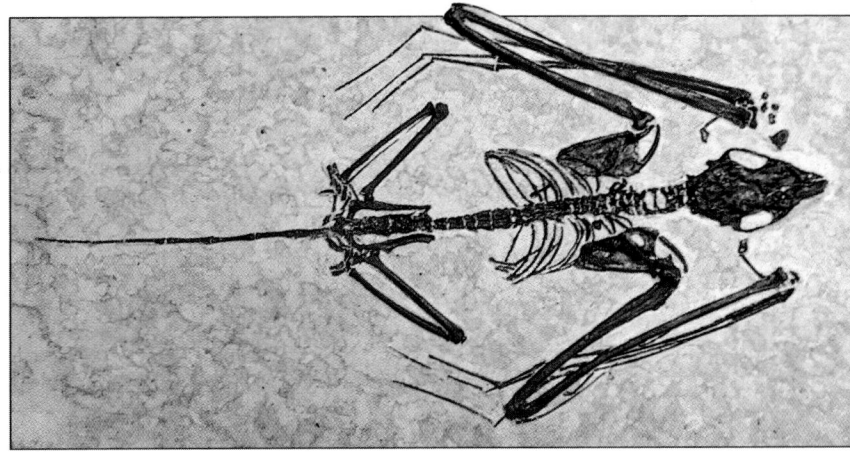
b

Figure 22.6 Some representative fossils. (**a**) Fossilized leaves from *Archeopteri*, which probably was on the evolutionary road that led to gymnosperms (such as pine trees) and angiosperms (flowering plants). (**b**) The complete skeleton of a bat that lived 50 million years ago. Fossils of this quality are extremely rare.

common fossils are bones, teeth, shells, spore capsules, seeds, and other hard parts of an organism. (Soft parts usually are the first to decompose when an organism dies.) **Fossilization** begins with burial in sediments or volcanic ash. Eventually, water infiltrates the organic remains, infusing them with dissolved metal ions and other inorganic compounds. As more and more sediments accumulate above the burial site, the remains are subjected to increasing pressure. Over great spans of time, the chemical changes and pressure transform them to stony hardness.

Preservation is favored when organisms are buried rapidly in the absence of oxygen. Gentle entombment by volcanic ash or anaerobic mud is best. Preservation also is favored when a burial site is undisturbed. Most often, though, erosion and other geologic insults have crushed, deformed, broken, or scattered the fossils.

Similar fossil-containing layers of sedimentary rock extend over vast areas, even on different continents. Such layers formed long ago, when silt, volcanic ash, and other materials were gradually deposited, one above the other. The layering of sedimentary deposits is called **stratification**. Most sedimentary layers must have formed horizontally. Where they are ruptured or tilted, this typically is evidence of later geologic disturbance.

Interpreting the Fossil Record

At present, we have fossils of about 250,000 species. However, judging from the current sweep of diversity, there must have been millions of ancient, now-extinct species, and we will not be able to recover fossils for most of them. This means that our "record" of past life is incomplete, with built-in biases.

Most important, large-scale movements in the Earth's crust have obliterated evidence from crucial periods in the history of life. Besides this, most members of ancient communities simply have not been preserved. For example, hard-shelled mollusks and bony fishes are well represented in the fossil record. Jellyfishes and soft-bodied worms are not, even though they may have been just as common or more so. Population density and body size further skew the record. A population of ancient plants may have produced millions of spores in a single growing season, whereas the earliest members of the human lineage lived in small groups and produced few offspring. Hence, the chance of finding a fossilized skeleton of an early human is small, compared to the chance of finding spores of plant species that lived at the same time.

The fossil record also is heavily biased toward certain environments. Most of the species represented lived on land or in shallow seas that, through geologic uplifting, became part of continents. We have few fossils from sediments beneath the ocean—which extends across three-fourths of the Earth's surface. Finally, most fossils have been found in the Northern Hemisphere, because that's where most geologists have lived.

Biogeographical patterns can provide clues to where a species arose. Correlated with evidence from plate tectonics, the patterns also can shed light on possible dispersal routes for some groups of organisms.

Fossils, the stone-hard physical evidence of ancient life, are present in layers of sedimentary rocks. The deeper the layers, the older the fossils.

The completeness of the fossil record varies as a function of the kinds of organisms represented, where they lived, and the stability of their burial sites.

COMPARISONS OF BODY FORM

More evidence for evolution comes from detailed comparisons of body form and structural patterns in major groups of organisms. Through **comparative morphology**, researchers reconstruct evolutionary history on the basis of information contained in the observed patterns of body form. This work involves detailed studies of embryos at different stages of development as well as the adult form.

Comparative Embryology

Vertebrates include fishes, amphibians, reptiles, birds, and mammals. Despite the diversity of these groups, however, comparisons of the ways in which their embryos develop provide compelling evidence of their evolutionary connection to one another.

Early in the program of development, the embryos of all the different vertebrate lineages proceed through strikingly similar stages (Figure 22.7). During vertebrate evolution, mutations that disrupted an early stage of development would have had devastating effects on the organized interactions required for later stages. Embryos of different groups remained similar because mutations that altered early steps in development were selected against.

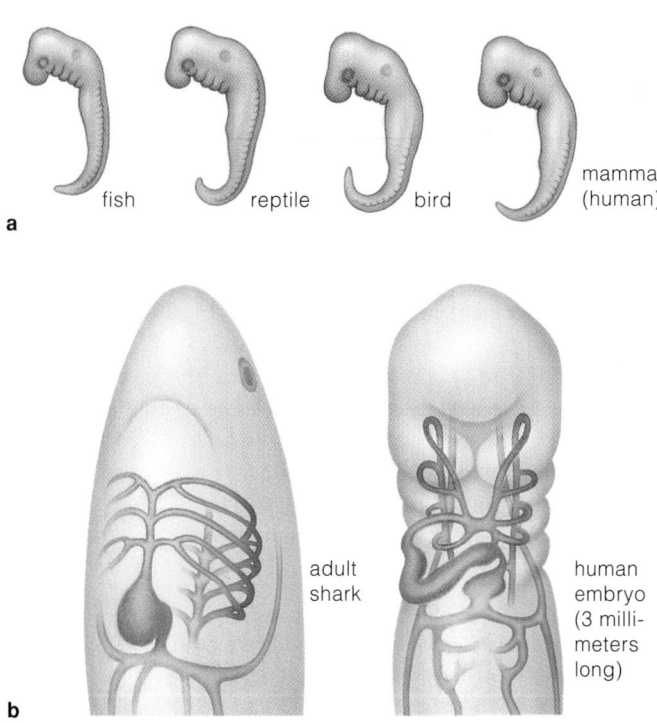

Figure 22.7 From comparative embryology, some evidence of evolutionary relationship among vertebrates. (**a**) Adult vertebrates show great diversity, yet the very early embryos retain striking similarities. This is evidence of change in a common program of development. (**b**) Fishlike structures still form in early embryos of reptiles, birds, and mammals. For example, a two-chambered heart (*orange*), certain veins (*blue*), and portions of arteries called aortic arches (*red*) develop in fish embryos and persist in adult fishes. The same structures form in an early human embryo.

So how did the *adults* of different vertebrate groups come to be so different? At least some differences resulted from mutations that altered the onset, rate, or time of completion of certain developmental steps. Such mutations would bring about changes in shape through increases or decreases in the size of body parts. They also could lead to adult forms that retain some juvenile features. Figure 22.8 illustrates how changes in the growth rate at a key developmental step may have caused differences in the proportions of chimpanzee and human skull bones, which are alike at birth. Later, they change dramatically for chimps but only slightly for humans. Chimps and humans arose from the same ancestral stock. *Their genes are nearly identical.* However, at some point on the separate evolutionary road leading to humans, some regulatory genes probably mutated in ways that proved adaptive. Since then, instead of promoting the rapid growth required for dramatic changes in skull bones, the mutated genes have blocked it.

Vestigial Structures

As Figure 22.9 indicates, the bodies of humans, pythons, and other organisms can have what seem like useless "vestigial" structures. Consider your own ear-wiggling muscles—which four-footed mammals (such as dogs) use

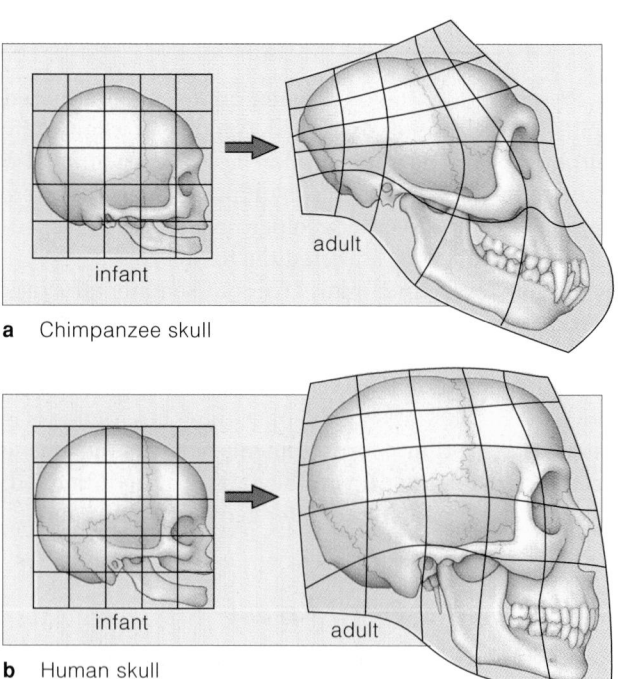

Figure 22.8 Comparison of proportional changes in a chimpanzee skull (**a**) and a human skull (**b**). Both skulls are quite similar in infants. Imagine that these representations of infant skulls are paintings on a blue rubber sheet divided into a grid. Stretching the sheet deforms the grid's squares. For the adult skulls, differences in size and shape within corresponding grid sections reflect differences in growth patterns.

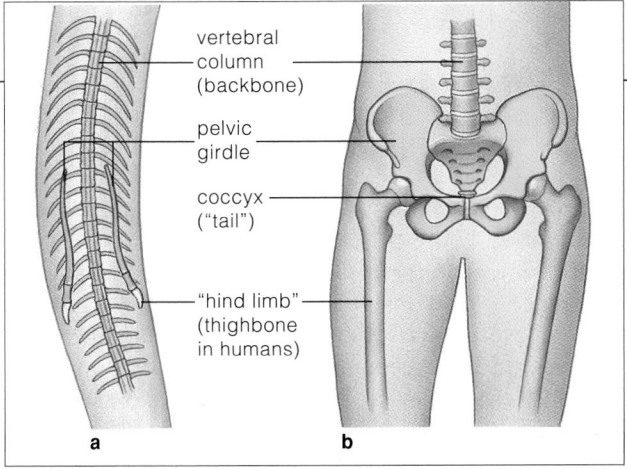

Figure 22.9 Bony parts of a python (**a**) correspond to the pelvic girdle of other vertebrates, including humans (**b**). A snake has small vestigial "hind limbs" on its underside—remnants from a limbed ancestor. The human coccyx is a similar vestige from an ancestral species that had a bony tail, although muscles still attach to it.

to orient their ears. In humans, such body parts are left over from a time when more functional versions were important for an ancestor.

Homologous and Analogous Structures

When populations of a species branch out in different evolutionary directions, they diverge in appearance, functions, or both. Yet the related species also remain alike in many ways, for their evolution proceeds by modification of a shared body plan. In such species we see **homologous structures**. These are the same body parts modified in different ways in different lines of descent from a common ancestor (*homo-* means "same").

For example, the same ancestral organism probably gave rise to most land-dwelling vertebrates, which have homologous structures. Apparently their common ancestor had four five-toed limbs. The limbs diverged in form and became wings in pterosaurs, birds, and bats (Figure 22.10). All these wings are homologous—they have the same parts. The five-toed limb also evolved into the flippers of porpoises and the anatomy of your own forearms and fingers.

Sometimes body parts in organisms that do *not* have a recent common ancestor come to resemble one another in form and function. These **analogous structures** (from *analogos*, meaning similar to one another) arise when different lineages evolve in the same or similar environments. Different body parts, which were put to similar uses, became modified through natural selection and ended up resembling one another. This pattern of change is called **morphological convergence**. For example, a dolphin, a fast-swimming marine mammal, has a sleek, torpedo-shaped torso—and so does a tuna, a fast-swimming fish.

Comparative morphology gleans evolutionary information from observed patterns of body form.

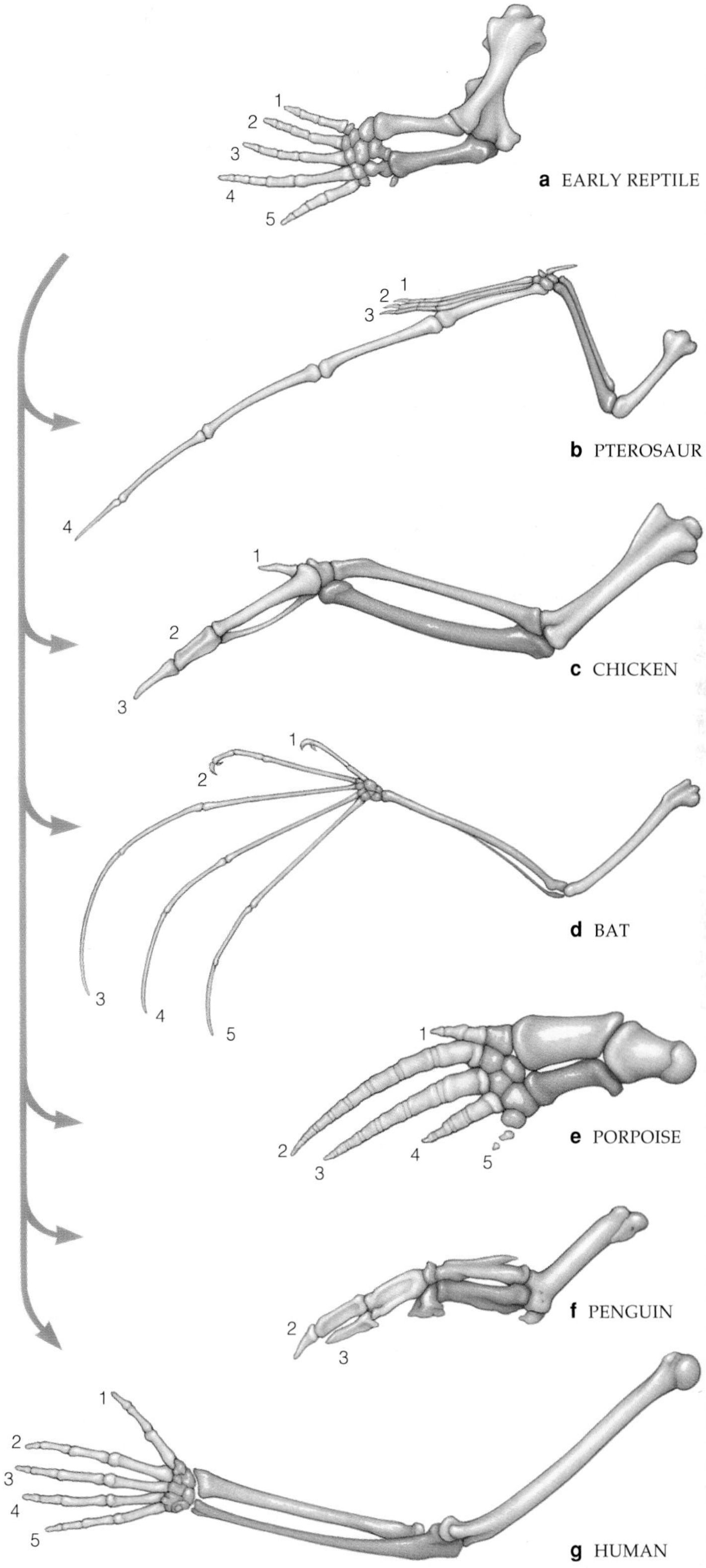

---

a EARLY REPTILE

b PTEROSAUR

c CHICKEN

d BAT

e PORPOISE

f PENGUIN

g HUMAN

Figure 22.10 Morphological divergence in the vertebrate forelimb, starting with a generalized form of ancestral early reptiles. Diverse forms evolved even as similarities in the number and position of bones were preserved. The drawings are not to the same scale.

COMPARING BIOCHEMISTRY

The genes and gene products (proteins) of different species contain information about evolutionary relationships. Recall from Section 21.4 that gel electrophoresis of proteins can be used to identify even a single amino acid substitution in the polypeptide chain of a protein. Genes dictate the amino acid sequence, so if the sequence is the same (or nearly so) in comparable proteins from two different species, the genes coding for those proteins point to a shared ancestor. The closer the match, the more recently the two species shared that common ancestor. The more remote the match, the more distantly the two species are related.

For example, organisms ranging from aerobic bacteria to corn plants to humans produce cytochrome *c*, a protein of electron transport chains. From various kinds of studies we know that the gene that codes for the protein has changed very little over huge spans of time. The cytochrome *c* sequence of 104 amino acids is identical in humans and chimps. In rhesus monkeys it is the same except for one amino acid. (On the basis of such data, would you assume that humans are more closely related to a chimpanzee or to a rhesus monkey?) By contrast, the cytochrome *c* of chickens differs from that of humans by 18 amino acids, and that of yeasts differs by 56 amino acids.

Nucleotide sequences of DNA are riddled with structural changes caused by mutation. This suggests that the extent to which a DNA strand from one species will base-pair with a comparable strand from another species is a rough measure of the evolutionary distance between them.

Some nucleic acid comparisons are based on **DNA-DNA hybridization**. In this method, the two strands of the double-stranded DNA from two species are induced to unwind from each other. Then the single strands are allowed to recombine into "hybrid" molecules. Heating breaks the hydrogen bonds between the two hybridized strands. The heat energy required to pull them apart is a measure of the bonding between the strands and therefore of the similarity between them. Among other discoveries related to DNA-DNA hybridization, we now know that the endangered giant panda is truly a member of the bear family, whereas the red panda (despite its name) belongs to the raccoon family.

Other methods used to compare DNA include PCR, gene sequencing, and restriction mapping—using the restriction enzymes described in Chapter 21 to obtain DNA fragments from different species. Using these methods, researchers can analyze genetic similarities and differences among species.

Comparative biochemistry analyzes genes and gene products (proteins) to help determine evolutionary relationships.

EVOLUTIONARY TREES AND THEIR BRANCHINGS

The fossil record, comparative morphology, and comparative biochemistry each provide insight into "who came from whom." Typically, information about this continuity of relationship among species is shown in the form of **evolutionary trees**. In these treelike diagrams, branches represent separate lines of descent from a common ancestor. As Figure 22.11 indicates, each branch point in a tree represents a time of divergence and speciation.

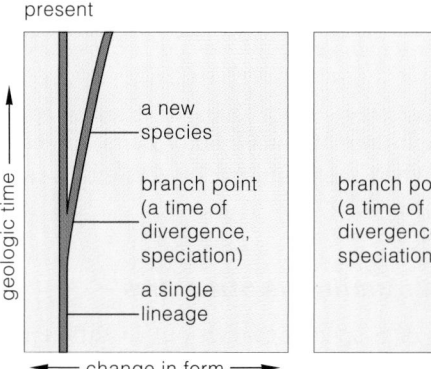
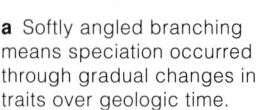
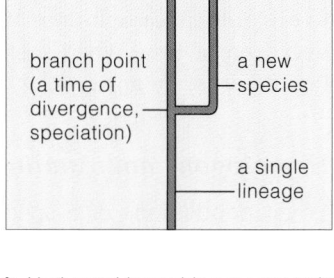

a Softly angled branching means speciation occurred through gradual changes in traits over geologic time.

b Horizontal branching means traits changed rapidly around the time of speciation. Vertical continuation of a branch means traits of the new species did not change much thereafter.

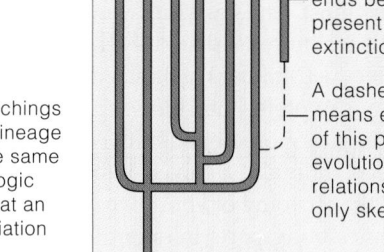

c Many branchings of the same lineage at or near the same point in geologic time mean that an adaptive radiation occurred.

A branch that ends before the present means extinction

A dashed line means evidence of this presumed evolutionary relationship is only sketchy

Figure 22.11 How to read evolutionary tree diagrams. The branch points represent times of divergence and speciation as brought about by microevolutionary processes.

Do changes from the original species form proceed slowly or rapidly? Many small changes may accumulate gradually within a lineage over long spans of time. (This is the premise of the *gradual* model of rates of change in an evolving lineage.) Conversely, species may originate through rapid spurts of change over hundreds or thousands of years, then change little for the next 2–6 million years. (This is the premise of the *punctuation* model, which has most changes occurring about the time of speciation.)

Extinctions and Adaptive Radiations

Inevitably, a lineage loses a number of species as local conditions change. The rather steady rate of their disappearance over time is "background extinction." Regional extinctions occur when the background rate is exceeded in a local area. By contrast, a **mass extinction** is an abrupt, widespread rise in extinction rates above the background level. It is a catastrophic, global event in which major groups are wiped out simultaneously. Dinosaurs and many marine groups died out during a mass extinction that occurred about 65 million years ago—possibly the result of environmental changes that occurred after one or more large meteorites struck the Earth during a short period of time.

In an **adaptive radiation**, new species of a lineage fill a wide range of habitats during bursts of microevolutionary activity. Many adaptive radiations have occurred during the first few million years after a mass extinction. For instance, fossil evidence reveals that this happened after dinosaurs became extinct. Many new species of mammals arose and radiated into the habitats the dinosaurs vacated.

In the last few hundred years human activities have caused the extinction of thousands of species. The extinction rate is accelerating as we cut down forests, fill in wetlands, and otherwise destroy the habitats of other animals and plants with which we share the Earth. The *Focus on Our Environment* essay in Section 22.8 considers some current efforts to rescue endangered species, and we will return to this topic in Chapters 24 and 25.

Organizing the Evidence: Classification

How can we organize the hundreds of thousands of known species of organisms in a meaningful way? For centuries, naturalists (and later, biologists) grouped organisms based on their morphological similarities (similar body structures). Modern evolutionary biologists also consider morphological similarities, along with other information. Such resemblances can be an important part of the evidence for constructing a **phylogeny**—a branching classification scheme that bases groupings on presumed evolutionary relatedness.

Classification requires the identification of each new species, which is assigned a unique name. (This work is called *taxonomy*.) In a *binomial system* devised two centuries ago by Carolus Linnaeus, a two-part name is used. The first part is the genus name (plural: genera). All species that are similar to one another and distinct from others in certain traits are grouped in the same genus. The second part of the name designates the particular species within the genus.

Table 22.2 Classification of Humans

Kingdom	Animalia
Phylum	Chordata
Class	Mammalia
Order	Primates
Family	Hominidae
Genus	*Homo*
Species	*sapiens* (only living species of this genus)

Current **classification systems** organize species into a series of ever more inclusive groupings, called *taxa*, beginning with the species. Typically these groupings include the following taxa:

kingdom
　　phylum (or division, in the case of plants)
　　　class
　　　　order
　　　　　family
　　　　　　genus
　　　　　　　species

Table 22.2 shows how these groupings are used in the classification of humans.

The branches of an evolutionary tree represent separate lines of descent from a common ancestor.

As lineages evolve, they may undergo adaptive radiation, in which new species rapidly (on a geological time scale) fill a range of habitats. Lineages also lose species to extinction.

A classification system organizes species into ever more inclusive groupings. The most inclusive grouping is the kingdom; the species level is the least inclusive.

GENES OF ENDANGERED SPECIES

Barbara Durrant and many other biologists at zoos, private conservation centers, and wildlife refuges are scrambling to buy time for the world's most *endangered species*. Populations of such species are so small that they are at the brink of extinction. The biologists face daunting problems, including an absence of genetic diversity.

Consider the cheetah (Figure 22.12). Only 20,000 of these sleek, swift cats have survived to the present. It seems that cheetahs went through a severe bottleneck about 10,000 years ago. Parents mated with their own offspring when no other mates were available. As a result of inbreeding among the survivors and their descendants, all cheetahs now carry strikingly similar alleles. They are so genetically uniform that a patch of skin from one can be successfully grafted to another. This seldom works with other mammals.

Among their shared alleles are previously rare harmful ones. Some of the harmful, mutated alleles affect fertility. Typically, a male cheetah's sperm count is low, and 70 percent of the sperm that do form are abnormal. Other alleles are responsible for weakened resistance to disease. Infections that are seldom life-threatening to other cats can reach epidemic proportions among cheetahs. During one outbreak of *feline infectious peritonitis* in a wild animal park, the infectious agent (a virus) had little effect on captive lions, but it killed every one of the cheetahs.

Intense hunting and other human activities have endangered another big cat, the Florida panther. With fewer than 50 individuals, the last remaining population is highly inbred. Researchers at the National Zoo retrieve DNA from unfertilized eggs of "road-killed" female panthers. With genetic engineering, the DNA may find use in captive breeding programs.

Figure 22.12 A few of the world's remaining cheetahs, all of which carry some harmful alleles that made it through a severe bottleneck at some time in the species' history.

EVOLUTION AND EARTH HISTORY

When all of the diverse kinds of evidence for evolution are carefully pieced together, an important fact emerges: *The evolution of life has been linked, from its origin to the present, to the physical and chemical evolution of the Earth.*

Origin of the Earth

Cloudlike remnants of stars are mostly hydrogen gas. But they also contain water, iron, silicates, hydrogen cyanide, methane, ammonia, formaldehyde, and many other simple inorganic and organic substances. Most likely, the contracting cloud from which our solar system evolved was similar in composition. Between 4.6 and 4.5 billion years ago, the outer regions of the cloud cooled. Swarms of mineral grains and ice orbited around the sun. By electrostatic attraction, by gravitational pull, they started clumping together. In time, the larger, faster clumps started colliding and shattering. Some of the larger ones became more massive by sweeping up asteroids, meteorites, and other rocky remnants of collisions—and gradually they evolved into planets.

While the early Earth was forming, much of its inner rocky material melted. Most likely, stupendous asteroid impacts generated the required heat, although heat released during the radioactive decay of minerals contributed to it. As rocks melted, nickel, iron, and other heavy materials moved toward the interior, and lighter minerals floated toward the surface. This process of differentiation resulted in the formation of a crust of basalt, granite, and other types of low-density rock, a rocky region of intermediate density (the mantle), and a high-density, partially molten core of nickel and iron.

Four billion years ago, the Earth was a thin-crusted inferno (Figure 22.13). Yet within 200 million years, life had originated on its surface! We have no record of the event. As far as we know, movements in the mantle and crust, volcanic activity, and erosion obliterated all traces of it. Still, we can put together a plausible explanation of how life originated by considering three questions:

1. What were the prevailing physical and chemical conditions on Earth at the time of life's origin?

2. Based on physical, chemical, and evolutionary principles, could large organic molecules have formed spontaneously and then evolved into molecular systems displaying the characteristics of life?

3. Can we devise experiments to test whether living systems could have emerged by chemical evolution?

We begin with a brief look at Earth's early atmosphere, which must have laid the physical and chemical foundation for life's beginnings. Section 22.10 then considers plausible scenarios for the emergence of living cells.

Figure 22.13 Representation of the primordial Earth, about 4 billion years ago. Within another 500 million years, diverse types of living cells would be present on the surface.

Conditions on the Early Earth

When patches of crust were forming, heat and gases blanketed the Earth. This first atmosphere probably consisted of gaseous hydrogen (H_2), nitrogen (N_2), carbon monoxide (CO), and carbon dioxide (CO_2). Were gaseous oxygen (O_2) and water also present? Probably not. Rocks release very little oxygen during volcanic eruptions. Even if oxygen were released, those small amounts would have reacted at once with other elements, and any water would have evaporated because of the intense heat.

When the crust finally cooled and solidified, water condensed into clouds and the rains began. For millions of years, runoff from rains stripped mineral salts and other compounds from the Earth's parched rocks. Salt-laden waters collected in depressions in the crust and formed the early seas.

What we are describing here are events that were crucial to the beginning of life. Without an oxygen-free atmosphere, the organic compounds that started the story of life never would have formed on their own.

Without liquid water, cell membranes never would have formed. Cells, recall, are the basic units of life. Each has a capacity for independent existence. Cells could not simply have appeared one day on the early Earth. Their emergence required the existence of biological molecules built from organic compounds. It also required metabolic pathways, which could be organized and controlled within the confines of a cell membrane. These essential events are our next topic.

From its origin to the present, the evolution of life on Earth has been linked to physical and chemical evolution of the Earth.

The oxygen-free atmosphere of the early Earth provided necessary conditions for the formation of organic compounds that would become the raw material for the first biological molecules. The accumulation of liquid water provided conditions in which the first cells could have formed.

Figure 22.14 shows fossil cells discovered in the Warrawoona rock formation in Western Australia—a fossil that has been dated at 3.5 billion years old! The first living cells probably emerged between 4 billion and 3.8 billion years ago. Before something as complex as a cell was possible, however, biological molecules must have come about through *chemical evolution*. In fact, researchers have been able to put together a reasonable scenario for the emergence of life.

Synthesis of Biological Molecules

Rocks collected from Mars, meteorites, and the Earth's moon—which all formed at the same time as the Earth, from the same cosmic cloud—contain precursors of biological molecules. Possibly, sunlight, lightning, or heat escaping from the Earth's crust supplied enough energy to drive condensation of precursors into more complex organic molecules. In the early 1950s a chemistry graduate student named Stanley Miller took the likely constituents of the early atmosphere—hydrogen, methane, ammonia, and water—and mixed them in a reaction chamber (Figure 22.15). He recirculated the mixture and for several days bombarded it with a spark discharge to simulate lightning. Within a week, many amino acids and other organic compounds had formed.

In other experiments that simulated conditions on the early Earth, glucose, ribose, deoxyribose, and other sugars were produced from formaldehyde. Adenine was produced from hydrogen cyanide. Adenine plus ribose occur in ATP, NAD, and other nucleotides vital to cells.

If *complex* organic compounds formed in the seas, they would not have lasted long. In water, the spontaneous direction of the required reactions would have been toward hydrolysis, not condensation.

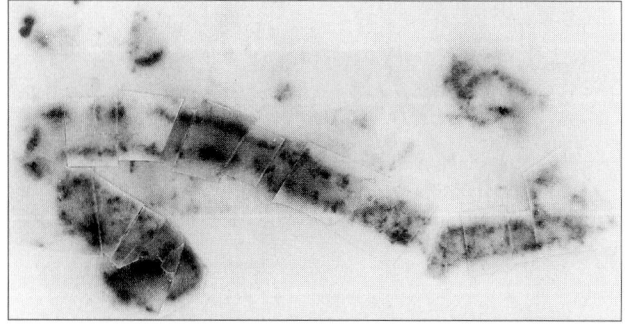

Figure 22.14 One of the oldest known fossils, dated at 3.5 billion years old. It is a string of walled cells unearthed in the Warrawoona rocks of Western Australia.

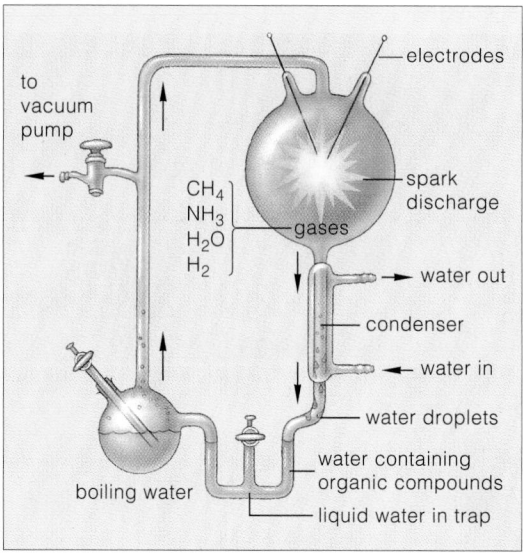

Figure 22.15 Stanley Miller's apparatus used in studying the synthesis of organic compounds under conditions believed to have been present on the early Earth. (The condenser cools circulating steam and causes water to condense into droplets.)

Perhaps more lasting bonds formed at the margins of seas. By one scenario, clay in the rhythmically drained muck of tidal flats and estuaries served as templates (structural patterns) for the spontaneous assembly of proteins and other complex organic compounds. Clay consists of thin, stacked layers of aluminosilicates with metal ions at its surface. Clay and metal ions attract amino acids. From experiments we know that when clay is first warmed by sunlight, then alternately dried out and moistened, it actually promotes condensation reactions that produce complex organic compounds.

Proteins that formed on some clay templates may have had the shape and chemical behavior needed to function as weak enzymes in hastening bonds between amino acids. If certain templates promoted such bonds, they would have had selective advantage over other templates in the chemical competition for available amino acids. Perhaps selection was at work before the origin of cells, favoring the chemical evolution of enzymes.

The First Metabolic Pathways

A defining characteristic of life is metabolism. The term refers to the reactions by which cells harness energy and use it to drive their activities, including biosynthesis. During the first 600 million years of Earth's history, enzymes, ATP, and other molecules could have assembled spontaneously in places where they were in close physical proximity. If so, their close association would have promoted chemical interactions—and the beginning of metabolic pathways.

Origin of Self-Replicating Systems

Another defining characteristic of life is a capacity for reproduction, which now starts with protein-building instructions in DNA. The DNA molecule is fairly stable, and it is easily replicated before each cell division. As you know from Chapter 19, arrays of enzymes and RNA molecules operate together to carry out DNA's encoded instructions.

Most existing enzymes are assisted by the small organic molecules or metal ions called coenzymes. Intriguingly, some types of coenzymes have a structure that is identical to that of the RNA nucleotides. And if you heat nucleotide precursors and short chains of phosphate groups together, they will self-assemble into strands of RNA. On the early Earth, energy from the sun or geothermal vents would have been sufficient to drive the spontaneous formation of RNA molecules.

Simple, self-replicating systems of RNA, enzymes, and coenzymes have been created in some laboratories. Did RNA later beome the information-storing templates for protein synthesis? Perhaps it initially did so, although existing RNA molecules are just too chemically fragile for such a role. Yet the use of RNA templates may have set up an **RNA world** that preceded DNA's dominance as the main informational molecule. Whatever the case, DNA eventually assumed this function, probably because it can form long nucleotide chains in more stable fashion.

We still don't know how DNA entered the picture. Until we identify the likely chemical ancestors of RNA and DNA, the story of life's origin will be incomplete.

Enter the First Cells

Before cells as we know them existed, there must have been "proto-cells." These structures may have been as simple as membrane-bound sacs that sheltered information-storing templates and various metabolic agents, such as amino acids, from the environment. Experiments show that such sacs can form spontaneously. In one classic study, Sidney Fox heated amino acids until they formed protein chains, which he placed in hot water. After the chains cooled, they self-assembled into small, stable *microspheres*. Like cell membranes today, the spheres were selectively permeable to different substances. They also picked up lipids from the water, and a lipid–protein film formed at their surface. In other experiments, lipids self-assembled into small, water-filled sacs that had many properties of cell membranes.

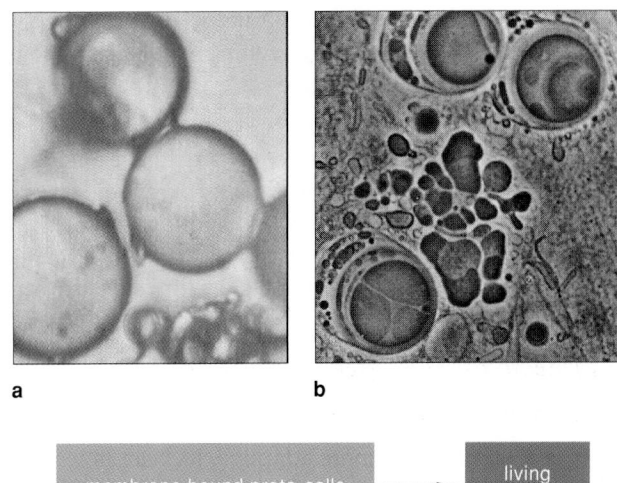

a b

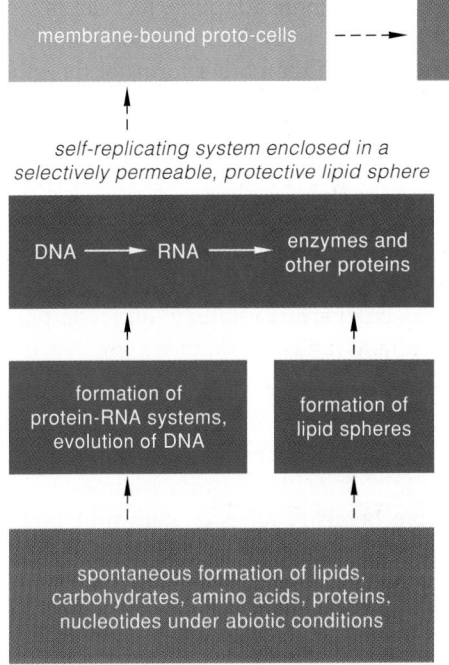

c

Figure 22.16 Microspheres of (**a**) proteins and (**b**) lipids that self-assembled under abiotic (nonliving) conditions. (**c**) One possible sequence of events that led to the first self-replicating systems, then to the first living cells.

Despite major gaps in our knowledge of life's origins, there is also good experimental evidence that chemical evolution could have given rise to the molecules and structures characteristic of life. Figure 22.16 summarizes some key events in that chemical evolution, which possibly led to the first cells.

Many experiments provide indirect evidence that the complex organic molecules characteristic of life could have formed under conditions that existed on the early Earth.

Experiments suggest that chemical and molecular evolution gave rise to proto-cells.

SUMMARY

1. Evolution proceeds by modifications of already existing species. Hence, there is a continuity of relationship among all species, past and present.

2. In a population, genetic variability gives rise to variations in traits. The frequencies of different genes (alleles) change as a result of four processes of microevolution: mutation, genetic drift, gene flow, and natural selection. The large-scale patterns, trends, and rates of change among groups of species (higher taxa) over time are called macroevolution.

3. Natural selection is a difference in survival and reproduction among members of a population that vary in one or more traits. That is, under prevailing conditions, one form of a trait may be more adaptive. Its bearers tend to survive and therefore to reproduce more often, so it becomes more common than other forms of the trait.

4. Speciation results when the differences between isolated populations become so great that their members are not able to interbreed successfully under natural conditions.

5. Evidence of evolutionary relationships comes from the fossil record and Earth history, radioactive dating methods, comparative morphology, and comparative biochemistry.

6. Comparative morphology often reveals similarities in embryonic development that indicate an evolutionary relationship. They may reveal homologous structures, shared as a result of descent from a common ancestor. Comparative biochemistry relies on gene mutations that have accumulated in different species. Methods such as DNA-DNA hybridization reveal the degree of similarity among genes and gene products of different species.

7. Family trees may be constructed to show evolutionary relationships. The branches of such trees are lines of descent (lineages). The branch points are speciation events brought about by mutation and by natural selection and other microevolutionary processes.

8. A mass extinction is a catastrophic event in which major lineages perish abruptly. An adaptive radiation is a burst of evolutionary activity; a lineage rapidly produces many new species. Both events have changed the course of biological evolution many times.

9. Every element of the solar system, the planet Earth, and life itself are products of the physical and chemical evolution of the universe and its stars.

10. Four billion years ago, the Earth had a high-density core, a mantle of intermediate density, and a thin, extremely unstable crust of low-density rocks. Probably gaseous hydrogen, nitrogen, carbon monoxide, and carbon dioxide made up the first atmosphere.

11. Many studies and experiments provide indirect evidence that life originated under conditions that presumably existed on the early Earth.

    a. Comparative analyses of the composition of cosmic clouds and of rocks from other planets and the Earth's moon suggest that precursors of the complex molecules associated with life were available.

    b. In laboratory tests that simulated primordial conditions, including the absence of free oxygen, the precursors spontaneously assembled into sugars, amino acids, and other organic compounds.

    c. Metabolic pathways could have evolved as a result of chemical competition for the limited supplies of organic molecules that had accumulated in the seas.

    d. Self-replicating systems of RNA, enzymes, and coenzymes have been synthesized in the laboratory. How DNA entered the picture is not yet understood.

    e. Lipid and lipid–protein membranes that show properties of cell membranes form spontaneously under the prescribed conditions. Life originated about 3.8 billion years ago.

Review Questions

1. What is the difference between microevolution and macroevolution? *459*

2. Explain the difference between:
    a. homologous and analogous structures *465*
    b. divergence and convergence *461, 465*

3. Fill in the blanks in this classification chart for humans: *467*

Classification of Humans	
Kingdom	Animalia
_____	Chordata
_____	Mammalia
_____	Primates
_____	Hominidae
_____	*Homo*
_____	*sapiens* (only living species of this genus)

4. Describe the chemical and physical characteristics of the Earth 4 billion years ago. How do we know what it was like? *468–471*

Critical Thinking: You Decide *(Key in Appendix V)*

1. Humans can inherit various alleles for the liver enzyme ADH (alcohol dehydrogenase), which breaks down ingested alcohol. People of Italian and Jewish descent commonly have a form of ADH that detoxifies alcohol very rapidly. People of northern European descent have forms of ADH that are moderately effective in alcohol breakdown, while people of Asian descent typically have ADH that is less efficient at processing alcohol. Explain why researchers have been able to use this information to help trace the origin of human use of alcoholic beverages.

2. The Atlantic Ocean is widening, and the Pacific and Indian Oceans are closing. Many millions of years from now, the continents will collide and form a second Pangea. Write a short essay on what environmental conditions might be like on that future supercontinent and on what types of species might survive there.

Self-Quiz *(Answers in Appendix IV)*

1. Large-scale patterns, trends, and rates of changes in *groups of species* over time are called _____ .

2. The rate of evolution has been _____ in some lineages, but in others it has been much more rapid, with _____ of speciation events followed by lengthy periods with little evidence of change.

3. The fossil record of evolution correlates with evidence from _____ .

    a. the geologic record
    b. radioactive dating
    c. comparative morphology
    d. comparative biochemistry
    e. all of the above

4. Comparative biochemistry _____ .
    a. is based mainly on the fossil record
    b. often reveals similarities in embryonic development stages that indicate evolutionary relationships
    c. is based on mutations that have accumulated in the DNA of different species
    d. compares the proteins and the DNA from different species to reveal relationships
    e. both c and d are correct

5. _____ reveals mutations that have accumulated in the DNA of different species.
    a. Fossil evidence   c. DNA hybridization
    b. Embryonic development   d. Comparative morphology

6. Comparative morphology _____ .
    a. is based mainly on the fossil record
    b. often reveals similarities in embryonic development stages that indicate an evolutionary relationship
    c. shows evidence of divergences and convergences in body parts among certain major groups
    d. compares the proteins and the DNA from different species to reveal relationships
    e. both b and c are correct

7. The frequencies of different genes (alleles) change as a result of four processes of microevolution: _____ , _____ , _____ , and _____ .

8. A difference in survival and reproduction among members of a population that vary in one or more traits is called _____ .

Selected Key Terms

adaptation *460*
adaptive radiation *467*
analogous structures *465*
biogeography *462*
classification system *467*
comparative morphology *464*
DNA-DNA hybridization *466*
evolution *459*
evolutionary tree *466*
fossil *462*
gene flow *460*
genetic drift *460*
homologous structure *465*

macroevolution *459*
mass extinction *467*
microevolution *459*
morphological convergence *465*
natural selection *460*
phylogeny *467*
population *459*
reproductive isolating
    mechanism *461*
RNA world *471*
speciation *461*
species *461*

Readings

"All the Way with RNA." January 1993. *Discover*. Intriguing non-technical discussion of RNA as the probable enzyme catalyst for Earth's first proteins.

de Duve, C. September-October 1995. "The Beginnings of Life on Earth." *American Scientist*.

————. April 1996. "The Birth of Complex Cells." *Scientific American*.

Hartman, W., and Chris Impey. 1994. *Astronomy: The Cosmic Journey*. 5th ed. Belmont, Calif.: Wadsworth.

Horgan, J. February 1991. "Trends in Evolution: In the Beginning . . ." *Scientific American*.

Mayr, E. 1976. *Evolution and the Diversity of Life*. Cambridge, Mass.: Belknap Press of Harvard University Press. Collection of the author's essays on major issues in evolutionary biology, revised to reflect his present views.

Otte, D., and J. Endler, eds. 1989. *Speciation and Its Consequences*. Sunderland, Mass.: Sinauer. Thought-provoking essays by evolutionists who specialize in a broad range of subjects. Paperback.

Ridley, M. 1993. *Evolution*. Boston: Blackwell.

Udall, J. R. September/October 1991. "Launching the Natural Ark." *Sierra*. Thoughtful article on ways humans can help save other species from extinction.

 For additional readings, go to InfoTrac College Edition, your online library at:

http://www.infotrac-college.com/wadsworth

23 HUMAN EVOLUTION

The Cave at Lascaux and the Hands of Gargas

Half a century ago, on a warm autumn day, four boys out for a romp stumbled into a cave near Lascaux, a town in the Perigord region of France. What they found inside that intricately tunneled cave stunned the world.

Magnificent sketches, engravings, and paintings swept out across the cave walls (Figure 23.1a). The red, yellow, purple, and brown pigments were as vivid as if they had been painted only recently. Yet radioisotope measurements revealed that they were between 17,000 and 20,000 years old. The prehistoric artists had worked deep within the cave, where sunlight could not fade the images, and winds and water could not wear them away. By the light of crude oil lamps, they had captured the graceful, dynamic lines of bison, stags, horses, ibexes, lions, a rhinoceros, and a large heifer, now known as the Great Black Cow (Figure 23.1b).

Caves throughout southern France, northern Spain, and Africa hold treasures from even earlier times. About 25,000 years ago, for example, prehistoric peoples carefully committed more than 150 imprints and outlines of their hands to walls in the cave of Gargas, in the Pyrenees.

Who were the people who did this? From their fossilized remains, we know that they were anatomically like us. From the way they planned and executed their art, we sense a level of abstract thinking that is unique to humans.

The quality of "humanness" may have emerged in Africa (Figure 23.2). Yet the story of the human species began more than 60 million years ago, with the origin of primates. In turn, the primate story began more than 250 million years ago, with the origin of mammals.

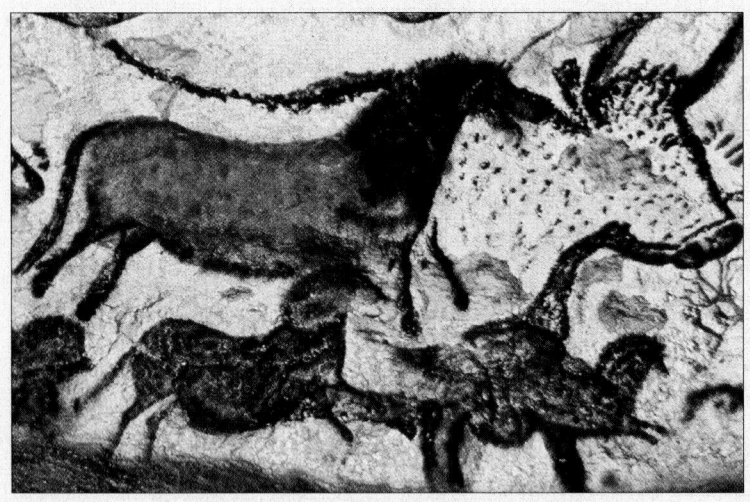

a

Figure 23.1 Part of the human cultural heritage—prehistoric cave paintings, a unique outcome of a long history of biological evolution.

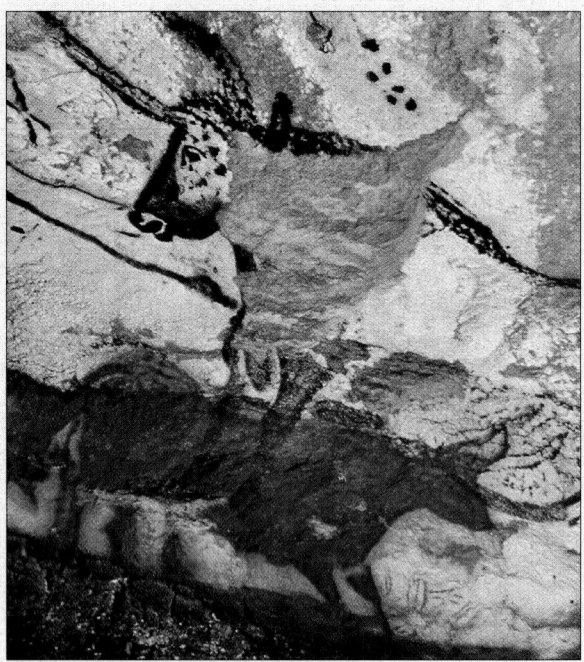

b

Figure 23.2 East Africa's Rift Valley, 3,200 kilometers long. Somewhere in this immense valley, sparsely wooded grasslands were the birthplace of the human lineage.

The story extends back to the origin of animals at some time before 750 million years ago—and so on back in time, to the origin of the first living cells.

As we sort through the branches of our family tree in this chapter, keep this greater evolutionary story in mind. *Our "uniquely" human traits emerged through modification of traits that had already evolved in ancestral forms.* At each branch in the family tree, certain mutations produced workable changes in traits, which proved useful in prevailing environments.

From this perspective, "ancient" cave paintings are the legacy of individuals who departed only yesterday, so to speak. The artists of Lascaux and Gargas are not remote from us. They *are* us.

KEY CONCEPTS

1. The primate branch of the mammalian lineage includes prosimians, tarsioids, and anthropoids, which encompass the monkeys, apes, humans, and ancestral humanlike forms. Humans and their humanlike ancestors are further classified as hominids.

2. Many early hominids became adapted to a wide range of challenges in complex, unpredictable environments.

3. The human capacity to make responses that work in different environments is the result of trends that occurred in certain primate lineages. These trends involved changes in bones, muscles, teeth, sensory systems, and the brain.

4. In ancestral primates, the skeleton became modified in ways that led to an upright stance. This change freed forelimbs and hands for new functions, including a capacity to hold, carry, use, and make objects.

5. Teeth became less specialized, allowing a greater variety of food sources to be tapped.

6. Reorganization of the skull and eye sockets led to increased reliance on daytime vision.

7. The brain enlarged and became more complex. Its evolution was interlocked with cultural evolution.

CHAPTER AT A GLANCE

23.1 Human Evolution in Perspective
23.2 From Primate to Human: Key Evolutionary Trends
23.3 The First Hominids
23.4 Emergence of Early Humans
23.5 *Out of Africa—Once, Twice, or . . .*

HUMAN EVOLUTION IN PERSPECTIVE

In the preceding chapters, we likened the history of life to a great tree. Each branch of the tree represents a line of descent, and the branch points represent a divergence leading to new species. There are more than 47,000 existing species of fishes, amphibians, reptiles, birds, and mammals on the vertebrate branch of that tree. When we turn to the evolution of any one of those species—as we do here—it helps to keep a key point in mind. At each crossroad leading to a new species, complex traits were already in place and functioning—*and new traits emerged only through modification of traits that already were in place*. The evolution of the human species speaks eloquently of this characteristic of life.

The Mammalian Heritage

Humans are members of the class Mammalia. Like other vertebrates, mammals have an internal skeleton with two key features: a nerve cord within a column of bones (backbone) and a skull that houses sense organs and a three-part brain (hindbrain, midbrain, and forebrain).

Mammals typically have hair. And of all vertebrates, only female mammals nourish their young with milk produced by mammary glands. Adult mammals feed and protect their young for an extended period and serve as models for their behavior. The young have an inborn capacity to learn and repeat a set of behaviors that have survival value. Mammals also show behavioral flexibility. They can expand on the basics with novel forms of behavior.

Mammals also are known by their **dentition**—the type, number, and size of teeth. As we will see, fossil remnants of teeth and jaws from human ancestors serve as a window on their diet and lifestyles.

Primates

Humans are **primates** (Figure 23.3). This order includes prosimians, tarsioids, and anthropoids. Figure 23.4 shows a few of these distinctive mammals.

Monkeys, apes, and humans are **anthropoids**. In their genes and body structure, apes are far more similar to humans than to monkeys. That is why apes, humans, and extinct species of the human lineage are classified together, as **hominoids**. A divergence from their common ancestor began many millions of years ago. All species on the separate evolutionary road leading to humans are further classified as **hominids**.

Most primates live in tropical or subtropical forests, woodlands, or savannas (open grasslands with a few stands of trees). Like their ancient ancestors, the vast majority are tree dwellers. Yet no one major feature sets

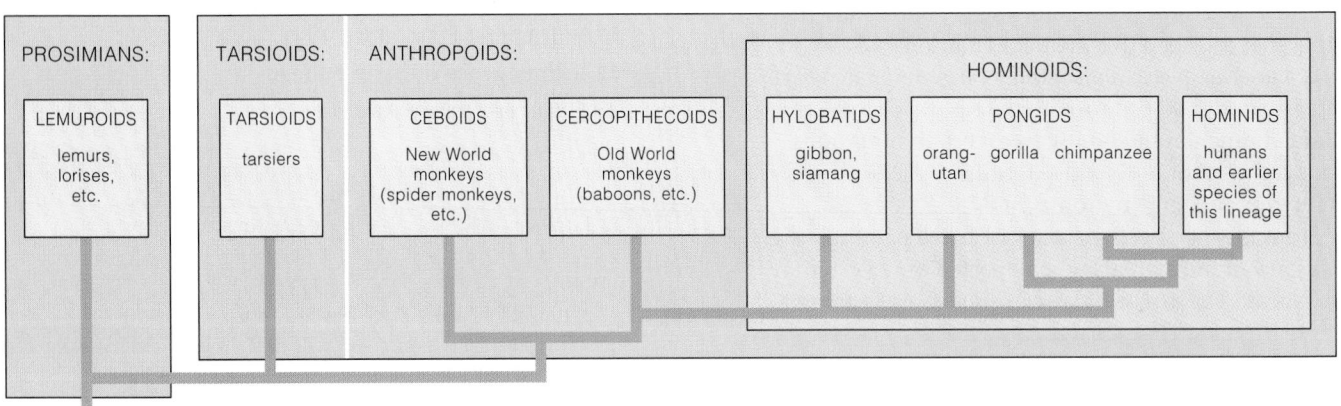

tree-dwelling, rodentlike
primate of Paleocene

Figure 23.3 Simplified family tree for primates. The "branch points" are junctures where lines diverged, a process that always begins with speciation. Major groups of living primates are indicated by the *gold* boxes.

a

b

c

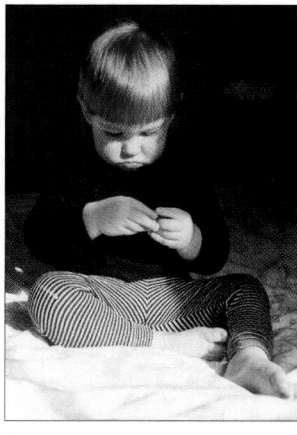

d

Figure 23.4 Representative primates. Gibbons (**a**) have limbs and a body adapted for brachiation (swinging arm over arm through the trees). Monkeys are quadrupedal (four-legged) climbers, leapers, and runners, as the spider monkey in (**b**) demonstrates. Tarsiers (**c**) are vertical clingers and leapers. (**d**) The most familiar primate of all.

"the primates" apart from other mammals. Each primate lineage evolved in a distinct way and has its own defining traits. Five trends define the human lineage, which is our focus here. They were set in motion by selection pressures that operated when primates first started adapting to life in the trees, and they contributed to the emergence of modern humans.

1. Skeletal changes led to upright walking, which freed the hands for new functions.

2. Changes in bones and muscles led to refined hand movements.

3. There was less reliance on the sense of smell and more on daytime vision.

4. Changes led to fewer, less specialized teeth.

5. Brain elaboration and changes in the skull led to speech. These developments became interlocked with each other and with cultural evolution.

In the next section we will explore these trends in more detail and then briefly consider how various selection pressures may have influenced the evolution of key primate—and ultimately, human—traits.

As members of the class **Mammalia**, humans have the defining mammalian traits of body hair, mammary glands (in females) that produce milk for young, an extended period of parental care of young, and notable behavioral flexibility. Humans also have characteristic mammalian dentition.

Humans also are primates, belonging to a primate subgroup known as the anthropoids. Apes and all species of the human lineage are classified as hominoids; only species of the lineage leading to humans are hominids.

Modern humans emerged as ancestral groups evolved adaptations to particular aspects of their environment. This evolutionary journey is marked by clear anatomical, physiological, and behavioral trends.

FROM PRIMATE TO HUMAN: KEY EVOLUTIONARY TRENDS

Upright Walking

Of all primates, only humans can stride freely on two legs for long periods of time. This habitual two-legged gait, called **bipedalism**, emerged through skeletal reorganizations in our primate ancestors. Compared with monkeys and apes, humans have a shorter, S-shaped, and somewhat flexible backbone. The position and shape of the human backbone, knee and ankle joints, and pelvic girdle are the basis of bipedalism (Figure 23.5). By current thinking, the evolution of bipedalism was the pivotal modification in the origin of hominids.

Precision Grip and Power Grip

The first mammals spread their toes apart to help support the body as they walked or ran on four legs. Primates still spread their toes or fingers. Many also make cupping motions, as when monkeys lift food to the mouth. Two other hand movements developed in ancient tree-dwelling primates. Through alterations in handbones, fingers could be wrapped around objects (prehensile movements), and the thumb and tip of each finger could touch (opposable movements).

In time, hands began to be freed from load-bearing functions. Later, when hominids evolved, refinements in hand movements led to the precision grip and power grip:

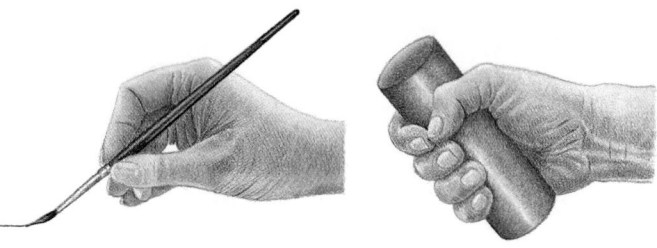

These hand positions gave early humans the capacity to make and use tools. They were a foundation for unique technologies and cultural development.

Enhanced Daytime Vision

Early primates had an eye on each side of the head. Later ones had forward-directed eyes, an arrangement that is better for sampling shapes and movements in three dimensions. Further modifications allowed the eyes to respond to variations in color and light intensity (dim to bright). Responsiveness to such complex visual stimuli is advantageous to life in the trees.

Changes in Dentition

Before hominids evolved, modifications in primate jaws and teeth accompanied a shift from eating insects, to fruits and leaves, and on to a mixed diet. Later, rectangular jaws and long canine teeth came to be further defining features of monkeys and apes. Along the road leading to humans, a bow-shaped jaw and teeth that were smaller and all about the same length evolved.

Figure 23.5 Comparison of the skeletal organization and stance of modern monkeys, apes (the gorilla is shown here), and humans. Modifications of the basic mammalian plan have allowed three distinct modes of locomotion. The quadrupedal monkeys climb and leap, and apes climb and swing by their forelimbs. Both modes of locomotion are well suited for life in the trees. On the ground, apes engage in "knuckle-walking." Humans are habitual two-legged walkers.

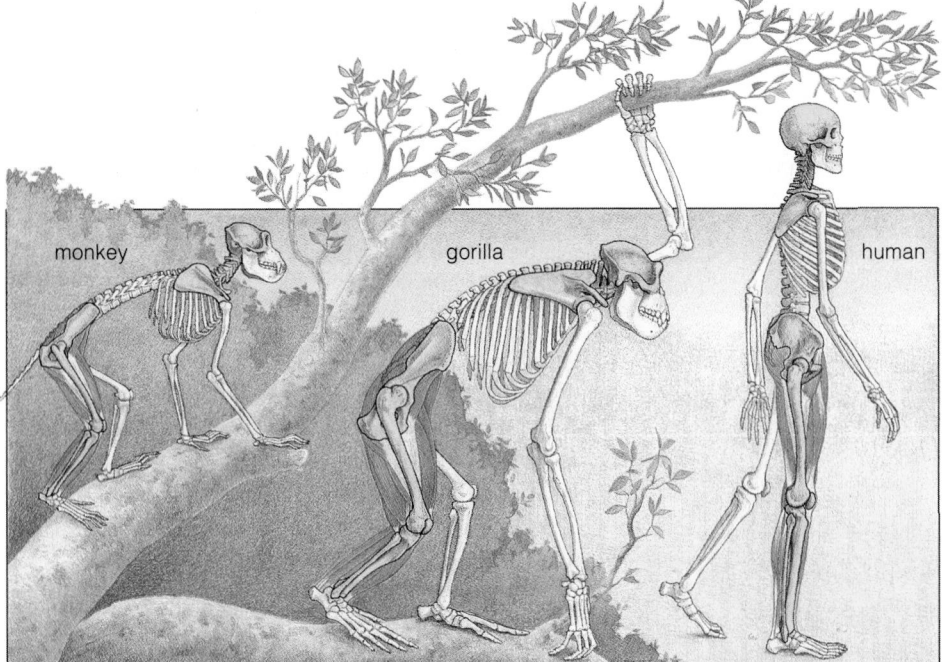

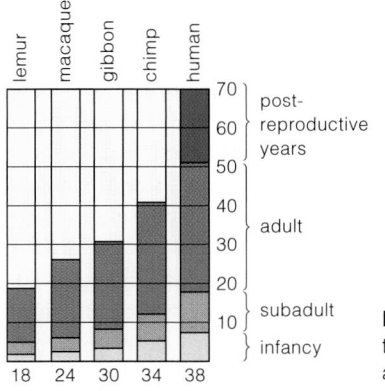

Figure 23.6 Trend toward longer life spans and longer dependency among primates.

Changes in the Brain and Behavior

Living on tree branches favored shifts in reproductive and social behavior. Imagine the advantages of single births over litters, for example, or of clinging longer to the mother. In many lineages, parents started to invest more effort in fewer offspring. They formed strong bonds with their young, maternal care became intense, and the learning period grew longer (Figure 23.6).

Brain regions concerned with encoding and processing information started expanding greatly. New behavior stimulated development of those regions—which in turn stimulated more new behavior. And so brain modifications and behavioral complexity became highly interlocked. The interlocking is most evident in the parallel evolution of the human brain and culture. **Culture** is the sum total of behavior patterns of a social group, passed between generations by learning and by symbolic behavior—especially language. The capacity for language arose among ancestral humans. And it arose

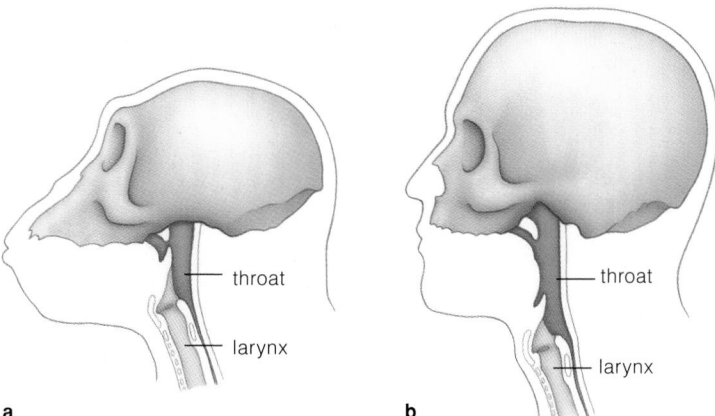

Figure 23.7 Structural basis of speech. (**a**) Like modern chimpanzees, early humans had a skull with a flattened base. Their larynx (a tube leading to the lungs) was not far below the skull, so the throat volume was small. (**b**) In modern humans, the skull's base angles down sharply as it develops. This moves the larynx down also. The result is an increase in the volume of the throat—the area in which sounds are produced.

through changes in the skull bones and expansion of parts of the brain (Figure 23.7).

Primate Origins

Primates evolved from ancestral mammals more than 60 million years ago. The first ones resembled small rodents or tarsiers (Figure 23.4c). Like tarsiers, they may have had huge appetites and foraged at night for insects, seeds, buds, and eggs near the forest floor. They had a long snout and a good sense of smell, useful for detecting predators or food. They could claw their way up through the shrubbery, although not with much speed.

Between 54 and 38 million years ago (the Eocene epoch), some primates were staying in the trees. Fossils give evidence of increased brain size, a shorter snout, enhanced daytime vision, and refined grasping movements. How did these traits evolve?

Consider the tree. Trees offered food and safety from ground-dwelling predators. They also were a habitat of uncompromising selection. Imagine dappled sunlight, boughs swaying in the wind, colorful fruit hidden among the leaves, perhaps predatory birds. A long, odor-sensitive snout would not have been of much use in the trees, where air currents disperse odors. But a brain that assessed depth, shape, movement, and color would have been a definite plus. So would a brain that worked fast when its owner was running and leaping (especially!) from branch to branch. Distance, body weight, winds, and suitability of the destination had to be estimated, and adjustments for miscalculations had to be quick. By 36 million years ago (before the dawn of the Oligocene), tree-dwelling anthropoids had evolved in tropical forests. One form, the squirrel-sized *Catopithecus*, was on or very close to the evolutionary road that led to monkeys, apes, and humans.

During the Miocene (25 to 5 million years ago), land masses were drifting into their current positions, and colossal uplifting produced high mountain ranges. The global climate became cooler and drier, and an adaptive radiation of apelike forms—the first hominoids—took place. By 13 million years ago, hominoids had spread into Africa, Europe, and southern Asia. Most became extinct by 7 to 8 million years ago. But genetic studies point to three divergences that occurred between 10 million and 5 million years ago. Two branchings of the Miocene apes led to gorillas and chimpanzees. The third gave rise to hominids, including the ancestors of humans.

These features emerged along the evolutionary road leading to humans: upright walking, refined hand movements, generalized dentition, refined vision, and the interlocked elaboration of brain regions and cultural behavior.

THE FIRST HOMINIDS

Most of the early hominids we know about lived in the East African Rift Valley and in southern Africa. By 5 to 10 million years ago (mya), once-vast tropical forests, with their bounty of edible soft fruits, leaves, and insects, gave way to dry woodlands and grasslands as the climate became more seasonal—and, accordingly, with seasonal shifts in water and food supplies. The African savanna had emerged. The new foods became increasingly hard to find, and early hominids had two options: Move into the new adaptive zone, or die out.

Fossil fragments 4 to 4.5 million years old indicate that this was a "bushy" period of evolution; a confounding variety of hominids appeared (Figures 23.8 and 23.9). At present it is impossible to figure out their family tree, so they are simply grouped as **australopiths** (meaning southern apes). *Australopithecus anamensis* is the oldest of these. Like *A. afarensis* and *A. africanus* it was gracile (slight of build). The forms called *A. boisei* and *A. robustus* were robust (muscular and heavily built).

We don't know how the australopiths were related or which ones may have been ancestral to us (Figure 23.10). But all shared three features. *First*, they walked upright, which meant their hands were free for new functions. *Second*, their jaws and large, thickly enameled teeth accommodated a variety of foods. *Third*, with an average volume of 400 cubic centimeters, their brain was not

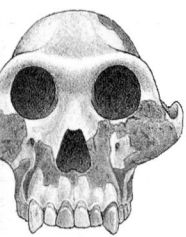
A. afarensis

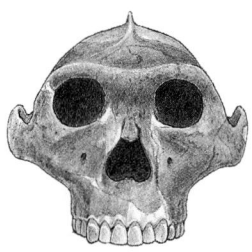
A. africanus

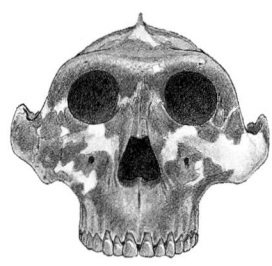
A. robustus

A. boisei

Figure 23.8 Skulls of australopiths, which are apelike in proportion and size. These reconstructions are based on fossilized fragments. The skulls of *A. afarensis*, *A. robustus*, and *A. boisei* males were crested, like a gorilla's, and provided attachment sites for large jaw muscles. Like apes, *A. afarensis* had a large face and projecting jaws. But its canines and molars were humanlike.

a　　　　　　b　　　　　　　　　　c　　　　　　　　　　　d

Figure 23.9 (**a**) Remains of "Lucy" (*Australopithecus afarensis*), who lived 3.2 million years ago. (**b**) At Laetoli, Tanzania, Mary Leakey found these footprints, made in soft, damp volcanic ash 3.7 million years ago. (**c,d**) The arch, big toe, and heel marks of such footprints are signs of bipedal hominids. Unlike apes, the early hominids did not have a splayed big toe, as the chimpanzee in (**c**) obligingly demonstrates.

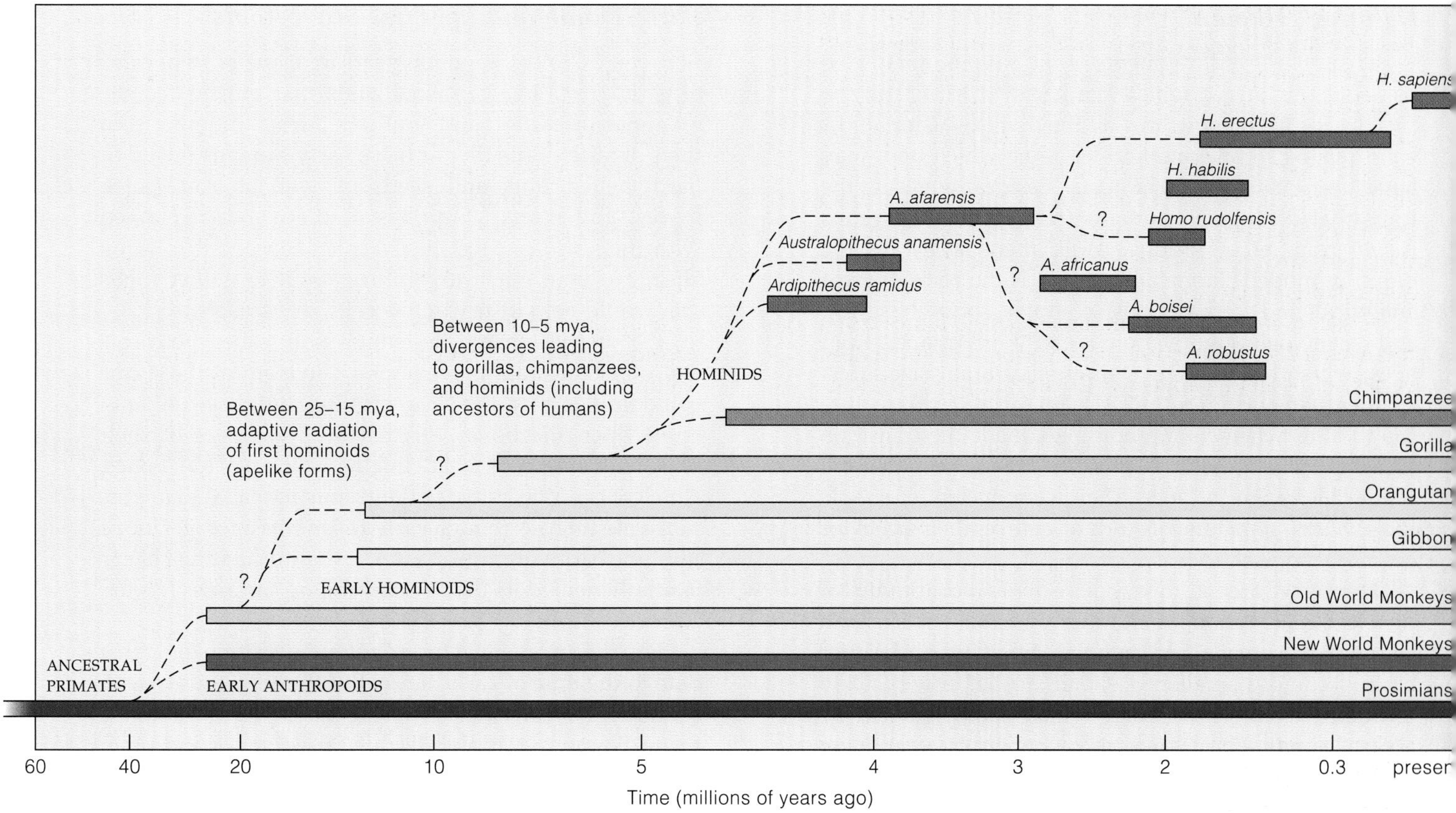

Figure 23.10 One proposed evolutionary tree for primates.

especially large, but it probably was more complex than that of their forerunners. At the very least, these hominids had to be able to think ahead, to plan when and where to get different foods when seasonal supplies ran out. All three features resulted from modifications of traits that we see among other primates. In other words, *they were based on the primate heritage.*

As one example, consider the *A. afarensis* female dubbed Lucy who lived 3.2 million years ago (Figure 23.9a). She was only about 3½ feet tall and in some respects was chimplike. Yet her foot bones and legbones had muscle attachment sites similar to those in your own body. Her thighbones angled inward so that her body weight had to be centered directly beneath her pelvis—a sure sign of bipedalism. And the most telling sign—australopiths left footprints (Figure 23.9b).

By 36 million years ago primate descendants included ancestors of monkeys, apes, and the hominids.

Between 10 million and 5 million years ago, a divergence led to the australopiths, the first hominids. They were apelike in some ways, humanlike in others. Unlike earlier forms, they were bipedal, they had smaller canines and a more flattened face, and their brain probably was more complex.

EMERGENCE OF EARLY HUMANS

Defining "Human"

What can fossilized fragments of the early hominids tell us about our own origins? The fossil record is still too sketchy for us to know how all the diverse australopiths were related to one another, let alone which ones may have been ancestral to humans. Besides, what *are* the traits that distinguish **humans** (members of the genus *Homo*) from other groups? Well, there is always the brain. In modern humans the brain is the basis of great analytical and verbal skills, complex social behavior, and technological innovation. It easily sets us apart from apes, which have a skull volume and brain size much smaller than ours (Figure 23.11). Yet by itself, this feature can't tell us when certain hominids made the transition to being human, because their brain size probably fell within the range for apes. They were makers of simple tools, but so are chimps. And, of course, their behavior did not lend itself to fossilization.

Hence, we are left to speculate about a continuum of physical traits among a number of fossils—a skeleton adapted for bipedalism, manual dexterity, and larger brain volume; a smaller face; and smaller, more thickly enameled teeth. These traits, which had originated by about 2.5 million years ago, were evident in what many consider to be the earliest humans, *Homo habilis* (meaning "handy man").

Early *Homo* and the First Stone Tools

Between 2.5 and 1.6 million years ago, one or two forms of early *Homo* lived in woodlands that punctuated the savannas of eastern and southern Africa. Judging from their dentition, the diet of these early humans included hard-shelled nuts and seeds as well as soft fruits, leaves, and insects. Supplies of these foods changed with the seasons. Most likely, early *Homo* had to think ahead, to plan when to venture about to gather foods that would help it survive the cold, dry seasons.

Early *Homo* shared its habitat with saber-tooth cats and other formidable predators. The teeth of those cats could impale prey and shear off flesh, but they could not crush open the bones for marrow. Carcasses of their kills, with shreds of meat clinging to the marrow bones, were concentrated stores of nutrients in places that were nutrient-stingy. Although early *Homo* was a forager, not a full-time carnivore (meat-eater), when it had the chance it supplemented its diet by scavenging carcasses (Figure 23.11c).

Fossil hunters have found numerous stone tools that date to the time of early *Homo*. But they cannot say with certainty that early *Homo* was the only species that made them. Possibly australopiths also used sticks and other perishable tools before then, as modern apes do, but we have no way of knowing.

Maybe individuals on the road to modern humans started down a toolmaking road by picking up rocks to crack marrow bones. Maybe they started to scrape flesh from bones with small, sharp flakes that had fractured naturally from rocks. Eventually, early humans started *shaping* stone implements. Paleoanthropologist Mary Leakey was the first to discover evidence of toolmaking at Africa's Olduvai Gorge, which cuts through a great sequence of sedimentary rock layers. The most ancient tools at this site are crudely chipped pebbles that were buried in the deepest layers. They may have been used to smash marrow bones, dig for roots, and poke insects from tree bark. More recent layers have more complex tools. Large numbers of bones and tools have been found along the shores of ancient lakes that would have beckoned plenty of thirsty animals.

At such sites we occasionally find fossils of one form of *H. habilis* that was twice as brainy as the australopiths and that obviously ate well. Although there apparently was selection pressure for more creativity in securing food resources, the stone tools of *H. habilis* did not change much for the next 500,000 years.

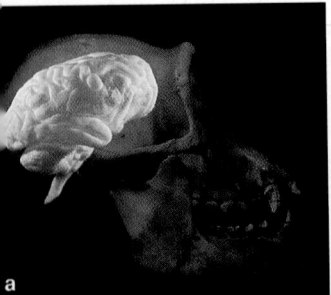
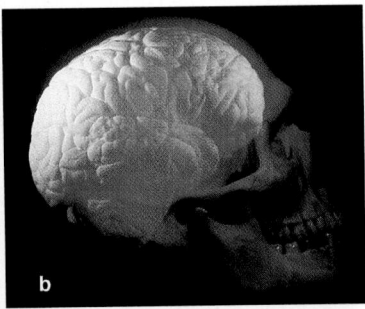

a   b

c

Figure 23.11 (**a**) Image of the brain of a modern chimpanzee, superimposed on a chimp skull. The brain of Lucy, one of the australopiths (*A. afarensis*), was only a bit larger than this and much smaller than (**b**) the modern human brain. (**c**) Artist's rendition of *Homo habilis* males in an East African woodland.

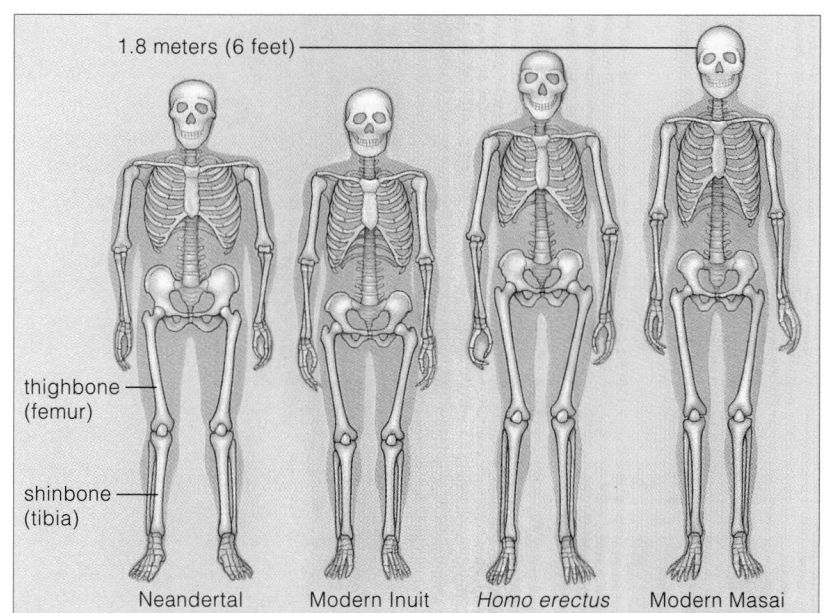

Figure 23.12 Climate and body build. Humans adapted to cold climates have a heat-conserving body—stockier, with shorter legs, compared to humans adapted to hot climates.

1.8 meters (6 feet)

thighbone (femur)

shinbone (tibia)

Neandertal   Modern Inuit   *Homo erectus*   Modern Masai

From Homo erectus *to* H. sapiens

The ancestors of modern humans apparently stayed put in Africa until about 2 million years ago. At that time, genetic divergence from early members of *Homo* led to **Homo erectus**, a human species that the fossil record places on the evolutionary road to modern humans. Its name means "upright human." Although its forerunners also were upright, two-legged walkers, *H. erectus* did the name justice; its populations walked out of Africa, and turned left into Europe or right into Asia. It took some of them a long time to walk the 14,000 kilometers to China. *H. erectus* fossils from Southeast Asia and the Republic of Georgia are 1.8 million and 1.6 million years old. The treks were strenuous. More than once, immense glaciers advanced down through northern Europe and parts of Asia.

Whatever selection pressures triggered the adaptive radiations, this was a time of physical changes, as in skull size and leg length (Figure 23.12). It also was the time of cultural liftoff for the human lineage. *H. erectus* had a larger brain, it was a more creative toolmaker, and its social organization and communication skills must have been well developed to make such successful treks. From southern Africa to England, different populations were using the same variety of hand axes and other tools designed to pound, scrape, shred, cut, and whittle. They withstood environmental challenges by building fires and using furs for clothing. Remains of fire use date from an ice age that occurred about a million years ago.

Judging from fossils in the Middle East, **Homo sapiens** had evolved by 100,000 years ago. The origin

and radiations of early *H. sapiens* are hotly debated topics (Section 23.5). Early *H. sapiens* had smaller teeth and jaws than *H. erectus*. Its facial bones were smaller, the skull was higher and rounder, and the brain was larger. Analysis of fossils indicates that these early forms may have had the capacity for complex language.

One group of early humans, the Neandertals, lived in Europe and the Near East from 150,000 to 35,000 years ago. Massively built and large brained, some of their populations were the first to adapt to the coldest regions (Figure 23.12). Their disappearance coincided with the appearance of anatomically modern humans in the same regions 40,000 to 30,000 years ago. There is no evidence that they warred or interbred with these later arrivals. We don't yet know what happened to them.

From 40,000 years ago to today, human evolution has been almost entirely cultural, not biological—and so we leave the story with these conclusions: Humans spread through the world by rapidly devising *cultural* means to deal with a broad range of environments. Compared with their predecessors, they developed rich and varied cultures. Even though hunters and gatherers persist in parts of the world, others moved from "stone-age" technology to the age of "high tech," attesting to the great plasticity and depth of human adaptations.

Cultural evolution has outpaced the biological evolution of the only remaining human species, *Homo sapiens*. Today, humans everywhere rely on cultural innovation to adapt rapidly to a broad range of environmental challenges.

OUT OF AFRICA—ONCE, TWICE, OR . . .

If researchers are interpreting the fossil record of human evolution correctly, then it would seem that Africa was the cradle for us all. *H. erectus* coexisted for a time with earlier humans (*H. habilis*) before dispersing from the African savannas to the cooler grasslands, forests, and mountains of Europe and Asia. They apparently left Africa in waves between about 2 million and 500,000 years ago. Judging from recent examination of *H. erectus* fossils from Java, some populations may have survived, in relative isolation, until 27,000 to 53,000 years ago.

Where, on the larger geologic stage, do we place the origin of *H. sapiens*? Here we find a good example of how the same body of evidence can be interpreted in different

ways. These interpretations are respectively called the multiregional model and the African emergence model for modern human origins. Both attempt to explain the world distribution of fossils of particular ages and the measured genetic distances among modern, existing human populations. For example, Figure 23.13 includes locations of *H. sapiens* fossils that have been dated to particular times. As another example, evidence from biochemical studies suggests the greatest genetic distance separates *H. sapiens* populations native to Africa from populations everywhere else; the next greatest distance separates Southeast Asia (and Australia) from everywhere else (Figure 23.14).

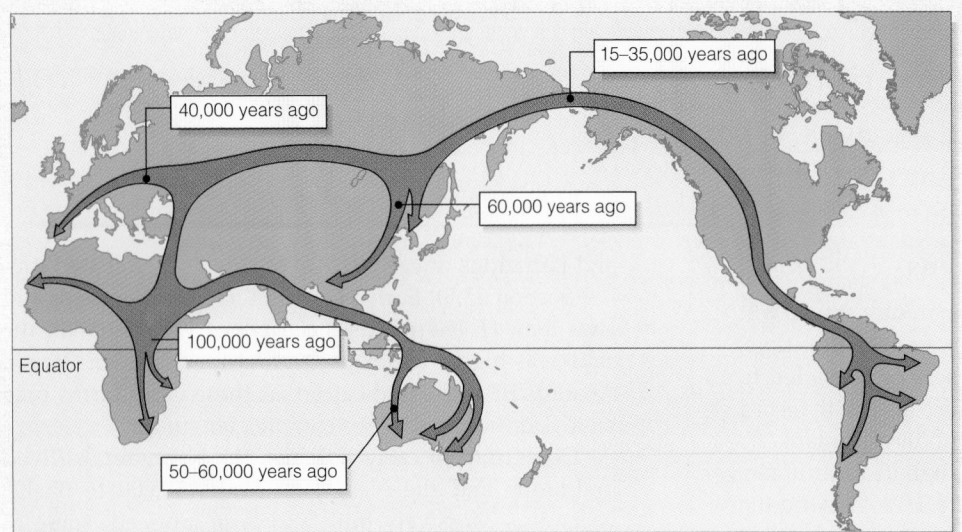

Figure 23.13 Dates when early *H. sapiens* populations were colonizing different parts of the world, based on fossil evidence. As shown here, the presumed dispersal routes (*brown* arrows) seem to support the African emergence model.

MULTIREGIONAL MODEL   By this model, *H. erectus* populations had spread through much of the world by 1 million years ago. Those geographically separated groups were subject to different selection pressures, so their traits evolved in regionally distinctive ways. Then subpopulations ("races") of *H. sapiens* evolved from them in different places. Although they differed phenotypically, they did not evolve into separate species because gene flow continued among them, even to the present day. (For example, while the armies of Alexander the Great were sweeping eastward, they also contributed blue-eye genes from the Greeks to the allele pool of generally brown-eyed subpopulations in Africa, the Near East, and Asia.)

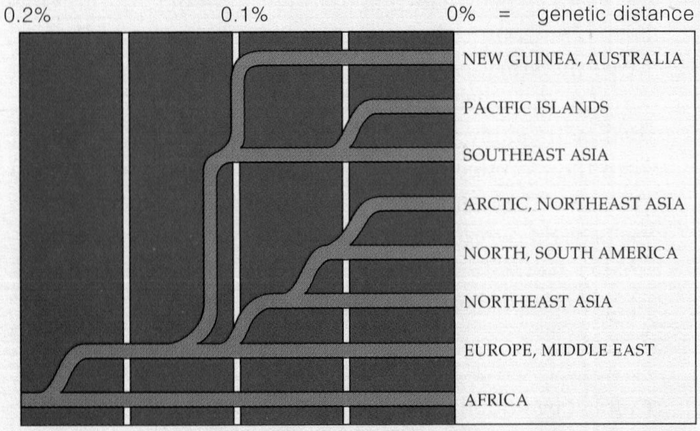

AFRICAN EMERGENCE MODEL   This model does not dispute fossil evidence that populations of *H. erectus* evolved in distinctive ways in different regions. But it holds that *H. sapiens*—modern humans—originated in sub-Saharan Africa somewhere between 200,000 and 100,000 years ago. Only later did *H. sapiens* populations move out of Africa, then into other regions along routes indicated by the Figure 23.13 map. In each region that *H. sapiens* populations settled, they replaced the archaic *H. erectus* populations that had preceded them. Only then did regional phenotypic differences become superimposed on the original *H. sapiens* body plan.

In support of this model, the oldest known *H. sapiens* fossils are indeed from Africa. Also, in Zaire, the discovery of a finely wrought barbed-bone harpoon and other tools suggests the African populations were as skilled at making tools as *Homo* populations known earlier from Europe.

Figure 23.14 One proposed family tree for populations of modern humans (*Homo sapiens*) that are native to different regions of the world. Branch points indicate presumed genetic divergences.

SUMMARY

1. Primates include prosimians (lemurs and related forms) and anthropoids (including monkeys, apes, and humans). Apes and humans are hominoids; only modern humans and others of their lineage (from *A. afarensis* to *H. erectus*) are further classified as hominids.

2. Important evolutionary trends in the primate lineage leading to *Homo sapiens* include: (1) a transition from a four-legged gait to bipedalism in the hominids, with related changes in the skeleton; (2) increased manipulative skills due to modification of the hands, which began to be freed from their locomotion function among tree-dwelling primates; (3) less reliance on the sense of smell and more reliance on enhanced daytime vision, including color vision and depth perception; (4) transition from specialized to omnivorous eating habits; and (5) brain expansion and reorganization. Larger, more complex brains are correlated with increasingly sophisticated technology (from simple to refined tools) and with social and linguistic development in the human lineage.

3. The first hominids (australopiths) emerged between 10 and 5 million years ago. All were bipedal.

4. Fossils of the first known representative of the human genus date from 2.5 million years ago. Early *Homo* was omnivorous, was larger-brained and taller than its predecessors, and used simple tools.

5. *Homo erectus* fossils associated with abundant cultural artifacts date from about 1.9 million years ago. This form was adapted to a wide range of habitats. Fully human forms (*H. sapiens*) emerged by 100,000 years ago, from *H. erectus* stock. By 40,000 years ago, modern forms had evolved. Ever since, cultural evolution has outstripped biological evolution of *Homo sapiens*.

Review Questions

1. What are the general evolutionary trends that occurred among the primates as a group? What way of life apparently was the foundation for these trends? 477–481

2. What is the difference between "hominoid" and "hominid"? Are we hominoids, hominids, or both? 476

Critical Thinking: You Decide (Key in Appendix V)

1. In 1992 the frozen body of a Stone Age man was discovered in the Austrian Alps. Although this "Iceman" died about 5,300 years ago, his body is amazingly intact and researchers have begun to analyze DNA extracted from bits of his tissue. Can these studies tell us something about early human evolution? Explain your reasoning.

2. Fossil evidence suggests that Neandertals and archaic human populations coexisted in the same regions for 25,000 to 50,000 years. By one hypothesis, the Neandertals evolved in Europe, then migrated to the Middle East, while the archaic human populations migrated out of Africa. Both apparently used the same kinds of tools, and the fact of their extended coexistence suggests they did not compete with each other. However, the Neandertals became extinct, and the other group gave rise to populations of fully modern humans. Does this scenario fit better with the multiregional model or the African emergence model of modern human origins?

Self-Quiz *(Answers in Appendix IV)*

1. Primates include _____.
   a. lemurs   c. apes   e. all of the above
   b. monkeys   d. humans

2. _____ are hominoids; only _____ are hominids.
   a. Lemurs; monkeys and their immediate ancestors
   b. Apes and humans; humans and their recent ancestors
   c. Monkeys; apes and their recent ancestors
   d. Monkeys, apes, and humans; apes and their recent ancestors

3. The pivotal modification in the origin of hominids was _____.
   a. transition to bipedalism
   b. hand modification that increased manipulative skills
   c. shift from omnivorous to specialized eating habits
   d. less reliance on smell, more on vision
   e. brain expansion and reorganization

4. Early hominids displayed great plasticity, meaning that _____.
   a. they were adapted to a wide range of demands in complex environments
   b. they had flexible bones that cracked easily
   c. they were limber enough to swing through the trees
   d. they were adapted for a narrow range of demands in complex environments

5. Primates evolved from ancestral mammals about _____.
   a. 80 million years ago
   b. 60 million years ago
   c. 120 million years ago
   d. 50,000 years ago

6. The first known hominids are generally classified as _____.
   a. *Homo*   c. cercopiths
   b. dryopiths   d. australopiths

7. Fossils of early *Homo*, the first known representative of the human line, date from _____ million years ago.
   a. 8   c. 4
   b. 6   d. 2.5

Selected Key Terms

anthropoid *476*	hominid *476*
australopith *480*	hominoid *476*
bipedalism *478*	*Homo erectus 483*
culture *479*	*Homo sapiens 483*
dentition *476*	human *482*

Readings

Gould, S. J., general editor. 1993. *The Book of Life.* New York: Norton. Splendid, easy-to-read essays, gorgeous illustrations.

Milton, K. August 1993. "Diet and Primate Evolution." *Scientific American.*

"Shaking the Tree." December 1992. *Scientific American.* How DNA analysis may help pinpoint the locale—or locales—where modern humans arose.

Weiss, M., and A. Mann. 1990. *Human Biology and Behavior.* 5th ed. New York: HarperCollins.

 For additional readings, go to InfoTrac College Edition, your online library at:

http://www.infotrac-college.com/wadsworth

24

ECOSYSTEMS

Crêpes for Breakfast, Pancake Ice for Dessert

Think of Antarctica, and you think of ice. Mile-thick slabs of the stuff hide all but a small fraction of a continent whipped by fierce winds and kept frozen by murderously low temperatures, on the order of –100°F. And yet, on patches of exposed rocky soil and on nearby islands, mosses and lichens grow. There, during the breeding season, penguins as well as seals form great noisy congregations, then reproduce and raise offspring. They cruise offshore or venture out in the open ocean in pursuit of food—krill, fishes, and squids.

The first explorers of Antarctica called it The Last Place on Earth. For all its harshness, though, the eerie, unspoiled beauty fired the imaginations of those first travelers and others who followed them.

In 1961, 38 nations signed a treaty to set aside Antarctica as a reserve for scientific research. This was the start of scientific outposts—and of the trashing of Antarctica. At first, researchers discarded a few oil drums and old tires. Then prefabricated villages went up. Onto the ice or into the water went garbage and used equipment. Just offshore, marine life became exposed to sewage and chemical wastes.

Antarctica also became a destination of cruise ships. Each summer thousands of tourists, fortified by three sumptuous meals a day plus snacks, are ferried from ship to land across channels glistening with pancake ice (Figure 24.1). They trample vegetation and weave among penguins, cameras clicking. Some scientists wonder whether there is a connection between this activity and the fact that at the communal breeding ground at Cape Royds, the penguin population has declined by half.

Meanwhile, nations looking for new sources of food started licking their chops over Antarctica's krill. Word also got out about potentially rich deposits of uranium, oil, and gold—and by 1988, treaty nations were poised to authorize digs and drillings. Of course, oil spills or krill harvesting could destroy the fragile ecosystem.

a

For example, the tiny, shrimplike krill are all that Adélie penguins eat. Krill are a key food for baleen whales and other marine animals that are, in turn, food for still others.

In 1991, the treaty nations thought about all of this and imposed a 50-year ban on mineral exploration. Research stations started burying or incinerating wastes, treating sewage, and taking other measures to curb pollution. Tour operators promised to supervise tourists. Krill are not being harvested on a massive scale—yet.

Because Antarctica seems so remote from the rest of the world, it is easier to see how life is interconnected there. We can ask whether harm to one species or one habitat will lead to collapse of the whole, and be fairly sure of the answer. What about places not as sharply defined? Does interconnectedness prevail there, also? Are other places as vulnerable to disturbance or more resilient? These topics will now occupy our attention.

b

Figure 24.1 (**a**) A boatload of tourists crossing a channel of "pancake ice" off the coast of Antarctica. (**b**) On the shore of an island near Antarctica, tourists get close to nature—maybe too close.

KEY CONCEPTS

1. An ecosystem is an association of organisms and their physical environment, linked by a flow of energy and a cycling of materials through it.

2. Every ecosystem is an open system, with inputs and outputs of both energy and nutrients.

3. Over time, energy flows in one direction through an ecosystem. Most commonly, the flow begins when photosynthesizing organisms (mainly green plants) harness sunlight energy and convert it to forms that they and other organisms of the ecosystem can use. Photo-synthesizers are "primary producer" organisms for the ecosystem.

4. Energy-rich organic compounds that the primary producers manufacture become incorporated in their body tissues. They are stored forms of energy, and they serve as the foundation for the ecosystem's food webs. Such webs consist of a number of interconnected food chains.

5. Each chain in a food web extends in a straight-line sequence from producers through arrays of consumers, decomposers, and detritivores ("detritus eaters"). Each chain also connects with other chains at different feeding levels.

6. Water and major nutrients move from the physical environment, through organisms, then back to the environment. Their movements, called biogeochemical cycles, take place on a global scale.

7. Human activities are disrupting the natural cycles of nature and so are endangering ecosystems.

CHAPTER AT A GLANCE

24.1 Introduction to Principles of Ecology

24.2 Ecosystem Organization

24.3 Energy Flow through Ecosystems

24.4 Case Study: Energy Flow at Silver Springs, Florida

24.5 Biogeochemical Cycles—An Overview

24.6 The Hydrologic Cycle

24.7 The Carbon Cycle

24.8 *From Greenhouse Gases to a Warmer Planet*

24.9 The Nitrogen Cycle

24.10 The Phosphorus Cycle

24.11 *Transfer of Harmful Compounds through Ecosystems*

INTRODUCTION TO PRINCIPLES OF ECOLOGY

The living world encompasses the **biosphere**—those regions of the Earth's crust, waters, and atmosphere in which organisms live. The Earth's surface is remarkably diverse. In climate, soils, vegetation, and animal life, its deserts differ from hardwood forests, which differ from tropical rain forests, prairies, and arctic tundra. Oceans, lakes, and rivers differ in their physical properties and arrays of organisms (Figure 24.2).

Ecology is the study of the interactions of organisms with one another and with the physical environment. The general type of place in which a species normally lives is its **habitat**. For example, muskrats live in a stream habitat, damselfish in a coral reef habitat. The habitat of any organism characteristically has certain physical and chemical features. Every species also interacts with others that occupy the same habitat. Humans live in "disturbed" habitats, which have been deliberately altered for agriculture, urban development, and other purposes. In any given habitat the populations of all species directly or indirectly associate with one another as a **community**.

What is an "ecological niche"? The **niche** ("nitch") of a species consists of all the physical, chemical, and biological conditions the species needs to live and reproduce in an ecosystem. Examples of those conditions include the amount of water, oxygen, and other nutrients a species needs, the ranges of temperature it can tolerate, the places it finds food, and the type of food it consumes. *Specialist* species have narrow niches. They may be able to use only one or a few types of food or live only in one type of habitat. For example, the red-cockaded woodpecker builds its nest mainly in longleaf pines that must be at least 75 years old. Humans and houseflies are examples of *generalist* species with broad niches. Both can live in a range of habitats and eat many types of food.

An **ecosystem** consists of one or more communities of organisms interacting with one another *and* with the physical environment through a flow of energy and a cycling of materials. Figure 24.3 shows some of the interacting organisms in one type of ecosystem, arctic tundra.

Communities of organisms make up the *biotic*, or living, portions of an ecosystem. New communities may arise in habitats initially devoid of life or in disturbed but previously inhabited areas such as abandoned pastures. Immature or disturbed areas are ecologically unstable. Through a process called **succession**, the first species that thrive in a habitat are then replaced by others, which are replaced by others in orderly progression until the composition of species becomes steady as long as other conditions remain the same. This more or less stable array of species is a "climax community."

In *primary succession*, changes begin when pioneer species colonize a newly available habitat, such as a

a

b

Figure 24.2 Major types of ecosystems on Earth are called "biomes." They include land biomes such as the warm desert near Tucson, Arizona, shown in (**a**), and aquatic realms such as the mountain lake in (**b**), which lies in the Canadian Rockies.

recently deglaciated region (Figure 24.4). In *secondary succession*, a community develops toward the climax state after parts of the habitat have been disturbed. For example, this pattern occurs in abandoned fields, where wild grasses and other plants quickly take hold when cultivation stops. Natural disasters (such as forest fires caused by lightning), changing climate, and other factors often interfere with succession toward climax conditions. Hence, truly stable climax conditions are rarely achieved.

The general type of place where a species normally lives is its habitat. A community encompasses the populations of all species in a given habitat.

An ecosystem consists of one or more communities of organisms interacting with one another and with the physical environment through a flow of energy and a cycling of materials.

Figure 24.3 Some of the producers, consumers, and decomposers of the arctic tundra: sedges, mosses, and other plants, along with the lemming (eater of plant parts); the snowy owl (eater of lemmings); and a fungal decomposer. As in all ecosystems, these organisms interact with one another and with the physical environment through a one-way flow of energy and a cycling of materials.

a

b

Figure 24.4 Primary succession in the Glacier Bay region of Alaska (**a**), where glaciers are receding and changes in newly deglaciated regions have been well documented. (**b**) The first plants here were small flowering plants that can grow and spread over glacial till. Within 20 years, young alders begin to flourish (**c**). After 80 years, a climax community of spruces crowds out the mature alders (**d**).

c

d

ECOSYSTEM ORGANIZATION

Although there are many different types of ecosystems, they are all alike in many aspects of their structure and function. Nearly every ecosystem runs on energy from the sun. Plants and other photosynthetic organisms are **autotrophs** (self-feeders). They capture sunlight energy and convert it to forms they can use to build organic compounds from simple inorganic substances. By securing energy from the physical environment, autotrophs serve as the **producers** for the entire system (Figure 24.5).

All other organisms in the system are **heterotrophs**, "other-feeders." They depend directly or indirectly on energy stored in the tissues of producers. Some heterotrophs are **consumers**, which feed on the tissues of other organisms. Within this category, *herbivores* such as grazing animals and insects eat plants, *carnivores* such as lions and snakes eat animals, and *parasites* reside in or on living hosts and extract energy from them. Most humans are *omnivores*—they feed on a variety of food types. Other heterotrophs, the **detritivores**, get energy from partly decomposed particles of organic matter. Crabs, nematodes, and earthworms are examples. Still other heterotrophs, the **decomposers**, include fungi and bacteria that extract energy from the remains or products of organisms.

Autotrophs secure nutrients as well as energy for the entire system. As they grow, they take up water and carbon dioxide (as sources of oxygen, carbon, and hydrogen) along with dissolved minerals, including nitrogen and phosphorus. Such materials are building blocks for carbohydrates, lipids, proteins, and nucleic acids. When detritivores and decomposers get their turn at this organic matter, they can break it down completely to inorganic bits. If those bits are not washed away or otherwise removed from the system, autotrophs can use them again as nutrients.

Keep in mind that ecosystems are *open* systems. They are not self-sustaining. Ecosystems require a continual *energy input* (as from the sun) and often *nutrient inputs* (as from minerals carried by erosion into a lake). Ecosystems also have *energy output* and *nutrient outputs*. Energy cannot be recycled; in time, most of the energy originally fixed by autotrophs is lost to the environment in the form of metabolic heat. Nutrients typically are cycled, but some are still lost (as through soil leaching).

Feeding Relationships in Ecosystems

Each species in an ecosystem fits somewhere in a hierarchy of feeding relationships called **trophic levels** (from *troph*, meaning nourishment). A key aspect of ecosystem functioning is the transfer of energy from one trophic level to another.

Primary producers, which gain energy directly from sunlight, make up the first trophic level. Photosynthetic autotrophs in a lake (including some bacteria and aquatic

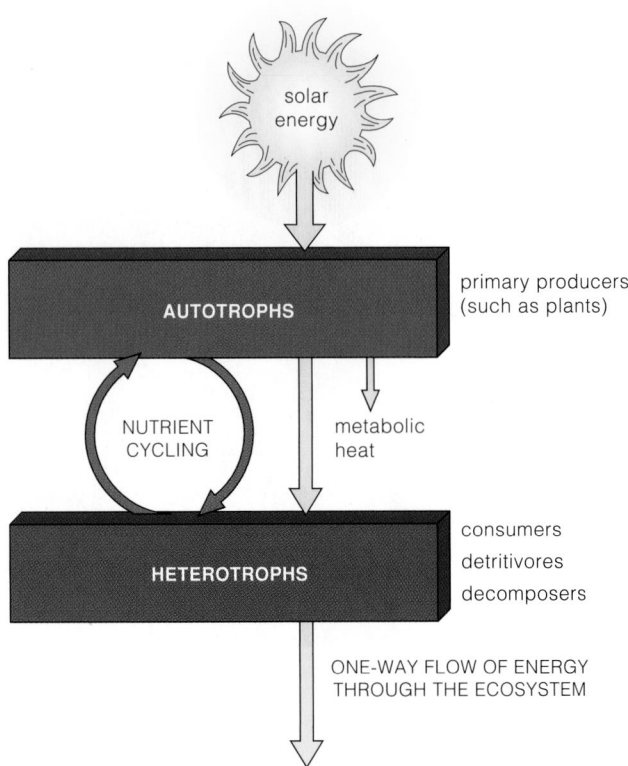

Figure 24.5 Generalized model of the one-way flow of energy and the cycling of materials through ecosystems.

plants) are examples. Snails and other herbivores feeding directly on the producers are at the next trophic level. Birds and other primary carnivores that prey directly on the herbivores form a third level. A hawk that eats a snake is a secondary carnivore. Decomposers, humans, and many other organisms can obtain energy from more than one source. They cannot be assigned to a single trophic level and are more like "trophic groups."

Food Webs

A straight-line sequence of who eats whom in an ecosystem is sometimes called a **food chain**. However, you will have a hard time finding such a simple, isolated sequence. Typically, the same food resource is part of more than one chain. This is especially true of plants and other resources at low trophic levels. It is more accurate to view food chains as cross-connecting with one another in **food webs**. Figure 24.6 shows an Antarctic food web.

A food web is a network of crossing, interlinked food chains encompassing primary producers and an array of consumers and decomposers.

By way of food webs, different species in an ecosystem are profoundly interconnected.

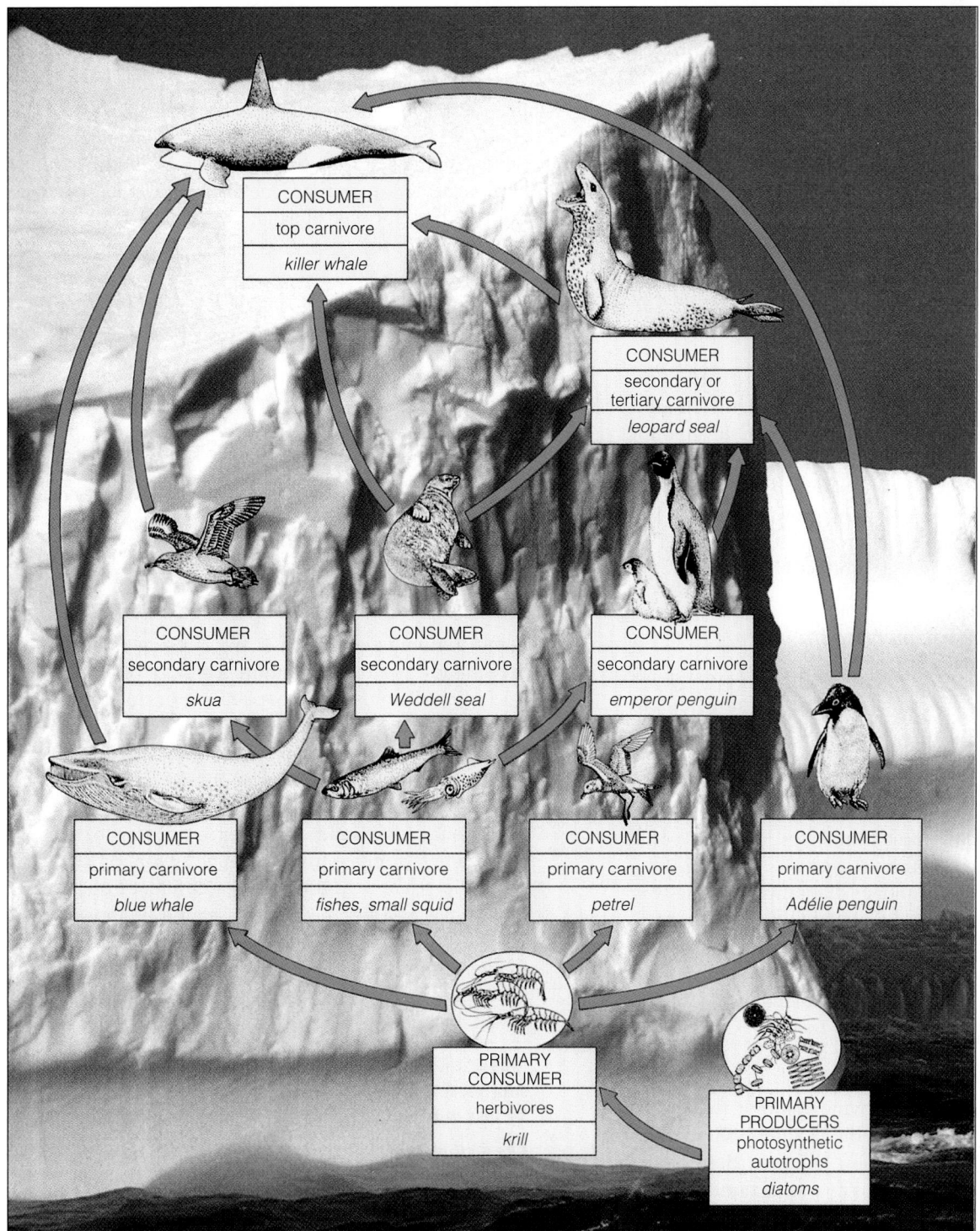

Figure 24.6 Simplified picture of a food web in the Antarctic. There are actually many more participants, including an array of decomposer organisms.

ENERGY FLOW THROUGH ECOSYSTEMS

Primary Productivity

To get an idea of how energy flow is studied, consider a land ecosystem for which multicellular plants are the primary producers. The plants trap sunlight energy and, through photosynthesis, convert it into the chemical energy of organic compounds in their tissues. The rate at which primary producers capture and store energy is an ecosystem's *gross* **primary productivity**. How much energy is stored depends on how many plants are present and on the balance between photosynthesis and energy used in aerobic respiration by the plants. This energy "bottom line" is *net* primary productivity.

Other factors affect the amount of net primary productivity, its seasonal patterns, and its distribution through the habitat. This is true of ecosystems on land and in the seas (Figure 24.7a). For example, the amount of new plant growth can vary depending on the availability of minerals, the temperature range, and the amount of sunlight and rainfall during the growing season. The harsher the environment, the fewer plant shoots will be produced—and the lower the productivity.

Major Pathways of Energy Flow

What is the direction of energy flow through ecosystems on land? Only a small part of the energy from sunlight becomes fixed in plants. The plants themselves use up to half of what they fix, and they lose metabolically generated heat. Other organisms tap into the energy stored in plant tissues, remains, or wastes. They, too, lose heat to the environment. All of these heat losses represent a one-way flow of energy out of the ecosystem.

Figure 24.7 (parts b and c) diagrams this one-way flow of energy through two kinds of food webs. In **grazing food webs**, the energy flows from plants to herbivores, then through an array of carnivores. In **detrital food webs**, it flows mainly from plants through detritivores and decomposers. Usually the two kinds cross-connect, as when a crab of a detrital food web is eaten by a herring gull of a grazing food web.

The amount of energy moving through food webs differs from one ecosystem to the next and often varies with the seasons. In most cases, however, the greatest portion of net primary productivity passes through detrital food webs. For instance, when cattle graze heavily on pasture plants, about half the net primary production enters a grazing food web. But cattle don't use all the stored energy they consume. Quantities of undigested plant parts and other material in cattle feces become available for decomposers and detritivores. In marshes, most of the stored energy is not used until plant parts die and become available for detrital food webs.

Ecological Pyramids

Often the trophic structure of an ecosystem is represented as an **ecological pyramid**, in which producers form a base for successive tiers of consumers above them. Some pyramids are based on biomass (the weight of all the individuals at each trophic level). Here is a pyramid of biomass (grams/square meter) for the aquatic ecosystem at Silver Springs, Florida:

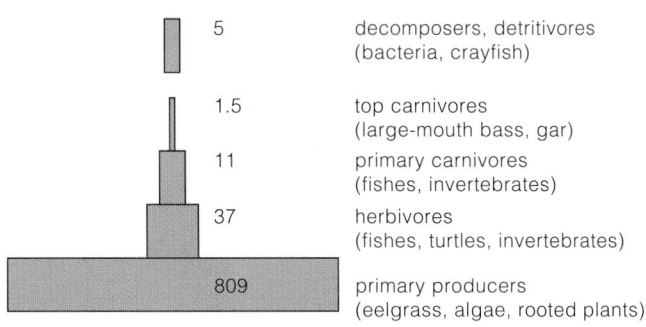

5	decomposers, detritivores (bacteria, crayfish)
1.5	top carnivores (large-mouth bass, gar)
11	primary carnivores (fishes, invertebrates)
37	herbivores (fishes, turtles, invertebrates)
809	primary producers (eelgrass, algae, rooted plants)

Some pyramids of biomass are "upside down," with the smallest tier on the bottom. Think of phytoplankton (including diatoms) and zooplankton (including rotifers) in a pond. A small biomass of rapidly growing and reproducing phytoplankton may support a greater biomass of larger zooplankton that grow more slowly and consume less energy per unit of weight.

An energy pyramid is another useful way to depict an ecosystem's trophic structure. Such pyramids show the energy losses at each transfer to a different trophic level. They have a large energy base at the bottom and are always "right-side up." They give a more accurate picture of ever diminishing amounts of energy flowing through successive trophic levels of the ecosystem.

Energy flows into food webs of ecosystems from an outside source—usually, the sun. Energy leaves ecosystems mainly by loss of metabolic heat, which each organism generates.

Gross primary productivity is the total energy stored by an ecosystem's photosynthesizers in a specified interval. Net primary productivity is the rate of energy storage in plant tissues minus energy used in the plants' aerobic respiration.

Living tissues of photosynthesizers are the basis of grazing food webs. The nonliving remains of photosynthesizers and consumers are the basis of detrital food webs.

The amount of useful energy flowing through consumer trophic levels declines at each energy transfer as metabolic heat is lost and as some food energy is shunted into organic wastes.

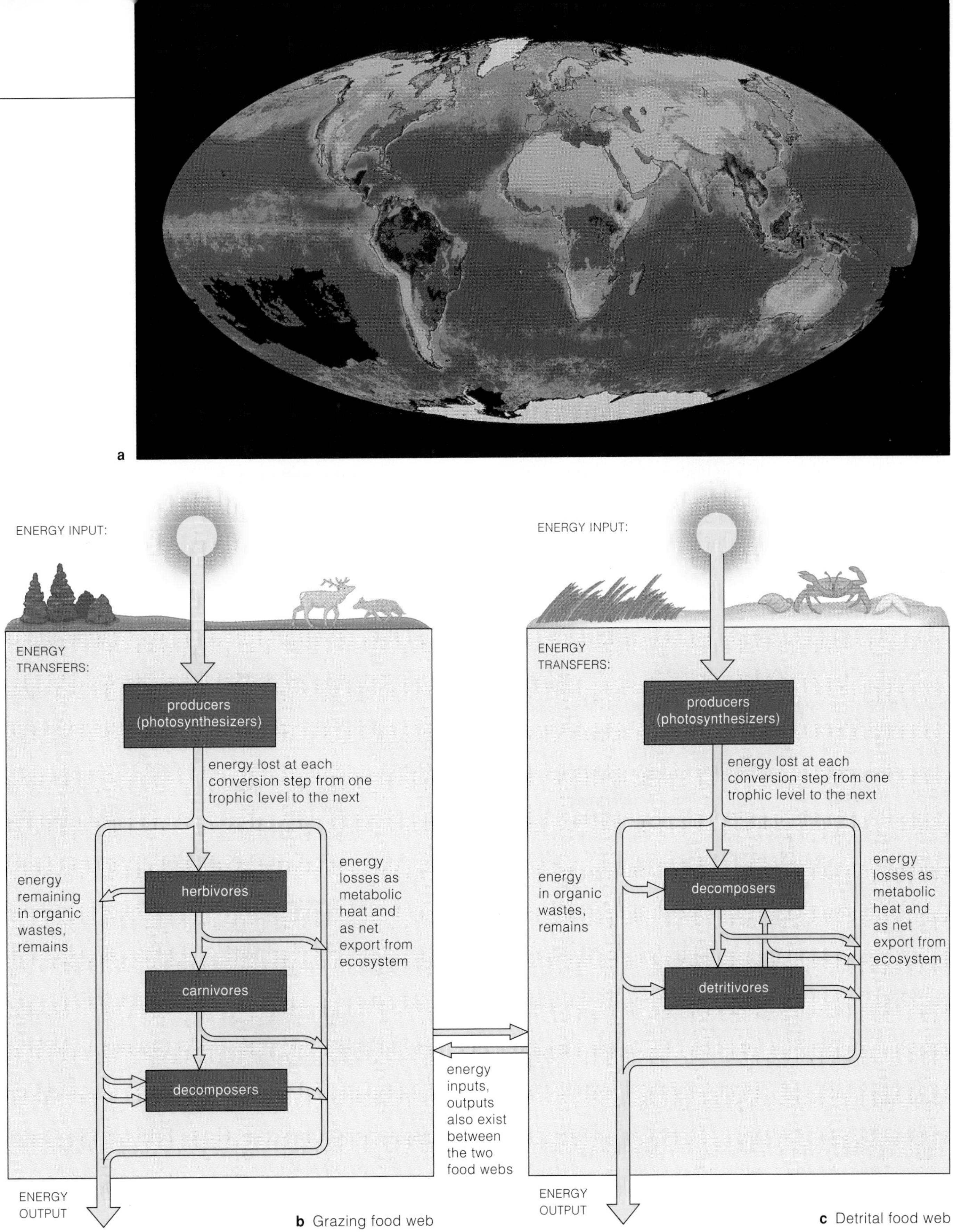

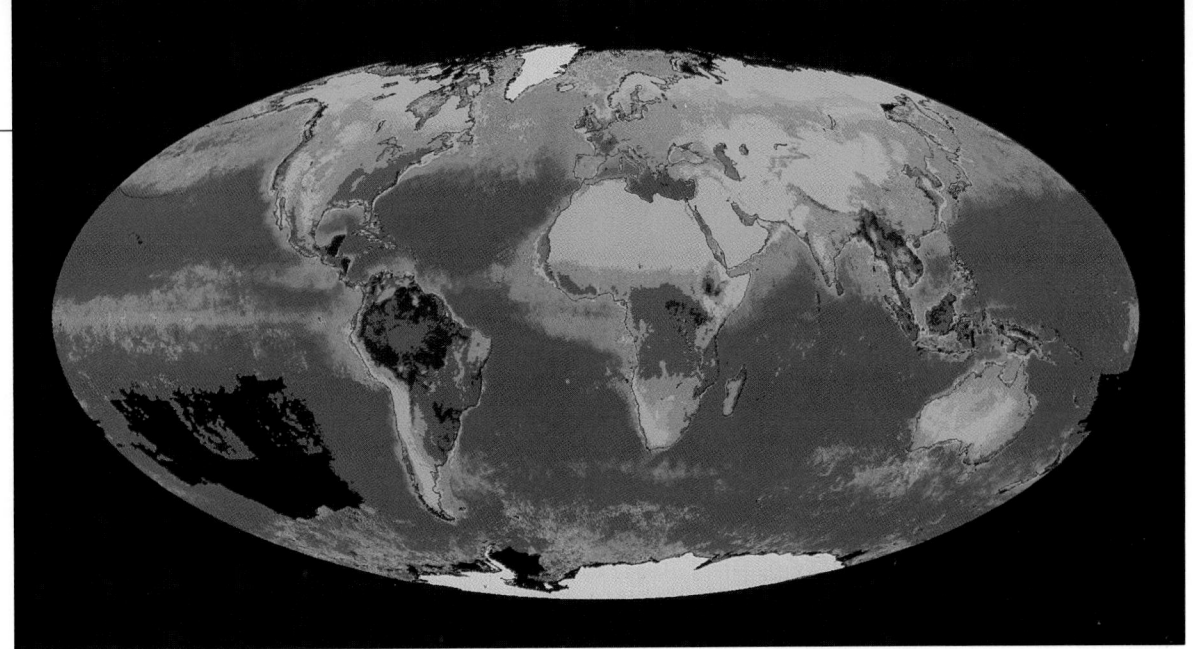
a

ENERGY INPUT:

ENERGY
TRANSFERS:

producers
(photosynthesizers)

energy lost at each
conversion step from one
trophic level to the next

energy
remaining
in organic
wastes,
remains

herbivores

energy
losses as
metabolic
heat and
as net
export from
ecosystem

carnivores

decomposers

ENERGY
OUTPUT

b Grazing food web

energy
inputs,
outputs
also exist
between
the two
food webs

ENERGY INPUT:

ENERGY
TRANSFERS:

producers
(photosynthesizers)

energy lost at each
conversion step from one
trophic level to the next

energy
in organic
wastes,
remains

decomposers

energy
losses as
metabolic
heat and
as net
export from
ecosystem

detritivores

ENERGY
OUTPUT

c Detrital food web

Figure 24.7 Energy bases and energy flow in ecosystems. (**a**) Summary of 3 years of satellite data on the Earth's primary productivity. *Dark green* denotes rain forests and other highly productive regions. *Yellow* denotes deserts (low productivity). Productivity in oceans, ranging from high to low, is coded *red* down through *orange, yellow, green,* and *blue.* (**b,c**) The one-way flow of energy through two kinds of cross-connected food webs in ecosystems.

CASE STUDY: ENERGY FLOW AT SILVER SPRINGS, FLORIDA

To gather data for an energy pyramid, researchers measure the energy that each type of individual in the ecosystem takes in, burns during metabolism, and stores in body tissues. They also measure the energy remaining in its waste products. They then multiply the energy per individual by population size. Energy inputs and outputs are calculated so that energy flow can be expressed per unit of land (or water) per unit of time.

The pyramid in Figure 24.8a is based on a long-term study of a grazing food web in an aquatic ecosystem in Florida. Figure 24.8b shows some of the calculations used to construct it. Such studies tell us that, given the metabolic demands of organisms and the amount of energy shunted into organic wastes, only about 6 to 16 percent of the energy entering one trophic level becomes available to organisms at the next level. Because the efficiency of these energy transfers is so low, ecosystems generally have no more than four consumer trophic levels.

Figure 24.8 Measurements of energy flow through an aquatic ecosystem—Silver Springs, Florida.

(**a**) The pyramid of energy flow, as calculated for 1 year. The numerals represent the number of individuals in each category.

(**b**) Annual energy flow, in kilocalories/square meter/year. Producers in this small spring are mostly aquatic plants. Carnivores are insects and small fishes; top carnivores are larger fishes. The energy source (sunlight) is available all year. Detritivores and decomposers cycle organic compounds from other trophic levels.

The producers trap 1.2 percent of incoming solar energy, and only a little more than a third of this amount is fixed in new plant biomass (4,245 + 3,368). The producers use more than 63 percent of the fixed energy for their own metabolism. About 16 percent of it is transferred to herbivores, and most of this is used for metabolism or transferred to detritivores and decomposers. Only 11.4 percent of the energy transferred to herbivores reaches the next trophic level (carnivores), and the carnivores use all but about 5.5 percent, which is transferred to top carnivores. In time, all of the 5,060 kilocalories transferred through the system appear as metabolically generated heat.

This diagram is oversimplified, for no community is isolated from others. Organisms and materials constantly drop into the springs. Organisms and materials are slowly lost by way of a stream that leaves the springs.

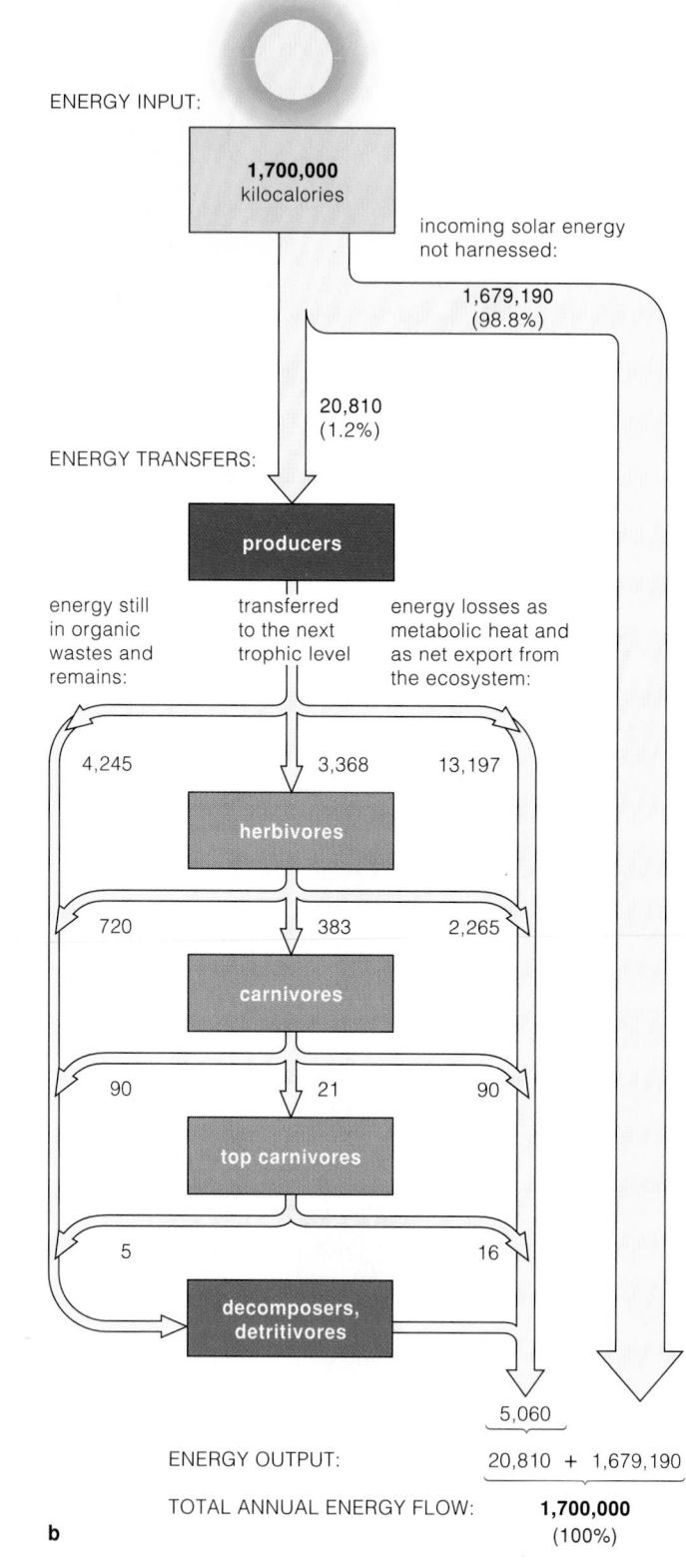

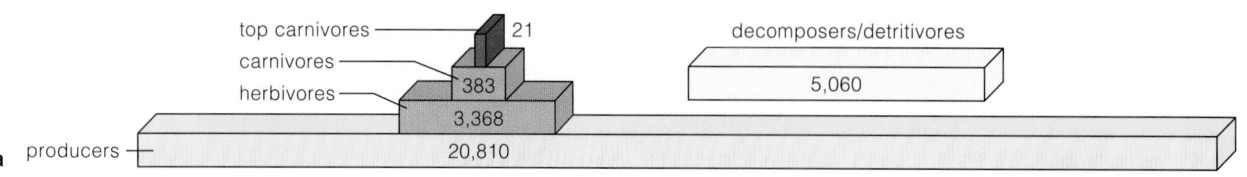

BIOGEOCHEMICAL CYCLES—AN OVERVIEW

Availability of *nutrients* as well as energy profoundly influences the structure of ecosystems. Photosynthetic producers require carbon, oxygen, and hydrogen, which they get from water and air. They require nitrogen, phosphorus, and other minerals. Because mineral deficiencies lower primary productivity, they affect the ecosystem at large.

Nutrients move in **biogeochemical cycles**. In such cycles, ions or molecules of a nutrient are transferred from the environment to organisms, then back to the environment—part of which serves as a reservoir for them. They generally move slowly through the reservoir, compared to their rapid exchange between organisms and the environment.

Figure 24.9 is a model of the relationship between geochemical cycles and most ecosystems. This model is based on four factors:

1. Elements used as nutrients usually become available to producer organisms as mineral ions, such as ammonium (NH_4^+).

2. Inputs from the physical environment and the cycling activities of decomposers and detritivores maintain an ecosystem's nutrient reserves.

3. The amount of a nutrient being cycled through most major ecosystems is greater than the amount entering or leaving in a given year.

4. Inputs to an ecosystem's nutrient reserves occur by rainfall or snowfall, metabolism (such as nitrogen fixation), and weathering of rocks. Outputs for land ecosystems include losses by runoff.

There are three types of biogeochemical cycles. In the *hydrologic* (water) cycle, oxygen and hydrogen move in the form of water molecules. In *atmospheric* cycles, a large portion of a nutrient is in the form of an atmospheric gas. This is true of carbon and nitrogen, for example. In *sedimentary* cycles, the nutrient moves from land to the seafloor and "returns" to land only through geological uplifting, which may take millions of years. The Earth's crust is the main storehouse for such nutrients, which include phosphorus.

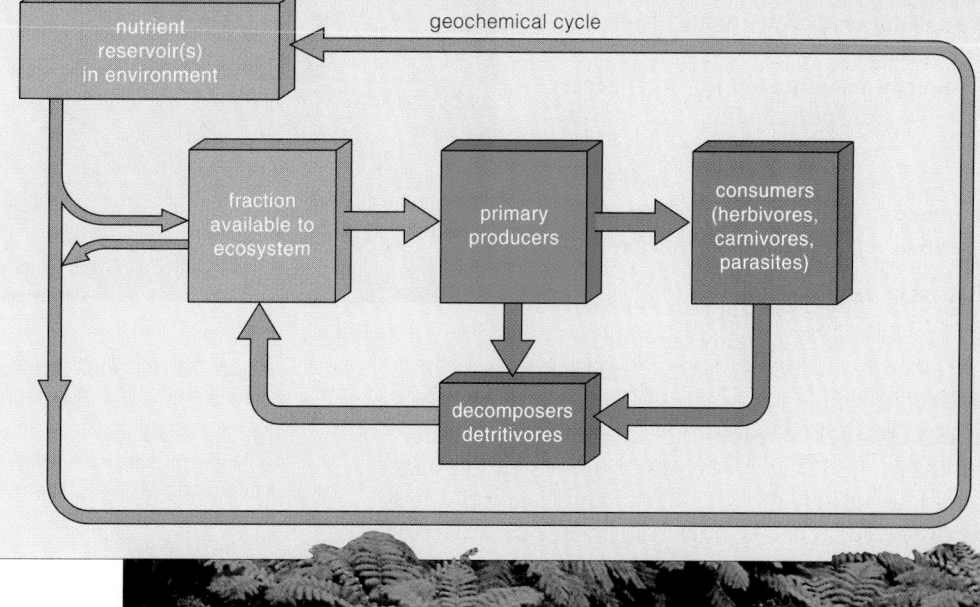

Figure 24.9 Generalized model of nutrient flow through a land ecosystem. The overall movement of nutrients from the physical environment, through organisms, and back to the environment constitutes a biogeochemical cycle.

In a biogeochemical cycle, ions or molecules of a nutrient move from the environment to organisms, then back to the environmental reservoir for them.

Nutrients generally move slowly through the environment but rapidly between organisms and the environment.

THE HYDROLOGIC CYCLE

Driven by solar energy, the waters of the Earth move slowly and on a vast scale from the ocean, into the atmosphere, to the land, and back to the ocean. Figure 24.10a gives an idea of the volume of water that moves in this global **hydrologic cycle**. Water evaporating into the atmosphere remains aloft as vapor, clouds, and ice crystals. It falls to Earth as precipitation—rain and snow, mostly. Ocean currents and wind patterns play roles in the cycle (Figure 24.10b).

Airborne water molecules stay aloft for no more than 10 days, on average. Water reaching the land as rain or snow stays there for about 10 to 120 days, depending on the season and the location, then evaporates or flows to the ocean. Water evaporates mainly from the world's oceans.

Water in itself is vital for organisms. It also is a transport medium that moves nutrients into and out of ecosystems. This became clear through studies in **watersheds**, where all precipitation in a specified region becomes funneled into a single stream or river. Watersheds may be any size. The Mississippi River watershed extends across roughly one-third of the United States. Watersheds at Hubbard Brook Valley in the White Mountains of New Hampshire average 14.6 hectares (36 acres). Most of the water entering a watershed seeps into soil or becomes surface runoff that moves into streams (Figure 24.11). Plants take up water and dissolved minerals from soil, then lose water by transpiration.

Measurements of water inputs and outputs at watersheds have practical application. Cities that depend on surface water supplies in watersheds can adjust water usage on the basis of seasonal variations.

Watershed studies also showed that plants greatly influence the movement of nutrients through the ecosystem phase of biogeochemical cycles. For example, you might think that water draining a watershed would rapidly leach calcium ions and other minerals. Yet in studies of young, undisturbed forests in Hubbard Brook watersheds, each hectare lost only about 8 kilograms of calcium, and rainfall and the weathering of rocks brought in calcium replacements. Tree roots were also "mining" the soil, so calcium was being stored in a growing biomass of tree tissues.

Nutrient outputs changed when experimental watersheds were stripped of vegetation. The soil was left undisturbed, and herbicides were applied for 3 years to prevent regrowth. Compared to undisturbed watersheds,

	Annual Volume (10^{18} grams)
Ocean	1,380,000
Sedimentary layers	210,000
Evaporation from ocean	319
Precipitation over ocean	283
Precipitation over land	95
Evaporation from land	59
Runoff and groundwater	36
Atmospheric water vapor	13

a

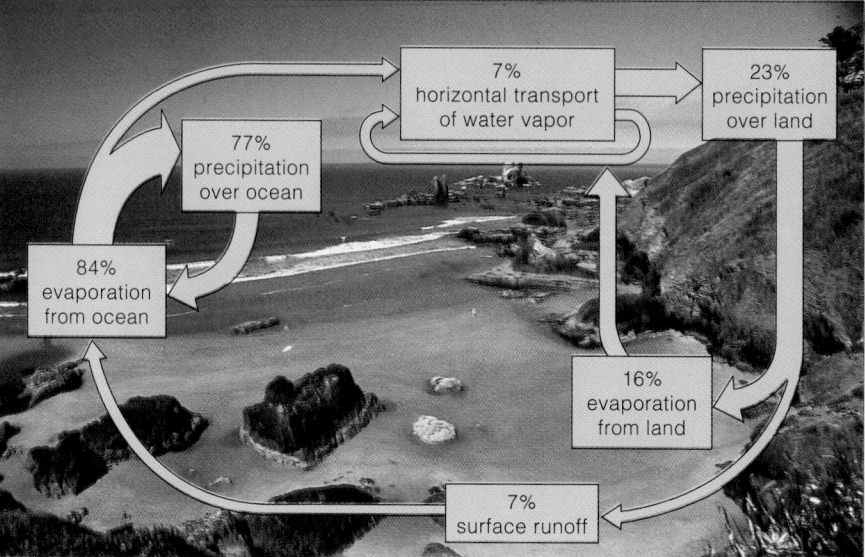

b

Figure 24.10 (a) Global water budget. Values reflect the annual movement of water into and out of the atmosphere.

(b) The hydrologic cycle. Values in boxes indicate percentages of water present at a given time in the atmosphere, in the ocean, and on land. There is an annual net rate of transfer of 37.3×10^3 cubic kilometers of water from the atmosphere to the land. Balancing this is a comparable net loss from the ocean to the atmosphere and a net gain of that amount by the ocean (through runoff).

six times as much calcium was lost by way of stream outflow (Figure 24.12c).

Because calcium and other nutrients move so slowly through geochemical cycles, deforestation may disrupt nutrient availability for an entire ecosystem. This is true of forests that cannot rapidly regenerate themselves. The conifer forests of the Pacific Northwest, which require decades to grow to maturity, are like this.

In the hydrologic cycle, water slowly moves on a global scale from the ocean reservoir, through the atmosphere, onto land, then back to the ocean.

By stabilizing the soil and absorbing dissolved minerals, plants minimize the loss of soil nutrients in runoff from land.

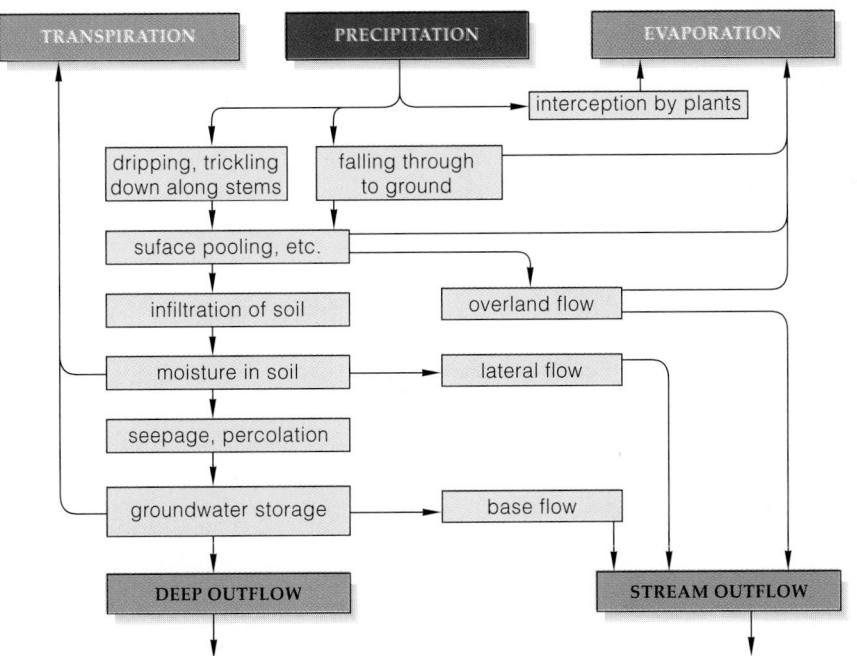

Figure 24.11 Model of the movement of water through a watershed. *Dark blue* indicates inputs to the watershed; *light blue*, distribution within the watershed; and *medium blue*, outputs from it. The process of transpiration noted at the upper left is the evaporation of water from plants. The plants take up water in the soil.

a

Figure 24.12 Studies of disturbances to a forest ecosystem.

(**a**) In experimental watersheds in the Hubbard Brook Valley, researchers studied the effects of deforestation. (**b**) All water that drains from an area being studied must flow over the V-notched concrete structure, where measurements are taken. This area was deforested; then herbicides were applied to prevent regrowth for 3 years.

(**c**) The *green* arrow indicates the time of deforestation. The concentrations of calcium ions and other mineral ions in the water passing through the catchment were compared with those in water passing through a control catchment in an undisturbed part of the same forest. The concentration of calcium in outflowing water increased sixfold. In addition, the volume of water flowing out of the deforested watershed also increased, so that total calcium loss was even greater.

b

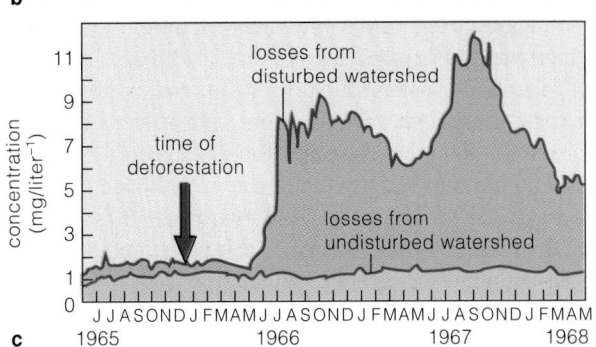

c

THE CARBON CYCLE

In one of the most important of all atmospheric cycles, carbon moves from reservoirs in the atmosphere and oceans, through organisms, then back to reservoirs. Figure 24.13 shows this global movement, which is called the **carbon cycle.**

Carbon enters the atmosphere via aerobic respiration, fossil fuel burning, and volcanic eruptions that release carbon from rocks deep in the Earth's crust. Most of the carbon is dissolved in the ocean (Figure 24.13b). Soil, the atmosphere, and plant biomass represent other large "holding stations" for carbon. Most carbon in the atmosphere is in the form of carbon dioxide (CO_2).

Through carbon dioxide fixation, photosynthesizers lock up billions of metric tons of carbon atoms into organic compounds every year. However, the average length of time that a carbon atom is held in any given ecosystem varies greatly. In tropical forests, leaves decompose rapidly, so not much carbon is tied up in litter on the soil surface. In marshes, bogs, and other anaerobic areas, organic compounds cannot be broken down completely, so carbon slowly accumulates in compressed organic matter, such as peat.

In aquatic food webs, carbon becomes incorporated into shells and other hard parts. When shelled organisms die, they sink and become buried in sediments at different depths. Carbon in deep sediments may stay buried for millions of years, until tectonic movements bring it to the surface. Other carbon is slowly converted to long-standing reserves of gas, petroleum, and coal, which we tap for use as fossil fuels.

Human activities, including the worldwide burning of fossil fuels, are putting more carbon into the atmosphere than can be cycled to the ocean reservoir and to holding stations. This adds to the greenhouse effect and so may help bring on global warming. *Science Comes to Life* in Section 24.8 describes this effect and some possible outcomes of increases in it.

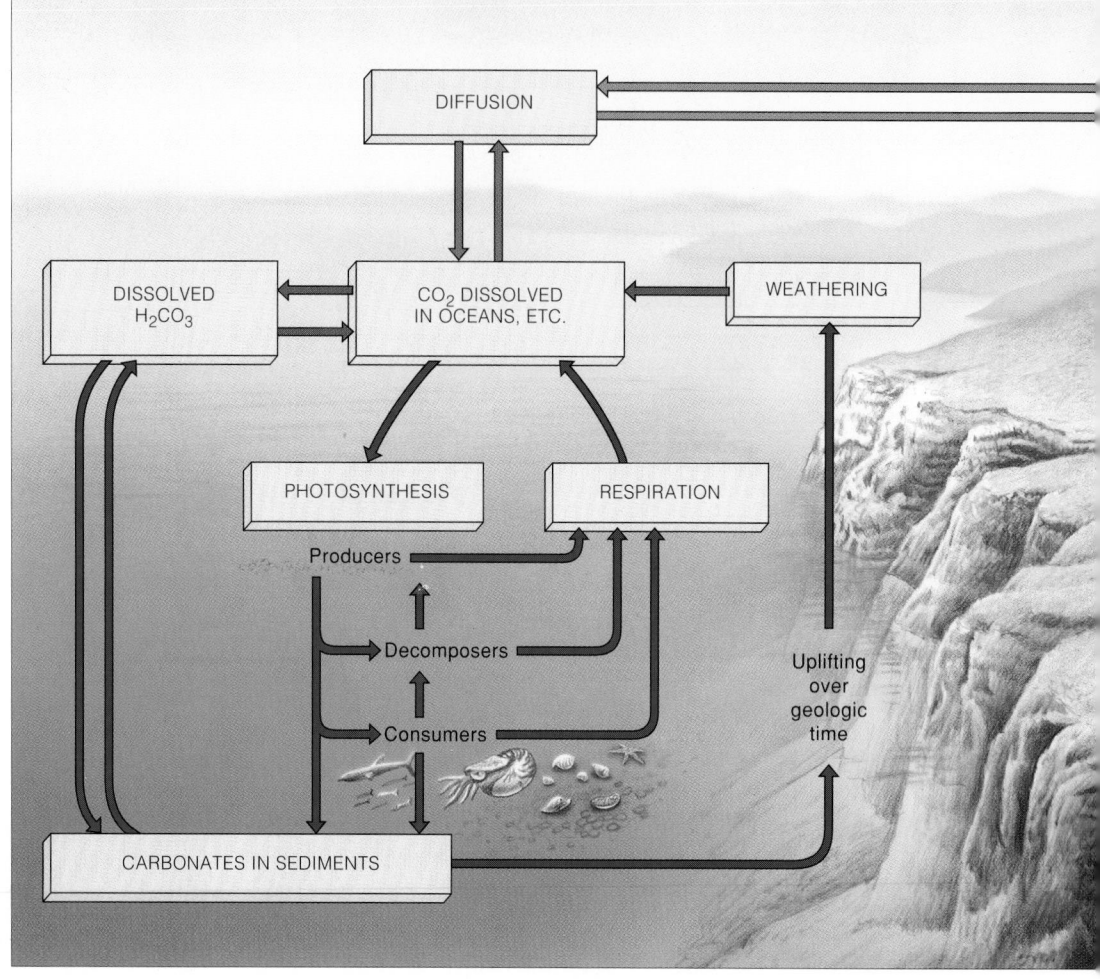

a

Figure 24.13 (**a**) Global carbon cycle. The portion of the diagram on this page illustrates the movement of carbon through marine ecosystems. The portion of the diagram on the facing page illustrates the movement of carbon through ecosystems on land.

(**b**) The present-day global carbon budget. The photograph shows a Los Angeles freeway system under its self-generated blanket of smog at twilight. The exhaust of vehicles as well as fossil fuel burning by industries and homes adds carbon and other substances to the atmosphere.

Data in (**b**) from W. Schlesinger, *Biogeochemistry: An Analysis of Global Change* (1991). New York: Academic Press.

Fossil fuel burning and other human activities may be contributing to imbalances in the global carbon budget.

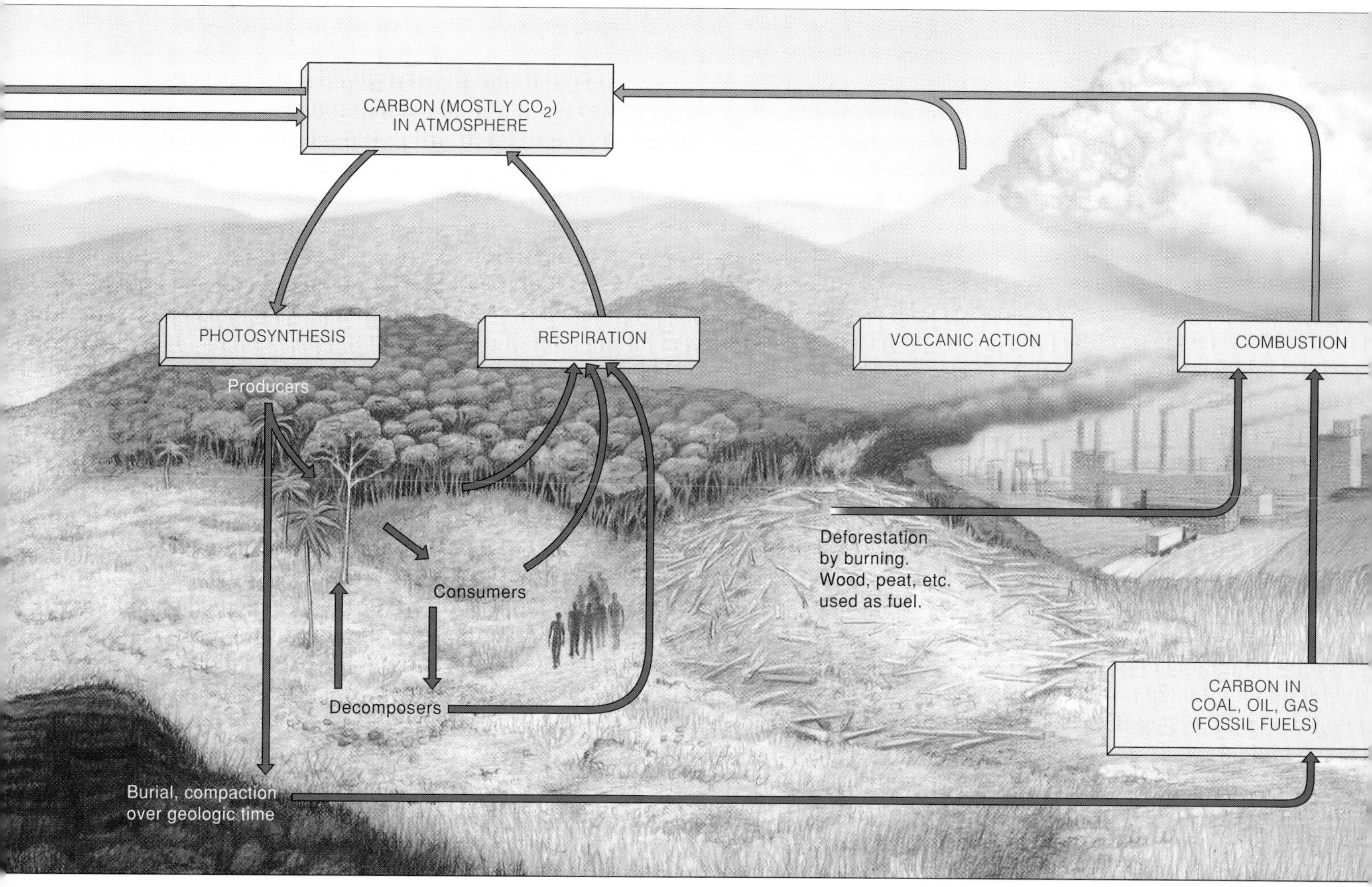

The following diagram shows the carbon cycle:

- CARBON (MOSTLY CO₂) IN ATMOSPHERE
- PHOTOSYNTHESIS — Producers
- RESPIRATION — Consumers, Decomposers
- VOLCANIC ACTION
- COMBUSTION
- Deforestation by burning. Wood, peat, etc. used as fuel.
- CARBON IN COAL, OIL, GAS (FOSSIL FUELS)
- Burial, compaction over geologic time

	Amount per Year (10^15 grams)
Carbon reservoirs and holding stations:	
Dissolved in ocean	38,000
Present in soil	1,500
Present in atmosphere	720
Plant biomass	560
Annual fluxes:	
From atmosphere to plants (carbon fixation)	120
From atmosphere to ocean	107
To atmosphere from ocean	105
To atmosphere from plants	60
To atmosphere from soil	60
To atmosphere from fossil fuel burning	5
To atmosphere from net destruction of plants	2
To ocean from runoff	0.4
Burial in ocean sediments	0.1

b

FROM GREENHOUSE GASES TO A WARMER PLANET

Near the Earth's surface, atmospheric concentrations of gaseous molecules play a profound role in shaping the average global temperature, which in turn has enormous effect on the global climate. Molecules of carbon dioxide, water, ozone, methane, nitrous oxide, and chlorofluorocarbons are the key players. Collectively, they act somewhat like a pane of glass in a greenhouse (hence their name, "greenhouse gases"). They let wavelengths of visible light reach the Earth's surface, but they impede the escape of longer, infrared wavelengths—that is, heat—from the Earth into space. They absorb infrared wavelengths, and much of the energy inherent in those wavelengths is reradiated back toward the Earth (Figure 24.14a). In short, greenhouse gases

cause heat to build up in the lower atmosphere, a warming action known as the **greenhouse effect**.

Without greenhouse gases, the Earth would be cold and lifeless. But there can be too much of a good thing. Largely as a result of human activities, greenhouse gases are building up to higher levels in the atmosphere (Figure 24.14b). The increase may be contributing to an alarming trend toward global warming.

What is so alarming about a warmer planet? Suppose the temperature of the lower atmosphere were to rise by only 4°C (7°F). Sea levels could rise by about 2 feet, or 0.6 meter. Why? The warming would increase ocean surface temperatures—and water expands when heated. Global

1. Sunlight penetrates the atmosphere and warms the Earth's surface.

2. The Earth's surface radiates heat (infrared wavelengths) to the atmosphere. Some heat escapes into space. Greenhouse gases and water vapor absorb some infrared wavelengths and reradiate a portion back toward the Earth.

3. When greenhouse gases build up in the atmosphere, more heat is trapped near the Earth's surface. Ocean surface temperatures rise, more water vapor enters the atmosphere, and the Earth's surface temperature rises.

a

Figure 24.14 (**a**) The greenhouse effect. (**b**) Shifts in atmospheric concentrations of carbon dioxide, correlated with the most recent glaciation and interglacial period during the past 160,000 years. (**c**) Recently documented increases in atmospheric concentrations of four greenhouse gases.

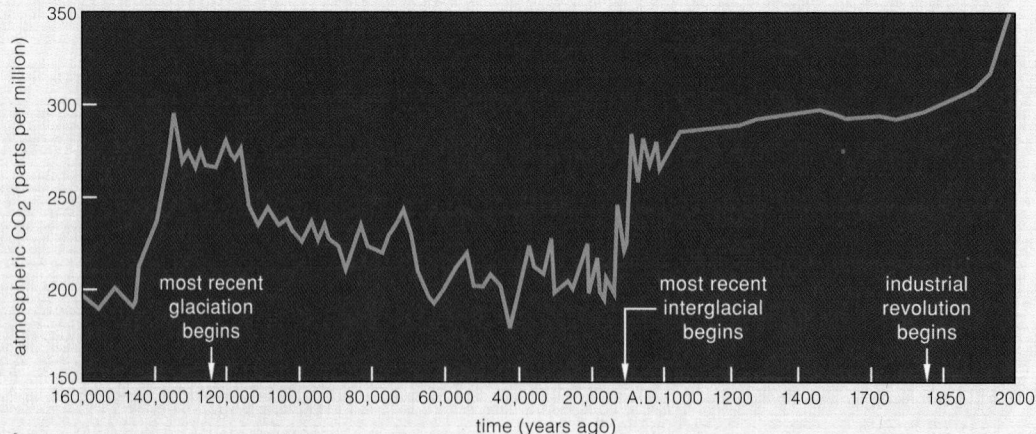

b

warming also could make glaciers and the Antarctic ice sheet melt faster, so low coastal regions could flood.

Think of a long-term rise in sea level, combined with high tides and storm waves. Waterfronts of Vancouver, Boston, San Diego, Galveston, and other coastal cities would be submerged. So would agricultural lowlands and deltas in India, China, and Bangladesh, where much of the world's rice is grown. Huge tracts of Florida and Louisiana would face saltwater intrusions. Besides this, global warming could disturb regional patterns of precipitation and temperature. Crop yields would decline in currently productive regions, including parts of Canada and the United States, and increase in others.

In the late 1950s, researchers on a mountaintop in the Hawaiian Islands started measuring concentrations of different greenhouse gases, and their monitoring activities are still going on. They chose the remote site because it was free of local contamination and reflected average conditions for the Northern Hemisphere.

Consider what they found out about carbon dioxide alone. Atmospheric levels of carbon dioxide follow the annual cycle of plant growth in the Northern Hemisphere. They are lower in summer, when photosynthesis rates are highest. They are higher in winter, when aerobic respiration continues and photosynthesis slows. The red line in

Figure 24.14c (part 1) traces the midline of the highs and lows in the cycle. As you can see, this midline steadily increased from 1958 to 1992. For the first time, scientists saw the integrated effects of the carbon balances for the land and water ecosystems of an entire hemisphere. Many scientists take this as evidence of a buildup of carbon dioxide that may intensify the greenhouse effect over the next century.

The global burning of fossil fuels is probably contributing most to increasing carbon dioxide levels. Deforestation adds to it; carbon is released when wood burns. Today, vast tracts of forests throughout the world are being cleared and burned at a rapid rate (refer to Figure 25.15). More important, the plant biomass is plummeting—and this affects global absorption of carbon dioxide in photosynthesis.

Many scientists wonder whether atmospheric levels of greenhouse gases will continue to increase until the middle of the 21st century—and whether the global temperature will rise by several degrees. If this is indeed a trend already in motion, we will not be able to reverse it now by stopping fossil fuel burning and deforestation. So there is widespread agreement that we should begin preparing for the consequences. For example, we might step up genetic engineering studies to develop drought-resistant and salt-resistant plants. Such plants may prove crucial in regions of saltwater intrusions and climatic change.

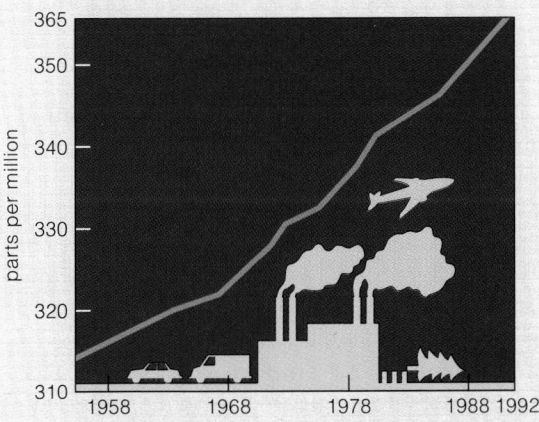

1. Carbon Dioxide (CO_2). Fossil fuel burning, factory emissions, car exhaust, and deforestation are contributing to the increased atmospheric concentration.

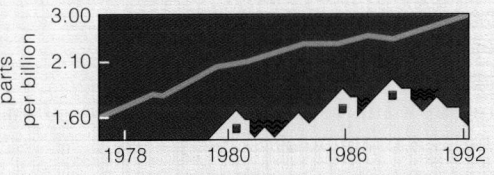

2. Chlorofluorocarbons (CFCs). These are used in plastic foams, air conditioners, refrigerators, and industrial solvents.

3. Methane (CH_4). This is produced by anaerobic bacteria in swamps, landfills, and termite activities. It is produced also by bacteria in the digestive tract of cattle and other ruminants.

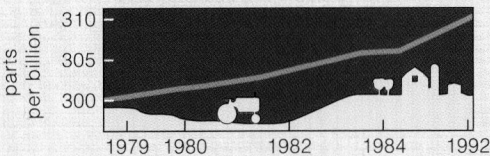

4. Nitrous Oxide (N_2O). This is a natural by-product of denitrifying bacteria. It also is released in great amounts from fertilizers and animal wastes, as in livestock feedlots.

c

Since the time of life's origin, the atmosphere and oceans have contained nitrogen. This component of all proteins and nucleic acids moves in an atmospheric cycle called the **nitrogen cycle**. Gaseous nitrogen (N_2) makes up about 80 percent of the atmosphere, the largest nitrogen reservoir. Triple covalent bonds hold its two atoms together ($N\equiv N$), and few organisms can break them. Only certain bacteria, volcanic action, and lightning convert N_2 into forms that enter food webs.

Of all nutrients required for plant growth, nitrogen often is the scarcest. Today, nearly all nitrogen in soils has been put there by nitrogen-fixing organisms. Ecosystems lose it through the activities of bacteria that "unfix" the fixed nitrogen. Land ecosystems lose more through leaching of soils, although this provides nitrogen inputs to aquatic ecosystems such as streams, lakes, and the oceans (Figure 24.15).

Cycling Processes

Let's follow nitrogen atoms through the ecosystem part of the nitrogen cycle. They move through organisms by way of processes of nitrogen fixation, assimilation and biosynthesis, decomposition, ammonification, and nitrification.

In **nitrogen fixation**, a few kinds of bacteria convert N_2 to ammonia (NH_3), which dissolves quickly in the cytoplasm to form ammonium (NH_4^+). Certain microorganisms (cyanobacteria) are nitrogen fixers of aquatic ecosystems. Others fix nitrogen in many land ecosystems. Collectively these organisms fix about 200 million metric tons of nitrogen each year! Plants assimilate and use this nitrogen in the biosynthesis of amino acids, proteins, and nucleic acids. Plant tissues serve as the only nitrogen source for animals, including humans, which feed directly or indirectly on plants.

Through **decomposition** and **ammonification**, bacteria and fungi break down nitrogen-containing wastes and remains of organisms. The decomposers use part of the released proteins and amino acids for their own metabolism. But most of the nitrogen is still in the decay products, in the form of ammonia or ammonium, which plants take up. Nitrifying bacteria also act on ammonia or ammonium. In **nitrification**, they strip these compounds of electrons, and nitrite (NO_2^-) is the result. Other bacteria use the nitrite in metabolism and produce nitrate (NO_3^-), which plants take up.

Certain plants are better than others at securing nitrogen. For example, peas, beans, clover, and other legumes have mutually beneficial associations with nitrogen-fixing bacteria. Also, most land plants have similar associations with fungi, forming specialized roots known as *mycorrhizae* ("fungus roots") that incorporate fungal tissues. The mycorrhizae enhance nitrogen uptake by the plant.

Nitrogen Scarcity

You would think that land plants could get enough nitrogen, given the cycling processes. However, the ammonium, nitrite, and nitrate that form during the cycle are highly vulnerable to leaching and runoff. With leaching, recall, soil water moves out of an area, which thereby loses the nutrients dissolved in the water.

Also, some nitrogen is lost to the air by **denitrification**. Here, bacteria convert nitrate or nitrite to N_2 and a bit of nitrous oxide (N_2O). Ordinarily, most denitrifying bacteria rely on aerobic respiration. When soil is waterlogged and poorly aerated, they switch to anaerobic pathways and use nitrate, nitrite, or nitrous oxide as the final electron acceptor instead of oxygen. In these reactions, fixed nitrogen is converted to N_2, much of which escapes into the atmosphere.

Besides this, nitrogen fixation comes at high metabolic cost to plants that are associated with nitrogen fixers. In exchange for nitrogen, they give up sugars and other substances. Such plants do have the competitive edge in nitrogen-poor soil. In nitrogen-rich soil, however, species that do not have to pay the metabolic price often displace them.

Human Impacts on the Nitrogen Cycle

Humans are altering the cycling of nitrogen in natural ecosystems. Consider how air pollutants contribute to changes in soil. Vehicles, power plants that burn fossil fuels, and nitrogenous fertilizers are sources of pollutants, including oxides of nitrogen. These contribute to soil acidity—which reduces the amounts of magnesium, calcium, and potassium that plants can take up. Figure 24.15 merely hints at the ion interactions in soil water.

And what about those nitrogen fertilizers? To be sure, nitrogen losses from soil are enormous in agricultural regions. With each harvest, nitrogen departs from the fields (in tissues of harvested plants). Soil erosion and leaching remove more. In Europe and North America, farmers traditionally have rotated crops, as when they alternate wheat with legumes. Along with other conservation practices, crop rotation has helped keep soils stable and productive, sometimes for thousands of years. Today, however, intensive agriculture is based on the application of nitrogen-rich fertilizers. New strains of crop plants are bred for an ability to take up fertilizers, and crop yields per hectare have doubled and

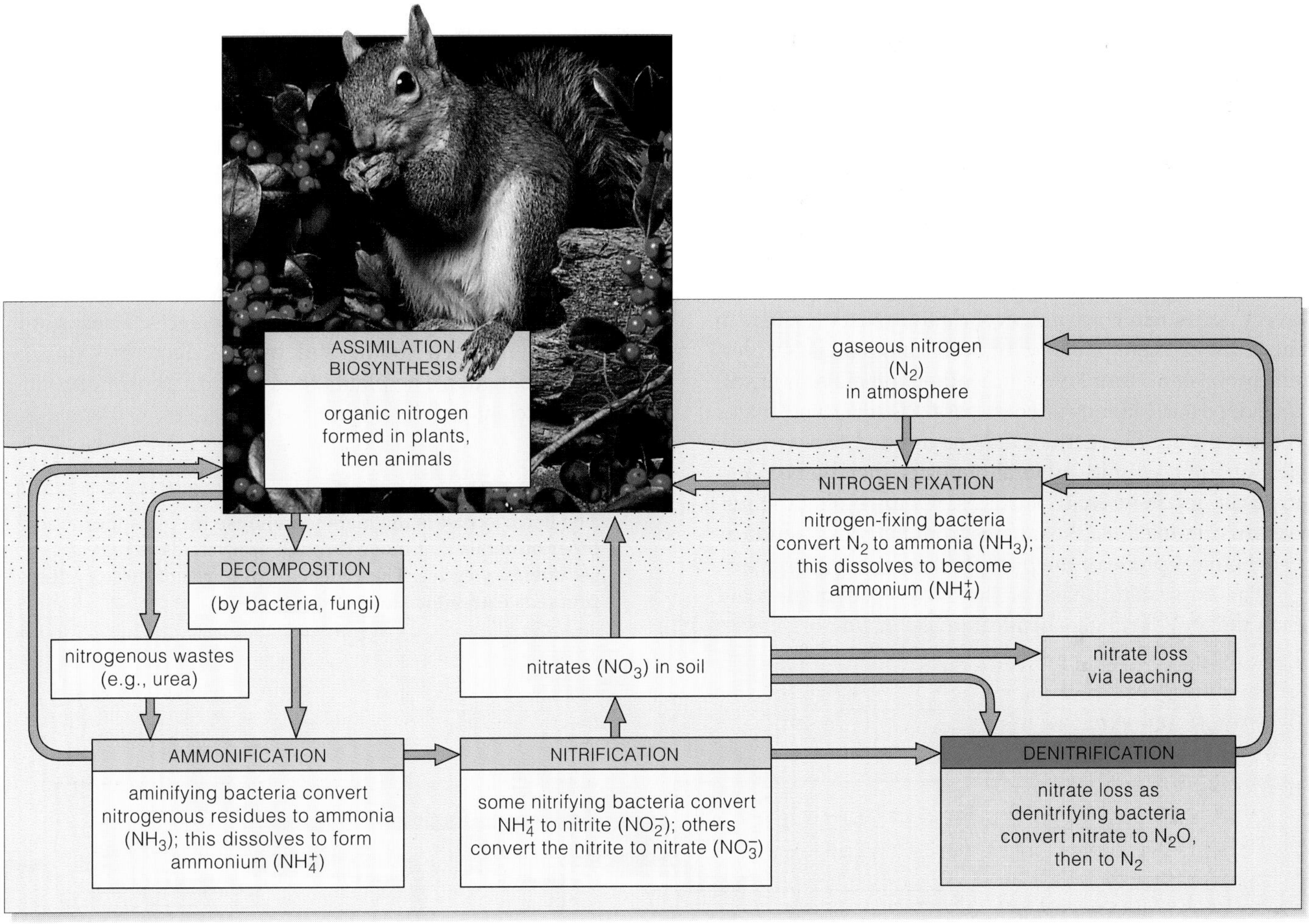

Figure 24.15 The nitrogen cycle.

The nitrogen cycle diagram contains the following labeled boxes:

- **ASSIMILATION BIOSYNTHESIS:** organic nitrogen formed in plants, then animals
- **gaseous nitrogen (N_2) in atmosphere**
- **NITROGEN FIXATION:** nitrogen-fixing bacteria convert N_2 to ammonia (NH_3); this dissolves to become ammonium (NH_4^+)
- **DECOMPOSITION** (by bacteria, fungi)
- **nitrogenous wastes** (e.g., urea)
- **nitrates (NO_3^-) in soil**
- **nitrate loss via leaching**
- **AMMONIFICATION:** aminifying bacteria convert nitrogenous residues to ammonia (NH_3); this dissolves to form ammonium (NH_4^+)
- **NITRIFICATION:** some nitrifying bacteria convert NH_4^+ to nitrite (NO_2^-); others convert the nitrite to nitrate (NO_3^-)
- **DENITRIFICATION:** nitrate loss as denitrifying bacteria convert nitrate to N_2O, then to N_2

even quadrupled over the past 40 years. Whether pest control and soil management technologies can sustain high yields indefinitely remains uncertain, for reasons that will become apparent in Chapter 25.

We can't get something for nothing. Fertilizer production requires huge amounts of energy from fossil fuels—not from free, unending sunlight. Few once believed that fossil fuel supplies might run out, so there was little concern about fertilizer costs. It still is common to pour more energy into soil (as in fertilizers) than we get out of it (in the form of food). As long as the human population continues to grow exponentially, farmers will be engaged in a race to grow as much food as they can for as many people as possible. Soil enrichment with nitrogen-containing fertilizers is part of the race, as it is now being run.

Nitrogen, a component of all proteins and of nucleic acids, moves in an atmospheric cycle.

Nitrogen atoms move through ecosystems by nitrogen fixation, assimilation and biosynthesis, decomposition, ammonification, and nitrification.

The cycling of nitrogen in natural ecosystems depends on the activity of nitrogen-fixing bacteria and on mycorrhizae. Human disruptions to that cycling may harm the ecosystems.

THE PHOSPHORUS CYCLE

We conclude our look at biogeochemical cycling with an example of a sedimentary cycle. In the **phosphorus cycle**, phosphorus moves from land to sediments in the seas and then back to the land (Figure 24.16). The Earth's crust is the main storehouse for this mineral and for others, including calcium and potassium.

Phosphorus is typically present in rock formations on land, in the form of phosphates. Through the natural processes of weathering and erosion, phosphates enter rivers and streams, which eventually transport them to the ocean. There, mainly on the continental shelves, phosphorus accumulates with other minerals as insoluble deposits. Millions of years pass. Where crustal plates collide, part of the seafloor may be uplifted and drained. The seafloor, with its mineral deposits, thereby becomes exposed as new land surfaces. Over geologic time, weathering releases phosphates from the rocks—and the geochemical phase of the phosphorus cycle begins again.

The ecosystem phase of the cycle is far more rapid than the long-term geochemical phase. Living organisms require small amounts of phosphorus. It is a key component of ATP, NADPH, phospholipids, nucleic acids, and other organic compounds. Plants have the metabolic means to take up dissolved, ionized forms of phosphorus. Actually, they do this so rapidly and efficiently that they often reduce soil concentrations of phosphorus to extremely low levels. Herbivores obtain phosphorus only by eating the plants; carnivores obtain it by eating herbivores. Herbivores and carnivores excrete phosphorus as a waste product in urine and feces. Phosphorus is also released to the soil by the decomposition of organic matter. The plants then take up phosphorus and so recycle it rapidly within the ecosystem.

The Earth's crust is the main storehouse for phosphorus and other minerals that move through ecosystems as part of sedimentary cycles. The geochemical phase of these cycles proceeds extremely slowly.

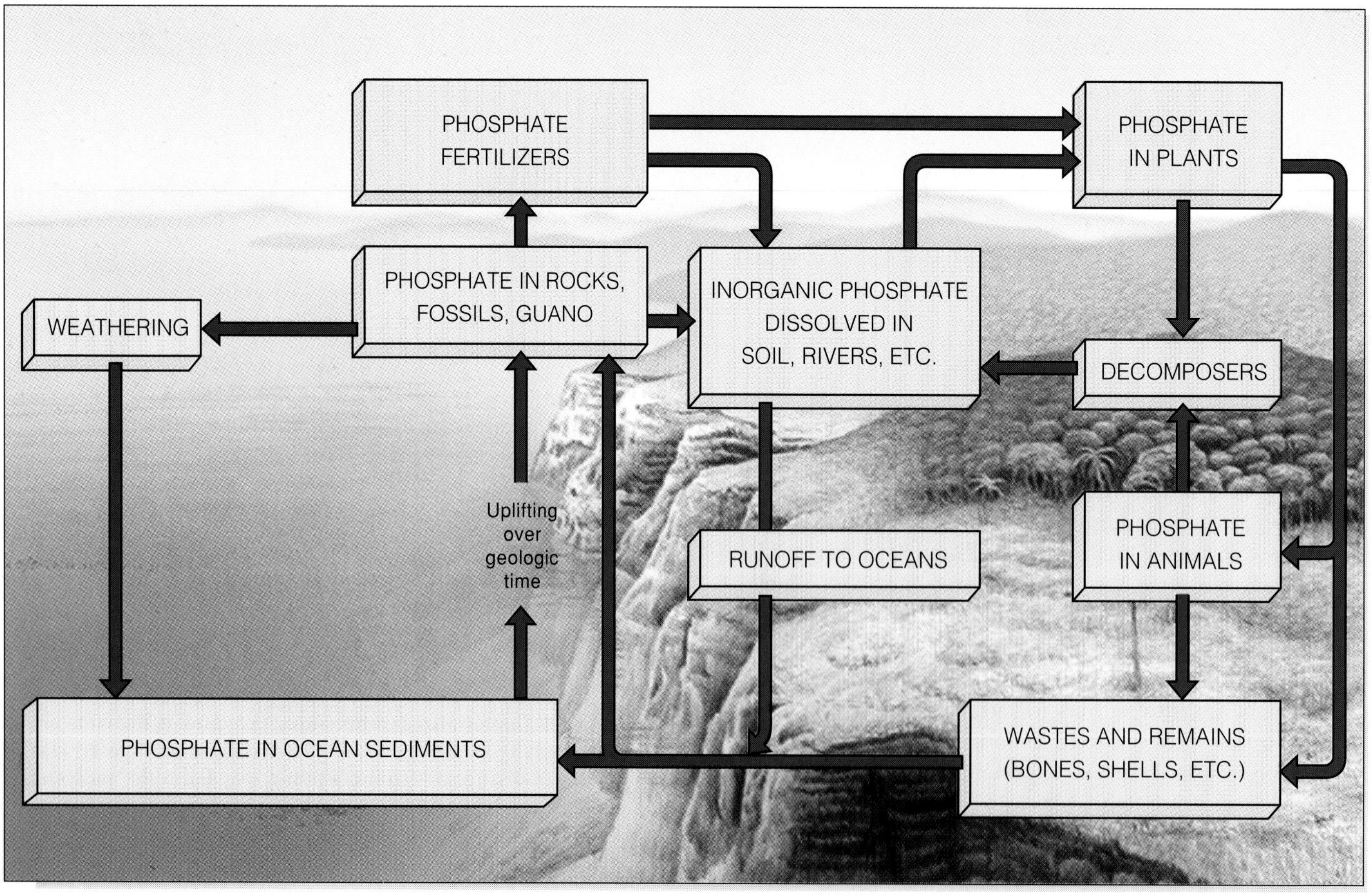

Figure 24.16 The phosphorus cycle. This is an example of a sedimentary cycle.

TRANSFER OF HARMFUL COMPOUNDS THROUGH ECOSYSTEMS

DDT, the first of the synthetic organic pesticides, was first used during World War II. In mosquito-infested regions of the tropical Pacific, people were vulnerable to a dangerous disease, malaria. DDT helped control the mosquitoes that were transmitting the sporozoan disease agents (*Plasmodium japonicum*). In war-ravaged cities of Europe, people were suffering from the crushing headaches, fevers, and rashes associated with typhus. DDT helped control the body lice that were transmitting *Rickettsia rickettsii*, the bacterial agent of this terrible disease. After the war, it seemed like a good idea to use DDT against insects that were agricultural or forest pests, transmitters of pathogens, or merely nuisances in homes and gardens.

DDT is a relatively stable hydrocarbon compound. It is nearly insoluble in water, so you might think that it would act only where applied. But winds can carry DDT in vapor form; water can transport fine particles of it. DDT also is highly soluble in fats, so it accumulates in the tissues of organisms. Thus, as we now know, DDT can show **biological magnification**. The term refers to an increase in concentration of a nondegradable (or slowly degradable) substance in organisms as it is passed upward through food chains (Figure 24.17). Most of the DDT from all the organisms that a consumer eats during its lifetime will become concentrated in its tissues. Besides this, many organisms have the means to partially metabolize DDT to DDE and other modified compounds with different but still disruptive effects. Both DDT and the modified compounds are toxic or physiologically disruptive to *many* aquatic and terrestrial animals.

After the war, DDT began to move through the global environment, infiltrate food webs, and affect organisms in ways that no one had predicted. In cities where DDT was sprayed to control Dutch elm disease, songbirds started dying. In streams flowing through forests where DDT was sprayed to control spruce budworms, salmon started dying. In croplands sprayed to control one kind of pest, new kinds of pests moved in. DDT was indiscriminately killing off the natural predators that had been keeping pest populations in check! It took no great leap of the imagination to make the connection. All of those organisms were dying at the same time and the same places as the DDT applications.

Then side effects of biological magnification started showing up in places far removed from the areas of DDT application—and much later in time. Most devastated were species at the end of food chains, including bald eagles, peregrine falcons, ospreys, and brown pelicans. One product of DDT breakdown interferes with physiological processes. As one consequence, birds produced eggs with brittle shells—and many of the chick embryos didn't make it to hatching time. Some species were at the brink of extinction.

Since the 1970s, DDT has been banned in the United States, except for restricted applications where public health is endangered. Many hard-hit species have partially recovered their numbers. Even today, however, some birds lay thin-shelled eggs. They pick up DDT at their winter ranges in Latin America. As recently as 1990, the California State Department of Health recommended that a fishery off the coast of Los Angeles be closed. DDT from industrial waste discharges that ended 20 years before is still contaminating that ecosystem.

Figure 24.17 An example of biological magnification. In 1987, scientists discovered a variety of toxic industrial chemicals in the tissues of a coelacanth (**top**), a "living fossil" that inhabits extremely deep waters off the west coast of Africa. The huge fish must have ingested other organisms that had been exposed to the chemical barrage.

SUMMARY

1. An ecosystem is an entire complex of producers, consumers, detritivores, and decomposers and their physical environment, all interacting through a flow of energy and a cycling of materials.

2. Ecosystems are open systems, with inputs and outputs of energy and nutrients.

   a. With few exceptions, sunlight is the source of energy, and photosynthesizing autotrophs are the primary producers. They convert the energy of sunlight to ATP and other forms that can be used to synthesize large organic compounds from simple inorganic substances.

   b. Primary producers also assimilate many of the nutrients that are eventually transferred to other members of the system.

3. Directly or indirectly, primary producers nourish an array of heterotrophs in an ecosystem.

   a. The heterotrophs include consumers: herbivores that feed on algae and plants, carnivores that feed on animals, and omnivores that have eclectic diets.

   b. The heterotrophs also include decomposers (mainly certain fungi and bacteria that digest organic substances) and detritivores (such as crabs and earthworms, which feed on particles of dead or decomposing material).

4. Feeding relationships within ecosystems are structured as trophic levels: a hierarchy of energy transfers, sometimes referred to as "who eats whom."

   a. Primary producers make up the first trophic level, herbivores make up the next, carnivores the next, and so on.

   b. Decomposers, humans, and many other organisms get energy from more than one source and cannot be assigned to a single trophic level.

5. An isolated food chain (a straight-line sequence of who eats whom in an ecosystem) is rare in nature. Instead, food chains cross-connect with one another, forming food webs.

6. Ecosystems generally are most open for inputs and outputs of energy, water, and carbon.

   a. The rate at which primary producers capture and store a given amount of energy in a given time interval is called the primary productivity. (The total rate is called the *gross* primary productivity. The rate of energy storage in plants in excess of the rate of aerobic metabolism by the plants is the *net* primary productivity.)

   b. Most mineral nutrients are cycled within a natural ecosystem.

7. Energy fixed by photosynthesizers passes through grazing food webs and detrital food webs. Both types of

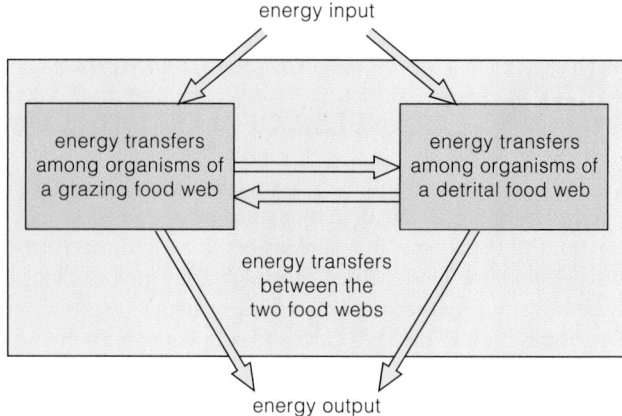

Figure 24.18 Summary of the one-way flow of energy through two kinds of cross-connected food webs in ecosystems.

food webs typically are interconnected in the same ecosystem (Figure 24.18).

   a. Both types of food webs lose energy (as heat) through metabolism of all organisms in the ecosystem.

   b. In either type of food web, the amount of useful energy flowing through consumer levels declines at each energy transfer. It declines through the loss of metabolically generated heat and as food energy is shunted into organic wastes.

8. In biogeochemical cycles, substances move from the physical environment, to organisms, then back to the environment.

   a. Water moves through the hydrologic cycle. Substances that exist primarily in gaseous phases move through atmospheric cycles. Phosphorus and other minerals move through sedimentary cycles.

   b. Nutrients generally move slowly through the geochemical phase of these cycles but rapidly between organisms and the environment.

9. Ecosystems on land have predictable rates of nutrient losses that generally increase when the land is cleared or otherwise disturbed.

10. Fossil fuel burning and conversion of natural ecosystems to cropland or grazing land are contributing to increased atmospheric concentrations of carbon dioxide. The increase may be contributing to a global warming trend.

11. Nitrogen availability is often a limiting factor for the total net primary productivity of land ecosystems. Gaseous nitrogen is abundant in the atmosphere, but it must be converted to ammonia and to nitrates that primary producers can use. Some bacteria as well as volcanic action and lightning can cause the conversion.

Review Questions

1. Define ecosystem, and name the central roles that autotrophs play in all ecosystems. *488, 490*

2. Define and give examples of trophic levels in ecosystems. *490–491*

3. Distinguish between food chain and food web. Can you imagine an extreme situation whereby you would be a participant in a food chain? *490*

4. If you were growing a vegetable garden, what variables might affect its net primary production? *492*

5. Characterize grazing and detrital food webs. Indicate how energy leaves each one and how the two are interconnected. *492*

6. Describe the greenhouse effect. Make a list of 20 agricultural products and manufactured goods that you depend on. Are any implicated in the amplification of the greenhouse effect? *500–501*

7. Describe the reservoirs and organisms involved in one of the biogeochemical cycles. *495–504*

8. Define nitrogen fixation, nitrification, ammonification, and denitrification. *502*

Self-Quiz *(Answers in Appendix IV)*

1. _____ can be thought of as an ecosystem.
   a. A freshwater spring   c. A city
   b. Antarctica   d. All of the above

2. Ecosystems have _____.
   a. energy inputs and outputs   c. one trophic level
   b. nutrient cycling but not outputs   d. a and b

3. Trophic levels can be described as _____.
   a. structured feeding relationships
   b. who eats whom in an ecosystem
   c. a hierarchy of energy transfers
   d. all of the above

4. A feeding relationship that proceeds from algae to a fish, then to a fisherman, and then to a shark is _____.
   a. a food chain   c. a and b
   b. a food web

5. Primary productivity is affected by _____.
   a. photosynthesis and respiration by plants
   b. how many plants are neither eaten nor decomposed
   c. rainfall
   d. temperatures
   e. all of the above

6. Deforestation of watersheds _____ most nutrient outputs.
   a. lessens   c. increases
   b. equalizes   d. stabilizes

7. Match the ecosystem terms with the suitable description.
   ____ primary producers   a. herbivores, carnivores, omnivores, parasites
   ____ consumers   b. feed on partly decomposed organic particles
   ____ decomposers
   ____ detritivores   c. break down remains or products of other organisms
        d. photoautotrophs

8. Match the ecosystem terms with the suitable description.
   ____ nitrogen availability   a. movement of water or nutrients from the environment, to organisms, and back
   ____ ecosystem components   b. form of nitrogen plants can take up
   ____ phosphorus   c. key limiting factor for net primary production in land ecosystems
   ____ ammonium
   ____ biogeochemical cycle   d. producers, consumers, detritivores, decomposers
        e. moves through a sedimentary cycle

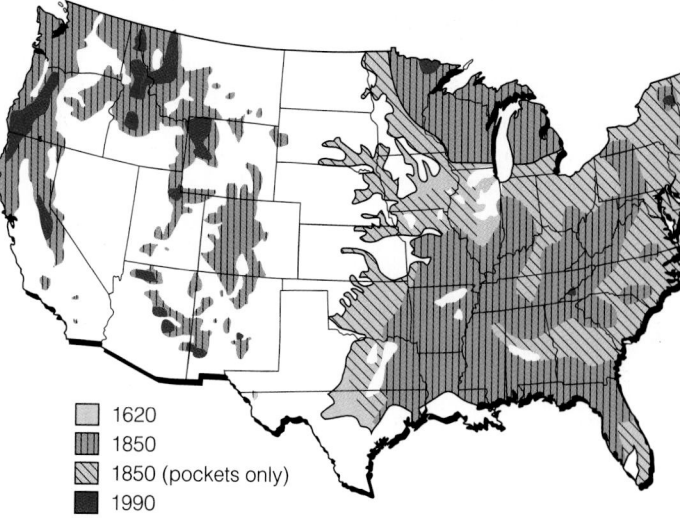

1620
1850
1850 (pockets only)
1990

Figure 24.19 Map showing the extent of deforestation in the United States from 1620 through 1990.

Critical Thinking

1. In 1995 biologists Reed Noss, J. Michael Scott, and Edward LaRoe issued a report on endangered ecosystems in the United States, including the once-vast forests (Figure 24.19). The report has prompted researchers and policymakers to ask whether we as a nation should protect entire ecosystems, rather than just endangered species. One objection is that doing so might endanger property values. Would you favor setting aside large tracts of land for ecological restoration? Or would you consider such efforts too intrusive on individual rights of property ownership?

Selected Key Terms

ammonification *502*	food web *490*
autotrophs *490*	grazing food web *492*
biogeochemical cycle *495*	greenhouse effect *500*
biological magnification *505*	habitat *488*
biosphere *488*	heterotroph *490*
carbon cycle *498*	hydrologic cycle *496*
community *488*	niche *488*
consumer *490*	nitrification *502*
decomposer *490*	nitrogen cycle *502*
decomposition *502*	nitrogen fixation *502*
denitrification *502*	phosphorus cycle *504*
detrital food web *492*	primary productivity *492*
detritivore *490*	producer *490*
ecological pyramid *492*	succession *488*
ecology *488*	trophic level *490*
ecosystem *488*	watershed *496*
food chain *490*	

Readings

Krebs, C. 1994. *Ecology.* 4th ed. New York: HarperCollins.

For additional readings, go to InfoTrac College Edition, your online library at:

http://www.infotrac-college.com/wadsworth

25   IMPACTS OF THE HUMAN POPULATION

Tales of Nightmare Numbers

Suppose that this year the U.S. Congress passes a law designed to stem the growth of the nation's population. With this law, a family can have a maximum of three children. After the third child is born, the father must be sterilized, regardless of his wishes in the matter. *It would never happen here*, you may be thinking. Such an invasion of privacy would never be tolerated in our society. And besides, family size is not much of an issue in North America, where the rates of food production and the standards of hygiene and medical care are among the world's highest.

Elsewhere in the world, especially where living conditions are already marginal, many local populations are growing at at rates that alarm most of the world's biologists—and a lot of other people as well. Consider India (Figure 25.1). At this writing, there are about 6 billion people on Earth. Of those, roughly 1 billion are living in India. Each year, India's huge population, which surpasses that of North and South America combined, grows by about 2 percent. A large percentage of the nation's inhabitants lack adequate food, shelter, and medical care. Each *week*, at least 125,000 of them enter the job market, with little hope for gainful employment. Each *day*, 100 acres of croplands that provide food for the population are lost to agriculture. Why? Among several reasons, too many salts have built up in the intensively irrigated soil; India does not get enough rain to flush them out.

Figure 25.1 This crowded scene in India shows a sampling of the more than 5.7 billion humans on Earth. In this chapter we turn to the principles governing the growth and sustainability of populations, including our own.

Recognizing that India's huge and rapidly growing population creates social problems and severely strains national resources, for decades the Indian government has sponsored family-planning programs. With what results? To begin with, administering such programs has been difficult, because many millions of people live in remote villages. Because of widespread illiteracy, information often must be conveyed by word of mouth. And because disease kills many children, villagers often resist limiting family size. Without a large family, they ask, who would survive and help tend fields? Who will go to the cities and earn money to send back home? Who will care for parents when they are too old to work? How can a father otherwise know that he will be survived by a son? This last bit of security is important because, by Hindu tradition, a man's soul will rest in peace only if his son conducts the last rites.

In 1976, out of desperation, government officials subjected some men to compulsory vasectomies. Public outrage over the policy contributed to the eventual downfall of the then-ruling party, and the law was rescinded.

Is there a way out of such dilemmas? Should the wealthier, less densely populated nations that now use most of the world's resources learn to get by more efficiently, on less? Should they donate surplus food to less fortunate nations? Would such donations help, or would they encourage dependency and even greater population growth? And what would happen if the benefactor nations suffered severe droughts year after year and had trouble meeting their own resource requirements?

For humans and all other organisms on Earth, we can be sure of one fact: *certain ecological principles govern the growth and sustainability of populations over time*. This chapter describes these principles, then shows how they apply to the past, present, and future growth of the human population.

KEY CONCEPTS

1. Ecological principles govern the growth and sustainability of populations over time. For example, all populations face limits to growth, for no environment can indefinitely sustain a continuously increasing number of individuals.

2. Population growth generally follows certain patterns. When a population grows exponentially, it increases in size by ever larger amounts per unit of time.

3. Now and in the foreseeable future, we have attained the population size, the technology, and the cultural inclination to use energy and modify our collective environment at astonishing rates.

4. The accumulation of pollutants generated by human activity is disrupting complex interactions among the atmosphere, oceans, and land in ways that may have the most serious kinds of consequences in the near future.

CHAPTER AT A GLANCE

25.1 Introduction to Population Dynamics

25.2 A Closer Look at Growth Patterns

25.3 Human Population Growth

25.4 Air Pollution

25.5 A Global Water Crisis

25.6 Coping with Solid Wastes and Problems of Land Use

25.7 Destruction of Forests

25.8 Concerns about Energy Use

25.9 *A Planetary Emergency: Loss of Biodiversity*

25.10 The Idea of Sustainable Living

Characteristics of Human Populations

Populations as a whole display certain characteristics, including density, dispersion, and age structure. Population *density* is the number of individuals per unit area or volume, such as the number of minnows in each liter of water in a small stream or the number of people living within a hectare of land. *Dispersion* refers to the general pattern in which the population's members are distributed through its habitat. We humans are social animals. Our populations tend to be clumped in villages, towns, and cities, where we interact with one another and have access to jobs and other resources.

A population's **age structure** is the relative proportion of individuals of each age. These are often divided into prereproductive, reproductive, and postreproductive age categories. The middle category represents the **reproductive base** for the population. Figure 25.2 graphs the age structure diagrams for populations growing at different rates. In these graphs, ages 15 to 44 are the average range of childbearing years (for both sexes). The population of the United States has a narrow base and is an example of slow growth. Figure 25.3 tracks its 78 million *baby boomers*, a group that formed in 1946 after soldiers returned home from World War II and settled down to start families. By contrast, age structure diagrams for a rapidly growing population have a broad base.

A country's *prereproductive base*—the proportion of individuals younger than 15—is the most telling statistic, because it reflects both recent population growth *and* potential future growth when those individuals mature. Kenya's prereproductive base is broad, for example; not only is that nation's population growing rapidly now, but we can also expect such growth to continue.

Population Size and Patterns of Growth

Over a given time span, the numerical size of a population depends on two factors: how many individuals enter it by birth or immigration and how many leave it by death or emigration:

$$\begin{pmatrix} \text{amount of} \\ \text{population} \\ \text{growth} \end{pmatrix} = \begin{pmatrix} \text{births} \\ + \\ \text{immigration} \end{pmatrix} - \begin{pmatrix} \text{deaths} \\ + \\ \text{emigration} \end{pmatrix}$$

If immigration and emigration are balanced, then population size is stable when the birth rate is balanced over the long term by the death rate. When this balance exists, there is **zero population growth**.

In reality, immigration may be a major contributor to a nation's population growth. In the United States, for example, immigration (legal and otherwise) accounts for more than 30 percent of annual population growth.

Populations increase in size when the number of births exceeds the number of deaths. But how fast and how much can they increase in a given period? Imagine a population that in a given year has 1,000 members ($N = 1,000$).

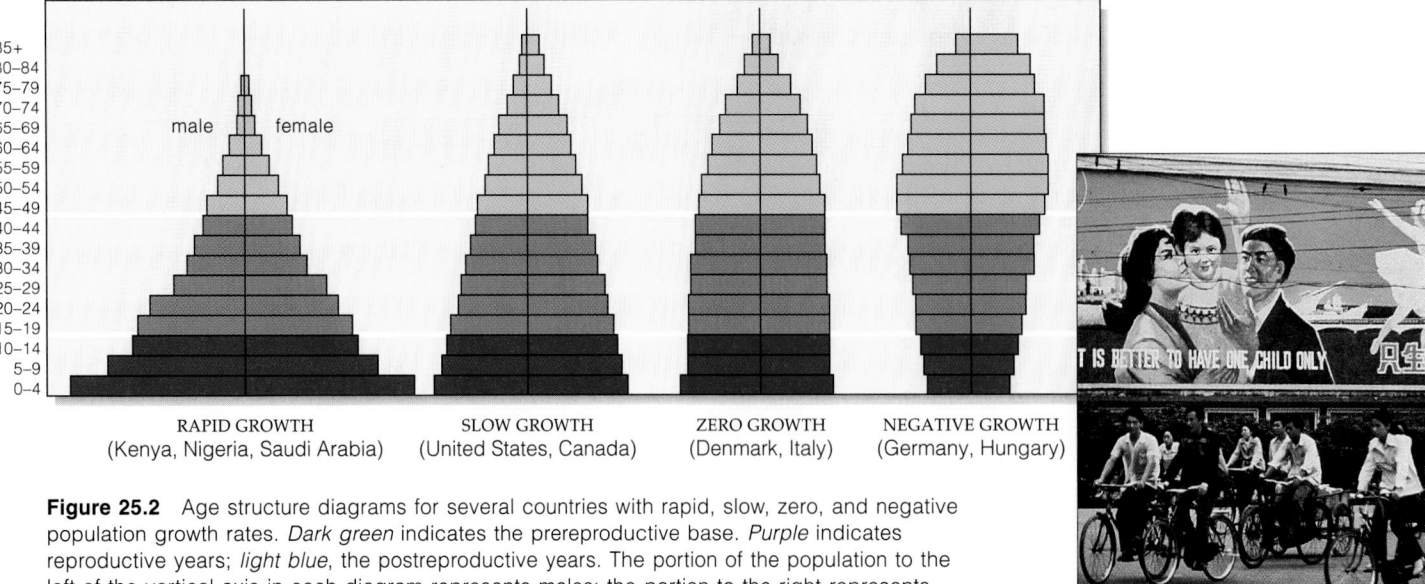

Figure 25.2 Age structure diagrams for several countries with rapid, slow, zero, and negative population growth rates. *Dark green* indicates the prereproductive base. *Purple* indicates reproductive years; *light blue*, the postreproductive years. The portion of the population to the left of the vertical axis in each diagram represents males; the portion to the right represents females. The photograph shows a family-planning billboard in China.

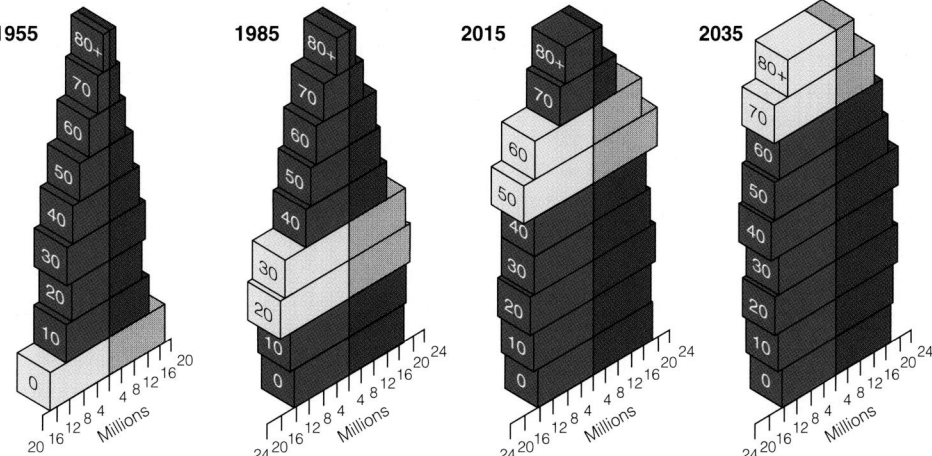

Figure 25.3 Age structure diagrams for the United States population. *Gold* bars track the baby-boom generation.

1955 1985 2015 2035

In each succeeding year, for each 1,000 members, 10 members die but 50 are born. The population's rate of increase (*r*) would be:

$$r = \frac{\text{births} - \text{deaths}}{N} = \frac{50 - 10}{1,000} = 0.04 \text{ or } 4\% \text{ (here, 40) per year}$$

As long as *r* holds constant, any population will show **exponential growth**. The number of its individuals increases in *doubling* increments—from 2 to 4, then 8, 16, 32, 64, and so on. For a population with a 4 percent rate of increase, this doubling time is 17.5 years.

We can observe a limited period of exponential growth in the laboratory by putting a single bacterium in a culture flask with a supply of nutrients. After about 30 minutes the bacterium divides into two, and 30 minutes later the two divide into four. If no cells die between divisions, the population will double every 30 minutes. The larger the population becomes, the more bacteria there are to divide. After only 9½ hours (19 doublings), the population will exceed 500,000. After 10 hours (20 doublings) it will soar past 1 million!

In a population undergoing unrestricted, exponential growth, a J-shaped curve results when population size is plotted against time (Figure 25.4). Populations of this type increase in size by ever larger amounts per unit of time. When nothing stops its growth, such a population will grow exponentially, even when the birth rate only slightly exceeds the death rate. Only the time scale changes. Doubling still occurs—it simply takes longer.

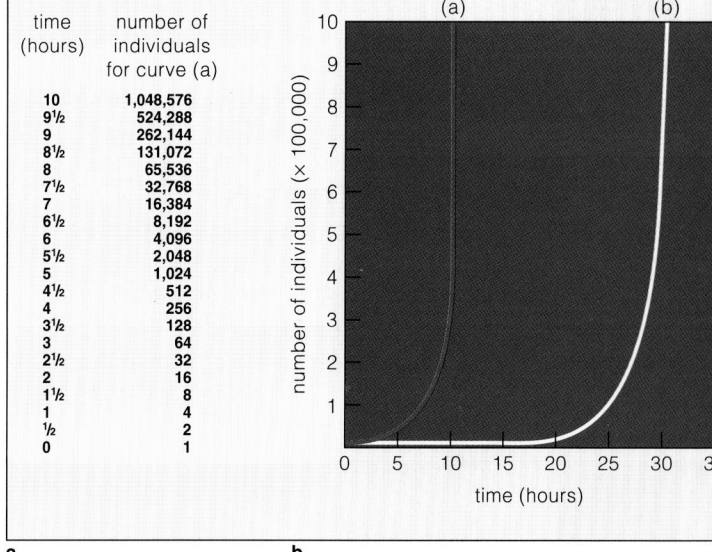

time (hours)	number of individuals for curve (a)
10	1,048,576
9½	524,288
9	262,144
8½	131,072
8	65,536
7½	32,768
7	16,384
6½	8,192
6	4,096
5½	2,048
5	1,024
4½	512
4	256
3½	128
3	64
2½	32
2	16
1½	8
1	4
½	2
0	1

a   b

Figure 25.4 (**a**) Exponential growth for a bacterial population that is dividing by fission every half hour. (**b**) Exponential growth of the population when division occurs every half hour but when 25 percent die between divisions. Although deaths slow the rate of increase, in themselves they are not enough to stop exponential growth.

As long as the birth rate remains even slightly above the death rate, a population will grow. If the rates remain constant, it will grow exponentially.

A CLOSER LOOK AT GROWTH PATTERNS

Biotic Potential

The **biotic potential** of a population is its *maximum* rate of growth under ideal conditions. It is the rate that might be achieved when food and living space are abundant, when other organisms aren't interfering with access to those resources, and when no predators or disease agents are present.

Biotic potential is not the same for every species. For many bacteria, it is 100 percent every half hour; for humans and other large mammals, it is between 2 and 5 percent per year. The differences arise through variations in (1) how soon individuals start reproducing, (2) how often reproduction occurs, and (3) how many offspring are born each time.

A population may grow exponentially even when it is not expressing its full biotic potential. It is biologically possible for human females to bear 20 children or more, although few have done so. Yet the human population has been growing exponentially since the mid-18th century. At any given time, the *actual* rate of increase is influenced by environmental circumstances affecting human society.

Limiting Factors and Carrying Capacity

Most often, environmental circumstances prevent a population from reaching its full biotic potential. For instance, when an essential resource such as food or water is in short supply, it becomes a **limiting factor** on population growth. Predation (as by disease organisms), competition for living space, and buildup of toxins are other examples of limiting factors. The number of such factors can be enormous, and their effects can vary.

The concept of limiting factors is important because it defines the **carrying capacity**—the number of individuals of a given species that can be sustained indefinitely by the resources in a given area. Some experts believe that Earth has the resources to support from 7 to 12 billion humans, with a reasonable standard of living for many. Others believe that the current human population of 5.7 billion is already exceeding its carrying capacity through soil erosion, deforestation, pollution of groundwater supplies, and changes in global climate. All these viewpoints share the premise that *overpopulation is the root of many, if not most, of the environmental problems the world now faces.*

A low-density population starts growing slowly, then goes through a rapid growth phase, and then growth levels off once the carrying capacity is reached. This pattern is called **logistic growth**. A plot of logistic growth gives us an S-shaped curve (Figure 25.5). This curve is only a simple approximation of what goes on in nature, however. Because environmental conditions vary, carrying capacity also can vary over time.

Checks on Population Growth

When a growing population's density increases, high density and overcrowding result in competition for resources. They also put individuals at increased risk of being killed by infectious diseases and parasites, which are more easily spread in crowded living conditions. These are **density-dependent controls** on population growth. Once such factors take their toll on a population and its density decreases, the pressures ease and the population may grow once more.

A classic example is the *bubonic plague* that killed 25 million Europeans—about one third of the population—during the 14th century. *Yersinia pestis*, the bacterium responsible, normally lives in wild rodents; fleas transmit it to new hosts. It spread like wildfire through the cities of medieval Europe because human dwellings were crowded together, sanitary conditions were poor, and rats were abundant. In 1994, bubonic plague and a related disease, pneumonic plague, raced through rat-infested cities in India where garbage and animal carcasses had piled up for months in the streets. Some fleeing residents carried the diseases as far away as London, England. Only concerted efforts by public health officials averted a pandemic.

Density-independent controls can also operate. These are events such as floods, earthquakes, or other natural disasters that cause deaths or births regardless of whether the members of a population are crowded or not.

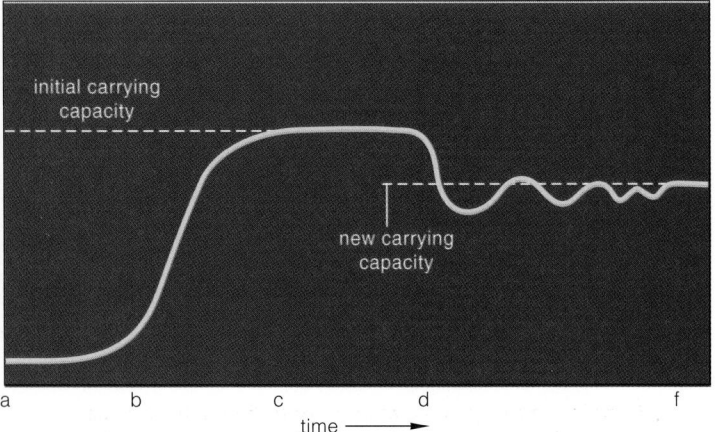

Figure 25.5 Idealized S-shaped curve characteristic of logistic growth. Following a rapid growth phase, growth slows and the curve flattens out as the carrying capacity is reached. Variations can occur in S-shaped growth curves, as when changed environmental conditions bring about a decrease in the carrying capacity. This happened to the human population in Ireland in the late 19th century, when a fungus wiped out the potatoes that were the mainstay of the diet.

Table 25.1 Life Table for the U.S. Human Population, 1989*

Age Interval (category for individuals between the two ages listed)	Survivorship (number alive at start of age interval, per 100,000 individuals)	Mortality (number dying during the age interval)	Life Expectancy (average lifetime remaining at start of age interval)	Number of Reported Live Births for Total U.S. Population
0–1	100,000	986	75.3	
1–5	99,104	192	75.0	
5–10	98,912	117	71.1	
10–15	98,975	132	66.2	11,486
15–20	98,663	429	61.3	506,503
20–25	98,234	551	56.6	1,077,598
25–30	97,683	606	51.9	1,263,098
30–35	97,077	737	47.2	842,395
35–40	96,340	936	42.5	253,878
40–45	95,404	1,220	37.9	44,401
45–50	94,184	1,766	33.4	1,599
50–55	92,148	2,727	28.9	
55–60	89,691	4,234	24.7	
60–65	85,357	6,211	20.8	
65–70	79,146	8,477	17.2	
70–75	70,669	11,470	13.9	
75–80	59,199	14,598	10.9	
80–85	44,601	17,448	8.3	
85 +	27,153	27,153	6.2	

Total: 4,040,958

*Compiled by Marion Hansen, based on data from U.S. Bureau of the Census, *Statistical Abstract of the United States, 1992* (edition 112).

Life History Patterns

Like all species, humans have a characteristic life span, although few people reach the maximum age possible. Death is more probable at some ages and less so at others. Also, individuals are more likely to reproduce at some ages than at others, and these ages vary from one species to the next. The study of such age-specific patterns is the subject of *demography*.

A **life table** (Table 25.1) is a summary of the age-specific patterns of birth and death for a given population in a given area. Such tables were originally developed by insurance companies and are used to help set the price of life or health insurance for people of different ages.

The biotic potential of a population is its maximum rate of increase under ideal conditions. For any population, the actual rate of increase is influenced by environmental circumstances.

Carrying capacity is the number of individuals of a species that can be sustained indefinitely by resources in a given area.

Factors that limit human population growth include availability of various types of resources, predation by disease organisms, and effects of pollution.

Because these factors vary in their effects over time, both the carrying capacity and the population size can fluctuate.

HUMAN POPULATION GROWTH

In 1997, the human population totaled over 5.8 billion (Figure 25.6). In that one year almost 100 million individuals were born—an average of 1.9 million per week, 273,000 per day. This growth is the consequence of advances in agriculture, industrialization, sanitation, and health care. It took *2 million years* for the human population to reach the first billion. It took only 130 years to reach the second billion, 30 years to reach the third, 15 years to reach the fourth, *and only 12 years to reach the fifth.* By 2000, some 6.2 billion humans will inhabit the Earth.

The **demographic transition model** (Figure 25.7) links changes in population growth with changes that unfold

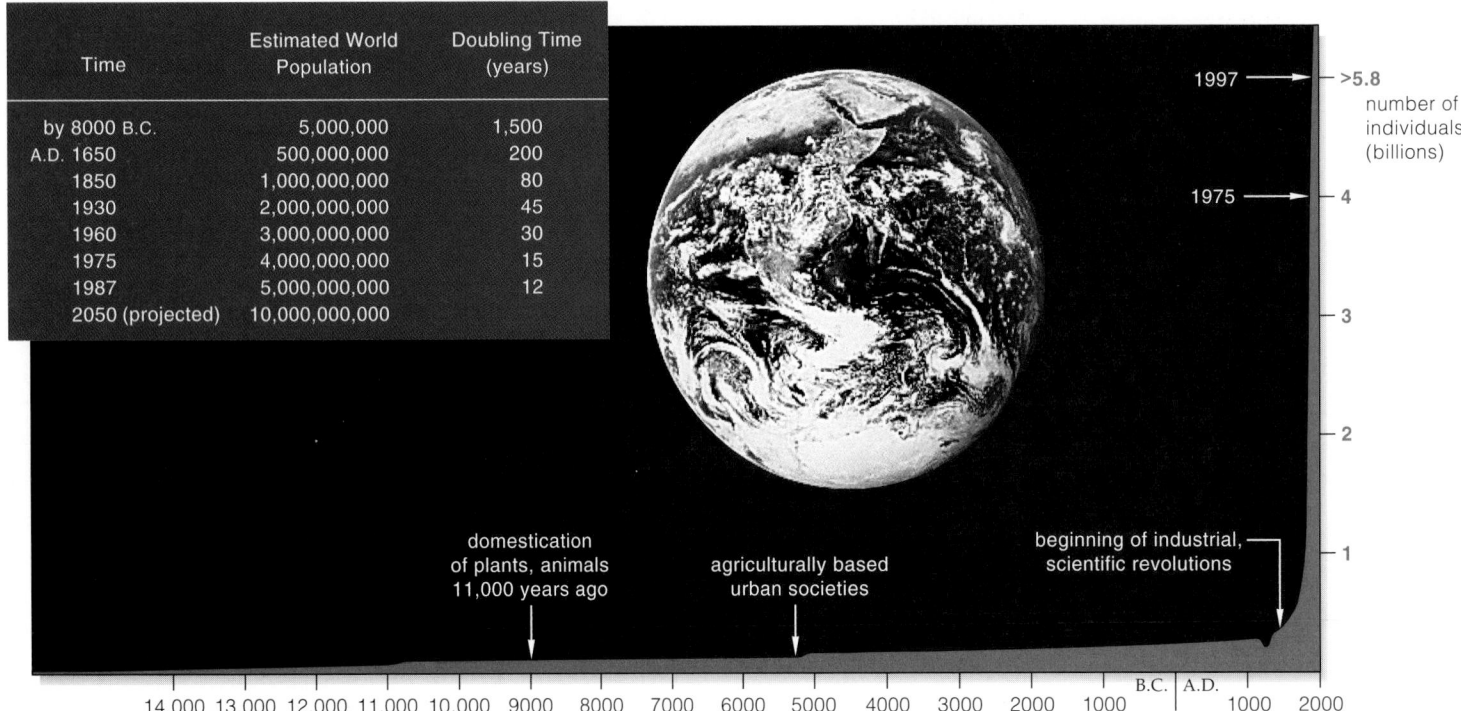

Time	Estimated World Population	Doubling Time (years)
by 8000 B.C.	5,000,000	1,500
A.D. 1650	500,000,000	200
1850	1,000,000,000	80
1930	2,000,000,000	45
1960	3,000,000,000	30
1975	4,000,000,000	15
1987	5,000,000,000	12
2050 (projected)	10,000,000,000	

domestication of plants, animals 11,000 years ago

agriculturally based urban societies

beginning of industrial, scientific revolutions

Figure 25.6 (**above**) Growth curve (*red*) of the human population. The diagram's vertical axis represents world population, in billions. (The slight dip between the years 1347 and 1351 is the time when 25 million people died in Europe as a result of bubonic plague.) The growth pattern over the past two centuries has been exponential, sustained by agricultural revolutions, industrialization, and improvements in health care. The list in the *blue* box tells us how long it took for the human population to double in size at different times in its history.

Figure 25.7 (**right**) The demographic transition model of changes in population size as correlated with changes in economic development.

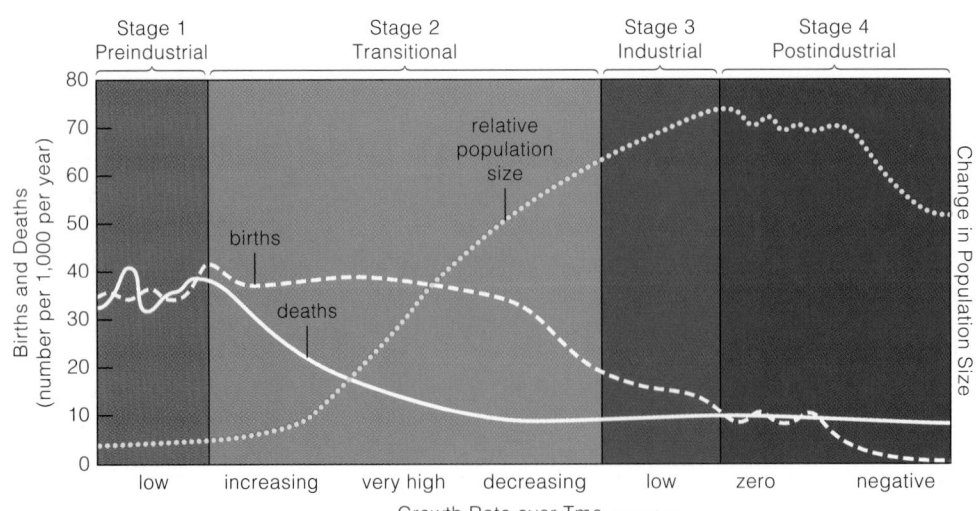

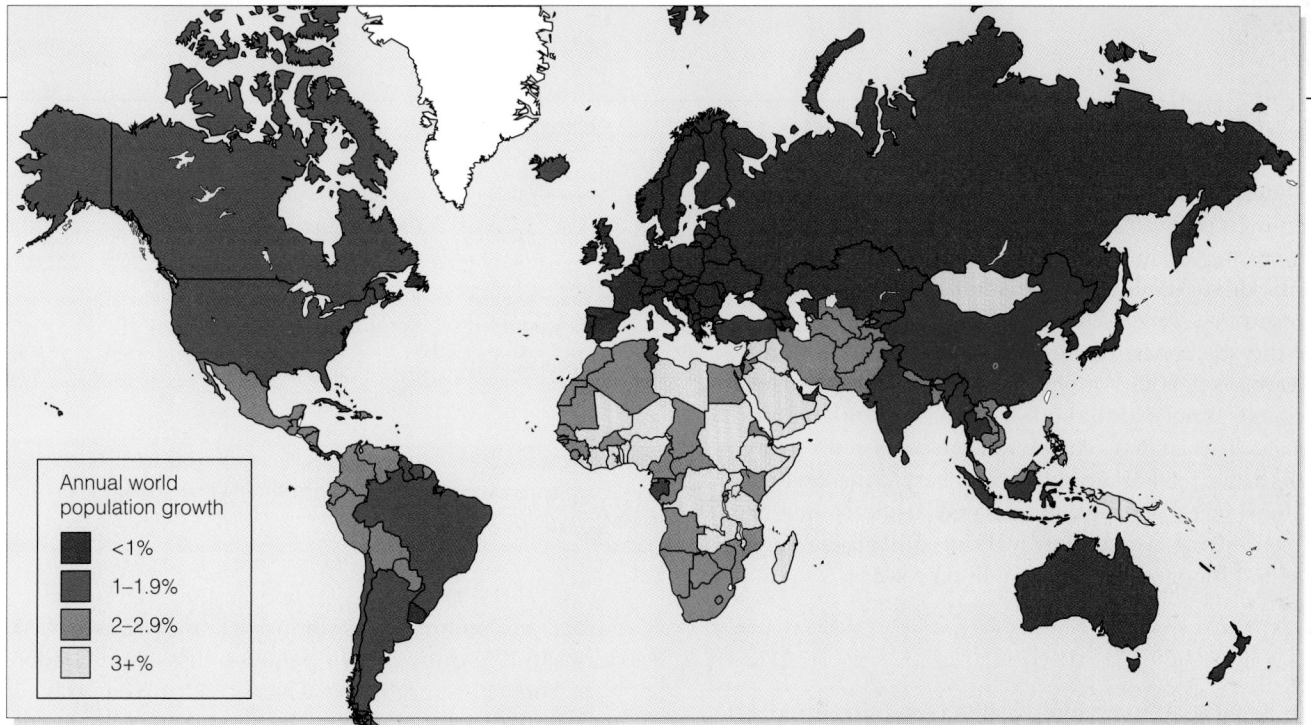

a

Figure 25.8 (**a**) The average annual population growth rate in different regions of the world. (**b**) Population sizes in 1996 (*orange* bars) and projected for 2025 (*blue* bars).

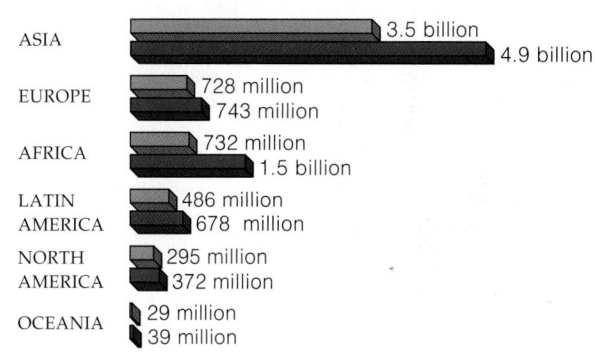

ASIA	3.5 billion
	4.9 billion
EUROPE	728 million
	743 million
AFRICA	732 million
	1.5 billion
LATIN AMERICA	486 million
	678 million
NORTH AMERICA	295 million
	372 million
OCEANIA	29 million
	39 million

b

during four stages of economic development. In the *preindustrial stage,* there is little population growth because living conditions are harsh, and birth and death rates are both high. In the *transitional stage,* industrialization begins, food production rises, and sanitation and health care improve. Death rates drop, but birth rates remain high, so the population grows rapidly over a long period. Then growth starts to level off as living conditions improve. In the *industrial stage*—when industrialization is in full swing—population growth slows, mostly because urban couples regulate family size. Many decide that raising children is expensive and that having too many puts them at an economic disadvantage. In the *postindustrial stage,* zero population growth is reached. Then the birth rate falls below the death rate, and the population slowly decreases in size.

Today, the United States, Canada, Australia, Japan, countries of the former Soviet Union, New Zealand, and most countries of western Europe are in the industrial stage, and their growth is slowly decreasing. Eighteen countries, including Sweden, the United Kingdom, Germany, and Hungary, are close to, at, or slightly below zero population growth.

Mexico and other developing countries are in the transitional stage. They may not stay there, however. In many countries in this stage, population growth exceeds economic growth, and it is difficult if not impossible for such nations to afford the costs of fossil fuels and other resources that drive industrialization. Some transitional countries and their populations may return to the harsh conditions of the preceding stage.

Figure 25.8a shows the average annual growth rate for different parts of the world in 1994. If that rate is maintained, we can expect the population to approach 8.5 billion by 2025 (Figure 25.8b). It may prove impossible to achieve corresponding increases in food production, drinkable water, energy reserves, and all the wood, steel, and other materials we use to meet everyone's basic needs—something we are not doing even now. As we see in the next section, there is evidence that harmful by-products of our activities—pollutants—are changing the land, seas, and the atmosphere in ominous ways. From what we know of the principles governing population growth, unless there are spectacular technological breakthroughs, it is realistic to expect an increase in human death rates. Although our stupendously accelerated growth continues, it cannot be sustained indefinitely.

Differences in population growth among countries correlate with levels of economic development. Globally, the population will reach 8.5 billion by 2025, a level that will severely strain supplies of food, energy, and other resources.

AIR POLLUTION

Let's start this survey of the impact of human activities by defining pollution, which is central to our problems. **Pollutants** are substances that adversely affect the health, activities, or survival of a population. Air pollutants are prime examples. As you can see in Table 25.2, they include carbon dioxide, oxides of nitrogen and sulfur, and chlorofluorocarbons (CFCs). Among them also are photochemical oxidants formed as the sun's rays interact with certain chemicals. The United States alone releases more than 700,000 metric tons of air pollutants every day. Whether these substances remain concentrated at the source or dispersed during a given interval depends on the local climate and topography, as you will now see.

Smog

During a **thermal inversion**, weather conditions trap a layer of cool, dense air under a layer of warm air. If the trapped air contains pollutants, winds cannot disperse them, and they may accumulate to dangerous levels. Thermal inversions have been key factors in some of the worst air pollution disasters, because they intensify an atmospheric condition called smog (Figure 25.9).

Two types of smog form in major cities. Where winters are cold and wet, **industrial smog** forms as a gray haze over industrialized cities that burn coal and other fossil fuels for manufacturing, heating, and generating electric power. The burning releases airborne dust, smoke, ashes, soot, asbestos, oil, bits of lead and other heavy

Table 25.2 Major Classes of Air Pollutants

Carbon oxides	Carbon monoxide (CO), carbon dioxide (CO_2)
Sulfur oxides	Sulfur dioxide (SO_2), sulfur trioxide (SO_3)
Nitrogen oxides	Nitric oxide (NO), nitrogen dioxide (NO_2), nitrous oxide (N_2O)
Volatile organic compounds	Methane (CH_4), benzene (C_6H_6), chlorofluorocarbons (CFCs)
Photochemical oxidants	Ozone (O_3), peroxyacyl nitrates (PANs), hydrogen peroxide (H_2O_2)
Suspended particles	Solid particles (dust, soot, asbestos, lead, etc.), liquid droplets (sulfuric acid, oils, dioxins, pesticides)

metals, and sulfur oxides. Industrial smog caused 4,000 deaths in 1952 during an air pollution disaster in London. New York, Pittsburgh, and Chicago once were gray-air cities until they restricted coal burning. Now, most industrial smog forms in cities of developing countries, including China and India, as well as in countries of eastern Europe.

In warm climates, **photochemical smog** forms as a brown, smelly haze over large cities. It becomes concentrated when the surrounding land forms a natural basin, as in Los Angeles and Mexico City. The key culprit is nitric oxide. After it is released from vehicles, nitric oxide reacts with oxygen in the air to form nitrogen dioxide. When exposed to sunlight, nitrogen dioxide can react with hydrocarbons (such as spilled or partly burned gasoline) to form photochemical oxidants. The main oxidants in smog are ozone and PANs (peroxyacyl nitrates). PANs resemble tear gas; even traces can sting the eyes, irritate lungs, and damage crops.

Acid Deposition

Oxides of sulfur and nitrogen are among the worst air pollutants. Coal-burning power plants, factories, and metal smelters emit most sulfur dioxides. Motor vehicles, power plants that burn gas and oil, and nitrogen-rich fertilizers produce nitrogen oxides. In dry weather, fine particles of oxides may briefly stay airborne, then fall to Earth as **dry acid deposition.** When dissolved in atmospheric water, they form weak solutions of sulfuric and nitric acids. Winds may disperse them over great distances. If they fall to Earth in rain and snow, we call this wet acid deposition or **acid rain.** Acid rain can be much more acidic than normal rainwater, sometimes becoming as acidic as lemon juice (pH 2.3). The deposited acids eat away at marble, metals, mortar, plastic, even nylon stockings. They also seriously disrupt the physiology of organisms and the chemistry of ecosystems.

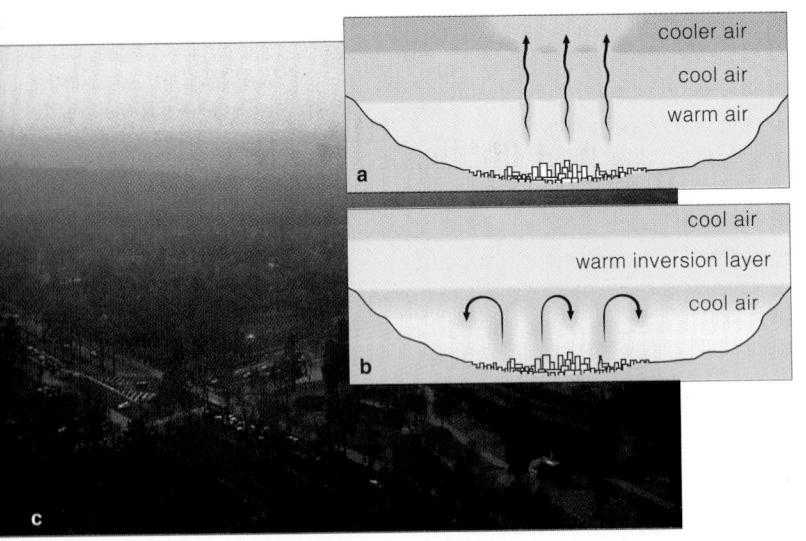

Figure 25.9 (a) Normal pattern of air circulation in smog-forming regions. (b) Air pollution trapped under a thermal inversion layer. (c) Mexico City on a sunny morning. Topography, huge numbers of people and vehicles, and industry combine to make its air among the world's smoggiest.

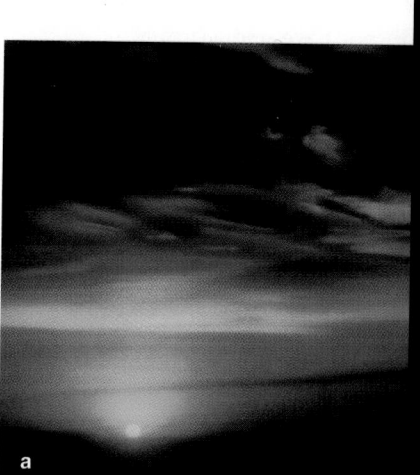

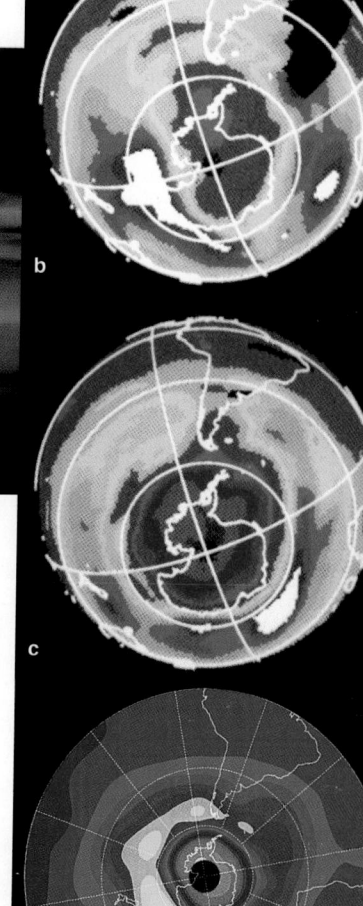

Figure 25.10 (**a**) Ice clouds above Antarctica that have a role in the ozone thinning each spring. Seasonal ozone thinning above Antarctica in (**b**) 1979, (**c**) 1991, and (**d**) 1996. Lowest ozone values are coded *magenta* and *purple*.

The precipitation in much of eastern North America is 30 to 40 times more acidic than it was several decades ago, and croplands and forests are suffering. Fish have vanished from more than 200 lakes of the Adirondack Mountains of New York. Acidic pollutants originating in the industrial regions of England and Germany are key factors in the destruction of large tracts of forests in northern Europe. Such pollutants also are a serious problem in heavily industrialized parts of Asia, Latin America, and Africa.

Damage to the Ozone Layer

Ozone is a molecule of three oxygen atoms (O_3). It occurs in two regions of Earth's atmosphere. In the troposphere, the region closest to the Earth's surface, ozone is part of smog and can damage the respiratory system as well as other organisms. However, ozone in the *next* atmospheric layer—the stratosphere (17–48 kilometers, or 11–30 miles above Earth)—intercepts harmful ultraviolet radiation that can cause skin cancer and eye cataracts. As you probably are aware, this protective screen has been thinning. Each year, from early September through mid-October, an ozone "hole" appears over the Antarctic, extending over an area about the size of the continental United States. Satellites and high-altitude planes have been monitoring the ozone hole (Figure 25.10). Since 1987 the ozone layer over the Antarctic has been thinning by about half every year. And since 1969 it has steadily decreased over some populated regions. By 2050 the ozone layer over North America, Europe, and Asia may thin by 10–25 percent.

As the ozone layer thins, more harmful ultraviolet radiation reaches Earth's surface. Already, the incidence of skin cancer is increasing dramatically. In addition, ultraviolet radiation can weaken the immune system, making people more vulnerable to infections. Reducing the ozone layer also may damage phytoplankton, microscopic photosynthetic organisms that are the basis of food webs in freshwater and marine ecosystems. Phytoplankton are also a factor in maintaining the composition of the atmosphere because they act to remove carbon dioxide from surface waters and release oxygen.

Chlorofluorocarbons (CFCs) are the chief culprits in ozone depletion. These compounds of chlorine, fluorine, and carbon are widely used as propellants in aerosol sprays, as coolants in refrigerators and air conditioners, and for other industrial and commercial uses. They are also released during volcanic activity.

When a CFC in the stratosphere absorbs ultraviolet light, it gives up a chlorine atom. The chlorine can react with ozone (O_3), yielding oxygen (O_2) and chlorine monoxide. Reactions between oxygen and chlorine monoxide release more chlorine atoms—each one able to destroy up to 10,000 molecules of ozone!

Today, officials from many countries are working with governments, industry, and environmental organizations, devising effective means for phasing out ozone-depleting substances. CFCs and several other substances now are banned in some places. Phase-out schedules for other offending chemicals have been moved up because of new evidence that in some regions ozone depletion is occurring more rapidly than suspected. At the same time, efforts to develop substitutes for CFCs are being speeded up. Even so, we do well to keep in mind that up to 95 percent of the CFCs released since 1955 are still making their way up to the stratosphere. For decades to come, all of us, our children, and our grandchildren will be living with their destructive effects.

A pollutant is a substance that adversely affects the health, activities, or survival of a population. Serious air pollutants include carbon dioxide, oxides of sulfur and nitrogen, and chlorofluorocarbons.

A GLOBAL WATER CRISIS

Three of every four humans do not have enough clean water to meet basic needs. Most of Earth's water is salty (in oceans). Of every million liters of water on our planet, only 6 liters are readily usable for human activities. As our population grows exponentially, so do demands and impacts on this limited water supply.

Impact of Large-Scale Irrigation

About a third of the world's food grows on irrigated land (Figure 25.11). Water is piped into agricultural fields from groundwater or diverted from lakes or rivers. Irrigation

Figure 25.11 Irrigation-dependent crops growing in the Sahara Desert of Algeria.

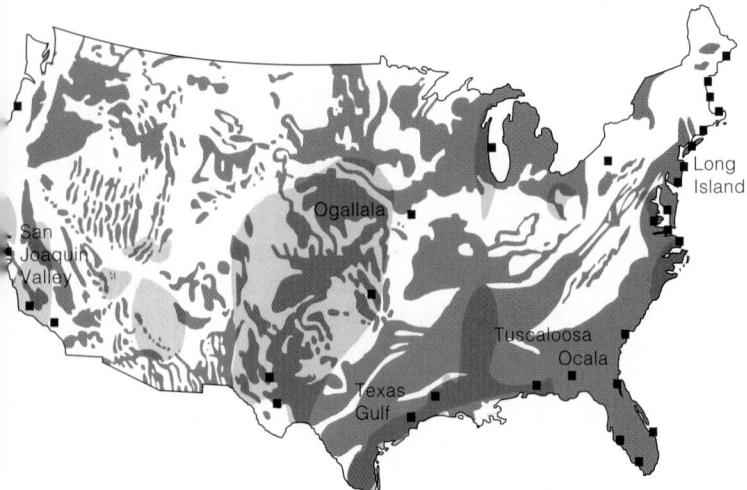

Figure 25.12 Major underground aquifers (shown in *blue*) containing 95 percent of all freshwater in the United States. These aquifers are being depleted in many areas (*gold*) and contaminated elsewhere through pollution and saltwater intrusion. The black boxes indicate areas of saltwater intrusion.

water often contains large amounts of mineral salts, and where soil drainage is poor, evaporation may cause salt buildup or **salinization**. Soil salinization can decrease yields and eventually kill crops. Worldwide, salinization is estimated to have reduced yields on 25 percent of all irrigated cropland. Improperly drained irrigated lands also can become waterlogged, so that soil around plant roots becomes saturated with toxic saline water.

Large-scale irrigation is also a major factor in groundwater depletion. For instance, farmers from South Dakota to Texas withdraw so much water from the Ogallala aquifer (Figure 25.12) that the annual overdraft (the amount not replenished) nearly equals the yearly flow of the Colorado River! The already-low water tables in much of the region are dropping rapidly, and stream and underground spring flows are dwindling. *Subsidence*—the sinking of land when groundwater is withdrawn—is increasing. In coastal areas, overuse of groundwater can cause saltwater intrusion into human water supplies.

Problems with Water Quality

In many regions, agricultural runoff pollutes public water sources with sediments, pesticides, and fertilizers. Power plants and factories pollute water with chemicals (including known carcinogens), radioactive materials, and excess heat (thermal pollution). Such pollutants may accumulate in lakes, rivers, and bays before reaching their ultimate destination, the oceans. Contaminants from human activities have begun to turn up even in supposedly "pure" water in underground aquifers.

Many people view the oceans as convenient refuse dumps. Cities throughout the world dump untreated sewage, garbage, and other noxious debris into coastal waters. Cities along rivers and harbors maintain shipping channels by dredging the polluted muck and barging it out to sea. We do not yet know the full impact of such practices on fisheries from which humans obtain food.

The United States has facilities for treating the liquid wastes from about 70 percent of the population and 87,000 industries. The remaining wastes, mostly from suburban and rural populations, are treated in lagoons or septic tanks or discharged—untreated—directly into waterways. Studies show that water pollution intensifies as rivers flow toward the oceans and that water treatment often does not remove toxic wastes from industries upstream from a city. Do you know where your own water supply has been?

Worldwide, human water supplies are threatened by overuse and by pollution with agricultural and industrial wastes. Parts of the world oceans also are becoming contaminated by wastes and toxins produced by human activities.

COPING WITH SOLID WASTES AND PROBLEMS OF LAND USE

Solid Wastes

Billions of metric tons of solid wastes are dumped, burned, and buried annually in the United States alone. This includes 50 billion nonreturnable cans and bottles. Paper products make up fully one-half of the total volume of solid wastes.

Paper is relatively easily recycled. In the Netherlands, 53 percent of wastepaper is recycled, and in Japan the figure is 50 percent. In 1992, almost 40 percent of the wastepaper in the United States was recycled, and efforts by individuals and businesses are improving. Simply recycling the nation's Sunday newspapers would save 500,000 trees per week. Using recycled paper also reduces air pollution that results from paper manufacturing by 95 percent, lowers water pollution by 35 percent, and takes 30–50 percent less energy than making new paper.

In natural ecosystems, solid wastes are recycled, but we bury them in landfills or burn them in incinerators. Incinerators can add heavy metals and other toxic pollutants to the air and leave a highly toxic ash that must be disposed of safely. Land available and acceptable for landfills is scarce and becoming scarcer. All landfills eventually "leak," posing a threat to groundwater supplies. That is one reason why communities increasingly take the "not in my backyard" approach to landfills. More and more people now participate in recycling programs, but it seems that few want to take responsibility for the nonrecyclable garbage they generate.

A transition from a throwaway mentality to one based on recycling and reuse is affordable, technically feasible, and environmentally essential. Consumers can play a role by refusing to buy goods that are lavishly wrapped and boxed, packaged in indestructible containers, or designed for one-time use. Individuals can ask the local post office to turn off their daily flow of junk mail. They can also urge local governments to develop resource recovery centers by which existing dumps and landfills become urban "mines" from which usable materials can be recovered. Several such centers now operate successfully in different parts of the United States.

Conversion of Marginal Lands for Agriculture

Today, human population growth is forcing expansion of agriculture onto marginally productive land. Almost 21 percent of the Earth's land is now being used for agriculture. Another 28 percent may be suitable for cropland or grazing land, but its potential productivity is so low that conversion may not be worth the cost.

Valiant efforts have been made to improve crop production on existing land. Under the banner of the **green revolution**, research has been directed toward improving the varieties of crop plants for higher yields and exporting modern agricultural practices and equipment to developing countries. Unfortunately, the green revolution involves intensive, mechanized agriculture. It is based on massive inputs of fertilizers and pesticides and ample irrigation to sustain high-yield crops. It is based also on fossil fuel energy to drive farm machines. Crop yields *are* four times as high as from traditional methods. But the modern practices use up a hundred times more energy and mineral resources, which makes them and the food they produce too expensive for both farmers and consumers in developing countries.

Overgrazing of livestock on marginal lands is a prime cause of **desertification**—the conversion of grasslands, rain-fed cropland, or irrigated cropland to a less productive, desertlike state (Figure 25.13). It occurs as vegetation is removed, leaving few or no plant roots to prevent wind and water from eroding the soil away. The subsoil then bakes rock-hard, and plants can no longer grow there. Worldwide, about 9 million square kilometers have become desertified over the past 50 years, and the trend is continuing.

Figure 25.13 Desertification in the Sahel, a region of West Africa that forms a belt between the dry Sahara Desert and tropical forests. This savanna country is rapidly undergoing desertification as a result of overgrazing and overfarming.

DESTRUCTION OF FORESTS

The world's great forests profoundly influence the biosphere. Like giant sponges, forested watersheds absorb, hold, and gradually release water. By intervening in the downstream flow of water, forests help control soil erosion, flooding, and sediment buildup in rivers, lakes, and reservoirs.

Deforestation is the removal of all trees from large tracts of land for logging, agricultural, or grazing operations. The loss of vegetation exposes the soil, and this promotes leaching of nutrients and erosion, especially on steep slopes. Figures 25.14 and 25.15 provide close-up and panoramic views of deforestation in South America's Amazon basin.

In the tropics, clearing forests for agriculture leads to a long-term loss in productivity. The irony is that tropical forests are one of the worst places to grow crops and raise pasture animals. The high temperatures and heavy, frequent rainfall favor decomposition. In intact forests, organic remains and wastes decompose too fast for litter

to build up. The forest trees and other plants absorb and assimilate nutrients as they become available.

Shifting cultivation (once called slash-and-burn agriculture) disrupts the forest ecosystem. People cut and burn trees, then till the ashes into the soil. The nutrient-rich ashes sustain crops for one to several seasons. Then, as a result of leaching, the cleared plots become infertile and are abandoned. Shifting cultivation on small, widely scattered plots may not damage forest ecosystems much, but fertility plummets when large areas are cleared and when plots are cleared again at shorter intervals.

Shifting rates of evaporation, transpiration, and runoff may even disrupt regional patterns of rainfall. Between 50 and 80 percent of the water vapor above tropical forests alone is released from the trees. Without trees, annual precipitation declines. Rain rapidly runs off the bare soil. As the region gets hotter and drier, soil fertility and moisture decline even more. In time, sparse

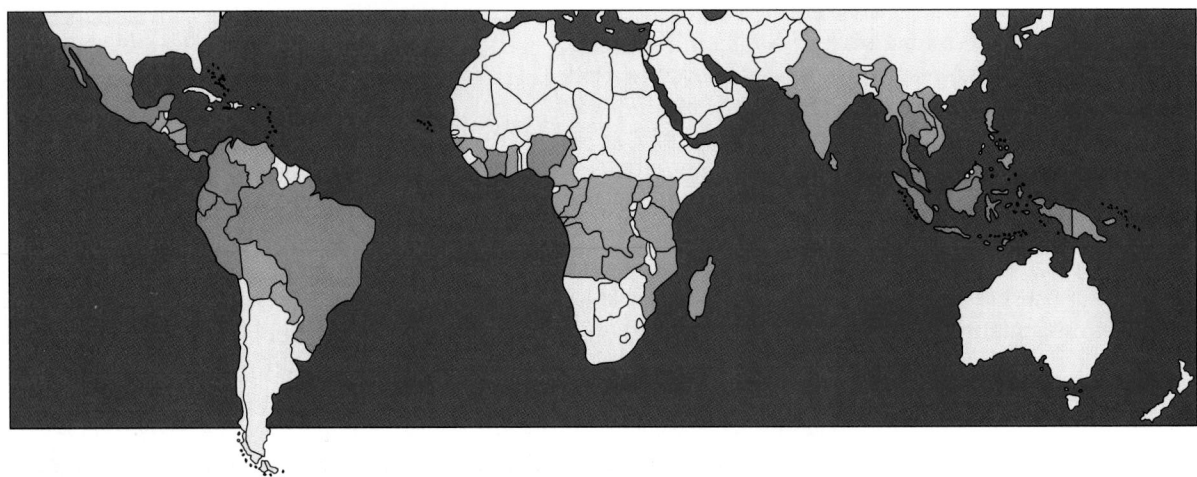

Figure 25.14 Countries permitting the greatest destruction of tropical forests. *Red* shading denotes where 2,000 to 14,800 square kilometers are deforested annually. *Orange* denotes "moderate" deforestation (100 to 1,900 square kilometers). The photograph shows the "slash-and-burn" clearing of a parcel of tropical forest.

Figure 25.15 The vast Amazon River basin of South America in September 1988. Its features were completely obscured by smoke from fires that had been set to clear tropical forests, pasturelands, and croplands during the dry season.

Smoke extends to the Andes Mountains on the western horizon, just over 1,000 kilometers (650 miles) away. The smoke cover was the largest that astronauts had ever before observed. It extended almost 175 million square kilometers (1,044,000 square miles).

The smoke plume near the center of the photograph alone covered an area comparable to the huge forest fire in Yellowstone National Park in that same year.

Massive deforestation is not confined to equatorial regions. For example, in the past century, 2 million acres of redwood forests along California's coast have been logged over. Most of the destruction of such temperate forests has been occurring since 1950, owing to the widespread use of chain saws and tractors and the practice of exporting the logs to lumbermills overseas, where wages are low.

ANDES

SMOKE FROM FIRES

grassland or even desertlike conditions prevail instead of a rich tropical forest.

Widespread tropical forest destruction may have global repercussions. These forests absorb much of the sunlight reaching equatorial regions of the Earth's surface. When the forests are cleared, the land becomes shinier, so to speak, and reflects more incoming energy back into space. Also, by their photosynthetic activity, the great numbers of trees in these vast forest biomes help sustain the global cycling of carbon and oxygen. When trees are harvested or burned, carbon stored in their biomass is released to the atmosphere in the form of carbon dioxide—and this may be amplifying the greenhouse effect.

About half the world's tropical forests have been destroyed for cropland, grazing land, timber, and fuel-

wood. Deforestation is greatest in Brazil, Indonesia, Colombia, and Mexico. If clearing and destruction continue at present rates, only Brazil and Zaire will have large tropical forests in 2010. By 2035, most of their forests will be gone.

Because forests profoundly influence the global cycling of carbon and oxygen, soil productivity, and the functioning of watersheds, large-scale deforestation can have serious repercussions on the biosphere.

Both tropical forests and temperate forests are rapidly being destroyed as trees are cut for timber and fuel and to clear land for agriculture and livestock grazing.

Paralleling the J-shaped curve of human population growth is a steep rise in total and per capita energy consumption. It is due to increased numbers of energy users and to extravagant consumption and waste. For example, in one of the most pleasant of all climates, a major university constructed seven- and eight-story buildings with narrow, sealed windows. The windows can't be opened to catch prevailing ocean breezes. The buildings and windows were not designed or aligned to use sunlight for passive solar heating and breezes for passive cooling. Massive energy-demanding cooling and heating systems were installed.

When you hear talk of abundant energy supplies, keep in mind that there is a huge difference between the *total* and the *net* amounts available. Net energy is that left over after subtracting the energy used to locate, extract, transport, store, and deliver energy to consumers. Some energy sources, such as direct solar energy, are renewable. Others, such as coal and petroleum, are not. Currently, 83 percent of the energy stores being tapped are in the second category (Figure 25.16).

Fossil Fuels

Forests that existed hundreds of millions of years ago gave us **fossil fuels**. Their carbon-containing remains became buried and compressed in sediments, then were transformed into coal, petroleum (oil), and natural gas. They are nonrenewable resources.

Even with strict conservation, known petroleum and natural gas reserves may be used up in the next century. As known reserves run out in accessible areas, we explore wilderness areas in Alaska and other fragile environments, such as continental shelves. Net energy declines as costs of extraction and transportation to and from remote areas increase. Environmental costs of extraction and transportation escalate. The long-term impact of the 11-million-gallon spill from the tanker *Valdez* off Alaska's coast is still not understood.

What about coal? In theory, world reserves can meet the energy needs of the human population for at least several centuries. But coal burning has been the main source of air pollution. Most coal reserves contain low-quality, high-sulfur material. Unless sulfur is removed before or after burning, sulfur dioxides are released into the air. They add to the global problem of acid deposition. Fossil fuel burning also releases carbon dioxide and adds to the greenhouse effect.

Extensive strip mining of coal reserves close to the Earth's surface carries its own problems. It removes land from agriculture, grazing, and wildlife. Restoration is difficult and expensive in arid and semiarid lands, where much of the strip mining is proceeding.

Nuclear Energy

Today, nuclear power plants dot the landscape. At one time, proponents of nuclear power predicted that by 2000 it would meet the energy needs of 70 percent of the world's people. Although industrialized nations that are poor in energy resources now depend heavily on nuclear power, in most countries plans to develop more nuclear energy have been delayed or canceled.

Questions surround nuclear energy's costs, efficiency, safety record, and environmental impact. By 1990 in the United States, it cost slightly more to generate electricity by nuclear energy than by using coal, and significantly more than by using natural gas, hydroelectric power, and most other technologies. By 2000, solar energy with natural gas backup also should cost less.

The greatest hazard associated with operation of a nuclear reactor is the potential danger of a **meltdown** if the reactor core becomes overheated. As happened at Chernobyl in Ukraine in 1986, the reactor core could melt through its thick concrete containment slab and into the Earth, contaminating groundwater and releasing lethal amounts of radiation. Thirty-one people died immediately, and several hundred others died later of radiation sickness. Inhabitants of entire villages were relocated; their former homes were bulldozed under. The fallout of

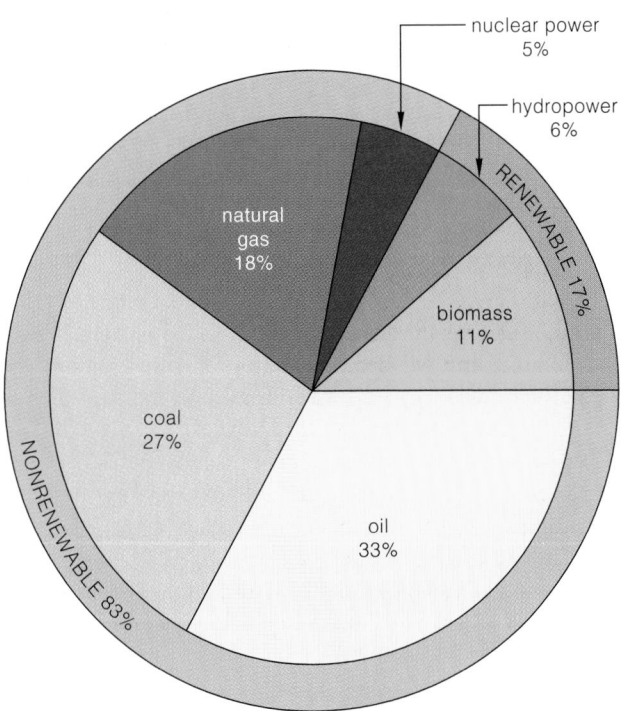

Figure 25.16 World consumption of nonrenewable and renewable energy sources in 1991.

Figure 25.17 Harnessing solar energy. (**a**) Electricity-producing photovoltaic cells in panels that collect sunlight energy. (**b**) Wind turbines, which exploit the air circulation patterns that arise from latitudinal variations in the intensity of incoming sunlight.

radioactive material put an estimated 300 to 400 million people throughout Europe at increased risk of leukemia and other radiation-induced disorders. In 1979 a nuclear accident occurred in a commercial nuclear facility at Three Mile Island near Harrisburg, Pennsylvania. Through mechanical failures and human error, one reactor came within 30 to 60 minutes of a total meltdown.

Unlike coal, nuclear fuel cannot be burned to harmless ashes. The fuel elements are spent after about 3 years, but they still contain uranium fuel and hundreds of new radioisotopes formed during the reactions. The wastes are extremely radioactive and dangerous. As they undergo radioactive decay, they become extremely hot. They are plunged at once into water-filled pools. The water cools them and keeps radioactive material from escaping. After being stored for several months, the remaining isotopes are still lethal. Some must be isolated for at least 10,000 years. If one kind of plutonium isotope (239^{Pu}) is not removed, the wastes must be kept isolated for a quarter of a million years!

After nearly 50 years of research, scientists still cannot agree on the best way to store high-level radioactive wastes. Even if they could do so, there is no politically acceptable solution. No one wants radioactive wastes anywhere near where they live.

Alternative Energy Sources

Various alternative energy sources are on the drawing boards. For example, in theory, **breeder reactors** may "breed" fuel by converting an isotope of uranium into a fissionable isotope of plutonium. However, unlike a conventional nuclear reactor (which cannot explode like an atomic bomb), breeder reactors could produce a small nuclear explosion. Also, it may cost several times more to put them into operation, and it may take 100 to 200 years to produce enough plutonium to fuel them.

Fusion power is another alternative. The idea is to fuse hydrogen atoms to form helium atoms, an event that releases considerable energy. There are great scientific, technological, and economic obstacles, however, and without major breakthroughs, fusion power is not expected to produce electricity on a commercial basis until the last half of the next century, if ever.

We might also develop low-cost solar cells that can harness sunlight energy, then use it to generate the electricity required to produce hydrogen gas (Figure 25.17). **Solar-hydrogen energy** would assure the human population of an affordable, renewable energy source, and the cleansing of the environment could begin in earnest.

Researchers in Germany, Japan, and elsewhere have been working intensively to develop the required technologies for hydrogen-fueled vehicles, factories, and homes. Prototype photovoltaic cells, which convert sunlight energy into electricity, are in place in 30,000 homes around the world and even in oil drilling platforms. Solar-hydrogen systems that will meet the heating, cooling, and other electrical needs of homes may soon be on the production line in Germany. Several automakers have built hydrogen-powered prototype vehicles.

With serious commitment, we could expect solar cells to be supplying as much energy as nuclear power plants by the year 2010. By 2050, solar cells could be satisfying half the energy requirements of the United States, at lower cost and at much lower risk than the existing alternatives. However, cuts in the federal budget have resulted in a decline in research efforts and applications in the United States.

What are the best options? For now and the near future, we must reduce consumption by using more energy-efficient vehicles, architecture, heating and cooling systems, and appliances. For the long term, some argue for hydrogen fuel systems based on energy from renewable resources such as the sun, wind, flowing water, and heated steam being vented from the Earth.

Increasing demand for safe, renewable, cost-efficient energy supplies is focusing attention on alternatives to fossil fuels and nuclear power.

25.9 A PLANETARY EMERGENCY: LOSS OF BIODIVERSITY

On March 24, 1900, in Ohio, a young boy shot the last known wild passenger pigeon. More recently, fisheries biologist John Musick of the Virginia Institute of Marine Science reported his latest findings on the decline in populations of the dusky shark. Over the past 25 years, Musick has tracked the numbers of the dusky and other shark species that spend much of their lives off the mid-Atlantic coast of the United States and have become popular as human food. Musick's numbers show that, like codfish in parts of the North Atlantic, today the dusky shark has all but disappeared. Based on similar evidence from many scientists, the National Marine Fisheries Service in 1995 imposed catch restrictions on commercial shark fishing operations, over the protests of fleet operators. Since then, additional shark species have been recognized as threatened, along with many other fish species.

Sooner or later all species become extinct, but humans have become a primary factor in the premature extinction of more and more of them. Already our actions are leading to the premature extinction of at least six species per hour.

Many biologists consider this epidemic of extinction an even more serious problem than depletion of stratospheric ozone or global warming because it is happening faster and is irreversible. Such rapid extinction cannot be balanced by speciation because it takes many thousands of generations for new species to evolve.

Globally, tropical deforestation is the greatest killer of species, followed by destruction of coral reefs. However, plant extinctions can be more important ecologically than animal extinctions since most animal species depend directly or indirectly on plants for food, and often for shelter as well. Humans also have traditionally depended on plants as sources of medicines. Indeed, 40 percent of prescription drug sales in the United States involve natural plant products. One striking example is the rosy periwinkle of Madagascar (Figure 25.18), from which we derive the anticancer drugs vincristine and vinblastine. As it happens, humans have destroyed 90 percent of the vegetation on Madagascar, so we may never know if it was home to plant sources of other life-saving drugs as well. Meanwhile, in the United States we are going about the business of habitat destruction at a dizzying pace. Among other things, we have cut down 92 percent of old-growth forests and drained half the wetlands, which

Figure 25.18 The rosy periwinkle found in the threatened tropical forests of Madagascar. The plant is a source of two anticancer drugs.

filter human water supplies and provide homes for waterfowl and juvenile fishes. At least 500 species native to these areas have been driven to extinction, and dozens more are endangered.

The underlying causes of wildlife extinction are human population growth and economic policies that fail to value the environment. Instead, they promote unsustainable exploitation. As our population grows we clear, occupy, and damage more land to supply food, fuel, timber, and other resources. In wealthy countries, affluence leads to greater than average resource use per person. In less-affluent nations, the combination of rapid population growth and poverty push the poor to cut forests, grow crops on marginal land, overgraze grasslands, and poach endangered animals.

How can we help stem the tide of extinctions? Among other strategies, individuals can support efforts to reduce deforestation, projected global warming, ozone depletion, and poverty—the greatest threats to Earth's wildlife *and* to the human species.

THE IDEA OF SUSTAINABLE LIVING

You may have read or heard the expressions "sustainable agriculture" or "sustainable development." A **sustainable society** is one that manages its population growth and economic activities in ways that prevent serious damage to the environment. Every life form must obtain energy and other resources from its environment in order to grow and reproduce. The key to sustainable living is simple: We must not take or use resources in a manner or amount that irreparably harms ecosystems—or causes them to collapse altogether.

Elsewhere in this chapter you have read of human activities that in the long run are unsustainable. Our heavy dependence on fossil fuels is an obvious one, because we know that the Earth has a limited supply of fossil fuels and eventually the supply will run out. Most likely, we also cannot long continue widespread deforestation, overfishing of the oceans, and pollution of the atmosphere with greenhouse gases and ozone-destroying chemicals (among other activities). Environmental experts disagree about the precise level of degradation the Earth's land and water ecosystems and its atmosphere can sustain. Most agree, however, that without significant changes in human behavior and technologies, our planet will become unfit for human habitation.

The large and growing numbers of humans and our disregard for the limits of ecosystems have put our species at grave risk. The final straw could be a prolonged global food shortage, irrevocable poisoning of water supplies, passing a critical threshold for ozone depletion, or some other nightmarish disaster. However, one reason we include a chapter such as this in a human biology text is that we firmly believe that *it is not too late*. Humans have the capacity to anticipate events *before* they happen. We are not locked into responding only after irreversible change has begun. We all have the capacity for adapting to a future that we can partly shape. We can, for example, learn to live with less. Far from being a return to primitive simplicity, this could be one of the most intelligent behaviors our species ever engaged in.

It is probably not too extreme to say that our species' survival depends on coming up with some alternatives for the future (Figure 25.19). It depends on designing and constructing human ecosystems that are in harmony not only with what we define as basic human values, but also with biological reality. Our values and expectations must be adapted to the altered world in which we now live, because the principles of energy flow and resource use that govern the survival of all systems of life do not change.

In the final analysis, we must come to terms with these principles and ask ourselves what will be our long-term contribution to the world of life. ✦

Figure 25.19 A biogas digester in a village in India. An example of an *appropriate technology*, this device produces methane gas. Its raw material is animal dung mixed with anaerobic bacteria. The microorganisms derive nutrients from the dung and produce methane gas as a by-product. The methane can be used as a household fuel. In addition to more than 750,000 of the devices operating in India, another 6 million provide usable fuel in rural areas of China. Farmers use solid residues of the process as fertilizer on food crops or trees.

SUMMARY

1. The growth rate of a population depends on the birth rate, the death rate, and the rates of immigration and emigration. If the birth rate exceeds the death rate by a constant amount, the population will grow exponentially (assuming immigration and emigration remain zero).

2. Carrying capacity is the number of individuals of a species that can be sustained indefinitely by the available resources in a given area.

3. Population size is determined by carrying capacity, competition, and other factors that limit population growth. Limiting factors vary in their relative effects and over time, so population size also can change over time.

4. Some limiting factors, such as competition for resources, disease, and predation, are density-dependent.

5. Currently, human population growth varies from zero in some of the more-developed countries to more than 4 percent per year in some of the less-developed countries.

6. Rapid growth of the human population during the past two centuries was possible largely because of our capacity to expand into new environments and because of agricultural and technological developments that increased the carrying capacity.

7. Accompanying the exponential growth of the human population are increased energy demands and environmental pollution.

8. Many pollutants are an outcome of human activities, and they adversely affect the health, activities, or survival of human populations.

9. Industrial smog and photochemical smog are examples of local air pollution. Acid rain, thinning of the ozone layer, and possible enhancement of the greenhouse effect are examples of global air pollution.

10. Global freshwater supplies are limited, yet they are being polluted by agricultural runoff (which includes sediments as well as pesticides and fertilizers), industrial wastes, and human sewage.

11. Human populations are damaging land surfaces by accumulation of vast amounts of solid wastes and by conversion of marginal lands for agriculture. Widespread desertification and the destruction of tropical and temperate forests may be altering regional soils and patterns of rainfall.

12. Energy supplies in the form of fossil fuels are non-renewable, dwindling, and environmentally costly to extract and use. Nuclear energy in itself is less polluting, but the costs and risks associated with fuel containment and with storing radioactive wastes are enormous. The challenge is to develop affordable alternatives based on renewable resources.

Review Questions

1. Why do populations that are not restricted in some way tend to grow exponentially? *510–511*

2. If the birth rate equals the death rate, what happens to the growth rate of a population? If the birth rate remains slightly higher than the death rate, what happens? *510*

3. At current growth rates, how many years will elapse before another billion people are added to the human population? *515*

4. Write a short essay about a hypothetical population that shows one of the following age structures. What might happen to younger and older age groups when members move into new categories? *510*

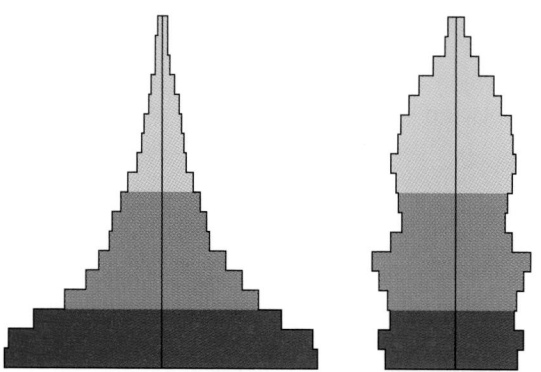

5. If a third of the world population is now below age 15, what effect will this age distribution have on the growth rate of the human population? What sorts of humane recommendations would you make that would encourage this age group to limit the number of children they plan to have? *510*

Critical Thinking: You Decide *(Key in Appendix V)*

1. How have humans increased the carrying capacity of their environments? Have they avoided some of the limiting factors on population growth, or is the avoidance an illusion?

2. The poet T. S. Eliot once wrote that the world would end "not with a bang but a whimper." Applying this grim thought to the topics of this chapter, design a 10-point plan for proving Eliot wrong.

3. Human populations (like those of other species) use resources and produce wastes. Use Figure 25.20 as a starting point for a brief essay on the accumulation and uses of energy and materials, including wastes, in a major city.

Self-Quiz *(Answers in Appendix IV)*

1. _____ is the study of how organisms interact with one another as well as with their physical and chemical environment.

2. The rate at which a population grows or declines depends on the _____.
   a. birth rate
   b. death rate
   c. immigration rate
   d. emigration rate
   e. all of the above

Figure 25.20 City as ecosystem.

3. Populations grow exponentially when _____.
   a. birth rate remains above death rate and neither changes
   b. death rate remains above birth rate
   c. immigration and emigration rates are equal (a zero value)
   d. emigration rates exceed immigration rates
   e. both a and c combined are correct

4. The number of individuals of a species that can be sustained indefinitely by the resources in a given region is the _____.
   a. biotic potential
   b. carrying capacity
   c. environmental resistance
   d. density control

5. Which of the following factors does *not* affect sustainable population size?
   a. predation
   b. competition
   c. available resources
   d. pollution
   e. each of the above can affect population size

6. Population growth controls such as resource competition and disease are said to be _____.
   a. density independent
   b. population sustaining
   c. population dynamics
   d. density dependent

7. During the past two centuries, rapid growth of the human population has occurred largely because of _____.
   a. increased birth rate worldwide
   b. increased death rate worldwide
   c. carrying capacity reduction
   d. carrying capacity expansion

8. Match the following population ecology terms.
   _____ carrying capacity
   _____ exponential growth
   _____ population growth rate
   _____ density-dependent controls

   a. examples are diseases and predation
   b. depends on birth, death, immigration, and emigration rates
   c. number of individuals of a given species that can be sustained indefinitely by the resources in a given area
   d. increases in population size by ever larger amounts per unit of time

Selected Key Terms

acid rain *516*
age structure *510*
biotic potential *512*
carrying capacity *512*
deforestation *520*
demographic transition model *514*
density-dependent control *512*
density-independent control *512*
desertification *519*
dry acid deposition *516*
exponential growth *511*
fossil fuel *522*
green revolution *519*

industrial smog *516*
life table *513*
limiting factor *512*
logistic growth *512*
meltdown *522*
photochemical smog *516*
pollutant *516*
reproductive base *510*
salinization *518*
shifting cultivation *520*
sustainable society *525*
thermal inversion *516*
zero population growth *510*

Readings

Cohen, J. November 1992. "How Many People Can the Earth Hold?" *Discover.*

Collins, M. 1990. *The Last Rain Forests.* New York: Oxford University Press.

Dixon, T. F., J. Boutwell, and G. Rathjens. February 1993. "Environmental Change and Violent Conflict." *Scientific American.*

Mohnen, V. August 1988. "The Challenge of Acid Rain." *Scientific American.*

Stolzenburg, W. July/August 1996. "10 Things (at Least) You Can Do to Save Life's Diversity." *Nature Conservancy.*

"Ten Years After." January 1997. Survey article by the editors of *Discover* on the lasting damage done by the nuclear meltdown at Chernobyl.

Western, D., and M. Pearl. 1989. *Conservation for the Twenty-First Century.* New York: Oxford University Press.

Wilson, E. 1988. *Biodiversity.* Washington, D.C.: National Academy of Sciences.

 For additional readings, go to InfoTrac College Edition, your online library at:

http://www.infotrac-college.com/wadsworth

APPENDIX I. PERIODIC TABLE OF THE ELEMENTS

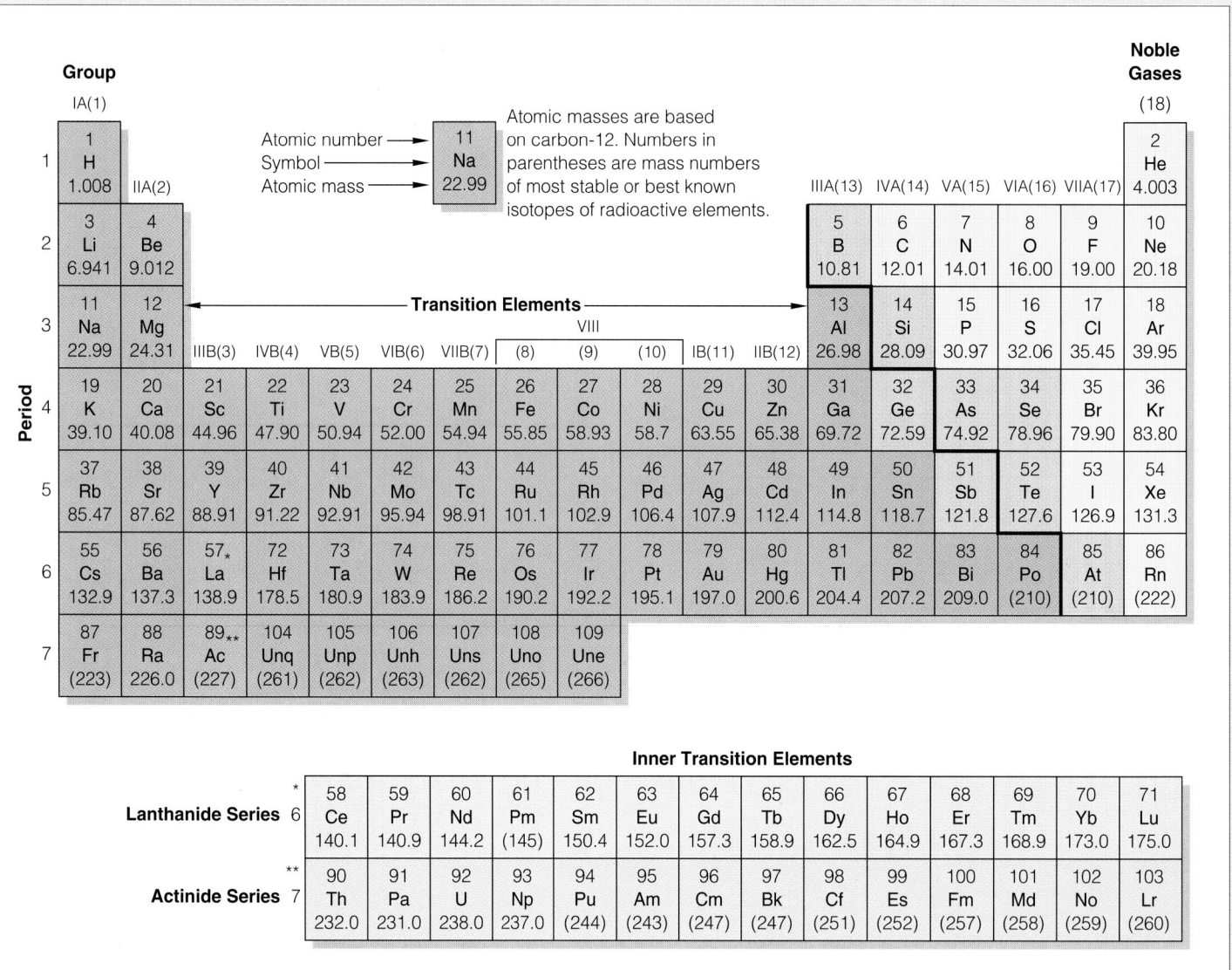

Group

Noble Gases (18)

Atomic number → 11
Symbol → Na
Atomic mass → 22.99

Atomic masses are based on carbon-12. Numbers in parentheses are mass numbers of most stable or best known isotopes of radioactive elements.

Transition Elements

Period	IA(1)	IIA(2)	IIIB(3)	IVB(4)	VB(5)	VIB(6)	VIIB(7)	VIII (8)	(9)	(10)	IB(11)	IIB(12)	IIIA(13)	IVA(14)	VA(15)	VIA(16)	VIIA(17)	(18)
1	1 H 1.008																	2 He 4.003
2	3 Li 6.941	4 Be 9.012											5 B 10.81	6 C 12.01	7 N 14.01	8 O 16.00	9 F 19.00	10 Ne 20.18
3	11 Na 22.99	12 Mg 24.31											13 Al 26.98	14 Si 28.09	15 P 30.97	16 S 32.06	17 Cl 35.45	18 Ar 39.95
4	19 K 39.10	20 Ca 40.08	21 Sc 44.96	22 Ti 47.90	23 V 50.94	24 Cr 52.00	25 Mn 54.94	26 Fe 55.85	27 Co 58.93	28 Ni 58.7	29 Cu 63.55	30 Zn 65.38	31 Ga 69.72	32 Ge 72.59	33 As 74.92	34 Se 78.96	35 Br 79.90	36 Kr 83.80
5	37 Rb 85.47	38 Sr 87.62	39 Y 88.91	40 Zr 91.22	41 Nb 92.91	42 Mo 95.94	43 Tc 98.91	44 Ru 101.1	45 Rh 102.9	46 Pd 106.4	47 Ag 107.9	48 Cd 112.4	49 In 114.8	50 Sn 118.7	51 Sb 121.8	52 Te 127.6	53 I 126.9	54 Xe 131.3
6	55 Cs 132.9	56 Ba 137.3	57* La 138.9	72 Hf 178.5	73 Ta 180.9	74 W 183.9	75 Re 186.2	76 Os 190.2	77 Ir 192.2	78 Pt 195.1	79 Au 197.0	80 Hg 200.6	81 Tl 204.4	82 Pb 207.2	83 Bi 209.0	84 Po (210)	85 At (210)	86 Rn (222)
7	87 Fr (223)	88 Ra 226.0	89** Ac (227)	104 Unq (261)	105 Unp (262)	106 Unh (263)	107 Uns (262)	108 Uno (265)	109 Une (266)									

Inner Transition Elements

Lanthanide Series 6*

58 Ce 140.1	59 Pr 140.9	60 Nd 144.2	61 Pm (145)	62 Sm 150.4	63 Eu 152.0	64 Gd 157.3	65 Tb 158.9	66 Dy 162.5	67 Ho 164.9	68 Er 167.3	69 Tm 168.9	70 Yb 173.0	71 Lu 175.0

Actinide Series 7**

90 Th 232.0	91 Pa 231.0	92 U 238.0	93 Np 237.0	94 Pu (244)	95 Am (243)	96 Cm (247)	97 Bk (247)	98 Cf (251)	99 Es (252)	100 Fm (257)	101 Md (258)	102 No (259)	103 Lr (260)

Metric-English Conversions

Length

English		Metric
inch	=	2.54 centimeters
foot	=	0.30 meter
yard	=	0.91 meter
mile (5,280 feet)	=	1.61 kilometer

To convert	multiply by	to obtain
inches	2.54	centimeters
foot	30.00	centimeters
centimeters	0.39	inches
millimeters	0.039	inches

Weight

English		Metric
grain	=	64.80 milligrams
ounce	=	28.35 grams
pound	=	453.60 grams
ton (short) (2,000 pounds)	=	0.91 metric ton

To convert	multiply by	to obtain
ounces	28.3	grams
pounds	453.6	grams
pounds	0.45	kilograms
grams	0.035	ounces
kilograms	2.2	pounds

Volume

English		Metric
cubic inch	=	16.39 cubic centimeters
cubic foot	=	0.03 cubic meter
cubic yard	=	0.765 cubic meters
ounce	=	0.03 liter
pint	=	0.47 liter
quart	=	0.95 liter
gallon	=	3.79 liters

To convert	multiply by	to obtain
fluid ounces	30.00	milliliters
quart	0.95	liters
milliliters	0.03	fluid ounces
liters	1.06	quarts

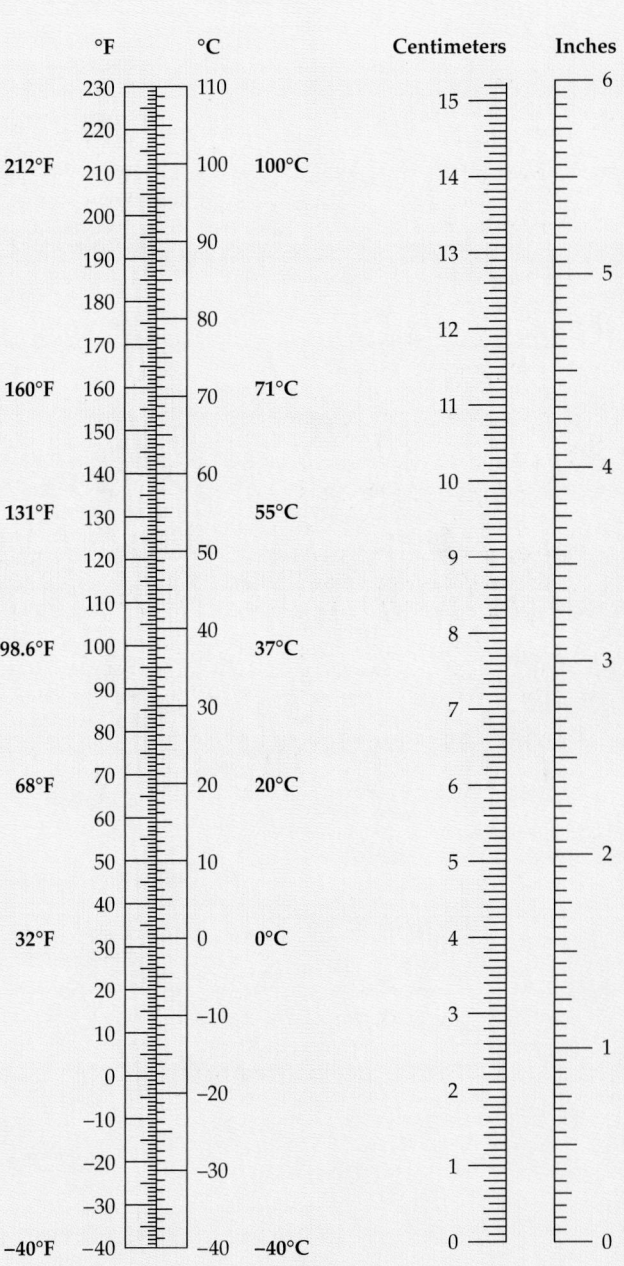

To convert temperature scales:

Fahrenheit to Celsius: $°C = 5/9(°F - 32)$

Celsius to Fahrenheit: $°F = 9/5(°C) + 32$

APPENDIX III. ANSWERS TO GENETIC PROBLEMS

CHAPTER 17

1. a: *AB*

   b: *AB* and *aB*

   c: *Ab* and *ab*

   d: *AB, aB, Ab,* and *ab*

2. a: *AaBB* will occur in all the offspring.

   b: 25% *AABB*; 25% *AaBB*; 25% *AABb*; 25% *AaBb*

   c: 25% *AaBb*; 25% *Aabb*; 25% *aaBb*; 25% *aabb*

   d: 1/16 *AABB* (6.25%)
   1/8 *AaBB* (12.5%)
   1/16 *aaBB* (6.25%)
   1/8 *AABb* (12.5%)
   1/4 *AaBb* (25%)
   1/8 *aaBb* (12.5%)
   1/16 *AAbb* (6.25%)
   1/8 *Aabb* (12.5%)
   1/16 *aabb* (6.25%)

3. a: Mother must be heterozygous for both genes; father is homozygous recessive for both genes. The first child is also homozygous recessive for both genes.

   b: The probability that the second child will not be able to roll the tongue and will have detached earlobes is 1/4 (25%).

4. a: ABC

   b: aBc

   c: ABC, ABc, aBc, aBC

   d: ABC, abc, aBC, AbC, Abc, ABc, abC, aBc

5. Because Molly does not exhibit the recessive hip disorder, she must be either homozygous dominant (HH) for this trait, or heterozygous (Hh). If the father is homozygous dominant (HH), then he and Molly cannot produce gametes that are homozygous recessive (hh), and so none of their offspring will have the undesirable phenotype. However, if Molly is a heterozygote for the trait, notice that the probability is 1/2 (50%) that a puppy will be heterozygous (Hh) and so carry the trait.

6. a: The mother must be heterozygous (*I*^A*i*). The man having type B blood could have fathered the child if he were also heterozygous. (*I*^B*i*).

   b: If the man is heterozygous, then he *could be* the father. However, because any other type B heterozygous male could also be the father, one cannot say that this particular man absolutely must be. Actually, any male who could contribute an O allele (*i*) could have fathered the child. This would include males with type O blood (*ii*) or type A blood who are heterozygous.

7. The probability is 1/2 (50%) that a child of this couple will be a heterozygote and have sickle cell trait. The probability is 1/4 (25%) that a child will be homozygous for the sickling allele and so will have sickle cell anemia.

8. For these ten traits, all the man's sperm will carry identical genes. He cannot produce genotypically different sperm. The woman can produce eggs with four genotypes. This example underscores the fact that the more heterozygous gene pairs that are present, the more genetically different gametes are possible.

9. The mating is *Ll* x *Ll*.

   Progeny genotypes: 1/4 *LL* + 1/2 *Ll* + 1/4 *ll*.

   Phenotypes: 1/4 homozyous survivors (*LL*)

   1/2 heterozygous survivors (*Ll*)

   1/4 lethal (*ll*) nonsurvivors

10. Bill's genotype: *Aa, Ss, Ee*
    Marie's genotype: *AA, SS, EE*

No matter how many children Bill and Marie have, the probability is 100% that each child will have the parents' phenotype. Because Marie can produce only dominant alleles, there is no way that a child could inherit a *pair* of recessive alleles for any of these three traits—and that is what would be required in order for the child to show the recessive phenotype. Thus the probability is zero that a child will have short lashes, high arches, and no achoo syndrome.

11. The white-furred parent's genotype is *bb*; the black-furred parent must be *Bb*; because if it were *BB* all offspring would be heterozygotes (*Bb*) and would have black fur. A monohybrid cross between a heterozygote and a homozygote typically yields a 1:1 phenotype ratio. 4 black and 3 white guinea pigs is close to a 1:1 ratio.

CHAPTER 18

1. Key: Among other complicated issues, here you must consider whether (in your view) a person's gender is determined by "genetic sex" or by reproductive structures. Also, review the discussion of sex determination at the beginning of Chapter 14. In addition, note that the secondary sex characteristics of normal males, which are due to the sensitivity of cells to testosterone, include greater muscle mass—which is why males typically are physically stronger than females.

2. The probability is the same for each child: 1/2, or 50%.

3. a: From his mother.

   b: A male can produce two types of gametes with respect to an X-linked gene. One type will possess only a Y chromosome and so lack this gene; the other type will have an X chromosome and will have the X-linked gene.

   c: A female homozygous for an X-linked gene will produce just one type of gamete containing an X chromosome with the gene.

   d: A female heterozygous for an X-linked gene will produce two types of gametes. One will contain an X chromosome with the dominant alleles, and the other type will contain an X chromosome with the recessive allele.

4. Two polled cattle who are both heterozygous for the recessive allele that causes the horned phenotype may mate. If they do, there is a 25% chance that any one of their offspring will be horned.

5. Most of chromosome 21 has been translocated to chromosome 14. While this individual has 46 chromosomes, there are in fact three copies of chromosome 21. The third copy of chromosome 21 is attached to chromosome 14.

6. 50%. The woman must be heterozygous for the allele. There is a 50% chance that any son will inherit the allele, regardless of the father's genotype. Since the father carries the allele, there is a 50% chance that any daughter will be homozygous and express the ocular albinism phenotype. No.

7. No. Many traits are sex-limited and influenced by genes on autosomes.

8. 50%. His mother is heterozygous for the allele, so there is a 50% chance that any male offspring will inherit the allele. Since males do not inherit an X chromosome from their fathers, the genotype of the father is irrelevant to this question.

9. The allele for cystic fibrosis is recessive. Most of the carriers are heterozygous for the allele and do not have the cystic fibrosis phenotype.

10. The parent with typical pigmentation is heterozygous for the albinism allele. The probability that any one child will have the albinism phenotype is 50%. However, with a small sample of only four offspring, there is a high probability of a deviation from a 1:1 ratio due to the random mixes of alleles that occur during meiosis and fertilization (discussed in Chapter 17).

APPENDIX IV. ANSWERS TO SELF-QUIZZES

INTRODUCTION: 1. DNA, 2. Metabolism, 3. Homeostasis, receptors, 4. adaptive, 5. mutations, 6. d, 7. c, 8. c

CHAPTER 1: 1. carbon, 2. a, 3. c, 4. c, 5. d, 6. b, 7. b, 8. c, 9. c, e, b, d, a, 10. a (plus R group interactions)

CHAPTER 2: 1. e, 2. cytoskeleton, 3. c, 4. a, 5. c, 6. b, g, f, d, c, a, e , 7. d, 8. d, 9. c, e, d, a, b, 10. d, 11. b, 12. b, a, c

CHAPTER 3: 1. epithelial, connective, muscle, nerve 2. All answers are activities for which the human body is structurally and functionally adapted. 3. d, 4. c, 5. d, 6. b, 7. a, 8. d, large food particles are first mechanically broken down by the digestive system (Chapter 5), 9. c, 10. a, 11. Receptors, integrator, effectors, 12. d, e, c, b, a

CHAPTER 4: 1. Skeletal and muscular, 2. skeletal, smooth, cardiac, 3. d, 4. b, 5. b, 6. c, 7. d, e, b, a, c

CHAPTER 5: 1. digestive, circulatory, respiratory, urinary, 2. digesting, absorbing, eliminating, 3. caloric, energy, 4. carbohydrates, 5. essential amino acids, essential fatty acids, 6. b, 7. c, 8. d, 9. c, 10. e, d, a, c, b, 11. e, 12. e

CHAPTER 6: 1. circulatory system, lymphatic system, 2. Arteries, veins; capillaries and venules; arterioles, 3. d, 4. d, 5. c, 6. a, 7. d, 8. b, 9. d, b, a, c , 10. c, d, a, b

CHAPTER 7: 1. a, 2. d, 3. c. , 4. d, 5. a, 6. d, 7. d, 8. b, 9. a, 10. c, b, a, e, d, f

CHAPTER 8: 1. b, 2. b, 3. d, 4. b, 5. a, 6. e, 7. c, 8. d, 9. d, 10. d, 11. f, a, e, d, b, c

CHAPTER 9: 1. c, 2. d, 3. d, 4. b, 5. a, 6. c, 7. d, 8. a, 9. d, 10. b, d, e, c, a

CHAPTER 10: 1. stimuli, neurons, 2. action potentials, 3. neurotransmitter, 4. integration, 5. d, 6. c, 7. b, 8. c, 9. e, d, b, c, a

CHAPTER 11: 1. stimulus, 2. sensation, 3. Perception, 4. d, 5. b, 6. d, 7. c, 8. c, 9. e, 10. b, 11. c, 12. e, c, d, a, b

CHAPTER 12: 1. f, 2. d, 3. b, 4. d , 5. e, 6. b, 7. b, 8. b, 9. b, e, g, d, f, a, c, 10. d, a, b, e, c

CHAPTER 13: 1. hypothalamus, 2. b, 3. d, 4. c, 5. b

CHAPTER 14: 1. a, 2. a, 3. c, 4. morphogenesis, 5. c, 6. b, 7. d, a, f, e, c, b, 8. a, 9. c, 10. b, 11. c

CHAPTER 15: 1. b, 2. a, 3. unsafe sexual contact; blood, semen, vaginal secretions, breast milk, 4. false, 5. d, 6. c, 7. a, 8. b, 9. d, e, b, a, c, g, f

CHAPTER 16: 1. mitosis; meiosis, 2. chromosomes; DNA, 3. c, 4. diploid, 5. c, 6. a, 7. b, 8. b, 9. c, 10. d, 11. c, 12. d, b, c, a

CHAPTER 17: 1. a, 2. c, 3. a, 4. c, 5. b, 6. d, 7. b, 8. d, 9. c, a, d, b

CHAPTER 18: 1. c, 2. e, 3. e, 4. c, 5. e, 6. c, 7. d, 8. d, 9. c, e, d, b, a

CHAPTER 19: 1. three, 2. e, 3. b, 4. c, 5. c, 6. a, 7. c, 8. b, 9. a, 10. c, 11. a, 12. e, c, d, a, b

CHAPTER 20: 1. c, 2. e, 3. tumor suppressor gene, 4. b, 5. d, 6. a, 7. c, 8. d, 9. b, 10. c

CHAPTER 21: 1. variation, 2. genetic engineering, 3. Plasmids, 4. vectors, 5. c, 6. cloned DNA, 7. d, 8. b, 9. c, 10. restriction enzymes

CHAPTER 22: 1. macroevolution, 2. gradual; bursts, 3. e, 4. e, 5. c, 6. e, 7. mutation, genetic drift, gene flow, and natural selection, 8. natural selection

CHAPTER 23: 1. e, 2. b, 3. a, 4. a, 5. b, 6. d, 7. d

CHAPTER 24: 1. d, 2. d, 3. d, 4. a, 5. e, 6. c, 7. d, a, c, b, 8. c, d, e, b, a

CHAPTER 25: 1. Ecology, 2. e, 3. e, 4. b, 5. e, 6. d, 7. a, 8. c, d, b, a

APPENDIX V. KEY TO CRITICAL THINKING QUESTIONS

INTRODUCTION

1. Key: The different approaches to "truth" inherent in scientific inquiry and in subjective spiritual, moral, and ethical approaches to issues. Another problem you might consider: Based on discussions in this chapter, can anyone—including scientists—ever claim to know the "whole" truth about a particular question?

2 and 3. Key: Consider experimental design utilizing the scientific method of investigation.

CHAPTER 1

1. Key: Understanding the chemical nature of carbohydrates. Such basic knowledge is very useful when it comes to evaluating health claims of diet books and other commercial products.

2. Key: The mathematical relationship between values on the pH scale.

3. Key: Knowing that cream is mainly animal fat, syrup is mainly water and a sugar such as fructose.

4. Review Section 1.6.

CHAPTER 2

1. See *Science Comes to Life*, page 48.

2. Key: The biochemical limits on mechanisms by which cells, especially muscle cells, can obtain energy to do work. In explaining the second observation, think of a "nonaerobic" exercise as one that, unlike aerobic exercise, does not require the body to make adjustments (as in breathing rate) in order to supply cells with the oxygen needed to keep aerobic cellular respiration going.

CHAPTER 3

1. Key: Review the basic functions of epidermis.

2. Key: Knowing the roles collagen plays in the body.

CHAPTER 4

1. Key: Review the various types of muscle fibers, and relate their characteristics to the athletic requirements of sprinting versus the prolonged activity required of a marathoner's leg muscles.

2. Key: Production of GH and the maximum length of an individual's long bones are both genetically determined.

3. Key: Knowing the components of *living* bone.

CHAPTER 5

1. Key: Consider the health impacts of an active lifestyle and low-fat diet.

2 and 3. Key: Consider the different caloric demands of an older, less active person compared to those of a normally-active child. Also, keep in mind that developing youngsters require ample protein.

4. Key: Consider how diarrhea affects the body (Section 5.6).

CHAPTER 6

1. Key: The relationship between blood pressure and blood volume, cross-sectional area of systemic vessels, and cardiac output.

2. Key: Review the function(s) of each type of formed element in blood.

CHAPTER 7

1. Key: The function of macrophages, compared to the function of T cells.

2. Key: Review mechanism by which memory cells recognize invaders.

CHAPTER 8

1. Key: Review health effects of second-hand smoke; effects of smoking on developing fetus.

2. Key: Hemoglobin binds much more readily with carbon monoxide than it does with oxygen. Also: consider the partial pressure of oxygen in air versus that in a tank of pure oxygen.

3. Key: Knowing how pressure gradients change during the respiratory cycle.

CHAPTER 9

1. Key: See *Focus on Your Health*, page 207.

2. Key: Review chapter section on filtration, reabsorption.

3. Key: Review the response of evaporative heat loss.

CHAPTER 10

1. Key: Review the various ways neurotransmitters can affect post-synaptic neurons.

2. Key: Consider the blood-brain barrier, and the functions of cerebrospinal fluid.

CHAPTER 11

1. Key: The role of olfaction in taste perception.

2. Key: Review the discussion of the sense of balance.

3. Key: Middle-ear anatomy and its role in sound transmission.

CHAPTER 12

1. Key: The biological functions of melatonin.

2. Key: Review the roles of anterior lobe secretions, including effects on target cells in other endocrine glands.

3. Key: Review diabetes discussion in Section 12.6.

CHAPTER 13

1. Key: What chemical conditions in the vagina are most conducive to fertilization?

2. Key: Review the role of estrogen at ovulation.

3. Key: Figure 13.8.

4. Key: Figures 12.10 and 13.8.

CHAPTER 14

1. Key: Take into account respiration, circulation, nutrient intake, waste excretion, immune functions, and functions of other basic body systems.

2. Key: What is the immediate outcome of gastrulation? Your answer to this question should also consider the long-term developmental consequences if gastrulation goes awry.

CHAPTER 15

1. Key: Compare modes of transmission and relative "contagiousness" of HIV, hepatitis B.

2. Key: Review discussions of HIV transmission and prevention of infection.

CHAPTER 16

1. Key: Review stages of meiosis and discussions of sources of genetic variation in offspring (such as crossing over and random alignment of chromosomes at metaphase; also, keep in mind that in each new generation, offspring receive half their chromosomes from each parent.

2. Key: Review what happens to a cell's DNA at different points in the cell cycle.

CHAPTER 17

See Answers to Genetics Problems.

CHAPTER 18

See Answers to Genetics Problems.

CHAPTER 19

1. Key: Consider the links between DNA, mutation, and inheritance.

2. Key: Review discussion of mutations.

3. Key: Review events of mitosis/meiosis (Chapter 16).

CHAPTER 20

1. Key: Review the types of cellular changes currently known to be associated with the onset of cancer.

2. Key: Consider current hypotheses about the roles of immune system cells in recognizing and eliminating cancerous cells.

3. Key: Review kidney functions that help maintain homeostasis (Chapter 9), as well as lung functions (Chapter 8) and liver functions (Chapter 5).

CHAPTER 21

1. Key: Genetic engineering and recombinant DNA technology are most applicable to plant and animal breeding, commercial and industrial microbiology (including use of bacteria and yeast in food processing), and to medical fields such as gene therapy and cancer treatments. You may also wish to relate potential new benefits or problems of genetic engineering to methods that have been or are being developed for manipulating human gametes and embryos.

2. Key: Review genetic engineering and recombinant DNA methods described in this chapter; also, keep in mind that pigs are mammals and so their growth and development entails many of the same gene-guided events—including production of specific hormones and other proteins—that occur in humans.

CHAPTER 22

1. Key: Consider that (1) proteins, including enzymes, are encoded by genes; (2) different populations of a species may show differing gene (allele) frequencies, and (3) various lines of evidence shed light on the geographical movements of early human populations.

2. Key: Review discussions of biogeography and evolutionary processes such as gene flow.

CHAPTER 23

1. Key: Review the time frame within which human evolution has occurred.

2. Key: Review the criteria for animal species designation given in Chapter 22. What general sort of evidence would be needed in order to say definitively that the two groups were the same species?

CHAPTER 24

1. Key: Although your personal views will shape your answer, you may find it useful to review the discussion of ecosystems in Section 24.1.

CHAPTER 25

1. Key: Consider changes in health care, agriculture, and public sanitation in the past few decades and centuries; for example, before penicillin was discovered in the 1940s, people often died of infections we would today consider minor. Also: review the discussion of limits on population growth.

2. Key: To survive, our species must grapple with environmental problems with energy and imagination. Although different people will focus their efforts on different aspects of our current predicament, it is generally agreed that a key element of sustaining a livable Earth is controlling human population growth.

3. Key: Figure 25.20.

GLOSSARY OF BIOLOGICAL TERMS

ABO blood typing Method of characterizing an individual's blood according to whether one or both of two protein markers, A and B, are present at the surface of red blood cells. The O signifies that neither marker is present.

abortion Spontaneous or induced expulsion of the embryo or fetus from the uterus.

absorption The movement of nutrients, fluid, and ions across the gastrointestinal tract lining and into the internal environment.

accommodation In the eye, adjustments of the lens position that move the focal point forward or back so that incoming light rays are properly focused on the retina.

acid A substance that releases hydrogen ions in water.

acrosome An enzyme-containing cap that covers most of the head of a sperm and helps the sperm penetrate an egg at fertilization.

actin (AK-tin) A globular contractile protein. In muscle cells, actin interacts with another protein, myosin, to bring about contraction.

action potential An abrupt, brief reversal in the steady voltage difference (resting membrane potential) across the plasma membrane of a neuron.

active site A crevice on the surface of an enzyme molecule where a specific reaction is catalyzed.

active transport The pumping of one or more specific solutes through a transport protein that spans the lipid bilayer of a cell membrane. Most often, the solute is transported against its concentration gradient. The protein is activated by an energy boost, as from ATP.

adaptation [L. *adaptare*, to fit] In evolutionary biology, the process of becoming adapted (or more adapted) to a given set of environmental conditions. Of sensory neurons, a decrease in the frequency of action potentials (or their cessation) even when a stimulus is maintained at constant strength.

adaptive behavior A behavior that promotes the propagation of an individual's genes and that tends to increase in frequency in a population over time.

adaptive radiation A burst of speciation events, with lineages branching away from one another as they partition the existing environment or invade new ones.

adaptive trait Any aspect of form, function, or behavior that helps an organism survive and reproduce under a given set of environmental conditions.

ADH Antidiuretic hormone. Produced by the hypothalamus and released by the posterior pituitary, it stimulates reabsorption in the kidneys and so reduces urine volume.

adipose tissue A type of connective tissue having an abundance of fat-storing cells and blood vessels for transporting fats.

ADP/ATP cycle In cells, a mechanism of ATP renewal. When ATP donates a phosphate group to other molecules (and so energizes them), it reverts to ADP, then forms again by phosphorylation of ADP.

adrenal cortex (ah-DREE-nul) Outer portion of the adrenal gland; its hormones have roles in metabolism, inflammation, maintaining extracellular fluid volume, and other functions.

adrenal medulla Inner region of the adrenal gland; its hormones help control blood circulation and carbohydrate metabolism.

aerobic respiration (air-OH-bik) [Gk. *aer*, air, and *bios*, life] The main energy-releasing metabolic pathway of ATP formation, in which oxygen is the final acceptor of electrons stripped from glucose or some other organic compound. The pathway proceeds from glycolysis through the Krebs cycle and electron transport phosphorylation. A typical net yield is 36 ATP for each glucose molecule.

age structure Of a population, the number of individuals in each of several or many age categories.

agglutination (ah-glue-tin-AY-shun) The clumping together of foreign cells that have invaded the body (as pathogens or in tissue grafts or transplants). Clumping is induced by cross-linking between antibody molecules that have already bound antigen at the surface of the foreign cells.

aging A range of processes, including the breakdown of cell structure and function, by which the body gradually deteriorates. All organisms showing extensive cell differentiation undergo aging.

agranulocyte Class of white blood cells that lack granular material in the cytoplasm; includes the precursors of macrophages (monocytes) and lymphocytes.

AIDS Acquired immunodeficiency syndrome. A set of chronic disorders following infection by the human immunodeficiency virus (HIV), which destroys key cells of the immune system.

alcoholic fermentation Anaerobic pathway of ATP formation in which pyruvate from glycolysis is broken down to acetaldehyde, which accepts electrons from NADH to become ethanol, and NAD is regenerated. Its net yield is two ATP.

aldosterone (al-DOSS-tuh-roan) Hormone secreted by the adrenal cortex that helps regulate sodium reabsorption.

allantois (ah-LAN-twahz) [Gk. *allas*, sausage] One of four extraembryonic membranes that form during embryonic development. In humans, it functions in early blood formation and development of the urinary bladder.

allele (uh-LEEL) For a given location on a chromosome, one of two or more slightly different molecular forms of a gene that code for different versions of the same trait.

allele frequency Of a given gene locus, the relative abundances of each kind of allele carried by the individuals of a population.

allergy An immune response made against a normally harmless substance.

all-or-none principle Principle that states that individual cells in a muscle's motor units always contract fully in response to proper stimulation. If the stimulus is below a certain threshold, the cells do not respond at all.

alveolar sac (al-VEE-uh-lar) Any of the pouchlike clusters of alveoli in the lungs; the major sites of gas exchange.

alveolus (ahl-VEE-uh-lus), plural alveoli [L. *alveus*, small cavity] Any of the many cup-shaped, thin-walled outpouchings of respiratory bronchioles. A site where oxygen diffuses from air in the lungs to the blood, and carbon dioxide diffuses from blood to the lungs.

amine hormone A hormone derived from the amino acid tyrosine.

amino acid (uh-MEE-no) A small organic molecule having a hydrogen atom, an amino group, an acid group, and an R group covalently bonded to a central carbon atom. The subunit of polypeptide chains, which represent the primary structure of proteins.

ammonification (uh-MOAN-ih-fih-KAY-shun) Together with decomposition, a process by which certain bacteria and fungi break down nitrogen-containing wastes and remains of other organisms.

amnion (am-NEE-on) Of land vertebrates, one of four extraembryonic membranes. It becomes a fluid-filled sac in which the embryo (and fetus) can grow, move freely, and be protected from sudden temperature shifts and impacts.

anaerobic pathway (AN-uh-row-bik) [Gk. *an*, without, and *aer*, air] Metabolic pathway in which a substance other than oxygen serves as the final acceptor of electrons that have been stripped from substrates.

analogous structures Body parts, once different in separate lineages, that were put to comparable uses in similar environments and that came to resemble one another in form and function. They are evidence of morphological convergence.

anaphase (AN-uh-faze) The stage at which microtubules of a spindle apparatus separate sister chromatids of each chromosome and move them to opposite spindle poles. During anaphase I of meiosis, the two members of each pair of homologous chromosomes separate. During anaphase II, sister chromatids of each chromosome separate.

aneuploidy (AN-yoo-ploy-dee) A change in the chromosome number following inheritance of one extra or one fewer chromosome.

antibiotic [Gk. *anti*, against] A normal metabolic product of certain microorganisms that kills or inhibits the growth of other microorganisms.

antibody Any of a variety of Y-shaped receptor molecules with binding sites for specific antigens. Only B cells produce antibodies, then position them at their surface or secrete them.

anticodon In a tRNA molecule, a sequence of three nucleotide bases that can pair with an mRNA codon.

antigen (AN-tih-jen) [Gk. *anti*, against, and *genos*, race, kind] Any molecular configuration that is recognized as foreign to the body and that triggers an immune response. Most antibodies are protein molecules at the surface of infectious agents or tumor cells.

antigen-MHC complex Unit consisting of fragments of an antigen molecule, bound to MHC proteins. MHC complexes displayed at the surface of an antigen-presenting cell such as a macrophage promote an immune response by lymphocytes.

antigen-presenting cell A macrophage or other cell that displays antigen-MHC complexes at its surface and so promotes an immune response by lymphocytes.

anus Terminal opening of the gastrointestinal tract.

aorta (ay-OR-tah) [Gk. *airein*, to lift, heave] Main artery of systemic circulation; carries oxygenated blood away from the heart to all body regions except the lungs.

aortic body Any of several receptors in artery walls near the heart that respond to changes in levels of carbon dioxide (and to some degree, oxygen) in arterial blood.

appendicular skeleton (ap-en-DIK-yoo-lahr) Bones of the limbs, hips, and shoulders.

appendix A slender projection from the cup-shaped pouch (cecum) at the start of the colon.

arrhythmia Irregular or abnormal heart rhythm, sometimes caused by stress, drug effects, or coronary artery disease.

arteriole (ar-TEER-ee-ole) Any of the blood vessels between arteries and capillaries. They are control points where the volume of blood delivered to different body regions can be adjusted.

artery Any of the large-diameter blood vessels that conduct deoxygenated blood to the lungs and oxygenated blood to all body tissues. The thick, muscular artery wall allows arteries to smooth out pulsations in blood pressure caused by heart contractions.

asexual reproduction Mode of reproduction by which offspring arise from a single parent and inherit the genes of that parent only.

atherosclerosis Condition in which an artery's wall thickens and loses its elasticity, and the vessel becomes clogged with lipid deposits. In the artery wall, abnormal smooth muscle cells accumulate, and there is an increase in connective tissue. Plaques consisting of lipids, calcium salts, and fibrous material extend into the artery lumen, disrupting blood flow.

atmosphere A region of gases, airborne particles, and water vapor enveloping the Earth; 80 percent of its mass is distributed within 17 miles of the Earth's surface.

atmospheric cycle A biogeochemical cycle in which the atmosphere is the largest reservoir of an element. The carbon and nitrogen cycles are examples.

atom The smallest unit of matter that is unique to a particular element.

atomic number The number of protons in the nucleus of each atom of an element; it differs for each element.

ATP Adenosine triphosphate (ah-DEN-uh-seen try-FOSS-fate) A nucleotide composed of adenine, ribose, and three phosphate groups. As the main energy carrier in cells, it directly or indirectly delivers energy to or picks up energy from nearly all metabolic pathways.

atrioventricular node (AV node) In the septum dividing the heart atria, a site that contains bundles of conducting fibers. Stimuli arriving at the AV node from the cardiac pacemaker (sinoatrial node) pass along the bundles and continue on via Purkinje fibers to contractile muscle cells in the ventricles.

atrioventricular valve One-way flow valve between the atrium and ventricle in each half of the heart.

atrium (AY-tree-um) Upper chamber in each half of the heart; the right atrium receives deoxygenated blood (from tissues) entering the pulmonary circuit of blood flow, and the left atrium receives oxygenated blood from pulmonary veins.

australopith (oss-TRAH-low-pith) [L. *australis*, southern, and Gk. *pithekos*, ape] Any of the earliest known species of hominids; that is, the first species of the evolutionary branch leading to humans.

autoimmune response Misdirected immune response in which lymphocytes mount an attack against normal body cells.

autonomic nervous system (ah-toe-NOM-ik) Those nerves leading from the central nervous system to the smooth muscle, cardiac muscle, and glands of internal organs and structures—that is, to the visceral portion of the body.

autosomal dominant inheritance Condition arising from the presence of a dominant allele on an autosome (not a sex chromosome). The allele is always expressed to some extent, even in heterozygotes.

autosomal recessive inheritance Condition arising from a recessive allele on an autosome (not a sex chromosome). Only recessive homozygotes show the resulting phenotype.

autosome Any of the chromosomes that are of the same number and kind in both males and females of the species.

autotroph (AH-toe-trofe) [Gk. *autos*, self, and *trophos*, feeder] An organism able to build its own large organic molecules by using carbon dioxide and energy from the physical environment. Compare heterotroph.

axial skeleton (AX-ee-uhl) In vertebrates, the skull, backbone, ribs, and breastbone (sternum).

axon Of a neuron, a long, cylindrical extension from the cell body, with finely branched endings. Action potentials move rapidly, without alteration, along an axon; their arrival at axon endings may trigger the release of neurotransmitter molecules that influence an adjacent cell.

B lymphocyte, or B cell The only white blood cell that produces antibodies, then positions them at the cell surface or secretes them as weapons in immune responses.

bacterial conjugation The transfer of plasmid DNA from one bacterial cell to another.

bacterial flagellum Of many bacterial cells, a whiplike motile structure that does not contain a core of microtubules.

bacteriophage (bak-TEER-ee-oh-fahj) [Gk. *baktērion*, small staff, rod, and *phagein*, to eat] Category of viruses that infect bacterial cells.

Barr body In the cells of female mammals, a condensed X chromosome that was inactivated during early embryonic development.

basal body A centriole that, after having given rise to the microtubules of a

flagellum or cilium, remains attached to its base in the cytoplasm.

basal metabolic rate Amount of energy required to sustain body functions when a person is resting, awake, and has not eaten for 12–18 hours.

base A substance that releases OH⁻ in water.

base pair A pair of hydrogen-bonded nucleotide bases in two strands of nucleic acids. In a DNA double helix, adenine pairs with thymine, and guanine with cytosine. When an mRNA strand forms on a DNA strand during transcription, uracil (U) pairs with the DNA's adenine.

base sequence The particular order in which one nucleotide base follows the next in a strand of DNA or RNA.

basement membrane Noncellular layer of mostly proteins and polysaccharides that is sandwiched in between an epithelium and underlying connective tissue.

basophil Fast-acting white blood cells that secrete histamine and other substances during inflammation.

biogeochemical cycle The movement of an element such as carbon or nitrogen from the environment to organisms, then back to the environment.

biogeography [Gk. *bios*, life, and *geographein*, to describe the surface of the Earth] The study of major land regions, each having distinguishing types of plants and animals and generally retaining its identity because of climate and geographic barriers to gene flow.

biological clock Internal time-measuring mechanism that has a role in adjusting an organism's daily activities, seasonal activities, or both in response to environmental cues.

biological magnification The increasing concentration of a nondegradable or slowly degradable substance in body tissues as it is passed along food chains.

biomass The combined weight of all the organisms at a particular trophic (feeding) level in an ecosystem.

biome A broad, vegetational subdivision of a biogeographic realm shaped by climate, topography, and composition of regional soils.

biopsy Diagnostic procedure in which a small piece of tissue is removed from the body through a hollow needle or exploratory surgery, and then examined

for signs of a particular disease (often cancer).

biosphere [Gk. *bios*, life, and *sphaira*, globe] All regions of the Earth's waters, crust, and atmosphere in which organisms live.

biosynthetic pathway A metabolic pathway in which small molecules are assembled into lipids, proteins, and other large organic molecules.

biotic potential Of a population, the maximum rate of increase per individual under ideal conditions.

bipedalism A habitual standing and walking on two feet, as by ostriches and humans.

blastocyst (BLASS-tuh-sist) [Gk. *blastos*, sprout, and *kystis*, pouch] In mammalian development, a blastula stage consisting of a hollow ball of surface cells and an inner cell mass.

blastula (BLASS-chew-lah) An embryonic stage consisting of a ball of cells produced by cleavage.

blood A fluid connective tissue composed of water, solutes, and formed elements (blood cells and platelets); it carries substances to and from cells and helps maintain an internal environment that is favorable for cell activities.

blood pressure Fluid pressure, generated by heart contractions, that keeps blood circulating.

blood-brain barrier Set of mechanisms that helps control which blood-borne substances reach neurons in the brain.

bone The mineral-hardened connective tissue of bones. *Bones* are organs that function in movement and locomotion, protection of other organs, mineral storage, and (in some bones) blood cell production.

brain Organ that receives, integrates, stores, and retrieves information, and coordinates appropriate responses by stimulating and inhibiting the activities of different body parts.

brain case The eight bones that together surround and protect the brain.

brainstem The vertebrate midbrain, pons, and medulla oblongata, the core of which contains the reticular formation that helps govern activity of the nervous system as a whole.

bronchiole A component of the finely branched bronchial tree inside each lung.

bronchus, plural bronchi (BRONG-cuss, BRONG-kee) [Gk. *bronchos*, windpipe] Tubelike branchings of the trachea that lead into the lungs.

buffer A substance that can combine with hydrogen ions, release them, or both. Buffers help stabilize the pH of blood and other fluids.

bulk A volume of fiber and other undigested material that absorption processes in the colon cannot decrease.

cancer A type of malignant tumor, the cells of which show profound abnormalities in the plasma membrane and cytoplasm, abnormal growth and division, and weakened capacity for adhesion within the parent tissue (leading to metastasis). Unless eradicated, cancer is lethal.

capillary [L. *capillus*, hair] A thin-walled blood vessel that functions in the exchange of gases and other substances between blood and interstitial fluid.

capillary bed Dense capillary networks containing true capillaries where exchanges occur between blood and tissues, and also thoroughfare channels that link arterioles and venules.

carbaminohemoglobin A hemoglobin molecule that has carbon dioxide bound to it; $HbCO_2$.

carbohydrate [L. *carbo*, charcoal, and *hydro*, water] A simple sugar or large molecule composed of sugar units. All cells use carbohydrates as structural materials, energy stores, and transportable forms of energy. The three classes of carbohydrates include monosaccharides, oligosaccharides, and polysaccharides.

carbon cycle A biogeochemical cycle in which carbon moves from its largest reservoir in the atmosphere, through oceans and organisms, then back to the atmosphere.

carbonic anhydrase Enzyme in red blood cells that catalyzes the conversion of unbound carbon dioxide to carbonic acid and its dissociation products, thereby helping maintain the gradient that keeps carbon dioxide diffusing from interstitial fluid into the blood.

carcinogen (kar-SIN-uh-jen) An environmental agent or substance, such as ultraviolet radiation, that can trigger cancer.

carcinogenesis The transformation of a normal cell into a cancerous one.

cardiac conduction system Set of non-contractile cells in heart muscle that spontaneously produce and conduct the electrical events that stimulate heart muscle contractions.

cardiac cycle (KAR-dee-ak) [Gk. *kardia*, heart, and *kyklos*, circle] The sequence of muscle contraction and relaxation constituting one heartbeat.

cardiac pacemaker Sinoatrial (SA) node; the basis of the normal rate of heartbeat. The self-excitatory cardiac muscle cells that spontaneously generate rhythmic waves of excitation over the heart chambers.

cardiovascular system Organ system that is composed of blood, the heart, and blood vessels and that functions in the rapid transport of substances to and from cells.

carotid body Any of several sensory receptors that monitor oxygen (and to some extent, carbon dioxide) levels in blood; located at the point where carotid arteries branch to the brain.

carrier protein Type of transport protein that binds specific substances and changes shape in ways that shunt the substances across a plasma membrane. Some carrier proteins function passively, others require an energy input.

carrying capacity The maximum number of individuals in a population (or species) that can be sustained indefinitely by a given environment.

cartilage A type of connective tissue with solid yet pliable intercellular material that resists compression.

cartilaginous joint Type of joint in which cartilage fills the space between adjoining bones; only slight movement is possible.

cDNA Any DNA molecule copied from a mature mRNA transcript by way of reverse transcription.

cell [L. *cella*, small room] The smallest living unit; an organized unit that can survive and reproduce on its own, given DNA instructions and suitable environmental conditions, including appropriate sources of energy and raw materials.

cell count The number of cells of a given type in a microliter of blood.

cell cycle Events during which a cell increases in mass, roughly doubles its number of cytoplasmic components, duplicates its DNA, then undergoes

nuclear and cytoplasmic division. It extends from the time a new cell is produced until it completes its own division.

cell differentiation The gene-guided process by which cells in different locations in the embryo become specialized.

cell-to-cell junction Of multicelled organisms, a point of contact that physically links two cells or that provides functional links between their cytoplasm.

cell theory A theory in biology, the key points of which are that (1) all organisms are composed of one or more cells, (2) the cell is the smallest unit that still retains a capacity for independent life, and (3) all cells arise from preexisting cells.

cell wall A rigid or semirigid wall outside the plasma membrane that supports a cell and imparts shape to it; a cellular feature of plants, fungi, protista, and most bacteria.

central nervous system The brain and spinal cord.

centriole (SEN-tree-ohl) A cylinder of triplet microtubules that gives rise to the microtubules of cilia and flagella.

centromere (SEN-troh-meer) [Gk. *kentron*, center, and *meros*, a part] A small, constricted region of a chromosome having attachment sites for microtubules that help move the chromosome during nuclear division.

cerebellum (ser-ah-BELL-um) [L. diminutive of *cerebrum*, brain] Hindbrain region with reflex centers for maintaining posture and refining limb movements.

cerebral cortex Thin surface layer of the cerebral hemispheres. Some regions of the cortex receive sensory input, others integrate information and coordinate appropriate motor responses.

cerebrospinal fluid Clear extracellular fluid that surrounds and cushions the brain and spinal cord.

cerebrum (suh-REE-bruhm) Part of the forebrain; the most complex integrating center.

chancroid Clinical term for the sexually transmitted disease caused by infection with the bacterium *Haemophilis ducreyi*. Infection results in soft, painful ulcers on the external genitals and can spread to pelvic tissues.

channel protein Type of transport protein that serves as a pore through which ions or other water-soluble substances move across the plasma membrane. Some channels remain open, while others are gated and open and close in controlled ways.

chemical bond A union between the electron structures of two or more atoms or ions.

chemical synapse (SIN-aps) [Gk. *synapsis*, union] A small gap, the synaptic cleft, that separates two neurons (or a neuron and a muscle cell or gland cell) and that is bridged by neurotransmitter molecules released from the presynaptic neuron.

chemiosmotic theory (keem-ee-oz-MOT-ik) Theory that an electrochemical gradient across a cell membrane drives ATP formation. Metabolic reactions cause hydrogen ions (H$^+$) to accumulate in a compartment formed by the membrane. The combined force of the resulting concentration and electric gradients propels hydrogen ions down the gradient, through channel proteins. Through enzyme action at these proteins, ADP and inorganic phosphate combine to form ATP.

chemoreceptor (KEE-moe-ree-sep-tur) Sensory receptor that detects chemical energy (ions or molecules) dissolved in the surrounding fluid.

chemotherapy The use of therapeutic drugs to kill cancer cells.

chlamydia Sexually transmitted disease caused by infection by the bacteria *Chlamydia trachomatis*. The bacterium infects cells of the genital organs and urinary tract.

chlorofluorocarbon (klore-oh-FLOOR-oh-car-bun), or CFC. One of a variety of odorless, invisible compounds of chlorine, fluorine, and carbon, widely used in commercial products, that are contributing to the destruction of the ozone layer above the Earth's surface.

chorion (CORE-ee-on) Of placental mammals, one of four extraembryonic membranes; it becomes a major component of the placenta. Absorptive structures (villi) that develop at its surface are crucial for the transfer of substances between the embryo and mother.

chromatid Of a duplicated eukaryotic chromosome, one of two DNA molecules and its associated proteins. One chromatid remains attached to its "sister" chromatid at the centromere until they are separated from each other during a nuclear division; then each is a separate chromosome.

chromosome (CROW-moe-soam) [Gk. *chroma*, color, and *soma*, body] Of eukaryotes, a DNA molecule with many associated proteins. A bacterial chromosome does not have a comparable profusion of proteins associated with the DNA.

chromosome number Of eukaryotic species, the number of each type of chromosome in all cells except dividing germ cells or gametes.

cilium (SILL-ee-um), plural cilia [L. *cilium*, eyelid] Of eukaryotic cells, a short, hairlike projection that contains a regular array of microtubules. Cilia serve as motile structures, help create currents of fluids, or are part of sensory structures. They typically are more profuse than flagella.

circadian rhythm (ser-KAYD-ee-un) [L. *circa*, about, and *dies*, day] Of many organisms, a cycle of physiological events that is completed every 24 hours or so, even when environmental conditions remain constant.

circulatory system An organ system consisting of the heart, blood vessels, and blood; the system transports materials to and from cells and helps stabilize body temperature and pH.

cladistics An approach to biological systematics in which organisms are grouped according to similarities that are derived from a common ancestor.

cladogram Branching diagram that represents patterns of relative relationships among organisms based on discrete morphological, physiological, and behavioral traits that vary among taxa being studied.

clavicle Long, slender "collarbone" that connects the pectoral girdle with the sternum (breastbone).

cleavage Stage of development when mitotic cell divisions convert a zygote to a ball of cells, the blastula.

cleavage furrow Of a cell undergoing cytoplasmic division, a shallow, ringlike depression that forms at the cell surface as contractile microfilaments pull the plasma membrane inward. It defines where the cytoplasm will be cut in two.

climate Prevailing weather conditions for an ecosystem, including temperature, humidity, wind speed, cloud cover, and rainfall.

climax community Following primary or secondary succession, the array of species that remains more or less steady under prevailing conditions.

clonal selection hypothesis Hypothesis that suggests that lymphocytes activated by a specific antigen will rapidly multiply and differentiate into huge subpopulations of cells, all having the parent cell's specificity against that antigen.

cloned DNA Multiple, identical copies of DNA fragments that have been inserted into plasmids or some other cloning vector.

cochlea Coiled, fluid-filled chamber of the inner ear. Sound waves striking the ear drum become converted to pressure waves in the cochlear fluid, and the pressure waves ultimately cause a membrane to vibrate and bend sensory hair cells. Signals from bent hair cells travel to the brain, where they may be interpreted as sound.

codominance Condition in which a pair of nonidentical alleles are both expressed, even though they specify two different phenotypes.

codon One of a series of base triplets in an mRNA molecule, most of which code for a sequence of amino acids of a specific polypeptide chain. (Of 64 codons, 61 specify different amino acids and three of these also serve as start signals for translation; one other serves only as a stop signal for translation.)

coenzyme A type of nucleotide that transfers hydrogen atoms and electrons from one reaction site to another. NAD$^+$ is an example.

coevolution The joint evolution of two or more closely interacting species; when one species evolves, the change affects selection pressures operating between the two species, so the other also evolves.

cofactor A metal ion or coenzyme; it helps catalyze a reaction or serves briefly as an agent that transfers electrons, atoms, or functional groups from one substrate to another.

colon (CO-lun) The large intestine.

community The populations of all species occupying a habitat; also applied to groups of organisms with similar lifestyles in a habitat.

compact bone Type of dense bone tissue that makes up the shafts of long bones and outer regions of all bones. Narrow channels in compact bone contain blood vessels and nerves.

comparative morphology [Gk. *morph*, form] Anatomical comparisons of major lineages.

complement system A set of about 20 proteins circulating in blood plasma with roles in nonspecific defenses and in immune responses. Some induce lysis of pathogens, others promote inflammation, and others stimulate phagocytes to engulf pathogens.

compound A substance in which the relative proportions of two or more elements never vary. Organic compounds have a backbone of carbon atoms arranged as a chain or ring structure. The simpler, inorganic compounds do not have comparable backbones.

concentration gradient A difference in the number of molecules (or ions) of a substance between two adjacent regions, as in a volume of fluid.

condensation Enzyme-mediated reaction leading to the covalent linkage of small molecules and, often, the formation of water as a by-product.

cone cell In the retina, a type of photoreceptor that responds to intense light and contributes to sharp daytime vision and color perception.

connective tissue A category of animal tissues, all having mostly the same components but in different proportions. These tissues contain fibroblasts and other cells, the secretions of which form fibers (of collagen and elastin) and a ground substance (of modified polysaccharides).

consumer [L. *consumere*, to take completely] Of ecosystems, a heterotrophic organism that obtains energy and raw materials by feeding on the tissues of other organisms. Herbivores, carnivores, omnivores, and parasites are examples.

continuous variation A more or less continuous range of small differences in a given trait among all the individuals of a population.

control group In a scientific experiment, a group used to evaluate possible side effects of a test involving an experimental group. Ideally, the control group should differ from the experimental group only with respect to the variable being studied.

core temperature The body's internal temperature, as opposed to temperatures of the tissues near its surface. Normal human core temperature is about 37°C (98.6°F).

cornea Transparent tissue in the outer layer of the eye, which causes incoming light rays to bend.

coronary artery Either of two arteries leading to capillaries that service cardiac muscle.

corpus callosum (CORE-pus ka-LOW-sum) A band of 200 million axons that functionally link the two cerebral hemispheres.

corpus luteum (CORE-pus LOO-tee-um) A glandular structure; it develops from cells of a ruptured ovarian follicle and secretes progesterone and some estrogen, both of which maintain the lining of the uterus (endometrium).

cortex [L. *cortex*, bark] In general, a rindlike layer; the kidney cortex is an example.

covalent bond (koe-VAY-lunt) [L. *con*, together, and *valere*, to be strong] A sharing of one or more electrons between atoms or groups of atoms. When electrons are shared equally, the bond is nonpolar. When electrons are shared unequally, the bond is polar—slightly positive at one end and slightly negative at the other.

cross-bridge The interaction between actin and myosin filaments that is the basis of muscle cell contraction.

crossing over During prophase I of meiosis, an interaction between a pair of homologous chromosomes. Their nonsister chromatids break at the same place along their length and exchange corresponding segments at the break points. Crossing over breaks up old combinations of alleles and puts new ones together in chromosomes.

culture The sum total of behavior patterns of a social group, passed between generations by learning and by symbolic behavior, especially language.

cyclic AMP Cyclic adenosine monophosphate. A nucleotide that has roles in intercellular communication, as when it serves as a second messenger (a cytoplasmic mediator of a cell's response to signaling molecules).

cytokinesis (sigh-toe-kih-NEE-sis) [Gk. *kinesis*, motion] Cytoplasmic division; the splitting of a parental cell into daughter cells.

cytomembrane system [Gk. *kytos*, hollow vessel] Organelles, functioning as a system to modify, package, and distribute newly formed proteins and lipids. Endoplasmic reticulum, Golgi bodies, lysosomes, and a variety of vesicles are its components.

cytoplasm (SIGH-toe-plaz-um) [Gk. *plassein*, to mold] All cellular parts, particles, and semifluid substances enclosed by the plasma membrane except for the nucleus.

cytosine (SIGH-toe-seen) A pyrimidine; one of the nitrogen-containing bases in nucleotides.

cytoskeleton A cell's internal "skeleton." Its microtubules and other components structurally support the cell and organize and move its internal components.

cytotoxic T cell Type of T lymphocyte that directly kills infected body cells and tumor cells by lysis.

decomposer [L. *de-*, down, away, and *companere*, to put together] Of ecosystems, a heterotroph that obtains energy by chemically breaking down the remains, products, or wastes of other organisms. Decomposer activities help cycle nutrients back to producers. Certain fungi and bacteria are examples.

deforestation The removal of all trees from a large tract of land, such as the Amazon Basin or the Pacific Northwest.

degradative pathway A metabolic pathway by which molecules are broken down in stepwise reactions that lead to products of lower energy.

deletion (chromosomal) A change in a chromosome's structure after one of its regions is lost as a result of irradiation, viral attack, chemical action, or some other factor.

demographic transition model Model of human population growth in which changes in the growth pattern correspond to different stages of economic development. These are a preindustrial stage, when birth and death rates are both high; a transitional stage; an industrial stage; and a postindustrial stage, when the death rate exceeds the birth rate.

denaturation (deh-nay-chur-AY-shun) Of any molecule, the loss of three-dimensional shape following disruption of hydrogen bonds and other weak bonds.

dendrite (DEN-drite) [Gk. *dendron*, tree] A short, slender extension from the cell body of a neuron.

denitrification (dee-nite-rih-fih-KAY-shun) The conversion of nitrate or nitrite by certain bacteria to gaseous nitrogen (N_2) and a small amount of nitrous oxide (N_2O).

density-dependent controls Factors, such as predation, parasitism, disease, and competition for resources, that limit population growth by reducing the birth rate, increasing the rates of death and dispersal, or all of these.

density-independent controls Factors such as storms or floods that increase a population's death rate more or less independently of its density.

dentition The type, size, and number of an animal's teeth.

dermis The layer of skin underlying the epidermis, consisting mostly of dense connective tissue.

desertification (dez-urt-ih-fih-KAY-shun) The conversion of grasslands, rain-fed cropland, or irrigated cropland to desertlike conditions, with a drop in agricultural productivity of 10 percent or more.

detrital food web Of most ecosystems, the flow of energy mainly from plants through detritivores and decomposers.

detritivores (dih-TRY-tih-vorz) [L. *detritus*; after *deterere*, to wear down] Heterotrophs that consume dead or decomposing particles of organic matter. Earthworms, crabs, and nematodes are examples.

diaphragm (DIE-uh-fram) [Gk. *dia-phragma*, to partition] Muscular partition between the thoracic and abdominal cavities, the contraction and relaxation of which contributes to breathing. Also, a contraceptive device used temporarily to prevent sperm from entering the uterus during sexual intercourse.

diastole Relaxation phase of the cardiac cycle.

diffusion Net movement of like molecules (or ions) down their concentration gradient. In the absence of other forces, molecular motion and random collisions cause their net outward movement from one region into a neighboring region where they are less concentrated (because collisions are more frequent where the molecules are most crowded together).

digestion The breakdown of food particles, then into nutient molecules small enough to be absorbed.

digestive system An internal tube from which ingested food is absorbed into the internal environment; divided into regions specialized for food transport, processing, and storage.

dihybrid cross An experimental cross in which offspring inherit two gene pairs, each consisting of two nonidentical alleles.

diploid number (DIP-loyd) For many sexually reproducing species, the chromosome number of somatic cells and of germ cells prior to meiosis. Such cells have two chromosomes of each type (that is, pairs of homologous chromosomes). Compare haploid number.

directional selection Of a population, a shift in allele frequencies in a steady, consistent direction in response to a new environment or to a directional change in the old one. The outcome is that forms of traits at one end of the range of phenotypic variation become more common than the intermediate forms.

disaccharide (die-SAK-uh-ride) [Gk. *di*, two, and *sakcharon*, sugar] A type of simple carbohydrate, of the class called oligosaccharides; two monosaccharides covalently bonded.

disruptive selection Of a population, a shift in allele frequencies to forms of traits at both ends of a range of phenotypic variation and away from intermediate forms.

distal tubule The tubular section of a nephron most distant from the glomerulus; a major site of water and sodium reabsorption.

divergence Accumulation of differences in allele frequencies between populations that have become reproductively isolated from one another.

DNA Deoxyribonucleic acid (dee-OX-ee-rye-bow-new-CLAY-ik) For all cells (and many viruses), the molecule of inheritance. A category of nucleic acids, each usually consisting of two nucleotide strands twisted together helically and held together by hydrogen bonds. The nucleotide sequence encodes the instructions for assembling proteins, and, ultimately, a new individual.

DNA-DNA hybridization *See* nucleic acid hybridization.

DNA fingerprint Of each individual, a unique array of RFLPs, resulting from

the DNA sequences inherited (in a Mendelian pattern) from each parent.

DNA library A collection of DNA fragments produced by restriction enzymes and incorporated into plasmids.

DNA ligase (LYE-gase) Enzyme that seals together the new base-pairings during DNA replication; also used by recombinant DNA technologists to seal base-pairings between DNA fragments and cut plasmid DNA.

DNA polymerase (poe-LIM-uh-rase) Enzyme that assembles a new strand on a parent DNA strand during replication; also takes part in DNA repair.

DNA repair Following an alteration in the base sequence of a DNA strand, a process that restores the original sequence, as carried out by DNA polymerases, DNA ligases, and other enzymes.

DNA replication Of cells, the process by which the hereditary material is duplicated for distribution to daughter nuclei. An example is the duplication of eukaryotic chromosomes during interphase, prior to mitosis.

dominant allele In a diploid cell, an allele that masks the expression of its partner on the homologous chromosome.

dry acid deposition The falling to Earth of sulfur and nitrogen oxides; called *wet acid deposition* when it occurs in rain or snow.

dryopith A type of hominoid, one of the first to appear during the Miocene about the time of the divergences that led to gorillas, chimpanzees, and humans.

duplication A change in a chromosome's structure resulting in the repeated appearance of the same gene sequence.

dysplasia An abnormal change in the sizes, shapes, and organization of cells in a tissue.

early *Homo* A type of early hominid that may have been the maker of stone tools that date from about 2.5 million years ago.

ecology [Gk. *oikos*, home, and *logos*, reason] Study of the interactions of organisms with one another and with their physical and chemical environment.

ecosystem [Gk. *oikos*, home] An array of organisms and their physical environment, all of which interact through a flow of energy and a cycling of materials.

ecosystem analysis Method of predicting unforeseen effects of disturbances to an ecosystem, based on computer programs and models.

ectoderm [Gk. *ecto*, outside, and *derma*, skin] The outermost primary tissue layer (germ layer) of an embryo, which gives rise to the outer layer of the integument and to tissues of the nervous system.

effector A muscle (or gland) that responds to signals from an integrator (such as the brain) by producing movement (or chemical change) that helps adjust the body to changing conditions.

effector cell Of the differentiated subpopulations of lymphocytes that form during an immune response, the type of cell that engages and destroys the antigen-bearing agent that triggered the response.

egg A mature female gamete; also called an ovum.

electron Negatively charged unit of matter, with both particulate and wavelike properties, that occupies one of the orbitals around the atomic nucleus. Atoms can gain, lose, or share electrons with other atoms.

electron transport system An organized array of enzymes and cofactors, bound in a cell membrane, that accept and donate electrons in sequence. When such systems operate, hydrogen ions (H^+) flow across the membrane, and the flow drives ATP formation and other reactions.

element Any substance that cannot be decomposed into substances with different properties.

embryo (EM-bree-oh) [Gk. *en*, in, and probably *bryein*, to swell] Of animals generally, the stage formed by way of cleavage, gastrulation, and other early developmental events.

embryonic disk In early development, the oval, flattened cell mass that gives rise to the embryo shortly after implantation.

endocrine gland A ductless gland that secretes hormones, which usually enter interstitial fluid and then the bloodstream.

endocrine system System of cells, tissues, and organs that is functionally linked to the nervous system and that exerts control by way of its hormones and other chemical secretions.

endocytosis (en-doe-sigh-TOE-sis) Movement of a substance into cells; the substance becomes enclosed by a patch of plasma membrane that sinks into the cytoplasm, then forms a vesicle around it. Phagocytic cells also engulf pathogens or prey in this manner.

endoderm [Gk. *endon*, within, and *derma*, skin] The inner primary tissue layer, or germ layer, of an embryo, which gives rise to the inner lining of the gut and organs derived from it.

endometrium (en-doh-MEET-ree-um) [Gk. *metrios*, of the womb] Inner lining of the uterus consisting of connective tissues, glands, and blood vessels.

endoplasmic reticulum or ER (en-doe-PLAZ-mik reh-TIK-yoo-lum) An organelle that begins at the nucleus and curves through the cytoplasm. In rough ER (which has many ribosomes on its cytoplasmic side), many new polypeptide chains acquire specialized side chains. In many cells, smooth ER (with no attached ribosomes) is the main site of lipid synthesis.

endotherm Animal such as a human that maintains body temperature from within, generally by metabolic activity and controls over heat conservation and dissipation.

energy The capacity to do work.

energy carrier A molecule that delivers energy from one metabolic reaction site to another. ATP is the most widely traveled of these; it readily donates energy to nearly all metabolic reactions.

energy pyramid A pyramid-shaped representation of an ecosystem's trophic structure, illustrating the energy losses at each transfer to a different trophic level.

enzyme (EN-zime) One of a class of proteins that greatly speed up (catalyze) reactions between specific substances. The substances that each type of enzyme acts upon are called its substrates.

eosinophil Fast-acting, phagocytic white blood cell that takes part in inflammation but not in immune responses.

epidermis The outermost tissue layer of a multicelled animal.

epiglottis A flaplike structure at the start of the larynx, the position of which directs the movement of air into the trachea or food into the esophagus.

epiphyseal plate Region of cartilage that covers either end of a growing long bone, permitting the bone to lengthen. The epiphyseal plate is replaced by bone when growth stops in late adolescence.

epithelium (ep-ih-THEE-lee-um) A tissue consisting of one or more layers of adhering cells that covers the body's external surfaces and lines its internal cavities and tubes. Epithelium has one free surface; the opposite surface rests on a basement membrane between it and an underlying connective tissue. Epidermis or skin is an example.

erythrocyte (eh-RITH-row-site) [Gk. *erythros*, red, and *kytos*, vessel] Red blood cell.

esophagus (ee-SOF-uh-gus) Tubular portion of the digestive system that receives swallowed food and leads to the stomach.

essential amino acid Any of eight amino acids that the body cannot synthesize and must be obtained from food.

essential fatty acid Any of the fatty acids that the body cannot synthesize and must be obtained from food.

estrogen (ESS-tro-jen) A sex hormone that helps oocytes mature, induces changes in the uterine lining during the menstrual cycle and pregnancy, and maintains secondary sexual traits; also influences body growth and development.

eukaryotic cell (yoo-carry-AH-tic) [Gk. *eu*, good, and *karyon*, kernel] A type of cell that has a "true nucleus" and other distinguishing membrane-bound organelles. Compare prokaryotic cell.

evolution, biological [L. *evolutio*, act of unrolling] Change within a line of descent over time. A population is evolving when some forms of a trait are becoming more or less common relative to the other kinds of traits. The shifts are evidence of changes in the relative abundances of alleles for that trait, as brought about by mutation, natural selection, genetic drift, and gene flow.

evolutionary tree A treelike diagram in which branches represent separate lines of descent from a common ancestor.

excitatory postsynaptic potential or EPSP One of two competing signals at an input zone of a neuron; a graded potential that brings the neuron's plasma membrane closer to threshold.

excretion Any of several processes by which excess water, excess or harmful solutes, or waste materials leave the body by way of the urinary system or certain glands.

exocrine gland (EK-suh-krin) [Gk. *es*, out of, and *krinein*, to separate] Glandular structure that secretes products, usually through ducts or tubes, to a free epithelial surface.

exocytosis (ek-so-sigh-TOE-sis) Movement of a substance out of a cell by means of a transport vesicle, the membrane of which fuses with the plasma membrane, so that the vesicle's contents are released outside.

exon Of eukaryotic cells, any of the nucleotide sequences of a pre-mRNA molecule that are spliced together to form the mature mRNA transcript and are ultimately translated into protein.

experiment A test in which some phenomenon in the natural world is manipulated in controlled ways to gain insight into its function, structure, operation, or behavior.

expiration Expelling air from the lungs; exhaling.

exponential growth (ex-po-NEN-shul) Pattern of population growth in which greater and greater numbers of individuals are produced during the successive doubling times; the pattern that emerges when the per capita birth rate remains even slightly above the per capita death rate, putting aside the effects of immigration and emigration.

external respiration Movement of oxygen from alveoli into the blood, and of carbon dioxide from the blood into alveoli.

extinction, background. A steady rate of species turnover that characterizes lineages through most of their histories. *Mass extinction* is an abrupt increase in the rate at which major taxa disappear, with several taxa being affected simultaneously.

extracellular fluid In animals generally, all the fluid not inside cells; includes plasma (the liquid portion of blood) and interstitial fluid (which occupies the spaces between cells and tissues).

extracellular matrix A material, largely secreted, that helps hold many animal tissues together in certain shapes; it

consists of fibrous proteins and other components in a ground substance.

extraembryonic membranes Membranes that form along with a developing embryo, including the yolk sac, amnion, allantois, and chorion.

F_1 (first filial generation) The offspring of an initial genetic cross.

F_2 (second filial generation) The offspring of parents who are the first filial generation from a genetic cross.

FAD Flavin adenine dinucleotide, a nucleotide coenzyme. When delivering electrons and unbound protons (H^+) from one reaction to another, it is abbreviated $FADH_2$.

family pedigree A chart of genetic relationships of the individuals in a family through successive generations.

fat A lipid with a glycerol head and one, two, or three fatty acid tails. The tails of saturated fats have only single bonds between carbon atoms and hydrogen atoms attached to all other bonding sites. Tails of unsaturated fats additionally have one or more double bonds between certain carbon atoms.

fatty acid A long, flexible hydrocarbon chain with a —COOH group at one end.

feedback inhibition Of cells, a control mechanism by which the production (or secretion) of a substance triggers a change in some activity that in turn shuts down further production of the substance.

femur Thigh bone; longest and strongest bone of the body.

fermentation [L. *fermentum*, yeast] A type of anaerobic pathway of ATP formation; it starts with glycolysis, ends when electrons are transferred back to one of the breakdown products or intermediates, and regenerates the NAD^+ required for the reaction. Its net yield is two ATP per glucose molecule broken down.

fertilization [L. *fertilis*, to carry, to bear] Fusion of a sperm nucleus with the nucleus of an egg, which thereupon becomes a zygote.

fetus Term applied to an embryo after it reaches the age of eight weeks.

fever Body temperature that has climbed above the normal set point, usually in response to infection. Mild fever promotes an increase in body defense activities.

fibrous joint Type of joint in which fibrous connective tissue unites the adjoining bones and no cavity is present.

filtration In urine formation, the process by which blood pressure forces water and solutes out of glomerular capillaries and into the cupped portion of a nephron wall (glomerular capsule).

flagellum (fluh-JELL-um), plural flagella [L., *whip*] Tail-like motile structure of many free-living eukaryotic cells; it has a distinctive 9 + 2 array of microtubules.

fluid mosaic model Model of membrane structure in which proteins are embedded in a lipid bilayer or attached to one of its surfaces. The lipid molecules give the membrane its basic structure, impermeability to water-soluble molecules, and (through packing variations and movements) fluidity. Proteins carry out most membrane functions, such as transport, enzyme action, and reception of signals or substances.

follicle (FOLL-ih-kul) In an ovary, a primary oocyte (immature egg) together with the surrounding layer of cells.

food chain A straight-line sequence of who eats whom in an ecosystem.

food web A network of cross-connecting, interlinked food chains encompassing primary producers and an array of consumers, detritivores, and decomposers.

forebrain Brain region that includes the cerebrum and cerebral cortex, the olfactory lobes, and the hypothalamus.

fossil Recognizable evidence of an organism that lived in the distant past. Most fossils are skeletons, shells, leaves, seeds, and tracks that were buried in rock layers before they decomposed.

FSH Follicle-stimulating hormone. The name comes from its function in females, in whom FSH helps stimulate follicle development in ovaries. In males, it acts in the testes as part of a sequence of events that trigger sperm production.

functional group An atom or group of atoms that is covalently bonded to the carbon backbone of an organic compound and that influences its behavior.

gallbladder Organ of the digestive system that stores bile secreted from the liver.

gamete (GAM-eet) [Gk. *gametēs*, husband, and *gametē*, wife] A haploid cell that functions in sexual reproduction. Sperm and eggs are examples.

ganglion (GANG-lee-un), plural ganglia [Gk. *ganglion*, a swelling] A distinct clustering of cell bodies of neurons in regions other than the brain or spinal cord.

gastrointestinal (GI) tract The digestive tube, extending from the mouth to the anus and including the stomach, small and large intestines, and other specialized regions with roles in food transport and digestion.

gastrulation (gas-tru-LAY-shun) The stage of embryonic development in which cells become arranged into two or three primary tissue layers (germ layers); in humans, the layers are an inner endoderm, an intermediate mesoderm, and a surface ectoderm.

gene A unit of information about a heritable trait that is passed on from parents to offspring. Each gene has a specific location on a chromosome.

gene flow A microevolutionary process; a physical movement of alleles out of a population as individuals leave (emigrate) or enter (immigrate); allele frequencies change as a result.

gene frequency More precisely, allele frequency: the relative abundances of all the different alleles for a trait that are carried by the individuals of a population.

gene locus A given gene's particular location on a chromosome.

gene pair In diploid cells, the two alleles at a given locus on a pair of homologous chromosomes.

gene pool Sum total of all genotypes in a population. More accurately, allele pool.

gene therapy Generally, the transfer of one or more normal genes into the body cells of an organism in order to correct a genetic defect.

genetic code [After L. *genesis*, to be born] The correspondence between nucleotide triplets in DNA (then in mRNA) and specific sequences of amino acids in the resulting polypeptide chains; the basic language of protein synthesis.

genetic disorder An inherited condition that results in mild to severe medical problems.

genetic drift A microevolutionary process; a change in allele frequencies over the generations due to chance events alone.

genetic engineering Altering the information content of DNA through use of recombinant DNA technology.

genetic recombination Presence of a new combination of alleles in a DNA molecule compared to the parental genotype; the result of processes such as crossing over at meiosis, chromosome rearrangements, gene mutation, and recombinant DNA technology.

genital herpes Infection of tissues in the genital area by a herpes simplex virus; an extremely contagious sexually transmitted disease.

genome All the DNA in a haploid number of chromosomes of a given species.

genotype (JEEN-oh-type) Genetic constitution of an individual. Can mean a single gene pair or the sum total of the individual's genes. Compare phenotype.

genus, plural genera (JEEN-us, JEN-er-ah) [L. *genus*, race, origin] A taxon into which all species exhibiting certain phenotypic similarities and evolutionary relationship are grouped.

germ cell One of a cell lineage set aside for sexual reproduction; germ cells give rise to gametes. Compare somatic cell.

germ layer One of two or three primary tissue layers that forms during gastrulation and that gives rise to certain tissues of the adult body. Compare ectoderm; endoderm; mesoderm.

gland A secretory cell or multicelled structure derived from epithelium and often connected to it.

glomerular capillaries The set of blood capillaries inside the glomerular capsule of the nephron.

glomerular (Bowman's) capsule The ballooned region of a nephron wall that forms a cup for water and other solutes filtered from the clustered blood capillaries of the glomerulus.

glomerulus (glow-MARE-you-luss) [L. *glomus*, ball] The first portion of the nephron, where water and solutes are filtered from blood.

glucagon (GLUE-kuh-gone) Hormone that stimulates conversion of glycogen and amino acids to glucose; secreted by alpha cells of the pancreas when the flow of glucose decreases.

glucocorticoid Hormone secreted by the adrenal cortex that influences metabolic reactions that help maintain the blood glucose level.

glyceride (GLISS-er-eyed) One of the molecules, commonly called fats and oils, that has one, two, or three fatty acid tails attached to a glycerol backbone. They are the body's most abundant lipids and its richest source of energy.

glycerol (GLISS-er-all) [Gk. *glykys*, sweet, and L. *oleum*, oil] A three-carbon molecule with three hydroxyl groups attached; together with fatty acids, a component of fats and oils.

glycogen (GLY-kuh-jen) In animals, a storage polysaccharide that is a main food reserve; can be readily broken down into glucose subunits.

glycolysis (gly-CALL-ih-sis) [Gk. *glykys*, sweet, and *lysis*, loosening or breaking apart] Initial reactions of both aerobic and anaerobic pathways by which glucose (or some other organic compound) is partially broken down to pyruvate with a net yield of two ATP. Glycolysis proceeds in the cytoplasm of all cells, and oxygen has no role in it.

glycoprotein A protein having linear or branched oligosaccharides covalently bonded to it. Most human cell surface proteins and many proteins circulating in blood are glycoproteins.

Golgi body (GOHL-gee) Organelle in which newly synthesized polypeptide chains as well as lipids are modified and packaged in vesicles for export or for transport to specific locations within the cytoplasm.

gonad (GO-nad) Primary reproductive organ in which gametes are produced.

gonorrhea Clinical term for the sexually transmitted disease caused by the bacterium *Neisseria gonorrhoeae*. This bacterium can infect epithelial cells of the genital tract, rectum, eye membranes, and the throat.

graded potential Of neurons, a local signal that slightly changes the voltage difference across a small patch of the plasma membrane. Such signals vary in magnitude, depending on the stimulus. With prolonged or intense stimulation, they may spread to a trigger zone of the membrane and initiate an action potential.

granulocyte Class of white blood cells that have a lobed nucleus and various types of granules in the cytoplasm;

includes neutrophils, eosinophils, and basophils.

gray matter The dendrites, neuron cell bodies, and neuroglial cells of the spinal cord and cerebral cortex.

grazing food web Of most ecosystems, the flow of energy from plants to herbivores, then through an array of carnivores.

green revolution In developing countries, the use of improved crop varieties, modern agricultural practices (including massive inputs of fertilizers and pesticides), and equipment to increase crop yields.

greenhouse effect Warming of the lower atmosphere due to the presence of greenhouse gases: carbon dioxide, methane, nitrous oxide, ozone, water vapor, and chlorofluorocarbons.

ground substance The intercellular material made up of cell secretions and other noncellular components.

growth factor A type of signaling molecule that can influence growth by regulating the rate at which target cells divide.

guanine A nitrogen-containing base; present in one of the four nucleotide building blocks of DNA and RNA.

habitat [L. *habitare*, to live in] The type of place where an organism normally lives, characterized by physical features, chemical features, and the presence of certain other species.

hair cell Type of mechanoreceptor that may give rise to action potentials when bent or tilted.

haploid number (HAP-loyd) The chromosome number of a gamete that, as an outcome of meiosis, is only half that of the parent germ cell (it has only one of each pair of homologous chromosomes). Compare diploid number.

HCG Human chorionic gonadotropin. A hormone that helps maintain the lining of the uterus during the menstrual cycle and during the first trimester of pregnancy.

heart Muscular pump that keeps blood circulating through the animal body.

helper T cell Type of T lymphocyte that produces and secretes chemicals that promote formation of large effector and memory cell populations.

hemoglobin (HEEM-oh-glow-bin) [Gk. *haima*, blood, and L. *globus*, ball] Iron-containing, oxygen-transporting protein that gives red blood cells their color.

hemostasis (hee-mow-STAY-sis) [Gk. *haima*, blood, and *stasis*, standing] Stopping of blood loss from a damaged blood vessel through coagulation, blood vessel spasm, platelet plug formation, and other mechanisms.

hepatic portal vein Vessel that receives nutrient-laden blood from villi of the small intestine and transports it to the liver, where excess glucose is removed. The blood then returns to the general circulation via a hepatic vein.

hepatitis B An extremely contagious, sexually transmitted disease caused by infection by the hepatitis B virus. Chronic hepatitis can lead to liver cirrhosis or cancer.

herbivore [L. *herba*, grass, and *vovare*, to devour] Plant-eating animal.

heterotroph (HET-er-oh-trofe) [Gk. *heteros*, other, and *trophos*, feeder] Organism that cannot synthesize its own organic compounds and must obtain nourishment by feeding on autotrophs, each other, or organic wastes. Animals, fungi, many protista, and most bacteria are heterotrophs. Compare autotroph.

heterozygous (het-er-oh-ZYE-guss) [Gk. *zygoun*, join together] For a given trait, having nonidentical alleles at a particular locus on a pair of homologous chromosomes.

hindbrain One of the three divisions of the brain; the medulla oblongata, cerebellum, and pons; includes reflex centers for respiration, blood circulation, and other basic functions; also coordinates motor responses and many complex reflexes.

histone Any of a class of proteins that are intimately associated with DNA and that are largely responsible for its structural (and possibly functional) organization in eukaryotic chromosomes.

homeostasis (hoe-me-oh-STAY-sis) [Gk. *homo*, same, and *stasis*, standing] A physiological state in which the physical and chemical conditions of the internal environment are being maintained within tolerable ranges.

homeostatic feedback loop An interaction in which an organ (or structure) stimulates or inhibits the output of another organ, then shuts down or increases this activity when it detects that the output has exceeded or fallen below a set point.

hominid [L. *homo*, man] All species on the evolutionary branch leading to modern humans. *Homo sapiens* is the only living representative.

hominoid Apes, humans, and their recent ancestors.

Homo erectus A hominid lineage that emerged between 1.5 million and 300,000 years ago and that may include the direct ancestors of modern humans.

Homo sapiens The hominid lineage of modern humans that emerged between 300,000 and 200,000 years ago.

homologous chromosome (huh-MOLL-uh-gus) [Gk. *homologia*, correspondence] One of a pair of chromosomes that resemble each other in size, shape, and the genes they carry, and that line up with each other at meiosis I. The X and Y chromosomes differ in these respects but still function as homologues.

homologous structure The same body part, modified in different ways, in different lines of descent from a common ancestor.

homozygous (hoe-moe-ZYE-guss) Having two identical alleles at a given locus (on a pair of homologous chromosomes).

homozygous dominant condition Having two dominant alleles at a given locus (on a pair of homologous chromosomes).

homozygous recessive condition Having two recessive alleles at a given gene locus (on a pair of homologous chromosomes).

hormone [Gk. *hormon*, to stir up, set in motion] Any of the signaling molecules secreted from endocrine glands, endocrine cells, and some neurons that the bloodstream distributes to nonadjacent target cells (any cell having receptors for that hormone).

human genome project A basic research project in which researchers throughout the world are working together to sequence the estimated 3 billion nucleotides present in the DNA of human chromosomes.

human immunodeficiency virus (HIV) A retrovirus; the pathogen that causes AIDS (acquired immune deficiency syndrome).

human papillomavirus (HPV) Virus that causes genital warts; HPV infection is suspected of having a role in the development of some cases of cervical cancer.

humerus The long bone of the upper arm.

hydrocarbon A molecule having only hydrogen atoms attached to a carbon backbone.

hydrogen bond Type of chemical bond in which an atom of a molecule interacts weakly with a neighboring atom that is already taking part in a polar covalent bond.

hydrogen ion A free (unbound) proton; a hydrogen atom that has lost its electron and so bears a positive charge (H^+).

hydrologic cycle A biogeochemical cycle, driven by solar energy, in which water moves slowly through the atmosphere, on or through surface layers of land masses, to the ocean and back again.

hydrolysis (high-DRAWL-ih-sis) [L. *hydro*, water, and Gk. *lysis*, loosening or breaking apart] Enzyme-mediated reaction in which covalent bonds break, splitting a molecule into two or more parts, and H^+ and OH^- (derived from a water molecule) become attached to the exposed bonding sites.

hydrophilic [Gk. *philos*, loving] A polar substance that is attracted to the polar water molecule and so dissolves easily in water. Sugars are examples.

hydrophobic [Gk. *phobos*, dreading] A nonpolar substance that is repelled by the polar water molecule and so does not readily dissolve in water. Oil is an example.

hydrosphere All liquid or frozen water on or near the Earth's surface.

hypodermis A subcutaneous layer having stored fat that helps insulate the body; although not part of skin, it anchors skin while allowing it some freedom of movement.

hypothalamus [Gk. *hypo*, under, and *thalamos*, inner chamber, or possibly *tholos*, rotunda] A brain center that monitors visceral activities (such as salt–water balance, temperature control, and reproduction) and that influences related forms of behavior (as in hunger, thirst, and sex).

hypothesis A possible explanation of a specific phenomenon.

immune response A series of events by which B and T lymphocytes recog-

nize a specific antigen, undergo repeated cell divisions that form huge lymphocyte populations, and differentiate into subpopulations of effector and memory cells. Effector cells engage and destroy antigen-bearing agents. Memory cells enter a resting phase and are activated during subsequent encounters with the same antigen.

immune therapy In cancer treatment, the use of specific chemicals that trigger a strong immune response in the patient's body against cancer cells. In general, immune therapies are still experimental.

immunization Various processes, including vaccination, that promote increased immunity against specific diseases.

immunoglobulin Any of the five classes of antibodies. Different immunoglobulins participate in specific ways in defense and immune responses. Examples are IgM antibodies (first to be secreted during immune responses) and IgG antibodies (which activate complement proteins and neutralize many toxins).

immunotherapy Procedures that enhance a person's immunological defenses against tumors or certain pathogens.

implantation Series of events in which a trophoblast (pre-embryo) invades the endometrium (lining of the uterus) and becomes embedded there.

incomplete dominance Of heterozygotes, the appearance of a version of a trait that is somewhere between the homozygous dominant and recessive conditions.

independent assortment Mendelian principle that each gene pair tends to assort into gametes independently of other gene pairs located on nonhomologous chromosomes.

induced-fit model Model of enzyme action whereby a bound substrate induces changes in the shape of the enzyme's active site, resulting in a more precise molecular fit between the enzyme and its substrate.

industrial smog A type of gray-air smog that develops in industrialized regions when winters are cold and wet.

inflammation, acute In response to tissue damage or irritation, phagocytes and plasma proteins, including complement proteins, leave the bloodstream, then defend and help repair the tissue.

Proceeds during both nonspecific and specific (immune) defense responses.

inheritance The transmission, from parents to offspring, of structural and functional patterns that have a genetic basis and are characteristic of each species.

inhibiting hormone A signaling molecule produced and secreted by the hypothalamus that controls secretions by the anterior lobe of the pituitary gland.

inhibitor A substance that can bind with an enzyme and interfere with its functioning.

inhibitory postsynaptic potential, or IPSP. Of neurons, one of two competing types of graded potentials at an input zone; tends to drive the resting membrane potential away from threshold.

inner cell mass In early development, a clump of cells in the blastocyst that will give rise to the embryonic disk.

insertion The end of a muscle attached to the bone that moves most when the muscle contracts.

inspiration The drawing of air into the lungs; inhaling.

integration, neural [L. *integrare*, to coordinate] Moment-by-moment summation of all excitatory and inhibitory synapses acting on a neuron; occurs at each level of synapsing in a nervous system.

integrator Of homeostatic systems, a control point where different bits of information are pulled together in the selection of a response. The brain is an example.

integument A protective body covering such as skin.

interferon Protein produced by T cells and that interferes with viral replication. Some interferons also stimulate the tumor-killing activity of macrophages.

interleukin One of a variety of communication signals, secreted by macrophages and by helper T cells, that drive immune responses.

intermediate filament A cytoskeletal component that consists of different proteins in different types of animal cells.

internal respiration Movement of oxygen into tissues from the blood, and of carbon dioxide from tissues into the blood.

interneuron Any of the neurons in the brain and spinal cord that integrate information arriving from sensory neurons and that influence other neurons in turn.

interphase Of cell cycles, the time interval between nuclear divisions in which a cell increases its mass, roughly doubles the number of its cytoplasmic components, and finally duplicates its chromosomes (replicates its DNA).

interstitial fluid (in-ter-STISH-ul) [L. *interstitus*, to stand in the middle of something] That portion of the extracellular fluid occupying spaces between cells and tissues.

intervertebral disk One of a number of disk-shaped structures containing cartilage that serve as shock absorbers and flex points between bony segments of the vertebral column.

intron A noncoding portion of a newly formed mRNA molecule.

inversion A change in a chromosome's structure after a segment separated from it was then inserted at the same place, but in reverse. The reversal alters the position and order of the chromosome's genes.

ion (EYE-on) An atom or a compound that has gained or lost one or more electrons and hence has acquired an overall negative or positive charge.

ionic bond An association between ions of opposite charge.

iris Of the eye, a circular pigmented region behind the cornea with a "hole" in its center (the pupil) through which incoming light enters.

isotonic Equality in the relative concentrations of solutes in two fluids; for two fluids separated by a cell membrane, there is no net osmotic (water) movement across the membrane.

isotope (EYE-so-tope) For a given element, an atom with the same number of protons as the other atoms but with a different number of neutrons.

J-shaped curve A curve, obtained when population size is plotted against time, that is characteristic of unrestricted, exponential growth.

joint An area of contact or near-contact between bones.

juxtaglomerular apparatus In kidney nephrons, a region of contact between the arterioles of the glomerulus and the distal tubule. Cells in this region secrete renin, which triggers hormonal events that stimulate increased reabsorption of sodium.

karyotype (CARRY-oh-type) Of eukaryotic individuals (or species), the number of metaphase chromosomes in somatic cells and their defining characteristics.

keratin A tough, water-insoluble protein manufactured by most epidermal cells.

keratinization (care-at-in-iz-AY-shun) Process by which keratin-producing epidermal cells of skin die and collect at the skin surface as keratinized "bags" that form a barrier against dehydration, bacteria, and many toxic substances.

kidney One of a pair of organs that filter mineral ions, organic wastes, and other substances from the blood and help regulate the volume and solute concentrations of extracellular fluid.

kilocalorie 1,000 calories of heat energy, or the amount of energy needed to raise the temperature of 1 kilogram of water by 1°C; the unit of measure for the caloric value of foods.

kinetochore A specialized group of proteins and DNA at the centromere of a chromosome that serves as an attachment point for several spindle microtubules during mitosis or meiosis. Each chromatid of a duplicated chromosome has its own kinetochore.

Krebs cycle Together with a few conversion steps that precede it, the stage of aerobic respiration in which pyruvate is completely broken down to carbon dioxide and water. Coenzymes accept the unbound protons (H^+) and electrons stripped from intermediates during the reactions and deliver them to the next stage.

lactate fermentation Anaerobic pathway of ATP formation in which pyruvate from glycolysis is converted to the three-carbon compound lactate, and NAD^+ (a coenzyme used in the reactions) is regenerated. Its net yield is two ATP.

lactation The production of milk by hormone-primed mammary glands.

lacteal Small lymph vessel in villi of the small intestine that receives absorbed triglycerides. Triglycerides move from the lymphatic system to the general circulation.

large intestine The colon; a region of the GI tract that receives unabsorbed food residues from the small intestine and concentrates and stores feces until they are expelled from the body.

larynx (LARE-inks) A tubular airway that leads to the lungs. It contains vocal cords, where sound waves used in speech are produced.

latency Of viruses, a period of time during which viral genes remain inactive inside the host cell.

lens Of the eye, a saucer-shaped region behind the iris containing multiple layers of transparent proteins. Ligaments can move the lens, which functions to focus incoming light onto photoreceptors in the retina.

Leydig cell In testes, cells in connective tissue around the seminiferous tubules that secrete testosterone and other signaling molecules.

LH Luteinizing hormone, secreted by the anterior lobe of the pituitary gland. In males it acts on interstitial cells of the testes and prompts them to secrete testosterone. In females, LH stimulates follicle development in the ovaries.

ligament A strap of dense, regular connective tissue that connects two bones at a joint.

limbic system Brain regions that, along with the cerebral cortex, collectively govern emotions.

limiting factor Any essential resource that is in short supply and so limits population growth.

lineage (LIN-ee-age) A line of descent.

linkage The tendency of genes located on the same chromosome to end up in the same gamete. For any two of those genes, the probability that crossing over will disrupt the linkage is proportional to the distance separating them.

lipid A greasy or oily compound of mostly carbon and hydrogen that shows little tendency to dissolve in water, but that dissolves in nonpolar solvents (such as ether). Cells use lipids as energy stores and structural materials, especially in membranes.

lipid bilayer The structural basis of cell membranes, consisting of two layers of mostly phospholipid molecules. Hydrophilic heads force all fatty acid tails of the lipids to become sandwiched between the hydrophilic heads.

liver Glandular organ with roles in storing and interconverting carbohydrates, lipids, and proteins absorbed from the gut; maintaining blood; disposing of nitrogen-containing wastes; and other tasks.

local signaling molecule Cellular secretion that alters chemical conditions in the immediate vicinity where it is secreted, then is swiftly degraded.

locus (LOW-cuss) The specific location of a particular gene on a chromosome.

logistic growth (low-JIS-tik) Pattern of population growth in which a low-density population slowly increases in size, goes through a rapid growth phase, then levels off once the carrying capacity is reached.

loop of Henle The hairpin-shaped, tubular region of a nephron that functions in reabsorption of water and solutes.

lung Saclike organ that serves as an internal respiratory surface.

lymph (limf) [L. *lympha*, water] Tissue fluid that has moved into the vessels of the lymphatic system.

lymph capillary A small-diameter vessel of the lymph vascular system that has no obvious entrance; tissue fluid moves inward by passing between overlapping endothelial cells at the vessel's tip.

lymph node A lymphoid organ that serves as a battleground of the immune system; each lymph node is packed with organized arrays of macrophages and lymphocytes that cleanse lymph of pathogens before it reaches the blood.

lymph vascular system [L. *lympha*, water, and *vasculum*, a small vessel] The vessels of the lymphatic system, which take up and transport excess tissue fluid and reclaimable solutes as well as fats absorbed from the digestive tract.

lymphatic system An organ system that supplements the circulatory system. Its vessels take up fluid and solutes from interstitial fluid and deliver them to the bloodstream; its lymphoid organs have roles in immunity.

lymphocyte Any of various white blood cells that take part in nonspecific and specific (immune) defense responses.

lymphoid organ The lymph nodes, spleen, thymus, tonsils, adenoids, and other organs with roles in immunity.

lysis [Gk. *lysis*, a loosening] Gross structural disruption of a plasma membrane that leads to cell death.

lysosome (LYE-so-sohm) The main organelle of digestion, with enzymes that can break down polysaccharides, proteins, nucleic acids, and some lipids.

lysozyme Infection-fighting enzyme that attacks and destroys various types of bacteria by digesting the bacterial cell wall. Present in mucous membranes that line the body's surfaces.

macroevolution The large-scale patterns, trends, and rates of change among groups of species.

macrophage One of the phagocytic white blood cells. It engulfs anything detected as foreign. Some also become antigen-presenting cells that serve as the trigger for immune responses by T and B lymphocytes. Compare antigen-presenting cell.

malnutrition A state in which body functions or development suffers due to inadequate or unbalanced food intake.

mammal A type of vertebrate; the only animal having offspring that are nourished by milk produced by mammary glands of females.

mandible The lower jaw, the largest facial bone.

mass extinction An abrupt rise in extinction rates above the background level; a catastrophic, global event in which major taxa are wiped out simultaneously.

mass number The total number of protons and neutrons in an atom's nucleus. The relative masses of atoms are also called atomic weights.

maternal chromosome One of the chromosomes bearing the alleles that are inherited from a female parent.

mechanoreceptor Sensory cell or cell part that detects mechanical energy associated with changes in pressure, position, or acceleration.

medulla oblongata Part of the brainstem with reflex centers for respiration, blood circulation, and other vital functions.

meiosis (my-OH-sis) [Gk. *meioun*, to diminish] Two-stage nuclear division process in which the chromosome number of a germ cell is reduced by half, to the haploid number. (Each daughter nucleus ends up with one of each type of chromosome.) Meiosis is the basis of gamete formation.

memory The storage and retrieval of information about previous experiences; underlies the capacity for learning.

memory cell Any of the various B or T lymphocytes of the immune system that are formed in response to invasion by a foreign agent and that circulate for some period, available to mount a rapid attack if the same type of invader reappears.

meninges Membranes of connective tissue that are layered between the skull bones and the brain and cover and protect the neurons and blood vessels that service the brain tissue.

menopause (MEN-uh-pozz) [L. *mensis*, month, and *pausa*, stop] End of the period of a human female's reproductive potential.

menstrual cycle The cyclic release of oocytes and priming of the endometrium (lining of the uterus) to receive a fertilized egg; the complete cycle averages about 28 days in female humans.

menstruation Periodic sloughing of the blood-enriched lining of the uterus when pregnancy does not occur.

mesoderm (MEH-so-derm) [Gk. *mesos*, middle, and *derm*, skin] In an embryo, a primary tissue layer (germ layer) between ectoderm and endoderm. Gives rise to muscle; organs of circulation, reproduction, and excretion; most of the internal skeleton (when present); and connective tissue layers of the gastrointestinal tract and body covering.

messenger RNA A linear sequence of ribonucleotides transcribed from DNA and translated into a polypeptide chain; the only type of RNA that carries protein-building instructions.

metabolic pathway One of many orderly sequences of enzyme-mediated reactions by which cells normally maintain, increase, or decrease the concentrations of substances. Different pathways are linear or circular, and often they interconnect.

metabolism (meh-TAB-oh-lizm) [Gk. *meta*, change] All controlled, enzyme-mediated chemical reactions by which cells acquire and use energy. Through these reactions, cells synthesize, store, break apart, and eliminate substances in ways that contribute to growth, survival, and reproduction.

metaphase Of mitosis or meiosis II, the stage when each duplicated chromosome has become positioned at the midpoint of the microtubular spindle, with its two sister chromatids attached to microtubules from opposite spindle poles. Of meiosis I, the stage when all pairs of homologous chromosomes are positioned at the spindle's midpoint, with the two members of each pair attached to opposite spindle poles.

metastasis The process in which cancer cells break away from a primary tumor and migrate (via blood or lymphatic tissues) to other locations, where they establish new cancer sites.

MHC marker Any of a variety of proteins that are self-markers. Some occur on all body cells of an individual; others are unique to the macrophages and lymphocytes.

microevolution Changes in allele frequencies brought about by mutation, genetic drift, gene flow, and natural selection.

microfilament [Gk. *mikros*, small, and L. *filum*, thread] One of a variety of cytoskeletal components. Actin and myosin filaments are examples.

microtubular spindle A bipolar structure composed of organized arrays of microtubules that forms during nuclear division and that moves the chromosomes.

microtubule Hollow cylinder of mainly tubulin subunits; a cytoskeletal element with roles in cell shape, motion, and growth and in the structure of cilia and flagella.

microvillus (my-crow-VILL-us) [L. *villus*, shaggy hair] A slender, cylindrical extension of the animal cell surface that functions in absorption or secretion.

midbrain A brain region that evolved as a coordination center for reflex responses to visual and auditory input; together with the pons and medulla oblongata, part of the brainstem, which includes the reticular formation.

mineral An inorganic substance required for the normal functioning of body cells.

mineralocorticoid Type of hormone, secreted by the adrenal cortex, that mainly regulates the concentrations of mineral salts in extracellular fluid.

mitochondrion (my-toe-KON-dree-on), plural mitochondria. Organelle that specializes in ATP formation; it is the site of the second and third stages of aerobic respiration, an oxygen-requiring pathway.

mitosis (my-TOE-sis) [Gk. *mitos*, thread] Type of nuclear division that maintains the parental chromosome number for daughter cells. It is the basis of bodily growth and, in many eukaryotic species, asexual reproduction.

molecule A unit of matter in which chemical bonding holds together two or more atoms of the same or different elements.

monoclonal antibody Antibody produced in the laboratory by a population of genetically identical cells that are clones of a single "parent" antibody-producing cell.

monohybrid cross [Gk. *monos*, alone] An experimental cross in which offspring inherit a pair of nonidentical alleles for a single trait being studied, so that they are heterozygous.

monomer A small molecule that is commonly a subunit of polymers, such as the sugar monomers of starch.

monosaccharide (mon-oh-SAK-ah-ride) [Gk. *monos*, alone, single, and *sakharon*, sugar] The simplest carbohydrate, with only one sugar unit. Glucose is an example.

monosomy Abnormal condition in which one chromosome of diploid cells has no homologue.

morphogenesis (more-foe-JEN-ih-sis) [Gk. *morphe*, form, and *genesis*, origin] Processes by which differentiated cells in an embryo become organized into tissues and organs, under genetic controls and environmental influences.

morphological convergence Lineages only remotely related; evolved in response to similar environmental pressures, they become similar in appearance, functions, or both. Analogous structures are evidence of this evolutionary pattern.

motor neuron A type of neuron; it delivers signals from the brain and spinal cord that can stimulate or inhibit the body's effectors (muscles, glands, or both).

motor unit A motor neuron and the muscle cells under its control.

mouth The oral cavity; in digestion, the site where polysaccharide breakdown begins.

multiple allele system Three or more different molecular forms of the same gene (alleles) that exist in a population.

muscle fatigue A decline in tension of a muscle that has been kept in a state of tetanic contraction as a result of continuous, high-frequency stimulation.

muscle tension A mechanical force, exerted by a contracting muscle, that resists opposing forces such as gravity and the weight of objects being lifted.

muscle tissue Tissue having cells able to contract in response to stimulation, then passively lengthen and so return to their resting stage.

muscle twitch Muscle response in which the muscle contracts briefly, then relaxes, when a brief stimulus activates a motor unit.

mutagen (MEW-tuh-jen) An environmental agent that can permanently modify the structure of a DNA molecule. Certain viruses and ultraviolet radiation are examples.

mutation [L. *mutatus*, a change] A heritable change in DNA due to the deletion, addition, or substitution of one to several bases in the nucleotide sequence.

myelin sheath Of many sensory and motor neurons, an axonal sheath that affects how fast action potentials travel; formed from the plasma membranes of Schwann cells that are wrapped repeatedly around the axon and are separated from each other by a small node.

myocardium The cardiac muscle tissue.

myofibril (MY-oh-fy-brill) One of many threadlike structures inside a muscle cell; each is functionally divided into sarcomeres, the basic units of contraction.

myosin (MY-uh-sin) A type of protein with a head and long tail. In muscle cells, it interacts with actin, another protein, to bring about contraction.

NAD$^+$ Nicotinamide adenine dinucleotide; a nucleotide coenzyme. When carrying electrons and unbound protons (H$^+$) between reaction sites, it is abbreviated NADH.

NADP Nicotinamide adenine dinucleotide phosphate; a phosphorylated nucleotide coenzyme. When carrying electrons and unbound protons (H$^+$) between reaction sites, it is abbreviated NADPH$_2$.

nasal cavity The region of the respiratory system where air is warmed, moistened, and filtered of airborne particles and dust.

natural killer cell Cell of the immune system, possibly a type of lymphocyte, that kills tumor cells (by lysis) or infected cells identified as abnormal.

natural selection A microevolutionary process; a difference in survival and reproduction among members of a population that vary in one or more traits.

negative feedback mechanism A homeostatic feedback mechanism in which an activity changes some condition in the internal environment and so triggers a response that reverses the changed condition.

nephron (NEFF-ron) [Gk. *nephros*, kidney] Of the vertebrate kidney, a slender tubule in which water and solutes filtered from blood are selectively reabsorbed and in which urine forms.

nerve Cordlike communication line of the peripheral nervous system, composed of axons of sensory neurons, motor neurons, or both packed within connective tissue. In the brain and spinal cord, similar cordlike bundles are called nerve pathways or tracts.

nerve tract A bundle of myelinated axons of interneurons inside the spinal cord and brain.

nervous system System of neurons oriented relative to one another in precise message-conducting and information-processing pathways.

nerve tissue A type of tissue composed of neurons.

net energy Of energy resources available to the human population, the amount of energy that is left over after subtracting the energy used to locate, extract, transport, store, and deliver energy to consumers.

neuroendocrine control center The portions of the hypothalamus and pituitary gland that interact to control many body functions.

neuroglial cell (nur-OH-glee-uhl) One of the cells that provide structural and metabolic support for neurons and that collectively represent about half the volume of the nervous system.

neuromodulator Type of signaling molecule that influences the effects of transmitter substances by enhancing or reducing membrane responses in target neurons.

neuromuscular junction Chemical synapses between axon terminals of a motor neuron and a muscle cell.

neuron A nerve cell; the basic unit of communication in nervous systems. Neurons collectively sense environmental change, integrate sensory inputs, then activate muscles or glands that initiate or carry out responses.

neurotransmitter Any of the class of signaling molecules that are secreted from neurons, act on immediately adjacent cells, and are then rapidly degraded or recycled.

neural mutation A gene mutation that has neither harmful nor helpful effects on the individual's ability to survive and reproduce.

neutron Unit of matter, one or more of which occupies the atomic nucleus, that has mass but no electric charge.

neutrophil Phagocytic white blood cell that takes part in inflammatory responses against bacteria.

niche (nitch) [L. *nidas*, nest] Of a species, the full range of physical and biological conditions under which its members can live and reproduce.

nitrification (nye-trih-fih-KAY-shun) A chemosynthetic process in which certain bacteria strip electrons from ammonia or ammonium present in soil. The end product, nitrite (NO_2^-), is broken down to nitrate (NO_3^-) by different bacteria.

nitrogen cycle Biogeochemical cycle in which the atmosphere is the largest reservoir of nitrogen.

nitrogen fixation Process by which a few kinds of bacteria convert gaseous nitrogen (N_2) to ammonia. This dissolves rapidly in their cytoplasm to form ammonium, which can be used in biosynthetic pathways.

NK cell Natural killer cell, possibly of the lymphocyte lineage, that reconnoiters and kills tumor cells and infected body cells.

nociceptor A receptor, such as a free nerve ending, that detects any stimulus causing tissue damage.

nondisjunction Failure of one or more chromosomes to separate properly during mitosis or meiosis.

nongonococcal urethritis (NGU) An inflammation of the urethra; often caused by infection by the bacterium that causes chlamydia and considered a sexually transmitted disease.

nonsteroid hormone A type of water-soluble hormone, such as a protein hormone, that cannot cross the lipid bilayer of a target cell. These hormones enter the cell by receptor-mediated endocytosis, or they bind to receptors that activate membrane proteins or second messengers within the cell.

nuclear envelope A double membrane (two lipid bilayers and associated proteins) that is the outermost portion of a cell nucleus.

nucleic acid (new-CLAY-ik) A long, single- or double-stranded chain of four different kinds of nucleotides joined one after the other at their phosphate groups. They differ in which nucleotide base follows the next in sequence. DNA and RNA are examples.

nucleic acid hybridization The base-pairing of nucleotide sequences from different sources, as used in genetics, genetic engineering, and studies of evolutionary relationship based on similarities and differences in the DNA or RNA of different species.

nucleoid Of bacteria, a region in which DNA is physically organized apart from other cytoplasmic components.

nucleolus (new-KLEE-oh-lus) [L. *nucleolus*, a little kernel] Within the nucleus of a nondividing cell, a site where the protein and RNA subunits of ribosomes are assembled.

nucleosome (new-KLEE-oh-sohm) Of chromosomes, one of many organizational units, each consisting of a small stretch of DNA looped twice around a "spool" of histone molecules, which another histone molecule stabilizes.

nucleotide (new-KLEE-oh-tide) A small organic compound having a five-carbon sugar (deoxyribose), nitrogen-containing base, and phosphate group. Nucleotides are the structural units of adenosine phosphates, nucleotide coenzymes, and nucleic acids.

nucleotide coenzyme A protein that transports hydrogen atoms (free protons) and electrons from one reaction site to another in cells.

nucleus (NEW-klee-us) [L. *nucleus*, a kernel] Of atoms, the central core of one or more positively charged protons and (in all but hydrogen) electrically neutral neutrons. In cells, a membranous organelle that physically isolates and organizes the DNA, out of the way of cytoplasmic machinery.

nutrition All those processes by which food is selectively ingested, digested, absorbed, and later converted to the body's own organic compounds.

obesity An excess of fat in the body's adipose tissues, caused by imbalances between caloric intake and energy output.

oligosaccharide A carbohydrate consisting of a short chain of two or more covalently bonded sugar units. One subclass, disaccharides, has two sugar units. Compare monosaccharide; polysaccharide.

omnivore [L. *omnis*, all, and *vovare*, to devour] An organism able to obtain energy from more than one source rather than being limited to one trophic level.

oncogene (ON-coe-jeen) Any gene having the potential to induce cancerous transformations in a cell.

oocyte An immature egg.

oogenesis (oo-oh-JEN-uh-sis) Formation of a female gamete, from a germ cell to a mature haploid ovum (egg).

orbital Volume of space around the nucleus of an atom in which electrons are likely to be at any instant.

organ A structure of definite form and function that is composed of more than one tissue.

organ of Corti Membrane region of the inner ear that contains the sensory hair cells involved in hearing.

organ system Two or more organs that interact chemically, physically, or both in performing a common task.

organelle Of cells, an internal, membrane-bounded sac or compartment that has a specific, specialized metabolic function.

organic compound In biology, a compound assembled in cells and having a carbon backbone, often with carbon atoms arranged as a chain or ring structure.

organogenesis Stage of development in which primary tissue layers (germ layers) split into subpopulations of cells, and different lines of cells become unique in structure and function; foundation for growth and tissue specialization, when organs acquire specialized chemical and physical properties.

origin The end of a muscle that is attached to the bone that remains relatively stationary when the muscle contracts.

osmosis (oss-MOE-sis) [Gk. *osmos*, act of pushing] Of cells, the tendency of water to move through channel proteins that span a membrane in response to a concentration gradient, fluid pressure, or both. Hydrogen bonds among water molecules prevent water itself from becoming more or less concentrated; but a gradient may exist when the water on either side of the membrane has more substances dissolved in it.

osteocyte A living bone cell.

osteon A set of thin, concentric layers of compact bone tissue surrounding a narrow canal carrying blood vessels and nerves; arrays of osteons make up compact bone.

ovary (OH-vuh-ree) In females, the primary reproductive organ in which eggs form.

oviduct (OH-vih-dukt) Duct through which eggs travel from the ovary to the uterus. Formerly called Fallopian tube.

ovulation (ahv-you-LAY-shun) During each turn of the menstrual cycle, the release of a secondary oocyte (immature egg) from an ovary.

ovum (OH-vum) A mature female gamete (egg).

oxidation-reduction reaction An electron transfer from one atom or molecule to another. Often hydrogen is transferred along with the electron or electrons.

oxidative phosphorylation (foss-for-ih-LAY-shun) Final stage of aerobic respiration, in which ATP forms after hydrogen ions and electrons (from the Krebs cycle) are sent through a transport system that gives up the electrons to oxygen.

oxyhemoglobin A hemoglobin molecule that has oxygen bound to it; HbO_2.

ozone hole A pronounced seasonal thinning of the ozone layer in the lower stratosphere above Antarctica.

palate Structure that separates the nasal cavity from the oral cavity. The bone-reinforced hard palate serves as a hard surface against which the tongue can press food as it mixes it with saliva.

pancreas (PAN-cree-us) Gland that secretes enzymes and bicarbonate into the small intestine during digestion, and

that also secretes the hormones insulin and glucagon.

pancreatic islets Any of the two million clusters of endocrine cells in the pancreas, including alpha cells, beta cells, and delta cells.

parasite [Gk. *para*, alongside, and *sitos*, food] An organism that obtains nutrients directly from the tissues of a living host, which it lives on or in and may or may not kill.

parasympathetic nerve Of the autonomic nervous system, any of the nerves carrying signals that tend to slow the body down overall and divert energy to basic tasks; also work continually in opposition with sympathetic nerves to bring about minor adjustments in internal organs.

parathyroid glands (pare-uh-THY-royd) Endocrine glands embedded in the thyroid gland that secrete parathyroid hormone, which helps restore blood calcium levels.

parturition Birth.

passive immunity Temporary immunity conferred by deliberately introducing antibodies into the body.

passive transport Diffusion of a solute through a channel or carrier protein that spans the lipid bilayer of a cell membrane. Its passage does not require an energy input; the protein passively allows the solute to follow its concentration gradient.

paternal chromosome One of the chromosomes bearing alleles that are inherited from a male parent.

pathogen (PATH-oh-jen) [Gk. *pathos*, suffering, and *-genēs*, origin] An infectious, disease-causing agent, such as a virus or bacterium.

pattern formation Mechanisms responsible for specialization and positioning of tissues during embryonic development.

pectoral girdle Set of bones, including the scapula (shoulder blade) and clavicle (collarbone), to which the long bone of each arm attaches. The pectoral girdles form the upper part of the appendicular skeleton and are only loosely attached to the rest of the body by muscles.

pedigree A chart of genetic connections among individuals, as constructed according to standardized methods.

pelvic girdle Set of bones including coxal bones that form an open basin, the pelvis; the lower part of the appendicular skeleton. The upper portions of the two coxal bones are the "hipbones"; the thighbones (femurs) join the coxal bones at hip joints. The pelvic girdle bears the body's weight when a person stands and is much more massive than the pectoral girdle.

pelvic inflammatory disease (PID) Generally, a bacterially caused inflammation of the uterus, oviducts, and ovaries. Often a complication of gonorrhea, chlamydia, or some other sexually transmitted disease.

penetrance In a given population, the percentage of individuals in which a particular genotype is expressed (that is, the percentage of individuals who have the genotype and also exhibit the corresponding phenotype).

penis Male organ that deposits sperm into the female reproductive tract.

pepsin Any of several digestive enzymes that are part of gastric fluid in the stomach.

peptide hormone A hormone that consists of a chain of specific amino acids.

perception The conscious interpretation of some aspect of the external world created by the brain from nerve impulses generated by sensory receptors.

perforin A type of protein secreted by a natural killer cell of the immune system, and which creates holes (pores) in the plasma membrane of a target cell.

peripheral nervous system (per-IF-ur-uhl) [Gk. *peripherein*, to carry around] The nerves leading into and out from the spinal cord and brain and the ganglia along those communication lines.

peristalsis (pare-ih-STAL-sis) A rhythmic contraction of muscles that moves food forward through the gastrointestinal tract.

peritoneum A lining of the coelom that also covers and helps maintain the position of internal organs.

peritubular capillaries The set of blood capillaries that threads around the tubular parts of a nephron; they function in reabsorption of water and solutes back into the body and in secretion of hydrogen ions and some other substances in the forming urine.

peroxisome Enzyme-filled vesicle in which fatty acids and amino acids are digested first into hydrogen peroxide (which is toxic), then to harmless products.

PGA Phosphoglycerate (foss-foe-GLISS-er-ate). A key intermediate in glycolysis.

PGAL Phosphoglyceraldehyde. A key intermediate in glycolysis.

pH scale A scale used to measure the concentration of free hydrogen ions in blood, water, and other solutions; pH 0 is the most acidic, 14 the most basic, and 7, neutral.

phagocyte (FAYG-uh-sight) [Gk. *phagein*, to eat, and -*kytos*, hollow vessel] A macrophage or certain other white blood cells that engulf and destroy foreign agents.

phagocytosis (fayg-uh-sigh-TOE-sis) [Gk. *phagein*, to eat, and -*kytos*, hollow vessel] Engulfment of foreign cells or substances by amoebas and some white blood cells by means of endocytosis.

pharynx (FARE-inks) A muscular tube by which food enters the gastrointestinal tract; the dual entrance for the tubular part of the digestive tract and windpipe (trachea).

phenotype (FEE-no-type) [Gk. *phainein*, to show, and -*typos*, image] Observable trait or traits of an individual; arises from interactions between genes, and between genes and the environment.

pheromone (FARE-oh-moan) [Gk. *phero*, to carry, and -*mone*, as in hormone] A type of signaling molecule secreted by exocrine glands that serves as a communication signal between individuals of the same species.

phospholipid A type of lipid that is the main structural component of cell membranes. Each has a hydrophobic tail (of two fatty acids) and a hydro-philic head that incorporates glycerol and a phosphate group.

phosphorus cycle Movement of phosphorus from rock or soil through organisms, then back to soil.

phosphorylation (foss-for-ih-LAY-shun) The attachment of unbound (inorganic) phosphate to a molecule; also the transfer of a phosphate group from one molecule to another, as when ATP phosphorylates glucose.

photochemical smog A brown-air smog that develops over large cities when the surrounding land forms a natural basin.

photoreceptor A light-sensitive sensory cell.

phylogeny Evolutionary relationships among species, starting with most ancestral forms and including the branches leading to their descendants.

pigment A light-absorbing molecule.

pilomotor response Contraction of smooth muscle controlling the erection of body hair when outside temperature drops. This creates a layer of still air that reduces heat losses from the body. (It is most effective in mammals that have more body hair than humans do.)

pineal gland (py-NEEL) A light-sensitive endocrine gland that secretes melatonin, a hormone that influences reproductive cycles and the development of reproductive organs.

pioneer species Typically small plants with short life cycles that are adapted to growing in exposed, often windy areas with intense sunlight, wide swings in air temperature, and soils deficient in nitrogen and other nutrients. By improving conditions in areas they colonize, pioneers invite their own replacement by other species.

pituitary gland An endocrine gland that interacts with the hypothalamus to coordinate and control many physiological functions, including the activity of many other endocrine glands. Its posterior lobe stores and secretes hypothalamic hormones; the anterior lobe produces and secretes its own hormones.

placenta (pluh-SEN-tuh) Of the uterus, an organ composed of maternal tissues and extraembryonic membranes (the chorion especially); it delivers nutrients to the fetus and accepts wastes from it, yet allows the fetal circulatory system to develop separately from the mother's.

plasma (PLAZ-muh) Liquid component of blood; consists of water, various proteins, ions, sugars, dissolved gases, and other substances.

plasma cell Of the immune system, any of the antibody-secreting daughter cells of a rapidly dividing population of B cells.

plasma membrane Of cells, the outermost membrane. Its lipid bilayer structure and proteins carry out most functions, including transport across the membrane and reception of extracellular signals.

plasmid Of many bacteria, a small, circular molecule of extra DNA that carries only a few genes and replicates independently of the bacterial chromosome.

plasticity Of the human species, the ability to remain flexible and adapt to a wide range of environments.

plate tectonics Arrangement of the Earth's outer layer (lithosphere) in slablike plates, all in motion and floating on a hot, plastic layer of the underlying mantle.

platelet (PLAYT-let) Any of the cell fragments in blood that release substances necessary for clot formation.

pleiotropy (pleye-ah-troe-pee) [Gk. *pleon*, more, and *trope*, direction] A type of gene interaction in which a single gene exerts multiple effects on seemingly unrelated aspects of an individual's phenotype.

pleura Thin, double membrane surrounding each lung.

polar body Any of three cells that form during the meiotic cell division of an oocyte; the division also forms the mature egg, or ovum.

pollutant Any substance with which an ecosystem has had no prior evolutionary experience in terms of kinds or amounts, and that can accumulate to disruptive or harmful levels. Can be naturally occurring or synthetic.

polymer (PAH-lih-mur) [Gk. *polus*, many, and *meris*, part] A molecule composed of three to millions of small subunits that may or may not be identical.

polymerase chain reaction (PCR). DNA amplification method; DNA containing a gene of interest is split into single strands, which enzymes (polymerases) copy; the enzymes also act on the accumulating copies, multiplying the gene sequence by the millions.

polymorphism (poly-MORE-fizz-um) [Gk. *polus*, many, and *morphe*, form] Of a population, the persistence through the generations of two or more forms of a trait, at a frequency greater than can be maintained by new mutations alone.

polypeptide chain Three or more amino acids joined by peptide bonds.

polyploidy (PAHL-ee-ployd-ee) A change in the chromosome number following inheritance of three or more of each type of chromosome.

polysaccharide [Gk. *polus*, many, and *sakharon*, sugar] A straight or branched chain of hundreds of thousands of covalently linked sugar units of the same or different kinds. The most common polysaccharides are cellulose, starch, and glycogen.

polysome Of protein synthesis, several ribosomes all translating the same messenger RNA molecule, one after the other.

population A group of individuals of the same species occupying a given area.

population density The number of individuals of a population that are living in a specified area or volume.

population distribution The general pattern of dispersion of individuals of a population throughout their habitat.

population size The number of individuals that make up the gene pool of a population.

positive feedback mechanism Homeostatic mechanism by which a chain of events is set in motion that intensifies a change from an original condition; after a limited time, the intensification reverses the change.

post-translational controls Of eukaryotes, controls that govern modification of newly formed polypeptide chains into functional enzymes and other proteins.

prediction A claim about what you can expect to observe in nature if a theory or hypothesis is correct.

primary immune response Actions by white blood cells and their products elicited by a first-time encounter with an antigen; includes both antibody-mediated and cell-mediated responses.

primary productivity Of ecosystems, *gross* primary productivity is the rate at which the producer organisms capture and store a given amount of energy during a specified interval. *Net* primary productivity is the rate of energy storage in the tissues of producers in excess of their rate of aerobic respiration.

primate The mammalian lineage that includes prosimians, tarsioids, and anthropoids (monkeys, apes, and humans).

probability With respect to any chance event, the most likely number of times it will turn out a certain way, divided by the total number of all possible outcomes.

producer, primary. Of ecosystems, any of the organisms that secure energy from the physical environment, as by photosynthesis or chemosynthesis.

progesterone (pro-JESS-tuh-rown) Female sex hormone secreted by the ovaries.

prokaryotic cell (pro-carry-OH-tic) [L. *pro*, before, and Gk. *karyon*, kernel] A bacterium; a single-celled organism that has no nucleus or any of the other membrane-bound organelles characteristic of eukaryotic cells.

promoter Of transcription, a base sequence that signals the start of a gene; the site where RNA polymerase initially binds.

prophase Of mitosis, the stage when each duplicated chromosome starts to condense, microtubules form a spindle apparatus, and the nuclear envelope starts to break up.

prophase I Of meiosis, the stage at which the microtubular spindle starts to form, the nuclear envelope starts to break up, and each duplicated chromosome also condenses and pairs with its homologous partner. At this time, their sister chromatids typically undergo crossing over and geneticize recombination.

prophase II Of meiosis, a brief stage after interkinesis during which each chromosome still consists of two chromatids.

prostaglandin Any of various lipids present in tissues throughout the body and that can act as local signaling molecules. Prostaglandins typically cause smooth muscle to contract or relax, as in blood vessels, the uterus, and respiratory airways.

prostate gland Gland in males that wraps around the urethra and ejaculatory ducts; its secretions become part of semen.

protein Large organic compound composed of one or more chains of amino acids held together by peptide bonds. Proteins have unique sequences of different kinds of amino acids in their polypeptide chains; such sequences are the basis of a protein's three-dimensional structure and chemical behavior.

proto-oncogene A gene sequence similar to an oncogene but that codes for a protein required in normal cell function; may trigger cancer, generally when specific mutations alter its structure or function.

proton Positively charged particle, one or more of which is present in the atomic nucleus.

proximal tubule Of a nephron, the tubular region that receives water and solutes filtered from the blood.

pulmonary circuit Blood circulation route leading to and from the lungs.

pulse Rhythmic pressure surge of blood flowing in an artery, created during each cardiac cycle when a ventricle contracts.

Punnett-square method A method to predict the possible outcome of a mating or an experimental cross in simple diagrammatic form.

purine Nucleotide base having a double ring structure. Adenine and guanine are examples.

pyrimidine (pie-RIM-ih-deen) Nucleotide base having a single ring structure. Cytosine and thymine are examples.

pyruvate (pie-ROO-vate) A compound with a backbone of three carbon atoms. Two pyruvate molecules are the end products of glycolysis.

r Designates net population growth rate; the birth and death rates are assumed to remain constant and so are combined into this one variable for population growth equations.

radioisotope An unstable atom that has dissimilar numbers of protons and neutrons and that spontaneously decays (emits electrons and energy) to a new, stable atom that is not radioactive.

radius One of two long bones of the forearm that extend from the humerus (at the elbow joint) to the wrist. The radius runs along the "thumb side" of the forearm, parallel to the ulna.

reabsorption In the kidney, the diffusion or active transport of water and usable solutes out of a nephron and into capillaries leading back to the general circulation; regulated by ADH and aldosterone.

receptor Of cells, a molecule at the surface of the plasma membrane or in the cytoplasm that binds molecules present in the extracellular environment. The binding triggers changes in cellular activities. Of nervous systems, a sensory cell or cell part that may be activated by a specific stimulus.

receptor protein Protein that binds a signaling molecule such as a hormone,

then triggers alterations in cell behavior or metabolism.

recessive (allele or trait) [L. *recedere,* to recede] In heterozygotes, an allele whose expression is fully or partially masked by expression of its partner; fully expressed only in the homozygous recessive condition.

recognition protein Protein at cell surface recognized by cells of like type; helps guide the ordering of cells into tissues during development and functions in cell-to-cell interactions.

recombinant DNA technology Procedures by which DNA (genes) from different species may be isolated, cut, spliced together, and the new recombinant molecules multiplied in quantity in a population of rapidly dividing cells such as bacteria.

rectum Final region of the gastrointestinal tract, which receives and temporarily stores undigested food residues (feces).

red blood cell Erythrocyte; an oxygen-transporting cell in blood.

red marrow A substance in the spongy tissue of many bones that serves as a major site of blood cell formation.

reductional division Mode of cell division represented by meiosis, in which daughter cells end up with one-half the normal diploid number of chromosomes.

reflex [L. *reflectere,* to bend back] A simple, stereotyped movement elicited directly by sensory stimulation.

reflex arc [L. *reflectere,* to bend back] Type of neural pathway in which signals from sensory neurons directly stimulate or inhibit motor neurons without intervention by interneurons.

refractory period Of neurons, the period following an action potential at a given patch of membrane when sodium gates are shut and potassium gates are open, so that the patch is insensitive to stimulation.

regulatory protein A protein that enhances or suppresses the rate at which a gene is transcribed.

releasing hormone A hypothalamic signaling molecule that stimulates or slows down secretion by target cells in the anterior lobe of the pituitary gland.

renal corpuscle In a kidney, the site where the nephron wall balloons around the glomerulus (a cluster of

capillaries). Compare glomerular capsule.

repressor protein Regulatory protein that provides negative control of gene activity by preventing RNA polymerase from binding to DNA.

reproduction In biology, processes by which a new generation of cells or multicelled individuals is produced. Sexual reproduction requires meiosis, formation of gametes, and fertilization. Asexual reproduction refers to the production of new individuals by any mode that does not involve gametes.

reproduction, sexual Mode of reproduction that begins with meiosis, proceeds through gamete formation, and ends at fertilization.

reproductive isolating mechanism Any aspect of structure, functioning, or behavior that restricts gene flow between two populations.

reproductive isolation An absence of gene flow between populations.

reproductive success The survival and production of the offspring of an individual.

respiration [L. *respirare,* to breathe] The overall exchange of oxygen from the environment for carbon dioxide wastes from cells by way of circulating blood. Compare aerobic respiration.

respiratory bronchiole Smallest airway in the respiratory system; opens onto alveoli.

respiratory surface In alveoli of the lungs, the thin, moist membrane across which gases diffuse.

respiratory system An organ system that functions in respiration.

resting membrane potential Of neurons and other excitable cells that are not being stimulated, the steady voltage difference across the plasma membrane.

restriction enzymes Class of bacterial enzymes that cut apart foreign DNA injected into them, as by viruses; also used in recombinant DNA technology.

restriction fragment A piece of DNA that has been spliced out of a chromosome by restriction enzymes.

reticular activating system A branch of the brain's reticular formation that controls the changing levels of consciousness by sending signals to the spinal cord, cerebellum, and cerebrum, as well as back to itself.

reticular formation Of the brainstem, a major network of interneurons that helps govern activity of the whole nervous system.

reverse transcriptase Viral enzyme required for reverse transcription of mRNA into DNA; used in recombinant DNA technology.

reverse transcription Assembly of DNA on a single-stranded mRNA molecule by viral enzymes.

RFLP Restriction fragment length polymorphism. Of DNA samples from different individuals, slight but unique differences in the banding pattern of fragments of the DNA that have been cut with restriction enzymes.

Rh blood typing A method of characterizing red blood cells on the basis of a protein that serves as a self-marker at their surface; Rh⁺ signifies its presence and Rh⁻, its absence.

rhodopsin Substance in rod cells of the eye consisting of the protein opsin and a side group, cis-retinal. When the side group absorbs incoming light energy, a series of chemical events follows that result in action potentials in associated neurons.

ribosomal RNA (rRNA) Type of RNA molecule that combines with proteins to form ribosomes, on which the polypeptide chains of proteins are assembled.

ribosome In all cells, the structure at which amino acids are strung together in specified sequence to form the polypeptide chains of proteins. An intact ribosome consists of two subunits, each composed of ribosomal RNA and protein molecules.

RNA Ribonucleic acid. A category of single-stranded nucleic acids that function in processes by which genetic instructions are used to build proteins.

rod cell Of the retina, a photoreceptor sensitive to very dim light and that contributes to coarse perception of movement.

S-shaped curve A curve, obtained when population size is plotted against time, that is characteristic of logistic growth.

salinization A salt buildup in soil as a result of evaporation, poor drainage, and often the importation of mineral salts in irrigation water.

salivary amylase Starch-degrading enzyme in saliva.

salivary gland Any of the glands that secrete saliva, a fluid that initially mixes with food in the mouth and starts the breakdown of starch.

salt An ionic compound formed when an acid reacts with a base.

saltatory conduction In myelinated neurons, rapid, node-to-node hopping of action potentials.

sarcomere (SAR-koe-meer) The basic unit of muscle contraction; a region of myosin and actin filaments organized in parallel between two Z lines of a myofibril inside a muscle cell.

sarcoplasmic reticulum (sar-koe-PLAZ-mik reh-TIK-you-lum) In muscle cells, a membrane system that takes up, stores, and releases the calcium ions required for cross-bridge formation in sarcomeres, hence for contraction.

scapula Flat, triangular bone on either side of the pectoral girdle; the scapulae form the shoulder blades.

Schwann cell A specialized neuroglial cell that grows around a neuron axon, forming a myelin sheath.

second messenger A molecule inside a cell that mediates and generally triggers amplified response to a hormone.

secondary immune response Rapid, prolonged response by white blood cells, memory cells especially, to a previously encountered antigen.

secondary oocyte An oocyte (unfertilized egg cell) that has completed meiosis I; it is this haploid cell that is released at ovulation.

secondary sexual trait Trait associated with maleness or femaleness, but not directly involved with reproduction. Beard growth in males and breast development in females are examples.

secretion Generally, the release of a substance for use by the organism producing it. (Not the same as excretion, the expulsion of excess or waste material.) Of kidneys, a regulated stage in urine formation in which ions and other substances move from capillaries into nephrons.

segregation, Mendelian principle of [L. *se-*, apart, and *grex*, herd] The principle that diploid organisms inherit a pair of genes for each trait (on a pair of homologous chromosomes) and that the two genes segregate during meiosis and end up in separate gametes.

selective gene expression Activation or suppression of a fraction of the genes in unique ways in different cells, leading to pronounced differences in structure and function among different cell lineages.

semen [L. *serere*, to sow] Sperm-bearing fluid expelled from the penis during male orgasm.

semicircular canals Fluid-filled canals positioned at different angles within the vestibular apparatus of the inner ear and that contain sensory receptors that detect head movements, deceleration, and acceleration.

semiconservative replication [Gk. *hēmi*, half, and L. *conservare*, to keep] Reproduction of a DNA molecule when a complementary strand forms on each of the unzipping strands of an existing DNA double helix, the outcome being two "half-old, half-new" molecules.

semilunar valve In each half of the heart; during each heartbeat, it opens and closes in ways that keep blood flowing in one direction from the ventricle to the arteries leading away from it.

seminiferous tubules Coiled tubes inside the testes where sperm develop.

senescence (sen-ESS-cents) [L. *senescere*, to grow old] Sum total of processes leading to the natural death of an organism or some of its parts.

sensation The conscious awareness of a stimulus.

sensory neuron Any of the nerve cells that act as sensory receptors, detecting specific stimuli (such as light energy) and relaying signals to the brain and spinal cord.

sensory receptor Of the nervous system, a sensory cell or specialized cell adjacent to it that can detect a particular stimulus.

sensory system Element of the nervous system consisting of sensory receptors (such as photoreceptors), nerve pathways from the receptors to the brain, and brain regions that process sensory information.

septum Of the heart, a thick wall that divides the heart into right and left halves.

Sertoli cell Any of the cells in seminiferous tubules that nourish and otherwise aid the development of sperm.

sex chromosome A chromosome whose presence determines a new individual's gender. Compare autosomes.

shifting cultivation The cutting and burning of trees, followed by tilling of ashes into the soil; once called "slash-and-burn agriculture."

sinoatrial node Region of conducting cells in the upper wall of the right atrium that generate periodic waves of excitation that stimulate the atria to contract.

sinus In the skull, an air-filled space lined with mucous membrane and that functions to lighten the skull.

sister chromatid Of a duplicated chromosome, one of two DNA molecules (and associated proteins) that remain attached at their centromere only during nuclear division. Each ends up in a separate daughter nucleus.

skeletal muscle Type of muscle that interacts with the skeleton to bring about body movements. A skeletal muscle typically consists of bundles of many long cylindrical cells encapsulated by connective tissue.

skull Bony structure that includes more than two dozen bones, including bones of the brain case and facial bones.

sliding filament model Model of muscle contraction in which myosin filaments physically slide along and pull two sets of actin filaments toward the center of the sarcomere, which shortens. The sliding requires ATP energy and cross-bridge formation between the actin and myosin.

small intestine The portion of the digestive system where digestion is completed and most nutrients absorbed.

smog, industrial Gray-colored air pollution that predominates in industrialized cities with cold, wet winters.

smog, photochemical Form of brown, smelly air pollution occurring in large cities with warm climates.

sodium-potassium pump A transport protein spanning the lipid bilayer of the plasma membrane. When activated by ATP, its shape changes and it selectively transports sodium ions out of the cell and potassium ions in.

solute (SOL-yoot) [L. *solvere*, to loosen] Any substance dissolved in a solution. In water, this means spheres of hydration surround the charged parts of individual ions or molecules and keep them dispersed.

solvent Fluid in which one or more substances is dissolved.

somatic cell (so-MAT-ik) [Gk. *somā*, body] Any body cell that is not a germ cell (which gives rise to gametes).

somatic nervous system Those nerves leading from the central nervous system to skeletal muscles.

somatic sensation Awareness of touch, pressure, heat, cold, pain, and limb movement.

speciation (spee-cee-AY-shun) The evolutionary process by which species originate. One speciation route starts with divergence of two reproductively isolated populations of a species. They become separate species when accumulated differences in allele frequencies prevent them from interbreeding successfully under natural conditions.

species (SPEE-ceez) [L. *species*, a kind] Of sexually reproducing organisms, a unit consisting of one or more populations of individuals that can interbreed under natural conditions to produce fertile offspring that are reproductively isolated from other such units.

sperm [Gk. *sperma*, seed] A type of mature male gamete.

spermatogenesis (sperm-at-oh-JEN-ih-sis) Formation of a mature sperm from a germ cell.

sphere of hydration Through positive or negative interactions, a clustering of water molecules around the individual molecules of a substance placed in water. Compare solute.

sphincter (SFINK-tur) Ring of muscle between regions of a tubelike system (as between the stomach and small intestine).

spinal cord Of the central nervous system, the portion threading through a canal inside the vertebral column and providing direct reflex connections between sensory and motor neurons as well as communication lines to and from the brain.

spindle apparatus A type of bipolar structure that forms during mitosis or meiosis and that moves the chromosomes. It consists of two sets of microtubules that extend from the opposite poles and that overlap at the spindle's equator.

spleen One of the lymphoid organs; it is a filtering station for blood, a reservoir of red blood cells, and a reservoir of macrophages.

spongy bone Type of bone tissue in which hard, needlelike struts separate large spaces filled with marrow. Spongy bone occurs at the ends of long bones and within the breastbone, pelvis, and bones of the skull.

stabilizing selection Of a population, a persistence over time of the alleles responsible for the most common phenotypes.

start codon Of protein synthesis, a base triplet in a strand of mRNA that serves as the start signal for mRNA translation.

stem cell Unspecialized cell that can give rise to descendants that differentiate into specialized cells.

sternum Elongated flat bone (also called the breastbone) to which the upper ribs attach and so form the rib cage.

steroid (STAIR-oid) A lipid with a backbone of four carbon rings and with no fatty acid tails. Steroids differ in their functional groups. Different types have roles in metabolism, intercellular communication, and cell membranes.

steroid hormone A type of lipid-soluble hormone synthesized from cholesterol. Many steroid hormones move into the nucleus and bind to a receptor for it there; others bind to a receptor in the cytoplasm, and the entire complex moves into the nucleus.

sterol A type of lipid with a rigid backbone of four fused carbon rings. Sterols occur in cell membranes; cholesterol is the main type in human tissues.

stimulus [L. *stimulus*, goad] A specific change in the environment, such as a variation in light, heat, or mechanical pressure, that the body can detect through sensory receptors.

stomach A muscular, stretchable sac that receives ingested food; the organ between the esophagus and intestine in which considerable protein digestion occurs.

stop codon Of protein synthesis, a base triplet in a strand of mRNA that serves as the stop signal for translation, so that no more amino acids are added to the polypeptide chain.

stromatolite Of shallow seas, layered structures formed from sediments and large mats of the slowly accumulated remains of photosynthetic populations.

substrate A reactant or precursor molecule for a metabolic reaction; a specific molecule or molecules that an enzyme can chemically recognize, bind briefly to itself, and modify in a specific way.

substrate-level phosphorylation The direct, enzyme-mediated transfer of a phosphate group from the substrate of a reaction to another molecule. An example is the transfer of phosphate from an intermediate of glycolysis to ADP, forming ATP.

succession, primary (suk-SESH-un) [L. *succedere,* to follow after] Orderly changes from the time pioneer species colonize a barren habitat through replacements by various species until the climax community, when the composition of species remains steady under prevailing conditions.

succession, secondary Orderly changes in a community or patch of habitat toward the climax state after having been disturbed, as by fire.

surface-to-volume ratio A mathematical relationship in which volume increases with the cube of the diameter, but surface area increases only with the square. Of growing cells, the volume of cytoplasm increases more rapidly than the surface area of the plasma membrane that must service the cytoplasm. Because of this constraint, cells generally remain small or elongated, or have elaborate membrane foldings.

survivorship curve A plot of the agespecific survival of a group of individuals in a given environment, from the time of their birth until the last one dies.

sympathetic nerve Of the autonomic nervous system, any of the nerves generally concerned with increasing overall body activities during times of heightened awareness, excitement, or danger; also work continually in opposition with parasympathetic nerves to bring about minor adjustments in internal organs.

synaptic integration (sin-AP-tik) The moment-by-moment combining of excitatory and inhibitory signals arriving at a trigger zone of a neuron.

synovial joint Freely movable joint in which adjoining bones are separated by a fluid-filled cavity and stabilized by straplike ligaments. An example is the ball-and-socket joint at the hip.

syphilis Clinical term for the sexually transmitted disease caused by infection by the spirochete bacterium *Treponema pallidum.* Untreated syphilis can lead to lesions in mucous membranes, the eyes, bones, skin, liver, and central nervous system.

systemic circuit (sis-TEM-ik) Circulation route in which oxygenated blood flows from the lungs to the left half of the heart, through the rest of the body (where it gives up oxygen and takes on carbon dioxide), then back to the right side of the heart.

systole Contraction phase of the cardiac cycle.

T lymphocyte A white blood cell with roles in immune responses.

target cell Any cell that has receptors for a specific signaling molecule and that may alter its behavior in response to the molecule.

tectorial membrane Inner ear structure against which sensory hair cells are bent, producing action potentials that travel to the brain via the auditory nerve.

telophase (TEE-low-faze) Of mitosis, the final stage when chromosomes decondense into threadlike structures and two daughter nuclei form. Of meiosis I, the stage when one of each pair of homologous chromosomes has arrived at one or the other end of the spindle pole. At telophase II, chromosomes decondense and four daughter nuclei form.

telophase II Of meiosis, final stage when four daughter nuclei form.

temperate pathway A viral infection that enters a latent period; the host is not killed outright.

temporal summation The adding together (summing) of several muscle contractions, resulting in a single, stronger contraction, when stimulatory signals arrive in rapid succession.

tendon A cord or strap of dense, regular connective tissue that attaches a muscle to bone or to another muscle.

test An attempt to produce actual observations that match predicted or expected observations.

testcross Experimental cross to reveal whether an organism is homozygous dominant or heterozygous for a trait. The organism showing dominance is crossed to an individual known to be homozygous recessive for the same trait.

testis, plural testes. Male gonad; primary reproductive organ in which male gametes and sex hormones are produced.

testosterone (tess-TOSS-tuh-rown) In males, a major sex hormone that helps control reproductive functions.

tetany Condition in which a muscle motor unit is maintained in a state of contraction for an extended period.

thalamus Of the forebrain, a coordinating center for sensory input and a relay station for signals to the cerebrum.

theory A testable explanation of a broad range of related phenomena. In modern science, only explanations that have been extensively tested and can be relied on with a very high degree of confidence are accorded the status of theory.

thermal inversion Situation in which a layer of dense, cool air becomes trapped beneath a layer of warm air; can cause air pollutants to accumulate to dangerous levels close to the ground.

thermoreceptor Sensory cell that can detect radiant energy associated with temperature.

thirst center Cluster of nerve cells in the hypothalamus that can inhibit saliva production, resulting in mouth dryness that the brain interprets as thirst and leading a person to seek out drinking fluids.

threshold Of neurons and other excitable cells, a certain minimum amount by which the voltage difference across the plasma membrane must change to produce an action potential.

thymine Nitrogen-containing base in some nucleotides.

thymus A lymphoid organ with endocrine functions; lymphocytes of the immune system multiply, differentiate, and mature in its tissues, and its hormone secretions affect their functions.

thyroid gland Of the endocrine system, a gland that produces hormones that affect overall metabolic rates, growth, and development.

tidal volume Volume of air, about 500 milliliters, that enters or leaves the lungs in a normal breath.

tissue A group of cells and intercellular substances that function together in one or more specialized tasks.

T lymphocyte One of a class of white blood cells that carry out immune responses. The helper T and cytotoxic T cells are examples.

tonicity The relative concentrations of solutes in two fluids, such as inside and outside a cell. When solute concentrations are isotonic (equal in both fluids), water shows no net osmotic movement in either direction. When one fluid is hypotonic (has less solutes than the other), the other is hypertonic (has more solutes) and is the direction in which water tends to move.

tracer A radioisotope used to label a substance so that its pathway or destination in a cell, organism, ecosystem, or some other system can be tracked, as by scintillation counters that detect its emissions.

trachea (TRAY-kee-uh) The windpipe, which carries air between the larynx and bronchi.

transcript-processing controls Controls that govern modification of new mRNA molecules into mature transcripts before shipment from the nucleus.

transcription [L. *trans*, across, and *scribere*, to write] Of protein synthesis, the assembly of an RNA strand on one of the two strands of a DNA double helix; the base sequence of the resulting transcript is complementary to the DNA region on which it was assembled.

transcriptional controls Controls influencing when and to what degree a particular gene will be transcribed.

transfer RNA (tRNA) Of protein synthesis, any of the type of RNA molecules that bind and deliver specific amino acids to ribosomes and pair with mRNA code words for those amino acids.

translation In protein synthesis, the conversion of the coded sequence of information in mRNA into a particular sequence of amino acids to form a polypeptide chain; depends on interactions of rRNA, tRNA, and mRNA.

translational controls Controls governing the rates at which mRNA transcripts that reach the cytoplasm will be translated into polypeptide chains at ribosomes.

translocation Of cells, a change in a chromosome's structure following the insertion of part of a nonhomologous chromosome into it.

transposable element DNA element that can spontaneously "jump" to new locations in the same DNA molecule or a different one. Such elements often inactivate the genes into which they become inserted and give rise to observable changes in phenotype.

triglyceride (neutral fat) A lipid having three fatty acid tails attached to a glycerol backbone. Triglycerides are the body's most abundant lipids and richest energy source.

trisomy (TRY-so-mee) Of diploid cells, the abnormal presence of three of one type of chromosome.

trophic level (TROE-fik) [Gk. *trophos*, feeder] All the organisms in an ecosystem that are the same number of transfer steps away from the energy input into the system.

trophoblast Surface layer of cells of the blastocyst that secrete enzymes that break down the uterine lining where the forthcoming embryo will implant.

tumor A tissue mass composed of cells that are dividing at an abnormally high rate.

tumor marker A substance that is produced by a specific type of cancer cell or by normal cells in response to cancer.

tumor suppressor gene A gene whose protein product operates to keep cell growth and division within normal bounds, or whose product has a role in keeping cells anchored in place within a tissue.

tympanic membrane The eardrum, which vibrates when struck by sound waves.

ulna One of two long bones of the forearm; the ulna extends along the "little finger" side of the forearm, parallel to the radius on the "thumb" side.

umbilical cord Structure containing blood vessels that connect a fetus to its mother's circulatory system.

uracil (YUR-uh-sill) Nitrogen-containing base found in RNA molecules; can base-pair with adenine.

ureter Tubular channel that carries urine from each kidney to the urinary bladder.

urethra Tube that carries urine from the bladder to the body surface.

urinary bladder Storage organ for urine.

urinary excretion A mechanism by which excess water and solutes are removed by way of the urinary system.

urinary system An organ system that adjusts the volume and composition of blood and so helps maintain extracellular fluid.

urine Fluid formed by filtration, reabsorption, and secretion in kidneys; consists of wastes, excess water, and solutes.

uterus (YOU-tur-us) [L. *uterus*, womb] Chamber in which the developing embryo is contained and nurtured during pregnancy.

vaccine Antigen-containing preparation injected into the body or taken orally; it elicits an immune response leading to the proliferation of memory cells that offer long-lasting protection against a particular pathogen.

vagina Part of a female reproductive system that receives sperm, forms part of the birth canal, and channels menstrual flow to the exterior.

variable Of a scientific experiment, the only factor that is not exactly the same in the experimental group as it is in the control group.

vas deferens Tube leading to the ejaculatory duct; one of several tubes through which sperm move after they leave the testes just prior to ejaculation.

vasoconstriction Decrease in the diameter of an arteriole, so that blood pressure rises; may be triggered by the hormones epinephrine and angiotensin.

vasodilation Enlargement of arteriole diameter, so that blood pressure falls; may be triggered by hormones including epinephrine and angiotensin.

vein Of the circulatory system, any of the large-diameter vessels that lead back to the heart.

ventricle (VEN-tri-kul) Of the heart, one of two chambers from which blood is pumped out. Compare atrium.

venule Small blood vessel that receives blood from tissue capillaries and merges into larger-diameter veins; a limited amount of diffusion occurs across venule walls.

vertebra, plural vertebrae. One of a series of hard bones arranged with intervertebral disks into a backbone.

vertebrate Animal having a backbone of bony segments, the vertebrae.

vesicle (VESS-ih-kul) [L. *vesicula*, little bladder] Within the cytoplasm of cells, one of a variety of small membrane-bound sacs that function in the transport, storage, or digestion of substances or in some other activity.

vestibular apparatus A closed system of fluid-filled canals and sacs in the inner ear that functions in the sense of balance. Compare semicircular canals.

villus (VIL-us), plural villi. Any of several types of absorptive structures projecting from the free surface of an epithelium.

virus A noncellular infectious agent consisting of DNA or RNA and a protein coat; can replicate only after its genetic material enters a host cell and subverts its metabolic machinery.

vision Precise light focusing onto a layer of photoreceptive cells that is dense enough to sample details concerning a given light stimulus, followed by image formation in the brain.

vital capacity Maximum volume of air that can move out of the lungs after a person inhales as deeply as possible.

vitamin Any of more than a dozen organic substances that the body requires in small amounts for normal cell metabolism but generally cannot synthesize for itself.

vocal cords A pair of elastic ligaments on either side of the larynx wall. Air forced between them causes the cords to vibrate and produce sounds.

water table The upper limit at which the ground in a specified region is fully saturated with water.

watershed Any specified region in which all precipitation drains into a single stream or river.

wax A type of lipid with long-chain fatty acid tails that help form protective, lubricating, or water-repellent coatings.

white blood cell Leukocyte; any of the macrophages, eosinophils, neutrophils, and other cells that, together with their products, make up the immune system.

white matter Of the spinal cord, major nerve tracts so named because of the glistening myelin sheaths of their axons.

X chromosome A sex chromosome with genes that cause an embryo to develop into a female, provided that it inherits a pair of these.

X-linked gene Any gene on an X chromosome.

X-linked recessive inheritance Recessive condition in which the responsible, mutated gene occurs on the X chromosome.

Y chromosome A sex chromosome with genes that cause the embryo that inherited it to develop into a male.

Y-linked gene Any gene on a Y chromosome.

yellow marrow Bone marrow that consists mainly of fat and hence appears yellow. It can convert to red marrow and produce red blood cells if the need arises.

yolk sac Of land vertebrates, one of four extraembryonic membranes. In humans, part becomes a site of blood cell formation and some of its cells give rise to the forerunners of gametes.

zero population growth A population for which the number of births is balanced by the number of deaths over a specified period, assuming immigration and emigration also are balanced.

zygote (ZYE-goat) The first cell of a new individual, formed by the fusion of a sperm nucleus with the nucleus of an egg (fertilization).

CREDITS AND ACKNOWLEDGMENTS

Page 1 Eric Gravé/Science Source/Photo Researchers

INTRODUCTION I.1 Frank Kaczmarek / I.2 Art by Carlyn Iverson / I.3 Rich Buzzelli/Tom Stack & Associates / **Page 5** Art by Precision Graphics / I.4 Daniel McDonald/The Stock Shop / I.5 Jon Feingersh/Tom Stack & Associates / I.9 (above) Lennart Nilsson © Boehringer Ingelheim GmbH; (below) photograph Ken Greer/Visuals Unlimited / I.10 Jan Halaska/Photo Researchers

CHAPTER 1 1.1 (a) Martin Rogers/FPG; (b) © Alan Craft/Photophile / 1.2 Photograph Jack Carey / 1.5 (b) Hank Morgan/Rainbow; (d) Dr. Harry T. Chugani, M.D., UCLA School of Medicine / 1.9 Photograph Richard Riley/FPG / 1.10 Art by Raychel Ciemma / 1.12 Michael Grecco/Picture Group / 1.13 Art by Precision Graphics / 1.18 (b) Micrograph Biophoto Associates/SPL/Photo Researchers; 1.19–1.22 Art by Precision Graphics / 1.24 Art by Palay/ Beaubois / 1.25 Micrograph CNRI/SPL/Photo Researchers; art by Robert Demarest / 1.27 Model A. Lesk/SPL/Photo Researchers / 1.28 © David R. Frazier Photolibrary

CHAPTER 2 2.1 (a) (left) National Library of Medicine; (right) Armed Forces Institute of Pathology; (b) Corbis-Bettmann / 2.2 (a) Photograph Ed Reschke; (b) photograph Carolina Biological Supply Company / 2.3, 2.4, 2.5 Art by Raychel Ciemma / 2.6 Art by Precision Graphics after Stephen L. Wolfe, *Molecular and Cellular Biology*, Wadsworth, 1993 / 2.7 Micrographs M. Sheetz, R. Painter, and S. Singer, *Journal of Cell Biology*, 70:193 (1976), by copyright permission of The Rockefeller University Press / 2.8 Art by Leonard Morgan; (above) after B. Alberts et al., *Molecular Biology of the Cell*, Second edition, Garland Publishing Co., 1989 / 2.9 Frank L. Lambrecht/Visuals Unlimited / 2.10 Art by Raychel Ciemma and American Composition and Graphics / 2.11 Micrograph G. L. Decker; art by Raychel Ciemma and American Composition and Graphics / 2.12 George Musil/Visuals Unlimited / 2.13 (a) Lennart Nilsson from *Behold Man*, © 1974 by Albert Bonniers Forlag and Little, Brown and Company, Boston; (b) David M. Phillips/Visuals Unlimited; (c) CNRI/SPL/Photo Researchers / 2.14 D. W. Fawcett, *The Cell*, Philadelphia: W. B. Saunders Co. / 2.15 Art by Raychel Ciemma and American Composition and Graphics / 2.16 (a), (b) Micrographs Don W. Fawcett/Visuals Unlimited; (below) art by Robert Demarest / 2.17 (above) Micrograph Gary W. Grimes; (below) art by Robert Demarest after a model by J. Kephart / 2.18 Micrograph Keith R. Porter / 2.19 (a) Art by Precision Graphics; (b) Art by Precision Graphics after Stephen L. Wolfe, *Molecular and Cellular Biology*, Wadsworth, 1993; micrograph J. Victor Small and Gottfried Rinnerthaler / 2.21 (a), (c) Thomas A. Steitz; art by Palay/Beaubois / 2.24 Art by Raychel Ciemma / 2.25 Photograph Gary Head

Pages 64–65 CNRI/SPL/Photo Researchers

CHAPTER 3 3.1 © Richard B. Levine / 3.2 (a) Photograph Focus on Sports; (inset) Manfred Kage/Bruce Coleman Ltd.; art by Palay/Beaubois; photographs (b) Lennart Nilsson from *Behold Man*, © 1974 by Albert Bonniers Forlag and Little, Brown and Company, Boston; (c) Manfred Kage/Bruce Coleman Ltd.; (d) Ed Reschke/ Peter Arnold, Inc. / 3.3 Photograph NASA/Johnson Space Center / 3.4 Simon Fraser/Science Photo Library / 3.5 Art by Raychel Ciemma / 3.6 Photographs by Ed Reschke / 3.7 Fred Hossler/Visuals Unlimited / 3.8 (left) Art by Joel Ito; (right) art by L. Calver / 3.9 Photograph Ed Reschke / 3.10 Photograph Ed Reschke; art by Precision Graphics / 3.11 Photograph Ed Reschke / 3.12 Lennart Nilsson from *Behold Man*, © 1974 Albert Bonniers Forlag and Little, Brown and Company, Boston / 3.13 Art by L. Calver / 3.14 Art by Palay/Beaubois / 3.15 (a) Art by Robert Demarest; (b) Ed Reschke / 3.16 CNRI/SPL/Photo Researchers / **Page 81** PhotoDisc / **Page 84** (a) Manfred Kage/ Bruce Coleman Ltd.; (b-d) Ed Reschke; inset Lennart Nilsson from *Behold Man*, © 1974 Albert

Bonniers Forlag and Little, Brown and Company, Boston

CHAPTER 4 4.1 © Ken Regan/Camera 5 / 4.3 (left) Art by L. Calver; (center and below) art by Joel Ito; micrograph Ed Reschke / 4.4 Art by K. Kasnot / 4.5 National Osteoporosis Foundation / 4.6 Art by Raychel Ciemma / 4.7 Art by John W. Karapelou / 4.8 Art by Ron Ervin / 4.9, 4.10 Art by John W. Karapelou / 4.11 Art by Precision Graphics / 4.12 Mike Devlin/SPL/Photo Researchers / 4.13 (a) Photograph C. Yokochi and J. Rohen, *Photographic Anatomy of the Human Body*, Second edition, Igaku-Shoin Ltd., 1979 / 4.14 Art by Robert Demarest / 4.15 Art by Raychel Ciemma / 4.16 art by Robert Demarest; (b) micrograph John D. Cunningham; (c) micrograph D. W. Fawcett, *The Cell*, Philadelphia: W. B. Saunders Co., 1966 / 4.17 Art by Nadine Sokol / 4.18 Art by Robert Demarest / 4.19 After Stephen L. Wolfe, *Molecular and Cellular Biology*, Wadsworth, 1993 / 4.21 (a) Art by Kevin Somerville; (b) Ed Reschke / 4.24 Photograph Michael Neveux / 4.23 Tim Davis/Photo Researchers

CHAPTER 5 5.1 Nancy J. Pierce/Photo Researchers / 5.3 Art by Kevin Somerville / 5.5 After A. Vander et al., *Human Physiology: Mechanisms of Body Function*, Fifth edition, McGraw-Hill, 1990. Used by permission.; (c) Redrawn from page 763 of *Human Anatomy and Physiology*, Fourth edition, by A. Spence and E. Mason. Copyright © 1992 by West Publishing Company. All rights reserved. / 5.6 (a),(b) Art by Carlyn Iverson; (c) art by Precision Graphics / 5.7, 5.9 Art by Carlyn Iverson / 5.10 (a) Art by Raychel Ciemma; (b), (c) (right) Lennart Nilsson © Boehringer Ingelheim International GmbH; (c) (left) Biophoto Associates/SPL/Photo Researchers; art by Robert Demarest / 5.11 Art by Raychel Ciemma / 5.12, 5.13 Art by Carlyn Iverson / 5.15 Modified after A. Vander et al. *Human Physiology: Mechanisms of Body Function*, Fourth edition, McGraw-Hill, 1985; photograph Ralph Pleasant/ FPG / 5.18 Photograph Gary Head / 5.19 Wide World Photos

CHAPTER 6 6.1 (a) From A. D. Waller, *Physiology, The Servant of Medicine*, Hitchcock Lectures, University of London Press, 1910; (b) photograph courtesy of The New York Academy of Medicine Library / 6.2 Art by Palay/Beaubois / 6.3 (a) Lennart Nilsson from *Behold Man*, copyright 1974 by Albert Bonniers Forlag and Little, Brown and Company, Boston; art by Raychel Ciemma / 6.4 Art by Palay/Beaubois / 6.5 Art by John W. Karapelou / 6.6 (a) After F. Ayala and J. Kiger, *Modern Genetics*, © 1980 Benjamin-Cummings; (b) Lester V. Bergman & Associates, Inc. / 6.7 After Gerard J. Tortora and Nicholas P. Anagnostakos, *Principles of Anatomy and Physiology*, Sixth edition, 1990. © 1990 by Biological Sciences Textbooks, Inc., A & P Textbooks, Inc., and Elia-Sparta, Inc. Reprinted by permission of Harper-Collins Publishers. / 6.9 Art by Kevin Somerville / 6.10 (a) C. Yokochi and J. Rohen, *Photographic Anatomy of the Human Body*, Second edition, Igaku-Shoin Ltd, 1979; (b), (c) Art by Raychel Ciemma / 6.11, 6.12 Art by John Karapelou / **Page 145** CNRI/ SPL/Photo Researchers / 6.13 Art by Kevin Somerville / 6.14 Art by John Karapelou / 6.15 Micrograph Ed Reschke / 6.16 Art by Raychel Ciemma / 6.17 Sheila Terry/SPL/Photo Researchers / 6.19 Art by Robert Demarest based on A. Spence, *Basic Human Anatomy*, Benjamin-Cummings, 1982 / 6.20 Art by John W. Karapelou / 6.21 Art by Kevin Somerville / 6.22 Art by Raychel Ciemma / 6.23 Photograph Lennart Nilsson, © Boehringer Ingelheim International GmbH / 6.24 Lewis L. Lainey / 6.25 (a) Ed Reschke / 6.27, 6.28 Art by Raychel Ciemma

CHAPTER 7 7.1 (a) The Granger Collection, New York; (b) Lennart Nilsson © Boehringer Ingelheim International GmbH / 7.2 Art by Nadine Sokol / 7.3 Robert R. Dourmashkin, courtesy of Clinical Research Centre, Harrow, England / 7.4 Lennart Nilsson © Boehringer Ingelheim International GmbH / 7.5, 7.6, 7.8 Art by Raychel Ciemma / 7.9 (a) Art by Raychel Ciemma; (b) micrograph Morton

H. Nielsen and Ole Werdlin, University of Copenhagen / 7.10 Photograph courtesy Don C. Wiley, Harvard University / 7.12, 7.13 Art by Palay/ Beaubois after B. Alberts et al., *Molecular Biology of the Cell*, Garland Publishing Company, 1983 / 7.15 (left) Lowell Georgia/Science Source/Photo Researchers; (right) Matt Meadows/Peter Arnold, Inc. / 7.18 Kent Wood/Photo Researchers / 7.19 (a) Oliver Meckes/Ottawa/Photo Researchers; (b) CAMR/A.B. Dowsett/Science Photo Library

CHAPTER 8 8.1 Galen Rowell/Peter Arnold, Inc. / 8.2 Art by Kevin Somerville / 8.3 CNRI/ SPL/Photo Researchers / 8.4 After A. Vander et al., *Human Physiology: Mechanisms of Body Function*, Sixth edition, McGraw-Hill, 1994. Used by permission of McGraw-Hill, Inc. / 8.5 Steve Lissau/ Rainbow / 8.6 O. Auerbach/Visuals Unlimited / 8.7 Larry Mulvehill/Photo Researchers / 8.8 SIU/ Visuals Unlimited; art by K. Kosnot / 8.9 From L. G. Mitchell, J. A. Mutchmor, and W. D. Dolphin, *Zoology*, © 1988, Benjamin-Cummings Publishing Company / 8.10 Modified from A. Spence and E. Mason, *Human Anatomy and Physiology*, Fourth edition, 1992, West Publishing Company / 8.11 Art by Carlyn Iverson / 8.12 Art by Leonard Morgan / 8.13 (a) (below) and (b) Art by Carlyn Iverson / 8.14 Lennart Nilsson from *Behold Man*, © 1974 by Albert Bonniers Forlag and Little Brown and Company, Boston

CHAPTER 9 9.1 (left) National Park Service; (right) © Greg Mancuso/Jeroboam / 9.3 U.S. Department of Agriculture / 9.4–9.6 Art by Robert Demarest / 9.7 Art by Precision Graphics / 9.10 Art by Joel Ito / 9.11 Art by Precision Graphics / 9.12 VU/SIU/Visuals Unlimited / 9.13 Photograph Colin Monteath, Hedgehog House, New Zealand; art by Kevin Somerville; art (left) Robert Demarest, (right) Kevin Somerville / **Page 217** Evan Cerasoli

CHAPTER 10 10.1 (left) Comstock, Inc.; (right) Manfred Kage/Peter Arnold, Inc. / 10.3, 10.5 Art by Raychel Ciemma / 10.7 (f) © C Raines/Visuals Unlimited / 10.8 Art re-drawn by Precision Graphics based on Sherwood, *Human Physiology*, 3rd edition, Wadsworth Publishing Company / 10.9 (b) Micrograph Dr. Constantino Sotelo; (d), (e) art by D. & V. Hennings / 10.11 Micrograph from *Tissues and Organs: A Text-Atlas of Scanning Electron Microscopy*, by R. G. Kessel and R. H. Kardon. Copyright © 1979 by W. H. Freeman and Company. Reprinted with permission. Art by Robert Demarest / 10.15 Art by Robert Demarest / 10.16 Art by Robert Demarest; micrograph Manfred Kage/Peter Arnold, Inc. / 10.17 (a) Colin Chumbley/Science Source/Photo Researchers; (b) C. Yokochi and J. Rohen, *Photographic Anatomy of the Human Body*, Second edition, Igaku-Shoin Ltd., 1979 / 10.18 Art by Kevin Somerville / **Page 235** (right) © Wiley/ Wales/ProFiles West / 10.19 (a) From James W. Kalat, *Introduction to Psychology*, Third edition, Brooks/Cole Publishing Company, 1993; (b) Marcus Raichle, Washington University School of Medicine / 10.20 (left) Art by Palay/Beaubois after Wilder Penfield and Theodore Rasmussen, *The Cerebral Cortex of Man*, © 1950 Macmillan Publishing Company; copyright renewed © 1978 by Theodore Rasmussen. Reprinted with permission of Macmillan Publishing; photograph Colin Chumbley/ Science Source/Photo Researchers / 10.22 Art by Robert Demarest / 10.23 Art by Palay/Beaubois / 10.26 M. Gadomski/Photo Researchers / 10.27 (a), (b) From Edythe D. London et al., *Archives of General Psychiatry*, 47:567-574 (1990); (right) photograph Ogden Gigli/Photo Researchers

CHAPTER 11 11.1 F. Wood/Superstock / 11.2 Art by Kevin Somerville / 11.3 From Hensel and Bowman, *Journal of Physiology*, 23:564-568, 1960 / 11.4 Art by Palay/ Beaubois after Wilder Penfield and Theodore Rasmussen, *The Cerebral Cortex of Man*, © 1950 Macmillan Publishing Company; copyright renewed © 1978 by Theodore Rasmussen. Reprinted with permission of Macmillan Publishing; photograph Colin Chumbley/Science Source/ Photo Researchers / 11.5 Art by Raychel Ciemma /

11.6 Art by D. & V. Hennings / 11.7 Art by Robert Demarest; micrograph Omikron/SPL/Photo Researchers / 11.9 Art by Robert Demarest / 11.10 (a) Photograph Edward R. Bower © 1991 TIB/West; (left) art by Kevin Somerville; (b) art by Betsy Palay/Artemis / 11.11 Robert E. Preston, courtesy Joseph E. Hawkins, Kresge Hearing Research Institute, University of Michigan Medical School / 11.12 Art by Robert Demarest / 11.13–11.14 Art by Kevin Somerville / 11.15 Photographs Gerry Ellis/The Wildlife Collection; art by Kevin Somerville / 11.16 Micrograph Lennart Nilsson © Boehringer Ingelheim International GmbH / 11.18 After S. Kuffler and J. Nicholls, *From Neuron to Brain*, Sinauer, 1977

CHAPTER 12   12.1 Evan Cerasoli / 12.2 Art by Kevin Somerville / 12.5, 12.6 Art by Robert Demarest / 12.7 (a) Mitchell Layton; (b) Syndication International (1986) Ltd. / 12.8 Photographs courtesy of Dr. William H. Daughaday, Washington University School of Medicine. From A. I. Mendelhoff and D. E. Smith, eds., *American Journal of Medicine*, 20:133 (1956) / 12.9 Art by Leonard Morgan / 12.11 The Mark Shaw Collection/Photo Researchers / 12.12 Art by Raychel Ciemma / 12.13 Corbis-Bettmann / 12.14 Biophoto Associates/SPL/Photo Researchers / 12.15 Thomas Zimmermann/FPG

CHAPTER 13   13.1 (left) Evan Cerasoli; (right) Art by Robert Demarest after Patten, Carlson, and others / 13.2 Art by Raychel Ciemma / 13.3 Art by Raychel Ciemma; (b) Micrograph Ed Reschke / 13.5, 13.6 Art by Raychel Ciemma / 13.7 Photograph Lennart Nilsson from *A Child Is Born*, © 1966, 1977 Dell Publishing Company. Inc. / 13.8 Art by Robert Demarest / 13.9 © Lennart Nilsson / 13.11 Sandy Roessler/FPG / 13.12 David Frazier/Photo Researchers

CHAPTER 14   14.1 Lennart Nilsson from *A Child Is Born*, © 1966, 1977 Dell Publishing Company. Inc. / 14.3 Art by Palay/Beaubois adapted from R. G. Ham and M. J. Veomett, *Mechanisms of Development*, St. Louis: C. V. Mosby Co., 1980 / 14.4–14.8 Art by Raychel Ciemma / 14.9 Sketches after B. Burnside, *Developmental Biology*, 26:416-441, 1971 / 14.10, 14.11 Photographs from Lennart Nilsson, *A Child Is Born*, © 1966, 1977 Dell Publishing Company, Inc.; art by Raychel Ciemma / 14.12 Art by Carlyn Iverson / 14.13 Art by Robert Demarest / 14.14 Art by Raychel Ciemma / 14.15 Art by Raychel Ciemma modified after Keith L. Moore, *The Developing Human: Clinically Oriented Embryology*, Fourth edition, W. B. Saunders Co., 1988 / 14.16 Photograph James W. Hanson, M.D. / 14.17 (a) Art by Palay/Beaubois; (b) from Fran Heyl Associates, © Jaques Cohen, Computer enhanced by © Pix Elation / 14.18 Art by Raychel Ciemma adapted from L. B. Arey, *Developmental Anatomy*, Philadelphia, W. B. Saunders Co., 1965 / 14.19 Photograph Gary Head / 14.20 Jose Carillo/Photophile / 14.21 (a) G. Musil/Visuals Unlimited; (b) Cecil Fox/Science Source/Photo Researchers

CHAPTER 15   15.1 (a) Allen Russell/ProFiles West; (b) Tony Brain and David Parker/SPL/Photo Researchers / 15.2 After Stephen L. Wolfe, *Molecular and Cellular Biology*, Wadsworth, 1993 / 15.3 Art by Palay/Beaubois / 15.4, 15.5 Art by Raychel Ciemma / 15.7 Lennart Nilsson © Boehringer Ingelheim International GmbH / 15.8 (left) After Stephen L. Wolfe, *Molecular and Cellular Biology*, Wadsworth, 1993 / 15.10 Z. Salahuddin, National Institutes of Health / 15.11 CNRI/SPL/Photo Researchers / 15.12 (a) Biophoto Associates/Photo Researchers; (b) Ken Greer/Visuals Unlimited; (c) Science VU/Visuals Unlimited; (d) Joel B. Baseman / 15.13 David M. Phillips/Visuals Unlimited / 15.14 George Musil/Visuals Unlimited / 15.15 E. Grau/SPL/Photo Researchers / 15.16 David M. Phillips/Visuals Unlimited / 15.17 © M. Richards/PhotoEdit

Pages 348–349 David M. Phillips/Science Source/Photo Researchers

CHAPTER 16   16.1 (a-c),(e) Lennart Nilsson from *A Child Is Born* © 1966, 1967 Dell Publishing Company, Inc.; (d) Lennart Nilsson from *Behold Man*,

© 1974 by Albert Bonniers Forlag and Little, Brown and Company, Boston / 16.2 C. J. Harrison et al., *Cytogenetics and Cell Genetics* 35:21-27, copyright 1983 S. Karger A.G., Basel / 16.3 CNRI/Photo Researchers / 16.4 Andrew S. Bajer, University of Oregon / 16.6 Micrographs Ed Reschke; art by Raychel Ciemma / 16.9 Micrographs H. Beams and R. G. Kessel, *American Scientist*, 64:279-290, 1976 / 16.11 Art by Raychel Ciemma / 16.12 Art by Raychel Ciemma and American Composition and Graphics / 16.15 Micrograph David M. Phillips/Visuals Unlimited / 16.17 Art by Raychel Ciemma

CHAPTER 17   17.1 (a) Frank Trapper/Sygma; (b) Focus on Sports; (c) Fabian/Sygma; (d) Moravian Museum, Brno / 17.3 Richard Corman/Outline / 17.9 Art by Raychel Ciemma / 17.12 Micrographs Stanley Flegler/Visuals Unlimited / 17.13 Jim Stevenson/SPL/Photo Researchers / 17.14 (top to bottom) Frank Cezus; Frank Cezus; Michael Keller; Ted Beaudin; Stan Sholik/all FPG / 17.15 Dan Fairbanks, Brigham Young University / Page 385 Evan Cerasoli

CHAPTER 18   18.1 Cystic Fibrosis Foundation / 18.2 Art by Palay/Beaubois; photograph Omikron/Photo Researchers / 18.3 (b) Redrawn by Robert Demarest with permission from page 126 of Michael Cummings, *Human Heredity: Principles and Issues*, Third edition. Copyright © 1994 by West Publishing Company. All rights reserved. / 18.4 (a), (b) Stuart Kenter Associates / 18.7 Photograph Dr. Victor A. McCusick / Page 393 Fran Heyl Associates, © Jacques Cohen, computer enhanced by © Pix Elation / 18.10 Steve Uzzell / 18.11 Giraudon/Art Resource / 18.13 After Victor A. McKusick, *Human Genetics*, Second edition, copyright 1969. Reprinted by permission of Prentice-Hall, Inc., Englewood Cliffs, NJ; (b) photograph Corbis-Bettmann / 18.14 W.M. Carpenter / 18.15 Rivera Collection/Superstock, Inc. / 18.17 Courtesy of G. H. Valentine / 18.21 Art by Raychel Ciemma / 18.22 (a) Cytogenetics Laboratory, University of California, San Francisco; (b) after Collman and Stoller, *American Journal of Public Health*, 52, 1962; (c) courtesy of Peninsula Association for Retarded Children and Adults, San Mateo Special Olympics, Burlingame, CA

CHAPTER 19   19.1 Photograph A. C. Barrington Brown © 1968 J. D. Watson; model A. Lesk/SPL/Photo Researchers / 19.6 (a) (left) C. J. Harrison et al., *Cytogenetics and Cell Genetics* 35:21-27, copyright 1983 S. Karger A.G., Basel; (b) micrograph B. Hamkalo; (c) micrograph O. L. Miller, Jr. and Steve L. McKnight; (b-d) art by Nadine Sokol / 19.7 Art by Hans & Cassady, Inc. / 19.11 (a) Courtesy of Thomas Steitz from *Science*, 246:1135-1142, December 1, 1989 / 19.13 Art by Raychel Ciemma and American Composition and Graphics / 19.14 Micrograph Dr. John E. Heuser, Washington University School of Medicine, St. Louis; art by Palay/Beaubois / Page 417 Micrograph C.J. Harrison / 19.17 Art by Raychel Ciemma

CHAPTER 20   20.1 Billy E. Barnes/Jeroboam / 20.2 Betsy Palay/Artemis / 20.3 Lennart Nilsson © Boehringer Ingelheim GmbH / 20.4 Betsy Palay/Artemis / 20.6 © Tony Freeman/PhotoEdit / 20.7 Paul Shambroom/Photo Researchers / 20.8 Michael Newman/PhotoEdit / 20.10 Photograph Mills-Peninsula Hospitals / 20.12 (a) John D. Cunningham/Visuals Unlimited; (b) Alfred Pasieka/Peter Arnold, Inc. / 20.13 Art by Betsy Palay/Artemis; (a) Biophoto Associates/Science Source/Photo Researchers; (b) James Stevenson/SPL/Photo Researchers; (c) Biophoto Associates/Science Source/Photo Researchers

CHAPTER 21   21.1 (a) Mark Clarke/SPL/Photo Researchers; (b) NCI/Science Source/Photo Researchers; (b) (a) Stanley N. Cohen/Science Source/Photo Researchers; (b) Dr. Huntington Potter and Dr. David Dressler / 21.4 After Stephen L. Wolfe, *Molecular and Cellular Biology*, Wadsworth, 1993 / 21.6 Cellmark Diagnostics, Abingdon, U.K. / 21.7 Art by Precision Graphics 21.9 (a), (b) Monsanto Company; (c), (d) Calgene, Inc. / 21.10 Cour-

tesy of Exxon Corporation / 21.11 (a) R. Brinster and R. E. Hammer, School of Veterinary Medicine, University of Pennsylvania; (b) Courtesy of Tufts University School of Veterinary Medicine/Genzyme Corporation / 21.12 Courtesy of Victor A. McKusick and Joanna Strayer Amberger, Johns Hopkins University / 21.13 Ted Thai/Time Magazine / 21.14 Michael Maloney/San Francisco Chronicle

Pages 454–455   Photograph NASA

CHAPTER 22   22.1 (left) Vatican Museums; (right) Martin Dohrn/SPL/Photo Researchers / 22.2 (a) Christopher Ralling; (b) (left) Courtesy George P. Darwin, Darwin Museum, Down House; (right) Heather Angel / 22.3 Alan Solem / 22.4 After F. Ayala and J. Valentine, *Evolving*, Benjamin-Cummings, 1979 / 22.5 (above) Art by Leonard Morgan; (below) art by Lloyd K. Townsend / 22.6 (a) Patricia G. Gensel; (b) Donald Baird, Princeton Museum of Natural History / 22.7 (below) From T. Storer et al., *General Zoology*, Sixth edition, McGraw-Hill, 1979. Reproduced by permission of McGraw-Hill, Inc. / 22.8 Art by Victor Royer / 22.10 Art by Raychel Ciemma / Page 467 Photograph Bruce Coleman Ltd. / 22.12 Kjell B. Sandved/Visuals Unlimited / 22.13 Chesley Bonestell / 22.14 Stanley W. Awramik / 22.16 (a) Sidney W. Fox; (b) W. Hargreaves and D. Deamer

CHAPTER 23   23.1 (left) FPG; (right) Douglas Mazonowicz/Gallery of Prehistoric Art / 23.2 Roger Burnard / 23.4 (a) Bruce Coleman Ltd.; (b) Tom McHugh/Photo Researchers; (c) Larry Burrows/Aspect Picture Library; (d) Stephen Rapley / 23.5 Art by D. & V. Hennings / 23.9 (a) Louise M. Robbins; (b) Dr. Donald Johanson, Institute of Human Origins; (c, d) Kenneth Garrett/National Geographic Image Collection / 23.8 Art by D. & V. Hennings / 23.11 (a, b) Kenneth Garrett/National Geographic Image Collection; (c) Jean Paul Tibbles / 23.12 Art by Raychel Ciemma

CHAPTER 24   24.1 Wolfgang Kaehler / 24.2 (a) Harlo H. Hadow; (b) Jack Carey / 24.3 Roger K. Burnard / 24.4 (a), (b) Roger K. Burnard; (c), (d) E. R. Degginger / 24.6 Photograph Sharon R. Chester / 24.7 (a) Gene C. Feldman and Compton J. Tucker/NASA, Goddard Space Flight Center / 24.9 Gerry Ellis/The Wildlife Collection / 24.10 Photograph © Gary Braasch / 24.12 (a) Photograph by Gene E. Likens from G. E. Likens and F. H. Bormann, *Proceedings First International Congress of Ecology*, pp. 330-335, September 1974, Centre Agric. Publ. Doc. Wagenigen, The Hague, The Netherlands; (b) Photograph by Gene E. Likens from G. E. Likens et al., *Ecology Monograph*, 40(1):23-47, 1970; (c) after G. E. Likens and F. H. Bormann, "An Experimental Approach to New England Landscapes" in A. D. Hasler (ed.), *Coupling of Land and Water Systems*, Chapman & Hall, 1975 / 24.13 (b) Photograph John Lawler/FPG / 24.14 Art by Precision Graphics; (b) after W. Dansgaard et al., *Nature*, 364:218-220, 15 July 1993; D. Raymond et al., *Science*, 259:926-933, February 1993; W. Post, *American Scientist*, 78:310-326, July–August 1990 / 24.15 Photograph William J. Weber/Visuals Unlimited / 24.17 (above) Peter Scoones/Planet Earth Pictures1; (below) Dennis Brokaw

CHAPTER 25   25.1 Antoinette Jongen/FPG / 25.2, 25.3 Data from Population Reference Bureau after G. T. Miller, Jr., *Living in the Environment*, Eighth edition, Wadsworth, Inc. 1993; photographs United Nations / 25.6 Photograph NASA / 25.9 Photograph United Nations / 25.10 (a) Photograph National Science Foundation; (b, c, d) Photographs NASA / 25.11 Dr. Charles Henneghiem/Bruce Coleman Ltd. / 25.12 From Water Resources Council / 25.13 Agency for International Development / 25.14 (below) R. Bieregaard/Photo Researchers; (below) after G. T. Miller, Jr., *Living in the Environment*, Eighth edition, Wadsworth, Inc., 1993 / 25.15 Photograph NASA / 25.16 Data from G. T. Miller, Jr. / 25.17 Alex MacLean/Landslides / 25.18 Richard Parker/Photo Researchers / 25.19 World Bank / 25.20 © 1983 Billy Grimes

A

Abdomen, *115*, 139, 191; fat storage, 75; muscle groups, *99*

Abdominal aorta, *143*

Abdominal cavity, *79*, 115, 121; blood flow, *143*, *146*; defined, 78; respiratory system, 186, 187; urinary system, *204*

ABO blood typing, 140, 141, 381

Abortion, 400; debate, 301, *301*

Abrasion: adhering junctions, *72*; human skin, 80, 81, 84

Absorption, 263; acid-base balance, 212; calcium, *126*; disruptive disorders, 122–123; homeostasis, 111; mechanics of, 118–119, *119*, 132; mineral ions, 281; of nutrients, 113, 122; simple epithelium and, 68, *69*; solute gains, 202, 216; water gained by, 202, 216

Absorption routes, 119, *119*

Abstinence, sexual, 298; to prevent disease transmission, 345, 346

Abstract thought, *233*, 234, 237, 474

Acceleration, straight-line, 248, 256, 257, 264

Accessory glands/ducts, reproductive system, 287; male reproductive tract, *288*, 289, 302

Accessory organs, digestive system, 111, *112*; enzyme secretion, 113, 116, 117, 132

Accommodation, *259*; defined, 259

Acetylcholine (ACh), 102, 103, 108, *123*, 226, 231; Alzheimer's disease and, 326; breakdown, 227; in brain tissue, *225*; as neurotransmitter, 77; as stimulant, 242

Acetyl cholinesterase, 227

Acetal CoA, 57, *57*, 59, 60, *61*

Achondroplasia, 395, *395*, 403

Achoo syndrome, *403*

Acid, 23, *23*, 36, 114; defined, 22; stomach, secretion, 22, 122; water interactions, 22

Acid-base balance, 216; defined, 212

Acid deposition, 516–517, 522

Acid group (carboxyl), 30, *30*, 31

Acidity: hormone secretions and, 116; oxygen/hemoglobin binding and, 138

Acidosis, defined, 23

Acid rain, 7, 23, 516–517, 526; defined, 516

Acne, 81, 107; scars, 73

Acquired immunodeficiency syndrome (AIDS), 45, *197*; cases by mode of transmission, *337*; confidential testing, 339; cumulative totals of cases, *346*; defined, 179; as global epidemic, 335, *335*; immune system and, *178*; indicator diseases, 336, 338; molecular structure of HIV, 8; onset, 338; as pandemic, 180; prevention, 339; as sexually transmitted disease, 330, 334, *334*, 340, 343, 346; transfusable blood and, 71; treatment plan, 181; vaccine, 446; as viral infection, *163*, 334, 438, *450*

Acromegaly, 274, *274*

Acrosome, 308, *308*; defined, 290

Actin, 76, 101, *101*, 102; defined, 100; in cytokinesis, 360; filaments, 101, 102, *102*, *103*, 108

Action potential, 103, 219; cardiac conduction system, 148; defined, 220; from graded, local signals, 247; in resting neurons, 221, *221*, *223*, *223*; in sensory system, 252, 253, 263; on

auditory nerve, 254, 255, *255*; on vestibular nerve, 257; photoreceptor system, 258, 259; propagation of, *222*, 223, *223*, 224, *224*, 227, *231*; rate of propagation, 228, 244; self-propagate, 248; sensory pathways and, 248, 249, *249*; spike, 222, 223, *223*, 227, 244

Active immunization, 176, 182

Active sites, 55; defined, 54

Active transport, 44, 62, 72, 216; defined, 44; of nutrients, 119, *119*; of sodium in urinary system, 121, 209, *209*, 210, 216

Active transport mechanism, maintains gradients, 221, *221*

Activity levels, hormonal adjustment, 267

Acute inflammation, defined, 166

Acyclovir, 181, 343

Adam's apple, 187, *280*

Adaptation, 464, 472, 475, 480, 483, *483*; defined, 460

Adaptive radiation, 466, 467, 472, 483; defined, 467

Adaptive trait, defined, 6

Addiction, 242, 243, 321; defined, 241

Addison's disease, 278, *279*

Adductor longus, *99*

Adenine (A), 56, 470; in DNA nucleotide bases, 406, *406*, 410, 418; in RNA nucleotide bases, 410, 411, *411*, 412, 417, 418

Adenine plus ribose, 470

Adenocarcinoma, 431, 434, *434*, 436

Adenosine deaminase (ADA), *450*, 451; deficiency, *450*, 451; genes, 451

Adenosine diphosphate (ADP): final stage of aerobic respiration, 58, *58*, 59, *59*; in reaction steps of glycolysis, *56*, 57

Adenosine diphosphate (ADP)/ATP cycle, 101, *103*; coenzyme role, *127*; defined, 59

Adenosine phosphates, 37

Adenosine triphosphate (ATP), 25, 37, *413*, 470–471, 504; conversion to cyclic AMP, 271; defined, 34; as energy carrier, 4, *5*, 52, 54, 55; as energy source, 221; functional groups, 25; hormonal triggers, *270*; metabolic pathways, *103*; minerals and, *127*; synthesis of, 52, *52*, *61*; use of glucose, 60

Adenosine triphosphate energy, for muscle action, *100*, 102, 105

Adenosine triphosphate formation, levels of exercise and, 103, *103*; vitamins and, *126*

Adenosine triphosphate-forming machinery, 58, 60

Adenosine triphosphate molecules, 46; transporting energy, 39, 44, 55, 56–57, *56*, 62

Adenosine triphosphate production, 52, 52, 55, 57, *61*, 62, 66; death and, 101; during aerobic respiration, 71; fatty acids for, 122; in muscle organelles, 87, 98; third stage, 58–59, *58*, 59

Adenyl cyclase, converting ATP to cyclic AMP, *270*, 271

Adhering junctions, *72*; defined, 72

Adhesion proteins, 41, 62; defined, 41; migration of cells, 313

Adhesive cues, 313

Adipose tissue, 28, 60, 73, *73*, 75; brown, 214; defined, 75; fat storage, 122, 125, 129; glucose uptake, 276; as hormone target, *275*; in breast, *319*; methyl mercury in, 235

Adjuvant therapy, 430

Adolescence: body development, 305, 323; diabetes onset, 276; remodeling, 90, 108

Adrenal androgens, 278

Adrenal cortex, 211, *269*, 278, 279; defined, 278; disorders, 278; hormonal sources/effects of, *275*; as hormone target, 272, *273*; mesoderm, *306*

Adrenal glands, 68, *83*; autonomic nervous system and, *230*; cortex, *269*; hormonal cascade, 318; hormones of, 242, 278–279; medulla, *269*, 279; nicotine and, 154; overview, *269*; steroid hormones, 272

Adrenalin, *275*, 279

Adrenal medulla, *269*; defined, 279; ectoderm, *306*; hormonal source/effects of, *275*

Adult life, 323, *323*

Aerobic exercise, 106

Aerobic metabolism, waste transportation, 138

Aerobic pathway, 56, 57, 59–62, *61*, 103, 105; entry point, *60*

Aerobic respiration, 31, 58–59, 62, 71, 492, 501, 506; ATP energy production, 87, 103, *103*; cellular, 185, 192, 193, 198; defined, 56; in carbon cycle, 498, *498*, 499; in nitrogen cycle, 502; oxygen transport, 136; summary of, *59*

Aerosol inhalants, 189. *See also* Ozone

Afferent arteriole, 205, *205*, 206, 211, *211*

Afferent nerves, 230

Aflatoxin, 426, 427

African emergence model, 484, *484*; defined, 484

Afterbirth, 318, *318*, 328

Age, energy output and, 128, 129

Age interval, 513

Age structure, *510*, *513*; defined, 510

Age structure diagrams, 510, *510*, *511*

Agglutination, *140*, 141; defined, 140

Agglutination response, *140*, 141

Aging, phenomenon of, 323, *324*, 328; antioxidants and, 21; bone formation and, 325; bone mass, 323; cartilage and, 96; causes of, 324; eye disorders, 260, 261; heart disease risk, 154, *155*; loss of muscle tissue, 106; menopause and, 293; physiological changes, 305, *327*, 328; weight gain and, 129

Aging population, demand for blood products, 71

Agranulocytes, *137*; defined, 137

Agricultural chemicals, and cancer, 428, *428*, 430

Agricultural runoff, 14, 45, 203

Agriculture, 508, 514, *514*, 515

Agrobacterium tumefaciens, 446

AIDS. *See* Acquired immunodeficiency syndrome

Air composition, 188, 198

Air flow rate, with blood flow rate, 194, 195

Air pollution, 135, *197*, 516–517, *516*, 517, 522, 525–526; breathing and, 189

Air pressure, 255; in eustachian tube, 255

Alanine, *412*

Albinism, 403

Albumin, 33, 136; in plasma, *136*; in urine, 207

Alcohol, 29, *37*, 157, 210, *241*, 243; bladder infections, 207; blood-brain barrier, 235; breakdown of, 51; con-

sumption, and cancer, 430, 434; defined, 242; dependence or habituation, 116, 120, 240, *240*, 345; drug action and interaction, 241, 242; functional groups, 24, *24*, 25, 26; HDL levels, 154; as legal drug, 218, 219; rate of absorption, 116; taste of, 253

Alcoholism, genetic links, 383

Alcohol use, prenatal risk, 320–321, *320*, 328

Aldehyde group, 25, 26, *26*, 37

Alders, *489*

Aldosterone, *208*, 210–211, 216, 271, 278; defined, 210; primary effects, *269*, 275

Algae, 492, 506

Alkaline solutions, 22, 212

Allantois, *310*, 311, 312, *317*, 328; defined, 310

Allele frequency, 459–461, *460*, 472

Allele pool, 484

Alleles, 364, 365, 368, 370, 449, 457, *459*; autosomal location, 392, 394–395; chromosomal combination of, 387; codominant, 373, 381; cystic fibrosis (CF), 386, 393; defined, 364, 374, 388, 459; dominant, 375, *375*, *380*, 384, 392–395, 402; dominant, in Mendel's single gene model, 372, 373, 381; dominant, penetrance and, 382, *382*; events contributing to mix of, 459; harmful, 468, *468*; harmful, recessive, 393, 394, 399; in cancer, 426, 429; microevolution and, 460, *460*, 461; as molecular form of gene, 372, 373, *374*; mutant, 381, 397, 460; nonidentical, 373; phenotype expression of, 383; rearrangements, 387; recessive, 375, *380*, 384; recessive, in genetic inheritance, 392, 394, 395, 402; recessive, in Mendel's single gene model, 372, 373, 381; recessive, penetrance of, 382; sex chromosome location, 392; shared, 468; X-linked, *398*; X-linked dominant, 396; X-linked dominant mutant, 397; X-linked recessive, 396–397, *396*

Allergens, 178, *179*; antibiotics as, 180; breathing and, 189; common, 178

Allergic response, 178–179, *179*, 182; aggravation of, 282; eosinophils and, 137, 166; glucocorticoid use, 278; pesticides and, 14

Allergic rhinitis, *197*

Allergies, *179*, 186; defined, 178; drug action and interaction, 241; food, 123; IgE antibodies and, 173, 177; sources of, 189

All-or-none principle, 108; defined, 105

All-or-nothing spike, 222, 223

Alpha-carotene, 21

Alpha cells, 276, *277*

Alpha globulins, 136

Alpha-ketoglutarate, 57

Alpha rhythm, *237*

Altitude sickness, 184

Aluminosilicates, 470

Aluminum, 16

Alu transposable element, 416

Alveolar duct, *186*

Alveolar sac, *186*, 193; air pressure in, 190, 191; defined, 187

Alveolus, *186*, 190, 191, 195; aging effects on, 325; breakdown of, 189, *197*; defined, 187; gas exchange in, 192, *192*, 193, 198; location, *187*; stretch receptors, 194, 195

Alzheimer's disease (AD), *225*, 227, 326, *326*, 449; aging and, 324; brain regions, 239; early-onset, 449; malfunctioning proteins and, 32

Amacrine cells, 262, *262*

American Cancer Society, 430, 432, 433, *433*, 434

American diet, 111

Amine hormones, 284; defined, 271, *271*

Amino acids, 60, *60*, 61, *119*, 132, 470–472, 502; blood transport, 139; breakdown of, 120, 122, 125; chemistry of life, 32, 36, *37*; coenzymes and metabolism, *126*; deamination, 202; defined, 30; as enzyme substrate, 54; essential, 125, *125*, 132; free, *412*; functional groups, 24–25, *25*, 30–31, *30*; genetic material and, 404; hormonal chemical families, 271; hormone stimulus and, 273; in plasma, 136, *136*, 160; linkage into polypeptide chain, 405, 418; negative charges, 408; as nutrient molecules, 110, 111, 119; peptide bonds, 414–415, *415*, 418; peroxisomes, 51; protein digestion, 117, *117*, 119, *123*; reabsorption, 206, 208; sequence, 405, *408*, 412–413, *412*, *416*, 418, 466; substitution, 416, *416*; taste of, 253; transport proteins and, 44; tRNA binding, *413*, 414, *414*

Amino group (—NH₃), 60

Aminopeptidase, 117, *117*

Ammonia (NH₃), 81, 120, 202, 203, 212, 212

Ammonification, 502, 503, *503*; defined, 502

Ammonite shells, fossilized, *457*

Ammonium, 495, 502

Amnesia, 239

Amniocentesis, 322, *322*; prenatal screening option, 393

Amnion, *310*, 311–312, *311*, *315*, 322, 328; defined, 310

Amniotic cavity, 310, *310*, 312

Amniotic fluid, 322, *322*; early formation, 310; HIV infection, 336

Amniotic sac, *314*, 318

Amoeba, *3*

Amoebic dysentery, 180, *181*

Amphetamines, *241*; defined, 242

Amphibians, 464, 476

Amplitude, of action potential of sensory neurons, 248, 249, *249*, 264; of wave cycles, 254, 255

Amygdala, 237, *237*; fact memory, 239, *239*

Amyl nitrate, 243

Amyloidosis, cerebroarterial, Dutch type, *449*

Amylose, 27

Amyotrophic lateral sclerosis (ALS), *403*, 449, *449*

Anabolic steroids, 105, 106–107

Anabolism, defined, 54

Anaerobic bacteria, 525

Anaerobic mud, 463

Anaerobic pathway, 56, 59, 62, 103, 502; ATP from, 58

Anaerobic respiration, 498, *501*

Anal canal, 112, 121

Analgesics, *241*, 243

Analogous structures, defined, 465

Anaphase, defined, 356; during mitosis, 354, *354*, *355*, 357–360, *357*, *359*, *369*, 370

Anaphase I, chromosome nonseparation, *400*; during meiosis, *361*, *362*, 365, 370; homologous chromosome separation in, 375

Anaphase II, chromosome nondisjunction at, *400*; during meiosis, *361*, *363*, 365, *369*, 370

Anaphylactic shock, 178

Anatomy, *457*; comparative, *458*; defined, 67

Androgen insensitivity, 397

Androgens, 397; in developing fetus, 278; primary effects, *269*, *275*

Anemia, 116, 131, 139, *155*, *446*; from ionizing radiation exposure, 367; vitamins and, *126*. *See also* sickle-cell anemia

Anencephaly, 11, 319

Anesthesia, drug abuse, 242, 243

Aneuploidy, defined, 400

Angelmann syndrome (AS), 383

Angina pectoris, *155*, 156

Angiogenin, 425

Angiography, 156

Angiosperms, *463*

Angiotensin, *275*

Angiotensin I, 211

Angiotensin II, 211

Angiotensinogen, 211

Animalia, *467*

Animal protein, 125, *125*

Animals, 450; genetic engineering of, 448–449; origin, 475; transgenic, 448–449, *448*

Animal species, *3*, 7

Ankle bones, *91*, 95, 478

Antacids, 22

Antagonistic drug interaction, 241

Antagonistic movement, 98, 108

Antarctica, 486–487, *486*, *487*, 491, 500–501; ozone thinning, 517, *517*; trashing of, 486

Anterior part, *79*

Anterior pituitary, *83*

Anthrax, 162

Anthropoids, 475–477, *476*, 479, *481*, 485; defined, 476

Antianxiety agents, 242

Antibiotics, 164, 180–181, 201–202, 446; broad-spectrum, 181; cholera, 45; defined, 180; in organ transplantation, 173; resistance to, 181, 440; sickle-cell anemia, 381; side effects, 180; STD infections and, 340, 341, 343; vaginitis and, 344; viruses and 334, 343

Antibodies, *37*, 172–173, 182, 446; anti-HIV, 338, 339; bound, 172; defined, 140, 169; gene activation for, 417; in blood screening, 343; in immune response, 164, 165, *165*; maternal, 141, *141*, 320; monoclonal, 176, 177, *177*; production site, *158*; Rh, 141, *141*; surface, 173

Antibody mediated response, 169, *169*, 172–173, *172*

Antibody molecules, 169, *169*, 172; structure, 174, *174*; synthesizing, 172

Anticancer drugs, 430, 524, *524*

Anticlotting factors, 167

Anticodon, *413*; -codon interactions, 413–415, *414*, *415*, 418; defined, 412

Antidepressant drugs, 227

Antidiuretic hormone (ADH), *208*, 210–211, *210*, 216; defined, 210; hypothalamus/pituitary production, 273, *273*, 284; primary effects, *269*, *271*, 272, 272

Antigens, 165, *168*, 169, *169*, 182, 427, 446; bound, 171–172; defined, 140, 163, 168; HIV proteins, 338–339; immune response to, 334; unbound, 170, *172*; vaccine and, 176; viral, 339

Antigen-antibody complexes, 140, 179

Antigen-binding sites, 173, *174*

Antigenic variants, 339

Antigen-MHC complexes, *169*, 170–173, *171*, *172*, 182; defined, 168

Antigen presenting cell, *169*, 172, *172*, 182; defined, 168; as triggers for immune response, 168, *168*, 170, *170*

Antigen-presenting macrophage, *171*

Antigen protein, 446

Antigen receptors, 172–175, *172*

Antigen recognition, 168, 170, 173, 182

Antigen-specific receptors, 182; formation of, 174

Antihistamines, 178, 241

Anti-inflammatry drugs, 178

Antimicrobial secretion, 164, *164*

Antioxidants, 21, 127, 130, 430, 439; defined, 21

Anti-Rh gamma globulin, 141

Antitrypsin, 189

Antiviral drugs, 181, 449

Antrum, 294

Anus, *112*, 112, 113, 293; defined, 121; genital warts, 343

Anvil, 254, *254*

Aorta, *144*, 145, 146, 147, *147*, *317*; arch of, *144*, *464*; blood pressure in, 149; coronary bypass, 156, *156*; defined, 144; oxygenated blood, 150, *151*

Aortic bodies, *194*; defined, 195

Aortic valve, *155*

Apnea, 195

Apneic episode, 195

Apocrine glands, 68

Apoptosis, 173, 307, 313, 430; defined, 170

Appendicitis, 121, 342

Appendicular skeleton, *91*, 93–95, 108; defined, 90

Appendix, *120*, 121, *121*, *158*, 159; deferred pain, 251

Appetite: stimulate, 129; suppress, 120, 129

Apples, 428

Appropriate technology, 525, *525*

Apres, 475–482, *480*, 485

Aquatic plants, 490

Aquatic realms, *488*

Aqueous humor, 258, *258*, 261

Aquifers, pollution of, 203, 518, *518*

Arable land, 131

Archeopteri, *463*

Arctic tundra, 488, *489*

Ardipithecus ramidus, *481*

Arginine, 54, *54*, *412*

Armpits, sweat glands in, 81

Arms, 94–95, *94*, 146; muscle groups, *99*; versatility of movement, 94–95, 98, *98*

Arrhythmia, *155*, 156–157, 243; defined, 157

Arsenic, *428*, 434

Arterial duct, 317, *317*

Arterial walls, 150–151, *150*; LDLs enter, 154

Arteries, *143*, 146, 151, 157, 160, *464*; arteriosclerosis in, 156; blood pressure in, 149, 151; blood velocity in, 149; cholesterol plaque and, 154, 156, *156*, 325; defined, 142; disease, 29; graft, *156*; homocysteine damage, 156; pressure reservoirs, 150, 160; respiratory control and, 195; to bones, 89, 97; as transport tubes, 135

Arterioles, 118, *118*, *150*, 166, 167; ADH regulation, 272; adjustable diameters, 135; blood pressure in, 149–151, *151*, *152*, 160; blood velocity in, 149; defined, 142; dilation, 150; as hormonal target, *275*, 279; in kidneys, 205, *208*, 211; in placenta, 310

Arteriosclerosis, 156

Arthritis, 97, 179, 450; aging and, 324, 326, 327; osteoarthritis, 96; rheumatoid, 96

Arthropods, 344

Arthroscopy, 97

Artificial blood, 145

Artificial insemination by donor (AID), 300

Artificial joint, 96, *97*

Artificial selection, 440

Artificial sweeteners, 253

Asbestos, 426, *428*, 430, 434, *516*

Asbestosis, *197*

Ascending aorta, *143*

Ascending colon, of large intestine, *120*, 121, 132

Ascorbic acid, *126*

Asparagine, *412*

Aspartame, 253, 394

Aspartate, *412*

Aspirin, 153, 215, 282; excessive use, 116

Assimilation, 502–503, *503*

Asteroids, 468

Asthma, 14, 178, *189*, *197*; bronchioles and, 187, 189; cortisone drugs, 278; preventive drugs, 189

Astigmatism, 260

Astrocytes, 220

Atherosclerotic plaque, 154, 156–157, *156*

Atherosclerosis, 107, 149, 154–157, *155*; defined, 156; free radicals and, 21

Athletes, anabolic steroids, 106–107; drug testing, 206; elite, 86; euphoric "high", 227; knee damage, 95; in rigorous training, 105; resting cardiac rate, 157

Athlete's foot, 164

Atmosphere, 517, *517*; early, 469, 470, *470*; first, 472; oxygen-free, 469

Atmospheric cycles, 495, 498, *498*, 503, 506

Atmospheric pressure, gas exchange and, 188, 190, 191, *190*, 198

Atomic nucleus, 16, 17, 19, 36

Atomic number, *17*; defined, 16

Atomic structure, model of, *16*

Atomic weight, 16

Atoms, 4, 15; chemical bonds and, 19–21, 36; in combination, 17, 36; defined, 16, 17; enzymes and transfer of, 55; ionization, 20; structure of, 16–17, *19*, 36; variant forms of, 17, 36

Atom transports, 54

Atrial natriuretic factor, *446*

Atrial natriuretic peptide (ANP), 275, 281

Atrioventricular node (AV node), 148, *148*; defined, 148

Atrioventricular valve (AV valve), *144*, 145, 146–147, *147*; defined, 144

Atrium, *144*, 145, *145*, 146; defined, 144; in cardiac cycle, 147, *147*, 148, 157, 160; in fetal circulation, 317

Atrophy, 106

Attachment sites, 72

Auditory canal, 254, *254*, 255

Auditory nerve, 254, 255, *255*

Auditory receptors, 248

AUG start codon, 412, *412*, 414, *414*

Australopithecus afarensis, 480–481, *480*, *481*, *482*, 485

Australopithecus africanus, 480, *480*, 481

Australopithecus anamensis, 480, *481*

Australopithecus boisei, 480, *480*, 481

Australopithecus robustus, 480, *480*, 481

Australopiths, *480*, 481–482, *482*, 485; defined, 480

Autoimmune disease, 225, 438; age-related, 327

Autoimmune disorders, 280

Autoimmune response, 179, 182; defined, 178; in type 1 diabetes, 276

Automated DNA sequencers, 443, 449

Automobile exhaust, 188

Autonomic nerve, 219, 228, 230, 231, 244; defined, 230

Autonomic nervous system, 148

Autonomic reflexes, 232

Autosomal dominant inheritance, 394–395, 402, *403*

Autosomal inheritance patterns, 394–395, 397

Autosomal recessive inheritance, 394–395, 402, *403*

Autosomes, 402; changes in number, 400–401; defined, 388, 389; dominant male inheritance, 397; mutated gene on, 397; recessive female inheritance, 397; sex-influenced genes on, 398

Autotrophs, *490, 491*, 506; defined, 490

Axial skeleton, *91, 92, 95*, 108; defined, 90; from somites, *312*; function, 93

Axon hillock, 220, *220*

Axons, 77, 103, 220, *220*, 225; adhesion protein on, 313; chemical synapses, *226*, 244; defined, 220; of motor neuron, *104, 222, 223*; myelination, 225; as nervous system component, 228, *228*, 229; nodes, 224, 225, *224*; propagation along sheaths, 224, *224*; sensory, 248, 249, 258, 262, 263, 264

Azidothymidine (AZT), 181, 339

B

Baboons, *476*

Baby boomers, *511*; defined, 510

Baby formula, dilution of, 131

Backbone, 92–93, *93*, 108, *465*, 476, 478; aging, mass and, 90; muscle groups, 99

Bacillus anthracis, 162

Backflow, 113, 132, 151, *151*, 159, *159*

Bacteria, *3*, 22, 80, 116, 438, 440–443, *440*, 445, *445*, 452; aerobic, 466; antibiotics' effect on, 180; antibodies and, 140; attack, 81, 84; bacillus, *331, 333*; bioengineered, 446, 447; cell structure, *333*; chlamydia, 260–261; ciliated cells and, 358; cleaning up pollutants, 447; coccus, *333*; converting nitrogen, 502, 503, *503*, 506; cultured cells, 162; decomposers, *7, 490*, 492, *501*, 506; defense, *158*; discovery of, 162; fermentation pathways, 58; filter, *158*; "friendly," 164, *164*, 182; genetic susceptibility, 393; growth rate, 511–512, *511*; in feces, 121; infectious disease and, *181*; leukocytes and, 137, 160; macrophages and, 73; marine, 447; mercury-resistant, 447; mitochondria and, 52; mucus defense for, 187, 189; oil-eating, 447, *447*; on skin surface, 164; as pathogens, 164–165, *165, 166, 167,* 168, *172*; prokaryotic, 40; protein as weapon, 30; recombinant, 448; reproduction, 333, *333*; secreted toxins, 105, 334; sexually transmitted disease and, 330–334, *331*, 340–343, *340, 341*, 346; shapes, 333; spirillum, 333; surface antigen, *174*; tooth decay, 114; transgenic, 446–447; tuberculosis, 175; urinary tract infections, 207; use in nonpolluting fuel, 35; *Vibrio cholerae*, 45, *45*; waterborne, 175

Bacterial dysentery, 180; drug-resistance, 181

Bacterial infections, 92, 116, 121, *197, 225*; asymptomatic, 340; embryo risk of, *292, 293*, 295, 320; heart disease and, *155*, 156; kidney inflammation, 213

Bacterium flagellum, 333, *333*

Balance, sense of, 233, *248*, 254, 256–257, 264

Bald eagles, 279, 505

Baldness, 107

Ball-and-socket joints, 95, 96, 97

Balloon angioplasty, 156

Banding patterns, 444, *444*

Barbiturates, 241; blood-brain barrier, 235; defined, 243; prenatal risk, 321

Barging of wastes, 518

Barkley, Charles, 372

Barnyard biotechnology, 448–449, *448*

Barometer, 188

Baroreceptor, 245, *248*

Barr body, 390, *390*, 401

Basal body, 34, *34, 53*; defined, 53

Basal cell carcinoma, *10, 80*, 435, *435*

Basal ganglia, fact memory, 239, *239*

Basal layer, *80*

Basal metabolic rate (BMR), 128, 129; age-related change, 327; by sex, 128

Base, 36; defined, 22

Basement layer, of retina, 258, 262

Basement membrane, *69, 72*, 84, *150, 425*; alveoli and, 192; defined, 68

Base-pairing, 441, 442, *442, 443*, 450, 466

Base-pairing rule, 406, 407, 410, 411, 412, 418; A-T pairing rule, 406, 407, 410, 416; A-U pairing rule, 410, 411, 412, 418; G-C pairing rule, 406–407, 410–413, 418

Base pairs, *407, 415, 415*, 418; defined, 406; in DNA replication, 410; in transcription, 410, *411*; substitutions, 416, *416*

Base sequence, in DNA, 410, 411, *410, 411, 412*

Base triplets, 412, *412*, 413, *412*, 416, *416*, 418

Basic solution, 22, 23, *23*

Basilar membrane, 255, *255*

Bass, large-mouth, 492

Bat, *463, 465, 465*

Bayliss, W., 268

B cells. *See* B lymphocytes

Beans, 502

Bear, 466

Bed rest, 93, 106

Beef, 428

Behavior: developmental stages and changes in, 323; ethical, STD and, 345, 346; Genetic links in, 383; hormones and, 267, 268; hostile, violent, 242; psychoactive drugs and, 240, 241, 243; sexual, hormonal controls for, 290; STD risks and, 331, 339, 342

Behavioral flexibility, 476, 477, 478, 479

The Bends. *See* Decompression sickness

Benzene, 426, *428*, 430, 435, *516*

Beriberi, 131

Beta amyloid plaques, 326, *326*

Beta carotene, 21, 430

Beta cells, 276, 277

Beta-cryptoxanthin, 21

Beta-galactosidase, 445, *445*; gene fragments, 445; molecules, 445

Beta globulins, 136

Beta interferon, 177

Biases, 463

Bicarbonate, 23, *23*, 114, 116, 117; in blood plasma, 193, 198; kidney action and, 212

Bicarbonate-carbon dioxide buffer system, *212*; defined, 212

Bicarbonate ions, active transport in urinary system, *209*, 212

Biceps, 98, *98, 231*

Biceps brachii, 99

Biceps femoris, 99

Bicuspid valve, 144, *145*, 146

Bilateral symmetry, 312

Bile, 121; bilirubin and, 139; for emulsifying fat, 112, 117, *132*; formed by liver, *120*, 132; storage, *112*, 120, 132

Bile pigment, 117, 207

Bile salts, 29, 117, 119, *119*; synthesis, 125

Bilirubin, 139, 343

Binding sites, 442; antibodies/antigens, 172–174, *174*; calcium to troponin, 102, *103*; in immune system, *168, 171*, 182; on ribosomes, 413, *414*

Binomial system, defined, 467

Biochemical comparisons, 457

Biochemical testing, 394; markers for prematurity, 319

Biochemistry, 162, 466

Biodiversity, 524

Biogas digester, *525*

Biogeochemical cycles, 487, 495–496, *495*, 504, 506; defined, 495

Biogeographical patterns, 463

Biogeography, defined, 462

Biological clock, 266, 267, *269*

Biological magnification, 505, *505*

Biological molecules, *18, 23*, 36; bonds, 24; carbohydrates as, 26; covalent bonds, 20; damage to, 21; five classes of reactions, 25; ionic bonds, 20; proteins as, 30–31; synthesis of, 470; types of, *15, 17*

Biology, as science, 3

Biomass, plant, 492, *494*, 496, 498, *499, 501*; for energy, 522, *522*

Biomes, 488; forest, 521

Biopsy, 429, *432*, 433, 436; defined, 429

Bioreactor, grown cells, 70

Bioremediation, 447, *447*; defined, 446

Biorhythms, *275*

Biosphere, *7*, 520, 521; defined, *7*, 488

Biosynthesis, 470, 502, 503, *503*; patterns in interphase, 354; proteins as enzymes in, 410

Biosynthetic pathways, 54, 60, 62

Biotechnological society issues, 452

Biotechnology, 141, 439

Biotic portions of ecosystem, 488

Biotic potential, 513; defined, 512

Biotin, *126*

Bipedalism, *480*, 481–482, 485; defined, 478

Bipolar cells, 220

Bipolar interneurons, 262, *262*

Bipolar spindle, 357, *357*

Birds, 464–465, *464*, 476, 479, 490, 505

Birth: body development and, 305; complications, 318–319; HIV transmission at, 336, 346; lungs functional at, 317, 318; premature, 316, 318–319; process of, 318, *318*

Birth canal, 292, *292, 293*

Birth control: biological bases, 298–299, *298*; global, 301

Birth defects, 203, 428; nondisjunction, 365; prematurity and, 319; prenatal diagnosis, 322; prevention, 319, 320; screening, 311; untreated STD and, 330–331, 346

Birthing, 479

Birth rate, 510–511, *513*, 515, 526

Birth weight, low, 131, *196*, 321

Bison, 474

Bladder: autonomic nervous system and, *230*; location, 78; smooth muscle tissue, 76

Bladder cancer, 423, 426, *428*, 431, 434, 436

Blastocyst, *293, 307, 309*, 310, 323; defined, 309

Blastomere, 306

Blindness, 260, 261; herpes and, 343, 346; prematurity and, 319; vitamins and, *126*, 131

"Blind spot," 258, *258*

Blistering, 81

Blood, 134–160; balance of water/solutes, 201, 202; calcium levels, 280; cholesterol level, 125, 395; components of, *75, 136, 137*, 142, 160; as connective tissue, 73, *73*, 75, *75*, 84; defined, 75, 136; deoxygenated, 135, *143*, 146, 160; DNA fingerprinting, 444, *444*; drug presence, 240; estrogen levels in, 294–295, *296*; FSH levels, *296*; functions of, 75, 135, 160; HCG detection, 309; HIV transmission, *178*, 336, 338; internal environment and, 5; LH levels, *296*;

oxygenated, 135, *143*, 144, 146–147, 150, 160; oxygen-carrying capacity of, 139, 145, 192–193, 198, 203; pH of, 23, *23*, 36, 193; phosphate levels, 280; protein in, 32, 33; pumping of, *76*, 78, *78*; red blood cells in, 139, *139*; sodium levels in, 278; testosterone levels in, 291, *291*; tissue fluid in, *79*; as transport medium, 135, *142*, 145; water content of, 22; white blood cells in, 166; whole, 145

Blood alcohol concentration (BAC), 242, 243

Blood bank, 145

Blood-brain barrier, defined, 235

Blood cell production, 74, *78*

Blood cells: early formation of, 310; marrow production, 88–89, 108

Blood circulation, 67, 78, 142, 146–147; fetal, *311*, 316–317, 328; improving, 106; interruption, 154; lymphatic system and, 158; maternal, *311*, 316–317, *317*, 328

Blood clotting, 54, 75, 167, 448, *448*, arterial, 156, *197*; cancer therapy and, 145; fibrinogen, 136; mechanism, 153; minerals and, *127*; oral contraceptives and, 299; plaques and heart attack, *154, 155*; platelets and, 137, 160; protein, *164, 182*; PTH and, 281; vitamins and, *126*. *See also* Clotting factor VIII, IX

Blood-clotting functions, X-linked gene, 388, 396

Blood donors, 71, 145; incompatibility, 140

"Blood doping," 139

Blood filtration, 206–207

Blood flow: core temperature and, 214; disruption, 381; diversion of, 151; peripheral, 214, 215; prostaglandins aid, 282; pulmonary circuit, 317, *317*; rate of, 142; rate/air flow rate, 194–195, 198; reduction of, *155*; resistance to, 150–151, 157; systemic circuit, 317, *317*; to skin, 83

Blood glucose levels, *275, 276, 277*, 279; set point, *277*, 278

Blood loss, 116, 153, 210; thirst, 211

Blood plasma, 136–137, 192; carbon dioxide ions dissolved in, 193; in extracellular fluid, 202, 216; PDGF in, 283

Blood plasma proteins, 120

Blood pressure, 154; arterial, 207; capillary, 152, *152*, 160; control, by vessels, 151; defined, 149; drop, 215; drug effect on, 242–243; high, 127, *127*, 149, 325; hormonal control, 272, *275*, 276, 278, 281; hydrostatic, 43; in capillaries, 150, *152*; in disease, 45; kidneys and, 206–207, 210–211, 216; loss of, 178; measuring, *149*; reduction in, *155*; resting, 149; set point, 272; surges, 150; vessel structure and, 150

Blood products: contamination of, 145; HIV-infected, 336–337

Blood reservoir, 80, 84

Bloodstream, 66, 78, 81; carbon dioxide transport, 57, 185, 192, 198; fatty acid delivery, 103; fluid exchange in, 82; glucose transport, 60, 103, *103*; hormone release, 117, 122, 129; hormone transport, 55, 68, 266, 268, 270, 272, 279; lymph vessels draining, 170; mineral salts concentration, 279; nutrient transport, 112, 113; oxygen transport, 185, 192, 198; reabsorption, 205; as toxin transport, 120, 334

Blood substitutes, 71, 145

Blood sugar levels, 5, 27, *275*. *See also* Glucose

Blood supply, contaminated, 336

Blood tests, 429, 436
Blood transfusions, 71, 140, 145; demand, 71; for sickle-cell anemia, 381; hepatitis B contamination, 343; HIV-tainted blood supply, 337; Rh antibodies, 141
Blood typing, 140–141, 140, 145, 460; in organ transplantation, 173. See also ABO typing
Blood velocity, 149
Blood vessel network, 64–65
Blood vessels, 75, 138, 160, 425, 425, 434–435; artificially grown cells and, 70; as cardiovascular system component, 142; carrying nutients, 88; carrying waste, 88; collagen in, 32; connective tissue, 73, 74; early formation, 312; endothelium lining, 144; enlarged rectal, 121; fetal, 311, 311, 316, 317; hormonal effects on, 275; in adipose tissue, 75; increased flow, 87; in skin, 80, 80, 81; in small intestine wall, 117, 118, 118; leaky, 178; maternal, 310, 311, 311, 316; monosaccharides enter, 118, 119, 119; network of, 135; pressure in, 149; smooth muscle tissue, 76, 83; structure, 150–151, 150; to bone, 88, 89, 92, 108; to teeth, 114; of umbilical cord, 311, 316, 328; wall lining, 68, 68, 69; at wrist, 94
Blood vessel spasm, 153, 153
Blood volume, 66, 136, 142; maintaining, 152; reduced, 155; veins as reservoirs of, 150
Blue light wavelengths, 247, 263, 383
Blue-white screening, 445, 445; defined, 445
B lymphocytes, 137, 137, 167, 171, 176, 182, 451; antigen-specific receptors, 174–175, 175, 182; autoimmune disorder, 178; clonal selection, 175, 177; defined, 168, 169; DNA, 174–175; as effector cells, 169, 169, 172, 179, 182; as memory cells, 169, 171–172, 172, 175, 175; specific immune response, 164, 168, 168; as virgin cells, 169, 170, 172, 172, 175, 175
Body, sense of natural position, 250, 251, 256, 257, 264
Body build, 483
Body builder, 107
Body builder's psychosis, 107
Body fat, 129; distribution, X-linked and, 287, 388; heart's health, 154
Body fluid exchange, 339
Body proportions, 323, 323
Body regions, in somatosensory cortex, 250
Body size, 463
Body temperature, 54; blood maintaining, 75, 135; feedback mechanisms and, 82, 83; fever and, 417; hormones and, 266; integumentary system, 78, 80; internal (core), 201, 214; maintenance mechanisms, 201, 215, 216; premature birth, 316; regulation of, 226–227, 325
Body weight: assessing, 128–129; excessive, 128; "ideal," 128–129, 128; low birth, 131, 196, 321; maintenance, 111, 132; set point, 129
Bolus, 115, 115; defined, 114
Bone, 78, 84, 463, 465, 475, 477, 481; around teeth, 114, 114; articulating ends, 87, 108; artificial, 70; basic structure, 88–89, 108; blood cell production, 87, 88, 89, 108; blood flow to, 147; as body transport, 88, 89, 91, 108; calcium loss, 325; collagen in, 32, 37; connective tissue, 73–74, 73; defined, 74, 88; development of, 89, 108; function of, 74, 87, 88, 89, 91, 108; from mesoderm, 306; from somites, 312; as hormonal targets, 273, 274, 275, 281; long, 86–89, 94,

108, 274; loss, 90, 127; mass, 87, 323; mineralization, 126; mineral storage, 87, 88, 89, 108; movement and, 87, 88, 89, 108; as organs, 108; regeneration of broken, 283; remodeling, 89, 90, 108; shaft, 89, 89, 108; skeletal locations, 91; strength, 89; structural protein, 30, 32; -to-bone attachment, 73; turnover, 90, 108; vitamin D and, 29
Bone cancer, 426, 429, 434, 436; from ionizing radiation exposure, 367; osteosarcoma, 383
Bone collar, 89, 89
Bone growth: growth hormone, 273; vitamins and, 126
Bone marrow, 74–75, 430–431, 435, 482; B cell formation, 172, 182; cell division, 355; growth of lymphocytes in, 166, 178; as hormonal target, 275; lymphatic system and, 158, 159; NK cell production, 171; stem cells, 137, 137, 139; T cell production, 170, 171, 172, 182
Bone marrow cancer, from ionizing radiation exposure, 367
Bone tissue, 88, 108; defined, 88; mineral storage, 88, 89, 108; replacement, 70; types of, 74
Booster injection, 176
Bottle feeding, 131
Botulism, 58
Bovine spongiform encephalitis, 71
Brachial artery, 143
Bradycardia, 157, 157
Brain, 2, 77, 224, 476–477, 479–482, 485; ACh and, 103; ATP use, 60; blood flow, 92, 143, 147; breathing regulation, 195, 198; capillaries, structure, 234–235; cavities and canals, 234–235; as central nervous system component, 77–78, 219–220, 228, 228, 244; components of, 219, 234; controlling blood pressure, 151; defined, 232; enlargement, 475, 478–479, 482–483, 485; function, 232, 233, 244; higher function, 11; information flow, 229, 231, 244, 247; as integrator, 82, 84; location of cerebrospinal fluid, 235, 235; monitors digestion, 122; nonspecific defense mechanisms, 166; neural tube and, 312, 312; neuronal interaction, 229–230, 266, 326; poor nutrition and, 320; protection, 88, 91, 234–235; protein deficiency, 125; sensory reception, 247, 249, 252, 258, 264; sensory systems and, 233, 248, 255, 259; signals secretory cells, 116; use of PET and, 18; viral infection, 341
Brain cancer, 424, 426, 429, 431, 450, 451
Brain case, defined, 92
Brain damage, 188, 225; drug abuse, 242–243; irreversible, 190; sickle cells, 381
Brain disorders, Huntington's disease, 394
Brain scan, radioisotopes and, 18
Brain stem, 11, 228, 230, 233, 234, 244; defined, 232; in breathing regulation, 194, 194, 195; reflex centers, 257
Brain tumor, 424
Branches, 467, 472, 476
Branchings, 466–467, 466
Branch points, 466, 472, 476, 476, 484
Breastbone, 89, 91, 92, 94, 96, 108, 154; red bone marrow in, 139; thymus location, 281
Breast cancer, 289, 299, 319, 423, 426, 429, 430–434, 432, 436; from ionizing radiation exposure, 367
Breast milk, 130; HIV transmission, 178, 336; hormones and secretion,

268, 269; immunity transfer, 173; "let-down," 272, 272; methyl mercury in, 235
Breasts, 292, 319; as hormonal target, 275
Breast self-examination, 429, 432, 432
Breathing, 82, 103, 115, 187–188, 188; controls over, 185, 194, 194, 198; cycles of, 190, 191, 198; disorders, 189; magnitude, 194, 195, 198; moving lymph, 159; rhythm, 194–195, 198; rib cage and, 91, 93
Breech position, 318
Breeder reactors, defined, 523
Breeding, 486
Breeding ground, communal, 486
Breeds, 440
Bronchial tree, 186
Bronchiole, 186, 190, 195, 196, 198; asthma attack and, 189; defined, 187; local signaling molecules and, 268, 279
Bronchitis, 189, 196, 197
Bronchus, 159, 187, 198; asthma and, 189; autonomic nervous system and, 230; infections, 387, 387; squamous epithelium in, 434
Bruising, 155
"Brush border," 118, 119
Brushing, of teeth, 114
Bubonic plague, 180, 512, 514
Budding, 332, 336, 337, 338, 338
Buffering mechanisms, hemoglobin bound hydrogen ions, 193, 212
Buffers, 24, 36, 112, 114, 132, 212; bicarbonate as, 117; defined, 23
Bulb of Krause, 250, 251
Bulbourethral glands, 290; defined, 289; function, 288, 289
Burkitt's lymphoma, 178, 434
Burnett, Macfarlane, 174
Burns: from ionizing radiation, 367; growing skin, 70; inflammatory response, 155; reduced blood pressure/volume, 155; regeneration, 283; skin mending, 80; thirst, 211
Bursae, 97
Butyrate, use in sickle-cell anemia, 381

C

Caffeine, 116, 157, 210, 253; across blood-brain barrier, 235; bladder infections, 207; defined, 242; dependence or habituation, 240
CAG repeats, 417
Calcification, 89
Calcitonin, 90, 275, 280, 281
Calcium, 16, 16, 126, 127, 206; absorption, 126, 281; deficiencies, 90; deposits, 325; excretion, 281; hormonal effect on blood levels, 275, 280, 281; in teeth, 114; metabolism, 80; reabsorption, 275; sarcoplasmic reticulum membranes and, 102, 102; storage, 74, 78, 88, 90, 108
Calcium carbonate, otoliths, 257
Calcium channels, 103
Calcium ions, 50, 275, 496, 497, 504; in chemical synapses, 226; myosin/actin binding and, 101, 102, 102; triggered release, 10, 108
Calcium phosphate, 90
Calcium pump, 44, 62
Calcium salts, 74, 88, 108, 206, 213; in atherosclerotic plaque, 154, 154
Calories, 124, 124, 131; expended in common activities, 129; intake by age, 327; intake by gender, 128; total intake, 124, 125, 128, 129, 132
Campodactyly, 382
Canaliculi, 88, 89, 108
Cancer, 107, 154, 179, 422–436, 438, 446, 449–451; advanced, 450; age

factor, 324, 423; artificial blood use, 71, 145; bone marrow production, 155; causes, 203, 423, 428, 428; childhood, 383; chromosomal translocation, 399; chronic hepatitis, 343; defined, 424; diagnostic tests, 289, 426, 429, 429, 436; from ionizing radiation exposure, 367; gender preference, 422, 423, 431, 436; heredity role, 423, 426–427, 427, 429, 432, 436; immunity breakdowns, 178, 427; malignant, 425, 426; monoclonal antibodies, 176, 177, 177; morbidity rate, 422; mortality rate, 422, 431; prevention, 430; research, 10; risk, personal, 336, 423, 429, 430, 436; treatments, 18, 177, 299, 423, 430, 436; triggers, 81, 334, 426–427; types, 189, 196, 196, 431; urinalysis, 207; warning signs, 429, 429, 434
Cancer cells, 10, 163, 424–425, 424, 425, 429–430, 433; rampant division, 355; treatments, 436; triggers, 427
Candida albicans, 344, 346
Candidiasis, 344
Canines, 114
Cannabis sativa, 243
Carotenoids, 21
Capacitation, 308
Capillaries, 106, 146, 150, 151, 160; blood flow resistance, 150; blood pressure in, 152, 160; blood velocity in, 149; defined, 142; in hand musculature, 105; inflammation and, 166; in kidney, 205, 205, 206, 216; obesity and, 154; oxyhemoglobin of red blood cells, 138; partial pressure gradients in, 193, 198; red blood cells in, 137, 137; true, 146, 147; as tubes for diffusion, 135, 152
Capillaries, lung, 187, 187, 188, 192; gas exchange in, 192, 193, 195, 198
Capillary beds, 142, 144, 146, 147, 166; blood/lymph capillaries in, 159; defined, 146; as diffusion zones, 150, 152, 152, 159, 160
Capillary permeability, in inflammation, 166, 167, 167, 182
Capillary walls, diffusion, 152; in inflammation, 166, 167; reabsorption, 167
Capsid proteins, 332–333
Captive breeding programs, 468
Carbaminohemoglobin, defined, 193
Carbohydrate chains, 50, 62
Carbohydrate molecules: binds complement proteins, 165; in whole blood substitutes, 71
Carbohydrates, 15, 15, 17, 25, 26–27, 37, 490; assembled in G1 phase, 354; cholesterol synthesized from, 125; complex, 60, 60, 61, 111, 124, 125, 132; defined, 26; digestion in small intestine, 116, 117, 117, 119; digestion of, 110, 116, 120, 122, 123, 124; digestive enzyme defect, 387; as energy source, 28; in bones, 88, 108; in cellular metabolism, 54, 57; making ATP, 56; metabolism, 112, 126, 233, 275, 279, 352; soluble fiber, 121; storage, 75; use in making nonpolluting fuels, 35
Carbon, 14, 15, 16, 16, 24, 36
Carbon atoms, 15, 15, 24, 28, 29, 30; atomic number, 17; atomic symbol, 17; decay of, 17; electrons in outer shell, 17, 19; in glycolysis, 56, 56, 57, 60; in substitute blood, 71; in sugar molecule, 34, 37, 56, 56; isotopes of, 17; mass number, 17; organic compounds and, 24–25; simple sugars, 26, 26, 27, 35
Carbon-based molecules, 34
Carbon bonding, 24
Carbon-bound hydrogen atoms, 60

Carbon compounds, 24; summary of, 37

Carbon cycle, 498–499, *498, 499*, 506; defined, 498

Carbon dioxide, 23, 52, 145, 516–517, *516*, 521; as air component, 188, 198; arterial level set point, 195, 198; as blood component, 202; blood levels of, 151, 185, 188, 194, *194*, 195, 198; blood transported, 136, *136*, 146, 192, *192*, 193, 198; breathing reflex and, 318; capillary exchange, 151; dissolved in extracellular fluid, *248* ; in gas exchange, 68, 185, 188, 192, 193, 195, 198; hemoglobin/oxygen binding and, 138, 160; in aerobic respiration, 58, *61*; in Krebs cycle, 57, *57*; in urea, 203; as metabolic waste, *79*, 202; placental diffusion, 311; receptors, *194*; respiratory release of, *123*, 135, *142*, 186, 187, 192, 198; selective permeability of, 42

Carbon dioxide fixation, 498, 499, 500, *500, 501, 501*, 506

Carbon dioxide ion, 212

Carbonic acid, 23, *23*, *212*; in blood plasma, 193

Carbonic anhydrase, defined, 193

Carbon monoxide, 154, *516*

Carbon monoxide poisoning, 188

Carbon oxides, 516, *516*

Carbon rings, 28, 29, 36, *37*; double, 34; single, 34

Carboxyl group, 28, 30, *30*, 31; as functional group, *25*

Carboxypeptidase, 117, *117*

Carcinogenesis, 426, *427*, 436; defined, 426

Carcinogens, 154, 423, 426–427, 518; chemical, 426, *427*, 436; defined, 10, 426

Carcinomas, 431, 434, *434*, 436

Cardiac conduction system, 148; defined, 148

Cardiac cycle, *147*, 148, 150, 156; defined, 147; heartbeat, 147

Cardiac failure, muscular dystrophy and, 397

Cardiac muscle, *76*, 108; autonomic nerves, 230; blood flow to, *147*; capillary beds and, 144; cells, 144, 148, *148, 157*; contraction, 87, 98; damage regeneration, 283; defined, 76; heart tissue, 87, 98; involuntary, 87, 98; striation, 100, *100*; tissue, 134, 144, 148, *148*, 149, 157

Cardiac pacemaker, 157; defined, 148

Cardiac vein, *145*

Cardiovascular circuits, 146–147; pulmonary circuit, 146, *146*; systemic circuit, 146, *146*

Cardiovascular disease, 107

Cardiovascular disorders, 29, 144, 154–155; aging and, 324, 325; risk factors, 154, *155*

Cardiovascular system, 135; basic role, 142; blood movement by, 75; components, 142, *143*, 160; defined, 142; from mesoderm, *306*; overview, 142–143

Caries, 114

Carnivore, 482, 492, 493, 494, 495, 504, 506; defined, 490; primary, 490, *491*, 492; secondary, 490, *491*; tertiary, *491*; top, *491*, 492, *494*

Carotenes, 80, 127

Carotid arteries, *143*; defined, 149

Carotid bodies, *194, 248*; defined, 195

Carotid sinus, *248*

Carpals, *91*, 94, *94*, 95, *96*

Carpal tunnel syndrome, 94

Carrier designation, 392, *394*; genetic screening, 393; Queen Victoria, 396, *396, 397*

Carrier proteins, 221, *221*

Carrying capacity, *512*, 513, 526; defined, 512

Cartilage, *37*, 70, 84, 88, 90, 451; cells, 74; as connective tissue, 73, *73*, 74; defined, 74; from mesoderm, *306*; functions of, 74; growth hormone and, 273, 274; inflammation of, 325; shaft, 89; vocal cords, 191, *191*

Cartilage model, 89, *89*; in skeletal system, *91*, 108

Cartilaginous joints, 93, 97, 108; defined, 96

Cartilaginous plates, 323

Cascading reactions, 165, *165*

Catabolism, defined, 54

Catalysts, enzymes as, 54

Catalytic molecules, 54, 56, 62

Cataracts, 261, 367, 517

Catopithecus, 479

Cattle, 492, 501

Caudate nucleus, skill memory, *239*

Cave at Lascaux, 474, *474*

Cave of Gargas, 474

Cave paintings, prehistoric, 474, *474*, 475

CCR5, 338

CD4 cells, *178*, 338, 346

cDNA, 445, *445*, 449; defined, 445

Ceboids, *476*

Cecum, 121, *121*

Cell adhesion, 426

Cell bodies, *220*, 228, *229*, 230, *230*; of neurons, 77, *77*, 232, 233

Cell culture, 355

Cell cycle, 360, 370; defined, 354; duration, 355; four phases of, 354, 355, *355, 356*, 370

Cell cytoplasm, 449

Cell death, 58; apoptosis, 170, *171*, 173, 313; viral infection, 333, 334

Cell determination, defined, 306

Cell differentiation, 168, 172, 314; defined, 306; as process of development, 306, 307, 323, 328

Cell division/differentiation, 173, *179*, 182; into armies, 168, 169, 170, 171, *171, 172*

Cell division mechanisms, 352–353; cytokinesis, 360; meiotic, 290, 351, 352, *362–363*, 368, *368*, 370; mitotic, 290, 351, 352–360, 368, *368*, 370

Cell junctions, 72, 76, *76*, 80; examples, 72

Cell-mediated responses, 170–171

Cell membrane, 22, 27, 28, 59, 469, 471, 472; diffusion, 42, 44, *44*, 62; free radicals and, 71; functions, 39, *39*, 41, 62; lipids in, 29, *29*, *37*, 125; selective permeability, 42, 43, 62; structure, 39, 40–41; transport proteins and, 30, 54, 62

Cell metabolism, 426

Cell reproduction, 39, 333, 350–370, 471; aging and, 324, 325; cancerous, 355, 425, 426, 433, 436; chromosome number, 352–353, 354–355; division of the cytoplasm, 360; DNA as basis of, 15, 49; DNA replication and, 405, 406; early development and, 304, *307*, 308, 313, 328; hormonal stimulus, 273, 282; ionizing radiation, 367; localized, 307; meiosis and life cycle, 366–367; meiosis and mitosis compared, 368, *368–369*; meiosis I, 364–365; meiosis overview, 361, *362–363*; mitosis, 354–359; of mitochondria, 52; protein synthesis and, 417

Cells, 16, 38–62, 75, 139, 469; ATP role, 34; as basic life units, 4–5, 6, 7; basic structure, 40–41, 46, *46*, 53, *123*, 267; carbohydrates and, 26; chemistry of, *15*, 36; cholesterol and, 29, *37*; coordinated activity, 67, 84; cutaway diagram, *46*; cytomembrane system, 50, *50*; damage, 51, 367;

defined, 4; energy-demanding, 58, 60, 279; eukaryotic, 39, 40, 41, 46, 53, 62; function, 40–41, 46; infection-fighting (macrophage), 73; living, 475; metabolism, 54–55; migrations, 53, *312*, 313; normal aging, 21; organelles, 46, *46*; oxidation and, 21, 36; pH of, 23, 36; protein functions, 22, 30, 31, *37*; shape, 40, 46, 53, *69*, 270, 313; signaling molecules, 268; solute concentrations, 43, 202; surface receptors, 5, 450; target, 68; transport mechanisms and, 43, 46; using organic compounds, 25, 36; use of microscopes, 40, 46, 48; water content of, 22

Cell survival, 39, *39*, 45, 82; hormonal secretion, 284; nutrient exchange, 135

Cell theory, defined, 38

Cell-to-cell adhesion, 425

Cell-to-cell recognition, 41, 62, 72

Cellular debris, 75, *75*, 136, 137, 165; blood transport, 160; lymph transported, 159

Cellulose, 27, *27*, *37*; dietary, 121; as nonpolluting fuel source, 35, *35*; structural formula, 27

Cell walls, bacterial, 333, *333*; early detection, 38, *38*; structural material, 27

Centers for Disease Control and Prevention, 335

Central canal, 232, 234, *235*

Central nervous system (CNS), 77, 78, 232–233, *306*; analgesics in, 243; cell division "brakes," 354; components of, 219, 228, *228*, 244; defined, 228; digestion and, *122*; drug abuse, 242; function, 219, 220; neurons, 232; no regeneration, 225; oligodendrocytes, 224, *225*; psychoactive drugs in, 240–241; structure, 234–235

Centrioles, *46*; defined, 53, 358; in meiotic cell division, *362, 363, 369*; in mitotic cell division, *356*, 358

Centromere, in meiosis, 361, 364; in mitosis, 352, *352*, 358, *358*

Cercopithecoids, *476*

Cerebellum, *233, 234*, 236, 244; coordinates motor activity, 239; defined, 233

Cerebral cortex, 233, 236–237, *236*, 244; defined, 234, 236; information storage, 239, *239*, 244; olfactory bulbs in, 252, 264; sensory information in, 250, 252, 264; somatosensory cortex, 236, *236*; stimulant use, 242

Cerebral hemisphere, 237, 250, 263; four lobes of, 236, 237; left, 234, *234*, 235, 237, 238, *238*; right, 234, *234*, 235, 237, 238, *238*

Cerebral palsy, prematurity and, 319

Cerebral ventricles, *235*

Cerebrospinal fluid, 232, 235, *235*; defined, 234; HIV infection, 336

Cerebrum, *233, 234*, 236, 244; defined, 233

Cervical canal, 295

Cervical cancer, 343, 426, *429*, 433

Cervical cap, 298

Cervical nerves, *228, 230*

Cervical vertebra, 92, 93, *93*

Cervix, 292, *292, 293*, 295, 322; dilation, 318, *318*, 319, 328; fertilization stage, 308, *308*; infections of, 343; secretions, 297, 298

Cesarean section, herpes and, 343

CFTR gene, 386, *450*

CGG repeats, 417

Chain building, 24, *24*, 26, *26*, 27; branched, 33; carbohydrate, 50; fatty acid, 28, 29; linear, 33, 36, *37*. *See also* Polypeptide chains

Chancre ulcer, 340, *341*

Chancroid, 346; defined, 34[...] 336

Channel proteins, 221, *221*; ga[...] genetic defect in, 386; trigger o[...] ings, 271

Cheekbones, 92

Cheetahs, 468, *468*

Chemical bonds, 17, 36; defined, 19; types of, 15

Chemical digestion of food, 115, 116, 132

Chemical equations, *17*

Chemical insults, 114, 279, 398

Chemical messages, psychoactive drugs and, 240

Chemical notation, 17, *17*

Chemical synapse, 225–226, *225*, 244; defined, 225

Chemical weapons, in immune responses, *164*, 167, 177, 182

Chemokine receptor, 338, *338*, 339

Chemoreceptor, 248, *248*, 252, 264

Chemotherapy, 434, 435, 436; defined, 430; side effects, 430

Chemotoxins, 167

Chen, Joan, 372, *372*

Chernobyl meltdown, 367, 522

Chest cavity, 93, 115, 187, 195; blood route, 146; respiratory system and, 187, 191, 198

Chewing, 114–115

Chiasmata, 364, *364*

Chickens, 465, 466

Childbearing, 510; function of, 95

Childbirth, 83, 96, 97; Rh blood typing and, 140–141

Childhood, 97; body development, 305; developmental stages, 323, *323*; diabetes onset, 276; remodeling, 90

Children: AIDS and, *337*; diarrhea in, 164; immunization schedule, *176*; undernutrition and, 130, 131

Chimpanzee, 461, *476, 479, 480, 481*, 482, *482*; comparative biochemistry, 466; comparative morphology, 464, *464*

China, *525*

China White, 243

Chinese hamster ovary cells, 448

Chlamydia, 207, 261, 342, 343; defined, 341

Chlamydia trachomatis, 341, *341*, 342

Chlamydial infections, 341, 342, 346

Chloride, 127

Chloride ions, 20, *20*, 22, *22*, 206, 208, *208*; active transport, 209, *209*

Chlorine, 16, *16*, 20, 517; atomic number, *17*; atomic symbol, *17*; electrons in outer shell, 17; mass number, *17*

Chlorine ion, extracellular pH, 136

Chlorine monoxide, 517

Chlorofluorocarbons (CFCs), 500, *501*, 516, *516*, 517

Choking, 115, 187

Cholecystokinin (CCK), 122

Cholera, 45, 180

Cholesterol, *37*, 117, 119, 271; arterial deposits, 154, *154, 156*; blood levels, 395; blood transported, 136, 154, *155*; excretion, 117; in cell membrane, 41, *41*; intake level, 111, 156; as lipid, 32, 136; sterol, 29; synthesized in liver, 125

Cholesterol plaques, 325

Chondroblasts, 74

Chondrocytes, 74, *74*

Chordae tendineae, 144

Chordata, 467

Chorion, 310, 311, *311*, 312, 328; defined, 310

Chorionic cavity, *310, 312*

Chorionic villi, 310, 310, 311, *311*

Chorionic villus sampling (CVS), 311, 322; prenatal screening option, 393

Choroid, 258, *258, 259*, 261, 262

Chromate, 430

Chromatids: genetic recombination, *391*; in meiotic cell division, *363*; mitotic cell division, 352, *352, 353, 358, 358, 359*

Chromation, 49, 352, *352*; defined, 49

Chromium, *428*

Chromosomal DNA, 440, *441*

Chromosomal partitioning: by meiosis, 352, 365, 370; by mitosis, 352, 354, 355, *355, 356–357*, 370

Chromosome bands, *449*

Chromosome mapping, defined, 391

Chromosome number, 353, *356*, 359, 370; change, improper separation and, 387, 388; change in, 400–401, *400, 401*, 402, 403; defined, 352; diploid number, 374

Chromosome pairs, 372, 373; as autosomes, 387, 388, 389

Chromosomes, 49, 286, *287*, 373, 426, 446; abnormal rearrangement, 388, 395; Alzheimer's disease link, 326; artificial, 448; at metaphase, 388, *389*; behavior during meiosis, 387, 388; decondensed, *356*, 370; defined, 49, 352; detecting defects, 387, 393, 395; diploid number, 290, *291*; DNA component, 405, *406*, 409, *409*; fragmented, 367; full complement, 308, 309; gene loci, 351, 387, 388; genetic birth defects, 322, *322*; haploid number, 290, *291*, 294; in unduplicated state, 361, *363*, 365, *391*, 405; ionizing radiation and, 367; lower arm, *449*; meiotic condensed, *362, 364*, 368; meiotic duplicated state, *362, 363, 364, 364*, 365, 368, 370, 406; meiotic movement, 361, *362*, 368, *369*, 370, 373, 379; meiotic shuffling, 364, 365; mitotic condensed, 252, 354, *354, 356–357*, 358, *369*, 370; mitotic duplicated state, *353, 355, 358, 359*, 368, *369*, 370; mitotic movement, *356–357*, 357, 370; physical organization, 352, 353; recombination, 440, 441, 459; segment change, 387, 388; sex chromosomes, 290, 387, 389, 390; sister chromatids, 406; structural change, 387, 388, 398–399, 402, *403*; structure of, 388; *449, 449*; upper arm, *449*

Chromosome variations, 386–402; chromosomal basis of inheritance, 387, 388–389; chromosome number changes, 387, 400–401; chromosome structure changes, 387, 398–399; crossing over, 387, 391; genetic screening, 387, 393; human genetic analysis, 387, 392–393; linkage groups, 387, 391; patterns of autosomal inheritance, 387, 394–395; patterns of X-linked inheritance, 387, 396–397; sex determination and X inactivation, 387, 390; sex-influenced inheritance, 387, 388

Chromosome Y, 448–449

Chronic bronchitis, *196*, 197

Chylomicrons, *119*

Chyme, 117, 118, 119, 120; defined, 116

Chymotrypsin, 117, *117*

Cigarette smoking. *See* Smoking

Cilia, *53*, 68; cleansing action of, 387; defined, 53; in immune responses, *164*; in olfactory epithelium, 246; in oviduct, 295; on respiratory epithelium, 187, 189, 196

Ciliary body, 258, *258*

Ciliary muscle, *258*, 259, 260

Ciliated cells, centrioles as, 358

Circadian rhythms, 233, *233*

Circulatory shock, 155

Circulatory system, 2, 66, 78, 120, 134–160, *151*, 425; central role of transportation, *142*, cholesterol damage, 125; collapse, 178, *197*;

fetal bypass vessels, 316–317, *317*, 318; forerunners, 305; gas exchange in lungs, 185; in embryonic period, 314; intense exercise and, 103; pumping blood, 76; reflex center, 233; regulation, 279

Circumduction, 97

Cirrhosis, 120, 207; alcohol and, 242; type B viral hepatitis, 343, 346

Cis-retinal, 263

Citrate, 57

CLA, 10

Classification systems, *467*; defined, 467

Clavicle, *91*, 94; defined, 94

Clay, 470

Cleavage, 36, *306, 307, 307*, 323, 328; defined, 25, 306; rotational pattern of, 306, *307*; as second life cycle stage, 305, 308, 309, *309*

Cleavage furrow, *360*; defined, 360

Climate, 483, *483*, 488, 492, 501, 512, 516

Climax community, 488, *489*

Clinical depression, 131

Clitoris, 107, *286, 293*, 297; defined, 292

Clonal selection hypothesis, 175, *175*; defined, 174

Clone, 170, 439, *441*, 448, 453; antibody-producing hybridoma cells, *177*; human cloning efforts, 319; use in immune response, 173, 174, 175, 182

Cloned DNA, *441*; defined, 442

Cloned human gene products, *446*

Clotting, 136, *136*; of blood, 153

Clotting factors, 167, *167*, 439; gene encoded, 353

Clotting factor VIII, 396

Clotting factor IX, 396

Clouds, cosmic, 472

Clover, 502

Clumping response, 140

Coagulation, of blood, 153

Coal, 522, *522*; as fossil fuel, 23, 35

Coal tar, 196, 426

Cobalt, *18*

Cocaine, 157, 219, 227, 241, 242; crack, 218, *241*, 242; defined, 242; drug dependence, *240*; granular, 242; prenatal risk, 321

Coccygeal nerves, 228

Coccyx, 93, *93*, 465

Cochlea, *254, 255*, 257; defined, 254

Cochlear duct, 254, 255, *255*

Cochlear fluid, 255

Codfish, 524

Codominance, 373, 384; defined, 381

Codon, *412*, 413, *415*, 418; anticodon interactions, 413, *413*, 414, *414*, 415, *415*, 418; defined, 412

Coelacanth, *505*

Coelum, 312

Coenzyme A (CoA), 57

Coenzymes, 57, 58, 59, *59*, 60, 62, 471, 472; defined, 34, 55; NAD⁺, 57, 62

Cofactors, 55, 62; defined, 54

Coffee beans, 428

Cohen, Stanley, 282

Cohesion, of water molecules, 190

Coitus, defined, 297

Cold, sensations of, 250, 251

Cold sores, 81, 181, *334*

Cold stress, 214, *214*; physiological response, 214, 215, *215*

"Coliform bacteria," 121

Collagen, 32, 37, 425; aging and, 324, 325; in designer skin, 70

Collagen fibers, 73, *73*, 90, *150*; in artery walls, 150; in cartilage/bone, 74, 88, 89, 108; production interference, *196*

Collagen implants, 73

Collagen synthesis, vitamins and, *126*

Collarbone, *91*, 94; breaks, 94

Collecting ducts, 204, 205, *205, 208, 209*, 216; reabsorption and, 210, *210*, 211

Colon, *112*, 132; autonomic nervous system and, 230; bacteria in, 126; deferred pain, 251; defined, 121

Colon cancer, 423, 426, 429, *429*, 430, *431*, 434–436; diet and, 111, 121, 125

Color, ability to distinguish, 247, 260, 262, 263, 478, 485

Color blindness, 260, 397, *403*

Colorectal cancer, 426, *429*, 431, 436

Colostrum, 319

Columnar epithelium, 68, *68*, 69

Coma, 185, 236, *237*, 244; drug abuse, 242; hypothermia and, 214; parathyroid disorders, 281

Common cold, *180*, 186, 334; chronic, 189

Communication cells, 219, 220

Communication junctions, 148

Communication molecules, 136, 283

Communication signals, 167; in immune response, *164*

Communication system, hormone-based, 2

Community, 7, *489*; defined, 488

Compact bone, 74, *88*, 89, 108; defined, 74

Comparative biochemistry, 472

Comparative embryology, 464, *464*

Comparative morphology, 465, 466, 467, 472; defined, 464

Complement protein cascades, 173

Complement proteins, 164, 165, *165*, 166, 167, 182; activated, 173, 179; C1, 165; in specific response, *164*, 165, *165*; as nonspecific response, *164*, 165, *165*, 182

Complement system, 165; defined, 165

Complex carbohydrates, 111, 132; as glucose source, 111, 124, 125

Compound light microscope, 48

Compounds, 15; complex organic, 470; defined, 17; identifying composition of, *17*

Computer-aided design (CAD), 70

Computerized tomography, 429

Conception, 286, 287; events of, 297; external, 300

Concussion, 225

Condensation, 25, 36, 470; defined, 25

Condoms, 298, *298*, 299, 340; latex, 299, 339, 345

Conducting cells, 148, *148*

Cone cells, 248, 260, 262, 263, 264; blue, 263; defined, 262; green, 263; red, 263; red/green color blindness, 260, 397, *403*

C1 protein, 165

Conjunctiva, 261

Connecting stalk, *310*, 312, *314, 315*

Connective tissue, 67, 68, 69, *69*; around muscle, 76, *100*; blood as, 135; collagen in, 324, 325; defined, 73; fibroblasts in, 283; fibrous, 98, 108; from mesoderm, *306*; growth hormone and, 274; heart, 144; hypodermis, 80; implantation stage, 309; in bones, 88, *88*, 89, 108; in dermis, 73, 81; in digestive tube, 113, 132; in embryo, 77; at joints, 96; ligaments as, 90; types of, 73, *73*, 84; vitamins and, *126*; white blood cells in, 166; wrapping nerves, 227, *229*

Connective tissue cancer, *428*, 431, 436

Connective tissue proper, 73, *73*, 84

Consciousness, loss of, 154, 157

Consciousness, states of, 236, 237, 244; psychoactive drugs and, 240, 241; two spheres of, 238

Constipation, 121

Consumers, 7, *7*, 487, *489, 490, 491, 492*; defined, 490; natural resources and, *495, 498, 499*, 505–506

Contact inhibition, 425

Contact lenses, 261

Contagion, 180

Continental shelves, 522

Continuous variation, *382*; defined, 383

Contraceptive injection, 299, 301

Contraceptive sponge, 298, *298*

Contractile activity, 103, 106, 108; fatty acids as fuel source, 103

Contractile proteins, 76, 84, 100

Contractile tissue, 76, 78, 148, *291*

Contraction: ATP formation and, 103; of heart muscle, 148, *148*, 149; muscle, 87, 98; sliding-filament model, 100, 101, *101*, 108; uterine, 308, 318, *318*, 319, 328

Contraction phase, 147, *147*

Control group, 9; defined, 8

Control mechanisms, 98, 116; for enzymes, 55; in cell division, 354, 355; of morphogenesis, 313; urine concentration, 201

Convergent circuits, 229

Coolant, water as, 22

Copper, *127*

Coral reefs, 524

Core, planetary, 468, 472

Core temperature, 54, 201. *See also* Body temperature defined, 214; fertile period and, 298; set point, 214, 215; sperm production and, 288

Cork tissue, 38, *38*

Cornea, 259, *259*, 260, 261; artificial, 261; collagen in, 32; defined, 258, *258*; transplants, 173

Corneal transplant surgery, 173, 261

Corn plants, 466

Coronary arteries, *143, 145, 156*; defined, 144; plaque clogging, 144, *155*, 156

Coronary artery disease, 155, 157, *196*; free radicals and, 21

Coronary bypass surgery, 156, *156*

"Coronary circulation," 144

Corpus callosum, 234, *234*, 238, *238*; neural bridge, 238

Corpus luteum, 282, *293, 294*; in menstrual cycle, 292, 295, 296, 302; pregnancy and, 309

Corpus striatum, skill memory, 239, *239*

Cortical granules, in egg cytoplasm, *308*

Corticotropin (ACTH), 273; hypothalamus/pituitary production, *273*, 278, 284; primary effects, 269, 272, *272*

Cortisol, 271, 275, 278; deficiency, 278; molecular receptor for, 270; primary effects, 269

Cosmic rays, 367

Cotton plants, 446, *447*

Coughing, 181, *197*; as immune response, *164*; reflex center, 233

Covalent bonds, 15, *24*, 25–26, *25*, 30, *34*, 36; defined, 20; double, 20, 24, 28, *28*, 34; making ATP, 56; nonpolar, 20, 21, 22, 37, 44; nucleotide double helix, 406; polar, 20–22, 24, 29, 36, 37, 42, 44; polypeptide patterns, 31, 33; single, 20, 24, 28, 34; triple, 28

Cowpox, 162

Cow's blood, 145

Coxal bones, 95, *95*

Crab lice, 344, *344*

Crabs, 490, 506

Crack cocaine, 218, *241*; dependence or habituation, 240; prenatal risk, 321
Cramping, 299
Cranial cavity, 78, *79*, *269*; defined, 78
Cranial nerves, 228, *228*; function, 230
Cranial vault, 92
Cranium, 92, *93*; bones of, *91*, 92, 232
Crayfish, 492
Creatine, *103*
Creatine phosphate, dephosphorylation of, 103, *103*
Creatinine, 203, *206*
Cretinism, 280
Creutzfeldt-Jakob disease (CJD), 334
CRH releasing hormone, 278
Crick, Francis, *404*, 405, 406
Cri-du-chat syndrome, 398, *398*, *403*
Criminal investigation, blood groups, 141
Crista, 52, *52*
Crohn's disease, 123
Croplands, 508
Crop plants, 446–447, *447*, 452, *452*, 502–503, 506, 519
Crop rotation, 502
Crop yields, 501, 502–503, 519
Cross breeding, of plants, 373
Cross-bridges, *101*, 103, 104; defined, 101; myosin, 102, *103*, 108
Crossing over, 368, *368*, 370, 390, 402, 459; alleles exchange places, 387, *391*; defined, 364, 388; during meiosis, *362*, *364*, 365; recombination in nature, 439, 440, 453
Cross-matching, 141
Crown, 114, *114*
Crude oil, 447
Crust, 468, 469, 470, 472, 495, 498, 504
Cryptosporidiosis, *181*
Cryptosporidium, 45
Cuboidal epithelium, 68, *68*, *69*
Culture, 479, 483, 485; defined, 479
Cupula, 256, *297*
Curare, 103
Cushions, fat as, 125
Cuspids, 114
Cuticle, *33*
Cuts, 182; bacterial invasion, 168; skin mending, 80; stopping blood loss, 153, *153*
Cyanobacteria, 502
Cyclic adenosine monophosphate (cyclic AMP) (cAMP), 34; formation, *270*, 284; as second messenger, *270*, 271
Cycling of materials, 7, 488, *489*, 490, *490*, 502
Cysteine, *412*, *419*
Cystic fibrosis (CF), 122, 376, 444, 449, 450, *450*, 451; as autosomal recessive condition, 394, *403*; CF phenotype, 386, 393; as genetic disorder, 386–387, *387*, 393
Cystitis, 207, 342
Cytochrome *c*, 466; sequence, *466*
Cytochromes, 55, *127*
Cytokines, 425, 451; defined, 177
Cytokinesis, 360, 365, 370; defined, 360
Cytomegalovirus, *155*
Cytomembrane system, 50–51, *50*, 62; defined, 50; polypeptide chains in, 414
Cytoplasm, 39, 40–41, *41*, 43, 49, *49*, 68; blastomeres and, 306, *307*; defined, 40; diffusion through, 42, *44*, 50, *50*, 62; of epithelial cell, 118; extensions of macrophages, *166*; gap junctions and, 72, *72*; glycolysis in, 56, 57, 58, *59*, 62; granulocytes in, *137*; hemoglobin in, *139*; hormonal diffusion and, *270*, 271; in cell death, 170; increase during interphase, 354, *354*, *356–357*, 370;

in cytomembrane system, 51; meiotic division of, 361, *362*, *363*, 366, *366*, 368; mitotic division of, 360; of neurons, 77, 226; platelet formation and, *137*; pseudopods, 313; secondary oocyte and, 367; translation in, 414–415, *414*, 418, *419*
Cytoplasmic concentration, hormonal-initiated change, 271
Cytoplasmic division: daughter cells, 352, 355, *355*, *357*, 359, 360, 370; meiosis, *351*, *369*; mitosis, *351*, 359, 360, *369*, 370
Cytoplasmic fluid, negative charge, 220, 221, 223
Cytosine (C), in DNA nucleotide bases, 406, *406*, 410, 418; in RNA nucleotide bases, 410, 411, 412, 417, 418
Cytoskeletal proteins, *41*
Cytoskeleton, 53, *53*, 313, 424; adhering junctions and, 72, *72*; defined, 53; main function, 46, *46*; microtubules, 358
Cytotoxic cells, 171, 182
Cytotoxic T cells, 169, 170, *171*, 177, 182, *182*, 430; defined, 168; effector, *169*, *171*; HIV antigenic protein, 338; immunity breakdown, 427, *427*; thymus function, 281; virgin, 169
Czarina Alexandra, *397*
Czar Nikolas II, 397

D

Dairy products, 428
Damselfish, 488
Darwin, Charles, 8, 458, *458*, 460, 462
Daughter cells, *307*, 308, 333, 358, 424, *461*; cell cycle and, 354, *355*, *356–357*; cytoplasmic machinery in, 352, 353, 360; ionizing radiation exposure, 367; diploid (2*n*), 353, *357*, *363*, 365, 370, *370*; haploid (*n*), 361, 365, *369*, 370, *370*; mitotic chromosome distribution, 351, 353, *356–357*, 359, *369*; sister chromatids, 406
Daughter nuclei, *363*, *369*, 370
da Vinci, Leonardo, 7
Daytime vision, 262, 263, 475
DDE, 505
DDT, 279, 505
Deafness, prenatal risk, 321
Deamination reactions, 202
Death, 163; abnormal chromosomal rearrangement, 388; agglutination, 140; AIDS-related, *181*, *346*; ATP production, 101; by infectious diseases, *181*; cardiovascular disorders as cause of, 154; defining, 11; drug abuse, 242; early infant, 319, 320; fetal, 141; from ionizing radiation, 367; hypothermia, 214; leading causes, 330; legal, 11; life cycle end, 305; programmed, 307; sudden, *155*; untreated STD and, 330, 331
Death rate, 510, 511, 513, 515, 526; children, 131
Death signal, 372
Deceleration, straight-line, 256, 257, 264
Decibels, 257
Decomposers, 495, *495*, 498, *498*, 499, 502, *504*, 506; defined, 490; and ecological balance, 487, *489*, 490, *490*, *491*; energy flow through ecosystem, 492, *493*, 494; organization and interdependence, 7, *7*
Decomposition, 463, 503, *503*, 504, 520; defined, 502
Decompression sickness, 188
Defecation, 121

Defender protein, 165
Defense, proteins used in, *136*
Defense mechanisms: cellular, 163, 164; chemical, 163, 164, 166; macrophages as part, 73; physical, 163, 164
Defibrillating drugs, 157
Deforestation, 519–521, *519*, *520*, *521*, 524–526; acid rain, 517; defined, 520; ecosystem disruption, 496, 497, *499*, 501, *501*; overpopulation and, 512; recycling, 519
Degenerative diseases, 334
Deglaciated regions, 488, *489*
Degradation, of red blood cells, *120*
Degradative pathways, 54, 62
Dehydration, 78, 80, 84, 164, 210, 211; adrenal cortex disorders, 278; as insulin deficiency and, 276; thirst, 211; vitamins and, *126*
Dehydration synthesis, defined, 25
Deletion, altering chromosome structure, 398–399, *398*, 402; defined, 398
Deliriants, 243
Delirium, 243
Delta cells, 276
Deltoid, *99*
de Medici, Maria, *280*
Dementia, 326
Demographic transtion model, *514*, defined, 514
Demography, defined, 513
Denaturation, defined, 33
Dendrites, 77, 220, 229, 232, 326; defined, 220
Denitrification, *503*; defined, 502
Dense, irregular connective tissue, *73*; defined, 73
Dense, regular connective tissue, 73; *73*; defined, 73
Density, 512; defined, 510
Density-dependent control, 512, 526; defined, 512
Density-independent controls, 512; defined, 512
Dentate nucleus, 236
Dentin, 114, *114*
Dentition, 476, 477, 478, 479, 482; defined, 476
Deoxyribonucleic acid. *See* DNA
Deoxyribose, 26, 406, 470; in DNA nucleotides, 406, *406*, 410
Depolarizing signals, 227
Depo-Provera, 299, 301
Depression, 218, 267; drug use/abuse, 242, 243; vitamins and, *126*
Depth perception, 479, 485
Dermatitis, *126*
Dermis, 70, *80*, 81, 250, *251*, 325; connective tissue, 73, 81; defined, 80; from somites, *312*
Descending colon of large intestine, *120*, *132*
Desensitization program, 178
Desertification, *519*, 526; defined, 519
Deserts, 488, *488*, 493
Designer drug, 243
Designer skin, 70
Desmosomes, 72, *72*
Detoxification, in liver, *120*
Detrital food webs, *493*, 506, *506*; defined, 492
Detritivores, 487, *493*, 494, 495, *495*, 506; defined, 490; energy flow through ecosystem, 490, *490*, 492
Development, 304–328; early embryo, 312–313; early events in, 308–309; extraembryonic membrane, 310–311; feature differentiation, 314–315; fetal development, 316–317; from birth onward, 318–319; hormonal adjustments, 267, 269, 280, 281, 284; mother's lifestyle, 320; postnatal, 323, *323*; prenatal, 323, *323*; stages of, 306–307, *307*, 323, *323*, 328; steroid

hormones and, 29; target cell response, 275, 285
Diabetes, 70, 207, *438*; blood glucose regulation and, 276; cataracts, 261; diet and, 111, 124, 276; kidney damage, 213; obesity and, 128; type 1, 276, 339, 446, *446*; type 2 diabetes, 276
Diabetes mellitus, 154, 276; heart attack and, *155*
Dialysis, 213
Diaphragm, *144*, *186*, 190, *190*, 191, 298, *298*; blood delivery to, *143*, *146*; contraction in breathing, 194, *194*, 195; deferred pain, *251*; defined, 187; muscle, 78, 115
Diarrhea, 45, 181; allergic reaction, 123; causes in colon, 121, 122; as immune response, 164, *164*; thirst, 211; vitamins and, *126*
Diastole, *147*, 150; defined, 147
Diastolic pressure, 149
Diatoms, *491*, 492
Dideoxycytidine (ddC), 339
Dideoxyinosine (ddI), 339
Diet, 111, 478–482, 485; aging effects and, 325; average American, 125; bulk in, 121; cancer risk, 423, 430, 434, 436; diabetes control, 276; gender and, 124; heart attack and, *155*, 156; hormonal adjustment, 267; iron-poor, 139; low-fat, 325; low-fiber, high-fat, 124, 125; low sodium, 208; PKU control, 394; prenatal, 320; well-balanced, *124*, 125, 127
Dieting, 129
Diffusion, 39, 42–43, *42*, 44, *44*, 62; at cell junctions, 72, *72*; blood velocity and, 149; defined, 42, 43; during immune reaction, 165; of hormones, *270*, 271, 272; of hydrogen ions, 116; in capillary beds, 150, 152; of ions across neuronal plasma membrane, 221, 225; of kidney filtrate, 206, 209; of neurotransmitter molecules, 227, *231*, 268; of nutrients into blood, 119, *119*; of oxygen/carbon dioxide, 185, 188, 192, *192*, 193; of pepsins, 116; placental, 311, *311*; simple epithelium and, 68, *69*, 81
Digestion, 110–132; of bilirubin, 139; chemical, 114; controls, 122–123; disruptive disorders, 123; nutrient turnover, 122–123; parasympathetic nerves, 231; process in small intestine, *119*
Digestive enzymes, breakdown products, 117, 132; cystic fibrosis and, 386–387; as exocrine secretion, 68; in specific immune response, 168, *171*
Digestive system, 2, 66, *78*, 110, 111, 434, 435; absorption, 113, 132; blood pressure and, 151; circulatory system link, *142*, 147; defined, 112; elimination, 111, 113, 132; functions, *112*, 132; major components, *112*; mechanical processing/motility, 113, 132; membrane lining, 69, 72; mucous membranes, 170, 173; overview, 112–113, 132; secretion, 113, 132
digestive tract, cystic fibrosis and, 386–387
Digestive tube, 122, 132; early development, 310; muscle layers, 113
Digital rectal examination, *429*, 433
Dihybrid cross, 379, 380, *380*, 384; defined, 378
Dihydroxytestosterone (DHT), 397
Dinosaurs, 426, 467
Dioxins, 279, *516*
2,3-diphosphoglycerate (PPG), 193
Diphtheria, 176, *176*, 181
Diploid cells (2*n*), 351, 352, *352*; Alu transposable element and, 416;

chromosomal components, 387, 388, 402; daughter cells, 353, *357, 363,* 365, 370, *370;* defined, 352; germ cells, *390;* homologous chromosomes contained, 374, *374, 375, 391,* 399; in meiotic division, *362,* 365, 366, *366,* 367, *368;* in mitotic division, 353, *356,* 361, *362, 368, 369,* 370; segregation, theory of, 375, *379, 384*

Diploid chromosome number, 290, *291,* 351, 367, *369,* 370, *370;* reduction, 375

Directional terms, *79*

Disability, cardiovascular disorders and, 154

Disaccharidase, *117*

Disaccharides, 26, *26, 37, 117*

Discoverer XVII satellite, 355

Disease, 180; artificial tissue and, 70; defenses against, 15; degenerative, 334; diagnosis of, *18;* homeostasis and, 44, *45;* immunization, 176; transmission, 71; viral, 71; water pollution and, 45. *See also* Infection

Disease agents, 145, 160, 331, 512, 513, 526; animal virus for STD, *334;* genetic susceptibility, 393

Disinfectants, HIV virus and, 337

Disjunction, 365

Disks, cartilage cushions, 74

Dislocation, shoulder, 94

Disorientation, 266

Dispersal routes, 462, 463

Dispersion, 510; defined, 510

Distal tubules, *79,* 205, *205, 208,* 209, 216; aldosterone action, 278; permeability, 210, *210,* 211, *211*

Distension, as a trigger, 122

Disulfide bonds, 32, *33*

Diuretic, 210

Dive reflex, defined, 214

Divergence, *461,* 466, *466,* 476, 479, 483, 484; defined, 461

Divergent circuits, 229

Diversity, 460, 463, 464, *464,* 465; genetic, 468

DNA (deoxyribonucleic acid), 3–6, *4, 12, 15, 37,* 40; autoimmune disease and, 179; chromosomal organization, 405, 409; consists of, 26, 388; defined, 34; duplication disorders, 400; FISH diagnosis, 322; foreign sequence, 448, 450, 453; free-radical damage, 21, 324; from mRNA to proteins, 405, 412–413; function, 405, 419, 420; gene region of, 406, *410, 411,* 417, 418, *419;* gene segment, 246, 374, 387, 388, 426, 450; gene/protein connection, 405, 408, 410–411, 412–413, *412,* 418, *419;* gene therapy and, 394; genomic imprinting and, 383; hereditary instruction, 6, 12, 348, 351–353, *353,* 370, 405; HIV gene incorporation, 339; HIV virus, 336; hormonal binding, 270, 271, 284; hydrogen bonds and, 21; in criminal investigation, 141; in head of sperm cell, 290, *291;* in prokaryotic fission, 333, *333;* into RNA, 405, 410–411, *411,* 418; location in cell, 46, *46, 47,* 49, *49,* 62; meiotic replication, 361, *361, 362, 363, 368;* of mitochondria, 52; mitotic duplication, 354, 355, *355, 356, 356,* 370; model of, *34;* mutagens and, 416, 418; mutation, 405, 416; nucleic acid digestion, *117;* nucleotide repeats, 405, 417; nucelotide sequence, 412, *412,* 413, 416, *416,* 418; ob gene, 129; proto-oncogenes, 81, 426; recombinant, 174, 175, 448; recombinant technology, 452, 453; reductional division, 361, 365; regulating gene action, 405, 417, 418; repair mechanisms,

324, 407, 416, *416,* 436; Sry gene, 287; strands, 445, *445,* 466; subunits, 404, 405, *406;* translations, 405, 414–415, *419;* of zygote, 308

DNA analysis, in detecting genetic disorders, 393

DNA databases, 444

DNA-DNA hybridization, 472; defined, 466

DNA fingerprint, 397, 444, *444;* defined, 444

DNA fragments, 170, *171,* 441, 444, 446, 453, 466; amplification procedures, 442; sorting out, 442–443

DNA library, 441, 442; defined, 441

DNA ligase, 407, 441; defined, 441

DNA mass screening, 444

DNA molecules, 438–444, *443,* 449–453, 466, 468, 471–472, *471;* blue-white screening, 445; chromation, 49; chromosome as, 352; double-stranded, *442;* during replication, 405; as master molecule of life, 404; per chromosome, 409; reverse transcriptase and, 333, *337*

DNA polymerases, 442, *443,* 553; defined, 407

DNA probe, defined, 429

DNA replication, 410, 430, 441; at interphase, *391;* before cell division, 405, *406;* conserved strand, 407; error in, 407, 418; mechanisms, 368; nucleotide strand separation, 406, 407, *407;* procedure, 405, 406–407; self replication, 405; semiconservative nature, 407, *407*

DNA structure, 352, 404–419; model of structure, *404*

DNA template, 406, 410, *410,* 411, *411, 412,* 418

DNA virus, 332, *332, 334,* 337, 343, 426, 435–436

Dogs, 464–465

Dolphins, 465

Dominant allele, 372, 373, 375, *375,* 402; defined, 374; for a sex-influenced trait, *398;* genetic analysis of transmission, 392, *392,* 393, 394–395, *394;* independent assortment, 378, 380, *380;* in Mendel's single-gene model, 381; in monohybrid cross, *377;* in Punnett square method, 376, *376;* in testcross, 378, *378;* penetrance, 382, *382,* 384

Dominant-to-recessive ratio, 377

Dopamine, 227; as signaling molecule, 242

Double helix, 352, 405; DNA structure as, 406, 407, *407, 410, 411, 412*

Double signal, 170, *171*

Double-stranded cDNA, 445, *445*

Double-stranded DNA, 445

Douching, 298, *298;* PID and, 342

Douglas, Michael, *375*

Downers, 243

Down syndrome, 400, *401, 403,* 435, *449;* frequency/maternal age relationship, *401*

Drainage vessels, 142, 158, *159*

Dreams, 237

Drinking, 218; drug dependence and, *240*

Drinking water, contamination of, 45, *45,* 203

Drought, 45, 439, 446–447

Drug abuse, 241, 242, 243; defined, 240

Drugs, 218; addiction to, 241; analgesics, 243; antagonistic interaction, 241; antiparasitic, 344; brain activity reduction, 242–243; deliriants, 243; dependence warning signs, *240;* dosage, 241; effects of, 240; hallucinogens, 243; inactivation of, 50, *120;* nonprescription overuse, 153; overdose, *225;* physical depen-

dence, 240; placental transfer, *320, 321,* 328; potentiating interaction, 241; psychedelics, 243; recreational drug use, *241;* SLE and, 179; STD protection, 345; stimulants, 242; synergistic interaction, 241; tachycardia, 157; testing, 206; therapeutic use, 106; tolerance, 240; withdrawal, *240,* 243

Drug therapies, 425, 430; "cocktail," 339; immune system suppression, 173; side effects, 339

Dry acid deposition, 516; defined, 516

Duchenne muscular dystrophy (DMD), 396–397, *403;* embryo screening, 393

Duckbilled platypus, 462

Duct system, embryonic, sex determination and, *390*

Ductus arteriosus, 317, *317*

Ductus deferens, 288

Ductus venosus, 317, *317*

Duodenum, 112, *113,* 116, 117, *132*

Duplication, altering chromosome structure, 398, 399, *399,* 402; defined, 399

Dura mater, 232

Durrant, Barbara, 468

Dust, *516*

Dutch elm disease, 505

Dye molecules, diffusion of, 42, *42*

Dyes, 426; granulocytes, 166; "staining," 48, 137; testing, 156

Dysplasia, defined, 424

Dystrophin, 397

E

Ear, 249, 465; components of, 254, *254, 255;* embryonic, *315;* protection, 257; sensory receptors in, 256

Ear canal, 29, 92

Eardrum, 29, 254, *254, 255*

Ear lobes: attached/detached traits, 372, *373;* attached gene, 364, 372–373; detached gene, 364, 372–373

Earth, 472

Earth history, 468–469, 472; early, 469, 470, *470,* 471, 472; origin, 458, 468; primordial, *469*

Earthquakes, 512

Earthworms, 490, 506

Earwax, 29, *37;* blockage, 29; as exocrine secretion, 68

Eating, 202, 237; psychoactive drugs and, 240

Eating behavior, 82

Ecological balance, 452

Ecological pyramid, defined, 492

Ecology, defined, 488; population growth, 509; principles, 488–489; threat to, 23

Economic development, *514,* 515

Ecosystem, 7, 486–506, *488, 489, 493, 506;* acid rain effect, 516; aquatic, 492, 494, *494, 498,* 501, 502; defined, 487, 488; forest, *497,* 520; human impacts, 454; land, 495, *495,* 501, 502; limits of, 525; organization, 490–491, *490;* types of, 488, 490

Ectoderm, *306,* 312; defined, 306; embryonic tissue layer, 305, 313

Ectoparasites, 344

Ectopic (tubal) pregnancy, 309, 342

Edema, 130; defined, *152;* localized, in acute inflammation, 167, *167;* plasma fluid, 185

Eelgrass, 492

Effector B cells, 169, *169, 172,* 173, *179,* 182

Effector cells, 169, 175, *175,* 176, 177; defined, 168; subpopulations, 170, 182

Effectors, *82, 83,* 84, 229, 230, 244; defined, 82

Effector T cells, 170; cytotoxic, *169, 171,* 182; helper, *169,* 171, 182

Efferent arteriole, 205, *205,* 206, 211

Efferent nerves, 230

Egg, 6, 286, 287, 290, 426, 452; development of, 292, *293;* early cleavage, *306, 309;* as female gametes, 374, 376, *376;* fertilization, 299, *305,* 306, *307,* 309, 348, 350; formation, hormone stimulus, *272;* gamete formation, 352, 355, 361, 365, *366,* 370; genetic engineering, 448, *448;* implantation, 309; production of, *79;* sex determination in, 390, *390*

Ejaculation, 289, *289,* 290, 297, 308

Ejaculatory duct, 288, *288, 289*

Elastic cartilage, 74

Elastic fibers, 73, *73*

Elastin, 73, 81, 88, 90, 150

Elastin fibers in bones, 88

Elbow, formation, *315*

Elbow joint, 94, 96, 98, *98*

Electrical charges, 17, 36; across neuronal plasma membrane, 220; at axon hillocks, 220; cardiac conduction system, 148; diffusion and, 42; in hydrogen bond, 21; ionization and, 15, 20, 22, 36; pattern of, 134

Electrical storms, 238

Electric gradient, 43, 58; defined, 42

Electricity, in creating nonpolluting fuel, 35

Electric shock, 157

Electrocardiogram (ECG), 134, 148, *157*

Electrochemical gradient, 221; across neuronal plasma membrane, 221, *221,* 223, 226

Electroencephalograms (EEG), 236, *237*

Electrolytes, 123; defined, 22

Electron microscope, 48, *48*

Electron microscopy, 268

Electrons, 15, *16,* 17, 36; coenzymes and, 34; defined, 16; distribution in atoms, 19, *19;* energy levels, 19; in chemical bonds, 19, 20, 21, 25; use in scanning electron microscope, 48

Electron transfer, 19, *20,* 20, 35, 36, *37,* 57; defined, 25; enzymes and, 55; in oxidation-reduction reaction, 54

Electron transport chains, 466; iron in, 126; minerals and, *127*

Electron transport phosphorylation, 56, *59*

Electron transports, 54, *126;* ATP production, 58, 59, *59,* 62

Electroporation, defined, 450

Electrostatic attraction, 468

Elements, 15, *16,* 36; atomic number of, 16, *17;* atomic weight, 16, *17;* atoms of, 17, 36; defined, 16; mass number, 16, *17;* molecules of, 17; symbol, *17*

Elimination, control of, 224; fecal, 202; water loss, 202, 216

"Elixir of life," 145

Elongation, 414, *414,* 418

Embolus, 156

Embryo, 2, 77, 287, 464, *464,* 472; cardiac tissue and, 134; cartilage model in, 89, *89,* 90; death, 279; germ layers, *306, 307,* 310; hyaline cartilage in, 74; measurement, *316;* NGF and developing, 282; oogenesis, 367; readying uterus for, 292, *293, 293,* 295; reproductive system, *79;* spleen in, 159; umbilical cord and, 311, *311;* viable, ethics and, 300; XX, 286, 292; XY, 286, 288

Embryo implants, 300

Embryonic development, 307, 308, 309, *309,* 323, 328; critical periods, 320–321, *320;* early, 304, 305, 306, 314, *314,* 448; organs of support,

293; period of, 312, *312*, 314, *314*, *315*; tissue layer formation, 305, *307*
Embryonic disc, *310*, 311, 312, *312*; defined, 310
Embryoscopy, 322
Embryo screening, 322, 393
Embryo transfer, 300
Emigration, 460, 510, 526
Emotional states, 243; in hippocampus, 239; limbic system, 252; regulation of, 227; weight gain and, 128, 129
Emotions, brain centers, 237, 244
Emphysema, 189, *189*, 196, *196*, *197*
Emulsification, 117, *119*, 132
Emulsion, 117, *119*
Enamel, tooth, 114, *114*
Encapsulated receptors, defined, 250
Encephalitis, 225
Endangered species, 467, 468, 524
Endemic disease, 180
Endocarditis, *155*
Endocardium, 144
Endocrine cells, *269*, *275*, 281; hormone secretion, 268, *275*; in heart, *269*, *275*, 281; in small intestine, 122; in stomach lining, 122
Endocrine disrupters, 279, *279*
Endocrine gland, 281, 284; pancreas as, 276
Endocrine system, 66, *78*, 268–269, 428; components of, *269*; control blood pressure, 150, 151; controls over digestion, 122, 123, 132; defined, 68, 268; disruption, 279; homeostasis and, 83, 84; integration and control, 266–284; kidney volume flow control, 207
Endocytic vesicle, *50*, *171*, 172
Endocytosis, 44, 152; defined, 44; hormone-receptor complex and, 271
Endoderm, *306*, 312; embryonic tissue layer, 305; folding, *313*
Endogenous pyrogen, 167
Endometrial cancer, *429*, 430, 433
Endometriosis, 299
Endometrium, *293*, 310, 311, 328, *429*; defined, 292; in menstrual cycle, *292*, 293, 295, *295*, *296*, 302; site of implantation, *293*, 295, 309, *309*, 310, *310*
Endoplasmic reticulum (ER), 49, 62, 102; main function, 46, *47*; rough, *46*, 50, *50*, *51*; smooth, *46*, 50, *50*, *51*
Endorphins, 227, 243
Endothelial cells, *150*; brain capillary walls, 235; edema and, 185; in blood vessels walls, 150, 152, *156*; in capillaries, 159, 166; in heart, 144, *144*; in inflammation, 166
Endotherms, 216; defined, 214
End products, defined, 54
Endurance, 106
Endurance sports, 87
Energy, 3, 5, *5*, 12, 470; cellular use of, 39, 54, 55; conversion in cytoplasm, 39; "cost" of movement, 87; defined, 4; demands for, 59, 60, 167, 191, 279; extraction, 4, *5*, 52, 131; intake/energy output, 111, 128, 129, 132; levels, electrons and, 19; measurement, 124; net, 522; output factors, 128; release, 52, 56; transfers of, 4, 7; transportable packets of, 25, 26, 27
Energy boost, in transport, 44, *44*, 62
Energy carriers, *37*, 52, *52*; ATP as, 34, 56, 62; defined, 54; mitochondria as, 52, *52*
Energy consumption, 509, 522, *522*, 525, 526
Energy, of ecosystem, *489*, 490–495, *490*, *493*, *494*, 506, *506*; flow in biosphere, 7, *7*; nonpolluting fuel, 34, 35; solar, 35; stored forms of, 487–488
Energy pyramid, 492, 494, *494*

Energy-releasing pathways, 60, 62, 101
Energy sources, 15, 25, 26, *26*, 27, *37*; alternative, 60, *60*; fat as, 123; fatty acids as, 103; foods as, 56, *61*, 124; glucose as main, 111, 124, 132; nutrients as, 110, 122; nonpolluting fuels, 34; triglycerides as, 28, 29
Energy storage, 7, 26, *26*, 27, 28, *37*, 75; fats as reserves, 125, 129
Engorgement mechanism, 297
Enkephalins, 243
Environment, 2, 3, 477, 512, 513; adaptive traits and, 6; atomic structure and, 16; bacteria to clean up pollutants, 447; cancer factors, 423; cell response to, 5; external, neuron function and, 77, *77*; in lungs, 138; nonpolluting fuels, 34, 35; recycling energy and, 7, *7*; support of populations, 458
Environmental biotechnology, 446
Environmental cues, 267, *275*, 281
Environmental pollutants, blood-brain barrier, 235
Environmental Protection Agency (U.S.), 279, 434, 447
Environmental reservoir, 495, *495*
Enzyme-coding gene sequence, 445
Enzyme-mediated reactions, 25, 36, 39, 58, 60, 62; in carbon dioxide transport, 193
Enzymes, 15, 27, 36, *37*, 139, *471*, *472*; of accessory organs, 111, 113, 132; action in condensation, 25, 36; action in hydrolysis, 25, 36; assembling DNA fragments, *443*, 445, *445*; blood transport, 75; bone tissue breakdown, 90; carbon compounds, *37*, 193; controlling, 55, 62; cut DNA, 439, 442; defined, 25, 54; digestive, 22, *112*, 114, 116, 122, 125, 132; DNA replication and, 405, 406, 407; DNA scanning, 354; from cancer cells, *425*; function, temperature and, 214; gene coding and, 382, 447, 448, 449; gene therapy and, *450*, 451; in ER system, 51; in Golgi bodies, 50; in immune responses, 164, 168; polysaccharide synthesis, 381; proteins as, 30, *30*, 31; splitting ATP, 101; substrates, 54, *54*, 55, *55*; synthesis, 55; use in nonpolluting fuel, 35; weak, 470–471
Eocene epoch, 479
Eosinophils: blood bone, *136*, 137, *137*; defined, 166; in nonspecific response, *164*, 166, 167, 182
Epidemic, 180; HIV as global, 335; sexually transmitted disease, 330
Epidermal cells, 81
Epidermal growth factor (EGF), 282, 283
Epidermis, 70, 80, *80*, 81; cell division, 325; defined, 80; free nerve endings on, 250, *251*; free surface, 80; from ectoderm, *306*
Epididymides, 288, 289, *290*, 297; function, *288*, *289*
Epiglottis, 74, 115, *115*, *186*, 187, *191*, 280
Epilepsy, 225, 449; barbiturates, 243; progressive myoclonus, *449*; split-brain and, 238
Epinephrine, *271*, 282; blood vessels and, 154; primary effects, *269*, *275*, 279
Epiphyseal plate, *89*, 108; defined, 89
Epiphyses, *89*
Epiphysis, 89, 108
Epithelial cells, *69*, 80, *118*, *119*, 120; radiation destruction of, 367; in the heart, 144; in urinary system, 204, 208; tight junctions, 72, *72*; villus covering, 118, *119*, 132
Epithelial linings cancer, *428*, 430, 431, 436

Epithelial tissue, 67, 68–69, *69*, 84; in embryo, 77; in endometrium, 292; taste buds in, *252*
Epithelium, bronchial, 189, 196; cell-to-cell contact, 72, 84; connective tissue in, 73, *73*; defined, 68, 69; free surface of, 68, 69, *69*, 72, *72*, 84; from endoderm, *306*; in blood vessel walls, 88, 108; in implementation stage, 309; mucosa, 113; on respiratory surface, 185, 186, 187, 188, 192, 198; pigmented, in retina, 258, 262, *262*; sensory neurons in, 77; types of, 68, *68*
Epstein-Barr virus, *155*, *178*, 434
Equilibrium, organs of, 256, 264. *See also* Balance
Equilibrium position, 256
ER, 414
Erectile tissue, *289*
Erection, 297, 327
Erosion, 468, 490, 502, 504, 512, 519, 520
Erythrocytes, 136; as blood component, *137*, *139*
Erythropoietin, 139, *275*, *446*
Escherichia coli (E. coli), 121
Esophageal cancer, 196, 430, *431*; diet and, 111
Esophagus, 112, *112*, 113, 115, *115*, *116*, 132; defined, 114; type of epithelium, *68*
Essential amino acids, *125*, 439, 446–447; defined, 125
Essential fatty acids, defined, 125
Estrogen, 29, 156, 268, *271*, 430, 432, 433; adrenal cortex production, 278; control fertility cycle, 287, 295, *296*, 302; defined, 292; environmental, 279; function, *293*; in pregnancy, 309, 310, 318, 319, 328; menopausal levels, 293, 327; primary effects, *269*, *275*; synthetic, 299
Estuaries, 470
Ether, *428*
Ethical problems, 298, 301, *301*, 345, 346, 452, 453; in vitro fertilization, 393
Ethmoid bone, 92, *92*
Ethmoid sinus, *93*
Ethyl group, 24
Eugenic engineering, 452
Eukaryotic cells, 39, 41, 46, 53, 62, 333; defined, 40; cell division, 351, 352; crossovers and, 391; DNA organization levels, *409*; mitochondrion in, 52
Eukaryotic species, 440
Euphoria; drug abuse, 242, 243
Eustachian tube, *254*, 255
Evaporation, *496*, *497*, 518, 520; water loss by, 202, *203*, 216
Evaporative heat loss, defined, 215
Evolution, 456–472, *461*; adaptive traits and, 6; biological, 2, 457, 472; chemical, 457, 468, 469, 470, 471, 472; cultural, 2, 475, 477, 479, 483, 485; defined, 459; history of, 2, 458; human, 93, 95, 474–485; molecular, 471; physical, 457, 472; rates of change, 461; sketchy relationship, *466*; vitamins and, 126
Evolution by natural selection, theory of, 8, 12, 459, 460
Evolutionary modifications, 364
Evolutionary trees, 466–467, *481*; defined, 466; diagrams, 466, *466*
Excitation, waves of, 148
Excitatory postsynaptic potentials (EPSP), 227, *227*
Excitatory signals, 226, 227, 229, 230, 236, 244; by acetylcholine (ACh), 102, *103*; drug abuse, 242
Excretion, *208*; in urine, *120*, 203; of waste, *79*; of water/solutes, 209, 210

Excretory system, *306*
Exercise, 90, 130; aging and level of, *325*; blood circulation during, 151, *152*; body functions and, 66, 67, 80; calories expended, 129, *129*; core temperature and, 214; defined, 106; dieting and, 129; energy for breathing during, 191; fluid balance, 210; HDL levels and, 154; heart rate and, 5; lack of, 121, 154, *155*; levels of/ATP formation, 103, *103*; quick energy source, 27; strenuous, 87, *103*, 122, 157, 191; stress electrocardiogram, 156; triglyceride use, 60
Exercise physiologist, 66, 86
Exhalation, *190*, 191, 198, 212; of carbon dioxide, 202; nervous system control, 194, 194, 195
Exocrine glands, compound, 68; defined, 68; genetic disorders, 386; pancreas as, 276; pheromone secretion, 268; secretions of, 68, 253, 276; simple, 68
Exocytic vesicles, *50*
Exocytosis, 44, 50, 119, 152; defined, 44
Exons, *411*, 418, 445; defined, 411
Experiment, 10, 12; defined, 8
Expiration, 191, 198; active, 191; defined, 190
Expiratory reserve volume, 191, *191*
Exploratory surgery, 18
Exponential growth, *511*, 526; defined, 511
Expressed sequence tags (EST) database, 449
External auditory meatus, 92
External environment, 84; cues, 281; decoding signals from, 5, 77, *78*, 247; homeostasis, 83; interstitial fluid as, 142; metabolic waste disposal, 201
External genitals, female, 292, *293*; STD symptoms, 343
External oblique, *99*
External respiration, defined, 192
External urethra sphincter, 206
Extinction, 466, 468, 505; background, 467; mass, 467; plant, 524; regional, 467
Extracellular fluid (ECF), *79*, 119, *172*, 202–203, 209, 284; acid-base balance, 211, *211*; calcium ions from, 103; cushion brain, 234, *235*; defined, 82, 202, 216; hormones in, 68; potassium concentration, 278; sodium concentration, 278; stability, 111, 202, 203, 213, 216; substances dissolved in, *248*; volume, 136, *136*, *152*; volume/composition balance, 201, 206, 207, 210, 211
Extracellular matrix, 313; defined, 73
Extracellular pathogens, *169*, 173
Extraembryonic membranes, *309*, 310–311, *310*, 328; defined, 310
Extrinsic clotting mechanism, 153
Eyeball, 258, *258*, 259, 260, 262; fat storage, 75
Eye cataracts, 261, 367, 517
Eye diseases, 260–261
Eye disorders, 260–261; new technology and, 261
Eyeglasses, 259, 261
Eyelids, 250, 274
Eyes, *238*, 249, 259, 264; autonomic nervous system and, *230*; color, continuous variation and, 382, 383; cones of, *248*; defined, 258; early development, 305; fat cushioning, 125; infections, 260–261, 341; injuries, 261; pigmented epithelium, 68; rods of, *248*; sensory neurons in, 77; sensory receptors in, 256, 257; structure of, 258, *258*
Eye sockets, 92, 475
Eye tissue transplantation, 173

F

Face, 481, 482, 483
Facial bones, 91, 92, 93
Facilitated diffusion, 44, 44, 119
Fact memory, 239, 239
Factor VIII, 446, 448
Factor IX, 446
FADH$_2$, 55, 57, 57, 59, 61
Fainting, 215
"Fake fats," 125
Fallopian tube, 292, 300
Falls, bone breakage, 94
Familial hypercholesterolemia, 154, 155, 395, 403
Family, 467, 467
Family pedigrees, in genetic analysis, 392, 393, 395, 396, 396, 397, 402
Family-planning programs, 301, 509, 510
Family size, 508, 509, 515
Family tree, 472, 475, 476, 481, 484
Farsightedness, vision, 259, 260, 260, 261, 327
Fascicles, 76
Fast-acting muscle cells, 105, 106
Fasting, 122
Fats, 15, 37, 119, 130; absorption, 123; aerobic exercise and, 106; carbon compounds, 24; deposited secondary sexual trait, 292; deposited in blood vessel wall, 154, 155, 156; deposition and aging, 323, 324, 325; digestion, 117, 117, 120, 120, 123, 123, 132; energy source, 60, 61; enter lymph vessels, 118; formation by coenzyme, 126, 132; functional groups, 25; gastrointestinal hormones and, 122; genetic predisposition, 129; hormonal breakdown of, 273, 276, 277, 278; level in diet, 111, 125, 132; linoleic acid and, 10; as lipid, 136; lymph transport, 159, 160; metabolism of, 29, 112; neutral, 28–29; nutrient turnover, 122; physical fitness and, 66; polyunsaturated, 125; saturated, 125, 154, 156; storage, 28, 120, 123, 132; storage in adipose tissue, 75, 75, 80; stored glycogen, 60, 60; subcutaneous, 325; substitutes, 125; unsaturated, 154
Fatigue, 106
Fat molecule, 46, 60
Fatty acids, 25, 29, 29, 36, 37, 60; called prostaglandins, 282; conversions, 60, 61; defined, 28; deliveries from bloodstream, 103; E. coli and, 121; free, 117, 117, 119, 119; hormonal stimulus, 273, 275, 278; "hydrogenated," 28; in interstitial fluid, 142; as nutrient molecules, 110, 111, 122, 132; peroxisomes and, 51; polyunsaturated, 28; structural formulas for, 28; synthesis by coenzyme, 126
Faulty enamel trait, 397, 397, 403
Fecal occult blood test, 429
Feces, 112, 121, 132; water component of, 200, 202, 203, 216
Feedback mechanisms, 82, 83, 84, 139, 281, 302; negative, 291, 291, 295; negative, hormone secretion and, 275, 278; positive, 295
Feet, 94, 99; soles of, 80, 81
Feline infectious peritonitis, 468
Femaleness, 287
Female reproductive disorders, 299
Female reproductive function, 292–295, 297, 302; components of, 292, 292–293; contractions in, 289; cyclic changes in ovary, 294; cyclic changes in uterus, 295; fertility cycle, 287, 293; from embryonic duct system, 390; gamete forma-

tion, 307; primary organ, 286, 287, 302; sex drive, 278
Female reproductive tract, 282
"Feminization" of males, 279
Femoral artery, 143
Femoral vein, 143
Femur, 86, 88, 91, 95, 97, 97, 483; defined, 95
Fermentation pathways, 56, 58, 62
Ferrous iron, 55
Fertile period, 298, 308
Fertility, 293; options, 298–299
"Fertility awareness," 298
Fertility, control of, 298–299, 300, 302; dilemmas of, 298, 301, 302; future options, 299; global perspective, 301; reliability, 298, 299; reversible methods, 299; side effects, 298–299
Fertilization, 302, 307, 307, 310, 318, 328, 459; artificial methods, 300; as chance event, 376; chromosome number change during, 400; defined, 297, 306; diploid chromosome, 374; as first life cycle stage, 308, 308, 309, 309; first trimester after, 312, 313, 314, 321; haploid gamete union, 287, 290, 351, 367, 368; hormones and, 293, 295; of ovulated oocyte, 304, 305; self, 374; sex determination at, 390; site of, 292, 292, 293, 294
Fertilizers, 501, 502–503, 504, 516, 518–519, 525, 526; biomediation and, 447; water pollution, 14, 45, 203
Fetal alcohol syndrome (FAS), 218, 321, 321
Fetal development, 316–317; critical periods, 320–324, 320; events, 305; malnutrition and, 131; organs of support, 293, 294; stages, 323
Fetal period, 316
Fetus, 79, 83, 96, 425; alcohol use and, 218, 242, 321, 321; at risk, 218; cretinism in, 280; defined, 314; IgG antibody protection, 173; Rh blood typing, 140, 141, 141; sex hormones, 278
Fever, 215, 417; defined, 167; enzyme structure and, 54; -reducing drugs, 215
Fiber, 124; insoluble, 121; level in diet, 27, 111, 121; protein, 31; soluble, 121
Fibrin, 153; blood protein, 75; fiber net, 153, 154
Fibrinogen, 153, 153; in blood clotting, 136; in plasma, 136
Fibroblast growth factor (FGF), 283
Fibroblasts, 73, 73; defined, 283
Fibrocartilage, 74, 97
Fibrosarcoma, 423
Fibrous joint, 97, 108; defined, 96
Fibula, 91, 95, 95
Fight-flight response, 279; defined, 231
Filtrate, kidney, 205, 206, 210, 212; reabsorption, 208, 209, 209, 216
Filtration, 208; defined, 206; rate, 207, 207
Fimbriae, 294
Finger bones, 91, 94
Fingernails, 81
Fingers, 250, 465, 478; formation of, 314, 315
Fingertips, sensory receptors in, 250
Fire, 483
First filial generation, 376; gene linkage crossovers and, 391
First polar body, in meiotic cell division, 366
Fishes, 476, 486, 491, 492, 494, 517, 524; bony fossils, 463; comparative body forms, 464, 464; reproductive isolation of, 461; transgenic, 452
Flagellum, 53, 290, 291; defined, 53
Flatulence, 122

FAD (flavin adenine dinucleotide), 34, 57, 57, 58, 59; defined, 55
Flavor, 252, 253
Fleas, 512
Flexibility, 93
Flex points, 93, 108
Flippers, 465
Floods, 501, 512, 520
Florida panther, 468
Flossing, 114
Fluid balance, maintaining, 152, 152
Fluid mosaic model, 41; defined, 41
Fluid pressure, 149, 151, 152, 152, 256, 257, 261; in inflammation, 166, 167, 167; osmosis and, 43
Fluorescent in situ hybridizing (FISH), 322
Fluorine, 127
Fluorine atoms, in substitute blood, 71
Fluorouracil, 430
Focal point, 259, 259, 260, 261
Focusing mechanisms, 259, 259, 260
Folate, 126
Folic acid, 126, 156; birth defect prevention, 320
Follicle, 292, 294, 295, 295, 296, 302; defined, 294; fertilization stage, 308
Follicle-stimulating hormone (FSH), 271, 273, 284, 291; control female reproductive function, 294, 295, 295, 296, 299; control male reproductive function, 287, 290, 291, 291; defined, 290; hypothalamus/pituitary production, 273; infertility methods, 300; menopause and, 327; primary effects, 269, 272, 272
Fontanels, 97
Food: additives, 10; nutritional value of, 110, 110, 125
Food allergies, 123
Food and Drug Administration, 428
Food chains, 487, 505, 506; defined, 490
Food guide pyramid, 124
Food supply, 509, 525; contaminated, 45, 120, 121, 180; global, 130, 131
Food webs, 487, 491, 502, 505, 506, 506, 517; aquatic, 498; defined, 490; detrital, 492, 493; grazing, 492, 493, 494
Foot, bones of, 95, 95
Foot plate, 315
Foragers, 482
Foramen magnum, 92, 92
Foramen ovale, 317, 317, 318
Forceps, 318
Forearm, 98, 99, 465; muscles of, 105
Forearm bone, 91
Forebrain, 219, 232, 233, 233, 244, 314, 476; defined, 233; hypothalamus location, 272
Forehead, 81, 92
Foreign agents, 159, 167, 168, 171, 173, 182, 431; "complement coated," 165, 165
Forelimbs, 465
Forensic medicine, 444
Forensic scientists, 95
Foreskin tissue, 70
Forests: conifer, 496; hardwood, 488; tropical rain, 488, 498, 519, 520–521, 520, 521, 526; tropical rain forest as energy flow, 493
Formaldehyde, 470
Formulas, 17; structural, 20
Fossil fuels, 35, 498, 503, 522, 523, 525, 526; burning, 498, 498, 499, 501–502, 501, 506; defined, 522; green revolution and, 519
Fossilization, defined, 463
Fossil record, 461, 466, 472, 482, 483, 484; interpretation, 463
Fossils, 426, 456–458, 463, 467, 476, 479–480, 485; cells, 470, 470; defined, 462; first hominids, 480;

harmful compounds traced, 505; human evolution, 482, 483–484, 484
Fossil sequences, 457
Founder effect, 460
Fovea, 258; defined, 263
Fox, Sidney, 471
Fractures, 94, 97; bone remodeling, 90
Fragile mental retardation 1 (FMR1), 417
Fragile X syndrome, 403, 417, 444
Frameshift mutation, 416, 416, 418
Franklin, Rosalind, 405
Fraternal twins, 308
Free nerve endings, 250, 251; defined, 250
Free nucleotides, 442
Free radicals, 126, 127; defined, 21; as mutagen, 416; scavengers, 21; tissue damage, 324
Free surface, of a villus, 118
Frequency distribution, for eye color trait, 382
Frequency of occurrence, gene crossover, 391, 391, 402; genetic nondisjunction, 401
Frequency: of signal on sensory pathway, 248, 249, 249, 264; of wave cycles, 254, 255, 255
Friction, blood flow and, 149
Friction reduction, 74
Frontal bone, 92, 92
Frontal lobe, 236
Frontal plane, 79
Frontal sinus, 93
Frostbite, 214
Fructose, 26, 26; in semen formation, 288, 289
Fruit flies, 391, 392
Fruit pectin, 121
Fuel, nonpolluting, 35, 35
Fumarate, 57
Functional groups, 25, 36; defined, 24; energy carriers and, 54, 55, 62
Functional group transfer, 36; defined, 25
Fungal spores, 162
Fungal infection, 197, 260
Fungicides, 15, 428
Fungus, 3, 446, 489, 490, 502, 506, 512; antibiotics, 180; carcinogens produced by, 426–427; as decomposers, 7; as pathogens, 164; sexually transmitted disease, 331, 344, 346
Fusion power, defined, 523

G

Galactose, 26
Gallbladder, 112, 113, 120, 120, 122, 132; deferred pain, 251; defined, 117; enzyme secretion, 117, 132; as hormonal target, 275
Gallbladder cancer, 430
Gallstones, 117
Gamete formation: allele crossover, 387, 388; chromosome types, 390; during meiosis, 368, 370, 390; ionizing radiation and, 367, meiosis, 350, 351, 352, 353, 361, 363, 365; nondisjunction during, 400, 400, 401, 401
Gamete intrafallopian transfer (GIFT), 300
Gametes, 290, 294, 307, 328, 459; chromosome number change, 400, 401; defined, 306; early body development, 305, 307, 310; female, 287, 289; gene pairs in, 373; haploid number in, 374, 375; independent assortment of genes, 375, 378, 379, 379, 391, 391; inheriting recessive alleles, 392, 394, 396; in oogenesis, 366, 367; in spermagenesis, 366, 367; linked alleles in, 388, 402; male,

287, 289, 290; mutations in, 416; one chromosome set, 351, 361; paternal/maternal mix, 368; segregation at meiosis, 376, 384

Gamma aminobutyric acid (GABA), 227

Gamma globulins, *136*; in immune response, 136

Gamma interferon, 177

Gamma rays, 427

Gancyclovir, *450*

Ganglia, 228, *232*; autonomic system, 230, *230*

Ganglion cells, 262, *262*, 263

Ganglioside, 283

Gap junctions, 72, *72*, 76, 148, *148*; defined, 72

Gar, 492

Garden pea plant, 374, 379, 384

Gargas, 474, 475

Garrod, Archibald, 408

Gas exchange, 185, 325; at birth, 317; between intracellular and extracellular, 202; controls over, 194–195, *194*, 198; factors influencing, 188; in capillary beds, 150; inhalants and, 243; in lungs, 146, 147; rates of, 188, 191; respiratory disease and, 189; sites of, 187, 198; transport and, 192–193, 198

Gas exchange mechanisms, 195, 198

Gasoline, 35, 516

Gastric emptying, rate of, 116

Gastric fluid, 72, *112*; immune response, 164, 182

Gastric juice, 132; defined, 116

Gastric mucosal barrier, 116, 164, 173

Gastrin, 116, 122, 275

Gastrocnemius, *99*

Gastrointestinal hormones, 122

Gastrointestinal tract (GI), 110,112, 113, *113*, 117, 118; blood flow, path of, 147; coelum, 312; defined, 112; diarrhea as immune response, 164; endocrine cells in, 276, 281; lipoproteins and, 32; mucosal lining and defense, 116, 164, 173; parts of, 121, 132; sickle-cell damage, *381*; water gain in, 202, 216

Gastrulation, 307, *307*, 312, *312*; defined, 306; end of, 313, *313*; as third life cycle stage, 305

Gated calcium channels, 225, *225*

Gated sodium channels, 221, *222*, 223, *223*, *224*, 244

GDP, 57

Gel electrophoresis, *408*, 442, 443, *443*, 444, 466; defined, 408

Gender, 286; alcohol intoxication, 242; diet and, 124, 128; fetal determination, 314; lung capacity and, 191; sex chromosomes govern, 389

Gene activation, in sex chromosomes, 287, 398

Gene coding, 406, 417, 418; in stages of development, 306, 307, 313

Gene flow, *460*, 461, 472, 484; defined, 460

Gene linkage, 388, 402; crossover and, 391, *391*, 402

Gene locus, 351, 373, 374, *374*, 380, *380*, 388, 402; crossing over frequency and, 391, *391*, 402

Gene mutation, 324, 338, 339, 386–387, *386*, 388; genetic disorders and, 393, 402, 405, 407

Gene pairs, 372, 373, 374, *374*, 384; independent assortment of, 378, 379

General anesthesia, 236

Generalist species, 488

Gene regulation, 417, 418

Genes, 438–452, *440*, *445*, *449*, 459–460, 466, 472; altering expression of, 449; basic function of, 406, 408, 418; cancerous changes, 423, 424, 426,

428, 429, 432; defined, 373, 374, 384, 406, 418; deletion, 398; de-tox, 447; of endangered species, 468; evolution and, 457; expression, 382–383; "foreign," 159, *167*, 168, 171, 173, 182, 439; for human somatatropin, *448*; HIV, 339; independent assortment, 378, 384; inactivation, 416; interactions, 382, 383, 384; ionizing radiation altering, 367; meiotic shuffling, 364, 370, 373; Mendel's single gene model, 373, 374, 381; mutation, 439, 440, 444, 460, 472; parental combination, 351, 370; pedigree analysis, 392, 393; primary structure of protein, 31; protein synthesis and, 416, *416*, 417, 418; regulation, 450; segregation, principle of, 375, *375*; sequence, 174, 175, 388, 417; similarity, 464; species gene flow, 484; steroid hormonal targets, 267, 271, 284; strategies for transferring, 450; trait encoded, *81*, 86, 353, 370, 388, 402; transcription, *270*, 405, 410, *410*; translation, 405, 410; as units of inheritance, 375; virus, 176; weight set point and, 129

Gene segment, *410*, 417; DNA and, 174, 374; as template, 410, *411*

Gene sequencing, 445, 466

Gene therapy, 449; defined, 449; examples, 451; methods and prospects, 450–451, *450*; and your future, 451

Gene therapy, for cystic fibrosis (CF), 394

Gene therapy trials, *450*, 451

Gene transfer, 440

Genetic abnormality, 393, 402, *403*; defined, 393

Genetically altered bacteria, 439

Genetically identical copies, 368

Genetic analysis, 392–393, 397

Genetic bioengineering, 438–439

Genetic code, 408, *412*, 413, 416, 418; defined, 412

Genetic counseling, 393

Genetic crosses, 376, *376*; Punnett square method, 376, *376*

Genetic damage, ionizing radiation and, 367; smoking effects, 196

Genetic defects, 393, 449, 451

Genetic determination, surface proteins, 140, 141, *141*

Genetic disease, 383

Genetic disorders, 154, 382, 449; abnormal chromosome rearrangement, 389, 399; carriers and, 392, 393, *394*, 396, *396*, 397; common, 402, *403*; cretinism, 280; cystic fibrosis, 122, 386–387, 393, 394; defined, 393; detection, 393; following nondisjunction, 400, 401, *401*; gene mutation, 389, 393, 396, 397, 402, 407, 416, 417; hemophilia as, 6; immunodeficiency, *178*; prenatal diagnosis, 322, 400; probability of transmission, 376, 393; recessive transmission, 393; spontaneous abortion, 314; X-linked genes and, 388

Genetic drift, *460*, 461, 472; defined, 460

Genetic engineering, 71, 339, 438–453, 468, 501; of animals, 448–449; bacteria, *438*, 446, 447; defined, 440; gene products, 466, 472; interferon, 177; virus, 176

Genetic heritage: blood groups, 141; fat storage, 75, 129

Genetic information, 246; in DNA, 34, 37, 49, 62

Genetic predisposition: Alzheimer's disease, *225*, 326; hyperthyroidism, 280; peptic ulcers, 116; rheumatoid

arthritis, 178; to allergies, 178; type 1 diabetes, 276

Genetic recombination, 174, *364*, 368, 370, 402, 441; defined, 364, 388; following crossover event, *391*; correlated with natural selection, 460

Genetics, 272, 383, 458; origins of, 274; variability, 459, 472

Genetic screening, 452; moral/ethical dilemma, 393

Genetic traits, variations in, 351, 365, 368, 370

Genital ducts, male, *68*

Genital epithelial cells, STD infection, 340, 341, 342

Genital herpes, 181, 331, 334, 342, *342*, 343, 346; defined, 342

Genitals, 250; autonomic nervous system and, *230*; development, hormone effect on, 275; in fetus, 314; sexual arousal, 297

Genital warts, *334*, 343, 346, 433

Genome, 383; mapping, 449; sequencing, 449

Genomic imprinting, 382, 383

Genotype, 376, 382; defined, 374; genetic counseling and, 393; in sex-influenced inheritance, 398; of sex chromosome abnormality, 401, *401*; probability of, 376, 377, *377*, 380, *380*; testcross for unknown, 378, *378*; vs. phenotype, *374*

Genus, 467, *467*

Geochemical cycle, 495, *495*, 496, 506

Geological uplifting, 495

Geologic record, 457

Geologic time, 457

Geology, 457, 458

Geothermal energy, 471, 523

German measles, *176*, 321

Germ cell, 328, 353, 426; at interphase, *362*; defined, 352; diploid, 366, *390*; early formation, 310; homologue separation, 375; ionizing radiation and, 367; meiosis and, 361, *361*, 365–366, 368, 370

Germ layers, *306*, 312; defined, 306

Germline, mutations in, 416

Gey, George, 355

Gey, Margaret, 355

GH, anterior pituitary secretion, 273

Gibbons, *476*, *477*, *481*

Gigantism, 274, *274*

Gingiva, *114*

Gingivitis, 114

Glaciation, *500*

Glaciers, 483, *489*, 500–501

Glands, 68, 69, 69, 113; classifying, 68; defined, 68; as effectors, 82, 84; nervous system and, 77, *77*

Glandular cells, 446; autonomic nerves, 230, 231, 244; secretion from, 182, 268

Glandular epithelium, 116, 132

Glans penis, 297

Glaucoma, 261

Glia, 77

Global carbon budget, 498, *498*, 499

Global cycling, 521

Global warming, 7, 498, 500–501, 506, 524

Globin, 32, 138

Globular protein, 100

Globulins, 136; in plasma, *136*

Glomerular (Bowman's capsule), 205, 206, *208*, 216; defined, 205

Glomerular capillaries, 205, 206, *208*, 213, 216; permeability of, 206, 207; sodium reabsorption and, *211*

Glomerulonephritis, 213

Glomerulus, 205, 206, 207, 208, *208*, 210; defined, 205; in juxtaglomerular apparatus, 211, *211*; kidney disorders, 213

Glottis, *191*; defined, 191

Glucagon, 117, *275*; defined, 276; hormones and level of, 268, *270*, 277

Glucocorticoids, *275*, 279; defined, 278

Gluconeogenesis, 278

Glucose, 26, 27, 37, 59, *59*, 62, 470; absorption, 202, 206; blood-brain barrier, 235; blood levels, 5, 135, *275*, 276, 277, 278, 279; breakdown of, 56–57, *56*, 446; carbohydrate digestion and, *117*, 119, 122, 124; coenzymes and, 55; depleted stores of, 122, 123; free, 60; hormones and level of, 268, *269*, 272, 273, 277; in ATP formation, 103, *103*; in urine, 207; as main energy source, 26, *26*, 111, 122, 124; model of, *55*; monomers, 27; as nonpolluting fuel, 35; reabsorption, 206, 207, 208; selective permeability of, 42; as simple sugar, 56, 60, *61*; storage of, 120, *277*; as substrate for fermentation pathways, 58; tissue absorption/PET and, *18*; transport proteins and, 44

Glucose insulinotropic peptide (GIP), 122

Glucose metabolism, coenzyme, *126*; hormonal effects on, *275*, 276, 277

Glucose molecule, energy-releasing pathways and, 60, 62; reaction steps of glycolysis, 56, *56*, 58

Glucose-6-phosphate, 60, *61*

"Glucose sparing," 278

Glutamate, 408, *412*, 419; GAA, GAG code words, 412, *412*; negative charge, 408, *408*

Glutamine, *412*, 419

Gluten intolerance, 123

Gluteus maximus, *99*

Glycerides, 37

Glycerol, 26, 28, *28*, 29, 37, 60, *61*; in interstitial fluid, 142; nutrient turnover, 122

Glycerol heads, 60

Glycine, *54*, *412*, 419; as enzyme substrate, 54

Glycogen, *26*, 27, *61*; carbohydrate conversion, 122, 124; conversion to glucose, *277*; in ATP formation, 103, *103*; molecule, 5; storage form, 60, 120, 122, 277; synthesis, 276

Glycogen formation, coenzyme, *126*

Glycolipids, 41

Glycolysis, 57, *60*, *61*, 62, 352; defined, 56; energy-releasing steps, 56, *56*, 58, 59, *59*, 60, *60*; energy-requiring step, 56, *56*; fermentation pathway and, 58; inhibited, 60; muscle cells and, 103, *103*, 105, 106; reaction steps of, *56*

Glycoproteins, 32, 271; carbohydrate chain alteration, 50; defined, 33; in ovarian cycle, 294; as peptide hormones, 271, 284

GM₁, 283

Goblet cells, 68

Golgi body, *46*, 47, 50, 62; main function, 46; micrograph of, *51*

Golgi membrane, *50*, 51

Gonadotropin, 272, 273

Gonadotropin-releasing hormone (GnRH), 273, 291, *291*, 294, *295*, 296, 302

Gonads, development, 314; female, *269*, 287, 293, 314; germ cells, meisosi of, 370; hormonal source/effects of, *275*; as hormone target, 273, 281; male, *269*, 287, 314; mutations in, 416

Gonococcus bacterium, 340, *340*

Gonorrhea, 207, 331, 333, 341, 342, 343, 346; defined, 340; drug-resistance, 181; prevention, 345

Gorilla, *476*, *478*, *479*, *480*, *481*

Gout, 202
Government regulations, 446
gp120, 338
Graafian follicle, 294
Graded, local signals, 222, 247, 248
Gradualism, defined, 461
Gradual model, defined, 466
Grand mal seizures, 225
Granulocytes, 137, 166; defined, 137
Granulosa cell, defined, 294
Grapes, 428
Graves' disease, 280
Gravitational pull, 468; blood flow and, 151
Gray matter, 232, 232, 233, 234, 235, 236; somatosensory cortex of, 250
Grazing food webs, 493, 506, 506; defined, 492
Great Black Cow, 474, 474
Great Deluge, 456, 456
Great Lakes, 279
Greenhouse effect, 498, 500–501, 500, 521, 522, 526; defined, 500
Greenhouse gases, 500–501, 500, 525; defined, 500
Green light wavelengths, 247, 260
Green revolution, defined, 519
Grip, power, 478; precision, 478
Ground substance, 73, 74; in bones, 88, 89, 108
Groundwater depletion, 518, 518, 519, 525; contamination, 203, 522
Groundwater supplies, 447
Groupings, inclusive, 467
Growth, hormonal adjustment, 267, 269, 275, 280, 281; mitosis and, 351, 351, 353, 368, 370, 370; steroid hormones and, 29
Growth and tissue specialization, 323; defined, 306; as fifth life cycle stage, 305; primary tissue, 306, 307
Growth factor, 283, 425, 426; defined, 282
Growth hormone (GH), 89, 106, 271, 284; abnormal production, 274, 274; hypothalamus/pituitary production, 273, 276; primary effects, 269, 272, 272, 273; rat, 448; as regulatory agent, 417
Growth retardation, 125
Growth spurts, 323
GTP, 57
Guanine (G), in DNA nucleotide bases, 406, 406, 417, 418; in RNA nucleotide bases, 410, 411, 413, 418
Guano, 504
Guilleman, Roger, 274
Gummas, 341
Gums, 114, 114
Gustatation, 252, 253
Gut: acid-base balance, 212, lumen, 117, 122; wall, 122
Gymnosperms, 463

H

Habitat, 467, 485, 487, 488, 492; defined, 488; destruction, 524; disturbed, 488
Habituation, 240, 241, 243
Haemanthus, 354
Haemophilus ducreyi, 343
Hair, 80, 81, 81, 430, 476, 477; color, continuous variation of, 383; growth/density, 81; loss of body, 325; root, 81; sex differentiation, 287, 290; shaft, 81, 81; structural protein, 30, 32, 33, 33, 37
Hair cells, sensory, 255, 256, 257, 348; damage to, 257, 257; defined, 255
Hair follicles, 80, 81, 251; death of, 325; thermoreceptor for, 250
Half-life, defined, 18
Hallucinogens, 241, 243

Hammer, 254, 254
Hamstring muscle, 99
Hand, 94, 94, 95; development of, 305, 350–351; movements, refined, 477, 478, 479, 482; muscles in, 105; palms of, 80, 81; pressure receptors in, 249, 250
Handbones, 478, 485
Hand plates, 315
Haploid cells, 363, 369, 370
Haploid gametes, 361, 363, 366, 367, 369, 370
Haploid number (n), 362, 363, 365, 367, 368, 370, 375; defined, 361; of gamete, 351; in telophase I and II, 290, 291
Haploid spermatids, 366, 366
Hardening of the arteries, 154, 157
Hard palate, 92, 92, 115
Harvard University study, 128, 128, 129
Hashish, 241
Haversian canals, 89
Haversian system, 88, 89
Hawks, 490
Hay fever, 178, 197
Hazardous waste, dumps, 203
HbA molecule, 408, 408
HbS molecule, 408, 408
Head, 79; blood route, 143, 146, 146; embryonic period growth, 314, 314; gravitational orientation of, 257; muscle groups, 99; rotational movement, 256, 256, 257; somatic nerves and, 230
Head cold, 92
Health care, 514, 514, 515
Health care delivery system, 330
Hearing, sense of, 92, 236, 237, 247, 254–255, 264; memory circuits, 239
Hearing loss, risk of, 257
Heart, 66, 67, 75, 78, 144–145, 160, 464; artificial replacement cells, 70; autonomic nerves, 219, 230, 244; bone protection to, 93; cardiac muscle, 98; cell junctions in, 69; contractions, 206; deferred pain, 251; defined, 144; early development, 305, 314; endocrine cells of, 269, 275, 281; enlarged, 157, 325; function of, 142; in fetal circulation, 317, 317, 318; location, 78, 187; membrane lining, 69; pulmonary circuit, 146; structure, 144, 145, 160; systemic circuit, 146
Heart abnormalities, 242; prenatal smoking, 321
Heart attack, 135, 154, 154, 155, 156, 446, 448; atherosclerosis, 154; growth factor and, 282; obesity and, 128, 129; referred pain, 251
Heartbeat, 134, 147, 148, 154, 157; fetal, 314; increased, 154
"Heartburn," 116
Heart damage: alcohol abuse, 242; sickle cells, 381
Heart defects, change in autosome number, 400
Heart disease: antioxidants and, 21; cholesterol-related, 395; diet and, 111, 124, 125; excessive weight and, 128, 128; familial history, 129, 154, 155; genetic disorders and, 416; post menopausal women, 327; risk factors, 154, 155, 157; trans fatty acids and, 29
Heart failure, 127, 152, 155, 157, 197; drug abuse, 242, 243; insulin deficiency and, 276; sickle-cell damage, 381
Heart monitoring, 135
Heart muscle: aging and, 325; damage/death, 154
Heart rate: age-related changes, 327; controls over, 148; drug effect on, 240, 241, 242, 243; hormonal effects on, 275, 279; nerves and, 5; resting,

135; sympathetic/parasympathetic integration, 231, 233
"Heartstrings," 144
Heart valves, 145; collagen in, 32; inflammation of, 155
Heat, sensations of, 250, 251
Heat conservation, 83, 83, 214, 214, 215, 215, 216
Heat dissipation, 135, 214, 214, 215, 215, 216; body temperature regulation, 200, 201
Heat energy, 7; enzymes and, 54; from blood flow friction, 149; measurement, 124; water properties and, 22
Heat exhaustion, 215
Heat production, 78, 214, 215, 215, 216; as metabolic byproduct, 201, 214, 215, 216
Heat shock factor (HSF), 417
"Heat shock" genes, 417
Heat stress, 214–215, 214; physiological response, 214, 215, 215
Heat stroke, 215
Heavy metals, toxic, 235, 447
Heel, 95
Heifer, 474
Height, continuous variation, 382, 383
Heimlich maneuver, 115, 115
HeLa cells, 355
Helicobacter pylori (H. pylori), 116
Helium: atomic structure model, 16; electrons in outer shell, 19
Helper T cells, 169, 170, 171, 173, 179, 182; defined, 168; effector, 169, 171; HIV and, 336, 336, 338, 339, 346; thymus function, 281; virgin, 169
Hematocrit, 75
Heme group, 32, 33, 138, 138, 139
Hemodialysis, 213
Hemoglobin (Hb), 32–33, 37, 75, 136, 155, 439; abnormal, 381, 381, 408; alpha polypeptide chains, 408, 408; beta polypeptide chains, 408, 408, 416; breakdown, 206, 343; carbon monoxide and, 154; defined, 32; DNA mutations and, 6; fetal, 316, 381; from donated blood, 145; in carbon dioxide transport, 193; iron in, 126, 127; macrophages and, 139; molecule structure, 138, 138; oxygenated, 138, 160, 188, 192, 193, 198; precursors and, 417; synthesis, 127, 139; tasks of, 71; as transport pigment, 80, 188, 192, 193; two gene codes, 408
Hemolytic anemia, 449
Hemolytic disease of the newborn, 141
Hemophilia, 403, 450, 460; DNA mutation and, 6; embryo screening, 393; X-linked recessive inheritance, 396
Hemophilia A, 446, 448; X-linked recessive inheritance, 396, 396, 397
Hemophilia B, 446, 450; X-linked recessive inheritance, 396
Hemophiliacs, 336, 337
Hemophilia influenzae, 176
Hemorrhage, 126, 210; anemia from, 139; intestinal, 367
Hemorrhoids, 121
Hemostasis, defined, 153
Hepadnaviruses, 334
Hepatic artery, 147
Hepatic portal vein, 120, 143, 147, 317; defined, 120
Hepatic vein, 120, 120, 147, 317
Hepatitis, 120, 207, 243, 332; immunization, 176, 176
Hepatitis A, 120, 180, 181
Hepatitis B, 120, 342, 346, 334, 434, 446; defined, 343; as global health threat, 181; virus, 343
Hepatitis C, 145, 177, 181
Hepatitis D, 181
Hepatitis E, 181
Herbicides, 15, 428, 452, 496, 497

Herbivores, 491, 492, 493, 494, 495, 504, 506; defined, 490
Hereditary information: DNA transfer, 49, 404, 407; heritable traits, 6, 12; nucleotide encoding, 405; on chromosomes, 351, 353; parental transmission, 372, 373
Heritable alteration, risk of, 416
Herniated disks, 93
Heroin, 241; blood-brain barrier, 235
Herpes simplex virus (HSV), 81, 225, 260; infection, 449; transmission, 342; type 1, 334, 342, 342; type 2, 334, 342, 342
Herpesviruses, 334, 336, 342, 342, 345
Herring gull, 492
Heslow, John, 458
Heterosexual contact, HIV spread, 337
Heterotrophs, 490, 506; defined, 490
Heterozygotes, 381, 394, 395, 398; dominance relations, 381, 384
Heterozygous condition, 374, 375, 376, 376, 378, 378, 379, 380; defined, 374
Heterozygous females, X-linked inheritance of dominant mutant alleles, 397
Heterozygous individuals, 394, 395, 398; autosomal inheritance patterns, 394, 395, 402; X-linked recessive inheritance, 396
Heterozygous recessive individual: autosomal recessive inheritance, 394, 394, 395; as carrier, 392, 393
Hexokinase, model of, 55
Hexosaminidase A, 394
High blood pressure, 127, 145, 154, 446; blood substitute side effects, 145; chloride and, 127
High-density lipoproteins, 154; in plasma, 136
High tides, 501
Hindbrain, 219, 232, 233, 233, 244, 476; defined, 232
Hind limb, 465
Hinge joints, 96
Hip bones, 89, 93, 95, 95; aging, mass and, 90
Hippocampus, 237, 237; fact memory, 239, 239
Hippocrates, 8
Hips, 95, 96, 97; fat storage, 75; flexibility of, 95; muscle groups, 99
Histadine, 408, 412
Histamine, 178, 179, 182, 197, 250; defined, 166; inflammatory response, 166, 167, 167
Histones, 409; defined, 409
Histoplasmosis, 197, 260
HIV. See Human immunodeficiency virus
Hives, 14
HMS Beagle, 458, 458
Hoarseness, 191
Hodgkin disease, 434
Holding stations, 498, 499
Holocrine glands, 68
Homeostasis, 2, 3, 5, 12, 84; acid-base balance, 212; blood function of, 75; blood glucose levels, 277; concept of, 66, 67; defined, 5, 82; digestive system's role in, 111; electrolyte imbalance and, 276; endocrine cells and, 276; feedback loops, 122, 129, 281, 284; gas exchange regulation, 194, 195; glucose sparing, 278; hemoglobin/oxygen binding, 138; maintenance of, 4; maintenance of blood pressure, 151; maintaining body temperature, 201, 215; maintaining calcium, 108, 281; mechanisms of, 15, 39, 60, 82–83, 83; negative feedback loops, 195, 275, 278; neuron function and, 77; parathyroid regulated balance, 280; undernutrition and, 130; water/ion balance, 200, 201, 327

Home pregnancy tests, 177
Hominidae, *467*
Hominids, 475–482, *476, 480, 481,* 485; defined, 476
Hominoids, 476–477, *476,* 479, *481,* 485; defined, 476
Homo, 465, *467*
Homocysteine, 156
Homocystinuria, *449*
Homo erectus, 481, 483–485, *483*
Homo genus, *481, 482, 482,* 485
Homo habilis, 481, 482, 482, 484
Homologous chromosome pairs, 373, *374, 379,* 384
Homologous chromosomes, *362,* 364, *364, 365, 365,* 370, 383; allele cross over, 387, 388; carrying alleles at a locus, 388, 399; defined, 353, 388, 402; deleted segment and, 399; in duplicated state, *391;* separation of, 375, 384; unduplicated, *374*
Homologous structures, 472; defined, 465
Homologues, 353, 361, 368, *368,* 375, 384
Homo rudofensis, 481
Homo sapiens, 2, *481,* 483, 484, *484,* 485; defined, 483
Homosexual contact, male, HIV spread by, *337, 337*
Homozygous condition, 375, 376, 378, 382, 384; defined, 374
Homozygous dominant individuals, *374, 374, 375, 377, 378,* 384, 395
Homozygous individuals, 381; autosomal inheritance patterns, 394, 395, 402
Homozygous recessive individuals, *374, 374, 375, 377, 378,* 384, 392; autosomal inheritance patterns, 394, 395; in pleiotropy, 381, *381;* sex-linked inheritance, *398,* 402
H1 histone, *409*
Hooke, Robert, 38, *38*
Hookworm, 181
Horizontal canal, *256*
Horizontal cells, 262, *262*
Hormonal controls, 90, 98, 108, 267; of blood pressure, 151; of female reproductive cycle, 294–295, *294, 296;* of male reproductive function, 290; over body functions, *78, 79;* over heat produced and lost, 214, 215, *215,* 216; over reabsorption, 210–211, 216; over urine concentration, 201, 207, *208,* 209, 216; as regulatory agents, 55, 417, 418
Hormonal imbalance, 300
Hormone activity, energy output and, 128, 129, 132
Hormone-based therapy, 129
Hormone interactions, 284; opposing, 268; permissive, 268; synergistic, 268
"Hormone-modulators," 279
Hormone-receptor complex, *270,* 271, 284
Hormone replacement therapy (HRT), 106, 327
Hormones, 41, *81,* 448; blood transported, 75, *88,* 135, 136, *136,* 160; cancer produced, 425, 426, 432; chemical families, 271, *271;* concentration, 275, 284; defined, 268; development of sex traits, 390; discovery of, 268; endocrine gland secretions, 67, 68, 266, 267, 268; gastrointestinal, 122; inactivation, *120;* interactions, 275; negative feedback mechanisms, 275; obesity and, 129; secretion, 5, 116, *123;* as signaling molecules, 117, 266, 267, 268, 270, 271, 284; sources/effects of, 268, *269,* 270, 275, *275;* synthetic, 105, 106
Horses, 474

Host cell, 332–333, *332,* 337, *337, 338, 441,* 450; genome of, 450; helper T cell, 336, 338; mode of attachment, *340;* modified, 445
Hosts, 490, 512
"Hot flashes," 293, 327
Houseflies, 488
H. pylori infection, 16
Hubel, David, *263*
Human biology: cornerstones of, 2; scientific method and, 8–9
Human cDNAs, 449
Human cells, 450; gene therapy use, 450
Human chorionic gonadotropin (HCG), 309, 310, 425, 429
Human chromosomes map, 448
Human DNA complement, 448
Human genes, 438, 445–448, *448*
Human gene sequences, 449
Human gene therapy, 439
Human Genome Project, 448, 449
Human genome sequence, 449
Human growth hormone, 446, *446,* 448, *448*
Human immunodeficiency virus (HIV), 179, *181, 197, 332,* 434, 438, 448; as cause of AIDS, 330, *334;* cumulative infection, 335, 338–339, *346;* defined, 335; genetic strains, 336; immune system and, *178;* infection, *225, 337,* 338–339, 346; method of infection, 333, 337; non-progressors, 338; rate of new infection, 330, 331, 335, 337; replication, 336, 338, *338,* 339; screening, 336, 339; structure, 8, 336, *337;* transfusable blood contamination, 71, 145; transmission, 336–337, 343, 345; treatment, 338, 339
HIV-1, 335
HIV-2, 335
HIV protease, 339
Humanness, 474
Human papillomavirus (HPV), *334,* 343, 346; defined, 343
Human proteins, 438–439, 446
Humans, 485, 506; biotic potential of, 512; classification of, *467;* comparative morphology, 464–465, *464, 465;* defined, 482; ecology and, 488; evolution of, 465–466, 474–483, *476, 477, 478, 479, 482;* traits, 482; vestigial structures, 464–465, *465*
Human species, 474, 485
Humerus, *91, 94, 94;* defined, 94
Hunger, 233
Hunting, 468
Huntington gene region, 417
Huntington's disease, 394–395, *395, 403,* 417, 444
Hyaline cartilage, 74
Hybrid molecules, 466
Hybrid offspring, 461
Hybridoma cells, 177, *177*
Hybrids, 440
Hydrocarbons, 28, *37,* 60, 426, *428,* 447, 516; chains, *24;* defined, 24
Hydrochloric acid (HCl), 22, *22, 23,* 112, 116, 117, *132;* formation, minerals and, *127*
hydroelectric power, 522, *522*
Hydrogen, 15, 24, 36; gas, 468; molecular, 20; as nonpolluting fuel, 35, *35;* solid metallic compounds, 35
Hydrogen atoms, 16, 17, 24, *24, 25, 28,* 30; atomic number, *17;* atomic structure model, *16;* atomic symbol, *17;* coenzymes and, 34, 55, 57; electron distribution, *17, 19, 19,* 36; glucose component, 56; in amino acid, 30; mass number, *17*
Hydrogen-based systems, 35
Hydrogen bonds, 15, *21;* defined, 21; DNA bases, 405; in nucleotides, *34;* in polypeptide chain, 31, *31,* 32, 33;

link base pairs, 406, 407; water molecules and, 22, 36
Hydrogen cyanide, 470
Hydrogen fuel, 523
Hydrogen ions, 22, 23, *23,* 24, *25,* 36, *37;* and blood pH, 193; coenzymes and, 55, 57; diffusion in stomach, 116; and electron transport systems, 58, *58,* 59, *59,* 62; excretion of, 206, *208,* 211, 212; extracellular pH, 136; in sensory reception, 253; levels in exercise, 151
Hydrogen peroxide, 51, *516*
Hydrogen revolution, 35
Hydrologic (water) cycle, 495, 496–497, *496,* 506; defined, 496
Hydrolysis, *25,* 27, 36, 470; defined, 25
Hydrophilic molecule, 29, 40, 41
Hydrophobic, 22, 29, 40, 41, 119; defined, 22
Hydrophobic compounds, 29
Hydropower, 523
Hydrostatic pressure, 43, 152, *152,* 206, 207; in kidneys, 206, 207, 213
Hydroxide ion, 22, 23, 36
Hydroxyl groups, 24, 25, *25,* 26, 27, *37*
Hydroxyl ions, 22, *25*
Hygiene, 181, 207
Hylobatids, *476*
Hyoid bone, *280*
Hyperopia, 260
Hyperparathyroidism, 281
Hyperpolarizing signal, 227
Hypertension, 154, 207, 211; blood substitute side effects, 145; heart disease risk, 154, *155,* 156–157; kidney damage, 213
Hyperthermia, 215
Hyperthyroidism, 157, 280
Hypertonic solution, 43, *43, 210*
Hyperventilation, 188
Hypnosis, 242
Hypnotics, 242
Hypodermis, 80, *80;* aging and, 325
Hypoglycemia, 278
Hypothalamic osmoreceptor, 248
Hypothalamus, 210–211, *233, 234,* 244, 269, *272,* 279; as body's thermostat, *83,* 214, 215, *215;* defined, 233, 272; fact memory, *239;* hormonal cascade, 318; as hormonal target, *275;* hormone production, 274, 276, 278, 284; hormones control reproduction, 291, 294, 295, *295, 296,* 302; in limbic system, 237, *237;* overview, *269;* psychoactive drug effects on, 240
Hypothalamus/pituitary interaction, 267, 272–273; anterior lobe, *273,* 284; feedback loops, 291, *291,* 302; hormone production, 273, *277;* posterior lobe, *273,* 284; reproductive hormones, 287, 291, *291, 295,* 302
Hypothermia, 214
Hypothesis, 9, *9;* alternative, 10; defined, 8, 12
Hypothyroidism, 280
Hypotonic solution, 43, *43*
Hypoxia, 188

I

Ibexes, 474
Ibuprofen, 215
Ice age, 483
Ice-minus bacteria, 452, *452*
Identical twins, 308, 351; behavior, genetic link in, 383
Ieucine, *408*
Ileum, *112,* 117, *132*
Iliac arteries, *143*
Iliac region, 95, *95*
Iliac veins, *143*

Illegal drugs, 242; prenatal risk, 321
Illness, diet and, 111
Immigration, 460, 510, 526
Immune-privileged tissues, 173
Immune response, 71, 163, *165, 168,* 427, 430; abnormal, 177, 178–179, 182; cell-mediated, *169;* classes of antibody molecules, 173; control of, 169; defined, 164; deficient, 178–179; fever, 215; gamma globulin function, 136; hormonal role, *275;* key players in, 168–169; kidney damage, 213; liver, *120;* lymphocytes in, 137; lymphoid organs and, 170; non-specific, 164, *164,* 165, 166–167, 182; organ transplantation, *169,* 170, 171, 173; primary, 175, *175,* 176, 179, 182; secondary, 175, *175,* 176, *179,* 182; specific, 164, *164,* 174–175, 176, 182; T lymphocyte functions, 170, *171;* to HIV infection, 338, 346; to viral infection, 334
Immune specificity, 164, *164,* 174–175, 176, 182
Immune system, 2, 51, 168–169, 446, *450,* 451, 517; allergies, 123; antibodies, 140; bacterial response, 334; cancer and, 10, 424–425, 427–428, *427,* 430, 434, 436; chronic infection, 343, 346; defined, 168; disorders, 96, 178–179; HIV target, 334, 336, 338, 339, 346; ionizing radiation exposure and, 367; rejection, 70; stimulants and, 242; suppression of, 81, 173, *173,* 278; thymus function, 281; undernutrition and, 130, 131
Immunity, 75, *79,* 162–182; acquired, 173; age-related changes in, 327; impaired, *126;* to genetically engineered virus, 394
Immunization, 162, *176,* 177, *177;* active, 176, 182; defined, 176; passive, 176, 177, 182
Immunodeficiency (non-HIV), *178,* 179, 182
Immunofluorescence, 340
Immunoglobins (Igs), 173; defined, 173; IgA antibodies, 173, *182;* IgD antibodies, 173, *182;* IgE antibodies, 173, 178, *179, 182;* IgG antibodies, 173, 178, *182;* IgM antibodies, 173, *182*
Immunological memory, 168, 174, 175, *175*
Immunological specificity, 168
Immunologists, *163*
Immunotherapy, 430, 436; defined, 430
Impetigo, 180
Implantation, 295, 299, 308–309, *309,* 310, 311; as critical period of development, 320; genetic screening and, 393
Impotence, 106
Impurity removal, in liver, 147
Inadequate blood flow to organs, visceral pain and, 251
Inbreeding, 468
Incinerators, 519
Incisors, 114, *114*
Incus, 254, *254*
Independent assortment, 379, *379,* 380, *380,* 384; defined, 378; of genes into gametes, 391, *391*
India, 508, *508,* 509, *525*
Indicator diseases, 336, 338
Industrial chemicals, 428, 430, 434, 435; hormonelike effects, 278–279
Industrialization, 426, 514, *514,* 515
Industrial pollution, 45, 203; airborne, pH of, 23; heavy metal mercury, 235
Industrial smog, *516,* 526; defined, 516
Industrial stage, *514,* 515
Infancy: acquired immunity, 173; development stages, 323, *323*
Infanta Margarita Teresa, *394*

Infant respiratory distress syndrome, 192

Infections, 438; birth control risk of, 299; cystic fibrosis and, 387, 387; defenses, 121; immuno system suppression and, 173; inflammatory response, 166; lymph nodes and, 159; scavenger cells, 135; susceptibility to, 336; undernutrition and, 130; uterine, 319; white blood cells' role in, 137

Infectious diseases, 180–181, 182, 334, 446, antibiotics, 180–181; global threats, 181; inoculation for, 162; modes of transmission, 180, 181; patterns of ocurrence, 180, 181; virulence, 180. See also specific disease

Infectious disease specialists, 181

Infectious mononucleosis, 155

Inferior part, 79

Inferior vena cava, 120, 143, 144, 145, 146, 147; flow velocity, 151

Infertility, 107, 300, 300, 302; ectopic (tubal) pregnancy, 309; genetic disorders and, 401; PID cause, 342

Inflammation, 116, 169, 173, 178; allergens and, 178; basophils and, 137; of bronchial walls, 189; of heart muscle, 155; of reproductive structures, 334, 340, 342, 346; signs and causes, 166; white blood cells and, 136

Inflammatory bowel disease, 434

Inflammatory disorders, 278

Inflammatory response, 163, 166–167, 167, 182, 342; blocking, 178; controlled, 334; glucocorticoid suppression of, 278; non-specific, 164, 164, 165, 166–167, 182; to viral infection, 334, 340

Influenza, 180, 197, 334; virus, 332, 446

Information flows, 229, 229; from eye to brain, 262; memory circuits, 239, 244

Infrared wavelengths, 248, 500, 500

Ingram, Vernon, 408

Inhalants, 241; defined, 243

Inhalation, 190, 198; fetal carbon dioxide buildup, 318; nervous system control, 194, 195

Inheritance, 15, 457, 460; chromosomal base, 388–389; defined, 6

Inheritance, patterns of, 372–384, 388; figuring probabilities, 376–377; genetic crosses, 376–377; genetic disorders and, 260, 387, 394, 397, 402; genomic imprinting, 383; independent assortment, 378–380; less predictable variation in traits, 382–383; Mendel's theory of segregation, 375, 381; testcross, 378

Inhibin, 291, 291

Inhibitor hormones, 102, 284; defined, 273; inhibiting signals, 226, 227, 229, 230, 236, 244; testosterone as, 291

Inhibitory postsynaptic potential (IPSP), 227, 227

Inipol™, 447

Initiation, 414, 414, 418

Initiation complex, 414, 414, 418

Initiator, defined, 426

Initiator tRNA, 414, 414, 415, 415, 418

Inner cell mass, 304, 309, 310, 323; defined, 309

Inner ear, 92, 233, 257, 264; damage to, 255; from endoderm, 306; sense of balance and, 256, 256; sense of hearing and, 254, 254, 255

Inorganic chemicals, water pollution, 203

Inorganic compounds, fossilization, 463

Insanity, syphilitic, 341

Insecticides, 428, 446

Insect resistance, 452

Insects, 3, 490, 494

Insertion, 98, 108; defined, 98

Inspiration, 191, 191, 198; defined, 190

Insulation, 77; stored fat as, 75, 80, 125

Insulin, 37, 117, 122, 268, 271, 284; deficiency, 276; defined, 276; genetically engineered, 438, 439, 446, 446; secretion of, 5, 277; source of, 275

Insulin-dependent diabetes, 276

Insulin-like growth factor (IGF), 275, 282, 283

Insurance charts, 128

Intensity, of stimulus, 249

Integument, defined, 80

Integumentary system, 78, 78, 84

Integrator, 82, 83, 84; defined, 82

Interbreeding, 461, 472

Intercalated disks, 76, 148, 148

Intercostal muscles, 186; external, 190; internal, 191

Intercostal nerve, 194

Interdependency, of organisms, 7

Interferons, 430, 436; beta, 177; defined, 177; gamma, 177; genetically engineered, 438, 446, 446

Interglacial period, 500

Interior vena cava, 317

Interkinesis, 365

Interleukins, 177, 430, 436, 451; defined, 167; from helper T-cells, 170, 171, 171, 172; in immune responses, 164; as signaling molecules, 177, 182, 182

Interleukin-1, 167, 451

Interleukin-2, 425, 430, 446, 450

Intermediate filaments, 53; defined, 53

Intermediate host, 180

Intermediates, 56, 58, 60, 60; defined, 54

Internal bleeding, 145

Internal chemical sense, 248

Internal environment, 2, 5, 6, 66, 200–216; awareness of change in, 247; decoding signals from, 247; digestion, 117, 118, 119, 119, 122; neuron functions, 77, 78; nutrient absorption, 112; oxygen diffusion in, 186; stability, 67, 82–84, 83, 135, 142; target cell responses, 284

Internal respiration, defined, 192

Internal transport, cardiovascular system and, 160

Internal urethra sphincter, 206

Interneurons, 229, 229, 233, 236, 244; bipolar, 262; defined, 220; in sensory pathway, 248, 249, 263

Interphase: DNA replication, 391; G1 period, 354, 355, 355, 370; G2 period, 354, 355, 355, 370; S period, 354, 355, 355, 370; as stage of meiosis, 361, 361, 362, 365, 368, 370; as stage of mitosis, 354–355, 354, 355, 356–357, 368, 370

Interstitial cells, in testes, 290

Interstitial fluid, 82, 90, 166; blood relationship with, 136, 142, 150, 152, 152, 160, 193; as "external environment," 142; hydrostatic pressure of, 152, 152; inflammation and, 166, 167, 168; in urinary system, 200, 202, 208–210, 208, 209, 212, 212; lung capillaries/alveoli, 192, 192, 198; lymph in, 159; relationship with blood, 158; neuronal membrane and, 220, 223

Interventions, behavioral, 298–299, 302

Intervertebral disk, 108; rupture of, 93

Intestinal bacteria, 180

Intestinal glandular cells, 118

Intestinal lining, 118, 119, 132; damage to, 122, 123; hormone secretion, 117

Intestinal nucleases, 117

Intestinal ulcers, 116

Intestines, 98; capillary bed of, 146; location, 78; macrophages in, 73; types of epithelium, 68

Intoxication, 240, 243

Intracellular digestion, 51, 51

Intracellular fluid, gas exchange in, 202

Intracellular pathogens, 169, 170

Intrafallopian transfer, 300

Intrapleural fluid, 187, 190

Intrapleural pressure, 190

Intrapleural space, 187

Intrapulmonary pressure, 190, 190

Intrauterine device (IUD), 298, 299

Intravenous drug abusers, 181; clean needle strategies, 339; HIV spread, 337, 337

Intrinsic clotting mechanisms, 153, 153

Intrinsic factor, 116

Introns, 411, 418, 445; defined, 411

Inuit, modern, 483

Inversion, altering chromosome structure, 398, 399, 402; defined, 399

Invertebral disks, 91, 232; defined, 93

Invertebrates, 426,492

In vitro fertilization, 300, 322; moral/ethical dilemmas, 393

Involuntary control, of respiratory system, 195

Involuntary muscles, in digestive system, 115, 115

Iodide deficiency, 131

Iodine, 16, 18, 127; radioactive, 280; thyroid function, 280, 280

Ion channels, 222

Ion diffusion, 152, 224; at concentration gradients, 221, 221, 222, 223; at electrochemical gradient, 221, 221, 222, 223, 226

Ionic bonds, 15, 21, 36; defined, 20, 20

Ionization, 15, 20

Ionizing radiation, 367; as mutagen, 416

Ion pumps, 221, 221

Ions, 21, 22, 22, 36, 66, 75; concentration homeostasis, 82; defined, 20; in ECF, 202; in membrane transfer, 39, 41, 42, 43, 49; transfer, at gap junctions, 72, 72

Ip, Clement, 10

IRAP gene, 451

IRAP protein, 451

Iris, 258, 258, 259; color and, 382, 383; defined, 258

Iron, 16, 127; deficiency, 131; in heme groups, 32, 32, 138, 138, 139, 160; in hemoglobin, 71, 126, 127, 131; -rich proteins, 105, 136

Iron deficiency anemia, 127, 139, 155

Irradiation, loss of chromosome region, 398, 399

Irrigation, 518, 518, 519

Isocitrate, 57

Isolated food chain, 506

Isolated populations, 472

Isoleucine, 125, 125, 412, 419

Isotonic solution, 43, 43, 210

Isotope, 36; defined, 17

Itano, Harvey, 408

Jaundice, 120, 343; vitamins and, 126

Jaws, 478, 480, 480, 483

Jejunum, 112, 117, 132

Jellyfishes, 463

Jenner, Edward, 162

Jet lag, 266

Jimmie the bulldog, 134, 135

Jogging, 90, 97, 106, 129

Joint, 74, 90, 98, 105, 256; arthritic, 96, 178, 179; artificial, 96, 97; ball and socket, 95, 96; defined, 87; elbow, 94, 96, 98; fibrous, 96, 97, 108; hinge, 96; injuries, 95, 96, 381; knee, 91, 95, 96; mechanoreceptors in, 250; nociceptors in, 251; range of movements, 96, 96, 108; synovial, 95, 108; uric acid crystals, 202

Joint breakdown, aging and, 325

Joint replacement, 71

Jugular foramen, 92

Jugular vein, 92, 143

Junk mail, 519

Juvenile-onset diabetes, 276

Juxta glomerular apparatus, 211; defined, 211

K

Kaposi's sarcoma, 338

Karotype analysis, 322

Karotype arrangement, 353; of chromosome 21, 400; of sex chromosome, 388, 389, 389

Kennedy, John F., 279

Keratin, 32, 33, 33, 37, 80, 81

Keratinocytes, 80

Ketone group, 25, 26, 26, 37; as functional group, 25

Ketones, insulin deficiency and, 276

Kidney, 66, 146; age-related change, 327; ATP use, 60; autonomic nervous system and, 230; blood balance of water/solutes, 201, 203, 208, 208, 216; blood-filtering, 201, 203, 204, 205, 205, 207, 207; blood flow to, 147, 151; connective tissue, 73; deferred pain, 251; defined, 204; endocrine cells of, 269, 275; epithelial lining, 69; excretion from, 120; fat storage, 75, 125; hormone production, 139, 282; as hormone target, 272, 272, 273, 275, 281; internal structure, 204, 216; nonspecific response in, 166; peroxisomes, 51; transplantation, 213

Kidney cancer, 177, 430, 431, 434, 450

Kidney cortex, 204, 204, 205, 208, 209, 210, 210

Kidney damage, 107; blood substitute side effects, 145; drug abuse and, 243; sickle-cell, 381

Kidney dialysis machine, 213

Kidney disease, blood pressure and, 149

Kidney disorders, 203, 207, 213

Kidney medulla, 204, 204, 205, 208, 209, 210, 210

Kidney stones, 127, 206, 207, 213, 281

Kilocalories, 124, 125, 128, 129, 130, 214; defined, 124

Kinetochore, 363; defined, 358; in meiotic cell division, 362, 363; in mitotic cell division, 358, 358, 359

Kingdom, 467, 467

Klinefelter syndrome, 401, 401, 403

Knee, 91, 451, 478; blood delivery to, 143; cartilage cushions, 74

Kneecap, 95

Knee joint, 91, 97; artificial, 97; damage to, 95, 97; range of movement, 96, 96, 99

Knuckles, 94

Koch, Robert, 162

Krebs, Sir Hans, 57

Krebs Cycle, 57, 59, 60, 61, 62, 352; defined, 57

Krill, 486, 487, 491

Kubicek, Mary, 355

Kwashiorkor, 130, 130, 131

L

Labia majora, 292, 293

Labia minora, 292, 293

Labor, 318; stages of, 318, 319, 328

Lacks, Henrietta, 355

Lacrimal bone, 92, 92

Lactase, 122

Lactate, in vagina, 164

Lactate fermentation, 103; defined, 58
Lactation, defined, 319
Lacteal, defined, 119
Lactic acid, 58
Lactobacillus, 164
Lactoferrin, 167
Lactose, 26, 123
Lactose intolerance, 122
Lacunae, 74, *74*, 88, *88*, 108
Lagoons, 518
Lakes, 488, 502
Land biomes, 488
Landfills, 519
Language, 479, 483, 485
Lanugo, 316
Laparoscope, 300
Large-cell carcinomas, 434, *434*
Large intestine, 111, 112, *112*, 121, 132; ascending portion, *112*, 121, *121*, *132*; defined, 121; descending portion, *112*, 121, *132*; transverse portion, *112*, 121, *132*
Laryngeal cancer, *196*, 430, *431*
Laryngitis, 191
Larynx, 74, *115*, *186*, 187, 191, 198, *479*; abnormal chromosome arrangment, 398; autonomic nervous system and, *230*; defined, 186; from pharyngeal arches, *312*
Lascaux, 474, *474*, 475
Laser angioplasty, 156
Laser coagulation, 261
Last Place on Earth, The, 486
Latency, 340; defined, 333
Lateral geniculate nucleus, *262*, 263
Lateral menisci, *97*
Latissimus dorsi, *99*
Lawn chemicals, 428
Leaching, 496, 502, *503*, 520
Lead, 203, *516*; water pollution, 203
Leakey, Mary, *480*, 482
Learn, inborn capacity to, 476
Learning disabilities, 218
Lecithin, 117, 119, 125
Left brain hemisphere, 232
Legal drugs, 218, 219
Legs, 95, 483; bones of, *95*
Legumes, 502
Lemming, *489*
Lemuroids, *476*
Lemurs, *476*, 485
Lens, *258*, *259*, *259*, 260, 261; age-related changes, 327; embryonic, *314*; defined, 258
Lentiform nucleus, skill memory, *239*
Leopard seal, *491*
Leptin, 129
Lettuce, 428
Leucine, 125, *125*, *412*
Leukemia, *155*, 423, 426, 434–436, 451, 523; acute myeloid, *449*; causes, 367, *428*; as major types of cancer, *431*; treatment, 430
Leukocyte adhesion deficiency, *449*
Leukocytes, 136, 137, *137*
Levi-Montalcini, Rita, 282
Levonorgestrel, 299
Leydig cells, *290*, 291, *291*; defined, 290
Lichens, 486
Life, artificial prolonging of, 11; ATP energy for, 56; atomic structures and, 16; cell theory, 38; characteristics of, 3, *3*, 4, 6; defined, 2; evolution of, 468–469; "meaningful," 301; molecules of, 25; origin of, 457, 468–472; water as essential to, 15, 21, 22, 45
Life cycle, 305, 352, *363*; meiosis, 366–367, *367*
Life expectancy, *196*, 323, 328, *479*, 513, *513*; genetic disorders and, 401; smoking and, *196*
Lifestyle, choices, 423, 429, 430, 436; life expectancy and, 323; heart dis-

ease risk, 154; pregnancy and, 320–321; sedentary, 66, 90, 137
Life-support systems, 2, 11
Life table, *513*; defined, 513
Ligaments, 73, 96, 97, *97*; at wrist joint, 94; defined, 90; of eye lens, 258, 259, *259*; mechanoreceptors in, 250; pelvic girdle/axial skeleton attachment, 95; vocal cord, 191; umbilical, *317*
Light, awareness of, 262, 263, 264
Light energy, 258, *258*; photoreceptors detect, 77, 247, 249, 263
Light intensity, 478; as sensory signal, 266, *266*, 267, 275, 281
Light microscope, 48, *48*, 74, 358
Lightning, 470, 488, 502, 506
Light wavelength, *248*, 258, *258*, 259, *259*; trajectories, 259, 261
Limb bones, 74, *74*, 95, 108; skeletal muscle movement, 87, 108
Limb buds, 305, 307, *314*
Limbic system, 233, *237*, 244; defined, 237; fact memory, 239; sensory information to, 252
Limbs, motion of, 233, 250, 251, 264; formation of, 314, *315*, 316, *320*, 321; somatic nerves and, 230, 244; valved veins in, 151
Limiting factor, 526; defined, 512
Lineage, 457, 460, 465–467; *466*, 472, 476; human, *475*, 477, 483; mammalian, 475
Linkage, 402, 405; crossing over and, 391, *391*, 402; defined, 391
Linoleic acid (LA), 10, 125
Linolenic acid, *28*
Lions, 474, 490
Lipase, *117*
Lipid bilayer, 42, 49, 62, 119, 165, 172, 235; basic cell structure, *40*, *41*; defined, 41; diffusion and, 44, *44*; hormonal diffusion, 271; of neuronal plasma membrane, 221, *221*
Lipid envelope, 332, *332*, 336, *337*
Lipid metabolism, 394
Lipid molecules, breakdown of, 214
Lipid-protein film, 471
Lipid-protein membranes, 472
Lipids, 28–29, 36, 37, 41, 51, 125, 354; as "biological molecule," 15, 17, 25; blood-brain barrier and, 235; blood transported, 135, 136; defined, 28; digestion of, 56, 110, 116, 119, 122, *123*; digestive enzyme defect, 387; ecosystem organization and, 490; envelopes, 165, *165*; free-radical damage, 324; in arterial wall, 156; in diet, 125; in plasma, 136, *136*; in plasma membrane, *39*, 41, 43, 62; as "proto" cell, 471, *471*, 472; synthesizing, 46, 50, *50*, *51*, 62, 277
Lipofuscins, 324
Lipoplexes, 451
Lipoplex therapy, 451
Lipoproteins, 32, 33, 136, 154
Lips, 250, 274
Liquid nitrogen freezing, 343
Lithotripsy, 206
Liver, 50, 52, 60, *112*, 113, 118, *306*; artificial replacement cells, 70; autonomic nervous system and, 230; bilirubin in, 139; blood flow to, *143*, 147; capillary bed of, *120*, *146*, 147; cells, 40, *47*, 51, 355; deferred pain, *251*; drug abuse and, 242, 243; drug detoxification, 240; endocrine cells of, *269*, 275, 282; fetal bypass, 316, 317, *317*; as hormonal target, 275, 277; lipid transportation, 136; location, 78; nonspecific response in, *164*, 166, 182; vitamins and, 127

Liver function, 26, 27, *120*, *123*, 203, 207; digestion in, 120, 125, 132, 154, 278; enzyme secretion, 117, 132; glucose conversion in, 5, *277*, 278; storage and, 122
Liver transplantation, 120
Living cells, 468, *469*
Living wills, 11
Local chemical controls, 195, 198, 207, 267
Local, graded potential, 248, 263
Local signaling molecules, 222, 244, 284; defined, 268, 282
Logging over, *521*
Logistic growth, *512*; defined, 512
Long-term adjustment, hormones induce, 267
Long-term memory, 239, *239*, 244
Loop of Henle, 205, *205*, 209, 210, *210*, 216
Loose connected tissue, 73; defined, 73
Lordosis, 90
Lorises, *476*
Lossa ovalis, *317*
Low-density lipoproteins (LDLs), *136*, 154, 395
Lower back, 93
Lower esophageal sphincter, *115*, 116
Lower extremities, *91*, 95; blood flow, *146*
Lower jaw, 92, 114, *114*
"Lucy," *480*, 481, *482*
Lumbar nerve, 228
Lumbar vertebrae, 93, *93*
Lumens, 156, *212*; of blood vessels, 142, 150, 154, *156*; of GI tract, 112, 113, *113*, 116, 122, *122*, 132; intestinal, 117, 118, *118*, *119*, 121
Lumpectomy, 432
Lung cancer, 189, *196*, 197, 242, 367, *431*, *434*; gene therapy, *450*, 451; smoke as carcinogen, *422*, 423, 426, *428*, 433–436
Lungs, 66, 74, 78, *144*, 160, 281, *479*; age and function, 325; airways to, 114, 186–187; alveoli, 68; artificial replacement cells, 70; autonomic nerves, 219, *230*, 244; bone protection to, 88, 93; breathing process, 190–191; capillary bed for, *146*; ciliated cells in, 358; collapsed, *197*; cystic fibrosis and, 386–387, *387*; deferred pain, 251; defined, 187; diseases of, 23, 123; fetal bypass, 316, 317, 318; gas exchange, *146*, 147, 150; lobes of, 187; local signaling molecules in, 268; membrane lining, 69, *69*; nonspecific defense mechanisms, 164, 166; pulmonary circuit, 57, *123*, 135, 138; respiratory system, component, 185, *186*, *194*; sickle-cell damage, *381*; smoke/smog damage, 21; substitute blood and, 71; volume, 191, *191*, 198, *327*
Luteinizing hormone (LH), 271, 273, *273*, 284; control female reproductive function, 294, 295, *295*, *296*, 299, 302; control male reproductive function, 287, 290, 291, *291*, 302; defined, 290; menopause and, 327; midcycle surge, 294, 295, *295*, *296*, 302; primary effects, 269, 272, *272*
Lymph, 113, *159*, 166, 170; defined, 158
Lymphatic system, 79, 135, 158–159, *306*, 425, *425*, 434–435; cardiovascular system link, 142, *152*; components of, *158*; defense, 159
Lymph nodes, *158*, *159*, 164, 170, 431–435; defined, 159; growth of lymphocytes in, *178*; HIV infection and, 338; immune response, 159, 160; nonspecific response in, *164*, 166, 182; specific response in, 168, *170*; STD spread, 341, 343

Lymphoblasts, 451
Lymphocytes, *159*, 160, 164, 171, 177, 427, 430; abnormal rsponse, 178, 179, 182; activated, 173, 417; antigen specificity, 174, 182; armies of, 168, *171*; blood-borne, *136*; foreign agents and, 159; as hormonal target, *275*; specific immune response and, 137, 164, 168, *168*, 169, *169*, 170, *170*; stem cells, 451
Lymphoid organs, 158; defined, 170
Lymphoid tissues, 158, *158*, 431, 434, 436
Lymphomas, *178*, 431, *431*, 434, 435, 436
Lymph vascular system, 159, *159*, 160, 170
Lymph vessels, 113, 118–119, *118*, 132, 158, *158*, 168; cancer cells in, *425*; defined, 159; in dermis, 81
Lysergic acid diethylamide (LSD), 243
Lysine, 125, *125*, *412*
Lysis, 165, 182, 333; defined, 165
Lysosomes, *46*, *47*, 51; defined, 50
Lysozyme, 165; defined, 164

M

McCagg, Betsy, 86, *87*
McCagg, Mary, 86, *87*
Macroevolution, 462–463, 472; defined, 457, 459
Macrofibrils, 33
Macrophages, 73, *171*, 177, 179, 220; antigen presenting, 338, 346; blood-borne, *136*, *137*; in rheumatoid arthritis, 178; as nonspecific response, 164, 182; phagocytic functions of, 137, 139, *164*, 166–168, *166*, *167*; spleen storage, 159
Maculae adherens junctions, 72
Mad cow disease, 71
"Magic bullets," 177, 430
Magnesium, *16*, 126, *127*; atomic number, *17*; atomic symbol, *17*; electrons in outer shell, *17*; mass number, *17*
Magnesium ions, 22
Magnetic resonance imaging (MRI), 429, *429*, 436
Malabsorption disorder, defined, 122
Malaria, 180, *181*, 434, 505; adaptive traits and, 6; drug-resistance, 181
Malate, 57
Maleness, 287
Male reproductive system, 288, 307, *390*; components of, *288*; defective hormone receptors in, 397; hormonal controls, 290–291, *291*; primary organ, 286, 287, 289, 302; reproductive tract, 290, *291*; sex determination, 287, 388
Malleus, 254, *254*
Malnourishment, 242
Malnutrition, 122, 130–131, 387; defined, 130
Malthus, Thomas, 458
Maltose, 26
Mammalia, *467*, 476–477
Mammals, 2, *3*, 464–465, *464*, 467–468, 474–479; biotic potential of, 512
Mammary ducts, 319
Mammary glands, *306*, 448, 476, 477; acquired immunity and, 173; as hormone target, 272, *272*, *273*; hormone triggers, 417; lactation, 68, *319*, 328; tumors in, 10, 279
Mammogram, *429*, 432, *432*
Mandible, 92; defined, 92
Manic behavior, 107
Mantle, 468, 472
Manual dexterity, 482
Mapping, chromosomal, 391
Marasmus, 131

Margarine, 28
Marginally productive land, 519
Marijuana, *241*, 243
Marine life, 458, 461, 486, 487
Markers, 449; MHC, 168, *168*, 170–172, *171*, *172*, 173, 182; nonself, 427; self, 140, 141, 168, *168*, 171, 182, 327
Marrow. *See also* Bone marrow
red, 88, 108; yellow, *88*, 108
Marrow cavity, 89, *89*
Masai, modern, *483*
Mass extinction, 472; defined, 467
Mass number, 17; defined, 16
Mast cells, *167*, 173, 178, *179*, 182; defined, 166
Mastectomy, modified radical, 432–433
"Master gene," 287
Master molecule of life, 404
Maternal chromosome, 383, 388; as carrier, *394*, *396*; in meiosis, *364*, 365, 368, *368–369*; in mitosis, *356*, *368–369*; patterns of autosomal inheritance, 394–395, *394*; patterns of X-linked inheritance, *396*, *396*, 397; transmission of, *373*, *374*
Maternity cases, 444
Mating rituals, 461
Matrix, 52, 57, *58*; cartilage, 74; osteoblasts and, 89
Matter, 3, 12; chemistry of, 15, 16
Maturity-onset diabetes, 276
Maxillary bones, 92, *92*
Maxillary sinus, *93*
Measles, 176, *176*, *181*
Mechanical breakup of food, 114, 115
Mechanical energy, 248, 254, 264
Mechanical insults, 81
Mechanical pressure, *248*, 249, 264
Mechanical stimulation, 297
Mechanical stresses, *91*, 114; bone strength and, 89, 90, 108
Mechanoreceptors, 248, *248*, 249, 264; near body surface, 250, 251; vibration-sensitive, 254, 255
Medial menisci, *97*
Medical genetics, 386
Medical imaging, 429, *429*, 436
Medulla, breathing rhythm and, 194, *194*, 195
Medulla oblongata, 148, 149, 151, *230*, *233*, 244; defined, 232–233; stimulant use, 242
Megakaryocytes, 75, 137, *137*
Meiosis, 290, *291*, 294, *370*, 406; abnormal events and, 388; basic function, *370*; chromosomal behavior during, 387, 388, *394*, *396*, 402; crossover event during, 387, *391*; defined, 352; independent assortment, 378, 379, *379*, 384, 459; making haploid gametes, 361, 366, *366*, 367, 370, 375; mechanism in gamete-formation, 350–351, *351*, 353, 361, 367, 370; nondisjunction during, 387, *400*, 400, 401, *401*, 402; segregation at, *375*, 376; stages of, 361, *362–363*, 370; two divisions, 361, *362–363*, 370, *370*; vs. mitosis, 368, *368*, *369*
Meiosis I, 290, 308, *366*, *368*, 370; egg formation, *294*, 294, 367; in monohybrid cross, *375*; key events, *364*, 364; overview, 361, *361*; sperm formation, *291*; stages of, *362*, 363
Meiosis II, 290, 294, 365, 366, 367, *369*; in monohybrid cross, *375*; key events, *370*; overview, 361, *361*; sperm formation, *291*; stages, *363*
Meiotic cell division, 290, *291*, 294
Meissner's corpuscle, 250, *251*
Melanin, 80, 81, *382*, 383; synthesis, *127*
Melanocytes, *80*, 435, *435*; defined, 80
Melanoma, malignant, 430, *431*, 435, *435*, *450*, 451

Melatonin, 21, 266, *266*, 267, 281; primary effects, *269*, *275*
Meltdown, 367, 522, 523; defined, 522
Membrane attach complexes, *165*, 271; defined, 165
Memory, 233, *237*; circuits, 239, *239*, 244; defined, 239; fact, 239, *239*; immune, 174–175; long-term, 239, *239*, 244; nervous system and, *219*, 244; short-term, 218, 239, *239*, 244, 326; stages of processing, 239
Memory cells, 169, 170, 176, 177; B cells, *169*, 172, *172*, 173, 175, *175*; defined, 168; IgG, 178; subpopulations of, 170, 182; T-cells, *169*, *171*, 175, *175*
Menarche, 293
Mendel, Gregor, 372, 373, *373*, 374, 383; deviation from single-gene model, 381; hybridization studies, 375; independent assortment, 378; inheritance patterns, 376, 392, 393, 444; segregation, theory of, 375, 384
Meninges, *232*; defined, 232
Meningitis, 225
Menisci, 97
Menopause, 157, 327, 433; physiological changes, 293; symptoms, 106
Menstrual cycle: birth control methods and, 298, 299; defined, 293; events of, *292*, 293, 294, *294*, 295, *295*, 302; follicular phase, *292*, *296*; luteal phase, *292*, *296*; ovulation, *292*, 298; visual summary, 296, *296*
Menstrual period, 253, 312, 313, 318, 327
Menstruation, 282, *292*, 295, 302; absence of pregnancy, 309; chronic problems, 342
Mental illness, genetic links, 383, 417
Mental retardation, 280, 321, 343; abnormal chromosome rearrangement, 388, *394*, 398, 400, 401, *403*
Mercury, 45, 188, 447; water pollution, 203
Mercury poisoning, 235
Merocrine glands, 68
Messenger RNA (mRNA), *411*, 412–413, *412*, *413*, 418; defined, 410; sequences, 412, 413, *412*, 416, *416*, 449; transcripts, 270, *417*, *419*, 445, *445*; transcripts translated into polypeptide chain, *412*, 414, *414*, 415, *415*, 418
Mesoderm, *306*, 310, 312, *312*; defined, 306; embryonic tissue layer, 305
Metabolic acidosis, 276
Metabolic disorders, 408
Metabolic pathways, *54*, 55, 59, 62, 469–472; ATP, *103*; blockage, 408; defined, 54
Metabolic wastes, 136, 146; blood transported, 135, 138, 142; disposal of, 142, 201, 202; removal, *80*, 81, 82
Metabolism, 25, 39, 47, 80, 212; cellular, 54–55, 62, 82, 83; core body temperature, 201, 214, 216; defined, 4, 39, 54; of glucose, 37; heat production, 201, 490, 490, 492, *493*, 494–495, *494*; homeostasis, 66, 67; hormonal regulation, 273, *275*, 277, 278, 280, 281; inactivation of harmful byproducts, 50; metabolic activity, 5, *5*, 12, 15, 111, 470–471; nitrogen cycle, 502; nucleotide role in, 34, 37; oxygen role in, 138; rates, 129, 280, 281; water by-product, 202, 216
Metacarpals, 91, 94, *94*, 95
Metafemale, *401*
Metal ions, 463, 470; cofactors as, 54, 55
Metal hydrides, 35
Metaphase, 388, *389*; defined, 356; during mitosis, 354, *354*, *355*, 356, 357, *357*, 369, 370

Metaphase I, 379, *379*; chromosomal movement, 388; during meiosis, *361*, *362*, 365, *365*, 368, *368*, 370
Metaphase II, 379, *379*; chromosome nondisjunction at, *400*; during meiosis, *361*, *363*, 365, *369*, 370; oogenesis arrested, 367
Metastasis, *424*, 425, *427*, 435, *435*, 436
Metatarsals, *91*, 95, *95*
Meteorites, 467, 468
Methamphetamine, *241*
Methane, *24*, 500, *501*, *516*, 525
Methaqualone, 243
3-methylfentanyl, 243
Methionine, 125, *125*, 156, *412*, *419*
Methyl group, *24*
Methyl mercury, 235
Metronidazole, 344
MHC markers, *168*, 170, *171*, *172*, *172*, 182; defined, 168; in tissue rejection, 173
Micelles, 119, *119*
Microbes, 114, 446, 447; defense, *78*, 116, 132, 137; recombinant, 452
Microevolution, *466*, 467; defined, 457, 459; processes of, 460–461, *460*
Microfilaments, *46*, 53, 313, *313*, 360, *360*; defined, 53
Microglia, 220
Micrographs, 48
Micro-injection, 448
Microorganism, 58, 121; antibiotics and, 180; defense mechanisms, 162, 163, 165
Microsatellites, 444
Microscopes, 38, *38*, 40, 162, 232; comparisons of, *48*; use with cell study, 40, 46, 48
Microscopy, *48*
Microspheres, *471*; defined, 471
Microtubular, spindle, 352, 388; during meiosis, 361, *362*, *363*, 365, 370; during mitosis, *354*, *357*, 356–358, 359–360, *359*, 370
Microtubule organizing centers (MTOC), 53, 358; defined, 53
Microtubules, *46*, *291*, 358, 433; defined, 53; form neural plate, 313, *313*
Microvillus, 68, *69*, *118*, 119, 132; defined, 118
Micturition, 206
Midbrain, 219, *230*, 232, 233, *233*, 244, 476; defined, 233
Middle ear, 92, 254, 255, 264; bones of, 254, *254*, *255*
Midline, 313
Midsagittal plane, *79*
Migraine headache, *225*
Milk, 68, 417; acquired immunity and, 173
Milk of magnesia, 22, *23*
Miller, Stanley, 470, *470*
Millivolts, 220
Mineral ions, 495, 496, *497*, 506; absorption of, *112*, 119, *132*, 202; in plasma, 136, *136*, 160; stored in bone tissue, 88
Mineralization, process of, 88, 90, 108
Mineralocorticoids, *275*, 279; defined, 278
Minerals, 110–111, 124, 132, 492, 496, 502, 504; deficiencies, *127*, 130, 131, 202, 495; defined, 126; functions, *127*; sources, *127*, 138; supplements, *127*, 139
Mineral salts, 278
Minisatellites, 444
Minnows, 510
Miocene epoch, 479
Miscarriage, 314, 340, 367, 400; risk of, 321, 322
Mitochondria, *46*, 51, 52, 54–55, 106, *179*; defined, 52; formation of ATP, 56, 57, *58*, 62; in muscles, 87, *100*,

105; main function, 46, *47*, 52; membrane system, 52, 58, *58*, 59, *59*; mutations in, 416; of sperm midpiece, 290, *291*; stages, 56, 57, *59*
Mitosis, 290, *291*, 425, 430; abnormal events and, 387, 388, 402; basics of, *370*; cell cycle and, 354–355, *355*; cell division, mechanism, 350, *350*, *351*, 353, 368, *370*; chromosome number and, 352–353; cytokinesis, 360, *360*, 370; defined, 352; DNA replication, 406; four stages of, 354, *354*, 356–357, *356–357*, 358–359, *369*, 370; immune responses and, *169*, *171*, *172*; vs. meiosis, 368, *368*, 369
Mitotic cell division, 307, 308, *308*, 400
Mitral valve, 144, *145*
Mixture, defined, 17
Mobility, 97
Molars, 114, *114*
Molecular biology, 163
Molecular genetics, 405
Molecules, 22; defined, 17; radioactively labeled, 18, *18*; rogue, 21; writing chemical reactions, 17
Molecules, biological, 6, 7, 17, 54, 136, 267, 313; complex, 5, 472; diffusion of, 42, 43; in membrane transfer, 39, 41; net surface charge, 408
Moles, 424
Mollusks, hard-shelled, 463
Monerans, *3*
Monkeys, 475–476, *476*, 477, 478, *478*, 485; primate evolution, 479, 481, *481*; Rhesus, 140, 466
Monoclonal antibodies, 177, 446; defined, 176, 177; radioactively labeled, 429, 430, 434–435, 436
Monocytes, 166, 167; blood-borne, *136*, 137, *137*
Monogamous behavior, 345, 346
Monoglycerides, 117, *117*, 119, *119*
Monohybrid cross, 375, 376, *377*, 378, 384; defined, 375
Monomers, 26, 27; defined, 25
Mononucleosis, 434
Monosaccharides, 26, *26*, 27, 37, 117, 119, *119*, 132; defined, 26; entering blood vessels, *118*
Monosomic, defined, 400
Montagu, Mary, 162
Montremes, 462
Morning-after pill, 299
Morphogenesis, 306, 307, 312, 313, *313*, 314, 328; defined, 307
Morphological abnormalities, 320
Morphological convergence, defined, 465
Morphological divergence, *465*
Mortality, *513*
Morula, *307*, 309, *309*, *323*; defined, 306
Mosses, 486, *489*
Motion sickness, 257
Motor coordination, drugs effect on, 240, 242
Motor cortex, 236, 244; skill memory, *239*; primary, *236*, 237
Motor end plate, *226*
Motor neurons, 40, 77, 102, *102*, 103, 215, 219, 449; action potentials in, 224, *224*; axons, 225, *225*, 228, 231; defined, 220; information flow, 229, *229*, 244; input zone, 222, 231; in sensory pathways, 248; loss of, 325; as part of motor unit, 104, *104*, 105, 108; reflex movement and, 232; resting potential, 221; trigger region, *222*
Motor proteins, in meiotic cell division, *362*, *363*
Motor unit, 104–105, *104*, 108; defined, 104
Mouth, 111, 112, *112*, 113, 114, 132, 312; carbohydrate digestion, 116, 132; dryness, 211; enzyme activity, *117*;

taste buds in, 252; type of epithelium, *68*

Mouth cancer, *196*, 430, *431*, 434

Movement, 39, 53, 78; cell triggers, 15; coordinated, 233; fetal limb, 316, *316*; involuntary, 76; muscle tissue and, 76, *76*, 78, *78*, 80, 84; perception of, 262, 263, 264; repetitive, 94; skeletal muscle, 87, 88, 98, 108; voluntary, *76*

mRNA-cDNA hybrid molecule, 445, *445*

Mucins, 114

Mucosa, 113, *113*, 117, 132; alcohol absorption, 242; "friendly" bacteria on, 164; intestinal, 117, 118, *118*, 119, 121

Mucous glands, 69

Mucous membranes, 182, *197*; age-related changes, 327; allergens and, 178; bacterial attachment to, 333; defined, 69; form vocal cords, 191; HIV point of entry, 334, 336; in sinuses, 92; list of body systems, 170; STD infection and, 340; white blood cell location, 164, *164*, 166

Mucus, 53, 68, 69, 196; cystic fibrosis and, 386–387, *387*, 394; immune response and, 164, 173; in cervix, *292*, *293*, 295; in digestive system, 112, 114, 116, *132*; in semen, *288*, 289, *289*; on respiratory lining, 186, 187, 189

Mucus epithelium, 112

Multicellular organism, 7

Multiple allele system, defined, 381

Multiple myeloma, *431*

Multiple sclerosis, 177, 224, *225*

Multipolar cells, *220*

Multiregional model, 484; defined, 484

Mumps, *176*

Muscle cells, 103–106, 121, 277; cellular "work," 4; circular, 113, *113*, 117; energy demanding, 59, 60; longitudinal, 113, *113*, 117; mitochondria in, 52, 87; specialization, 98; use of fermentation pathway, 58

Muscle contraction, 44, 50, 83, 103, 281, 289; chemical synapses and, 226, *226*, 244; during digestion, 122, 132; sarcomeres, 100, *100*, 101, *101*; sliding filament model of, 100, 101, *101*; strength of, 105, 108; tetanic, *105*

Muscle groups, 98, 108

Muscles, 2, 66, *306*, 464–465, *465*, 475, 477; /bone interaction, 87–88, 90, 93–94, *97*, 98, 98, 104; degenerative disorders, 396–397; as effectors, 82, 84; fatigue, 105, 251; function, 100–101; glucose uptake, 276, 277; mass, 129, 323, 325; nervous system and, 77, *77*; properties of whole, 104–105; sickle-cell damage, *381*; spindles, *231*, 233, *248*; strength, *327*; structural protein, 30, 31; structure, 76, 98, 100–101; work capacity of, 106

Muscle sense, 247, 250, 264

Muscle system, 78, *78*; overview, 87

Muscle tension, 106; defined, 104; isometrically contracting, 104; isotonically contracting, 104; lengthening contracting, 104

Muscle tissue, 26, 27, 78, 82, 103; contraction, 90, 95, 98; defined, 76; in embryo, 77; types of, 76, *76*

Muscle to bone attachment, 73, 74, *78*

Muscle twitch, *105*; defined, 105

Muscular dystrophy, 396

Musculoskeletal system, 86–109, 98; defined, 87

Music, 234, 237

Musick, John, 524

Muskrats, 488

Mutagens, 418, 423; defined, 416

Mutations, 12, 386–388, *386*, 439–440, 466, 472; beneficial, 461; cancer and, 423, 426–429, *427*, 432, 436; DNA, 6; frameshift, 416, 418; gene alteration, 405, 407; genetic disorders and, 393, 394, 396, 397, 402; genetic engineering, 445–446, 450, 453, 459–460, *460*, 464; harmful, *461*; mutagen induced, 416, 418; neutral, 461; spontaneous, 416, 418; substitution, 416, *416*, 418; of viruses, 452

Myasthenia gravis, 226

Mycobacterium tuberculosis, (TB), 180, *181*, 197

Mycorrhizae, 502, 503

Myelin sheath, *224*, 225, *226*, 228, *229*, 232; age-related changes, 326; defined, 224; destruction, 225

Myeloma cells, 177, *177*

Myocardial infarction, *155*

Myocarditis, *155*

Myocardium, *144*; defined, 144

Myofibrils, 100, 102, *102*, 106, 108; defined, 100

Myoglobin, 105, 106

Myometrium, 292, *293*

Myopia, 260

Myosin, 76, 101, *101*, 102, *102*, 103, 108; defined, 100

Myotonic muscular dystrophy, 393

N

NADH, 55, *56*, 57–58, *57*, *58*, 59, 61

NADP⁺, *37*

NADPH, 504

Nails, *37*

Nasal cavity, 92, *92*, *186*, 198, 246; from pharyngeal arches, *312*

Nasal epithelium cancer, *428*

Nasal passages, *68*, 92

Nasal septum, 92

National Academy of Sciences, 428

National Aeronautics and Space Administration (NASA), 70

National Marine Fisheries Service, 524

National resources, 509, 512

Native Americans, blood type, 460

Natural disasters, 488, 512

Natural gas, 522, *522*

"Natural history," 8

Naturalists, 458, 462

Natural killer (NK) cells, 177, *182*, 427, *427*, 430; defined, 171

Natural resources, 512–513, 515, 519, 522, *522*, 525–526; reserves, 522

Natural selection, 457–459, *460*, 461, 465, 472; defined, 460

Nausea, 120, 121, *126*

Navel, 115, 121, 317, 318

Neanderthal, 483, *483*

Nearsightedness vision, 259, 260, *260*, 261

Neck muscles, 93, 190

Needle contamination, 336, 337; "clean" strategies, 339

Negative feedback mechanisms, 82, 84, 195, *295*; defined, 82; hormone secretion and, 275, 278

Neisseria gonorrhoeae, 340, *340*

Nematodes, 490

Nembutal, 241

Neonate, 323, *323*

Neoplasm, *424*; defined, 424

Nephritis, 213

Nephrons, 205, 208, *208*, 209, 272, 278; age-related changes, 327; defined, 204; function, 282; functional regions, *205*, 211, 216; kidney disorders, 213; permeability characteristics, 210, *210*, 212

Nephron tubule, 211, 212, *212*

Nerve cells, *40*, 113; potassium, 126; sodium, 126

Nerve cord, 476

Nerve fascicle, 229

Nerve growth factor (NGF), 225, 282, 283, *283*

Nerve network, 122, *122*

Nerve pathways, from sensory receptor to brain, 247, 248, 249, *249*, 251

Nerve pathways/tracts, sensory, 248–249, *249*, 252, *253*, 262, 264

Nerves, 94, *114*, 219; compression of, 93; connective tissues, 73, 74; damage, 106; defined, 77, 228; function, 121; structure of, *229*; to bone, 88, 89, 92

Nerve tissue, 67, 77, 84, 234; defined, 77; in bones, 88, 108; in embryo, 77

Nerve tract, 229, 234, 244; defined, 232; sensory, 233, 248

Nervous system, 6, 43, 67, 78, 218–244, 305; age-related changes, 326–327; blood pressure controls, 151; calcium and, 90, 102; chemistry of, 22; control musculoskeletal movement, 87, 98, 102, *102*, 103, 105, 108; control of defecation, 121; controls kidney flow volume, 207; controls over digestion, 122, *122*, 123, 132; disease, 177; drug disruption, 218, 240–243; electrolyte imbalance, 123; flow of signals through, 219, 227, 244, 281; genetic degeneration of, 394–395, *395*; heart rate controls, 5, 148, 150, 157; homeostasis, 83, 84; interaction with adrenal glands, *269*; maturation, 323; metabolic heat output, 215, 216; neurulation, 313; overview, 228–229, *228*; respiratory controls, 188, 194, 195, 198; sensory systems and, 247; under attack, *225*, 235; vocal cords, 191, *191*

Nervous system disorders, 428

Net energy yield, of ATP, 56–59, *56*, *59*, 62

Net surface charge, of a molecule, 408

Neural-endocrine control center, 272, 278

Neuralgia, 225, 227, 232

Neural plate, 313, *313*

Neural signals, 267, 279

Neural tube, *312*, 313, *313*, *314*; defined, 312; disorders, 320

Neurofibrillary tangles, 326, *326*

Neurofibromatosis, 416

Neuroglia, 77

Neuroglial cells, 224, 228; defined, 220

Neuromodulators, defined, 227

Neuromuscular junctions, 102–103, 104, 108, 226, *226*; defined, 103

Neuronal architecture, 262

Neuronal damage, 235, 282, 283; regeneration, 282, 283

Neuronal plasma membrane, 221–224, *221*, *223*, 244

Neuron injuries, repair of, 225

Neurons, 77, 220–221, 229, 255, 268, 313; breathing rhythm and, 194, 198; cell bodies of, 232; cell division "brakes," 354; classification of, *220*; conducting zones, 220, *220*; death of, 283; defined, 220; electrical charge, 219, 244; functional zones of, 220, *220*; input zones, 220, *220*, 222, *222*, 225, 227, *231*; output zones, 220, *220*, 226, 244; structure/function, 77; trigger zones, 220, *220*, *222*, 223, 227

Neurotransmitter molecules, 77, 225, *225*, 227; psychoactive drugs and, 240; removal from synaptic cleft, 227

Neurotransmitters, *225*, 226–227, *231*, 244, 336; acetylcholine as, 102, 108;

Nephron tubule, 211, 212, *212*

Nerve cells, *40*, 113; potassium, 126; sodium, 126

age-related changes in, 326; defined, 225, 268; in sensory pathways, 248, 253, 255; as signaling molecule, 268, 279, 284

Neurulation, 313

Neutrons, 15, *16*, 17, 36; defined, 16

Neutrophils: blood-borne, *136*, 137, *137*; defined, 166; in nonspecific response, 164, 167, *167*, 182

Newspaper, as cellulose source, 35, *35*

Niacin, 126

Niche, defined, 488

Nickel, *428*, 430

Nicotinamide adenine dinucleotide (NAD⁺), 34, 37, 59, 62, 470; as cofactor helpers, *56*, 57–58, *57*, *58*; defined, 55

Nicotine, 154, 157, *240*, *241*; blood-brain barrier, 235; defined, 242; dependence or habituation, 240

Night blindness, *126*

Night sweats, *155*

Nipples, 250, *319*

Nitrate, 203, 502, *503*, 506

Nitric acid, 516

Nitric oxide, 516, *516*

Nitrification, 503, *503*; defined, 502

Nitrite, 502, *503*

Nitrogen, 14, 16, *16*, 36, 37, 60, 123; blood transported, 136; containing wastes, *120*, 200, 202, 203; decay and, 17; gaseous, 188; in adenine, 56; as nucleotide component, 34, *34*; scarcity of, 502, 503, 506

Nitrogen atoms, 31; atomic number, *17*; atomic symbol, *17*; electrons in outer shell, 17, 19; mass number, *17*; molecular, 17

Nitrogen-containing bases, in DNA, 405, 406, *406*; in RNA, 410

Nitrogen cycle, 502–503, *504*; defined, 502

Nitrogen dioxides, 516, *516*

Nitrogen fixation, 495, 502, 503, *503*; defined, 502

Nitrogen narcosis, 188

Nitrogen oxides, 516, *516*, 517

Nitrous oxide, 500, *501*, 502, *516*

Nits, 344

Nobel Prize, 1962 for medicine, 405

Nociceptor, 248, *248*, 250, 264; somatic pain, 251

Noise pollution, 257

Noncoding sequences, 417

Nondisjunction, 365, *400*, 401, *401*, 402; defined, 400

Nongonococcal urethritis (NGU), 343, 346; defined, 342

Non-Hodgkin lymphoma, 434

Nonoxynol-9, 345

"Nonself" markers, 140, 168, 169, 182, 427; immunity breakdown, 427

Nonshivering heat production, 214, 215; defined, 214

Nonsister chromatids, 364, *364*; meiotic crossover event, *391*

Nonspecific defense responses, 163, 164, *164*, 165, *165*

Nonsteroid hormones, 284; main categories, 271, *271*

Nonsurgical sterilization, 299

Nonverbal skills, right hemisphere, 234

Noradrenalin, 279

Norepinephrine, 231, *271*, 275, 282; primary effects, *269*, 275, 279; as signaling molecule, 242

Norplant, *298*, 299, 301

Nose, 92, 115, 186, 274; cartilage, 74; olfactory receptors, 248, 252, *253*; sensory neurons in, 77

Notochord, 312

Nuclear energy, 522–523, *522*, 526

Nuclear envelope, 46, 47, 49, 53, 332; defined, 49; in meiotic cell division,

362, 364, *364*, 365; in mitotic cell division, *356–357*, 358, 359, 370
Nuclear medicine, *18*
Nuclear power, 522–523, *522*, 526
Nuclear power plants, 367
Nuclear reactors, 522, 523
Nucleic acid, *34*, 36, 37, *127*, 132, 490, 502–504; altering gene expression, 449; breakdown of, 51, 202; chains, 410; coenzyme, *126*; defined, 34; digestion in small intestine, 116, 117, *117*; as DNA component, 15, *15*, 17; nucleotides and, 406; sequences, 34; viral, 332
Nucleolus, *46*, 47, 49; defined, 49
Nucleoplasm, *46*, 49
Nucleosomes, *409*; defined, 409
Nucleotides, 25, 36, *117*, *123*, *443*, 449, 470–471; ATP as, 44, 56; bases, 405, 412, *412*, 416, 418; change and mutation, 388, 395; coenzymes, *37*; defined, 34, 406; as DNA subunit, 405–407, *406*, *407*; free, 406–407; in RNA, 410, 411, *411*, 412; kinds of, 406, *406*, 407; precursors, 471; repeats, 417, 418; structure of, *34*; sugar/phosphate backbone, 410, *410*
Nucleotide sequence, 412–413, *412*, 416, *416*, 418, 426, 441–443, *442*, *443*, 448–449, 466
Nucleus, 41, *46*, 47, 50, 62, 233, 448; cytoskeleton and, 53; defined, 40, 49; division to haploid nuclei, 361, *362*, *363*; main function, 46; mitotic division, 351, *354*, *356–357*, 358–360, 370; multinuclear arrangement, 76, *76*; of neurons, 77; number of chromosomes, 352; pores, 49, *49*; of pseudostratified epithelium cells, 68; of red blood cells, 75, 139, *139*; steroid hormonal binding, *270*, 271
Nursing, 319, 328
Nutrient cycling, *7*
Nutrient deficiencies, 131
Nutrient molecules, 72, 110, *110*; adsorption of, 111–113, 116, 118–119, *119*, 121; bile processing, 120; organic metabolism, 122
Nutrients, 77–78, 82, 106, 520, *525*; basic types, 124; blood transported, 75, 81, 87–89, 135, *142*, 147; capillary exchanges, 150; of ecosystem, 487, 490, *490*, 495–496, *495*, 502, 506; placental diffusion, 310–311, *311*, 316, *317*; reserves, 495
Nutrient turnover, 122
Nutrition, *81*, 110–132; age-related changes in, 327; maternal, 313, 320, 321
Nutritionist, 124, 131

O

Oak Ridge National Laboratory, 367
Obesity, 129, *155*; defined, 128; diabetes and, 276; dietary habits, 111; heart disease risk, *155*; venous valves and, 151
ob genes, 129
Observable evidence, 372, 373, *373*
Occipital bone, 92, *92*
Occipital lobe, 236
Ocean reservoir, 498
Oceans, 488, 496, *496*, 499, 500, *500*, 502; currents, 496
Octet rule, 19
Odor molecules, 77, 246, *248*, 252
Oil gland, *80*, 81
Oils, 28, 29, *37*, *516*, 522, *522*; carbon compounds, *24*; as exocrine secretion, 68; as fossil fuel, 35; nondigestible, 125; as nonpolar, 22
Oil spills, 486, 522

Olduvai Gorge, 482
Oleic acid, *28*
Olestra, 125
Olfaction, 252–253
Olfactory bulbs, 252, *253*
Olfactory epithelium, 246, 252, 253
Olfactory lobes, 233, *233*, 237
Olfactory neurons, 252, *253*
Olfactory receptors, *248*, 252, 253, *253*; defined, 252
Oligocene, 479
Oligodendrocytes, 224, 228
Oligosaccharides, 33, *37*, *41*, *117*; defined, 26–27
Olympic Games, 86, 87
Omnivores, 485, 506; defined, 490
Oncogenes, 427, 436, 449; defined, 423, 426
One-child limit, 301
"One gene, one enzyme" hypothesis, 408
Oocyte, *293*, 294, 300, 302, 304, 308; defined, 292; in menstrual cycle, *292*, 293, *295*; primary, 294; secondary, *292*, *293*; X chromosome only, 390
Oogenesis, *366*, 367; defined, 366
Oognia, 352, 361, *366*
Open systems, 490, 506
Opioids, 243
Opium, 243
"Opportunistic" infection, 338
Opposable movements, 478
Opsin protein, 397; in rhodopsin molecule, 263
Optic chiasm, *238*, 269
Optic disk, 258
Optic nerve, *230*, 238, 249, 258, *258*, 262, 263
Oral cancer, *431*, 434, 435; tobacco use, 242
Oral cavity, 112, *112*, 132, *186*; defined, 114
Oral contraceptives, 180, *298*, 299, 342
Oral sex, 345
Orangutan, *476*, 481
Orbitals, 36; defined, 19
Order (classification scheme), 467, *467*
Organelles, *39*, 43, 46, 50, *50*; in cell death, 170, *171*; in cytoplasm division, 360; defined, 46; in red blood cells, 75; kinds of, 46, 62; mitochondria as, 52, *52*; producing ATP energy, 87, 100
Organ formation, as fourth life cycle stage, 305, *307*
Organic chemicals, water pollution, 203
Organic compounds, 15, 24–25, 36, 469–470, *470*, 472, 504; and ATP formation, 103; defined, 24; energy-rich, 487, 490
Organic metabolism, 122, *123*
Organic molecules, 60, 62, 468; cofactors as, 54; large, digestion of, 117, 132; synthesizing, 59
Organisms, 7; interdependency of, *3*, 7, 12
Organoapatites, 70
Organ of corti, 255, *255*; defined, 255
Organogenesis, 312, 328; defined, 306
Organs, 1, 67, 69, 77, 82, 116; accessory, 111; artificially grown, 70; beginning formation, 312, 313; bones as, 88, 89; connective tissue and, 73; damage from untreated STD, 330, 331; defined, 67, 84; maturation, 316, 317, 318; placenta as, 310
Organ systems, 7, 66, 77, 78–79, *78*, *79*; beginning development, 312; defined, 67, 84; maturation, 316, 317, 318; stable internal environment and, 82, 83
Organ transplant, 11, 171, 173, 319; cell-mediated immune response,

169, 170, 182; HIV transmission, 336; immunodeficiency (non HIV), *178*
Organ transplant recipient, rejection, 173
Orgasm, defined, 297
Origin, *98*, 108; defined, 98
Osmoreceptor, 248, *248*, 264
Osmosis, 42, 62, 152; defined, 43; water transport by, 119, 121
Osmotic gradient, 208, 209, *209*, 216
Osmotic imbalance, 130
Osmotic pressure, 43, 136, 152, *152*, 167
Ospreys, 505
Osteoarthritis, 96, 325
Osteoblasts, 88, 89, *89*, 90, 108
Osteoclasts, 90, 108
Osteocytes, 74, *88*, 89, 108; defined, 88
Osteogenesis imperfecta, 381
Osteon, *88*, 108; defined, 89
Osteoporosis, 90, *90*, 325; vitamin K, 127; postmenopausal women, 327
Osteosarcoma, 383
Otoliths, *256*, 257
Otters, 462
Outer ear, 254, *254*, 255, 264
Oval window, 254, *254*, 255
Ovarian cancer, 299, 423, *429*, 433, *450*, 451; incidence, *431*; treatment, 430
Ovarian hormones, 292
Ovaries, 68, 282, 286, 292, *293*, 302, 314; absence of Y chromosome, 390, *390*; cyclic changes, *292*, 293–296, *294*, *295*, *296*; deferred pain, *251*; defined, 287; fertilization stage, *308*, *309*; germ cells in, 361, *361*, 370; as hormone target, *272*; meiosis in, 352; overview, *269*, *275*; PID infection, 342; testicular feminizing syndrome, 397
Overdraft, 518
Overfarming, 519, *519*, 521, 524
Overfishing, 525
Overgrazing, 519, *519*, 521, 524
Overheating, 83
Overpopulation, 512
Oviducts, *292*, 293, *293*, 294, *294*, 297; bacterial infection, 340, 342; birth control and, 299; defined, 292; egg fertilization, 308; fertilization, 304, *305*, 308; scarring, 340, 342, 346; site of fertilization, 308, *309*
Ovulation, 282, *294*, 297, 367; birth control methods, 298, 299; fertilization stage and, *308*, 308; hormone stimulation, *272*; infertility and, 300; suppression, 299; triggers, 294, 295, *295*, *296*, 302
Ovum, mature, 294, *308*; corpus luteum and, 282; defined, 308; meiotic cell division, *366*, 367
Oxaloacetate, 57
Oxidation, 21
Oxidation-reduction reactions, defined, 54
Oxidative damage, aging and, 324
Oxidative phosphorylation, *61*, 62; defined, 58
Oxygen, 14–16, *16*, 24, 27, 36, 106, 469; in aerobic respiration, 56–59, 62, 186–187, 198; as air component, 188; binding factors, 138; blood levels of, 151, 194, *194*, 195, 198; blood transported, 135–136, *136*, 142, *142*, 157, 160; as byproduct, 51; capillary exchange, 152; cellular demand for, 66, 71, 139, 151; diffusion of, 68; free radicals, 21; in disease, 45; in plasma, 138, 160, 202; lungs as entry point, 135, 160; metabolic pathways and, *103*, 105, 106; mitochondria system and, 52; placental diffusion, 310–311, *311*, 316, *317*, 328

Oxygen atom, 17, 25; atomic number, *17*; atomic symbol, *17*; electrons in outer shell, *17*, 19, 20; mass number, *17*
Oxygen debt, 184, 188; defined, 103
Oxygen diffusion, in gas exchange, 185, *186*, 188, 192, *193*, 198
Oxygen-free atmosphere, 472
Oxygen molecule, 32, *32*, 192, *192*, 194; blood-borne, 146; dissolved in extracellular fluid, *248*; selective permeability, 42, 44
Oxygen transport, *37*, 75, 79, 192, *192*
Oxyhemoglobin, 138, *193*; defined, 192
Oxytocin, 83, 268, *271*, 273, 282; hypothalamus/pituitary production, *273*, 284; primary effects, *269*, *272*, *272*, 284; uterine contraction, 318, 319
Ozone, 500, 516, *516*, 517, *517*, 524–526
Ozone layer depletion, 7, 8, 81

P

Pacemaker, 134, 135; artificial, *18*, 148; self excitatory cells, 148
Pacific yew tree, 433
Pacinian corpuscle, *248*, 250, *251*
Pain, 81, 93, *251*, 330, 342; defined, 250; detecting tissue damage, 248, 250; disk ruptures and, 93; in heart attack, 154, *155*; perception of, 227, 243, 251, 264; receptors, 247, *248*, 250, 264; receptors, inflammation and, *166*, 167, *167*, 182; types of, 251
Painkillers, 93, 227, *241*
Palate, defined, 114
Palatine bones, 92, *92*
Paleontologist, 95
Palm bones, *91*
Pancake ice, 486, *486*, 500–501
Pancreas, *112*, 113, 118, *120*, *300*, 446; autonomic nervous system and, *230*; bicarbonate secretion, 122; deferred pain, *251*; defined, 117; disease, 123; endocrine function, 276; enzyme secretion, 5, 116–117, *117*, *119*, 120, 122–123, 132; ER cells, 50; exocrine function, 276; genetic disorder, 386; as hormonal target, *275*; role in glucose metabolism, 277
Pancreatic amylase, *117*
Pancreatic cancer, 196, *431*, 434
Pancreatic enzymes, 5, *119*, 120, 123, 132; secretion of, 116–117, *117*, 122
Pancreatic islets, 276–277, 280; defined, 276; overview, *269*, *275*
"Pancreatic juice," 117, *119*; triggers, 268
Pancreatic nucleases, *117*
Panda, giant, 466
Panda, red, 466
Pandemic, 180, 512
Pangaea, 462, *464*
Panic attacks, 243
Pantothenic acid, *126*
Paper, recycling, 519
Papillary muscles, 144, *252*
Papovaviruses, *334*
Pap test, 343, *429*
Paraganglioma, 383
Paralysis, 224, 283, *283*; decompression sickness, 182; sickle-cell damage, 381; syphilitic, 341
Paranasal sinuses, 186
Paranoia, 243
Paranoid delusions, 243
Parasites, 180, 338, *344*, 346, 490, 495; defined, 330; eosinophils and, 137, *137*; STD infection, 344
Parasitic worms, as pathogens, 164–166, 173, *181*, 182

Parasympathetic nerves, 148, *228*, 230–231, *230*; defined, 231
Parathyroid gland, *280*, 281, 284, *306*; defined, 280; disorders, 281; overview, *269*, *275*
Parathyroid hormone (PTH), 90; primary effects, *269*, *275*, 280–281
Parental care of young, 477, *479*
Parents at risk, 393, 394, 395
Parent species, *461*
Parietal bones, 92, *92*
Parietal lobe, 236
Pariza, Michael, 10
Parkinson's disease, 227, 239
Parotid gland, 114, *114*
Partial pressure gradient, of carbon dioxide, 188, 192, 193, *193*, 195, 198; of oxygen, 188, 192, 193, *193*, 198
Parturition, defined, 318
Passenger pigeon, 524
Passive immunization, 176, 177, 182
Passive transport, *44*, 62; defined, 44
Pasteurization, 162
Pasteur, Louis, 162
Patella, *91*, 95, *95*, *97*
Paternal chromosomes: behavior during meiosis, 388; as carrier, *394*; genomic imprinting, 383; in meiosis, *364*, *365*, *368*; in mitosis, *356*; patterns of autosomal inheritance, 394–395, *394*; patterns of X-linked inheritance, 396, *396*, 397; sex determination, 390; transmission of, *373*, *374*
Paternity cases, 141, 444
Pathogens, *112*, 163–164, *164*, *165*, 174–175, 182; broad spectrum of, 338; defined, 164; extracellular, *169*; immunological specificity, 168, 170; infectious disease and, 180, 181; inflammatory response to, 334; intracellular, *169*; in vaccines, 176; macrophages and, 167; of sexually transmitted disease, 330, 334, 345, *345*; viruses as, 334
Pattern baldness, 398, *398*, *403*
Pauling, Linus, 405, 408
PCBs, 279
Peanut butter, raw, 427
Peas, 502
Pectoral girdle, *91*, 94, *94*, 95, 108; defined, 94
Pectoralis major, *99*
Pediatricians, 176
Pedigrees, 397, 402; defined, 392; family, 393, *395*, 396, *396*; symbols used, *392*
Pelicans, brown, 505
Pellagra, vitamins and, *126*
Pelvic cavity, *79*, 288, 292; defined, 78
Pelvic examination, *429*
Pelvic girdle, *91*, 93, 108, *465*, 478, 481; bones of, *95*; defined, 95
Pelvic inflammatory disease (PID), 340, 342, 343, 346; defined, 342
Pelvic nerve, *230*
Pelvic organs, blood delivery to, *143*
Pelvis, 90, 95
Penetrance, *308*, 384; defined, 382
Penguins, *465*, 486; Adélie, *487*, *491*; emperor, *491*
Penicillin, 180, 206, 341, 446; antibiotic-resistant strains, 340
Penis, *286*, 288, *290*, 292, 390; age-related changes, 327; condom over, 299; discharge from, 341, 344, 345; function, *288*, *289*, 297; genital warts, 343; HIV point of entry, 336
Pepsin, 117, *117*; defined, 116
Pepsinogens, 116
Peptic ulcer, 116
Peptide bond, *30*, 31, 414, 415, *415*, 418; HCl secretion and, 116; in enzyme substrates, 54
Peptide, digestion of, 122
Peptide fragments, 117

Peptide hormones, *117*, 271, 284; defined, 271; initiating change, *270*
Peptidoglycan, 333
Perception, 251, 252, 262; defined, 248
Percussion therapy, *387*
Peregrine falcons, 505
Perfluorocarbon (PFC), 71, 145
Perforins, *171*; defined, 170
Pericardium, *144*; defined, 144
Periodic table of elements, 16
Periodontal disease, 114
Periodontal membrane, 114, *114*
Periosteum, *88*, 89
Peripheral nerves, *228*, 233
Peripheral nervous system (PNS), 77, *225*, *228*, 230–232, 244, *306*; axon regeneration, 225; components of, 219, 228, *228*, 244; defined, 228
Peripheral vasoconstriction, 214, *214*, 215
Peripheral vasodilation, 214, 215; defined, 214
Peripheral vision, 261
Peristalsis, 115, *115*, 116, 121; defined, 113
Peristaltic movement, 113, *113*
Peritoneal dialysis, 213
Peritoneum, *204*, 213
Peritonitis, 121
Peritubular capillaries, *205*, *208*, *209*, 212, 216; defined, 205
Periwinkle, rosy, 430, 435, 524, *524*
Permutation, 417
Pernicious anemia, 139
Peroxisomes, 51; defined, 51
Peroxyacyl nitrates (PANs), 516, *516*
Pest control, 502–503, 505
Pesticides, *15*, 206, *516*, 518, 519, 526; banning of, 505; as carcinogen, 426, 428, *428*, 430; exposure to, 14, *14*; genetic engineering, 446–447; synthetic organic, 505; water pollution, 203
Pests, 439
Petit mal seizures, *225*
Petrel, *491*
Petroleum, 522, *522*
p53/antisense *ras* RNA, *450*
p53 tumor suppressor gene, 426
pH, 24, 33, 36, 138, 193; cellular, 43; of cerebrospinal fluid, *194*, 195; enzyme activity and, 54, 55; extracellular, 136, 212, *212*; hyperventilating and, 188; levels in blood, 135, 193, 195, 198, 276; of mouth, 114; regulation, 75, *78*, *79*; scale, 22, *23*; scale, defined, 23; vagina, 164, 289
Phagocytes, 81, 139, 163, *165*, 173; blood transported, 135, 137, 160; effects of smoking on, 196; role in inflammation, 166, *166*, *167*, 182
Phagocytosis, 44, 51, *136*, 163
Phalanges, *91*, 94, *94*, 95, *95*
Pharmaceutical drugs, 448, 449
Pharyngeal arches, 312, *312*, *314*
Pharynx, 112, *112*, 114–115, *115*, 132, 186, 198; defined, 185; eustachian tube, 255; from pharyngeal arches, *312*
Phase-out schedules, 517
Phencyclidine (PCP), 243
Phenotypes, 381–384, 386, 392–393, *394*, 397, 459; defined, 374; gene mutation and, 388; in a testcross, 378, *378*; nondisjunction outcomes, 401; probability of, 377, *377*, 379, 380, *380*; sex-influenced inheritance, 398, *398*; vs. genotype, *374*
Phenylalanine, 125, *125*, 394; UUU, UUC triplets as, *412*
Phenylketonuria (PKU), 394, *403*
Phenylpyruvic acid, 394
Pheromones, 253, 282, 284; defined, 268; as signalling molecules, 268

Phosphate, 101, 103, *103*, 212, 504, *504*; blood levels of, 280; storage, *78*, 88, 90; unbound, *58*, *59*
Phosphate groups, 29, 34, *34*, *37*, 59, 443, 471; ATP and, 56, 101; as functional group, *25*; in DNA nucleotide, 406, *406*; in RNA nucleotide, 410
Phosphate ions, 212, *275*, 281
Phosphatidylcholine, *37*
Phosphofructokinase deficiency, *449*
Phosphoglyceraldehyde (PGAL), *56*, 57, *59*, 60, *60*, 61; defined, 56
Phospholipids, 28, 32, 117, 125, *127*, 136, 504; as carbon compound, *37*; defined, 29; formation of micelles, *119*; in cell membrane, 39, 40, *40*, 41, 62; structural formula for, 29, *29*
Phosphoric acid, 203
Phosphorus, 16, *16*, *127*; atomic number, *17*; atomic symbol, *17*; electrons in outer shell, *17*; mass number, *17*
Phosphorus cycle, 504, *504*, 506; defined, 504
Phosphorylations, *56*; defined, 56
Photochemical oxidants, 516, *516*
Photochemical smog, 526; defined, 516
Photons, absorption of, 263
Photoreceptor, 248, *248*, 249, *258*, 260, 264; detecting light energy, 247; in visual pathway, 262, *262*, 263; organs, *258*, 259
Photosensitive organ, third eye as, 281
Photosynthesis, 492, *498*, 499, 501
Photosynthesizing organisms, 487, 490, *491*, 493, 495, 498, 506
Photovoltaic cells, 523
Phrenic nerve, *194*
Phylogeny, 467; defined, 467
Phylum, 467, *467*
Physical drug dependence, 240, *240*
Physical fitness, 67; defined, 66
Physical therapy, cystic fibrosis and, 387
Physiological mechanisms, 82
Physiologists, 134
Physiology, defined, 67
Phytoplankton, 492, 517
Pigs, as, transgenic, 173
Pili, 333, *340*
Pilomotor response, 215; defined, 214
Pineal gland, *233*, *234*; defined, 281; melatonin secretion, 266, 267; overview, *269*, *275*
Pinna, *254*
"Pinky finger," 94
Pinna, *254*
Pioneer species, 488
Pitch, 255
Pituitary dwarfism, 274, *274*, 446
Pituitary gland, 68, *233*, 275, *306*, 318, 319; abnormal output, 274, *274*; anterior lobe, *269*, 272–273, *272*, 284; defined, 272; location, *269*, 272; overview, 269; posterior lobe, *269*, 272, *272*, 284; regulatory hormones, 55, 89, 280, 417; reproductive hormones, 293, 294, *295*
Pituitary hormone, anterior lobe, 290–291, *291*, 294, *295*, *296*, 299, 302
Pituitary/hypothalamus interaction, 267, 272–273, 284; anterior lobe, 273, 284; hormone productions, *273*, *277*, 284; posterior lobe, 273, 284; reproduction, 287, 291, *291*, *295*
Pituitary stalk, *269*, 272
Placenta, *293*, *310*, *311*, 314, *316*; as afterbirth, 318, *318*, 319; as barrier, 328; defined, 310; fetal circulation and, 316–317; IgG antibodies crossing, 173; umbilical cord and, 310, 311, *317*
Planes of space, three, 256
Planes of symmetry, *79*
Planets, 468

Plant bioengineering, 439, 452
Plant protein, 125, *125*
Plants, *3*, 490, 492, *497*, 499, 502, 506; aquatic, 494; bioengineered, 439, 452; cancer in, 426; carcinogens produced by, 426; deglaciated regions, *489*; as food producers, 7; genetic modification of, 446–447, *447*; rooted, 492; tissues, 446; transgenic, 446–447, *447*
Plaques, 72; in blood vessels, 154
Plasma, 75, *75*, *136*, 160; cells, 173, *179*, 182; defined, 136; extracellular fluid, 82; in edema, 185; in inflammation, *166*, 167; solute concentration, 152
Plasma membrane, 39, *39*, *50*, 51, 282, *306*–307, 424; adhering junctions and, 72, *72*; bacterial, 333, *333*; cytoskeleton and, 53, *53*; defined, 40; diffusion, 42–44, *44*, 62, 270, 271; during cell division, *356*, *360*, *360*, *362*; of epithelial cell, 118, 119, *119*; fluid mosaic model of, *41*; function, 41, *47*, 62; Golgi body location, 50; in cardiac muscle cells, 76, *76*; in platelets, 137; intercalated discs, 148; keratin fibers in, 80; membrane attack complexes, 165, *165*, 170, *171*; muscle cells, 102, *102*, 103, 231; proteins in, 168, 172, 327; receptors located in, 271, 284; red blood cells, 145; selective permeability, 43, 62, 267; structure, 40–41, 43, *46*, *47*; viral budding, 332, 336, *337*, 338, *338*; voltage, 263
Plasma proteins, 153, 193; blood transported, 135, 136, 160; in immune response, 163, 164, 166, *166*, *167*, 182; in lymph fluid, 159, 160
Plasmids, 439, *440*, 441–442, 446, 448, 450, 453; defined, 440; DNA, 440, 441; nonrecombinant, 445; recombinant, 445, *445*, 447
Plasmodium falciparum, 180
Plasmodium japonicum, 505
Plastic surgeons, collagen implants, 73
Platelet-derived growth factor (PDGF), 283
Platelets, 75, *75*, *136*, 156, *158*, 160; blood-borne, *136*, *137*, 283; defined, 137; freezing, 145; plug formation, 153, *153*
Plate tectonics, *462*, *463*; defined, 462
Pleasure center, 242
Pleiotropy, defined, 381
Pleurae, 187; defined, 187
Pleural membrane, *186*, 187, *197*
Pleural sac, 187, 190; pressure, 190
Pleurisy, 187
Plutonium, 18
Pneumocystis carinii, 338
Pneumonia, 180, *181*, 187, 196, *197*; cystic fibrosis and, 387; drug-resistance, 181; as opportunisitic disease, 336, 338; sickle-cell damage, 381; STD and, 336, 341
Pneumonic plague, 512
Pneumothorax injury, *197*
Podophyllin, 343
Point mutations, 418; definition, 416
Poisoning, *155*
Polar body, 294, *294*, 295, 308, *308*, *309*; meiotic cell division, *366*, 367
Polarity: of bonds in biological molecules, 20, 21; of water molecules, 22
Polarity of charge, 219
Polio, 176
Poliovirus, 176; vaccine, 176
Pollen allergies, 92
Pollutants, 447, 526; defined, 516; generated by humans, 509, 512–513, 515–519, *516*, *517*, *518*
Pollution, 14, 487. *See also* Agricultural runoff, Industrial waste, Pollutants, Soil salination, Water pollution

air, 502; environmental estrogens, 279; making nonpolluting fuel, 35; noise, 257
Polydactyly, 382, *382, 403*; pedigree for, *392*
Polygenic traits, defined, 383
Polymer, defined, 25
Polymerase chain reaction, *442, 443, 444, 453, 466*; defined, 442
Polypeptide chain, 37, 46, 117, 324, 466; alpha chain, 32; amino acid sequence, 405, 412, *412*, 413, 418; of antibody molecule, 174, *174*; beta chain, 32; building, 30, 31, *31, 32, 32*; cytomembrane system and, 50, *50*; defined, 31; formation, 50, *50, 419*; gene coding for, 406, 418; in hemoglobin, 32, *32*, 33, *33*, 138, *138*; in RNA translation, *412*, 414, *414, 415, 415, 415*; RNA templates, 410
Polypeptides hormone, *270*, 271, 284
Polyploidy, defined, 400
Polysaccharides, 36, *37*, *117, 132*, 276, 333; defined, 26, 27; digestion, *112*, 114; in basement membrane, 68; in collagen, 150; in ground substance, 73; red blood cells and, 381
Polysomes, *415*; defined, 415
Polyunsaturated fat, 125
Pongids, *476*
Pons, *194*, 195, *233*, 244; defined, 233
Poppers, 243
Population, 457, 458, 463, 472, 484, *484*, 488; defined, 459; evolving, 460; food crisis and, 131; growth and sustainability, 508–526; interdependent organisms and, 7; speciation, 461, *461*; variation in, 459, 460
Population density, 463, 512; defined, 510
Population dynamics, 510–511, *510*
Population growth, *510*, 512, *513, 514–515, 514*, 524, 526; annual, 510, 515, *515*; rate of increase, 301, 511, 513
Population size, 510, 511, *511*, 513–515, *514, 515*, 526
Porpoises, 465, *465*
Positive feedback mechanism, 83–84, 222, 295; defined, 83
Positron-emission-tomography (PET), *18*, 236
Posterior pituitary gland, 210
Posterior section, *79*
Post ganglionic neurons, 230
Postindustrial stage, *514*, 515
Postreproductive years, *510*
Postsynaptic cell, 225, *225*, 227, *227*, 244
Postural muscles, 105
Potassium, 16, *16*, 126, *127, 206*, 504; atomic number, *17*; atomic symbol, *17*; electrons in outer shell, *17*; extracellular concentration, 278; mass number, *17*
Potassium gates, 223, *223*
Potassium ion channels, 253
Potassium ions, 22, 151; electrically charged, 221, *221*, 223, 224, 244; excretion of, 206, 278; extracellular pH, 136
Potatoes, 428, 447, *447, 512*; as fuel source, 35
Potentiating drug interaction, 241
Poverty, 524
Power stroke, 101, *101*, 108
Prader-Willi syndrome (PWS), 383
Prairies, 488
Pre-carcinogens, 427
Precipitation, 492, 495–496, *496, 497*, 501, 520, 526; acid rain effect, 517
Precursor, 54, 76, 417; defined, 26
Predation, 458
Predators, 479, 482, 505, 512, 513, 526
Prediction, *9*, 10; defined, 8, 12
Pre-erythrocyte, nucleated, *139*

Prefrontal cortex, 236, *236*; fact memory, 239, *239*
Preganglionic neurons, 230
Pregnancy, 81, 151, 268, 282, 318, 425; food risks, 235; genetic counseling and, 393; hormonal secretion and, *275*; IgG antibodies, 173; infertility and, 300; myths and, 297; preparation for, 287, 293, *293, 294*, 295, *295*; prescription drugs and, 299, 321; Rh blood typing and, 140–141; risk of, 298, *298*; risky behavior during, *196*, 218, 336, *337*, 340, 342, 346; termination of, 301; unplanned, *298*, 302
Pregnancy test, 299; at-home, 309
Pregnancy vaccine, 299
Prehensile movements, 478
Preimplantation diagnosis, 322
Preindustrial stage, *514*, 515
Premature birth, 316, 318–319; causes of, 319; smoking and, 321; immature lungs, 192; saving babies, 319
Premolars, 114, *114*
Premotor cortex, skill memory, *239*
Prenatal diagnosis, 400
Prenatal risk, 320–321
Prepregnancy stage, 322
Prereproductive base, *510*; defined, 510
Prescription drugs: the Pill, 299; prenatal risk, 321
Preservation, 463
Pressure: changes in, 248, 250, 251, *251*; fluid, *256*, 257, 261; sense of, 247, 248, *248*, 264; sustained, *251*
Pressure gradients, 62, 190; breathing and partial, 188, 191, 192, 198; carbon dioxide and, 185, 192, 193; defined, 43; negative, 190; oxygen diffusion, 185
Pressure reservoirs, 150
Pressure surge, 150
Pressure waves, 254, 255, *255, 256*
Presynaptic cell, 225, *225*, 227, 244
Primary immune response, 175, *175*, 176, 179
Primary oocytes, 294, *294, 361*, 366, *366*; prophase I arrested, 367
Primary producers, 487, 490, *490, 491, 493, 495*
Primary productivity: defined, 492; gross, 492, 506; net, 492, 506
Primary reproductive organs, 286, 287
Primary spermatocytes, 290, *291, 361*, 366, *366*
Primary succession, 488, *489*
Primary teeth, 114
Primate heritage, 481
Primates, 2; classification, *467*; defined, 476; evolution of, 474–479, *476, 477, 479*, 481, *481*, 485; origins, 479
Primers, 442, *443*
"Primitive streak," 312, *312*
Primordial follicle, *294*
Prions, defined, 334
Probability, 377, *379*, 384; calculating, *377*, 380, *380*; as constant, 376, *379*; defined, 376; genetic analysis and, 392, 393; monohybrid cross and, 377, *377*; Punnett square method, 376, *376*
Processing, of food, 111
Producers, *489*, 493, 494, 495, *495, 498*; defined, 490; organization and interdependency, 7, *7*; primary, 487, 490, *490, 491*, 492, *495*, 506
Product molecule, *55*
Progesterone, 268, *271, 275, 293*, 295, *295, 296*; adrenal cortex production, 278; controlling fertility cycle, 287, 302; defined, 292, 299; in pregnancy, 309, 310, 319, 328; post-menopausal women, 327; primary effects, 269, *275*; synthetic, 299

Programmed death, 307
Prokaryotes, 30; bacteria as, 333
Prokaryotic fission, 333, *333*
Prolactin (PRL), 268, *271*, 273, *273*, 284; primary effects, *269*, 272, *272*; as regulatory agent, 417; stimulates lactation, 319
Proline, 408, *412*; CCC, CCA, CCG triplets as, 413
Prometaphase, 358
Promoter, defined, 410
Promoters, cancer, 427
Prophase, defined, 356; during mitosis, 354, *354, 355, 356*, 357–358, *369*, 370
Prophase I: arrested stage, 367; crossover event, 391; during meiosis, *361, 362*, 364, *364*, 365, 368, 370
Prophase II, during meiosis, *361, 363*, 365, *369*
Proprioception, 243
Prosimians, 475, 476, *476, 481*, 485
Prostaglandins, 167, 178; defined, 282; reproductive signals, 289, *289*, 295; targets of, 268, 318
Prostate cancer, 207, 289, *429*, 430, *431*, 433, 436; screening for, 177
Prostate gland, 207, 289, *290*, 297, 433; defined, 289; enlarged, 327; function, *288*, 289; inflammation of, 342
Protease inhibitor, 339
Protein channels, *41*, 72, 248
Protein deficiency, 125, *125*, 130–131, *178*
Protein-energy malnutrition, 130
Protein metabolism, 56, *112*, 122, 123, 125, 132, 213; byproducts of, 202, 203; hormonal effects on, *275*, 278
Protein molecules: DNA instructions for, 49, *49*; osmosis and, *42*, 43; synthesis of, 106
Protein-RNA systems, *471*
Proteins, 27, 46, 108, 354, 414; actin filaments and, 102; antibodies as, 172, 177, 182; as biological molecule, 15, 17, 25, 36; cancer and, 424, 426, 430, 436; carbohydrate building blocks, 490; coenzymes and, 55; in collagen, 150; complete/incomplete, 125, *125*; cytoskeletal, *41*, 53; defined, 30; dietary intake, *125*, 130, 131; digestion location, 110, 112, 116–117, *117, 119*, 132; DNA structure and, 405, 409, 418; as energy source, 60, *61*; as enzymes, 54, 352, 354, 470; excess, 90, 132; fibrin, 75; fibrous, 32, 33, *37*, 80; functional groups, 24, *25*, 405, 450; as gene product, 352, 404, 466; gene regulation, 417, 418; genetic engineering, 439, 446, 447, 448, 451; globular, 31, 32, *37*, 100, 132; hormone, 30, 33, *270, 271*; immune system and, 166, 168; keratin as, 80; malfunctioning, 32, 387; markers, 140, 141; mucins as, 114; in nitrogen fixation, 502, 503; nonfunctional, 450; in plasma, *136*, 137, 142, 159; in plasma basement membrane, 68; in plasma membrane, 39, *39*, 41, 168; polypeptide chains, 406, 418; 471; receptor, *41*, 174, 246; in red blood cells, 139; self-replicating, 471; storage, 60, *60*, 75; tight junctions, 72, *72*; transport, *41*, 42–43, 47, 54, 58, 62, 71; transport across cell membrane, 44, *44*. *See also* Regulatory proteins
Protein-specific antigen (PSA) test, 433
Protein structure, 70, 73, *123*, 352, 354; mutational alteration, 405; primary, 31, 33; quaternary, 32, *32*, 33; secondary, 31; tertiary, 31
Protein synthesis, 30, 31, 39, 49, 139, 449, 471; amino acid substitution, 125, 416; blockage, 180; control of, *120, 123*; cytomembrane system

and, 50, *50*, 51, *51*, 62; genetic code as basis of, 412, 418; hormonal targets, 267; minerals and, *127*; mutation, 405, 416; pancreatic secretions and, 276; regulatory hormones, 417, 418; steroid influence on, 270, 271, *272, 273*, 284; transcription, 412–413, 418, *419*; translation, 414–415, *414*, 418, *419*; two steps, 410–411
Prothrombin, *153*
Protistans, *3*
Proto-cells, 471, *471*
Protons, 15, *16*, 17, 36, *37*; defined, 16; "free," 23; in a chemical bond, 19, 20, 22, 37
Proto-oncogenes, 81, 426, 427, *427*, 436; defined, 426
Protozoa, *181*, 338, 344, *344*, 346; as pathogens, 164
Provirus, 333, 336
Proximal plane, *79*
Proximal tubule, 205, *205*, 206, 216; defined, 208; reabsorption in, 208–209, *208, 209*, 210, *210*
Prozac, 227
Pseudopodia, 425
Pseudopods, 313
Pseudostratified epithelium, 68, *68*
Psychedelics, 243
Psychoactive drugs, 237; abuse, 242–243; action and interaction, 241; defined, 240; dependence, 240; effects of, 240–241; examples of, *240*, 242; habituation, 240
Psychological drug dependence, 240
Psychosis, 243
Pterosaurs, 465, *465*
Puberty, 30, 323, *323*; genetic disorders and, 401; meiosis resumes, 367; onset, 281; sperm production, 287, 290, 291, *291*
Pubescent, *323*
Pubic arch, 95, *95*
Pubic hair, 323, 344
Pubic lice, 344, 346
Pulmonary arteries, *143, 144, 145*, 146–147, *146, 147*, 317; deoxygenated blood, 150; partial pressure gradients and, *193*
Pulmonary circuit, 135, *146*, 147, 150, 160; at birth, 317, *317*; defined, 146
Pulmonary embolism, 197
Pulmonary surfactant, 192, *192, 197*
Pulmonary trunk, *144, 146*
Pulmonary veins, *143, 144, 145*, 146, *146, 317*; partial pressure gradients and, *193*
Pulp, 114
Pulp cavity, *114*
Pulse, defined, 150
Pump, heart role as, 144–145, *146*
Pumping iron, 90
Punctuated equilibrium, defined, 461
Punctuation model, defined, 466
Punctures, inflammatory reponse, 166
Punnett square, *376*, 380, *380*; defined, 376
Pupil, *238*, 258, *258*
Purkinje fibers, 148, *148, 157*
P wave, *157*
Pyelonephritis, 207
Pyloric sphincter, *113*, 116, *116*
Pyruvate, 56, 57, *57*, 60, *61*; defined, 56; fermentation pathway, 58, *59*
Python, 465

Q

QRS wave, complex, *157*
Quaaludes, *241*, 243
Quadriceps femoris, *99*
Queen Victoria, 396, *396, 397*
Quinine, taste of, 253

R

Rabies, 446; virus, 180
Raccoon, 466
Races, 484
Radial keratomy, 261
Radial muscles, in eye pupil, 258
Radiant energy, from heat source, 247, 248, 248, 264
Radiation, 71; background, 427; as cancer cause, 423, 427, 427, 430, 434–436; as cancer therapy, 18, 139, 430, 433, 434; as therapy, immune system suppression, 173, 178
Radiation sickness, 522–523
Radioactive dating, 457, 472
Radioactive decay, 468, 523; defined, 17
Radioactive iodine, 280
Radioactive isotopes, diagnostic tests, 367
Radioactive materials, 518, 523, 526
Radioactive radon gas, 367
Radioactivity, exposure to, 197
Radioisotopes, 428, 523; defined, 17; measurements of, 474; uses of, 17, 18
Radium, 18
Radius, 91; defined, 94
Rage, 237
Rapid Eye Movement (REM), 237
Rashes, 14
Rat gene, 448
Rats, 512; atherosclerosis and, 154
Raw materials, 111
RBCs, 155
Reabsorption, 121, 208, 209; across capillary walls, 167; defined, 206; of urea, 203, 204; of water/solutes, 205, 207, 207, 209, 216
Reactants, 17, 34, 54
Reaction sequences, 54; of metabolic pathways, 54
Rearrangement, defined, 25
Reasoning, 219
Rebound effect, 231
Receptive fields, 263, 263
Receptor mediated endocytosis, 271
Receptor molecules, antigen-binding, 168, 171; CD4 cells, 338; chemokine, 338, 339; hormonal diffusion and, 270, 275; in cytoplasm, 271, 284; lymphocytes, 168; on plasma membrane, 270, 271, 284; on Sertoli cells, 291, 291
Receptor protein, 31, 41, 62; defined, 41; diversity, 174; for hormones, 55, 62; in a chemical synapse, 226, 226; on immune-privileged tissue, 173
Recessive alleles, 372, 373, 375, 384; defined, 374; for a sex-influenced trait, 398; genetic analysis of transmission, 392–395, 392, 394, 395, 402; in Mendel's single gene model, 381; in monohybrid cross, 377; in Punnett square method, 376, 380; in testcross, 378, 378; penetrance, 382
Recessive trait, red-green color blindness, 397
Reciprocal innervation, 98
Recognition protein, 41, 62, 168; defined, 41
Recombinant DNA, 322, 438–453; methods, 451
Recombinant DNA technology, 439; defined, 440
Recombinant molecules, 441
Recombinant organisms, 452
Recombinant plasmids, 440, 441, 441
Recombinate, 448
Recombination, 439, 445, 453; viral, 440
Recreational drugs, 241, 241
Rectum, 112, 112, 120, 121, 132; autonomic nervous system and, 230;

bacterial infection, 340; HIV point of entry, 336; location, 78
Rectus abdominus, 99
Recycling, 519; interdependency of organisms and, 7, 7
Red blood cells, 40, 43, 75, 120, 126, 417; abnormal hemoglobin, 408; as blood component, 136, 136, 137, 160; chief function, 75; defined, 136; destruction of, 155; disposal of, 120, 158; fetal, 140, 141, 141; formation, 158, 275; freezing, 145; hemoglobin breakdown, 343; hemoglobin transport, 32, 192, 192; in capillaries, 137, 150; in urine, 207; iron in, 126; life cycle, 139, 139; micrographs of, 48, 48; sickling, 381, 381; spleen storage, 159; substitute blood, 71; typing of, 140, 141
Red/green color blindness, 260, 397
Red light wavelengths, 247, 260
Red marrow, 88; blood cell production in, 89, 108, 139, 139; defined, 89
"Red" muscle, 105
Redox reactions, 54
Red pulp, 159
Reductional division, 362, 370, 370; defined, 361
Referred pain, 251, 251
Reflex action, fetal, 316
Reflex arc, 230, 231, 244; defined, 229
Reflexes, 244, 297, 327; control center, 232, 233, 233, 244; defined, 229
Reflex pathways, 233; local, 122; long-distance, 122
Regulatory proteins, 30, 417, 418, 426; defined, 417
Rejection, immune system and, 70
Relaxation phase, 147, 147
Release factors, 414, 415, 418
Releasing hormones, 274, 278, 284; defined, 273; GnRH, 291, 291
Religious beliefs, 11; blood transfusions and, 71
Remodeling, 89, 108; defined, 90
Renal artery, 143, 146, 204, 208
Renal capsule, 204, 204, 313
Renal corpuscle, defined, 205, 205
Renal pelvis, 204, 204, 206, 208, 213
Renal vein, 143, 204
Renewable resources, 35
Renin, 211, 211
Repair enzymes, 416, 416
Repeated cell divisions, 168
Replication, 407, 442, 443; defined, 406; DNA mechanism of, 406–407; enzymes, 440, 442; errors, 416, 418; viral, 177
Reproduction, 5, 12, 457, 479, 512, 513; of cells, 39, 39; cycles, 267, 268, 269, 272, 281, 285; cytoplasmic machinery, 352; defined, 6; failure, 400; meaning of, 352, 353; of mitochondria, 52; natural selection, 460, 460, 472; prostaglandins and, 282; self-replicating, 471, 472; steroid hormones and, 29
Reproductive base, 510; defined, 510
Reproductive isolating mechanisms, defined, 461
Reproductive organs, 2, 78, 287, 323; female, 292–293; male, 288, 288–289; primary female, 286, 293, 302; primary male, 286, 289, 289, 302; synthesizing hormones, 55, 278
Reproductive systems, 79, 286–302, 306, 433; age-related changes, 327; in embryonic period, 314; membrane lining, 69, 170, 173
Reptiles, 464, 464, 465, 476
Residual volume, 191, 191
Respiration, 185, 186; control H+ concentration, 212; defined, 186; drugs' effect on, 240, 241, 243; reflex center, 233; solute gains, 202, 216
Respirators, 11

Respiratory bronchioles, defined, 187. See also Bronchiolus
Respiratory cycle, 190, 191; defined, 190; regulation mechanisms, 195, 198
Respiratory distress syndrome, 197, 316
Respiratory failure, 387; muscular dystrophy, 397
Respiratory surface, ciliated cells and, 53; defined, 188; membrane lining, 69, 69; mucous membranes, 170, 173
Respiratory system, 2, 66, 79, 112, 132, 184–198; airways, 186, 187, 189, 190; components/functions, 186, 198; defined, 186; from endoderm, 306; intense exercise and, 103; local chemical control, 195, 198; macrophages in, 73; neural control, 188, 194, 195, 198; systemal links, 111, 142; transporting gases, 138; viral attack, 171, 332
Resting blood pressure, 149, 151, 157
Resting cardiac rate, 157
Resting membrane potential, 223, 223, 227, 227, 244; defined, 220; recording, 236, 236; restoring and maintaining, 221
Restriction enzymes, 440, 441, 442, 444, 446; defined, 441
Restriction fragment length polymorphism (RFLPs), 443; analysis, 444; defined, 444; production of, 441
Restriction mapping, 466
Reticular activating system (RAS), 236, 237, 244
Reticular formation, 194, 194, 195, 236, 237, 244; defined, 233; drug use and, 242
Retina, 77, 238, 258, 259, 259, 264; defined, 258; disorders/disease, 260, 261; embryonic pigment, 315; organization of, 262, 262
Retinal detachment, 261
Retinoblastoma, 423, 426
Retinoic acid (Retin-A), 321; prenatal risk, 321
Retrovirus, 333, 334, 336, 337, 426, 350; recombinant, 451
Reverberating circuits, 229
Reversals, 219
Reverse transcriptase, 333, 336, 337, 445, 450; action blocking, 339
Reverse transcription, defined, 445
R group, 30, 30, 31
Rh blood typing, 140–141; defined, 140
Rhesus monkeys, 140, 466
Rheumatism, sickle-cell damage, 381
Rheumatoid arthritis, 96, 178, 451; age-related, 327; free radicals and, 21
Rhinoceros, 474
Rhodopsin, defined, 263
RhoGam, 141
Rhythm method, 298, 298
Rib cage, 93, 159; bones of, 91; Heimlich maneuver, 115; organ protection, 93; osteoporosis and, 90; role in breathing cycle, 190, 190, 194, 194
Rib muscle, 195
Riboflavin, 125, 126
Ribonucleic acid. See RNA
Ribose, 26, 56, 410, 470
Ribosomal RNA (rRNA), 413, 418; defined, 410
Ribosomes, 46, 51, 333, 410, 411, 412; defined, 413; main function, 46; of mitochondria, 52; protein subunits, 49, 49; red blood cells, 139; RNA subunits, 49, 49; subunits, 413–415, 413, 414, 415, 418, 419
Ribozymes, defined, 449
Ribs, 89, 91, 92, 96, 108; cartilage, 74; defined, 93. See also Rib cage

Rickets, 131, 281, 281; vitamins and, 126
Rickettsia rickettsii, 505
Rift Valley, East Africa, 475, 480
Right brain hemisphere, 232
Right lymphatic duct, 158
Right semilunar valve, 144, 146
Rigor mortis, 101
Ring building, 24, 24, 26; carbon, 28, 29, 36
Ringworm, 180
Rivers, 488, 496, 504, 504; water pH, 23
RNA (ribonucleic acid), 15, 26, 37, 419, 445, 471; antisense molecules, 449; defined, 34, 410; DNA into, 405, 410–411, 410, 418; functions, 34, 54; gene coding, 406, 418; HIV virus, 336; nucleic acid digestion, 117; ribosomal subunits, 49, 49; synthesis, 410, 410, 418; transcript, 411, 411
RNA molecules, 471, 471, 472
RNA nucleotides, 471
RNA polymerases, 410, 411, 416; defined, 410
RNA template, 410, 411, 418, 471
RNA viruses, 332, 333, 334, 450
RNA world, defined, 471
Rod cells, 248, 262, 263, 264; defined, 262
Rodents, 479, 512
Roe v. Wade, 301
Roid rage, 107
Romanov, Alexis, 397
Romanov, Anastasia, 397
Root, 114, 114
Root canal, 114
Rotational cleavage, 306, 307
Rotational movement, 256, 256, 257
Rotifers, 492
Round window, 254, 255, 255
rRNA, roles of, 412–413; transcript/translation summary, 419
RU-486, 299
Rubella, 176
Ruffini endings, 250, 251
Rugae, 116, 116
Runoff, 469, 495–496, 496, 499, 502, 504, 520; agricultural, 518, 526
Ruptures, stopping blood loss, 153, 153
Russian royal family, 397

S

Saber-tooth cats, 482
Sabin polio vaccine, 176
Saccharin, 253
Saccule, 256, 257
Saccule, 256, 257
Sacral nerve, 228, 230
Sacrum, 93, 93, 95
Safer sex, 331, 339; guidelines, 345
Salinization, defined, 518
Saliva, 112, 114, 115, 132; ABO markers in, 141; dissolved substances in, 248; exocrine secretion, 68; hepatitis B transmission, 343; HIV infection and, 336; immune response, 164, 164, 182; pH value, 23; production, 211; taste receptors and, 252
Salivary amylase, 117; defined, 114
Salivary glands, 68, 69, 112, 113, 114, 116, 132; autonomic nervous system and, 230; cancer of, 434; defined, 114; enzyme source, 117; location, 114
Salmon, 505
Salmonella, 180; infection, 133
Saltatory conduction, 224
Salts: absorption, 118; defined, 22; in soil, 508; level in diet, 111; reabsorption of, 121; water interaction, 22
Salt/water balance, 211

Saltwater intrusion, 518, *518*

Salty taste class, 246, 252, 253

Sanger method of DNA sequencing, 443, *443*

Sanitation, 514, 515; public, *45*

SA nodes, 157

Sarcomas, 436; defined, 431

Sarcomeres, defined, 100; in muscle contraction, 100–101, *100, 101*, 108

Sarcoplasmic reticulum (SR), 50, *102*, 108; defined, 102

Sartorius, *99*

Saturated fat, 125, 154

Saturated fatty acid, 28, *28, 29*; defined, 28

Scabies, 344, 346

Scala tympani, 254, *255*

Scala vestibuli, 254, *255*

Scalp, *81*

Scanning electron microscope (SEM), 48, *48, 81*

Scapula, *91, 94*; defined, 94

Scar tissue, 97; bronchial, *197*; emphysema, 189; hearing loss and, *257*; of liver, 120, 242; navel, 318; of oviducts, 340, 342, 346; smallpox, 162; stretch marks as, 81; TB, 181; of vas deferens, 340; vasectomy, 299

Scavenger cells, 97; fighting infections, 135; for dead/worn-out cells, 137; tissue debris, 135, 160

Schally, Andrew, 274

Schistosomiasis, *181*

Schizophrenia, chronic, *449*

Schwann cells, 220, *224*, 225, 228, 313; defined, 224

Sciatic nerve, *228*

Scientific methods, 8–9, 12, 372; example of, *9*, 10

Scientific theory, 3, 9, *9*; defined, 8, 12

Sclera, 258, *258*, 259

Scrotum, 288, *288*, 299, *433*; cancer of, 426

Scurvy, 131; vitamins and, *126*

Seals, 486; Weddell, *491*

"Search and destroy" cells, 137; neutrophils as, 137, *137*

Seasonal affective disorder (SAD), 266, 267

Sebaceous glands, 68, 81

Seconal, 241

Secondary bone, 89

Secondary immune response, 175, *175*, 176, *179*

Secondary infections, 116, 261

Secondary oocyte, *292*, 294, 295–296, *295, 296*, 302; defined, 294; fertilization, *308*; meiotic cell division, *366*, 367

Secondary sexual traits, *269*, 271, 302, 323, *323*; defined, 287; female, 292; genetic disorders and, 401; male, 290

Secondary spermatocyte, 290, *291*

Secondary succession, 488

Second filial generation, 376, 378

Secondhand smoke, 196, *196, 197*; during pregnancy, 195, *196*

Second messengers, *270*; amplified response, 271, 284; defined, 271

Secret handshake, 338

Secretin, 122, 268, *275*

Secretion, 207, 216; ADH, 210; defined, 206; simple epithelium and, 68, *69*; tubular, *208*

Secretory cells, 116; glands as, 68, 69

Sedatives, *241*; abuse, 242, 243; prenatal risk, 321

Sedentary lifestyle, 90, 106, 137

Sedges, *489*

Sedimentary cycles, 495, 504, *504*, 506

Sedimentary rocks, 458, 463; layers, 482, *496*, 498, *498, 499*

Sediments, 463, 518, 520, 522, 526

Segmentation, *113*, 118; defined, 113

Segregation, principle of, 384; defined, 375

Seizures, 238, 243

Selective permeability, *42, 43*, 62; defined, 42

Self-gratifying behavior, 237

"Self" markers, 140, 141, 168, *168*, 171, 182; age-related changes, 327

Self-replicating systems, 405, 471, *471*, 472

Semen, *288, 289*, 297, 299, 300; ABO markers in, 141; composition of, 282; defined, 288; formation, 288–289; hepatitis B transmission, 343; HIV transmission, *178*, 336

Semicircular canals, 254, *256*, 257; defined, 256

Semiconservative replication, 407, *407*

Semilunar valve, *144, 145*, 146, *147*; defined, 144

Seminal vesicle, 288, *288, 289, 290*, 297; prostaglandin secretion, 289

Seminiferous tubules, 290, *290, 291*; defined, 288

Sensations, 264; compound, 248; defined, 247, 248

Sensory adaptation, defined, 249

Sensory axons, 248, 249

Sensory cortex, fact memory, 239

Sensory epithelium, *306*

Sensory hair cells, 255, *255, 256*; damage to, 257, *257*

Sensory input, 233, 244, 252; longterm storage, 239, *239*; short-term storage, 239, *239*

Sensory neurons, 227, *228*, 247–249, 252–253, *252*, 256, 264; action potentials in, 224; defined, 220; free nerve endings of, 250; in dermis, *80*, 81; in eye, 77, 259, 262; input zone, 222; long reflex pathway, *122*; reflex movement and, 232

Sensory organs, *91*, 249, 264; agerelated changes, 327; eyes as, 259; muscle spindles as, *231*; taste buds as, 253

Sensory pathways, *122*, 248–249, 253, 262; chemoreceptors and, 252, 264; example of, *249*; information flows, 229, 244

Sensory perception, 226

Sensory processing, psychoactive drugs and, 240, 241, 243

Sensory reception, 246–264

Sensory receptors, 80, 82, *82, 83*, 84, 246, *249*; defined, 248; for carbon dioxide/oxygen gas concentration, *194*, 195; function, 77; in cardiac muscle tissue, 149; in ear, *255*; in eye, 260; in glans penis, 297; in kidney blood vessels, 210–211; in pharynx wall, 115; in semicircular canals, 256; in stomach lining, 116; major categories of, 247, 248, *248*, 264; near body surface, 250; on muscle cell membranes, 103; short reflex pathways, 122, *122*

Sensory stimuli, 266; information storage, 239, *239*

Sensory systems, 249, 264, 475; defined, 247, 248; maturation, 323

Septic tanks, 518

Septum, *144*, 148, 186; defined, 144

Serine, *412*

Serotonin, 153, 243; blood clotting and, 153; as neurotransmitter, 226, 227, 236

Serosa, 113, *113*, *117*, 132

Serous membranes, 69

Serratus anterior, *99*

Sertolic cells, 291, *291*; defined, 290

Serum albumin, 447

Set point, 82, *83*, 129; arterial carbon dioxide level, 195; of blood glucose levels, *277, 278*; of blood pressure, 272; testosterone level, 291

Severe combined immune deficiencies (SCIDS), 179, *450*, 451

Sewage, 486, 487; human, 45, *45*

Sex, 30

Sex chromosomes, 287, 290, 353, *364*, 370, 398; abnormalities, 401; at metaphase, 388, *389*; change in number of, 401; defined, 388; as type of chromosome, 387, 388, 389; X-linked gene, 286, 389; Y-linked, 286, 389

Sex determination, 286, 287, *287*, 390; pattern of, *390*

Sex differentiation, 95

Sex hormones, 30, *37*, 106, 286, 287; deficiencies, 90; embryonic period, 314; female, 292, *292*; in developing fetus, 278; male, *288*, 397; maturation and, 323; organ development, 267, *269*; synthesis, 55

Sex-influenced inheritance, 398, *398, 403*

Sexual behavior, 233, 237–244; hormonal controls, 290; number of partners, 339, 340, 433; psychoactive drugs and, 240; sympathetic nerves, 231

Sexual intercourse, 287, 288, 297; birth control during, 298–299; female organ of, *292, 293*; male organs of, *288, 289*; STD and, 331, 340, 344, 346

Sexuality, age-related changes, 327

Sexually transmitted disease (STD), 120, 177, 179, 180, 330–346; animal viruses causing, *334*; chlamydia, 261; complications of untreated, 330, 331; defined, 330; diagnosing, 340; ethical obligation, 345, 346; latex condoms and, 299; prevention, 299–331, *331*, 345, *345*; recurrence, 342, 343, 345; transmission, 340; untreated, 330–331, *334*, 340, 342, 345–346; urinary tract infections, 207

Sexual nose, 253

Sexual orientation, genetic links, 383

Sexual reproduction: mechanisms, 374; meiosis, *351*, 352, 353, *363*, 368

Sexual stimulation, 81, 292, *293*; female arousal, 297; male arousal, 288, 289, *289, 290*, 297

Sexual traits, hormonal effect on, 275

Shark, *464*; dusky, 524

Shark fishing, 524

Shell model, *19*

Shells, 459, 463

Shifting cultivation, 520–521, *520, 521*, 524, 526; defined, 520

Shigella, 180

Shinbone, 95, *99, 483*

Shivering, *83*, 214, 215, *215*; defined, 214

Shock, *127*, 278

Shock absorbers, 93, 108

Short-term changes, hormonal adjustment to, 267

Short-term memory, 239, *239*, 244

Shoulder, 95, *99*; dislocation of, 94; range of movement, 96, *96, 97*, 98

Shoulderblade, *91*, 94, *99*

Siamang, *476*

Sickle-cell anemia, 376, *381, 403*, 405, 444; DNA mutation and, 6; embryo screening, 393; genetic disorder, 416; one gene, one enzyme hypothesis, 408; trait for, 381

Sigmoid colon, 121

Sigmoidoscopy, *429*

Signaling molecules, 55, 227, 231, 270–271, 282; *cis*-retinal, 263; effects of, 270; for pleasure center, 242, 253; hormones as, 254, 267, 284, 289;

lipid soluble, 271; water soluble, 271

"Silent killer," 157

Silicon, *16*; atomic number, *17*; atomic symbol, *17*; mass number, *17*

Silicosis, *197*

Silt, 463

Simple epithelium, 68, *68*, 69, *69*

Simultaneous inputs, 248

Single-celled organisms, *3*

Single-gene defects, 376

Sinoatrial (SA) node, *148*; defined, 148

Sinuses, 92; allergens and, 178; defined, 92; type of epithelium, *68*

Sinusitis, 92

Sister chromatids, 352, *353*; defined, 352; in meiosis, 362–363, 364, *364*, 365, *368, 369*, 370; in meiosis, DNA replication, 406; in meiosis II, 361, *361*; in mitosis, 357, *356–357, 358*, 359, *359, 369*, 370; in mitosis, DNA replication, 406

Skeletal muscles, 56, 106, 108, *191*, 206, 244; atrophy, 325; blood flow to, *147*, 151, *151*, 207; contraction, 87–88, 98, 102–103, *102*; defined, 98; equalize body temperature, 135; external urethra sphincter and, 206; from somites, 312, *312*; function of, 87, 88; glucose use, 278; heat production, 214, *215*; in breathing rhythm, 194; major groups, *99*; move lymph, 159; muscle spindle in, 248; nociceptors in, 251; organization of nerves in, 219, 230, 231, *231*, 250; sex differences in, 287; structural features, 100–101, *100*; tension, 104; types of, 105; voluntary, 87, 98

Skeletal muscle tissue, 76, 83, 148; defined, 76; smooth ER and, 50; striation, 100, *100*

Skeletal system, 78, *78*, 88, *91*; overview, 87

Skeleton, 2, 475–477, 482, 485; abnormal development, 400; function, 74; number of bones, 90, 108; organ protection, *91*, 93; structural support, 87–88, *91*, 93

Skills memory, 239, *239*

Skin, 68, 80, 88; as barrier to invasion, 164, *164*, 182; blood flow to, 83, *83*, *147*; collagen in, 32; color phenotype, 80, 383; designer skin, 70; epithelium, 72, *73*, 274; eruptions, 81; fat storage, 75, 80, 125; hair cells and, *33*; as integumentary system, 78, 80–81, 84; keratinized, *68*; mechanical insults, 81; mechanoreceptors in, 250; neuron receptors in, 77, 81; nociceptors in, 251; sensory axons and, 230, 244; structure, 80–81, *80*; wrinkling, 73, 323, 324, 325

Skin cancer, 423, 435, *435*, 517; basal cell carcinoma, *10*, 80, 435, *435*; carcinogens, 10, *10*, *428*; from ionizing radiation exposure, 367, 427; gene defects and, 407, 426; gene therapy, 449, 451; malignant melanomas, *431*, 435, *435*; squamous cell carcinoma, 434, 435, *435*; ultraviolet radiation, 427, 430, 517

Skin diseases, herpes simplex as, 260

Skin tests: allergens, 178; TB, 175

Skull, 90, *91*, 92, 93, 108; bones, *91*, 464, *464*; defined, 92; evolutionary development, 475–477, 479, *479*, *480*, 482, 483; fibrous joints, 96; red bone marrow in, 139

Skunks, 462

Slash-and-burn agriculture, 520, *520*

Sleeping, 30, 236, *237*; behavior, 266, 267; braincells governing, 226;

medulla and, 233; sedation, 242; slow-wave pattern, 237; /wake cycles, 195, 266, 281

Sliding-filament model, 100, 101, 108; defined, 101

Slipped disk, 93

Slow-acting muscle cells, 105

Slow-wave sleep pattern, 237

Small-cell carcinomas, 434

Small intestine, 111–112, 112, 120, 121, 132; absorption in, 118–119, 119, 121, 132; alcohol absorption, 242; autonomic nervous system and, 230; blood delivery to, 143; deferred pain, 251; digestion in, 116, 117, 119, 122, 132; endocrine cells of, 269, 275; enzyme activity, 50, 117, 132; lining, 132; lymphoid tissue in, 158, 159

Smallpox, 162, 164

Smell, sense of, 252–253, 264, 477, 479, 485; brain region for, 233, 237, 237; food and, 116; memory circuits, 239; odor molecules and, 77

Smell genes, 246

Smell receptors, 246, 247, 248

Smog, 498, 499, 516, 516, 517, 526

Smoke and smoking, 90, 116, 129, 155, 156, 240, 521; addiction, 242; aging and, 325; associated risks, 196; benefits of quitting, 196; breathing and, 185, 188, 189, 196; cancers and, 422, 423, 426, 430, 433–434, 436; carbon monoxide, 154, 188; free radicals and, 21; marijuana, 241; oral contraceptives and, 299; prenatal risk, 195, 321, 328; secondhand, 434; SIDS and, 195

Smoker's cough, 196

Smooth muscle tissue, 40, 76, 80, 83, 108; asthma attack and, 189; autonomic nerves, 230; contraction mechanism, 87, 282, 288; defined, 76; in arterioles, 215; in blood vessel walls, 87, 150, 150, 151, 153; in digestive tract, 113, 122, 132; in internal organ walls, 87, 98; in intestinal wall, 118; in male reproductive tract, 297; in respiratory system, 189, 195; in uterine wall, 292, 318; involuntary, 87; local signaling molecules and, 268; regeneration, 283; true capillaries, 147; vasodilation and, 166

Snack foods, 111

Snails, 490

Snake, 465, 490

Sneezing, 180, 181, 197; as allergic response, 178; as immune response, 164

Snout, 479

Snowfall, 495, 496

Snowy owl, 489

Social behavior, integrating, 268

Social group, 479

Social problems, 509

Sodium, 16, 16, 20, 126, 127, 201; active transport of, 209, 209; atomic number, 17; atomic symbol, 17; electrons in outer shell, 17; excretion, 206, 206, 275; extracellular concentration, 278; hormonal conservation of, 269; mass number, 17; reabsorption, 208–211, 208, 211

Sodium bicarbonate, as pancreatic secretion, 276

Sodium chloride, 22, 22, 210; blood pressure and, 211; body temperature and, 215

Sodium fluoride, 20

Sodium hydroxide, 22

Sodium ion channels, in sensory reception, 253

Sodium ions, 20, 20, 22, 22; active transport, 209, 209, 211; electrically

charged, 221, 221, 222, 222, 244; extracellular pH, 136; GI tract absorption, 202; in sensory reception, 253; reabsorption, 121, 206, 207, 208, 208, 275, 278

Sodium-potassium pump, defined, 221, 221, 223, 223

Sodium pumps, 208, 208, 209, 209

"Soft spots," 97

Soil fertility, 520, 521, 526

Soil leaching, 490

Soil management technologies, 502–503

Soil pollution, 14

Soil salinization, 518

Solar energy, 471, 522, 523, 523

Solar-hydrogen energy, defined, 523

Solar system, 457, 468, 472

Sole bones, 91

Solenoid, 409

Soluble fiber, 121

Solutes, 44, 142, 167, 267; blood transported, 135, 142, 152, 159; capillary exchange, 152, 152, 159; concentration gradients, 43, 62; concentration of, 209, 213, 216, 248, 248, 264; defined, 22; gains and losses, 202, 216; kidney filtrate, 205; net osmotic movement of, 42, 43, 62; reabsorption, 209, 209; transport mechanisms and, 43, 44, 62; urine composition, 206, 208, 216; /water balance, 201, 216, 276

Solvent, water as, 22

Somatic cells, 353–354, 368, 368, 370, 370, 389, 447; defined, 352; DNA in, 409; ionizing radiation effects on, 367

Somatic nerves, 219, 228, 230, 244; defined, 230; motor axons, 230, 231; sensory axons, 230

Somatic sensations, 247, 250–251, 264; defined, 249

Somatosensory cortex, 236, 236, 250; defined, 250

Somatostatin, 122, 273, 275, 276

Somatotropin (STH), 272, 272, 273, 273

Somites, 312, 314; defined, 312

Soot, 516

Sound, 254; decibels, 257; frequencies, 254, 255; perceived tone (pitch), 255; production, 191, 191; stimulus, 248, 264; waves, 254, 255, 255, 264

Sour taste class, 246, 252, 253

Southern blot method, 444

Spatial ability, 227, 234, 237

Specialist species, 488

Special senses, 247, 252, 264

Speciation, 461, 466, 466, 472, 476, 524; defined, 461

Species, 438–441, 459, 465–466, 467, 472, 476, 484; ancestral, 462; biotic potential of, 512, 513; defined, 457; ecosystems and survival of, 487, 488, 490; extinction rate, 467, 524; of fossils, 463; origin, 458; of population, 7, 459; rates of change, 459, 466; splitting of, 460, 461; survival, 525

Specificity, 174–176, 182; of response, 163–165, 164, 165; viral, 332

Speech, 477; structural basis of, 479, 479;

"Speed," 242

Sperm, 38, 38, 39, 286–287, 426, 452; defective, 21, 300; defined, 290; fertilization and, 305, 306, 307, 308–309, 348; flagellum and, 53, 53; formation, 272, 275, 279, 288, 288, 290–291, 291; gamete formation, 352, 361, 365–367, 366, 370; head, 290, 291; human reproduction and, 6; in fertilization, 297–300; as male gametes, 374, 376, 376; midpiece,

290, 291; motility, 289, 292; nucleus in fusion, 306, 307, 308, 308; rapid transport of, 288, 289; reproductive system, 79; sex determination, 390, 390; storage, 288, 289; tail, 290, 291, 308; travel medium, 295

Spermatids, 290, 291, 361; haploid, 366, 366

Spermatogenesis, 288, 291, 291, 366, 367; defined, 366

Spermatogonia, 352, 361, 366; diploid, 290, 291; haploid, 290

Spermatozoa, 290, 291

Sperm bank, 300

Sperm count, 279, 468; low, 299, 300

Spermicides, 345; foam, 298, 298; jelly, 298, 298; nonoxynol-9, 339

Sperm-producing tubules, 289

Sperry, Roger, 238

S phase, 430

Sphenoid bone, 92, 92

Sphenoid sinus, 93

Sphincters, 113, 113, 115, 115, 117, 121, 132; male bladder, 297; urinary, 327

Sphingolipids metabolism, 394

Spike. See Action potential

Spina bifida, 313

Spinal cavity, 79, 93; defined, 78

Spinal column, 92; collapsing, 90; disks, 74

Spinal cord, 77, 92, 93, 95; ACh and, 103; centers for heart control, 148; central nervous system component, 219–220, 228, 228, 244; controls over breathing, 194; defects, 313; defined, 232; glial cell sheaths, 224; in central nervous system, 77, 78; information flow, 229, 231, 244; interneurons on, 233; motor axons and, 104, 230, 230, 231; neural tube, 312, 312; neuronal interaction, 229; organization of, 232, 233; as pathway for nervous system signals, 102; protection, 93, 234, 235, 235; reciprocal innervation, 98; reflexes, 232; sensory systems and, 248, 249; viral infection, 341

Spinal cord injury, 155, 282; growth factor and, 282, 283, 283

Spinal nerves, 228, 228; function, 230

Spindle apparatus, 361, 367, 370; in cell division, 354, 356; defined, 356

Spindle equater: in meiotic cell division, 362, 363, 365, 368, 369, 370; in mitotic cell division, 354, 357, 357, 360, 369, 370

Spindle microtubules, 363, 365, 370

Spindle poles: independent assortment, 379, 379; in meiotic cell division, 362, 363, 365, 368, 368, 370; in mitotic cell division, 357, 357, 358–359, 359, 369, 370

Spindles: in meiotic cell division, 362, 364, 365, 368, 369; in mitotic cell division, 354, 357, 356–357, 358–359, 369

Spirochete bacterium, 333, 340

Spleen, 120, 139, 158, 434; autonomic nervous system and, 230; B cells, 176; defined, 159; from mesoderm, 306; lymphatic growth, 178; nonspecific response in, 164, 166, 170, 170, 182; sickle cells in, 381

Split-brain experiments, 238, 238

Spoken language skills, left hemisphere, 234

Spongy bone tissue, 74, 74, 88, 89, 108

Spontaneous abortion, 314

Spontaneous mechanisms, 98

Spontaneous mutation, 416, 418

Sporadic disease, 180

Spore capsules, 463

Sprain, 97

Spraying programs, community, 428

Sprinters, 105

Spruce budworms, 505

Spruces, 489

Squamous cell carcinomas, 434, 435, 435

Squamous epithelium, 68, 68, 69, 80

Squids, 486, 491

SRY gene, as male-determining gene, 287, 314, 390

Stags, 474

Staining, 48, 137. See also Dyes

Stamina, 87, 324

Standard of living, 512

Stapes, 254, 254

Staphylococcus aureus (Staph A), 81, 181

Starch, 27, 27, 37, 124; digestion of, 114, 132; as nonpolluting fuel, 35; undernutrition and, 130, 131

Starfish, 458

Starling, E., 268

Stars, 457, 472; remnants of, 468

Start codon (AUG), 412, 412, 414, 414

Starting molecules, 60, 60

Starvation, 129, 130, 458: response, 129

Stearic acid, 28

Stem cells, 75, 155, 430, 431, 435, 451; ABO blood typing in, 381; B cell production, 172; as blood component, 137, 137; defined, 139; gamete producing, 350; NK cell production, 171; platelet formation, 137; red blood cells formation, 139, 139; T cell production, 170, 172; white blood cell production, 166

Sterility, 300; genetic disorder and, 401; STD infections, 330, 334, 340, 342, 346

Sterilization, 508

Sternum, 89, 91, 92, 93, 94, 190; defined, 93; red bone marrow in, 139; thymus location, 281

Steroid hormones, 29, 37, 123, 284; defined, 271; initiating change, 270; main categories, 271, 271; synthesis, 125, 126; use/abuse, 106–107

Sterols, 28, 29, 29; defined, 29

Sticky ends, 441, 441; defined, 441

Stillbirth, 196, 203, 321; genetic disorders and, 395; STD and, 340

Stimulants, 241, 242

Stimulation, 76; ongoing, 250

Stimulus, 82, 83, 84, 249; artificial, 242; chemical, 218, 244; defined, 82, 248; as form of energy, 248, 264; light intensity as, 259, 263, 263, 264; long and short reflex pathways, 122; motor neurons on muscle cells, 87, 98, 102, 103; sensory receptors and, 247; visual, 258

Stirrup, 254, 254, 255

Stomach, 74, 111–112, 112, 114, 115, 118, 120; alcohol absorption, 242; autonomic nervous system and, 230; cell division in, 355; cell types, 40; controls over digestion, 122; cutaway view, 116; deferred pain, 251; defined, 116; endocrine cells of, 275; enzyme activity, 117; functions of, 116, 132; as hormonal target, 275; lining, 116, 117; location, 78; peristalsis, 113; protein digestion in, 116–117, 132; smooth muscle tissue, 76, 76

Stomach acid, 23

Stomach cancer, 111, 431, 434

Stomach epithelium, 68, 72; mucussecreting cells, 68

Stop codon, 412, 412, 414–415, 415, 418

Stop signals, UAA, UAG, UGA codons as, 412, 412

Storage, of food, 111, 116

Stored memory, 252

Storm waves, 501

Strata, 456, 457

Stratification, defined, 463

Stratified epithelium, 68, 68, 69, 69, 80

Stratified rock layers, 456–457
Stratosphere, 517
Stratum basales, *80*
Stratum corneum, 80, *80*
Strawberry plants, 447, 452, *452*
Streams, 496, 502, 504, 510
Strength, 87; of stimulus, 249, *249*
Strength training, 106
Strep infections, 181
Streptomycin, 180; prenatal risk, 321
Stress, 81, *157*; allergies and emotional, 178; chronic, 116; colon problems and, 121; environmentally induced, 323; heart attacks and, 155; hormonal secretions and, 269, 278, 279; immunosuppressive effects, *178*, 345
Stress electrocardiogram, 156
Stretching, 81, 121
Stretch marks, 81
Stretch receptors, 77, 98, *248*; in skeletal muscle, 249, 250; in alveoli, *194*, 195
Striated muscle cells, 76, *76*
Strip mining, 522
Stroke, 154, 157, *446*, 448
Structural formulas, 24, *24*
Strychnine, 253
Subatomic particles, 16, 17
Subcellular components, 7
Subcutaneous layer, 80, *80*
Sublingual gland, 114, *114*
Submandibular glands, 114, *114*
Submucosa, 113, *113*, 117, *118*, 132
Subpopulations, 484; of cells, *169*, 170, 182; of free nerve endings, 250
Subsidence, defined, 518
Substance abuse, 219
Substance P, 227
Substitution mutation, 416, *416*
Substrate-level phosphorylation, 56, *56*, *59*; defined, 56
Substrates, 55, *55*, 57, 62; defined, 54; of enzymes, *117*; glucose as, 103
Succession: defined, 488; primary, 488, *489*; secondary, 488
Succinate, 57
Succinyl-CoA, *57*
Sucking reflex, 316. *See also* Nursing
Sucrose, 37, 125, 253; bacteria and, 114; chemical makeup, 17, 26, *26*; tonicity, *43*
Sucrose polyester, 125
Sudden death, *155*
Sudden infant death syndrome (SIDS), 195
Suffocation, 189
Sugar, 27, 36, 470, 472, 502; bacteria and, 114; calories, *124*; carbohydrates as simple, 26, *26*, 27; as carbon compounds, 37, 56; complex, breakdown of, 51; as component of nucleotides, 34, *34*, 37; 5-carbon, 406, *406*; functional groups, 24; glucose, *18*, 35; hormonal effects on blood levels, *275*; in semen formation, 288; in urine, 270; lactose, 123; molecular structure, 22 25, 26, *26*; monomers, 26; physical fitness and, 66; red blood cells and, 381; refined, 125; ribose, 410; simple, 56, 60, *61*, 110, 120, *124*; simple, in plasma, 136, *136*, 160; table, 26; taste of, 253
Sugar beets, 26
Sugar cane, 26
Suicide, 170
Suicide tags, 451
Sulfur, 16, *16*, *127*; electrons in outer shell, *17*
Sulfur atoms, in disulfide bridges, 32
Sulfur dioxide, 23, *23*, 516, *516*, 522
Sulfuric acid, 203, 516, *516*
Sulfur oxides, 516, *516*, 517, 522
Sulfur trioxide, *516*
Summation, 227
Sunburns, 81

Sunlight energy, 487, 490, *490*, 492, *494*, 496, 506; harnessing solar, 522, 523, *523*
Sunlight exposure, 81; causing cancer, 423, 427, 430, 435, 436
Suntanning, 80, 81
Supercontinent, 462, *462*
Superior canal, *256*
Superior part, *79*
Superior vena cava, *143*, *144*, *145*, 146, *147*, *317*; flow velocity, *151*
Supernatural belief, 9
Superweeds, 452
Supreme Court rulings, 301
Surface barriers to invasion, 163, 164, *164*, 170
Surface markers, 140; in immune response, 164, 171; Rh, 141
Surface tension, 192
Surface tissue, somatic sensations in, 250
Surgery, 93, 430, 432, 433, 435, 436; axon regeneration after, 225; blood transfusion, 71, 336; exploratory, *18*; for arterial blockage, 156; knee joint, 97, *97*; patients, and contaminated blood supply, 336
Surgical-wound infections, 181
Surrogate mother, 448
Survival, 457, 460, 472; adaptive traits and, 6, 12; of cells, 39, *39*; cellular oxygen demand, 71; coordinated activity and, 67, 84; digestive system and, 110, *110*; early detection, 289; homeostasis and, 5, 84; interdependent organisms and, 7; ion concentrations and, 82; nutrient exchange and, 135; organ systems and, 78; of organ transplant recipients, 173; premature birth and, 316, 318–319; secondary immune response, 175; water pollution, 45
Survival of the fittest, 460
Survivorship, 513
Suspended particles, *516*
Sustainable society, 526; defined, 525
Sutures, *92*, 97
Swallowing, 74, 114–115, *115*, 187; reflex, 115
Sweat, 83, 203; exocrine gland, 386; pheromones in female, 253
Sweat glands, 22, 68, *69*, *80*, 81; ducts of, *68*; evaporative heat loss, 215, *215*; homeostatic control of inner temperature, 83, *83*; loss of, 325
Sweating, 22; ADH regulates, 272; as menopausal symptom, 293, 327; as temperature mechanism, 200, 211, 215, *215*; water loss, 202, 216
Sweet taste class, 246, 252, 253
Sympathetic nerves, *215*, *228*, 230–231, *230*, *269*, 279; defined, 231; heart rate control and, 148
Sympto-thermal method, 298
Synapse, 103, 108, 226–227, 229–230, *231*; neurotransmitter diffusion, 268; sensory nerves, *256*
Synapsis, 364
Synaptic cleft, 225, *225*, *231*, 242, 244; removing neurotransmitters, 227
Synaptic integration, 227, 229, 231, 244; defined, 227; visual pathway, 262
Synaptic vesicles, 226, 327
Synergistic drug interaction, 241
Synergistic movement, 98, 108
Synovial fluid, 96, 97
Synovial joint, *96*, 97, 108, 451; defined, 96
Synovial membrane cells, 451
Synthetic hormones, 105, 106
Syphilis, 331, 343, 345, 346; defined, 340; primary stage, 340, *341*; secondary stage, 340, *341*; as STD, 336; tertiary stage, 340–341, *341*

Systemic circuit, 135, *146*, 147, 149, 160; at birth, 317, *317*; defined, 146
Systemic lupus erythematosus (SLE), 179
Systole, *147*, 150, *157*; defined, 147
Systolic pressure, 149

T

Table salt, 127, 211; iodized, 280; ionic bonding, 20, *20*, 22
Tachycardia, 157, *157*
Tactile receptor, *251*
TADH, 273
Tail, *465*; embryonic, *314*, *315*
Tailbone, 93
Tanning, 80, 81
Taq DNA polymerase, defined, 442
Target cells, 273, 291, 450, 451; of hormones, 68, 266, 267, 269, 270, 272, *277*; nonresponse, 271, 275, 276; priming, 268; response, 275, 276, 285; of signaling molecules, 268, 269, 282, 284
Tarsals, *91*, 95, *95*
Tarsiers, 476, *477*, 479
Tarsioids, 475, 476, *476*
Tar smoke, *428*
Tastant molecules, 253
Tastants, 253
Taste buds, 246, 252, *252*, 253; taste receptors in, *252*
Taste classes, 252, 253
Taste receptors, 264; location of, *252*; on tongue, *248*, 252; sensitivity of, 253
Taste, sense of, 77, 247, *248*, 252–253, 264; food and, 116; olfactory element of, 252, 253
Taxa, 472; defined, 467
Taxoids, 433
Taxol, 433, 435
Taxonomy, defined, 467
Tay-Sachs disease, 394, *403*; embryo screening, 393
T cells. *See* T lymphocytes
Tear glands, 186
Tears: HIV infection, 336; as immune response, 164, *164*, 182
Tectonic movements, 498
Tectorial membrane, 255; defined, 255
Tectum, 233, *233*
Teeth, *91*, *92*, 114–115, 463, 475–477, 480, 482–483; anatomy of, *114*; canine, 478, *480*, 481; molars, *480*; types of, *114*; vitamin D and, 29
Telophase: defined, 356; during mitosis, 354, *354*, 355, 357, *357*, 359–360, *369*, 370
Telophase I, during meiosis, 361, *362*, 365
Telophase II, during meiosis, 361, *363*, 365, *369*
Temperature, allergies and air, 178; body set point, 167; enzymes and body, 54, 55; external environment, 214,*214*; hydrogen bonds and, 33, 36; internal regulation, *78*, 80, 81, 83; oxygen/hemoglobin binding and, 138; oxygen transport and, 193, 196; sense of, 5, 247, *248*, *248*, 264; sperm production and, 288, 298, *298*; stabilization of, 15, 22, 23
Templates, 406, 410, 418; strands, *443*, *445*
Temporal bones, *92*, *92*
Temporal lobe, 236
Temporal summation, 227; defined, 105
Tendons, *97*, 98; at wrist, 94; connective tissue, 73, *73*; damage, 94; defined, 90; mechanoreceptors in, 250; nociceptors in, 251; sensory axons and, 230
Teratogen, *320*

Termination, 414, *415*, 418
Testcross technique, *378*; defined, 378
Testes, 286, 289, 291, *291*, 299, 433; defective hormone receptors in, 397; defined, 287; formation, 288, 292; function, *288*, *289*, 302; germ cells in, 361, *361*; hormonal control, 290, 291, *291*; meiosis in, 352; overview of, *269*, *275*; Sry chromosome trigger, 314; steroid use and, 107; testicular inflammation, 334; testosterone secretion, 271; Y chromosome, 390, *390*
Testicles, 433, *433*; inflammation, 340, 342; self-examination, *429*, 433, *433*; tissue transplantation, 173
Testicular cancer, 107, 279, 289, *429*, 433
Testicular feminizing syndrome, 397, *403*
Testosterone, 29, *275*, 286–287, *289*, 291, *291*, 302; defective receptors for, 397; defined, 290; feminine development and, 314; hormonal controls on production, 290, 291; Klinefelter syndrome and, 410; primary effects, *269*, 271, *271*, *275*; synthetic, 106
Tetanus, 58, 105, *105*, 176, *176*
Tetany, defined, 105
Tetracycline, 180; prenatal risk, 321; STD infections, 341, 342
Tetrahydrocannibol (THC), 243
Tetramune, *176*
TGF-alpha, 283
TGF-beta, 283
Thalamus, 233, *234*, 237, *237*, 244; awareness of pain in, 250; defined, 233; fact memory, 239, *239*; optic nerve to, 258; sensory information in, 252
Thalidomide, 321
Therapeutic drugs, 106
Therapeutic waste, blood-borne, 136
Thermal inversion, *516*; defined, 516
Thermal pollution, 518
Thermometer, 298
Thermoreceptors, 214, *215*, *248*, *248*, 250, 264; detect heat energy, 247
Thiamine, *126*
Thigh, *99*, 105; blood delivery to, *143*; fat storage, 75; muscle groups, *99*
Thighbone, 88, *91*, 95, *465*, *483*; bone turnover and, 90
"Third eye," 281
Thirst, 82, 200, 202, 211, 233, 244; insulin deficiency and, 276
Thirst center, defined, 211
Thoracic cavity, *79*, 186, 187; blood flow, *146*; defined, 78; membrane lining, 69, *69*; muscle groups, *99*; respiratory system and, 186, 187, 190, 191
Thoracic duct, *158*
Thoracic nerves, *228*, 230
Thoracic vertebrae, 93, *93*
"Thoroughfare channels," 146, *147*
Three Mile Island, 523
Threonine, 125, *408*, *412*
Threshold level, 223, 226, 227, *227*, 244; defined, 222
Throat, 114, *115*, *186*, 479; bacterial infection, 340; eustachian tube to, 255; types of epithelium, *68*
Throat cancer, 434
Thrombin, 54; formation, 153, *153*
Thrombocytes, 137
Thrombus, 156
Thumb, *91*, 94, 250
Thymidine kinase, *450*
Thymine, 406, *406*, 410, 418
Thymosins, *275*; primary effects, *269*, *275*
Thymus gland, *158*, 434; defined, 159, 281; from endoderm, *306*; overview, *269*, *275*; T cell maturation, 170, *170*, 171, 174, 182

Thyroid cancer, 367
Thyroid cartilage, 187, *280*
Thyroid gland, 68, *83*, 271, *280*, 281; abnormal output, 280; defined, 280; excess hormone, 157; from endoderm, *306*; hormone synthesis, 89, 280, 284; as hormone target, *272*, *273*; overview, *269*, 275; radioisotopes and, *18*
Thyroid hormones, 131; formation, *127*
Thyroid-stimulating hormone, 280
Thyrotropin (TSH), *271*, 273, 274, 280, 284; hypothalamus/pituitary production, 273, *273*; primary effects, *269*, *272*, 272
Thyroxine, *275*, 280, 281; primary effects, *269*, 275
Tibia, *91*, 95, *95*, 97, *97*, *483*
Tibialis anterior, *99*
Tidal flats, 470
Tidal volume, *191*; defined, 191
Tight junctions, *72*; defined, 72
Time scale, 511
Ti plasmid, 446
Tissue, *37*, 66, 110; artificially grown, 70, 447; defense mechanisms, 163; defined, 67, 84; fat storage, 28; folding, *312*, 313; four basic classes of, 67; growth/specialization, 307, *307*; inflammation, 282; organism interdependence and, 7; primary, 306, *307*, 312, *315*; radioisotopes effect on, *18*; rapid water loss, 45; stable internal environment, 82; stretching, 73; sugar storage in, 26; systemic path of blood flow, 147
Tissue damage, 264, *248*; agglutination, 140; defense mechanisms, 163, 164, 166, 167, 182; H. pylori infection, 116; regeneration, 283
Tissue engineers, 70
Tissue fluids, pH of, 23, 36
Tissue plasminogen activator (tPA), *446*, 448, *448*
Tissue rejection: cytotoxic T cells and, 170; strategies for prevention, 173
Tissue repair, 110, 167; mitosis, 351, *351*, 353, 368, 370, *370*
T lymphocytes, 137, *137*, 167, *168*, 169, 281, 451; antigen-specific receptors, 174, 175, *175*, 182; autoimmune disorder, 178; as cytotoxic cells, 168–170, *171*, 176, 182; defined, 168; DNA, 174, 175; as effector cells, 170–171, *171*, 182; formation, 170, 171, 182; as helper cells, 168–171, *169*, *171*, *172*, 173, *178*, 179; as helper cells, defensive role of, 182, *182*; as memory cells, *169*, *171*; and receptor molecules, 170, 171, 172, *172*; specific immune response, *164*, 168, *168*, *171*, 179; thymus function and, 281; as virgin cells, 170, 175, *175*
Tobacco, 434, 436; smoke, as mutagen, 416; as stimulant, 242; use, 154, *196*, *197*. *See also* Smoke and smoking
Tobacco plants, 447
Toe bones, *91*
Toenails, 81
Toes, 95; formation of, 314
Tolerance, 240, 241, 242
Tomatoes, 428
Tongue, 114, 115, *115*, 191; cancer of the, 434; epithelium, 274; muscle, 114; tip, sensory receptors of, 250
Tonicity, *43*; defined, 43
Tonsils, *158*, 170, *170*; from endoderm, *306*
Tools, 482, 483, 484, 485
Tooth decay, 114, *127*
Tooth enamel, *306*
Tooth sockets, 92
Topography, 516, *516*
Torso, 93; cardiovascular system and, 146

Touch, sense of, *237*, 250, 264; memory circuits, *239*
Touch-kill mechanism, 168, *169*, 170, 171, *171*
Touch receptor, 247, 248, *249*, 250, 251, *251*
Toxic-dump sites, 447
Toxic psychosis, 243
Toxins: antibody-mediated response, 173; bacterial, 121, 176; inactivated, in vaccine, 176, 182; removal, in liver, 120
Trace elements, 16
Tracers, defined, *18*
Trachea, *68*, 114–115, *115*, *186*, 189, *197*, 198; cartilage, 74; defined, 187; mucus-secreting cells, 68; thyroid location, 280, *280*
Trachoma, 260–261
Training regimen, 86
Traits (phenotypes), 459–460, *459*, *466*, 467, 475, 477, 481–482; behavioral, 459; beneficial, 460; codominance, 373, 381, 384; complex, 476; crossbreeding for, 373; dominant forms, 373; evolution and, 457; genetic basis, 392, 440; governed by one gene, *373*, 377; heritable, 6, 12, 351, 353, 404, 461; inheritance patterns, 392, 394–395, *396*, 402; morphological, 459; mutations and differentiation, 405; physiological, 459; polygenic, 383, 384; recessive form, 373; two different genes, 378, 379, 380, *380*; X-linked, 388, 390, 398; X-linked recessive, 396, 397
Trait variations, 364–365, 368, 373–374, 382–383, 458–459, 472; allele rearrangement, 387, 388; change in chromosome number, 387, 388; change in chromosome structure, 387, 388
Tranquilizers, prenatal risk, 321
Transcription, *332*, 336, *337*, 405, 445, 449; defined, 410; hormonal trigger of, *270*; in protein synthesis, 410–411, *410*, *411*, 417–418; in RNA, *412*, 414, *416*, 417; summary, *419*
Trans fatty acids, 28, 29
Transfection, defined, 450
Transfer RNA (tRNA), *418*; defined, 410; as free molecules, 412; as initiator, 414, *414*, 415, *415*, 418; molecule model, *413*; roles of, 412–413, 418; transcription/translation summary, *419*
Transforming growth factors (TGF), 283
Transfusable blood, 145; shelf life, 145
Transfusion, 381
Transgenic organisms, 173, 452; defined, 446
Transitional stage, *514*, 515
Translation, 405, 445; AUG codon, 412, *412*; defined, 410; mRNA into polypeptide chain, 412, 413, *416*, 417; stages of, 414–415, 418; summary, *419*
Translocation, 398, *399*, 402, 426; defined, 399
Transmission electron microscope, 48, *48*, *100*
Transpiration, 496, *497*, 520
Transplants, 171; corneal, 173; kidney, 213; liver, 120. *See also* Organ transplant
Transport pigments, 188
Transport proteins, *41*, 42, 43, 47, 54, 62; activate, 271; ATP production, 57, 58, *59*; defined, 41; hemoglobin, 71, 75; in proximal tubule, 208; on neuronal plasma membrane, 227, 235; passive transport, 44, *44*
Transport systems, 44, 46, 57, 58, 59, 62; by blood, 71, 75, *78*; by respiratory system, *79*
Transposable elements, 416, 418

Trans-retinal, 263
Transverse colon, 121, *132*
Transverse plane, *79*, *232*
Trapezius, *99*
Tree dwellers, 476, 479, 485
Trends, 472, 477, 485
Treponema pallidum, 340, 341, *341*
TRH, *271*, 273
Triceps, 98, *98*
Triceps brachii, *99*
Trichomonas vaginalis, 344, *344*
Tricuspid valve, 144, *145*
Triglycerides, *28*, 29, 60, *117*, 119, *119*, 132; defined, 28; in LDL, 154; lipoproteins and, 32
Triiodothyronine (T$_3$), *275*, 280, 281; primary effects, *269*, 275
Trimesters: first, 301, 312, 314, 321; second, 301, 316, 317; third, 301, 316, 317, 320
Trinucleotide repeat expansion, 417
Trisomic, defined, 400
Trisomy 21, 400, *401*
Trophic levels, 492, *493*, 494, *494*, 506; defined, 490
Trophoblast, *309*, 310
Tropomyosin, 102, *103*
Troponin, 102, *103*
Troposphere, 517
"True capillaries," 146, 147
Trunk, *79*, 230; positioning, 233
Trypsin, 117, *117*
Tryptophan, 125, *125*, 412
Tubal ligation, 299, *298*
Tubercles, 181
Tuberculosis, *181*, *197*; bacterium, 175; drug-resistant strain, 181, 336, 338; screening programs, 181; skin test, 175
Tubulin, 53, *53*; subunits, 358
Tumor, *178*, 446, 451; benign, 424, *424*; cancerous, 424–425, *424*, 427, 429, 430, *432*; defined, 424; genetic disorders and, 407; growth, 425, *425*; in lung, *197*; irradiation, 430, 436; malignant, 424, *424*; mass, 424, *424*; primary, 424, 425
Tumor angiogenesis factor (TAF), 283
Tumor cells, 170, 177, 451; antigen, 168, 182, 427, 429; touch-killing, 168, *169*, 171
Tumor-inducing genes, 446
Tumor-infiltrating lymphocytes (TILs), 451
Tumor markers, 436; defined, 429
Tumor necrosis factor (TNF), 177, *446*, 450, 451
Tumor suppressor genes, 423, 426, 427, *427*, 436; defined, 426; FHIT, 426; p53, 426; *Rb*, 426
Tuna, 465
Tunics, 259; fibrous, 258, *258*; sensory, 258, *258*; vascular, 258, *258*
Turner syndrome, 401, *401*, *403*
Turnover, 90, *90*, 108
Turtles, 492
Tympanic membrane, *254*, *255*; defined, 254
Type A hepatitis, 120
Type B hepatitis, 120, 342, 343, 346
Typhoid fever, 180
Typhus, 505
Tyrosine, 271, 394, *412*
T wave, 157

U

Ulcers, 435; corneal, 260; growing skin, 70; STD produced, 340, *340*, 343
Ulna, *91*; defined, 94
Ulnar nerve, *228*
Ultrafiltration, 167

Ultrasound, *248*, 429; pregnancy and, 314, 322
Ultraviolet light, 322; genetic disorders, 407; as mutagen, 416
Ultraviolet (UV) radiation, 80–81, 84, 423, 427, 435, *435*, 517; free radicals and, 21; UV-B, 81
Umbilical arteries, 316, *317*, 328
Umbilical cord, 311, *311*, 314, *315*; at birth, 318, *318*; blood, 451; defined, 310; fetal circulation and, *317*
Umbilical ligaments, *317*
Umbilical vein, 316, *317*, 328
Umbilicus, *317*
Unbound phosphate, *58*, *59*
Unconsciousness, 236
Undernutrition, 125, 130–131
Underwater pressure, 188
Unipolar cells, 220
United Nations, 131
U.S. Environmental Protection Agency, 235
United States national rowing team, 86
Universe, 457, 472
Unsaturated fats, 154
Unsaturated fatty acid, *29*; defined, 28
Upper arm bone, *91*, 94
Upper extremities, blood flow, *146*
Upper nasal passage, 253
Upper respiratory tract, 92
Upright posture, 108
Uracil, 411, 412, 418; defined, 410
Uranium fuel, 523
Urea, 120, 203, 206, *206*, 207, 209; amino group conversion, 60, *61*; formation, *120*, *123*; in nitrogen cycle, *503*
Ureter, *204*, 205, 216; deferred pain, *251*; defined, 204
Urethra, *204*, 205–207, 216, *306*; defined, 204; female, 292, *293*; male, 342, 344; use in male reproduction, 288, 289, *289*, 290, 297
Uric acid, 202, 203, 206, 213
Urinalysis, 206; defined, 207
Urinary bladder, *204*, 205, 206, 216, *289*, *293*; deferred pain, *251*; defined, 204; fetal, *306*, *317*; infections, 327
Urinary excretion, 209; controls hydrogen ion concentration, 212; defined, 202; from urethra, 288, *289*, 297; water loss, 202, 216
Urinary frequency, 327
Urinary incontinence, 327
Urinary system, 2, 66, 78, 111, 203, 204–205; acid-base balance, 212; age-related changes in, 327; circulatory system link, *142*; components and function, *204*, 216; defined, 202; macrophages in, 73; water/solute balance, 201
Urinary system cancers, 434–435
Urinary "toilet training," 206
Urinary tract infection, 207, 327, 341; drug-resistance, 181
Urination, 82, *164*, 206; control of, 224
Urine, 60, *120*, *123*; concentration/dilution, 209, 210, 211, 216; defined, 206; excess water/salts in, 272; formation of, 206–207, 213, 216; HCG detection, 309; HIV infection, 336; in male urinary tract, 292; pH value, 23, *164*, *164*; sugar in, 276; water/solute excretion, 200–203, *203*, 211
Uterine cancer, 429, 430, *431*, 433, 436; hormone replacement therapy and, 327
Uterine cavity, *309*
Uterine contractions, 308, 318, *318*, 319, 328; hormone-induced, 272, *272*, 282
Uterine lining, 425, 433; fertility cycle and, 287, *293*; pregnancy, 309, 310

Uterus, *292*, 293, *293*, *294*, 300, 390; autonomic nervous system and, *230*; birth control in, 299; defined, 292; fertility cycle and, 287, 295, *295*, 296, *296*, 302; fertilization stage, 304, *305*, *308*; fetal development, 316; as hormone target, 272, *272*, *273*, *275*, 282; implantation, 309, *309*; infection, 319; perforation of, 299; PID infection, 342; placenta and, *311*; pregnancy and, 268; smooth muscle tissue, 76; testicular feminizing syndrome and, 397; type of epithelium, *68*; walls of, 83

Utricle, *256*, 257

UV-B, 81

V

Vaccination, 162; for German measles, 321; for Hepatitis B, 343

Vaccines, 182, 430, 438; against HIV, 336; bioengineered, 446; combined injected, *176*; defined, *176*; oral, *176*; viral, 339

Vagina, 292, *292*, 293, *293*, *294*, 295; artificial insemination, 300; birth, *318*; discharge, 340, 341, 342, 345; epithelium, 344; female reproductive system component, *390*; fertilization stage, 308, *308*; as HIV point of entry, 334, 336; HIV transmission, 336; infections, 164, 344, *344*, 346; nonkeratinized, *68*; pH, 289, 344; sexual penetration, 297, 298

Vaginal canal, 293; birth through, 318

Vaginal opening, *286*, 292

Vaginitis, 344, 346

Vagus nerve, *194*, 230

Valine, 125, *125*, 408, *412*, 419; net charge, 408, *408*

Valves, 144–145, *150*, 159, *159*; in veins in limbs, 151, *151*

van Leeuwenhoek, Antony, 38, *38*

Variable, defined, 8

Variable expressivity, 382

Variation, *459*; heritable, 460

Varicose veins, 151

Vascular circuit, 142, 154

Vascular system, 150–151

Vas deferens, scarring, 340

Vas deferentia, *288*, *289*, *290*, 297, 299; defined, 288

Vasectomy, *298*, 299; compulsory, 509

Vasoconstriction, 214, *214*, *215*; defined, 151; during intercourse, 297

Vasoconstrictors, prostaglandins as, 282

Vasodilation, 166, *166*, 207, 214, *214*, *215*; defined, 151; during intercourse, 297; menopausal symptom, 293

Vasodilators, prostaglandins as, 282

Vasopressin, 272

Vectors, 180

Vegetarians, 125, *125*; vitamin/mineral supplements, 127

Veins, *143*, *145*, 146, *150*, 160, *464*; blood pressure in, 150–151, *151*; blood velocity in, 149, *151*; defined, 142; lymph draining into, 159; obstruction, *152*; partial pressure gradients in, *173*; to bone, 89, *97*; as transport tubes, 135; valved in limbs, 151

Velazquez, *394*

Venous duct, 317, *317*

Ventilation, 188, 190, 192; rate of, 191, 195

Ventricle, *144*, 145, *145*, 146, 157, 160; defined, 144; in brain, 234; in cardiac cycle, 147, *147*, *157*; in fetal circulation, 317

Ventricular contraction, 160

Ventricular fibrillation, 157, *157*

Ventricular pressure, 147, 150

Venule, 118, *118*, 146–147, *151*, 160, 205; blood pressure in, 150–151, *152*; defined, 142; in placenta, 310; as tubes for diffusion, 135

Vernix caseosa, 316

Vertebrae, *91*, 96, 232; number of, 93; red bone marrow in, 139

Vertebral column, *91*, 92, 93, *204*, 232, *232*, 465; from notochord, 312; ribs and, 93

Vertebral disk, 74; deterioration, 325

Vertebrates, 2, *3*, 77, 426, *464*, 465, 476; body plan, 312, 313, 314, *314*, *315*; land-dwelling, 465; lineages, 464; "third eye," 281

Vertical axis, 92, 93, 108

Very low-density lipoproteins (VLDL), 154

Vesicles, 44, *44*, *46*, 62; budding, 50, *50*, 51, *51*; endocytic, *50*; exocytic, *50*; in axon endings, 103; main function, 46, *51*; in metaphase of mitosis, 358, 359; variety of, 50–51

Vestibular apparatus, *256*, 257; defined, 256

Vestibular nerve, 257

Vestigial structures, 464–465, *465*

Vibration, 255, *255*; defined, 254

Vibration stimulus, *248*; Meissner's corpuscle, 250, *251*

Vibrio cholerae bacterium, 45, *45*

Villi, *118*, 119, 120, 132; defined, 118

Vinblastine, 435, 524

Vincristine, 430, 435, 524

Vinyl chloride, 426, 428, 430

Viral attack, 447

Viral disease, 71, 177, 179, *181*; HIV retrovirus, 335–339, *336*, *337*, *338*; sexually transmitted disease (STD), 330–339, *332*, *334*, 342, 345, 346

Viral DNA, 332, *332*, 337, 426

Viral glycoprotein, 338

Viral infections, *197*, 225, 440, *446*; and cancer, 423, 426, *427*, 436; heart disease and, *155*, 156; influenza, 180, *181*; prenatal risk, 320–321, *320*; type 1 diabetes and, 276

Viral protein, *117*; synthesis, 332, *332*, 336, 337

Viral RNA, 332, *332*, 333, 336, 337

Virgin B cells, *169*, 170, 172, *172*, 175, *175*; IgD antibodies, 173; IgM antibodies, 173

Virgin T cells, 170, 175, *175*

Virulence, 180

Viruses, *163*, 170, 171, *171*, 468; altered, 339; antibiotics and, 181; antibodies and, 140; combatting, 81; components, 332; defined, 332; genetically engineered, 176, 394, 438, 449–452; genetic susceptibility, 393; gene transfer and recombination, 440; leukocytes and, 137, 160; liver cell injury, 120; loss of chromosome region, 398, 399; multiplication cycle, 332, *332*, 333; as pathogens, 164, 165, 168, 182; specificity, 332; surface antigen, *174*; surface markers, 163; transgenic, 452

Visceral pain, 251

Visible light, 248

Vision, daytime, 262–263, 475, 477–479, 485

Vision, sense of, 116, 236, 237, 258, 264; memory circuits, *239*; overview, 258–259

Visual acuity, *258*, 263

Visual cortex, 236, *238*, 262, *262*, 263, *263*; defined, 258

Visual field, *238*, 263, *263*; defined, 236; peripheral, 261

Visual information, 262, 264

Visual receptors, 247, *248*

Visual stimulus, 238, *238*, *239*, 244

Vital capacity, *191*; defined, 191

Vitamin A, *126*, 127, 131, 263; in colostrum, 319

Vitamin B, 156

Vitamin B$_1$, *126*; deficiency, 131

Vitamin B$_2$, 125, *126*

Vitamin B$_6$, *126*

Vitamin B$_{12}$, 116, *125*, *126*; deficiency, 139

Vitamin C, 21, 26, 81, *126*, 127; deficiency, 131; depletion, *196*

Vitamin D, 29, 37, 80, *126*, 127, 281, 325; deficiency, 131

Vitamin D$_3$, *275*

Vitamin E, 21, *126*, 127

Vitamin K, *126*, 127; *E. coli* and, 121

Vitamins, 110, 111, 124, 130, 430; absorption, 118, 132; blood transported, 135, 136; as coenzymes, 34, 55; deficiencies, *126*, 130, 131, *155*; defined, *126*; excess, *126*; fat-soluble, 125, *126*; functions, *126*; in plasma, 136, *136*, 160; reabsorption, 206; sources, *126*; water-soluble, *126*

Vitamin supplements, 127; prenatal, 320

Vitreous body, 258, *258*, 261

VLDL, in plasma, *136*

Vocal cord, 115, *115*, *191*; defined, 191

Voicebox, 115, *186*

Volatile organic compounds, *516*

Volatile substances, 252

Volcanic activity, 468

Volcanic ash, 463, *480*

Volcanic eruptions, 469, 498, *499*, 502, 506, 517

Voltage-sensitive gates, 221, 222–224, *223*, 227, 244

Voluntary muscle movement, 115, *115*, 233, 244, 319; loss of control, 227

Vomer bone, 92, *92*

Vomeronasal organ, 253, 268

Vomiting reflex, 257; allergic reaction, 123; appendicitis, 121

Vulva, 292, 297, 344

W

Waking, 30, 236, 237; medulla and, 233

Walking, 106, *129*; brisk, 90; upright, 475, 477–480, 483

Waller, Augustus, 134

Warrawoona rock formation, 470, *470*

Warts, 334, *334*. *See also* Genital warts

Waste, 487, 492; capillary exchanges, 150, 152, 160; carbon dioxide as, 193; carried by blood vessels, 87, 89; cellular, 43, 60; chemical, 486; greenhouse gas release and, *501*; industrial discharges, 505, *505*; metabolic, 136, 138, 146; nitrogen cycle and, 502, *503*; organic, 492, *493*, 494, *494*, *504*, 506; radioactive, 523, 526; solid, 519, 526; treatment of, 518, 526

Waste-burning, controls over, 235

Waste dumps, toxic, *203*, 447

Waste products, kidneys and, 203, 207; placental diffusion of, 310, 311, 316, 328; urea as, 120, *120*. *See also* Metabolic waste

Waste removal, 77; by blood supply, 106; disposal of, 111; organ systems and, 66, 67, 75, 78, *78*, *79*, 82

Wasting, 106

Water, 24, *30*, 78, 469–471, *470*, 500, 506; as byproduct, 51, 57, *57*, 58, *61*, 202; as compound, 17; diarrhea and, 121; evaporation, 83; excess loss, 276, 282; excreted in urine, *208*, 209, 211, 216; extracellular fluid (ECF) component, 202; forced into interstitial fluid, 142, 152, *152*, 159; as fuel byproduct, 35; gains and losses, 202, *203*; global crisis, 518, 526; hydrogen ion concentration, 23, *23*; imbalance, 201, *201*, 202; in plasma, 136, *136*; "insensible" losses, 202; interactions, 22–23, 28, 36, *37*; metabolism and, 36; osmotic flow of, 43, 130, 209; properties of, 15, 22, *22*, 36; repellants, 28, 29; saliva component, 114; as solvent, 136, 200; surface tension of, 192; sweat component, 81

Water absorption, *112*, 118, 119

Waterborne bacteria, 175

Water conservation: ADH regulation, 272, *272*; global crisis, 518, 526; mechanisms, 81, 200, 201, 215

Water–hydrogen bonds, functional groups, 24

Water/ion balance, 327

Water molecules, 25, *25*; cell exchange, 39, *39*, 42–43; cohesiveness, 190; diffusion of, 42, *42*, 43, *44*; structure of, 20, 21; tonicity, 43

Water pollution, 14, 35, 120, 518, *518*, 519, 525, 526; disease and, 45, *45*; of drinking, 203; fecal contamination, 121; of water supply, 180

Water reabsorption, 121, 207–211, *207*, 209, 216; ADH regulation, 272; sodium reabsorption and, 278

Water/salt balance, hormonal effects on, *275*

Watersheds, *497*, 520, 521; defined, 496

Water-solute balance, 201, 216, 276

Water treatment, 518

Water vapor, 23

Watson, James, *404*, 405, 406

Wave cycles: amplitude, 254; frequency, 254

Waxes, *24*, 28, 29, *37*; defined, 29

Weaning, protein deficiency and, 130

Weasels, 462

Weathering of rocks, 495, 496, *498*, 504, *504*

Weight control, 125, 129, 132; aging and, 325, *325*; diabetes and, 276; during pregnancy, 320; insulin deficiency and, 276; loss, 129. *See also* Body weight

Weight guidelines by sex, *128*

Wet acid deposition, 516

Wexler, Nancy, 395

Whales, 487, *491*

Wheat, 502

White blood cells, 51, 136, 160, *425*, 435, 451; antibody activators, 417; as blood component, 75, *75*, *136*, *137*; defined, 136; in immune responses, 163, *163*, 168, 182; in inflammation, 166; in urine, 207; leukemia, *155*; a line of defense, 164, *164*; lymphatic system and, 158, 159; types of, 137

White matter, 232, *232*, 234

"White" muscle, 105

Whole blood, 145

Whooping cough, *176*, 180, *181*

Wiesel, Torsten, *263*

Wilkins, Maurice, 405

Wilms tumor, 434

Wind energy, 523, *523*

Wind patterns, 496

Windpipe, 114, *186*, 187

Winter blues, *266*, 267

Withdrawal, 298, *298*
"Wobble effect," 413
Woodpecker, red-cockaded, 488
World Health Organization, 335, *335*
World Wide Web, 449
Worm attacks, 446, *447*
Worms, 446, *447*, 463
Wounds, mitotic healing, 351
Wrist, 150; formation, *315*
Wrist bones, *91*, 94, 95
Wrist joint, 94

X

X chromosome, 286, 314, 353; condensed to Barr bodies, 401; homologous region, 388; in meiotic cell division, 364, *364*; mutated gene on, 397; nondisjunction, 401, *401*; as sex chromosome, 387–390, *390*, 402

Xenotransplantation, 173
Xeroderma pigmentosum, 407
Xeropthalmia, 131
X-inactivation, defined, 390
X-linked anhidrotic ectodermal dysplasia, *403*
X-linked genes, 353, 389, 390, 402; androgen insensitivity, 397; defined, 388; dominant inheritance, 397, *403*; recessive inheritance, 396–397, *396*, *403*
X-linked trait, red-green color blindness, 260
XO chromosomes, 401, *401*
X ray diffraction, DNA analysis, 405
X rays, 156, 427, 429, *429*, 432; chest, for TB, 181; diagnostic, 367; exposure, dental, 427
XX chromosomes, 314; as female, 387–390, *389*, *390*
XX embryo, 286
XXX females, 401, *401*
XXY chromosomes, 401, *401*

XY chromosomes, as male, 387–390, *389*, *390*; testicular feminizing syndrome, 397
XY embryos, 286
XYY chromosomes, 401, *403*

Y

Y chromosome, 286, *286*, 287, 314, 353; homologous region, 388; in meiotic cell division, 364, *364*; as sex chromosome, 387–390, *390*, 401, 402; Sry, 314
Yeast infections, 336, 338; genital, 164, 344
Yeasts, 442, 443, 446, 466
Yersinia pestis, 512
Y-linked gene, 353, 389; defined, 388
Yolk sac, *310*, 311, 312, *312*, *314*, 328; defined, 310

Z

Zalcitabine, 339
Zero population growth, 515, 526; defined, 510
Zinc, *127*
Z-line, *100*, *101*, *102*
Zona pellucida, *294*; defined, 294; fertilization stage, 308, *308*; implantation stage, 309
Zonula adherens junctions, 72
Zooplankton, 492
Zovirax, 343
Zygomatic bones, 92, *92*
Zygote, *323*, 328, 461; defined, 306; formation, 307–309, *307*, *308*, *309*; in infertility treatments, 300; X inactivation in, 390
Zygote intrafallopian transfer (ZIFT), 300

A

ABO blood typing, 140, 141, 381
Abortion, 400; debate, 301, *301*
Abrasion: adhering junctions, *72*; human skin, 80, 81, 84
Achondroplasia, 395, *395*, 403
Achoo syndrome, *403*
Acidosis, defined, 23
Acid rain, 7, 23, 516–517, 526; defined, 516
Acne, 81, 107; scars, 73
Acquired immunodeficiency syndrome (AIDS), 45, *197*; cases by mode of transmission, *337*; confidential testing, 339; cumulative totals of cases, *346*; defined, 179; as global epidemic, 335, *335*; immune system and, *178*; indicator diseases, 336, 338; molecular structure of HIV, 8; onset, 338; as pandemic, 180; prevention, 339; as sexually transmitted disease, 330, 334, *334*, 340, 343, 346; transfusable blood and, 71; treatment plan, 181; vaccine, 446; as viral infection, *163*, 334, 438, *450*
Acromegaly, 274, *274*
Acute inflammation, defined, 166
Addiction, 242, 243, 321; defined, 241
Addison's disease, 278, *279*
Adenocarcinoma, 431, 434, *434*, 436
Adipose tissue, methyl mercury in, 235
Adjuvant therapy, 430
Aerosol inhalants, 189, *See also* Ozone
Aging, phenomenon of, 323, *324*, 328; antioxidants and, 21; bone formation and, 325; bone mass, 323; cartilage and, 96; causes of, 324; eye disorders, 260, 261; heart disease risk, 154, *155*; loss of muscle tissue, 106; menopause and, 293; physiological changes, 305, 327, 328; weight gain and, 129
Agricultural chemicals, and cancer, 428, *428*, 430
Agricultural runoff, 14, 45, 203
AIDS. *See* Acquired immunodeficiency syndrome
Air pollution, 135, *197*, 516–517, *516*, *517*, 522, 525–526; breathing and, 189
Albinism, 403
Alcohol, 29, *37*, 157, 210, *241*, 243; bladder infections, 207; blood-brain barrier, 235; breakdown of, 51; consumption, and cancer, 430, 434; defined, 242; dependence or habituation, 116, 120, 240, *240*, 345; drug action and interaction, 241, 242; functional groups, 24, *24*, 25, 26; HDL levels, 154; as legal drug, 218, 219; rate of absorption, 116; taste of, 253
Alcoholism, genetic links, 383
Alcohol use, prenatal risk, 320–321, *320*, 328
Allergens, 178, *179*; antibiotics as, 180; breathing and, 189; common, 178
Allergic response, 178–179, *179*, 182; aggravation of, 282; eosinophils and, 137, 166; glucocorticoid use, 278; pesticides and, 14
Allergic rhinitis, *197*
Allergies, *179*, 186; defined, 178; drug action and interaction, 241; food, 123; IgE antibodies and, 173, 177; sources of, 189
Altitude sickness, 184
Alzheimer's disease (AD), 225, 227, 326, *326*, 449; aging and, 324; brain

regions, 239; early-onset, 449; malfunctioning proteins and, 32
American Cancer Society, 430, 432, 433, *433*, 434
Amnesia, 239
Amniocentesis, 322, *322*; prenatal screening option, 393
Amoebic dysentery, 180, *181*
Amphetamines, *241*; defined, 242
Amyloidosis, cerebroarterial, Dutch type, *449*
Amyotrophic lateral sclerosis (ALS), *403*, 449, *449*
Anabolic steroids, 105, 106–107
Anaphylactic shock, 178
Anemia, 116, 131, 139, *155*, *446*; from ionizing radiation exposure, 367; vitamins and, *126*. *See also* sickle-cell anemia
Anesthesia, drug abuse, 242, 243
Angelmann syndrome (AS), 383
Angina pectoris, *155*, 156
Antacids, 22
Antagonistic drug interaction, 241
Anthrax, 162
Antianxiety agents, 242
Antibiotics, 164, 180–181, 201–202, 446; broad-spectrum, 181; cholera, 45; defined, 180; resistance to, 181, 440; in organ transplantation, 173; sickle-cell anemia, 381; side effects, 180; STD infections and, 340, 341, 343; vaginitis and, 344; viruses and 334, 343
Antibodies, *37*, 172–173, 182, 446; anti-HIV, 338, 339; bound, 172; defined, 140, 169; gene activation for, 417; in blood screening, 343; in immune response, 164, 165, *165*; maternal, 141, *141*, 320; monoclonal, 176, *177*, *177*; production site, *158*; Rh, 141, *141*; surface, 173
Anticancer drugs, 430, 524, *524*
Anticlotting factors, 167
Antidepressant drugs, 227
Antihistamines, 178, 241
Anti-inflammatory drugs, 178
Antioxidants, 21, 127, 130, 430, 439; defined, 21
Antiviral drugs, 181, 449
Apnea, 195
Apneic episode, 195
Apoptosis, 173, 307, 313, 430; defined, 170
Appendicitis, 121, 342
Appetite: stimulate, 129; suppress, 120, 129
Aquifers, pollution of, 203, 518, *518*
Arable land, 131
Arrhythmia, *155*, 156–157, 243; defined, 157
Arteriosclerosis, 156
Arthritis, 97, 179, 450; aging and, 324, 326, 327; osteoarthritis, 96; rheumatoid, 96
Arthroscopy, 97
Artificial blood, 145
Artificial insemination by donor (AID), 300
Artificial joint, 96, 97
Artificial selection, 440
Artificial sweeteners, 253
Asbestosis, *197*
Aspirin, 153, 215, 282; excessive use, 116
Asthma, 14, 178, *189*, *197*; bronchioles and, 187, 189; cortisone drugs, 278; preventive drugs, 189
Astigmatism, 260
Atherosclerosis, 107, 149, 154–157, *155*; defined, 156; free radicals and, 21

Atherosclerotic plaque, 154, 156–157, *156*
Athletes, anabolic steroids, 106–107; drug testing, 206; elite, 86; euphoric "high", 227; knee damage, 95; in rigorous training, 105; resting cardiac rate, 157
Athlete's foot, 164
Atrial natriuretic factor, *446*
Atrophy, 106
Autoimmune disease, 225, 438; age-related, 227
Autoimmune disorders, 280
Automobile exhaust, 188

B

Baby formula, dilution of, 131
Bacillus anthracis, 162
Bacterial dysentery, 180; drug-resistance, 181
Bacterial infections, 92, 116, 121, *197*, 225; asymptomatic, 340; embryo risk of, *292*, *293*, 295, 320; heart disease and, *155*, 156; kidney inflammation, 213
Baldness, 107
Balloon angioplasty, 156
Barbiturates, *241*; blood-brain barrier, 235; defined, 243; prenatal risk, 321
Barging of wastes, 518
Barnyard biotechnology, 448–449, *448*
Basal cell carcinoma, *10*, *80*, 435, *435*
Basal metabolic rate (BMR), 128, 129; age-related change, 327; by sex, 128
Behavior: developmental stages and changes in, 323; ethical, STD and, 345, 346; genetic links in, 383; hormones and, 267, 268; hostile, violent, 242; psychoactive drugs and, 240, 241, 243; sexual, hormonal controls for, 290; STD risks and, 331, 339, 342
The Bends. *See* Decompression sickness
Beriberi, 131
Biochemical testing, 394; markers for prematurity, 319
Biogas digester, *525*
Biogeography, defined, 462
Biological clock, 266, 267, 269
Biological magnification, 505, *505*
Biopsy, 429, *432*, 433, 436; defined, 429
Bioreacter, grown cells, 70
Bioremediation, 447, *447*; defined, 446
Biorhythms, 275
Biotechnological society issues, 452
Biotechnology, 141, 439
Birth control: biological bases, 298–299, *298*; global, 301
Birth defects, 203, 428; nondisjunction, 365; prematurity and, 319; prenatal diagnosis, 322; prevention, 319, 320; screening, 311; untreated STD and, 330–331, 346
Birth weight, low, 131, *196*, 321
Bladder cancer, 423, 426, *428*, *431*, 434, 436
Blindness, 260, 261; herpes and, 343, 346; prematurity and, 319; vitamins and, *126*, 131
Blistering, 81
Blood alcohol concentration (BAC), 242, 243
Blood bank, 145
Blood clotting, 54, 75, 167, 448, *448*; arterial, 156, *197*; cancer therapy and, 145; fibrinogen, 136; mechanism, 153; minerals and, *127*; oral

contraceptives and, 299; plaques and heart attack, 154, *155*; platelets and, 137, 160; protein, *164*, 182; PTH and, 281; vitamins and, *126*. *See also* Clotting factor VIII, IX
Blood-clotting functions, X-linked gene, 388, 396
"Blood doping," 139
Blood filtration, 206–207
Blood products: contamination of, 145; HIV-infected, 336–337
Blood substitutes, 71, 145
Blood supply, contaminated, 336
Blood tests, 429, 436
Blood transfusions, 71, 140, 145; demand, 71; for sickle-cell anemia, 381; hepatitis B contamination, 343; HIV-tainted blood supply, *337*; Rh antibodies, 141
Blood typing, 140–141, *140*, 145, 460; in organ transplantation, 173. *See also* ABO typing
Blood velocity, 149
Blood vessels, artificially grown cells and, 70
Blue-white screening, 445, *445*; defined, 445
Body builder's psychosis, 107
Body fat, *129*; distribution, X-linked and, 287, 388; heart's health, 154
Bone cancer, 426, 429, 434, 436; from ionizing radiation exposure, 367; osteosarcoma, 383
Bone marrow cancer, from ionizing radiation exposure, 367
Booster injection, 176
Bottle feeding, 131
Botulism, 58
Bovine spongiform encephalitis, 71
Bradycardia, 157, *157*
Brain, use of PET and, *18*; viral infection, 341
Brain cancer, 424, 426, 429, *431*, *450*, 451
Brain damage, 188, 225; drug abuse, 242–243; irreversible, 190; sickle cells, *381*
Brain disorders, Huntington's disease, 394
Brain scan, radioisotopes and, *18*
Brain tumor, 424
Breast cancer, 289, 299, *319*, 423, 426, *429*, 430–434, *434*, 436; from ionizing radiation exposure, 367
Breast milk, 130; HIV transmission, *178*, 336; hormones and secretion, 268, *269*; immunity transfer, 173; "let-down," 272, *272*; methyl mercury in, 235
Breast self-examination, *429*, 432, *432*
Breech position, 318
Bronchitis, 189, 196, *197*
Bruising, *155*
Bubonic plague, 180, 512, *514*
Burkitt's lymphoma, *178*, 434
Burns: from ionizing radiation, 367; growing skin, 70; inflammatory response, 166; reduced blood pressure/volume, *155*; regeneration, 283; skin mending, 80; thirst, 211

C

Caffeine, 116, 157, 210, 253; across blood-brain barrier, 235; bladder infections, 207; defined, 242; dependence or habituation, 240
Calcium salts, in atherosclerotic plaque, 154, *154*
Campodactyly, 382

Cancer, 107, 154, 179, 422–436, 438, 446, 449–451; advanced, 450; age factor, 324, 423; artificial blood use, 71, 145; bone marrow production, 155; causes, 203, 423, 428, 428; childhood, 383; chromosomal translocation, 399; chronic hepatitis, 343; defined, 424; diagnostic tests, 289, 426, 429, 429, 436; from ionizing radiation exposure, 367; gender preference, 422, 423, 431, 436; heredity role, 423, 426–427, 427, 429, 432, 436; immunity breakdowns, 178, 427; malignant, 425, 426; monoclonal antibodies, 176, 177, 177; morbidity rate, 422; mortality rate, 422, 431; prevention, 430; research, 10; risk, personal, 336, 423, 429, 430, 436; treatments, 18, 177, 299, 423, 430, 436; triggers, 81, 334, 426–427; types, 189, 196, 196, 431; urinalysis, 207; warning signs, 429, 429, 434

Candidiasis, 344

Captive breeding programs, 468

Carbohydrates, digestive enzyme defect, 387; use in making nonpolluting fuels, 35

Carbon atoms, in substitute blood, 71

Carbon monoxide poisoning, 188

Carcinogenesis, 426, 427, 436; defined, 426

Carcinogens, 154, 423, 426–427, 518; chemical, 426, 427, 436; defined, 10, 426

Carcinomas, 431, 434, 434, 436

Cardiac failure, muscular dystrophy and, 397

Cardiac pacemaker, 157; defined, 148

Cardiovascular disease, 107

Cardiovascular disorders, 29, 144, 154–155; aging and, 324, 325; risk factors, 154, 155

Caries, 114

Carnivore, 482, 492, 493, 494, 495, 504, 506; defined, 490; primary, 490, 491, 492; secondary, 490, 491; tertiary, 491; top, 491, 492, 494

Carotenes, 80, 127

Carotid arteries, 143; defined, 149

Carotid bodies, 194, 248; defined, 195

Carotid sinus, 248

Carpals, 91, 94, 94, 95, 96

Carpal tunnel syndrome, 94

Carrier designation, 392, 394; genetic screening, 393; Queen Victoria, 396, 396, 397

Cartilage, inflammation of, 325

Cataracts, 261, 367, 517

Cell death, 58; apoptosis, 170, 171, 173, 313; viral infection, 333, 334

Cell reproduction, ionizing radiation, 367

Cellulose, as nonpolluting fuel source, 35, 35

Centers for Disease Control and Prevention, 335

Central nervous system (CNS), drug abuse, 242; psychoactive drugs in, 240–241

Cerebral palsy, prematurity and, 319

Cerebrospinal fluid, HIV infection, 336

Cervical cancer, 343, 426, 429, 433

Cesarean section, herpes and, 343

Chancre ulcer, 340, 341

Chancroid, 346; defined, 343; as STD, 336

Channel proteins, 221, 221; gated, 41; genetic defect in, 386; trigger openings, 271

Chemical insults, 114, 279, 398

Chemical weapons, in immune responses, 164, 167, 177, 182

Chemotherapy, 434, 435, 436; defined, 430; side effects, 430

Chemotoxins, 167

Chernobyl meltdown, 367, 522

Childbirth, 83, 96, 97; Rh blood typing and, 140–141

Children: AIDS and, 337; diarrhea in, 164; immunization schedule, 176; undernutrition and, 130, 131

China White, 243

Chlamydia, 207, 261, 342, 343; defined, 341

Chlamydial infections, 341, 342, 346

Chlamydia trachomatis, 341, 341, 342

Choking, 115, 187

Cholera, 45, 180

Cholesterol, arterial deposits, 154, 154, 156

Cholesterol plaques, 325

Chorionic villus sampling (CVS), 311, 322; prenatal screening option, 393

Chromosomes, abnormal rearrangement, 388, 395; Alzheimer's disease link, 326; detecting defects, 387, 393, 395; genetic birth defects, 322, 322; ionizing radiation and, 367; segment change, 387, 388

Chronic bronchitis, 196, 197

Circadian rhythms, 233, 233

Circulatory shock, 155

Circulatory system, cholesterol damage, 125; collapse, 178, 197

Cirrhosis, 120, 207; alcohol and, 242; type B viral hepatitis, 343, 346

Classification systems, 467; defined, 467

Clinical depression, 131

Clonal selection hypothesis, 175, 175; defined, 174

Clone, 170, 439, 441, 448, 453; antibody-producing hybridoma cells, 177; human cloning efforts, 319; use in immune response, 173, 174, 175, 182

Cloned DNA, 441; defined, 442

Cloned human gene products, 446

Clotting, 136, 136; of blood, 153

Clotting factors, 167, 167, 439; gene encoded, 353

Clotting factor VIII, 396

Clotting factor IX, 396

Clumping response, 140

Coal, 522, 522; as fossil fuel, 23, 35

Coal tar, 196, 426

Cocaine, 157, 219, 227, 241, 242; crack, 218, 241, 242; defined, 242; drug dependence, 240; granular, 242; prenatal risk, 321

Cold sores, 81, 181, 334

Cold stress, 214, 214; physiological response, 214, 215, 215

"Coliform bacteria," 121

Collagen, 32, 37, 425; aging and, 324, 325; in designer skin, 70

Collagen implants, 73

Colon cancer, 423, 426, 429, 429, 430, 431, 434–436; diet and, 111, 121, 125

Color blindness, 260, 397, 403

Colorectal cancer, 426, 429, 431, 436

Coma, 185, 236, 237, 244; drug abuse, 242; hypothermia and, 214; parathyroid disorders, 281

Common cold, 180, 186, 334; chronic, 189

Compound light microscope, 48

Computer-aided design (CAD), 70

Computerized tomography, 429

Concussion, 225

Condoms, 298, 298, 299, 340; latex, 299, 339, 345

Cone cells, 248, 260, 262, 263, 264; blue, 263; defined, 262; green, 263; red, 263; red/green color blindness, 260, 397, 403

Connective tissue cancer, 428, 431, 436

Connective tissue, implantation stage, 309

Consciousness, loss of, 154, 157

Consciousness, states of, 236, 237, 244; psychoactive drugs and, 240, 241; two spheres of, 238

Constipation, 121

Contact inhibition, 425

Contact lenses, 261

Contagion, 180

Contraceptive injection, 299, 301

Contraceptive sponge, 298, 298

Cornea, artificial, 261; collagen in, 32; defined, 258, 258; transplants, 173

Corneal transplant surgery, 173, 261

Coronary arteries, plaque clogging, 144, 155, 156

Coronary artery disease, 155, 157, 196; free radicals and, 21

Coronary bypass surgery, 156, 156

Cosmic rays, 367

Coughing, 181, 197; as immune response, 164; reflex center, 233

Cowpox, 162

Cow's blood, 145

Crab lice, 344, 344

Crack cocaine, 218, 241; dependence or habituation, 240; prenatal risk, 321

Cramping, 299

Cretinism, 280

Creutzfeldt-Jakob disease (CJD), 334

Cri-du-chat syndrome, 398, 398, 403

Criminal investigation, blood groups, 141

Crohn's disease, 123

Crop rotation, 502

Crop yields, 501, 502–503, 519

Cross breeding, of plants, 373

Cross-matching, 141

Cryptosporidiosis, 181

Curare, 103

Cuts, 182; bacterial invasion, 168; skin mending, 80; stopping blood loss, 153, 153

Cystic fibrosis (CF), 122, 376, 444, 449, 450, 450, 451; as autosomal recessive condition, 394, 403; CF phenotype, 386, 393; as genetic disorder, 386–387, 387, 393

Cystitis, 207, 342

Cytomegalovirus, 155

D

Daughter cells, ionizing radiation exposure, 367

Daytime vision, 262, 263, 475

DDT, 279, 505

Deafness, prenatal risk, 321

Death, 163; abnormal chromosomal rearrangment, 388; agglutination, 140; AIDS-related, 181, 346; ATP production, 101; by infectious diseases, 181; cardiovascular disorders as cause of, 154; defining, 11; drug abuse, 242; early infant, 319, 320; fetal, 141; from ionizing radiation, 367; hypothermia, 214; leading causes, 330; legal, 11; life cycle end, 305; programmed, 307; sudden, 155; untreated STD and, 330, 331

Death rate, 510, 511, 513, 515, 526; children, 131

Death signal, 372

Decomposers, 495, 495, 498, 498, 499, 502, 504, 506; defined, 490; and ecological balance, 487, 489, 490, 490, 491; energy flow through ecosystem, 492, 493, 494; organization and interdependence, 7, 7

Decomposition, 463, 503, 503, 504, 520; defined, 502

Decompression sickness, 188

Deforestation, 519–521, 519, 520, 521, 524–526; acid rain, 517; defined, 520; ecosystem disruption, 496, 497, 499, 501, 501; overpopulation and, 512; recycling, 519

Degenerative diseases, 334

Dehydration, 78, 80, 84, 164, 210, 211; adrenal cortex disorders, 278; insulin deficiency and, 276

Dehydration synthesis, defined, 25

Deliriants, 243

Delirium, 243

Dementia, 326

Depo-Provera, 299, 301

Depression, 218, 267; drug use/abuse, 242, 243; vitamins and, 126

Dermatitis, 126

Desensitization program, 178

Desertification, 519, 526; defined, 519

Designer drug, 243

Designer skin, 70

Detoxification, in liver, 120

Diabetes, 70, 207, 438; blood glucose regulation and, 276; cataracts, 261; diet and, 111, 124, 276; kidney damage, 213; obesity and, 128; type 1, 276, 339, 446, 446; type 2 diabetes, 276

Diabetes mellitus, 154, 276; heart attack and, 155

Dialysis, 213

Diarrhea, 45, 181; allergic reaction, 123; causes in colon, 121, 122; as immune response, 164, 164; thirst, 211; vitamins and, 126

Diet, 111, 478–482, 485; aging effects and, 325; average American, 125; bulk in, 121; cancer risk, 423, 430, 434, 436; diabetes control, 276; gender and, 124; heart attack and, 155, 156; hormonal adjustment, 267; iron-poor, 139; low-fat, 325; low-fiber, high-fat, 124, 125; low sodium, 208; PKU control, 394; prenatal, 320; well-balanced, 124, 125, 127

Dieting, 129

Digestive enzymes, cystic fibrosis and, 386–387

Digestive tract, cystic fibrosis and, 386–387

Digital rectal examination, 429, 433

Dihybrid cross, 379, 380, 380, 384; defined, 378

Dioxins, 279, 516

Diphtheria, 176, 176, 181

Disability, cardiovascular disorders and, 154

Discoverer XVII satellite, 355

Disease, 180; artificial tissue and, 70; defenses against, 15; degenerative, 334; diagnosis of, 18; homeostasis and, 44, 45; immunization, 176; transmission, 71; viral, 71; water pollution and, 45. See also Infection

Disease agents, 145, 160, 331, 512, 513, 526; animal virus for STD, 334; genetic susceptibility, 393

Disinfectants, HIV virus and, 337

Dislocation, shoulder, 94

Disorientation, 266

Diuretic, 210

Dive reflex, defined, 214

DNA (deoxyribonucleic acid), autoimmune disease and, 179; duplication disorders, 400; FISH diagnosis, 322; free-radical damage, 21, 324; gene therapy and, 394; HIV gene incorporation, 339; HIV virus, 336; in criminal investigation, 141; mutagens and, 416, 418; mutation, 405, 416; recombinant technology, 452, 453

DNA analysis, in detecting genetic disorders, 393

DNA databases, 444

DNA-DNA hybridization, 472; defined, 466

DNA fingerprint, 397, 444, 444; defined, 444

DNA library, *441*, 442; defined, 441
DNA mass screening, 444
DNA virus, 332, *332*, 334, 337, 343, 426, 435–436
Douching, 298, *298*; PID and, 342
Downers, 243
Down syndrome, 400, *401*, *403*, 435, *449*; frequency/maternal age relationship, *401*
Dreams, 237
Drinking, 218; drug dependence and, *240*
Drinking water, contamination of, 45, *45*, 203
Drought, 45, 439, 446–447
Drug abuse, 241, 242, 243; defined, 240
Drugs, 218; addiction to, 241; analgesics, 243; antagonistic interaction, 241; antiparasitic, 344; brain activity reduction, 242–243; deliriants, 243; dependence warning signs, 240; dosage, 241; effects of, 240; hallucinogens, 243; inactivation of, 50, 120; nonprescription overuse, 153; overdose, *225*; physical dependence, 240; placental transfer, *320*, 321, 328; potentiating interaction, 241; psychedelics, 243; recreational drug use, *241*; SLE and, 179; STD protection, 345; stimulants, 242; synergistic interaction, 241; tachycardia, 157; testing, 206; therapeutic use, 106; tolerance, 240; withdrawal, *240*, 243
Drug therapies, 425, 430; "cocktail," 339; immune system suppression, 173; side effects, 339
Duchenne muscular dystrophy (DMD), 396–397, *403*; embryo screening, 393
Dutch elm disease, 505
Dyes, 426; granulocytes, 166; "staining," 48, 137; testing, 156
Dysplasia, defined, 424

E

Earwax, 29, 37; blockage, 29; as exocrine secretion, 68
Eating, 202, 237; psychoactive drugs and, 240
Ecology, defined, 488; population growth, 509; principles, 488–489; threat to, 23
Economic development, 514, 515
Ecosystem, 7, 486–506, *488*, *493*, *506*; acid rain effect, 516; aquatic, 492, 494, *494*, 498, 501, 502; defined, 487, 488; forest, *497*, 520; human impacts, 454; land, 495, *495*, 501, 502; limits of, 525; organization, 490–491, *490*; types of, 488, 490
Ectoparasites, 344
Ectopic (tubal) pregnancy, 309, 342
Edema, 130; defined, *152*; localized, in acute inflammation, 167, *167*; plasma fluid, 185
Egg, genetic engineering, 448, *448*; implantation, 309
Electrical storms, 238
Electricity, in creating nonpolluting fuel, 35
Electric shock, 157
Electrocardiogram (ECG), 134, 148, 157
Electroencephalograms (EEG), 236, 237
Electron microscope, 48, *48*
Electron microscopy, 268
Electrons, use in scanning electron microscope, 48
Electroporation, defined, 450

Embolus, 156
Embryo implants, 300
Embryoscopy, 322
Embryo screening, 322, 393
Embryo transfer, 300
Emotional states, 243; in hippocampus, 239; limbic system, 252; regulation of, 227; weight gain and, 128, 129
Emphysema, 189, *189*, 196, *196*, *197*
Emulsification, 117, *119*, 132
Encephalitis, 225
Endangered species, 467, 468, 524
Endemic disease, 180
Endocarditis, *155*
Endometrial cancer, *429*, 430, 433
Endometriosis, 299
Endorphins, 227, 243
Endurance, 106
Energy consumption, 509, 522, *522*, 525, 526
Energy, of ecosystem, *489*, 490–495, *490*, *493*, *494*, 506, *506*; flow in biosphere, 7, *7*; nonpolluting fuel, 34, 35; solar, 35; stored forms of, 487–488
Energy pyramid, 492, 494, *494*
Energy sources, 15, 25, 26, *26*, 27, *37*; alternative, 60, *60*; fat as, 123; fatty acids as, 103; foods as, 56, *61*, 124; glucose as main, 111, 124, 132; ; nonpolluting fuels, 34; nutrients as, 110, 122triglycerides as, 28, 29
Environment, bacteria to clean up pollutants, 447; cancer factors, 423; nonpolluting fuels, 34, 35; recycling energy and, 7, *7*
Environmental biotechnology, 446
Environmental pollutants, blood-brain barrier, 235
Environmental Protection Agency (U.S.), 279, 434, 447
Enzymes, from cancer cells, 425; gene therapy and, *450*, 451; use in nonpolluting fuel, 35
Epidemic, 180; HIV as global, 335; sexually transmitted disease, 330
Epilepsy, 225, 449; barbiturates, 243; progressive myoclonus, *449*; split-brain and, 238
Epithelial cells, *69*, 80, *118*, *119*, 120; radiation destruction of, 367
Epithelial linings cancer, *428*, 430, 431, 436
Epstein-Barr virus, *155*, *178*, 434
Erosion, 468, 490, 502, 504, 512, 519, 520
Erythropoietin, 139, *275*, *446*
Esophageal cancer, 196, 430, *431*; diet and, 111
Estrogen, environmental, 279; menopausal levels, 293, 327; synthetic, 299
Ethical problems, 298, 301, *301*, 345, 346, 452, 453; in vitro fertilization, 393
Eugenic engineering, 452
Euphoria, 242; drug abuse, 242, 243
Evolution, 456–472, *461*; adaptive traits and, 6; biological, 2, 457, 472; chemical, 457, 468, 469, 470, 471, 472; cultural, 2, 475, 477, 479, 483, 485; defined, 459; history of, 2, 458; human, 93, 95, 474–485; molecular, 471; physical, 457, 472; rates of change, 461; sketchy relationship, *466*; vitamins and, 126
Evolutionary modifications, 364
Evolutionary trees, 466–467, *481*; defined, 466; diagrams, 466, *466*
Evolution by natural selection, theory of, 8, 12, 459, 460
Excitatory signals, 226, 227, 229, 230, 236, 244; by acetylcholine (ACh), 102, 103; drug abuse, 242

Exercise, 90, 130; aging and level of, *325*; blood circulation during, 151, *152*; body functions and, 66, 67, 80; calories expended, 129, *129*; core temperature and, 214; defined, 106; dieting and, 129; energy for breathing during, 191; fluid balance, 210; HDL levels and, 154; heart rate and, 5; lack of, 121, 154, *155*; levels of/ATP formation, 103, *103*; quick energy source, 27; strenuous, 87, *103*, 122, 157, 191; stress electrocardiogram, 156; triglyceride use, 60
Exercise physiologist, 66, 86
Exocrine glands, genetic disorders, 386
Exploratory surgery, 18
Expressed sequence tags (EST) database, 449
External environment, 84; cues, 281; decoding signals from, 5, 77, *78*, 247; homeostasis, 83; interstitial fluid as, 142; metabolic waste disposal, 201
External genitals, STD symptoms, 343
External oblique, 99
Extinction, *466*, 468, 505; background, 467; mass, 467; plant, 524; regional, 467
Extrinsic clotting mechanism, 153
Eye cataracts, 261, 367, 517
Eye diseases, 260–261
Eye disorders, 260–261; new technology and, 261
Eyeglasses, 259, 261
Eyes: infections, 341; injuries, 261
Eye tissue transplantation, 173

F

Factor VIII, *446*, 448
Factor IX, *446*
Fainting, 215
"Fake fats," 125
Falls, bone breakage, 94
Familial hypercholesterolemia, 154, *155*, 395, *403*
Family pedigrees, in genetic analysis, 392, 393, *395*, 396, *396*, 397, 402
Family-planning programs, 301, 509, *510*
Family tree, 472, 475, *476*, *481*, 484
Fasting, 122
Fatigue, 106
Fats: aerobic exercise and, 106; deposited secondary sexual trait, 292; deposited in blood vessel wall, *154*, *155*, 156; deposition and aging, 323, 324, 325; genetic predisposition, 129; level in diet, 111, 125, 132; physical fitness and, 66; substitutes, 125
Fecal occult blood test, *429*
Feline infectious peritonitis, 468
Femaleness, 287
Female reproductive disorders, 299
"Feminization" of males, 279
Fertility, 293; options, 298–299
Fertility, control of, 298–299, 300, 302; dilemmas of, 298, 301, 302; future options, 299; global perspective, 301; reliability, *298*, 299; reversible methods, 299; side effects, 298–299
Fertilization: artificial methods, 300
Fertilizers, *501*, 502–503, *504*, 516, 518–519, *525*, 526; biomediation and, 447; water pollution, 14, 45, 203
Fetal alcohol syndrome (FAS), 218, 321, *321*
Fetal development, 316–317; critical periods, 320–324, *320*; malnutrition and, 131

Fetus: alcohol use and, 218, 242, 321, *321*; at risk, 218; cretinism in, 280; Rh blood typing, 140, 141
Fever, 215, 417; defined, 167; enzyme structure and, 54; -reducing drugs, 215
Fiber, level in diet, 27, 111, 121
Fibroblast growth factor (FGF), 283
Fibrosarcoma, 423
Fight-flight response, 279; defined, 231
First filial generation, 376; gene linkage crossovers and, *391*
Fish, transgenic, 452
Flatulence, 122
Flossing, 114
Fluorescent in situ hybridizing (FISH), 322
Fluorine atoms, in substitute blood, 71
Food: additives, 10; nutritional value of, 110, *110*, 125
Food allergies, 123
Food and Drug Administration, 428
Food guide pyramid, *124*
Food supply, 509, 525; contaminated, 45, 120, 121, 180; global, 130, 131
Foragers, 482
Foreign agents, 159, *167*, 168, 171, 173, 182, 439; "complement coated," 165, *165*
Forensic medicine, 444
Forensic scientists, 95
Forests: tropical rain, 488, 498, *519*, 520–521, *520*, *521*, 526; tropical rain forest as energy flow, *493*
Fossil fuels, 35, 498, 503, *522*, 523, 525, 526; burning, 498, *498*, *499*, 501–502, *501*, 506; defined, *522*; green revolution and, 519
Fossilization, defined, 463
Fossil record, 461, 466, 472, 482, 483, 484; interpretation, 463
Fossils, 426, 456–458, *463*, 467, 476, 479–480, 485; cells, 470, *470*; defined, 462; first hominids, *480*; harmful compounds traced, *505*; human evolution, 482, 483–484, *484*
Fossil sequences, 457
Fractures, 94, 97; bone remodeling, 90
Fragile mental retardation 1 (FMR1), 417
Fragile X syndrome, *403*, 417, 444
Frameshift mutation, 416, *416*, 418
Frostbite, 214
Fuel, nonpolluting, 35, *35*
Fumarate, 57
Fungal infection, *197*, 260
Fungicides, 15, 428
Fungus, 3, 446, *489*, 490, 502, 506, *512*; antibiotics, 180; carcinogens produced by, 426–427; as decomposers, 7; as pathogens, 164; sexually transmitted disease, 331, 344, 346
Fusion power, defined, 523

G

Gallbladder, deferred pain, *251*
Gallbladder cancer, 430
Gallstones, 117
Gamete formation: ionizing radiation and, 367; nondisjunction during, 400, *400*, 401, *401*
Gametes: mutations in, 416
Gamma interferon, 177
Gamma rays, 427
Ganglioside, 283
Gastrointestinal tract (GI):diarrhea as immune response, 164; sickle-cell damage, *381*; water gain in, 202, 216
Gel electrophoresis *408*, 442, 443, *443*, 444, 466; defined, 408

Gender, alcohol intoxication, 242
Gene mutation, 324, 338, 339, 386–387, *386*, 388; genetic disorders and, 393, 402, 405, 407
General anesthesia, 236
Genes, cancerous changes, 423, 424, 426, 428, 429, 432; de-tox, 447; of endangered species, 468; evolution and, 457; HIV, 339; independent assortment, 378, 384; inactivation, 416; ionizing radiation altering, 367; meiotic shuffling, 364, 370, 373; mutation, 439, 440, 444, 460, 472; pedigree analysis, 392, 393; virus, 176
Gene therapy, 449; for cystic fibrosis (CF), 394; defined, 449; examples, 451; methods and prospects, 450–451, *450*; and your future, 451
Gene therapy trials, *450*, 451
Genetic abnormality, 393, 402, *403*; defined, 393
Genetically altered bacteria, 439
Genetic analysis, 392–393, 397
Genetic bioengineering, 438–439
Genetic counseling, 393
Genetic damage, ionizing radiation and, 367; smoking effects, 196
Genetic defects, 393, 449, 451
Genetic disease, 383
Genetic disorders, 154, 382, 449; abnormal chromosome rearrangement, 389, 399; carriers and, 392, 393, *394*, 396, *396*, *397*; common, 402, *403*; cretinism, 280; cystic fibrosis, 122, 386–387, 393, 394; defined, 393; detection, 393; following nondisjunction, 400, 401, *401*; gene mutation, 389, 393, 396, 397, 402, 407, 416, 417; hemophilia as, 6; immunodeficiency, *178*; prenatal diagnosis, 322, 400; probability of transmission, 376, 393; recessive transmission, 393; spontaneous abortion, 314; X-linked genes and, 388
Genetic engineering, 71, 339, 438–453, 468, 501; of animals, 448–449; bacteria, *438*, 446, 447; defined, 440; gene products, 466, 472; interferon, 177; virus, 176
Genetic predisposition: Alzheimer's disease, *225*, 326; hyperthyroidism, 280; peptic ulcers, 116; rheumatoid arthritis, 178; to allergies, 178; type 1 diabetes, 276
Genetic screening, 452; moral/ethical dilemma, 393
Genital epithelial cells, STD infection, 340, 341, 342
Genital herpes, 181, 331, 334, 342, *342*, 343, 346; defined, 342
Genital warts, *334*, 343, 346, 433
Genotype, 376, 382; defined, 374; genetic counseling and, 393; in sex-influenced inheritance, 398; of sex chromosome abnormality, 401, *401*; probability of, 376, 377, *377*, 380, *380*; testcross for unknown, 378, *378*; vs. phenotype, *374*
Geothermal energy, 471, 523
German measles, *176*, 321
Germ cell, ionizing radiation and, 367
Germline, mutations in, 416
Gigantism, 274, *274*
Gingivitis, 114
Glaucoma, 261
Glia, 77
Global carbon budget, 498, *498*, *499*
Global cycling, 521
Global warming, 7, 498, 500–501, 506, 524
Glomerulonephritis, 213
Gluten intolerance, 123

Goiter, *127*, 131, 280, *280*
Gonads, mutations in, 416
Gonorrhea, 207, 331, 333, 341, 342, 343, 346; defined, 340; drug-resistance, 181; prevention, 345
Gout, 202
Government regulations, 446
Grand mal seizures, *225*
Graves' disease, 280
Greenhouse effect, 498, 500–501, *500*, 521, 522, 526; defined, 500
Greenhouse gases, 500–501, *500*, 525; defined, 500
Green revolution, defined, 519
Groundwater depletion, 518, *518*, 519, 525; contamination, 203, 522
Growth hormone (GH), abnormal production, 274, *274*

H

Habituation, 240, 241, 243
Hair cells, sensory, 255, *256*, 257, *348*; damage to, 257, *257*; defined, 255
Hallucinogens, *241*, 243
Hardening of the arteries, 154, 157
Harvard University study, 128, *128*, 129
Hashish, *241*
Hay fever, 178, *197*
Hazardous waste, dumps, *203*
Head cold, 92
Health care, 514, *514*, 515
Health care delivery system, 330
Hearing loss, risk of, 257
Heart abnormalities, 242; prenatal smoking, 321
Heart: artificial replacement cells, 70; deferred pain, *251*; enlarged, 157, 325
Heart attack, 135, 154, *154*, *155*, 156, *446*, 448; atherosclerosis, 154; growth factor and, 282; obesity and, 128, 129; referred pain, 251
Heart damage: alcohol abuse, 242; sickle cells, *381*
Heart defects, change in autosome number, 400
Heart disease: antioxidants and, 21; cholesterol-related, 395; diet and, 111, 124, 125; excessive weight and, 128, *128*; familial history, 129, 154, *155*; genetic disorders and, 416; post-menopausal women, 327; risk factors, 154, *155*, 157; trans fatty acids and, 29
Heart failure, *127*, *152*, *155*, 157, *197*; drug abuse, 242, 243; insulin deficiency and, 276; sickle-cell damage, 381
Heart monitoring, 135
Heart muscle: aging and, 325; damage/death, 154
Heart rate: age-related changes, *327*; controls over, 148; drug effect on, 240, 241, 242, 243; hormonal effects on, *275*, 279; nerves and, 5; resting, 135; sympathetic/parasympathetic integration, 231, 233
Heat exhaustion, 215
Heat production, 78, *214*, 215, *215*, 216; as metabolic byproduct, 201, 214, *215*, 216
Heat shock factor (HSF), 417
"Heat shock" genes, 417
Heat stress, 214–215, *214*; physiological response, 214, 215, *215*
Heat stroke, 215
Heimlich maneuver, 115, *115*
HeLa cells, 355
Hemodialysis, 213

Hemoglobin (Hb): abnormal, 381, *381*, 408; DNA mutations and, 6; from donated blood, 145
Hemolytic anemia, *449*
Hemolytic disease of the newborn, 141
Hemophilia, *403*, 450, 460; DNA mutation and, 6; embryo screening, 393; X-linked recessive inheritance, 396
Hemophilia A, *446*, 448; X-linked recessive inheritance, 396, *396*, 397
Hemophilia B, *446*, *450*; X-linked recessive inheritance, 396
Hemophiliacs, 336, 337
Hemophilia influenzae, 176
Hemorrhage, *126*, 210; anemia from, 139; intestinal, 367
Hemorrhoids, 121
Hemostasis, defined, 153
Hepadnaviruses, *334*
Hepatitis, 120, 207, 243, 332; immunization, *176*, *176*
Hepatitis A, 120, 180, *181*
Hepatitis B, 120, 342, 346, *334*, 434, *446*; defined, 343; as global health threat, *181*; virus, 343
Hepatitis C, 145, 177, *181*
Hepatitis D, *181*
Hepatitis E, *181*
Herbicides, *15*, 428, 452, 496, *497*
Heritable alteration, risk of, 416
Herniated disks, 93
Heroin, *241*; blood-brain barrier, 235
Herpes simplex virus (HSV), 81, *225*, 260; infection, 449; transmission, 342; type 1, *334*, 342, *342*; type 2, *334*, 342, *342*
Herpesviruses, *334*, 336, 342, *342*, 345
Heterosexual contact, HIV spread, 337
High blood pressure, 127, 145, 154, *446*; blood substitute side effects, 145; chloride and, *127*
Histoplasmosis, *197*, 260
HIV. *See* Human immunodeficiency virus
Hives, 14
Hoarseness, 191
Hodgkin disease, 434
Home pregnancy tests, 177
Homocystinuria, *449*
Homosexual contact, male, HIV spread by, 337, *337*
Hookworm, 181
Hormonal imbalance, 300
Hormone-based therapy, 129
Hormone replacement therapy (HRT), 106, 327
Hormones: cancer produced, 425, 426, 432; obesity and, 129; synthetic, 105, 106
"Hot flashes," 293, 327
H. pylori infection, 16
Human cells, 450; gene therapy use, 450
Human gene therapy, 439
Human Genome Project, 448, 449
Human growth hormone, 446, *446*, 448, *448*
Human immunodeficiency virus (HIV), 179, *181*, *197*, 332, 434, 438, 448; as cause of AIDS, 330, *334*; cumulative infection, 335, 338–339, 346; defined, 335; genetic strains, 336; immune system and, *178*; infection, *225*, 337, 338–339, 346; method of infection, 333, 337; non-progressors, 338; rate of new infection, 330, 331, 335, 337; replication, 336, 338, *338*, 339; screening, 336, 339; structure, 8, 336, *337*; transfusable blood contamination, 71, 145; transmission, 336–337, 343, 345; treatment, 338, 339
HIV-1, 335
HIV-2, 335

HIV protease, 339
Human papillomavirus (HPV), *334*, 343, 346; defined, 343
Hunger, 233
Huntington gene region, 417
Huntington's disease, 394–395, *395*, *403*, 417, 444
Hybrid offspring, 461
Hybridoma cells, 177, *177*
Hybrids, 440
Hydroelectric power, 522, *522*
Hydrogen, as nonpolluting fuel, 35, *35*
Hydrogen fuel, 523
Hydropower, 523
Hyperopia, 260
Hyperparathyroidism, 281
Hypertension, 154, 207, 211; blood substitute side effects, 145; heart disease risk, 154, *155*, 156–157; kidney damage, 213
Hyperthermia, 215
Hyperthyroidism, 157, 280
Hyperventilation, 188
Hypnosis, 242
Hypnotics, 242
Hypoglycemia, 278
Hypothalamus, psychoactive drug effects on, 240
Hypothermia, 214
Hypothesis, 9, *9*; alternative, 10; defined, 8, 12
Hypothyroidism, 280
Hypoxia, 188

I

Ice-minus bacteria, 452, *452*
Illegal drugs, 242; prenatal risk, 321
Illness, diet and, 111
Immune response: abnormal, 177, 178–179, 182; deficient, 178–179; fever, 215; organ transplantation, *169*, 170, 171, 173; specific, *164*, *164*, 174–175, 176, 182; to HIV infection, 338, 346; to viral infection, 334
Immune specificity, 164, *164*, 174–175, 176, 182
Immune system: allergies, 123; antibodies, 140; bacterial response, 334; cancer and, 10, 424–425, 427–428, *427*, 430, 434, 436; chronic infection, 343, 346; defined, 168; disorders, 96, 178–179; HIV target, 334, 336, 338, 339, 346; ionizing radiation exposure and, 367; rejection, 70; stimulants and, 242; suppression of, 81, 173, *173*, 278; thymus function, 281; undernutrition and, 130, 131
Immunity, 75, *79*, 162–182; acquired, 173; age-related changes in, 327; impaired, *126*; to genetically engineered virus, 394
Immunization, 162, *176*, 177, *177*; active, 176, 182; defined, 176; passive, 176, 177, 182
Immunodeficiency (non-HIV), *178*, 179, 182
Immunofluorescence, 340
Immunological memory, 168, 174, 175, *175*
Immunological specificity, 168
Immunologists, *163*
Immunotherapy, 430, 436; defined, 430
Impetigo, 180
Implantation, 295, 299, 308–309, *309*, 310, 311; as critical period of development, 320; genetic screening and, 393
Impotence, 106
Inadequate blood flow to organs, visceral pain and, 251
Inbreeding, 468

Incinerators, 519
Indicator diseases, 336, 338
Industrial chemicals, 428, 430, 434, 435; hormonelike effects, 278–279
Industrial pollution, 45, 203; airborne, pH of, 23; heavy metal mercury, 235
Industrial smog, 516, 526; defined, 516
Infant respiratory distress syndrome, 192
Infections, 438; birth control risk of, 299; cystic fibrosis and, 387, 387; defenses, 121; immuno system suppression and, 173; inflammatory response, 166; lymph nodes and, 159; scavenger cells, 135; susceptibility to, 336; undernutrition and, 130; uterine, 319; white blood cells' role in, 137
Infectious diseases, 180–181, 182, 334, 446; antibiotics, 180–181; global threats, 181; inoculation for, 162; modes of transmission, 180, 181; patterns of ocurrence, 180, 181; virulence, 180. *See also* specific disease
Infectious disease specialists, 181
Infectious mononucleosis, 155
Infertility, 107, 300, 300, 302; ectopic (tubal) pregnancy, 309; genetic disorders and, 401; PID cause, 342
Inflammation, 116, 169, 173, 178; allergens and, 178; basophils and, 137; of bronchial walls, 189; of heart muscle, 155; of reproductive structures, 334, 340, 342, 346; signs and causes, 166; white blood cells and, 136
Inflammatory bowel disease, 434
Inflammatory disorders, 278
Inflammatory response, 163, 166–167, 167, 182, 342; blocking, 178; controlled, 334; glucocorticoid suppression of, 278; non-specific, 164, 164, 165, 166–167, 182; to viral infection, 334, 340
Influenza, 180, 197, 334; virus, 332, 446
Inheritance, patterns of, 372–384, 388; genetic crosses, 376–377; genetic disorders and, 260, 387, 394, 397, 402; genomic imprinting, 383; Mendel's theory of segregation, 375, 381; testcross, 378
Inorganic chemicals, water pollution, 203
Inorganic compounds, fossilization, 463
Insanity, syphilitic, 341
Insecticides, 428, 446
Insect resistance, 452
Insulin-dependent diabetes, 276
Insulin-like growth factor (IGF), 275, 282, 283
Insurance charts, 128
Interferons, 430, 436; beta, 177; defined, 177; gamma, 177; genetically engineered, 438, 446, 446
Interleukins, 177, 430, 436, 451; defined, 167; from helper T-cells, 170, 171, 171, 172; in immune responses, 164; as signaling molecules, 177, 182, 182
Interleukin-1, 167, 451
Interleukin-2, 425, 430, 446, 450
Internal bleeding, 145
Interventions, behavioral, 298–299, 302
Intervertebral disk, 108; rupture of, 93
Intestinal ulcers, 116
Intoxication, 240, 243
Intrauterine device (IUD), 298, 299
Intravenous drug abusers, 181; clean needle strategies, 339; HIV spread, 337, 337
Intrinsic clotting mechanisms, 153, 153
In vitro fertilization, 300, 322; moral/ethical dilemmas, 393
Iodide deficiency, 131

Ionizing radiation, 367; as mutagen, 416
Iron deficiency anemia, 127, 139, 155
Irradiation, loss of chromosome region, 398, 399
Irrigation, 518, 518, 519

J

Jaundice, 120, 343; vitamins and, 126
Joint: arthritic, 96, 178, 179; artificial, 96, 97
Joint breakdown, aging and, 325
Joint replacement, 71
Juvenile-onset diabetes, 276

K

Kaposi's sarcoma, 338
Ketones, insulin deficiency and, 276
Kidney: age-related change, 327; transplantation, 213
Kidney cancer, 177, 430, 431, 434, 450
Kidney damage, 107; blood substitute side effects, 145; drug abuse and, 243; sickle-cell, 381
Kidney dialysis machine, 213
Kidney disease, blood pressure and, 149
Kidney disorders, 203, 207, 213
Kidney stones, 127, 206, 207, 213, 281
Klinefelter syndrome, 401, 401, 403
Knee joint, 91, 97; artificial, 97; damage to, 95, 97

L

Labor, 318; stages of, 318, 319, 328
Lactose intolerance, 122
Landfills, 519
Laparoscope, 300
Large-cell carcinomas, 434, 434
Laryngeal cancer, 196, 430, 431
Laryngitis, 191
Laser angioplasty, 156
Laser coagulation, 261
Last Place on Earth, The, 486
Lawn chemicals, 428
Lead, 203, 516; water pollution, 203
Learn, inborn capacity to, 476
Learning disabilities, 218
Legal drugs, 218, 219
Lens, age-related changes, 327
Leukemia, 155, 423, 426, 434–436, 451, 523; acute myeloid, 449; causes, 367, 428; as major types of cancer, 431, 431; treatment, 430
Leukocyte adhesion deficiency, 449
Life, artificial prolonging of, 11; "meaningful," 301
Life expectancy, 196, 323, 328, 479, 513, 513; genetic disorders and, 401; smoking and, 196
Lifestyle, choices, 423, 429, 430, 436; life expectancy and, 323; heart disease risk, 154; pregnancy and, 320–321; sedentary, 66, 90, 137
Life-support systems, 2, 11
Light microscope, 48, 48, 74, 358
Lipids: blood-brain barrier and, 235; digestive enzyme defect, 387; ecosystem organization and, 490; free-radical damage, 324; in arterial wall, 156
Lipoplexes, 451
Lipoplex therapy, 451
Liquid nitrogen freezing, 343
Liver: artificial replacement cells, 70; drug abuse and, 242, 243; drug detoxification, 240

Liver cancer, 426–427, 428, 430, 431, 434; type B viral hepatitis link, 332, 346
Liver transplantation, 120
Living wills, 11
Local chemical controls, 195, 198, 207, 267
Logging over, 521
Lordosis, 90
"Lucy," 480, 481, 482
Lumpectomy, 432
Lung cancer, 189, 196, 197, 242, 367, 431, 434; gene therapy, 450, 451; smoke as carcinogen, 422, 423, 426, 428, 433–436
Lungs, age and function, 325; artificial replacement cells, 70; collapsed, 197; cystic fibrosis and, 386–387, 387; deferred pain, 251; diseases of, 23, 123; sickle-cell damage, 381; smoke/smog damage, 21; substitute blood and, 71
Lymph nodes, HIV infection and, 338; immune response, 159, 160; non-specific response in, 164, 166, 182; specific response in, 168, 170; STD spread, 341, 343
Lymphocytes: abnormal response, 178, 179, 182; foreign agents and, 159
Lymphomas, 178, 431, 431, 434, 435, 436
Lymph vessels, cancer cells in, 425
Lysergic acid diethylamide (LSD), 243

M

Macroevolution, 462–463, 472; defined, 457, 459
Macrophages, in rheumatoid arthritis, 178
Mad cow disease, 71
"Magic bullets," 177, 430
Magnetic resonance imaging (MRI), 429, 429, 436
Malabsorption disorder, defined, 122
Malaria, 180, 181, 434, 505; adaptive traits and, 6; drug-resistance, 181
Maleness, 287
Male reproductive system, defective hormone receptors in, 397
Malnourishment, 242
Malnutrition, 122, 130–131, 387; defined, 130
Mammary glands: acquired immunity and, 173; tumors in, 10, 279
Mammogram, 429, 432, 432
Manic behavior, 107
Manual dexterity, 482
Marasmus, 131
Marginally productive land, 519
Marijuana, 241, 243
Markers, 449; MHC, 168, 168, 170–172, 171, 172, 173, 182; nonself, 427; self, 140, 141, 168, 168, 171, 182, 327
Mass extinction, 472; defined, 467
Mastectomy, modified radical, 432–433
"Master gene," 287
Master molecule of life, 404
Maternal chromosome, 383, 388; as carrier, 394, 396
Maternity cases, 444
Mating rituals, 461
Maturity-onset diabetes, 276
Measles, 176, 176, 181
Medical genetics, 386
Medical imaging, 429, 429, 436
Medulla, breathing rhythm and, 194, 194, 195
Medulla oblongata, stimulant use, 242
Melanoma, malignant, 430, 431, 435, 435, 450, 451
Meltdown, 367, 522, 523; defined, 522
Menarche, 293

Mendel, Gregor, hybridization studies, 375; independent assortment, 378; inheritance patterns, 376, 392, 393, 444; segregation, theory of, 375, 384
Meningitis, 225
Menopause, 157, 327, 433; physiological changes, 293; symptoms, 106
Menstrual cycle, birth control methods and, 298, 299
Menstruation, chronic problems, 342
Mental illness, genetic links, 383, 417
Mental retardation, 280, 321, 343; abnormal chromosome rearrangement, 388, 394, 398, 400, 401, 403
Mercury, 45, 188, 447; water pollution, 203
Mercury poisoning, 235
Metabolic acidosis, 276
Metabolic disorders, 408
Metabolic pathways, blockage, 408
Metabolism, inactivation of harmful byproducts, 50
Methamphetamine, 241
Methane, 24, 500, 501, 516, 525
Methaqualone, 243
MHC markers, 168, 170, 171, 172, 172, 182; defined, 168; in tissue rejection, 173
Microevolution, 466, 467; defined, 457, 459; processes of, 460–461, 460
Micrographs, 48
Micro-injection, 448
Microorganism, 58, 121; antibiotics and, 180; defense mechanisms, 162, 163, 165
Microsatellites, 444
Microscopes, 38, 38, 40, 162, 232; comparisons of, 48; use with cell study, 40, 46, 48
Microscopy, 48
Migraine headache, 225
Milk, 68, 417; acquired immunity and, 173
Minerals, deficiencies, 127, 130, 131, 202, 495; supplements, 127, 139
Minisatellites, 444
Molecular genetics, 405
Molecules, rogue, 21
Monoclonal antibodies, radioactively labeled, 429, 430, 434–435, 436
Monogamous behavior, 345, 346
Monohybrid cross, 375, 376, 377, 378, 384; defined, 375
Mononucleosis, 434
Morning-after pill, 299
Mortality, 513
Motion sickness, 257
Motor coordination, drugs effect on, 240, 242
Motor neurons, loss of, 325
Mouth, dryness, 211
Mouth cancer, 196, 430, 431, 434
Mucous membranes: age-related changes, 327; allergens and, 178; HIV point of entry, 334, 336; STD infection and, 340
Multiple myeloma, 431
Multiple sclerosis, 177, 224, 225
Multiregional model, 484; defined, 484
Mumps, 176
Muscles, degenerative disorders, 396–397; fatigue, 105, 251
Muscular dystrophy, 396
Mutagens, 418, 423; defined, 416
Mutations, 12, 386–388, 386, 439–440, 466, 472; beneficial, 461; cancer and, 423, 426–429, 427, 432, 436; DNA, 6; frameshift, 416, 418; gene alteration, 405, 407; genetic disorders and, 393, 394, 396, 397, 402; genetic engineering, 445–446, 450, 453, 459–460, 460, 464; harmful, 461; mutagen induced, 416, 418; neutral,

461; spontaneous, 416, 418; substi-
tution, 416, *416*, 418; of viruses, 452
Myasthenia gravis, 226
Mycobacterium tuberculosis, (TB),
180, *181*, 197
Myocardial infarction, *155*
Myocarditis, *155*
Myopia, 260
Myotonic muscular dystrophy, 393

N

Nasal epithelium cancer, *428*
National Marine Fisheries Service, 524
Naturalists, 458, 462
Natural selection, 457–459, *460*, 461,
465, 472; defined, 460
Nausea, 120, 121, *126*
Nearsightedness vision, 259, 260, *260*,
261
Needle contamination, 336, 337;
"clean" strategies, 339
Nephritis, 213
Nephrons, kidney disorders, 213
Nervous system: age-related changes,
326–327; disease, 177; drug disrup-
tion, 218, 240–243; genetic degener-
ation of, 394–395, *395*; under attack,
225, 235
Nervous system disorders, 428
Neuralgia, 225, 227, 232
Neural tube, disorders, 320
Neurofibrillary tangles, 326, *326*
Neurofibromatosis, 416
Neuronal damage, 235, 282, 283;
regeneration, 282, 283
Neuron injuries, repair of, 225
Neurotransmitter molecules, psy-
choactive drugs and, 240
Neurotransmitters, age-related
changes in, 326
Nicotine, 154, 157, *240*, *241*; blood-
brain barrier, 235; defined, 242;
dependence or habituation, 240
Night blindness, *126*
Night sweats, *155*
Nitrification, 503, *503*; defined, 502
Nitrogen narcosis, 188
Nits, 344
Nobel Prize, 1962 for medicine, 405
Noise pollution, 257
Nongonococcal urethritis (NGU), 343,
346; defined, 342
Non-Hodgkin lymphoma, 434
"Nonself" markers, 140, 168, 169, 182,
427; immunity breakdown, 427
Nonshivering heat production, *214*,
215; defined, 214
Nonspecific defense responses, 163,
164, *164*, 165, *165*
Nonsurgical sterilization, 299
Nonverbal skills, right hemisphere,
234
Norplant, *298*, 299, 301
Nuclear energy, 522–523, *522*, 526
Nuclear medicine, *18*
Nuclear power, 522–523, *522*, 526
Nuclear power plants, 367
Nuclear reactors, 522, 523
Nutrient deficiencies, 131
Nutrition, *81*, 110–132; age-related
changes in, 327; maternal, 313, 320,
321
Nutritionist, 124, 131

O

Obesity, 129, *155*; defined, 128; diabetes and, 276; dietary habits, 111;
heart disease risk, *155*; venous
valves and, 151

Oils, as fossil fuel, 35
Olestra, 125
Opioids, 243
Opium, 243
"Opportunistic" infection, 338
Oral cancer, *431*, 434, 435; tobacco use,
242
Oral contraceptives, 180, *298*, 299,
342
Organic chemicals, water pollution,
203
Organ transplant, 11, 171, 173,319;
cell-mediated immune response,
169, 170, 182; HIV transmission,
336; immunodeficiency (non HIV),
178
Organ transplant recipient, rejection,
173
Osteoarthritis, 96, 325
Osteogenesis imperfecta, 381
Osteoporosis, 90, *90*, 325; vitamin K,
127; postmenopausal women, 327
Osteosarcoma, 383
Ovarian cancer, 299, 423, *429*, 433,
450, 451; incidence, *431*; treatment,
430
Ovaries, deferred pain, *251*; PID infec-
tion, 342; testicular feminizing syn-
drome, 397
Overdraft, 518
Overfarming, 519, *519*, 521, 524
Overfishing, 525
Overgrazing, 519, *519*, 521, 524
Overheating, 83
Overpopulation, 512
Oviducts, bacterial infection, 340, 342;
birth control and, 299; scarring, 340,
342, 346
Ovulation: birth control methods, 298,
299; infertility and, 300
Oxygen debt, 184, 188; defined, 103
Ozone, 500, 516, *516*, 517, *517*, 524–526
Ozone layer depletion, 7, 8, 81

P

Pacemaker, 134, 135; artificial, *18*, 148;
self excitatory cells, 148
Pain, 81, 93, *251*, 330, 342; defined,
250; detecting tissue damage, 248,
250; disk ruptures and, 93; in heart
attack, 154, *155*; perception of, 227,
243, 251, 264; receptors, 247, *248*,
250, 264; receptors, inflammation
and, *166*, 167, *167*, 182; types of,
251
Painkillers, 93, 227, *241*
Pancreas, deferred pain, *251*; disease,
123; genetic disorder, 386
Pancreatic cancer, 196, *431*, 434
Pandemic, 180, 512
Panic attacks, 243
Paper, recycling, 519
Papovaviruses, *334*
Pap test, 343, *429*
Paraganglioma, 383
Paralysis, 224, 283, *283*; decompres-
sion sickness, 182; sickle-cell dam-
age, 381; syphilitic, 341
Paranoia, 241
Paranoid delusions, 243
Parasites, 180, 338, *344*, 346, 490, *495*;
defined, 330; eosinophils and, 137,
137; STD infection, 344
Parasitic worms, as pathogens,
164–166, 173, *181*, 182
Parathyroid gland, disorders, 281
Parkinson's disease, 227, 239
Passive immunization, 176, 177, 182
Pasteurization, 162
Paternal chromosome, as carrier, *394*;
patterns of autosomal inheritance,
394–395, *394*; patterns of X-linked

inheritance, 396, *396*, 397; sex deter-
mination, 390
Paternity cases, 141, 444
Pathogens, infectious disease and,
180, 181; inflammatory response to,
334; of sexually transmitted disease,
330, 334, 345, *345*; viruses as, 334
Pattern baldness, 398, *398*, *403*
PCBs, 279
Pedigrees, 397, 402; defined, 392; fam-
ily, 393, *395*, 396, *396*; symbols used,
392
Pellagra, vitamins and, *126*
Pelvic examination, *429*
Pelvic inflammatory disease (PID),
340, 342, 343, 346; defined, 342
Penicillin, 180, 206, 341, 446; antibi-
otic-resistant strains, 340
Penis, *286*, 288, *290*, 292, 390; age-
related changes, 327; condom over,
299; discharge from, 341, 344, 345;
function, *288*, *289*, 297; genital
warts, 343; HIV point of entry, 336
Peptic ulcer, 116
Percussion therapy, *387*
Periodontal disease, 114
Peritoneal dialysis, 213
Peritonitis, 121
Pernicious anemia, 139
"Persistent vegetative state," 11
Pest control, 502–503, 505
Pesticides, *15*, 206, *516*, 518, 519, 526;
banning of, 505; as carcinogen, 426,
428, *428*, 430; exposure to, 14, *14*;
genetic engineering, 446–447; syn-
thetic organic, 505; water pollution,
203
Petit mal seizures, *225*
p53/antisense *ras* RNA, *450*
p53 tumor suppressor gene, 426
Phagocytes, effects of smoking on,
196
Phagocytosis, 44, 51, *136*, 163
Phalanges, *91*, 94, *94*, 95, *95*
Pharmaceutical drugs, 448, 449
Phencyclidine (PCP), 243
Phenotypes, gene mutation and, 388;
in a testcross, 378, *378*; nondisjunc-
tion outcomes, 401; sex-influenced
inheritance, 398, *398*
Phenylketonuria (PKU), 394, *403*
Phosphofructokinase deficiency, *449*
Photochemical oxidants, 516, *516*
Photochemical smog, 526; defined,
516
Photosensitive organ, third eye as,
281
Photosynthesis, 492, *498*, *499*, 501
Photosynthesizing organisms, 487,
490, *491*, 493, 495, 498, 506
Phylogeny, 467; defined, 467
Physical drug dependence, 240, *240*
Physical therapy, cystic fibrosis and,
387
Physiological mechanisms, 82
Physiologists, 134
Pioneer species, 488
Pituitary dwarfism, 274, *274*, 446
Placenta, as barrier, 328; IgG antibod-
ies crossing, 173
Plant bioengineering, 439, 452
Plants, cancer in, 426; carcinogens
produced by, 426; transgenic,
446–447, *447*
Plastic surgeons, collagen implants,
73
Platelets, freezing, 145
Plate tectonics, *462*, 463; defined, 462
Pleasure center, 242
Pleurisy, 187
Pneumonia, 180, *181*, 187, *196*, *197*;
cystic fibrosis and, 387; drug-resis-
tance, 181; as opportunisitic dis-
ease, 336, 338; sickle-cell damage,
381; STD and, 336, 341

Pneumonic plague, 512
Pneumothorax injury, *197*
Point mutations, 418; definiti[on]
Poisoning, *155*
Polio, 176
Poliovirus, 176; vaccine, 176
Pollen allergies, 92
Pollutants, 447, 526; defined, 516; gen-
erated by humans, 509, 512–513,
515–519, *516*, *517*, *518*
Pollution, 14, 487. *See also* Agricultural
runoff, Industrial waste, Pollutants,
Pollution, Soil salination
air, 502; environmental estrogens, 279;
making nonpolluting fuel, 35;
noise, 257
Polydactyly, 382, *382*, *403*; pedigree
for, *392*
Polygenic traits, defined, 383
Polyploidy, defined, 400
Poppers, 243
Population density, 463, 512; defined,
510
Population dynamics, 510–511, *510*
Population, food crisis and, 131;
growth and sustainability, 508–526
Population growth, *510*, 512, *513*,
514–515, *514*, 524, 526; annual, 510,
515, *515*; rate of increase, 301, 511,
513
Population size, 510, 511, *511*, 513–515,
514, 515, 526
Positron-emission-tomography (PET),
18, 236
Potatoes, as fuel source, 35
Potentiating drug interaction, 241
Prader-Willi syndrome (PWS), 383
Pre-carcinogens, 427
Precipitation, acid rain effect, 517
Predators, 479, 482, 505, 512, 513, 526
Pregnancy, 81, 151, 268, 282, 318, 425;
food risks, 235; genetic counseling
and, 393; hormonal secretion and,
275; IgG antibodies, 173; infertility
and, 300; myths and, 297; prepara-
tion for, 287, 293, *293*, *294*, 295, 295;
prescription drugs and, 299, 321;
Rh blood typing and, 140–141; risk
of, 298, *298*; risky behavior during,
196, 218, 336, *337*, 340, 342, 346; termination of, 301; unplanned, *298*,
302
Pregnancy test, 299; at-home, 309
Pregnancy vaccine, 299
Preimplantation diagnosis, 322
Premature birth, 316, 318–319; causes
of, 319; smoking and, 321; immature lungs of baby, 192; saving
babies, 319
Prenatal diagnosis, 400
Prenatal risk, 320–321
Prescription drugs: the Pill, 299; prenatal risk, 321
Preservation, 463
Primary immune response, 175, *175*,
176, *179*
Probability, 377, *379*, 384; calculating,
377, 380, *380*; as constant, 376, *379*;
defined, 376; genetic analysis and,
392, 393; monohybrid cross and,
377, *377*; Punnett square method,
376, *376*
Processing, of food, 111
Progesterone, post-menopausal
women, 327; synthetic, 299
Programmed death, 307
Promoters, cancer, 427
Prostate cancer, 207, 289, *429*, 430, *431*,
433, 436; screening for, 177
Protein deficiency, 125, *125*, 130–131,
178
Protein-energy malnutrition, 130
Proteins, cancer and, 424, 426, 430, 436
Protein-specific antigen (PSA) test, 433
Proto-cells, 471, *471*

427,

n. 416

...al drug dependence, 240
..sychosis, 243
Puberty, genetic disorders and, 401
Pubic lice, 344, 346
Pulmonary embolism, 197
Pumping iron, 90
Punctures, inflammatory response, 166
Punnett square, 375, 380, 380; defined, 376
Pyelonephritis, 207

Q

Quaaludes, 241, 243

R

Rabies, 446; virus, 180
Radial keratomy, 261
Radial muscles, in eye pupil, 258
Radiation, 71; background, 427; as cancer cause, 423, 427, 427, 430, 434–436; as cancer therapy, 18, 139, 430, 433, 434; as therapy, immune system suppression, 173, 178
Radiation sickness, 522–523
Radioactive dating, 457, 472
Radioactive decay, 468, 523; defined, 17
Radioactive isotopes, diagnostic tests, 367
Radioactivity, exposure to, 197
Radioisotopes, 428, 523; defined, 17; diagnostic tests, 367; measurements, 474; uses of, 17, 18
Rage, 237
Rapid Eye Movement (REM), 237
Rashes, 14
Rats, 512; atherosclerosis and, 154
Receptor mediated endocytosis, 271
Recessive alleles, 372, 373, 375, 384; defined, 374; for a sex-influenced trait, 398; genetic analysis of transmission, 392–395, 392, 394, 395, 402; in Mendel's single gene model, 381; in monohybrid cross, 377; in Punnett square method, 376, 380; in testcross, 378, 378; penetrance, 382
Recognition protein, 41, 62, 168; defined, 41
Recombinant DNA, 322, 438–453; methods, 451
Recombinant DNA technology, 439; defined, 440
Recombinate, 448
Recombination, 439, 445, 453; viral, 440
Recreational drugs, 241, 241
Rectum, bacterial infection, 340; HIV point of entry, 336
Recycling, 519; interdependency of organisms and, 7, 7
Red blood cells, abnormal hemoglobin, 408; freezing, 145; micrographs of, 48, 48; sickling, 381, 381; substitute blood, 71; typing of, 140, 141
Red/green color blindness, 260, 397

Referred pain, 251, 251
Rejection, immune system and, 70
Remodeling, 89, 108; defined, 90
Replication, 407, 442, 443; defined, 406; DNA mechanism of, 406–407; enzymes, 440, 442; errors, 416, 418; viral, 177
Reproductive systems, age-related changes, 327
Respiration, drugs' effect on, 240, 241, 243
Respirators, 11
Respiratory distress syndrome, 197, 316
Respiratory failure, 387; muscular dystrophy, 397
Respiratory system, intense exercise and, 103; viral attack, 171, 332
Resting blood pressure, 149, 151, 157
Resting cardiac rate, 157
Reticular formation, drug use and, 242
Retina, disorders/disease, 260, 261
Retinal detachment, 261
Retinoblastoma, 423, 426
Retinoic acid (Retin-A), 321; prenatal risk, 321
Retrovirus, 333, 334, 336, 337, 426, 350; recombinant, 451
Reverse transcriptase, 333, 336, 337, 445, 450; action blocking, 339
Reverse transcription, defined, 445
Rh blood typing, 140–141; defined, 140
Rheumatism, sickle-cell damage, 381
Rheumatoid arthritis, 96, 178, 451; age-related, 327; free radicals and, 21
Rhythm method, 298, 298
Rib cage, Heimlich maneuver, 115; organ protection, 93; osteoporosis and, 90
Rickets, 131, 281, 281; vitamins and, 126
Rigor mortis, 101
Ringworm, 180
Roe v. Wade, 301
Roid rage, 107
Rubella, 176
Runoff, 469, 495–496, 496, 499, 502, 504, 520; agricultural, 518, 526
Ruptures, stopping blood loss, 153, 153

S

Sabin polio vaccine, 176
Saccharin, 253
Salinization, defined, 518
Saliva, ABO markers in, 141; hepatitis B transmission, 343; HIV infection and, 336; immune response, 164, 164, 182
Salivary glands, cancer of, 434
Salmonella, 180; infection, 133
Salts: absorption, 118; defined, 22; in soil, 508; level in diet, 111; reabsorption of, 121; water interaction, 22
Salt/water balance, 211
Salt/water intrusion, 518, 518
Salty taste class, 246, 252, 253
Sanger method of DNA sequencing, 443, 443
Sanitation, 514, 515; public, 45
Sarcomas, 436; defined, 431
Sarcomeres, defined, 100; in muscle contraction, 100–101, 100, 101, 108
Saturated fat, 125, 154
Saturated fatty acid, 28, 28, 29; defined, 28
Scabies, 344, 346
Scanning electron microscope (SEM), 48, 48, 81

Scar tissue, 97; bronchial, 197; emphysema, 189; hearing loss and, 257; of liver, 120, 242; navel, 318; of oviducts, 340, 342, 346; smallpox, 162; stretch marks as, 81; TB, 181; of vas deferens, 340; vasectomy, 299
Schistosomiasis, 181
Schizophrenia, chronic, 449
Scientific methods, 8–9, 12, 372; example of, 9, 10
Scientific theory, 3, 9, 9; defined, 8, 12
Scrotum cancer, 426
Scurvy, 131; vitamins and, 126
"Search and destroy" cells, 137; neutrophils as, 137, 137
Seasonal affective disorder (SAD), 266, 267
Seconal, 241
Secondary immune response, 175, 175, 176, 179
Secondary infections, 116, 261
Secondary sexual traits, genetic disorders and, 401
Second filial generation, 376, 378
Secondhand smoke, 196, 196, 197; during pregnancy, 195, 196
Sedatives, 241; abuse, 242, 243; prenatal risk, 321
Sedentary lifestyle, 90, 106, 137
Seizures, 238, 243
Self-gratifying behavior, 237
"Self" markers, 140, 141, 168, 168, 171, 182; age-related changes, 327
Self-replicating systems, 405, 471, 471, 472
Semen, ABO markers in, 141; hepatitis B transmission, 343; HIV transmission, 178, 336
Sensory processing, psychoactive drugs and, 240, 241, 243
Severe combined immune deficiencies (SCIDS), 179, 450, 451
Sewage, 486, 487; human, 45, 45
Sex chromosomes, abnormalities, 401
Sex hormones, deficiencies, 90
Sex-influenced inheritance, 398, 398, 403
Sexual behavior, 233, 237–244; hormonal controls, 290; number of partners, 339, 340, 433; psychoactive drugs and, 240; sympathetic nerves, 231
Sexual intercourse, 287, 288, 297; birth control during, 298–299; female organ of, 292, 293; male organs of, 288, 289; STD and, 331, 340, 344, 346
Sexuality, age-related changes, 327
Sexually transmitted disease (STD), 120, 177, 179, 180, 330–346; animal viruses causing, 334; chlamydia, 261; complications of untreated, 330, 331; defined, 330; diagnosing, 340; ethical obligation, 345, 346; latex condoms and, 299; prevention, 299–331, 331, 345, 345; recurrence, 342, 343, 345; transmission, 340; untreated, 330–331, 334, 340, 342, 345–346; urinary tract infections, 207
Sexual orientation, genetic links, 383
Sexual stimulation, 81, 292, 293; female arousal, 297; male arousal, 288, 289, 289, 290, 297
Shifting cultivation, 520–521, 520, 521, 524, 526; defined, 520
Shigella, 180
Shivering, 83, 214, 215, 215; defined, 214
Short-term memory, 239, 239, 244
Sickle-cell anemia, 376, 381, 403, 405, 444; DNA mutation and, 6; embryo screening, 393; genetic disorder,

416; one gene, one enzyme hypothesis, 408; trait for, 381
Sigmoidoscopy, 429
"Silent killer," 157
Silicosis, 197
Silt, 463
Single-gene defects, 376
Sinuses, 92; allergens and, 178; defined, 92; type of epithelium, 68
Sinusitis, 92
Skeleton, abnormal development, 400
Skills memory, 239, 239
Skin, as barrier to invasion, 164, 164, 182; color phenotype, 80, 383; designer skin, 70; eruptions, 81; mechanical insults, 81; wrinkling, 73, 323, 324, 325
Skin cancer, 423, 435, 435, 517; basal cell carcinoma, 10, 80, 435, 435; carcinogens, 10, 10, 428; from ionizing radiation exposure, 367, 427; gene defects and, 407, 426; gene therapy, 449, 451; malignant melanomas, 431, 435, 435; squamous cell carcinoma, 434, 435, 435; ultraviolet radiation, 427, 430, 517
Skin diseases, herpes simplex as, 260
Skin tests: allergens, 178; TB, 175
Slash-and-burn agriculture, 520, 520
Sleeping, 30, 236, 237; behavior, 266, 267; braincells governing, 226; medulla and, 233; sedation, 242; slow-wave pattern, 237; /wake cycles, 195, 266, 281
Slipped disk, 93
Slow-wave sleep pattern, 237
Small-cell carcinomas, 434
Smallpox, 162, 164
Smog, 498, 499, 516, 516, 517, 526
Smoke and smoking, 90, 116, 129, 155, 156, 240, 521; addiction, 242; aging and, 325; associated risks, 196; benefits of quitting, 196; breathing and, 185, 188, 189, 196; cancers and, 422, 423, 426, 430, 433–434, 436; carbon monoxide, 154, 188; free radicals and, 21; marijuana, 241; oral contraceptives and, 299; prenatal risk, 195, 321, 328; secondhand, 434; SIDS and, 195
Smoker's cough, 196
Sneezing, 180, 181, 197; as allergic response, 178; as immune response, 164
Soil leaching, 490
Soil management technologies, 502–503
Soil pollution, 14
Soil salinization, 518
Solar energy, 471, 522, 523, 523
Solar-hydrogen energy, defined, 523
Solar system, 457, 468, 472
Sole bones, 91
Solenoid, 409
Soluble fiber, 121
Southern blot method, 444
"Speed," 242
Sperm bank, 300
Sperm count, 279, 468; low, 299, 300
Sperm, defective, 21, 300
Spermicides, 345; foam, 298, 298; jelly, 298, 298; nonoxynol-9, 339
Spina bifida, 313
Spinal cord injury, 155, 282; growth factor and, 282, 283, 283
Spinal cord, viral infection, 341
Spinal nerves, 228, 228; function, 230
Spleen, sickle cells in, 381
Split-brain experiments, 238, 238
Spoken language skills, left hemisphere, 234
Spontaneous abortion, 314
Spontaneous mutation, 416, 418

Sporadic disease, 180
Sprain, 97
Spraying programs, community, 428
Squamous cell carcinomas, 434, 435, *435*
Staining, 48, 137. *See also* Dyes
Starvation, 129, 130, 458: response, 129
Sterility, 300; genetic disorder and, 401; STD infections, 330, 334, 340, 342, 346
Sterilization, 508
Steroid hormones, use/abuse, 106–107
Stillbirth, *196*, 203, 321; genetic disorders and, 395; STD and, 340
Stimulants, *241*, 242
Stomach cancer, 111, *431*, 434
Strep infections, 181
Streptomycin, 180; prenatal risk, 321
Stress, 81, *157*; allergies and emotional, 178; chronic, 116; color problems and, 121; environmentally induced, 323; heart attacks and, *155*; hormonal secretions and, 269, 278, 279; immunosuppressive effects, *178*, 345
Stress electrocardiogram, 156
Strip mining, 522
Stroke, 154, 157, *446*, 448
Substance abuse, 219
Substitution mutation, 416, *416*
Sudden death, *155*
Sudden infant death syndrome (SIDS), 195
Suffocation, 189
Suicide, 170
Suicide tags, 451
Sunburns, 81
Sunlight energy, 487, 490, *490*, 492, *494*, 496, 506; harnessing solar, 522, 523, *523*
Sunlight exposure, 81; causing cancer, 423, 427, 430, 435, 436
Suntanning, 80, 81
Supernatural belief, 9
Superweeds, 452
Supreme Court rulings, 301
Surgery, 93, 430, 432, 433, 435, 436; axon regeneration after, 225; blood transfusion, 71, 336; exploratory, *18*; for arterial blockage, 156; knee joint, 97, *97*; patients, and contaminated blood supply, 336
Surgical-wound infections, 181
Surrogate mother, 448
Survival of the fittest, 460
Survivorship, *513*
Sustainable society, 526; defined, 525
Synergistic drug interaction, 241
Synergistic movement, 98, 108
Synthetic hormones, 105, 106
Syphilis, 331, 343, 345, 346; defined, 340; primary stage, 340, *341*; secondary stage, 340, *341*; as STD, 336; tertiary stage, 340–341, *341*
Systemic lupus erythematosus (SLE), 179

T

Tachycardia, 157, *157*
Tanning, 80, 81
Tay-Sachs disease, 394, *403*; embryo screening, 393
Tears: HIV infection, 336; as immune response, 164, *164*, 182
Termination, 414, *415*, 418
Testcross technique, 378, *378*
Testicles, 433, *433*; inflammation, 340, 342; self-examination, *429*, 433, *433*; tissue transplantation, 173
Type A hepatitis, 120

Testicular cancer, 107, 279, 289, *429*, 433
Testicular feminizing syndrome, 397, *403*
Tetanus, 58, 105, *105*, 176, *176*
Tetracycline, 180; prenatal risk, 321; STD infections, 341, 342
Tetrahydrocannibol (THC), 243
Thalidomide, 321
Therapeutic drugs, 106
Thermal pollution, 518
"Third eye," 281
Three Mile Island, 523
Threonine, 125, *125*, *408*, *412*
Threshold level, *223*, 226, 227, *227*, 244; defined, 222
Throat, bacterial infection, 340
Throat cancer, 434
Thrombus, 156
Thyroid cancer, 367
Tissue damage, 264, *298*; agglutination, 140; defense mechanisms, 163, 164, 166, 167, 182; H. pylori infection, 116; regeneration, 283
Tissue engineers, 70
Tissue plasminogen activator (tPA), *446*, 448, *448*
Tissue rejection: cytotoxic T cells and, 170; strategies for prevention, 173
Tobacco, 434, 436; smoke, as mutagen, 416; as stimulant, 242; use, 154, *196*, *197*. *See also* Smoke and smoking
Tongue, cancer of the, 434
Tooth decay, 114, *127*
Touch-kill mechanism, 168, *169*, 170, 171, *171*
Toxic-dump sites, 447
Toxic psychosis, 243
Toxins: antibody-mediated response, 173; bacterial, 121, 176; inactivated, in vaccine, 176, 182; removal, in liver, 120
Tranquilizers, 321
Transforming growth factors (TGF), 283
Transfusable blood, 145; shelf life, 145
Transgenic organisms, 173, 452; defined, 446
Transmission electron microscope, 48, *48*, *100*
Transplants, 171; corneal, 173; kidney, 213; liver, 120. *See also* Organ transplant
Tubal ligation, 299, *298*
Tubercles, 181
Tuberculosis, *181*, *197*; bacterium, 175; drug-resistant strain, 181, 336, 338; screening programs, 181; skin test, 175
Tumor, *178*, 446, 451; benign, 424, *424*; cancerous, 424–425, *424*, 427, 429, 430, *432*; defined, 424; genetic disorders and, 407; growth, 425, *425*; in lung, *197*; irradiation, 430, 436; malignant, 424, *424*; mass, 424, *424*; primary, 424, 425
Tumor angiogenesis factor (TAF), 283
Tumor cells, 170, 177, 451; antigen, 168, 182, 427, 429; touch-killing, 168, *169*, 171
Tumor-inducing genes, 446
Tumor-infiltrating lymphocytes (TILs), 451
Tumor markers, 436; defined, 429
Tumor necrosis factor (TNF), 177, *446*, *450*, 451
Tumor suppressor genes, 423, 426, 427, *427*, 436; defined, 426; FHIT, 426; p53, 426; *Rb*, 426
Turner syndrome, 401, *401*, *403*

Type B hepatitis, 120, 342, 343, 346
Typhoid fever, 180
Typhus, 505

U

Ulcers, 435; corneal, 260; growing skin, 70; STD produced, 340, *340*, 343
Ultrafiltration, 167
Ultrasound, *248*, 429; pregnancy and, 314, 322
Ultraviolet light, 322; genetic disorders, 407; as mutagen, 416
Ultraviolet (UV) radiation, 80–81, 84, 423, 427, 435, *435*, 517; free radicals and, 21; UV-B, 81
Umbilical cord, blood, 451
Unconsciousness, 236
Undernutrition, 125, 130–131
Underwater pressure, 188
U.S. Environmental Protection Agency, 235
Unsaturated fats, 154
Unsaturated fatty acid, *29*; defined, 28
Uranium fuel, 523
Urinalysis, 206; defined, 207
Urinary bladder, infections, 327
Urinary frequency, 327
Urinary incontinence, 327
Urinary system, age-related changes in, 327
Urinary system cancers, 434–435
Urinary "toilet training," 206
Urinary tract infection, 207, 327, 341; drug-resistance, 181
Urine, excess water/salts in, 272; HIV infection, 336; sugar in, 276
Uterine cancer, *429*, 430, *431*, 433, 436; hormone replacement therapy and, 327
Uterus, implantation, 309, *309*; infection, 319; perforation of, 299; PID infection, 342; placenta and, *311*; pregnancy and, 268; testicular feminizing syndrome and, 397
UV-B, 81

V

Vaccination, 162; for German measles, 321; for Hepatitis B, 343
Vaccines, 182, 430, 438; against HIV, 336; bioengineered, 446; combined injected, *176*; defined, 176; oral, *176*; viral, 339
Vagina, artificial insemination, 300; discharge, 340, 341, 342, 345; as HIV point of entry, 334, 336; HIV transmission, 336; infections, 164, 344, *344*, 346; sexual penetration, 297, 298
Vaginitis, 344, 346
Variable expressivity, 382
Variation, *459*; heritable, 460
Varicose veins, 151
Vas deferens, scarring, 340
Vasectomy, *298*, 299; compulsory, 509
Vasoconstriction, during intercourse, 297
Vasodilation, during intercourse, 297; menopausal symptom, 293
Vegetarians, 125, *125*; vitamin/mineral supplements, 127
Ventricular fibrillation, 157, *157*
Vertebrates, "third eye," 281
Vestigial structures, 464–465, *465*
Vibration, 255, *255*; defined, 254

Vibration stimulus, *248*; Meissner's corpuscle, 250, *251*
Vinblastine, 435, 524
Vincristine, 430, 435, 524
Vinyl chloride, 426, 428, 430
Viral attack, *447*
Viral disease, 71, 177, 179, *181*; HIV retrovirus, 335–339, *336*, *337*, *338*; sexually transmitted disease (STD), 330–339, *332*, *334*, 342, 345, 346
Viral infections, *197*, 225, 440, *446*; and cancer, 423, 426, 427, 436; heart disease and, *155*, 156; influenza, 180, *181*; prenatal risk, 320–321, *320*; type 1 diabetes and, 276
Virulence, 180
Viruses, *163*, 170, 171, *171*, 468; altered, 339; antibiotics and, 181; antibodies and, 140; combatting, 81; components, 332; defined, 332; genetically engineered, 176, 394, 438, 449–452; genetic susceptibility, 393; gene transfer and recombination, 440; leukocytes and, 137, 160; liver cell injury, 120; loss of chromosome region, 398, 399; multiplication cycle, 332, *332*, 333; as pathogens, 164, 165, 168, 182; specificity, 332; surface antigen, *174*; surface markers, 163; transgenic, 452
Visceral pain, 251
Vision, daytime, 262–263, 475, 477–479, 485
Visual information, 262, 264
Vitamins, deficiencies, *126*, 130, 131, *155*
Vitamin supplements, 127; prenatal, 320
Volatile organic compounds, *516*
Volatile substances, 252
Volcanic activity, 468
Volcanic eruptions, 469, 498, *499*, 502, 506, 517
Vomiting reflex, 257; allergic reaction, 123; appendicitis, 121

W

Waking, 30, 236, 237; medulla and, 233
Walking, 106, *129*; brisk, 90; upright, 475, 477–480, 483
Warts, 334, *334*. *See also* Genital warts
Waste-burning, controls over, 235
Waste dumps, toxic, *203*, 447
Waterborne bacteria, 175
Water conservation: ADH regulation, 272, *272*; global crisis, 518, 526; mechanisms, 81, 200, 201, 215
Water, global crisis, 518, 526
Water pollution, 14, 35, 120, 518, *518*, 519, 525, 526; disease and, 45, *45*; of drinking, 203; fecal contamination, 121; of water supply, 180
Water treatment, 518
Weaning, protein deficiency and, 130
Weight control, 125, 129, 132; aging and, 325, *325*; diabetes and, 276; during pregnancy, 320; loss, insulin deficiency and, 276. *See also* Body weight
Weight guidelines by sex, *128*
Whales, 487, *491*
Wheat, 502
White blood cells: in inflammation, 166; leukemia, *155*; a line of defense, 164, *164*
Whooping cough, *176*, 180, *181*
Wilms tumor, 434
Wind energy, 523, *523*
Winter blues, 266, 267

Withdrawal, 298, *298*
World Health Organization, 335, *335*
World Wide Web, 449

X

Xenotransplantation, 173
Xeroderma pigmentosum, 407
Xeropthalmia, 131
X-inactivation, defined, 390

X-linked anhidrotic ectodermal dysplasia, *403*
X-linked genes, androgen insensitivity, 397
X-linked trait, red-green color blindness, 260
X ray diffraction, DNA analysis, 405
X rays, 156, 427, 429, *429*, 432; chest, for TB, 181; diagnostic, 367; exposure, dental, 427
XXX females, 401, *401*
XXY chromosomes, 401, *401*

XY chromosomes, as male, 387–390, *389*, *390*; testicular feminizing syndrome, 397
XYY chromosomes, 401, *403*

Y

Yeast infections, 336, 338; genital, 164, 344
Yeasts, 442, 443, 446, 466

Z

Zalcitabine, 339
Zero population growth, 515, 526; defined, 510
Zovirax, 343
Zygomatic bones, 92, *92*
Zygote: in infertility treatments, 300; X inactivation in, 390
Zygote intrafallopian transfer (ZIFT), 300